MACHINE DESIGN
An Integrated Approach

Robert L. Norton

Worcester Polytechnic Institute

Worcester, Massachusetts

Prentice-Hall
Upper Saddle River
New Jersey 07458

Library of Congress Cataloging-in-Publication Data
Norton, Robert L.
  Machine Design   An Integrated Approach /
  Robert L. Norton
      p.   cm.
  Includes bibliographical references and index.
  ISBN: 0-13-55575-6 (hard cover)
  1. Machine design.   I. Title
  TJ230.N64   1996
  621.8'15--dc20   96-40909   CIP

Acquisitions editor: Bill Stenquist
Production editor: Rose Kernan
Copy editor: David George
Book designer: AGT Inc.
Cover designer: Koren Pacheco
Manufacturing buyer: Donna Sullivan
Editorial assistant: Meg Weist

This book is dedicated to:

Donald N. Zwiep

A gentleman and a leader,
without whose faith and foresight,
this book would never have been written.

and to:

Kayla, Nick, and Sarah

The next generation

The artwork was drawn, equations set, and the book electronically typeset by the author on a *Macintosh* computer using *Pagemaker 5.0* and *Freehand 3.1* by Aldus, *Word 5.0* and *Excel 4.0* by Microsoft, *Adobe Photoshop 2.5, Mathtype 3.0* by Design Science, *Endnote Plus 1.2* by Niles Associates, and *TKSolver* by UTS. Body text is set in Times Roman and headings are in Syntax font.

© 1996 by Prentice-Hall Inc.
Simon & Schuster / A Viacom Company
Upper Saddle River, N.J. 07458

Printed in the United States of America

10 9 8 7 6 5 4 3 2 1

Prentice-Hall International (UK) Limited, *London*
Prentice-Hall of Australia Pty. Limited, *Sydney*
Prentice-Hall Canada Inc., *Toronto*
Prentice-Hall Hispanoamericana, S.A., *Mexico*
Prentice-Hall of India Private Limited, *New Delhi*
Prentice-Hall of Japan, Inc., *Tokyo*
Prentice-Hall of Southeast Asia Pte. Ltd., *Singapore*
Editora Prentice-Hall do Brasil, Ltda., *Rio de Janeiro*

ISBN 0-13-555475-6

Contents

PART II   MACHINE DESIGN   535

CHAPTER 8   DESIGN CASE STUDIES _____ 537

Preface

Introduction

This text is intended for the *Design of Machine Elements* courses typically given in the junior year of most mechanical engineering curricula. The usual prerequisites are a first course in *Statics and Dynamics*, and one in *Strength of Materials*. The purpose of this book is to present the subject matter in an up-to-date manner with a strong design emphasis. The level is aimed at junior-senior mechanical engineering students. A primary goal was to write a text that is very easy to read and that students will enjoy despite the inherent dryness of the subject matter. The book makes extensive use of over 200 *TKSolver* computer-model files included on CD-ROM that support the examples and case studies to aid in the learning process. The student edition of the *TKSolver* computer program and its manual is also packaged with the text as an option.

This textbook is designed to be an improvement over those currently available and to provide methods and techniques that take full advantage of the graphics microcomputer. It emphasizes design and synthesis as well as analysis. Example problems, case studies, and solution techniques are spelled out in detail: verbally, graphically, and within the *TKSolver* files. All the illustrations are done in two colors with computer-drawing packages. Scanned images of photos are also included. Short problems are provided in each chapter and, where appropriate, longer unstructured design-project assignments are given.

While this book attempts to be thorough and complete on the engineering-mechanics topics of failure theory and analysis, it also emphasizes the synthesis and design aspects of the subject to a greater degree than most other texts in print on this subject. It points out the commonality of the analytical approaches needed to design a wide variety of elements and emphasizes the use of computer-aided engineering as an approach to the design and analysis of these classes of problems. The author's approach to this course is based on over 30 year's experience in mechanical engineering design, both in industry and as a consultant. He has taught mechanical engineering design at the university level for about 20 of those years as well.

Philosophy

This is often the first course that mechanical engineering students see that presents them with design challenges rather than set-piece problems. Nevertheless, the type of design addressed in this course is that of *detailed design,* which is only one part of the entire design-process spectrum. In *detailed design*, the general concept, application, and even general shape of the required device are typically known at the outset. We are not trying to invent a new device so much as define the shape, size, and material of a particular machine element such that it will not fail under the loading and environmental conditions expected in service.

The traditional approach to the teaching of the *Elements* course has been to emphasize the design of individual machine parts, or elements, such as gears, springs, shafts, etc. One criticism that is sometimes directed at the *Elements* course (or textbook) is that it can easily become a "cookbook" collection of disparate topics that does not prepare the student to solve other types of problems not found in the recipes presented. There is a risk of this happening. It is relatively easy for the instructor (or author) to allow the course (or text) to degenerate into the mode "Well it's Tuesday, let's design springs—on Friday, we'll do gears." If this happens, it may do the student a disservice because it doesn't necessarily develop a fundamental understanding of the practical application of the underlying theories to design problems.

However, many of the machine elements typically addressed in this course provide superb examples of the underlying theory. If viewed in that light, and if presented in a general context, they can be an excellent vehicle for the development of student understanding of complex and important engineering theories. For example, the topic of preloaded bolts is a perfect vehicle to introduce the concept of prestressing used as a foil against fatigue loading. The student may never be called upon in practice to design a preloaded bolt, but he or she may well utilize the understanding of prestressing gained from the experience. The design of helical gears to withstand time-varying loads provides an excellent vehicle to develop the student's understanding of combined stresses, Hertzian stresses, and fatigue failure. Thus the *elements* approach is a valid and defensible one as long as the approach taken in the text is sufficiently global. That is, it should not be allowed to degenerate into a collection of apparently unrelated exercises, but rather provide an integrated approach.

Another area in which the author has found existing texts (and *Machine Elements* courses) to be deficient is the lack of connection made between the dynamics of a system and the stress analysis of that system. Typically, these texts present their machine elements with (magically) predefined forces on them. The student is then shown how to determine the stresses and deflections caused by those forces. In real machine design, the forces are not always predefined and can, in large part, be due to the accelerations of the masses of the moving parts. However, the masses cannot be accurately determined until a stress analysis is done to determine the strength of the assumed part. Thus an impasse exists that is broken only by iteration, i.e., assume a part geometry and define its geometric and mass properties, calculate the dynamic loads due in part to the material and geometry of the part. Then calculate the stresses and deflections resulting from those forces, find out it fails, redesign, and repeat.

An Integrated Approach

The text is divided into two parts. The first part presents the fundamentals of stress, strain, deflection, materials properties, failure theories, fatigue phenomena, fracture mechanics, etc. These theoretical aspects are presented in similar fashion to other texts. The second part presents treatments of specific, common design elements used as examples of applications of the theory but attempts to avoid presenting a string of disparate topics in favor of an integrated approach that ties the various topics together via *case studies*.

Most *Elements* texts contain many more topics and more content than can possibly be covered in a one-semester course. A survey questionnaire was sent to 200 U.S. university instructors of this course to solicit their opinions on the relative importance and desirability of the typical set of topics in an *Elements* text. The results from the 52 responses were analyzed and used to influence the structure and content of this book. One of the strongest desires expressed by the respondents was for *case studies* that present realistic design problems.

We have attempted to accomplish this goal by structuring the text around a series of nine case studies. These case studies present different aspects of the same design problem in successive chapters, for example, defining the static or dynamic loads on the device in Chapter 3, calculating the stresses due to the static loads in Chapter 4, and applying the appropriate failure theory to determine its safety factor in Chapter 5. Later chapters present more complex case studies, which have more design content. The case study in Chapter 6 on fatigue design is one such example.

The case studies provide a series of machine design projects throughout the book that contain various combinations of the elements normally dealt with in this type of text. The assemblies contain some collection of elements such as links subjected to combined axial and bending loads, column members, shafts in combined bending and torsion, gearsets under alternating loads, return springs, fasteners under fatigue loading, rolling element bearings, etc. This integrated approach has several advantages. It presents the student with a generic design problem in context rather than as a set of disparate, unrelated entities. The student can then see the interrelationships and the rationales for the design decisions that affect the individual elements. These more comprehensive case studies are in Part II of the text. The case studies in Part I are more limited in scope and directed to the engineering mechanics topics of the chapter. In addition to the case studies, each chapter has a selection of worked-out examples to reinforce particular topics.

Chapter 8, Design Case Studies, is devoted to the setup of three design case studies that are used in the following chapters to reinforce the concepts behind the design and analysis of shafts, springs, gears, fasteners, etc. Not all aspects of these design case studies are addressed as worked-out examples since another purpose is to provide material for student-project assignments. The author has used these case study topics as multi-week or term-long project assignments for groups or individual students with good success. Assigning open-ended project assignments serves to reinforce the design and analysis aspects of the course much better than set-piece homework assignments.

Many of the problem sets are integrated by being built upon in succeeding chapters. These linked problems have the same dash number in each chapter and are marked with a † to indicate their commonality among chapters. For example, Problem 3-4 asks for a static force analysis of a trailer hitch. Problem 4-4 requests a stress analysis of the same hitch based on the forces calculated in Problem 3-4. Problem 5-4 asks for the static safety factor for the hitch using the stresses calculated in Problem 4-4. Problem 6-4 requests a fatigue-failure analysis of the same hitch, and Problem 7-4 requires a surface stress analysis, etc. Thus the complexity of the underlying design problem is unfolded as new topics are introduced. An instructor who wishes to use this approach may want to assign problems with the same dash number in succeeding chapters.

Text Arrangement

Chapter 1 provides an introduction to the design process, problem formulation, safety factors, and units. An introduction to the use of *TKSolver* is also provided in the form of 10 simple examples. Material properties are reviewed in Chapter 2 since even the student who has had a first course in material science or metallurgy typically has but a superficial understanding of the wide spectrum of engineering material properties needed for machine design. Chapter 3 presents a review of static and dynamic loading analysis, including beam, vibration, and impact loading, and sets up a series of case studies that are used in later chapters to illustrate the stress and deflection analysis topics with some continuity.

The *Design of Machine Elements* course, at its core, is really an intermediate-level, applied stress-analysis course. Accordingly, a review of the fundamentals of stress and deflection analysis is presented in Chapter 4. Static failure theories are presented in detail in Chapter 5 since the students have typically not yet fully digested these concepts from their first stress-analysis course. Fracture-mechanics analysis for static loads is also introduced.

The *Elements* course is typically the student's first exposure to fatigue analysis since most introductory stress-analysis courses deal only with statically loaded problems. Accordingly, fatigue-failure theory is presented at length in Chapter 6 with the emphasis on stress-life approaches to high-cycle fatigue design, which is commonly used in the design of rotating machinery. Fracture-mechanics theory is further discussed with regard to crack propagation under cyclic loading. Strain-based methods for low-cycle fatigue analysis are not presented but their application and purpose are introduced to the reader and bibliographic references are provided for further study. Residual stresses are also addressed. Chapter 7 presents a thorough discussion of the phenomena of wear mechanisms, surface contact stresses, and surface fatigue. These seven chapters comprise Part I of the text and lay the analytical foundation needed for design of machine elements. They are arranged to be taken up in the order presented and build upon each other.

Part II of the text presents the design of machine elements in context as parts of a whole machine. The chapters in Part II are essentially independent of one another and can be taken (or skipped) in any order that the instructor desires (except that Chapter 11 on spur gears should be studied before Chapter 12 on helical, bevel, and worm gears). It is unlikely that all topics in the book can be covered in a one-term course. Chapter 8 presents a set of design case studies to be used as assignments and as example case studies in the following chapters. Chapter 8 also provides a set of suggested design project assignments in addition to the detailed case studies as described above. Chapter 9 investigates shaft design using the fatigue-analysis techniques developed in Chapter 6. Chapter 10 discusses fluid-film and rolling-element bearing theory and application. Chapter 11 gives a thorough introduction to the kinematics, design and stress analysis of spur gears using the latest AGMA recommended procedures. Chapter 12 extends gear design to helical, bevel, and worm gearing. Chapter 13 covers spring design including helical compression, extension and torsion springs, as well as a thorough treatment of Belleville springs. Chapter 14 deals with screws and fasteners including power screws and preloaded fasteners. Chapter 15 presents an introduction to the design and specification of disk and drum clutches and brakes. The appendices contain geometric formulas, extensive material-strength data, beam tables, and stress-concentration factors, as well as answers to selected problems and a list of included software.

The program *TKSolver* is used to accomplish most of the computational burden associated with the topics in this text. A special edition of the *TKSolver* program is included with each text, along with numerous *TKSolver* files that solve all the examples and case studies and some general problems. In addition, several custom-written programs for kinematic, dynamic, and stress analysis that do not use *TKSolver* are provided. Appendix H contains a list of included software. The fully capable student edition of *TKSolver* and its student manual is also available as an option, packaged with the text.

Supplements

A solutions manual is available that uses *TKSolver* to solve the problem sets in the text. Over 1100 *TKSolver* problem-solution files are made available with the solutions manual. This computerized approach to the solutions manual has significant advantages to the instructor as he or she can easily change any assigned problem's data and instantly solve it. Thus an essentially infinite supply of problem sets is available, going far beyond those defined in the text. The instructor also can easily prepare and solve exam problems by changing data in the supplied *TKSolver* files.

Acknowledgments

The author would like to express his sincere appreciation to all those who reviewed this text in its various stages of development including Professors J. E. Beard, Michigan Technological University; J. M. Henderson, University of California, Davis; L. R. Koval, University of Missouri at Rolla; S. N. Kramer, University of Toledo; L. D. Mitchell, Virginia Polytechnic Institute; G. R. Pennock, Purdue University; D. A. Wilson, Tennessee Technical University; Mr. John Lothrop; and Professor J. Ari-Gur, Western Michigan University, who also taught from a class-test version of the book. Robert Herrmann (WPI-ME '94) provided some problems and Charles Gillis (WPI-ME '96) solved most of the problem sets. Professors John R. Steffen of Valparaiso University, R. Jay Conant of Montana State University, Norman E. Dowling of Virginia Polytechnic, and Francis E. Kennedy of Dartmouth College made many useful suggestions for improvement and caught many errors. Special thanks go to Professor Hartley T. Grandin of WPI, who provided much encouragement and many good suggestions and ideas throughout the book's gestation, and also taught from the various class-test versions.

Two (former and present) Prentice-Hall editors deserve special mention for their efforts in developing this book: Doug Humphrey, who wouldn't take no for an answer in persuading me to write it, and Bill Stenquist who usually said yes to my requests and expertly shepherded the book through to completion. Finally, my infinitely patient wife, Nancy, deserves kudos for her unfailing support and encouragement during four summers of "book widowhood." The author takes full responsibility for all errors that remain and will appreciate them being called to his attention so that the book can be improved in future printings and editions.

Robert L. Norton,
Mattapoisett, Mass.
August 1, 1995

FUNDAMENTALS

1 INTRODUCTION TO DESIGN

Learning without thought is labour lost;
thought without learning is perilous.
CONFUCIUS, 6TH CENTURY B.C.

1.1   DESIGN

What is design? Wallpaper is designed. You may be wearing "designer" clothes. Automobiles are "designed" in terms of their external appearance. The term *design* clearly encompasses a wide range of meaning. In the above examples, design refers primarily to the object's aesthetic appearance. In the case of the automobile, all of its other aspects also involve design. Its mechanical internals (engine, brakes, suspension, etc.) must be designed, more likely by engineers than by artists, though even the engineer gets to exhibit some artistry when designing machinery.

The word design is from the Latin word *designare* meaning, "*to designate, or mark out..*" Webster's dictionary gives several definitions of the word **design**, the most applicable of which is "*to outline, plot, or plan as action or work . . . to conceive, invent, contrive.*" We are more concerned here with engineering design than with artistic design. **Engineering design** can be defined as "*The process of applying the various techniques and scientific principles for the purpose of defining a device, a process or a system in sufficient detail to permit it's realization.*"

Machine Design

This text is concerned with one aspect of engineering design   **machine design**. Machine design deals with the creation of machinery that works safely, reliably, and well. A **machine** can be defined in many ways. The Random House dictionary[1] lists twelve definitions, among which are these two:

Machine

1. An apparatus consisting of interrelated units, or
2. A device that modifies force or motion.

The **interrelated parts** referred to in the definition are also sometimes called **machine elements** in this context. The notion of **useful work** is basic to a machine's function as there is almost always some energy transfer involved. The mention of **forces** and **motion** is also critical to our concerns as, in converting energy from one form to another, machines **create motion** and **develop forces**. It is the engineer's task to define and calculate those motions, forces, and changes in energy in order to determine the sizes, shapes, and materials needed for each of the interrelated parts in the machine. This is the essence of **machine design**.

While one must, of necessity, design a machine one part at a time, it is crucial to recognize that each part's function and performance (and thus its design) are dependent on many other interrelated parts within the same machine. Thus, we are going to attempt to "design the whole machine" here, rather than simply designing individual elements in isolation from one another. To do this we must draw upon a common body of engineering knowledge encountered in previous courses, e.g., statics, dynamics, mechanics of materials (stress analysis) and material properties. Brief reviews and examples of these topics are included in the early chapters of this book.

The ultimate goal in machine design is to size and shape the parts (machine elements) and choose appropriate materials and manufacturing processes so that the resulting machine can be expected to perform its intended function without failure. This requires that the engineer be able to calculate and predict the mode and conditions of failure for each element and then design it to prevent that failure. This requires that a **stress and deflection analysis** be done for each part. Since stresses are a function of the applied and inertial loads, and of the part's geometry, an analysis of the forces, moments, torques, and the dynamics of the system must be done before the stresses and deflections can be completely calculated.

If the "machine" in question has no moving parts, then the design task becomes much simpler because only a static force analysis is required. But if the machine has no moving parts, it is not much of a machine (and doesn't meet the definition above); it is then a **structure**. Structures also need to be designed against failure, and, in fact, large external structures (bridges, buildings, etc.) are also subjected to dynamic loads from wind, earthquakes, traffic, etc., and thus must also be designed for these conditions. Structural dynamics is an interesting subject but one which we will not address in this text. We will concern ourselves with the problems associated with machines that move. If the machine's motions are very slow and the accelerations negligible, then a static force analysis will suffice. But if the machine has significant accelerations within it, then a dynamic force analysis is needed and the accelerating parts become "victims of their own mass."

Title-page photograph courtesy of Boeing Airplane Co. Inc., Seattle, Wash.

In a static structure, such as a building's floor, designed to support a particular weight, the safety factor of the structure can be increased by adding appropriately distributed material to its structural parts. Though it will be heavier (more "dead" weight), if properly designed it may nevertheless carry more "live" weight (payload) than it did before, still without failure. In a dynamic machine, adding weight (mass) to moving parts may have the opposite effect, reducing the machine's safety factor, its allowable speed, or its payload capacity. This is because some of the loading that creates stresses in the moving parts is due to the inertial forces predicted by **Newton's second law**, $F = ma$. Since the accelerations of the moving parts in the machine are dictated by its kinematic design and by its running speed, adding mass to moving parts will increase the inertial loads on those same parts unless their kinematic accelerations are reduced by slowing its operation. Even though the added mass may increase the strength of the part, that benefit may be reduced or cancelled by the resultant increases in inertial forces.

Iteration

Thus, we face a dilemma at the initial stages of machine design. Generally, before reaching the stage of sizing the parts, the kinematic motions of the machine will have already been defined. External forces provided by the "outside world" on the machine are also often known. Note that in some cases, the external loads on the machine will be very difficult to predict, for example, the loads on a moving automobile. The designer cannot predict with accuracy what environmental loads the user will subject the machine to (potholes, hard cornering, etc.) In such cases, statistical analysis of empirical data gathered from actual testing can provide some information for design purposes.

What remain to be defined are the inertial forces that will be generated by the known kinematic accelerations acting on the as yet undefined masses of the moving parts. The dilemma can be resolved only by **iteration**, which means *to repeat, or to return to a previous state*. We must assume some trial configuration for each part, use the mass properties (mass, *CG* location, and mass moment of inertia) of that trial configuration in a dynamic force analysis to determine the forces, moments, and torques acting on the part, and then use the cross-sectional geometry of the trial design to calculate the resulting stresses. In general, accurately determining all the loads on a machine is the most difficult task in the design process. If the loads are known, the stresses can be calculated.

Most likely, on the first trial, we will find that our design fails because the materials cannot stand the levels of stress presented. We must then redesign the parts (iterate) by changing shapes, sizes, materials, manufacturing processes, or other factors in order to reach an acceptable design. It is generally not possible to achieve a successful result without making several iterations through this design process. Note also that a change to the mass of one part will also affect the forces applied to parts connected to it and thus require their redesign also. It is truly the design of **interrelated parts**.

TABLE 1-1   A Design Process

1	Identification of need
2	Background research
3	Goal statement
4	Task specifications
5	Synthesis
6	Analysis
7	Selection
8	Detailed design
9	Prototyping and testing
10	Production

1.2   A DESIGN PROCESS*

The process of design is essentially an exercise in applied creativity. Various "design processes" have been defined to help organize the attack upon the "unstructured problem," i.e., one for which the problem definition is vague and for which many possible solutions exist. Some of these design process definitions contain only a few steps and others a detailed list of 25 steps. One version of a design process is shown in Table 1-1, which lists ten steps. The initial step, **Identification of Need** usually consists of an ill-defined and vague problem statement. The development of **Background Research** information (step 2) is necessary to fully define and understand the problem, after which it is possible to restate the **Goal** (step 3) in a more reasonable and realistic way than in the original problem statement.

Step 4 calls for the creation of a detailed set of **Task Specifications** which bound the problem and limit its scope. The **Synthesis** step (5) is one in which as many alternative possible design approaches are sought, usually without regard (at this stage) for their value or quality. This is also sometimes called the **ideation and invention** step in which the largest possible number of creative solutions is generated.

In step 6, the possible solutions from the previous step are **Analyzed** and either accepted, rejected or modified. The most promising solution is **Selected** at Step 7. Once an acceptable design is selected, the **Detailed Design** (step 8) can be done in which all the loose ends are tied up, complete engineering drawings made, vendors identified, manufacturing specifications defined, etc. The actual construction of the working design is first done as a **Prototype** in step 9 and finally in quantity in **Production** at step 10. A more complete discussion of this design process can be found in reference 2, and a number of references on the topics of creativity and design are provided in the bibliography at the end of this chapter.

---

*   Adapted from Norton, *Design of Machinery*, McGraw-Hill, New York, 1992, with the publisher's permission.

The above description may give an erroneous impression that this process can be accomplished in a linear fashion as listed. On the contrary, **iteration is required within the entire process**, moving from any step back to any previous step, in all possible combinations, and doing this repeatedly. The best ideas generated at step 5 will invariably be discovered to be flawed when later analyzed. Thus a return to at least the Ideation step will be necessary in order to generate more solutions. Perhaps a return to the Background Research phase may be necessary to gather more information. The Task Specifications may need to be revised if it turns out that they were unrealistic. In other words, anything is "fair game" in the design process, including a redefinition of the problem, if necessary. One cannot design in a linear fashion. It's three steps forward and two (or more) back, until you finally emerge with a working solution.

Theoretically, we could continue this iteration on a given design problem forever, continually creating small improvements. Inevitably, the incremental gains in function or reductions in cost will tend toward zero with time. At some point, we must declare the design "good enough" and ship it. Often someone else (most likely, the boss) will snatch it from our grasp and ship it over our protests that it isn't yet "perfect." Machines that have been around a long time and that have been improved by many designers, reach a level of "perfection" that makes them difficult to improve upon. One example is the ordinary bicycle. Though inventors continue to attempt to improve this machine, the basic design has become fairly static after more than a century of development.

In machine design, the early design process steps usually involve the **Type Synthesis** of suitable kinematic configurations which can provide the necessary motions. Type synthesis involves the choice of the *type of mechanism best suited to the problem*. This is a difficult task for the student as it requires some experience and knowledge of the various types of mechanisms that exist and that might be feasible from a performance and manufacturing standpoint. As an example, assume that the task is to design a device to track the constant-speed, straight-line motion of a part on a conveyor belt, and attach a second part to it as it passes by. This has to be done with good accuracy and repeatability and must be reliable and inexpensive. You might not be aware that this task could be accomplished by any of the following devices:

- a straight-line linkage
- a cam and follower
- an air cylinder
- a hydraulic cylinder
- a robot
- a solenoid

Each of these solutions, while possible, may not be optimal or even practical. Each has good and bad points. The straight-line linkage is large and may have undesirable accelerations, the cam and follower is expensive, but is accurate and repeatable. The air cylinder is inexpensive, but noisy and unreliable. The hydraulic cylinder and the robot are more expensive. The inexpensive solenoid has high impact loads and velocities. So, the choice of device type can have a big effect on design quality. A bad choice at the type synthesis stage can create major problems later on. The design might have to be changed after completion, at great expense. Design is essentially an exercise in trade-offs. There is usually no clear-cut solution to a real engineering design problem.

Once the type of required mechanism is defined, its detailed kinematics must be synthesized and analyzed. The motions of all moving parts and their time derivatives through acceleration must be calculated in order to be able to determine the dynamic forces on the system. (See reference 2 for more information on this aspect of machine design.)

In the context of machine design addressed in this text, we will not exercise the entire design process as described in Table 1-1. Rather, we will propose examples, problems, and case studies that already have had steps 1-4 defined. The type synthesis and kinematic analysis will already be done, or at least set up, and the problems will be structured to that degree. The tasks remaining will largely involve steps 5 through 8, with a concentration on **synthesis** (step 5) and **analysis** (step 6).

Synthesis and analysis are the "two faces" of machine design, like two sides of the same coin. **Synthesis** means *putting together* and **analysis** means *to decompose, to take apart, to resolve into its constituent parts.* Thus they are opposites, but they are symbiotic. We cannot take apart "nothing," thus we must first synthesize something in order to analyze it. When we analyze it we will probably find it lacking, requiring further synthesis, and then further analysis *ad nauseam*, finally iterating to a better solution. You will need to draw heavily upon your understanding of statics, dynamics, and mechanics of materials to accomplish this.

1.3   PROBLEM FORMULATION AND CALCULATION

It is extremely important for every engineer to develop good and careful computational habits. Solving complicated problems requires an organized approach. Design problems also require good record keeping and documentation habits in order to record the many assumptions and design decisions made along the way so that the designer's thought process can be later reconstructed if redesign is necessary.

A suggested procedure for the designer is shown in Table 1-2, which lists a set of subtasks appropriate to most machine-design problems of this type. These steps should be documented for each problem in a neat fashion, preferably in a bound notebook in order to maintain their chronological order.*

---

* If there is a possibility of a patentable invention resulting from the design, then the notebook should be permanently bound (not loose-leaf) and its pages should be consecutively numbered, dated, and witnessed by someone who understands the technical content.

Table 1-2   Problem Formulation and Calculation

1	Define the problem	
2	State the givens	— Definition stage
3	Make appropriate assumptions	
4	Preliminary design decisions	
5	Design sketches	— Preliminary design stage
6	Mathematical models	
7	Analysis of the design	— Detailed design stage
8	Evaluation	
9	Document results	— Documentation stage

Definition Stage

In your design notebook, first **Define the Problem** clearly in a concise statement. The "**givens**" for the particular task should be clearly listed, followed by a record of the **assumptions** made by the designer about the problem. Assumptions expand upon the given (known) information to further constrain the problem. For example, one might assume the effects of friction to be negligible in a particular case, or assume that the weight of the part can be ignored because it will be small compared to the applied or dynamic loads expected.

Preliminary Design Stage

Once the general constraints are defined, some **Preliminary Design Decisions** must be made in order to proceed. The reasons and justifications for these decisions should be documented. For example, we might decide to try a solid, rectangular cross section for a connecting link, and choose aluminum as a trial material. On the other hand, if we recognized from our understanding of the problem that this link would be subjected to significant accelerations of a time-varying nature that would repeat for millions of cycles, a better design decision might be to use a hollow or I-beam section in order to reduce its mass and also to choose steel for its infinite fatigue life. Thus, these design decisions can have significant effect on the results and will often have to be changed or abandoned as we iterate through the design process. It has often been noted that 90% of a design's characteristics may be determined in the first 10% of the total project time, during which these preliminary design decisions are made. If they are bad decisions, it may not be possible to save the bad design through later modifications without essentially starting over. The preliminary design concept should be documented at this stage with clearly drawn and labeled **Design Sketches** that will be understandable to another engineer or even to oneself after some time has passed.

Detailed Design Stage

With a tentative design direction established we can create one or more **engineering** (mathematical) **models** of the element or system in order to analyze it. These models will usually include a loading model consisting of free-body diagrams which show all forces, moments, and torques on the element or system and the appropriate equations for their calculation. Models of the stress and deflection states expected at locations of anticipated failure are then defined with appropriate stress and deflection equations.

Analysis of the design is then done using these models and the safety or failure of the design determined. The results are **evaluated** in conjunction with the properties of the chosen **engineering materials** and a decision made whether to proceed with this design or iterate to a better solution by returning to an earlier step of the process.

Documentation Stage

Once sufficient iteration through this process provides satisfactory results, the **documentation** of the element's or system's design should be completed in the form of detailed engineering drawings, material and manufacturing specifications, etc. If properly approached, a great deal of the documentation task can be accomplished concurrent with the earlier stages simply by keeping accurate and neat records of all assumptions, computations, and design decisions made throughout the process.

1.4   THE ENGINEERING MODEL

The success of any design is highly dependent on the validity and appropriateness of the engineering models used to predict and analyze its behavior in advance of building any hardware. Creating a useful engineering model of a design is probably the most difficult and challenging part of the whole process. Its success depends a great deal on experience as well as skill. Most important is a thorough understanding of the first principals and fundamentals of engineering. The engineering model that we are describing here is an amorphous thing which may consist of some sketches of the geometric configuration and some equations that describe its behavior. It is a mathematical model that describes the physical behavior of the system. This engineering model invariably requires the use of computers to exercise it. Using computer tools for analyzing engineering models is discussed in the next section. A physical model or prototype usually comes later in the process and is necessary to prove the validity of the engineering model through experiments.

Estimation and First-Order Analysis

The value of making even very simplistic engineering models of your preliminary designs cannot be overemphasized. Often, at the outset of a design, the problem is so loosely and poorly defined that it is difficult to develop a comprehensive and thorough model in the form of equations that fully describe the system. The engineering student is used to problems that are fully structured, of a form such as *"Given A, B, and C, find D."* If one can identify the appropriate equations (model) to apply to such a problem, it is relatively easy to determine an answer (which might even match the one in the back of the book).

Real-life engineering design problems are not of this type. They are very **unstructured** and must be structured by you before they can be solved. Also, there is no *"back of the book"* to refer to for the answer.* This situation makes most students and beginning engineers very nervous. They face the "blank paper syndrome," not knowing where to begin. A useful strategy is to recognize that

1   You must begin somewhere.

2   Wherever you begin, it will probably not be the "best" place to do so.

3   The magic of iteration will allow you to back up, improve your design, and eventually succeed.

With this strategy in mind, you can feel free to make some estimation of a design configuration at the outset, assume whatever limiting conditions you think appropriate, and do a "first-order analysis," one that will be only an estimate of the system's behavior. These results will allow you to identify ways to improve the design. Remember that it is preferable to get a reasonably approximate but quick answer that tells you whether the design does or doesn't work than to spend more time getting the same result to more decimal places. With each succeeding iteration, you will improve your understanding of the problem, the accuracy of your assumptions, the complexity of your model, and the quality of your design decisions. Eventually, you will be able to refine your model to include all relevant factors (or identify them as irrelevant) and obtain a higher-order, final analysis in which you have more confidence.

The Engineering Sketch

A sketch of the concept is often the starting point for a design. This may be a freehand sketch, but should always be reasonably to scale in order to show realistic geometric proportions. This sketch often serves the primary purpose of communicating the concept to other engineers and even to yourself. It is one thing to have a vague concept in mind and quite another to define it in a sketch. This sketch should, at a minimum, contain three or more orthographic views, aligned according to proper drafting convention, and may also include an isometric or trimetric view. Figure 1-1 shows a freehand sketch of a simple design for one subassembly of a trailer hitch for a tractor. While often incomplete in terms of detail needed for manufacture, the engineering sketch should contain enough information to allow the development of an engineering model for design and analysis. This may include critical, if approximate, dimensional information, some material assumptions and any other data germane to its function that is needed for further analysis. The engineering sketch captures some of the givens and assumptions made, even implicitly, at the outset of the design process.

---

*   A student once commented that *"Life is an odd-numbered problem."* This (slow) author had to ask for an explanation, which was: *"The answer is not in the back of the book."*

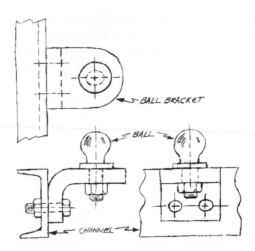

FIGURE 1-1

A Freehand Sketch of a Trailer Hitch Assembly for a Tractor

1.5   COMPUTER-AIDED DESIGN AND ENGINEERING

The computer has created a true revolution in engineering design and analysis. Problems whose solution methods have literally been known for centuries but that only a generation ago were practically unsolvable due to their high computational demands can now be solved in minutes on inexpensive microcomputers. Tedious graphical solution methods were developed in the past to circumvent the lack of computational power available from slide rules. Some of these graphical solution methods still have value in that they can show the results in an understandable form. But one can no longer "do engineering" without using its latest and most powerful tool, the computer.

Computer-Aided Design (CAD)

As the design progresses, the crude freehand sketches made at the earliest stages will be supplanted by formal drawings made either with conventional drafting equipment or, as is increasingly common, with computer-aided design or computer-aided drafting software. If the distinction between these two terms (both of which share the acronym CAD) was ever clear (a subject for debate which will be avoided here), then that distinction is fading as more sophisticated CAD software becomes available. The original CAD systems of a generation ago were essentially drafting tools that allowed the creation of computer-generated multiview drawings similar to those done for centuries before by hand on a drafting board. The data stored in these early CAD systems were strictly two-dimensional representations of the orthographic projections of the part's true 3-D geometry. Only the edges of the part are defined in the database. This is called a **wireframe model**. Some 3-D CAD packages use wireframe representation as well.

Present versions of most CAD software packages allow (and sometimes require) that the geometry of the parts be encoded in a 3-D data base as **solid models**. In a solid model the edges and the faces of the part are defined. From this 3-D information, the conventional 2-D orthographic views can be automatically generated if desired. The major advantage of creating a 3-D solid-model geometric data base for any design is that its mass-property information can be rapidly calculated. (This is not possible in a 2-D or 3-D wireframe model.) For example, in designing a machine part, we need to determine the location of its center of gravity (CG), its mass, its mass moment of inertia and its cross-sectional geometries at various locations. Determining this information from a 2-D model must be done outside the CAD package. That is tedious to do and can only be approximate when the geometry is complex. But, if the part is designed in a solids-modeling CAD system such as *CadKey,*[3], *Aries,* [4] or one of many others, the mass properties can be instantly calculated for the most complicated part geometries.

Solid modeling systems usually provide an interface to one or more Finite Element Analysis (FEA) programs and allow direct transfer of the model's geometry to the FEA package for stress, vibration, and heat transfer analysis. Some CAD systems include a mesh-generation feature which creates the FEA mesh automatically before sending the data to the FEA software. This combination of tools provides an extremely powerful means to obtain superior designs whose stresses are more accurately known than would be possible by conventional analysis techniques when the geometry is complex.

While it is highly likely that the students reading this textbook will be using CAD tools including finite element or boundary element analysis (BEA) methods in their professional practice, it is still necessary that the fundamentals of applied stress analysis be thoroughly understood. That is the purpose of this text. FEA techniques will be discussed briefly in Chapter 4 but will not be used in this text. Rather we will concentrate on the classical stress analysis techniques in order to lay the foundation for a thorough understanding of the fundamentals and their application to machine design.

FEA and BEA methods are rapidly becoming the methods of choice for the solution of complicated stress analysis problems. However, there is danger in using those techniques without a solid understanding of the theory behind them. These methods will always give *some* results. Unfortunately, those results can be incorrect if the problem was not well formulated and well meshed. Being able to recognize incorrect results from a computer-aided solution is extremely important to the success of any design. The student may want to take courses in FEA and BEA to become familiar with these tools.

Figure 1-2 shows a solid model of the ball bracket from Figure 1-1 that was created in the *Aries Solid modeling* CAD software package. The shaded, isometric view in the upper right corner shows that the solid volume of the part is defined. The other three views show orthographic projections of the part. Figure 1-3 shows the mass properties data which are calculated by the *Aries* software. Figure 1-4 shows a wireframe rendering of the same part generated from the solids geometry data base. A wireframe version is used principally to speed up the screen-drawing time when working on the model. There is much less wireframe display information to calculate than for the solid rendering of Figure 1-2.

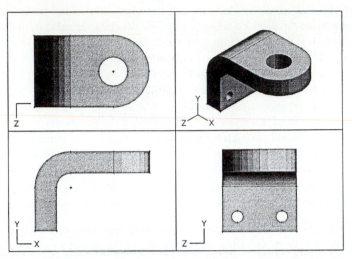

FIGURE 1-2

A CAD Solid Model of the Ball Bracket from the Trailer Hitch Assembly of Figure 1-1

Figure 1-5 shows a fully dimensioned, orthographic, multiview drawing of the ball bracket that was generated in the *Computervision CADDS5* software package. Another major advantage of creating a solid model of a part is that the dimensional and tool-path information needed for its manufacture can be generated in the CAD system and sent

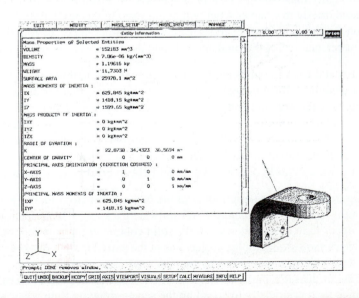

FIGURE 1-3

Mass Properties of the Ball Bracket Calculated Within the *Aries* CAD System from its Solid Model

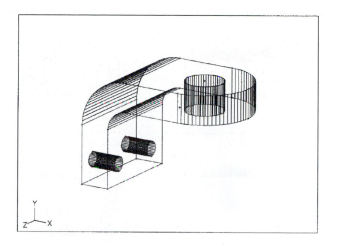

FIGURE 1-4

A Wireframe Representation of the Ball Bracket Generated from its Solid Model in the *Aries* CAD

over a network to a computer-controlled machine on the manufacturing floor. This feature allows the production of parts without the need for paper drawings such as Figure 1-5. Figure 1-6 shows the same part after a finite element mesh was applied to it by the *Aries* software before sending it to the FEA software for stress analysis.

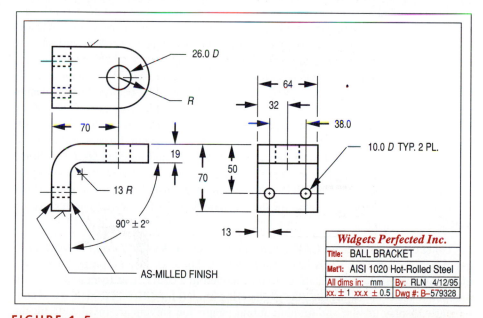

FIGURE 1-5

A Dimensioned, 3-View Orthographic Drawing Done in a 2-D CAD Drawing Package

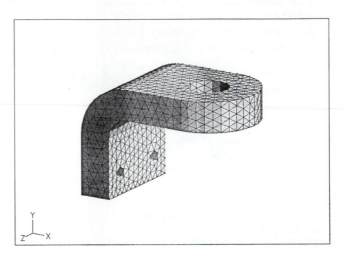

FIGURE 1-6

An FEM Mesh Applied to the Solid Model of the Ball Bracket in the *Aries* CAD System

Computer-Aided Engineering (CAE)

The techniques generally referred to above as CAD are a subset of the more general topic of computer-aided engineering (CAE), which term implies that more than just the geometry of the parts is being dealt with. However, the distinctions between CAD and CAE continue to blur as more sophisticated software packages become available, In fact, the description of the use of a solid modeling CAD system and an FEA package together as described in the previous section is an example of CAE. When some analysis of forces, stresses, deflections, or other aspects of the physical behavior of the design is included, with or without the solid geometry aspects, the process is called CAE. Many commercial software packages do one or more aspects of CAE. The FEA and BEA software packages mentioned above are in this category. Dynamic force simulations of mechanisms can be done with such packages as *ADAMS*[5] and *Working Model.*[6] Some software packages such as *ProEngineer*[7] combine aspects of a CAD system with general analysis capabilities. These constraint-based programs allow constraints to be applied to the design which can control the part geometry as the design parameters are varied.

Other classes of tools for CAE are equation solvers such as *TKSolver,*[8] *MathCad,*[9] and spreadsheets such as *Lotus 123*[10] or *Excel*[11]. These are general-purpose tools that allow any combination of equations to be encoded in a convenient form and then will manipulate the equation set (i.e., the engineering model) for different trial data and conveniently display tabular and graphic output. Equation solvers are invaluable for the solution of force, stress, and deflection equations in machine design problems because they allow rapid "what-if" calculations to be done. The effects of dimensional or material changes on the stresses and deflections in the part can be seen instantly. In the absence of a true solid modeling system, an equation solver also can be used to approximate the part's mass properties while iterating the geometry and material properties of trial part designs. Rapid iteration to an acceptable solution is thus enhanced.

We make extensive use of these software tools in this text. The CD-ROM included with the text contains the program *TKSolver* and a large number of *TK* data files that support the examples and case studies presented in the text. A brief introduction to *TKSolver* is provided in a later section of this chapter along with some examples of its use. In addition, some custom-written computer programs are provided on the CD-ROM to aid in the calculation of dynamic loads when solving the open-ended design problems assigned. See Appendix H for a listing of the computer programs and *TKSolver* files supplied with this text.

However, one must be aware that these computer tools are just tools and are not a substitute for the human brain. Without a thorough understanding of the engineering fundamentals on the part of the user, the computer will not give good results. Garbage in, garbage out. *Caveat Lector*.

Computational Accuracy

Computers and calculators make it very easy to obtain numerical answers having many significant figures. Before writing down all those digits, you are advised to recall the accuracy of your initial assumptions and given data. If, for example, your applied loads were known to only two significant figures, it is incorrect and misleading to express the calculated stresses to more significant figures than your input data possessed. However, it is valid and appropriate to make all intermediate calculations to the greatest accuracy available in your computational tools. This will minimize computational round-off errors. But, when done, round off the results to a level consistent with your known or assumed data.

1.6   THE ENGINEERING REPORT[*]

Communication of your ideas and results is a very important aspect of engineering. Many engineering students picture themselves in professional practice spending most of their time doing calculations of a nature similar to those they have done as students. Fortunately, this is seldom the case, as it would be very boring. Actually, engineers spend the largest percentage of their time communicating with others, either orally or in writing. Engineers write proposals and technical reports, give presentations, and interact with support personnel. When your design is done, it is usually necessary to present the results to your client, peers, or employer. The usual form of presentation is a formal engineering report. In addition to a written description of the design, these reports will usually contain engineering drawings or sketches as described earlier, as well as tables and graphs of data calculated from the engineering model.

---

[*]   Excerpted from Norton, *Design of Machinery*, McGraw-Hill, New York, 1992, with the publisher's permission.

It is very important for engineering students to develop their communication skills. *You may be the cleverest person in the world, but no one will know that if you cannot communicate your ideas clearly and concisely.* In fact, if you cannot explain what you have done, you probably don't understand it yourself. The design project assignments in Chapter 8 are intended to be written up in formal engineering reports to give you some experience in this important skill of technical communication, Information on writing engineering reports can be found in the suggested readings listed in the bibliography.

1.7   FACTORS OF SAFETY AND DESIGN CODES

The quality of a design can be measured by many criteria. It is always necessary to calculate one or more factors of safety to estimate the likelihood of failure. There may be legislated, or generally accepted, design codes which must be adhered to as well.

Factor of Safety[†]

A factor of safety or safety factor can be expressed in many ways. It is typically a ratio of two quantities that have the same units, such as strength/stress, critical load/applied load, load to fail part/expected service overload, maximum cycles/applied cycles, or maximum safe speed/operating speed. A safety factor is always unitless.

The form of expression for a safety factor can usually be chosen based on the character of loading on the part. For example, consider the loading on the side wall of a cylindrical water tower that can never be "more than full" of a liquid of known density within a known temperature range. Since this loading is highly predictable over time, a comparison of the strength of the material to the stress in the wall of a full tank might be appropriate as a safety factor. Note in this example that the possibility of rust reducing the thickness of the wall over time must be considered. (See Section 4.16 for a discussion of stresses in cylinder walls and Section 7.7 for a discussion of corrosion.)

If this cylindrical water tower is standing on legs loaded as columns, then a safety factor for the legs based on a ratio of the column's critical buckling load over the applied load from a full water tower would be appropriate. (See section 4.15 for a discussion of column buckling.)

If a part is subjected to loading that varies cyclically with time, it may experience fatigue failure. The resistance of a material to some types of fatigue loading can be expressed as a maximum number of cycles of stress reversal at a given stress level. In such cases, it may be appropriate to express the safety factor as a ratio of the maximum number of cycles to expected material failure over the number of cycles applied to the part in service for its desired life. (See Chapter 6 for a discussion of fatigue failure phenomena and several approaches to the calculation of safety factors in such situations.)

[†] Also called *safety factor* . We will use both terms interchangeably in this text.

The safety factor of a part such as a rotating sheave (pulley) or flywheel is often expressed as a ratio of its maximum safe speed over the highest expected speed in service. In general, if the stresses in the parts are a linear function of the applied service loads and those loads are predictable, then a safety factor expressed as strength/stress or failure load/applied load will give the same result. Not all situations fit these criteria. A column does not because its stresses are a nonlinear function of the loading (see Section 4.15). Thus a critical (failure) load for the particular column must be calculated for comparison to the applied load.

Another complicating factor is introduced when the magnitudes of the expected applied loads are not accurately predictable. This can be true in virtually any application in which the use (and thus the loading) of the part or device is controlled by humans. For example, there is really no way to prevent someone from attempting to lift a 10-ton truck with a jack designed to lift a 2-ton automobile. When the jack fails, the manufacturer (and designer) may be blamed even though the failure was probably due more to the "nut behind the jack handle." In situations where the user may subject the device to overloading conditions, an assumed overload may have to be used to calculate a safety factor based on a ratio of the load that causes failure over the assumed service overload. Labels warning against inappropriate use may be needed in these situations as well.

Since there may be more than one potential mode of failure for any machine element, it can have more than one value of safety factor N. The smallest value of N for any part is of greatest concern since it predicts the most likely mode of failure. When N becomes reduced to 1, the stress in the part is equal to the strength of the material (or the applied load is equal to the load that fails it, etc.) and failure occurs. Therefore, we desire N to be always greater than 1.

Choosing a Safety Factor

Choosing a safety factor is often a confusing proposition for the beginning designer. The safety factor can be thought of as a measure of the designer's uncertainty in the analytical models, failure theories, and material-property data used, and should be chosen accordingly. How much greater than one N must be depends on many factors, including our level of confidence in the model on which the calculations are based, our knowledge of the range of possible in-service loading conditions, and our confidence in the available material strength information. If we have done extensive testing on physical prototypes of our design to prove the validity of our engineering model and of the design, and have generated test data on the particular material's strengths, then we can afford to use a smaller safety factor. If our model is less well proven or the material property information is less reliable, a larger N is in order. In the absence of any design codes that may specify N for particular cases, the choice of factor of safety involves engineering judgment. A reasonable approach is to determine the largest loads expected in service (including possible overloads) and the minimum expected material strengths and base the safety factors on these data. The safety factor then becomes a reasonable measure of uncertainty.

If you fly, it may not give you great comfort to know that safety factors for commercial aircraft are in the range of 1.2 to 1.5. Military aircraft can have $N < 1.1$, but their crews wear parachutes. (Test pilots deserve their high salaries.) Missiles have $N = 1$ but have no crew and aren't expected to return anyway. These small factors of safety in aircraft are necessary to keep weight low and are justified by sophisticated analytical modeling (usually involving FEA), testing of the actual materials used, extensive testing of prototype designs, and rigorous in-service inspections for incipient failures of the equipment. The opening photograph of this chapter shows an elaborate test rig used by the Boeing Aircraft Co. to mechanically test the airframe of full-scale prototype or production aircraft by applying dynamic forces and measuring their effects.

It can be difficult to predict the kinds of loads that an assembly will experience in service, especially if those loads are under the control of the end-user, or Mother Nature. For example, what loads will the wheel and frame of a bicycle experience? It depends greatly on the age, weight, and recklessness of the rider, whether used on- or off-road, etc. The same problem of load uncertainty exists with all transportation equipment, ships, aircraft, automobiles, etc. Manufacturers of these devices engage in extensive test programs to measure typical service loads. See Figures 3-16 and 6-7 for examples of such service-load data.

Some guidelines for the choice of a safety factor in machine design can be defined based on the quality and appropriateness of the material-property data available, the expected environmental conditions compared to those under which the material test data were obtained, and the accuracy of the loading and stress analysis models developed for the analyses. Table 1-3 shows a set of factors for ductile materials which can be chosen in each of the three categories listed based on the designer's knowledge or judgment of the quality of information used. The overall safety factor is then taken as the largest of the three factors chosen. Given the uncertainties involved, a safety factor typically should not be taken to more than 1 decimal place accuracy.

$$N_{ductile} \cong MAX(F1, F2, F3) \qquad (1.1a)$$

The ductility or brittleness of the material is also a concern. Brittle materials are designed against the ultimate strength, so failure means fracture. Ductile materials under static loads are designed against the yield strength and are expected to give some visible warning of failure before fracture unless cracks indicate the possibility of a fracture-mechanics failure (see Sections 5-3 and 6-3). For these reasons, the safety factor for brittle materials is often made twice that which would be used for a ductile material in the same situation:

$$N_{brittle} \cong 2 * MAX(F1, F2, F3) \qquad (1.1b)$$

This method of determining a safety factor is only a guideline to obtain a starting point and is obviously subject to the judgment of the designer in selecting fac-

Table 1-3   Factors Used to Determine a Safety Factor for Ductile Materials

Information	Quality of Information	Factor
		F1
	The actual material used was tested	1.3
Material property data	Representative material test data are available	2
available from tests	Fairly representative material test data are available	3
	Poorly representative material test data are available	5+
		F2
	Are identical to material test conditions	1.3
Environmental conditions	Essentially room-ambient environment	2
in which it will be used	Moderately challenging environment	3
	Extremely challenging environment	5+
		F3
	Models have been tested against experiments	1.3
Analytical models for	Models accurately represent system	2
loading and stress	Models approximately represent system	3
	Models are crude approximations	5+

tors in each category. **The designer has the ultimate responsibility to ensure that the design is safe. A larger safety factor than any shown in Table 1-3 may be appropriate in some circumstances.**

Design and Safety Codes

Many engineering societies and government agencies have developed codes for specific areas of engineering design. Most are only recommendations, but some have the force of law. The ASME provides recommended guidelines for safety factors to be used in particular applications such as steam boilers and pressure vessels. Building codes are legislated in most U.S.A states and cities and usually deal with publicly accessible structures or their components, such as elevators and escalators. Safety factors are sometimes specified in these codes and may be quite high. (The code for escalators in one state called for a factor of safety of 14.) Clearly, where human safety is involved, high values of N are justified. However, they come with a weight and cost penalty, as parts must often be made heavier to achieve large values of N. The design engineer must always be aware of these codes and standards and adhere to them where applicable.

The following is a partial list of engineering societies and governmental, industrial, and international organizations that publish standards and codes of potential interest to the mechanical engineer. Addresses and data on their publications can be obtained in any technical library.

American Gear Manufacturers Association (AGMA)

American Institute of Steel Construction (AISC)

American Iron and Steel Institute (AISI)

American National Standards Institute (ANSI)

American Society for Metals (ASM)

American Society of Mechanical Engineers (ASME)

American Society of Testing and Materials (ASTM)

American Welding Society (AWS)

Anti-Friction Bearing Manufacturers Association (AFBMA)

International Standards Organization (ISO)

National Institute for Standards and Technology (NIST)[*]

Society of Automotive Engineers (SAE)

Society of Plastics Engineers (SPE)

Underwriters Laboratories (UL)

1.8   STATISTICAL CONSIDERATIONS

Nothing is absolute in engineering any more than in any other endeavor. The strengths of materials will vary from sample to sample. The actual size of different examples of the "same" part made in quantity will vary due to manufacturing tolerances. As a result, we should take the statistical distributions of these properties into account in our calculations. The published data on the strengths of materials may be stated either as minimum values or as average values of tests made on many specimens. If it is an average value, there is a 50% chance that a randomly chosen sample of that material will be weaker or stronger than the published average value. To guard against failure, we can reduce the material strength value that we will use in our calculations to a level that will include a larger percentage of the population. To do this requires some understanding of statistical phenomena and their calculation. All engineers should have this understanding and should include a statistics course in their curriculum. We briefly discuss some of the fundamental aspects of statistics in Chapter 2.

1.9   UNITS[†]

Several different systems of units are used in engineering. The most common in the United States are the **U.S. foot-pound-second system (*fps*)**, the **U.S. inch-pound-sec-**

---

[*] Formerly the *National Bureau of Standards* (NBS).

ond system (*ips*), and the **System International** (*SI*). The metric **centimeter, gram, second** (*cgs*) system is also sometimes used. All systems are created from the choice of three of the quantities in the general expression of Newton's second law

$$F = \frac{mL}{t^2} \tag{1.2a}$$

where F is force, m is mass, L is length, and t is time. The units for any three of these variables can be chosen and the other is then derived in terms of the chosen units. The three chosen units are called *base units*, and the remaining one is a *derived unit*.

Most of the confusion that surrounds the conversion of computations between either one of the U.S. systems and the *SI* system is due to the fact that the *SI* system uses a different set of base units than the U.S. systems. Both U.S. systems choose *force*, *length* and *time* as the base units. Mass is then a derived unit in the U.S. systems which are referred to as *gravitational systems* because the value of mass is dependent on the local gravitational constant. The *SI* system chooses **mass,** *length,* and *time* as the base units and force is the derived unit. *SI* is then referred to as an *absolute system* since the mass is a base unit whose value is not dependent on local gravity.

The **U.S.** *foot-pound-second* (*fps*) system requires that all lengths be measured in feet (ft), forces in pounds (lb), and time in seconds (sec). Mass is then derived from Newton's law as

$$m = \frac{Ft^2}{L} \tag{1.2b}$$

and its units are pounds seconds squared per **foot** (lb sec^2/ft) = **slugs**

The **U.S.** *inch-pound-second* (*ips*) system requires that all lengths be measured in inches (in), forces in pounds (lb) and time in seconds (sec). Mass is still derived from Newton's law, equation 1.2b, but the units are now

pounds seconds squared per **inch** (lb sec^2/in) = **blobs**[*]

This mass unit is not slugs! It is worth twelve slugs or one "blob"!

Weight is defined as the force exerted on an object by gravity. Probably the most common units error that students make is to mix up these two unit systems (*fps* and *ips*) when converting weight units (which are pounds force) to mass units. Note that the gravitational acceleration constant (g or g_c) on earth at sea level is approximately 32.17 **feet** per second squared which is equivalent to 386 **inches** per second squared. The relationship between mass and weight is

mass = weight / gravitational acceleration

$$m = \frac{W}{g_c} \tag{1.3}$$

[†] Excerpted from Norton, *Design of Machinery*, 1992 McGraw-Hill, New York, with the publisher's permission.

---

[*] It is unfortunate that the mass unit in the *ips* system has never officially been given a name such as the term *slug* used for mass in the *fps* system. The author boldly suggests (with tongue only slightly in cheek) that this unit of mass in the *ips* system be called a *blob* (bl) to distinguish it more clearly from the *slug* (sl), and to help the student avoid some of the common errors listed above. Twelve *slugs* = one *blob*. *Blob* does not sound any sillier than slug, is easy to remember, implies mass, and has a convenient abbreviation (bl) which is an anagram for the abbreviation for pound (lb). Besides, if you have ever seen a garden slug, you know it looks just like a *"little blob."*

It should be obvious that if you measure all your lengths in **inches** and then use $g = g_c = 32.17$ **feet**/sec^2 to compute mass, you will have an error of a *factor of 12* in your results. This is a serious error, large enough to crash the airplane you designed. Even worse off is the student who neglects to convert weight to mass *at all*. The results of this calulation will have an error of either 32 or 386, which is enough to sink the ship!

The value of mass is needed in Newton's second-law equation to determine forces due to accelerations.

$$F = ma \qquad\qquad (1.4a)$$

The units of mass in this equation are either *g, kg, slugs,* or *blobs* depending on the units system used. Thus, in either English system, the weight W (lb$_f$) must be divided by the acceleration due to gravity g_c as indicated in equation 1.3 to get the proper mass quantity for equation 1.4a.

Adding further to the confusion is the common use of the unit of ***pounds mass*** (lb$_m$). This unit is often used in fluid dynamics and thermodynamics, and comes about through the use of a slightly different form of Newton's equation:

$$F = \frac{ma}{g_c} \qquad\qquad (1.4b)$$

where m = mass in lb$_m$, a = acceleration and g_c = the gravitational constant. On earth, the value of the **mass** of an object measured in ***pounds mass*** (lb$_m$) is *numerically equal* to its **weight** in ***pounds force*** (lb$_f$). However, the student *must remember to divide* the value of m in lb$_m$ by g_c when using this form of Newton's equation. Thus the lb$_m$ will be divided either by 32.17 or by 386 when calculating the dynamic force. The result will be the same as when the mass is expressed in either slugs or blobs in the $F = ma$ form of the equation. Remember that in round numbers at sea level on earth

$$1 \text{ lb}_m = 1 \text{ lb}_f \qquad\qquad 1 \text{ slug} = 32.17 \text{ lb}_f \qquad\qquad 1 \text{ blob} = 386 \text{ lb}_f$$

The *SI system* requires that lengths be measured in meters (m), mass in kilograms (kg), and time in seconds (sec). This is sometimes also referred to as the *mks* system. Force is derived from Newton's law and the units are:

$$\text{kg m/sec}^2 = \text{newtons}$$

In the *SI* system there are distinct names for mass and force which helps alleviate confusion. When converting between *SI* and U.S. systems, be alert to the fact that mass converts from kilograms (kg) to either slugs (sl) or blobs (bl), and force converts from newtons (N) to pounds (lb). The gravitational constant (g_c) in the *SI* system is approximately 9.81 m/sec^2.

The *cgs* system requires that lengths be measured in centimeters (cm), mass in grams (g), and time in seconds (sec). Force is measured in dynes. The *SI* system is generally preferred over the *cgs* system.

The systems of units used in this textbook are the U.S. *ips* system and the *SI* system. Much of machine design in the United States is still done in the *ips* system though the *SI* system is becoming more common. Table 1-4 shows some of the variables used in this text and their units. Table 1-5 shows a number of conversion factors between commonly used units. The student is cautioned always to check the units in any equation written for a problem solution, whether in school or in professional practice. If properly written, an equation should cancel all units across the equal sign. If it does not, then you can be *absolutely sure it is **incorrect**.* Unfortunately, a unit balance in an equation does not guarantee that it is correct, as many other errors are possible. Always double-check your results. You might save a life.

EXAMPLE 1-1

Units Conversion

Problem   The weight of an automobile is known in lb_f. Convert it to mass units in the *SI*, *cgs*, *fps*, and *ips* systems. Also convert it to lb_m

Given   The weight = 4500 lb_f.

Assumptions The automobile is on earth at sea level.

Solution

1   Equation 1.4*a* is valid for the first four systems listed.

For the *fps* system:

$$m = \frac{W}{g} = \frac{4\,500 \text{ lb}_f}{32.17 \text{ ft}/\text{sec}^2} = 139.9 \frac{\text{lb}_f - \text{sec}^2}{\text{ft}} = 139.9 \text{ slugs} \qquad (a)$$

For the *ips* system:

$$m = \frac{W}{g} = \frac{4\,500 \text{ lb}_f}{386 \text{ in}/\text{sec}^2} = 11.66 \frac{\text{lb}_f - \text{sec}^2}{\text{in}} = 11.66 \text{ blobs} \qquad (b)$$

For the *SI* system:

$$W = 4\,500 \text{ lb} \frac{4.448 \text{ N}}{\text{lb}} = 20\,016 \text{ N}$$

$$m = \frac{W}{g} = \frac{20\,016 \text{ N}}{9.81 \text{ m}/\text{sec}^2} = 2\,040 \frac{\text{N} - \text{sec}^2}{\text{m}} = 2\,040 \text{ kg} \qquad (c)$$

For the *cgs* system:

Table 1-4   Variables and Units
Base Units in Boldface - Abbreviations in ( )

Variable	Symbol	ips unit	fps unit	SI unit
Force	F	**pounds (lb)**	**pounds (lb)**	newtons (N)
Length	l	**inches (in)**	**feet (ft)**	**meters (m)**
Time	t	**seconds (sec)**	**seconds (sec)**	**seconds (sec)**
Mass	m	lb-sec^2/in (bl)	lb-sec^2/ft (sl)	**kilograms (kg)**
Weight	W	pounds (lb)	pounds (lb)	newtons (N)
Pressure	p	psi	psf	N/m^2 = Pa
Velocity	v	in/sec	ft/sec	m/sec
Acceleration	a	in/sec^2	ft/sec^2	m/sec^2
Stress	σ, τ	psi	psf	N/m^2 = Pa
Angle	θ	degrees (deg)	degrees (deg)	degrees (deg)
Angular velocity	ω	radians/sec	radians/sec	radians/sec
Angular acceleration	α	radians/sec^2	radians/sec^2	radians/sec^2
Torque	T	lb-in	lb-ft	N-m
Mass moment of inertia	I	lb-in-sec^2	lb-ft-sec^2	kg-m^2
Area moment of inertia	I	in^4	ft^4	m^4
Energy	E	in-lb	ft-lb	joules = N-m
Power	P	in-lb/sec	ft-lb/sec	N-m/sec = watt
Volume	V	in^3	ft^3	m^3
Specific weight	ν	lb/in^3	lb/ft^3	N/m^3
Mass density	ρ	bl/in^3	sl/ft^3	kg/m^3

$$W = 4\ 500\ \text{lb}\ \frac{4.448E5\ \text{dynes}}{\text{lb}} = 2.002E9\ \text{dynes}$$

$$m = \frac{W}{g} = \frac{2.002E9\ \text{dynes}}{981\ \text{cm}/\text{sec}^2} = 2.04E6\ \frac{\text{dynes} - \text{sec}^2}{\text{cm}} = 2.04E6\ g \qquad (d)$$

2   For mass expressed in lb$_m$, equation 1.4b must be used.

$$m = W\frac{g_c}{g} = 4\ 500\ \text{lb}_\text{f}\ \frac{386\ \text{in}/\text{sec}^2}{386\ \text{in}/\text{sec}^2} = 4\ 500\ \text{lb}_\text{m} \qquad (e)$$

Note that lb$_m$ is numerically equal to lb$_f$ and so must not be used as a mass unit unless you are using the form of Newton's law expressed as equation 1.4b.

Table 1-5   Selected Units Conversion Factors

Note That These Conversion Factors (and Others) are Built Into the *TKSolver* Files UNITMAST and STUDENT

Multiply this	by	this	to get	this		Multiply this	by	this	to get	this
acceleration						**mass moment of inertia**				
in/sec^2	x	0.0254	=	m/sec^2		$lb\text{-}in\text{-}sec^2$	x	0.1138	=	$N\text{-}m\text{-}sec^2$
ft/sec^2	x	12	=	in/sec^2		**moments and energy**				
angles						in-lb	x	0.1138	=	N-m
radian	x	57.2958	=	deg		ft-lb	x	12	=	in-lb
						N-m	x	8.7873	=	in-lb
area						N-m	x	0.7323	=	ft-lb
in^2	x	645.16	=	mm^2		**power**				
ft^2	x	144	=	in^2		HP	x	550	=	ft-lb/sec
area moment of inertia						HP	x	33 000	=	ft-lb/min
in^4	x	416 231	=	mm^4		HP	x	6 600	=	in-lb/sec
in^4	x	4.162E–07	=	m^4		HP	x	745.7	=	watts
m^4	x	1.0E+12	=	mm^4		N-m/sec	x	8.7873	=	in-lb/sec
m^4	x	1.0E+08	=	cm^4		**pressure and stress**				
ft^4	x	20 736	=	in^4		psi	x	6 894.8	=	Pa
density						psi	x	6.895E-3	=	MPa
lb/in^3	x	27.6805	=	g/cc		psi	x	144	=	psf
g/cc	x	0.001	=	g/mm^3		kpsi	x	1 000	=	psi
lb/ft^3	x	1 728	=	lb/in^3		N/m^2	x	1	=	Pa
kg/m^3	x	1.0E–06	=	g/mm^3		N/mm^2	x	1	=	MPa
force						**spring rate**				
lb	x	4.448	=	N		lb/in	x	175.126	=	N/m
N	x	1.0E+05	=	dyne		lb/ft	x	0.08333	=	lb/in
ton (short)	x	2 000	=	lb		**stress intensity**				
length						$MPa\text{-}m^{0.5}$	x	0.909	=	$ksi\text{-}in^{0.5}$
in	x	25.4	=	mm		**velocity**				
ft	x	12	=	in		in/sec	x	0.0254	=	m/sec
mass						ft/sec	x	12	=	in/sec
blob	x	386.4	=	lb		rad/sec	x	9.5493	=	rpm
slug	x	32.2	=	lb		**volume**				
blob	x	12	=	slug		in^3	x	16 387.2	=	mm^3
kg	x	2.205	=	lb		ft^3	x	1 728	=	in^3
kg	x	9.8083	=	N		cm^3	x	0.061023	=	in^3
kg	x	1 000	=	g		m^3	x	1.0E+9	=	mm^3

1.10   INTRODUCTION TO TKSOLVER

The *TKSolver* program was chosen to accompany this textbook because it possesses some unique features that are extremely useful for the solution of design problems in general and for machine design problems in particular. These features include the ability to backsolve for any variable by simply switching it from an input column to an output column and then moving some other variable from output to input. In addition the program will automatically invoke a root-finding algorithm applying user-supplied guess values for the unknown parameters. It can thus solve equations in which the unknown parameters are implicit (i.e., appear more than once in the equation). Systems of simultaneous nonlinear equations can also be solved by its root-finding algorithm. It comes with an extensive library of prewritten functions that do many mathematical tasks such as numerical integration, numerical solution of differential equations, etc.

Design problems typically contain more variables than we have equations for and thus cannot be directly solved. Many assumptions (read intelligent guesses) must then be made to obtain a trial solution from which some insight is gained into the problem. The assumed values are then changed and the model is recalculated to obtain a better solution. *TKSolver* allows this iteration process to proceed in an easy and rapid manner.

TKSolver is a member of the class of programs known as **equation solvers** which typically allow the typing in of equations in a line-by-line fashion as you would do with any programming language. Unlike programming languages and most other equation solvers, which require the equation to be put in explicit form (i.e., the one unknown variable isolated on the left of the equal sign and all other terms on the right side), *TK* allows the equation to be input in **free form**. This may seem like a minor convenience, but is really quite powerful as it allows implicit variables to be solved as described above. In fact, there is really no distinction between input and output variables within the equations of a *TK* model. Any variable can take on either input or output characteristics at different times. This is what allows "instant" back-solving. This means that you can, for example, specify the safety factor as an input variable, assign a desired value, and solve for any one of the part dimensions that affect the safety factor, no matter how complicated the equations or how many times that dimension's variable appears in the collection of equations that define the engineering model. Once the model is calculated, output of the data in table or graph form is simple to accomplish.

There are many more useful features of this program but rather than list them, it will be more productive to present a few simple examples of its use to show what it can do. The *Student Manual* included with the software provides additional instruction for, and examples of, its use. Context-sensitive, on-line help is built into the program as well.

We will create a simple design problem to use as an example and gradually increase its complexity in succeeding examples as additional topics and features of the program are introduced. Before presenting the examples, we must define some of the terminology associated with this program.

Terminology

The *TKSolver* program contains a number of "sheets" which can be thought of as a paradigm for the sheets of paper that clutter the typical engineer's desk. One of these sheets, the **rule sheet,** contains the equations or "rules" that define the model. Another sheet, the **variable sheet,** contains the values of the known input parameters, and, after the rules are solved will also contain the output values. One of the advantages of *TKSolver* is its clean separation of rules (equations) and variables (data). The rules can (and should) be kept in strictly symbolic terms and the data changed on the variable sheet to recalculate or iterate the problem.

Beneath the variable sheet (and other sheets) are additional "subsheets" that contain more detailed information on each variable. These subsheets can be accessed from a pull-down menu or with a keystroke combination which differs depending on the version of *TKSolver* used.[*] This is referred to as "**diving**" down to the subsheet, as in rummaging beneath the papers on your desk to find that sheet with additional data on it. When the program is first run, both the rule and variable sheets are open and visible on the screen. Other sheets (**list**, **function**, **plot**, **table**, **unit**, **format**, and **comment**) are accessible from the WINDOW pull-down menu. The use of these sheets will be introduced in later examples.

Simple Models, Rule and Variable Sheets, Subsheets

We wish to design a 1-liter-capacity cooking pot, as shown in Figure 1-7, to be made of 1-mm-thick stainless steel. Ultimately we would like to minimize the amount of stainless steel used. We will start with the simplest possible model.

EXAMPLE 1-2

Simple Equation-Solving Using TKSolver - Part A

Problem   **Find the height and empty weight of an open-top, cylindrical container (a cooking pot) for a desired volume.**

Given   **The volume = 1 liter, inside diameter = 12 cm, wall thickness = 0.1 cm.**

Assumptions   **The material is stainless steel with a mass density of 7.75 g/cc.**

Solution   **See Figure 1-7, Tables 1-6 and 1-7, and the *TKSolver* file EX01-02.**

1   An equation for the volume of a cylinder can be expressed as

$$vol = basearea \cdot H \qquad\qquad (a)$$

where H is the height of the cylinder.

2   The area of the base can be found from

* In the *Windows* version, a "dive" to a subsheet is done with *Alt* + *Enter* or by clicking the right mouse button with the cursor on the item to be dived into. In the *DOS* version, use *Shift* + > to dive down and *Shift* + < to dive back up. In the *Macintosh* use *Command* + *mouse-click*. The *Windows* and *Macintosh* versions also provide pull-down menu picks for opening subsheets. These examples will refer to the *Windows* versions of these (and other) commands. See your manual for equivalents in other versions.

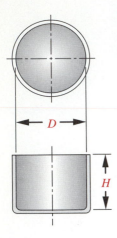

$$basearea = \frac{\pi D^2}{4} \qquad\qquad (b)$$

where D is the inside diameter of the base.

3 Since we want to know the weight of the cooking pot, we need its total surface area, which can be found from

$$surface = basearea + \pi DH \qquad\qquad (c)$$

4 The weight is then

$$surface \cdot thick \cdot dens = weight \qquad\qquad (d)$$

5 These equations are encoded into the rule sheet as shown in Table 1-6. Though this model is provided on disk, it will be of more value if you create this model (and those in later examples) from scratch as described here. If you do that, you will notice that as soon as you hit ENTER after typing each equation, the variable names in that equation automatically appear on the variable sheet as shown in Table 1-7.

6 To solve the model we must provide sufficient input data to constrain it. The data given in this example are typed into the input column of the variable sheet for *vol, thick, dens,* and *D*. The function key F9 will solve the model and the results should appear in the output column beside the variables as seen in Table 1-7. Try it.

7 Comments may be typed into the comment column or not as you wish. They serve the useful purpose of documenting the model and that should always be done. The comments do not enter into the solution, however. The labels in the units column have special meanings when used in conjunction with a **units sheet**. We will introduce that powerful feature of *TK* in a later example. If you are typing in this model from scratch, then it is better to leave the units labels out for now.

8 The only caution that ALWAYS applies, whether using a units sheet or not, is that when first entering data to the variable sheet you must use only data that are **defined in a consistent unit system**. We have chosen *cgs* units in this example. Having made that choice we had to be consistent in using only data measured in that system,

FIGURE 1-7

A Stainless Steel Cylindrical Container (Cooking Pot)

Table 1-6   **TKSolver Rule Sheet for Example 1-2 from File EX01-02**

; 1- calculate the volume of the cylinder   (Eq. *a*)
    *vol = basearea * H*
; 2 - calculate the base area of the cylinder   (Eq. *b*)
    *basearea = PI() * D^2 / 4*
; 3 - calculate the total surface area of the cylinder   (Eq. *c*)
    *surface = basearea + PI() * D * H*
; 4 - calculate the weight of the cylinder   (Eq. *d*)
    *surface * thick * dens = weight*

Table 1-7   TKSolver Variable Sheet for Example 1-2 from File EX01-02

St	Input	Variable	Output	Unit	Comments
	1 000	*vol*		cm^3	volume of container
	0.10	*thick*		cm	thickness of wall
	7.75	*dens*		g/cc	density of material
	12	*D*		cm	diameter of base of cylinder
		H	8.8	cm	height of side of cylinder
		basearea	113.1	cm^2	area of base
		surface	446.4	cm^2	total surface area
		weight	346	g	weight of empty cylinder

i.e., cm, g, and cc in this case. We will have a great deal of freedom to change the units of any variable after its first entry, and with a proper units sheet, *TK* will automatically convert them! But that's getting ahead of the story. (See Example 1-10 for a discussion of units conversion.)

Let's examine this model more closely. Note in Table 1-6 that equation (*a*) computes volume as a function of the base area. But the base area is not calculated until step 2. This arrangement of equations would not work in any conventional programming language or in most other equation solvers. They require any variable used on the right side of an equation to be defined numerically in advance of the execution of the equation in which it appears. All other programming languages and equation solvers are **procedural solvers**, meaning that they *proceed linearly through a set of instructions, from top to bottom, evaluating each expression as they go* (unless redirected by control statements, such as loops, if statements, etc.).

TKSolver operates differently. It is a **declarative solver**, which in simple terms means that it is *capable of sorting out the order in which a set of equations must be solved and reordering them* (in its "head") to solve them (if possible) no matter what order you present them to it. We will see in a later example that it is possible to make *TK* behave in a procedural manner when we wish to, as when controlling a loop structure to manipulate an array for example. Thus, *TK* is able to figure out that it needs to evaluate the second equation for *basearea* before it can solve the first equation for *vol*. It does this by making multiple passes through the rule sheet until it sorts out these hierarchies and will ultimately solve the model if enough data have been supplied to constrain it.

Equations (*b*) and (*c*) in Table 1-6 use a built-in function *PI( )* which returns the value of π to as many significant figures as the computer is capable of carrying (typically >15). The ( ) are required for proper syntax and, for most functions, the ( ) will contain one or more arguments to be passed to the function (e.g., *SIN(x)*); but *PI( )* does not need an argument since it just returns a constant. Using this function is preferable to typing 3.1416 for π as it gives more accuracy.

Look at equation (*d*) in Table 1-6. It is written in *implicit* form in that the left side has more than one term. This is not allowed in any procedural solver or programming language. That simple equation could easily have been written with the single term *weight* on the left side, which would make it acceptable to a procedural solver. However, one often encounters complicated equations that either cannot be put into explicit form for a particular variable or in which to do so would involve a great deal of algebraic manipulation. In such instances this ability of *TK* to handle an implicit form is very useful. It was done in this example only to make this point.

Switching Variables from Input to Output and Vice Versa

We now redo the previous example with a different parameter specified as the input.

EXAMPLE 1-3

Simple Equation Solving Using TKSolver - Part B

Problem	**Find the diameter and weight of an open-top, cylindrical container (a cooking pot) for a desired volume.**
Given	**The volume = 1 liter, inside height = 10 cm, wall thickness = 0.1 cm.**
Assumptions	**The material is stainless steel with a mass density of 7.75 g/cc.**
Solution	**See Figure 1-7, Table 1-8, and the *TKSolver* file EX01-03.**

1 The only difference between this example and the previous one is that we have specified the pot's height and asked for the diameter needed to match a given volume. The rule sheet is thus the same as before. The model is unchanged, but we want to solve for a different variable.

2 Using *TKSolver,* this becomes a trivial issue. Table 1-8 shows the variable sheet for this example. Note that the value for H is now defined in the input column and the value of D has been computed in the output column. This switch was accomplished by simply typing the letter I (for input) in the leftmost column (labeled *St* for status) opposite the variable H. The I does not show, but it pulls the value into the input column. Then type a number 10 in the input column for H. Typing a letter O in the *St* column opposite D pushes its value to the output column. We have thus switched the status of these two variables. (Note that simply typing the number in the input column of H will change its status to input without requiring an *I* in the *St* column.)

3 Solving the model with function key F9 gives the results shown in Table 1-8.

Table 1-8   TKSolver Variable Sheet for Example 1-3 from File EX01-03

St	Input	Variable	Output	Unit	Comments
	1 000	vol		cm^3	volume of container
	0.10	thick		cm	thickness of wall
	7.75	dens		g/cc	density of material
O*		D	11.3	cm	diameter of base of cylinder
I*	10	H		cm	height of side of cylinder
		basearea	100	cm^2	area of base
		surface	454.5	cm^2	total surface area
		weight	352.3	g	weight of empty cylinder

* These letters are not visible on the variable sheet, but change the status of the variable to input (I) or output (O).

Note the ease with which variables can be switched from input to output status or vice versa. This allows any variable to be solved for quickly and easily.

Using Iteration and Root Finding

We now complicate the problem by introducing a **height to diameter ratio** constraint (*H/D*) which will then require an iterative root-finding solution to a set of simultaneous equations.

EXAMPLE 1-4

Simultaneous Equation-Solving Using TKSolver

Problem   **Find the diameter, height, and weight of an open-top, cylindrical container (a cooking pot) for a desired volume and height/diameter ratio (*H/D*).**

Given   **The volume = 1 liter, *H/D* = 0.6, wall thickness = 0.1 cm.**

Assumptions   **The material is stainless steel with a mass density of 7.75 g/cc.**

Solution   **See Figure 1-7, Tables 1-9, 1-10, and 1-11, and the *TKSolver* file EX01-04.**

1   This example introduces a new constraint to the problem posed in the previous two examples, but is otherwise the same. Instead of specifying either the diameter or height, we now specify the aspect ratio between those two parameters, and want to find some combination of height and diameter that satisfies the two constraints of volume and aspect ratio. Define an additional rule that involves the aspect ratio.

$$ratio = \frac{H}{D} \qquad (e)$$

2 Add this equation to the rule sheet as shown in Table 1-9 and push both H and D to the output column by typing an O in the status column as described in the previous example. Type the given value of 0.6 in the input column of the variable *ratio*.

3 Use Function key F9 to try solving the model now. You will find that it does not solve because there are now two unknowns, H and D. An unsolved model is evidenced by the presence of asterisks to the left of the rules. When a rule is satisfied, the asterisk is erased.

4 We have two equations in these two unknowns, so they can be solved simultaneously. Equations (*a*) and (*b*) could be combined to give one expression for volume as $f(H, D)$, and equation (*e*) expresses the ratio as $f(H, D)$. Note that it is not necessary to rewrite the rule sheet to algebraically combine equations (*a*) and (*b*). The solver will "combine" them numerically.

5 To solve simultaneous, nonlinear equations it is necessary to provide guess values for one or more variables so that the Newton-Raphson root-finding algorithm[*] built into *TKSolver* can iterate to a solution. There are two ways to provide a guess value for any variable in *TK*. If you type a G in the *St* column next to the variable, it will use whatever number you type in the input column as an initial guess. If that guess is close enough to one of the roots of the equation system, it will converge to that root. Be aware, however, that nonlinear equations can have multiple roots, meaning that your solution can be different for different guess values.

6 The second (and preferred) way to provide a guess value for a variable is to type it on the variable's subsheet. To access the subsheet, place the cursor on the line containing the variable, pull down the WINDOW menu and select DISPLAY SUBSHEET. (This selection has the keyboard equivalent of *Alt + Enter.*) Once the variable's subsheet is open, any desired guess value can be typed on the line so labeled. The difference between these two methods of establishing a guess value is that a guess

* For information on root-finding algorithms, see *Numerical Recipes* by Press et al, Cambridge University Press.

Table 1-9   TKSolver Rule Sheet for Example 1-4 from File EX01-04

; 1- calculate the volume of the cylinder   (EQ. *a*)
     vol = basearea * H

; 2 - calculate the base area of the cylinder   (EQ. *b*)
     basearea = PI() * D^2 / 4

; 3 - calculate the total surface area of the cylinder   (EQ. *c*)
     surface = basearea + PI() * D * H

; 4 - calculate the weight of the cylinder   (EQ. *d*)
     surface * thick * dens = weight

; 5 - calculate the ratio of cylinder height to diameter   (EQ. *e*)
     ratio = H / D

Table 1-10   TKSolver Variable Sheet for Example 1-4 - Before Solving

St	Input	Variable	Output	Unit	Comments
	1 000	vol		cm^3	volume of container
	0.10	thick		cm	thickness of wall
	7.75	dens		g/cc	density of material
	0.6	ratio			height to diameter ratio
G	2	D		cm	diameter of base of cylinder
		H		crn	height of side of cylinder
		baseare		cm^2	area of base
		surface		cm^2	total surface area
		wt		g	weight of empty cylinder

placed on the variable's subsheet will remain there through multiple solves of the model. But a guess value placed in the input column of the variable sheet along with a G in the *St* column will convert to an output value after solving and both the G and the guess value will have to be retyped for each subsequent solve.

7   Use either of these methods to define a guess value (the supplied example file has a guess value of 2 on *D*'s subsheet) and solve the model with F9. It will automatically iterate to the solution. Table 1-10 shows the variable sheet for this example with the G shown in the *St* column and the guess value of 2 in the input column *before solution*. Table 1-11 shows the same variable sheet after solution. The solver has iterated to the simultaneous solution that satisfies the constraints on both volume and ratio to give the values of *H* and *D* shown in the output column.

Table 1-11   TKSolver Variable Sheet for Example 1-4 - After Solving

St	Input	Variable	Output	Unit	Comments
	1 000	vol		cm^3	volume of container
	0.10	thick		cm	thickness of wall
	7.75	dens		g/cc	density of material
	0.6	ratio			height to diameter ratio
		D	12.9	cm	diameter of base of cylinder
		H	7.7	cm	height of side of cylinder
		basearea	129.7	cm^2	area of base
		surface	441	cm^2	total surface area
		wt	341.8	g	weight of empty cylinder

The previous three examples show how *TKSolver* can quickly solve individual or simultaneous equations. Variables can be switched from input to output status allowing a model to be solved for any parameter present in its rules without requiring any rewriting of those rules. Iterative root finding is automatically invoked if the model cannot be directly solved and if a sufficient number of guesses for the unknown variable values have been provided, either on the variable sheet, or on the variables' subsheets.

Lists, Tables, Plots, Optimization, and Built-in Functions

The next two examples introduce the use of lists (arrays) in *TKSolver* and show how an optimum solution to a problem can be quickly found. Once lists of variables are created, plots and tables of the model's parameters can also be quickly generated.

EXAMPLE 1-5

List-Solving and Plotting Using TKSolver

Problem	Find the height/diameter ratio and dimensions of an open-top, cylindrical container that will minimize its weight for a given volume. Plot the variation of weight with the height/diameter ratio.
Given	Volume = 1 liter, wall thickness = 0.1 cm.
Assumptions	The material is stainless steel with a mass density of 7.75 g/cc. The ratio *H/D* will be varied from 0.1 to 2.0 in increments of 0.1.
Solution	See Figures 1-7 and 1-8, Tables 1-12, 1-13, and 1-14, and the *TKSolver* file EX01-05.

1 There should be an optimum *H/D* ratio for this container that will minimize its surface area and weight. Since the material is sold by weight, its cost will then be minimized as well. This simple example could be optimized by writing an expression for the weight as a function of *H/D* ratio, differentiating it with respect to that ratio, setting the differential equal to zero, and solving for the ratio. However, we can also obtain a numerical approximation to that optimum ratio from the existing *TKSolver* model by creating lists of variables and *list solving* the model for all values in the input list.

2 A variable can be made into a list simply by typing a letter L in the status column of that variable as shown in Table 1-12. The variables ratio, *D*, *H*, and weight have been so designated.

3 At least one list must be declared as an input list and a set of input values provided to it. The variable *ratio* is made to be the input list simply by the fact that it has a value in its input column on the variable sheet. Note that all the other list variables' values are in the output column.

Table 1-12   TKSolver Variable Sheet for Example 1-5 from File EX01-05

St	Input	Variable	Output	Unit	Comments
	1 000	*vol*		cm^3	volume of container
	0.10	*thick*		cm	thickness of wall
	7.75	*dens*		g/cc	density of material
L	0.6	*ratio*			height to diameter ratio
L		*D*	12.9	cm	diameter of base of cylinder
L		*H*	7.7	cm	height of side of cylinder
		basearea	129.7	cm^2	area of base
		surface	441	cm^2	total surface area
L		*weight*	341.8	g	weight of empty cylinder

4   To put values in the input list, pull down the COMMANDS menu and select LIST FILL. The resulting dialog box allows the desired first, last, and step values for the list to be specified, and fills the list. To view its contents, place the cursor on the variable-sheet line containing that variable and use *Alt + Enter* twice in succession to "dive" down two levels into the sub-subsheet for that variable. This will expose the list in its own window. The list for the variable *ratio* was filled by this technique with 20 values from 0.1 to 2.0 in steps of 0.1 and is shown in Table 1-13.

5   Use the function key F10 to do a **list solve**. This performs a separate solution of the rules for each value in the input list. In this example it solved the rules 20 times. The output lists *D, H,* and *weight* are now filled with 20 values apiece, each one corresponding to a solution for one of the values in the input list *ratio*. In terms of a conventional programming language such as BASIC, Fortran, Pascal or C, the list-solve operation has, in effect, executed a loop which repeated the rule sheet calculations while incrementing through all the values in the input list.

6   To see the results in convenient form, we can create an **interactive table**. Open the **table sheet** from the WINDOW pull-down menu. Type any name you wish in the column labeled "name." A title for the table also can be typed in the title column if desired, or it can be left blank. With the cursor sitting on the line in the table sheet containing the name you just typed, "dive" (*Alt + Enter*) to go to the subsheet for this table. In the subsheet column headed "List," type the name of any list variable that you want to put in the table. In this example, we typed *ratio, D, H,* and *weight,* each on a new line in the List column. To display the table "dive" again using *Alt + Enter.* The **interactive table** will now be visible in its own window and should look like Table 1-14. It can be scrolled through on-screen and can be printed. Since it is an "interactive" table, any changes typed into it will automatically change the parent lists. For example, if you wanted to change the last value of the input list *ratio* from 2.0 to 3.0, typing the new value in the interactive table will have the same effect as if you typed it directly into the list. They are "hot-linked."

Table 1-13

TKSolver List for the Variable *ratio*

Element	Value
1	0.1
2	0.2
3	0.3
4	0.4
5	0.5
6	0.6
7	0.7
8	0.8
9	0.9
10	1.0
11	1.1
12	1.2
13	1.3
14	1.4
15	1.5
16	1.6
17	1.7
18	1.8
19	1.9
20	2.0

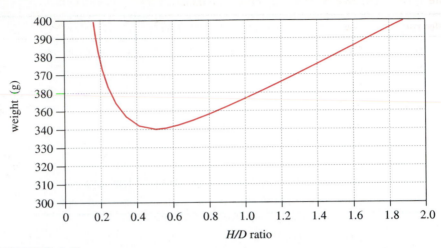

FIGURE 1-8

Variation of Weight with Height to Diameter Ratio in Example 1-5

7   A plot of any list versus any other list (or lists) can also be quickly created. Open the plot sheet from the WINDOWS menu. Type any name you wish in the column labeled "name." A title for the plot also can be typed in the title column if desired, or it can be left blank. With the cursor sitting on the line in the plot sheet containing the name you just typed, "dive" (*Alt + Enter*) to go to the subsheet for this plot. The cursor will be on a field labeled "X-Axis List." Type the name of the list that you want plotted on the *x*-axis as the independent variable. For our example this is the list *ratio*. Then place the cursor on one of the fields in the column under the label "Y-Axis" and type the name of a list that you want plotted as a dependent variable on the *y*-axis. For our example this could be the list *weight*. Use the function key F7 to display the plot. It should look like Figure 1-8.

8   The optimum solution is obvious from Figure 1-8. The weight of the container is a minimum at an H/D ratio of about 0.5. Table 1-14 shows this weight to be 340.5 g.

The preceding example shows how easy it is to obtain plots and tables of solution data and from them determine an approximate optimum design. To do this without a tool such as an equation solver would be tedious indeed. Even to write a custom computer program in Pascal, BASIC, or C would be more time-consuming than this approach. An equation solver eliminates the "programming overhead" associated with input and output of data, table and plot generation, etc. Solutions to even simple problems like this example can be obtained in less time than with most other methods. When the problem is more complex, an equation solver becomes an indispensable tool for the engineer.

Table 1-14   TKSolver Interactive Table for Example 1-5

Element	ratio	D	H	weight
1	0.1	23.4	2.3	464.7
2	0.2	18.5	3.7	376.4
3	0.3	16.2	4.9	351.1
4	0.4	14.7	5.9	342.5
5	0.5	13.7	6.8	340.5
6	0.6	12.9	7.7	341.8
7	0.7	12.2	8.5	344.7
8	0.8	11.7	9.3	348.5
9	0.9	11.2	10.1	352.9
10	1.0	10.8	10.8	357.5
11	1.1	10.5	11.5	362.4
12	1.2	10.2	12.2	367.3
13	1.3	9.9	12.9	372.2
14	1.4	9.7	13.6	377.1
15	1.5	9.5	14.2	382.0
16	1.6	9.3	14.8	386.8
17	1.7	9.1	15.4	391.6
18	1.8	8.9	16.0	396.3
19	1.9	8.8	16.6	400.9
20	2.0	8.6	17.2	405.4

Using Built-in Functions

We now modify the *TKSolver* file from the preceding example to add the ability to extract the optimum solution values from the lists without the need for a table or plot
(though both of those displays will still be of value in visualizing the solution.) In the
process, we also increase the size of the lists from their present 20 to 80 elements in order
to get a closer approximation to the optimum solution.

EXAMPLE 1-6

List Solving and Built-in Functions in TKSolver

Problem   Find the height/diameter ratio and dimensions of an open-top, cylindrical container that will minimize its weight for a given volume. Plot the
variation of weight with the height/diameter ratio. Extract the optimum values from the lists of solution variables.

Given   Volume = 1 liter, wall thickness = 0.1 cm.

Assumptions   The material is stainless steel with a mass density of 7.75 g/cc. The ratio *H/D* will be varied from 0.1 to 2.0 in increments of 0.025.

Solution   See Tables 1-15 and 1-16 and the *TKSolver* file EX01-06.

1 Fill the input list *ratio* with 77 elements with values from 0.1 to 2.0 in increments of 0.025. See step 4 in Example 1-5 for the procedure.

2 *TKSolver* has about 100 built-in functions to perform common mathematical operations such as square roots, logarithms, trigonometric operations, etc. Their syntax can be viewed with the on-line help in *TKSolver*, either from the HELP pull-down menu, or by using the function key F1. We will use three of these functions in this example to demonstrate function use in general and also to show how to extract information from lists.

3 Table 1-15 shows the rule sheet for this example. Three rules have been added to those of Example 1-5. At line 6, the *MIN(x)* function is used to return the smallest value in the list '*weight*. The single quote in front of *weight* indicates that it is the name of a list rather than a variable for which a single value exists on the variable sheet. Equation (*f*) in Table 1-15 places the minimum value returned from this function in the variable *minwt*. The variable sheet in Table 1-16 shows the value it returned to *minwt* after solving. Note a subtlety here. A list solve (F10) must first be done in order to fill the list '*weight* with its calculated values. Then a direct solve

---

Table 1-15   TKSolver Rule Sheet for Example 1-6 from File EX01-06

; 1- calculate the volume of the cylinder	(Eq. *a*)
*vol = basearea * H*	
; 2 - calculate the base area of the cylinder	(Eq. *b*)
*basearea = PI() * D^2 / 4*	
; 3 - calculate the total surface area of the cylinder	(Eq. *c*)
*surface = basearea + PI() * D * H*	
; 4 - calculate the weight of the cylinder	(Eq. *d*)
*surface * thick * dens = weight*	
; 5 - calculate the ratio of cylinder height to diameter	(Eq. *e*)
ratio = H / D	
; 6 - find the minimum value in the list 'weight	(Eq. *f*)
minwt = MIN('weight)	
; 7 - find the location (subscript) of the min value in the list 'weight	(Eq. *g*)
locmin = MEMBER(minwt, 'weight)	
; 8 - find the value in the list 'ratio with the subscript locmin	(Eq. *h*)
minratio = ELEMENT('ratio, locmin)	

(F9) also must be done to cause equation (*f*) to extract the minimum value from that list.

4   We now have found an approximation of the lowest weight for our design, but we really need to know the value of the *H/D* ratio that is responsible for that minimum weight. The function *MEMBER*(*minwt*, '*weight*) at line 7 returns the location (or subscript) in the array '*weight* of the element that has the value *minwt*. This value is shown as *locmin* in Table 1-16.

5   We can now use the function *ELEMENT*('*ratio*, *locmin*) at line 8 to return the value of '*ratio*[*locmin*]. This is the desired solution, i.e., the value of the *H/D* ratio that will give the minimum weight container.*

6   Compare these results with Table 1-14 and the plot in Figure 1-8.

It is worth the effort to spend a little time looking at the collection of built-in functions provided with the software. Their use can save programming time. See the *TKSolver* manual and its on-line help for more information.

User-Defined Functions

One of the most powerful features of *TKSolver* is its ability to accommodate user-defined functions. Three types of user-functions are available: **rule functions**, **procedure functions** and **list functions**. Each has a different purpose and set of applications. We present an example of each in the order listed above.

Table 1-16   TKSolver Variable Sheet for Example 1-6 from File EX01-06

St	Input	Variable	Output	Unit	Comments
	1 000	*vol*		cm^3	volume of container
	0.10	*thick*		cm	thickness of wall
	7.75	*dens*		g/cc	density of material
L	0.6	*ratio*			height to diameter ratio
L		*D*	12.9	cm	diameter of base of cylinder
L		*H*	7.7	cm	height of side of cylinder
		basearea	129.7	cm^2	area of base
		surface	441	cm^2	total surface area
L		*weight*	341.8	g	weight of empty cylinder
		minwt	340.5	g	minimum weight for ratio
		locmin	17		array location of minwt
		minratio	0.5		optimum *H/D* ratio for min wt

* Note that this problem could have been solved quickly and easily for the exact minimum weight using the calculus. We use a less-exact numerical method here to show the process. A numerical method is often the only possible solution to more complicated problems.

RULE FUNCTIONS   A rule function performs like the rule sheet but allows a set of rules to be isolated within a callable function that has its own set of local variables. Its local variables are isolated from those on the variable sheet. Values are passed into the rule function as **argument variables** when it is called either from the rule sheet or from another function. Its computed values are passed back to the calling program as result variables. Within a rule function, the declarative solver is in operation, so the order of equations is not critical and implicit rules are acceptable. Anything that can be done on a **rule sheet** can be done in a **rule function**.

PROCEDURE FUNCTIONS   A procedure function is very similar to a procedure in PASCAL or a subroutine in FORTRAN. It has local variables, input and output variables, and is activated with a CALL statement. Its execution is procedural, meaning its statements are executed in linear fashion, top to bottom, unless redirected with control statements. It behaves like any procedural programming language. All its statements must be explicit with only one variable on the left side. Its purpose is to allow looping through lists and/or execution of control structures that are not possible within the declarative solver environment.

LIST FUNCTIONS   are table-look-up functions that allow several types of mapping between their input and output lists, including two forms of interpolation. List functions prove to be extremely useful in design problems where much information needed for the solution is discrete in nature, such as material strengths for various alloys, or stock sizes available for fasteners, wire, or I-beams, etc. Empirical data generated from tests can be encoded as list functions and looked-up "on-the-fly" while a model's rules are iterating to a solution using guess values for other continuous variables.

Rule Functions

We now modify the previous example to introduce the use of a **rule function**. Subsequent examples will use procedure and list functions as well.

EXAMPLE 1-7

Using Rule Functions in TKSolver

Problem	Use a rule function to find the height/diameter ratio and dimensions of an open-top, cylindrical container that will minimize its weight for a given volume. Plot the variation of weight with height/diameter ratio.
Given	Volume = 1 liter, wall thickness = 0.1 cm.
Assumptions	The material is stainless steel with a mass density of 7.75 g/cc. The ratio H/D will be varied from 0.1 to 2.0.
Solution	See Tables 1-17, 1-18, and 1-19 and the *TKSolver* file EX01-07.

1   Table 1-17 shows the rule function *cyl_rule** which was created to solve the first four equations (*a - d*) in the rule sheet for Example 1-5. To do this, open the **function sheet** from the WINDOW menu and type the desired function name (*cyl_rule*) in the column labeled "Name." In the column labeled "Type," put the letter *R* to identify it as a **rule function**. A comment is optional. Dive on this line to open the actual rule function subsheet shown in Table 1-17.

2   Type the desired equations into the rule function as they would appear on a rule sheet. In fact, the equations from an existing rule sheet such as that of Example 1-5 can be copied to the clipboard and pasted into the rule function to save typing time. Note in Table 1-17 that different variable names have been used for some of the parameters. This is arbitrary and was done to make the point that these are local variables, different from those on the rule sheet.

3   At the top of the rule function are three lines labeled **parameter variables**, **argument variables**, and **result variables**. **Parameter variables** have common values among the variable sheet and all functions in which they appear. Anything in this parameter list becomes a **global variable**, and loses its local character. We do not need any of these in this function, so the line is left blank.

4   **Argument variables** are the input variables to the rule function. Type all of the variable names needed as input to solve the rules in the function on this line, separated by commas. In this example, only *V* and *ratio* are required as input. (Note that these are the only input values on the variable sheet of Table 1-12 that are used in these rules.)

5   **Result variables** are the output from the function. Type any of the variables from the equations in the function for which you wish to return values to the calling program. Note that it is not necessary to put all of the equations' variables in these lists. Any variables not listed in one of these three lines (such as *basearea* in this example) will remain as local variables known only to this rule function.

* Note that the use of the underscore (_) in the name of the function is NOT required. Any word can be used for a function name as long as it does not contain certain characters that have mathematical meaning such as +, -, etc. The underscore is used here simply to increase readability of the function name. Variable and function names in *TKSolver* are "case sensitive" meaning that capitalization of any letter(s) changes the name, i.e., Cyl_rule is a different name than cyl_rule. This case sensitivity can cause some consternation when writing and debugging models as one's eye may not notice that the "same" variable name has been inadvertently typed in two places with different case. *TKSolver* will consider them different variables and the model will probably not solve because only one of them has a value associated with it on the variable sheet. This is one of the most common errors to check for when your model refuses to solve. See the *TKSolver* manual for complete information on naming functions and variables.

Table 1-17   TKSolver Rule Function for Example 1-7 from File EX01-07

Rule Function cyl_rule

Comment:	Calculates volume and area of cylinder
Parameter variables:	
Argument variables:	V, ratio
Result variables:	dia, height, A
S Rule	

$$; \text{Rule Function to calculate surface area and volume of a cylindrical container}$$

; to use: CALL cyl_rule (V, ratio ; dia, height , A)

   $ratio = height / dia$
   $basearea = PI() * dia^2 / 4$
   $A = basearea + PI() * dia * height$
   $V = basearea * height$

Table 1-18   TKSolver Rule Sheet for Example 1-7 from File EX01-07

; 1 - calculate the dimensions and area of the cylinder using the rule function "cyl_rule"

CALL cyl_rule (vol, ratio ; D, H , surface)

; 2 - calculate the weight of the cylinder

weight = surface * thick * dens

---

6   The comments within the rule function preceded by a semicolon[*] are optional and serve to document the function. It is a good idea to document the required CALL statement that will be used in any rule sheet or other function to activate this function. The format of the CALL statement is as shown in Tables 1-17 and 1-18. Note that the argument variables (inputs), separated by commas, are listed first and are separated from the list of comma-delimited result variables with a semicolon.

7   The rule statement is now reduced to two statements as shown in Table 1-18. The first is the CALL statement (without the preceding ; or " so it will execute). Note that the names of the variables passed to the rule function do not have to be the same as the names used in the rule function. The variable names in the rule function are "dummies," each of which takes on the label of whatever is passed to it in the CALL statement. The mapping between the actual and dummy variables is by their **order** in the argument list, first to first, second to second, etc. In this case *vol* in the rule sheet (Table 1-18) becomes *V* in the rule function (Table 1-17), *ratio* is *ratio*, *dia* becomes *D*, etc. The second rule-sheet equation calculates the weight of the cylinder. This equation could have been placed inside the rule function if desired, which would then require *thick* and *dens* to be added to its argument list.

8   When this model is solved, it gives the same solution as in Example 1-5. The variable sheet is shown in Table 1-19. Note that the dummy variables *V*, *dia*, *height*, *A*, and *basearea* do not appear on the variable sheet. Only the global variables created in the rule sheet appear there.

Table 1-19   TKSolver Variable Sheet for Example 1-7 from File EX01-07

St	Input	Variable	Output	Unit	Comments
	1 000	*vol*		cm^3	volume of container
	0.10	*thick*		cm	thickness of wall
	7.75	*dens*		g/cc	density of material
L	0.6	*ratio*			height to diameter ratio
L		*D*	12.9	cm	diameter of base of cylinder
L		*H*	7.7	cm	height of side of cylinder
		surface	441	cm^2	total surface area
L		*weight*	341.8	g	weight of empty cylinder

[*] The semicolon is used to denote a comment line in the Windows and DOS versions of TKSolver. The Macintosh version uses a quote (") instead of a semicolon for the same purpose.

9   The principal advantage of using rule (or other) functions is to encapsulate rule sets that may be useful in more than one program into a form that is easily transferred from one model to another. The localization of variables inside the function means that you can call the variables by different names in different models and still use the same rule function. Another advantage is that of modularizing the model, which means breaking it down into more tractable pieces that can be debugged and proven independently of the rest of the model. The value of this approach increases as your models get more complicated.

Procedure Functions

We will now introduce the use of procedure functions to control access to lists in the following example.

EXAMPLE 1-8

Using Procedure Functions in TKSolver

Problem   Create a procedure function to load the input list *ratio* in Example 1-7 so that the range of values for calculation can be defined by the user at run time. Use the rule function *cyl_rule* to find the height/diameter ratio and dimensions of an open-top, cylindrical container that will minimize its weight for a given volume.

Given   Volume = 1 liter, wall thickness = 0.1 cm.

Assumptions   The material is stainless steel with a mass density of 7.75 g/cc. The ratio *H/D* will be varied from 0.1 to 2.0 using 40 data points.

Solution   See Tables 1-20, 1-21, and 1-22 and the *TKSolver* file EX01-08.

1   The task here is to automatically fill a list with *N* values that start at a user defined minimum value (*min*) and end at a user-defined maximum value (*max*). The value of *N* will be user-specified as well. We cannot do this on the rule sheet or in a rule function because the declarative solver will control the order in which the statements are executed. We need to control a loop structure and cause the desired numbers to be placed in the list one by one in order. This requires the sequential solving of a procedure function.

2   To create a procedure function, open the **function sheet** from the WINDOW menu and type the desired function name (here *filalist*) in the column labeled "Name." In the column labeled "Type," put the letter *P* to identify it as a procedure function. A comment is optional. Dive on this line to open the procedure-function subsheet.

3   There are two built-in functions that will help with this task. The BLANK (*listname*) function blanks the list whose '*name* is provided in the variable *listname*. The

---

Table 1-20   TKSolver Procedure Function for Example 1-8 from File EX01-08

Procedure filalist

Comment: Fills a list with range of values

Parameter Variables:

Input Variables: N,min,max,name

Output Variables:

S   Statement

    ; N is the number of data in the list

    ; min is first value

    ; max is last value

    ; name is a string variable containing the name of the list

    ; to use: CALL filalist (N, min, max, 'listname)

    ; first blank the list

        CALL BLANK(name)

        T = min

        delta = (max - min) / (N - 1)

        FOR I = 1 to N

            PLACE (name, I) = T

            T = T + delta

        NEXT I

PLACE (*listname, I*) = x function places the value of x in the Ith element of the list whose '*name* is provided on the variable sheet in the variable *listname*.

4   The finished procedure function is shown in Table 1-20. The input variables are listed on the line with that label. Their meanings are defined in the comment lines. The heart of this procedure is the FOR-NEXT loop which runs its index I from 1 to N and uses I as an argument in the built-in function PLACE(*name,I*) = T. This statement inserts the value of T into the Ith element of the list "name." The value of T is initialized to the value *min* and increased on each pass through the loop by *delta*. *Delta* is calculated from the values of *min*, *max* and N. Note the similarity of this procedure to a BASIC computer program; the loop structure is identical.

5   The rule sheet that calls the procedure is shown in Table 1-21. The CALL to *filalist* passes the list name 'ratio for loading with N values between *min* and *max*. This will have the same effect as manually loading the list with the LISTFILL menu command, but is now automated. Open the list sheet to see that the list 'ratio now has N values from *min* to *max* in it after solving the model. The rest of the rule sheet is identical to that of Example 1-7.

6   The variable sheet for this model is shown in Table 1-22. The values of N *min,* and *max* are the only additions to the model of Example 1-7 shown in Table 1-19.

Table 1-21   TKSolver Rule Sheet for Example 1-8 from File EX01-08

; 1 - fill the input list called 'ratio with desired values
     CALL filalist (N, min, max, 'ratio)
; 2 - calculate the area and volume of the cylinder using the rule function ;cyl_rule;
     CALL cyl_rule (vol, ratio ; D, H , surface)
; 3 - calculate the weight of the cylinder
     *weight = surface * thick * dens*

This is a somewhat trivial example of the use of a procedure function[*] but it does illustrate its ability to manipulate lists (arrays) in conventional programming fashion. Other applications of procedure functions can be found in examples and case studies presented in later chapters.

List Functions

We now add to our example a list function to select the density of the material based on the input of an alphameric code to the variable sheet. Again, this is a somewhat trivial application of the power of list functions but be assured that these functions have great value in automatically selecting tabular data while calculating a model. We will use them extensively in later chapters' examples and case studies.

Table 1-22   TKSolver Variable Sheet for Example 1-8 from File EX01-08

St	Input	Variable	Output	Unit	Comments
	40	N			number of data points desired
	0.1	min			minimum value for input list
	2	max			maximum value for input list
L	0.5	ratio			height to diameter ratio
	1 000	vol		cm^3	volume of container
	0.1	thick		cm	thickness of wall
	7.75	dens		g/cc	density of material
L		D	13.7	cm	diameter of base of cylinder
L		H	6.8	cm	height of side of cylinder
		surface	439.4	cm^2	total surface area
L		weight	340.5	g	weight of empty cylinder

[*] It is also a crude implementation in that the CALL to the procedure *filalist* will be repeated (unnecessarily) each time the list solver performs an execution of the rule sheet. The only penalty in this case is a waste of execution time. The process could be speeded up by using an IF statement to limit the execution of *filalist* to the first pass through the rule sheet only. This was not done in this example in order to adhere to the principle of KISS (Keep It Simple, Stupid!).

1

EXAMPLE 1-9

Using List Functions in TKSolver

Problem	Create a list function to return the density of a selected material. Use that density value in combination with the rule function *cyl_rule* to find the height/diameter ratio and dimensions of an open-top, cylindrical container that will minimize its weight for a given volume.
Given	Volume = 1 liter, wall thickness = 0.1 cm.
Assumptions	The material is aluminum with a mass density of 2.77 g/cc. The height to diameter ratio *H/D* will be varied from 0.1 to 2.0.
Solution	See Tables 1-23, 1-24, and 1-25 and the *TKSolver* file EX01-09.

1 A list function relates the contents of two lists. The input list is called the **domain** of the function and the output list is called its **range**. The input value is passed to the list function as an argument and it returns the corresponding output value.

2 To create a list function, open the **function sheet** from the WINDOW menu and type the desired function name (here *get_dens*) in the column labeled "Name." In the column labeled "Type," put the letter *L* to identify it as a list function. A comment is optional. Dive on this line to open the list-function subsheet.

3 Table 1-23 shows this list function. When making your own, start by typing, on the line beside the label "Domain List," the name of the list that you want to use as the input. This list name can be one that already exists on the list sheet or can be a new list, still to be filled. We called the domain list *matl* for this example.

4 For this example we want **Table** mapping, which creates a one-for-one correspondence between input and output variables with no interpolation. Other possible mappings are **step**, **linear interpolation,** and **cubic interpolation**, See the *TKSolver* manual for further information on these.

Table 1-23   TKSolver List Function for Example 1-9 from File EX01-09

List Function: get_dens

Comment:	Returns the weight density of a material
Domain List:	material
Mapping:	Table
Range List:	density

Element	Domain	Range
1	'alum	2.76805
2	'steel	7.75054
3	'copper	8.58096

Table 1-24   TKSolver Rule Sheet for Example 1-9 from File EX01-09

; 1 - fill the input list called *'ratio* with desired values
> CALL *filalist (N, min, max, 'ratio)*

; 2 - retrieve the density of the chosen material from the list function *"get_rule"*
> *dens = get_dens (material)*

; 3 - calculate the dimensions and area of the cylinder using the rule function *"cyl_rule"*
> CALL *cyl_rule (vol, ratio ; D, H , surface)*

; 4 - calculate the weight of the cylinder
> *weight = surface * thick * dens*

5   Type the name of the desired output list on the line labeled "Range List." Here we used the list name *density*.

6   If these lists have already been defined on the list sheet and filled with data, they will immediately appear in the columns headed Domain and Range. If they have not yet been filled, you can fill them from this sheet. Simply type the appropriate words or numbers on each element's line. In this simple example we want some convenient labels for the materials in the domain list (*matl*) and the corresponding densities of those materials in the range list (*density*). To create your list function, type the data as shown in Table 1-23, using the single quote on the names to designate them as alphameric data.

7   The rule sheet is shown in Table 1-24. At line 2, the list function is called by using it in the assignment statement:

$$dens = get\_dens(material) \qquad (f)$$

8   Note on the variable sheet in Table 1-25 that the variable *material* is set to the value 'alum which is one of the labels in the list *matl*. When the rule (*f*) is evaluated, it passes the value 'alum to the list function *get_dens* which scans its domain list for that value. If it finds an 'alum in its domain, it returns the corresponding value from the range list, (2.76805) and places it in the variable *dens* on the variable sheet. Note that neither of the list variables *matl* or *density* that belong to the list function appears on the variable sheet.

9   The solution then proceeds as before using the returned value of *dens* which is now an output variable.

Units and Formatting

TKSolver allows automatic unit conversions within any model that possesses a **units sheet**. The units sheet must be created by the user. A fairly extensive one is supplied on disk with this text as the file UNITMAST. A small portion of this unit sheet is shown

Table 1-25   TKSolver Variable Sheet for Example 1-9 from File EX01-09

St	Input	Variable	Output	Unit	Comments
	40	N			number of data points desired
	0.1	min			minimum value for input list
	2	max			maximum value for input list
L	0.5	ratio			height to diameter ratio
	1 000.	vol		cm^3	volume of container
	0.1	thick		cm	thickness of wall
	'alum	material			one of 'alum, 'steel, or 'copper
		dens	2.77	g/cc	density of material
L		D	13.7	cm	diameter of base of cylinder
L		H	6.8	cm	height of side of cylinder
		surface	439.4	cm^2	total surface area
L		weight	121.6	g	weight of empty cylinder

as Table 1-26. This file can be merged into any *TK* model by using the MERGE command on the FILE pull-down menu. Once merged into the model, typing any of this sheet's unit abbreviations in the units column of the variable sheet will invoke conversion of that variable's units provided that a conversion factor exists on the units sheet. Formatting of variables is also possible by using the **format sheet**. The use of both units and format sheets will be demonstrated in an example.

EXAMPLE 1-10

Using Unit Sheets and Format Sheets in TKSolver

Problem   Convert the *cgs* units used in the previous example to the *SI* system.

Solution   See Tables 1-26, 1-27, and 1-28 and the *TKSolver* file EX01-10.

1   The previous examples were all calculated in the *cgs* units system. The *SI* system is generally preferred for metric units. With the units sheet shown in Table 1-26 present in the *TK* file, the conversions to *SI* units are accomplished simply by typing the desired unit symbols from Table 1-26 into the units column of any variable to be converted. The result is shown in Table 1-27. If a unit symbol that does not exist on the units sheet is typed, it will not convert the variable's value but will instead put a question mark in front of its value to indicate that the conversion is not possible with this unit sheet. If more than one symbol is desired for a given variable, such as cc and cm^3 (or cm3) for cubic centimeters, this can be accomplished by providing a unity conversion factor between the equivalent symbols as was done in Table 1-26.

2   If any variable's subsheet is opened (by diving on its line in the variable sheet), it will look like that shown in Table 1-28 for the variable *D*. Detailed information about the variable is recorded here. In this case, the status line indicates *D* is a list variable. It has a guess value of 0.02 assigned to it which will be used whenever the **direct solver** cannot resolve the rules and must invoke the **iterative solver**. Putting a guess value here eliminates the need to repeatedly put a G in the status column on the variable sheet when solving. Since this is an output variable at present, the input value line is blank and its calculated value appears on the output value line.

3   Note the two lines labeled **Display Unit** and **Calculation Unit**. The first of these defines the units currently displayed on the variable sheet and the second defines the unit in which all calculations of this variable are done. **It is critical that all variables in the model have calculation units consistent with one of the standard units systems**. In this case, the calculation unit is cm because that was the unit label first typed into the variable sheet in Example 1-2. **Whatever unit is first typed on the variable sheet becomes the calculation unit and will remain so unless changed on each variable's subsheet. Any subsequent changes to the unit label on the variable sheet change only the display unit**. This is why it is so important to be consistent with one units system when first applying unit labels to a variable sheet. To not do so may lead to erroneous numerical results. You can have any mix of display units on the variable sheet and this will not affect the correctness of the results as long as the calculation units for all variables are in a consistent units system.

Table 1-26   Part of the TKSolver Units Sheet from the File UNITMAST

From	To	Multiply By	Add Offset	Comment
lb	N	4.448 2		force
N	dyne	100 000		force
m	cm	100		length
in	cm	2.54		length
in^2	mm^2	645.162 6		area
m^2	mm^2	1 000 000		area
cm^2	mm^2	100		area
m^3	mm^3	1E+09		volume
cm^3	mm^3	1 000		volume
cm^3	cc			volume
g/cc	g/mm^3	0.001		density
kg/m^3	g/mm^3	0.000 001		density
lb/in^3	g/cc	27.680 5		density
lb/ft^3	lb/in^3	1 728		density
C	F	1.8	32	temperature

Table 1-27   TKSolver Variable Sheet for Example 1-10 from File EX01-10

St	Input	Variable	Output	Unit	Comments
	40	N			number of data points desired
	0.1	min			minimum value for input list 'ratio
	2	max			maximum value for input list 'ratio
	0.001	vol		m^3	volume of container
	0.001	thick		m	thickness of wall
	alum	material			one of 'alum, 'steel, or 'copper
		dens	2 768.	kg/m^3	density of material
L		D	0.137	m	diameter of base of cylinder
L		H	0.068	m	height of side of cylinder
L	0.5	ratio			height to diameter ratio
		surface	0.044	m^2	total surface area
L		weight	0.122	kg	weight of empty cylinder

4   The **Numeric Format** line allows specification of a format for this variable provided that the format code (*d3*) is defined on the format sheet. Table 1-29 shows the format subsheet for this *d3* specification. The user can make up any set of format codes and define their attributes on this sheet. The *TK* file FORMATS supplied on disk has a format sheet that defines useful decimal and exponential formats. This file can be merged into any *TK* model. A default global format for the entire variable sheet can also be defined from the SETTINGS menu found under the FORMAT pull-down menu. The file STUDENT contains both the units sheet from UNITMAST and the format sheet from FORMATS.

Table 1-28   TKSolver Variable Subsheet for Example 1-10 from File EX01-10

	Variable: D
Status:	L
First Guess:	0.02
Associated List:	D
Input Value:	
Output Value:	0.137
Numeric Format:	d3
Display Unit:	m
Calculation Unit:	cm
Comment:	Diameter of base of cylinder

Table 1-29   TKSolver Format Sheet for Example 1-10 from File EX01-10

Format: d3	
Comment:	3 decimal places
Numeric Notation:	Decimal
Significant Digits:	18
Decimal Places:	3
Padding:	Zero
Decimal Point Symbol:	.
Digit Grouping Symbol:	,
Zero Representation:	0.000
+/− Notation:	– Only
Prefix:	
Suffix:	
Justification:	Right
Left Margin Width:	0
Right Margin Width:	0

1.11   SUMMARY

Design can be fun and frustrating at the same time. Because design problems are very unstructured, a large part of the task is creating sufficient structure to make it solvable. This naturally leads to multiple solutions. To students used to seeking an answer that matches the one in the "back of the book" this exercise can be frustrating. There is no "one right answer" to a design problem, only answers that are arguably better or worse than others. The marketplace has many examples of this phenomenon. How many different makes and models of new automobiles are available? Don't they all do more or less the same task? But you probably have your own opinion about which ones do the task better than others. Moreover, the task definition is not exactly the same for all examples. A four-wheel-drive automobile is designed for a slightly different problem definition than is a two-seat sports car (though some examples incorporate both those features).

The message to the beginning designer then is to be open-minded about the design problems posed. Don't approach design problems with the attitude of trying to find "the right answer" as there is none. Rather, be daring! Try something radical. Then test it with analysis. When you find it doesn't work, don't be disappointed; instead realize that you have learned something about the problem you didn't know before. Negative results are still results! We learn from our mistakes and can then design a better solution the next time. This is why *iteration* is so crucial to successful design.

The computer is a necessary tool to the solution of contemporary engineering problems. Problems can be solved more quickly and more accurately with proper use of computer-aided engineering (CAE) software. However, the results are only as good as the quality of the engineering models and data used. The engineer should not rely on computer-generated solutions without also developing and applying a thorough understanding of the fundamentals on which the model and the CAE tools are based.

Important Equations Used in this Chapter

See the referenced sections for information on the proper use of these equations.

Mass (see Section 1.9):

$$m = \frac{W}{g_c} \tag{1.3}$$

Dynamic Force - for use with standard mass units (kg, slugs, blobs) (See Section 1.9):

$$F = ma \tag{1.4a}$$

Dynamic Force - for use with mass in $lb_m = lb_f$ (see Section 1.9):

$$F = \frac{ma}{g_c} \tag{1.4b}$$

1.12   REFERENCES

1 *Random House Dictionary of the English Language*. 2nd ed. unabridged, S. B. Flexner, ed, Random House: New York, 1987, pp. 1151.

2 **R. L. Norton**, *Design of Machinery: An Introduction to the Synthesis and Analysis of Mechanisms and Machines*. McGraw-Hill: New York, 1992, pp. 6-13.

3 **Cadkey 6**, Cadkey Inc., Windsor, Ct.

4 **Aries ConceptStation 5.1**, MSC Aries Technologies Inc., Lowell, MA.

5 **ADAMS**, Mechanical Dynamics, Ann Arbor, Mich.

6 **Working Model** 2.0, Knowledge Revolution, San Francisco, CA.

7 **ProEngineer**, Parametric Technology Corp., Waltham, Mass.

8 **TKSolver 2.0**, Universal Technical Systems, Rockford, Il.

9 **Mathcad 4.0**, Mathsoft Inc., Cambridge, Mass.

10 **Lotus 123**, Lotus Development Corp., Cambridge, Mass.

11 **Excel 4.0**, Microsoft Corp., Redmond, Wash.

1.13   BIBLIOGRAPHY

For information on creativity and the design process, the following are recommended:

J. L. Adams, *The Care and Feeding of Ideas*. 3rd ed. Addison Wesley: Reading Mass, 1986.

J. L. Adams, *Conceptual Blockbusting*. 3rd ed. Addison Wesley: Reading, Mass, 1986.

J. R. M. Alger and C. V. Hays, *Creative Synthesis in Design*. Prentice-Hall: Englewood Cliffs, N.J., 1964.

M. S. Allen, *Morphological Creativity*. Prentice-Hall: Englewood Cliffs, N.J., 1962.

H. R. Buhl, *Creative Engineering Design*. Iowa State University Press: Ames, Iowa, 1960.

W. J. J. Gordon, *Synectics*. Harper and Row: New York, 1962.

J. W. Haefele, *Creativity and Innovation*. Reinhold: New York, 1962.

L. Harrisberger, *Engineersmanship*. 2nd ed. Brooks/Cole: Monterey, Calif, 1982.

D. A. Norman, *The Psychology of Everyday Things*. Basic Books: New York, 1986.

A. F. Osborne, *Applied Imagination*. Scribners: New York, 1963.

C. W. Taylor, *Widening Horizons in Creativity*. John Wiley: New York, 1964.

E. K. Von Fange, *Professional Creativity*. Prentice-Hall: Englewood Cliffs, N.J., 1959.

For information on writing engineering reports, the following are recommended:

R. Barrass, *Scientists Must Write*. Chapman and Hall: New York, 1978.

W. G. Crouch and R. L. Zetler, *A Guide to Technical Writing*. 3rd ed. The Ronald Press Co.: New York, 1964.

D. S. Davis, *Elements of Engineering Reports*. Chemical Publishing Co.: New York, 1963.

D. E. Gray, *So You Have to Write a Technical Report*. Information Resources Press: Washington, D.C., 1970.

H. B. Michaelson, *How to Write and Publish Engineering Papers and Reports*. ISI: Philadelphia, Pa., 1982.

J. R. Nelson, *Writing the Technical Report*. 3rd ed. McGraw-Hill: New York, 1952.

1.14   PROBLEMS

1-1   It is often said, *"Build a better mousetrap and the world will beat a path to your door."* Consider this problem and write a goal statement and a set of at least 12 task specifications that you would apply to its solution. Then suggest 3 possible concepts to achieve the goal. Make annotated, freehand sketches of the concepts.

1-2   A bowling machine is desired to allow quadriplegic youths, who can only move a joystick, to engage in the sport of bowling at a conventional bowling alley. Consider the factors involved, write a goal statement, and develop a set of at least 12 task specifications that constrain this problem. Then suggest 3 possible concepts to achieve the goal. Make annotated, freehand sketches of the concepts.

1-3   A quadriplegic needs an automated page-turner to allow her to read books without assistance. Consider the factors involved, write a goal statement, and develop a set of at least 12 task specifications that constrain this problem. Then suggest 3 possible concepts to achieve the goal. Make annotated, freehand sketches of the concepts.

*1-4   Convert a mass of 1 000 lb_m to (a) lbf, (b) slugs, (c) blobs, (d) kg.

*1-5   A 250-lb_m mass is accelerated at 40 in/sec^2. Find the force in lb needed for this acceleration.

*1-6   Express a 100 kg mass in units of slugs, blobs, and lb_m. How much does this mass weigh in lb_f and in N? Check your answer with the *TKSolver* units sheet provided.

1-7   Write a *TKSolver* program to calculate the cross-sectional properties for all the shapes shown in Appendix A.

1-8   Convert the program in Problem 1-7 to a set of rule functions, one for each shape in Appendix A. Each function should return the area and second moments of area for one section shape.

1-9   Write a *TKSolver* program to calculate the mass properties for all the solids shown in Appendix B.

1-10   Convert the program in Problem 1-9 to a set of rule functions, one for each solid in Appendix B. Each function should return the volume and second moments of mass for one section shape.

---

*   Answers to these problems are provided in Appendix G.

MATERIALS AND PROCESSES

*There is no subject so old that something
new cannot be said about it.*

DOSTOEVSKY

2.0 INTRODUCTION

Whatever you design, you must make it out of some material and be able to manufacture it. A thorough understanding of material properties, treatments and manufacturing processes is essential to good machine design. It is assumed that the reader has had a first course in material science. This chapter presents a brief review of some basic metallurgical concepts and a short summary of engineering material properties to serve as background for what follows. This is not intended as a substitute for a text on material science and the reader is encouraged to review references such as those listed in the bibliography of this chapter for more detailed information. Later chapters of this text will explore some of the common material-failure modes in more detail.

Table 2-0 shows the variables used in this chapter and references the equations, figures, or sections in which they are used. At the end of the chapter, a summary section is provided which groups the significant equations from this chapter for easy reference and identifies the chapter section in which they are discussed.

2.1 MATERIAL PROPERTY DEFINITIONS

Mechanical properties of a material are generally determined through destructive testing of samples under controlled loading conditions. The test loadings do not accurately duplicate actual service loadings experienced by machine parts except in certain special cases. Also, there is no guarantee that the particular piece of material you purchase

2

Table 2-0   Variables Used in This Chapter

Symbol	Variable	ips units	SI units	See
A	area	in^2	m^2	Sect. 2.1
A_0	original area, test specimen	in^2	m^2	Eq. 2.1a
E	Young's modulus	psi	Pa	Eq. 2.2
el	elastic limit	psi	Pa	Figure 2-2
f	fracture point	none	none	Figure 2-2
G	shear modulus, modulus of rigidity	psi	Pa	Eq. 2.4
HB	Brinell hardness	none	none	Eq. 2.10
HRB	Rockwell B hardness	none	none	Sect. 2.4
HRC	Rockwell C hardness	none	none	Sect. 2.4
HV	Vickers hardness	none	none	Sect. 2.4
J	polar second moment of area	in^4	m^4	Eq. 2.5
K	stress intensity	$kpsi\text{-}in^{0.5}$	$MPa\text{-}m^{0.5}$	Sect. 2.1
K_c	fracture toughness	$kpsi\text{-}in^{0.5}$	$MPa\text{-}m^{0.5}$	Sect. 2.1
l_0	gage length, test specimen	in	m	Eq. 2.3
N	number of cycles	none	none	Figure 2-10
P	force or load	lb	N	Sect. 2.1
pl	proportional limit	psi	Pa	Figure 2-2
r	radius	in	m	Eq. 2.5a
S_d	standard deviation	any	any	Eq. 2.9
S_e	endurance limit	psi	Pa	Figure 2-10
S_{el}	strength at elastic limit	psi	Pa	Eq. 2.7
S_f	fatigue strength	psi	Pa	Figure 2-10
S_{us}	ultimate shear strength	psi	Pa	Eq. 2.5
S_{ut}	ultimate tensile strength	psi	Pa	Figure 2-2
S_y	tensile yield strength	psi	Pa	Figure 2-2
S_{ys}	shear yield strength	psi	Pa	Eq. 2.6
T	torque	lb-in	N-m	Sect. 2.1
U_R	modulus of resilience	psi	Pa	Eq. 2.7
U_T	modulus of toughness	psi	Pa	Eq. 2.8
y	yield point	none	none	Figure 2-2
ε	strain	none	none	Eq. 2.1b
σ	tensile stress	psi	Pa	Sect. 2.1
τ	shear stress	psi	Pa	Eq. 2.3
θ	angular deflection	rad	rad	Eq. 2.3
μ	arithmetic mean value	any	any	Eq. 2.9b
ν	Poisson's ratio	none	none	Eq. 2.4

2

for your part will exhibit the same strength properties as the samples of similar materials tested previously. There will be some statistical variation in the strength of any particular sample compared to the average tested properties for that material. For this reason, much of the published strength data are given as minimum values. It is with these caveats that we must view all published material-property data, as it is the engineer's responsibility to ensure the safety of his or her design.

The best material-property data will be obtained from destructive or nondestructive testing under actual service loadings of prototypes of your actual design, made from the actual materials by the actual manufacturing process. This is typically done only when the economic and safety risks are high. Manufacturers of aircraft, automobiles, motorcycles, snowmobiles, farm equipment, and other products regularly instrument and test finished assemblies under real or simulated service conditions.

In the absence of such specific test data, the engineer must adapt and apply published material property data from standard tests to the particular situation. The *American Society for Testing and Materials* (ASTM) defines standards for test specimens and test procedures for a variety of material property measurements.[*] The most common material test used is the tensile test.

The Tensile Test

A typical tensile test specimen is shown in Figure 2-1. This tensile bar is machined from the material to be tested in one of several standard diameters d_o and gage lengths l_o. The gage length is an arbitrary length defined along the small-diameter portion of the specimen by two indentations so that its increase can be measured during the test. The larger-diameter ends of the bar are threaded for insertion into a tensile test machine which is capable of applying either controlled loads or controlled deflections to the ends of the bar, and the gage-length portion is mirror polished to eliminate stress concentrations from surface defects. The bar is stretched slowly in tension until it breaks, while the load and the distance across the gage length (or alternatively the strain) are continuously monitored. The result is a stress-strain plot of the material's behavior under load as shown in Figure 2-2a which depicts a curve for a low-carbon or "mild" steel.

STRESS AND STRAIN   Note that the parameters measured are load and deflection, but those plotted are stress and strain. **Stress** (σ) is defined as *load per unit area* (or *unit load*) and for the tensile specimen is calculated from

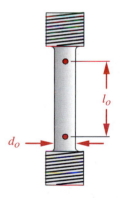

FIGURE 2-1

A Tensile-Test Specimen

$$\sigma = \frac{P}{A_o} \qquad (2.1a)$$

where P is the applied load at any instant and A_o is the original cross-sectional area of the specimen. The stress is assumed to be uniformly distributed across the cross section. The stress units are psi or Pa.

* ASTM, 1994 Annual Book of ASTM Standards, Vol. 03.01, Am. Soc. for Testing and Materials, Philadelphia, PA.

Strain is the *change in length per unit length* and is calculated from

$$\varepsilon = \frac{l - l_o}{l_o} \qquad (2.1b)$$

where l_o is the original gage length and l is the gage length at any load P. The strain is unitless, being length divided by length.

MODULUS OF ELASTICITY   This tensile stress-strain curve provides us with a number of useful material parameters. Point *pl* in Figure 2-2*a* is the **proportional limit** below which the stress is proportional to the strain as expressed by **Hooke's law**:

$$E = \frac{\sigma}{\varepsilon} \qquad (2.2)$$

where E defines the slope of the stress-strain curve up to the proportional limit and is called **Young's modulus** or the **modulus of elasticity** of the material. E is a measure of the stiffness of the material in its elastic range and has the units of stress. Most metals exhibit this linear stiffness behavior and also have elastic moduli that vary very little with heat treatment or with the addition of alloying elements. For example, the highest strength steel has the same E as the lowest strength steel at about 30 Mpsi (207 GPa). For most ductile materials (defined below), the modulus of elasticity in compression is the same as in tension. This is not true for cast irons and other brittle materials (defined below) or for magnesium.

ELASTIC LIMIT   The point labeled *el* in Figure 2-2*a* is the **elastic limit**, or the point beyond which the material will take a permanent set, or plastic deformation. The elastic limit marks the boundary between the **elastic-behavior** and **plastic-behavior** regions of the material. Points *el* and *pl* are typically so close together that they are often considered to be the same.

YIELD STRENGTH   At a point *y* slightly above the elastic limit, the material begins to yield more readily to the applied stress and its rate of deformation increases (note the lower slope). This is called the **yield point** and the value of stress at that point defines the **yield strength** S_y of the material.

Materials that are very ductile, such as low-carbon steels, will sometimes show an apparent drop in stress just beyond the yield point, as shown in Figure 2-2*a*. Many less ductile materials, such as aluminum and medium- to high-carbon steels will not exhibit this apparent drop in stress and will look more like Figure 2-2*b*. The yield strength of a material that does not exhibit a clear yield point has to be defined with an offset line, drawn parallel to the elastic curve and offset some small percentage along the strain axis. An offset of 0.2% strain is most often used. The yield strength is then taken at the intersection of the stress-strain curve and the offset line as shown in Figure 2-2*b*.

ULTIMATE TENSILE STRENGTH   The stress in the specimen continues to increase non-linearly to a peak or **ultimate tensile strength** value S_{ut} at point *u*. This is considered to be the largest tensile stress the material can sustain before breaking. However, for the ductile steel curve shown, the stress appears to fall off to a smaller value at the fracture point f. The drop in apparent stress before the fracture point (from u to f in Figure 2-2a) is an artifact caused by the "necking down" or reduction in area of the ductile specimen. The reduction of cross-sectional area is nonuniform along the length of the specimen as can be seen in Figure 2-3.

Because the stress is calculated using the original area A_o in equation 2.1a, it understates the true value of stress after point u. It is difficult to accurately monitor the dynamic change in cross-sectional area during the test, so these errors are accepted. The strengths of different materials can still be compared on this basis. When based on the uncorrected area A_o this is called the **engineering stress-strain curve,** as shown in Figure 2-2.

The stress at fracture is actually larger than shown. Figure 2-2 also shows the **true stress-strain curve** that would result if the change in area were accounted for. The **engineering stress-strain** data from Figure 2-2 are typically used in practice. The most commonly used strength values for **static loading** are the yield strength S_y and the ultimate tensile strength S_{ut}. The material stiffness is defined by Young's modulus, E.

In comparing the properties of different materials it is quite useful to express those properties normalized to the material's density. Since light weight is nearly always a goal in design, we seek the lightest material that has sufficient strength and stiffness to withstand the applied loads. The **specific strength** of a material is defined as *the strength divided by the density*. Unless otherwise specified, strength in this case is assumed to mean ultimate tensile strength, though any strength criterion can be so normalized. The **strength-to-weight ratio** (SWR) is another way to express the specific strength. **Specific stiffness** is the *Young's modulus divided by material density*.

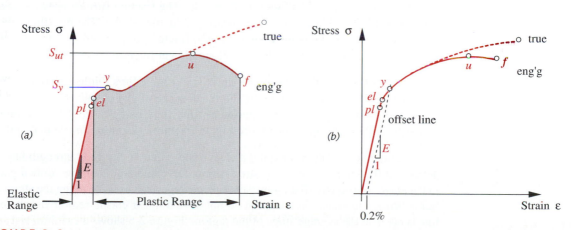

FIGURE 2-2

Engineering and True Stress-Strain Curves for Ductile Materials: (a) Low-Carbon Steel (b) Annealed High-Carbon Steel

FIGURE 2-3

A Tensile Test Specimen of Mild, Ductile Steel After Fracture

Ductility and Brittleness

The tendency for a material to deform significantly before fracturing is a measure of its ductility. The absence of significant deformation before fracture is called brittleness.

DUCTILITY   The stress-strain curve in Figure 2-2a is of a ductile material, mild steel. Take a common paper clip made of mild-steel wire. Straighten it out with your fingers. Bend it into some new shape. You are yielding this ductile steel wire but not fracturing it. You are operating between point *y* and point *f* on the stress-strain curve of Figure 2-2a. The presence of a significant plastic region on the stress-strain curve is evidence of ductility.

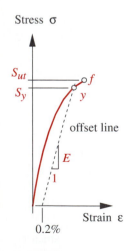

FIGURE 2-4

Stress-Strain Curve of a Brittle Material

Figure 2-3 shows a test specimen of ductile steel after fracture. The distortion called *necking-down* can clearly be seen at the break. The fracture surface appears torn and is laced with hills and valleys, also indicating a ductile failure. The **ductility** of a material is measured by its percent elongation to fracture, or percent reduction in area at fracture. Materials with more than 5% elongation at fracture are considered ductile.

BRITTLENESS   Figure 2-4 shows a stress-strain curve for a brittle material. Note the lack of a clearly defined yield point and the absence of any plastic range before fracture. Repeat your paper-clip experiment, this time using a wooden toothpick or matchstick. Any attempt to bend it results in fracture. Wood is a brittle material.

Brittle materials do not exhibit a clear yield point, so the yield strength has to be defined at the intersection of the stress-strain curve and an offset line, drawn parallel to the elastic curve and offset some small percentage such as 0.2% along the strain axis. Some brittle materials like cast iron do not have a linear elastic region and the offset line is taken at the average slope of the region. Figure 2-5 shows a cast iron test specimen after fracture. The break shows no evidence of necking and has the finer surface contours typical of a brittle fracture.

FIGURE 2-5

A Tensile Test Specimen of Brittle Cast Iron After Fracture

The same metals can be either ductile or brittle depending on the way they are manufactured, worked, and heat treated. Metals that are **wrought** (meaning drawn or pressed into shape in a solid form while either hot or cold) can be more ductile than metals that are cast by pouring molten metal into a mold or form. There are many exceptions to this broad statement, however. The cold working of metal (discussed below) tends to reduce its ductility and increase its brittleness. Heat treatment (discussed below) also has a marked effect on the ductility of steels. Thus it is difficult to generalize about the relative ductility or brittleness of various materials. A careful look at all the mechanical properties of a given material will tell the story.

The Compression Test

The tensile test machine can be run in reverse to apply a compressive load to a specimen that is a constant diameter cylinder as shown in Figure 2-6. It is difficult to obtain a useful stress-strain curve from this test because a ductile material will yield and increase its cross-sectional area, as shown in Figure 2-6a, eventually stalling the test machine. The ductile sample will not fracture in compression. If enough force were avail-

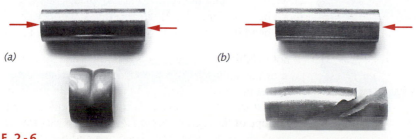

(a)   (b)

FIGURE 2-6

Compression Test Specimens After Failure *(a)* Ductile Steel *(b)* Brittle Cast Iron

2

able from the machine, it could be crushed into a pancake shape. Most ductile materials have compressive strengths similar to their tensile strengths and the tensile stress-strain curve is used to represent their compressive behavior as well. A material that has essentially equal tensile and compressive strengths is called an **even material**.

Brittle materials will fracture when compressed. A failed specimen of brittle cast iron is shown in Figure 2-6b. Note the rough, angled fracture surface. The reason for the failure on an angled plane is discussed in Chapter 4. Brittle materials generally have much greater strength in compression than in tension. Compressive stress-strain curves can be generated since the material fractures rather than crushes and the cross-sectional area doesn't change appreciably. A material that has different tensile and compressive strengths is called an **uneven material**.

The Bending Test

A thin rod, as shown in Figure 2-7, is simply supported at each end as a beam and loaded transversely in the center of its length until it fails. If the material is ductile, failure is by yielding, as shown in Figure 2-7a. If the material is brittle, the beam fractures as shown in Figure 2-7b. Stress-strain curves are not generated from this test because the stress distribution across the cross section is not uniform. The tensile test's σ–ε curve is used to predict failure in bending since the bending stresses are tensile on the convex side and compressive on the concave side of the beam.

The Torsion Test

The shear properties of a material are more difficult to determine than its tensile properties. A specimen similar to the tensile test specimen is made with noncircular details on its ends so that it can be twisted axially to failure. Figure 2-8 shows two such samples, one of ductile steel and one of brittle cast iron. Note the painted lines along their lengths. The lines were originally straight in both cases. The helical twist in the ductile specimen's line after failure shows that it wound up for several revolutions be-

(a)   (b)

FIGURE 2-7

Bending Test Specimens After Failure (a) Ductile Steel (b) Brittle Cast Iron

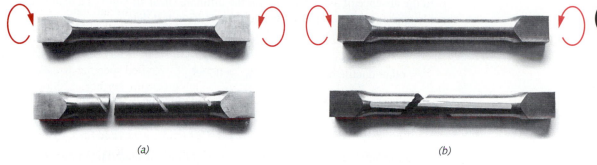

FIGURE 2-8

Torsion Test Specimens After Failure (a) Ductile Steel (b) Brittle Cast Iron

fore breaking. The brittle, torsion-test specimen's line is still straight after failure as there was no significant plastic distortion before fracture.

MODULUS OF RIGIDITY The stress-strain relation for pure torsion is defined by

$$\tau = \frac{Gr\theta}{l_o} \tag{2.3}$$

where τ is the shear stress, r is the radius of the specimen, l_o is the gage length, θ is the angular twist in radians, and G is the shear modulus or **modulus of rigidity**. G can be defined in terms of Young's modulus E and Poisson's ratio ν:

$$G = \frac{E}{2(1+\nu)} \tag{2.4}$$

Poisson's ratio (ν) is the ratio between lateral and longitudinal strain and for most metals is around 0.3 as shown in Table 2-1.

ULTIMATE SHEAR STRENGTH The breaking strength in torsion is called the ultimate shear strength or modulus of rupture S_{us} and is calculated from

$$S_{us} = \frac{Tr}{J} \tag{2.5a}$$

where T is the applied torque necessary to break the specimen, r is the radius of the specimen, and J is the polar second moment of area of the cross section. The distribution of stress across the section loaded in torsion is not uniform. It is zero at the center and maximum at the outer radius. Thus the outer portions have already plastically yielded while the inner portions are still below the yield point. This nonuniform stress distribution in the torsion test (unlike the uniform distribution in the tension test) is the reason for calling the measured value at failure of a solid bar in torsion a *modulus of rupture*. A thin-walled tube is a better torsion-test specimen than a solid bar for this reason and can give a better measure of the ultimate shear strength.

Table 2-1

Poisson's Ratio ν

Material	ν
Aluminum	0.34
Copper	0.35
Iron	0.28
Steel	0.28
Magnesium	0.33
Titanium	0.34

2

In the absence of available data for the ultimate shear strength of a material, a reasonable approximation can be obtained from tension test data:

steels : $S_{us} \cong 0.80 S_{ut}$

                                                                                                      (2.5b)

other ductile metals : $S_{us} \cong 0.75 S_{ut}$

Note that the shear yield strength has a different relationship to the tensile yield strength:

$$S_{sy} \cong 0.58 S_y \qquad\qquad (2.6)$$

This relationship is derived in Chapter 5 where failure of materials under static loading is discussed in more detail.

Fatigue Strength and Endurance Limit

The tensile test and the torsion test both apply loads slowly and only once to the specimen. These are static tests and measure static strengths. While some machine parts may see only static loads in their lifetime, most will see loads and stresses that vary with time. Materials behave very differently in response to loads that come and go (called **fatigue loads**) than they do to loads that remain static. Most of machine design deals with the design of parts for time-varying loads, so we need to know the **fatigue strength** of materials under these loading conditions.

One test for fatigue strength is the R. R. Moore rotating-beam test in which a similar, but slightly smaller, test specimen than that shown in Figure 2-1 is loaded as a beam in bending while being rotated by a motor. Recall from your first course in strength of materials that a bending load causes tension on one side of a beam and compression on the other. (See Section 4-10 for a review of beams in bending.) The rotation of the beam causes any one point on the surface to go from compression to tension to compression each cycle. This creates a load-time curve as shown in Figure 2-9.

The test is continued at a particular stress level until the part fractures and the number of cycles N is then noted. Many samples of the same material are tested at various stress levels S until a curve similar to Figure 2-10 is generated. This is called a Wohler strength-life diagram or an S-N diagram. It depicts the breaking strength of a particular material at various numbers of repeated cycles of fully-reversed stress.

Note in Figure 2-10 that the **fatigue strength** S_f at one cycle is the same as the static strength S_{ut}, and it decreases steadily with increasing numbers of cycles N (on a log-log plot) until reaching a plateau at about 10^6 cycles. This plateau in fatigue strength exists only for certain metals (notably steels and some titanium alloys) and is called the endurance limit S_e'. Fatigue strengths of other materials keep falling beyond that point. While there is considerable variation among materials, their raw (or uncorrected) fatigue strengths at about $N = 10^6$ cycles tend to be no more than about 40-50% of their static tensile strength S_{ut}. This is a significant reduction and, as we will learn in Chapter 6,

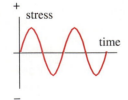

FIGURE 2-9

Time-Varying Loading

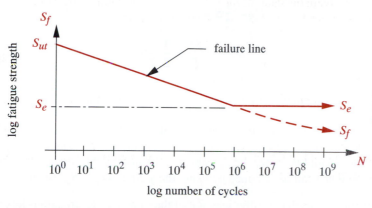

An endurance limit S_e exists for some ferrous metals and titanium alloys. Other materials show no endurance limit.

FIGURE 2-10

Wohler Strength-Life or *S-N* Diagram Plots Fatigue Strength Against Number of Fully Reversed Stress Cycles

further reductions in the fatigue strengths of materials will be necessary due to other factors such as surface finish and type of loading.

It is important at this stage to remember that the tensile stress-strain test does not tell the whole story and that a material's static strength properties are seldom adequate by themselves to predict failure in a machine-design application. This topic of fatigue strength and endurance limit is so important and fundamental to machine design that we devote Chapter 6 exclusively to a study of fatigue failure.

The rotating-beam test is now being supplanted by axial-tension tests performed on modern test machines which can apply time-varying loads of any desired character to the axial-test specimen. This approach provides more testing flexibility and more accurate data because of the uniform stress distribution in the tensile specimen. The results are consistent with (but slightly lower-valued than) the historical rotating-beam test data for the same materials.

Impact Resistance

The stress-strain test is done at very low, controlled strain rates, allowing the material to accommodate itself to the changing load. If the load is suddenly applied, the energy absorption capacity of the material becomes important. The energy in question is its **strain energy**, or the area under the stress-strain curve. The **resilience** and **toughness** of the material are functions of this area.

RESILIENCE   The ability of a material to absorb energy without permanent deformation is called its **resilience** U_R (also called **modulus of resilience**) and is proportional to the area under the stress-strain curve up to the elastic limit, shown as the color-shaded area in Figure 2-2*a*. Resilience is defined as:

2

$$U_R = \frac{1}{2} S_{el} \varepsilon_{el}$$

$$= \frac{1}{2} S_{el} \frac{S_{el}}{E} = \frac{1}{2} \frac{S_{el}^2}{E} \qquad (2.7)$$

$$U_R \cong \frac{1}{2} \frac{S_y^2}{E}$$

where S_{el} and ε_{el} represent, respectively, the strength and strain at the elastic limit. Substitution of Hooke's law from equation 2.2 expresses the relationship in terms of strength and Young's modulus. Since the S_{el} value is seldom available, a reasonable approximation of resilience can be obtained by using the yield strength S_y instead.

This relationship shows that a stiffer material of the same elastic strength is less resilient than a more compliant one. A rubber ball can absorb more energy without permanent deformation than one made of glass.

TOUGHNESS   The ability of a material to absorb energy without fracture is called its **toughness** U_T (also called **modulus of toughness**) and is proportional to the area under the stress-strain curve up to the fracture point, shown as the entire shaded area in Figure 2-2a. Toughness is defined as:

$$U_T = \int_0^{S_{ut}} \sigma d\varepsilon$$

$$\cong \left(\frac{S_y + S_{ut}}{2} \right) \varepsilon_f \qquad (2.8)$$

where S_{ut} and ε_f represent, respectively, the ultimate tensile strength and the strain at fracture. Since an analytical expression for the stress-strain curve is seldom available for actual integration, an approximation of toughness can be obtained by using the average of the yield and ultimate strengths and the strain at fracture to calculate an area. The units of toughness and resilience are energy per unit volume (in-lb/in^3 or joules/m^3). Note that these units are numerically equivalent to psi or Pa.

A ductile material of similar ultimate strength to a brittle one will be much more tough. A sheet-metal automobile body will absorb more energy from a collision through plastic deformation than will a brittle, fiberglass body.[*]

IMPACT TESTING   Various tests have been devised to measure the ability of materials to withstand impact loading. The **Izod** and the **Charpy** tests are two such procedures which involve striking a notched specimen with a pendulum and recording the kinetic energy needed to break the specimen at a particular temperature. While these data do not directly correlate with the area under the stress-strain curve, they nevertheless provide a means to compare the energy absorption capacity of various materials under controlled conditions. Materials handbooks such as those listed in this chapter's bibliography give data on the impact resistance of various materials.

[*] It is interesting to note that one of the toughest and strongest materials known is that of *spider webs*! These tiny arachnids spin a monofilament that has an ultimate tensile strength of 200 to 300 kpsi (1380 to 2070 MPa) and 35% elongation to fracture! It also can absorb more energy without rupture than any fiber known, absorbing 3 times as much energy as *Kevlar*, the man-made fiber used for bullet-proof vests.

Fracture Toughness

Fracture toughness K_c (not to be confused with the modulus of toughness defined above) is *a material property that defines its ability to resist stress at the tip of a crack.* The fracture toughness of a material is measured by subjecting a standardized, pre-cracked test specimen to cyclical tensile loads until it breaks. Cracks create very high local stress concentrations which cause local yielding (see Section 4.13). The effect of the crack on the local stress is measured by a **stress intensity factor** K which is defined in Section 5.3. When the stress intensity K reaches the fracture toughness K_c, a sudden fracture occurs with no warning. The study of this failure phenomenon is called **fracture mechanics** and it is discussed in more detail in Chapters 5 and 6.

Creep and Temperature Effects

The tensile test, while slow, does not last long compared to the length of time an actual machine part may be subjected to constant loading. All materials will, under the right environmental conditions (particularly elevated temperatures), slowly creep (deform) under stress loadings well below the level (yield point) deemed safe in the tensile test. Ferrous metals tend to have negligible creep at room temperature or below. Their creep rates increase with increasing ambient temperature, usually becoming significant around 30-60% of the material's absolute melting temperature.

Low-melt-temperature metals such as lead, and many polymers, can exhibit significant creep at room temperature as well as increasing creep rates at higher temperatures. Creep data for engineering materials are quite sparse due to the expense and time required to develop the experimental data. The machine designer needs to be aware of the creep phenomenon and obtain the latest manufacturer's data on the selected materials if high ambient temperatures are anticipated or if polymers are specified. The creep phenomenon is more complex than this simple description implies. See the bibliography to this chapter for more complete and detailed information on creep in materials.

It is also important to understand that all material properties are a function of temperature and published test data are usually generated at room temperature. Increased temperature usually reduces strength. Many materials that are ductile at room temperature can behave as brittle materials at low temperatures. Thus, if your application involves either elevated or low temperatures, you need to seek out relevant material property data for your operating environment. Material manufacturers are the best source of up-to-date information. Most manufacturers of polymers publish creep data for their materials at various temperatures.

2.2 THE STATISTICAL NATURE OF MATERIAL PROPERTIES

Some published data for material properties represent average values of many samples tested. (Other data are stated as minimum values.) The range of variation of the published test data is sometimes stated, sometimes not. Most material properties will vary about the average or mean value according to some statistical distribution such as the *Gaussian* or *normal distribution* shown in Figure 2-11. This curve is defined in terms of two parameters, the *arithmetic mean* μ and the *standard deviation* S_d. The equation of the Gaussian distribution curve is

$$f(x) = \frac{1}{\sqrt{2\pi}S_d} \exp\left[-\frac{(x-\mu)^2}{2S_d^2}\right], \qquad -\infty \le x \le \infty \qquad (2.9a)$$

where *x* represents some material parameter, *f(x)* is the frequency with which that value of *x* occurs in the population, and μ and S_d are defined as

$$\mu = \frac{1}{n}\sum_{i=1}^{n} x_i \qquad (2.9b)$$

$$S_d = \sqrt{\frac{1}{n-1}\sum_{i=1}^{n}(x_i - \mu)^2} \qquad (2.9c)$$

The *mean* μ defines the most frequently occurring value of *x* at the peak of the curve and the *standard deviation* S_d is a measure of the "spread" of the curve about the *mean*. A small value of S_d relative to μ means that the entire population is clustered closely about the mean. A large S_d indicates that the population is widely disbursed about the mean. We can expect to find 68% of the population within μ ± 1S_d, 95% within μ ± 2S_d, and 99% within μ ± 3S_d.

There is considerable scatter in multiple tests of the same material under the same test conditions. Note that there is a 50% chance that the samples of any material that you buy will have a strength less than that material's published mean value. Thus, you may not want to use the mean value alone as a predictor of the strength of a randomly chosen sample of that material. If the standard deviation of the test data is published along with the mean, we can "factor it down" to a lower value that is predictive of some larger percentage of the population based on the ratios listed above. For example, if you want to have a 99% probability that all possible samples of material are stronger than your assumed material strength, you would subtract 3S_d from μ to get an allowable value for your design. This assumes that the material property's distribution is Gaussian and not skewed toward one end or the other of the spectrum. If a minimum value of the material property is given (and used), then its statistical distribution is not of concern.

Usually, no data are available on the standard deviation of the material samples tested. But you can still choose to reduce the published mean strength by a reliability factor based on an assumed S_d. One such approach assumes S_d to be some percentage of μ based on experience. Haugen and Wirsching[1] report that the standard deviations of strengths of steels seldom exceed 8% of their mean values. Table 2-2 shows reliability

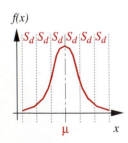

f(x)

μ   *x*

FIGURE 2-11

The Gaussian (Normal) Distribution

reduction factors based on an assumption of $S_d = 0.08\,\mu$ for various reliabilities. Note that a 50% reliability has a factor of 1 and the factor reduces as you choose higher reliability. The reduction factor is multiplied by the mean value of the relevant material property. For example, if you wish 99.99% of your samples to meet or exceed the assumed strength, multiply the mean strength value by 0.702.

In summary, the safest approach is to develop your own material property data for the particular materials and loading conditions relevant to your design. Since this approach is usually prohibitively expensive in both time and money, the engineer often must rely on published material property data. Some published strength data are expressed as the minimum strength to be expected in a statistical sample, but other data may be given as the average value for the samples tested. In that case, some of the tested material samples failed at stresses lower than the average value and your design strength may need to be reduced accordingly.

Table 2-2
Reliability Factors
for $S_d = 0.08\,\mu$

Reliability %	Factor
50	1.000
90	0.897
99	0.814
99.9	0.753
99.99	0.702
99.999	0.659

2.3   HOMOGENEITY AND ISOTROPY

All discussion of material properties so far has assumed that the material is homogeneous and isotropic. **Homogeneous** means that the *material properties are uniform throughout its continuum*, e.g., they are not a function of position. This ideal state is seldom attained in real materials, many of which are subject to the inclusion of discontinuities, precipitates, voids, or bits of foreign matter from their manufacturing process. However, most metals and some nonmetals can be considered, for engineering purposes, to be macroscopically homogeneous despite their microscopic deviations from this ideal.

An **isotropic** material is one whose *properties are independent of orientation or direction*. That is, the strengths across the width and thickness are the same as along the length of the part, for example. Most metals and some nonmetals can be considered to be macroscopically isotropic. Other materials are **anisotropic**, meaning that *there is no plane of material property symmetry*. **Orthotropic** materials *have three mutually perpendicular planes of property symmetry and can have different material properties along each axis*. Wood, plywood, fiberglass, and some cold-rolled sheet metals are orthotropic.

One large class of materials that is distinctly nonhomogeneous (i.e., heterogeneous) and nonisotropic is that of **composites** (also see below). Most composites are man-made, but some, such as wood, occur naturally. Wood is a composite of long fibers held together in a resinous matrix of lignin. You know from experience that it is easy to split wood along the grain (fiber) lines and nearly impossible to do so across the grain. Its strength is a function of both orientation and position. The matrix is weaker than the fibers and it always splits between fibers.

2

2.4   HARDNESS

The hardness of a material can be an indicator of its resistance to wear (but is not a guarantee of wear resistance). The strengths of some materials such as steels are also closely correlated to their hardness. Various treatments are applied to steels and other metals to increase hardness and strength. These are discussed below.

Hardness is most often measured on one of three scales: *Brinell, Rockwell,* or *Vickers.* These hardness tests all involve the forced impression of a small probe into the surface of the material being tested. The **Brinell test** uses a 10-mm hardened steel or tungsten-carbide[*] ball impressed with either a 500- or 3000-kg load depending on the range of hardness of the material. The diameter of the resulting indent is measured under a microscope and used to calculate the Brinell hardness number. The **Vickers test** uses a diamond-pyramid indenter and measures the width of the indent under the microscope. The **Rockwell test** uses either a 1/16-in ball or a 120° cone-shaped diamond indenter and measures the depth of penetration. Hardness is indicated by a number followed by the letter H, followed by letter(s) to identify the method used, e.g., 375 HB or 396 HV. Several lettered scales (A, B, C, D, F, G) are used for materials in different Rockwell hardness ranges and it is necessary to specify both the letter and number of a Rockwell reading such as 60 HRC.

All these tests are nondestructive in the sense that the sample remains intact. However, the indentation can present a problem if the surface finish is critical or if the section is thin, so they are actually considered destructive tests. The Vickers test has the advantage of having only one test setup for all materials. Both the Brinell and Rockwell tests require selection of the tip size or indentation load, or both, to match the material tested. The Rockwell test is favored for its lack of operator error since no microscope reading is required and the indentation tends to be smaller. But, the Brinell hardness number provides a very convenient way to quickly estimate the ultimate tensile strength (S_{ut}) of the material from the relationship

$$S_{ut} \cong 500 H_B \pm 30 H_B \quad \text{psi}$$

$$S_{ut} \cong 3.45 H_B \pm 0.2 H_B \quad \text{MPa}$$

(2.10)

where H_B = the Brinell hardness number. This gives a convenient way to obtain a rough experimental measure of the strength of any low- or medium-strength carbon or alloy steel sample, even one that has already been placed in service and cannot be truly destructively tested.

Microhardness tests use a low force on a small diamond indenter and can provide a profile of microhardness as a function of depth across a sectioned sample. The hardness is computed on an absolute scale by dividing the applied force by the area of the indent. The units of **absolute hardness** are kg_f/mm^2. Brinell and Vickers hardness numbers also have these hardness units, though the values measured on the same sample can differ with each method. For example, a Brinell hardness of 500 HB is about the same as a Rockwell C hardness of 52 HRC and an absolute hardness of 600 kg/mm^2.

---

[*] Tungsten-carbide is one of the hardest substances known.

Table 2-3   Approximate Equivalent Hardness Numbers and Ultimate Tensile Strengths for Steels

Brinell	Vickers	Rockwell		Ultimate, σ_u	
HB	HV	HRB	HRC	MPa	ksi
627	667	—	58.7	2393	347
578	615	—	56.0	2158	313
534	569	—	53.5	1986	288
495	528	—	51.0	1813	263
461	491	—	48.5	1669	242
429	455	—	45.7	1517	220
401	425	—	43.1	1393	202
375	396	—	40.4	1267	184
341	360	—	36.6	1131	164
311	328	—	33.1	1027	149
277	292	—	28.8	924	134
241	253	100	22.8	800	116
217	228	96.4	—	724	105
197	207	92.8	—	655	95
179	188	89.0	—	600	87
159	167	83.9	—	538	78
143	150	78.6	—	490	71
131	137	74.2	—	448	65
116	122	67.6	—	400	58

Note: Load 3000 kg for HB.

Source: Table 5-10, p.185, in N. E. Dowling, *Mechanical Behavior of Materials*, Prentice Hall, Englewood Cliffs, NJ, 1993, with permission.

Note that these scales are not linearly related so conversion is difficult. Table 2-3 shows approximate conversions between the Brinell, Vickers, and Rockwell hardness scales for steels and their approximate equivalent ultimate tensile strengths.

Heat Treatment

The steel heat-treatment process is quite complicated and is dealt with in detail in materials texts such as those listed in the bibliography at the end of this chapter. The reader is referred to such references for a more complete discussion. Only a brief review of some of the salient points is provided here.

The hardness and other characteristics of many steels and some nonferrous metals can be changed by heat treatment. Steel is an alloy of iron and carbon. The weight percent of carbon present affects the alloy's ability to be heat-treated. A low-carbon steel will have about 0.03 to 0.30% of carbon, a medium-carbon steel about 0.35 to 0.55% and a high-carbon steel about 0.60 to 1.50%. (Cast irons will have greater than 2% car-

2

bon.) Hardenability of steel increases with carbon content. Low-carbon steel has too little carbon for effective through-hardening so other surface-hardening methods must be used (see below). Medium- and high-carbon steels can be through-hardened by appropriate heat treatment. The depth of hardening will vary with alloy content.

QUENCHING   To harden a medium- or high-carbon steel, the part is heated above a critical temperature (about 1400°F {760°C}), allowed to equilibrate for some time, and then suddenly cooled to room temperature by immersion in a water or oil bath. The rapid cooling creates a supersaturated solution of iron and carbon called martensite which is extremely hard and much stronger than the original soft material. Unfortunately it is also very brittle. In effect, we have traded off the steel's ductility for its increased strength. The rapid cooling also introduces strains to the part. The change in the shape of the stress-strain curve as a result of quenching a ductile, medium-carbon steel is shown in Figure 2-12 (not to scale). While the increased strength is desirable, the severe brittleness of a fully quenched steel usually makes it unusable without tempering.

TEMPERING   Subsequent to quenching, the same part can be reheated to a lower temperature (400-1300°F {200-700°C}), heat-soaked, and then allowed to cool slowly. This will cause some of the martensite to convert to ferrite and cementite which reduces the strength somewhat but restores some ductility. A great deal of flexibility is possible in terms of tailoring the resulting combination of properties by varying time and temperature during the tempering process. The knowledgeable materials engineer or metallurgist can achieve a wide variety of properties to suit any application. Figure 2-12 also shows a stress-strain curve for the same steel after tempering.

ANNEALING   The quenching and tempering process is reversible by annealing. The part is heated above the critical temperature (as for quenching) but now allowed to cool slowly to room temperature. This restores the solution conditions and mechanical properties of the unhardened alloy. Annealing is often used even if no hardening has been previously done in order to eliminate any residual stresses and strains introduced by the forces applied in forming the part. It effectively puts the part back into a "relaxed" and soft state restoring its original stress-strain curve as shown in Figure 2-12.

NORMALIZING   Many tables of commercial steel data indicate that the steel has been normalized after rolling or forming into its stock shape. Normalizing is similar to annealing but involves a shorter soak time at elevated temperature and a more rapid cooling rate. The result is a somewhat stronger and harder steel than a fully annealed one but one that is closer to the annealed condition than to any tempered condition.

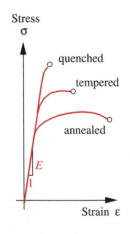

FIGURE 2-12

Stress-Strain Curves for Annealed, Quenched, and Tempered Steel

Surface (Case) Hardening

When a part is large or thick, it is difficult to obtain uniform hardness within its interior by through-hardening. An alternative is to harden only the surface, leaving the core soft. This also avoids the distortion associated with quenching a large, through-heated part. If the steel has sufficient carbon content, its surface can be heated, quenched, and tempered as would be done for through hardening. For low-carbon (mild) steels other techniques are needed to obtain a hardened condition. These involve heating the part in a special atmosphere rich in either carbon, nitrogen or both and then quenching it, a process called *carburizing*, *nitriding*, or *cyaniding*. In all situations, the result is a hard surface (i.e., *case*) on a soft core referred to as being **case-hardened.**

Carburizing heats low-carbon steel in a carbon monoxide gas atmosphere, causing the surface to take up carbon in solution. **Nitriding** heats low-carbon steel in a nitrogen-gas atmosphere and forms hard iron nitrides in the surface layers. **Cyaniding** heats the part in a cyanide salt bath at about 1500°F (800°C) and the low-carbon steel takes up both carbides and nitrides from the salt.

For medium- and high-carbon steels, no artificial atmosphere is needed as the steel has sufficient carbon for hardening. Two methods are in common use. **Flame-hardening** passes an oxyacetylene flame over the surface to be hardened and follows it with a water jet for quenching. This results in a somewhat deeper hardened case than obtainable from the artificial-atmosphere methods. **Induction hardening** uses electric coils to rapidly heat the part surface which is then quenched before the core can get hot.

Case hardening by any appropriate method is a very desirable hardening treatment for many applications. It is often advantageous to retain the full ductility (and thus the toughness) of the core material for better energy absorption capacity while also obtaining high hardness on the surface in order to reduce wear and increase surface strength. Large machine parts such as cams and gears are examples of elements that can benefit more from case hardening than from through-hardening as heat distortion is minimized and the tough, ductile core can better absorb impact energy.

Heat Treating Nonferrous Materials

Some nonferrous alloys are hardenable and others are not. Some of the aluminum alloys can be **precipitation hardened**, also called **age hardening**. An example is aluminum alloyed with up to about 4.5% copper. This material can be hot-worked (rolled, forged, etc.) at a particular temperature and then heated and held at a higher temperature to force a random dispersion of the copper in the solid solution. It is then quenched to capture the supersaturated solution at normal temperature. The part is subsequently reheated to a temperature below the quenching temperature and held for an extended period of time while some of the supersaturated solution precipitates out and increases the material's hardness.

Other aluminum alloys, magnesium, titanium, and a few copper alloys are amenable to similar heat treatment. The strengths of the hardened aluminum alloys approach those of medium-carbon steels. Since all aluminum is about 1/3 the density of steel, the stronger aluminum alloys can offer better strength-to-weight ratios than low-carbon (mild) steels.

2

Mechanical Forming and Hardening

COLD WORKING   The mechanical working of metals at room temperature to change their shape or size will also work-harden them and increase their strength at the expense of ductility. Cold working can result from the rolling process in which metal bars are progressively reduced in thickness by being squeezed between rollers, or from any operation that takes the ductile metal beyond the yield point and permanently deforms it. Figure 2-13 shows the process as it affects the material's stress-strain curve. As the load is increased from the origin at O beyond the yield point y to point B, a permanent set OA is introduced.

If the load is removed at that point, the stored elastic energy is recovered and the material returns to zero stress at point A along a new elastic line BA parallel to the original elastic slope E. If the load is now reapplied and brought to point C, again yielding the material, the new stress-strain curve is $ABCf$. Note that there is now a new yield point y' which is at a higher stress than before. The material has **strain-hardened**, increasing its yield strength and reducing its ductility. This process can be repeated until the material becomes brittle and fractures.

If significant plastic deformation is required for manufacture, such as in making deep-drawn metal pots or cylinders, it is necessary to cold form the material in stages and anneal the part between successive stages to avoid fracture. The annealing resets the material to more nearly the original ductile stress-strain curve, allowing further yielding without fracture.

HOT WORKING   All metals have a recrystallization temperature below which the effects of mechanical working will be as described above, e.g., cold-worked. If the material is mechanically worked above its recrystallization temperature (hot-working),

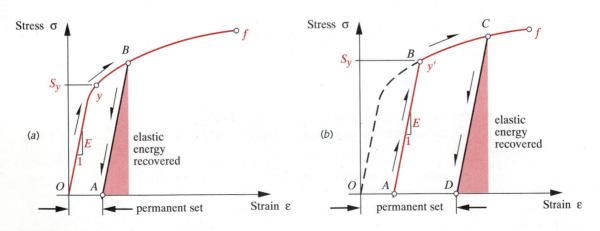

FIGURE 2-13

Strain Hardening a Ductile Material by Cold Working (a) First Working (b) Second Working

2

it will tend to at least partially anneal itself as it cools. Thus hot working reduces the strain-hardening problem but introduces its own problems of rapid oxidation of the surface due to the high temperatures. Hot-rolled metals tend to have higher ductility, lower strength and poorer surface finish than cold-worked metals of the same alloy. Hot working does not increase the hardness of the material appreciably, though it can increase the strength by improving grain structure and aligning the "grain" of the metal with the final contours of the part. This is particularly true of forged parts.

FORGING   is an automation of the ancient art of blacksmithing. The blacksmith heats the part red-hot in the forge, then beats it into shape with a hammer. When it cools too much for forming it is reheated and the process is repeated. Forging uses a series of hammer dies shaped to gradually form the hot metal into the final shape. Each stage's die shape represents an achievable change in shape from the original ingot form to the final desired part shape. The part is reheated between blows from the hammer dies which are mounted in a forging press. The large forces required to plastically deform the hot metal require massive presses for parts of medium to large size. Machining operations are required to remove the large "flash" belt at the die parting line and to machine holes, mounting surfaces, etc. The surface finish of a forging is as rough as any hot-rolled part due to oxidation and decarburization of the heated metal.

Virtually any wrought, ductile metal can be forged. Steel, aluminum, and titanium are commonly used. Forging has the advantage of creating stronger parts than casting or machining can. Casting alloys are inherently weaker in tension than wrought alloys. The hot forming of a wrought material into the final forged shape causes the material's

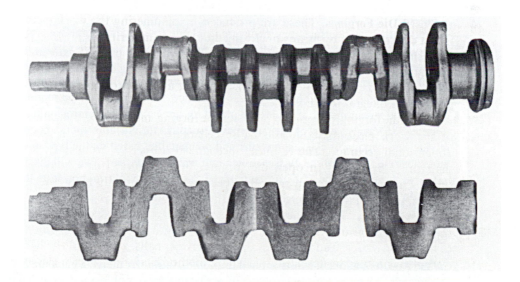

FIGURE 2-14

Forged Steel Crankshaft for a Diesel-Truck Engine - *Courtesy of Wyman-Gordon Corp, Grafton, MA*

2

Table 2-4

Galvanic Series
of Metals in Seawater

Least noble

    Magnesium

    Zinc

    Aluminum

    Cadmium

    Steel

    Cast Iron

    Stainless steel

    Lead

    Tin

    Nickel

    Brass

    Copper

    Bronze

    Monel

    Silver

    Titanium

    Graphite

    Gold

    Platinum

Most noble

internal "flow-lines" or "grain" to approximate the contours of the part, which can result in greater strength than if a stock shape's flow lines were severed by machining to the final contour. Forgings are used in highly stressed parts, such as aircraft wing and fuselage structures, engine crankshafts and connecting rods, and vehicle suspension links. Figure 2-14 shows a forged truck crankshaft. In the cross-section, the grain lines can be seen to follow the crankshaft's contours. The high cost of the multiple dies needed for forged shapes makes it an impractical choice unless production quantities are large enough to amortize the tooling cost.

EXTRUSION   is used principally for nonferrous metals (especially aluminum) as it typically uses steel dies. The usual die is a thick, hardened-tool-steel disk with a tapered "hole" or orifice ending in the cross-sectional shape of the finished part. A billet of the extrudate is heated to a soft state and then rammed at fairly high speed through the die which is clamped in the machine. The billet flows, or extrudes, into the die's shape. The process is similar to the making of macaroni. A long strand of the material in the desired cross section is extruded from the billet. The extrusion then passes through a water-spray cooling station. Extrusion is an economical way to obtain custom shapes of constant cross section since the dies are not very expensive to make. Dimensional control and surface finish are good. Extrusion is used to make aluminum mill shapes such as angles, channels, I-beams, and custom shapes for storm-door and -window frames, sliding-door frames, etc. The extrusions are cut and machined as necessary to assemble them into the finished product. Some extruded shapes are shown in Figure 2-15.

2.5   COATINGS AND SURFACE TREATMENTS

Many coatings and surface treatments are available for metals. Some have the prime purpose of inhibiting corrosion while others are intended to improve surface hardness and wear resistance. Coatings are also used to change dimensions (slightly) and to alter physical properties such as reflectance, color, and resistivity. For example, piston rings are chrome-plated to improve wear resistance, fasteners are plated to reduce corrosion, and automobile trim is chrome-plated for appearance and corrosion resistance. Figure 2-16 shows a chart of various types of coatings for machine applications. These divide into two major classes, metallic and nonmetallic, based on the type of coating, not substrate. Some of the classes divide into many subclasses . We will discuss only a few of these here. The reader is encouraged to seek more information from the references in the bibliography.

Galvanic Action

When a coating of one metal is applied to another dissimilar metal, a galvanic cell may be created. All metals are electrolytically active to a greater or lesser degree and if sufficiently different in their electrolytic potential will create a battery in the presence of a conductive electrolyte such as seawater or even tap water. Table 2-4 lists some common metals ordered in terms of their galvanic action potential from the least noble (most

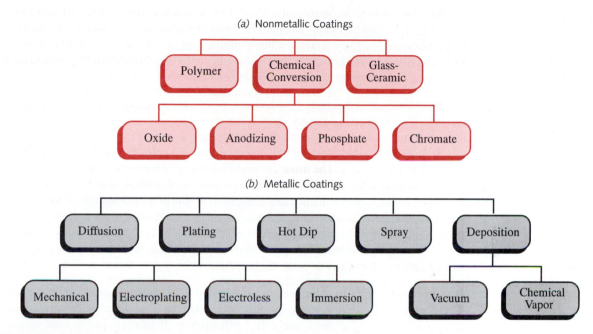

FIGURE 2-16

Coating Methods Available for Metals

electrolytically active) to the most noble (least active). Combinations of metals that are close to each other in the galvanic series, such as cast iron and steel, are relatively safe from galvanic corrosion. Combinations of metals far apart on this scale, such as aluminum and copper, will experience severe corrosion in an electrolyte or even in a moist environment.

In a conductive medium, the two metals become anode and cathode, with the less-noble metal acting as the anode. The self-generated electrical current flow causes a loss of material from the anode and a deposition of material on the cathode. The less-noble metal gradually disappears. This problem occurs whenever two metals sufficiently far apart in the galvanic series are present in an electrically conductive medium. Thus, not only coatings but fasteners and mating parts must be made of metal combinations that will not create this problem.

Electroplating

Electroplating involves the deliberate creation of a galvanic cell in which the part to be plated is the cathode and the plating material is the anode. The two metals are placed in an electrolyte bath and a direct current applied from anode to cathode. Ions of the plating material are driven to the plating substrate through the electrolyte and cover the part with a thin coating of the plating material. Allowance must be made for the plating thickness, which is controllable. Plating thickness is uniform except at sharp corners or in holes and crevices. The plating builds up on the outside corners and will not go into holes or narrow crevices. Thus grinding may be necessary after plating to restore dimensions. Worn parts (or mistakes) can sometimes be repaired by plating on a coating of suitable material, then regrinding to dimension.

Steels, nickel- and copper-based alloys, as well as other metals are readily electroplatable. Two approaches are possible. If a more noble (less active) metal is plated onto the substrate, it can reduce the tendency to oxidize as long as the plating remains intact to protect the substrate from the environment. Tin, nickel, and chromium are often used to electroplate steel for corrosion resistance. Chrome plating also offers an increase in surface hardness to HRC 70, which is above that obtainable from many hardened alloy steels.[*] Unfortunately, any disruptions or pits in the plating can provide nodes for galvanic action if conductive media (such as rainwater) are present. Because the substrate is less noble than the plating, it becomes the sacrificial anode and rapidly corrodes. Electroplating with metals more noble than the substrate is seldom used for parts that will be immersed in water or other electrolytes.

Alternatively, a less-noble metal can be plated onto the substrate to serve as a sacrificial anode which will corrode instead of the substrate. The most common example of this is zinc coating of steel, also called galvanizing. (Cadmium can be used instead of zinc and will last longer in saltwater or salt-air environments.) The zinc or cadmium coating will gradually corrode and protect the more noble steel substrate until the coating is used up, after which the steel will oxidize. Zinc coating can be applied by a process called "hot dipping" rather than by electroplating which will result in a thicker and more protective coating recognizable by its "mother-of-pearl" appearance. Galvanizing is often applied by manufacturers to automobile body panels to inhibit corrosion. Sacrificial zinc anodes are also attached to aluminum outboard motors and aluminum boat hulls to short-circuit corrosion of the aluminum in seawater.

A caution about electroplated coatings is that hydrogen embrittlement of the substrate can occur causing significant loss of strength. Electroplated finishes should not be used on parts that are fatigue loaded. Experience has shown that electroplating severely reduces the fatigue strength of metals and can cause early failure.

Electroless Plating

Electroless plating puts a coating of nickel on the substrate without any electric current needed. The substrate "cathode" in this case (there is no anode) acts as a catalyst to start a chemical reaction that causes nickel ions in the electrolyte solution to be reduced and deposited on the substrate. The nickel coating also acts as a catalyst and keeps the reaction going until the part is removed from the bath. Thus, relatively thick coatings can be developed. Coatings are typically between 0.001 in and 0.002 in thick. Unlike electroplating, the electroless nickel plate is completely uniform and will enter holes and crevices. The plate is dense and fairly hard at around 43 HRC. Other metals can also be electroless plated but nickel is most commonly used.

[*] It is interesting to note that chromium in the pure form is softer than hardened steel but when electroplated onto steel, it becomes harder than the steel substrate. Nickel and iron also increase their hardness when electroplated on metal substrates. The mechanism is not well understood, but it is believed that internal microstrains are developed in the plating process that harden the coating. The hardness of the plating can be controlled by changes in process conditions.

Anodizing

While aluminum can be electroplated (with difficulty), it is more common to treat it by anodizing. This process creates a very thin layer of aluminum oxide on the surface. The aluminum oxide coating is self-limiting in that it prevents atmospheric oxygen from further attacking the aluminum substrate in service. The anodized oxide coating is naturally colorless but dyes can be added to color the surface and provide a pleasing appearance in a variety of hues. This is a relatively inexpensive surface treatment with good corrosion resistance and negligible distortion. Titanium, magnesium, and zinc can also be anodized.

A variation on conventional anodizing of aluminum is so-called "hard anodizing." Since aluminum oxide is a ceramic material, it is naturally very hard and abrasion resistant. **Hard anodizing** provides a thicker (but not actually harder) coating than conventional anodizing and is often used to protect the relatively soft aluminum parts from wear in abrasive contact situations. The hardness of this surface treatment exceeds that of the hardest steel and hard-anodized aluminum parts can be run against hardened steel though the somewhat abrasive aluminum oxide surface is not kind to the steel.

Plasma-Sprayed Coatings

A variety of very hard ceramic coatings can be applied to steel and other metal parts by a plasma-spray technique. The application temperatures are high which limits the choice of substrate. The coatings as sprayed have a rough "orange-peel" surface finish which requires grinding or polishing to obtain a fine finish. The main advantage is a surface with extremely high hardness and chemical resistance. However, the ceramic coatings are brittle and subject to chipping under mechanical or thermal shock.

Chemical Coatings

The most common chemical treatments for metals range from a phosphoric acid wash on steel (or chromatic acid on aluminum) that provides limited and short-term oxidation resistance, to paints of various types designed to give more lasting corrosion protection. Paints are available in a large variety of formulations for different environments and substrates. One-part paints give somewhat less protection than two-part epoxy formulations, but all chemical coatings should be viewed as only temporary protection against corrosion, especially when used on corrosion-prone materials such as steel. Baked enamel and porcelain finishes on steel have longer lives in terms of corrosion resistance, though they suffer from brittleness. New formulations of paints and protective coatings are continually being developed. The latest and best information will be obtained from vendors of these products.

2

2.6   GENERAL PROPERTIES OF METALS

The large variety of useful engineering materials can be confusing to the beginning engineer. There is not space enough in this book to deal with the topic of material selection in complete detail. Several references are provided in this chapter's bibliography which the reader is encouraged to use. Tables of mechanical property data are also provided for a limited set of materials in Appendix C of this book. Figure 2-17 shows the Young's moduli for several engineering metals.

The following sections attempt to provide some general information and guidelines for the engineer to help identify what types of materials might be suitable in a given design situation. It is expected that the practicing engineer will rely heavily on the expertise and help available from materials manufacturers in selecting the optimum material for each design. Many references are also published which list detailed property data for most engineering materials. Some of these references are listed in the bibliography to this chapter.

Cast Iron

Cast irons constitute a whole family of materials. Their main advantages are relatively low cost and ease of fabrication. Some are weak in tension compared to steels but, like most cast materials, have high compressive strengths. Their densities are slightly lower than steel at about 0.25 lb/in³ (6 920 kg/m³). Most cast irons do not exhibit a linear stress-strain relationship below the elastic limit; they do not obey Hooke's law. Their modulus of elasticity E is estimated by drawing a line from the origin through a point on the curve at 1/4 the ultimate tensile strength and is in the range of 14-25 Mpsi (97-172 MPa). Cast iron's chemical composition differs from steel principally in its higher carbon content, being between 2 and 4.5%. The large amount of carbon, present in some cast irons as graphite, makes some of these alloys easy to pour as a casting liquid and also easy to machine as a solid. The most common means of fabrication is sand casting with subsequent machining operations. Cast irons are not easily welded, however.

WHITE CAST IRON is a very hard and brittle material. It is difficult to machine and has limited uses, such as in linings for cement mixers where its hardness is needed.

GRAY CAST IRON is the most commonly used form of cast iron. Its graphite flakes give it its gray appearance and name. The ASTM grades gray cast iron into seven classes based on the minimum tensile strength in kpsi. Class 20 has a minimum tensile strength of 20 kpsi (138 MPa). The class numbers of 20, 25, 30, 35, 40, 50, and 60 then represent the tensile strength in kpsi. Cost increases with increasing tensile strength. This alloy is easy to pour, easy to machine and offers good acoustical damping. This makes it the popular choice for machine frames, engine blocks, brake rotors and drums, etc. The graphite flakes also give it good lubricity and wear resistance. Its relatively low tensile strength recommends against its use in situations where large bending or fatigue loads are present, though it is sometimes used in low-cost engine crankshafts. It runs reasonably well against steel if lubricated.

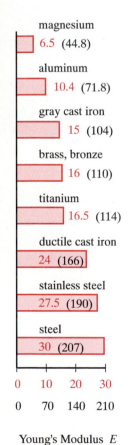

magnesium
6.5 (44.8)

aluminum
10.4 (71.8)

gray cast iron
15 (104)

brass, bronze
16 (110)

titanium
16.5 (114)

ductile cast iron
24 (166)

stainless steel
27.5 (190)

steel
30 (207)

| 0 | 10 | 20 | 30 |
| 0 | 70 | 140 | 210 |

Young's Modulus E

Mpsi (GPa)

FIGURE 2-17

Young's Moduli for Various Metals

MALLEABLE CAST IRON has superior tensile strength to gray cast iron but does not wear as well. The tensile strength can range from 50 to 120 kpsi (345 to 827 MPa) depending on formulation. It is often used in parts where bending stresses are present.

NODULAR (DUCTILE) CAST IRON has the highest tensile strength of the cast irons, ranging from about 70 to 135 kpsi (480 to 930 MPa). The name *nodular* comes from the fact that its graphite particles are spheroidal in shape. Ductile cast iron has a higher modulus of elasticity (about 25 Mpsi {172 GPa}) than gray cast iron and exhibits a linear stress-strain curve. It is tougher, stronger, more ductile, and less porous than gray cast iron. It is the cast iron of choice for fatigue-loaded parts such as crankshafts, pistons, and cams.

Cast Steels

Cast steel is similar to wrought steel in terms of its chemical content, i.e., it has much less carbon than cast iron. The mechanical properties of cast steel are superior to cast iron but inferior to wrought steel. Its principal advantage is ease of fabrication by sand or investment (lost wax) casting. Cast steel is classed according to its carbon content into low carbon (< 0.2%), medium carbon (0.2 - 0.5%) and high carbon (> 0.5%). Alloy cast steels are also made containing other elements for high strength and heat resistance. The tensile strengths of cast steel alloys range from about 65 to 200 kpsi (450 to 1380 MPa).

Wrought Steels

The term "wrought" refers to all processes that manipulate the shape of the material without melting it. Hot rolling and cold rolling are the two most common methods used though many variants exist, such as wire drawing, deep drawing, extrusion, and cold-heading. The common denominator is a deliberate yielding of the material to change its shape either at room or at elevated temperatures.

HOT-ROLLED STEEL is produced by forcing hot billets of steel through sets of rollers or dies which progressively change their shape into I-beams, channel sections, angle irons, flats, squares, rounds, tubes, sheets, plates, etc. The surface finish of hot-rolled shapes is rough due to oxidation at the elevated temperatures. The mechanical properties are also relatively low because the material ends up in an annealed or normalized state unless deliberately heat-treated later. This is the typical choice for low-carbon structural steel members used for building- and machine-frame construction. Hot-rolled material is also used for machine parts that will be subjected to extensive machining (gears, cams, etc.) where the initial finish of the stock is irrelevant and uniform, non-cold-worked material properties are desired in advance of a planned heat treatment. A wide variety of alloys and carbon contents are available in hot-rolled form.

COLD-ROLLED STEEL is produced from billets or hot-rolled shapes. The shape is brought to final form and size by rolling between hardened steel rollers or drawing

2

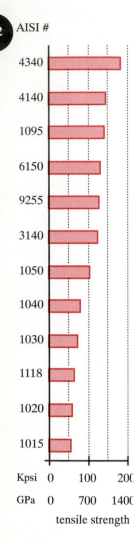

AISI #

Kpsi   0   100   200

GPa   0   700   1400

tensile strength

FIGURE 2-18

Approximate Ultimate
Tensile Strengths of
Some Normalized Steels

* ASTM is the American
Society for Testing and
Materials, AISI is the
American Iron and Steel
Institute, and SAE is the
Society of Automotive
Engineers. AISI and SAE
both use the same
designations for steels.

through dies at room temperature. The rolls or dies burnish the surface and cold work the material, increasing its strength and reducing its ductility as was described in the section on **Mechanical Forming and Hardening** above. The result is a material with good surface finish and accurate dimensions compared to hot-rolled material. Its strength and hardness are increased at the expense of significant built-in strains which can later be released during machining, welding, or heat treating, then causing distortion. Cold-rolled shapes commonly available are sheets, strips, plates, round and rectangular bars, tubes, etc. Structural shapes such as I-beams are typically available only as hot rolled.

Steel Numbering Systems

Several steel numbering systems are in general use. The ASTM, AISI, and SAE* have devised codes to define the alloying elements and carbon content of steels. Table 2-5 lists some of the AISI/SAE designations for commonly used steel alloys. The first two digits indicate the principal alloying elements. The last two digits indicate the amount of carbon present, expressed in hundreths of percent. ASTM and the SAE have developed a new Unified Numbering System for all metal alloys, which uses the prefix UNS followed by a letter and a 5-digit number. The letter defines the alloy category, F for cast iron, G for carbon and low-alloy steels, K for special-purpose steels, S for stainless steels, and T for tool steels. For the G series, the numbers are the same as the AISI/SAE designations in Table 2-5 with a trailing zero added. For example, SAE 4340 becomes UNS G43400. See reference 2 for more information on metal numbering systems. We will use the AISI/SAE designations for steels.

PLAIN CARBON STEEL is designated by a first digit of 1 and a second digit of 0 since no alloys other than carbon are present. The low-carbon steels are those numbered AISI 1005 to 1030, medium-carbon from 1035 to 1055, and high-carbon from 1060 to 1095. The AISI 11xx series adds sulphur, principally to improve machinability. These are called free-machining steels and are not considered alloy steels as the sulphur does not improve the mechanical properties and also makes it brittle. The ultimate tensile strength of plain carbon steel can vary from about 60 to 150 kpsi (414 to 1034 MPa) depending on heat treatment.

ALLOY STEELS have various elements added in small quantities to improve the material's strength, hardenability, temperature resistance, corrosion resistance, and other properties. Any level of carbon can be combined with these alloying elements. Chromium is added to improve strength, ductility, toughness, wear resistance, and hardenability. Nickel is added to improve strength without loss of ductility, and it also enhances case hardenability. Molybdenum, used in combination with nickel and/or chromium, adds hardness, reduces brittleness, and increases toughness. Many other alloys in various combinations, as shown in Table 2-5, are used to achieve specific properties. Specialty steel manufacturers are the best source of information and assistance for the engineer trying to find the best material for any application. The ultimate tensile strength of alloy steels can vary from about 80 to 300 kpsi (550 to 2070 MPa), depending on its

2

Table 2-5   AISI/SAE Designations of Steel Alloys
A partial list - other alloys are available - consult the manufacturers

Type	AISI/SAE Series	Principal Alloying Elements
Carbon Steels		
Plain	10xx	Carbon
Free-cutting	11xx	Carbon plus Sulphur (resulphurized)
Alloy Steels		
Manganese	13xx	1.75% Manganese
	15xx	1.00 to 1.65% Manganese
Nickel	23xx	3.50% Nickel
	25xx	5.00% Nickel
Nickel-Chrome	31xx	1.25% Nickel and 0.65 or 0.80% Chromium
	33xx	3.50% Nickel and 1.55% Chromium
Molybdenum	40xx	0.25% Molybdenum
	44xx	0.40 or 0.52% Molybdenum
Chrome-Moly	41xx	0.95% Chromium and 0.20% Molybdenum
Nickel-Chrome-Moly	43xx	1.82% Nickel, 0.50 or 0.80% Chromium, and 0.25% Molybdenum
	47xx	1.45% Nickel, 0.45% Chromium, and 0.20 or 0.35% Molybdenum
Nickel-Moly	46xx	0.82 or 1.82% Nickel and 0.25% Molybdenum
	48xx	3.50% Nickel and 0.25% Molybdenum
Chrome	50xx	0.27 to 0.65% Chromium
	51xx	0.80 to 1.05% Chromium
	52xx	1.45% Chromium
Chrome-Vanadium	61xx	0.60 to 0.95% Chromium and 0.10 to 0.15% Vanadium minimum

alloying elements and heat treatment. Appendix C contains tables of mechanical property data for a selection of carbon and alloy steels. Figure 2-18 shows approximate ultimate tensile strengths of some normalized carbon and alloy steels and Figure 2-19 shows engineering stress-strain curves from tensile tests of three steels.

TOOL STEELS are medium- to high-carbon alloy steels especially formulated to give very high hardness in combination with wear resistance and sufficient toughness to resist the shock loads experienced in service as cutting tools, dies and molds. There is a very large variety of tool steels available. Refer to the bibliography and to manufacturers' literature for more information.

STAINLESS STEELS are alloy steels containing at least 10% chromium and offer much improved corrosion resistance over plain or alloy steels, though their name should not be taken too literally. Stainless steels will stain and corrode (slowly) in severe environ-

2

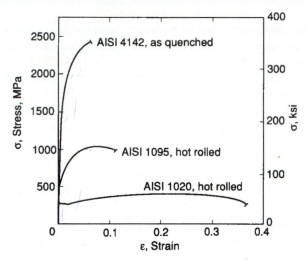

FIGURE 2-19

Tensile Test Stress-Strain Curves of Three Steel Alloys (From Fig. 5.16, p. 160, in N. E. Dowling, *Mechanical Behavior of Materials*, Prentice-Hall, Englewood Cliffs, N.J., 1993, with permission)

ments such as seawater. Some stainless-steel alloys have improved resistance to high temperature. There are four types of stainless steel, called **martensitic, ferritic, austenitic**, and **precipitation hardening**.

Martensitic stainless steel contains 11.5 to 15% Cr and 0.15 to 1.2% C, is magnetic, can be hardened by heat treatment, and is commonly used for cutlery. **Ferritic** stainless steel has over 16% Cr and a low carbon content, is magnetic, soft, and ductile, but is not heat treatable though its strength can be increased modestly by cold working. It is used for deep-drawn parts such as cookware and has better corrosion resistance than the martensitic SS. The ferritic and martensitic stainless steels are both called **400 series** stainless steel.

Austenitic stainless steel is alloyed with 17 to 25% chromium and 10 to 20% nickel. It has better corrosion resistance due to the nickel, is nonmagnetic, and has excellent ductility and toughness. It cannot be hardened except by cold-working. It is classed as **300 series** stainless steel.

Precipitation-hardening stainless steels are designated by their alloy percentages followed by the letters PH, as in 17-4 PH which contains 17% nickel and 4% copper. These alloys offer high strength, and high temperature and corrosion resistance.

The **300 series** stainless steels are very weldable but the 400 series are less so. All grades of stainless steel have poorer heat conductivity than regular steel and many of the stainless alloys are difficult to machine. All stainless steels are significantly more expensive than regular steel. See Appendix C for mechanical property data.

Aluminum

Aluminum is the most widely used nonferrous metal, being second only to steel in world consumption. Aluminum is produced in both "pure" and alloyed forms. Aluminum is commercially available up to 99.8% pure. The most common alloying elements are copper, silicon, magnesium, manganese, and zinc, in varying amounts up to about 5%. The principal advantages of aluminum are its low density, good strength-to-weight ratio (SWR), ductility, excellent workability, castability, and weldability, corrosion resistance, high conductivity, and reasonable cost. Compared to steel it is 1/3 as dense (0.10 lb/in^3 versus 0.28 lb/in^3), about 1/3 as stiff (E = 10.3 Mpsi {71 GPa} versus 30 Mpsi {207 GPa}), and generally less strong. If you compare the strengths of low-carbon steel and pure aluminum, the steel is about three times as strong. Thus the specific strength is approximately the same in that comparison. However, pure aluminum is seldom used in engineering applications. It is too soft and weak. Pure aluminum's principal advantages are its bright finish and good corrosion resistance. It is used mainly in decorative applications.

The aluminum alloys have significantly greater strengths than pure aluminum and are used extensively in engineering, with the aircraft and automotive industries among the largest users. The higher-strength aluminum alloys have tensile strengths in the 70 to 90 kpsi (480 to 620 MPa) range, and yield strengths about twice that of mild steel. They compare favorably to medium-carbon steels in specific strength. Aluminum competes successfully with steel in some applications, though few materials can beat steel if very high strength is needed. See Figure 2-20 for tensile strengths of some aluminum alloys. Figure 2-21 shows tensile-test engineering stress-strain curves for three aluminum alloys. Aluminum's strength is reduced at low temperatures as well as at elevated temperatures.

Some aluminum alloys are hardenable by heat treatment and others by strain hardening or precipitation and aging. High-strength aluminum alloys are about 1.5 times harder than soft steel and surface treatments such as *hard anodizing* can bring the surface to a condition harder than the hardest steel.

Aluminum is among the most easily worked of the engineering materials though it tends to work harden. It casts, machines, welds,[*] and hot and cold forms[†] easily. It can also be extruded. Alloys are specially formulated for both sand and die casting as well as for wrought and extruded shapes and for forged parts.

WROUGHT ALUMINUM ALLOYS are available in a wide variety of stock shapes such as I-beams, angles, channels, bars, strip, sheet, rounds, and tubes. Extrusion allows relatively inexpensive custom shapes as well. The Aluminum Association numbering system for alloys is shown in Table 2-6. The first digit indicates the principal alloying element and defines the series. Hardness is indicated by a suffix containing a letter and up to 3 numbers as defined in the table. The most commonly available and most-used aluminum alloys in machine design applications are the 2000 series and the 6000 series.

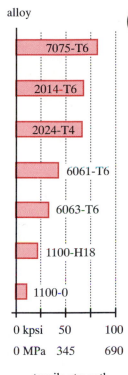

alloy

7075-T6
2014-T6
2024-T4
6061-T6
6063-T6
1100-H18
1100-0

0 kpsi   50   100

0 MPa   345   690

tensile strength

FIGURE 2-20

Ultimate Tensile Strengths of Some Aluminum Alloys

[*] The heat of welding causes localized annealing, which can remove the desirable strengthening effects of cold work or heat treatment in any metal.

[†] Some aluminum alloys will cold-work when formed to the degree that trying to bend them again (without first annealing) will cause fractures. Some bicycle racers prefer steel frames over aluminum despite their added weight because, once an aluminum frame is bent in a fall, it cannot be straightened without cracking. Damaged steel tube frames can be straightened and reused.

Table 2-6   Aluminum Association Designations of Aluminum Alloys

A partial list - other alloys are available - consult the manufacturers

Series	Major Alloying Elements	Secondary Alloys
1xxx	Commercially pure (99%)	None
2xxx	Copper (Cu)	Mg, Mn, Si
3xxx	Manganese (Mn)	Mg, Cu
4xxx	Silicon (Si)	None
5xxx	Magnesium (Mg)	Mn, Cr
6xxx	Magnesium and Silicon	Cu, Mn
7xxx	Zinc (Zn)	Mg, Cu, Cr

Hardness Designations

xxxx-F	As fabricated	
xxxx-O	Annealed	
xxxx-Hyyy	Work-hardened	
xxxx-Tyyy	Thermal/age-hardened	

The oldest aluminum alloy is 2024, which contains 4.5% copper, 1.5% magnesium, and 0.8% manganese. It is among the most machinable of the aluminum alloys and is heat treatable. In the higher tempers, such as -T3 and -T4, it has a tensile strength approaching 70 kpsi (483 MPa), which also makes it one of the strongest of the alumi-

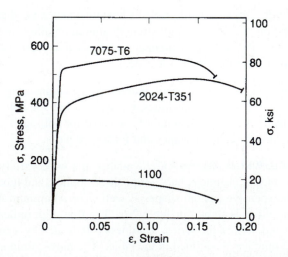

FIGURE 2-21

Tensile Test Stress-Strain Curves of Three Aluminum Alloys (From Fig. 5.17, p. 160, in N. E. Dowling, *Mechanical Behavior of Materials*, Prentice-Hall, Englewood Cliffs, NJ, 1993, with permission)

num alloys. It also has high fatigue strength. However, it has poor weldability and formability compared to the other aluminum alloys.

The 6061 alloy contains 0.6% silicon, 0.27% copper, 1.0% manganese, and 0.2% chromium. It is widely used in structural applications because of its excellent weldability. Its strength is about 40 to 45 kpsi (276 to 310 MPa) in the higher tempers. It has lower fatigue strength than 2024 aluminum. It is easily machined and is a popular alloy for extrusion which is a hot-forming process.

The 7000 series is called aircraft aluminum and is used mostly in airframes. These are the strongest alloys of aluminum with tensile strengths up to 98 kpsi (676 MPa) and the highest fatigue strength of about 22 kpsi (152 MPa) @ 10^8 cycles. Some alloys are also available in an *alclad* form which bonds a thin layer of pure aluminum to one or both sides to improve corrosion resistance.

CAST ALUMINUM ALLOYS are differently formulated than the wrought alloys. Some of these are hardenable but their strength and ductility are less than those of the wrought alloys. Alloys are available for sand casting, die casting, or investment casting. See Appendix C for mechanical properties of wrought and cast aluminum alloys.

Titanium

Though discovered as an element in 1791, commercially produced titanium has been available only since the 1940s, so it is among the newest of engineering metals. Titanium can be the answer to an engineer's prayer in some cases. It has an upper service-temperature limit of 1 200 to 1 400°F (650 to 750°C), weighs half as much as steel (0.16 lb/in^3 {4 429 kg/m^3}), and is as strong as a medium-strength steel (135 kpsi {930 MPa} typical). Its Young's modulus is 16 to 18 Mpsi (110 to 124 GPa), or about 60% that of steel. Its specific strength approaches that of the strongest alloy steels and exceeds that of medium-strength steels by a factor of 2. Its specific stiffness is greater than steel, making it as good or better in limiting deflections. It is also nonmagnetic.

Titanium is very corrosion resistant and is nontoxic, allowing its use in contact with acidic or alkaline foodstuffs and chemicals, and in the human body as replacement heart valves and hip joints, for example. Unfortunately, it is expensive compared to aluminum and steel. It finds much use in the aerospace industry, especially in military aircraft structures and in jet engines, where strength, light weight, and high temperature and corrosion resistance are all required.

Titanium is available both pure and alloyed with combinations of aluminum, vanadium, silicon, iron, chromium, and manganese. Its alloys can be hardened and anodized. Limited stock shapes are available commercially. It can be forged and wrought, though it is quite difficult to cast, machine, and cold form. Like steel and unlike most other metals, some titanium alloys exhibit a true endurance limit, or leveling off of the fatigue strength beyond about 10^6 cycles of repeated loading, as shown in Figure 2-10. See Appendix C for mechanical property data.

Magnesium

Magnesium is the lightest of commercial metals but is relatively weak. The tensile strengths of its alloys are between 10 and 50 kpsi (69 and 345 MPa). The most common alloying elements are aluminum, manganese, and zinc. Because of its low density (0.065 lb/in^3 {$1\ 800$ kg/m^3}), its specific strength approaches that of aluminum. Its Young's modulus is 6.5 Mpsi (45 GPa) and its specific stiffness exceeds those of aluminum and steel. It is very easy to cast and machine but is more brittle than aluminum and thus is difficult to cold form.

It is nonmagnetic and has fair corrosion resistance, better than steel, but not as good as aluminum. Some magnesium alloys are hardenable, and all can be anodized. It is the most active metal on the galvanic scale and cannot be combined with most other metals in a wet environment. It is also extremely flammable, especially in powder or chip form, and its flame cannot be doused with water. Machining requires flooding with oil coolant to prevent fire. It is roughly twice as costly per pound as aluminum. Magnesium is used where light weight is of paramount importance such as in castings for chain-saw housings and other hand-held items. See Appendix C for mechanical property data.

Copper Alloys

Pure copper is soft, weak, and malleable and is used primarily for piping, flashing, electrical conductors (wire) and motors. It cold works readily and can become brittle after forming, requiring annealing between successive draws.

Many alloys are possible with copper. The most common are brasses and bronzes which themselves are families of alloys. **Brasses**, in general, are alloys of copper and zinc in varying proportions and are used in many applications, from artillery shells and bullet shells to lamps and jewelry.

Bronzes were originally defined as alloys of copper and tin, but now also include alloys containing no tin, such as silicon bronze, and aluminum bronze, so the terminology is somewhat confusing. **Silicon bronze** is used in marine applications such as ship propellers.

Beryllium copper is neither brass nor bronze and is the strongest of the alloys, with strengths approaching those of alloy steels (200 kpsi {$1\ 380$ MPa}). It is often used in springs that must be nonmagnetic, carry electricity, or exist in corrosive environments. **Phosphor bronze** is also used for springs but unlike beryllium copper, it cannot be bent along the grain or heat treated.

Copper and its alloys have excellent corrosion resistance and are nonmagnetic. All copper alloys can be cast, hot- or cold-formed, and machined, but pure copper is difficult to machine. Some alloys are heat treatable and all will work harden. The Young's modulus of most copper alloys is about 17 Mpsi (117 GPa) and their weight density is slightly higher than that of steel at 0.31 lb/in^3 (8 580 kg/m^3). Copper alloys are expensive compared to other structural metals. See Appendix C for mechanical property data.

2.7   GENERAL PROPERTIES OF NONMETALS

The use of nonmetallic materials has increased greatly in the last 50 years. The usual advantages sought are light weight, corrosion resistance, temperature resistance, dielectric strength, and ease of manufacture. Cost can range from low to high compared to metals depending on the particular nonmetallic material. There are three general categories of nonmetals of general engineering interest: **polymers** (plastics), **ceramics,** and **composites**.

   Polymers have a wide variety of properties, principally low weight, relatively low strength and stiffness, good corrosion and electrical resistance, and relatively low cost per unit volume. **Ceramics** can have extremely high compressive (but not tensile) strengths, high stiffness, high temperature resistance, high dielectric strength (resistance to electrical current), high hardness, and relatively low cost per unit volume. **Composites** can have almost any combination of properties you want to build into them, including the highest specific strengths obtainable from any materials. Composites can be low or very high in cost. We briefly discuss nonmetals and some of their applications. Space

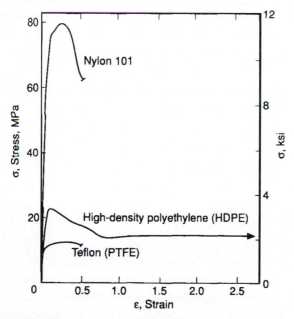

FIGURE 2-22

Tensile Test Stress-Strain Curves of Three Thermoplastic Polymers (From Fig. 5.18, p. 161 in Dowling, N. E., *Mechanical Behavior of Materials*, Prentice-Hall, 1993, with permission)

does not permit a complete treatment of these important classes of materials. The reader is directed to the bibliography for further information. Appendix C also provides some mechanical property data for polymers.

Polymers

The word polymers comes from **poly** = *many* and **mers** = *molecules*. Polymers are long-chain molecules of organic materials or carbon-based compounds. (There is also a family of silicon-based polymeric compounds.) The source of most polymers is oil or coal, which contains the carbon or hydrocarbons necessary to create the polymers. While there are many natural polymer compounds (wax, rubber, proteins, ...), most polymers used in engineering applications are man-made. Their properties can be tailored over a wide range by copolymerization with other compounds or by alloying two or more polymers together. Mixtures of polymers and inorganic materials such as talc or glass fiber are also common.

Because of their variety, it is difficult to generalize about the mechanical properties of polymers, but compared to metals they have low density, low strength, low stiffness, nonlinear elastic stress-strain curves as shown in Figure 2-22 (with a few exceptions), low hardness, excellent electrical and corrosion resistance, and ease of fabrication. Their apparent moduli of elasticity vary widely from about 10 kpsi (69 MPa) to about 400 kpsi (2.8 GPa), all much less stiff than any metals. Their ultimate tensile strengths range from about 4 kpsi (28 MPa) for the weakest unfilled polymer to about 22 kpsi (152 MPa) for the strongest glass-filled polymers. The specific gravities of most polymers range from about 0.95 to 1.8 compared to about 2 for magnesium, 3 for aluminum, 8 for steel, and 13 for lead. So, even though the absolute strengths of polymers are low, their specific strengths are respectable due to their low densities.

Polymers are divided into two classes, **thermoplastic** and **thermosets**. **Thermoplastic polymers** can be repeatedly melted and solidified, though their properties can degrade due to the high melt temperatures. Thermoplastics are easy to mold and their rejects or leftovers can be reground and remolded. **Thermosetting polymers** become cross-linked when first heated and will burn, not melt, on reheating. Cross-linking creates connections (like the rungs of a ladder) between the long-chain molecules which wind and twist through a polymer. These cross-connections add strength and stiffness.

Another division among polymers can be made between filled and unfilled compounds. The fillers are usually inorganic materials, such as carbon black, graphite, talc, chopped glass fibers, and metal powders. Fillers are added to both thermoplastic and thermosetting resins, though they are more frequently used in the latter. These filled compounds have superior strength, stiffness, and temperature resistance over that of the raw polymers but are more difficult to mold and to fabricate.

A confusing array of polymers is available commercially. The confusion is increased by a proliferation of brand names for similar compounds made by different manufacturers. The generic chemical names of polymers tend to be long, complex, and

Table 2-7
Families of Polymers

Thermoplastics
- Cellulosics
- Ethylenics
- Polyamides
- Polyacetals
- Polycarbonates
- Polyphenyline oxides
- Polysulfones

Thermosets
- Aminos
- Elastomers
- Epoxies
- Phenolics
- Polyesters
- Silicones
- Urethanes

hard to remember. In some cases a particular polymer brand name has been so widely used that it has become generic. Nylon, plexiglass, and fiberglass are examples. Learning the generic chemical names and associated brand names of the main families of engineering polymers will eliminate some of the confusion. Table 2-7 shows a number of important polymer families. The mechanical properties of a few of these that have significant engineering applications are included in Appendix C.

Ceramics

Ceramic materials are finding increasing application in engineering and a great deal of effort is being devoted to the development of new ceramic compounds. Ceramics are among the oldest known engineering materials; clay bricks are ceramic materials. Though still widely used in building, clay is not now considered an engineering ceramic. Engineering ceramics are typically compounds of metallic and nonmetallic elements. They may be single oxides of a metal, mixtures of metallic oxides, carbides, borides, nitrides, or other compounds such as Al_2O_3, MgO, SiC, and Si_3N_4, for example. The principal properties of ceramic materials are high hardness and brittleness, high temperature and chemical resistance, high compressive strength, high dielectric strength, and potentially low cost and weight. Ceramic materials are too hard to be machined by conventional techniques and are usually formed by compaction of powder then fired or sintered to form bonds between particles and increase their strength. The powder compaction can be done in dies or by hydrostatic pressure. Sometimes, glass powder is mixed with the ceramic and the result is fired to melt the glass and fuse the two together. Attempts are being made to replace traditional metals with ceramics in such applications as cast engine blocks, pistons, and other engine parts. The low tensile strength, porosity, and low fracture toughness of most ceramics can be problems in these applications. Plasma-sprayed ceramic compounds are often used as hard coatings on metal substrates to provide wear and corrosion-resistant surfaces.

Composites

Most composites are man-made, but some, such as wood, occur naturally. Wood is a composite of long cellulose fibers held together in a resinous matrix of lignin. Man-made composites are typically a combination of some strong, fibrous material such as glass, carbon or boron fibers glued together in a matrix of resin such as epoxy or polyester. The fiberglass material used in boats and other vehicles is a common example of a glass-fiber reinforced polyester (GFRP) composite. The directional material properties of a composite can be tailored to the application by arranging the fibers in different juxtapositions such as parallel, interwoven at random or particular angles, or wound around a mandrel. Custom composites are finding increased use in highly stressed applications such as airframes due to their superior strength-to-weight ratios compared to the common structural metals. Temperature and corrosion resistance can also be designed into some composite materials. These composites are typically neither homogeneous nor isotropic as was discussed in Section 2.3.

2

Table 2-8
Iron and Steel Strengths

Form	S_{ut} kpsi (MPa)
Theoretical	2900 (20E3)
Whisker	1800 (12E3)
Fine wire	1400 (10E3)
Mild steel	60 (414)
Cast iron	40 (276)

It is interesting to note that if one calculates the theoretical strength of any "pure" elemental crystalline material based on the interatomic bonds of the element, the predicted strengths are orders of magnitude larger that those seen in any test of a "real" material, as seen in Table 2-8. The huge differences in actual versus theoretical strengths are attributed to disruptions of the atomic bonds due to crystal defects in the real material. That is, it is considered impossible to manufacture "pure anything" on any realistic superatomic scale. It is presumed that if we could make a "wire" of pure iron only one atom in diameter, it would exhibit its theoretical "super strength." Crystal "whiskers" have been successfully made of some elemental materials and exhibit very high tensile strengths which approach their theoretical values (Table 2-8).

Other empirical evidence for this theory comes from the fact that fibers of any material made in very small diameters exhibit much higher tensile strengths than would be expected from stress-strain tests of larger samples of the same material. Presumably, the very small cross-sections are approaching a "purer" material state. For example, it is well-known that glass has poor tensile strength. However, small-diameter glass fibers show much larger tensile strength than sheet glass, making them a practical (and inexpensive) fiber for use in boat hulls, which are subjected to large tensile stresses in use. Small-diameter fibers of carbon and born exhibit even higher tensile strengths than glass fiber, which explains their use in composites for spacecraft and military aircraft applications, where their relatively high cost is not a barrier.

2.8   SUMMARY

There are many different kinds of material strengths. It is important to understand which ones are important in particular loading situations. The most commonly measured and reported strengths are the **ultimate tensile strength** S_{ut} and the **tensile yield strength** S_y. The S_{ut} indicates the largest stress that the material will accept before fracture, and S_y indicates the stress beyond which the material will take a permanent set. Many materials have **compressive strengths** about equal to their tensile strengths and are called **even materials**. Most wrought metals are in the *even* category. Some materials have significantly different compressive and tensile strengths and these are called **uneven materials**. Cast metals are usually in the *uneven* category, with compressive strengths much greater than their tensile strengths. The **shear strengths** of even materials tend to be about half their tensile strengths, while shear strengths of uneven materials tend to be between their tensile and compressive strengths.

One or more of these strengths may be of interest when the loading is static. If the material is ductile, then S_y is the usual criterion of failure, as a ductile material is capable of significant distortion before fracture. If the material is brittle, as are most cast materials, then the S_{ut} is a more interesting parameter, because the material will fracture before any significant yielding distortion takes place. Yield strength values are nevertheless reported for brittle materials, but are usually calculated based on an arbitrary, small value of strain rather than on any measured yielding of the specimen. Chapter 5 deals with the mechanisms of material failure for both ductile and brittle materials in more detail than does this chapter.

The **tensile test** is the most common measure of these static strength parameters. The **stress strain curve** (σ–ε) generated in this test is shown in Figure 2-2. The so-called **engineering σ–ε curve** differs from the **true σ–ε curve** due to the reduction in area of a ductile test specimen during the failure process. Nevertheless, the **engineering σ–ε curve** is the standard used to compare materials since the true σ–ε curve is more difficult to generate.

The slope of the σ–ε curve in the elastic range, called **Young's modulus** or the **modulus of elasticity** E is a very important parameter as it defines the material's stiffness or resistance to elastic deflection under load. If you are designing to control deflections as well as stresses, the value of E may be of more interest than the material's strength. While various alloys of a given base material may vary markedly in terms of their strengths, they will have essentially the same E. If deflection is the prime concern, a low-strength alloy is as good as a high-strength one of the same base material.

When the loading on the part varies with time it is called **dynamic** or **fatigue loading**. Then the static strengths do not give a good indication of failure. Instead, the **fatigue strength** is of more interest. This strength parameter is measured by subjecting a specimen to dynamic loading until it fails. Both the magnitude of the stress and the number of cycles of stress at failure are reported as the strength criterion. The fatigue strength of a given material will always be lower than its static strength, and often is less than half its S_{ut}. Chapter 6 deals with the phenomenon of fatigue failure of materials in more detail than does this chapter.

Other material parameters of interest to the machine designer are **resilience**, which is the ability to absorb energy without permanent deformation, and **toughness** or the ability to absorb energy without fracturing (but *with* permanent deformation). **Homogeneity** is the uniformity of a material throughout its volume. Many engineering materials, especially metals, can be assumed to be macroscopically homogeneous even though at a microscopic level they are often heterogeneous. **Isotropism** means having properties that are the same regardless of direction within the material. Many engineering materials are reasonably isotropic in the macro and are assumed so for engineering purposes. However, other useful engineering materials such as wood and composites are neither homogeneous nor isotropic and their strengths must be measured separately in different directions. **Hardness** is important in wear resistance and is also related to strength. **Heat treatment**, both through and surface, as well as **cold working** can increase the hardness and strength of some materials.

Important Equations Used in this Chapter

See the referenced sections for information on the proper use of these equations.

Axial tensile stress (Section 2.1):

$$\sigma = \frac{P}{A_o} \qquad (2.1a)$$

Axial tensile strain (Section 2.1):

$$\varepsilon = \frac{l - l_o}{l_o} \qquad (2.1b)$$

Modulus of elasticity (Young's modulus) (Section 2.1).

$$E = \frac{\sigma}{\varepsilon} \qquad (2.2)$$

Modulus of rigidity (Section 2.1):

$$G = \frac{E}{2(1 + \nu)} \qquad (2.4)$$

Ultimate shear strength (Section 2.1):

$$
\begin{array}{ll}
\text{steels}: & S_{us} \cong 0.80 S_{ut} \\
\text{other ductile metals}: & S_{us} \cong 0.75 S_{ut}
\end{array}
\qquad (2.5b)
$$

Shear yield strength (Section 2.1):

$$S_{sy} \cong 0.58 S_y \qquad (2.6)$$

Modulus of resilience (Section 2.1):

$$U_R \cong \frac{1}{2}\frac{S_y^2}{E} \qquad (2.7)$$

Modulus of toughness (Section 2.1):

$$U_T \cong \left(\frac{S_y + S_{ut}}{2}\right)\varepsilon_f \qquad (2.8)$$

Arithmetic mean (Section 2.2):

$$\mu = \frac{1}{n}\sum_{i=1}^{n} x_i \qquad (2.9b)$$

Standard deviation (Section 2.2):

$$S_d = \sqrt{\frac{1}{n-1}\sum_{i=1}^{n}(x_i - \mu)^2} \qquad (2.9c)$$

Ultimate tensile strength as a function of Brinell hardness (Section 2.4):

$$S_{ut} \cong 500 H_B \pm 30 H_B \quad \text{psi}$$

$$S_{ut} \cong 3.45 H_B \pm 0.2 H_B \quad \text{MPa}$$

(2.10)

2.9   REFERENCES

1   **E. B. Haugen and P. H. Wirsching**, "Probabilistic Design." *Machine Design*, v. 47, nos. 10-14, Penton Publishing, Cleveland, Ohio, 1975.

2   **H. E. Boyer and T. L. Gall**, ed. *Metals Handbook*. Vol. 1. American Society for Metals: Metals Park, Ohio, 1985.

2.10   BIBLIOGRAPHY

For general information on materials, consult the following:

Metals & Alloys in the Unified Numbering System. 6th ed. ASTM/SAE: Philadelphia, Pa., 1994.

Brady, ed. *Materials Handbook*. 13th ed. McGraw-Hill: New York. 1992.

H. E. Boyer, ed. *Atlas of Stress-Strain Curves*. Amer. Soc. for Metals: Metals Park, Ohio, 1987.

K. Budinski, *Engineering Materials: Properties and Selection*. 4th ed. Reston-Prentice-Hall: Reston, Va., 1992.

M. M. Farag, *Selection of Materials and Manufacturing Processes for Engineering Design*. Prentice-Hall International: Hertfordshire, U.K., 1989.

I. Granet, *Modern Materials Science*. Reston-Prentice-Hall: Reston, Va., 1980.

H. W. Pollack, *Materials Science and Metallurgy*. 2nd ed. Reston-Prentice-Hall: Reston, Va., 1977.

S. P. Timoshenko, *History of Strength of Materials*. McGraw-Hill: New York, 1983.

L. H. V. Vlack, *Elements of Material Science and Engineering*. 6th ed. Addison-Wesley: Reading, Mass., 1989.

M. M. Schwartz, ed. *Handbook of Structural Ceramics*. McGraw-Hill: New York. 1984.

For specific information on material properties, consult the following:

H. E. Boyer and T. L. Gall, ed. *Metals Handbook*. Vol. 1. American Society for Metals: Metals Park, Ohio, 1985.

R. Juran, ed. *Modern Plastics Encyclopedia*. McGraw-Hill: New York. 1988.

2

J. D. Lubahn and R. P. Felgar, *Plasticity and Creep of Metals*. Wiley: New York, 1961.

For information on failure of materials, consult the following:

J. A. Collins, *Failure of Materials in Mechanical Design*. Wiley: New York, 1981.

N. E. Dowling, *Mechanical Behavior of Materials*. Prentice-Hall: Englewood Cliffs, N.J., 1992.

R. C. Juvinall, *Stress, Strain and Strength*. McGraw-Hill: New York, 1967.

For information on plastics and composites, consult the following:

ASM, *Engineered Materials Handbook: Composites*. Vol. 1. American Society for Metals: Metals Park, Ohio, 1987.

ASM, *Engineered Materials Handbook: Engineering Plastics*. Vol. 2. American Society for Metals: Metals Park, Ohio, 1988.

Harper, ed. *Handbook of Plastics, Elastomers and Composites*. 2nd ed. McGraw-Hill: New York, 1990.

J. E. Hauck, "Long-Term Performance of Plastics." *Materials in Design Engineering*, p. 113-128, November, 1965.

M. M. Schwartz, *Composite Materials Handbook*. McGraw-Hill: New York, 1984.

For information on manufacturing processes, see:

R. W. Bolz, *Production Processes: The Productivity Handbook*. Industrial Press: New York, 1974.

J. A. Schey, *Introduction to Manufacturing Processes*. McGraw-Hill: New York, 1977.

2.11   PROBLEMS

2-1   Figure P2-1 shows stress-strain curves for three failed tensile-test specimens. All are plotted on the same scale.

(a)   Characterize each material as brittle or ductile.
(b)   Which is the stiffest?
(c)   Which has the highest ultimate strength?
(d)   Which has the largest modulus of resilience?
(e)   Which has the largest modulus of toughness?

2-2   Determine an approximate ratio between the yield strength and ultimate strength for each material shown in Figure P2-1.

2-3   Which of the steel alloys shown in Figure 2-19 would you choose to obtain

(a)   Maximum strength
(b)   Maximum modulus of resilience
(c)   Maximum modulus of toughness
(d)   Maximum stiffness

* Answers to these problems are provided in Appendix G.

2-4 Which of the aluminum alloys shown in Figure 2-21 would you choose to obtain

(a)   Maximum strength
(b)   Maximum modulus of resilience
(c)   Maximum modulus of toughness
(d)   Maximum stiffness

2-5 Which of the thermoplastic polymers shown in Figure 2-22 would you choose in order to obtain

(a)   Maximum strength
(b)   Maximum modulus of resilience
(c)   Maximum modulus of toughness
(d)   Maximum stiffness

*2-6 A metal has a strength of 60 kpsi (414 MPa) at its elastic limit and the strain at that point is 0.002. What is its modulus of elasticity? What is the strain energy at the elastic limit? Can you define the type of metal based on the given data?

2-7 A metal has a strength of 41.2 kpsi (284 MPa) at its elastic limit and the strain at that point is 0.004. What is its modulus of elasticity? What is the strain energy at the elastic limit? Can you define the type of metal based on the given data?

*2-8 A metal has a strength of 19.5 kpsi (134 MPa) at its elastic limit and the strain at that point is 0.003. What is its modulus of elasticity? What is the strain energy at the elastic limit? Can you define the type of metal based on the given data?

2-9 A metal has a strength of 100 kpsi (689 MPa) at its elastic limit and the strain at that point is 0.006. What is its modulus of elasticity? What is the strain energy at the elastic limit? Can you define the type of metal based on the given data?

*2-10 A steel has a yield strength of 100 kpsi (689 MPa) at an offset of 0.6% strain. What is its modulus of resilience?

2-11 An aluminum has a yield strength of 60 kpsi (414 MPa) at an offset of 0.2% strain. What is its modulus of resilience?

*2-12 A steel has a yield strength of 60 kpsi (414 MPa), an ultimate tensile strength of 100 kpsi (689 MPa), and an elongation at fracture of 15%. What is its approximate modulus of toughness? What is its approximate modulus of resilience?

2-13 The Brinell hardness of a steel specimen was measured to be 250 HB. What is the material's approximate tensile strength? What is its hardness on the Vickers scale? The Rockwell scale?

*2-14 The Brinell hardness of a steel specimen was measured to be 340 HB. What is the material's approximate tensile strength? What is its hardness on the Vickers scale? The Rockwell scale?

2-15 What are the principal alloy elements of an AISI 4340 steel? How much carbon does it have? Is it hardenable? By what techniques?

*2-16 What are the principal alloy elements of an AISI 1095 steel? How much carbon does it have? Is it hardenable? By what techniques?

2-17 What are the principal alloy elements of an AISI 6180 steel? How much carbon does it have? Is it hardenable? By what techniques?

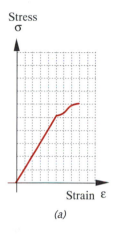

(a)

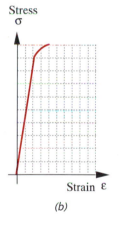

(b)

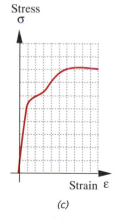

(c)

FIGURE P2-1

Stress-Strain Curves

2

2-18   Which of the steels in Problems 2-15, 2-16, and 2-17 is the stiffest?

2-19   Calculate the *specific strength* and *specific stiffness* of the following materials and pick one for use in an aircraft wing spar.

(a)   Steel S_{ut} = 80 kpsi (552 MPa)
(b)   Aluminum S_{ut} = 60 kpsi (414 MPa)
(c)   Titanium S_{ut} = 90 kpsi (621 MPa)

2-20   If maximum *impact resistance* were desired in a part, which material properties would you look for?

2-21   Write a *TKSolver* program that contains a List Function that returns the modulus of elasticity for steel, gray cast iron, aluminum, titanium, and magnesium in response to supplying it with one of the following arguments: 'st, 'ci, 'al, 'ti, 'mg. Print out a table of these values and labels.

2-22   Write a *TKSolver* program that contains a rule function that returns the approximate ultimate shear strength S_{us} and approximate shear yield strength S_{ys} of any material when it is given the ultimate tensile strength S_{ut} and tensile yield strength S_y of the material. Print out a table of these data for a material having S_{ut} = 400 MPa and S_y = 300 MPa.

2-23   Write a *TKSolver* program that contains a List Function that returns the density for steel, gray cast iron, aluminum, titanium, and magnesium in response to supplying it with one of the following arguments: 'st, 'ci, 'al, 'ti, 'mg. Print out a table of these values and labels.

2-24   Call your local steel and aluminum distributors (consult the Yellow Pages) and obtain current costs per pound for round stock of consistent size in low-carbon (SAE 1020) steel, SAE 4340 steel, 2024-T4 aluminum, and 6061-T6 aluminum. Calculate a strength/dollar ratio and a stiffness/dollar ratio for each alloy. Which would be your first choice on a cost-efficiency basis for an axial-tension-loaded round rod

(a)   If maximum strength were needed?
(b)   If maximum stiffness were needed?

2-25   Call your local plastic stock-shapes distributors (consult the Yellow Pages) and obtain current costs per pound for round rod or tubing of consistent size in plexiglass, acetal, nylon 6/6, and PVC. Calculate a strength/dollar ratio and a stiffness/dollar ratio for each alloy. Which would be your first choice on a cost-efficiency basis for an axial-tension-loaded round rod or tube of particular diameters. (Note: material parameters can be found in Appendix C.)

(a)   If maximum strength were needed?
(b)   If maximum stiffness were needed?

3

LOAD
DETERMINATION

*If a builder has built a house for a man and his
work is not strong and the house falls in and
kills the householder, that builder shall be slain*

FROM THE CODE OF HAMMURABI, 2150 BC

3.0   INTRODUCTION

This chapter provides a review of the fundamentals of static and dynamic force analysis, impact forces, and beam loading. The reader is assumed to have had first courses in statics and dynamics. Thus, this chapter presents only a brief, general overview of those topics but also provides more powerful solution techniques, such as the use of singularity functions for beam calculations. The Newtonian solution method of force analysis is reviewed and a number of case-study examples are presented to reinforce understanding of this subject. The case studies also set the stage for analysis of these same systems for stress, deflection, and failure modes in later chapters.

Table 3-0 shows the variables used in this chapter and references the equations, sections, or case studies in which they are used. At the end of the chapter, a summary section is provided which groups all the significant equations from this chapter for easy reference and identifies the chapter section in which their discussion can be found.

3.1   LOADING CLASSES

The type of loading on a system can be divided into several classes based on the character of the applied loads and the presence or absence of system motion. Once the general configuration of a mechanical system is defined and its kinematic motions calculated, the next task is to determine the magnitudes and directions of all the forces and couples present on the various elements. These loads may be constant or may be

Table 3-0   Variables Used in This Chapter

Symbol	Variable	ips units	SI units	See
a	distance to load	in	m	Sect. 3.9
b	distance to load	in	m	Sect. 3.9
d	damping	lb-sec/in	N-sec/m	Eq. 3.6
E	energy	in-lb	joules	Eq. 3.9, 3.10
F	force or load	lb	N	Sect. 3.1
f_d	damped natural frequency	Hz	Hz	Eq. 3.7
f_n	natural frequency	Hz	Hz	Eq. 3.4
g	gravitational acceleration	in/sec^2	m/sec^2	Eq 3.12
I_x	mass moment of inertia about x axis	lb-in-sec^2	kg-m^2	Sect. 3.1
I_y	mass moment of inertia about y axis	lb-in-sec^2	kg-m^2	Sect. 3.1
I_z	mass moment of inertia about z axis	lb-in-sec^2	kg-m^2	Sect. 3.1
k	spring rate or spring constant	lb/in	N/m	Eq. 3.5
l	length	in	m	Sect. 3.9
m	mass	lb-sec^2/in	kg	Sect. 3.1
N	normal force	in	m	Case 4a
M	moment, moment function	lb-in	N-m	Sect. 3.1, 3.9
q	beam loading function	lb	N	Sect. 3.9
$\mathbf{R}$	position vector	in	m	Sect 3.4
R	reaction force	lb	N	Sect. 3.9
v	linear velocity	in/sec	m/sec	Eq. 3.10
V	beam shear function	lb	N	Sect. 3.9
W	weight	lb	N	Eq. 3.14
x	generalized length variable	in	m	Sect. 3.9
y	displacement	in	m	Eq. 3.5, 3.8
δ	deflection	in	m	Eq. 3.5
η	correction factor	none	none	Eq 3.10
μ	coefficient of friction	none	none	Case 4a
ω	rotational or angular velocity	rad/sec	rad/sec	Case 5a
ω_d	damped natural frequency	rad/sec	rad/sec	Eq. 3.7
ω_n	natural frequency	rad/sec	rad/sec	Eq. 3.4

varying over time. The elements in the system may be stationary or moving. The most general class is that of a moving system with time-varying loads. The other combinations are subsets of the general class.

Table 3-1   Load Classes

	Constant Loads	Time-Varying Loads
Stationary Elements	Class 1	Class 2
Moving Elements	Class 3	Class 4

Table 3-1 shows the four possible classes. Class 1 is a stationary system with constant loads. One example of a Class 1 system is the base frame for an arbor press used in a machine shop. The base is required to support the dead weight of the arbor press which is essentially constant over time, and the base frame does not move. The parts brought to the arbor press (to have something pressed into them) temporarily add their weight to the load on the base, but this is usually a small percentage of the dead weight. A static load analysis is all that is necessary for a Class 1 system.

Class 2 describes a stationary system with time-varying loads. An example is a bridge which, though essentially stationary, is subjected to changing loads as vehicles drive over it and wind impinges on its structure. Class 3 defines a moving system with constant loads. Even though the applied external loads may be constant, any significant accelerations of the moving members can create time-varying reaction forces. An example might be a powered rotary lawn mower. Except for the case of mowing the occasional rock, the blades experience a nearly constant external load from mowing the grass. However, the accelerations of the spinning blades can create high loads at their fastenings. A dynamic load analysis is necessary for Classes 2 and 3.

Note however that, if the motions of a Class 3 system are so slow as to generate negligible accelerations on its members, it could qualify as a Class 1 system. An automobile scissors jack (see Figure 3-5) can be considered to be a Class 1 system since the external load (when used) is essentially constant, and the motions of the links are slow with negligible accelerations. The only complexity introduced by the motions of the elements in this example is that of determining in which position the internal loads on the jack's elements will be maximal since they vary as the jack is raised, despite the essentially constant external load.

Class 4 describes the general case of a rapidly moving system subjected to time-varying loads. Note that even if the applied external loads are essentially constant in a given case, the dynamic loads developed on the elements from their accelerations will still vary with time. Most machinery, especially if powered by a motor or engine, will be in Class 4. An example of such a system is the engine in your car. The internal parts (crankshaft, connecting rods, pistons, etc.) are subjected to time-varying loads from the gasoline explosions, and also experience time-varying inertial loads from their own accelerations. A dynamic load analysis is necessary for Class 4.

3.2   FREE-BODY DIAGRAMS

In order to correctly identify all potential forces and moments on a system, it is necessary to draw accurate free-body diagrams (FBDs) of each member of the system. These FBDs should show a general shape of the part and display all the forces and moments that are acting on it. There may be external forces and moments applied to the part from outside the system, and there will be interconnection forces and/or moments where each part joins or contacts adjacent parts in the assembly or system.

In addition to the known and unknown forces and couples shown on the FBD, the dimensions and angles of the elements in the system are defined with respect to local coordinate systems located at the **centers of gravity** (CG) of each element.[*] For a dynamic load analysis, the kinematic accelerations, both angular and linear (at the CG), need to be known or calculated for each element prior to doing the load analysis.

3.3   LOAD ANALYSIS

This section presents a brief review of Newton's laws and Euler's equations as applied to dynamically-loaded and statically-loaded systems in both 3-D and 2-D. The method of solution presented here may be somewhat different than that used in your previous statics and dynamics courses. The approach taken here in setting up the equations for force and moment analysis is designed to facilitate computer programming of the solution.

This approach assumes all *unknown* forces and moments on the system to be positive in sign, regardless of what one's intuition or an inspection of the free-body diagram might indicate as to their probable directions. However, all *known* force components are given their proper signs to define their directions. The simultaneous solution of the set of equations that results will cause all the unknown components to have the proper signs when the solution is complete. This is ultimately a simpler approach than the one often taught in statics and dynamics courses which requires that the student assume directions for all unknown forces and moments (a practice that does help the student develop some intuition, however). Even with that traditional approach, an incorrect assumption of direction results in a sign reversal on that component in the solution. Assuming all unknown forces and moments to be positive allows the resulting computer program to be simpler than would otherwise be the case. The simultaneous equation solution method used is extremely simple in concept, though it requires the aid of a computer to solve. Software is provided with the text to solve the simultaneous equations.

Real dynamic systems are three dimensional and thus must be analyzed as such. However, many 3-D systems can be analyzed by simpler 2-D methods. Accordingly, we will investigate both approaches.

[*] While it is not a requirement that the local coordinate system for each element be located at its CG, this approach provides consistency and simplifies the dynamic calculations. Further, most solids-modeling CAD/CAE systems will automatically calculate the mass properties of parts with respect to their CGs. The approach taken here is to apply a consistent method that works for both static and dynamic problems and that is also amenable to computer solution.

Three-Dimensional Analysis

Since three of the four cases potentially require dynamic load analysis, and because a static force analysis is really just a variation on the dynamic analysis, it makes sense to start with the dynamic case. Dynamic load analysis can be done by any of several methods, but the one that gives the most information about internal forces is the Newtonian approach based on Newton's laws.

NEWTON'S FIRST LAW   *A body at rest tends to remain at rest and a body in motion at constant velocity will tend to maintain that velocity unless acted upon by an external force.*

NEWTON'S SECOND LAW   *The time rate of change of momentum of a body is equal to the magnitude of the applied force and acts in the direction of the force.*

Newton's second law can be written for a rigid body in two forms, one for linear forces and one for moments or torques:

$$\sum \mathbf{F} = m\mathbf{a} \qquad\qquad \sum \mathbf{M}_G = \dot{\mathbf{H}}_G \qquad\qquad (3.1a)$$

where $\mathbf{F}$ = force, m = mass, $\mathbf{a}$ = acceleration, $\mathbf{M}_G$ = moment about the center of gravity, and $\dot{\mathbf{H}}_G$ = the time rate of change of the moment of momentum, or the angular momentum about the CG. The left sides of these equations respectively sum all the forces and moments that act on the body, whether from known applied forces or from interconnections with adjacent bodies in the system.

For a three-dimensional system of connected rigid bodies, this vector equation for the linear forces can be written as three scalar equations involving orthogonal components taken along a local x, y, z axis system with its origin at the CG of the body:

$$\sum F_x = ma_x \qquad \sum F_y = ma_y \qquad \sum F_z = ma_z \qquad (3.1b)$$

If the x, y, z axes are chosen coincident with the principal axes of inertia of the body,[*] the angular momentum of the body is defined as

$$\mathbf{H}_G = I_x \omega_x \hat{\mathbf{i}} + I_y \omega_y \hat{\mathbf{j}} + I_z \omega_z \hat{\mathbf{k}} \qquad\qquad (3.1c)$$

where I_x, I_y, and I_z are the principal centroidal mass moments of inertia (second moments of mass) about the principal axes. This vector equation can be substituted into equation 3.1a to yield the three scalar equations known as **Euler's equations**:

$$\sum M_x = I_x \alpha_x - \left(I_y - I_z\right)\omega_y \omega_z$$

$$\sum M_y = I_y \alpha_y - \left(I_z - I_x\right)\omega_z \omega_x \qquad\qquad (3.1d)$$

$$\sum M_z = I_z \alpha_z - \left(I_x - I_y\right)\omega_x \omega_y$$

[*] This is a convenient choice for symmetric bodies but may be less convenient for other shapes. See F. P. Beer and E. R. Johnson, *Vector Mechanics for Engineers*, 3rd ed., 1977, McGraw-Hill, New York, Chap. 18, "Kinetics of Rigid Bodies in Three Dimensions."

where M_x, M_y, M_z are moments about those axes and α_x, α_y, α_z are the angular accelerations about the axes. This assumes that the inertia terms remain constant with time, i.e., the mass distribution about the axes is constant.

NEWTON'S THIRD LAW   states that *when two particles interact, a pair of equal and opposite reaction forces will exist at their contact point. This force pair will have the same magnitude and act along the same direction line, but have opposite sense.*

We will need to apply this relationship as well as applying the second law in order to solve for the forces on assemblies of elements that act upon one another. The six equations in equations 3.1*b* and 3.1*d* can be written for each rigid body in a 3-D system. In addition, as many (third-law) reaction force equations as are necessary will be written and the resulting set of equations solved simultaneously for the forces and moments. The number of second-law equations will be up to six times the number of individual parts in a three-dimensional system (plus the reaction equations), meaning that even simple systems result in large sets of simultaneous equations. A computer is needed to solve these equations, though high-end pocket calculators will solve large sets of simultaneous equations also. The reaction (third-law) equations are often substituted into the second-law equations to reduce the total number of equations to be solved simultaneously.

Two-Dimensional Analysis

All real machines exist in three dimensions but many three-dimensional systems can be analyzed two dimensionally if their motions exist only in one plane or in parallel planes. **Euler's equations** 3.1*d* show that if the rotational motions (ω, α) and applied moments or couples exist about only one axis (say the *z* axis), then that set of three equations reduces to one equation,

$$\sum M_z = I_z \alpha_z \qquad (3.2a)$$

because the ω and α terms about the *x* and *y* axes are now zero. Equation 3.1*b* is reduced to

$$\sum F_x = ma_x \qquad \sum F_y = ma_y \qquad (3.2b)$$

Equations 3.2 can be written for all the connected bodies in a two-dimensional system and the entire set solved simultaneously for forces and moments. The number of second-law equations will now be up to three times the number of elements in the system plus the necessary reaction equations at connecting points, again resulting in large systems of equations for even simple systems. Note that even though all motion is about one (*z*) axis in a 2-D system, there may still be loading components in the *z* direction due to external forces or couples.

Static Load Analysis

The difference between a dynamic loading situation and a static one is the presence or absence of accelerations. If the accelerations in equations 3.1 and 3.2 are all zero, then for the three-dimensional case these equations reduce to

$$\sum F_x = 0 \qquad \sum F_y = 0 \qquad \sum F_z = 0$$

$$\sum M_x = 0 \qquad \sum M_y = 0 \qquad \sum M_z = 0 \qquad (3.3a)$$

and for the two-dimensional case,

$$\sum F_x = 0 \qquad \sum F_y = 0 \qquad \sum M_z = 0 \qquad (3.3b)$$

Thus, we can see that the static loading situation is just a special case of the dynamic loading one, in which the accelerations happen to be zero. A solution approach based on the dynamic case will then also satisfy the static one with appropriate substitutions of zero values for the absent accelerations.

3.4   TWO-DIMENSIONAL, STATIC LOADING CASE STUDIES

This section presents a series of three case studies of increasing complexity, all limited to two-dimensional static loading situations. A bicycle handbrake lever, a crimping tool, and a scissors jack are the systems analyzed. These case studies provide examples of the simplest form of force analysis, having no significant accelerations and forces acting in only two dimensions.

CASE STUDY 1A

Bicycle Brake Lever Loading Analysis

Problem: Determine the forces on the elements of the bicycle brake lever assembly shown in Figure 3-1 during braking.

Given: The geometry of each element is known. The average humans hand can develop a grip force of about 267 N (60 lb) in the lever position shown. A very strong hand can exert about 712 N (160 lb).

Assumptions: The accelerations are negligible. All forces are coplanar and two dimensional. A Class 1 load model is appropriate and a static analysis is acceptable. The higher applied load will be used as a worst case, assuming that it can be reached before bottoming the tip of the handle on the handgrip. If that occurs, it will change the beam's boundary conditions and the analysis.

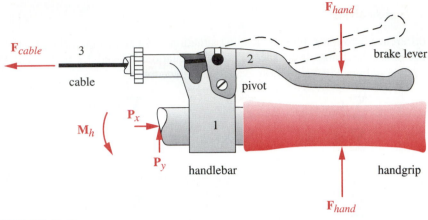

FIGURE 3-1

Bicycle Brake Lever Assembly

Solution: **See Figures 3-1, 3-2, Table 3-2, and TKSolver file CASE1A.**

1   Figure 3-1 shows the handbrake lever assembly, which consists of three subassemblies: the handlebar (1), the lever (2), and the cable (3). The lever is pivoted to the handlebar and the cable is connected to the lever. The cable runs within a plastic-lined sheath (for low friction) down to the brake caliper assembly at the bicycle's wheel rim. The user's hand applies equal and opposite forces at some point on the lever and handgrip. These forces are transformed to a larger force in the cable by reason of the lever ratio of part 2.

Figure 3-1 is a free-body diagram of the entire assembly since it shows all the forces and moments acting on it except for its weight which is small compared to the applied forces and is thus neglected for this analysis. The "broken away" portion of the handlebar provides x and y force components and a moment. These are arbitrarily shown as positive in sign. Their actual signs will "come out in the wash" in the calculations. The known applied forces are shown in their actual directions and senses.

2   Figure 3-2 shows the three subassembly elements separated and drawn as free-body diagrams with all relevant forces and moments applied to each element, again neglecting the weights of the parts. The lever (part 2) has three forces on it, $\mathbf{F}_{b2}$, $\mathbf{F}_{c2}$, and $\mathbf{F}_{12}$. The two-character subscript notation used here should be read as, force of element 1 **on** 2 ($\mathbf{F}_{12}$) or force **at** B on 2 ($\mathbf{F}_{b2}$), etc. This defines the source of the force (first subscript) and the element on which it acts (second subscript).

This notation will be used consistently throughout this text for both forces and position vectors such as $\mathbf{R}_{b2}$, $\mathbf{R}_{c2}$, and $\mathbf{R}_{12}$ in Figure 3-2 which serve to locate the above three forces in a local, nonrotating coordinate system whose origin is at the center of gravity (CG) of the element or subassembly being analyzed.[*]

On this brake lever, $\mathbf{F}_{b2}$ is an applied force whose magnitude and direction are known. $\mathbf{F}_{c2}$ is the force in the cable. Its direction is known but not its magnitude. Force $\mathbf{F}_{12}$

[*] Actually, for a simple static analysis such as the one in this example, any point (on or off the element) can be taken as the origin of the local coordinate system. However, in a dynamic force analysis it simplifies the analysis if the coordinate system is placed at the CG. So, for the sake of consistency, and to prepare for the more complicated dynamic analysis problems ahead, we will use the CG as the origin even in the static cases here.

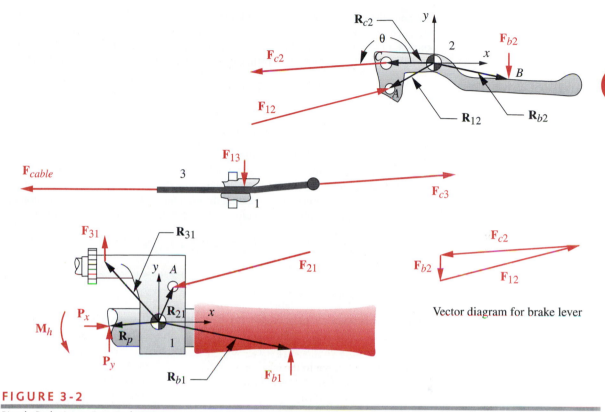

FIGURE 3-2

Bicycle Brake Lever Free-Body Diagrams

is provided by part 1 on part 2 at the pivot pin. Its magnitude and direction are both unknown. We can write equations 3.3b for this element to sum forces in the x and y directions and sum moments about the CG. Note that all unknown forces and moments are initially assumed positive in the equations. Their true signs will come out in the calculation.[†]

$$\sum F_x = F_{12x} + F_{b2x} + F_{c2x} = 0$$

$$\sum F_y = F_{12y} + F_{b2y} + F_{c2y} = 0 \qquad (a)$$

$$\sum M_z = \left(\mathbf{R}_{12} \times \mathbf{F}_{12}\right) + \left(\mathbf{R}_{b2} \times \mathbf{F}_{b2}\right) + \left(\mathbf{R}_{c2} \times \mathbf{F}_{c2}\right) = 0$$

The cross products in the moment equation represent the "turning forces" or moments created by the application of these forces at points remote from the CG of the element. Recall that these cross products can be expanded to

$$\sum M_z = \left(R_{12x}F_{12y} - R_{12y}F_{12x}\right) + \left(R_{b2x}F_{b2y} - R_{b2y}F_{b2x}\right)$$

$$+ \left(R_{c2x}F_{c2y} - R_{c2y}F_{c2x}\right) = 0 \qquad (b)$$

[†] You may not have done this in your statics class but this approach makes the problem more amenable to a computer solution. Note that regardless of the direction shown for any unknown force on the FBD, we will assume its components to be positive in the equations. The angles of the known (given) forces (or the signs of their components) do have to be correctly input to the equations, however.

We have three equations and four unknowns (F_{12x}, F_{12y}, F_{c2x}, F_{c2y}) at this point, so we need another equation. It is available from the fact that the direction of $\mathbf{F}_{c2}$ is known. (The cable can pull only along its axis). We can express one component of the cable force $\mathbf{F}_{c2}$ in terms of its other component and the known angle, θ of the cable.

$$F_{c2y} = F_{c2x} \tan\theta \qquad (c)$$

We could now solve the four unknowns for this element, but will wait to do so until the equations for the other two links are defined.

3   Part 3 in Figure 3-2 is the cable which passes through a hole in part 1. This hole is lined with a low friction material which allows us to assume no friction at the joint between parts 1 and 3. We will further assume that the three forces $\mathbf{F}_{13}$, $\mathbf{F}_{c3}$ and $\mathbf{F}_{cable}$ form a concurrent system of forces acting through the CG and thus create no moment. With this assumption only a summation of forces is necessary for this element.

$$\sum F_x = F_{cable_x} + F_{13x} + F_{c3x} = 0$$
$$\sum F_y = F_{cable_y} + F_{13y} + F_{c3y} = 0 \qquad (d)$$

4   The assembly of elements labeled part 1 in Figure 3-2 has both forces and moments on it (i.e., it is not a concurrent system), so the three equations 3.3b are needed.

$$\sum F_x = F_{21x} + F_{b1x} + F_{31x} + P_x = 0$$
$$\sum F_y = F_{21y} + F_{b1y} + F_{31y} + P_y = 0 \qquad (e)$$
$$\sum M_z = M_h + \left(\mathbf{R}_{21} \times \mathbf{F}_{21}\right) + \left(\mathbf{R}_{b1} \times \mathbf{F}_{b1}\right) + \left(\mathbf{R}_{31} \times \mathbf{F}_{31}\right) + \left(\mathbf{R}_p \times \mathbf{F}_p\right) = 0$$

Expanding cross products in the moment equation gives the moment magnitude as

$$\sum M_z = M_h + \left(R_{21x}F_{21y} - R_{21y}F_{21x}\right) + \left(R_{b1x}F_{b1y} - R_{b1y}F_{b1x}\right)$$
$$+ \left(R_{31x}F_{31y} - R_{31y}F_{31x}\right) + \left(R_{Px}F_{Py} - R_{Py}F_{Px}\right) = 0 \qquad (f)$$

5   The grand total of unknowns at this point (including those listed in step 2 above) is nineteen: F_{b1x}, F_{b1y}, F_{12x}, F_{12y}, F_{21x}, F_{21y}, F_{c2x}, F_{c2y}, F_{c3x}, F_{c3y}, F_{13x}, F_{13y}, F_{31x}, F_{31y}, F_{cablex}, F_{cabley}, P_x, P_y, and M_h. We have only nine equations so far, three in equation set (a), one in set (c), two in set (d) and three in set (e). We need ten more equations to solve this system. We can get six of them from the Newton's third-law relationships between contacting elements:

$$F_{c3x} = -F_{c2x} \qquad\qquad F_{c3y} = -F_{c2y}$$
$$F_{21x} = -F_{12x} \qquad\qquad F_{21y} = -F_{12y} \qquad (g)$$
$$F_{31x} = -F_{13x} \qquad\qquad F_{31y} = -F_{13y}$$

Two more equations come from the assumption (shown in Figure 3-1) that the two forces provided by the hand on the brake lever and handgrip are equal and opposite:

$$F_{b1x} = -F_{b2x}$$
$$F_{b1y} = -F_{b2y}$$

$$(h)$$

The remaining two equations come from the given geometry and the assumptions made about the system. The direction of the force F_{cable} is known to be in the same direction as that end of the cable. In the figure it is seen to be horizontal, so we can set

$$F_{cable_y} = 0$$

$$(i)$$

Table 3-2   **Case Study 1A – Brake Lever Force Analysis**
Part 1 of 3   Rule Sheet from *TKSolver* file CASE1A

Rules

   ; within-link relations - Newton's second law

    ; for link 2

      Fc2y = Fc2x * TAND(Angc2)
      F12x + Fb2x + Fc2x = 0
      F12y + Fb2y + Fc2y = 0
      Cross2 (R12x, R12y, F12x, F12y) + Cross2 (Rb2x, Rb2y, Fb2x, Fb2y) + Cross2 (Rc2x, Rc2y, Fc2x, Fc2y) = 0

    ; for link 3

      Fcablex + Fc3x + F13x = 0
      Fcabley + Fc3y + F13y = 0
      Fcabley = Fcablex * TAND (Angcable)

    ; for link 1

      F21x + Fb1x + F31x + Px = 0
      F21y + Fb1y + F31y + Py = 0
      Mz + Cross2 (R21x, R21y, F21x, F21y) + Cross2 (Rb1x, Rb1y, Fb1x, Fb1y) + Cross2 (R31x, R31y, F31x, F31y) = 0

   ; cross-link relations - Newton's third law

      Fb1x = – Fb2x
      Fb1y = – Fb2y
      F31x = – F13x
      F31y = – F13y
      F21x = – F12x
      F21y = – F12y
      Fc3x = – Fc2x
      Fc3y = – Fc2y

Functions

   ; Rule Function Cross2 to calculate a two-dimensional cross-product.
   ; Inputs are the Cartesian coordinates of the two vectors
   ; to use:
   ;   CALL Cross2 (Rx, Ry, Fx, Fy ; M)   or: M = Cross2 (Rx, Ry, Fx, Fy)

      M = Rx * Fy – Ry * Fx

Table 3–2   Case Study 1A – Brake Lever Force Analysis

Part 2 of 3   Given Data - See *TKSolver* File CASE1A

Input	Variable	Output	Unit	Comments
				Applied Force
0	*Fb2x*		N	component of handle force
–712	*Fb2y*		N	component of handle force
184	*Angc2*		deg	known angle of cable force F_{C2}
180	*Angcabl*		deg	known angle of force Fcable
				Link 2 Geometry
19	*Rb2x*		mm	component of position vector R_{b2}
–4	*Rb2y*		mm	component of position vector R_{b2}
–25	*Rc2x*		mm	component of position vector R_{c2}
0	*Rc2y*		mm	component of position vector R_{c2}
–12	*R12x*		mm	component of position vector R_{12}
–7	*R12y*		mm	component of position vector R_{12}
				Link 1 Geometry
7	*R21x*		mm	component of position vector R_{21}
19	*R21y*		mm	component of position vector R_{21}
70	*Rb1x*		mm	component of position vector R_{b1}
–14	*Rb1y*		mm	component of position vector R_{b1}
–27	*R31x*		mm	component of position vector R_{31}
30	*R31y*		mm	component of position vector R_{31}

Because of our no-friction assumption, the force F_{31} can be assumed to be normal to the surface of contact between the cable and the hole in part 1. This surface is horizontal in this example so F_{31} is vertical and

$$F_{31x} = 0 \qquad\qquad (j)$$

This completes the set of nineteen equations (equation sets *a, c, d, e, g, h, i,* and *j*) and they can be solved for the nineteen unknowns.

6   This set of equations can be solved simultaneously "as is," that is, all nineteen equations can be put into matrix form and solved with a **matrix-reduction computer program,** or equations *g, h, i,* and *j* can be manually substituted into the others to reduce them to a set of nine equations in nine unknowns.   In this case, since it is a simple system, the four equations in (*a*) and (*c*) can then be manually solved as noted in step 2 above. This will yield the values of F_{12x}, F_{12y}, F_{c2x}, and F_{c2y}. Equations (*d*) can then be solved for F_{13y} and F_{cablex}. Finally, these results can be used to solve equations (*e*) and (*f*) for the unknown forces and moments on part 1 (P_x, P_y, and M_h).

3

Table 3–2   Case Study 1A – Brake Lever Force Analysis
Part 3 of 3   Calculated Results - See *TKSolver* File CASE1A

Input	Variable	Output	Unit	Comments
				Link 2 Forces
	$Fc2x$	–2 851	N	force of cable on 2 in x - iterated
	$Fc2y$	–199	N	force of cable on 2 in y - iterated
	$F12x$	2 851	N	force of 1 on 2 in x - iterated
G*	$F12y$	911	N	force of 1 on 2 in y - guessed
				Link 3 Forces
	$Fc3x$	2 851	N	force of cable on 3 in x - iterated
	$Fc3y$	199	N	force of cable on 3 in y - iterated
0	$F13x$		N	force of 1 on 3 in x - assumed zero
	$F13y$	–199	N	force of 1 on 3 in y - iterated
	$Fcablex$	–2 851	N	force of cable in x - iterated
	$Fcabley$	0	N	force of cable in y - iterated
				Link 1 Forces
	$Fb1x$	0	N	force of hand on 1 in x - iterated
	$Fb1y$	712	N	force of hand on 1 in y - iterated
	$F31x$	0	N	force of 3 on 1 in x - iterated
	$F31y$	199	N	force of 3 on 1 in y - iterated
	$F21x$	–2 851	N	force of 2 on 1 in x - iterated
	$F21y$	–911	N	force of 2 on 1 in y - iterated
	Px	2 851	N	reaction force on handlebar - in x
	Py	0	N	reaction force on handlebar - in y
	Mh	–93	N-m	reaction moment on handlebar

* Indicates that a guess value was required to start the iteration.

7   Another approach to the solution of sets of simultaneous equations is to use an **iterative equation solver** such as *TKSolver,* a copy of which is included with this text. (See Section 1-10 and the *TKSolver* manual for instructions in using this software.) This case study (and most of the others in the text) was solved with this software and its data file is included on disk. You may input the file labeled CASE1A into the program to examine the equations and the solution. The 19 equations *a* through *i* have been typed into the *rule sheet* and the values of the known variables and the known geometry have been typed into the *variable sheet.* The cross products are computed with a function routine called *Cross2* written on the *function sheet.* You may change the values of the known data on the *variable sheet* to alter the problem as you wish. Since this software uses a Newton-Raphson root-finding method of solution, an **initial guess** value is required to begin the iteration. This initial guess is built-in to the *subsheet* for one of the unknowns (F_{12y}).* You need only to hit the **F9** (Function 9)

* See also Example 1-4.

3

key to do the calculation. See Section 1-10 and the *TKSolver* manual for further information on running the software.

8   Table 3-2 shows the solution to this problem for the dimensions in Figure 3-2, assuming a worst-case 712-N (160-lb) force applied by the person's hand normal to the brake lever. The force generated in the cable (F_{cable}) is then 2 851 N (641 lb), the reaction force against the handlebar (F_{21}) is 2 993 N (673 lb) at −162°, and a 93-N-m (818-lb-in) moment (M_h) is applied to the handlebar.

9   As an exercise, run the *TKSolver* model and move the point of application of the hand force along the lever by changing the values of $\mathbf{R}_{b2}$, recalculate and observe the changes to the forces and moments. Note that the *TK* model is set up to allow input of the forces and position vectors in the more convenient polar form. It then converts them to Cartesian coordinates for use in the equations.

CASE STUDY 2A

Hand-Operated Crimping-Tool Loading Analysis

Problem: Determine the forces on the elements of the crimping tool shown in Figure 3-3 during a crimp operation.

Given: The geometry is known and the tool develops a crimp force of 2 000 lb (8 896 N) at closure in the position shown.

Assumptions: The accelerations are negligible. All forces are coplanar and two dimensional. A Class 1 load model is appropriate and a static analysis acceptable.

Solution: See Figures 3-3 and 3-4, Table 3-3, and TKSolver file CASE2A.

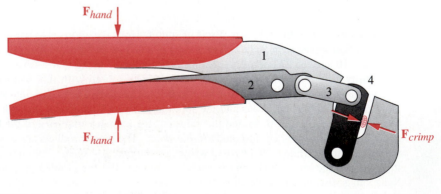

FIGURE 3-3

Wire Connector Crimping Tool

1   Figure 3-3 shows the tool in the closed position, in the process of crimping a metal connector onto a wire. The user's hand provides the input forces between links 1 and 2, shown as the reaction pair F_{hand}. The user can grip the handle anywhere along its length but we are assuming a nominal moment arm of R_{hand} for the application of the resultant of the user's grip force (see Figure 3-4). The high mechanical advantage of the tool transforms the grip force to a large force at the crimp.

Figure 3-3 is a free-body diagram of the entire assembly, neglecting the weight of the tool which is small compared to the crimp force. There are four elements, or links, in the assembly, all pinned together. Link 1 can be considered to be the "ground" link, with the other links moving with respect to it as the jaw is closed. The desired magnitude of the crimp force F_c is defined and its direction will be normal to the surfaces at the crimp. The third law relates the action-reaction pair acting on links 1 and 4:

$$F_{c1x} = -F_{c4x}$$
$$F_{c1y} = -F_{c4y}$$

(a)

2   Figure 3-4 shows the elements of the crimping tool assembly separated and drawn as free-body diagrams with all forces applied to each element, again neglecting their

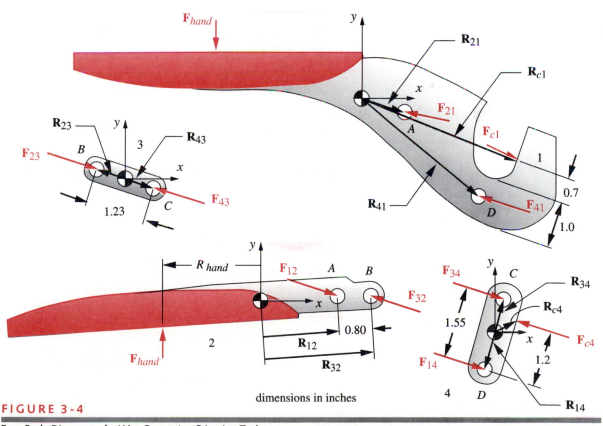

dimensions in inches

FIGURE 3-4

Free-Body Diagrams of a Wire Connector Crimping Tool

Table 3-3   **Case Study 2A – Crimping Tool Force Analysis**
Part 1 of 3   Rule Sheet from *TKSolver* file CASE2A

Rules

; within-link relations - Newton's second law

    ; for link 2

        F12x + F32x = 0

        F12y + F32y + Fhand= 0

        M12 + Cross2 (R12x, R12y, F12x, F12y) + Cross2 (R32x, R32y, F32x, F32y) = 0

    ; for link 3

        F43x + F23x = 0

        F43y + F23y = 0

        Cross2 (R43x, R43y, F43x, F43y) + Cross2 (R23x, R23y, F23x, F23y) = 0

    ; for link 4

        F34x + F14x + Px = 0

        F34y + F14y + Py = 0

        Cross2 (R34x, R34y, F34x, F34y) + Cross2 (R14x, R14y, F14x, F14y) + Cross2 (Rpx, Rpy, Px, Py) = 0

; cross-link relations - Newton's third law

        F34x = – F43x

        F34y = – F43y

        F32x = – F23x

        F32y = – F23y

        F21x = – F12x

        F21y = – F12y

        F41x = – F14x

        F41y = – F14y

        Fhand = M12/Rhand

---

weights as being insignificant compared to the applied forces. The centers of gravity of the respective elements are used as the origins of the local, nonrotating coordinate systems in which the points of application of all forces on the elements are located.[*]

3   We will consider link 1 to be the ground plane and analyze the remaining moving links. Note that all unknown forces and moments are initially assumed positive. Link 2 has three forces acting on it: $\mathbf{F}_{hand}$ is the unknown force from the hand, and $\mathbf{F}_{12}$ and $\mathbf{F}_{32}$ are the reaction forces from links 1 and 3, respectively. Force $\mathbf{F}_{12}$ is provided by part 1 on part 2 at the pivot pin and force $\mathbf{F}_{32}$ is provided by part 3 acting on part 2 at their pivot pin. The magnitudes and directions of both these pin forces are unknown. We can write equations 3.3*b* for this element to sum forces in the *x* and *y* directions and sum moments about the CG (with cross products expanded).

$$\sum F_x = F_{12x} + F_{32x} = 0$$

$$\sum F_y = F_{12y} + F_{32y} + F_{hand} = 0 \qquad\qquad (b)$$

$$\sum M_z = F_{hand}\left(R_{hand}\right) + \left(R_{12x}F_{12y} - R_{12y}F_{12x}\right) + \left(R_{32x}F_{32y} - R_{32y}F_{32x}\right) = 0$$

[*] Again, in a static analysis it is not necessary to take the CG as the coordinate system origin (any point can be used), but we do so to be consistent with the dynamic analysis approach in which it is quite useful to do so.

3

Table 3–3   **Case Study 2A – Crimping Tool Force Analysis**
Part 2 of 3   Given Data - See *TKSolver* File CASE2A

Input	Variable	Output	Unit	Comments
				Applied Force
−1 956.30	*Fc4x*		lb	component of applied force
415.82	*Fc4y*		lb	component of applied force
0.09	*Rc4x*		in	component of position vector
0.05	*Rc4y*		in	component of position vector
				Link 2 Geometry
1.40	*R12x*		in	component of position vector
0.05	*R12y*		in	component of position vector
2.20	*R32x*		in	component of position vector
0.08	*R32y*		in	component of position vector
−4.40	*Rhand*		in	radius to F_{hand} from CG
				Link 3 Geometry
−0.60	*R23x*		in	component of position vector
0.13	*R23y*		in	component of position vector
0.60	*R43x*		in	component of position vector
−0.13	*R43y*		in	component of position vector
				Link 4 Geometry
−0.16	*R14x*		in	component of position vector
−0.76	*R14y*		in	component of position vector
0.16	*R34x*		in	component of position vector
0.76	*R34y*		in	component of position vector

4   Link 3 has two forces on it, $\mathbf{F}_{23}$ and $\mathbf{F}_{43}$. Write equations 3.3b for this element:

$$\sum F_x = F_{23x} + F_{43x} = 0$$

$$\sum F_y = F_{23y} + F_{43y} = 0 \qquad\qquad (c)$$

$$\sum M_z = \left(R_{23x}F_{23y} - R_{23y}F_{23x}\right) + \left(R_{43x}F_{43y} - R_{43y}F_{43x}\right) = 0$$

5   Link 4 has three forces acting on it: $\mathbf{F}_{c4}$ is the known (desired) force at the crimp, and $\mathbf{F}_{14}$ and $\mathbf{F}_{34}$ are the reaction forces from links 1 and 3, respectively. The magnitudes and directions of both these pin forces are unknown. Write equations 3.3b for this element:

Table 3-3   Case Study 2A – Crimping Tool Force Analysis

Part 3 of 3   Calculated Results - See *TKSolver* File CASE2A

Input	Variable	Output	Unit	Comments
				Link 2 Forces
	F12x	1 514.6	lb	force of 1 on 2 in x - iterated
G*	*F12y*	–373.6	lb	force of 1 on 2 in y - guessed
	F32x	–1 514.6	lb	force of 3 on 2 in x - iterated
	F32y	321.9	lb	force of 3 on 2 in y - iterated
	Fhand	51.7	lb	force of hand on 2 in y - iterated
				Link 3 Forces
	F43x	–1 514.6	lb	force of 4 on 3 in x - iterated
	F43y	321.9	lb	force of 4 on 3 in y - iterated
	F23x	1 514.6	lb	force of 2 on 3 in x - iterated
G*	*F23y*	–321.9	lb	force of 2 on 3 in y - guessed
				Link 4 Forces
G*	*F34x*	1 514.6	lb	force of 3 on 4 in x - guessed
	F34y	–321.9	lb	force of 3 on 4 in y - iterated
	F14x	441.7	lb	force of 1 on 4 in x - iterated
	F14y	–93.9	lb	force of 1 on 4 in y - iterated
				Link 1 Forces
	F21x	–1 514.6	lb	force of 2 on 1 in x - from third law
	F21y	373.6	lb	force of 2 on 1 in y - from third law
	F41x	–441.7	lb	force of 4 on 1 in x - from third law
	F41y	93.9	lb	force of 4 on 1 in y - from third law
	Fhand	–51.7	lb	force of hand on 1 in y - third law

* Indicates that a guess value was required to start the iteration.

$$\sum F_x = F_{14x} + F_{34x} + F_{c4x} = 0$$

$$\sum F_y = F_{14y} + F_{34y} + F_{c4y} = 0 \qquad\qquad (d)$$

$$\sum M_z = \left(R_{14x}F_{14y} - R_{14y}F_{14x}\right) + \left(R_{34x}F_{34y} - R_{34y}F_{34x}\right)$$
$$+ \left(R_{c4x}F_{c4y} - R_{c4y}F_{c4x}\right) = 0$$

6   The nine equations in sets *b* through *d* have 13 unknowns: F_{12x}, F_{12y}, F_{32x}, F_{32y}, F_{23x}, F_{23y}, F_{43x}, F_{43y}, F_{14x}, F_{14y}, F_{34x}, F_{34y}, and F_{hand}. We can write the third-law relationships between action-reaction pairs at each of the joints to obtain the four additional equations needed:

$$F_{32x} = -F_{23x} \qquad\qquad F_{34x} = -F_{43x}$$

$$F_{32y} = -F_{23y} \qquad\qquad F_{34y} = -F_{43y}$$

(e)

3

7   Unlike Case Study 1A, this problem cannot be solved by direct substitution of equations even though substituting equation set e into sets b-d will reduce it to nine unknowns. The resulting nine equations can be solved simultaneously by matrix reduction or by iteration with a root-finding algorithm. This case study was solved by the latter method, using *TKSolver*. Its data file is included on disk. Input the file CASE2A into that program to examine the equations and the solution.

8   Table 3-3 shows the solution to this problem for the scaled dimensions in Figure 3-3 assuming a 2 000-lb (8 896-N) force applied at the crimp, normal to the crimp surface. The force generated in link 3 is 1 548 lb (6 888 N), the reaction force against link 1 by link 2 ($\mathbf{F}_{21}$) is 1 560 lb (6 939 N) at 166°, the reaction force against link 1 by link 4 ($\mathbf{F}_{41}$) is 452 lb (2 009 N) at 168°, and a −227 lb-in (−26-N-m) moment must be applied to the handles to generate the specified crimp force. This moment can be obtained with a 52-lb (230-N) force applied at mid-handle. This force is within the physiological grip-force capacity of the average human.

9   As an exercise, run the *TKSolver* model and move the point of application of the crimp force along the jaw, recalculate, and observe the changes to the forces and moments. The *TK* model is set up to allow input of the forces and position vectors in polar form and converts them to Cartesian coordinates for use in the equations.

CASE STUDY 3A

Automobile Scissors-Jack Loading Analysis

Problem: Determine the forces on the elements of the scissors jack in the position shown in Figure 3-5.

Given: The geometry is known and the jack supports a force of P = 1 000 lb (4 448 N) in the position shown.

Assumptions: The accelerations are negligible. The jack is on level ground. The angle of the elevated car chassis does not impart an overturning moment to the jack. All forces are coplanar and two dimensional. A Class 1 load model is appropriate and a static analysis is acceptable.

Solution: See Figures 3-5 through 3-8, Table 3-4, and TKSolver file CASE3A.

1   Figure 3-5 shows a schematic of a simple scissors jack used to raise a car. It consists of six links which are pivoted and/or geared together and a seventh link in the form of a lead screw which is turned to raise the jack. While this is clearly a three-dimensional device, it can be analyzed as a two-dimensional one if we assume that the ap-

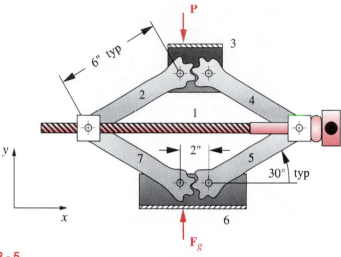

FIGURE 3-5

An Automobile Scissors Jack

plied load (from the car) and the jack are exactly vertical (in the z direction). If so, all forces will be in the xy plane. This assumption is valid if the car is jacked from a level surface. If not, then there will be some forces in the yz and xz planes as well. The jack designer needs to consider the more general case, but for our simple example we will initially assume two-dimensional loading. For the overall assembly as shown in Figure 3-5, we can solve for the reaction force $\mathbf{F}_g$, given force $\mathbf{P}$, by summing forces: $\mathbf{F}_g = -\mathbf{P}$.

2   Figure 3-6 shows a set of free-body diagrams for the entire jack. Each element or subassembly of interest has been separated from the others and the forces and moments shown acting on it (except for its weight which is small compared to the applied forces and is thus neglected for this analysis). The forces and moments can be either internal reactions at interconnections with other elements or external loads from the "outside world." The centers of gravity of the respective elements are used as the origins of the local, nonrotating coordinate systems in which the points of application of all forces on the elements are located. In this design, stability is achieved by the mating of two pairs of crude (noninvolute) gear segments acting between links 2 and 4 and between links 5 and 7. These interactions are modeled as forces acting along a *common normal* shared by the two teeth. This common normal is perpendicular to the common tangent at the contact point.

There are 3 second-law equations available for each of the seven elements allowing 21 unknowns. An additional 10 third-law equations will be needed for a total of 31. This is a cumbersome system to solve for such a simple device, but we can use its symmetry to advantage in order to simplify the problem.

3   Figure 3-7 shows the upper half of the jack assembly. Because of the mirror symmetry between the upper and lower portions, the lower half can be removed to simplify the analysis. The forces calculated for this half will be duplicated in the other. If we

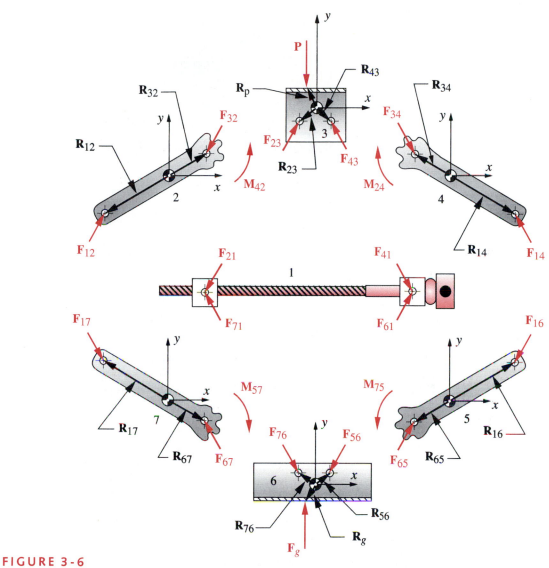

FIGURE 3-6

Free-Body Diagrams of the Complete Scissors Jack

wished, we could solve for the reaction forces at A and B using equations 3.3b from this free-body diagram of the half-jack assembly.

4   Figure 3-8a shows the free-body diagrams for the upper half of the jack assembly which are essentially the same as those of Figure 3-6. We now have four elements but can consider the subassembly labeled 1 to be the "ground," leaving three elements on which to apply equations 3.3. Note that all forces and moments are initially assumed positive in the equations.

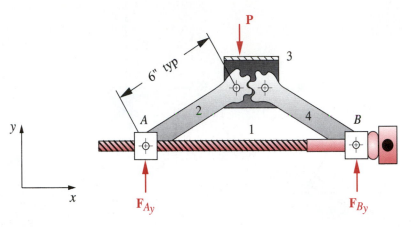

FIGURE 3-7

Free-Boby Diagram of thhe Symmetrical Upper Half of an Automobile Scissors Jack

5   Link 2 has three forces acting on it: $\mathbf{F}_{42}$ is the unknown force at the gear tooth contact with link 4; $\mathbf{F}_{12}$ and $\mathbf{F}_{32}$ are the unknown reaction forces from links 1 and 3 respectively. Force $\mathbf{F}_{12}$ is provided by part 1 on part 2 at the pivot pin, and force $\mathbf{F}_{32}$ is provided by part 3 acting on part 2 at their pivot pin. The magnitudes and directions of these pin forces and the magnitude of $\mathbf{F}_{42}$ are unknown. The direction of $\mathbf{F}_{42}$ is along the common normal shown in Figure 3-8b. Write equations 3.3b for this element to sum forces in the x and y directions and sum moments about the CG (with the cross products expanded)[*]:

$$\sum F_x = F_{12x} + F_{32x} + F_{42x} = 0$$

$$\sum F_y = F_{12y} + F_{32y} + F_{42y} = 0 \qquad\qquad (a)$$

$$\sum M_z = \left(R_{12x}F_{12y} - R_{12y}F_{12x}\right) + \left(R_{32x}F_{32y} - R_{32y}F_{32x}\right)$$
$$+ \left(R_{42x}F_{42y} - R_{42y}F_{42x}\right) = 0$$

6   Link 3 has three forces acting on it: the applied load $\mathbf{P}$, $\mathbf{F}_{23}$, and $\mathbf{F}_{43}$. Only $\mathbf{P}$ is known. Writing equations 3.3b for this element gives

$$\sum F_x = F_{23x} + F_{43x} + P_x = 0$$

$$\sum F_y = F_{23y} + F_{43y} + P_y = 0 \qquad\qquad (b)$$

$$\sum M_z = \left(R_{23x}F_{23y} - R_{23y}F_{23x}\right) + \left(R_{43x}F_{43y} - R_{43y}F_{43x}\right)$$
$$+ \left(R_{Px}P_y - R_{Py}P_x\right) = 0$$

7   Link 4 has three forces acting on it: $\mathbf{F}_{24}$ is the unknown force from link 2; $\mathbf{F}_{14}$ and $\mathbf{F}_{34}$ are the unknown reaction forces from links 1 and 3 respectively.

[*] Note the similarity to equations (b) in Case Study 2A. Only the subscript for the reaction moment is different because a different link is providing it. The consistent notation of this force analysis method makes it easy to write the equations for any system.

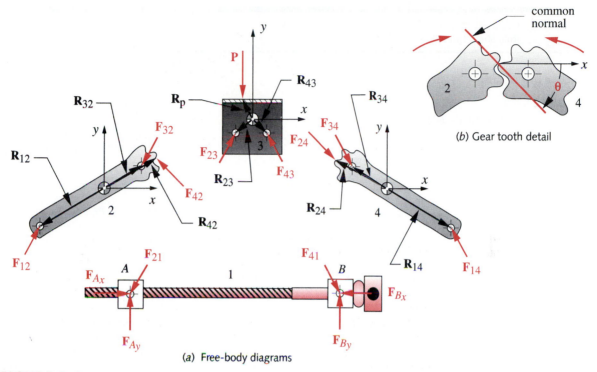

(a) Free-body diagrams

FIGURE 3-8

Free-Body Diagrams of Elements of the Half Scissors Jack

$$\sum F_x = F_{14x} + F_{24x} + F_{34x} = 0$$

$$\sum F_y = F_{14y} + F_{24y} + F_{34y} = 0 \qquad (c)$$

$$\sum M_z = \left(R_{14x}F_{14y} - R_{14y}F_{14x}\right) + \left(R_{24x}F_{24y} - R_{24y}F_{24x}\right)$$

$$+ \left(R_{34x}F_{34y} - R_{34y}F_{34x}\right) = 0$$

8   The nine equations in sets *a* through *c* have (16) unknowns in them, F_{12x}, F_{12y}, F_{32x}, F_{32y}, F_{23x}, F_{23y}, F_{43x}, F_{43y}, F_{14x}, F_{14y}, F_{34x}, F_{34y}, F_{24x}, F_{24y}, F_{42x}, and F_{42y}. We can write the third-law relationships between action-reaction pairs at each of the joints to obtain six of the seven additional equations needed:

$$
\begin{aligned}
F_{32x} &= -F_{23x} & F_{32y} &= -F_{23y} \\
F_{34x} &= -F_{43x} & F_{34y} &= -F_{43y} & \qquad (d)\\
F_{42x} &= -F_{24x} & F_{42y} &= -F_{24y}
\end{aligned}
$$

9   The last equation needed comes from the relationship between the *x* and *y* components of the force F_{24} (or F_{42}) at the tooth/tooth contact point. Such a contact (or half) joint can transmit force (excepting friction force) only along the **common normal**[4] which is perpendicular to the joint's common tangent as shown in Figure 3-8b. The common

Table 3-4   Case Study 2A – Scissors-Jack Force Analysis
Part 1 of 3   Rule Sheet from *TKSolver* file CASE3A

Rules

; within-link relations - Newton's second law

   ; for link 2

$$F12x + F32x + F42x = 0$$
$$F12y + F32y + F42y = 0$$
$$Cross2\ (\ R42x,\ R42y,\ F42x,\ F42y\ ) + Cross2\ (\ R12x,\ R12y,\ F12x,\ F12y\ ) + Cross2(\ R32x,\ R32y,\ F32x,\ F32y\ ) = 0$$

   ; for link 3

$$F43x + F23x + Px = 0$$
$$F43y + F23y + Py = 0$$
$$Cross2\ (\ R43x,\ R43y,\ F43x,\ F43y\ ) + Cross2\ (\ R23x,\ R23y,\ F23x,\ F23y\ ) + Cross2(\ Rpx,\ Rpy,\ Px,\ Py\ ) = 0$$

   ; for link 4

$$F34x + F14x + F24x = 0$$
$$F34y + F14y + F24y = 0$$
$$Cross2\ (\ R24x,\ R24y,\ F24x,\ F24y\ ) + Cross2\ (\ R34x,\ R34y,\ F34x,\ F34y\ ) + Cross2\ (\ R14x,\ R14y,\ F14x,\ F14y\ )\ = 0$$

; use the geometry at gear teeth

$$F24y = F24x * TAND\ (\ Theta\ )$$

; cross-link relations - Newton's third law

$$F34x = - F43x$$
$$F34y = - F43y$$
$$F32x = - F23x$$
$$F32y = - F23y$$
$$F24x = - F42x$$
$$F24y = - F42y$$

normal is also called the **axis of transmission**. The tangent of the angle of this common normal relates the two components of the force at the joint:

$$F_{24y} = F_{24x} \tan\theta \qquad (e)$$

10 Equations (*d*) and (*e*) can be substituted into equations (*a*) through (*c*) to create a set of ten simultaneous equations for solution either by matrix reduction or by iterative root finding methods. This case study was solved by the latter method, using *TKSolver*. Its data file is included on disk. Input the file labeled Case3A into that program to examine the equations and the solution.

11 Table 3-4 shows the solution to this problem for the scaled dimensions in Figure 3-3 assuming a vertical 1 000-lb (4 448-N) applied force **P**. Note that the forces on link 1 can also be found from Newton's third law.

$$F_{Ax} = -F_{21x} = F_{12x}$$
$$F_{Ay} = -F_{21y} = F_{12y}$$
$$F_{Bx} = -F_{41x} = F_{14x}$$
$$F_{By} = -F_{41y} = F_{14y} \qquad (f)$$

Table 3–4   Case Study 3A – Scissors-Jack Force Analysis

Part 2 of 3   Given Data - See *TKSolver* File CASE3A

Input	Variable	Output	Unit	Comments
				Applied Force
0.0	*Px*		lb	component of applied force
−1 000.0	*Py*		lb	component of applied force
−0.5	*Rpx*		in	component of its position vector
0.9	*Rpy*		in	component of its position vector
−45.0	*q*		deg	angle of common normal
				Link 2 Geometry
−3.1	*R12x*		in	component of position vector
−1.8	*R12y*		in	component of position vector
2.1	*R32x*		in	component of position vector
1.2	*R32y*		in	component of position vector
2.7	*R42x*		in	component of position vector
1.0	*R42y*		in	component of position vector
				Link 3 Geometry
−0.8	*R23x*		in	component of position vector
−0.8	*R23y*		in	component of position vector
0.8	*R43x*		in	component of position vector
−0.8	*R43y*		in	component of position vector
				Link 4 Geometry
3.1	*R14x*		in	component of position vector
−1.8	*R14y*		in	component of position vector
−2.6	*R24x*		in	component of position vector
1.0	*R24y*		in	component of position vector
−2.1	*R34x*		in	component of position vector
1.2	*R34y*		in	component of position vector

12   As an exercise, run the *TKSolver* model and move the point of application of **P** along the *x* direction, recalculate and observe the changes to the forces and moments on the links. What happens when the vertical force **P** is centered on link 3? Also, change the angle of the applied force **P** to create an *x* component and observe the effects on the forces and moments on the elements. Note that the *TK* model is set up to allow input of the forces and position vectors in polar form. It then converts them to Cartesian coordinates for use in the equations.

Table 3–4   Case Study 3A – Scissors-Jack Force Analysis

Part 3 of 3   Calculated Results - See *TKSolver* File CASE3A

Input	Variable	Output	Unit	Comments
				Link 2 Forces
	$F12x$	877.9	lb	force of 1 on 2 in x - iterated
G *	$F12y$	530.0	lb	force of 1 on 2 in y - guessed
	$F32x$	−586.5	lb	force of 3 on 2 in x - iterated
	$F32y$	−821.4	lb	force of 3 on 2 in y - iterated
	$F42x$	−291.4	lb	force of 4 on 2 in x - iterated
	$F42y$	291.4	lb	force of 4 on 2 in y - iterated
				Link 3 Forces
	$F23x$	586.5	lb	force of 2 on 3 in x - iterated
	$F23y$	821.4	lb	force of 2 on 3 in y - iterated
	$F43x$	−586.5	lb	force of 4 on 3 in x - iterated
	$F43y$	178.6	lb	force of 4 on 3 in y - iterated
				Link 4 Forces
	$F14x$	−877.9	lb	force of 1 on 4 in x - iterated
	$F14y$	470.0	lb	force of 1 on 4 in y - iterated
G *	$F24x$	291.4	lb	force of 2 on 4 in x - guessed
	$F24y$	−291.4	lb	force of 2 on 4 in y - iterated
G *	$F34x$	586.5	lb	force of 3 on 4 in x - guessed
	$F34y$	−178.6	lb	force of 3 on 4 in y - iterated

* Indicates that a guess value was required to start the iteration.

3.5   THREE-DIMENSIONAL, STATIC LOADING CASE STUDY

This section presents a case study involving three-dimensional static loading on a bicycle brake caliper assembly. The same techniques used for two-dimensional load analysis also work for the three-dimensional case. The third dimension requires more equations, which are available from the summation of forces in the z direction and the summation of moments about the x and y axes as defined in equations 3.1 and 3.3 for the dynamic and static cases, respectively. As an example, we will now analyze the bicycle brake arm that is actuated by the handbrake lever that was analyzed in Case Study 1A.

CASE STUDY 4A

Bicycle Brake Arm Loading Analysis

Problem: Determine the forces acting in three dimensions on the bicycle brake arm in its actuated position as shown in Figure 3-9. This brake arm has been failing in service and may need to be redesigned.

Given: The existing brake arm geometry is known and the arm is acted on by a cable force of 2 851 N in the position shown. (See also Case Study 1A.)

Assumptions: The accelerations are negligible. A Class 1 load model is appropriate and a static analysis is acceptable. The coefficient of friction between the brake pad and wheel rim has been measured and is 0.45 at room temperature and 0.40 at 150°F.

Solution: See Figures 3-9 and 3-10, Table 3-5, and TKSolver file CASE4A.

1   Figure 3-9 shows a center-pull brake arm assembly commonly used on bicycles. It consists of six elements or subassemblies, the frame and its pivot pins (1), the two brake arms (2 and 4), the cable spreader assembly (3), the brake pads (5), and the wheel rim (6). This is clearly a three-dimensional device, and must be analyzed as such.

2   The cable is the same one that is attached to the brake lever in Figure 3-1. The 712-N (160-lb) hand force is multiplied by the mechanical advantage of the hand lever and transmitted via this cable to the pair of brake arms as was calculated in Case Study 1A. We will assume no loss of force in the cable guides, thus the full 2 851-N (641-lb) cable force is available at this end.

3   The direction of the normal force between the brake pad and the wheel rim is shown in Figure 3-9 to be at 172° with respect to the positive x axis, and the friction force is directed along the z axis. (See Figures 3-9 and 3-10 for the xyz axis orientations.)

4   Figure 3-10 shows free-body diagrams of the arm, frame, and cable spreader assembly. We are principally interested in the forces acting on the brake arm. However, we first need to analyze the effect of the cable spreader geometry on the force applied to the arm at A. This analysis can be two dimensional since the cables are in a common z-plane. Note also that the cable subassembly (3) is a concurrent force system. Writing equations 3.3b for this subassembly, and noting the symmetry about point A, we can write from inspection of the FBD:

$$\sum F_x = F_{23x} + F_{43x} = 0$$

$$\sum F_y = F_{23y} + F_{43y} + F_{cable} = 0 \tag{a}$$

This equation set can be easily solved to yield

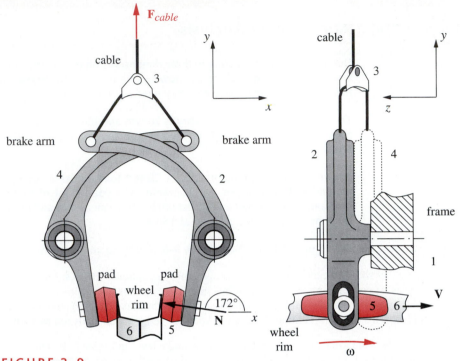

FIGURE 3-9

Center-Pull Bicycle Brake Arm Assembly

$$F_{23y} = F_{43y} = -\frac{F_{cable}}{2} = -\frac{2\,851}{2} = -1\,426 \text{ N}$$

$$F_{23x} = \frac{F_{23y}}{\tan(56°)} = -962 \text{ N} \qquad (b)$$

$$F_{43x} = -F_{23x} = 962 \text{ N}$$

Newton's third law relates these forces to their reactions on the brake arm at point A:

$$F_{32x} = -F_{23x} = 962 \text{ N}$$

$$F_{32y} = -F_{23y} = 1\,426 \text{ N} \qquad (c)$$

$$F_{32z} = 0$$

5   We can now write equations 3.3*a* for the arm (link 2).

For the forces:

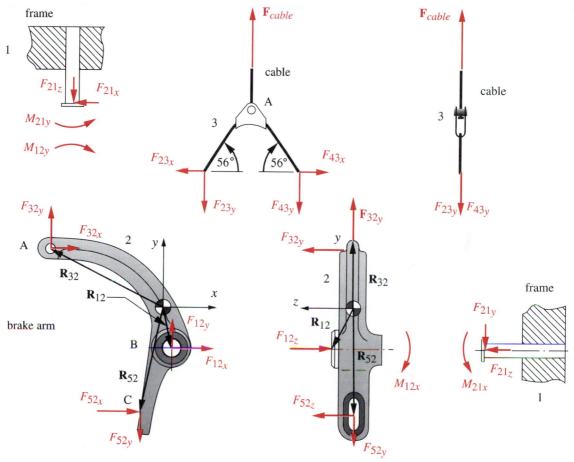

FIGURE 3-10

Brake Arm Free-Body Diagrams

$$\sum F_x = F_{12x} + F_{32x} + F_{52x} = 0$$

$$F_{12x} + F_{52x} = -962$$

$$\sum F_y = F_{12y} + F_{32y} + F_{52y} = 0$$

$$F_{12y} + F_{52y} = -1\,426 \qquad\qquad (d)$$

$$\sum F_z = F_{12z} + F_{32z} + F_{52z} = 0$$

$$F_{12z} + F_{52z} = 0$$

For the moments:

3

Table 3-5   **Case Study 4A – Bicycle Brake Arm Force Analysis**
Part 1 of 2   Rule Sheet from *TKSolver* file CASE4A

Rules

; summing forces
      F12x + F32x + F52x = 0
      F12y + F32y + F52y = 0
      F12z + F32z + F52z = 0

; summing moments about the CG
      M12x + Cross2 (R12y, R12z, F12y, F12z) + Cross2 (R32y, R32z, F32y, F32z) + Cross2 (R52y, R52z, F52y, F52z) = 0
      M12y + Cross2 (R12x, R12z, F12x, F12z) + Cross2 (R32x, R32z, F32x, F32z) + Cross2 (R52x, R52z, F52x, F52z) = 0
      M12z + Cross2 (R12x, R12y, F12x, F12y) + Cross2 (R32x ,R32y ,F32x ,F32y) + Cross2 (R52x, R52y, F52x, F52y) = 0

; find friction force
      *Ff = mu * N*

; find normal force
      *F52y = F52x * TAND (Angnorm)*
      F52z = –Ff
      N = SQRT (F52x^2 + F52y^2)

; find force in cable
      Fcable = SQRT (F32x^2 + F32y^2)

$$\sum M_x = M_{12x} + \left(R_{12y}F_{12z} - R_{12z}F_{12y}\right) + \left(R_{32y}F_{32z} - R_{32z}F_{32y}\right)$$
$$+\left(R_{52y}F_{52z} - R_{52z}F_{52y}\right) = 0$$

$$\sum M_y = M_{12y} + \left(R_{12z}F_{12x} - R_{12x}F_{12z}\right) + \left(R_{32z}F_{32x} - R_{32x}F_{32z}\right)$$
$$+\left(R_{52z}F_{52x} - R_{52x}F_{52z}\right) = 0 \qquad (e)$$

$$\sum M_z = \left(R_{12x}F_{12y} - R_{12y}F_{12x}\right) + \left(R_{32x}F_{32y} - R_{32y}F_{32x}\right)$$
$$+\left(R_{52x}F_{52y} - R_{52y}F_{52x}\right) = 0$$

Note that all unknown forces and moments are initially assumed positive in the equations, regardless of their apparent directions on the FBDs. The moments M_{12x} and M_{12y} are due to the fact that there is a moment joint between the arm (2) and the pivot pin (1) about the x and y axes. We assume negligible friction about the z axis, thus allowing M_{12z} to be zero.

6   The joint between the brake pad (5) and the wheel rim (6) transmits a force normal to the plane of contact. The friction force magnitude, F_f, in the contact plane is related to the normal force by the Coulomb friction equation,

$$F_f = \mu N \qquad (f)$$

where μ is the coefficient of friction and N is the normal force. The velocity of the point on the rim below the center of the brake pad is in the z direction. The force com-

Table 3-5   **Case Study 4A – Bicycle Brake Arm Force Analysis**

Part 2 of 2   See *TKSolver* File CASE4A

Input	Variable	Output	Unit	Comments
				Geometry
0.4	*m*		none	coefficient of friction
172	*Angnorm*		deg	angle of normal force
5.2	*R12x*		mm	x component of link 1 vs CG of 2
–27.2	*R12y*		mm	y component of link 1 vs CG of 2
23.1	*R12z*		mm	z component of link 1 vs CG of 2
–75.4	*R32x*		mm	x component of link 3 vs CG of 2
38.7	*R32y*		mm	y component of link 3 vs CG of 2
0	*R32z*		mm	z component of link 3 vs CG of 2
–12.9	*R52x*		mm	x component of link 5 vs CG of 2
–69.7	*R52y*		mm	y component of link 5 vs CG of 2
0	*R52z*		mm	z component of link 5 vs CG of 2
				Forces
962	*F32x*		N	known applied cable force in x
1 426	*F32y*		N	known applied cable force in y
0	*F32z*		N	known applied cable force in z
	F12x	–4 916	N	force of 1 on 2 in x - guessed
	F12y	–870	N	force of 1 on 2 in y - iterated
	F12z	1 597	N	force of 1 on 2 in z - iterated
	F52x	3 954	N	force of 5 on 2 in x - iterated
	F52y	–556	N	force of 5 on 2 in y - iterated
	F52z	–1 597	N	force of 5 on 2 in z - iterated
	M12x	–89	N–m	moment of 1 on 2 in x - iterated
	M12y	–144	N–m	moment of 1 on 2 in y - iterated
0	*M12z*		N–m	assumed zero
	N	3 993	N	normal force
	Ff	1 597	N	friction force

ponents F_{52x} and F_{52y} are due entirely to the normal force being transmitted through the pad to the arm and must be in the ratio dictated by the angle of the normal force:

$$N = \sqrt{F_{52x}{}^2 + F_{52y}{}^2}$$

$$F_{52y} = F_{52x} \tan 172° \qquad (g)$$

The direction of the friction force F_f must always oppose motion and thus it acts in the negative z direction on the wheel rim. Its reaction force on the arm has the opposite sense.

$$F_{52z} = -F_f \qquad\qquad (h)$$

7   We now have 10 equations (in the sets labeled d, e, f, g, and h) containing 10 unknowns: F_{12x}, F_{12y}, F_{12z}, F_{52x}, F_{52y}, F_{52z}, M_{12x}, M_{12y}, N, and F_f. Forces F_{32x}, F_{32y}, and F_{32z} are known from equations c. These (10) equations can be solved simultaneously either by matrix reduction or by iterative root finding methods. This case study was solved by the latter method, using *TKSolver*. Its data file is included on disk. Input the file labeled CASE4A into that program to examine the equations and the solution. The results are shown in Table 3-5.

3.6   DYNAMIC LOADING CASE STUDY

This section presents a case study involving two-dimensional dynamic loading on a four-bar linkage designed as a dynamic-load demonstration device. A photograph of this machine is shown in Figure 3-11. This machine can be analyzed in two dimensions since all elements move in parallel planes. The presence of significant accelerations on the moving elements in a system requires that a dynamic analysis be done with equations 3.1. The approach is identical to that used in the preceding static load analyses except for the need to include the $m\mathbf{A}$ and $I\alpha$ terms in the equations.

CASE STUDY 5A

Fourbar Linkage Loading Analysis

Problem: Determine the theoretical rigid-body forces acting in two dimensions on the fourbar linkage shown in Figure 3-11.

Given: The linkage geometry, masses and mass moments of inertia are known and the linkage is driven at up to 120 rpm by a speed-controlled electric motor.

Assumptions: The accelerations are significant. A Class 4 load model is appropriate and a dynamic analysis is required. There are no external loads on the system, all loads are due to the accelerations of the links. The weight forces are insignificant compared to the inertial forces and will be neglected. The links are assumed to be ideal rigid bodies. Friction and the effects of clearances in the pin joints also will be ignored.

Solution: See Figures 3-11 through 3-13, Table 3-6, and file CASE5A.

1 Figure 3-11 shows the fourbar linkage demonstrator model. It consists of three moving elements (links 2, 3, and 4) plus the frame or ground link (1). The motor drives link 2 through a gearbox. The two fixed pivots are instrumented with piezoelectric force transducers to measure the dynamic forces acting in x and y directions on the ground plane. A pair of accelerometers is mounted to a point on the floating coupler (link 3) to measure its accelerations.

2 Figure 3-12 shows a schematic of the linkage. The links are designed with lightening holes to reduce their masses and mass moments of inertia. The input to link 2 can be an angular acceleration (or a constant angular velocity) plus a torque. Link 2 rotates fully about its fixed pivot at O_2. Even though link 2 may have a zero angular acceleration α_2, if run at constant angular velocity ω_2, there will still be time-varying angular accelerations on links 3 and 4 since they oscillate back and forth. In any case, the CGs of the links will experience time-varying linear accelerations as the linkage moves. These angular and linear accelerations will generate inertia forces and torques as defined by Newton's second law. Thus, even with no external forces or torques applied to the links, the inertial forces will create reaction forces at the pins. It is these forces that we wish to calculate.

3 Figure 3-13 shows the free-body diagrams of the individual links. The local, nonrotating, coordinate system for each link is set up at its CG. The kinematic equations of motion must be solved to determine the linear accelerations of the CG of each link and the link's angular acceleration for every position of interest during the cycle. (See reference 1 for an explanation of this procedure.) These accelerations, $\mathbf{A}_{Gn}$ and α_n, are shown acting on each of the n links. The forces at each pin connection are shown as xy pairs, numbered as before, and are initially assumed to be positive.

4 Equations 3.1 can be written for each moving link in the system. The masses and the mass moments of inertia of each link about its CG must be calculated for use in these equations. In this case study, an Aries[*] solid modelling CAD system was used to design the links' geometries and to calculate their mass properties.

5 For link 2:

$$\sum F_x = F_{12x} + F_{32x} = m_2 A_{G2_x}$$

$$\sum F_y = F_{12y} + F_{32y} = m_2 A_{G2_y} \qquad (a)$$

$$\sum M_z = T_2 + \left(R_{12x}F_{12y} - R_{12y}F_{12x}\right) + \left(R_{32x}F_{32y} - R_{32y}F_{32x}\right) = I_{G2}\alpha_2$$

6 For link 3:

$$\sum F_x = F_{23x} + F_{43x} = m_3 A_{G3_x}$$

$$\sum F_y = F_{23y} + F_{43y} = m_3 A_{G3_y} \qquad (b)$$

$$\sum M_z = \left(R_{23x}F_{23y} - R_{23y}F_{23x}\right) + \left(R_{43x}F_{43y} - R_{43y}F_{43x}\right) = I_{G3}\alpha_3$$

[*] MSC Aries Technologies Inc., Suffolk St., Lowell, MA.

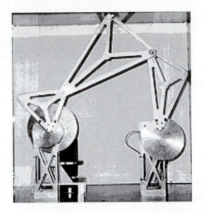

FIGURE 3-11

Fourbar Linkage Dynamic Model

7   For link 4:

$$\sum F_x = F_{14x} + F_{34x} = m_4 A_{G4_x}$$

$$\sum F_y = F_{14y} + F_{34y} = m_4 A_{G4_y} \qquad\qquad (c)$$

$$\sum M_z = \left(R_{14x}F_{14y} - R_{14y}F_{14x}\right) + \left(R_{34x}F_{34y} - R_{34y}F_{34x}\right) = I_{G4}\alpha_4$$

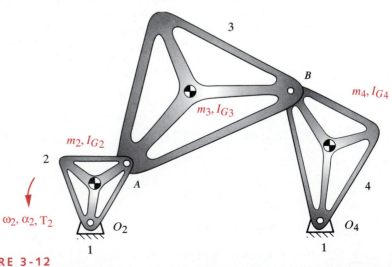

FIGURE 3-12

Fourbar Linkage Schematic

8   There are 13 unknowns in these nine equations: F_{12x}, F_{12y}, F_{32x}, F_{32y}, F_{23x}, F_{23y}, F_{43x}, F_{43y}, F_{14x}, F_{14y}, F_{34x}, F_{34y}, and T_2. Four third-law equations can be written to equate the action-reaction pairs at the joints.

$$F_{32x} = -F_{23x}$$
$$F_{32y} = -F_{23y}$$
$$F_{34x} = -F_{43x} \qquad\qquad (d)$$
$$F_{34y} = -F_{43y}$$

9   The set of thirteen equations in a through d can be solved simultaneously to determine the forces and driving torque either by matrix reduction or by iterative root-finding methods. This case study was solved by the latter method, using *TKSolver.* See Table 3-6. Its data file is included on disk. Input the file labeled CASE5A into that program to examine the equations and the solution. Note that the masses and mass moments of inertia of the links are constant with time and position, but the accelerations are time-varying. Thus, a complete analysis requires that equations a-d be solved for all positions or time steps of interest. This *TK* model uses *lists* to store the calculated values from equations a-d for 13 values of the input angle θ_2 of the driving link (0 to 360° by 30° increments). The model also calculates the kinematic accelerations of the links and their CGs which are needed for the force calculations. The largest and smallest forces present on each link during the cycle can then be determined for use in later stress and deflection analyses. The results of this force analysis for one crank position (30°) are shown in Table 3-6. Plots of the forces at the fixed pivots for one complete revolution of the crank are shown in Figure 3-14.

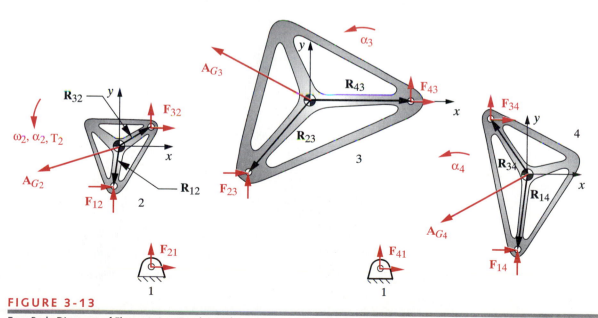

FIGURE 3-13

Free-Body Diagrams of Elements in a Fourbar Linkage

Table 3-6   **Case Study 5A – Fourbar Linkage Dynamic Force Analysis**

Part 1 of 3   Partial Rule Sheet - See the *TKSolver* file CASE5A for the Complete Set of Rules and Functions

Rules

; second-law relations

     ; for link 2

$$F12x + F32x = mass2 * Acg2x$$
$$F12y + F32y = mass2 * Acg2y$$
$$T12 + Cross2 (R12x, R12y, F12x, F12y) + Cross2 (R32x, R32y, F32x, F32y) = Icg2 * a2$$

     ; for link 3

$$F43x + F23x = mass3 * Acg3x$$
$$F43y + F23y = mass3 * Acg3y$$
$$Cross2 (R43x, R43y, F43x, F43y) + Cross2 (R23x, R23y, F23x, F23y) = Icg3 * a3$$

     ; for link 4

$$F34x + F14x + Px = mass4 * Acg4x$$
$$F34y + F14y + Py = mass4 * Acg4y$$
$$Cross2 (R34x, R34y, F34x, F34y) + Cross2 (R14x, R14y, F14x, F14y) + Cross2 (Rpx, Rpy, Px, Py) = Icg4$$

; third-law relations

$$F34x = - F43x$$
$$F34y = - F43y$$
$$F32x = - F23x$$
$$F32y = - F23y$$
$$F21x = - F12x$$
$$F21y = - F12y$$
$$F41x = - F14x$$
$$F41y = - F14y$$

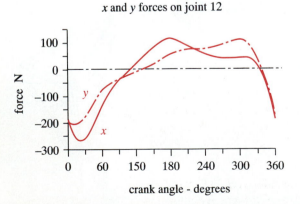

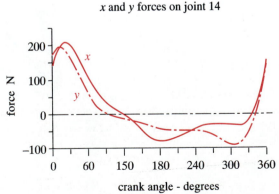

FIGURE 3-14

Calculated Rigid-Body Dynamic Forces in the Fourbar Linkage of Case Study 5A

Table 3–6   Case Study 5A – Fourbar Linkage Dynamic Force Analysis

Part 2 of 3   Given Data - See *TKSolver* File CASE5A

Input	Variable	Output	Unit	Comments
30.00	*theta2*		deg	link 2 angle
120.00	*rpm*		rpm	link 2 rpm (constant)
0.525	*mass2*		kg	mass of link 2
1.050	*mass3*		kg	mass of link 3
1.050	*mass4*		kg	mass of link 4
0.057	*Icg2*		kg–m^2	2nd mass moment of link 2
0.011	*Icg3*		kg–m^2	2nd mass moment of link 3
0.455	*Icg4*		kg–m^2	2nd mass moment of link 4
–46.9	*R12x*		mm	component of position vector
–71.3	*R12y*		mm	component of position vector
85.1	*R32x*		mm	component of position vector
4.9	*R32y*		mm	component of position vector
–150.7	*R23x*		mm	component of position vector
–177.6	*R23y*		mm	component of position vector
185.5	*R43x*		mm	component of position vector
50.8	*R43y*		mm	component of position vector
–21.5	*R14x*		mm	component of position vector
–100.6	*R14y*		mm	component of position vector
–10.6	*R34x*		mm	component of position vector
204.0	*R34y*		mm	component of position vector

3.7   VIBRATION LOADING

In systems that are dynamically loaded, there will usually be vibration loads superimposed on the theoretical loads predicted by the dynamic equations. These vibration loads can be due to a variety of causes. If the elements in the system were infinitely stiff, then vibrations would be eliminated. But all real elements, of any material, have elasticity and thus act as springs when subjected to forces. The resulting deflections can cause additional forces to be generated from the inertial forces associated with the vibratory movements of elements or if clearances allow contact of mating parts to generate impact (shock) loads (see below) during their vibrations.

A complete discussion of vibration phenomena is beyond the scope of this text and will not be attempted here. References are provided in the bibliography at the end of this chapter for further study. The topic is introduced here mainly to alert the machine designer to the need to consider vibration as a source of loading. Often the only way

Table 3–6   **Case Study 5A – Fourbar Linkage Dynamic Force Analysis**
Part 3 of 3   Calculated Results - See *TKSolver* File CASE5A

Input	Variable	Output	Unit	Comments
	$F12x$	−255.8	N	force of link 1 on 2 in x - iterated
G*	$F12y$	−178.1	N	force of link 1 on 2 in y - guessed
	$F32x$	252.0	N	force of link 3 on 2 in x - iterated
	$F32y$	172.2	N	force of link 3 on 2 in y - iterated
G*	$F34x$	−215.6	N	force of link 3 on 4 in x - guessed
	$F34y$	−163.9	N	force of link 3 on 4 in y - iterated
	$F14x$	201.0	N	force of link 1 on 4 in x - iterated
	$F14y$	167.0	N	force of link 1 on 4 in y - iterated
	$F43x$	215.6	N	force of link 4 on 3 in x - iterated
	$F43y$	163.9	N	force of link 4 on 3 in y - iterated
	$F23x$	−252.0	N	force of link 2 on 3 in x - iterated
	$F23y$	−172.2	N	force of link 2 on 3 in y - iterated
	$T12$	−3.55	N-m	torque of link 1 on 2 - iterated
	$a3$	56.7	rad/sec^2	angular acceleration (alpha) of link 3
	$a4$	138.0	rad/sec^2	angular acceleration (alpha) of link 4
	$Acg2x$	−7.4	m/sec^2	acceleration of link 2 CG
	$Acg2y$	−11.3	m/sec^2	acceleration of link 2 CG
	$Acg3x$	−34.6	m/sec^2	acceleration of link 3 CG
	$Acg3y$	−7.9	m/sec^2	acceleration of link 3 CG
	$Acg4x$	−13.9	m/sec^2	acceleration of link 4 CG
	$Acg4y$	2.9	m/sec^2	acceleration of link 4 CG

* Indicates that a guess value was required to start the iteration.

to get an accurate measure of the effects of vibration on a system is to do testing of prototypes or production systems under service conditions. The discussion of safety factors in Section 1.7 mentioned that many industries (automotive, aircraft, etc.) engage in extensive test programs to develop realistic loading models of their equipment. This topic will be discussed further in Section 6.4 when fatigue loading is introduced. Modern finite element (FEA) and boundary element (BEA) analysis techniques also allow vibration effects on a system or structure to be modeled and calculated. It is still difficult to obtain a computer model of a complex system that is as accurate as a real, instrumented prototype. This is especially true when clearances (gaps) between moving parts allow impacts to occur in the joints when loads reverse. Impacts create nonlinearities which are very difficult to model mathematically.

Natural Frequency

When designing machinery, it is desirable to determine the natural frequencies of the assembly or subassemblies in order to predict and avoid resonance problems in operation. Any real system can have an infinite number of natural frequencies at which it will readily vibrate. The number of natural frequencies that are necessary or desirable to calculate will vary with the situation. The most complete approach to the task is to use Finite Element Analysis (FEA) to break the assembly into a large number of discrete elements. The stresses, deflections, and number of natural frequencies that can be calculated by this technique are mainly limited by time and the computer resources available.

If not using FEA, we would like to determine, at a minimum, the system's lowest, or fundamental natural frequency since this frequency will usually create the largest magnitude of vibrations. The undamped fundamental natural frequency ω_n, with units of rad/sec, or f_n, with units of Hz, can be computed from the expressions

$$\omega_n = \sqrt{\frac{k}{m}}$$

$$f_n = \frac{1}{2\pi}\omega_n \tag{3.4}$$

where ω_n is the fundamental natural frequency, m is the moving mass of the system, and k is the effective spring constant of the system. (The period of the natural frequency is its reciprocal in seconds, $T_n = 1/f_n$.)

Equation 3.4 is based on a single degree of freedom, lumped model of the system. Figure 3-15 shows such a model of a simple cam-follower system consisting of a cam, a sliding follower, and a return spring. The simplest lumped model consists of a mass connected to ground through a single spring and a single damper. All the moving mass in the system (follower, spring) is contained in m and all the "spring" including the physical spring and the springiness of all other parts is lumped in the effective spring constant k.

SPRING CONSTANT   A spring constant k is an assumed linear relationship between the force, F, applied to an element and its resulting deflection δ (see Figure 3-17):

$$k = \frac{F}{\delta} \tag{3.5a}$$

If an expression for the deflection of an element can be found or derived, it will provide this spring constant relationship. This topic is revisited in the next chapter. In the example of Figure 3-15, the spring deflection δ is equal to the displacement y of the mass.

$$k = \frac{F}{y} \tag{3.5b}$$

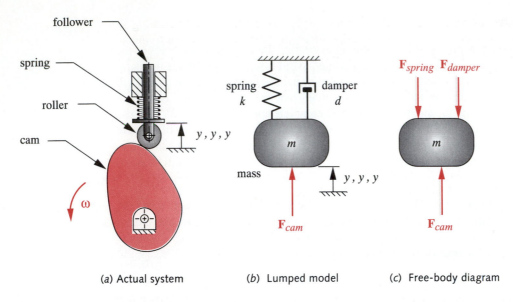

(a) Actual system   (b) Lumped model   (c) Free-body diagram

FIGURE 3-15

Lumped Model of a Cam-Follower Dynamic System

DAMPING   All the damping, or frictional losses are lumped in the damping coefficient d.   For this simple model, damping is assumed to be inversely proportional to the velocity y_{dot} of the mass.

$$d = \frac{F}{\dot{y}}$$ (3.6)

Equation 3.4 simplifies this model even further by assuming the damping d to be zero. If damping is included, the expressions for the fundamental, damped natural frequency ω_d with units of radians/sec, or f_d, with units of Hz, become

$$\omega_d = \sqrt{\frac{k}{m} - \left(\frac{d}{2m}\right)^2}$$

$$f_d = \frac{1}{2\pi}\omega_d$$ (3.7)

This damped frequency ω_d will be slightly smaller than the undamped frequency ω_n.

EFFECTIVE VALUES   Determining the effective mass for a lumped model is straightforward and requires only summing all the values of the connected, moving masses in appropriate mass units. Determining the values of the effective spring constant and the effective damping coefficient is more complicated and will not be addressed here. See reference 2 for an explanation.

RESONANCE   A condition called resonance can be experienced if the operating or forcing frequency applied to the system is the same as any one of its natural frequencies. That is, if the input angular velocity applied to a rotating system is the same as, or close to, ω_n, the vibratory response will be very large. This can create large forces and cause failure. Thus it is necessary to avoid operation at or near the natural frequencies if possible.

Dynamic Forces

If we write equation 3.1 for the simple, one DOF model of the dynamic system in Figure 3-15 and substitute equations 3.5 and 3.6, we get

$$\sum F_y = ma = m\ddot{y}$$
$$F_{cam} - F_{spring} - F_{damper} = m\ddot{y} \qquad\qquad (3.8)$$
$$F_{cam} = m\ddot{y} + d\dot{y} + ky$$

If the kinematic parameters of displacement, velocity, and acceleration are known for the system, this equation can be solved directly for the force on the cam as a function of time. If the cam force is known and the kinematic parameters are desired, then the well-known solution to this linear, constant coefficient differential equation can be applied. See reference 3 for a detailed derivation of that solution. Though the coordinate system used for a dynamic analysis can be arbitrarily chosen, it is important to note that both the kinematic parameters (displacement, velocity, and acceleration) and the forces in equation 3.8 must be defined in the **same** coordinate system.

As an example of the effect of vibration on the dynamic forces of a system, we now revisit the fourbar linkage of Case Study 5A and see the results of actual measurements of dynamic forces under operating conditions.

CASE STUDY 5B

Fourbar Linkage Dynamic Loading Measurement

Problem:	Determine the actual forces acting on the fixed pivots of the fourbar linkage in Figure 3-11 during one revolution of the input crank.
Given:	The linkage is driven at 60 rpm by a speed-controlled electric motor, and force transducers are placed between the fixed pivot bearings and the ground plane.
Assumptions:	There are no applied external loads on the system; all loads are due to the accelerations of the links. The weight forces are insignificant compared to the inertial forces and will be neglected. The force transducers measure only dynamic forces.

Solution: **See Figures 3-12 and 3-16.**

1   Figure 3-11 shows the fourbar linkage. It consists of three moving elements (links 2, 3 and 4) plus the frame or ground link (1). The shock-mounted motor drives link 2 through a gearbox and shaft coupling. The two fixed pivots are instrumented with piezoelectric force transducers to measure the dynamic forces acting in x and y directions on the ground plane.

2   Figure 3-16 shows an actual force and torque measured while the linkage was running at 60 rpm and compares them to the theoretical force and torque predicted by equations a-d in Case Study 5A.[5] Only the x component of force at the pivot between link 2 and the ground and the torque on link 2 are shown as examples. The other pin forces and components show similar deviations from their predicted theoretical values. Some of these deviations are due to variations in instantaneous angular velocity of the drive motor. The theoretical analysis assumes constant input shaft velocity. Vibrations and impacts account for other deviations.

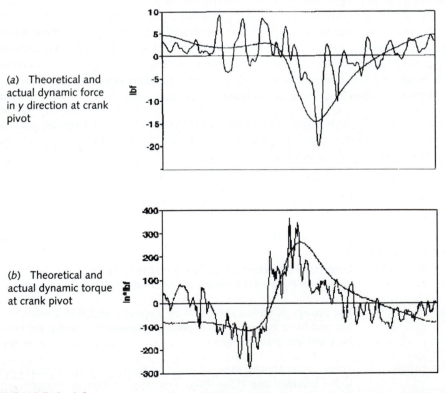

(a) Theoretical and actual dynamic force in y direction at crank pivot

(b) Theoretical and actual dynamic torque at crank pivot

FIGURE 3-16

Actual Measured Dynamic Forces and Torques in the Fourbar Linkage

This example of deviations from theoretical forces in a very simple dynamic system is presented to point out that our best calculations of forces (and thus the resulting stresses) on a system may be in error due to factors not included in a simplified force analysis. It is common for the theoretical predictions of forces in a dynamic system to understate reality, which is, of course, a nonconservative result. Wherever feasible, the testing of physical prototypes will give the most accurate and realistic results.

The effects of vibration on a system can cause significant loadings and are difficult to predict without test data of the sort shown in Figure 3-16, where the actual loads are seen to be double their predicted values; this will obviously double the stresses. A traditional, and somewhat crude, approach used in machine design has been to apply overload factors to the theoretical calculated loads based on experience with the same or similar equipment. As an example, see Table 11-17 in the chapter on spur-gear design. This table lists industry-recommended overload factors for gears subjected to various types of shock loading. These sorts of factors should be used only if one cannot develop more accurate test data of the type shown in Figure 3-16.

3.8   IMPACT LOADING

The loading considered so far has either been static or, if time-varying, has been assumed to be gradually and smoothly applied, with all mating parts continually in contact. Many machines have elements that are subjected to sudden loads or impacts. One example is the crank-slider mechanism which forms the heart of an automobile engine. The piston head is subjected to an explosive rise in pressure every two crank revolutions when the cylinder fires, and the clearance between the circumference of the piston and the cylinder wall can allow an impact of these surfaces as the load is reversed each cycle. A more extreme example is a jackhammer, whose purpose is to impact pavement and break it up. The loads that result from impact can be much greater than those that would result from the same elements contacting gradually. Imagine trying to drive a nail by gently placing the hammer head on the nail rather than by striking it.

What distinguishes impact loading from static loading is the time duration of the application of the load. If the load is applied slowly, it is considered static; if applied rapidly then it is impact. One criterion used to distinguish the two is to compare the time of load application t_l (defined as the time it takes the load to rise from zero to its peak value) to the period of the natural frequency T_n of the system. If t_l is less than half T_n, it is considered to be impact. If t_l is greater than three times T_n, it is considered static. Between those limits is a gray area in which either condition can exist.

Two general cases of impact loading are considered to exist, though we will see that one is just a limiting case of the other. Burr[6] calls these two cases *striking impact* and *force impact*. **Striking impact** refers to an actual collision of two bodies, such as in hammering or the taking up of clearance between mating parts. **Force impact** refers to a suddenly applied load with no velocity of collision, as in a weight suddenly being taken up by a support. This condition is common in friction clutches and brakes (see Chapter 15). These cases can occur independently or in combination.

Severe collisions between moving objects can result in permanent deformation of the colliding bodies as in an automobile accident. In such cases the permanent deformation is desirable in order to absorb the large amount of energy of the collision and protect the occupants from more severe harm. We are concerned here only with impacts that do not cause permanent deformation, that is, the stresses will remain in the elastic region. This is necessary to allow continued use of the component after impact.

If the mass of the striking object m is large compared to that of the struck object, m_b and if the striking object can be considered rigid, then the kinetic energy possessed by the striking object can be equated to the energy stored elastically in the struck object at its maximum deflection. This energy approach gives an **approximate value** for the impact loading. It is not exact because it assumes that the stresses throughout the impacted member reach peak values at the same time. However, waves of stress are set up in the struck body, travel through it at the speed of sound and reflect from the boundaries. Calculating the effects of these longitudinal waves on the stresses in elastic media gives exact results and is necessary when the ratio of mass of the striking object to that of the struck object is small. The wave method will not be discussed here. The reader is directed to reference 6 for further information.

Energy Method

The kinetic energy of the striking body will be converted to stored potential energy in the struck body, assuming that no energy is lost to heat. If we assume that all particles of the combined bodies come to rest at the same instant, then just before rebound, the force, stress, and deflection in the struck body will be maximal. The elastic energy stored in the struck body will be equal to the area under the force-deflection curve defined by its particular spring constant. A generalized force-deflection curve for a linear spring element is shown in Figure 3-17. The elastic energy stored is the area under the curve between zero and any combination of force and deflection. Because of the linear relationship, this is the area of a triangle, $A = 1/2bh$. Thus, the energy stored at the point of peak impact deflection, δ_i, is

$$E = \frac{1}{2} F_i \delta_i \tag{3.9a}$$

Substituting equation 3.5 gives

$$E = \frac{F_i^2}{2k} \tag{3.9b}$$

Figure 3-18a shows a mass about to impact the end of a horizontal rod. This device is sometimes called a *slide hammer* and is used to remove dents from automobile sheet metal among other uses. At the point of impact, the portion of the kinetic energy of the moving mass that is imparted to the struck mass is

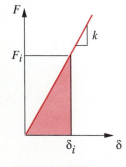

FIGURE 3-17

Energy Stored in a Spring

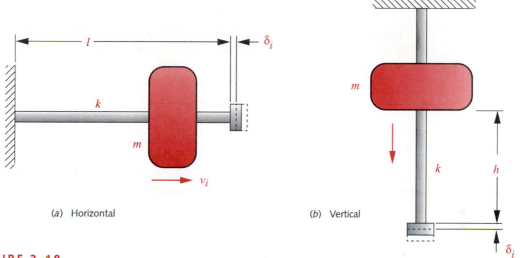

(a) Horizontal   (b) Vertical

FIGURE 3-18

Axial Impact on a Slender Rod

$$E = \eta\left(\frac{1}{2}mv_i^2\right) \qquad (3.10)$$

where m is its mass and v_i its velocity at impact. We need to modify the kinetic energy term by a correction factor η to account for the energy dissipation associated with the particular type of elastic member being struck. If the dissipation is negligible, η will be 1.

Assuming that all the kinetic energy transferred from the moving mass is converted to elastic energy stored in the struck member allows us to equate equations 3.9 and 3.10:

$$\frac{F_i^2}{2k} = \eta\frac{mv_i^2}{2}$$

$$F_i = v_i\sqrt{\eta mk} \qquad (3.11)$$

In part b of Figure 3-18, if the mass were allowed to statically load the struck member, its only effect would be its weight $W = mg$. The resulting static deflection would be $\delta_{st} = W/k$. Substituting these into equation 3.11 gives a ratio of dynamic force to static force or dynamic deflection to static deflection:

$$\frac{F_i}{W} = \frac{\delta_i}{\delta_{st}} = v_i\sqrt{\frac{\eta}{g\delta_{st}}} \qquad (3.12)$$

Thus if the static deflection can be calculated for the application of a force equal to the weight of the mass, an estimate of the dynamic force and dynamic deflection can be obtained. Methods for calculating the deflections of various cases are addressed in the next chapter.

For the case of a mass falling through a distance h onto a rod, as shown in Figure 3-18b, the potential energy given up by the mass on impact is

$$E = mg(\eta h + \delta_i)$$
$$= W(\eta h + \delta_i) \tag{3.13}$$

since the total vertical distance traveled by the mass includes the deflection due to impact.

Equating this potential energy to the elastic energy stored in the struck member, substituting $W = k\delta_{st}$, and solving for the dynamic force ratio gives

$$\frac{F_i}{W} = \frac{\delta_i}{\delta_{st}} = 1 + \sqrt{1 + \frac{2\eta h}{\delta_{st}}} \tag{3.14}$$

If the distance h to which the mass is raised is set to zero, Equation 3.14 becomes equal to 2. This says that if the mass is held **in contact** with the "struck" member (with the weight of the mass separately supported) and then allowed to suddenly impart its weight to that member, the dynamic force will be twice the weight. This is the case of "force impact" described earlier, in which there is no actual collision between the objects. A more accurate analysis, using wave methods predicts that the dynamic force will be more than doubled even in this noncollision case of sudden application of load.[6] Many designers use 3 or 4 as a more conservative estimate of this dynamic factor for the case of sudden load application. This is only a crude estimate however, and, if possible, experimental measurements or a wave-method analysis should be made to determine more suitable dynamic factors for any particular design.

Burr derives the correction factors η for several impact cases in reference 6. Roark and Young provide factors for additional cases in reference 7. For the case of a mass axially impacting a rod as shown in Figure 3-18, the correction factor is[6]

$$\eta = \frac{1}{1 + \dfrac{m_b}{3m}} \tag{3.15}$$

where m is the mass of the striking object, and m_b is the mass of the struck object. As the ratio of the striking mass to the struck mass increases, the correction factor η asymptotically approaches one. Figure 3-19 shows η as a function of mass ratio for three cases: an axial rod (equation 3.15) plotted in color, a simply supported beam struck at midspan (black-solid), and a cantilever beam struck at the free end (black-dotted).[6] The correction factor η is always less than one (and > 0.9 for mass ratios > 5) , so assuming it to be one is conservative. However, be aware that this energy method gives approximate and nonconservative results in general, and it needs to be used with larger-than-usual safety factors applied.

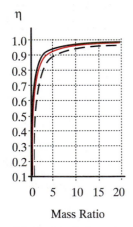

FIGURE 3-19

Correction Factor η as a Function of Mass Ratio

EXAMPLE 3-1

Impact Loading on an Axial Rod

Problem: The axial rod shown in Figure 3-18a is hit by a mass moving at 1 m/sec.

    a.   Determine the sensitivity of the impact force to the length/diameter ratio of the rod for a constant 1 kg moving mass.

    b.   Determine the sensitivity of the impact force to the ratio of moving mass to rod mass for a constant length/diameter ratio of 10.

Given: The round rod is 100 mm long. The rod and moving mass are steel with E = 207 Gpa and a mass density of 7.86 g/cm³.

Assumptions: An approximate energy method will be acceptable. The correction factor for energy dissipation will be applied.

Solution: See Figures 3-18a, 3-20, and 3-21, Table 3-7, and file EX03-01.

1 Figure 3-18a shows the system. The moving mass strikes the flange on the end of the rod with the stated velocity of 1 m/sec.

2 For part (a), we will keep the moving mass constant at 1 kg, and the rod length constant at 100 mm and vary the rod diameter to obtain l/d ratios in the range of 1 to 20. In the equations that follow we will show only one calculation made for the l/d ratio of 10. The *TKSolver* file EX03-01A computes all the values for a list of l/d ratios. Its output for l/d = 10 is shown in Table 3-7. The static deflection that would result from application of the weight force of the mass is calculated from the expression for the deflection of a bar in tension. (See equation 4.8 in the next chapter for the derivation.)

$$\delta_{st} = \frac{Wl}{AE} = \frac{9.81\,\text{N}\,(100\,\text{mm})}{78.54\,\text{mm}^2 \left(2.07E5\,\text{N}/\text{mm}^2\right)} = 0.06\,\mu\text{m} \qquad (a)$$

The correction factor η is calculated here for an assumed mass ratio of 16.2,

$$\eta = \frac{1}{1 + \dfrac{m_b}{3m}} = \frac{1}{1 + \dfrac{0.0617}{3(1)}} = 0.98 \qquad (b)$$

These values (calculated for each different rod diameter d) are then substituted into Equation 3.12 to find the force ratio F_i/W and the dynamic force F_i. For d = 10 mm

$$\frac{F_i}{W} = v_i\sqrt{\frac{\eta}{g\delta_{st}}} = 1\sqrt{\frac{0.98}{9.81(0.00006)}} = 1\,285.9$$

$$F_i = 1285.9(9.81) = 12\,612\,\text{N} \qquad (c)$$

The variation in force ratio with changes in l/d ratio for a constant amount of moving mass and a constant impact velocity (i.e., constant input energy) is shown in Figure 3-20. As the l/d ratio is reduced, the rod becomes much stiffer and generates

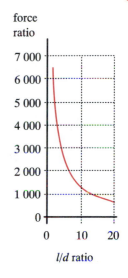

force ratio

l/d ratio

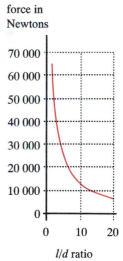

force in Newtons

l/d ratio

FIGURE 3-20

Dynamic Force and Force Ratio as a Function of *l/d* Ratio for the System in Example 3-1

3

much larger dynamic forces from the same impact energy. This clearly shows that impact forces can be reduced by increasing the compliance of the impacted system.

3   For part (*b*), we will keep the *l* / *d* ratio constant at 10 and vary the ratio between the moving mass and the rod mass over the range of 1 to 20. The *TKSolver* file EX03-01B computes all the values for a list of mass ratios. The results for a mass ratio of 16.2 are the same as in part (*a*) above. Figure 3-21*a* shows that the dynamic force ratio F_i / W varies inversely with the mass ratio. However, the value of the dynamic force is increasing with mass ratio as shown in Figure 3-21*b*, because the static force W is also increasing with mass ratio.

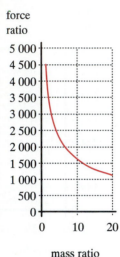

force ratio

mass ratio

force in Newtons

mass ratio

FIGURE 3-21

Dynamic Force and Force Ratio as a Function of Mass Ratio for the System in Example 3-1

Table 3-7   Example 3-1 – Impact Force Analysis

Varying Length to Diameter Ratio - See *TKSolver* File EX03-01

Input	Variable	Output	Unit	Comments
				Given Data
9.81	*g*		m/sec^2	gravitational constant
7.86	*density*		gm/cm^3	material density (steel)
207	*E*		Gpa	Young's modulus (steel)
1	*Vi*		m/sec	impact velocity
1	*m*		kg	moving mass
100	*l*		mm	length of rod
10	*loverd*			L / D Ratio
				Calculated Data
	d	10.00	mm	diameter of rod
	A	78.54	mm^2	area of rod
	mrod	0.06	kg	mass of rod
	Wrod	0.61	N	weight of rod
	W	9.81	N	weight of moving mass
	dst	0.06	µm	static deflection
	msratio	16.20		mass ratio
	n	0.98		correction factor
	Fi	12 612.53	N	dynamic force
	Fratio	1 285.91		ratio of dynamic to static force
	di	77.64	µm	dynamic deflection

3.9   BEAM LOADING

A beam is any element that carries loads transverse to its long axis and may carry loads in the axial direction as well. A beam supported on pins or narrow supports at each end is said to be **simply supported,** as shown in Figure 3-22a. A beam fixed at one end and unsupported at the other is a **cantilever beam** (Figure 3-22b). A simply supported beam that overhangs its supports at either end is an **overhung beam** (Figure 3-22c). If a beam has more supports than are necessary to provide kinematic stability (i.e., make the kinematic degree of freedom zero), then the beam is said to be overconstrained or indeterminate, as shown in Figure 3-22d. An **indeterminate beam** problem cannot be solved for its loads using only equations 3.3. Other techniques are necessary. This problem is addressed in the next chapter.

Beams are typically analyzed as static devices, though vibrations and accelerations can cause dynamic loading. A beam may carry loads in three dimensions in which case equations 3.3a apply. For the two-dimensional case, equations 3.3b suffice. The review examples used here are limited to 2-D cases for brevity.

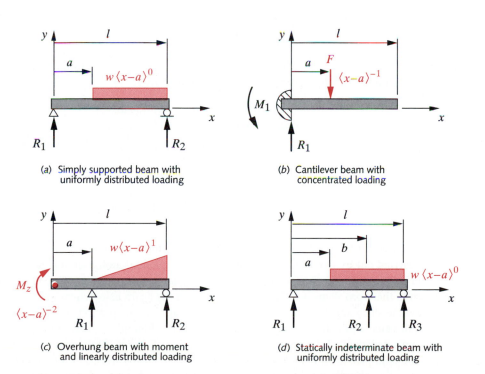

(a) Simply supported beam with uniformly distributed loading

(b) Cantilever beam with concentrated loading

(c) Overhung beam with moment and linearly distributed loading

(d) Statically indeterminate beam with uniformly distributed loading

FIGURE 3-22

Types of Beams and Beam Loadings

Shear and Moment

A beam may be loaded with some combination of distributed and/or concentrated forces or moments as in Figure 3-22. The applied forces will create both shearing forces and bending moments in the beam. A load analysis must find the magnitudes and spatial distributions of these shear forces and bending moments on the beam. The shear forces V and the moment M in a beam are related to the loading function $q(x)$ by

$$q(x) = \frac{dV}{dx} = \frac{d^2M}{dx^2} \qquad (3.16a)$$

The loading function $q(x)$ is typically known and the shear V and moment M distributions can be found by integrating equation 3.16a:

$$\int_{V_A}^{V_B} dV = \int_{x_A}^{x_B} q\,dx = V_B - V_A \qquad (3.16b)$$

Equation 3.16b shows that the difference in the shear forces between any two points, A and B, is equal to the area under the graph of the loading function, equation 3.16a.

Integrating the relationship between shear and moment gives

$$\int_{M_A}^{M_B} dM = \int_{V_A}^{V_B} V\,dx = M_B - M_A \qquad (3.16c)$$

showing that the difference in the moment between any two points, A and B, is equal to the area under the graph of the shear function, equation 3.16b.

SIGN CONVENTION   The usual (and arbitrary) sign convention used for beams is to consider a moment positive if it causes the beam to deflect concave downward (as if to collect water). This puts the top surface in compression and the bottom surface in tension. The shear force is considered positive if it causes a clockwise rotation of the section on which it acts. These conventions are shown in Figure 3-23 and result in positive moments being created by negative applied loads. All the applied loads shown in Figure 3-22 are negative. For example, in Figure 3-22a, the distributed load magnitude from a to l is $q = -w$.

EQUATION SOLUTION   The solution of equations 3.3 and 3.16 for any beam problem may be carried out by one of several approaches. Sequential and graphical solutions are described in many textbooks on statics and mechanics of materials. One classical approach to these problems is to find the reactions on the beam using equations 3.3 and then draw the shear and moment diagrams using a graphical integration approach combined with calculations for the significant values of the functions. This approach has value from a pedagogical standpoint as it is easily followed, but is cumbersome to implement. The approach most amenable to computer solution uses a class of mathematical functions called **singularity functions** to represent the loads on the beam. We present both the classical approach as a pedagogical reference and also introduce the use of singularity functions which offer some computational advantages. While this approach

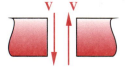

positive shear

positive moment

FIGURE 3-23

Beam Sign Convention

may be new to some students, when compared to the methods usually learned in other courses, it has significant advantages in computerizing the solution.

Singularity Functions

Because the loads on beams typically consist of collections of discrete entities, such as point loads or segments of distributed loads that can be discontinuous over the beam length, it is difficult to represent these discrete functions with equations that are valid over the entire continuum of beam length. A special class of functions called **singularity functions** was invented to deal with these mathematical situations. Singularity functions are often denoted by a binomial in angled brackets as shown in equations 3.17. The first quantity in the brackets is the variable of interest, in our case x, the distance along the beam length. The second quantity a is a user-defined parameter that denotes where in x the singularity function either acts or begins to act. For example, for a point load, the quantity a represents the particular value of x at which the load acts (see Figure 3-22b). The definition of this singularity function, called the **unit impulse** or **Dirac delta function**, is given in equation 3.17d. Note that all singularity functions involve a conditional constraint. The unit impulse evaluates to 1 *if* $x = a$ and is 0 at any other value of x. The **unit step** function (Eq. 3.17c) evaluates to 0 for all values of x less than a and to 1 for all other x.

Since these functions are defined to evaluate to unity, multiplying them by a coefficient creates any magnitude desired. Their application is shown in the following three examples and is explained in the most detail in Example 3-2B. If a loading function both starts and stops within the range of x desired, it needs two singularity functions to describe it. The first defines the value of a_1 at which the function begins to act and has a positive or negative coefficient as appropriate to its direction. The second defines the value a_2 at which the function ceases to act and has a coefficient of the same magnitude but opposite sign as the first. These two functions will cancel beyond a_2, making the load zero. Such a case is shown in Example 4-6 in the next chapter.

Quadratically distributed loads can be represented by a **unit parabolic function**,

$$\langle x - a \rangle^2 \tag{3.17a}$$

which is defined as 0 when $x \le a$, and equal to $(x - a)^2$ when $x > a$.

Linearly distributed loads can be represented by a **unit ramp function**,

$$\langle x - a \rangle^1 \tag{3.17b}$$

which is defined as 0 when $x \le a$, and equal to $(x - a)$ when $x > a$.

A uniformly distributed load over a portion of a beam can be represented mathematically by a **unit step** function,

$$\langle x - a \rangle^0 \tag{3.17c}$$

defined as 0 when $x < a$, unity when $x > a$, and is undefined at $x = a$.

A concentrated force can be represented by the **unit impulse** function,

$$\langle x - a \rangle^{-1} \tag{3.17d}$$

which is defined as 0 when $x < a$, ∞ when $x = a$, and 0 when $x > a$. Its integral evaluates to unity at a.

A concentrated moment can be represented by the **unit doublet** function,

$$\langle x - a \rangle^{-2} \tag{3.17e}$$

which is defined as 0 when $x < a$, indeterminate when $x = a$, and 0 when $x > a$. It generates a unit couple moment at a.

This process can be extended to obtain polynomial singularity functions of any order $\langle x - a \rangle^n$ to fit distributed loads of any shape. Four of the five singularity functions described here are shown in Figure 3-22, as applied to various beam types. Table 3-8 shows four of the singularity functions implemented as *TKSolver Rule Functions*. The so-called *argument variables* are the input values to the function and the *result variables* are returned from its execution. See Table 3-9 for examples of the statements that pass the argument variables to it for evaluation. (See also Example 1-7 on p. 42.)

The integrals of these singularity functions have special definitions that, in some cases, defy common sense but nevertheless provide the desired mathematical results. For example, the unit impulse function (Eq. 3.17d) is defined in the limit as having zero width and infinite magnitude, yet its area (integral) is defined as equal to one (Eq. 3.18d). (See reference 8 for a more complete discussion of singularity functions.) The integrals of the singularity functions in equations 3.17 are defined as

$$\int_{-\infty}^{x} \langle \lambda - a \rangle^2 d\lambda = \frac{\langle x - a \rangle^3}{3} \tag{3.18a}$$

$$\int_{-\infty}^{x} \langle \lambda - a \rangle^1 d\lambda = \frac{\langle x - a \rangle^2}{2} \tag{3.18b}$$

$$\int_{-\infty}^{x} \langle \lambda - a \rangle^0 d\lambda = \langle x - a \rangle^1 \tag{3.18c}$$

$$\int_{-\infty}^{x} \langle \lambda - a \rangle^{-1} d\lambda = \langle x - a \rangle^0 \tag{3.18d}$$

$$\int_{-\infty}^{x} \langle \lambda - a \rangle^{-2} d\lambda = \langle x - a \rangle^{-1} \tag{3.18e}$$

Where λ is just an integration variable running from $-\infty$ to x. These expressions can be used to evaluate the shear and moment functions that result from any loading function that is expressed as a combination of singularity functions.

Table 3-8   TKSolver Rule Functions to Evaluate Singularity Functions

; impulse singularity function - to use: y = PULSE (x, a, mag)

Argument Variables: x, a, mag

Result Variables: y

IF ABS (x - a) < .0001 THEN y = mag ELSE y = 0

; step singularity function - to use: y = STEPF (x, a, mag)

Argument Variables: x, a, mag

Result Variables: y

IF x < a THEN y = 0 ELSE y = mag

; ramp singularity function - to use: y = RAMP (x, a, mag)

Argument Variables: x, a, mag

Result Variables: y

*IF x <= a THEN y = 0 ELSE y = mag * (x - a)*

; parabolic singularity function - to use: y = PARA (x, a, mag)

Argument Variables: x, a, mag

Result Variables: y

*IF x <= a THEN y = 0 ELSE y = mag * (x - a) ^ 2*

EXAMPLE 3-2A

Shear and Moment Diagrams of a Simply Supported Beam Using a Graphical Method

Problem: Determine and plot the shear and moment functions for the simply supported beam with uniformly distributed load shown in Figure 3-22a.

Given: Beam length l = 10 in, and load location a = 4 in. The magnitude of the uniform force distribution is w = 10 lb/in.

Assumptions: The weight of the beam is negligible compared to the applied load and so can be ignored.

Solution: See Figures 3-22a and 3-24

1   Solve for the reaction forces using equations 3.3. Summing moments about the right end and summing forces in the y direction gives

$$\sum M_z = 0 = R_1 l - \frac{w(l-a)^2}{2}$$

$$R_1 = \frac{w(l-a)^2}{2l} = \frac{10(10-4)^2}{2(10)} = 18$$

(a)

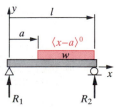

Simply supported beam with uniformly distributed loading

FIGURE 3-22a

Repeated

$$\sum F_y = 0 = R_1 - w(l-a) + R_2$$

$$R_2 = w(l-a) - R_1 = 10(10-4) - 18 = 42 \qquad (b)$$

2   The shape of the shear diagram can be sketched by graphically integrating the loading diagram shown in Figure 3-24a. As a "device" to visualize this graphical integration process, imagine that you walk backward across the loading diagram of the beam, starting from the left end and taking small steps of length dx. You will record on the shear diagram (Figure 3-24b) the area ($force \cdot dx$) of the loading diagram that you can see as you take each step. As you take the first step backward from $x = 0$, the shear diagram rises immediately to the value of R_1. As you walk from $x = 0$ to $x = a$, no change occurs since you see no additional forces. As you step beyond $x = a$, you begin to see strips of area equal to $-w \cdot dx$, which subtract from the value of R_1 on the shear diagram. When you reach $x = l$, the total area $w \cdot (l-a)$ will have taken the value of the shear diagram to $-R_2$. As you step backward off the beam's loading diagram (Figure 3-24a) and plummet downward, you can now see the reaction force R_2 which closes the shear diagram to zero. The largest value of the shear force in this case is then R_2 at $x = l$.

3   If your reflexes are quick enough, you should try to catch the shear diagram (Figure 3-24b) as you fall, climb onto it, and repeat this backward-walking trick across it to create the moment diagram which is the integral of the shear diagram. Note in Figure 3-24c that from $x = 0$ to $x = a$ this moment function is a straight line with slope = R_1. Beyond point a, the shear diagram is triangular, and so integrates to a parabola. The peak moment will occur where the shear diagram crosses zero (i.e., zero slope on the moment diagram). The value of x at $V = 0$ can be found with a little trigonometry, noting that the slope of the triangle is $-w$:

$$x_{@V=0} = a + \frac{R_1}{w} = 4 + \frac{18}{10} = 5.8 \qquad (c)$$

Positive shear area adds to the moment value and negative area subtracts. So the value of the peak moment can be found by adding the areas of the rectangular and triangular portions of the shear diagram from $x = 0$ to the point of zero shear at $x = 5.8$:

$$M_{@x=5.8} = R_1(a) + R_1\frac{1.8}{2} = 18(4) + 18\frac{1.8}{2} = 88.2 \qquad (d)$$

The above method gives the magnitudes and locations of the maximum shear and moment on the beam and is useful for a quick determination of those values. However, all that walking and falling can become tiresome, and it would be useful to have a method that can be conveniently computerized to give accurate and complete information on the shear and moment diagrams of any beam loading case. Such a method will also allow us to obtain the beam's deflection curve with little additional work. The simple method shown above is not as useful for determining deflection curves, as will be seen in the next chapter. We will now repeat this example using singularity functions to determine the loading, shear, and moment diagrams.

EXAMPLE 3-2B

Shear and Moment Diagrams of a Simply Supported Beam Using Singularity Functions

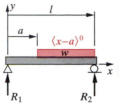

Problem: Determine and plot the shear and moment functions for the simply supported beam with uniformly distributed load shown in Figure 3-22a.

Given: Beam length l = 10 in, and load location a = 4 in. The magnitude of the uniform force distribution is w = 10 lb/in.

Assumptions: The weight of the beam is negligible compared to the applied load and so can be ignored.

Simply supported beam with uniformly distributed loading

FIGURE 3-22a

Repeated

Solution: See Figures 3-22a and 3-24 , and the *TKSolver* file EX03-02.

1   Write equations for the load function in terms of equations 3.17 and integrate the resulting function twice using equations 3.18 to obtain the shear and moment functions. For the beam in Figure 3-22a,

$$q = R_1 \langle x - 0 \rangle^{-1} - w \langle x - a \rangle^0 + R_2 \langle x - l \rangle^{-1} \qquad (a)$$

$$V = \int q\,dx = R_1 \langle x - 0 \rangle^0 - w \langle x - a \rangle^1 + R_2 \langle x - l \rangle^0 + C_1 \qquad (b)$$

$$M = \int V\,dx = R_1 \langle x - 0 \rangle^1 - \frac{w}{2} \langle x - a \rangle^2 + R_2 \langle x - l \rangle^1 + C_1 x + C_2 \qquad (c)$$

(a) Loading Diagram

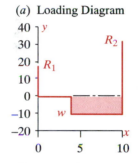

There are two reaction forces and two constants of integration to be found. We are integrating along a hypothetical infinite beam from $-\infty$ to x. The variable x can take on values both before and beyond the end of the beam. If we consider the conditions at a point infinitesimally to the left of $x = 0$ (denoted as $x = 0^-$), the shear and moment will both be zero there. The same conditions apply at a point infinitesimally to the right of $x = l$ (denoted as $x = l^+$). These observations provide the four boundary conditions needed to evaluate the four constants C_1, C_2, R_1, R_2: when $x = 0^-$, $V = 0$, $M = 0$; when $x = l^+$, $V = 0$, $M = 0$.

(b) Shear Diagram

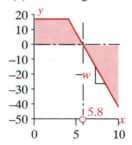

2   The constants C_1 and C_2 are found by substituting the boundary conditions $x = 0^-$, $V = 0$, and $x = 0^-$, $M = 0$ in equations (b) and (c), respectively:

$$V = 0 = R_1 \langle 0^- - 0 \rangle^0 - w \langle 0^- - a \rangle^1 + R_2 \langle 0^- - l \rangle^0 + C_1$$

$$C_1 = 0$$

$$\qquad (d)$$

$$M = 0 = R_1 \langle 0^- - 0 \rangle^1 - \frac{w}{2} \langle 0^- - a \rangle^2 + R_2 \langle 0^- - l \rangle^1 + C_1 (0^-) + C_2$$

$$C_2 = 0$$

(c) Moment Diagram

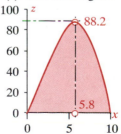

FIGURE 3-24

Example 3-2 Plots

Note that in general, the constants C_1 and C_2 will always be zero if the reaction forces and moments acting on the beam are included in the loading function, because the shear and moment diagrams must close to zero at each end of the beam.

3   The reaction forces R_1 and R_2 can be calculated from equations (c) and (b) respectively by substituting the boundary conditions $x = l^+$, $V = 0$, $M = 0$. Note that we can substitute l for l^+ since their difference is vanishingly small.

$$M = R_1 \langle l - 0 \rangle^1 - w \frac{\langle l - a \rangle^2}{2} + R_2 \langle l - l \rangle^1 = 0$$

$$0 = R_1 l - \frac{w(l - a)^2}{2} \tag{e}$$

$$R_1 = \frac{w(l - a)^2}{2l} = \frac{10(10 - 4)^2}{2(10)} = 18$$

$$V = R_1 \langle l \rangle^0 - w \langle l - a \rangle^1 + R_2 \langle l \rangle^0 = 0$$

$$0 = R_1 - w(l - a) + R_2 \tag{f}$$

$$R_2 = w(l - a) - R_1 = 10(10 - 4) - 18 = 42$$

Since w, l, and a are known from the given data, equation (e) can be solved for R_1, and this result substituted in equation (f) to find R_2. Note that equation (f) is just $\Sigma F = 0$, and equation (e) is the sum of moments taken about point l and set to 0.

4   To generate the shear and moment functions over the length of the beam, equations (b) and (c) must be evaluated for a range of values of x from 0 to l, after substituting the above values of C_1, C_2, R_1, and R_2 in them. This was done using the *TKSolver* file EX03-02, which is provided on the enclosed disk. A list of values for the independent variable x was run from 0 to $l = 10$ at 0.1 increments. The reactions, loading function, shear force function, and moment function were calculated from equations (a) through (f) above and are plotted in Figure 3-24.

5   The largest absolute values of the shear and moment functions are of interest for the calculation of stresses in the beam. The plots show that the shear force is largest at $x = l$ and the moment has a maximum M_{max} near the center. The value of x at M_{max} can be found by setting V to 0 in equation (b) and solving for x. (The shear function is the derivative of the moment function and so must be zero at each of its minima and maxima.) This gives $x = 5.8$ at M_{max}. The function values at these points of maxima or minima can then be calculated from equations b and c respectively by substituting the appropriate values of x and evaluating the singularity functions. For the maximum absolute value of shear force at $x = l$,

$$V_{max} = V_{@x=l^-} = R_1 \langle l^- - 0 \rangle^0 - w \langle l^- - a \rangle^1 + R_2 \langle l^- - l \rangle^0$$

$$= R_1 - w \left(l^- - a \right) + 0 \tag{g}$$

$$= 18 - 10(10 - 4) + 0 = -42$$

3

Table 3-9   **Example 3-2B – Uniformly Loaded, Simply Supported Beam**
Part 1 of 2   *TKSolver* Rule Sheet from File EX03-02

; find reactions
$$R1 = w * (l - a)\char`\^2 / (2 * l)$$
$$R2 = w * (l - a) - R1$$
; beam equations
$$q = PULSE\,(x, 0, R1) - STEPF\,(x, a, w) + PULSE\,(x, l, R2)$$
$$V = STEPF\,(x, 0, R1) - RAMP\,(x, a, w) + STEPF\,(x, l, R2)$$
$$M = RAMP\,(x, 0, R1) - 1/2 * PARA\,(x, a, w) + RAMP\,(x, l, R2)$$

Note that the first singularity term evaluates to 1 since $l > 0$ (see Eq. 3.17c), the second singularity term evaluates to $(l - a)$ because $l > a$ in this problem (see Eq. 3.17b), and the third singularity term evaluates to 0 as defined in equation 3.17c. The maximum moment is found in similar fashion:

$$M_{max} = M_{@x=5.8} = R_1\langle 5.8 - 0\rangle^1 - w\frac{\langle 5.8 - a\rangle^2}{2} + R_2\langle 5.8 - l\rangle^1$$

$$= R_1\langle 5.8\rangle^1 - w\frac{\langle 5.8 - 4\rangle^2}{2} + R_2\langle 5.8 - 10\rangle^1 \qquad (h)$$

$$= 18(5.8) - 10\frac{(5.8 - 4)^2}{2} + 0 = 88.2$$

Table 3–9   **Example 3-2B – Uniformly Loaded, Simply Supported Beam**
Part 2 of 2   *TKSolver* Variable Sheet from File EX03-02

	Input	Variable	Output	Unit	Comments
	10	l		in	length of beam
	4	a		in	distance to load w
L*	5.8	x		in	dimension along beam
	10	w		lb/in	applied load starts at point a
		$R1$	18	lb	left reaction force
		$R2$	42	lb	right reaction force
L		q	–10	lb	loading function at point x
L		V	0	lb	shear function at point x
L		M	88.2	lb-in	moment function at point x
		$Mmin$	0	lb-in	minimum moment
		$Mmax$	88.2	lb-in	maximum moment
		$Vmin$	–41	lb	minimum shear load
		$Vmax$	18	lb	maximum shear load
		$xmax$	5.8	in	x value at maximum moment

* Indicates that a List of values of this variable has been created for plotting.

3

The third singularity term evaluates to 0 because $5.8 < l$ (see Eq. 3.17b).

6   The results are

$$R_1 = 18 \qquad R_2 = 42 \qquad V_{max} = -42 \qquad M_{max} = 88.2 \qquad (i)$$

7   The *TKSolver* rule and variable sheets for this example problem are shown in Table 3-9. These models contain custom-written rule functions called PULSE, STEPF, RAMP, PARA, CUBIC, and QUARTIC (some of which are shown in Table 3-8), which evaluate the singularity functions. These are called from the rule statements that evaluate the loading function, shear function, and moment function. These functions have three arguments: x, a list of independent variable values; a, the position along the beam at which the load begins to act; and F or w, the magnitude of the load. Each rule function applies the appropriate definition of its singularity function to these argument values and returns its value corresponding to x. List-solving causes these calculations to be repeated for all listed values of x and stores the results in the lists q, V, and M. Note the similarity among the three rules for q, V, and M in Tables 3-9, 3-10, and 3-11. Each corresponding term has the same arguments and the singularity function changes with each integration. A PULSE function in the q equation becomes a STEPF function in V and a RAMP function in M, etc.

EXAMPLE 3-3A

Shear and Moment Diagrams of a Cantilever Beam Using a Graphical Method

Problem: Determine and plot the shear and moment functions for the cantilever beam with a concentrated load as shown in Figure 3-22b.

Given: Beam length l = 10 in, and load location a = 4 in. The magnitude of the applied force is F = 40 lb.

Assumptions: The weight of the beam is negligible compared to the applied load and so can be ignored.

Solution: See Figures 3-22b and 3-25

1   Solve for the reaction forces using equations 3.3. Summing moments about the left end and summing forces in the y direction gives

$$\sum M_z = 0 = Fa - M_1$$

$$M_1 = Fa = 40(4) = 160 \qquad (a)$$

$$\sum F_y = 0 = R_1 - F$$

$$R_2 = F_1 = 40 \qquad (b)$$

2 By the sign convention, the shear is positive and the moment is negative in this example. To graphically construct the shear and moment diagrams for a cantilever beam take an imaginary "backward walk" starting at the fixed end of the beam and moving toward the free end (from left to right in Figure 3-24). In this example, that results in the first observed force being the reaction force R_1 acting upward This shear force remains constant until the downward force F at $x = a$ is reached which closes the shear diagram to zero.

3 The moment diagram is the integral of the shear diagram, which in this case is a straight line of slope = 40.

4 Both the shear and moment are maximum at the wall in a cantilever beam. Their maximum magnitudes are as shown in equations (a) and (b) above.

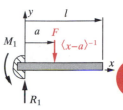

Cantilever beam with concentrated loading

FIGURE 3-22b

Repeated

This example will now be repeated using singularity functions.

EXAMPLE 3-3B

Shear and Moment Diagrams of a Cantilever Beam Using Singularity Functions

Problem: Determine and plot the shear and moment functions for the cantilever beam with a concentrated load as shown in Figure 3-22b.

Given: Beam length $l = 10$ in, and load location $a = 4$ in. The magnitude of the applied force is $F = 40$ lb.

Assumptions: The weight of the beam is negligible compared to the applied load and so can be ignored.

Solution: See Figures 3-22b, 3-25, Table 3-10, and *TKSolver* file EX03-03.

1 Write equations for the load function in terms of equations 3.17 and integrate the resulting function twice using equations 3.18 to obtain the shear and moment functions. Note the use of the unit doublet function to represent the moment at the wall. For the beam in Figure 3-22b,

$$q = M_1\langle x - 0\rangle^{-2} + R_1\langle x - 0\rangle^{-1} - F\langle x - a\rangle^{-1} \tag{a}$$

$$V = \int q\,dx = M_1\langle x - 0\rangle^{-1} + R_1\langle x - 0\rangle^{0} - F\langle x - a\rangle^{0} + C_1 \tag{b}$$

$$M = \int V\,dx = -M_1\langle x - 0\rangle^{0} + R_1\langle x - 0\rangle^{1} - F\langle x - a\rangle^{1} + C_1 x + C_2 \tag{c}$$

The reaction moment M_1 at the wall is in the z direction and the forces R_1 and F are in the y direction in equation (b). All moments in equation (c) are in the z direction.

(a) Loading Diagram

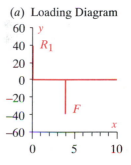

(b) Shear Diagram

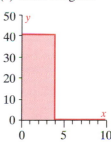

(c) Moment Diagram

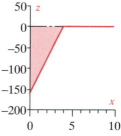

FIGURE 3-25

Example 3-3 Plots

2   Because the reactions have been included in the loading function, the shear and moment diagrams both close to zero at each end of the beam, making $C_1 = C_2 = 0$.

3   The reaction force R_1 and reaction moment M_1 are calculated from equations (b) and (c) respectively by substituting the boundary conditions $x = l^+$, $V = 0$, $M = 0$. Note that we can substitute l for l^+ since their difference is vanishingly small.

$$V_y = 0 = R_1 \langle l - 0 \rangle^0 - F \langle l - a \rangle^0$$
$$0 = R_1 - F \tag{d}$$
$$R_1 = F = 40 \text{ lb}$$

$$M_z = 0 = -M_1 \langle l - 0 \rangle^0 + R_1 \langle l - 0 \rangle^1 - F \langle l - a \rangle^1$$
$$0 = -M_1 + R_1(l) - F(l - a) \tag{e}$$
$$M_1 = R_1(l) - F(l - a) = 40(10) - 40(10 - 4) = 160 \text{ lb-in cw}$$

Since w, l, and a are known from the given data, equation (d) can be solved for R_1, and this result substituted in equation (e) to find M_1. Note that equation (d) is just $\Sigma F_y = 0$, and equation (e) is $\Sigma M_z = 0$. M_1 does not appear in equation (d) because it is in a different vector direction than the y forces.

4   To generate the shear and moment functions over the length of the beam, equations (b) and (c) must be evaluated for a range of values of x from 0 to l, after substituting the above values of C_1, C_2, R_1, and M_1 in them. This was done using the *TKSolver* file EX03-03 which is provided on the enclosed disk. A list of values for the independent variable x, was run from 0 to $l = 10$ at 0.1 increments. The reactions, loading function, shear force function and moment function were calculated from equations (a) through (e) above and are plotted in Figure 3-25.

5   The largest absolute values of the shear and moment functions are of interest for the calculation of stresses in the beam. The plots show that the shear force and the moment are both largest at $x = 0$. The function values at these points can be calculated from equations b and c respectively by substituting $x = 0$ and evaluating the singularity functions:

$$R_1 = 40 \qquad\qquad V_{max} = 40 \qquad\qquad |M_{max}| = 160 \tag{f}$$

Table 3-10   Example 3-3B – Cantilever Beam with Concentrated Load

Part 1 of 2   *TKSolver* Rule Sheet from File EX03-03

; find reactions

    R1 = F
    M1 = R1 * l − F * (l − a)

; beam equations

    q   =   PULSE (x, 0, R1) − PULSE (x, a, F)
    V =   STEPF (x, 0, R1)   − STEPF (x, a, F)
    M = RAMP (x, 0, R1) − RAMP (x, a, F) − STEPF (x, 0, M1)

Table 3–10   Example 3-3B – Cantilever Beam with Concentrated Load

Part 2 of 2   *TKSolver* Variable Sheet from File EX03-03

	Input	Variable	Output	Unit	Comments
	10	l		in	length of beam
	4	a		in	distance to load w
L*	2	x		in	dimension along beam
	40	F		lb	applied load at point a
		$R1$	40	lb	reaction force at wall
		$M1$	160	lb-in	reaction moment at wall
L		q	0	lb	loading function at point x
L		V	40	lb	shear function at point x
L		M	−80	lb-in	moment function at point x
		$Mmin$	−160	lb-in	minimum moment
		$Mmax$	0	lb-in	maximum moment
		$Vmin$	0	lb	minimum shear load
		$Vmax$	40	lb	maximum shear load

* Indicates that a list of values of this variable has been created for plotting.

6   The *TKSolver* rule and variable sheets for this example problem are shown in Table
     3-10. These models contain custom-written procedure functions called PULSE,
     STEPF, RAMP, PARA, CUBIC, and QUARTIC (some of which are shown in Table
     3-8), which evaluate the singularity functions. These are called from the rule
     statements that evaluate the loading function, shear function and moment function.
     These functions have three arguments: x, a list of independent variable values; a, the
     position along the beam at which the load begins to act; and F or w, the magnitude of
     the load. Each procedure applies the appropriate definition of its singularity function
     to these argument values and returns its value corresponding to x. The list-solving
     feature of *TKSolver* causes these calculations to be repeated for all listed values of x
     and stores the results in the lists q, V, and M.

EXAMPLE 3-4

Shear and Moment Diagrams of an Overhung Beam Using Singularity Functions

Problem: Determine and plot the shear and moment functions for the overhung
                   beam with an applied moment and ramp load as shown in Figure 3-22c.

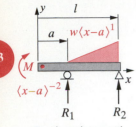

Overhung beam with
moment and linearly
distributed loading

FIGURE 3-22c

Repeated

Given: Beam length l = 10 in, and load location a = 4 in. The magnitude of the
applied moment M = 20 lb-in, and the slope of the force distribution is
w = 10 lb/in/in.

Assumptions: The weight of the beam is negligible compared to the applied load and
so can be ignored.

Solution: See Figures 3-22c and 3-26, Table 3-11, and File EX03-04.

1   Write equations for the load function in terms of equations 3.17 and integrate the
resulting function twice using equations 3.18 to obtain the shear and moment functions.
For the beam in Figure 3-22c,

$$q = M\langle x - 0\rangle^{-2} + R_1\langle x - a\rangle^{-1} - w\langle x - a\rangle^1 + R_2\langle x - l\rangle^{-1} \tag{a}$$

$$V = \int q\,dx = M\langle x - 0\rangle^{-1} + R_1\langle x - a\rangle^0 - \frac{w}{2}\langle x - a\rangle^2 + R_2\langle x - l\rangle^0 + C_1 \tag{b}$$

$$M = \int V\,dx = M\langle x - 0\rangle^0 + R_1\langle x - a\rangle^1 - \frac{w}{6}\langle x - a\rangle^3 + R_2\langle x - l\rangle^1 + C_1 x + C_2 \tag{c}$$

2   As demonstrated in the previous two examples, the constants of integration C_1 and
C_2 will always be zero if the reaction forces are included in the equations for shear
and moment. So we will set them to zero.

3   The reaction forces R_1 and R_2 can be calculated from equations (c) and (b) respectively by substituting the boundary conditions $x = l^+$, $V = 0$, $M = 0$. Note that we can
substitute l for l^+ since their difference is vanishingly small.

$$M = M_1\langle l\rangle^0 + R_1\langle l - a\rangle^1 - \frac{w}{6}\langle l - a\rangle^3 + R_2\langle l - l\rangle^1 = 0$$

$$0 = M_1 + R_1(l - a) - \frac{w}{6}(l - a)^3$$

$$R_1 = \frac{w}{6}(l - a)^2 - \frac{M_1}{(l - a)} \tag{d}$$

$$= \frac{10}{6}(10 - 4)^2 - \frac{20}{(10 - 4)} = 56.67 \text{ lb}$$

$$V = M\langle l\rangle^{-1} + R_1\langle l - a\rangle^0 - \frac{w}{2}\langle l - a\rangle^2 + R_2\langle l - l\rangle^0 = 0$$

$$0 = M(0) + R_1 - \frac{w}{2}(l - a)^2 + R_2 \tag{e}$$

$$R_2 = \frac{w}{2}(l - a)^2 - R_1 = \frac{10}{2}(10 - 4)^2 - 56.67 = 123.33 \text{ lb}$$

Note that equation (d) is just $\Sigma M_z = 0$, and equation (e) is $\Sigma F_y = 0$.

4   To generate the shear and moment functions over the length of the beam, equations
(b) and (c) must be evaluated for a range of values of x from 0 to l, after substituting

3

Table 3-11   Example 3-4 – Overhung Beam with Ramp Loading

Part 1 of 2   *TKSolver* Rule Sheet from File EX03-04

; find reactions

$R1 = w/6 * (l - a)\text{^}2 - M1 / (l - a)$

$R2 = w/2 * (l - a)\text{^}2 - R1$

; beam equations

$q = PULSE\ (x, a, R1) - RAMP\ (x, a, w) + PULSE\ (x, l, R2)$

$V = STEPF\ (x, a, R1) - 1/2 * PARA\ (x, a, w) + STEPF\ (x, l, R2)$

$M = RAMP\ (x, a, R1) - 1/6 * CUBIC\ (x, a, w) + RAMP\ (x, l, R2) + STEPF\ (x, 0, M1)$

---

the values of $C_1 = 0$, $C_2 = 0$, R_1, and R_2 in them. This was done using the *TKSolver* file EX03-04, which is provided on the enclosed disk. A list of values for the independent variable x was run from 0 to $l = 10$ at 0.1 increments. The reactions, loading function, shear force function, and moment function were calculated from equations (a) through (f) above and are plotted in Figure 3-26.

5   The largest absolute values of the shear and moment functions are of interest for the calculation of stresses in the beam. The plots show that the shear force is largest at $x = l$ and the moment has a maximum to the right of the beam center. The value of x at M_{max} can be found by setting V to 0 in equation (b) and solving for x. The shear function is the derivative of the moment function and so must be zero at each of its minima and maxima. This gives $x = 7.4$ at M_{max}. The function values at these points of maxima or minima can be calculated from equations b and c respectively by substituting the appropriate values of x and evaluating the singularity function:

$$R_1 = 56.7 \qquad R_2 = 123.3 \qquad V_{max} = -120 \qquad M_{max} = 147.2 \qquad (f)$$

6   The *TKSolver* rule and variable sheets for this example problem are shown in Table 3-11. These models contain custom-written procedure functions called PULSE, STEPF, RAMP, PARA, CUBIC, and QUARTIC (some of which are shown in Table 3-8), which evaluate the singularity functions. These are called from the rule statements that evaluate the loading function, shear function, and moment function. These functions have three arguments: x, a list of independent variable values; a, the position along the beam at which the load begins to act; and F or w, the magnitude of the load. Each procedure applies the appropriate definition of its singularity function to these argument values and returns its value corresponding to x. The list-solving feature of *TKSolver* causes these calculations to be repeated for all listed values of x and stores the results in the lists q, V, and M.

(a) Loading Diagram

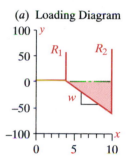

(b) Shear Diagram

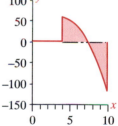

(c) Moment Diagram

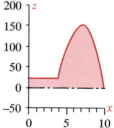

FIGURE 3-26

Example 3-4 Plots

Superposition

These examples of beam systems represent only a small fraction of all the possible combinations of beam loadings and constraints that one will encounter in practice. Rather

Table 3–11   Example 3-4 – Overhung Beam with Ramp and Concentrated Moment

Part 2 of 2   *TKSolver* Variable Sheet from File EX03-04

	Input	Variable	Output	Unit	Comments
	10	l		in	length of beam
	4	a		in	distance to load w
L*	8	x		in	dimension along beam
	20	$M1$		lb-in	applied moment
	10	w		lb/in/in	ramp magnitude
		$R1$	56.7	lb	left reaction force
		$R2$	123.3	lb	right reaction force
L		q	−40	lb	loading function at point x
L		V	−23.3	lb	shear function at point x
L		M	140	lb-in	moment function at point x
		$Mmin$	0	lb-in	minimum moment
		$Mmax$	147.2	lb-in	maximum moment
		$Vmin$	−117.4	lb	minimum shear load
		$Vmax$	56.7	lb	maximum shear load

* Indicates that a list of values of this variable has been created for plotting.

than having to write and integrate the loading functions for every new beam situation from scratch, the particular problem can often be solved by using superposition, which simply means adding the individual results together. For small deflections, it is safe to assume linearity in these problems and linearity is a requirement for superposition to be valid. For example, the load due to the weight of a beam (ignored in the above examples) can be accounted for by superposing a uniform load over the beam's entire length on whatever other applied loads may be present.

The effects on the shear and moment diagrams of multiple loads on a beam can also be determined by superposition of the individual loadings. If, for example, the beam of Example 3-3 had two point loads applied to it, each at a different distance a, their combined effect could be found by applying the equations of that example twice, once for each load and position, and then adding (superposing) the two results. Appendix D contains a collection of common beam loading situations solved for shear and moment functions giving their equations and plots. Their solutions are also supplied on disk as *TKSolver* files. These solutions can be combined by superposition to accommodate more complicated situations. If you use the supplied files that contain these beam cases as callable functions, they can be superposed within your model to obtain and plot the total shear and moment diagrams, their maxima and minima.

3.10   SUMMARY

Though the student learning about stress analysis for the first time may not think so, the subject of load analysis can often be more difficult and complicated than that of stress analysis. Ultimately, the accuracy of any stress analysis is limited by the quality of our knowledge about the loads on the system, since the stresses are proportional to the loads as will be discussed in Chapter 4.

This chapter has presented a review of Newtonian methods of force and moment analysis for both dynamically and statically loaded systems of a few types. It is by no means a complete treatment of the complex subject of load analysis, and the references in the bibliography of this chapter should be consulted for more detail and for cases not covered here.

The following factors should be kept in mind when attempting to determine loads on a system:

1 Determine the character of the loading in terms of its load class as defined in Section 3.1 in order to decide on whether a static or dynamic load analysis is in order.

2 Draw complete free-body diagrams (FBD) of the system and of as many subsystems within it as are necessary to define the loads acting on its elements. Include all applied moments and torques as well as forces. The importance of a carefully drawn FBD cannot be overemphasized. Most errors in force analysis occur at this step because the FBD is often incorrectly drawn.

3 Write the relevant equations using Newton's laws to define the unknown forces and moments acting on the system. The solution of these equations for most real problems requires some sort of computer tool such as an equation solver or spreadsheet to get satisfactory results in a reasonable time. This is especially true for dynamic systems, which must be solved for a multiplicity of positions in order to determine the maximum loads.

4 The presence of any impact forces can significantly increase the loads on any system. The accurate calculation of forces due to impact is quite difficult. The energy method for impact force estimation presented in this chapter is crude and should be considered an approximation. Detailed information about the deformations of the bodies at impact is needed for a more accurate result, and this may not be available without testing of the actual system under impact. More sophisticated analysis techniques exist for impact force analysis but are beyond the scope of this introductory design text. The reader is referred to the bibliography for more information.

5 Vibration loading can also severely increase the actual loading above the theoretically calculated levels as shown in Case Study 5B. Experimental measurements made under real loading conditions are the best way to develop information in these cases.

The case studies in this chapter are designed to set up realistic problems for stress and failure analysis in the following chapters. Though their complexity may be a bit daunting to the student at first encounter, much benefit can be gained from time spent studying them. This effort will be rewarded with a better understanding of the stress-analysis and failure-theory topics in succeeding chapters.

Important Equations Used in this Chapter

See the referenced sections for information on the proper use of these equations.

Newton's Second Law (Section 3.1):

$$\sum F_x = ma_x \qquad \sum F_y = ma_y \qquad \sum F_z = ma_z \qquad (3.1b)$$

Euler's Equations (Section 3.1):

$$\sum M_x = I_x \alpha_x - \left(I_y - I_z\right)\omega_y \omega_z$$

$$\sum M_y = I_y \alpha_y - \left(I_z - I_x\right)\omega_z \omega_x \qquad (3.1d)$$

$$\sum M_z = I_z \alpha_z - \left(I_x - I_y\right)\omega_x \omega_y$$

Static Loading (Section 3.1):

$$\sum F_x = 0 \qquad \sum F_y = 0 \qquad \sum F_z = 0$$

$$\sum M_x = 0 \qquad \sum M_y = 0 \qquad \sum M_z = 0 \qquad (3.3a)$$

Undamped Natural Frequency (Section 3.8):

$$\omega_n = \sqrt{\frac{k}{m}}$$

$$f_n = \frac{1}{2\pi}\omega_n \qquad (3.4)$$

Damped Natural Frequency (Section 3.8):

$$\omega_d = \sqrt{\frac{k}{m} - \left(\frac{d}{2m}\right)^2}$$

$$f_d = \frac{1}{2\pi}\omega_d \qquad (3.7)$$

Spring Constant (Section 3.8):

$$k = \frac{F}{y} \qquad (3.5b)$$

Viscous Damping (Section 3.8):

$$d = \frac{F}{\dot{y}}$$

(3.6)

Dynamic Force Ratio (Section 3.8):

$$\frac{F_i}{W} = \frac{\delta_i}{\delta_{st}} = 1 + \sqrt{1 + \frac{2\eta h}{\delta_{st}}}$$

(3.14)

Beam Loading, Shear and Moment Functions (Section 3.9):

$$q(x) = \frac{dV}{dx} = \frac{d^2 M}{dx^2}$$

(3.16a)

Integrals of Singularity Functions (Section 3.9):

$$\int_{-\infty}^{x} \langle \lambda - a \rangle^2 d\lambda = \frac{\langle x - a \rangle^3}{3}$$

(3.18a)

$$\int_{-\infty}^{x} \langle \lambda - a \rangle^1 d\lambda = \frac{\langle x - a \rangle^2}{2}$$

(3.18b)

$$\int_{-\infty}^{x} \langle \lambda - a \rangle^0 d\lambda = \langle x - a \rangle^1$$

(3.18c)

$$\int_{-\infty}^{x} \langle \lambda - a \rangle^{-1} d\lambda = \langle x - a \rangle^0$$

(3.18d)

$$\int_{-\infty}^{x} \langle \lambda - a \rangle^{-2} d\lambda = \langle x - a \rangle^{-1}$$

(3.18e)

3.11   REFERENCES

1   **R. L. Norton**, *Design of Machinery: An Introduction to the Synthesis and Analysis of Mechanisms and Machines*. McGraw-Hill: New York, pp. 108-139, 185-264, 1992.

2   *Ibid.*, pp. 623-637.

3   *Ibid.*, pp. 637-662.

4   *Ibid.*, pp. 199.

5   **R. L. Norton, et al.**, "Bearing Forces as a Function of Mechanical Stiffness and Vibration in a Fourbar Linkage," in *Effects of Mechanical Stiffness and Vibration on Wear*, R. G. Bayer, ed. American Society for Testing and Materials: Philadelphia, Pa., 1995.

6   **A. H. Burr and J. B. Cheatham**, *Mechanical Analysis and Design*. 2nd ed. Prentice-Hall: Englewood Cliffs, N.J., pp. 835-863, 1995.

7   **R. J. Roark and W. C. Young**, *Formulas for Stress and Strain*. 6th ed. McGraw-Hill: New York, 1989.

8   **C. R. Wylie and L. C. Barrett**, *Advanced Engineering Mathematics*. 5th ed. McGraw-Hill: New York, 1982.

3.12   BIBLIOGRAPHY

For further review of static and dynamic force analysis see:

R. C. Hibbler, *Engineering Mechanics: Statics*. 7th ed. Prentice-Hall: Englewood Cliffs, N.J., 1995.

R. C. Hibbler, *Engineering Mechanics: Dynamics*. 7th ed. Prentice-Hall: Englewood Cliffs, N.J., 1995.

I. H. Shames, *Engineering Mechanics: Statics and Dynamics*. 3rd ed. Prentice-Hall: Englewood Cliffs, N.J., 1980.

For further information on impact see:

A. H. Burr and **J. B. Cheatham**, *Mechanical Analysis and Design*. 2nd ed., Prentice-Hall, Englewood Cliffs, N.J., Chapter 14, 1995.

W. Goldsmith, *Impact*. Edward Arnold Ltd.: London, 1960.

H. Kolsky, *Stress Waves in Solids*. Dover Publications: New York, 1963.

For further information on vibrations see:

L. Meirovitch, *Elements of Vibration Analysis*. McGraw-Hill: New York, 1975.

For beam loading formulas and tables see:

R. J. Roark and **W. C. Young**, *Formulas for Stress and Strain*. 6th ed. McGraw-Hill: New York, 1989. (Also available as *TKSolver* files for use with that program from Universal Technical Systems, Rockford, Ill.)

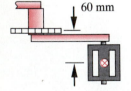

3.13   PROBLEMS

3-1   Which load class from Table 3-1 best suits these systems?

 (a) Bicycle frame   (b) Flagpole   (c) Boat oar
 (d) Diving board   (e) Pipe wrench   (f) Golf club

3-2   Draw free-body diagrams for the systems of Problem 3-1.

†3-3   Draw a free-body diagram of the pedal-arm assembly from a bicycle with the pedal-arms in the horizontal position and dimensions as shown in Figure P3-1. (Consider the two arms, pedals, and pivot as one piece.). Assuming a rider-applied force of 1500 N at the pedal, determine the torque applied to the chain sprocket and the maximum bending moment and torque in the pedal arm.

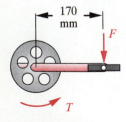

FIGURE P3-1

Problem 3-3

*†3-4   The trailer hitch from Figure 1-1 (p. 12) has loads applied as shown in Figure P3-2. The tongue weight of 100 kg acts downward and the pull force of 4 905 N acts horizontally. Using the dimensions of the ball bracket in Figure 1-5 (p. 15), draw a free-body diagram of the ball bracket and find the tensile and shear loads applied to the two bolts that attach the bracket to the channel in Figure 1-1.

†3-5   For the trailer hitch of Problem 3-4, determine the horizontal force that will result on the ball from accelerating a 2 000-kg trailer to 60 m/sec in 20 sec. Assume a constant acceleration.

*†3-6   For the trailer hitch of Problem 3-4, determine the horizontal force that will result on the ball from an impact between the ball and the tongue of the 2 000-kg trailer if the hitch deflects 2.8 mm dynamically on impact. The tractor weighs 1 000 kg. The velocity at impact is 0.3 m/sec.

*†3-7   The piston of an internal-combustion engine is connected to its connecting rod with a "wrist pin." Find the force on the wrist pin if the 0.5-kg piston has an acceleration of 2 500 g.

*†3-8   A paper mill processes rolls of paper having a density of 984 kg/m^3. The paper roll is 1.50 m outside dia (OD) by 0.22 m inside dia (ID) by 3.23 m long and is on a simply supported, hollow, steel shaft whose lateral spring rate is 1.717E4 N/mm. The deflection at the shaft center is 1 mm under the load of the paper. What is the approximate lateral natural frequency of the combined assembly of paper roll and shaft? *Hint: Assume that the paper roll contributes negligibly to the system stiffness and that the shaft contributes negligibly to the system mass.*

†3-9   A ViseGrip® plier-wrench is drawn to scale in Figure P3-3. Scale the drawing for dimensions. Find the forces acting on each pin and member of the assembly for an assumed clamping force of $P = 4\,000$ N in the position shown. What force F is required to keep it in the clamped position shown? *Note: A similar tool is probably available for inspection in your school's machine shop.*

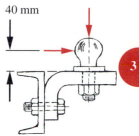

40 mm

FIGURE P3-2

Problems 3-4, 3-5, 3-6

*   Answers to these problems are provided in Appendix G.

†   These problems are extended with similar problems in later chapters with the same –number, e.g., Problem 4-4 is based on Problem 3-4, etc.

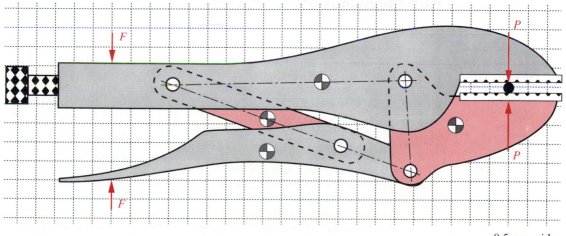

FIGURE P3-3

Problem 3-9

0.5-cm grid

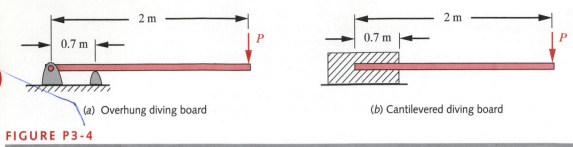

(a) Overhung diving board   (b) Cantilevered diving board

FIGURE P3-4

Problems 3-10 through 3-13

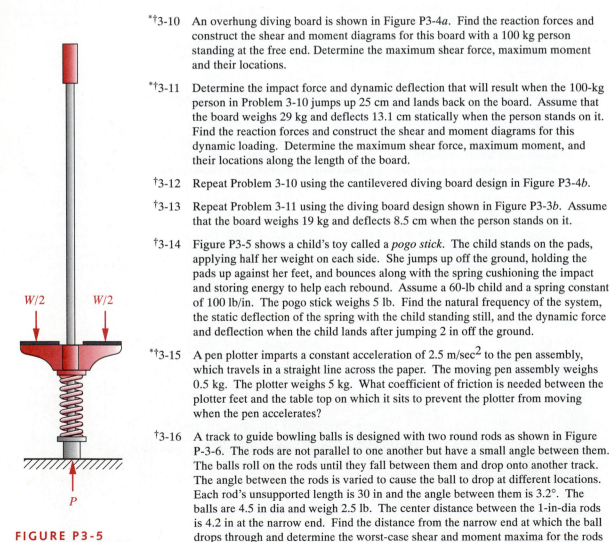

FIGURE P3-5

Problem 3-14

*†3-10   An overhung diving board is shown in Figure P3-4a. Find the reaction forces and construct the shear and moment diagrams for this board with a 100 kg person standing at the free end. Determine the maximum shear force, maximum moment and their locations.

*†3-11   Determine the impact force and dynamic deflection that will result when the 100-kg person in Problem 3-10 jumps up 25 cm and lands back on the board. Assume that the board weighs 29 kg and deflects 13.1 cm statically when the person stands on it. Find the reaction forces and construct the shear and moment diagrams for this dynamic loading. Determine the maximum shear force, maximum moment, and their locations along the length of the board.

†3-12   Repeat Problem 3-10 using the cantilevered diving board design in Figure P3-4b.

†3-13   Repeat Problem 3-11 using the diving board design shown in Figure P3-3b. Assume that the board weighs 19 kg and deflects 8.5 cm when the person stands on it.

†3-14   Figure P3-5 shows a child's toy called a *pogo stick*. The child stands on the pads, applying half her weight on each side. She jumps up off the ground, holding the pads up against her feet, and bounces along with the spring cushioning the impact and storing energy to help each rebound. Assume a 60-lb child and a spring constant of 100 lb/in. The pogo stick weighs 5 lb. Find the natural frequency of the system, the static deflection of the spring with the child standing still, and the dynamic force and deflection when the child lands after jumping 2 in off the ground.

*†3-15   A pen plotter imparts a constant acceleration of 2.5 m/sec^2 to the pen assembly, which travels in a straight line across the paper. The moving pen assembly weighs 0.5 kg. The plotter weighs 5 kg. What coefficient of friction is needed between the plotter feet and the table top on which it sits to prevent the plotter from moving when the pen accelerates?

†3-16   A track to guide bowling balls is designed with two round rods as shown in Figure P-3-6. The rods are not parallel to one another but have a small angle between them. The balls roll on the rods until they fall between them and drop onto another track. The angle between the rods is varied to cause the ball to drop at different locations. Each rod's unsupported length is 30 in and the angle between them is 3.2°. The balls are 4.5 in dia and weigh 2.5 lb. The center distance between the 1-in-dia rods is 4.2 in at the narrow end. Find the distance from the narrow end at which the ball drops through and determine the worst-case shear and moment maxima for the rods as the ball rolls along their length. Assume rods are simply supported at each end.

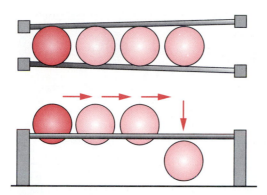

FIGURE P3-6

Problem 3-16

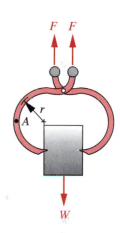

†3-17   A pair of ice tongs is shown in Figure P3-7. The ice weighs 50 lb and is 10 in wide across the tongs. The distance between the handles is 4 in, and the mean radius *r* of a tong is 6 in. Draw free-body diagrams of the two tongs and find all forces acting on them. Determine the bending moment at point *A*.

*3-18   A tractor-trailer tipped over while negotiating an on-ramp to the New York Thruway. The road has a 50-ft radius at that point and tilts 3° toward the outside of the curve. The 45-ft-long by 8 ft wide by 8.5-ft-high trailer box (13 ft from ground to top) was loaded with 44 415 lb of paper rolls in two rows by two high as shown in Figure P3-8. The rolls are 40 in dia by 38 in long and weigh about 900 lb each. They are wedged against backward rolling but not against sidewards sliding. The empty trailer weighed 14 000 lb. The driver claims that he was traveling at less than 15 mph and that the load of paper shifted inside the trailer, struck the trailer sidewall, and tipped the truck. The paper company that loaded the truck claims the load was

FIGURE P3-7

Problem 3-17

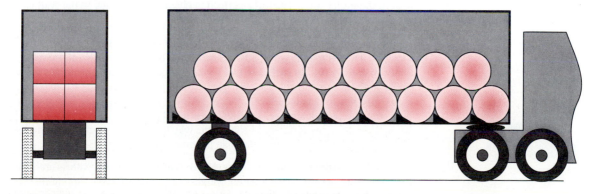

FIGURE P3-8

Problem 3-18

3

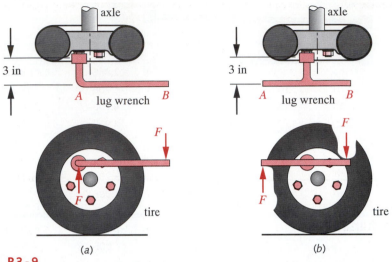

FIGURE P3-9

Problems 3-21 and 3-22

properly stowed and would not shift at that speed. Independent tests of the coefficient of friction between similar paper rolls and a similar trailer floor give a value of 0.43 ± 0.08. The composite center of gravity of the loaded trailer is estimated to be 7.5 ft above the road. Determine the truck speed that would cause the truck to just begin to tip and the speed at which the rolls will just begin to slide sideways. What do you think caused the accident?

3-19   Assume that the CG of the paper rolls in the truck of Problem 3-18 is 2.5 ft above the floor of the trailer. At what speed on the same curve will the pile of rolls tip over (not slide) with respect to the trailer.

3-20   Assume that the load of paper rolls in Problem 3-18 will slide sideways at a truck speed of 20 mph on the curve in question. Estimate the impact force of the cargo against the trailer wall. The force-deflection characteristic of the trailer wall has been measured as approximately 400 lb/in.

†3-21   Figure P3-9 shows an automobile wheel with two common styles of lug wrench being used to tighten the wheel nuts, a single-ended wrench in (a), and a double-ended wrench in (b). In each case two hands are required to provide forces respectively at A and B as shown. The distance between points A and B is 1 ft in both cases. The wheel nuts require a torque of 70 ft-lb. Draw free-body diagrams for both wrenches and determine the magnitudes of all forces and moments on each wrench. Is there any difference between the way these two wrenches perform their assigned task? Is one design better than the other? If so, why? Explain.

*†3-22   A roller-blade skate is shown in Figure P3-10. The polyurethane wheels are 72 mm dia. The skate-boot-foot combination weighs 2 kg. The effective "spring rate" of the person-skate system is 6000 N/m. Find the forces on the wheels' axles for a 100-kg person landing a 0.5-m jump on one foot. (a) Assume that all 4 wheels land simultaneously. (b) Assume that one wheel absorbs all the landing force.

FIGURE P3-10

Problem 3-22

Table P3-1   Data for Problems 3-23 through 3-26

Use only data relevant to the particular problem. Lengths in m, forces in N, I in m^4.

Row	l	a	b	w	F	I	c	E
a	1.00	0.40	0.60	200	500	2.85E–08	2.00E–02	steel
b	0.70	0.20	0.40	80	850	1.70E–08	1.00E–02	steel
c	0.30	0.10	0.20	500	450	4.70E–09	1.25E–02	steel
d	0.80	0.50	0.60	65	250	4.90E–09	1.10E–02	steel
e	0.85	0.35	0.50	96	750	1.80E–08	9.00E–03	steel
f	0.50	0.18	0.40	450	950	1.17E–08	1.00E–02	steel
g	0.60	0.28	0.50	250	250	3.20E–09	7.50E–03	steel
h	0.20	0.10	0.13	400	500	4.00E–09	5.00E–03	alum
i	0.40	0.15	0.30	50	200	2.75E–09	5.00E–03	alum
j	0.20	0.10	0.15	150	80	6.50E–10	5.50E–03	alum
k	0.40	0.16	0.30	70	880	4.30E–08	1.45E–02	alum
l	0.90	0.25	0.80	90	600	4.20E–08	7.50E–03	alum
m	0.70	0.10	0.60	80	500	2.10E–08	6.50E–03	alum
n	0.85	0.15	0.70	60	120	7.90E–09	1.00E–02	alum

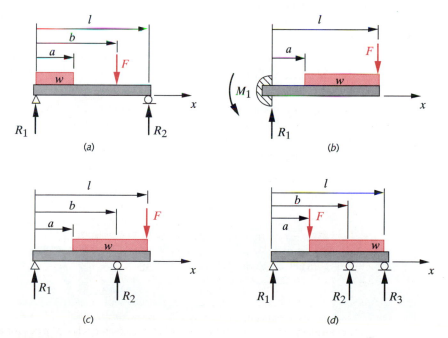

(a)

(b)

(c)

(d)

FIGURE P3-11

Beams and Beam Loadings for Problems 3-23 to 3-26 - see Table P3-1 for Data

* Answers to these problems are provided in Appendix G.

† These problems are extended with similar problems in later chapters with the same –number, e.g., Problem 4-4 is based on Problem 3-4, etc.

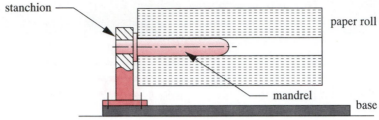

FIGURE P3-12

Problem 3-27

*†3-23   A beam is supported and loaded as shown in Figure P3-11*a*. Find the reactions, maximum shear, and maximum moment for the data given in the row(s) assigned from Table P3-1.

*†3-24   A beam is supported and loaded as shown in Figure P3-11*b*. Find the reactions, maximum shear, and maximum moment for the data given in the row(s) assigned from Table P3-1.

*†3-25   A beam is supported and loaded as shown in Figure P3-11*c*. Find the reactions, maximum shear, and maximum moment for the data given in the row(s) assigned from Table P3-1.

*†3-26   A beam is supported and loaded as shown in Figure P3-11*d*. Find the reactions, maximum shear, and maximum moment for the data given in the row(s) assigned from Table P3-1.

†3-27   A storage rack to hold the paper roll of Problem 3-8 is shown in Figure P3-12. Determine the reactions and draw the shear and moment diagrams for the mandrel that extends 50% into the roll.

†3-28   Figure P3-13 shows a forklift truck negotiating a 15° ramp to drive onto a 4-ft-high loading platform. The truck weighs 5000 lb and has a 42-in wheelbase. Determine the reactions and draw the shear and moment diagrams for the worst case of loading as the truck travels up the ramp.

*   Answers to these problems are provided in Appendix G.

† These problems are extended with similar problems in later chapters with the same –number, e.g., Problem 4-4 is based on Problem 3-4, etc.

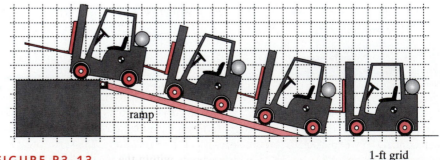

1-ft grid

FIGURE P3-13

Problem 3-28

STRESS, STRAIN, AND DEFLECTION

'Tis not knowing much, but what is useful that makes a wise man.

THOMAS FULLER, M.D.

4.0 INTRODUCTION

You have probably had a first course in stress analysis (perhaps called *Strength of Materials* or *Mechanics of Materials*) and thus should understand the fundamentals of that subject. Nevertheless, this chapter will present a review of the basics in order to set the stage for the topic of fatigue analysis in later chapters. Stress and strain were discussed in Chapter 2 on materials properties but were incompletely defined at that juncture. In this chapter we will present a more complete definition of what is meant by the terms stress, strain, and deflection.

Table 4-0 shows the variables used in this chapter and references the equations, tables, or sections in which they are used. At the end of the chapter, a summary section is provided that also groups all the significant equations from this chapter for easy reference and identifies the chapter section in which their discussion can be found.

4.1 STRESS

Stress was defined in Chapter 2 as force per unit area with units of psi or MPa. In a part subjected to some forces, stress is generally distributed as a continuously varying function within the continuum of material. Every infinitesimal element of the material can conceivably experience different stresses at the same time. Thus, we must look at stresses as acting on vanishingly small elements within the part. These infinitesimal elements are typically modeled as cubes, shown in Figure 4-2. The stress components

4

Table 4-0　Variables Used in this Chapter
Part 1 of 2

Symbol	Variable	ips units	SI units	See
A	area	in^2	m^2	Sect. 4.8, 4.10, 4.12
b	width of beam cross section	in	m	Sect. 4.10
c	distance to outer fiber - straight beam	in	m	Eq. 4.11b
c_i	distance to inner fiber - curved beam	in	m	Eq. 4.12
c_o	distance to outer fiber - curved beam	in	m	Eq. 4.12
d	diameter of cross section	in	m	Sect. 4.11
E	Young's modulus	psi	Pa	Sect. 4.8, 4.10, 4.12
e	eccentricity of a column	in	m	Sect. 4.14
e	shift of neutral axis - curved beam	in	m	Sect. 4.10, Eq. 4.12a
F	force or load	lb	N	Sect. 4.12
G	shear modulus, modulus of rigidity	psi	Pa	Sect. 4.11
h	depth of beam cross section	in	m	Sect. 4.10
I	second moment of area	in^4	m^4	Eq. 4.11a
K	geometry parameter - torsion	in^4	m^4	Eq. 4.26b, Table 4-7
k	radius of gyration	in	m	Sect. 4.14
K_t	geometric stress conc. factor - normal stress	none	none	Sect. 4.13
K_{ts}	geometric stress conc. factor - shear stress	none	none	Sect. 4.13
l	length	in	m	Sect. 4.8, 4.10, 4.12
M	moment, moment function	lb-in	N-m	Sect. 4.10
P	force or load	lb	N	Sect. 4.8
P_{cr}	critical column load	lb	N	Sect. 4.14
q	beam loading function	lb	N	Sect. 4.10
Q	integral of 1st moment of area - beam	in^3	m^3	Eq. 4.13
Q	geometry parameter - torsion	in^3	m^3	Eq. 4.26a, Table 4-7
r	radius - general	in	m	Sect. 4.10, Eq. 4.12
r_i	inside radius of curved beam	in	m	Eq. 4.12
r_o	outside radius of curved beam	in	m	Eq. 4.12
S_r	slenderness ratio - column	none	none	Sect. 4.14
S_y	yield strength	psi	Pa	Sect. 4.14
T	torque	lb-in	N-m	Sect. 4.12
V	beam shear function	lb	N	Sect. 4.10
x	generalized length variable	in	m	Sect. 4.10
y	distance from neutral axis - beam	in	m	Eq. 4.11a
y	deflection - general	in	m	Sect. 4.10, 4.12
Z	section modulus	in^3	m^3	Eq. 4.11d

Table 4-0   Variables Used in this Chapter

Part 2 of 2

Symbol	Variable	ips units	SI units	See
x, y, z	generalized coordinates	any	any	Sect. 4.1, 4.2
ε	strain	none	none	Sect. 4.2
θ	beam slope	rad	rad	Sect. 4.10
θ	angular deflection - torsion	rad	rad	Sect. 4.12
σ	normal stress	psi	Pa	Sect. 4.2
σ_1	principal stress	psi	Pa	Sect. 4.4
σ_2	principal stress	psi	Pa	Sect. 4.4
σ_3	principal stress	psi	Pa	Sect. 4.4
τ	shear stress	psi	Pa	Sect. 4.2
τ_{13}	maximum shear stress	psi	Pa	Sect. 4.4
τ_{21}	principal shear stress	psi	Pa	Sect. 4.4
τ_{32}	principal shear stress	psi	Pa	Sect. 4.4

are considered to be acting on the faces of these cubes in two different manners. **Normal stresses** act perpendicular (i.e., normal) to the face of the cube and tend to either pull it out (tensile normal stress) or push it in (compressive normal stress). **Shear stresses** act parallel to the faces of the cubes, in pairs (couples) on opposite faces, which tends to distort the cube into a rhomboidal shape. This is analogous to grabbing both pieces of bread of a peanut-butter sandwich and sliding them in opposite directions. The peanut butter will be sheared as a result. These normal and shear components of stress acting on an infinitesimal element make up the terms of a **tensor**.[*]

Stress is a tensor of order two[†] and thus requires nine values or components to describe it in three dimensions. The 3-D stress tensor can be expressed as the matrix:

$$\begin{bmatrix} \sigma_{xx} & \tau_{xy} & \tau_{xz} \\ \tau_{yx} & \sigma_{yy} & \tau_{yz} \\ \tau_{zx} & \tau_{zy} & \sigma_{zz} \end{bmatrix} \qquad (4.1a)$$

where the notation for each stress component contains three elements, a magnitude (either σ or τ), the direction of a normal to the reference surface (first subscript) and a direction of action (second subscript). We will use σ to refer to normal stresses and τ for shear stresses.

Many elements in machinery are subjected to three-dimensional stress states and thus require the stress tensor of equation 4.1a. There are some special cases, however, which can be treated as two-dimensional stress states.

[*] For a discussion of tensor notation, see C. R., Wylie and L. C. Barrett, *Advanced Engineering Mathematics*, 5th ed., McGraw-Hill, New York, 1982.

[†] It is more correctly a tensor for rectilinear Cartesian coordinates. The more general tensor notation for curvilinear coordinate systems will not be used here.

The stress tensor for 2-D is

$$\begin{bmatrix} \sigma_{xx} & \tau_{xy} \\ \tau_{yx} & \sigma_{yy} \end{bmatrix} \tag{4.1b}$$

Figure 4-1 shows an infinitesimal cube of material taken from within the material continuum of a part that is subjected to some 3-D stresses. The faces of this infinitesimal cube are made parallel to a set of xyz axes taken in some convenient orientation. The orientation of each face is defined by its surface normal vector[†] as shown in Figure 4-1a. The x face has its surface normal parallel to the x axis, etc. Note that there are thus two x faces, two y faces, and two z faces, one of each being positive and one negative as defined by the sense of its surface normal vector.

The nine stress components are shown acting on the surfaces of this infinitesimal element in Figure 4-1b and c. The components σ_{xx}, σ_{yy}, and σ_{zz} are the normal stresses, so-called because they act, respectively, in directions normal to the x, y, and z surfaces of the cube. The components τ_{xy} and τ_{xz}, for example, are shear stresses that act on the x face and whose directions of action are parallel to the y and z axes, respectively. The sign of any one of these components is defined as positive if the signs of its surface normal and its stress direction are the same, and as negative if they are different. Thus the components shown in Figure 4-1b are all positive because they are acting on the positive faces of the cube and their directions are also positive. The components shown in Figure 4-1c are all negative because they are acting on the positive faces of the cube and their directions are negative. This sign convention makes tensile normal stresses positive and compressive normal stresses negative.

For the 2-D case, only one face of the stress cube may be drawn. If the x and y directions are retained and z eliminated, we look normal to the xy plane of the cube of Figure 4-1 and see the stresses shown in Figure 4-2, acting on the unseen faces of the cube. The reader should confirm that the stress components shown in Figure 4-2 are all positive by the sign convention stated above.

Note that the definition of the dual subscript notation given above is consistent when applied to the normal stresses. For example, the normal stress σ_{xx} acts on the x face and is also in the x direction. Since the subscripts are simply repeated for normal stresses, it is common to eliminate one of them and refer to the normal components simply as σ_x, σ_y, and σ_z. Both subscripts are needed to define the shear stress components and they will be retained. It can also be shown[1] that the stress tensor is symmetric, which means that

$$\tau_{xy} = \tau_{yx}$$
$$\tau_{yz} = \tau_{zy} \tag{4.2}$$
$$\tau_{zx} = \tau_{xz}$$

This reduces the number of stress components to be calculated.

[†] A surface normal vector is defined as "growing out of the surface of the solid in a direction normal to that surface." Its sign is defined as the sense of this surface normal vector in the local coordinate system.

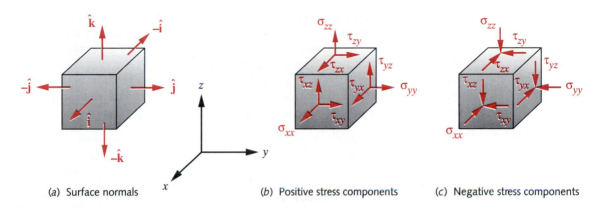

(a) Surface normals   (b) Positive stress components   (c) Negative stress components

FIGURE 4-1

The Stress Cube, its Surface Normals, and its Stress Components

4.2   STRAIN

Stress and strain are linearly related by Hooke's law in the elastic region of most engineering materials as discussed in Chapter 2. Strain is also a second-order tensor and can be expressed for the 3-D case as

$$\begin{bmatrix} \varepsilon_{xx} & \varepsilon_{xy} & \varepsilon_{xz} \\ \varepsilon_{yx} & \varepsilon_{yy} & \varepsilon_{yz} \\ \varepsilon_{zx} & \varepsilon_{zy} & \varepsilon_{zz} \end{bmatrix} \tag{4.3a}$$

and for the 2-D case as

$$\begin{bmatrix} \varepsilon_{xx} & \varepsilon_{xy} \\ \varepsilon_{yx} & \varepsilon_{yy} \end{bmatrix} \tag{4.3b}$$

where ε represents either a normal or a shear strain, the two being differentiated by their subscripts. We will also simplify the repeated subscripts for normal strains to ε_x, ε_y, and ε_z for convenience while retaining the dual subscripts to identify shear strains.

4.3   PRINCIPAL STRESSES

The axis systems taken in Figures 4-1 and 4-2 are arbitrary and are usually chosen for convenience in computing the applied stresses. For any particular combination of applied stresses, there will be a continuous distribution of the stress field around any point analyzed.[*] The normal and shear stresses at the point will vary with direction in any coordinate system chosen. There will always be planes on which the shear stress components are zero. The normal stresses acting on these planes are called the principal

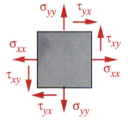

FIGURE 4-2

Two-Dimensional Stress Element

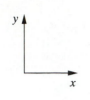

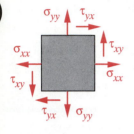

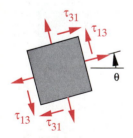

stresses. The planes on which these principal stresses act are called the **principal planes**. The directions of the surface normals to the principal planes are called the **principal axes** and the normal stresses acting in those directions are the **principal normal stresses**. There will also be another set of mutually perpendicular axes along which the shear stresses will be maximal. The **principal shear stresses** act on a set of planes that are at 45° angles to the planes of the principal normal stresses. The principal planes and principal stresses for the 2-D case of Figure 4-2 are shown in Figure 4-3.

Since, from an engineering standpoint, we are most concerned with designing our machine parts so that they will not fail, and since failure will occur if the stress at any point exceeds some safe value, we need to find the largest stresses (both normal and shear) that occur anywhere in the continuum of material that makes up our machine part. We may be less concerned with the directions of those stresses than with their magnitudes as long as the material can be considered to be at least macroscopically isotropic, thus having strength properties that are uniform in all directions. Most metals and many other engineering materials meet these criteria, although wood and composite materials are notable exceptions.

The expression relating the applied stresses to the principal stresses is

$$\begin{bmatrix} \sigma_x - \sigma & \tau_{xy} & \tau_{xz} \\ \tau_{yx} & \sigma_y - \sigma & \tau_{yz} \\ \tau_{zx} & \tau_{zy} & \sigma_z - \sigma \end{bmatrix} \begin{bmatrix} n_x \\ n_y \\ n_z \end{bmatrix} = 0 \qquad (4.4a)$$

where σ is the principal stress magnitude and n_x, n_y, and n_z are the direction cosines of the unit vector **n**, which is normal to the principal plane:

$$\hat{\mathbf{n}} \cdot \hat{\mathbf{n}} = 1$$
$$\hat{\mathbf{n}} = n_x\hat{\mathbf{i}} + n_y\hat{\mathbf{j}} + n_z\hat{\mathbf{k}} \qquad (4.4b)$$

For the solution of equation 4.4a to exist, the determinant of the coefficient matrix must be zero. Expanding this determinant and setting it to zero we obtain

$$\sigma^3 - C_2\sigma^2 - C_1\sigma - C_0 = 0$$

where

$$C_2 = \sigma_x + \sigma_y + \sigma_z$$
$$C_1 = \tau_{xy}^2 + \tau_{yz}^2 + \tau_{zx}^2 - \sigma_x\sigma_y - \sigma_y\sigma_z - \sigma_z\sigma_x \qquad (4.4c)$$
$$C_0 = \sigma_x\sigma_y\sigma_z + 2\tau_{xy}\tau_{yz}\tau_{zx} - \sigma_x\tau_{yz}^2 - \sigma_y\tau_{zx}^2 - \sigma_z\tau_{xy}^2$$

Equation 4.4c is a cubic polynomial in σ. The coefficients C_0, C_1, and C_2 are called the tensor invariants because they have the same values regardless of the initial choice of *xyz* axes in which the applied stresses were measured or calculated. The three principal (normal) stresses σ_1, σ_2, σ_3 are the three roots of this cubic polynomial The roots

(a) Applied stresses

(b) Principal normal stresses

(c) Principal shear stresses

FIGURE 4-3

Principal Stresses on a Two-Dimensional Stress Element

of this polynomial are always real[2] and are usually ordered such that $\sigma_1 > \sigma_2 > \sigma_3$. If needed, the directions of the principal stress vectors can be found by substituting each root of equation 4.4c into 4.4a and solving for n_x, n_y, and n_z for each of the three principal stresses. The directions of the three principal stresses are mutually orthogonal.

The principal shear stresses can be found from the values of the principal normal stresses using

$$\tau_{13} = \frac{|\sigma_1 - \sigma_3|}{2}$$

$$\tau_{21} = \frac{|\sigma_2 - \sigma_1|}{2} \qquad (4.5)$$

$$\tau_{32} = \frac{|\sigma_3 - \sigma_2|}{2}$$

If the principal normal stresses have been ordered as shown above, then $\tau_{max} = \tau_{13}$. The directions of the planes of the principal shear stresses are 45° from those of the principal normal stresses and are also mutually orthogonal.

The solution of equation 4.4c for its three roots can be done trigonometrically by Viete's method* or using an iterative root finding algorithm. The *TKSolver* file STRESS3D provided solves equation 4.4c and finds the three principal-stress roots by Viete's method and orders them by the above convention. STRESS3D also computes the stress function (Eq. 4.4c) for a list of user-defined values of σ and then plots that function. The root crossings can be seen on the plot. Figure 4-4 shows the stress function for an arbitrary set of applied stresses plotted over a range of values of σ that includes all three roots. Table 4-1 shows the *TKSolver* variable sheet for the computation.

For the special case of a two-dimensional stress state, the equations 4.4c for principal stress reduce to†

$$\sigma_a, \sigma_b = \frac{\sigma_x + \sigma_y}{2} \pm \sqrt{\left(\frac{\sigma_x - \sigma_y}{2}\right)^2 + \tau_{xy}^2}$$

$$\sigma_c = 0 \qquad (4.6a)$$

The two nonzero roots calculated from equation 4.6a are temporarily labeled σ_a and σ_b and the third root σ_c is always zero in the 2-D case. Depending on their resulting values, the three roots are then labeled according to the convention: *algebraically largest* = σ_1, *algebraically smallest* = σ_3, and *other* = σ_2. Using equation 4.6a to solve the example shown in Figure 4-4 would yield values of $\sigma_1 = \sigma_a$, $\sigma_3 = \sigma_b$, and $\sigma_2 = \sigma_c = 0$ as labeled in the figure.‡ Of course, equation 4.4c for the 3-D case can still be used to solve any two-dimensional case. One of the three principal stresses found will then be zero. The example in Figure 4-4 is of a two-dimensional case solved with equation 4.4c. Note the root at $\sigma = 0$.

* See *Numerical Recipes* by Press et al., Cambridge Univ. Press, 1986, p. 146, or *Standard Mathematical Tables*, CRC Press. 22nd Ed., 1974, p. 104, or any collection of standard mathematical formulas.

† Equations 4.6 can also be used when one principal stress is nonzero but is directed along one of the axes of the *xyz* coordinate system chosen for calculation. The stress cube of Figure 4-2 is then rotated about one principal axis to determine the angles of the other two principal planes.

‡ If the 3-D numbering convention is strictly observed in the 2-D case, then sometimes the two nonzero principal stresses will turn out to be σ_1 and σ_3 if they are opposite in sign (as in Example 4-1). Other times they will be σ_1 and σ_2 when they are both positive and the smallest (σ_3) is zero (as in Example 4-2). A third possibility is that both nonzero principal stresses are negative (compressive) and the algebraically largest of the set (σ_1) is then zero. Equation 4.6 arbitrarily calls the two nonzero 2-D principal stresses σ_a and σ_b with the remaining one (σ_c) reserved for the zero member of the trio. Application of the standard convention can result in σ_a and σ_b being called any of the possible combinations σ_1 and σ_2, σ_1 and σ_3, or σ_2 and σ_3 depending on their relative values. See Examples 4-1 and 4-2.

4

stress function

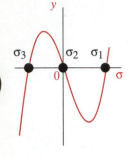

FIGURE 4-4

The Three Roots of the
Stress Function for a
Plane Stress Case

Table 4-1   Solution to the Cubic Stress Function for a Plane Stress Case
See *TKSolver* File STRESS3D.TK

Input	Variable	Output	Unit	Comments
1000	σ_{xx}		psi	applied normal stress in x direction
750	σ_{yy}		psi	applied normal stress in y direction
0	σ_{zz}		psi	applied normal stress in z direction
500	τ_{xy}		psi	applied shear stress in xy direction
0	τ_{yz}		psi	applied shear stress in yz direction
0	τ_{zx}		psi	applied shear stress in zx direction
	C_2	250		coefficient of σ^2 term
	C_1	1.0 E6		coefficient of σ^1 term
	C_0	0		coefficient of σ^0 term
G *	σ_1	1133	psi	principal stress root #1
G	σ_2	0	psi	principal stress root #2
G	σ_3	−883	psi	principal stress root #3

* Indicates that a guess value was required to start the iteration to the root of the equation.

Once the three principal stresses are found and ordered as described above, the maximum shear stress is found from equation 4.5:

$$\tau_{max} = \tau_{13} = \frac{|\sigma_1 - \sigma_3|}{2} \qquad (4.6b)$$

4.4   PLANE STRESS AND PLANE STRAIN

The general state of stress and strain is three-dimensional but there exist particular geometric configurations that can be treated differently.

Plane Stress

The two-dimensional, or biaxial, stress state is also called plane stress. **Plane stress** requires that one principal stress be zero. This condition is common in some applications. For example, a thin plate or shell may also have a state of plane stress away from its boundaries or points of attachment. These cases can be treated with the simpler approach of equations 4.6.

Plane Strain

There are principal strains associated with the principal stresses. If one of the principal strains (say ε_3) is zero, and if the remaining strains are independent of the dimension along its principal axis, $\mathbf{n}_3$, it is called **plane strain**. This condition occurs in particular geometries. For example, if a long, solid, prismatic bar is loaded only in the transverse direction, regions within the bar that are distant from any end constraints will see essentially zero strain in the direction along the axis of the bar and be in plane strain. (However, the stress is not zero in the zero-strain direction.)   A long, hydraulic dam can be considered to have a plane strain condition in regions well removed from its ends or base at which it is attached to surrounding structures.

4.5   MOHR'S CIRCLES

Mohr's circles[*] have long provided a means to do a graphical solution of equation 4.6 and find the principal stresses for the plane-stress case. Many textbooks on machine design present the Mohr's circle method as a primary solution technique for determining principal stresses. Before the advent of programmable calculators and computers, Mohr's graphical method was a reasonable and practical way to solve equation 4.6. Now, however, it is more practical to find the principal stresses numerically. Nevertheless, we present the graphical method for several reasons. It can serve as a quick check on a numerical solution, and it may be the only viable method if the power to your computer fails or your calculator's batteries go dead. It also serves the useful purpose of providing a visual presentation of the stress state at a point.

Mohr's circles exist for the three-dimensional stress case as well, but a graphical construction method is not available to create them directly from the applied-stress data except for the special case where one of the principal stresses is coincident with an axis of the *xyz* coordinate system chosen, i.e., where one plane is a plane of principal stress. However, once the principal stresses are calculated from equation 4.4c by a suitable root-finding technique, the 3-D Mohr's circles can be drawn by using the calculated principal stresses. A computer program called MOHR is provided on disk for that purpose. In the special 3-D stress case where one principal stress lies along a coordinate axis, the three Mohr's circles can be graphically constructed.

The Mohr plane, on which Mohr's circles are drawn, is arranged with its axes drawn mutually perpendicular, but the angle between them represents 180° in real space. All angles drawn on the Mohr plane are double their value in real space. The abscissa is the axis of all normal stresses. The applied normal stresses σ_x, σ_y, and σ_z are plotted along this axis and the principal stresses σ_1, σ_2, and σ_3 are also found on this axis. The ordinate is the axis of all shear stresses. It is used to plot the applied shear stresses τ_{xy}, τ_{yx}, and τ_{xz} and to find the maximum shear stress.[†] Mohr used a sign convention for shear stresses that makes *cw* shear couples positive, *which is **not consistent** with the now-standard right-hand rule*. Nevertheless, this left-handed convention is still used for his circles. The best way to demonstrate the use of Mohr's circle is with examples.

[*] Devised by the German engineer, Otto Mohr (1835-1918). His circles are also used for the coordinate transformation of strains, and area moments and products of inertia.

[†] The fact that Mohr used the same axes to plot more than one variable is one of the sources of confusion to students when they first encounter this method. Just remember that all σ's are plotted on the horizontal axis whether they are applied normal stresses (σ_x, σ_y, σ_z) or principal stresses (σ_1, σ_2, σ_3) and all τ's are plotted on the vertical axis whether they are applied shear stresses (τ_{xy}, etc.) or maximum shear stresses (τ_{12}, etc.). The Mohr axes are **not** conventional Cartesian axes.

4

EXAMPLE 4-1

Determining Principal Stresses Using Mohr's Circles

Problem: A biaxial stress element as shown in Figure 4-2 has σ_x = 40 000 psi, σ_y = –20 000 psi, and τ_{xy} = 30 000 psi *ccw*. Use Mohr's circles to determine the principal stresses. Check the result with a numerical method.

Solution: See Figures 4-2 and 4-5.

1   Construct the Mohr-plane axes as shown in Figure 4-5*b* and label them σ and τ.

2   Lay off the given applied stress σ_x (as line *OA*) to any convenient scale along the normal stress (horizontal) axis. Note that σ_x is a tensile (positive) stress in this example.

3   Lay off the given applied stress σ_y (as line *OB*) to scale along the normal stress axis. Note that σ_y is a compressive (negative) stress in this example.

4   Figure 4-2 shows that the pair of shear stresses τ_{xy} create a *ccw* couple on the element. This couple is balanced for equilibrium by the *cw* couple provided by the shear stresses τ_{yx}. Recall that both of these shear stresses, τ_{xy} and τ_{yx}, are equal in magnitude according to equation 4.2 and are positive according to the stress sign convention. But, instead of using the stress sign convention, they are plotted on the Mohr circle according to the rotation that they imply to the element, using Mohr's left-handed sign convention of *cw*+ and *ccw*–.

5   Draw a vertical line downward (*ccw*–) from the tip of σ_x (as line *AC*) to represent the scaled magnitude of τ_{xy}. Draw a vertical line upward (*cw*+) from the tip of σ_y (as line *BD*) to represent the scaled magnitude of τ_{yx}.

6   The diameter of one Mohr's circle is the distance from point *C* to point *D*. Line *AB* bisects *CD*. Draw the circle using this intersection as a center.

7   Two of the three principal normal stresses are then found at the two intersections that this Mohr's circle makes with the normal stress axis at points P_1 and P_3: σ_1 = 52 426 psi at P_1 and σ_3 = –32 426 psi at P_3.

8   Since there were no applied stresses in the *z* direction in this example, it is a 2-D stress state and the third principal stress, σ_2, is zero, located at point *O,* which is also labeled P_2.

9   There are still two other Mohr's circles to be drawn. The three Mohr's circles are defined by the diameters (σ_1– σ_3), (σ_1– σ_2), and (σ_2– σ_3), which are the lines P_1P_3, P_2P_1, and P_2P_3. The three circles are shown in Figure 4-5*c*.

10 Extend horizontal tangent lines from the top and bottom extremes of each Mohr's circle to intersect the shear (vertical) axis. This determines the values of the principal shear stresses associated with each pair of principal normal stresses: τ_{13} = 42 426, τ_{12} = 26 213, and τ_{23} = 16 213 psi. Note that despite having only two nonzero principal normal stresses, there are three nonzero principal shear stresses. However, only the largest of these, τ_{max} = τ_{13} = 42 426 psi is of interest for design purposes.

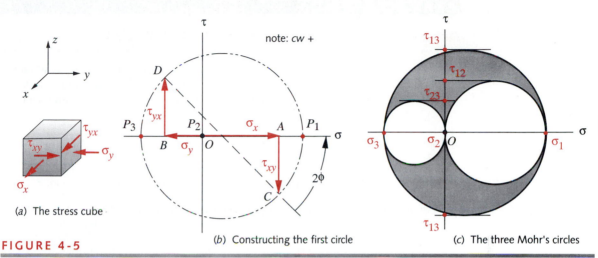

FIGURE 4-5

The Stress Cube and Mohr's Circles for Example 4-1

11 We can also determine the angles (with respect to our original *xyz* axes) of the principal normal and principal shear stresses from the Mohr's circle. These angles are only of academic interest if the material is homogeneous and isotropic. If it is not isotropic, its material properties are direction dependent and the directions of the principal stresses are then important. The angle $2\phi = -45°$ in Figure 4-5a represents the orientation of the principal normal stress with respect to the *x* axis of our original system. Note that the line *DC* on the Mohr plane is the *x* axis in real space and the angles are measured according to Mohr's left-handed convention (*cw*+). Since angles on the Mohr plane are double those in real space, the angle of the principal stress σ_1 with respect to the real space *x* axis, is $\phi = -22.5°$. The stress σ_3 will be 90° from σ_1 and the maximum shear stress τ_{13} will be 45° from the σ_1 axis in real space.

A computer program called MOHR has been written and is included, on disk, with this text. Program MOHR allows the input of any set of applied stresses and computes the principal normal and shear stresses using equations 4.4 and 4.5. The program then plots the Mohr's circles and will also display the stress function in the vicinity of the three principal stress roots. See the appendices for instructions on running the program. Data files that can be read into this program are also supplied. Input the file EX04-01.IPM to see the analytical solution to the example above. Alternatively, you may input the file EX04-01 to the *TKSolver* program supplied. This will calculate the principal stresses and plot the cubic stress function for Example 4-1.

We will now change the previous example only slightly to show the need for drawing all three Mohr's circles even in the plane stress case. The change of significance makes the applied stresses σ_x and σ_y both positive, instead of opposite in sign.

4

EXAMPLE 4-2

Determining Plane Stresses Using Mohr's Circles

Problem: A biaxial stress element as shown in Figure 4-2 has $\sigma_x = 40\ 000$ psi, $\sigma_y = 20\ 000$ psi, and $\tau_{xy} = 10\ 000$ psi *ccw*. Use Mohr's circles to determine the principal stresses. Check the result with a numerical method.

Solution: See Figures 4-2 and 4-6, *TKSolver* file EX04-02, and MOHR program file EX04-02.IPM.

1   Construct the Mohr-plane axes as shown in Figure 4-6 and label them σ and τ.

2   Lay off the given applied stress σ_x (as line *OA*) to scale along the normal stress (horizontal) axis. Note that σ_x is a tensile (positive) stress in this example.

3   Lay off the given applied stress σ_y (as line *OB*) to scale along the normal stress axis. Note that σ_y is also a tensile (positive) stress in this example and so lies in the same direction as σ_x along the σ axis.

4   Figure 4-2 shows that the shear stresses τ_{xy} create a *ccw* couple on the element. This couple is balanced for equilibrium by the *cw* couple provided by the shear stresses τ_{yx}. Recall that both of these shear stresses, τ_{xy} and τ_{yx}, are equal according to equation 4.2 and are positive according to the stress sign convention. But, instead of using the stress sign convention, they are plotted on the Mohr circle according to the rotation that they imply to the element, using Mohr's left-handed sign convention of *cw+* and *ccw−*.

5   Draw a vertical line downward (*ccw−*) from the tip of σ_x (as line *AC*) to represent the scaled magnitude of τ_{xy}. Draw a vertical line upward (*cw+*) from the tip of σ_y (as line *BD*) to represent the scaled magnitude of σ_{yx}.

6   The diameter of one Mohr's circle is the distance from point *C* to point *D*. Line *AB* bisects *CD*. Draw the circle using this intersection as a center.

7   Two of the three principal normal stresses are then found at the two intersections that this Mohr's circle makes with the normal stress axis at points P_1 and P_2: $\sigma_1 = 44\ 142$ and $\sigma_2 = 15\ 858$ psi. Note that if we stop at this point, the maximum shear stress appears to be $\tau_{12} = 14\ 142$ psi as defined by the projection of a horizontal tangent from the top of the one circle to the τ axis, as shown in Figure 4-6b.

8   Since there were no applied stresses in the z direction in this example, it is a 2-D stress state and the third principal stress, σ_3, is known to be zero, thus is located at point *O,* also labeled P_3.

9   There are still two other Mohr's circles to be drawn. The three Mohr's circles are defined by the diameters $(\sigma_1 - \sigma_3)$, $(\sigma_1 - \sigma_2)$, and $(\sigma_2 - \sigma_3)$, which in this case, are the lines P_1P_3, P_1P_2, and P_2P_3 as shown in Figure 4-6.

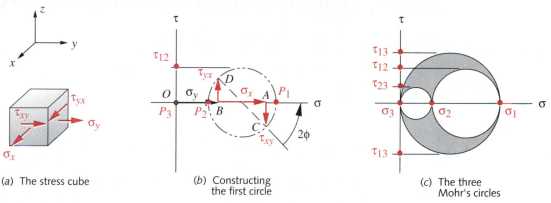

(a) The stress cube

(b) Constructing the first circle

(c) The three Mohr's circles

FIGURE 4-6

The Stress Cube and Mohr's Circles for Example 4-2

10   Extend horizontal tangent lines from the top and bottom extremes of each Mohr's circle to intersect the shear (vertical) axis. This determines the value of the principal shear stress associated with each pair of principal normal stresses: i.e., $\tau_{13} = 22\,071$, $\tau_{12} = 14\,142$, and $\tau_{23} = 7\,929$ psi. The largest of these is $\tau_{max} = 22\,071$, not the value $14\,142$ found in step 7.

11   Note that it is always the circle lying between the largest and smallest principal stresses that determines the maximum shear stress. In the previous example the zero principal stress was not the smallest of the three because one principal stress was negative. In the present example, the zero principal stress is the smallest. Thus, failing to draw all three circles would have led to a serious error in the value of τ_{max}.

The previous two examples point out some uses and limitations of the Mohr's circle approach to plane stress calculations. From a practical standpoint, as long as computational resources (at least in the form of a programmable pocket calculator) are available to the modern engineer, the analytical solution (Eq. 4.4c) is the preferred method of solution for determining principal stresses. It is universal (serving for plane stress, plane strain or any general stress case) and it yields all three principal stresses.

EXAMPLE 4-3

Determining 3-D Stresses Using Analytical Methods

Problem: A triaxial stress element as shown in Figure 4-1 has $\sigma_x = 40\,000$, $\sigma_y = -20\,000$, $\sigma_z = -10\,000$, $\tau_{xy} = 5\,000$, $\tau_{yz} = -1\,500$, $\tau_{zx} = 2\,500$ psi. Find the principal stresses using a numerical method and draw the resulting Mohr's circles.

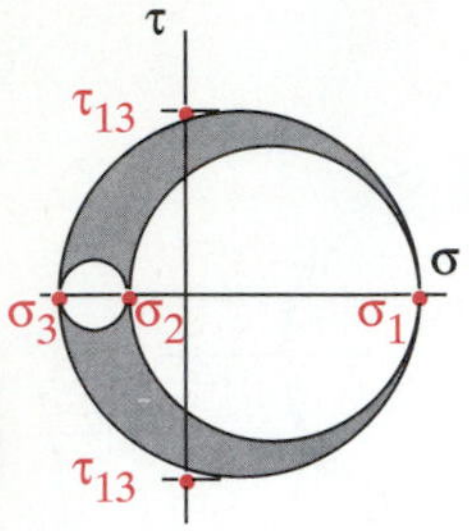

FIGURE 4-7

Mohr's Circles for
Example 4-3

Solution: See Figures 4-2 and 4-7, *TKSolver* file EX04-03, and MOHR file EX04-03.IPM.

1   Calculate the tensor invariants C_0, C_1, and C_2 from equation 4.4c.

$$C_2 = \sigma_x + \sigma_y + \sigma_z = 40\,000 - 20\,000 - 10\,000 = 10\,000 \qquad (a)$$

$$
\begin{aligned}
C_1 &= \tau_{xy}^2 + \tau_{yz}^2 + \tau_{zx}^2 - \sigma_x\sigma_y - \sigma_y\sigma_z - \sigma_z\sigma_x \\
&= (5\,000)^2 + (-1\,500)^2 + (2\,500)^2 - (40\,000)(-20\,000) \\
&\quad -(-20\,000)(-10\,000) - (-10\,000)(40\,000) = 10.335\ \text{E8}
\end{aligned}
\qquad (b)
$$

$$
\begin{aligned}
C_0 &= \sigma_x\sigma_y\sigma_z + 2\tau_{xy}\tau_{yz}\tau_{zx} - \sigma_x\tau_{yz}^2 - \sigma_y\tau_{zx}^2 - \sigma_z\tau_{xy}^2 \\
&= 40\,000(-20\,000)(-10\,000) + 2(5\,000)(-1\,500)(2\,500) \\
&\qquad -40\,000(-1\,500)^2 - (-20\,000)(2\,500)^2 \\
&\qquad -(-10\,000)(5\,000)^2 = 8.248\ \text{E12}
\end{aligned}
\qquad (c)
$$

2   Substitute the invariants into equation 4.4c and solve for its three roots using a numerical method.

$$\sigma^3 - C_2\sigma^2 - C_1\sigma - C_0 = 0$$

$$\sigma^3 - 10\,000\ \sigma^2 - 10.335\ \text{E8}\ \sigma - 8.248\ \text{E12} = 0 \qquad (d)$$

$$\sigma_1 = 40\,525; \qquad \sigma_2 = -9\,838; \qquad \sigma_3 = -20\,687$$

This was solved with the *TKSolver file* EX04-03, which is on the enclosed disk.

3   The principal shear stresses can now be found from equation 4.5.

$$\tau_{13} = \frac{|\sigma_1 - \sigma_3|}{2} = \frac{|40\,525 - (-20\,687)|}{2} = 30\,606$$

$$\tau_{21} = \frac{|\sigma_2 - \sigma_1|}{2} = \frac{|-9\,838 - 40\,525|}{2} = 25\,182 \qquad (e)$$

$$\tau_{32} = \frac{|\sigma_3 - \sigma_2|}{2} = \frac{|-20\,687 - (-9\,838)|}{2} = 5\,425$$

4.6   APPLIED VERSUS PRINCIPAL STRESSES

We now want to summarize the differences between the stresses **applied to an element** and the principal stresses that may occur on other planes as a result of the applied stresses. The **applied stresses** are the nine *components of the stress tensor* (Eq. 4.4*a*) that result from whatever loads are applied to the particular geometry of the object as defined in a coordinate system chosen for convenience. The **principal stresses** are the three *principal normal stresses* and the three *principal shear stresses* defined in Section 4.3. Of course, many of the applied stress terms may be zero in a given case. For example, in the tensile test specimen discussed in Chapter 2, the only nonzero applied stress is the σ_x term in equation 4.4*a*, which is unidirectional and normal. There are no applied shear stresses on the surfaces normal to the force axis in pure tensile loading. However, the principal stresses are **both normal and shear.**

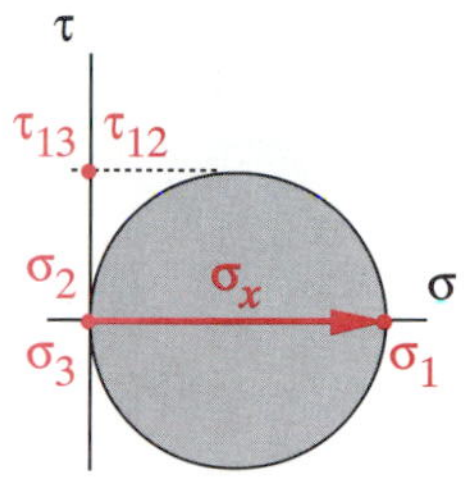

Figure 4-8 shows the Mohr's circle for a tensile-test specimen. In this case, the applied stress is pure tensile and the maximum principal normal stress is equal to it in magnitude and direction. But a principal shear stress of half the magnitude of the applied tensile stress acts on a plane 45° from the plane of the principal normal stress. Thus, the principal shear stresses will typically be nonzero even in the absence of any applied shear stress. This fact is important to an understanding of why parts fail and it will be discussed in more detail in Chapter 5. The examples in the previous section also reinforce this point. The most difficult task for the machine designer in this context is to correctly determine the locations, types, and magnitudes of all the applied stresses acting on the part. The calculation of the principal stresses is then *pro forma* using equations 4.4 to 4.6,

FIGURE 4-8

Mohr's Circles for Unidirectional Tensile Stress (two circles are coincident and the third is a point since $\sigma_2 = \sigma_3 = 0$).

4.7   AXIAL TENSION

Axial loading in tension (Figure 4-9) is one of the simplest types of loading that can be applied to an element. It is assumed that the load is applied through the area centroid of the element and that the two opposing forces are colinear along the *x* axis. At some distance away from the ends where the forces are applied, the stress distribution across the cross section of the element is essentially uniform as shown in Figure 4-10. This is one reason that this loading method is used to test material properties, as was described in Chapter 2. The applied normal stress for pure axial tension can be calculated from

$$\sigma_x = \frac{P}{A} \qquad (4.7)$$

where *P* is the applied force and *A* is the cross-sectional area at the point of interest. This is an applied normal stress. The principal normal stresses and the maximum shear stress can be found from equations 4-6. The Mohr's circle for this case was shown in Figure 4-8. The allowable load for any particular tension member can be determined by a comparison of the principal stresses with the appropriate strength of the material. For example, if the material is ductile, then the tensile yield strength, S_y, could be compared to the principal normal stress and the safety factor calculated as $N = S_y / \sigma_1$. Failure criteria will be dealt with in detail in Chapter 5.

FIGURE 4-9

A Bar in Axial Tension

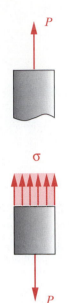

FIGURE 4-10

Stress Distribution Across
a Bar in Axial Tension

The change in length Δs of a member of uniform cross section loaded in pure axial tension is given by

$$\Delta s = \frac{Pl}{AE} \qquad (4.8)$$

where P is the applied force, A is the cross-sectional area, l is the loaded length, and E is Young's modulus for the material.

Tension loading is very common, occurring in cables, struts, bolts, and many other axially loaded elements. The designer needs to check carefully for the presence of other loads on the member that, if present in combination with the tensile load, will create a different stress state than the pure axial tension described here.

4.8   DIRECT SHEAR STRESS, BEARING STRESS, AND TEAROUT

These types of loading occur mainly in pin-jointed, bolted, or riveted connections. Possible modes of failure are direct shear of the connector (pin, rivet, or bolt), bearing failure of connector or surrounding material, or a tearing out of the material surrounding the connector. See the Case Studies later in this chapter for examples of the calculation of these types of stresses.

Direct Shear

Direct shear occurs in situations where there is no bending present. A pair of scissors (also called a *pair of shears*) is designed to produce direct shear on the material being cut. A poor-quality or worn-out pair of scissors will not cut well (even if sharp) if it allows a gap to exist between the two blades in a direction perpendicular to the blades' motion. Figure 4-11 shows a condition of direct shear and also one in which bending occurs instead. If the gap between the two shearing "blades" or surfaces can be kept close to zero, then a state of direct shear can be assumed and the resulting average stress on the shear face can be estimated from

$$\tau_{xy} = \frac{P}{A_{shear}} \qquad (4.9)$$

where P is the applied load and A_{shear} is the shear-area being cut, i.e., the cross-sectional area being sheared. The assumption here is that the shear stress is uniformly distributed over the cross section. This is not accurate since higher local stresses occur at the blade.

In Figure 4-11a the shear blade is tight against the jaws that hold the workpiece. Thus, the two forces P are in the same plane and do not create a couple. This provides a condition of direct shear with no bending. Figure 4-11b shows the same workpiece with a small gap (x) between the shear blade and the jaws. This creates a moment arm, turning the pair of forces P into a couple and thus bending, rather than directly shearing the part. Of course, there will still be significant shearing stresses developed in addition to the bending stresses in this case. Note that it is difficult to create situations in which pure direct shear is the only loading. Even the slight clearances necessary for function can superimpose bending stresses on the applied shear stresses. We will discuss stresses due to bending in the next section.

The situation depicted in Figures 4-11*a* and 4-12*a* is also called *single shear* because only one cross-sectional area of the part needs to be severed to break it. Figure 4-12*b* shows a pivot pin in *double shear*. Two areas must fail before it separates. This is called a *clevis-pin joint*, where the yoke-shaped link is the *clevis*. The area to be used in equation 4-9 is now $2A$. Double shear is preferred over single shear for pivot pin designs. Single-shear pivots should only be used where it is impossible to support both ends of the pin as in some linkage cranks, which must pass over adjacent links on one side. Bolted and riveted joints are in single shear when only two flat pieces are fastened together.

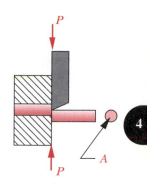

(a) Direct shear

Direct Bearing

A pivot pin in a hole such as is depicted in Figure 4-12 may fail in other ways than in direct shear. The surfaces of the pin and hole are subjected to a direct *bearing stress,* which is compressive in nature. Bearing stress occurs whenever two surfaces are pressed together. This stress tends to crush the hole or pin rather than to shear it. The bearing stress is normal, compressive and can be calculated from equation 4.7. The area used for this calculation is typically taken as the **projected area of contact** of pin and hole, **not the circumferential area**. That is,

$$A_{bearing} = l\,d \qquad (4.10)$$

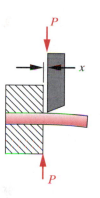

(b) Shear with bending

where l is the length of the bearing contact and d is the diameter of the hole or pin. Figure 4-13*a* shows the bearing areas for the clevis pin joint of Figure 4-12. Each of the two links joined must be checked separately for bearing failure as either can fail independent of the other. The length l (i.e., link thickness) as well as the pin diameter can be adjusted to create sufficient bearing area and avoid failure.

FIGURE 4-11

Shear Loading

Tearout Failure

Another possible mode of failure for pinned joints is tearout of the material surrounding the hole. This will occur if the hole is placed too close to the edge. This is a double-shear failure as it requires both sides of the hole to separate from the parent material. Equation 4.9 is applicable to this case provided that the correct shear area is used. Figure 4-13*b* shows the tearout areas for the clevis pin joint from Figure 4-12. It appears that the area could be calculated as the product of the link thickness and the distance from the center of the hole to the outer edge of the part, doubled to account for both sides of the hole. However, this assumption implies that the very thin wedge of material within

4

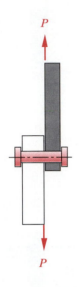

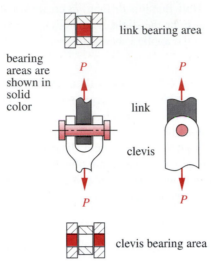

(a) Bearing-stress areas

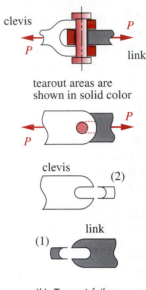

(b) Tearout failure

FIGURE 4-13

Bearing and Tearout Failures

(a) Pivot in single shear

the hole diameter adds significant shear strength. A more common and conservative assumption is to use twice the product of the total link thickness and the dimension from **the edge of the hole to the outside of the part** for the tearout area. It is a simple matter to provide sufficient material around holes to prevent tearout failure. A minimum of one pin-diameter of material between edge of hole and the part outer edge is a reasonable starting point for your design calculations.

4.9   BEAMS AND BENDING STRESSES

Beams are very common elements in structures and machines of all kinds. Any intermittently supported member subjected to loads transverse to its length will act as a beam. Structural floor joists, roof rafters, machinery shafts, springs, and frames are a few examples of elements that are often loaded as beams. Beams will usually have some combination of normal and shear stresses distributed over their cross sections. It is important for the designer to understand how stresses are distributed within beams in order to choose the correct locations for the calculation of maximum stresses. Memorization of the beam stress formulas, while useful, is not sufficient without also gaining an understanding of how and where to properly apply them.

(b) Pivot in double shear

FIGURE 4-12

Single and Double Shear

Beams in Pure Bending

While it is rare in practice to encounter a beam that is loaded strictly in "pure" bending, it is nevertheless useful to explore this simplest loading case as a means of developing the theory of stresses due to bending loads. Most real beams will also be subjected to shear loading in combination with the bending moment. That case will be addressed in the next section.

STRAIGHT BEAMS   As an example of a pure bending case, consider the simply supported, straight beam shown in Figure 4-14. Two identical, concentrated loads P are applied at points A and B, which are each the same distance from either end of the beam. The shear and bending moment diagrams for this loading show that the center section of the beam, between points A and B, has zero shear force and a constant bending moment of magnitude M. The absence of a shear force makes it pure bending.

Figure 4-15 shows a removed and enlarged segment of the beam taken between points A and B. The assumptions for the following analysis are as follows:

1   The segment analyzed is distant from applied loads or external constraints on the beam.

2   The beam is loaded in a plane of symmetry.

3   Cross sections of the beam remain plane and perpendicular to the neutral axis during bending.

4   The material of the beam is homogeneous, and obeys Hooke's law.

5   Stresses remain below the elastic limit and deflections are small.

6   The segment is subjected to pure bending with no axial or shear loads.

7   The beam is initially straight.

The unloaded segment in Figure 4-15a is straight but as the bending moment is applied in Figure 4-15b, the segment becomes curved (shown exaggerated). The line from N to N along the neutral axis does not change length, but all other lines along the x direction must either shorten or lengthen in order to keep all cross sections perpendicular to the neutral axis. The outer fibers of the beam at A-A are shortened, which puts them in compression and the outer fibers at B-B are lengthened and put in tension. This causes the bending stress distribution shown in Figure 4-15b. The bending stress magnitude is zero at the neutral axis and is linearly proportional to the distance y from the neutral axis. This relationship is expressed by the familiar bending stress equation:

$$\sigma_x = \frac{My}{I} \qquad (4.11a)$$

where M is the applied bending moment at the section in question, I is the second moment of area (area moment of inertia) of the beam cross section about the neutral plane (which passes through the centroid of the cross section of a straight beam), and y is the distance from the neutral plane to the point at which the stress is calculated.

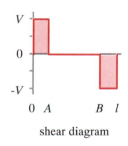

loading diagram

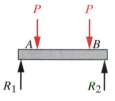

shear diagram

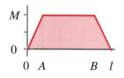

moment diagram

FIGURE 4-14

Pure Bending in a Beam

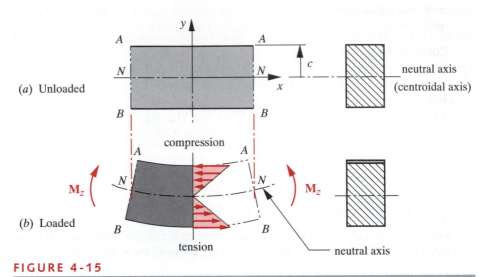

FIGURE 4-15

Segment of a Straight Beam in Pure Bending

The maximum bending stress occurs at the outer fibers, and is expressed as

$$\sigma_{max} = \frac{Mc}{I} \qquad (4.11b)$$

where c is the distance from the neutral plane to the outer fiber at either top or bottom of the beam. Note that these two distances will only be the same for sections that are symmetrical about the neutral plane. The value of c is usually taken as positive for both top and bottom surfaces and the proper sign then applied to the stress based on inspection of the beam loading to determine which surface is in compression (−) and which is in tension (+).

An alternate form of equation 4.11b is often used:

$$\sigma_{max} = \frac{M}{Z} \qquad (4.11c)$$

where Z is the beam's *section modulus*:

$$Z = \frac{I}{c} \qquad (4.11d)$$

These equations, though developed for the case of pure bending, are nevertheless applicable to cases, where shear strain is negligible, in which other loads in addition to the moment are applied to the beam. In such situations the effects of the combined loadings must be properly accounted for. This will be discussed in later sections. Formulas for the geometric properties *(A, I, Z)* of typical beam cross sections can be found in Appendix D and are also provided as *TKSolver* files on disk.

CURVED BEAMS Many machine parts such as crane hooks, C-clamps, punch-press frames, etc., are loaded as beams, but are not straight. They have a radius of curvature. The first six assumptions listed above for straight beams still apply. If a beam has significant curvature, then the neutral axis will no longer be coincident with the centroidal axis and equations 4.11 do not directly apply. The neutral axis shifts toward the center of curvature by a distance *e* as shown in Figure 4-16.

$$e = r_c - \frac{A}{\int \frac{dA}{r}} \qquad (4.12a)$$

where r_c is the radius of curvature of the centroidal axis of the curved beam, A is the cross-sectional area, and r is the radius from the beam's center of curvature to the differential area dA. Numerical evaluation of the integral can be done for complex shapes.*

The stress distribution across the section is no longer linear, but is now hyperbolic and is greatest on the inner surface of a rectangular cross section as shown in Figure 4-16. The convention is to define a positive moment as one that tends to straighten the beam. This creates tension on the inside and compression on the outside surface from a positive applied moment and vice versa. For pure bending loads, the expressions for the maximum stresses at inner and outer surfaces of a curved beam now become:

$$\sigma_i = +\frac{M}{eA}\left(\frac{c_i}{r_i}\right) \qquad (4.12b)$$

$$\sigma_o = -\frac{M}{eA}\left(\frac{c_o}{r_o}\right) \qquad (4.12c)$$

* Expressions for this integral for many common cross-sectional shapes can be found in reference [4]. For example, for a rectangular cross section, $e = r_c - (r_o - r_i) / \text{LN}(r_o / r_i)$

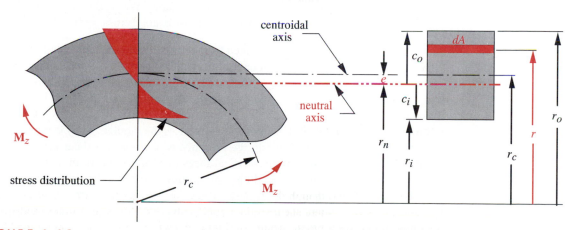

FIGURE 4-16

Segment of a Curved Beam in Pure Bending

4

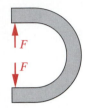

F

F

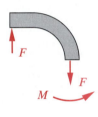

(a) Force-loaded
curved beam

where the subscript i denotes the inside surface and o the outside, M is the applied moment at the section in question, A is the cross-sectional area, and r_i and r_o are the radii of curvature of the inner and outer surfaces. These expressions contain the ratio c/r. If the radius of curvature r is large compared to c, then the beam looks more "straight" than "curved." When c/r becomes less than about 1:10, the stresses will be only about 10% greater than those of a straight beam of the same dimensions and loading. (Note that this is not a linear relationship as e is also a function of c and r.)

It is more common to have applied forces loading a curved beam as shown in Figure 4-17a. An example is a clamp or a hook. The free-body diagrams in Figure 4-17b show that there is now an axial force as well as a moment on the cut section. The equations for stress at the inside and outside of the beam become

$$\sigma_i = +\frac{M}{eA}\left(\frac{c_i}{r_i}\right) + \frac{F}{A} \tag{4.12d}$$

$$\sigma_o = -\frac{M}{eA}\left(\frac{c_o}{r_o}\right) + \frac{F}{A} \tag{4.12e}$$

The second terms in equations 4.12d and 4.12e represent the direct axial tensile stress on the midsection of the beam. The *TKSolver* file CURVBEAM provided computes equations 4.12 for five common shapes of curved beam cross sections, round, elliptical, trapezoidal, rectangular, and tee. This program uses numerical integration to evaluate equation 4.12a and to find the area and centroid of the cross section.

Shear Due to Transverse Loading

The more common condition in beam loading is a combination of both shear force and bending moment applied to a particular section. Figure 4-18 shows a point-loaded, simply supported beam and its shear and moment diagrams. We need to now consider how the shear loading affects the stress state within the beam's cross sections.

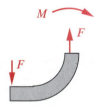

(b) Free-body diagrams

FIGURE 4-17

A Curved Beam with
Force Loading

Figure 4-19a shows a segment taken from the beam around point A in Figure 4-18. An element labeled P is shown cut out of the beam at point A. This element is dx wide and is cut in from the outer fiber at c to a depth y_1 from the neutral axis. Note that the magnitude of the moment $M(x_1)$ on the left-hand side of P (face b_1-c_1) is less than the moment $M(x_2)$ on the right-hand side (face b_2-c_2) and their difference is the differential moment dM. Figure 4-18 shows that at point A, the moment $M(x)$ is increasing as a function of beam length x, due to the presence of the nonzero shear force V at that point. The normal stresses on the vertical faces of P are found from equation 4.11a. Since the normal stress due to bending is proportional to $M(x)$, the stress σ on the left-hand face of P is less than on its right-hand face as shown in Figure 4-19b. For equilibrium, this stress imbalance must be counteracted by some other stress component, which is shown as the **shear stress** τ in Figure 4-19b.

The force acting on the left-hand face of P at any distance y from the neutral axis can be found by multiplying the stress by the differential area dA at that point.

$$\sigma dA = \frac{My}{I}dA \qquad (4.13a)$$

The total force acting on the left-hand face is found by integrating

$$F_{1x} = \int_{y_1}^{c} \frac{My}{I}dA \qquad (4.13b)$$

and similarly for the right-hand face:

$$F_{2x} = \int_{y_1}^{c} \frac{(M+dM)y}{I}dA \qquad (4.13c)$$

The shear force on the top face at distance y_1 from the neutral axis is found from

$$F_{xy} = \tau b\,dx \qquad (4.13d)$$

where the product bdx is the area of the top face of element P.

For equilibrium, the forces acting on P must sum to zero,

$$F_{xy} = F_{2x} - F_{1x}$$

$$\tau b\,dx = \int_{y_1}^{c} \frac{(M+dM)y}{I}dA - \int_{y_1}^{c} \frac{My}{I}dA \qquad (4.13e)$$

$$\tau = \frac{dM}{dx}\frac{1}{Ib}\int_{y_1}^{c} y\,dA$$

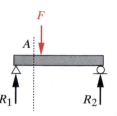

loading diagram

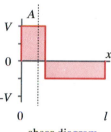

shear diagram

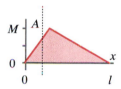

moment diagram

FIGURE 4-18

Shear Force and Bending Moment in a Beam

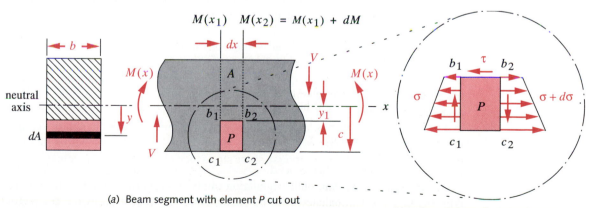

(a) Beam segment with element P cut out

(b) Enlarged view of removed element P

FIGURE 4-19

Segment of a Beam in Bending and Transverse Shear - Shown Removed at Point A in Figure 4-18

which gives an expression for the shear stress τ as a function of the change in moment with respect to x, the distance y from the neutral axis, the second moment of area I of the cross section, and the width b of the cross section at y. Equation 3.16a shows that the slope of the moment function dM/dx is equal to the magnitude of the shear function V at any point, so:

$$\tau_{xy} = \frac{V}{Ib} \int_{y_1}^{c} y \, dA \qquad (4.13f)$$

The integral in equation 4.13f represents the first moment about the neutral axis of that portion of the cross sectional area that exists outside of the value of y_1 for which the shear stress is being calculated. It is conventional to assign the variable Q to the value of this integral.

$$Q = \int_{y_1}^{c} y \, dA$$

and then $(4.13g)$

$$\tau_{xy} = \frac{VQ}{Ib}$$

The integral Q will obviously vary with the shape of the beam's cross section and also with the distance y_1 from the neutral axis. Thus, for any particular cross section, we should expect the shear stress to vary across the beam. It will always be zero at the outer fibers because Q vanishes when y_1 becomes equal to c. This makes sense as there is no material to shear against at the outer fiber. The shear stress due to transverse loading will be a maximum at the neutral axis. These results are very fortuitous since the normal stress due to bending is maximum at the outer fiber and zero at the neutral axis. Thus their combination on any particular element within the cross section seldom creates a worse stress state than exists at the outer fibers.

The shear stress due to transverse loading will be small compared to the bending stress Mc / I if the beam is long compared to its depth. The reason for this can be seen in equation 3.16 and in the shear and moment diagrams, one example of which is shown in Figure 4-17. Since the magnitude of the moment function is equal to the area under the shear function, for any given value of V in Figure 4-18, the area under the shear function and thus the maximum moment will increase with beam length. So, while the maximum shear stress magnitude remains constant, the bending stress increases with beam length, eventually dwarfing the shear stress. A commonly used rule of thumb says that the shear stress due to transverse loading in a beam will be small enough to ignore if the length-to-depth ratio of the beam is 10 or more. Short beams below that ratio should be investigated for transverse shear stress as well as for bending stress.

RECTANGULAR BEAMS   The calculation of shear stress due to transverse loading typically becomes an exercise in evaluating the integral Q for the particular beam cross section. Once that is done, the maximum value of τ is easily found. For a beam with a rectangular cross section of width b and depth h, $dA = b\,dy$ and $c = h / 2$.

$$Q = \int_{y_1}^{c} y\,dA = b\int_{y_1}^{c} y\,dy = \frac{b}{2}\left(\frac{h^2}{4} - y_1^2\right)$$

and   (4.14a)

$$\tau = \frac{V}{2I}\left(\frac{h^2}{4} - y_1^2\right)$$

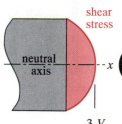

shear
stress

neutral
axis

- - - x

The shear stress varies parabolically across a rectangular beam as shown in Figure 4-19. When $y_1 = h/2$, $\tau = 0$ as expected. When $y_1 = 0$, $\tau_{max} = Vh^2/8I$. For a rectangle, $I = bh^3/12$, which gives

$$\tau_{max} = \frac{3}{2}\frac{V}{A}$$ (4.14b)

$$\tau_{max} = \frac{3}{2}\frac{V}{A}$$

This is valid **only for rectangular cross-sectional beams** and is shown in Figure 4-20a.

(a) rectangular beam

ROUND BEAMS   Equations 4.13g apply to any cross section. The integral Q for a circular cross section is

$$Q = \int_{y_1}^{c} y\,dA = 2\int_{y_1}^{c} y\sqrt{r^2 - y^2}\,dy = \frac{2}{3}\left(r^2 - y_1^2\right)^{\frac{3}{2}}$$ (4.15a)

and the shear stress distribution is

shear
stress

neutral
axis

- - - x

$$\tau = \frac{VQ}{bI} = \frac{V\left[\dfrac{2}{3}\left(r^2 - y_1^2\right)^{\frac{3}{2}}\right]}{2\sqrt{r^2 - y^2}\left(\dfrac{\pi r^4}{4}\right)} = \frac{4}{3}\frac{V}{\pi r^2}\left(1 - \frac{y_1^2}{r^2}\right)$$ (4.15b)

$$\tau_{max} = \frac{4}{3}\frac{V}{A}$$

(b) solid round beam

This is also a parabolic distribution but has a smaller peak value than the rectangular section as shown in Figure 4-20b. The maximum shear stress in a **solid, circular cross-sectional beam** is, at the neutral axis:

$$\tau_{max} = \frac{4}{3}\frac{V}{A}$$ (4.15c)

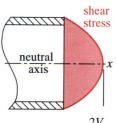

shear
stress

neutral
axis

- - - x

If the **round beam is hollow and thin-walled**, (wall thickness < about 1/10 the outside diameter) the maximum shear stress at the neutral axis will be approximately

$$\tau_{max} \cong \frac{2V}{A}$$

$$\tau_{max} \cong \frac{2V}{A}$$ (4.15d)

(c) hollow round beam

as shown in Figure 4-20c.

FIGURE 4-20

I-BEAMS   It can be shown mathematically that the I-beam configuration in Figure 4-21a is the optimal cross-sectional shape for a beam in terms of strength-to-weight

Shear-Stress Distribution and Maxima in Round, Round-Hollow, and Rectangular Beams

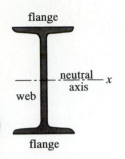

(a) I-beam shape

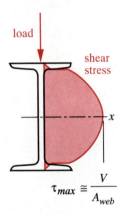

$$\tau_{max} \cong \frac{V}{A_{web}}$$

(b) Stress distribution

FIGURE 4-21

Shear-Stress Distribution and Maximum in an I-Beam

ratio. This explains why I-beams are commonly used as floor and roof beams in large structures. Their shape puts most of the material at the outer fibers where the bending stress is maximum. This gives a large area moment of inertia to resist the bending moment. As the shear stress is maximum at the neutral axis, the narrow web connecting the flanges (called the shear web) serves to resist the shear forces in the beam. In a long beam, the shear stresses due to bending are small compared to the bending stresses, which allows the web to be thin, reducing weight. An approximate expression for the maximum shear stress in an I-beam uses only the area of the web and ignores the flanges:

$$\tau_{max} \cong \frac{V}{A_{web}} \qquad (4.16)$$

Figure 4-21b shows the shear-stress distribution across the I-beam section depth. Note the discontinuities at the flange-web interfaces. The shear stress in the flange is small due to its large area. The shear stress jumps to a larger value on entering the web, then rises parabolically to a maximum at the neutral axis.

4.10   DEFLECTION IN BEAMS

In addition to the stresses in a beam, a designer also needs to be concerned with its deflections. Any applied bending load will cause a beam to deflect since it is made of an elastic material. If the deflection does not create strains in excess of the material's strain at its yield point, the beam will return to its undeflected state when the load is removed. If the strain exceeds that of the material's yield point, the beam will yield and "take a set" if ductile, or possibly fracture if brittle. If the beam is sized to prevent stresses that exceed the material's yield point (or other appropriate strength criterion) then no permanent set or fracture should occur. However, elastic deflections at stresses well below the material's failure levels may still cause serious problems in a machine.

Deflections can cause interferences between moving parts or misalignments that destroy the required accuracy of the device. In general, designing to minimize deflections will lead to larger beam cross sections than will designing only against stress failure. Even in static structures such as buildings, deflection can be the limiting criterion in sizing floor or roof beams. You have probably walked across a residential floor that bounced noticeably with each step. The floor was undoubtedly safe against collapse due to excessive stresses, but had not been designed stiff enough to prevent undesirable deflections under normal working loads.

The bending deflection of a beam is calculated by double integration of the beam equation,

$$\frac{M}{EI} = \frac{d^2 y}{dx^2} \qquad (4.17)$$

which relates the applied moment M, the material's modulus of elasticity E, and the cross section's area moment of inertia I to the second derivative of the beam deflection y. The independent variable x is the position along the beam length. Equation 4.17 is only valid for small deflections, which is not a limitation in most cases of beam design for machine or structural applications. Sometimes beams are used as springs and their deflections may then exceed the limitations of this equation. Spring design will be covered in a later chapter. Equation 4.17 also does not include the effects of deflection due to transverse shear loads. The transverse-shear component of deflection in long beams is small compared to that due to bending and is typically ignored unless the beam's length/depth ratio is < about 10.

Equation 4.17 can be differentiated twice and integrated twice to create the set of five equations 4.18 (including Eq. 4.17 repeated as Eq. 4.18c), which define beam behavior. Section 3.9 showed the relationship between the loading function $q(x)$, the shear function $V(x)$, and the moment function $M(x)$. V is the first derivative and q the second derivative of equation 4.17 with respect to x. Integrating equation 4.17 once gives the beam slope θ and integrating a second time gives the beam deflection y. These relationships form the following set of beam equations:

$$\frac{q}{EI} = \frac{d^4 y}{dx^4} \tag{4.18a}$$

$$\frac{V}{EI} = \frac{d^3 y}{dx^3} \tag{4.18b}$$

$$\frac{M}{EI} = \frac{d^2 y}{dx^2} \tag{4.18c}$$

$$\theta = \frac{dy}{dx} \tag{4.18d}$$

$$y = f(x) \tag{4.18e}$$

The only material parameter in these equations is Young's modulus E, which defines its stiffness. Since most alloys of a given base metal have essentially the same modulus of elasticity, equations 4.18 show why there is no advantage in using a stronger and more expensive alloy when designing to minimize deflection. Higher-strength alloys typically only provide higher yield or break strengths, and designing against a deflection criterion will usually result in relatively low stresses. This is the reason that I-beams and other structural-steel shapes are made primarily in low-strength, low-carbon steels.

Determining the deflection function of a beam is an exercise in integration. The loading function q is typically known and can be integrated by any one of several methods, analytical, graphical, or numerical. The constants of integration are evaluated from the boundary conditions of the particular beam configuration. Changes in section modulus across the beam require creating the M/EI function from the moment diagram before integrating for the beam slope. If the beam's area moment of inertia I and material E is uniform across its length, the moment function can just be divided by the constant EI. If the beam's cross section changes over its length, then the integration must be done piecewise to accommodate the changes in I. The integral forms of the beam equations are

$$V = \int q\, dx + C_1 \qquad\qquad 0 < x < l \qquad (4.19a)$$

$$M = \int V\, dx + C_1 x + C_2 \qquad\qquad 0 < x < l \qquad (4.19b)$$

$$\theta = \int \frac{M}{EI} dx + C_1 x^2 + C_2 x + C_3 \qquad\qquad 0 < x < l \qquad (4.19c)$$

$$y = \int \theta\, dx + C_1 x^3 + C_2 x^2 + C_3 x + C_4 \qquad\qquad 0 < x < l \qquad (4.19d)$$

The constants C_1 and C_2 can be found from boundary conditions on the shear and moment functions. For example, the moment will be zero at a simply supported beam end and either zero (or known if applied) at an unsupported free end of a beam. The shear force will be zero at an unloaded free end. Note that if the reaction forces are included in the loading function $q(x)$, then $C_1 = C_2 = 0$.

The constants C_3 and C_4 can be found from boundary conditions on the slope and deflection functions. For example, the deflection will be zero at any rigid support, and the beam slope will be zero at a moment joint. Substitute two known combinations of values of x and y or x and θ along with C_1 and C_2 in equations 4.19c and 4.19d and solve for C_3 and C_4. Many techniques for solution of these equations have been developed such as graphical integration, the area-moment method, energy methods, and singularity functions. We will explore the last two of these.

Deflection by Singularity Functions

Section 3.9 presented the use of singularity functions to represent loads on the beam. These functions make it relatively simple to perform the integration analytically and can easily be programmed for computer solution. Section 3.9 also applied this approach to obtain the shear and moment functions from the loading function. We will now extend that technique to develop the beam's slope and deflection functions. The best way to explore this method is by way of examples. Accordingly, we will calculate the beam shear, moment, slope, and deflection functions for the beams shown in Figure 4-22.

EXAMPLE 4-4

Finding Beam Slope and Deflection of a Simply Supported Beam Using Singularity Functions

Problem: Determine and plot the slope and deflection functions for the simply supported beam shown in Figure 4-22a.

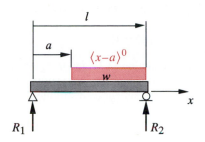

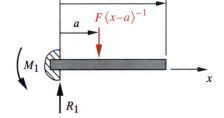

(a) Simply-supported beam with
uniformly distributed loading

(b) Cantilever beam with
concentrated loading

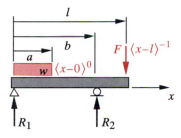

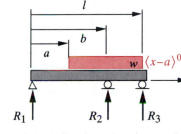

(c) Overhung beam with concentrated force
and uniformly distributed loading

(d) Statically indeterminate beam with
uniformly distributed loading

FIGURE 4-22

Various Beams and Beam Loadings

Given: The load is uniform over part of the beam length. Let beam length $l = 10$ in, and load location $a = 4$ in. The beam's $I = 0.163$ in^4 and $E = 30$ Mpsi. The distributed force is $w = 100$ lb/in.

Assumptions: The weight of the beam is negligible compared to the applied load and so can be ignored.

Solution: See Figures 4-22a and 4-23, and Table 4-2.

1 Solve for the reaction forces using equations 3.3. Summing moments about the right-hand end and summing forces in the y direction:

$$\sum M_z = 0 = R_1 l - \frac{w(l-a)^2}{2}$$

$$R_1 = \frac{w(l-a)^2}{2l} = \frac{100(10-4)^2}{2(10)} = 180 \tag{a}$$

Loading Diagram (lb)

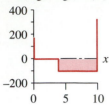

Shear Diagram (lb)

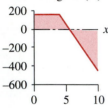

Moment Diagram (lb-in)

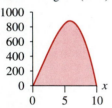

Slope Diagram (rad)

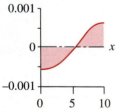

Deflection Diagram (in)

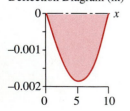

FIGURE 4-23

Example 4-4 Plots

$$\sum F_y = 0 = R_1 - w(l-a) + R_2 \tag{b}$$

$$R_2 = w(l-a) - R_1 = 100(10-4) - 180 = 420$$

2   Write equations for the load function in terms of equations 3.17 and integrate the resulting function four times using equations 3.18 to obtain the shear, moment, slope, and deflection functions. For the simply supported beam with a distributed load over part of its length:

$$q = R_1\langle x-0\rangle^{-1} - w\langle x-a\rangle^0 + R_2\langle x-l\rangle^{-1} \tag{c}$$

$$V = \int q\,dx = R_1\langle x-0\rangle^0 - w\langle x-a\rangle^1 + R_2\langle x-l\rangle^0 + C_1 \tag{d}$$

$$M = \int V\,dx = R_1\langle x-0\rangle^1 - \frac{w}{2}\langle x-a\rangle^2 + R_2\langle x-l\rangle^1 + C_1 x + C_2 \tag{e}$$

$$\theta = \int \frac{M}{EI}\,dx = \frac{1}{EI}\left(\begin{array}{c} \frac{R_1}{2}\langle x-0\rangle^2 - \frac{w}{6}\langle x-a\rangle^3 + \frac{R_2}{2}\langle x-l\rangle^2 \\[2mm] + \frac{C_1 x^2}{2} + C_2 x + C_3 \end{array}\right) \tag{f}$$

$$y = \int \theta\,dx = \frac{1}{EI}\left(\begin{array}{c} \frac{R_1}{6}\langle x-0\rangle^3 - \frac{w}{24}\langle x-a\rangle^4 + \frac{R_2}{6}\langle x-l\rangle^3 \\[2mm] + \frac{C_1 x^3}{6} + \frac{C_2 x^2}{2} + C_3 x + C_4 \end{array}\right) \tag{g}$$

3   There are four constants of integration to be found. The constants C_1 and C_2 are zero because the reaction forces and moments acting on the beam are included in the loading function. The deflection y is zero at the supports. The constants C_3 and C_4 are found by substituting the boundary conditions $x = 0$, $y = 0$ and $x = l$, $y = 0$ into equation (g).

$$y = 0 = \frac{1}{EI}\left(\frac{R_1}{6}\langle 0-0\rangle^3 - \frac{w}{24}\langle 0-a\rangle^4 + \frac{R_2}{6}\langle 0-l\rangle^3 + C_3(0) + C_4 \right)$$

$$C_4 = -\frac{R_1}{6}\langle 0-0\rangle^3 + \frac{w}{24}\langle 0-4\rangle^4 - \frac{R_2}{6}\langle 0-10\rangle^3 - C_3(0)$$

$$C_4 = -\frac{R_1}{6}(0) + \frac{w}{24}(0) - \frac{R_2}{6}(0) - C_3(0) = 0 \tag{h}$$

$$y = 0 = \frac{1}{EI}\left(\frac{R_1}{6}\langle l-0\rangle^3 - \frac{w}{24}\langle l-a\rangle^4 + \frac{R_2}{6}\langle l-l\rangle^3 + C_3 l + C_4 \right)$$

$$C_3 = \frac{w}{24l}\left[(l-a)^4 - 2l^2(l-a)^2 \right]$$

$$C_3 = \frac{100}{24(10)}\left[(10-4)^4 - 2(10)^2(10-4)^2 \right] = -2460 \tag{i}$$

4   Substitution of the values or expressions for C_1, C_2, C_3, C_4, R_1, and R_2 from equations (a), (b), (h), and (i) into equation (g) gives the resulting deflection equation for the beam in part (a) of Figure 4-22:

$$y = \frac{w}{24lEI}\left\{\left[2(l-a)^2\right]x^3 + \left[(l-a)^4 - 2l^2(l-a)^2\right]x - l\langle x-a\rangle^4\right\} \qquad (j)$$

5   The maximum deflection will occur at the point in x where the slope of the deflection curve is zero. Set the beam-slope equation (f) to zero and solve for x:

$$\theta = \frac{1}{EI}\left(\frac{R_1}{2}x^2 - \frac{w}{6}(x-a)^3 + \frac{R_2}{2}(x-l)^2 + C_3\right) = 0$$

$$0 = \frac{1}{3E7(0.163)}\left(90x^2 - 16.67(x-4)^3 + 210(x-10)^2 - 2460\right)$$

$$0 = -\frac{1}{14.65E5}\left(5x^3 - 150x^2 + 1500x - 5882\right)$$

$$x \cong 5.8 \qquad (k)$$

Note that either Viete's method or a numerical root-finding algorithm is needed to find the roots of this cubic equation.

6   Use this value of x in equation (g) to find the largest deflection.

$$y_{max} = \frac{100}{24(10)(4.883E6)}\left\{\begin{array}{l}\left[2(10-4)^2\right](5.8)^3 + \left[(10-4)^4 - 2(10)^2(10-4)^2\right](5.8) \\ -10(5.8-4)^4\end{array}\right\}$$

$$y_{max} = -0.00175 \text{ in} \qquad (l)$$

Table 4-2   Example 4-4 – Uniformly Loaded Simply Supported Beam
Part 1 of 2   *TKSolver Rule Sheet - See File EX04-04*

Rules

; 1 - find the reaction forces

        R1 = w * (l - a)^2 / (2 * l)
        R2 = w * (l - a) - R1

; 2 - solve for the nonzero constant of integration

        C3 = (-1/6 * CUBIC(l, 0, R1) + 1/24 * QUARTIC (l, a, w) – 1/6 * CUBIC(l, l, R2))/l

; 3 - solve the beam equations using singularity functions - see Function Sheet

        q = PULSE (x, 0, R1) – STEPF (x, a, w) + PULSE (x, l, R2)
        V = STEPF (x, 0, R1) – RAMP (x, a, w) + STEPF (x, l, R2)
        M = RAMP (x, 0, R1) – 1/2 * PARA (x, a, w) + RAMP (x, l, R2)
        θ = (1/2 * PARA (x, 0, R1) – 1/6 * CUBIC (x, a, w) + 1/2 * PARA (x, l, R2) + C3) /(E * I)
        y = (1/6 * CUBIC (x, 0, R1) – 1/24 * QUARTIC (x, a, w) + 1/6 * CUBIC (x, l, R2) + C3 * x) /(E * I)

Table 4-2   Example 4-4 – Uniformly Loaded Simply Supported Beam

Part 2 of 2   *TKSolver* Variable Sheet - See File EX04-04

	Input	Variable	Output	Unit	Comments
	3.0E+7	E		psi	modulus of elasticity
	10	l		in	length of beam
	100	w		lb/in	applied unit load
	4	a		in	distance to load w
L*	5.80	x		in	dimension along beam
		$xmax$	5.80	in	x at maximum deflection
		I	0.163 0	in^4	area moment of Inertia
L		q	–100	lb	loading function at point x
L		V	0	lb	shear function at point x
L		M	882	lb-in	moment function at point x
L		$theta$	0.000 1	rad	slope function at point x
L		y	–0.001 7	in	deflection function at point x
		$ymin$	–0.001 8	in	max negative deflection

* Indicates that a list of values of this variable has been created for plotting.

7 Plots of the loading, shear, moment, slope, and deflection functions for part *(a)* are
shown in Figure 4-23. The *TKSolver* rule and variable sheets for this example
problem are shown in Table 4-2. These *TKSolver* models contain custom-written
rule functions called PULSE, STEPF, RAMP, PARA, CUBIC, QUARTIC, and
QUINTIC which evaluate the singularity functions as shown in Table 3-8. These are
called from the rule statements that evaluate the loading, shear, moment, slope, and
deflection functions of the beam. These functions all have three arguments, x, the
independent variable representing distance along the beam, a, the position along the
beam at which the load begins to act, and F, the magnitude of the load. Each *rule
function* applies the appropriate definition of its singularity function to these
argument values and returns the singularity function's value corresponding to x. The
list-solving feature of *TKSolver* allows these calculations to be repeated for all listed
values of x and stores the results in the lists q, V, M, θ, and y. Note the similarity
among the rules for q, V, M, θ, and y in Table 4-2. Each corresponding term in the
beam equations has the same arguments but the singularity function changes with
each integration. A PULSE function in the q equation becomes a STEPF function in
V, a RAMP function in M, a PARA function in θ, and a CUBIC function in y, etc.
You may input the file EX04-04 to *TKSolver* to see more detail of the solutions and
to make larger-scale plots of the functions shown in Figure 4-23.

EXAMPLE 4-5

Finding Beam Slope and Deflection of a Cantilever Beam Using Singularity Functions

Problem: Determine and plot the slope and deflection functions for the beam shown in Figure 4-22b.

Given: The load is the concentrated force shown. Let beam length l = 10 in, and load location a = 4 in. The beams' I = 0.5 in⁴ and E = 30 Mpsi. The magnitude of the applied force is F = 400 lb.

Assumptions: Ignore the beam weight as negligible compared to the applied load.

Solution: See Figures 4-22b and 4-24, and Table 4-3.

1 Write equations for the load function in terms of equations 3.17 and integrate the resulting function twice using equations 3.18 to obtain the shear and moment functions. Note the use of the unit doublet function to represent the moment at the wall. For the beam in Figure 4-22b,

$$q = M_1\langle x - 0\rangle^{-2} + R_1\langle x - 0\rangle^{-1} - F\langle x - a\rangle^{-1} \qquad (a)$$

$$V = \int q\,dx = M_1\langle x - 0\rangle^{-1} + R_1\langle x - 0\rangle^{0} - F\langle x - a\rangle^{0} + C_1 \qquad (b)$$

$$M = \int V\,dx = -M_1\langle x - 0\rangle^{0} + R_1\langle x - 0\rangle^{1} - F\langle x - a\rangle^{1} + C_1 x + C_2 \qquad (c)$$

$$\theta = \int \frac{M}{EI}\,dx = \frac{1}{EI}\left(\begin{array}{c} -M_1\langle x - 0\rangle^{1} + \dfrac{R_1}{2}\langle x - 0\rangle^{2} - \dfrac{F}{2}\langle x - a\rangle^{2} \\[2mm] + \dfrac{C_1 x^2}{2} + C_2 x + C_3 \end{array} \right) \qquad (d)$$

$$y = \int \theta\,dx = \frac{1}{EI}\left(\begin{array}{c} -\dfrac{M_1}{2}\langle x - 0\rangle^{2} + \dfrac{R_1}{6}\langle x - 0\rangle^{3} - \dfrac{F}{6}\langle x - a\rangle^{3} \\[2mm] + \dfrac{C_1 x^3}{6} + \dfrac{C_2 x^2}{2} + C_3 x + C_4 \end{array} \right) \qquad (e)$$

The reaction moment M_1 at the wall is in the z direction and the forces R_1 and F are in the y direction in equation (b). All moments in equation (c) are in the z direction.

2 Because the reactions have been included in the loading function, the shear and moment diagrams both close to zero at each end of the beam, making $C_1 = C_2 = 0$.

3 The reaction force R_1 and reaction moment M_1 are calculated from equations (b) and (c), respectively, by substituting the boundary conditions $x = l^{+}$, $V = 0$, $M = 0$. Note that we can substitute l for l^{+} since their difference is vanishingly small.

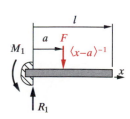

Cantilever beam with concentrated loading

FIGURE 4-22b

Repeated

4

Loading Diagram (lb)

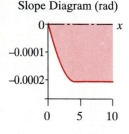

Shear Diagram (lb)

Moment Diagram (lb-in)

Slope Diagram (rad)

Deflection Diagram (in)

FIGURE 4-24

Example 4-5 Plots

$$V_y = 0 = R_1 \langle l - 0 \rangle^0 - F \langle l - a \rangle^0$$

$$0 = R_1 - F \tag{f}$$

$$R_1 = F = 400 \text{ lb}$$

$$M_z = 0 = -M_1 \langle l - 0 \rangle^0 + R_1 \langle l - 0 \rangle^1 - F \langle l - a \rangle^1$$

$$0 = -M_1 + R_1(l) - F(l - a) \tag{g}$$

$$M_1 = R_1(l) - F(l - a) = 400(10) - 400(10 - 4) = 1\,600 \text{ lb - in cw}$$

Since w, l, and a are known from the given data, equation (f) can be solved for R_1, and this result substituted in equation (g) to find M_1. Note that equation (f) is just $\Sigma F_y = 0$, and equation (g) is $\Sigma M_z = 0$. M_1 does not appear in equation (f) because it is in a different vector direction than the y forces.

4   Substitute $x = 0$, $\theta = 0$ and $x = 0$, $y = 0$ in (d) and (e) and solve for C_3 and C_4:

$$\theta = 0 = \frac{1}{EI}\left(-M_1 \langle 0 - 0 \rangle^1 + \frac{R_1}{2}\langle 0 - 0 \rangle^2 - \frac{F}{2}\langle 0 - a \rangle^2 + C_3\right)$$

$$C_3 = M_1 \langle 0 - 0 \rangle^1 - \frac{R_1}{2}\langle 0 - 0 \rangle^2 + \frac{F}{2}\langle 0 - 4 \rangle^2 = 0 \tag{h}$$

$$y = 0 = \frac{1}{EI}\left(-\frac{M_1}{2}\langle 0 - 0 \rangle^2 + \frac{R_1}{6}\langle 0 - 0 \rangle^3 - \frac{F}{6}\langle 0 - a \rangle^3 + C_3(0) + C_4\right)$$

$$C_4 = \frac{M_1}{2}\langle 0 - 0 \rangle^2 - \frac{R_1}{6}\langle 0 - 0 \rangle^3 + \frac{F}{6}\langle 0 - 4 \rangle^3 = 0 \tag{i}$$

5   Substitution of the expressions for C_3, C_4, R_1, and M_1 from (f), (g), (h), and (i) into equation (e) gives the deflection equation for the cantilever beam in Figure 4-22b:

Table 4-3   Example 4-5 – Cantilever Beam with Concentrated Load

Part 1 of 2   *TKSolver Rule Sheet - See File EX04-05*

Rules

; 1 - find the reaction force and moment

    R1 = F
    M1 = R1 * l + F * (l − a)

; 2 - solve the beam equations using singularity functions - see Function Sheet

    q = PULSE (x, 0, R1) − PULSE (x, a, F)
    V = STEPF (x, 0, R1) − STEPF (x, a, F)
    M = RAMP (x, 0, R1) − RAMP (x, a, F) − STEPF (x, 0, M1)
    θ = (1/2 * PARA (x, 0, R1) − 1/2 * PARA (x, a, F) − RAMP (x, 0, M1)) / (E * I)
    y = (1/6 * CUBIC (x, 0, R1) − 1/6 * CUBIC (x, a, F) − 1/2 * PARA (x, 0, M1) / (E * I)

$$y = \frac{F}{6EI}\left[x^3 - 3ax^2 - \langle x - a \rangle^3\right] \qquad (j)$$

6   The maximum deflection of a cantilever beam is at its free end, Substitute $x = l$ in equation (j) to find y_{max}.

$$y_{max} = \frac{F}{6EI}\left[l^3 - 3al^2 - (l - a)^3\right] = \frac{Fa^2}{6EI}(a - 3l)$$

$$y_{max} = \frac{400(4)^2}{6(3E7)(0.5)}\left[4 - 3(10)\right] = -0.001\,85 \text{ in} \qquad (k)$$

7   Plots of the loading, shear, moment, slope, and deflection functions are shown in Figure 4-24. Note that the beam slope is becoming increasingly negative for the portion of beam between the support and the load and then becomes constant to the right of the load. While not very apparent at the small scale of the figure, the beam deflection becomes a straight line to the right of the point of application of the load. The *TKSolver* rule and variable sheets for equations (*a -g*) are shown in Table 4-3. Some of the *rule functions'* definitions are shown in Table 3-8. Input the file EX04-05 to *TKSolver* to see more detail and make larger-scale plots of the functions.

Table 4-3   Example 4-5 – Cantilever Beam with Concentrated Load
Part 2 of 2   *TKSolver* Variable Sheet - See File EX04-05

	Input	Variable	Output	Unit	Comments
	10	l		in	length of beam
	4	a		in	distance to load F
	400	F		lb	applied load
	0.50	I		in^4	area moment of inertia
	3.0E+7	E		psi	modulus of elasticity
L*	0	x		in	dimension along beam
L		q	0	lb	loading function at point x
L		V	400	lb	shear function at point x
L		M	−1 600	lb-in	moment function at point x
L		*theta*	0	rad	beam slope at point x
L		y	0	in	deflection at point x
		ymin	−0.001 8	in	max negative deflection
		$R1$	400	lb	reaction force
		$M1$	1 600	lb	reaction moment

* Indicates that a list of values of this variable has been created for plotting.

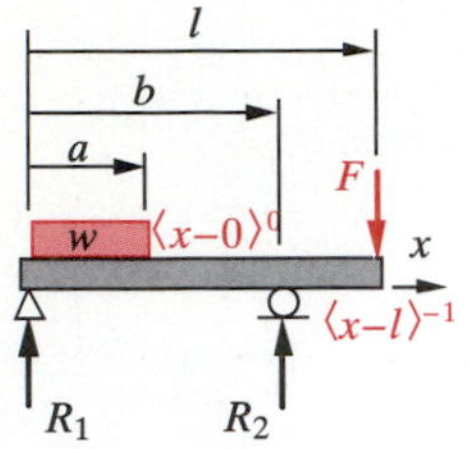

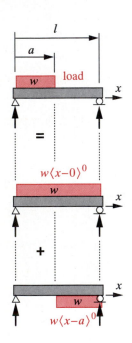

Overhung beam with
concentrated force
and uniformly
distributed loading

FIGURE 4-22c

Repeated

4

EXAMPLE 4-6

Finding Beam Slope and Deflection of an Overhung Beam Using Singularity Functions

Problem: Determine and plot the slope and deflection functions for an overhung beam with a uniformly distributed load over part of its length and a concentrated force at its end as shown in Figure 4-22c.

Given: The loads are as shown. Let beam length l = 10 in, and load locations a = 4 in and b = 7 in. The beams' I = 0.2 in⁴ and E = 30 Mpsi. The magnitude of the concentrated force is F = 200 lb and the distributed force is w = 100 lb/in.

Assumptions: The weight of the beam is negligible compared to the applied loads and so can be ignored.

Solution: See Figures 4-22c, 4-25, 4-26, and Table 4-4.

1 The distributed load does not extend over the entire length of this beam. All singularity functions extend from their initial point to the end of the beam. So, to terminate the uniform load's step function at some point short of the end of the beam it is necessary to apply another step function of equal amplitude and opposite sign in order to cancel it for all points beyond length a as shown in Figure 4-25. The sum of the two step functions of opposite sign is then zero to the right of distance a.

$$q = R_1\langle x-0\rangle^{-1} - w\langle x-0\rangle^0 + w\langle x-a\rangle^0 + R_2\langle x-b\rangle^{-1} - F\langle x-l\rangle^{-1} \qquad (a)$$

$$V = \int q\,dx = R_1\langle x-0\rangle^0 - w\langle x-0\rangle^1 + w\langle x-a\rangle^1 + R_2\langle x-b\rangle^0 - F\langle x-l\rangle^0 + C_1 \qquad (b)$$

$$M = \int V dx = \left(\begin{array}{c} R_1\langle x-0\rangle^1 - \dfrac{w}{2}\langle x-0\rangle^2 + \dfrac{w}{2}\langle x-a\rangle^2 + R_2\langle x-b\rangle^1 \\[2mm] - F\langle x-l\rangle^1 + C_1 x + C_2 \end{array} \right) \qquad (c)$$

$$\theta = \int \dfrac{M}{EI}\,dx = \dfrac{1}{EI}\left(\begin{array}{c} \dfrac{R_1}{2}\langle x-0\rangle^2 - \dfrac{w}{6}\langle x-0\rangle^3 + \dfrac{w}{6}\langle x-a\rangle^3 + \dfrac{R_2}{2}\langle x-b\rangle^2 \\[2mm] - \dfrac{F}{2}\langle x-l\rangle^2 + \dfrac{C_1}{2}x^2 + C_2 x + C_3 \end{array} \right) \qquad (d)$$

$$y = \int \theta\,dx = \dfrac{1}{EI}\left(\begin{array}{c} \dfrac{R_1}{6}\langle x-0\rangle^3 - \dfrac{w}{24}\langle x-0\rangle^4 + \dfrac{w}{24}\langle x-a\rangle^4 + \dfrac{R_2}{6}\langle x-b\rangle^3 \\[2mm] - \dfrac{F}{6}\langle x-l\rangle^3 + \dfrac{C_1}{6}x^3 + \dfrac{C_2}{2}x^2 + C_3 x + C_4 \end{array} \right) \qquad (e)$$

2 Because the reactions have been included in the loading function, the shear and moment diagrams both close to zero at each end of the beam, making $C_1 = C_2 = 0$.

---

load = $w\langle x-0\rangle^0$
         $+ w\langle x-a\rangle^0$

FIGURE 4-25

Interrrupted Singularity Functions are Formed by Combining Functions of Opposite Sign that Start at Different Points Along the Beam

3 Since both the shear and moment are zero at $x = l^+$, the reactions R_1 and R_2 can be calculated simultaneously from (b) and (c) with $x = l^+ = l$:

$$V = 0 = R_1\langle l-0\rangle^0 - w\langle l-0\rangle^1 + w\langle l-a\rangle^1 + R_2\langle l-b\rangle^0 - F\langle l-l\rangle^0$$

$$0 = R_1 - wl + w(l-a) + R_2 - F \tag{f}$$

$$R_2 = -R_1 + wl - w(l-a) + F = 400 \text{ lb}$$

$$M = 0 = \left(R_1\langle l\rangle^1 - \frac{w}{2}\langle l\rangle^2 + \frac{w}{2}\langle l-a\rangle^2 + R_2\langle l-b\rangle^1 - F\langle l-l\rangle^1\right)$$

$$0 = R_1 l - \frac{wl^2}{2} + \frac{w(l-a)^2}{2} + R_2(l-b) \tag{g}$$

$$R_1 = \frac{1}{l}\left[\frac{wl^2}{2} - \frac{w(l-a)^2}{2} - R_2(l-b)\right] = 200 \text{ lb}$$

Note that the equations (f) are just the sum of forces = 0, and the sum of moments taken about point l and set to 0.

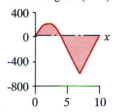

Loading Diagram (lb)

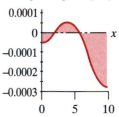

Shear Diagram (lb)

4 Substitute $x = 0$, $y = 0$ and $x = b$, $y = 0$ in equation (e) and solve for C_3 and C_4:

$$y = 0 = \frac{1}{EI}\left(\begin{array}{c}\dfrac{R_1}{6}\langle 0-0\rangle^3 - \dfrac{w}{24}\langle 0-0\rangle^4 + \dfrac{w}{24}\langle 0-a\rangle^4 + \dfrac{R_2}{6}\langle 0-b\rangle^3 \\[2mm] - F\langle 0-l\rangle^3 + \dfrac{C_1}{6}(0)^3 + \dfrac{C_2}{2}(0)^2 + C_3(0) + C_4\end{array}\right)$$

$$C_4 = 0 \tag{h}$$

$$y = 0 = \frac{1}{EI}\left(\begin{array}{c}\dfrac{R_1}{6}\langle b-0\rangle^3 - \dfrac{w}{24}\langle b-0\rangle^4 + \dfrac{w}{24}\langle b-a\rangle^4 + \dfrac{R_2}{6}\langle b-b\rangle^3 \\[2mm] - F\langle b-l\rangle^3 + \dfrac{C_1}{6}(b)^3 + \dfrac{C_2}{2}(b)^2 + C_3(b) + C_4\end{array}\right)$$

$$C_3 = \frac{1}{b}\left[-\frac{R_1}{6}b^3 + \frac{w}{24}b^4 - \frac{w}{24}\langle b-a\rangle^4\right]$$

$$= \frac{1}{7}\left[-\frac{200}{6}(7)^3 + \frac{100}{24}(7)^4 - \frac{100}{24}(7-4)^4\right] = -252.4 \tag{i}$$

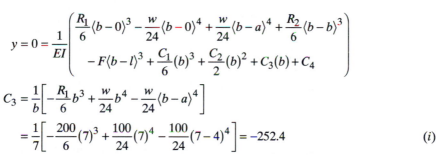

Moment Diagram (lb-in)

Slope Diagram (rad)

5 Substitution of the expressions for C_1, C_2, C_3, C_4, R_1, and R_2 from equations (f), (g), (h), and (i) into equation (e) gives the resulting deflection equation

$$y = \frac{1}{EI}\left(\begin{array}{c}\dfrac{R_1}{6}x^3 - \dfrac{w}{24}x^4 + \dfrac{w}{24}\langle x-a\rangle^4 + \dfrac{R_2}{6}\langle x-b\rangle^3 - \dfrac{F}{6}\langle x-l\rangle^3 \\[3mm] + \dfrac{1}{b}\left[-\dfrac{R_1}{6}(b)^3 + \dfrac{w}{24}(b)^4 - \dfrac{w}{24}\langle b-a\rangle^4\right]x\end{array}\right) \tag{j}$$

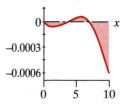

Deflection Diagram (in)

FIGURE 4-26

Example 4-6 Plots

Table 4-4   Example 4-6 – Overhung Beam with Uniformly Distributed Load

Part 1 of 2   *TKSolver* Rule Sheet - See File EX04-06

Rules

; 1 - find the reaction forces

$R1 = 1/ l *(w*l^2 / 2 - w * (l - a)^2 / 2 - R2 * (l - b))$

$R2 = -R1 + w*l - w * (l - a) + F$

; 2 - solve for the nonzero constant of integration

$C3 = 1/b * (-1/6 * CUBIC(b, 0, R1) + 1/24 * QUARTIC(b, 0, w) - 1/24 * QUARTIC (b, a, w) - 1/6 * CUBIC(b, b, R2)$
$+ 1/6 * CUBIC(b, l, F) - 1/6 * C1 * b^3 - .5 * C2 * b^2 - C4)$

; 3 - solve the beam equations using singularity functions - see Function Sheet

$q = PULSE(x, 0, R1) - STEPF(x, 0, w) + STEPF(x, a, w) + PULSE(x, b, R2) - PULSE(x, l, F)$

$V = STEPF(x, 0, R1) - RAMP(x, 0, w) + RAMP(x, a, w) + STEPF(x, b, R2) - STEPF(x, l, F)$

$M = RAMP(x, 0, R1) - 1/2 * PARA(x, 0, w) + 1/2 * PARA(x, a, w) + RAMP(x, b, R2) - RAMP(x, l, F)$

$\theta = 1/EI * (.5 * PARA(x, 0, R1) - 1/6 * CUBIC(x, 0, w) + 1/6 * CUBIC(x, a, w) + .5 * PARA(x, b, R2)$
$- .5 * PARA(x, l, F) + C3)$

$y = 1/EI * (1/6 * CUBIC(x, 0, R1) - 1/24 * QUARTIC(x, 0, w) + 1/24 * QUARTIC(x, a, w) + 1/6 * CUBIC(x, b, R2) - 1/6 *$
$CUBIC(x, l, F) + C3 * x)$

---

6   Since an overhung beam is a form of cantilever beam, the maximum deflection is most likely at the free end. Substitute $x = l$ in equation (f) to find y_{max}.

$$y_{max} = \frac{1}{EI}\left(\begin{array}{c} \frac{R_1}{6}l^3 - \frac{w}{24}l^4 + \frac{w}{24}\langle l - a \rangle^4 + \frac{R_2}{6}\langle l - b \rangle^3 - \frac{F}{6}\langle l - l \rangle^3 \\ + \frac{1}{b}\left[-\frac{R_1}{6}(b)^3 + \frac{w}{24}(b)^4 - \frac{w}{24}\langle b - a \rangle^4 \right]l \end{array} \right)$$

$$= \frac{1}{3E7(0.2)}\left(\begin{array}{c} \frac{200}{6}10^3 - \frac{100}{24}10^4 + \frac{100}{24}\langle 10 - 4 \rangle^4 + \frac{400}{6}(10 - 7)^3 \\ - \frac{200}{6}0^3 + \frac{1}{7}\left[-\frac{200}{6}(7)^3 + \frac{100}{24}(7)^4 - \frac{100}{24}(7 - 4)^4 \right]10 \end{array} \right)$$

$y_{max} = -0.0006 \text{ in}$ (k)

7   Plots of the loading, shear, moment, slope, and deflection functions for part (c) are shown in Figure 4-26. The rule sheet for equations (a - i) and the variable sheet are shown in Table 4-4. Some of the called *rule functions* are shown in Table 3-8. Input the file EX04-06 to *TKSolver* to examine the model and see larger plots of the functions in Figure 4-26.

Statically Indeterminate Beams

When a beam has redundant supports as shown in Figure 4-22d, it is said to be statically indeterminate. This example is also called a continuous beam and is quite com-

Table 4-4   Example 4-6 – Overhung Beam with Uniform Load and Conc. Force

Part 2 of 2   *TKSolver* Variable Sheet - See File EX04-06.TK

	Input	Variable	Output	Unit	Comments
	10	l		in	length
	4	a		in	dist over which load w applies
	7	b		in	distance to R_2
	200	F		lb	applied force
	100	w		lb/in	step magnitude
	0.20	I		in^4	area moment of inertia
	3.0E+7	E		psi	Youngs modulus
L*	10	x		in	dimension along beam
L		q	−200	lb	loading function at point x
L		V	0	lb	shear function at point x
L		M	0	lb/in	moment function at point x
L		theta	−0.000 3	rad	slope function at point x
L		y	−0.000 6	in	deflection function at point x
		ymin	−0.000 6	in	max negative deflection
		ymax	0.000 1	in	max positive deflection
		R1	200	lb	left
		R2	400	lb	right
		Vmin	−200	lb	minimum shear
		Vmax	200	lb	maximum shear
		Mmin	−600	lb/in	minimum moment
		Mmax	200	lb/in	maximum moment
		tmax	0.002 7	deg	minimum beam slope
		tmin	−0.014 5	deg	maximum beam slope

* Indicates that a list of values of this variable has been created for plotting.

mon. Supporting beams for buildings often have multiple columns distributed under a long beam span. The magnitudes of more than two reaction forces or moments cannot be found using only the two equations of static equilibrium, $\Sigma F = 0$ and $\Sigma M = 0$. To find more than two reactions requires additional equations and the deflection function can be used for this purpose. The deflection can be assumed to be zero at each simple support (as a first approximation) and the beam slope is known or can be closely estimated at a moment support.[*] These provide an additional boundary condition for each added reaction allowing the solution to be calculated.

* Since nothing is truly rigid, the beam's supports may deflect (compress) under the applied loads. However, if the supports are suitably stiff, this support movement will usually be small compared to the beam deflection and can be assumed to be zero for beam analysis.

SOLVING INDETERMINATE BEAMS WITH SINGULARITY FUNCTIONS   The singularity functions provide a convenient way to set up and evaluate the equations for the loading, shear, moment, slope, and deflection functions as was demonstrated in the previous example. This approach can also be used to solve the indeterminate beam problem and is best demonstrated by another example.

EXAMPLE 4-7

Finding Reactions and Deflection of Statically Indeterminate Beams Using Singularity Functions

Problem: Determine and plot the loading, shear, moment, slope, and deflection functions for the beam in Figure 4-22d. Find maximum deflection.

Given: The load is uniformly distributed over part of the beam as shown. Length l = 10 in, a = 4 in, and b = 7 in. The beams' I = 0.08 in^4 and E = 30 Mpsi. The magnitude of the distributed force is w = 500 in/lb.

Assumptions: Ignore the beam weight as negligible compared to the applied load.

Solution: See Figures 4-22d and 4-27, and Table 4-5.

1   Write an equation for the load function in terms of equations 3.17 and integrate the resulting function four times using equations 3.18 to obtain the shear, moment, slope, and deflection functions.

$$q = R_1\langle x - 0\rangle^{-1} - w\langle x - a\rangle^0 + R_2\langle x - b\rangle^{-1} + R_3\langle x - l\rangle^{-1} \qquad (a)$$

$$V = \int q\,dx = R_1\langle x - 0\rangle^0 - w\langle x - a\rangle^1 + R_2\langle x - b\rangle^0 + R_3\langle x - l\rangle^0 + C_1 \qquad (b)$$

$$M = \int V\,dx = R_1\langle x - 0\rangle^1 - \frac{w}{2}\langle x - a\rangle^2 + R_2\langle x - b\rangle^1 + R_3\langle x - l\rangle^1 + C_1 x + C_2 \qquad (c)$$

$$\theta = \int \frac{M}{EI}\,dx = \frac{1}{EI}\left(\begin{array}{c} \dfrac{R_1}{2}\langle x - 0\rangle^2 - \dfrac{w}{6}\langle x - a\rangle^3 + \dfrac{R_2}{2}\langle x - b\rangle^2 + \dfrac{R_3}{2}\langle x - l\rangle^2 \\ + \dfrac{C_1 x^2}{2} + C_2 x + C_3 \end{array}\right) \qquad (d)$$

$$y = \int \theta\,dx = \frac{1}{EI}\left(\begin{array}{c} \dfrac{R_1}{6}\langle x - 0\rangle^3 - \dfrac{w}{24}\langle x - a\rangle^4 + \dfrac{R_2}{6}\langle x - b\rangle^3 + \dfrac{R_3}{6}\langle x - l\rangle^3 \\ + \dfrac{C_1 x^3}{6} + \dfrac{C_2 x^2}{2} + C_3 x + C_4 \end{array}\right) \qquad (e)$$

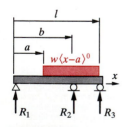

Indeterminate beam with uniformly distributed loading

FIGURE 4-22d

Repeated

2   There are 3 reaction forces and 4 constants of integration to be found. The constants C_1 and C_2 are zero because the reaction forces and moments acting on the beam are included in the loading function. This leaves 5 unknowns to be found.

3   If we consider the conditions at a point infinitesimally to the left of $x = 0$ (denoted as $x = 0^-$), the shear and moment will both be zero. The same conditions obtain at a point infinitesimally to the right of $x = l$ (denoted as $x = l^+$). Also, the deflection y must be zero at all three supports. These observations provide the 5 boundary conditions needed to evaluate the 3 reaction forces and 2 remaining integration constants: i.e., when $x = 0^-$, $V = 0$, $M = 0$; when $x = 0$, $y = 0$; when $x = b$, $y = 0$; when $x = l$, $y = 0$; when $x = l^+$, $V = 0$, $M = 0$.

4   Substitute the boundary conditions $x = 0$, $y = 0$, $x = b$, $y = 0$, and $x = l$, $y = 0$ into (e).

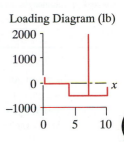

Loading Diagram (lb)

for $x = 0$:

$$y = 0 = \frac{1}{EI}\left(\frac{R_1}{6}\langle 0 - 0\rangle^3 - \frac{w}{24}\langle 0 - a\rangle^4 + \frac{R_2}{6}\langle 0 - b\rangle^3 + \frac{R_3}{6}\langle 0 - l\rangle^3 + C_3(0) + C_4\right)$$

$$C_4 = 0 \qquad\qquad (f)$$

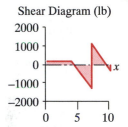

Shear Diagram (lb)

for $x = b$:

$$y = 0 = \frac{1}{EI}\left(\frac{R_1}{6}\langle b - 0\rangle^3 - \frac{w}{24}\langle b - a\rangle^4 + \frac{R_2}{6}\langle b - b\rangle^3 + \frac{R_3}{6}\langle b - l\rangle^3 + C_3 b + C_4\right)$$

$$C_3 = \frac{1}{b}\left(-\frac{R_1}{6}b^3 + \frac{w}{24}\langle b - a\rangle^4\right) = \frac{1}{7}\left(-\frac{R_1}{6}7^3 + 100(7 - 4)^4\right) = 385.7 - 8.17 R_1 \qquad (g)$$

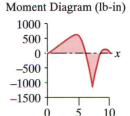

Moment Diagram (lb-in)

for $x = l$:

$$y = 0 = \frac{1}{EI}\left(\frac{R_1}{6}\langle l - 0\rangle^3 - \frac{w}{24}\langle l - a\rangle^4 + \frac{R_2}{6}\langle l - b\rangle^3 + \frac{R_3}{6}\langle l - l\rangle^3 + C_3 l + C_4\right)$$

$$C_3 = \frac{1}{l}\left(-\frac{R_1}{6}l^3 + \frac{w}{24}\langle l - a\rangle^4 - \frac{R_2}{6}\langle l - b\rangle^3\right)$$

$$C_3 = \frac{1}{10}\left(-\frac{R_1}{6}10^3 + \frac{100}{24}\langle 10 - 4\rangle^4 - \frac{R_2}{6}\langle 10 - 7\rangle^3\right) = 540 - 16.67 R_1 - 4.5 R_2 \qquad (h)$$

5   Two more equations can be written using equations (c) and (b) and noting that at a point l^+, infinitesimally beyond the right end of the beam, both V and M are zero. We can substitute l for l^+ since their difference is vanishingly small.

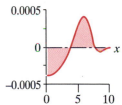

Slope Diagram (rad)

$$M = 0 = R_1\langle l - 0\rangle^1 - \frac{w}{2}\langle l - a\rangle^2 + R_2\langle l - b\rangle^1 + R_3\langle l - l\rangle^1$$

$$0 = R_1 l - \frac{w}{2}(l - a)^2 + R_2(l - b)$$

$$R_1 = \frac{1}{l}\left[\frac{w}{2}(l - a)^2 - R_2(l - b)\right]$$

$$R_1 = \frac{1}{10}\left[\frac{100}{2}(10 - 4)^2 - R_2(10 - 7)\right] = 180 - 0.3 R_2 \qquad (i)$$

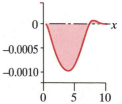

Deflection Diagram (in)

$$V = 0 = R_1\langle l\rangle^0 - w\langle l - a\rangle^1 + R_2\langle l - b\rangle^0 + R_3\langle l - l\rangle^0 = 0$$

$$0 = R_1 - w(l - a) + R_2 + R_3$$

$$R_2 = w(l - a) - R_1 - R_3 = 600 - R_1 - R_3 \qquad (j)$$

FIGURE 4-27

Example 4-7 Plots

Table 4-5   Example 4-7 – Indeterminate Beam with Uniform Load
Part 1 of 2   *TKSolver* Rule Sheet - See File EX04-07

Rules

; 1 - the beam equations using singularity functions - see Function Sheet

$q = PULSE(x, 0, R1) - STEPF(x, a, w) + PULSE(x, b, R2) + PULSE(x, l, R3)$

$V = STEPF(x, 0, R1) - RAMP(x, a, w) + STEPF(x, b, R2) + STEPF(x, l, R3)$

$M = RAMP(x, 0, R1) - .5 * PARA(x, a, w) + RAMP(x, b, R2) + RAMP(x, l, R3)$

$\theta = 1/EI * (.5 * PARA(x, 0, R1) - 1/6 * CUBIC(x, a, w) + .5 * PARA(x, b, R2) + .5 * PARA(x, l, R3) + C3)$

$y = 1/EI * (1/6 * CUBIC(x, 0, R1) - 1/24 * QUARTIC(x, a, w) + 1/6 * CUBIC(x, b, R2) + 1/6 * CUBIC(x, l, R3) + C3 * x)$

; 2 - the one nonzero constant of integration in this case

$C3 = 1/l * (-1/6 * CUBIC(l, 0, R1) + 1/24 * QUARTIC(l, a, w) - 1/6 * CUBIC(l, b, R2))$

; 3 - the reaction forces

$R1 = 1/l * (w * (l - a)^2 /2 - R2 * (l - b))$

$R2 = w * (l - a) - R1 - R3$

; 4 - note: the above set of equations must be solved simultaneously - R1 and R2 are guessed for iteration.

6   Equations *(f)* through *(j)* provide 5 equations in the 5 unknowns, R_1, R_2, R_3, C_3, C_4 and can be solved simultaneously. Table 4-5 shows the *TKSolver* rule sheet and variable sheets for this problem and also show the value of y_{max}. Guess values are provided for R_1 and R_2 to seed the calculation and the program iterates to the solution. The deflection function can be expressed in terms of the geometry plus the loading and reaction forces but, in this case, a simultaneous solution is necessary.

$$y = \frac{1}{EI}\left(\begin{array}{l} \dfrac{R_1}{6}x^3 + \dfrac{1}{b}\left(\dfrac{w}{24}\langle b - a\rangle^4 - \dfrac{R_1}{6}b^3\right)x - \dfrac{w}{24}\langle x - a\rangle^4 \\ \\ + \dfrac{R_2}{6}\langle x - b\rangle^3 + \dfrac{R_3}{6}\langle x - l\rangle^3 \end{array}\right) \qquad (k)$$

7   Input the file EX04-07 to *TKSolver* to see more detail of the solution and to make larger-scale plots of the loading, shear, moment, slope, and deflection functions shown in Figure 4-27.

This example shows that singularity functions provide a good way to solve beam problems for reactions and deflections simultaneously when there are redundant reactions present. The singularity functions allow the writing of a single expression for each function that applies across the entire beam. They also are inherently computerizable in conjunction with an equation solver that will solve simultaneous equations. The singularity function method presented here is universal and will solve any problems of the types presented.

There are also other techniques for the solution of deflection and redundant reaction problems. **Finite element analysis** (FEA) will solve these problems. The **area-**

Table 4-5   Example 4-7 – Indeterminate Beam with Uniform Load
Part 2 of 2   *TKSolver* Variable Sheet - See File EX04-07

Input	Variable	Output	Unit	Comments
500	w		lb	applied unit load
4	a		in	distance to load w
7	b		in	distance to reaction R_2
10	l		in	length of beam
7.0	x		in	dimension along beam
0.5	I		in^4	area moment of inertia
3.0E7	E		psi	Young's modulus
G *	$R1$	158.4	lb	left reaction force
G	$R2$	2 471.9	lb	center reaction force
	$R3$	369.6	lb	right reaction force
	$C3$	−1 052.7	lb	constant of integration
	q	−500.0	lb	loading function at point x
	$Vmax$	−1 291.6	lb	largest shear force
	$Mmax$	−1 141.1	lb-in	largest moment
	$qmax$	0.000 5	rad	largest slope
	$ymin$	−0.001 1	in	largest deflection

* Indicates that a guess value was required to start the iteration.

moment method treats the moment function as if it were a "loading" function and integrates twice to obtain the deflection function. The reader is referred to this chapter's bibliography for additional information on these topics. **Castigliano's method** uses strain energy equations to determine the deflection at any point.

4.11 CASTIGLIANO'S METHOD

Energy methods often provide simple and rapid solutions to problems. One such method useful for the solution of beam deflections is that of Castigliano. It can also provide a solution to indeterminate beam problems. When an elastic member is deflected by the application of a force, torque or moment, strain energy is stored in the member. For small deflections of most geometries, the relationship between the applied force, moment, or torque and the resulting deflection can be assumed to be linear as shown in Figure 3-17, repeated here. This relationship is often called the spring rate k of the system. The area under the load-deflection curve is the strain energy U stored in the part. For a linear relationship, this is the area of a triangle.

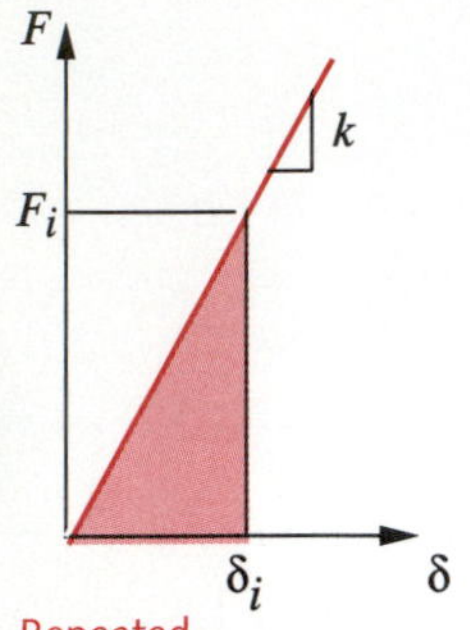

$$U = \frac{F\delta}{2} \tag{4.20}$$

where F is the applied load and δ is the deflection.

Castigliano observed that when a body is elastically deflected by any load, the deflection in the direction of that load is equal to the partial derivative of the strain energy with respect to that load. Letting Q represent a generalized load and Δ a generalized deflection

$$\Delta = \frac{\partial U}{\partial Q} \tag{4.21}$$

This relationship can be applied to any loading case, whether axial, bending, shear, or torsion. If more that one such loading case exists on the same part, their effects can be superposed using equation 4.21 for each case.

STRAIN ENERGY IN AXIAL LOADING   For axial loading, the strain energy is found by substituting the expression for axial deflection (Eq. 4.8) into equation 4.20:

$$U = \frac{1}{2}\frac{F^2 l}{EA} \tag{4.22a}$$

which is valid only if neither E nor A varies over the length l. If they do vary with the distance x along the member, then an integration is necessary:

$$U = \frac{1}{2}\int_0^l \frac{F^2}{EA}dx \tag{4.22b}$$

STRAIN ENERGY IN TORSIONAL LOADING   For torsional loading (see next section) the strain energy is

$$U = \frac{1}{2}\int_0^l \frac{T^2}{GK}dx \tag{4.22c}$$

where T is the applied torque, G is the modulus of rigidity, and K is a geometric property of the cross section as defined in Table 4-6.

STRAIN ENERGY IN BENDING LOADING   For bending, the strain energy is

$$U = \frac{1}{2}\int_0^l \frac{M^2}{EI}dx \tag{4.22d}$$

where M is the bending moment which may be a function of x.

STRAIN ENERGY IN TRANSVERSE SHEAR LOADING   For transverse shear loading in a beam, the strain energy will be a function of the cross-sectional shape as well as the load and length. For a rectangular cross-sectional beam the strain energy is

4

Repeated

FIGURE 3-17

Energy Stored in a Spring

$$U = \frac{3}{5} \int_0^l \frac{V^2}{GA} dx \qquad\qquad (4.22e)$$

where V is the shear loading, which may be a function of x. For cross sections other than rectangular, the fraction 3/5 will be different.

The effects of transverse shear loads on deflections in beams typically will be less than 1% of the effects due to bending moments when the beam's length to depth ratio is >10. Thus, only short beams will have significant transverse shear effects. For nonrectangular beam cross sections, 1/2 is often used in equation 4.22e rather than 3/5 to get a quick approximation of the strain energy due to transverse shear loads. Such a rough calculation will give an indication as to whether the order of magnitude of the deflection due to transverse shear is sufficient to justify a more accurate calculation.

Deflection by Castigliano's Method

This method is useful for calculating deflections at particular points on a system. Equation 4.21 relates force and deflection through the strain energy. For a system deflected by more than one type of load, the individual effects can be superposed using a combination of equations 4.21 and 4.22. When bending and torsional loads are present, their deflection components will often be significantly larger than those due to any axial loading present. For this reason the axial effects are sometimes ignored.

The deflection at points where no actual load is applied can be found by applying a "dummy load" at that point and solving equation 4.21 with the dummy load set to zero. The computation is made easier if the partial differentiation of equation 4.21 is done before performing the integration defined in equations 4.22.

To find the maximum deflection, some knowledge of its location along the beam is needed. The singularity function method, on the other hand, provides the deflection function over the entire beam from which the maxima and minima are easily found.

Finding Redundant Reactions with Castigliano's Method

Castigliano's method also provides a convenient means to solve statically indeterminate problems. For example, reaction forces at redundant supports on a beam can be found using equation 4.21 by setting the deflection at the support to zero and solving for the force.

4.12 TORSION

When members are loaded with a moment about a longitudinal axis, they are said to be in **torsion**, and the applied moment is then called a **torque**. This situation is common in shafts that transmit power, in screw fasteners, and in any situation where the applied moment vector is parallel to the long axis of a part rather than transverse to it as in the case of bending. Many machine parts are loaded with combinations of torques and bending moments and these situations will be dealt with in later chapters. Here we wish to consider only the simple case of **pure torsional loading**.

Figure 4-28*a* shows a straight bar having uniform circular cross section with a pure torque applied in such manner that no bending moment or other forces are present. This can be accomplished with a two-handled wrench such as a tap-handle, which allows a pure couple to be applied with no net transverse force. The fixed end of the bar is embedded in a rigid wall. The bar twists about its long axis and its free end deflects through an angle θ. The assumptions for this analysis are as follows:

1   The element analyzed is distant from applied loads or external constraints on the bar.

2   The bar is subjected to pure torsion in a plane normal to its axis and no axial, bending, or direct shear loads are present.

3   Cross sections of the bar remain plane and perpendicular to the axis.

4   The material of the bar is homogeneous, isotropic, and obeys Hooke's law.

5   Stresses remain below the elastic limit.

6   The bar is initially straight.

CIRCULAR SECTIONS   A differential element taken anywhere on the outer surface will be sheared as a result of the torque loading. The stress τ is pure shear and varies from zero at the center to a maximum at the outer radius, as shown in Figure 4-28*b*,

$$\tau = \frac{T\rho}{J} \qquad (4.23a)$$

where T = applied torque, ρ = radius to any point, and J = the polar area moment of inertia of the cross section. The stress is maximum at the outer surface, at radius r,

$$\tau_{max} = \frac{Tr}{J} \qquad (4.23b)$$

The angular deflection due to the applied torque is

$$\theta = \frac{Tl}{JG} \qquad (4.24)$$

where l is length of the bar and G is the shear modulus (modulus of rigidity) of the material as defined in equation 2.5.

Note that **equations 4.24 only apply to circular cross sections**. Any other cross-sectional shape will behave quite differently. The polar area moment of inertia of a solid circular cross section of diameter d is

$$J = \frac{\pi d^4}{32} \qquad (4.25a)$$

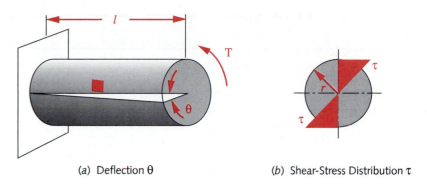

(a) Deflection θ   (b) Shear-Stress Distribution τ

FIGURE 4-28

A Round Bar in Pure Torsion

and for a hollow circular cross section of outside dia d_o and inside dia d_i is

$$J = \frac{\pi\left(d_o^4 - d_i^4\right)}{32} \qquad (4.25b)$$

The circular cross section is the optimum shape for any bar subjected to torsional loading and should be used in all torsion situations if possible.

NONCIRCULAR SECTIONS   In some cases, other shapes may be necessary for design reasons. Noncircular cross sections subjected to torsion exhibit behavior that violates some of the assumptions listed above. Sections do not remain plane, and will warp. Radial lines do not remain straight and the shear stress distribution is not necessarily linear across the section. A general expression for the **maximum shear stress due to torsion in noncircular sections** is

$$\tau_{max} = \frac{T}{Q} \qquad (4.26a)$$

where Q is a function of cross section geometry. The angular deflection is

$$\theta = \frac{Tl}{KG} \qquad (4.26b)$$

where K is a function of cross-sectional geometry. Note the similarity between this equation and equation 4.24. For a closed round section (only), the geometry factor K is the polar moment of inertia, J. For any closed cross-sectional shape other than round, the factor K will be less than J for the same section dimensions, which is an indication of the value of using a closed round section for torsional loading. This fact will be demonstrated in the next example.

Expressions for Q and K for various cross sections can be found in reference 3 as well as in other sources. Table 4-6 shows expressions for Q and K for a few common cross sections and also shows the locations of the maximum shear stress.

4

Table 4-6 Expressions for K and Q for Some Cross-Section Shapes in Torsion

The Black Dots Indicate Points of Maximum Shear Stress (Source: Ref. 3 with Permission)

Shape	K	Q
solid square 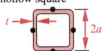	$K = 2.25a^4$	$Q = \dfrac{a^3}{0.6}$
hollow square 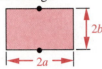	$K = \dfrac{2t^2(a-t)^4}{2at - 2t^2}$	$Q = 2t(a-t)^2$
solid rectangle 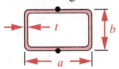	$K = ab^3\left[\dfrac{16}{3} - 3.36\dfrac{b}{a}\left(1 - \dfrac{b^4}{12a^4}\right)\right]$	$Q = \dfrac{8a^2b^2}{3a + 1.8b}$
hollow rectangle	$K = \dfrac{2t^2(a-t)^2(b-t)^2}{at + bt - 2t^2}$ inside corners may have higher stress if radius is small	$Q = 2t(a-t)(b-t)$
solid ellipse 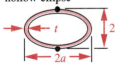	$K = \dfrac{\pi a^3 b^3}{a^2 + b^2}$	$Q = \dfrac{\pi ab^2}{2}$
hollow ellipse	$K = \dfrac{\pi a^3 b^3}{a^2 + b^2}\left[1 - \left(1 - \dfrac{t}{a}\right)^4\right]$	$Q = \dfrac{\pi ab^2}{2}\left[1 - \left(1 - \dfrac{t}{a}\right)^4\right]$
open circular tube	$K = \dfrac{2}{3}\pi rt^3; \quad t \ll r$	$Q = \dfrac{4\pi^2 r^2 t^2}{6\pi r + 1.8t}; \quad t \ll r$
open arbitrary shape	$K = \dfrac{1}{3}Ut^3; \quad t \ll U$	$Q = \dfrac{U^2 t^2}{3U + 1.8t}; \quad t \ll U$

U = length of median line

EXAMPLE 4-8

Design of a Torsion Bar

Problem: Determine the best cross-sectional shape for a hollow torsion bar to be made from a sheet of steel of known dimensions in order to withstand a pure torsional load with minimum angular deflection. Also find the maximum shear stress.

Given: The applied torque is 10 N-m. The sheet of steel has length $l = 1$ m, width $w = 100$ mm, and thickness $t = 1$ mm. Shear modulus $G = 80.8$ GPa.

Assumptions: Try four different cross-sectional shapes: unformed flat plate, open-circular section, closed-circular section, and closed-square section. The open-circular shape is rolled but is not welded at the seam. The closed shapes' seams are welded to create a continuous cross section.

Solution: See Figure 4-29 and Table 4-6.

1   Equations 4.26 will apply for all sections provided that we substitute J for K and J/r for Q in the case of the closed-circular section.

2   The unformed flat plate behaves as a solid rectangular section as shown in Figure 4-28a and Table 4-6. It has dimensions $a = w / 2 = 0.05$ m and $b = t / 2 = 0.000\,5$ m:

$$K = ab^3 \left[\frac{16}{3} - 3.36 \frac{b}{a} \left(1 - \frac{b^4}{12a^4} \right) \right]$$

$$= (0.05)(0.0005)^3 \left[5.333 - 3.36 \frac{0.0005}{0.05} \left(1 - \frac{(0.0005)^4}{12(0.05)^4} \right) \right] \qquad (a)$$

$$K = 3.312E - 11 \text{ m}^4 = 33.12 \text{ mm}^4$$

$$\theta = \frac{Tl}{GK} = \frac{(10)(1)}{(8.08E10)(3.312E - 11)} = 3.71 \text{ rad} = 212.6°$$

This is obviously a rather large angular deflection indicating that the flat plate has been wound into a corkscrew by this torsional load.

$$Q = \frac{8a^2 b^2}{3a + 1.8b}$$

$$= \frac{8(0.05)^2 (0.0005)^3}{3(0.05) + 1.8(0.0005)} \qquad (b)$$

$$Q = 3.313E - 8 \text{ m}^3 = 33.13 \text{ mm}^3$$

$$\tau = \frac{T}{Q} = \frac{10}{3.313E - 8} = 300 \text{ MPa} = 43\,460 \text{ psi}$$

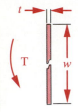

4

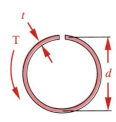

(a) Flat plate

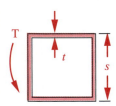

(b) Open-circular

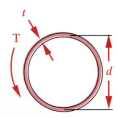

(c) Closed-square

(d) Closed-circular

FIGURE 4-29

Cross-sections for
Example 4-6

The maximum shear stress is 300 MPa which would require a material with a tensile yield strength of over 520 MPa (75 000 psi) in order not to yield and take a set. This requires a high-strength steel. (See Section 5.1 for discussion of this relationship between tensile and shear strength).

3 The open-circular shape is now formed into a 3.18 cm diameter tube but its longitudinal seam is left unwelded and open as shown in Figure 4-29b. The expressions for K and Q from Table 4-6 are

$$K = \frac{2}{3}\pi r t^3$$

$$= \frac{2}{3}\pi \frac{\left(\frac{w}{\pi} - t\right)}{2}t^3 = \frac{1}{3}\left(w - \frac{t}{2}\right)t^3 = \frac{1}{3}\left(0.1 - \frac{0.001}{2}\right)0.001^3 \qquad (c)$$

$$K = 3.33E-11 \text{ m}^4 = 33.33 \text{ mm}^4$$

$$\theta = \frac{Tl}{GK} = \frac{(10)(1)}{(8.08E10)(3.33E-11)} = 3.69 \text{ rad} = 211.2°$$

This is essentially as large an angular deflection as that of the flat plate.

$$Q = \frac{4\pi^2 r^2 t^2}{6\pi r + 1.8t}$$

$$= \frac{4\pi^2(0.016)^2(0.001)^3}{6\pi(0.016) + 1.8(0.001)} \qquad (d)$$

$$Q = 3.313E-8$$

$$\tau = \frac{T}{Q} = \frac{10}{3.313E-8} = 300 \text{ MPa} = 43\,460 \text{ psi}$$

The stress and deflection are unacceptable. It is just as bad a design as the flat plate.

4 The closed-square tube is formed by folding the sheet into a square section with side dimension $s = 2a = w/4$, and $a = w/8$. The seam is welded as shown in Figure 4-29c. From Table 4-6, K and Q, and the stress and deflection are now

$$K = \frac{2t^2(a - t)^4}{2at - 2t^2}$$

$$= \frac{2t^2\left(\frac{w}{8} - t\right)^4}{2\frac{w}{8}t - 2t^2} = \frac{2(0.001)^2\left(\frac{0.1}{8} - 0.001\right)^4}{2\left(\frac{0.1}{8}\right)(0.001) - 2(0.001)^2} \qquad (e)$$

$$K = 1.521E-9$$

$$\theta = \frac{Tl}{GK} = \frac{(10)(1)}{(8.08E10)(1.521E-9)} = 0.081 \text{ rad} = 4.63°$$

$$Q = 2t(s-t)^2$$

$$= 2t\left(\frac{w}{8} - t\right)^2 = 2(0.001)\left(\frac{0.1}{8} - 0.001\right)^2 \tag{f}$$

$$Q = 2.645E - 7$$

$$\tau = \frac{T}{Q} = \frac{10}{2.645E - 7} = 37.5 \text{ MPa} = 5\ 444 \text{ psi}$$

This angular deflection of the square tube is much less than that of either open section and the maximum shear stress is now much more reasonable.

5   The closed-circular shape is formed into a 3.18 cm diameter tube and its longitudinal seam is welded shut as shown in Figure 4-29d. We can either use equations 4.24 and 4.25 or use the general equations 4.26 involving K and Q, which are now a function of J for this circular shape. For the deflection,

$$K = J = \frac{\pi\left(d_o^4 - d_i^4\right)}{32}; \qquad d_0 = \frac{w}{\pi}, \qquad d_i = d_o - 2t$$

$$= \frac{\pi\left[\left(\dfrac{w}{\pi}\right)^4 - \left(\dfrac{w}{\pi} - 2t\right)^4\right]}{32} = \frac{\pi\left[\left(\dfrac{0.1}{\pi}\right)^4 - \left(\dfrac{0.1}{\pi} - 2\{0.001\}\right)^4\right]}{32} \tag{g}$$

$$K = 2.304E - 8$$

$$\theta = \frac{Tl}{GK} = \frac{(10)(1)}{(8.08E10)(2.304E - 8)} = 0.005 \text{ rad} = 0.306°$$

and for the maximum shear stress at the outer surface,

$$Q = \frac{J}{r} = \frac{\pi\left(d_o^4 - d_i^4\right)}{32r}$$

$$= \frac{\pi\left[\left(\dfrac{w}{\pi}\right)^4 - \left(\dfrac{w}{\pi} - 2t\right)^4\right]}{32\left(\dfrac{w}{2\pi}\right)} = \frac{\pi\left[\left(\dfrac{0.1}{\pi}\right)^4 - \left(\dfrac{0.1}{\pi} - 2\{0.001\}\right)^4\right]}{32\left(\dfrac{0.1}{2\pi}\right)} \tag{h}$$

$$Q = 1.448E - 6$$

$$\tau = \frac{T}{Q} = \frac{10}{1.448E - 6} = 6.86 \text{ MPa} = 995 \text{ psi}$$

Note the much smaller deflection and stress for this design. It is clearly the best choice of the four designs presented and is viable for any steel, though the wall thickness could be increased to further reduce stress and deflection if desired. It needs to be checked for possible torsional buckling as well. See the *TKSolver* file EX04-08.

The previous example points out the advantage of using circular sections whenever torsional loads are present. Remember that the amount of material and thus the weight is identical in all four designs in this example. The closed-square section has 15 times the angular deflection of the closed-circular section (tube). The flat plate has 696 times the deflection of the closed-circular tube. Note that the open-circular section is no better in torsion than the flat plate. It has 691 times the angular deflection of the closed tube. This kind of result is true of any open section in torsion, whether I-beam, channel, angle, square, circle, or arbitrary shape. **Any open section is generally no better in torsion than a flat plate of the same cross-sectional dimensions**. Obviously, **open sections should be avoided for all torsionally-loaded applications**. Even nonround-closed sections should be avoided as they are much less efficient in torsion than round-closed sections. **Only closed-circular sections, either hollow or solid, are recommended for torsional loading applications**.

4.13   COMBINED STRESSES

It is very common in machine parts to have combinations of loadings that create both normal and shear stresses on the same part. There may be locations within the part where these applied stresses must be combined to find the principal stresses and maximum shear stress. The best way to demonstrate this is with an example.

EXAMPLE 4-9

Combined Bending and Torsional Stresses

Problem	**Find the most highly stressed locations on the bracket shown in Figure 4-30 and determine the applied and principal stresses at those locations.**
Given	**The rod length l = 6 in and arm a = 8 in. The rod outside diameter d = 1.5 in. Load F = 1 000 lb.**
Assumptions	**The load is static and the assembly is at room temperature. Consider shear due to transverse loading as well as other stresses.**
Solution	**See Figures 4-30 to 4-33 and the *TKSolver* file EX04-09**

1 We will limit our investigation to the rod which is loaded in both bending (as a cantilever beam) and in torsion. (The arm would also need to be analyzed for a complete design). First, the load distributions over the rod's length need to be determined by drawing shear, moment, and torque diagrams for the rod.

2 The shear and moment diagrams will look similar to those for the cantilever beam in Example 4-5, the difference being that this force is at the end of the beam rather than at some intermediate point. Figure 4-31 shows that the shear force is uniform across

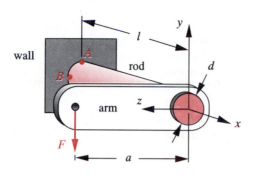

FIGURE 4-30

Bracket for Example 4-9

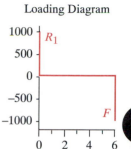

Loading Diagram

Shear Diagram

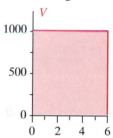

Moment Diagram

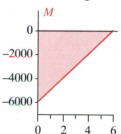

Torque Diagram

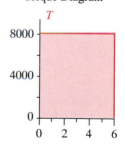

FIGURE 4-31

Loading Diagrams for Bending Shear, Moment, and Torque in Example 4-9

the beam length and its magnitude is equal to the applied load $V_{max} = F = 1\,000$ lb. The maximum moment occurs at the wall and its magnitude is $M_{max} = Fl = (1\,000)$ (6) = 6 000 lb-in. (See Example 4-5 for derivations).

The torque applied to the rod is due to the force F acting at the end of the 8 in arm and is $T_{max} = Fa = (1\,000)(8) = 8\,000$ lb-in. Note that this torque is uniform over the length of the rod as it can only be reacted against by the wall. Figure 4-31 shows all three of these loading functions. It is clear from these plots that the most heavily loaded cross section is at the wall where all three loads are maximum.

3   We will now take a section through the rod at the wall and examine the stress distributions within it due to these external loads. Figure 4-32a shows the distribution across the section of the normal bending stresses, which are a maximum (+/–) at the outer fibers and zero at the neutral axis. The shear stress due to transverse loading is a maximum at all points in the neutral (xz) plane and zero at the outer fibers (Figure 4-32b).

The shear stress due to torsion is proportional to the radius so is zero at the center and a maximum at all points on the outer surface as shown in Figure 4-32c. Note the differences between the distributions of the normal bending stress and the torsional shear stress. The bending stress magnitude is proportional to the distance y from the neutral plane and so is maximum at only the top and bottom of the section, whereas the torsional shear stress is maximal all around the perimeter.

4   We choose two points, A and B of Figure 4-30, to analyze (also shown in Figure 4-32a) because they have the worst combinations of stresses. The largest tensile bending stress will be in the top outer fiber at point A and it combines with the largest torsional shear stress that is all around the outer circumference of the rod. A differential element taken at point A is shown in Figure 4-33b. Note that the normal stress (σ_x) acts on the x face in the x direction and the torsional shear stress (τ_{xz}) acts on the x face in the +z direction.

At point B the torsional shear stress has the same magnitude as at point A, but the direction of the torsional shear stress (τ_{xy}) at point B is 90° different than at point A. The shear stress due to transverse loading (τ_{xy}) is a maximum at point B. Note that these shear stresses both act in the $-y$ direction on the x face at point B as shown in Figure 4-33c. The bending and torsional shear stresses then add at point B.

5   Find the normal bending stress and torsional shear stress on point A using equations 4.11b and 4.19b, respectively.

$$\sigma_x = \frac{Mc}{I} = \frac{(Fl)c}{I} = \frac{1\,000(6)(0.75)}{0.249} = 18\,108 \text{ psi} \tag{a}$$

$$\tau_{xz} = \frac{Tr}{J} = \frac{(Fa)r}{J} = \frac{1\,000(8)(0.75)}{0.497} = 12\,072 \text{ psi} \tag{b}$$

6   Find the maximum shear stress and principal stresses that result from this combination of applied stresses using equations 4.6.

$$\tau_{max} = \sqrt{\left(\frac{\sigma_x - \sigma_z}{2}\right)^2 + \tau_{xz}^2} = \sqrt{\left(\frac{18\,108 - 0}{2}\right)^2 + 12\,072^2} = 15\,090 \text{ psi}$$

$$\sigma_1 = \frac{\sigma_x + \sigma_z}{2} + \tau_{max} = \frac{18\,108}{2} + 15\,090 = 24\,144 \text{ psi}$$

$$\sigma_2 = 0 \tag{c}$$

$$\sigma_3 = \frac{\sigma_x + \sigma_z}{2} - \tau_{max} = \frac{18\,108}{2} - 15\,090 = -6\,036 \text{ psi}$$

7   Find the shear due to transverse loading at point B on the neutral axis. The maximum transverse shear stress at the neutral axis of a round rod was given as equation 4.15c.

$$\tau_{bending} = \frac{4V}{3A} = \frac{4(1\,000)}{3(1.767)} = 755 \text{ psi} \tag{d}$$

Point B is in pure shear. The total shear stress at point B is the algebraic sum of the transverse shear stress and the torsional shear stress, which both act on the same planes of the differential element.

$$\tau_{max} = \tau_{torsion} + \tau_{bending} = 12\,072 + 755 = 12\,827 \text{ psi} \tag{e}$$

which from equation 4.6 or the Mohr's circle can be shown to be equal to the largest principal stress for this point.

8   Point A has the larger principal stress in this case but note that the relative values of the applied torque and moment determine which of these two points will have the higher principal stress. Both points must then be checked.

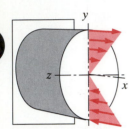

(a) Bending normal-stress distribution across section

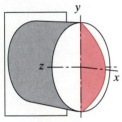

(b) Transverse shear-stress distribution across section

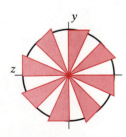

(c) Torsional shear-stress distribution across section

FIGURE 4-32

Cross Sections of Rod for Example 4-9

4.14 SPRING RATES

Every part made of material having an elastic range can behave as a spring. Some parts are designed to function as springs, giving a controlled and predictable deflection in response to an applied load or vice versa. The "springiness" of a part is defined by its spring rate k, which is the load per unit deflection. For rectilinear motion springs,

$$k = \frac{F}{y} \qquad (4.27a)$$

where F is the applied load and y is the resulting deflection. Typical units are lb/in or N/m. For angular motion springs the general expression is

$$k = \frac{T}{\theta} \qquad (4.27b)$$

where T is the applied torque and θ is the resulting angular deflection. Typical units are in-lb/rad or N-m/rad, or sometimes expressed as in-lb/rev or N-m/rev.

The spring rate equation for any part is easily obtained from the relevant deflection equation, which provides a relationship between force (or torque) and deflection. For example, for a uniform bar in axial tension, the deflection is given by equation 4.8, repeated here and rearranged to define its axial spring rate.

$$y = \frac{Fl}{AE}$$

$$k = \frac{F}{y} = \frac{AE}{l} \qquad (4.28)$$

This is a constant spring rate, dependent only on the bar's geometry and its material properties.

For a uniform-section round bar in pure torsion, the deflection is given by equation 4.24c, repeated here and rearranged to define its torsional spring rate:

$$\theta = \frac{Tl}{GJ}$$

$$k = \frac{T}{\theta} = \frac{GJ}{l} \qquad (4.29)$$

This is also a constant spring rate, dependent only on the bar's geometry and material properties.

For a cantilever beam with a concentrated point load as shown in Figure 4-22b, the deflection is given by equation (j) in Example 4-5, repeated here and rearranged to define the beam's spring rate with the force applied at the end of the beam ($a = l$):

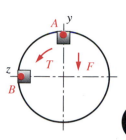

(a) Two points of interest for stress calculations

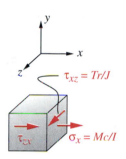

(b) Stress element at point A

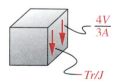

(c) Stress element at point B

FIGURE 4-33

Stress Elements at Points A and B within Cross Section of Rod for Example 4-9

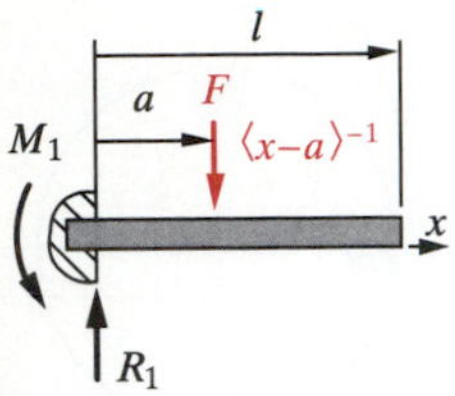

M_1

$\langle x-a \rangle^{-1}$

R_1

Cantilever beam with
concentrated loading

FIGURE 4-22b

Repeated

$$y = \frac{F}{6EI}\left[x^3 - 3ax^2 - \langle x-a \rangle^3\right]$$

but, when $a = l$, $y = \frac{F}{6EI}\left(x^3 - 3lx^2 - 0\right) = \frac{Fx^2}{6EI}(x - 3l)$

for F at $x = l$: $y = \frac{Fl^2}{6EI}(l - 3l) = -\frac{Fl^3}{3EI}$

then $k = \frac{F}{y} = \frac{3EI}{l^3}$ (4.30)

Note that the spring rate k of a beam is unique to its manner of support and its loading distribution since k depends on the particular beam's deflection equation and point of load application. We will investigate spring design in more detail in a later chapter.

4.15   STRESS CONCENTRATION

All the discussion of stress distributions within loaded members has, up to now, assumed that the members' cross sections were uniform throughout. However, most real machine parts will have varying cross sections. For example, shafts often are stepped to different diameters to accommodate bearings, gears, pulleys, etc. A shaft may have grooves for snap-rings or O-rings, or have keyways and holes for the attachment of other parts. Bolts are threaded and have heads bigger than their shank. Any one of these changes in cross-sectional geometry will cause localized stress concentrations.

Figure 4-34 shows the stress concentration introduced by notches and fillets in a flat bar subjected to a bending moment. The stress effects were measured using photoelastic techniques and the resulting "fringes" indicate the stress distribution in the part when loaded. Note that, at the right end of the part where the cross section is uniform, the fringe lines are straight, of uniform width and equispaced. This indicates a linear stress distribution across this portion of the bar. But, at the fillet where the width of the part is reduced from D to d, the fringe lines indicate a disruption and concentration of stresses at this sudden change in geometry. The same effect is seen near the left-hand end around the two notches. The nonlinear stress distribution across the section at the notches is plotted on the lower figure. This is experimental evidence of the existence of stress concentrations at any change in geometry. Such geometric changes in a part are often called "stress-raisers" and should be avoided or at least minimized as much as possible in design. Unfortunately, it is not practical to eliminate all such stress-raisers since such geometric details are needed to connect mating parts and provide functional part shapes.

The amount of stress concentration in any particular geometry is denoted by a geometric stress-concentration factor K_t for normal stresses, or as K_{ts} for shear stresses. The maximum stress at a local stress-raiser is then defined as

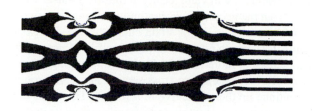

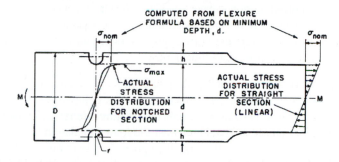

FIGURE 4-34

Photoelastic Measurement of Stress Concentration in a Flat, Stepped, Notched Bar in Bending
Reproduced from reference 4, Fig. 2, p. 3 reprinted by permission of John Wiley & Sons Inc.

$$\sigma_{max} = K_t\,\sigma_{nom}$$
$$\tau_{max} = K_{ts}\,\tau_{nom} \tag{4.31}$$

where σ_{nom} and τ_{nom} are the nominal stresses calculated for the particular applied loading and net cross section assuming a stress distribution across the section that would obtain for a uniform geometry. For example, in the beam of Figure 4-34, the nominal stress distribution is linear and at the outer fiber, $\sigma_{nom} = Mc\,/\,I$. The stress at the notches would then be $\sigma_{max} = K_t\,Mc\,/\,I$. In an axial tension case, the nominal stress distribution would be as defined in Figure 4-10 and for torsion as defined in Figure 4-28. Note that the nominal stresses are calculated using the **net cross section**, which is *reduced by the notch geometry,* i.e., using d instead of D as the width at the notches in Figure 4-34.

The factors K_t and K_{ts} only take the effects of part geometry into account and do not consider how the material behaves in the face of stress concentrations. The ductility or brittleness of the material and the type of loading, whether static or dynamic also affects how it responds to stress concentrations.

Stress Concentration Under Static Loading

The ductility or brittleness of the material has a pronounced effect on its response to stress concentrations under static loads. We will discuss each of these cases in turn.

4

DUCTILE MATERIALS   will yield locally at the stress-raiser while the lower-stressed material further from the geometric discontinuity remains below the yield point. When the material yields locally, its stress strain curve there becomes nonlinear and of low slope (see Figure 2-2), which prevents further significant increase in stress at that point. As the load is increased, more material is yielded, bringing more of the cross section to that stress. Only when the entire cross section has been brought to the yield point will the part continue up the $\sigma \angle \varepsilon$ curve to fracture. Thus, it is common to ignore the effects of geometric stress concentration in *ductile materials under static loading*. The stress for the net cross section is calculated as if the stress concentration were not there. However, the reduction in net cross-sectional area or in area moment of inertia due to the removed material **is** accounted for, thus producing higher stresses than for an un-notched part of the same overall dimensions.

BRITTLE MATERIALS   will not yield locally since they do not have a plastic range. Thus, stress concentrations do have an effect on their behavior even under static loads. Once the stress at the stress-raiser exceeds the fracture strength, a crack begins to form. This reduces the material available to resist the load and also increases the stress concentration in the narrow crack. The part then goes quickly to failure. So, for brittle materials under static loads, the stress-concentration factor should be applied to increase the apparent maximum stress according to equation 4.31.

The one exception to this is for brittle, cast materials that tend to contain many disruptions and discontinuities within their structure due to graphite flakes in the alloy, or air bubbles, foreign matter, sand particles, etc., which found their way into the molten material in the mold. These discontinuities within the material create many stress-raisers, which are also present in the test specimens used to establish the material's basic strengths. Thus the published strength data include the effects of stress concentration. Adding geometric stress-raisers to the part design, it is argued, adds little to the overall statistical effect of those already in the material. Thus, the geometric stress-concentration factor is often ignored for cast-brittle materials or any material with known defects distributed throughout its interior. But, it should be applied to stresses in other brittle materials.

Stress Concentration Under Dynamic Loading

Ductile materials under dynamic loading behave and fail as if they were brittle. So, regardless of the ductility or brittleness of the material, the stress-concentration factor should be applied when dynamic loads (fatigue or impact) are present. However, there are still material-related parameters to account for. While all materials are affected by stress concentrations under dynamic loads, some materials are more sensitive than others. A parameter called notch sensitivity q is defined for various materials and used to modify the geometric factors K_t and K_{ts} for a given material under dynamic loading. These procedures will be discussed in detail in Chapter 6.

Determining Geometric Stress-Concentration Factors

The theory of elasticity can be used to derive stress-concentration functions for some simple geometries. Figure 4-35 shows an elliptical hole in a semi-infinite plate subjected to axial tension. The hole is assumed to be small compared to the plate and far removed from the plate boundaries. The nominal stress is calculated based on the applied force and the total area, $\sigma_{nom} = P/A$. The theoretical stress-concentration factor at the edge of the hole was developed by Inglis in 1913[*] and is

$$K_t = 1 + 2\left(\frac{a}{c}\right) \tag{4.32a}$$

where a is the half-width of the ellipse and c is the half-height. Clearly, as the height of the elliptical hole approaches zero, creating a sharp-edged crack, the stress concentration goes to infinity. When the hole is a circle, $a = c$ and $k_t = 3$. Figure 4-34 also shows a plot of k_t as a function of c/a, the reciprocal of the ratio in equation 4.32a. The function is asymptotic to $k_t = 1$ at large values of c/a.

The theory of elasticity can provide stress-concentration values for some cases. Other stress-concentration factors have come from experimental investigations of parts under controlled loading. Experimental measurements can be made with strain gages, photoelastic techniques, laser holography, or other means. Finite element analysis (FEA) and boundary element analysis (BEA) techniques are increasingly being used to develop stress-concentration factors. When a stress analysis is done with these numerical techniques, the stress concentrations "come out in the wash" as long as the mesh is made sufficiently fine around areas of geometric stress risers. (See Figure 4-51.)

The best-known and most-referenced collection of stress-concentration factor data is in Peterson[3]. This book compiles the theoretical and experimental results of many researchers into useful design charts from which the values of K_t and K_{ts} for various geometric parameters and types of loading can be read. Roark and Young[4] also provide tables of stress-concentration factors for a number of cases.

Figure 4-36 and Appendix E contain stress-concentration functions and their plots based on data from the technical literature for a set of cases representing commonly encountered situations in machine design. In some cases, mathematical functions have been derived to fit the various empirical curves as closely as possible. In other cases *TKSolver* List Functions (table lookups) have been created that allow interpolation and automatic retrieval of the value of K_t in the process of a stress calculation. While these stress-concentration functions (SCF) are approximations of the data in the literature, they go beyond the originals in terms of usefulness since they can be incorporated into a mathematical model of a machine-design problem. These SCF are supplied with this text as *TKSolver* files, which can be merged with other *TKSolver* models or used as stand-alone tools to calculate K_t and K_{ts} for any supplied geometry. This is preferable to looking up data from charts for each calculation.

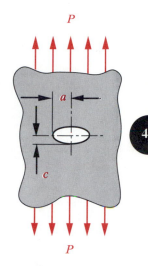

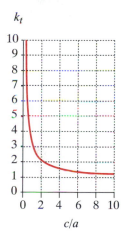

FIGURE 4-35

Stress Concentration at the Edge of an Elliptical Hole in a Plate

* Inglis, C. E., 1913, "Stresses in a Plate Due to the Presence of Cracks and Sharp Corners," *Engineering* (London), Vol. 95, p. 415.

As an example, Figure 4-36 shows the stress-concentration function for the case of a stepped, flat bar in bending. (It and other cases are also shown in Appendix E.) The reduction in width $(D-d)$ at the step creates a stress-raiser and the size of the fillet radius r is also a factor. These two geometric parameters are expressed as the dimensionless ratios r/d and D/d. The first of these is used as the independent variable in the equation and the second determines the member of the family of curves that result. This stress-concentration function is really a three-dimensional surface with the axes, r/d, D/d, and K_t. In Figure 4-36, we are looking at lines on that 3-D surface computed at different values of D/d and projected forward to the r/d–K_t plane. The geometry of the part and its stress equation are defined in the figure, as is the function that defines each stress-concentration curve. In Figure 4-36 it is an exponential function of the form

$$K_t = Ax^b \tag{4.32b}$$

where x represents the independent variable or r/d in this case. The values of the coefficient A and exponent b for any one value of D/d are determined by nonlinear regression on several data points taken from the experimental data. The resulting values of A and b for various magnitudes of the second independent variable D/d are given in the table within the figure. A and b for other values of D/d can be interpolated. The *TKSolver* file name that evaluates these functions and interpolates between them is also noted in this figure and in Appendix E for each of the 14 cases shown there.

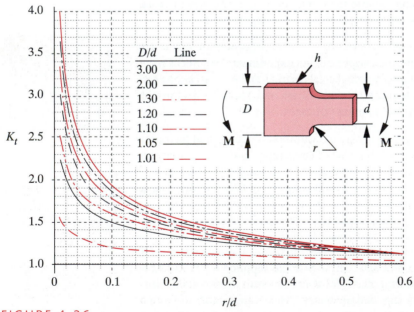

FIGURE 4-36

Geometric Stress-Concentration Factors and Functions for a Stepped Flat Bar in Bending - Also see the *TKSolver* File APP_E-10
Source: Fig. 73, p. 98, R. E. Peterson, Stress Concentration Factors, John Wiley & Sons, 1975, with the publisher's permission

The stress-concentration plots and functions provided in Appendix E along with their corresponding *TKSolver* files will prove useful in the design of machine parts throughout this text and in your practice of engineering. For loading and geometry cases not covered in Appendix E of this text, see references 3 and 4.

Designing to Avoid Stress Concentrations

Complicated geometry is often necessary for the proper function of machine parts. For example, a crankshaft must have particular contours for its purpose. The designer is always faced with the problem of stress concentrations at sections having abrupt changes of shape. The best that can be done is to minimize their effects. A study of the stress-concentration curves for various geometries in Appendix E will show that, in general, the sharper the corner and the larger in magnitude the change in contour, the worse will be the stress concentration. For the stepped bar in Figure 4-36, larger D/d ratios and smaller r/d ratios give worse stress concentration. From these observations, we can state some general guidelines for designing to minimize stress concentrations.

1 Avoid abrupt and/or large-magnitude changes in cross section if possible.

2 Avoid sharp corners completely and provide the largest possible transition radii between surfaces of different contours.

These guidelines are fine to state and better to observe, but practical design constraints often intervene to preclude strict adherence to them. Some examples of good and bad designs for stress concentration are shown in Figures 4-37 through 4-39 along with some common tricks used by experienced designers to improve the situation.

FORCE-FLOW ANALOGY Figure 4-37*a* shows a shaft with an abrupt step and a sharp corner, while Figure 4-37*b* shows the same step in a shaft with a large transition radius. A useful way to visualize the difference in the stress states in contoured parts such as these is to use a "force-flow" analogy, which considers the forces (and thus the stresses) to flow around contours in a way similar to the flow of an ideal incompressible fluid inside a pipe or duct of changing contour. (See also Figure 4-34.) A sudden narrowing of the pipe or duct causes an increase in fluid velocity at the neckdown to maintain constant flow. The velocity profile is then "concentrated" into a smaller region. Streamlined shapes are used in pipes and ducts (and on objects that are pushed through a fluid medium such as aircraft and boats) to reduce turbulence and resistance to flow. "Streamlining" our part contours (at least internally) can have similar beneficial effects in reducing stress concentrations. The force-flow contours at the abrupt step transition in Figure 4-37*a* are more concentrated than in the design of Figure 4-37*b*.

The example in Figure 4-38 is a stepped shaft to which a ball bearing is to be fitted. A step is needed to locate the bearing axially as well as radially on the shaft diameter. Commercial ball- and roller bearings have quite small radii on their corners, which forces the designer to create a fairly sharp corner at the shaft step. To reduce the stress concentration at the step (*a*), a larger radius is needed than the bearing will allow. Three

(a) Force flow around a sharp corner

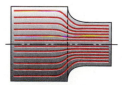

(b) Force flow around a radiused corner

FIGURE 4-37

The Force-Flow Analogy for Contoured Parts

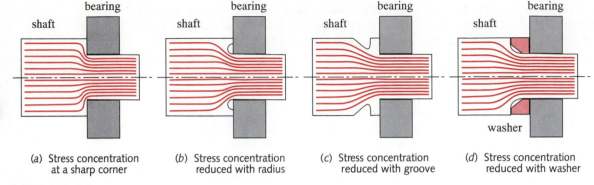

(a) Stress concentration
    at a sharp corner

(b) Stress concentration
    reduced with radius

(c) Stress concentration
    reduced with groove

(d) Stress concentration
    reduced with washer

FIGURE 4-38

Design Modifications to Reduce Stress Concentrations at a Sharp Corner

possible design modifications to create a better force flow around the step are shown in the figure. The first design (b) removes additional material at the corner in order to increase the radius and then "returns" the contour to provide the needed axial locating surface for the bearing. The second approach (c) removes material behind the step to improve the force streamlines. The third approach (d) provides a suitably large corner radius and adds a special washer that bridges the radius to provide the bearing seat. The stress concentration is reduced in each case versus the original sharp-cornered design.

A similar approach of removing material to improve the force flow is seen in Figure 4-39a, which shows a snap-ring groove in a shaft with additional relief grooves provided on each side to smooth the effective transition of the cross-sectional dimension. The effect on the force flow lines is similar to that shown in Figure 4-38c. Another common source of stress concentration is a key needed to torque-couple gears, pulleys, flywheels, etc., to a shaft. The keyway groove creates sharp corners at locations of maximum bending and torsional stresses. Different key styles are available, the most common being the square key and the circular-segment Woodruff key as shown in Figures 4-38b and 4-38c. See Chapter 9 for more information on keys and keyways.

Another example of removing material to reduce stress concentration (not shown) is the reduction of the unthreaded portion of a bolt shank's diameter to a dimension less than that of the root diameter of the thread. Since the thread contours create large stress concentrations, the strategy is to keep the force-flow lines within the solid (unthreaded) portion of the bolt.

These examples show the usefulness of the *force-flow analogy* in providing a means to qualitatively improve the design of machine parts for reduced stress concentration. The designer should attempt to minimize sharp changes in the contours of internal force-flow lines by suitable choice of part shape.

4.16   AXIAL COMPRESSION - COLUMNS

Section 4.7 discussed stress and deflection due to axial tension and developed equations for their calculation, which are repeated here for convenience.

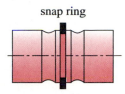

snap ring

(a) Reliefs can reduce stress concentration around a snap-ring groove

$$\sigma_x = \frac{P}{A} \tag{4.7}$$

$$\Delta s = \frac{Pl}{AE} \tag{4.8}$$

When the axial load direction is reversed so as to put the member in compression, equation 4.7 alone may not be sufficient to determine the safe load for the member. It is now a **column** and may fail by buckling rather than by compression. **Buckling** occurs suddenly and without warning, even in ductile materials, and as such is one of the more dangerous modes of failure. You can demonstrate buckling for yourself by taking a common rubber eraser between the palms of your two hands and gradually loading it in axial compression. It will resist the load until at some point it suddenly buckles into a bowed shape and collapses. (If you are feeling stronger, you can do the same with an aluminum beverage can.)

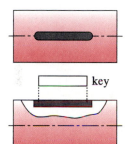

key

(b) Slot for square key creates stress concentrations

Slenderness Ratio

A **short column** *will fail in compression* as shown in Figure 2-6 (p. 63) and its compressive stress can be calculated from equation 4.7. An **intermediate** or a **long column** *will fail by buckling* when the applied axial load exceeds some critical value. The compressive stress can be **well below** the material's yield strength at the time of buckling. The factor that determines if a column is short or long is its **slenderness ratio** S_r,

$$S_r = \frac{l}{k} \tag{4.33}$$

where l is the length of the column and k is its *radius of gyration*. **Radius of gyration** is defined as

$$k = \sqrt{\frac{I}{A}} \tag{4.34}$$

where I is the smallest *area moment of inertia* (*second moment of area*) of the column's cross section (about any neutral axis), and A is its *area at the same cross section*.

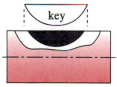

key

(c) Slot for Woodruff key creates stress concentrations

Short Columns

A short column is usually defined as one whose slenderness ratio is less than about 10. The material's yield strength in compression is then used as the limiting factor to compare to the stress calculated in equation 4.7.

FIGURE 4-39

Stress Concentrations in Shafts

Long Columns

A long column requires the calculation of its **critical load**. Figure 4-40 shows a slender column with rounded ends acted upon by compressive forces at either end, which are coaxial and initially act through the area centroid of the column. (A section has been removed to show a reaction force and moment within the column.) The column is shown slightly deflected in the positive y direction, which shifts its area centroid out of colinearity with the applied forces at its ends. This shift of the area centroid creates a moment arm for the force to act about and puts the member in bending as well as in compression. The bending moment tends to increase the lateral deflection, which then also increases the moment arm! Once a critical value of load P_{cr} is exceeded, the positive feedback of this mechanism causes a sudden, catastrophic buckling. There is no visible warning.

The lateral deflection due to the bending moment is given by

$$M = Py \tag{4.35}$$

For small deflections of a beam (Eq. 4.17 repeated),

$$\frac{M}{EI} = \frac{d^2 y}{dx^2} \tag{4.17}$$

Combining 4.35 and 4.17 yields a familiar differential equation:

$$\frac{d^2 y}{dx^2} + \frac{P}{EI} y = 0 \tag{4.36}$$

which has the well-known solution:

$$y = A \sin \sqrt{\frac{P}{EI}} x + B \cos \sqrt{\frac{P}{EI}} x \tag{4.37a}$$

where A and B are constants of integration that depend on the boundary conditions defined for the column of Figure 4-40 as $y = 0$ at $x = 0$; $y = 0$ at $x = l$. Substitution of these conditions shows that $B = 0$ and

$$A \sin \sqrt{\frac{P}{EI}} l = 0 \tag{4.37b}$$

This equation will hold if $A = 0$ but that is a null solution. Thus A must be nonzero and

$$\sin \sqrt{\frac{P}{EI}} l = 0 \tag{4.37c}$$

which will be true for

$$\sqrt{\frac{P}{EI}} l = n\pi; \quad n = 1, 2, 3, \ldots \tag{4.37d}$$

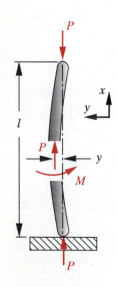

FIGURE 4-40

Buckling of an Euler Column

The first critical load will occur for $n = 1$, which gives

$$P_{cr} = \frac{\pi^2 EI}{l^2} \qquad (4.38a)$$

This is known as the **Euler-column formula** for rounded-end or pinned-end columns. Note that the critical load is a function only of the column's cross sectional geometry I, its length l, and the material's modulus of elasticity E. **The strength of the material is not a factor.** Using a stronger (higher yield strength) steel, for example, will not help matters because all steel alloys have essentially the same modulus of elasticity and will thus fail at the same critical load regardless of yield strength.

Substitute equation 4.33 and the expression $I = Ak^2$ from equation 4.34 into 4.38a:

$$P_{cr} = \frac{\pi^2 EAk^2}{l^2} = \frac{\pi^2 EA}{\left(\dfrac{l}{k}\right)^2} = \frac{\pi^2 EA}{S_r^2} \qquad (4.38b)$$

Normalizing equation 4.38b by the cross-sectional area of the column, we get an expression for the **critical unit load**,

$$\frac{P_{cr}}{A} = \frac{\pi^2 E}{S_r^2} \qquad (4.38c)$$

which has the same units as stress, or strength. It is the **load per unit area** of a rounded- (or pinned-) end column that will cause buckling to occur. Thus, it represents the strength of the particular column rather than the strength of the material from which it is made.

Substituting 4.38a in 4.37a gives the deflection curve for this column as

$$y = A \sin \frac{\pi x}{l} \qquad (4.39)$$

which is a half-period sine wave. Note that applying different boundary or **end conditions** will yield a different deflection curve and a different critical load.

End Conditions

Several possible end conditions for columns are shown in Figure 4-41. The *rounded-rounded* and *pinned-pinned* conditions of Figures 4-41a and 4-41b are essentially the same. They each allow forces but not moments to be supported at their ends. Their boundary conditions are identical, as described above. Their critical unit load is defined in equation 4.38c and their deflection in equation 4.39.

The *fixed-free* column of Figure 4-41c supports a moment and a force at its base and thus controls both the deflection, y and the slope, y' at that end, but it controls nei-

4

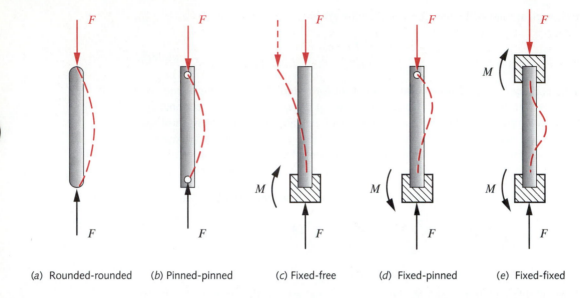

(a) Rounded-rounded   (b) Pinned-pinned   (c) Fixed-free   (d) Fixed-pinned   (e) Fixed-fixed

FIGURE 4-41

Various End Conditions for Columns and their Resultant Deflection Curves *(Applied Loads Shown in Color - Reactions in Black)*

ther x nor y movement at its tip. Its boundary conditions are $y = 0$ and $y' = 0$ at $x = 0$. Substitution of these conditions in equation 4.37a gives

$$\frac{P_{cr}}{A} = \frac{\pi^2 E}{4 S_r^2}$$
(4.40a)

$$y = A \sin \frac{\pi x}{2l}$$
(4.40b)

The deflection curve of a fixed-free Euler column is a quarter sine-wave making it effectively twice as long as a pinned-pinned column having the same cross section. This column can only support 1/4 the critical load of a pinned-pinned column. This reduction can be accounted for by using an effective length l_{eff} for a column with end conditions other than the pinned-pinned conditions used to derive the critical-load equations.

The *fixed-pinned* column (Figure 4-41d) has an $l_{eff} = 0.707l$ and the *fixed-fixed* column (Figure 4-41e) has an $l_{eff} = 0.5l$. The more rigid end constraints make these columns behave as if they were shorter (i.e., stiffer) than a pinned-pinned version and they will thus support more load. Substitute the appropriate effective length in equation 4.33 to obtain the proper slenderness ratio to use in any of the critical-load formulas:

$$S_r = \frac{l_{eff}}{k}$$
(4.41)

where l_{eff} takes the values shown in Table 4-7 for various end conditions. Note that the *fixed-pinned* and *fixed-fixed* conditions have theoretical l_{eff} values of $0.5l$ and $0.707l$, respectively, but these values are seldom used because it is very difficult to obtain a joint fixation that does not allow **any** change in slope at the column end. Welded joints will usually allow some angular deflection, which is dependent on the stiffness of the structure to which the column is welded.

Also, the theoretical analysis assumes that the loading is perfectly centered on the column axis. This condition is seldom realized in practice. Any loading eccentricity will cause a moment and create larger deflections than this model predicts. For these reasons, the AISC* suggests higher values for l_{eff} than the theoretical ones, and some designers use even more conservative values as shown in the third column of Table 4-7. The problem of eccentrically loaded columns is discussed in a later section.

Intermediate Columns

Equations 4.7 and 4.38c are plotted in Figure 4-42 as a function of slenderness ratio. The material's compressive yield strength, S_{yc}, is used as the value of σ_x in equation 4.7 and the critical unit load from equation 4.38c is plotted on the same axis as the material strength. The envelope *OABCO* defined by these two lines and the axes would seem to describe a safe region for column unit loads. However, experiments have demonstrated that columns loaded within this apparently safe envelope will sometimes fail. The problem occurs when the unit loads are in the region *ABDA* near the intersection of the two curves at point *B*. J. B. Johnson suggested fitting a parabolic curve between point *A* and a tangent point *D* on the Euler curve (Eq. 4.38c), which excluded the empirical failure zone. Point *D* is usually taken at the intersection of the Euler curve and a horizontal line at $S_{yc}/2$. The value of $(S_r)_D$ corresponding to this point can be found from equation 4.38c.

Table 4-7   Column End-Condition Effective Length Factors

End Conditions	Theoretical Value	AISC* Recommended	Conservative Value
Rounded-Rounded	$l_{eff} = l$	$l_{eff} = l$	$l_{eff} = l$
Pinned-Pinned	$l_{eff} = l$	$l_{eff} = l$	$l_{eff} = l$
Fixed-Free	$l_{eff} = 2l$	$l_{eff} = 2.1l$	$l_{eff} = 2.4l$
Fixed-Pinned	$l_{eff} = 0.707l$	$l_{eff} = 0.80l$	$l_{eff} = l$
Fixed-Fixed	$l_{eff} = 0.5l$	$l_{eff} = 0.65l$	$l_{eff} = l$

* The American Institute of Steel Construction, in their *Manual of Steel Construction*.

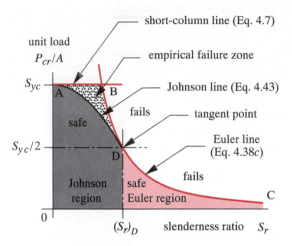

(a) Construction of column failure lines

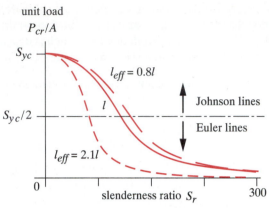

(b) Failure lines for different end conditions

FIGURE 4-42

Euler-, Johnson-, and Short-Column Failure Lines

$$\frac{S_y}{2} = \frac{\pi^2 E}{S_r^2}$$

$$\left(S_r\right)_D = \pi \sqrt{\frac{2E}{S_y}} \tag{4.42}$$

The equation of the parabola fitted between points A and D is

$$\frac{P_{cr}}{A} = S_y - \frac{1}{E}\left(\frac{S_y S_r}{2\pi}\right)^2 \tag{4.43}$$

Equations 4.38c and 4.43 taken together over their appropriate regions then provide a reasonable failure model for all concentrically loaded columns. If the slenderness ratio is $\leq (S_r)_D$, use equation 4.43, else use equation 4.38c. Note that equation 4.43 is both valid and conservative for short columns as well. Equations 4.38c and 4.43 predict failure at the calculated critical unit loads, so an appropriate safety factor must be applied to the result to reduce the allowable load accordingly.

The *TKSolver* file COLMPLOT will compute the critical load and plot the column failure curves of Figure 4-42 for any choices of S_{yc}, S_r, E, and end-condition factor. It can be used to check the design of any concentrically loaded column or to investigate trial designs. The reader is urged to experiment with this program file by changing the values of the above factors and observing the effects on the plotted curves.

EXAMPLE 4-10

Column Design for Concentric Loading

Problem: A beachfront house is to be jacked up 10 ft above grade and placed on a set of steel columns. The weight to be supported by each column is estimated to be 200 000 lb. Two designs are to be considered, one using square steel tubes and the other using round steel tubes.

Given: Design the columns using a safety factor of 4. Determine the columns' outer dimensions for each shape assuming a 0.5-in-thick tube wall in each case. The steel alloy has a compressive yield stress $S_{yc} = 60$ kpsi.

Assumptions: The loading is concentric and the columns are vertical. Their bases are set in concrete and their tops are free, creating a *fixed-free* end-constraint condition. Use AISC recommended end-condition factors.

Solution: See Tables 4-8 and 4-9.

1 This problem, as stated, requires an iterative solution because the allowable load is specified and the column cross-sectional dimensions are requested. If the reverse were desired, equations 4.38c, 4.42, and 4.43 could be solved directly to determine the allowable load for any chosen geometry.

2 To solve this problem using only a calculator requires choosing a trial cross-section dimension, such as the outside diameter, calculating its cross-sectional properties of area A, second moment of area I, radius of gyration k, and the slenderness ratio l_{eff}/k, then using these values in equations 4.38c, 4.42, and 4.43 to determine the allowable load after applying a safety factor. It is not known at the outset whether the column will turn out to be a Johnson or Euler one, so the slenderness ratio $(S_r)_D$ at the tangent point should be found from Eq 4.42 and compared to the column's actual S_r to decide whether Euler's or Johnson's equation is required.

3 Assume an 8-in outside diameter for a first-trial, round column. The area A, second moment of area I, and radius of gyration k for a 0.5-in wall-thickness round tube of that outside diameter are then

$$A = \frac{\pi\left(d_o^2 - d_i^2\right)}{4} = \frac{\pi(64 - 49)}{4} = 11.781 \text{ in}^2$$

$$I = \frac{\pi\left(d_o^4 - d_i^4\right)}{64} = \frac{\pi(4096 - 2401)}{64} = 83.203 \text{ in}^4 \qquad (a)$$

$$k = \sqrt{\frac{I}{A}} = \sqrt{\frac{83.203}{11.781}} = 2.658 \text{ in}$$

4 Calculate the slenderness ratio S_r for this column and compare it to the value of $(S_r)_D$ corresponding to the tangent point between the Euler and Johnson curves (Eq 4.42). Use the AISC recommended value (Table 4-7) for a *fixed-free* column of $l_{eff} = 2.1\ l$.

$$S_r = \frac{l_{eff}}{k} = \frac{120(2.1)}{2.658} = 94.825$$

$$(S_r)_D = \pi\sqrt{\frac{2E}{S_y}} = \pi\sqrt{\frac{2(30E6)}{60\,000}} = 99.346$$

(b)

5   This column's slenderness ratio is to the left of the tangent point and thus is in the Johnson region of Figure 4-42, so use equation 4.43 to find the critical load P_{cr} and apply the safety factor to determine the allowable load P_{allow}.

$$P_{cr} = A\left[S_y - \frac{1}{E}\left(\frac{S_y S_r}{2\pi}\right)^2\right] = 11.8\left\{6E4 - \frac{1}{3E7}\left[\frac{6E4(94.83)}{2\pi}\right]^2\right\} = 384\,866 \text{ lb}$$

(c)

$$P_{allow} = \frac{P_{cr}}{SF} = \frac{384\,866}{4} = 96\,217 \text{ lb}$$

6   This load is substantially below the required 200 000 lb so we must repeat the calculations in steps 3 - 5 using larger outside diameters (or wall thicknesses) until we obtain a suitable allowable load. The problem also requests the design of a square-section column, which will only change the equations (a) in step 3 above.

7   This is clearly a tedious solution process when using only a calculator and it cries out for a better approach. An equation solver or spreadsheet package can provide such a tool. Table 4-8 shows the rule sheets (equations) and the variable sheet for this problem as set up in *TKSolver*. The file EX04-10 is provided on the accompanying disk. The iterative-solving feature of *TKSolver* is well-suited to this problem as it allows us to specify the desired allowable load (*Allow*) and provide a guess value for the outside dimension of the column. Any wall thickness can be specified (set it equal to the outside radius for a solid column). The program then will iterate automatically until it converges on a value for the outside diameter (*Dout*) that provides the specified allowable load.

8   This program also provides for the solution of either a square or round cross section design. The user must type either *'circle* or *'square* (with the leading single quote) into the variable sheet as an input string for the variable *Shape*. The program then tests for this value and calls one of two functions that calculate the column geometry for step 3 above. A guess value has been placed into the subsheet (not shown in table) for the variable *Dout*. This guess value is used to start the iteration. Another test in the program chooses whether to execute the Euler or Johnson functions based on the relative values of the slenderness ratios calculated in step 4 above and then proceeds to step 5 above. The complete solution takes only a few seconds and shows that a round column of 11.3-in diameter and 0.5-in wall is adequate to carry the specified load. This is also a Johnson column with an effective slenderness ratio of 65.6 and it weighs 579 lb. Note that the Euler formula now predicts a critical load nearly 1.5 times that of the Johnson formula and this column would be in the "danger region" labeled *ABDA* in Figure 4-42 if the Euler formula were used.

9 If a 0.5-in wall, square-column cross section is chosen, the outside dimension will be 9.3 in for the same slenderness ratio and allowable load but the column will weigh more at 600 lb. A square column will always be stronger than a round one of the

---

Table 4-8   **Example 4-10 – Column Design**
Part 1 of 2   *TK Solver* Rule and Function Sheets – See *TKSolver* File EX04-10

Rules   **Euler and Johnson Column Formulas for a solid/hollow, round/square column**

; Calculate column's cross–sectional geometry

IF shape = 'circle   THEN CALL Circle(;Din,Dout,A,I,k)
IF shape = 'square THEN CALL Square(;Din,Dout,A,I,k)

; Calculate column's effective length and slenderness ratio

$L_{eff} = end * L$
$Sr = L_{eff} / k$

; Calculate slenderness ratio at tangency between Euler and Johnson

$Srd = PI() * SQRT((2 * E) / Sy)$

; Calculate critical load from either Euler or Johnson

IF Sr < Srd THEN CALL Johnson(Sr ; Unitload) ELSE CALL Euler(Sr ; Unitload)

; Calculate allowable load from critical load and safety factor

*Load = Unitload * A*
Allow = Load / FS

; Calculate both Euler and Johnson loads for comparison

CALL Euler(Sr ; Euler)
CALL Johnson(Sr ; Johnson)

Functions   **(E, Sy, and Wall are in Common to their respective functions)**

;SQUARE - to use: CALL Square (; Din, Dout, A, I, k)

$Din = Dout - 2 * Wall$
$A = Dout^2 - Din^2$
$I = (Dout^4 - Din^4) / 12$
$k = SQRT(I / A)$

;CIRCLE - to use: CALL Circle (; Din, Dout, A, I, k)

$Din = Dout - 2 * Wall$
$A = PI() * (Dout^2 - Din^2) / 4$
$I = PI() * (Dout^4 - Din^4) / 64$
$k = SQRT(I/A)$

;JOHNSON - to use: CALL Johnson (Sr ; Unitload)

$Unitload = Sy - 1/E * (Sy * Sr / (2 * PI()))^2$

;EULER - to use: CALL Euler (Sr ; Unitload)

$Unitload = PI()^2 * E / Sr^2$

Table 4-8   Example 4-10 – Column Design
Part 2 of 2   *TKSolver* Variable Sheet - See File EX04-10

Input	Variable	Output	Unit	Comment
				Input Data
'circle	*Shape*			column shape 'square or 'circle
120	*L*		in	Length of column
0.5	*Wall*		in	column wall thickness
2.1	*end*			AISC end condition factor
30E6	*E*		psi	Young's modulus
60 000	*Sy*		psi	compressive yield stress
4	*FS*			safety factor
200 000	*Allow*		lb	allowable load desired
				Output Data
G *	*Dout*	11.35	in	outside dim of column
	Leff	252	in	effective length of column
	Sr	65.60		slenderness ratio
	Srd	99.35		tangency point in S_r
	Load	46 921	lb	critical unit load
	Johnson	46 921	lb	Johnson unit load
	Euler	68 811	lb	Euler unit load
	Din	10.35	in	inside dim of column
	k	3.84	in	radius of gyration
	I	251.63	in^4	second moment of area
	A	17.05	in^2	area of cross section

* Indicates that a guess value is required to start the iteration.

same outside dimension and wall thickness because its area, second moment of area, and radius of gyration are larger due to the material in the corners being at a larger radius. The additional weight of material also makes it more expensive than a round column of the same strength.

10 Note that the structure of *TKSolver* allows any variable to be declared as either input or output (though some choices as output will require the specification of a guess value for other input parameters in order to start the iteration). Thus, we could choose to specify both the allowable load and an outside column diameter and ask for the column length as the output result. Many other combinations are possible with this program. Try it for yourself with the supplied file.

Eccentric Columns

The above discussion of column failure assumed that the applied load was concentric with the column and passed exactly through its centroid. Even though this condition is desirable, it is seldom achieved in practice as manufacturing tolerances will usually cause the load to be somewhat eccentric to the centroidal axis of the column. In other cases, the design may deliberately introduce an eccentricity e as shown in Figure 4-43. Whatever the cause, the eccentricity changes the loading situation significantly by superposing a bending moment Pe on the axial load P. The bending moment causes a lateral deflection y which in turn increases the moment arm to $e + y$. Summing moments about point A gives

$$\Sigma M_A = -M + Pe + Py = -M + P(e + y) = 0 \qquad (4.44a)$$

Substituting Equation 4.17 yields the differential equation:

$$\frac{d^2y}{dx^2} + \frac{P}{EI}y = -\frac{Pe}{EI} \qquad (4.44b)$$

The boundary conditions are $x = 0$, $y = 0$, and $x = l/2$, $dy/dx = 0$, which give the solution for the deflection at midspan as

$$y = e\left[\sec\left(\frac{l}{2}\sqrt{\frac{P}{EI}}\right) - 1\right] \qquad (4.45a)$$

and for the maximum bending moment as

$$M_{max} = -P(e + y) = -Pe\,\sec\left(\frac{l}{2}\sqrt{\frac{P}{EI}}\right) \qquad (4.45b)$$

The compressive stress is

$$\sigma_c = \frac{P}{A} - \frac{Mc}{I} = \frac{P}{A} - \frac{Mc}{Ak^2} \qquad (4.46a)$$

Substituting the expression for maximum moment from equation 4.45b:

$$\sigma_c = \frac{P}{A}\left[1 + \left(\frac{ec}{k^2}\right)\sec\left(\frac{l}{k}\sqrt{\frac{P}{4EA}}\right)\right] \qquad (4.46b)$$

Failure will occur at midspan when the maximum compressive stress exceeds the yield strength of the material if ductile, or its fracture strength if brittle. Setting σ_c equal to the compressive yield strength for a ductile material gives an expression for the critical unit load of an eccentric column:

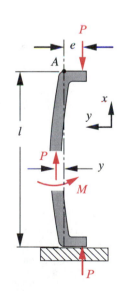

FIGURE 4-43

An Eccentrically-Loaded Column

$$\frac{P}{A} = \frac{S_{yc}}{1 + \left(\dfrac{ec}{k^2}\right) \sec\left(\dfrac{l_{eff}}{k} \sqrt{\dfrac{P}{4EA}}\right)} \qquad (4.46c)$$

This is called the **secant column formula**. The appropriate end-condition factor from Table 4-7 is used to obtain an effective length l_{eff}, which accounts for the column's boundary conditions. The radius of gyration k for equation 4.46c is taken with respect to the axis about which the applied bending moment acts. If the column cross section is asymmetrical and the bending moment does not act about the weakest axis, it must be checked for concentric-column failure about the axis having the smallest k as well as for failure due to eccentric loading in the bending plane.

The fraction ec / k^2 in equation 4.46c is called the **eccentricity ratio E_r** of the column. A 1933 study[*] concluded that assuming a value of 0.025 for the eccentricity ratio would account for typical variations in loading eccentricity of concentrically loaded Euler columns. However, if the column is in the Johnson range, Johnson's formula will apply for E_r's of less than about 0.1. (See Figure 4-44 and its discussion below.)

Equation 4.46c is a difficult function to evaluate. Not only does it require an iterative solution, but the secant function goes to $\pm \infty$ causing computation problems. It also yields incorrect results when the secant function goes negative. The *TKSolver* file SECANT computes and plots equation 4.46c (as well as the Euler and Johnson formulas) over a range of slenderness ratios for any choice of eccentricity ratio and round-column cross sectional parameters. Nonround columns can also be calculated in this program by declaring the area A and moment of inertia I to be input values instead of using the column's linear dimensions. When using this program, take care to plot the resulting function and note the regions, if any, in which the results are incorrect due to the secant's behavior. It will be obvious from the plots.

Figure 4-44 shows plots of equation 4.46c from SECANT (over valid ranges[†]) superimposed on the Euler- Johnson- and short-column plots from Figure 4-42. These curves are normalized to the compressive yield strength of the material. The curve shapes are the same for any material modulus of elasticity E; only the horizontal scale changes. The ratio of the S_r scales to the ratio of E values of different materials is given in the figure.

The secant curves are all asymptotic to the Euler curve at large S_r's. At an eccentricity ratio of zero, the secant curve becomes coincident with the Euler curve up to nearly the level of the short column line. When the eccentricity ratio becomes smaller than about 0.1, the secant functions protrude into the concentric-column empirical-failure region labeled *ABDA* in Figure 4-42, i.e., they move above the Johnson line. This indicates that **for eccentric intermediate columns with small eccentricity ratios**, *the Johnson concentric-column formula (rather than the secant formula) may be the failure criterion* and should also be computed.

[*] Report of a Special Committee on Steel Column Research, *Trans. Amer. Soc. Civil Engrs.*, 98 (1933).

[†] Note in Figure 4-39 that the secant curves for eccentricity ratios of 0.01, 0.05, and 0.1 abruptly end short of the Euler line. This is where the first discontinuity in the secant function occurs and the data beyond those points is invalid until the secant again goes positive. See the plots in file SECANT for further edification.

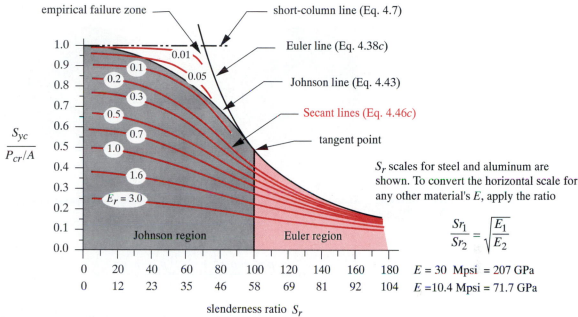

FIGURE 4-44

Secant Lines (in color) Superimposed on Euler-, Johnson-, and Short-Column Failure Lines

4.17   STRESSES IN CYLINDERS

Cylinders are often used as pressure vessels or pipelines and can be subjected to internal and/or external pressure as shown in Figure 4-45. Some common applications are air or hydraulic cylinders, fluid storage tanks and pipes, and gun barrels. Some of these devices are open-ended and some are closed-ended. If open-ended, a two-dimensional stress state will exist in the cylinder walls, with radial and tangential (hoop) stress components. If close-ended, a third-dimensional stress called longitudinal or axial will also be present. These three applied stresses are mutually orthogonal and are principal since there is no applied shear from the uniformly distributed pressure.

Thick-Walled Cylinders

In Figure 4-45, an annular differential element is shown at radius r. The radial and tangential stresses on that element for an open-ended cylinder are given by Lame's equation:

$$\sigma_t = \frac{p_i r_i^2 - p_o r_o^2 - r_i^2 r_o^2 (p_o - p_i)}{r^2 (r_o^2 - r_i^2)} \qquad (4.47a)$$

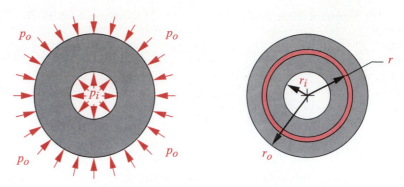

FIGURE 4-45

A Cylinder Subjected to Internal and External Pressure

$$\sigma_r = \frac{p_i r_i^2 - p_o r_o^2 + r_i^2 r_o^2 (p_o - p_i)}{r^2 (r_o^2 - r_i^2)} \tag{4.47b}$$

where r_i and r_o are the inside and outside radii, p_i and p_o are the internal and external pressures, respectively, and r is the radius to the point of interest. Note that these stresses vary nonlinearly throughout the wall thickness.

If the ends of the cylinder are closed, the axial stress in the walls is:

$$\sigma_a = \frac{p_i r_i^2 - p_o r_o^2}{r_o^2 - r_i^2} \tag{4.47c}$$

Note the absence of r in this equation as the axial stress is uniform throughout the wall thickness.

If the external pressure $p_o = 0$ then the equations reduce to

$$\sigma_t = \frac{r_i^2 p_i}{r_o^2 - r_i^2}\left(1 + \frac{r_o^2}{r^2}\right) \tag{4.48a}$$

$$\sigma_r = \frac{r_i^2 p_i}{r_o^2 - r_i^2}\left(1 - \frac{r_o^2}{r^2}\right) \tag{4.48b}$$

and if closed-ended:

$$\sigma_a = \frac{p_i r_i^2}{r_o^2 - r_i^2} \tag{4.48c}$$

The distributions of these stresses across the wall thickness for $p_o = 0$ are shown in Figure 4-46. Under internal pressure, both are maximum at the inside surface. The tangential (hoop) stress is tensile and the radial stress is compressive. When two parts are press- or shrink-fitted together with an interference, the stresses developed in the two parts are defined by equations 4.47. Their mutual elastic deflections create internal pressure on the outer part and external pressure on the inner part. Interference fits are discussed further in Section 9.12.

Thin-Walled Cylinders

When the wall thickness is less than about 1/10 of the radius, the cylinder can be considered thin-walled. The stress distribution across the thin wall can be approximated as uniform and the expressions for stress simplify to

$$\sigma_t = \frac{pr}{t} \tag{4.49a}$$

$$\sigma_r = 0 \tag{4.49b}$$

and if closed-ended:

$$\sigma_a = \frac{pr}{2t} \tag{4.49c}$$

All of these equations are valid only at locations removed from any local stress concentrations or changes in section. For true pressure-vessel design, consult the ASME Boiler Code for more complete information and guidelines to safe design. Pressure vessels can be extremely dangerous even at relatively low pressures if the stored volume is large and the pressurized medium is compressible. Large amounts of energy can be released suddenly at failure, possibly causing serious injury.

4.18   CASE STUDIES IN STATIC STRESS AND DEFLECTION ANALYSIS

We will now present some case studies that continue the design of devices whose forces were analyzed in the case studies of Chapter 3. The same case-study number will be retained for a given design throughout the text and successive installments will be designated by a series of letter suffixes. For example, Chapter 3 presented six case studies labeled 1A, 2A, 3A, 4A, 5A, and 5B. This chapter will continue case studies 1 through 4 as 1B, 2B, 3B, and 4B. Some of these will be continued in later chapters and given successive letter designators. Thus, the reader can review the earlier installments of any case study by referring to its common case number. See the list of case studies in the table of contents to locate each part.

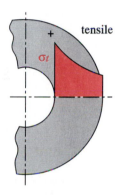

(a) Tangential stress

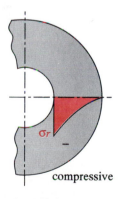

(b) Radial stress

FIGURE 4-46

Tangential and Radial Stress Distributions in Wall of Internally Pressurized Cylinder

Since stresses vary continuously over a part, we must make some engineering judgement as to where they will be the largest and calculate for those locations. We do not have time to calculate the stresses at an infinity of locations. Since the geometries of these parts are fairly complicated, short of doing a complete finite-element stress analysis, we must make some reasonable simplifications in order to model them. The aim is to quickly generate some information about the stress state of the design in order to determine its viability before investing more time in a complete analysis.

CASE STUDY 1B

Bicycle Brake Lever Stress and Deflection Analysis

Problem Determine the stresses and deflections at critical points in the brake lever shown in Figures 3-1 and 4-47.

Given The geometry and loading are known from the previous Case Study 1A (p. 107). The pivot pin is 8-mm dia. The average human's hand can develop a grip force of about 267 N (60 lb) in the lever position shown. A very strong hand can exert about 712 N (160 lb).

Assumptions The most likely failure points are the two holes where the pins insert and at the root of the cantilever-beam lever handle. The cross section of the lever handle is essentially circular.

Solution See Figures 4-47 to 4-48 and the TKSolver file CASE1B.

1 The 14.3-mm-diameter portion of the handle can be modeled as a cantilever beam with an intermediate concentrated load as shown in Figure 4-48 if we assume that the more massive block at its left end serves as a "ground plane." The location of likely failure is the root of the round-handle portion where the shear and moment are both maximum as shown in Figure 4-24 for this model that was analyzed for reactions, moments, and deflections in Examples 3-3 and 4-5. From $\Sigma F = 0$ and $\Sigma M = 0$, we find that $R_1 = 712$ N and $M_1 = 54.6$ N-m. (See Case Study 1-A on p. 107.) The tensile bending stress at the root of the cantilever is maximum at the outer fiber (at point P as shown in Figure 4-47) and is found from equation 4.11b:

$$\sigma_x = \frac{Mc}{I} = \frac{54.6 \text{ N-m}\left(\frac{0.0143}{2}\right)\text{m}}{\frac{\pi(0.0143)^4}{64}\text{m}^4} = 190 \text{ MPa} \qquad (a)$$

This is a relatively low stress for this material. There is some stress concentration due to the small radius at the root of the beam but since this is made of a marginally ductile cast material (5% elongation to fracture) we can ignore the stress concentrations on the basis that local yielding will relieve them.

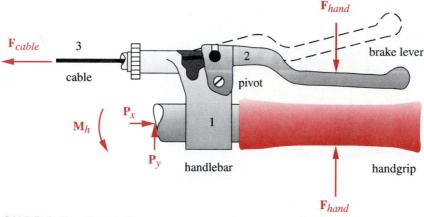

FIGURE 3-1   Repeated

Bicycle Brake Lever Assembly

2   The effective length-to-depth ratio of this beam is small at 76.2 / 14.3= 5.3. Since this ratio is less than 10, the shear stress due to transverse loading will be calculated. For this solid-circular section, it is found from equation 4.15c to be:

$$\tau_{xy} = \frac{4V}{3A} = \frac{4(712) \text{ N}}{\dfrac{3\pi(14.3)^2}{4} \text{ mm}^2} = 6 \text{ MPa} \qquad (b)$$

The shear stress is maximum at the neutral axis (point Q) and the normal bending stress is maximum at the outer fiber (point P). The largest principal stress at the top outer fiber is then $\sigma_1 = \sigma_x = 190$ MPa, $\sigma_2 = \sigma_3 = 0$, and $\tau_{max} = 95$ MPa. The Mohr circle for this stress element looks like the one in Figure 4-8 (repeated here for convenience).

3   The deflection calculation for the handle is complicated by its curved geometry and its slight taper from root to end. A first approximation of the deflection can be obtained by simplifying the model to a straight beam of constant cross section as shown in Figure 4-48b. The deflection due to transverse shear will also be neglected. This will be in slight error in a nonconservative direction but will nevertheless give an order-of-magnitude indication of the deflection. If this result shows a problem with excessive deflection it will be necessary to improve the model. Equation (i) from Example 4-5 provides the deflection equation for our simple model. In this case, $l = 127$ mm, $a = 76.2$ mm, and $x = l$ for the maximum deflection at the end of the beam.

$$y = \frac{F}{6EI}\left[x^3 - 3ax^2 - \langle x - a\rangle^3\right]$$

$$= \frac{712}{6(71.7E3)(2.04E3)}\left[127^3 - 3(76.2)127^2 - (127 - 76.2)^3\right] \qquad (c)$$

$$= -1.4 \text{ mm}$$

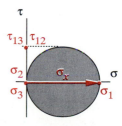

Repeated

FIGURE 4-8

Mohr's Circles for Unidirectional Tensile Stress (two circles are coincident and the third is a point since $\sigma_2 = \sigma_3 = 0$).

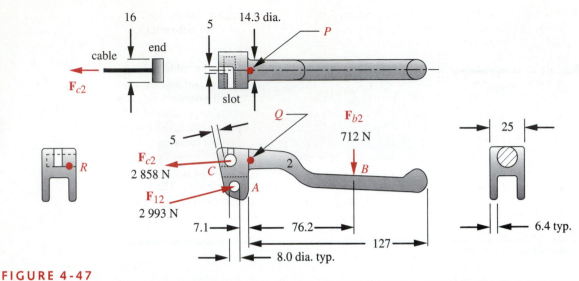

FIGURE 4-47

Bicycle Brake Lever Free-Body Diagram with Forces in N and Dimensions in mm

This is about a 1/16-in deflection at the handle-end, which is not considered excessive in this application. See Figure 4-24 for a plot of the general shape of this beam's deflection curve, though the values are different in that example.

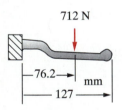

(a) The handle as a cantilever beam

4 Other locations of likely failure must also be checked. The material around the two holes may experience any of several modes of failure due to bearing stress, direct shear stress, or tearout. The hole at point A in Figure 4-47 contains a pivot pin which bears against the handle with the 2 993-N force shown. We will check this for the three modes listed above.

5 Bearing stress is compressive and is considered to act upon the projected area of the hole which, in this case, is the 8-mm hole diameter times the total length of the bearing (two 6.4-mm-thick flanges).

$$A_{bearing} = dia \cdot thickness = 8(2)(6.4) = 102 \text{ mm}^2$$

$$\sigma_{bearing} = \frac{F_{12}}{A_{bearing}} = \frac{2\,993}{102} = 30 \text{ MPa} \qquad (d)$$

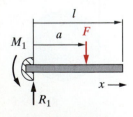

(b) The cantilever-beam model

FIGURE 4-48

Cantilever-Beam Model of Handle

6 Tearout in this case requires that (4) 6.4-mm-thick sections fail in shear through the 5-mm length of material between hole and edge. (See also Figure 4-13 for a definition of tearout area).

$$A_{tearout} = length \cdot thickness = 7.1(4)(6.4) = 181 \text{ mm}^2$$

$$\tau_{tearout} = \frac{F_{12}}{A_{tearout}} = \frac{2\,993}{181} = 17 \text{ MPa} \qquad (e)$$

7   These are very low stresses for the specified material but remember that the applied force used is based on typical human-hand force capability and does not anticipate abuse due to impact or other means.

8   The cable-end inserts in a blind hole, which is half-slotted to allow the cable to pass through at assembly as shown in Figure 4-47. This slot weakens the part and makes the section at C the most likely failure location at this joint. We will assume that failure of the open (slotted) half of the material around the hole is sufficient to disable the part since the cable-end could then slip out. The small section that retains the cable pin can be modeled to a first approximation as a cantilever beam with a cross-sectional width of $(25 - 5) / 2 = 10$ mm and a depth of 5 mm. This is a conservative assumption as it ignores the increase in depth due to the radius of the hole. The moment arm of the force will be assumed to be equal to the radius of the pin or 4 mm. The force on the slotted half of the width is taken as half of the total force of 2 858 N on the cable. The bending stress at the outer fiber at point C is then

$$\sigma_x = \frac{Mc}{I} = \frac{\dfrac{2\,858}{2}\left(\dfrac{5}{2}\right)(4)}{\dfrac{10(5)^3}{12}} = 137 \text{ MPa} \tag{f}$$

and the shear due to transverse loading at the neutral axis is (from equation 4.14b)

$$\tau_{xy} = \frac{3V}{2A} = \frac{3(2\,858)}{2(10)(5)} = 76 \text{ MPa} \tag{g}$$

9   The normal stress is principal here as shown in Figure 4-8 and the maximum shear stress is then half of the principal normal stress. These are the highest stresses found for the three sections checked. A failure analysis of this part will be done in the continuation of this case study in the next chapter.

10 A more complete analysis can also be done using finite element methods to determine the stresses and deformations at many other locations on the part. This preliminary analysis shows some of the areas that might benefit from further investigation. See the *TKSolver* file CASE1B.

CASE STUDY 2B

Crimping-Tool Stress and Deflection Analysis

Problem   Determine the stresses and deflections at critical points in the crimping-tool shown in Figures 3-3 and 4-48.

Given   The geometry and loading are known from the previous Case Study 2A on p. 113. The thickness of link 1 is 0.35 in, of link 3 is 0.125 in and of link 4, 0.313 in. All material is 1095 steel as-rolled with S_y = 83 kpsi and E = 30 Mpsi.

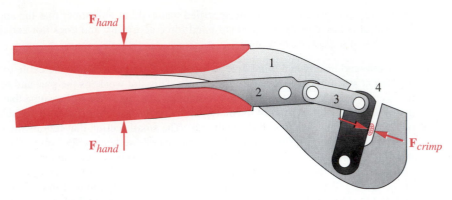

FIGURE 3-3 Repeated

Wire Connector Crimping Tool

Assumptions   The most likely failure points are link 3 as a column, the holes where the pins insert, the connecting pins in shear, and link 4 in bending. The number of cycles expected over the life of the tool is low so a static analysis is acceptable. Stress concentration can be ignored due to the material's ductility and the static loading assumption.

Solution   See Figures 3-3, 4-49 to 4-51 and the TKSolver files CASE2B-1, CASE2B-2, and CASE2B-3.

1   Link 3 is a pinned-pinned column loaded with $F_{43} = 1\,514$ lb as calculated in Case Study 2A (p. 113) and shown in Table 3-3 (p. 118). Note that $l_{eff} = l$ from Table 4-7. We need to first check its slenderness ratio (Eq 4.41). This requires the radius of gyration (Eq. 4.34) for the weakest buckling direction (the z direction in this case).[*]

$$k = \sqrt{\frac{I}{A}} = \sqrt{\frac{bh^3/12}{bh}} = \sqrt{\frac{h^2}{12}} = \sqrt{\frac{0.125^2}{12}} = 0.036 \text{ in} \qquad (a)$$

The slenderness ratio for z direction buckling is then

$$S_r = \frac{l_{eff}}{k} = \frac{1.228}{0.036} = 34 \qquad (b)$$

which is >10 making it other than a short column. Calculate the slenderness ratio of the tangent point between the Johnson and Euler lines of Figure 4-42.

$$(S_r)_D = \pi \sqrt{\frac{2E}{S_y}} = \pi \sqrt{\frac{2(30E6)}{83E3}} = 84.5 \qquad (c)$$

The slenderness ratio of this column is less than that of the tangent point between the Johnson and Euler lines shown in Figure 4-42. It is thus an intermediate-column and the Johnson-column formula (Eq. 4.43) should be used to find the critical load.

[*] Even a small amount of clearance in the holes will prevent the pins from acting as a moment joint along their axes, thus creating an effective pinned-pinned connection in two dimensions.

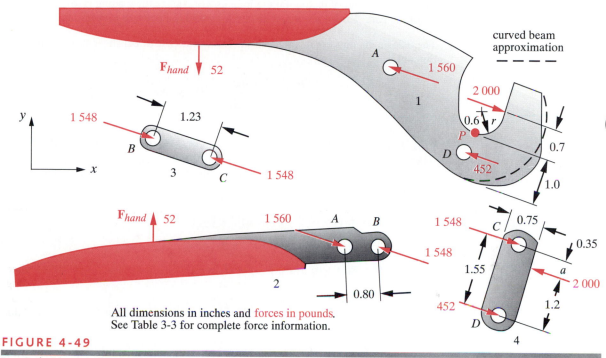

FIGURE 4-49

Free-Body Diagrams, Dimensions, and Force Magnitudes for a Wire Connector Crimping Tool

$$P_{cr} = A\left[S_y - \frac{1}{E}\left(\frac{S_y S_r}{2\pi}\right)^2\right]$$

$$(d)$$

$$= 0.125(.5)\left[83\,000 - \frac{1}{30E6}\left(\frac{83\,000(34)}{2\pi}\right)^2\right] = 4\,765\text{ lb}$$

The critical load is 3.08 times larger than the applied load. It is safe against buckling. Link 2 is a shorter, wider column than link 3 and has lower axial forces so can be assumed to be safe against buckling based on the link 3 calculations. See the *TKSolver* file CASE2B-1.

2   Since it does not buckle, the deflection of link 3 in axial compression is (from Eq. 4.7):

$$x = \frac{Pl}{AE} = \frac{1\,548(1.23)}{0.0625(30E6)} = 0.001\text{ in}$$

$$(e)$$

3   Any of the links could also fail in bearing in the 0.25-in dia holes. The largest force on any pin is 1 560 lb. This worst-case bearing stress (Eqs. 4.7 and 4.10) is then

$$\sigma_b = \frac{P}{A_{bearing}} = \frac{P}{length(dia)} = \frac{1\,560}{0.125(0.25)} = 49\,920\text{ psi}$$

$$(f)$$

There is no danger of tearout failure in links 2 or 3 since the loading is toward the center of the part. Link 1 has ample material around the holes to prevent tearout.

4   The 0.25-in dia pins are in single shear. The worst-case direct shear stress from equation 4.9 is:

$$\tau = \frac{P}{A_{shear}} = \frac{1\,560}{\dfrac{\pi(0.25)^2}{4}} = 31\,780 \text{ psi} \tag{g}$$

5   Link 4 is a 1.55-in-long beam, simply supported at the pins and loaded with the 2 000-lb crimp force at 0.35 in from point C. Write the equations for the load, shear, moment, slope, and deflection using singularity functions, noting that the integration constants C_1 and C_2 will be zero:

$$q = R_1\langle x - 0\rangle^{-1} - F\langle x - a\rangle^{-1} + R_2\langle x - l\rangle^{-1}$$
$$V = R_1\langle x - 0\rangle^{0} - F\langle x - a\rangle^{0} + R_2\langle x - l\rangle^{0} \tag{h}$$
$$M = R_1\langle x - 0\rangle^{1} - F\langle x - a\rangle^{1} + R_2\langle x - l\rangle^{1}$$

$$\theta = \frac{1}{EI}\left(\frac{R_1}{2}\langle x - 0\rangle^2 - \frac{F}{2}\langle x - a\rangle^2 + \frac{R_2}{2}\langle x - l\rangle^2 + C_3\right)$$
$$y = \frac{1}{EI}\left(\frac{R_1}{6}\langle x - 0\rangle^3 - \frac{F}{6}\langle x - a\rangle^3 + \frac{R_2}{6}\langle x - l\rangle^3 + C_3 x + C_4\right) \tag{i}$$

6   The reaction forces can be found from substitution of the boundary conditions $x = 0$, $M = 0$ and $x = l$, $M = 0$ in the moment equation.

$$0 = R_1\langle l - 0\rangle^{1} - F\langle l - a\rangle^{1} + R_2\langle l - l\rangle^{1}$$
$$R_1 = \frac{F(l - a)}{l} = \frac{2000(1.55 - 0.35)}{1.55} = 1\,548 \text{ lb} \tag{j}$$

$$0 = R_1\langle 0 - 0\rangle^{1} - F\langle 0 - a\rangle^{1} + R_2\langle 0 - l\rangle^{1}$$
$$R_2 = \frac{Fa}{l} = \frac{2000(0.35)}{1.55} = 452 \text{ lb} \tag{k}$$

Note that these forces are consistent with the data in Table 3-3. The maximum moment is $1\,548(0.35) = 541.8$ lb-in at the applied load. The shear and moment diagrams for link 4 are shown in Figure 4-50.

7   The beam depth at the point of maximum moment is 0.75 in and the thickness is 0.187. The bending stress is then

$$\sigma = \frac{Mc}{I} = \frac{541.8\left(\dfrac{0.75}{2}\right)}{\dfrac{0.187(0.75)^3}{12}} = 30\,905 \text{ psi} \tag{l}$$

8   The beam slope and deflection functions require calculation of the integration constants C_3 and C_4, which are found by substituting the boundary conditions $x = 0$, $y = 0$ and $x = l$, $y = 0$ in the deflection equation.

$$0 = \frac{1}{EI}\left(\frac{R_1}{6}\langle 0 - 0\rangle^3 - \frac{F}{6}\langle 0 - a\rangle^3 + \frac{R_2}{6}\langle 0 - l\rangle^3 + C_3(0) + C_4 \right)$$

$$C_4 = 0 \qquad\qquad\qquad\qquad\qquad\qquad\qquad\qquad\qquad\qquad\qquad (m)$$

$$0 = \frac{1}{EI}\left(\frac{R_1}{6}\langle l - 0\rangle^3 - \frac{F}{6}\langle l - a\rangle^3 + \frac{R_2}{6}\langle l - l\rangle^3 + C_3(l) \right)$$

$$\qquad\qquad\qquad\qquad\qquad\qquad\qquad\qquad\qquad\qquad\qquad\qquad (n)$$

$$C_3 = \frac{1}{6l}\left[F(l-a)^3 - R_1 l^3 \right] = \frac{1}{6(1.55)}\left[2000(1.55 - 0.35)^3 - 1548(1.55)^3 \right] = 2.143E6$$

9   The deflection equation is found by combining equations i, j, k, m, and n:

$$y = \frac{F}{6lEI}\left\{ (l-a)\left(x^3 + \left[(l-a)^2 - l^2 \right]x \right) - l\langle x - a\rangle^3 + a\langle x - l\rangle^3 \right\} \qquad (o)$$

and the maximum deflection at $x = a = 0.35$ is

$$y_{max} = \frac{Fa(l-a)}{6lEI}\left(a^2 + (l-a)^2 - l^2 \right)$$

$$\qquad\qquad\qquad\qquad\qquad\qquad\qquad\qquad\qquad\qquad\qquad\qquad (p)$$

$$= \frac{2\,000(0.35)(1.55 - 0.35)}{6(1.55)(30E6)(0.006\,6)}\left[0.35^2 + (1.55 - 0.35)^2 - 1.55^2 \right] = 0.000\,5 \text{ in}$$

Only a very small deflection is allowed here to guarantee the proper crimp stroke, and this amount is acceptable. The slope and deflection diagrams are shown in Figure 4-50. Also see the *TKSolver* file CASE2B-1.

10 Link 1 is relatively massive compared to the others and the only area of concern is the jaw, which is loaded by the 2 000-lb crimp force and has a hole in the cross section at its root. While the shape of this element is not exactly that of a curved beam with concentric inside and outside radii, this assumption will be acceptably conservative if we use an outer radius equal to the smallest section dimension as shown in Figure 4-49. This makes its inside radius 0.6 in and its approximate outside radius 1.6 in. The eccentricity e of the curved beam's neutral axis versus the beam's centroidal axis r_c is found from equation 4.12a.

$$e = r_c - \frac{A}{\int_0^{r_o} \frac{dA}{r}} = 1.1 - \frac{0.313(1)}{0.306} = 0.078 \qquad (q)$$

The radius to the neutral axis (r_n) and the distances (c_i and c_o) from the inner and outer fiber radii (r_i and r_o) to the neutral axis are then (see Figure 4-16)

$$r_n = r_c - e = 1.10 - .08 = 1.02$$

$$c_i = r_n - r_i = 1.02 - 0.60 = 0.42 \qquad (r)$$

$$c_o = r_o - r_n = 1.60 - 1.02 = 0.58$$

Loading Diagram (lb)

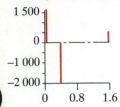

Shear Diagram (lb)

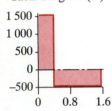

Moment Diagram (lb-in)

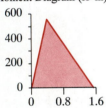

Slope Diagram (rad)

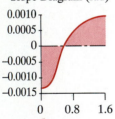

Deflection Diagram (in)

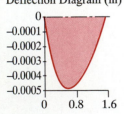

FIGURE 4-50

Link 4 Plots - Case 2B

11　The applied bending moment on the curved beam section is taken as the applied load times its distance to the beam's neutral axis.

$$M = Fl = 2\,000(0.7 - 0.6 + 1.02) = 2\,240 \text{ lb - in} \qquad (s)$$

12　Find stresses at the inner and outer fibers from equations 4.12b and 4.12c.

$$\sigma_i = +\frac{M}{eA}\left(\frac{c_i}{r_i}\right) = \frac{2\,240}{0.08(0.31)}\left(\frac{0.42}{0.60}\right) = 64\,200 \text{ psi} \qquad (t)$$

$$\sigma_o = -\frac{M}{eA}\left(\frac{c_o}{r_o}\right) = \frac{2\,240}{0.08(0.31)}\left(\frac{0.58}{1.60}\right) = 33\,220 \text{ psi}$$

13　There is also a direct axial tensile stress, which adds to the bending stress in the inner fiber at point P:

$$\sigma_a = \frac{F}{A} = \frac{2\,000}{(1.0 - 0.25)(0.3125)} = 8\,533 \text{ psi} \qquad (u)$$

$$\sigma_{max} = \sigma_a + \sigma_i = 72\,734 \text{ psi}$$

This is the principal stress for point P since there is no applied shear or other normal stress at this edge point. The maximum shear stress at point P is half this principal stress or 36 367 psi. The bending stress at the outer fiber is compressive and thus subtracts from the axial tensile stress for a net of $-33\,220 + 8\,533 = -24\,687$ psi.

14　The stress at the hole also needs to be checked as there is significant stress concentration with the axial tensile stress from equation u of 8 533 psi. The theoretical stress-concentration factor for the case of a circular hole in an infinite plate is $k_t = 3$ as defined in equation 4.32a and Figure 4-35. For a circular hole in a finite plate, k_t is a function of the ratio of the hole diameter to the plate width. Peterson gives a chart of stress-concentration factors for a round hole in a flat plate under tension[5] from which we find that $k_t = 2.42$ for a dia / width ratio = 1/4. The local principal stress at the hole is then 2.42(8 533) = 20 650 psi, which is still less than the stress at the inner fiber.

15　While this is far from a complete stress and deflection analysis of these parts, the calculations done address the areas judged to be most likely to fail or to have problem deflections. The stresses and deflections in link 1 were also computed using the ANSYS finite element analysis program, which gave an estimated maximum principal stress at point P of 66 248 psi compared to our estimate of 72 734 psi. The FEA mesh and stress distribution calculated by the ANSYS model is shown in Figure 4-51.

Our analysis simplified the part geometry in order to allow the use of a known closed-form model (the curved beam) whereas the ANSYS FEA model included all the material in the actual part but discretized its geometry. Both analyses should be recognized as only estimates of the stress states in the parts, not exact solutions.

16　Redesign may be needed to reduce these stresses and deflections, based on a failure analysis. This case study will be revisited in the next chapter after various failure theories are presented. See the *TKSolver* file CASE2B-2 for further information and plots.

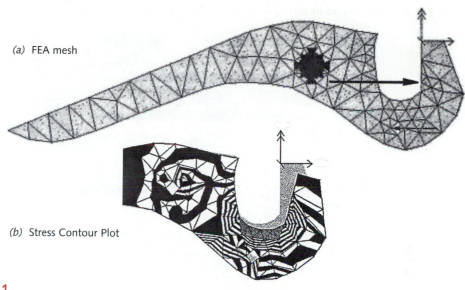

(a) FEA mesh

(b) Stress Contour Plot

FIGURE 4-51

Finite Element Analysis of the Stresses in the Crimping Tool of Case Study 2B[*]

CASE STUDY 3B

Automobile Scissors-Jack Stress and Deflection Analysis

Problem   Determine the stresses and deflections at critical points in the scissors jack assembly shown in Figures 3-5 and 4-52.

Given   The geometry and loading are known from the previous Case Study 3A (p. 118). The design load is 2 000 lb total or 1 000 lb per side. The width of the links is 1.032 in and their thickness is 0.15 in. The screw is a 1/2-13 UNC thread with root dia. = 0.406 in. The material of all parts is ductile steel with $E = 30E6$ psi and $S_y = 60\,000$ psi.

Assumptions   The most likely failure points are the links as columns, the holes in bearing where the pins insert, the connecting pins in shear, the gear teeth in bending, and the screw in tension. There are two sets of links, one set on each side, Assume the two sides share the load equally. The jack is typically used for very few cycles over its lifetime so a static analysis is appropriate.

Solution   See Figures 4-52 to 4-53 and the TKSolver files CASE3B-1 and CASE3B-2.

* The FEA analysis of the crimping tool was done by Stephen Zamarro as part of his senior design project at Worcester Polytechnic Institute.

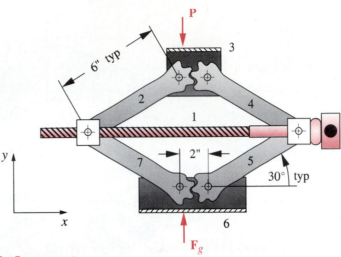

FIGURE 3-5 Repeated

An Automobile Scissors Jack

1   The forces on this jack assembly for the position shown were calculated in the previous installment of this case study (3A) in Chapter 3, (p. 118). Please see that section and Table 3-4 (p. 124) for additional force data. Also see the *TKSolver* file CASE3A for more complete force information than is provided in the table.

2   The jack screw feels the –878-lb component F_{21x} at point A and the +878-lb component F_{41x} at point B as defined in Table 3-4. These forces put the screw in axial tension. The tensile stress is found from equation 4.7 using the 0.406 in root diameter of the thread to calculate the cross-sectional area. This is a conservative assumption as we shall see when we analyze threaded fasteners in a Chapter 14.

$$\sigma_x = \frac{P}{A} = \frac{878}{\dfrac{\pi(0.406)^2}{4}} = \frac{878}{0.129} = 6\ 782 \text{ psi} \qquad (a)$$

The axial deflection of the screw is found from equation 4.8.

$$x = \frac{Pl}{AE} = \frac{878(12.55)}{0.129(30E6)} = 0.003 \text{ in} \qquad (b)$$

3   Link 2 is the most heavily loaded of the links due to the applied load P being slightly offset to the left of center, so we will calculate its stresses and deflections. This link is loaded as a beam-column with both an axial compressive force P between points C and D and a bending couple applied between D and E. Note that the force F_{12} is virtually colinear with the link axis. The axial load is equal to $F_{12} \cos(1°) = 1\ 026$ lb and the bending couple created by F_{42} acting about point D is $M = 412(0.9) = 371$ in-lb. This couple is equivalent to the axial load being eccentric at point D by distance $e = M/P = 371 / 1026 = 0.36$ in.

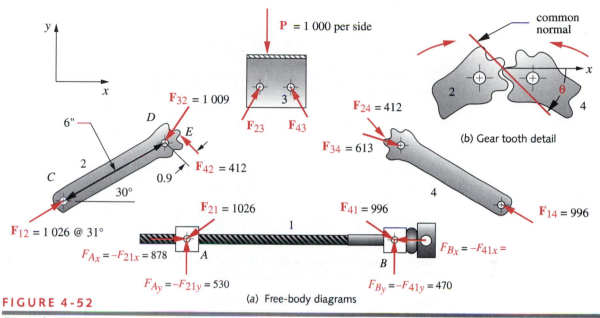

FIGURE 4-52

(a) Free-body diagrams

Free-Body Diagrams, Dimensions, and Forces for Elements of the Scissors Jack

The secant-column formula (Eq. 4.46c) can be used with this effective eccentricity e accounting for the applied couple in the plane of bending; c is 1/2 of the 1.032-in width of the link. Since it is a pinned-pinned column, $l_{eff} = l$ from Table 4-7. The radius of gyration k is taken in the xy plane of bending for this calculation (Eq. 4.34):

$$k = \sqrt{\frac{I}{A}} = \sqrt{\frac{bh^3}{12bh}} = \sqrt{\frac{0.15(1.032)^3}{12(0.15)(1.032)}} = 0.298 \qquad (c)$$

The slenderness ratio is $l_{eff}/k = 20.13$. The secant formula can now be applied and iterated for the value of P. (See Figure 4-53 and *TKSolver* file CASE3B-2.)

$$\frac{P}{A} = \frac{S_{yc}}{1 - \left(\dfrac{ec}{k^2}\right) \sec\left(\dfrac{l_{eff}}{k} \sqrt{\dfrac{P}{4EA}}\right)} = 18\,975 \text{ psi} \qquad (d)$$

$$P_{crit} = 0.155(18\,975) = 2\,937 \text{ lb}$$

The column also has to be checked for concentric-column buckling in the weaker (z) direction with $c = 0.15/2$. The radius of gyration in the z direction is found from

$$k = \sqrt{\frac{I}{A}} = \sqrt{\frac{bh^3}{12bh}} = \sqrt{\frac{1.032(0.15)^3}{12(1.032)(0.15)}} = 0.043 \qquad (e)$$

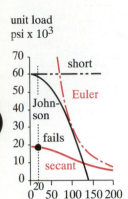

unit load
psi x 10^3

slenderness ratio

FIGURE 4-53

Solution to Eccentric
Column in Case Study 3B

The slenderness ratio in the z direction is

$$S_r = \frac{l_{eff}}{k} = \frac{6}{0.043} = 138.6 \qquad (f)$$

This needs to be compared to the slenderness ratio $(S_r)_D$ at the tangency between the Euler and Johnson lines to determine which buckling equation to use for this column:

$$(S_r)_D = \pi\sqrt{\frac{2E}{S_y}} = \pi\sqrt{\frac{2(30E6)}{60\,000}} = 99.3 \qquad (g)$$

The S_r for this column is greater than $(S_r)_D$, making it an Euler column (see Figure 4-53). The critical Euler load is then found from equation 4.38a.

$$P_{cr} = \frac{\pi^2 EI}{l^2} = \frac{\pi^2(30E6)(1.032)(0.15)^3}{12(6)^2} = 2\,387 \text{ lb} \qquad (h)$$

Thus it is more likely to buckle in the weaker z direction than in the plane of the applied moment. Its safety factor against buckling is 2.3.

4   The pins are all 0.437-in dia. The bearing stress in the most heavily loaded hole at C is

$$\sigma_{bearing} = \frac{P}{A_{bearing}} = \frac{1\,026}{0.15(0.437)} = 15\,652 \text{ psi} \qquad (i)$$

The pins are in single shear and their worst shear stress is

$$\tau = \frac{P}{A_{shear}} = \frac{1\,026}{\dfrac{\pi(0.437)^2}{4}} = 6\,841 \text{ psi} \qquad (j)$$

5   The gear tooth on link 2 is subjected to a force of 412 lb applied at a point 0.22 in from the root of the cantilevered tooth. The tooth is 0.44 in deep at the root and 0.15 in thick. The bending moment is 412(0.22) = 91 in-lb and the bending stress at the root is

$$\sigma = \frac{Mc}{I} = \frac{91(0.22)}{\dfrac{0.15(0.44)^3}{12}} = 18\,727 \text{ psi} \qquad (k)$$

6   This analysis could be continued, looking at other points in the assembly and, more importantly, at stresses when the jack is in different positions. We have used an arbitrary position for this case study but, as the jack moves to the lowered position, the link and pin forces will increase due to poorer transmission angles. A complete stress analysis should be done for multiple positions.

This case study will be revisited in the next chapter for the purpose of failure analysis. See the *TKSolver* files CASE3B-1 and CASE3B-2 for a more complete analysis and to see all the equations and data used.

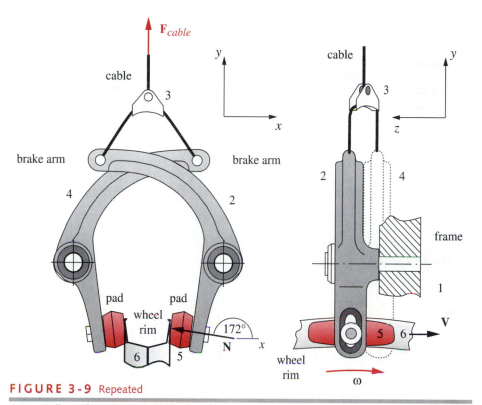

FIGURE 3-9 Repeated

Center-Pull Bicycle Brake Arm Assembly

CASE STUDY 4B

Bicycle Brake-Arm Stress Analysis

Problem   Determine the stresses at critical points in the bicycle brake arm shown in Figures 3-9 and 4-54.

Given   The geometry and loading are known from the previous Case Study 4A (p. 127) and are shown in Table 3-5. The cast-aluminum arm is a tee-section curved beam whose dimensions are shown in Figure 4-54. The pivot pin is ductile steel. The loading is three-dimensional.

Assumptions The most likely failure points are the arm as a double-cantilever beam (one end of which is curved), the hole in bearing, and the connecting pin in bending as a cantilever beam. Since this is a marginally ductile cast material (5% elongation to fracture) we can ignore the stress concentration on the basis that local yielding will relieve it.

Solution   See Figures 4-54 to 4-56 and the TKSolver file CASE4B.

1 The brake arm is a double-cantilever beam. Each end can be treated separately. The curved-beam portion has a tee-shaped cross section as shown in Section X-X of Figure 4-54. The neutral axis of a curved beam shifts from the centroidal axis toward the center of curvature a distance e as described in Section 4-9 and in equation 4.12a. To find e requires an integration of the beam cross section and knowledge of its centroidal radius. Figure 4-55 shows the tee-section broken into two rectangular segments, flange and web. The radius of the centroid of the tee is found by summing moments of area for each segment about the center of curvature:

$$\sum M = A_1 r_{c_1} + A_2 r_{c_2} = A_t r_{c_t}$$

$$r_{c_t} = \frac{A_1 r_{c_1} + A_2 r_{c_2}}{A_t} = \frac{A_1(r_i + y_1) + A_2(r_i + y_2)}{A_1 + A_2} \qquad (a)$$

$$r_{c_t} = \frac{(20)(7.5)(58 + 3.75) + (10)(7.5)(58 + 11.25)}{(20)(7.5) + (10)(7.5)} = 64.25 \text{ mm}$$

See Figures 4-53 and 4-54 for dimensions and variable names. The integral dA/r for equation 4.12a can be found in this case by adding the integrals for web and flange.

$$\int_0^{r_o} \frac{dA}{r} = \frac{A_1}{r_{c_1}} + \frac{A_2}{r_{c_2}} = \frac{(20)(7.5)}{58 + 3.75} + \frac{(10)(7.5)}{58 + 11.25} = 3.51 \text{ mm} \qquad (b)$$

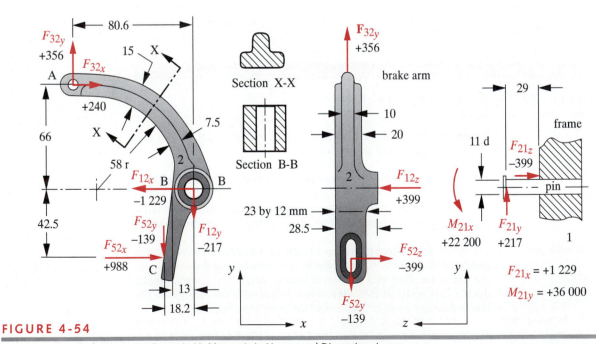

FIGURE 4-54

Brake-Arm Free-Body Diagrams, Forces in N, Moments in N-mm, and Dimensions in mm

The radius of the neutral axis and the distance e are then

$$r_n = \frac{A_t}{\int_0^{r_o} \frac{dA}{r}} = \frac{225}{3.51} = 64.06 \text{ mm}$$

$$(c)$$

$$e = r_c - r_n = 64.25 - 64.06 = 0.187 \text{ mm}$$

The bending moment acting on the curved section of the beam is found by taking the cross-product of the force F_{32} and its position vector R_{AB} referenced to the pivot at B in Figure 4-54.

$$M_{AB} = R_{AB_x}F_{32_y} - R_{AB_y}F_{32_x} = -80.6(356) - 66(240) = 44\,572 \text{ N} \cdot \text{mm} \qquad (d)$$

The stresses at the inner and outer fibers can now be found using equations 4.12b and 4.12c (with lengths in mm and moments in N-mm for proper unit balance):

$$\sigma_i = +\frac{M}{eA}\left(\frac{c_i}{r_i}\right) = \frac{44\,572(6.062)}{(0.187)(225)(58)} = 110.6 \text{ MPa}$$

$$(e)$$

$$\sigma_o = -\frac{M}{eA}\left(\frac{c_o}{r_o}\right) = \frac{44\,572(8.937)}{(0.187)(225)(73)} = -129.5 \text{ MPa}$$

2   The hub cross section, shown as Section B-B in Figure 4-54 is a possible location of failure since there is a combination of bending and axial tension stresses here and the pin hole removes substantial material. The bending stress is due to the moment acting on the curved beam and the tensile stress is due to the y-component of the force at A. There is also a shear stress due to transverse loading but this will be zero at the outer fiber where the sum of the bending and axial stresses are maximum. The area and area moment of inertia of the hub cross section are needed:

$$A_{hub} = length(d_{out} - d_{in}) = 28.5(25 - 11) = 399 \text{ mm}^2$$

$$(f)$$

$$I_{hub} = \frac{length(d_{out}^3 - d_{in}^3)}{12} = \frac{28.5(25^3 - 11^3)}{12} = 33\,948 \text{ mm}^4$$

The stress on the left half of Section B-B is the sum of the bending and axial stress:

$$\sigma_{hub} = \frac{Mc}{I_{hub}} + \frac{F_{32_y}}{A_{hub}} = \frac{44\,572(12.5)}{33\,948} + \frac{356}{399} = 17.3 \text{ MPa}$$

$$(g)$$

The stress on the right half of Section B-B is lower because the compression due to bending is reduced by the axial tension.

3   The straight portion of the brake arm is a cantilever beam loaded in two directions, in the xy plane and in the yz plane. The section moduli and moments are different in these bending directions. The z-moment in the xy plane is equal and opposite to the moment on the curved section. The cross section at the root of the cantilever is a rectangle of

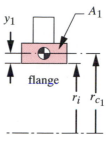

(a) Centroid of flange

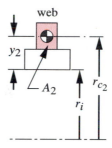

(b) Centroid of web

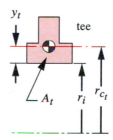

(c) Centroid of tee

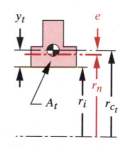

(d) Neutral axis of tee

FIGURE 4-55

Finding Neutral Axis of a Tee-Section Curved Beam - Case Study 4B

y

23 by 12 mm as shown in Figure 4-54. The bending stress at the outer fiber of the 23-mm side due to this moment is

$$\sigma_{y_1} = \frac{Mc}{I} = \frac{44\,572\left(\dfrac{12}{2}\right)}{\dfrac{23(12)^3}{12}} = 80.75 \text{ MPa} \qquad (h)$$

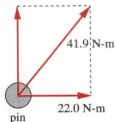

(a) Moment of couple
F_{12z} - F_{52z}

The x moment is due to force F_{52z} acting at the 42.5 radius, bending the link in the z-direction. The bending stress at the surface of the 12-mm side is

$$\sigma_{y_2} = \frac{Mc}{I} = \frac{16\,985\left(\dfrac{23}{2}\right)}{\dfrac{12(23)^3}{12}} = 16.05 \text{ MPa} \qquad (i)$$

These two y-direction normal stresses add at the corners of the two faces to give

$$\sigma_y = \sigma_{y_1} + \sigma_{y_2} = 80.75 + 16.05 = 96.8 \text{ MPa} \qquad (j)$$

4 Another possible failure point is the slot in the cantilever arm. Though the moment is zero there, the shear force is present and can cause tearout in the z direction. The tearout area is the shear area between the slot and edge.

$$A_{tearout} = thickness(width) = 8(4) = 32 \text{ mm}^2$$
$$\tau = \frac{F_{52_z}}{A_{tearout}} = \frac{399}{32} = 12.5 \text{ MPa} \qquad (k)$$

(b) Moment of force F_{21}

5 The pivot pin is subjected to force F_{21}, which has both x and y components and to a couple M_{21} due to the forces F_{12z} and F_{52z}. The force F_{21} creates a bending moment having components $F_{21x}l$ and $F_{21y}l$ in the xz and yz planes, respectively, where l is the length of the pin. Figure 4-56a shows the moment of the couple M_{21} and Figure 4-55b shows the moment of the force F_{21}. Their combination is shown in Figure 4-56c.

It is this combined moment of 72 N-m that creates the largest bending stresses in the pin at 36° and 216° around its circumference. The maximum bending stress in the pin (with lengths in mm and moments in N-mm for unit balance) is

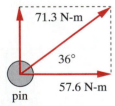

(c) Combined moments

$$\sigma_{pin} = \frac{M_{pin}c_{pin}}{I_{pin}} = \frac{71\,270\left(\dfrac{11}{2}\right)}{\dfrac{\pi(11)^4}{64}} = 545 \text{ MPa} \qquad (l)$$

6 A more complete analysis could be done using finite-element methods to determine the stresses and deformations at many other locations on the part. See the *TKSolver* file CASE4B for more detailed information on the equations and calculations.

FIGURE 4-56

Bending Moments on
Pivot Pin - Case Study 4B

4.19   SUMMARY

The equations used for stress analysis are relatively few and are fairly easy to remember. (See the equation summary later in this section.) The major source of confusion among students seems to be in understanding when to use which stress equation and how to determine where in the part's continuum to calculate the stresses, since they vary over the part's internal geometry.

There are two types of applied stresses of interest, **normal stress** σ, and **shear stress** τ. These each may be present on the same stress element and will combine to create a set of principal normal stresses and maximum shear stress, as evidenced on the Mohr's circle plane. It is ultimately these principal stresses that we need to find in order to determine the safety of the design. So, regardless of the source of loading or type of stress that may be applied to the part, you should always determine the principal stresses and maximum shear stress that result from their combination. (See Sections 4.3 and 4.5.)

There are only a few types of loading that commonly occur on machine parts, but they may occur in combination on the same part. Loading types that create **applied normal stresses** are **bending loads**, **axial loads,** and **bearing loads**. Bending loads will always create both tensile and compressive normal stresses at different locations within the part. Beams provide the most common example of bending loads. (See Section 4.9.) Axial loads create normal stresses that can be either tensile or compressive (but not both at once) depending on whether the axial load is in a tensile or compressive direction. (See Section 4.7.) Fasteners such as bolts often have significant axial tension loads. If the axial load is compressive then there may be a danger of **column buckling** and the equations of Section 4.16 must also be applied. Bearing loads create compressive normal stresses in the shaft and bushing (bearing).

Loading types that create **applied shear stresses** are **torsional loads**, **direct shear loads,** and **bending loads**. Torsional loads involve the twisting of a part around its long axis by application of a torque. A transmission shaft is a typical example of a torsionally loaded part. (Chapter 9 deals with the design of transmission shafts.)

Direct shear can be caused by loads that tend to slice the part transversely. Fasteners such as rivets or pins sometimes experience direct shear loads. A pin trying to tear its way out of its hole also causes direct shear on the tearout area. (See Section 4.8.) Bending loads also cause transverse shear stresses on the cross section of the beam. (See Section 4.9.)

Stresses can vary continuously over the internal continuum of a part's geometry and are calculated as acting at an infinitesimally small point within that continuum. To do a complete analysis of the stresses at all the infinite number of potential sites within the part would require infinite time, which we obviously do not have. So, we must intelligently select a few sites for our calculations such that they represent the worst-case situations.

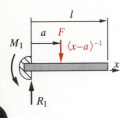

FIGURE 4-57

Cantilever Beam with
Concentrated Load

4

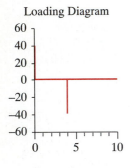

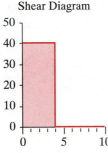

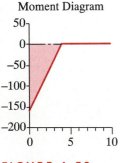

FIGURE 4-58

Load Distributions

The student needs to understand how the various stresses are distributed within the continuum of a loaded part. There are two aspects to determining appropriate locations at which to make the stress calculations on a given part. The first aspect concerns the load distribution over the part's geometry and the second concerns the stress distribution within the part's cross section. For example, consider a straight cantilever beam loaded with a single force at some point along its length as shown in Figure 4-57. The first aspect requires some knowledge of the way the loads on the beam are distributed in response to the applied force. This comes from an analysis of the beam's shear and moment diagrams as shown in Figure 4-58, which indicate that in this case, the section with greatest load is at the wall. We would then concentrate our attention on a vanishingly thin "bologna slice" taken from this beam at the wall. Note that the presence of stress concentrations at other locations having lower nominal stresses would require their investigation too.

The second aspect is then to determine where in this cross-sectional "bologna slice" the stresses will be greatest. Figures in the relevant sections of this chapter show the stress distributions across sections for various types of loadings. These stress-distribution diagrams are collected in Figure 4-59, which also shows the relevant stress equations for each case. Since the loading in this beam example creates bending stresses, we must understand that there will be a normal stress that is maximally compressive at one extreme fiber and maximally tensile at the other extreme fiber as shown in Figures 4-15 and 4-59c. Thus we would take a stress element at the outer fiber of this slice of the beam to calculate the worst-case bending normal stress from equation 4.11b.

Bending loads also cause shear stress but its distribution is maximum at the neutral plane and zero at the outer fiber as shown in Figures 4-19 and 4-59d. So, a different stress element is taken at the neutral plane of the cross-sectional slice for calculation of the shear stress due to transverse loading using the appropriate equation such as Eq. 4.14b for a rectangular cross section. Each of these two stress elements will have its own set of principal stresses and maximum shear stress, which can be calculated from equation 4.6 for this 2-D case.

More complicated loadings on more complicated geometries may have multiple stresses applied to the same infinitesimal stress element. It is very common in machine parts to have loadings that create both bending and torsion on the same part. Example 4-9 deals with such a case and should be studied carefully.

Stress is only one consideration in design. The deflections of parts must also be controlled for proper function. Often, a requirement for small deflections will dominate the design and require thicker sections than would be necessary to guard against excessive stress. The deflections of a designed part, as well as its stresses, should always be checked. Equations for deflection under various loadings are given in the relevant sections and are also collected in Appendix D for beams of various types and loadings.

Figure 4-60 shows a flow chart depicting a set of steps that can be followed to analyze stresses and deflections under static loading.

Static Loading
Stress Analysis
Materials are
assumed to be
Homogeneous
and Isotropic

↓

Find all applied forces,
moments, torques,
etc. and draw the
free-body diagrams to
show them applied to
the part's geometry.

↓

Based on the load
distributions over the
part's geometry,
determine what cross
sections of the part
are most heavily
loaded.

↓

Determine the stress
distributions within
the cross sections of
interest and identify
locations of the
highest applied and
combined stresses.

↓

Draw a 3-D stress
element for each of
the selected points of
interest within the
section and identify
the stresses acting on
it.

↓

Calculate the applied
stresses acting on each
face of every element
and then calculate the
principal stresses and
maximum shear stress
resulting therefrom.

↓

Calculate critical
deflections of the
parts.

FIGURE 4-60

Flow Chart for Static
Stress Analysis

(a) Uniaxial tension,
stress distribution
across section

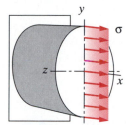

$$\sigma = \frac{P}{A}$$

Eq. 4.7

(b) Direct shear,
average-stress
distribution
across section

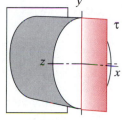

$$\tau = \frac{P}{A_{shear}}$$

Eq. 4.9

(c) Bending, normal-
stress distribution
across section

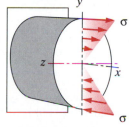

$$\sigma = \frac{My}{I}$$

Eq. 4.11a

(d) Bending, shear-
stress distribution
across section

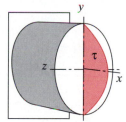

$$\tau = \frac{VQ}{Ib}$$

Eq. 4.13d

(e) Torsion, shear-
stress distribution
across section

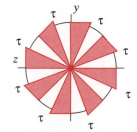

$$\tau = \frac{Tr}{J}$$

Eq. 4.23b

FIGURE 4-59

Distribution of Stresses Across a Cross Section Under Various Types of Loading

Important Equations Used in this Chapter

See the referenced sections for information on the proper use of these equations.

The Stress Cubic - its roots are the 3-D principal stresses (Section 4.3):

$$\sigma^3 - C_2\sigma^2 - C_1\sigma - C_0 = 0$$

where

$$C_2 = \sigma_x + \sigma_y + \sigma_z$$

$$C_1 = \tau_{xy}^2 + \tau_{yz}^2 + \tau_{zx}^2 - \sigma_x\sigma_y - \sigma_y\sigma_z - \sigma_z\sigma_x \qquad (4.4c)$$

$$C_0 = \sigma_x\sigma_y\sigma_z + 2\tau_{xy}\tau_{yz}\tau_{zx} - \sigma_x\tau_{yz}^2 - \sigma_y\tau_{zx}^2 - \sigma_z\tau_{xy}^2$$

Maximum Shear Stresses (Section 4.3):

$$\tau_{13} = \frac{|\sigma_1 - \sigma_3|}{2}$$

$$\tau_{21} = \frac{|\sigma_2 - \sigma_1|}{2} \qquad (4.5)$$

$$\tau_{32} = \frac{|\sigma_3 - \sigma_2|}{2}$$

Two-Dimensional Principal Stresses (Section 4.3):

$$\sigma_a, \sigma_b = \frac{\sigma_x + \sigma_y}{2} \pm \sqrt{\left(\frac{\sigma_x - \sigma_y}{2}\right)^2 + \tau_{xy}^2}$$

$$\sigma_c = 0 \qquad (4.6a)$$

$$\tau_{max} = \tau_{13} = \frac{|\sigma_1 - \sigma_3|}{2} \qquad (4.6b)$$

Axial Tension Stress (Section 4.7):

$$\sigma_x = \frac{P}{A} \qquad (4.7)$$

Axial Deflection (Section 4.7):

$$\Delta s = \frac{Pl}{AE} \qquad (4.8)$$

Direct Shear Stress (Section 4.8):

$$\tau_{xy} = \frac{P}{A_{shear}} \qquad (4.9)$$

Direct Bearing Area (Section 4.8):

$$A_{bearing} = l\,d \qquad (4.10)$$

Maximum Bending Stress - Straight Beams (Section 4.9):

$$\sigma_{max} = \frac{Mc}{I} \qquad (4.11b)$$

Maximum Bending Stress - Curved Beams (Section 4.9):

$$\sigma_i = + \frac{M}{eA}\left(\frac{c_i}{r_i}\right) \qquad (4.12b)$$

Transverse Shear Stress in Beams - General Formula (Section 4.9):

$$\tau_{xy} = \frac{V}{Ib}\int_{y_1}^{c} y\,dA \qquad (4.13f)$$

Maximum Transverse Shear Stress - Rectangular Beam (Section 4.9):

$$\tau_{max} = \frac{3}{2}\frac{V}{A} \qquad (4.14b)$$

Maximum Transverse Shear Stress - Round Beam (Section 4.9):

$$\tau_{max} = \frac{4}{3}\frac{V}{A} \qquad (4.15c)$$

Maximum Transverse Shear Stress - I-Beam (Section 4.9):

$$\tau_{max} \cong \frac{V}{A_{web}} \qquad (4.16)$$

The General Beam Equations (Section 4.9):

$$\frac{q}{EI} = \frac{d^4 y}{dx^4} \qquad (4.18a)$$

$$\frac{V}{EI} = \frac{d^3 y}{dx^3} \qquad (4.18b)$$

$$\frac{M}{EI} = \frac{d^2 y}{dx^2} \qquad (4.18c)$$

$$\theta = \frac{dy}{dx} \qquad (4.18d)$$

$$y = f(x) \qquad (4.18e)$$

Maximum Torsional Shear Stress - Round Section (Section 4.12):

$$\tau_{max} = \frac{Tr}{J}$$

(4.23b)

Maximum Torsional Deflection - Round Section (Section 4.12):

$$\theta = \frac{Tl}{JG}$$

(4.24)

Maximum Torsional Shear Stress - Nonround Section (Section 4.12):

$$\tau_{max} = \frac{T}{Q}$$

(4.26a)

Maximum Torsional Deflection - Nonround Section (Section 4.12):

$$\theta = \frac{Tl}{KG}$$

(4.26b)

Spring Rate or Spring Constant - linear (a), angular (b) (Section 4.14):

$$k = \frac{F}{y}$$

(4.27a)

$$k = \frac{T}{\theta}$$

(4.27b)

Stress with Stress Concentration (Section 4.15):

$$\sigma_{max} = K_t \, \sigma_{nom}$$

$$\tau_{max} = K_{ts} \, \tau_{nom}$$

(4.31)

Column Radius of Gyration (Section 4.16):

$$k = \sqrt{\frac{I}{A}}$$

(4.34)

Column Slenderness Ratio (Section 4.16):

$$S_r = \frac{l}{k}$$

(4.33)

Column Critical Unit Load - Euler Formula (Section 4.16):

$$\frac{P_{cr}}{A} = \frac{\pi^2 E}{S_r{}^2}$$

(4.38c)

Column Critical Unit Load - Johnson Formula (Section 4.16):

$$\frac{P_{cr}}{A} = S_y - \frac{1}{E}\left(\frac{S_y S_r}{2\pi}\right)^2 \tag{4.43}$$

Column Critical Unit Load - Secant Formula (Section 4.16):

$$\frac{P}{A} = \frac{S_{yc}}{1 + \left(\dfrac{ec}{k^2}\right)\sec\left(\dfrac{l_{eff}}{k}\sqrt{\dfrac{P}{4EA}}\right)} \tag{4.46c}$$

Pressurized Cylinder (Section 4.17)

$$\sigma_t = \frac{p_i r_i^2 - p_o r_o^2 - r_i^2 r_o^2 (p_o - p_i)}{r^2 (r_o^2 - r_i^2)} \tag{4.47a}$$

$$\sigma_r = \frac{p_i r_i^2 - p_o r_o^2 + r_i^2 r_o^2 (p_o - p_i)}{r^2 (r_o^2 - r_i^2)} \tag{4.47b}$$

$$\sigma_a = \frac{p_i r_i^2 - p_o r_o^2}{r_o^2 - r_i^2} \tag{4.47c}$$

4.20   REFERENCES

1   **I. H. Shames and C. L. Dym**, *Energy and Finite Element Methods in Structural Mechanics*. Hemisphere Publishing: New York, Sect. 1.6, 1985.

2   **I. H. Shames and F. A. Cossarelli**, *Elastic and Inelastic Stress Analysis*. Prentice-Hall: Englewood Cliffs, N.J., pp. 46-50, 1991.

3   **R. E. Peterson**, *Stress Concentration Factors*. John Wiley & Sons: New York, 1974.

4   **R. J. Roark and W. C. Young**, *Formulas for Stress and Strain*. 6th ed. McGraw-Hill: New York, 1989.

5   **R. E. Peterson**, *Stress Concentration Factors*. John Wiley & Sons: New York, p. 150, 1974.

4.21   BIBLIOGRAPHY

for general information on stress and deflection analysis see:

F. P. Beer and E. R. Johnston, *Mechanics of Materials*. 2nd ed. McGraw-Hill: New York, 1992.

J. P. D. Hartog, *Strength of Materials*. Dover: New York, 1961.

4

R. J. Roark and W. C. Young, *Formulas for Stress and Strain*. 6th ed. McGraw-Hill: New York, 1989.

I. H. Shames, *Introduction to Solid Mechanics*. Prentice-Hall: Englewood Cliffs, N.J., 1989.

I. H. Shames and F. A. Cossarelli, *Elastic and Inelastic Stress Analysis*. Prentice-Hall: Englewood Cliffs, N.J., 1991.

S. Timoshenko and D. H. Young, *Elements of Strength of Materials*. 5th ed. Van Nostrand: New York, 1968.

for information on the finite element method see:

K. J. Bathe, *Finite Element Procedures in Engineering Analysis*. Prentice-Hall: Englewood Cliffs, N.J., 1982.

H. T. Grandin, *Fundamentals of the Finite Element Method*. Waveland Press: Prospect Heights, Il., 1991.

D. L. Logan, *A First Course in the Finite Element Method*. 2nd ed. PWS Kent: Boston, 1992.

I. H. Shames and C. L. Dym, *Energy and Finite Element Methods in Structural Mechanics*. Hemisphere Publishing: New York, 1985.

E. Zahavi, *The Finite Element Method in Machine Design*. Prentice-Hall: Englewood Cliffs, N.J., 1992.

4.22   PROBLEMS

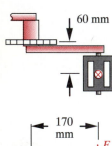

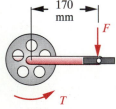

FIGURE P4-1

Problem 4-3

*4-1   A differential stress element has a set of applied stresses on it as indicated in each row of Table P4-1. For the row(s) assigned, draw the stress element showing the applied stresses, find the principal stresses and maximum shear stress analytically and check the results by drawing Mohr's circles for that stress state.

4-2   A 400-lb chandelier is to be hung from two 10-ft-long solid steel cables in tension. Choose a suitable diameter of cable which will not exceed an allowable stress of 5 000 psi. What will be the deflection of the cables? State all assumptions.

†4-3   For the bicycle pedal-arm assembly in Figure P4-1 with a rider-applied force of 1500 N at the pedal, determine the maximum principal stress in the pedal arm if its cross section is 15 mm in dia. The pedal attaches to the pedal arm with a 12-mm screw thread. What is the stress in the pedal screw?

*†4-4   The trailer hitch shown in Figure P4-2 and Figure 1-1 (p. 12) has loads applied as defined in Problem 3-4. The tongue weight of 100 kg acts downward and the pull force of 4 905 N acts horizontally. Using the dimensions of the ball bracket shown in Figure 1-5 (p. 15), determine:

(a) The principal stresses in the shank of the ball where it joins the ball bracket.
(b) The bearing stress in the ball bracket hole.
(c) The tearout stress in the ball bracket.

(d) The normal and shear stresses in the attachment bolts if they are 19-mm dia,
(e) The principal stresses in the ball bracket as a cantilever.

†4-5   Repeat Problem 4-4 for the loading conditions of Problem 3-5.

*†4-6   Repeat Problem 4-4 for the loading conditions of Problem 3-6.

*†4-7   Design the wrist pin of Problem 3-7 for a maximum allowable principal stress of 20 kpsi if the pin is hollow and loaded in double shear.

*†4-8   A paper mill processes rolls of paper having a density of 984 kg/m^3. The paper roll is 1.50-m outside dia (OD) x 22 cm inside dia (ID) x 3.23-m long and is on a simply supported, hollow, steel shaft. Find the shaft ID needed to obtain a maximum deflection at the center of 3 mm if the shaft OD is 22 cm.

†4-9   For the ViseGrip® plier-wrench drawn to scale in Figure P4-3, and for which the forces were analyzed in Problem 3-9, find the stresses in each pin for an assumed clamping force of $P = 4\,000$ N in the position shown. The pins are 8-mm dia and are all in double shear.

*†4-10 The overhung diving board of Problem 3-10 is shown in Figure P4-4a. Assume cross-section dimensions of 305 mm x 32 mm. The material has $E = 10.3$ GPa. Find the largest principal stress at any location in the board when a 100-kg person is standing at the free end. What is the maximum deflection?

*†4-11 Repeat Problem 4-10 using the loading conditions of Problem 3-11. Assume the board weighs 29 kg and deflects 13.1 cm statically when the person stands on it. Find the largest principal stress at any location in the board when the 100-kg person in Problem 4-10 jumps up 25 cm and lands back on the board. Find the maximum deflection.

†4-12   Repeat Problem 4-10 using the cantilevered diving board design in Figure P4-4b.

†4-13   Repeat Problem 4-11 using the diving board design shown in Figure P4-4b. Assume the board weighs 19 kg and deflects 8.5 cm statically when the person stands on it.

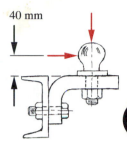

40 mm

FIGURE P4-2

Problems 4-4, 4-5, 4-6

4

Table P4-1   Data for Problem 4-1
Rows *a-g* are two-dimensional, others are 3-D problems

Row	σ_x	σ_y	σ_z	τ_{xy}	τ_{yz}	τ_{zx}
a	1 000	0	0	500	0	0
b	−1 000	0	0	750	0	0
c	500	−500	0	1 000	0	0
d	0	−1 500	0	750	0	0
e	750	250	0	500	0	0
f	−500	1 000	0	750	0	0
g	1 000	0	−750	0	0	250
h	750	500	250	500	0	0
i	1 000	−250	−750	250	500	750
j	−500	750	250	100	250	1 000

* Answers to these problems are provided in Appendix G.

† These problems are based on similar problems in previous chapters with the same −number, e.g., Problem 4-4 is based on Problem 3-4, etc. Problems in succeeding chapters may also continue and extend these problems.

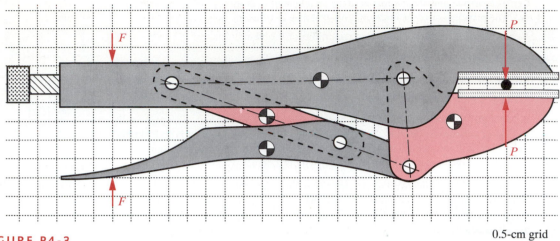

FIGURE P4-3

0.5-cm grid

Problem 4-9

†4-14   Figure P4-5 shows a child's toy called a *pogo stick*. The child stands on the pads,
applying half her weight on each side. She jumps up off the ground, holding the
pads up against her feet, and bounces along with the spring cushioning the impact
and storing energy to help each rebound. Assume a 60-lb child and a spring constant
of 100 lb/in. The pogo stick weighs 5 lb. Design the aluminum cantilever beam
sections on which she stands to survive jumping 2 in off the ground. Assume an
allowable stress of 20 kpsi. Define the beam shape and size it appropriately.

*4-15   Design a shear pin for the propeller shaft of an outboard motor if the shaft through
which the pin is placed is 25-mm diameter, the propeller is 20-cm diameter, and the
pin must fail when a force > 400 N is applied to the propeller tip. Assume an
ultimate shear strength for the pin material of 100 MPa.

†4-16   A track to guide bowling balls is designed with two round rods as shown in Figure
P-4-6. The rods are not parallel to one another but have a small angle between them.
The balls roll on the rods until they fall between them and drop onto another track.

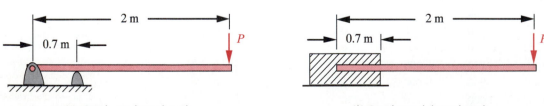

(*a*) Overhung diving board   (*b*) Cantilevered diving board

FIGURE P4-4

Problems 4-10 through 4-13

The angle between the rods is varied to cause the ball to drop at different locations. Each rod's unsupported length is 30 in and the angle between them is 3.2°. The balls are 4.5-in dia and weigh 2.5 lb. The center distance between the 1-in-dia rods is 4.2 in at the narrow end. Find the maximum stress and deflection in the rods.

(a) Assume rods are simply supported at each end.
(b) Assume rods are fixed at each end.

†4-17   A pair of ice tongs is shown in Figure P4-7. The ice weighs 50 lb and is 10 in wide across the tongs. The distance between the handles is 4 in, and the mean radius r of a tong is 6 in. The rectangular cross-sectional dimensions are 0.75 in x 0.312 in. Find the stress in the tongs.

*4-18   A set of steel reinforcing rods is to be stretched axially in tension to create a tensile stress of 30 kpsi prior to being cast in concrete to form a beam. Determine how much force will be required to stretch them the required amount and how much deflection is required. There are 10 rods; each is 0.75-in diameter and 30 ft long.

*4-19   The clamping fixture used to pull the rods in Problem 4-18 is connected to the hydraulic ram by a clevis like that shown in Figure P4-8. Determine the size of the clevis pin needed to withstand the applied force. Assume an allowable shear stress of 20 000 psi and an allowable normal stress of 40 000 psi. Determine the required outside radius of the clevis end to not exceed the above allowable stresses in either tearout or bearing if the clevis flanges are each 0.8 in thick.

4-20   Repeat Problem 4-19 for 12 rods each 1 cm in diameter and 10 m long. The desired rod stress is 200 MPa. The allowable normal stress in the clevis and pin is 280 MPa and their allowable shear stress is 140 MPa. Each clevis flange is 2 cm wide.

†4-21   Figure P4-9 shows an automobile wheel with two common styles of lug wrench being used to tighten the wheel nuts, a single ended wrench in (*a*), and a double-ended wrench in (*b*). In each case two hands are required to provide forces, respectively, at *A* and *B* as shown. The distance between points *A* and *B* is 1 ft in both cases and the handle diameter is 0.625 in. The wheel nuts require a torque of 70 ft-lb. Find the maximum principal stress and maximum deflection in each wrench design.

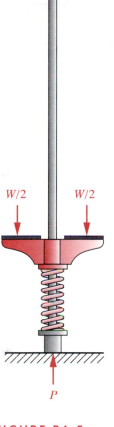

W/2   *W*/2

P

FIGURE P4-5

Problem 4-14

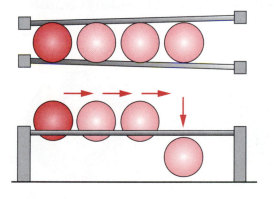

FIGURE P4-6

Problem 4-16

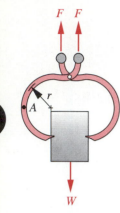

FIGURE P4-7

Problem 4-17

FIGURE P4-8

Problems 4-19 and 4-20

---

*†4-22   An inline "roller-blade" skate is shown in Figure P4-10. The polyurethane wheels are 72-mm dia and spaced on 104-mm centers. The skate-boot-foot combination weighs 2 kg. The effective "spring rate" of the person-skate system is 6 000 N/m. The axles are 10-mm-dia steel pins in double shear. Find the stress in the pins for a 100-kg person landing a 0.5-m jump on one foot. (a) Assume that all four wheels land simultaneously. (b) Assume that one wheel absorbs all the landing force.

*†4-23   A beam is supported and loaded as shown in Figure P4-11a Find the reactions, maximum shear, maximum moment, maximum slope, maximum bending stress, and maximum deflection for the data given in the assigned row(s) in Table P4-2.

*†4-24   A beam is supported and loaded as shown in Figure P4-11b, find the reactions, the maximum shear, maximum moment, maximum slope, maximum bending stress, and maximum deflection for the data given in the assigned row(s) in Table P4-2.

*†4-25   A beam is supported and loaded as shown in Figure P4-11c, find the reactions, the maximum shear, maximum moment, maximum slope, maximum bending stress, and maximum deflection for the data given in the assigned row(s) in Table P4-2.

*†4-26   A beam is supported and loaded as shown in Figure P4-11d. Find the reactions, maximum shear, maximum moment, maximum slope, maximum bending stress, and maximum deflection for the data given in the assigned row(s) in Table P4-2.

†4-27   A storage rack is to be designed to hold the paper roll of Problem 4-8 as shown in Figure P4-12. Determine suitable values for dimensions a and b in the figure. Consider bending, shear, and bearing stresses. Assume an allowable tensile/compressive stress of 100 MPa and an allowable shear stress of 50 MPa for both stanchion and mandrel, which are steel. The mandrel is solid and inserts halfway into the paper roll. Balance the design to use all of the material strength. Calculate the deflection at the end of the roll.

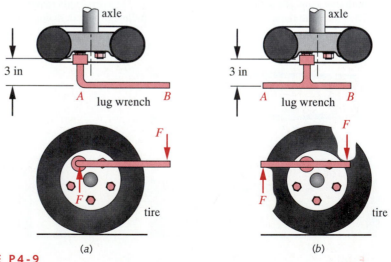

          (a)   (b)

FIGURE P4-9

Problem 4-21

Table P4-2 Data for Problems 4-23 through 4-26 and 4-29 through 4-32

Use only data relevant to the particular problem. Lengths in m, forces in N, I in m^4.

Row	l	a	b	w	F	I	c	E
a	1.00	0.40	0.60	200	500	2.85E–08	2.00E–02	steel
b	0.70	0.20	0.40	80	850	1.70E–08	1.00E–02	steel
c	0.30	0.10	0.20	500	450	4.70E–09	1.25E–02	steel
d	0.80	0.50	0.60	65	250	4.90E–09	1.10E–02	steel
e	0.85	0.35	0.50	96	750	1.80E–08	9.00E–03	steel
f	0.50	0.18	0.40	450	950	1.17E–08	1.00E–02	steel
g	0.60	0.28	0.50	250	250	3.20E–09	7.50E–03	steel
h	0.20	0.10	0.13	400	500	4.00E–09	5.00E–03	alum
i	0.40	0.15	0.30	50	200	2.75E–09	5.00E–03	alum
j	0.20	0.10	0.15	150	80	6.50E–10	5.50E–03	alum
k	0.40	0.16	0.30	70	880	4.30E–08	1.45E–02	alum
l	0.90	0.25	0.80	90	600	4.20E–08	7.50E–03	alum
m	0.70	0.10	0.60	80	500	2.10E–08	6.50E–03	alum
n	0.85	0.15	0.70	60	120	7.90E–09	1.00E–02	alum

FIGURE P4-10

Problem 4-22

4

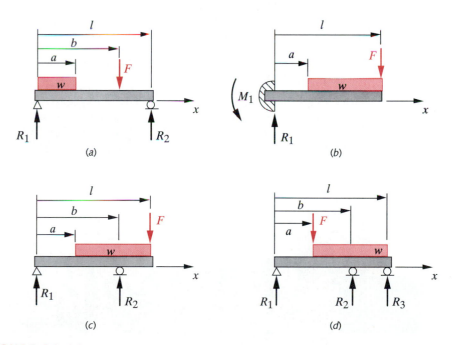

(a)

(b)

(c)

(d)

FIGURE P4-11

Beams and Beam Loadings for Problems 4-23 to 4-26 and 4-29 to 4-32 - See Table P4-2 for Data

4

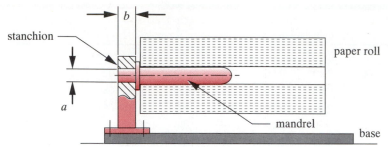

b

stanchion

paper roll

a

mandrel

base

FIGURE P4-12

Problem 4-27

†4-28   Figure P4-13 shows a forklift truck negotiating a 15° ramp to drive onto a 4-ft-high
         loading platform. The truck weighs 5 000 lb and has a 42-in wheelbase. Design two
         (one for each side) 1-ft-wide ramps of steel to have no more than 1-in deflection in
         the worst case of loading as the truck travels up them. Minimize the weight of the
         ramps by using a sensible cross-sectional geometry.

*4-29   Find the spring rate of the beam in Problem 4-23 at the applied concentrated load for
         the row(s) assigned in Table P4-2.

*4-30   Find the spring rate of the beam in Problem 4-24 at the applied concentrated load for
         the row(s) assigned in Table P4-2.

*4-31   Find the spring rate of the beam in Problem 4-25 at the applied concentrated load for
         the row(s) assigned in Table P4-2.

*4-32   Find the spring rate of the beam in Problem 4-26 at the applied concentrated load for
         the row(s) assigned in Table P4-2.

*4-33   For the bracket shown in Figure P4-14 and the data in the row(s) assigned from
         Table P4-3, determine the bending stress at point A and the shear stress due to
         transverse loading at point B. Also find the torsional shear stress at both points.
         Then determine the principal stresses at points A and B.

* Answers to these problems
are provided in Appendix G.

† These problems are based
on similar problems in
previous chapters with the
same –number, e.g.,
Problem 4-4 is based on
Problem 3-4 etc. Problems
in succeeding chapters may
also continue and extend
these problems.

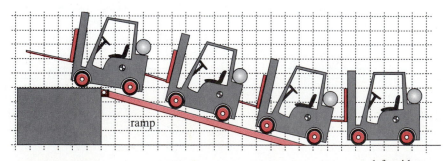

ramp

FIGURE P4-13   1-ft grid

Problem 4-28

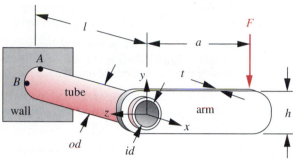

FIGURE P4-14

Problems 4-33 to 4-36

*4-34   For the bracket shown in Figure P4-14 and the data in the row(s) assigned from Table P4-3, determine the deflection at load F.

*4-35   For the bracket shown in Figure P4-14 and the data in the row(s) assigned from Table P4-3, determine the spring rate of the tube in bending, the spring rate of the arm in bending and the spring rate of the tube in torsion. Combine these into an overall spring rate in terms of the force F and the linear deflection at force F.

4-36   For the bracket shown in Figure P4-14 and the data in the row(s) assigned from Table P4-3, redo Problem 4-33 considering the stress concentration at points A and B. Assume a stress-concentration factor of 2.5 in both bending and torsion.

Table P4-3   Data for Problems 4-33 through 4-36 and 4-49 through 4-52

Use only data that are relevant to the particular problem. Lengths in mm, forces in N.

Row	l	a	t	h	F	OD	ID	E
a	100	400	10	20	50	20	14	steel
b	70	200	6	80	85	20	6	steel
c	300	100	4	50	95	25	17	steel
d	800	500	6	65	160	46	22	alum
e	85	350	5	96	900	55	24	alum
f	50	180	4	45	950	50	30	alum
g	160	280	5	25	850	45	19	steel
h	200	100	2	10	800	40	24	steel
i	400	150	3	50	950	65	37	steel
j	200	100	3	10	600	45	32	alum
k	120	180	3	70	880	60	47	alum
l	150	250	8	90	750	52	28	alum
m	70	100	6	80	500	36	30	steel
n	85	150	7	60	820	40	15	steel

* Answers to these problems are provided in Appendix G.

† These problems are based on similar problems in previous chapters with the same –number, e.g., Problem 4-4 is based on Problem 3-4 etc. Problems in succeeding chapters may also continue and extend these problems.

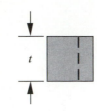

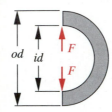

*4-37　A semicircular curved beam as shown in Figure P4-15 has od = 150 mm, id = 100 mm and t = 25 mm. For a load pair F = 14 kN applied along the diameter, find the eccentricity of the neutral axis and the stress at the inner and outer fibers.

4-38　Design a solid, straight, steel torsion bar to have a spring rate of 10 000 in-lb per radian per foot of length. Compare designs of solid round and solid square cross sections. Which is more efficient in terms of material use?

4-39　Design a 1-ft-long steel, end-loaded cantilever spring for a spring rate of 10 000 lb/in at the load. Compare designs of solid round and solid square cross sections. Which is more efficient in terms of material use?

4-40　Redesign the roll support of Problem 4-8 to be like that shown in Figure P4-16. The stub mandrels insert to 10% of the roll length at each end. Choose appropriate dimensions a and b to fully utilize the material's strength, which is the same as in Problem 4-27. See Problem 4-8 for additional data.

*4-41　A 10-mm ID steel tube carries liquid at 7 MPa. Determine the principal stresses in the wall if its thickness is: a) 1 mm, b) 5 mm.

4-42　A cylindrical tank with hemispherical ends is required to hold 150 psi of pressurized air at room temperature. Find the principal stresses in the 1-mm-thick wall if the tank diameter is 0.5 m and its length is 1 m.

4-43　Figure P4-17 shows an off-loading station at the end of a paper rolling machine. The finished paper rolls are 0.9-m OD by 0.22-m ID by 3.23-m long and have a density of 984 kg/m^3. The rolls are transferred from the machine conveyor (not shown) to the forklift truck by the V-linkage of the off-load station which is rotated through 90° by an air cylinder. The paper then rolls onto the waiting forks of the truck. The forks are 38-mm thick by 100-mm wide by 1.2-m long and are tipped at a 3° angle from the horizontal. Find the stresses in the two forks on the truck when the paper rolls onto it under two different conditions (state all assumptions):

(a) The two forks are unsupported at their free end.
(b) The two forks are contacting the table at point A.

4-44　Determine a suitable thickness for the V-links of the off-loading station of Figure P4-17 to limit their deflections at the tips to 10 mm in any position during their rotation. Assume that there are two V-links supporting the roll, arranged at the 1/4

FIGURE P4-15

Problem 4-37

---

* Answers to these problems are provided in Appendix G.

---

† These problems are based on similar problems in previous chapters with the same –number, e.g., Problem 4-4 is based on Problem 3-4 etc. Problems in succeeding chapters may also continue and extend these problems.

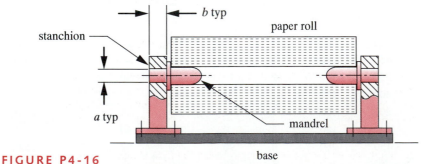

FIGURE P4-16

Problem 4-40

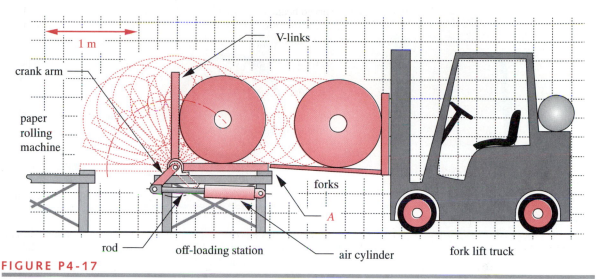

FIGURE P4-17

Problems 4-43 to 4-47

and 3/4 points along the roll's length and that each of the V arms is 10 cm wide by 1 m long. The V arms are welded to a steel tube that is rotated by the air cylinder. See Problem 4-43 for more information.

4-45   Determine the critical load on the air-cylinder rod in Figure P4-17 if the crank arm that it rotates is 0.3 m long and the rod has a maximum extension of 0.5 m. The 25 mm dia rod is solid steel with a yield strength of 400 MPa. State all assumptions.

4-46   The V-links of Figure P4-17 are rotated by the crank arm through a shaft that is 60 mm dia by 3.23 m long. Determine the maximum torque applied to this shaft during the motion of the V-linkage and find the maximum stress and deflection for the shaft. See problem 4-43 for more information.

4-47   Determine the maximum forces on the pins at each end of the air cylinder of Figure P4-17. Determine the stress in these pins if they are 30-mm dia and in single shear.

4-48   A 100 kg wheelchair marathon racer wants an exerciser that will allow indoor practicing in any weather. The design shown in Figure P4-18 is proposed. Two free-turning rollers on bearings support the rear wheels. A platform supports the front wheels. Design the 1-m-long rollers as hollow tubes of aluminum to minimize the height of the platform and also limit the roller deflections to 1 mm in the worst case. The wheelchair has 65-cm-dia drive wheels separated by a 70-cm track width. The flanges shown on the rollers limit the lateral movement of the chair while exercising and thus the wheels can be anywhere between those flanges. Specify suitable sized steel axles to support the tubes on bearings. Calculate all significant stresses.

[*]4-49   A hollow, square column has a length l, material E, as shown in the row(s) assigned in Table P4-3. Its cross-sectional dimensions are 4 mm outside and 3 mm inside.

4

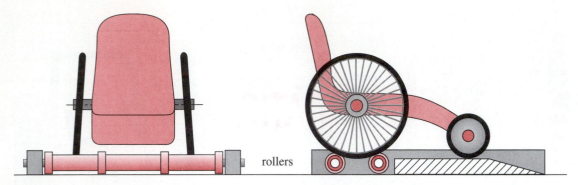

rollers

FIGURE P4-18

Problem 4-48

Use S_y = 150 MPa for aluminum and 300 MPa for steel. Determine if it is a Johnson or an Euler column and find the critical load:

(a) If its boundary conditions are pinned-pinned.
(b) If its boundary conditions are fixed-pinned.
(c) If its boundary conditions are fixed-fixed.
(d) If its boundary conditions are fixed-free.

*4-50   A hollow, round column has a length of 1.5 m, and material E, and cross-sectional dimensions *od* and *id* as shown in the row(s) assigned in Table P4-3. Use S_y = 150 MPa for aluminum and 300 MPa for steel. Determine if it is a Johnson or an Euler column and find the critical load:

(a) If its boundary conditions are pinned-pinned.
(b) If its boundary conditions are fixed-pinned.
(c) If its boundary conditions are fixed-fixed.
(d) If its boundary conditions are fixed-free.

*4-51   A solid, rectangular column has a length *l*, material E, and cross-sectional dimensions *h* and *t* as shown in the row(s) assigned in Table P4-3. Use S_y = 150 MPa for aluminum and 300 MPa for steel. Determine if it is a Johnson or an Euler column and find the critical load:

(a) If its boundary conditions are pinned-pinned.
(b) If its boundary conditions are fixed-pinned.
(c) If its boundary conditions are fixed-fixed.
(d) If its boundary conditions are fixed-free.

* Answers to these
problems are provided in
Appendix G.

† These problems are based
on similar problems in
previous chapters with the
same –number, e.g.,
Problem 4-4 is based on
Problem 3-4 etc. Problems
in succeeding chapters may
also continue and extend
these problems.

*4-52   A solid, circular column has a length *l*, material E, diameter *od,* and an eccentricity *t* as shown in the row(s) assigned in Table P4-3. Use S_y = 150 MPa for aluminum and 300 MPa for steel. Determine if it is a Johnson or an Euler column and find the critical load:

(a) If its boundary conditions are pinned-pinned.
(b) If its boundary conditions are fixed-pinned.
(c) If its boundary conditions are fixed-fixed.
(d) If its boundary conditions are fixed-free.

5

STATIC FAILURE THEORIES

The whole of science is nothing more than a refinement of everyday thinking.

ALBERT EINSTEIN

5.0   INTRODUCTION

Why do parts fail? This is a question that has occupied scientists and engineers for centuries. Much more is understood about various failure mechanisms today than was known even a few decades ago, largely due to improved testing and measuring techniques. If you were asked to respond to the above question based on what you have learned so far, you would probably say something like "parts fail because their stresses exceed their strength", and you would be right up to a point. The follow-up question is the critical one; what kind of stresses cause the failure: Tensile? Compressive? Shear? The answer to this one is the classic, "it depends." It depends on the material in question and its relative strengths in compression, tension, and shear. It also depends on the character of the loading (whether static or dynamic) and on the presence or absence of cracks in the material.

Table 5-0 shows the variables used in this chapter and references the equations or sections in which they are used. At the end of the chapter, a summary section is provided that also groups all the significant equations from this chapter for easy reference and identifies the chapter section in which their discussion can be found.

Figure 5-1a shows the Mohr's circle for the stress state in a tensile test specimen. The tensile test (see Section 2-1) slowly applies a pure tensile loading to the part and causes a tensile, normal stress. However, the Mohr's circle shows that a shear stress is also present, which happens to be exactly half as large as the normal stress. Which stress failed the part, the normal stress or the shear stress?

5

Symbol	Variable	ips units	SI units	See
a	half-width of crack	in	m	Sect. 5.3
b	half-width of cracked plate	in	m	Sect. 5.3
E	Young's modulus	psi	Pa	Sect. 5.1
K_c	stress intensity	kpsi-in$^{0.5}$	MPa-m$^{0.5}$	Sect. 5.3
K	fracture toughness	kpsi-in$^{0.5}$	MPa-m$^{0.5}$	Sect. 5.3
N	safety factor	none	none	Sect. 5.1
N_{fm}	safety factor for fracture mechanics failure	none	none	Sect. 5.3
S_{uc}	ultimate compressive strength	psi	Pa	Sect. 5.2
S_{ut}	ultimate tensile strength	psi	Pa	Sect. 5.2
S_y	tensile yield strength	psi	Pa	Eq. 5.8a, 5.9b
S_{ys}	shear yield strength	psi	Pa	Eq. 5.9b, 5.10
U	total strain energy	in-lb	Joules	Eq. 5.1
U_d	distortion strain energy	in-lb	Joules	Eq. 5.2
U_h	hydrostatic strain energy	in-lb	Joules	Eq. 5.2
β	stress-intensity geometry factor	none	none	Eq. 5.14c
ϵ	strain	none	none	Sect. 5.1
ν	Poisson's ratio	none	none	Sect. 5.1
σ_1	principal stress	psi	Pa	Sect. 5.1
σ_2	principal stress	psi	Pa	Sect. 5.1
σ_3	principal stress	psi	Pa	Sect. 5.1
$\tilde{\sigma}$	Modified-Mohr effective stress	psi	Pa	Eq. 5.12
σ'	von Mises effective stress	psi	Pa	Eq. 5.7

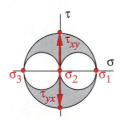

(a)

(b)

FIGURE 5-1

Mohr's Circles for
Unidirectional Tensile
Stress (a) and Pure
Torsion (b)

Figure 5-1b shows the Mohr's circle for the stress state in a torsion test specimen. The torsion test (see Section 2-1) slowly applies a pure torsion loading to the part and causes a shear stress. However, the Mohr's circle shows that a normal stress is also present, which happens to be exactly equal to the shear stress. Which stress failed the part, the normal stress or the shear stress?

In general, ductile materials in static tensile loading are limited by their shear strengths while brittle materials are limited by their tensile strengths (though there are exceptions to this rule when ductile materials can behave as if brittle). This situation requires that we have different failure theories for the two classes of materials, ductile and brittle. Recall from Chapter 2 that ductility can be defined in several ways, the most common being a material's percent elongation to fracture, which, if >5% is considered ductile. Most ductile metals have elongations to fracture >10%.

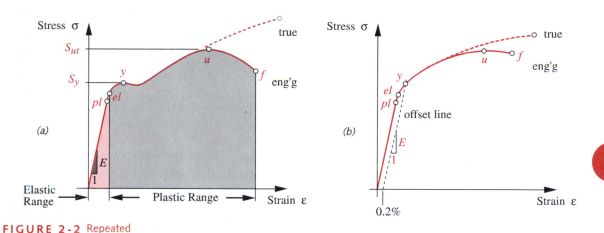

FIGURE 2-2 Repeated

Engineering and True Stress-Strain Curves for Ductile Materials: (a) Low-Carbon Steel   (b) Annealed High-Carbon Steel

Most importantly, we must carefully define what we mean by failure. A part may fail if it yields and distorts sufficiently to not function properly. Also, a part may fail by fracturing and separating. Either of these conditions is a failure, but the mechanisms causing them can be very different. Only ductile materials may yield significantly before fracturing. Brittle materials proceed to fracture without significant shape change. The stress-strain curves of each type of material reflect this difference as shown in Figures 2-2 and 2-4, which are reproduced here for your convenience. Note that if cracks are present in a ductile material, it can suddenly fracture at nominal stress levels well below the yield strength, even under static loads.

Another significant factor in failure is the character of the loading, whether it is static or dynamic. Static loads are slowly applied and remain essentially constant with time. Dynamic loads are either suddenly applied (impact loads) or repeatedly varied with time (fatigue loads), or both. The failure mechanisms are quite different in either case. Table 3-1 defined four classes of loading based on the motion of the loaded parts and the time-dependence of the loading. By that definition, only Class 1 loading is static. The other three classes are dynamic to a greater or lesser degree. When the loading is dynamic, the distinction between ductile and brittle materials' failure behavior blurs, and ductile materials fail in a "brittle" manner. Because of the significant differences in failure mechanisms under static and dynamic loading, we will consider them each separately, discussing failures due to static loading in this chapter and failures due to dynamic loading in the next chapter. For the static loading case (Class 1), we will consider the theories of failure separately for each type of material, ductile and brittle.

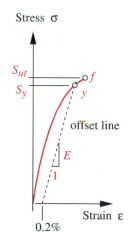

FIGURE 2-4

Stress-Strain Curve of a Brittle Material

Repeated

5.1   FAILURE OF DUCTILE MATERIALS UNDER STATIC LOADING

While ductile materials will fracture if statically stressed beyond their ultimate tensile strength, their failure in machine parts is generally considered to occur when they yield under static loading. The yield strength of a ductile material is appreciably less than its ultimate strength.

Historically, several theories have been formulated to explain this failure: *the maximum normal-stress theory, the maximum normal-strain theory, the total strain-energy theory, the distortion-energy (von Mises-Hencky) theory, and the maximum shear-stress theory*. Of these only the last two agree closely with experimental data for this case, and of those, the von Mises-Hencky theory is the most accurate. We will discuss only the last two in detail, starting with the most accurate (and preferred) approach.

The von Mises-Hencky or Distortion-Energy Theory

The microscopic yielding mechanism is now understood to be due to relative sliding of the material's atoms within their lattice structure. This sliding is caused by shear stress and is accompanied by distortion of the shape of the part. The energy stored in the part from this distortion is an indicator of the magnitude of the shear stress present.

TOTAL STRAIN ENERGY   It was once thought that the total strain energy stored in the material was the cause of yield failure, but experimental evidence did not bear this out. The strain energy U in a unit volume associated with any stress is the area under the stress strain curve up to the point of the applied stress as shown in Figure 5-2 for a unidirectional stress state. Assuming that the stress-strain curve is essentially linear up to the yield point, then we can express the total strain energy at any point in that range as

$$U = \frac{1}{2}\sigma\varepsilon \tag{5.1a}$$

Extending this to a three-dimensional stress state gives

$$U = \frac{1}{2}\left(\sigma_1\varepsilon_1 + \sigma_2\varepsilon_2 + \sigma_3\varepsilon_3\right) \tag{5.1b}$$

using the principal stresses and principal strains that act on planes of zero shear stress.

This expression can be put in terms of principal stresses alone by substituting the relationships

$$\varepsilon_1 = \frac{1}{E}\left(\sigma_1 - \nu\sigma_2 - \nu\sigma_3\right)$$

$$\varepsilon_2 = \frac{1}{E}\left(\sigma_2 - \nu\sigma_1 - \nu\sigma_3\right) \tag{5.1c}$$

$$\varepsilon_3 = \frac{1}{E}\left(\sigma_3 - \nu\sigma_1 - \nu\sigma_2\right)$$

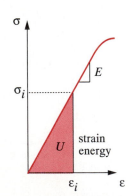

FIGURE 5-2

Internal Strain Energy Stored in a Deflected Part

where ν is Poisson's ratio, giving

$$U = \frac{1}{2E}\left[\sigma_1^2 + \sigma_2^2 + \sigma_3^2 - 2\nu\left(\sigma_1\sigma_2 + \sigma_2\sigma_3 + \sigma_1\sigma_3\right)\right] \qquad (5.1d)$$

HYDROSTATIC LOADING   Very large amounts of strain energy can be stored in materials without failure if they are hydrostatically loaded to create stresses that are uniform in all directions. This can be done in compression very easily by placing the specimen in a pressure chamber. Many experiments have shown that materials can be hydrostatically stressed to levels well beyond their ultimate strengths in compression without failure as this just reduces the volume of the specimen without changing its shape. P. W. Bridgman subjected water ice to 1 Mpsi hydrostatic compression with no failure. The explanation is that uniform stresses in all directions, while creating volume change and potentially large strain energies, cause no distortion of the part and thus no shear stress. Consider the Mohr's circle for a specimen subjected to $\sigma_x = \sigma_y = \sigma_z = 1$ Mpsi compressive stress. The Mohr's "circle" is a **point** on the σ axis at -1 Mpsi and $\sigma_1 = \sigma_2 = \sigma_3$. The shear stress is zero, so there is no distortion and no failure. This is true for ductile or brittle materials when the principal stresses are identical in magnitude and sign.

Den Hartog[1] describes the condition of rocks at great depth in the earth's crust where they withstand uniform, hydrostatic compressive stresses of 5 500 psi/mile of depth due to the weight of the rock above. This is in excess of their typical 3 000 psi ultimate compressive strength as measured in a compression test. While it is much more difficult to create hydrostatic tension, Den Hartog[1] also describes such an experiment done by the Russian scientist Joffe in which he slowly cooled a glass marble in liquid air, allowed it to equilibrate to a stress-free state at the low temperature then removed it to a warm room. As the marble warmed from the outside in, the temperature differential versus its cold core created uniform tensile stresses calculated to be well in excess of the material's tensile strength, but it did not crack. Thus, it appears that distortion is the culprit in tensile failure as well.

COMPONENTS OF STRAIN ENERGY   The total strain energy in a loaded part (Eq. 5.1d) can be considered to consist of two components one due to hydrostatic loading which changes its volume and one due to distortion, which changes its shape. If we separate the two components, the **distortion energy** portion will give a measure of the shear stress present. Let U_h represent the hydrostatic or volumetric component and U_d the **distortion-energy component**, then

$$U = U_h + U_d \qquad (5.2)$$

We can also express each of the principal stresses in terms of a hydrostatic (or volumetric) component σ_h that is common to each face and a distortion component σ_{id} that is unique to each face, where the subscript i represents the principal stress direction, 1, 2, or 3:

$$\sigma_1 = \sigma_h + \sigma_{1_d}$$
$$\sigma_2 = \sigma_h + \sigma_{2_d} \qquad (5.3a)$$
$$\sigma_3 = \sigma_h + \sigma_{3_d}$$

Adding the three principal stresses in equation 5.3a gives:

$$\sigma_1 + \sigma_2 + \sigma_3 = \sigma_h + \sigma_{1_d} + \sigma_h + \sigma_{2_d} + \sigma_h + \sigma_{3_d}$$

$$\sigma_1 + \sigma_2 + \sigma_3 = 3\sigma_h + \left(\sigma_{1_d} + \sigma_{2_d} + \sigma_{3_d}\right) \tag{5.3b}$$

$$3\sigma_h = \sigma_1 + \sigma_2 + \sigma_3 - \left(\sigma_{1_d} + \sigma_{2_d} + \sigma_{3_d}\right)$$

For a volumetric change with no distortion, the term in parentheses in equation 5.3b must be zero, giving an expression for the volumetric or hydrostatic component of stress σ_h:

$$\sigma_h = \frac{\sigma_1 + \sigma_2 + \sigma_3}{3} \tag{5.3c}$$

which you will note is merely the average of the three principal stresses.

Now, the strain energy U_h associated with the hydrostatic volume change can be found by replacing each principal stress in equation 5.1d with σ_h:

$$
\begin{aligned}
U_h &= \frac{1}{2E}\left[\sigma_h^2 + \sigma_h^2 + \sigma_h^2 - 2\nu\left(\sigma_h\sigma_h + \sigma_h\sigma_h + \sigma_h\sigma_h\right)\right] \\
&= \frac{1}{2E}\left[3\sigma_h^2 - 2\nu\left(3\sigma_h^2\right)\right] \\
U_h &= \frac{3}{2}\frac{(1-2\nu)}{E}\sigma_h^2
\end{aligned}
\tag{5.4a}
$$

and substituting equation 5.3c:

$$
\begin{aligned}
U_h &= \frac{3}{2}\frac{(1-2\nu)}{E}\left(\frac{\sigma_1 + \sigma_2 + \sigma_3}{3}\right)^2 \\
&= \frac{1-2\nu}{6E}\left[\sigma_1^2 + \sigma_2^2 + \sigma_3^2 - 2\left(\sigma_1\sigma_2 + \sigma_2\sigma_3 + \sigma_1\sigma_3\right)\right]
\end{aligned}
\tag{5.4b}
$$

DISTORTION ENERGY   The distortion energy U_d is now found by subtracting equation 5.4b from 5.1d in accordance with equation 5.2:

$$
\begin{aligned}
U_d &= U - U_h \\
&= \left\{\frac{1}{2E}\left[\sigma_1^2 + \sigma_2^2 + \sigma_3^2 - 2\nu\left(\sigma_1\sigma_2 + \sigma_2\sigma_3 + \sigma_1\sigma_3\right)\right]\right\} \\
&\qquad - \left\{\frac{1-2\nu}{6E}\left[\sigma_1^2 + \sigma_2^2 + \sigma_3^2 - 2\left(\sigma_1\sigma_2 + \sigma_2\sigma_3 + \sigma_1\sigma_3\right)\right]\right\} \\
U_d &= \frac{1+\nu}{3E}\left[\sigma_1^2 + \sigma_2^2 + \sigma_3^2 - \sigma_1\sigma_2 - \sigma_2\sigma_3 - \sigma_1\sigma_3\right]
\end{aligned}
\tag{5.5}
$$

To obtain a failure criterion, we will compare the distortion energy per unit volume given by equation 5.5 to the distortion energy per unit volume present in a tensile test specimen at failure, because the tensile test is our principal source of material-strength data. The failure stress of interest here is the yield strength S_y. The tensile test is a **uniaxial stress state** where, at yield, $\sigma_1 = S_y$ and $\sigma_2 = \sigma_3 = 0$. The distortion energy associated with yielding in the tensile test is found by substituting these values in equation 5.5:

$$U_d = \frac{1+v}{3E} S_y^2 \qquad (5.6a)$$

and the failure criterion is obtained by equating the general expression 5.5 with the specific failure expression 5.6a to get

$$\frac{1+v}{3E} S_y^2 = U_d = \frac{1+v}{3E} \left[\sigma_1^2 + \sigma_2^2 + \sigma_3^2 - \sigma_1\sigma_2 - \sigma_2\sigma_3 - \sigma_1\sigma_3 \right]$$

$$S_y^2 = \sigma_1^2 + \sigma_2^2 + \sigma_3^2 - \sigma_1\sigma_2 - \sigma_2\sigma_3 - \sigma_1\sigma_3 \qquad (5.6b)$$

$$S_y = \sqrt{\sigma_1^2 + \sigma_2^2 + \sigma_3^2 - \sigma_1\sigma_2 - \sigma_2\sigma_3 - \sigma_1\sigma_3}$$

which applies to the three-dimensional stress state.

For a two-dimensional stress state, $\sigma_2 = 0^*$ and equation 5.6b reduces to:

$$S_y = \sqrt{\sigma_1^2 - \sigma_1\sigma_3 + \sigma_3^2} \qquad (5.6c)$$

The two-dimensional distortion-energy equation 5.6c describes an ellipse, which when plotted on the σ_1, σ_3 axes is as shown in Figure 5-3. The interior of this ellipse defines the region of combined biaxial stresses safe against yielding under static loading.

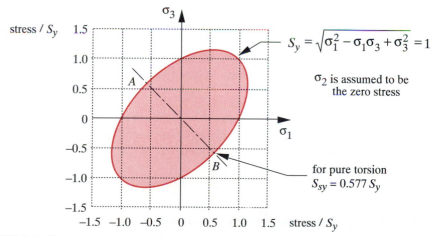

$$S_y = \sqrt{\sigma_1^2 - \sigma_1\sigma_3 + \sigma_3^2} = 1$$

σ_2 is assumed to be the zero stress

for pure torsion
$S_{sy} = 0.577\, S_y$

FIGURE 5-3

The 2-D Distortion-Energy Ellipse Normalized to the Yield Strength of the Material

* Note that this assumption will be consistent with the conventional ordering of principal stresses in the 3-D case ($\sigma_1 > \sigma_2 > \sigma_3$) only if $\sigma_3 < 0$. If both nonzero principal stresses are positive, then the assumption that $\sigma_2 = 0$ violates the ordering convention, Nevertheless we will use σ_1 and σ_3 to represent the two nonzero principal stresses in the 2-D case regardless of their signs in order to simplify their representation in figures and equations.

The three-dimensional distortion-energy equation 5.6b describes a circular cylinder, inclined to the σ_1, σ_2, σ_3 axes such that its intersection with any of the three principal planes is an ellipse as shown in Figure 5-3. The interior of this cylinder defines the region safe against yielding for combined stresses σ_1, σ_2, σ_3.

VON MISES EFFECTIVE STRESS   It is often convenient in situations involving combined tensile and shear stresses acting on the same point to define an effective stress that can be used to represent the stress combination. The distortion-energy approach provides a good means to do this for ductile materials. The **von Mises effective stress σ'** is defined as *the uniaxial tensile stress that would create the same distortion energy as is created by the actual combination of applied stresses*. This approach allows us to treat cases of combined, multiaxial tension and shear stresses as if they were due to pure tensile loading.

The von Mises effective stress σ' for the three-dimensional case is found from equation 5.6b:

$$\sigma' = \sqrt{\sigma_1^2 + \sigma_2^2 + \sigma_3^2 - \sigma_1\sigma_2 - \sigma_2\sigma_3 - \sigma_1\sigma_3} \qquad (5.7a)$$

This can also be expressed in terms of the applied stresses:

$$\sigma' = \sqrt{\frac{\left(\sigma_x - \sigma_y\right)^2 + \left(\sigma_y - \sigma_z\right)^2 + \left(\sigma_z - \sigma_x\right)^2 + 6\left(\tau_{xy}^2 + \tau_{yz}^2 + \tau_{zx}^2\right)}{2}} \qquad (5.7b)$$

and for the two-dimensional case from equation 5.6c (with $\sigma_2 = 0$):

$$\sigma' = \sqrt{\sigma_1^2 - \sigma_1\sigma_3 + \sigma_3^2} \qquad (5.7c)$$

and if expressed in terms of the applied stresses:

$$\sigma' = \sqrt{\sigma_x^2 + \sigma_y^2 - \sigma_x\sigma_y + 3\tau_{xy}^2} \qquad (5.7d)$$

Use these effective stresses for any combined stress situation. (See Example 5-1.) This von Mises effective stress will be revisited later when examples of combined stresses are encountered.

SAFETY FACTOR   Equations 5.6b and 5.6c define the conditions at failure. For design purposes it is convenient to include a chosen safety factor N in the calculation so that the stress state will be safely inside the failure-stress ellipse of Figure 5-3.

$$N = \frac{S_y}{\sigma'} \qquad (5.8a)$$

For the three-dimensional stress case this becomes

$$\frac{S_y}{N} = \sqrt{\sigma_1^2 + \sigma_2^2 + \sigma_3^2 - \sigma_1\sigma_2 - \sigma_2\sigma_3 - \sigma_1\sigma_3} \qquad (5.8b)$$

and for the two-dimensional stress case:

$$\frac{S_y}{N} = \sqrt{\sigma_1^2 - \sigma_1\sigma_3 + \sigma_3^2} \qquad (5.8c)$$

PURE SHEAR   For the case of pure shear as encountered in pure torsional loading the principal stresses become $\sigma_1 = \tau = -\sigma_3$ and $\sigma_2 = 0$ as shown in Figure 5-1b. Figure 5-3 also shows the pure torsional stress state plotted on the σ_1, σ_3 axes. The locus of pure torsional shear stress is a straight line through the origin at $-45°$. This line intersects the failure ellipse at two points, A and B. The absolute values of σ_1 and σ_3 at these points are found from equation 5.6c for the two-dimensional case.

$$S_y^2 = \sigma_1^2 + \sigma_1\sigma_1 + \sigma_1^2 = 3\sigma_1^2 = 3\tau_{max}^2$$

$$\sigma_1 = \frac{S_y}{\sqrt{3}} = 0.577\,S_y = \tau_{max} \qquad (5.9a)$$

This relationship defines the **shear yield strength** S_{ys} of any ductile material as a fraction of its yield strength in tension S_y determined from the tensile test.

$$S_{ys} = 0.577\,S_y \qquad (5.9b)$$

DUCTILE FAILURE THEORY   We can now answer the question posed in the first paragraph of this chapter as to whether the shear stress or the tensile stress was responsible for the failure of a ductile specimen in the tensile test. Based on experiments and the distortion energy theory, *failure in the case of ductile materials in static tensile loading is considered to be due to shear stress.*

AN HISTORICAL NOTE   The distortion-energy approach to failure analysis has many fathers. In fact, equation 5.6 can be derived by five different approaches[2]. The distortion-energy method presented here was originally proposed by James Clerk Maxwell[3] in 1856 but not further developed until additional contributions were made in 1904 by Hueber, in 1913 by von Mises, and in 1925 by Hencky. Today it is most often credited to von Mises and Hencky, and sometimes only to von Mises. The effective stress defined in equations 5.7 is usually referred to as the von Mises or just the Mises (pronounced meeses) stress. Eichinger (in 1926) and Nadai (in 1937) independently developed equation 5.6 by a different method involving **octahedral stresses**, and others have arrived at the same result by still different routes. The number of independent developments of this theory using different approaches, in combination with the very close correlation of experimental data to its predictions, make it the *best choice for predicting failure in the case of static loading of ductile materials in which the tensile and compressive strengths are equal.*

The Maximum Shear-Stress Theory

The role of shear stress in static failure was recognized prior to the development of the von Mises approach to the failure analysis of ductile materials under static loading. The maximum shear stress theory was first proposed by Coulomb (1736-1806) and later described by Tresca in an 1864 publication. Around the turn of the twentieth century, J. Guest performed experiments in England that confirmed the theory. It is sometimes called the Tresca-Guest theory.

The **maximum shear-stress theory** (or just **maximum shear theory**) states that *failure occurs when the maximum shear stress in a part exceeds the shear stress in a tensile specimen at yield (one-half of the tensile yield strength)*. This predicts that the shear-yield strength of a ductile material is

$$S_{ys} = 0.50\,S_y \qquad\qquad (5.10)$$

Note that this is a more conservative limit than that of the distortion-energy theory given in equation 5.9b.

Figure 5-4 shows the hexagonal failure envelope for the two-dimensional maximum shear theory superposed on the distortion-energy ellipse. It is inscribed within the ellipse and contacts it at six points. Combinations of principal stresses σ_1 and σ_2 that lie within this hexagon are considered safe, and failure is considered to occur when the combined stress state reaches the hexagonal boundary. This is obviously a more conservative failure theory than distortion energy as it is contained within the latter. The conditions for torsional (pure) shear are shown at points C and D.

To use this theory for either two- or three-dimensional static stress in homogeneous, isotropic, ductile materials, first compute the three principal normal stresses σ_1, σ_2, σ_3 (one of which will be zero for a 2-D case) and the **maximum shear stress**, τ_{13}, as defined in equation 4.5. Then compare the maximum shear stress to the failure criterion in equation 5.10. The **safety factor** for the **maximum shear-stress theory** is found from

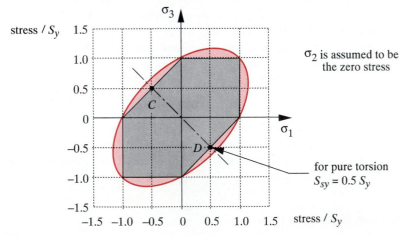

FIGURE 5-4

The 2-D Shear-Stress Theory Hexagon Inscribed Within the Distortion-Energy Ellipse

$$N = \frac{S_{ys}}{\tau_{max}} = \frac{0.50\,S_y}{\tau_{max}} = \frac{S_y/2}{(\sigma_1 - \sigma_3)/2} = \frac{S_y}{(\sigma_1 - \sigma_3)} \qquad (5.11)$$

where τ_{max} is the largest result from equations 4.5. Remember that even in a two-dimensional applied-stress case, there can be three principal shear stresses, the largest of which is τ_{max}.

Maximum Normal-Stress Theory

This theory is presented for historical interest and completeness **but it must be noted that it is not a safe theory to use for ductile materials**. Modifications of this theory will shortly be discussed that are valid and useful for brittle materials whose ultimate tensile strengths are lower than their shear and compressive strengths. The **maximum normal-stress theory** states that *failure will occur when the normal stress in the specimen reaches some limit on normal strength such as tensile yield strength or ultimate tensile strength*. For ductile materials, yield strength is the desired criterion.

Figure 5-5 shows the two-dimensional failure envelope for the maximum normal stress theory. It is a square. Compare this square envelope to those shown in Figure 5-4. In the first and third quadrants, the maximum normal-stress theory envelope is coincident with that of the maximum shear theory. But, in the second and fourth quadrants, the normal-stress theory envelope is well outside of both the distortion energy ellipse and its inscribed maximum-shear-theory hexagon. Since experiments show that ductile materials fail in static loading when their stress states are outside of the ellipse, the normal stress theory is an unsafe failure criterion in the second and fourth quadrants. **The wise designer will *avoid* using the normal-stress theory with ductile materials.**

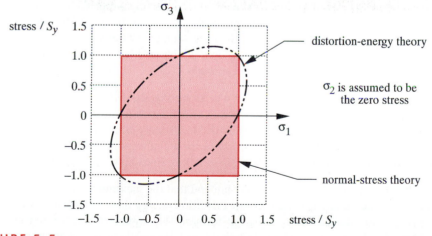

FIGURE 5-5

The Maximum Normal-Stress Theory - Incorrect for Ductile Materials in 2nd and 4th Quadrants

5

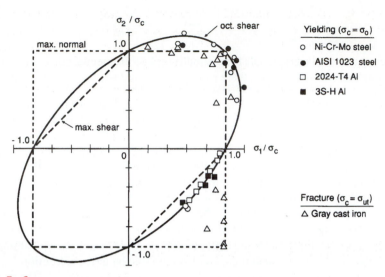

FIGURE 5-6

Experimental Data from Tensile Tests Superposed on Three Failure Theories *(Reproduced from Fig. 7.11, p. 252, in Mechanical Behavior of Materials by N. E. Dowling, Prentice-Hall, Englewood Cliffs, NJ, 1993)*

Comparison of Experimental Data with Failure Theories

Many tensile tests have been done on various materials. The data show statistical scatter but in the aggregate tend to fit the distortion-energy ellipse fairly well. Figure 5-6 shows experimental data for two ductile steels, two ductile aluminum alloys, and a brittle cast-iron superposed on the failure envelopes for the three failure theories discussed above. Note how the ductile-yield data cluster on or near the distortion-energy ellipse (labeled *oct shear*) with a few data points falling between the maximum shear-theory hexagon and the ellipse, both of which are normalized to the yield strength of the material. The brittle cast-iron fracture (not yield) data are seen to cluster more closely about the (square) maximum normal-stress envelope, which in this figure is normalized to the ultimate tensile strength, not the yield strength.

These data are typical. From them we can see that the distortion energy theory most closely approximates the ductile yield data and the maximum shear theory provides a more conservative criterion that is safely inside virtually all of the data points for yielding of the ductile materials. Since a safety factor will always be applied, the actual stress state can be expected to fall inside of these failure lines by some margin.

It was common in the past to recommend that the maximum shear theory be used in design rather than the more accurate distortion energy theory because it was considered easier to calculate results using the former. This argument may (or may not) have been justified in the days of slide rules but is not defensible in an age of programmable calculators and computers. The distortion-energy method is very easy to use, even with

only a pocket calculator, and provides a theoretically more accurate result. Nevertheless, since some experimental data fall inside the ellipse but outside the shear hexagon,
some designers prefer the more conservative approach of the maximum shear theory.
As the engineer in charge, the choice is ultimately your decision.

   *Both the distortion-energy theory and the maximum shear theory are acceptable as
failure criteria in the case of static loading of ductile, homogeneous, isotropic materials whose compressive and tensile strengths are of the same magnitude.* Most wrought
engineering metals and some polymers are in this category of so-called **even materials**. **Uneven materials** such as cast-brittle metals and composites that do not exhibit
these uniform properties require more complex failure theories, some of which are described in a later section and some in reference 4. See the next section for a discussion
of even and uneven materials.

EXAMPLE 5-1

Failure of Ductile Materials Under Static Loading

Problem   Determine the safety factors for the bracket rod shown in Figure 5-7
                 based on both the distortion-energy theory and the maximum shear
                 theory and compare them.

Given   The material is 2024-T4 aluminum with a yield strength of 47 000 psi.
                 The rod length l = 6 in and arm a = 8 in. The rod outside diameter
                 d = 1.5 in. Load F = 1 000 lb.

Assumptions The load is static and the assembly is at room temperature. Consider
                 shear due to transverse loading as well as other stresses.

Solution   See Figures 5-7 and 4-33 (repeated here) and the *TKSolver* file EX05-01.
                 Also see Example 4-9 for a more complete explanation of the stress
                 analysis for this problem.

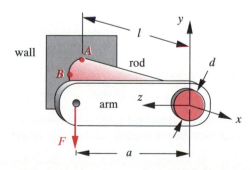

FIGURE 5-7

Bracket for Example 5-1 and 5-2

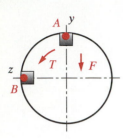

(a) Two points of
interest for
stress calculations

5

1 The rod is loaded in both bending (as a cantilever beam) and in torsion. The largest tensile bending stress will be in the top outer fiber at point A. The largest torsional shear stress will be all around the outer circumference of the rod. (See Example 4-9 for more detail.) First take a differential element at point A where both of these stresses combine as shown in Figure 4-33b. Find the normal bending stress and torsional shear stress on point A using equations 4.11b and 4.24b, respectively.

$$\sigma_x = \frac{Mc}{I} = \frac{(Fl)c}{I} = \frac{1\,000(6)(0.75)}{0.249} = 18\,108 \text{ psi} \qquad (a)$$

$$\tau_{xz} = \frac{Tr}{J} = \frac{(Fa)r}{J} = \frac{1\,000(8)(0.75)}{0.497} = 12\,072 \text{ psi} \qquad (b)$$

2 Find the maximum shear stress and principal stresses that result from this combination of applied stresses using equations 4.6.

$$\tau_{max} = \sqrt{\left(\frac{\sigma_x - \sigma_z}{2}\right)^2 + \tau_{xz}^2} = \sqrt{\left(\frac{18\,108 - 0}{2}\right)^2 + 12\,072^2} = 15\,090 \text{ psi}$$

$$\sigma_1 = \frac{\sigma_x + \sigma_z}{2} + \tau_{max} = \frac{18\,108}{2} + 15\,090 = 24\,144 \text{ psi}$$

$$\sigma_2 = 0 \qquad (c)$$

$$\sigma_3 = \frac{\sigma_x + \sigma_z}{2} - \tau_{max} = \frac{18\,108}{2} - 15\,090 = -6\,036 \text{ psi}$$

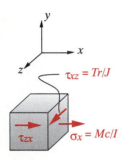

(b) Stress element
at point A

3 Find the von Mises effective stress from the principal stresses using equation 5.7a with $\sigma_2 = 0$.

$$\sigma' = \sqrt{\sigma_1^2 - \sigma_1\sigma_3 + \sigma_3^2}$$

$$\sigma' = \sqrt{24\,144^2 - 24\,144(-6\,036) + (-6\,036)^2} = 27\,661 \text{ psi} \qquad (d)$$

4 The safety factor using the distortion-energy theory can now be found using equation 5.8a.

$$N = \frac{S_y}{\sigma'} = \frac{47\,000}{27\,661} = 1.7 \qquad (e)$$

(c) Stress element
at point B

5 The safety factor using the maximum shear-stress theory can be found from equation 5.10.

$$N = \frac{0.50\,S_y}{\tau_{max}} = \frac{0.50(47\,000)}{15\,090} = 1.6 \qquad (f)$$

6 Comparing these two results shows the more conservative nature of the maximum shear-stress theory, which gives a slightly lower safety factor.

7 Since the rod is a short beam, we need to check the shear due to transverse loading at point B on the neutral axis. The maximum transverse shear stress at the neutral axis of a round rod was given as equation 4.15c.

FIGURE 4-33

Stress Elements at Points
A and B within Cross
Section of Rod for
Example 4-9
Repeated

$$\tau_{bending} = \frac{4V}{3A} = \frac{4(1\,000)}{3(1.767)} = 755 \text{ psi} \tag{g}$$

Point B is in pure shear. The total shear stress at point B is the algebraic sum of the transverse shear stress and the torsional shear stress, which both act on the same planes of the differential element in this case, in the same direction as shown in Figure 4-32c.

$$\tau_{max} = \tau_{torsion} + \tau_{bending} = 12\,072 + 755 = 12\,827 \text{ psi} \tag{h}$$

8   The safety factor for point B using the distortion energy theory for pure shear (Eq. 5.9b) is

$$N = \frac{0.577 S_y}{\tau_{max}} = \frac{0.577(47\,000)}{12\,827} = 2.1 \tag{i}$$

and for the maximum shear theory (Eq 5.10) is

$$N = \frac{0.50 S_y}{\tau_{max}} = \frac{0.50(47\,000)}{12\,827} = 1.8 \tag{j}$$

Again, the latter is more conservative.

5.2   FAILURE OF BRITTLE MATERIALS UNDER STATIC LOADING

Brittle materials fracture rather than yield. **Brittle fracture in tension** is *considered to be due to the normal tensile stress alone* and thus the maximum normal-stress theory is applicable in this case. **Brittle fracture in compression** *is due to some combination of normal compressive stress and shear stress* and requires a different theory of failure. To account for all loading conditions a combination of theories is used.

Even and Uneven Materials

Some wrought materials, such as fully hardened tool steel, can be brittle. These materials tend to have compressive strengths equal to their tensile strengths and so are called *even materials*. Many cast materials, such as gray cast iron, are brittle but have compressive strengths much greater than their tensile strengths. These are called *uneven materials*. Their low tensile strength is due to the presence of microscopic flaws in the casting, which when subjected to tensile loading, serve as nuclei for crack formation. But when subjected to compressive stress, these flaws are pressed together, increasing the resistance to slippage from the shear stress. Gray cast irons typically have compressive strengths 3 to 4 times their tensile strengths and ceramics have even larger ratios.

Another characteristic of **some cast, brittle materials** is that their *shear strength can be greater than their tensile strength*, falling between their compressive and ten-

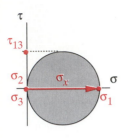

(a)

5

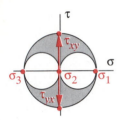

(b)

Repeated

FIGURE 5-1

Mohr's Circles for
Unidirectional Tensile
Stress (a) and Pure
Torsion (b)

sile values. This is quite different than ductile materials in which the shear strength is about one-half the tensile strength. The effects of the stronger shear strength in cast materials can be seen in their failure characteristics in the tension and torsion tests. Figure 2-3 (p. 62) shows a ductile-steel tensile specimen whose failure plane is at 45° to the applied tensile stress indicating a shear failure occurred, which we also know to be true from the distortion-energy theory. Figure 2-5 (p. 63) shows a brittle cast-iron tensile specimen whose failure plane is normal to the applied tensile stress indicating that a tensile failure occurred. The Mohr's circle for this stress state is shown in Figure 5-1*a* **and is the same for both specimens**. The different failure mode is due to the difference in relative shear and tensile strengths between the two materials.

Figure 2-8 (p. 65) shows two torsion-test specimens. The Mohr's circle for the stress state in both specimens is shown in Figure 5-1*b*. The ductile-steel specimen fails on a plane normal to the axis of the applied torque. The applied stress here is pure shear acting in a plane normal to the axis. The applied shear stress is also the maximum shear stress, and the failure is along the maximum shear plane because the ductile material is weakest in shear. The brittle, cast-iron specimen fails in a spiral fashion along planes inclined 45° to the specimen axis. The failure is on the planes of maximum (principal) normal stress because this material is weakest in tension.

Figure 5-8 shows Mohr's circles for both compression and tensile tests of an *even material* and an *uneven material*. The lines tangent to these circles constitute failure lines for all combinations of applied stresses between the two circles. The area enclosed by the circles and the failure lines represents a safe zone. In the case of the even material, the failure lines are independent of the normal stress and are defined by the maximum shear strength of the material. This is consistent with the maximum shear-stress theory for ductile materials (which tend also to be even materials). For the uneven

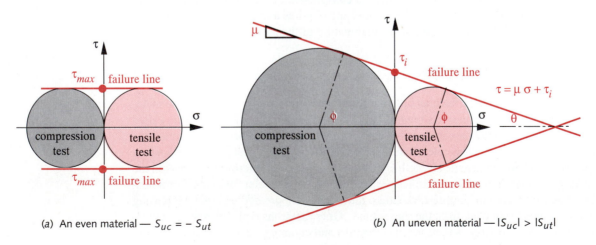

(a) An even material — $S_{uc} = -S_{ut}$ (b) An uneven material — $|S_{uc}| > |S_{ut}|$

FIGURE 5-8

Mohr's Circles for Both Compression and Tensile Tests Showing the Failure Envelopes for (a) *Even* and (b) *Uneven* Materials

material, the failure lines are a function of both the normal stress σ and the shear stress τ. For the compressive regime, as the normal compressive stress component becomes increasingly negative (i.e., more compression), the material's resistance to shear stress increases. This is consistent with the idea expressed above that compression makes it more difficult for shear slippage to occur along fault lines within the material's internal flaws. The equation of the failure line can be found for any material from the test data shown in the figure. The slope μ and the intercept τ_i can be found from geometry using only the radii of the Mohr's circles from the tensile and compression tests.

The interdependence between shear and normal stress shown in Figure 5-8*b* is confirmed by experiment for cases where the compressive stress is dominant, specifically where the principal stress having the largest absolute value is compressive. However, experiments also show that, in tensile-stress-dominated situations with uneven, brittle materials, failure is due to tensile stress alone. The shear stress appears not to be a factor in uneven materials if the principal stress with the largest absolute value is tensile.

THE COULOMB-MOHR THEORY

These observations lead to the Coulomb-Mohr theory of brittle failure, which is an adaptation of the maximum normal-stress theory. Figure 5-9 shows the two-dimensional case plotted on the σ_1, σ_3 axes and normalized to the ultimate tensile strength, S_{ut}. The maximum normal-stress theory is shown for an *even material* as the dotted square of half-dimensions $\pm S_{ut}$. This could be used as the failure criterion for a brittle material in static loading if its compressive and tensile strengths were equal (even material).

The maximum normal-stress theory envelope is also shown (gray-shaded) for an *uneven material* as the asymmetric square of half-dimensions S_{ut}, $-S_{uc}$. This failure

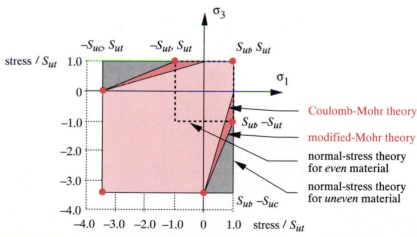

FIGURE 5-9

Coulomb-Mohr, Modified-Mohr, and Maximum Normal-Stress Theories for Uneven Brittle Materials

envelope is only valid in the first and third quadrants as it does not account for the interdependence of the normal and shear stresses shown in Figure 5-8, which affects the second and fourth quadrants. The Coulomb-Mohr envelope (light-color shaded area) attempts to account for the interdependence by connecting opposite corners of these two quadrants with diagonals. Note the similarity of the shape of the Coulomb-Mohr hexagon to the maximum shear-theory hexagon for ductile materials in Figure 5-4. The only differences are the Coulomb Mohr's asymmetry due to the uneven material properties and its use of ultimate (fracture) strengths instead of yield strengths.

Figure 5-10 shows some gray cast-iron experimental test data superposed on the theoretical failure envelopes. Note that the failures in the first quadrant fit the maximum normal-stress theory line (which is coincident with the other theories). The failures in the fourth quadrant fall **inside** the maximum normal-stress line (indicating its unsuitability) and also fall well outside the Coulomb-Mohr line (indicating its conservative nature). This observation leads to a modification of the Coulomb-Mohr theory to make it better fit the observed data.

THE MODIFIED-MOHR THEORY

The actual failure data in Figure 5-10 follow the even materials' maximum normal-stress theory envelope down to a point S_{ut}, $-S_{ut}$ below the σ_1 axis and then follow a straight line to 0, $-S_{uc}$. This set of lines, shown as the combined light-and-dark-color shaded portions of Figure 5-9 (also marked by colored dots), is the **modified-Mohr failure theory envelope**. *It is the preferred failure theory for uneven, brittle materials in static loading.*

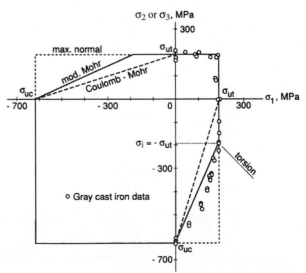

FIGURE 5-10

Biaxial Fracture Data of Gray Cast Iron Compared to Various Failure Criteria *(From Fig 7.13, p 255, in Mechanical Behavior of Materials by N. E. Dowling, Prentice-Hall, Englewood Cliffs NJ, 1993. Data from R. C. Grassi and I. Cornet, "Fracture of Gray Cast Iron Tubes under Biaxial Stresses," J. App. Mech, v. 16, p.178, 1949)*

If the 2-D principal stresses are ordered $\sigma_1 > \sigma_3$, $\sigma_2 = 0$, then only the first and fourth quadrants of Figure 5-10 need to be drawn as shown in Figure 5-11, which plots the stresses normalized by N/S_{ut}. Figure 5-11 also depicts three plane stress conditions labeled A, B, and C. Point A represents any stress state in which the two nonzero principal stresses, σ_1, σ_3 are positive. Failure will occur when the load line OA crosses the failure envelope at A'. The safety factor for this situation can be expressed as

$$N = \frac{S_{ut}}{\sigma_1} \tag{5.12a}$$

If the two nonzero principal stresses have opposite sign, then two possibilities exist for failure, as depicted by points B and C in Figure 5-11. The only difference between these two points is the relative values of their two stress components σ_1, σ_3. The load line OB exits the failure envelope at B' above the point $S_{ut}, -S_{ut}$ and the safety factor for this case is given by equation 5.12a above.

If the stress state is as depicted by point C, then the intersection of the load line OC and the failure envelope occurs at C' below point $S_{ut}, -S_{ut}$. The safety factor can be found by solving for the intersection between the load line OC and the failure line. Write the equations for these lines and solve simultaneously to get

$$N = \frac{S_{ut}S_{uc}}{S_{uc}\sigma_1 - S_{ut}(\sigma_1 + \sigma_3)} \tag{5.12b}$$

If the stress state is in the fourth quadrant, both equations 5.12a and 5.12b should be checked and the smaller resulting safety factor used.

To use this theory, it would be convenient to have expressions for an **effective stress** that would account for all the applied stresses and allow direct comparison to a material strength property, as was done for ductile materials with the von Mises stress. Dowling[5] develops a set of expressions for this effective stress involving the three principal stresses:[*]

$$C_1 = \frac{1}{2}\left[|\sigma_1 - \sigma_2| + \frac{S_{uc} + 2S_{ut}}{S_{uc}}(\sigma_1 + \sigma_2)\right]$$

$$C_2 = \frac{1}{2}\left[|\sigma_2 - \sigma_3| + \frac{S_{uc} + 2S_{ut}}{S_{uc}}(\sigma_2 + \sigma_3)\right] \tag{5.12c}$$

$$C_3 = \frac{1}{2}\left[|\sigma_3 - \sigma_1| + \frac{S_{uc} + 2S_{ut}}{S_{uc}}(\sigma_3 + \sigma_1)\right]$$

The largest of the set of six values (C_1, C_2, C_3, plus the three principal stresses) is the desired effective stress as suggested by Dowling:

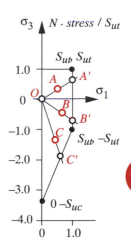

FIGURE 5-11

Modified-Mohr Failure Theory for Brittle Material

---

[*] See reference 5 for a complete derivation of the Coulomb-Mohr and modified-Mohr theories and the effective stress.

$$\tilde{\sigma} = \text{MAX}(C_1, C_2, C_3, \sigma_1, \sigma_2, \sigma_3)$$

$$\tilde{\sigma} = 0 \quad \text{if MAX} < 0 \tag{5.12d}$$

where the function MAX denotes the largest of the six supplied arguments. If all of the arguments are negative then the effective stress is zero.

This *modified-Mohr effective stress* can now be compared to the ultimate tensile strength of the material to determine a safety factor.

$$N = \frac{S_{ut}}{\tilde{\sigma}} \tag{5.12e}$$

This approach allows easy computerization of the process.

EXAMPLE 5-2

Failure of Brittle Materials Under Static Loading

Problem　　　Determine the safety factors for the bracket rod shown in Figure 5-7 based on the modified-Mohr theory.

Given　　　The material is class 50 gray cast iron with S_{ut} = 52 500 psi and S_{uc} = –164 000 psi. The rod length l = 6 in and arm a = 8 in. The rod outside diameter d = 1.5 in. Load F = 1 000 lb.

Assumptions　　　The load is static and the assembly is at room temperature. Consider shear due to transverse loading as well as other stresses.

Solution　　　See Figures 5-7 and 4-33 (repeated here) and the *TKSolver* file EX05-02. Also see Examples 4-9 and 5-1.

1　The rod of Figure 5-7 is loaded in both bending (as a cantilever beam) and in torsion. The largest tensile bending stress will be in the top outer fiber at point A. The largest torsional shear stress will be all around the outer circumference of the rod. First take a differential element at point A where both of these stresses combine. Find the normal bending stress and torsional shear stress on point A using equations 4.11b and 4.24b, respectively.

$$\sigma_x = \frac{Mc}{I} = \frac{(Fl)c}{I} = \frac{1\,000(6)(0.75)}{0.249} = 18\,108 \text{ psi} \tag{a}$$

$$\tau_{xz} = \frac{Tr}{J} = \frac{(Fa)r}{J} = \frac{1\,000(8)(0.75)}{0.497} = 12\,072 \text{ psi} \tag{b}$$

2　Find the maximum shear stress and principal stresses that result from this combination of applied stresses using equations 4.6.

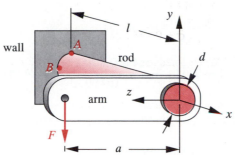

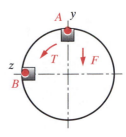

(a) Two points of interest for stress calculations

FIGURE 5-7   Repeated

Bracket for Example 5-1 and 5-2

$$\tau_{max} = \sqrt{\left(\frac{\sigma_x - \sigma_z}{2}\right)^2 + \tau_{xz}^2} = \sqrt{\left(\frac{18\,108 - 0}{2}\right)^2 + 12\,072^2} = 15\,090 \text{ psi}$$

$$\sigma_1 = \frac{\sigma_x + \sigma_z}{2} + \tau_{max} = \frac{18\,108}{2} + 15\,090 = 24\,144 \text{ psi}$$

$$\sigma_2 = 0 \tag{c}$$

$$\sigma_3 = \frac{\sigma_x + \sigma_z}{2} - \tau_{max} = \frac{18\,108}{2} - 15\,090 = -6\,036 \text{ psi}$$

Note that these stresses are identical to those of Example 5-1.

3   The principal stresses for point A can now be plotted on a modified-Mohr diagram as shown in Figure 5-12a. This shows that the load line crosses the failure envelope above the S_{ut},–S_{ut} point, making equation 5.12a appropriate for the safety factor calculation.

$$N = \frac{S_{ut}}{\sigma_1} = \frac{52\,400}{24\,144} = 2.2 \tag{d}$$

4   An alternative approach that does not require drawing the modified-Mohr diagram is to find the Dowling factors C_1, C_2, C_3 using equation 5.12b:

$$C_1 = \frac{1}{2}\left[|\sigma_1 - \sigma_2| + \frac{S_{uc} + 2S_{ut}}{S_{uc}}(\sigma_1 + \sigma_2)\right]$$

$$= \frac{1}{2}\left[|24\,144 - 0| + \frac{(-164\,000 + 2(52\,500))}{-164\,000}(24\,144 + 0)\right] = 16\,415 \text{ psi} \tag{e}$$

$$C_2 = \frac{1}{2}\left[|\sigma_2 - \sigma_3| + \frac{S_{uc} + 2S_{ut}}{S_{uc}}(\sigma_2 + \sigma_3)\right]$$

$$= \frac{1}{2}\left[|0 - (-6\,036)| + \frac{(-164\,000 + 2(52\,500))}{-164\,000}(0 - 6\,036)\right] = 1\,932 \text{ psi} \tag{f}$$

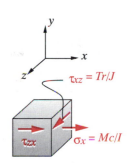

(b) Stress element at point A

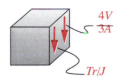

(c) Stress element at point B

FIGURE 4-33

Stress Elements at Points A and B within Cross Section of Rod for Example 4-9   Repeated

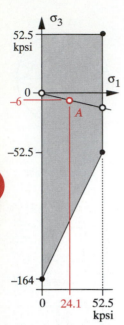

(a) Stresses at point A

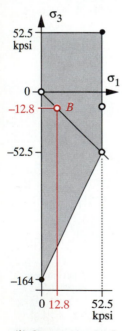

(b) Stresses at point B

FIGURE 5-12

Example 5-2

$$C_3 = \frac{1}{2}\left[|\sigma_3 - \sigma_1| + \frac{S_{uc} + 2S_{ut}}{S_{uc}}(\sigma_3 + \sigma_1)\right]$$

$$= \frac{1}{2}\left[|24\,144 - (-6\,036)| + \frac{(-164\,000 + 2(52\,500))}{-164\,000}(24\,144 - 6\,036)\right] = 18\,348 \quad (g)$$

5   Then find the largest of the six stresses $C_1, C_2, C_3, \sigma_1, \sigma_2, \sigma_3$:

$$\tilde{\sigma} = \text{MAX}(C_1, C_2, C_3, \sigma_1, \sigma_2, \sigma_3)$$

$$\tilde{\sigma} = \text{MAX}(16\,415, 1\,932, 18\,348, 24\,144, 0, -6\,036) = 24\,144$$

   (h)

   which is the modified-Mohr effective stress.

6   The safety factor for point A can now be found using equation 5.12d:

$$N = \frac{S_{ut}}{\tilde{\sigma}} = \frac{52\,500}{24\,144} = 2.2 \qquad (i)$$

   which is the same as was found in step 3.

7   Since the rod is a short beam, we need to check the shear stress due to transverse loading at point B on the neutral axis. The maximum transverse shear stress at the neutral axis of a solid round rod was given as equation 4.15c.

$$\tau_{bending} = \frac{4V}{3A} = \frac{4(1\,000)}{1.767} = 755 \text{ psi} \qquad (j)$$

   Point B is in pure shear. The total shear stress at point B is the algebraic sum of the transverse shear stress and the torsional shear stress which both act on the same planes of the differential element and, in this case, act in the same direction as shown in Figure 4-33b.

$$\tau_{max} = \tau_{torsion} + \tau_{bending} = 12\,072 + 755 = 12\,827 \text{ psi} \qquad (k)$$

8   Find the principal stresses for this pure shear loading:

$$\sigma_1 = \tau_{max} = 12\,827 \text{ psi}$$

$$\sigma_2 = 0 \qquad (l)$$

$$\sigma_3 = -\tau_{max} = -12\,827 \text{ psi}$$

9   These principal stresses for point B can now be plotted on a modified-Mohr diagram as shown in Figure 5-12b. Because this is a pure shear loading, the load line crosses the failure envelope at the $S_{ut}, -S_{ut}$ point, making equation 5.12a appropriate for the safety factor calculation.

$$N = \frac{S_{ut}}{\sigma_1} = \frac{52\,400}{12\,827} = 4.1 \qquad (m)$$

10 To avoid drawing the modified-Mohr diagram, find the Dowling factors C_1, C_2, C_3 using equations 5.12b:

$$C_1 = 8\ 721 \text{ psi}$$
$$C_2 = 0 \text{ psi} \tag{n}$$
$$C_3 = 8\ 721 \text{ psi}$$

11 And find the largest of the six stresses $C_1, C_2, C_3, \sigma_1, \sigma_2, \sigma_3$:

$$\tilde{\sigma} = 12\ 827 \text{ psi} \tag{o}$$

which is the modified-Mohr effective stress.

12 The safety factor for point B can now be found using equation 5.12d:

$$N = \frac{S_{ut}}{\tilde{\sigma}} = \frac{52\ 500}{12\ 827} = 4.1 \tag{p}$$

and it is the same as was found in step 9.

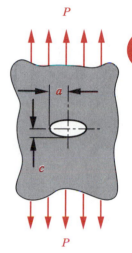

5.3   FRACTURE MECHANICS

The static failure theories discussed so far have all assumed that the material is perfectly homogeneous and isotropic, and thus free of any defects such as cracks, voids or inclusions, which could serve as stress-raisers. This is seldom true for real materials. Actually, all materials are considered to contain microcracks too small to be seen with the naked eye. Dolan[6] says that "... *every structure* contains small flaws whose size and distribution are dependent upon the material and its processing. These may vary from nonmetallic inclusions and microvoids to weld defects, grinding cracks, quench cracks, surface laps, etc." Scratches or gouges in the surface due to mishandling can also serve as incipient cracks. Functional geometric contours that are designed into the part may raise local stresses in predictable ways and can be taken into account in the stress calculations as was discussed in Chapter 4 (and will be further discussed in the next chapter). Cracks that occur spontaneously in service, due to damage or material flaws, are more difficult to predict and account for.

    The presence of a sharp crack in a stress field creates stress concentrations that theoretically approach infinity. See Figure 4-35 and equation 4.32a, which are repeated here for your convenience.

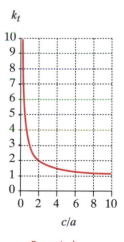

Repeated

FIGURE 4-35

Stress Concentration at the Edge of an Elliptical Hole in a Plate

$$K_t = 1 + 2\left(\frac{a}{c}\right) \tag{4.32a}$$

    Note that when the value of c approaches zero, the stress concentration, and thus the stress, approaches infinity. Since no material can sustain such high stresses, local yielding (for ductile materials), local microfracture (for brittle materials), or local crazing (for polymers) will occur at the crack tip[7]. If stresses are high enough at the tip of a crack of sufficient size, a sudden, "brittle-like" failure can result, even in ductile materials under static loads. The science of **fracture mechanics** has been developed to explain and predict this sudden-failure phenomenon.

Cracks commonly occur in welded structures, bridges, ships, aircraft, land vehicles, pressure vessels, etc. Many catastrophic failures of tankers and Liberty Ships built during World War II occurred.[8], [9] Twelve of these failures occurred shortly after the ships were launched and before they had sailed anywhere. They simply split in half while tied to the pier. One such ship is shown in Figure 5-13. The hull material was welded, ductile steel and the ship had not yet been dynamically loaded to any significant degree. The nominal stresses were well below the material's yield strength. Other examples of sudden failure at stresses below yield strength have occurred in this century, such as the Boston molasses-tank rupture in January 1919, which drowned 21 people and many horses under 2.3 million gallons of the sticky liquid.[10] A more recent example is the failure of a 22-ft dia rocket-motor case while being pressure tested by the manufacturer. Figure 5-14 shows the pieces of the rocket case after failure. It "… was designed to stand proof pressures of 960 psi (but) failed … at 542 psi."[11] These and other sudden "brittle-like" failures of ductile materials under static loading led researchers to seek better failure theories since the ones then available could not adequately explain the observed phenomena.

Where human life is at risk, as in bridges, aircraft, etc., periodic structural-safety inspections for cracks are required by law or government regulation. These inspections can be by X-ray, ultrasonic energy, or just be visual. When cracks are discovered, an engineering judgment must be made whether to repair or replace the flawed part, retire the assembly, or to continue it in service for a further time subject to more frequent inspection. (Many commercial aircraft currently flying contain structural cracks.) These decisions can now be made sensibly through the use of fracture-mechanics theory.

Fracture-Mechanics Theory

Fracture mechanics presumes the presence of a crack. The stress state in the region of the crack may be one of plane strain or plane stress (see Section 4.4, p. 182). If *the zone of yielding around the crack is small compared to the dimensions of the part*, then **linear-elastic fracture-mechanics** (LEFM) theory is applicable. LEFM assumes that the bulk of the material is behaving according to Hooke's law. However, if a significant portion of the bulk material is in the plastic region of its stress-strain behavior, then a more complicated approach is required than that described here. For the following discussion, we will assume that LEFM applies.

MODES OF CRACK DISPLACEMENT Depending on the orientation of the loading versus the crack, the applied loads may tend to pull the crack open in tension (Mode I), shear the crack in-plane (Mode II), or shear (tear) it out-of-plane (Mode III). Most of

FIGURE 5-13

WW II Tanker Cracked in Two While Berthed Prior to Being Placed in Service, Portland, Oregon, January 16, 1943 *(Courtesy of the Ship Structures Committee, U. S. Government)*

FIGURE 5-14

Failed Rocket-Motor Case *(Courtesy of NASA-Lewis Research Center)*

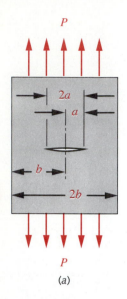

5

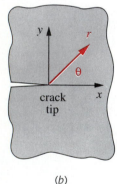

(a)

FIGURE 5-15

A Through-Crack in a
Plate in Tension

---

* The nominal stress for a
fracture-mechanics analysis
is calculated based on the
gross cross-sectional area,
without any reduction for
the crack area. Note that
this is different than the
procedure used for
calculation of nominal stress
when using stress-
concentration factors in a
regular stress analysis. Then,
the net cross section is used
to find the nominal stress.

the fracture-mechanics research and testing has been devoted to the tensile loading case (Mode I), and we will limit our discussion to it.

STRESS INTENSITY FACTOR K   Figure 5-15a shows a plate (not to scale) of width $2b$ under tension with a through crack of width $2a$ in the center. The crack is assumed to be sharp at its ends and b is much larger than a. The crack's cross section is in the xy plane. An r-θ polar coordinate system is also set up in the xy plane with its origin at the crack tip as shown in Figure 5-15b. From the *theory of linear elasticity*, for $b \gg a$ the stresses around the crack tip, expressed as a function of the polar coordinates, are

$$\sigma_x = \frac{K}{\sqrt{2\pi r}}\cos\frac{\theta}{2}\left[1 - \sin\frac{\theta}{2}\sin\frac{3\theta}{2}\right] + \ldots \qquad (5.13a)$$

$$\sigma_y = \frac{K}{\sqrt{2\pi r}}\cos\frac{\theta}{2}\left[1 + \sin\frac{\theta}{2}\sin\frac{3\theta}{2}\right] + \ldots \qquad (5.13b)$$

$$\tau_{xy} = \frac{K}{\sqrt{2\pi r}}\cos\frac{\theta}{2}\sin\frac{\theta}{2}\sin\frac{3\theta}{2} + \ldots \qquad (5.13c)$$

or

$$\sigma_z = 0 \qquad\qquad \text{for plane stress}$$
$$\qquad\qquad\qquad\qquad\qquad\qquad\qquad (5.13d)$$
$$\sigma_z = \nu\left(\sigma_x + \sigma_y\right) \qquad \text{for plane strain}$$

$$\tau_{yz} = \tau_{zx} = 0 \qquad\qquad\qquad\qquad\qquad (5.13e)$$

with higher-order terms of small value omitted. Note that, when the radius r is zero, the xy stresses are infinite, which is consistent with equation 4.32b. The stresses diminish rapidly as r increases. The angle θ defines the geometric distribution of the stresses around the crack tip at any radius. **The quantity K is called the *stress intensity factor*.** (A subscript can be added to designate the mode I, II, III of loading as in K_I, K_{II}, K_{III}. Since we are dealing only with mode I loading, we will eliminate the subscript and let $K = K_I$.)

If we take the plane-stress case and compute the von Mises stress σ' from the x, y, and shear components (Eqs. 5.13a, -b, -c), we can plot the distribution of σ' versus θ for any chosen r as shown in Figure 5-16a for $r = 10E{-}6$ in and $K = 1$. The maxima occur at about $\pm 81°$. If we set θ to that angle and compute the distribution of σ' as a function of r, it looks like Figure 5-16b, which plots r from $10E{-}6$ to 1 in on a log scale.

The high stresses near the crack tip cause local yielding and create a plastic zone of radius r_y as shown (not to scale) in Figure 5-16c. For any radius r and angle θ, the stress state in this plastic zone at the crack tip is directly proportional to the **stress intensity factor K**. If $b \gg a$ then K can be defined for the center-cracked plate as

$$K = \sigma_{nom}\sqrt{\pi a} \qquad\qquad a \ll b \qquad\qquad (5.14a)$$

where σ_{nom} is the nominal stress* in the absence of the crack, a is the crack half-width and b is the plate half-width (see Figure 5-15). This equation will be accurate within

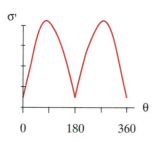

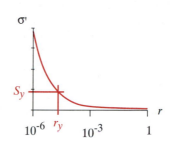

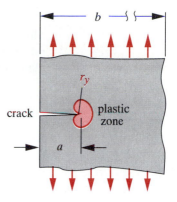

(a) Von Mises stress as a function of angle around crack tip ($r = 10E{-}6$ in)

(b) Von Mises stress as a function of distance from crack tip ($\theta = 81°$)

(c) Zone of plastic yielding around crack tip

FIGURE 5-16

Von Mises Plane Stress Field Around an Edge Crack in a Plate Subject to Axial Tension for Stress Intensity Factor $K = 1$

10% if $a/b \le 0.4$. Note that the stress intensity factor K is directly proportional to the applied nominal stress and proportional to the square root of the crack width. The units of K are either MPa-m$^{0.5}$ or kpsi-in$^{0.5}$.

If the crack-width a is not small compared to the plate-width b, and/or if the geometry of the part is more complicated than the simple cracked plate shown in Figures 5-15, then an additional factor β is needed to calculate K.

$$K = \beta \sigma_{nom} \sqrt{\pi a} \qquad (5.14b)$$

where β is a dimensionless quantity that depends on the part's geometry, the type of loading and the ratio a / b. Its value is also affected by the manner in which σ_{nom} is calculated. It is customary to use the gross stress for σ_{nom} calculated from the original section dimensions unreduced by the crack dimensions. Using the net stress would be more accurate but is less convenient to calculate and the difference can be accounted for in the determination of the geometry factor β. Values of β for various geometries and loadings can be found in handbooks, some of which are noted in the bibliography at the end of this chapter. For example, the value of β for the center-cracked plate of Figure 5-15a is

$$\beta = \sqrt{\sec\left(\frac{\pi a}{2b}\right)} \qquad (5.14c)$$

This asymptotically approaches 1 for small values of a / b and is ∞ for $a / b = 1$.

For example, if the crack is at the edge rather than in the center of the plate, as shown in Figure 5-16c, the factor $\beta = 1.12$:

$$K = 1.12\sigma_{nom}\sqrt{\pi a} \qquad a << b \qquad (5.14d)$$

This equation will be accurate within 10% if $a/b \leq 0.13$. This equation is also accurate within 10% for a plate cracked on both edges if $a/b \leq 0.6$, and for an edge-cracked plate in bending if $a/b \leq 0.4$.

Fracture Toughness K_c

As long as the stress intensity factor K is below a critical value called the **fracture toughness** K_c* (which is a property of the material) the crack can be considered to be in a *stable mode* (if the load is static *and* the environment is noncorrosive), in a *slow-growth mode* (if the load is time-varying *and* the environment is noncorrosive), or in a *fast-growth* mode (if the environment is corrosive).[13] **When K reaches K_c**, by reason of an increase in the nominal stress or by growth of the crack width, *the crack will propagate suddenly to failure*. The rate of this unstable crack propagation can be spectacular, reaching velocities as high as 1 mile/sec![14] The structure effectively "unzips."† The factor of safety for fracture-mechanics failure is defined as

$$N_{FM} = \frac{K_c}{K} \qquad (5.15)$$

Note that this can be a moving target if cracks are in a growth-mode because K is a function of crack width. If the current, or typical, crack-width is known for the part and the fracture toughness K_c is known for the material, then the maximum allowable nominal stress can be determined for any chosen safety factor or vice versa. The allowable stress for any chosen safety factor calculated from the appropriate version of equation 5.14c will typically be lower than that calculated based on yield strength using equation 5.8 or 5.11. The effect of time-varying (dynamic) stresses on the stress intensity factor K and on failure will be addressed in the next chapter.

To determine the fracture toughness, K_c, ASTM-standardized specimens,‡ containing a crack of defined dimensions, are tested to failure. For axial tests, the specimen is gripped in a servo-hydraulic testing machine such that it can be tensioned across the crack. (Bending tests place the crack on the tension side of the beam.) The specimen is loaded dynamically with increasing displacements and its load-displacement characteristic (effective spring rate) is monitored. The load-displacement function becomes nonlinear at the start of rapid crack growth. The fracture toughness K_c is measured at this point.

Fracture toughness K_c for engineering metals ranges between 20 to 200 MPa-m$^{0.5}$; engineering polymers' and ceramics' K_c ranges between 1 to 5 MPa-m$^{0.5}$.[15] *Fracture toughness generally parallels ductility* and increases substantially at high temperatures. Higher-strength steels tend to be less ductile and have lower K_c than lower-strength steels. Substitution of a high-strength steel for a low-strength steel has led to failures in some applications due to the reduction in fracture toughness that accompanied the material change.

---

* More correctly called K_{Ic} where the I refers to mode I loading. Fracture toughness values for the other modes of loading are designated K_{IIc} and K_{IIIc}. Since we are discussing only mode I loading here, it is shortened to K_c.

† On the occasion of the 75th anniversary of the Boston molasses tank rupture described earlier, a 91-year-old Boston resident was interviewed and described what he saw and heard as a 16-year-old boy in January 1919 when he witnessed the tank failure from atop Cobb's Hill in Boston's North End. He recalled a sudden popping sound, like machine-gun fire, followed quickly by a loud explosion. The "machine-gun sound" was quite likely the sound of the crack propagating across the tank wall at up to 1 mile/second and the loud explosion was probably that of the molasses pressure bursting and disintegrating the tank, large pieces of which landed on and destroyed houses hundreds of yards away.

‡ See ASTM E-399-83 "Standard Test Method for Plane Strain Fracture Toughness of Metallic Materials."

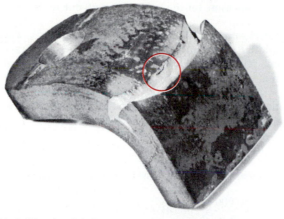

(a) Ball bracket failed suddenly while bending at red heat   (b) 12.5x magnification of preexisting crack within material

FIGURE 5-17

Ductile Steel Trailer Hitch Fractured While Being Bent at Red Heat. Note Pre-existing Crack and Sharpness of the Failure Edges. *(Courtesy of Steven Taylor, Mobile Logic Inc, Port Townsend, Wash.)*

Another example of a fracture-mechanics failure is shown in Figure 5-17*a* which is a photograph of the low-carbon, steel trailer hitch ball bracket of Figures 1-2 through 1-6. This part failed suddenly while being bent to shape at a red heat. The fracture surface can be seen to be relatively smooth and the edges of the crack are extremely sharp. Since elevated temperature increases both ductility and fracture toughness, a sudden brittle failure is unusual under these circumstances. A closer inspection of the failure surface (shown at 12.5x magnification in Figure 5-17*b*) shows a small crack that was apparently a flaw in the hot-rolled bar of steel. The stress intensity at this crack tip exceeded the fracture toughness of the material at its elevated temperature and a sudden brittle failure resulted.[*]

This brief discussion of fracture mechanics has barely *scratched the surface* of this complex topic. The reader is encouraged to read further on this subject. Sources for general information on fracture mechanics, stress intensity factors, and fracture-toughness properties of materials are noted in the bibliography of this chapter. Some fracture-toughness data is also provided in the material-property tables in Appendix C.

EXAMPLE 5-3

Failure of Cracked Materials Under Static Loading

Problem   A steel support strap designed to hold a 60 000 N static load in axial tension was accidently sawcut during production and now has an edge crack in it. Determine the safety factor of the original, uncracked strap based on yielding and its new "cracked" safety factor based on fracture

[*] Note that LEFM cannot be used to analyze this failure because it was not in the linear elastic range. The entire cross section was being plastically deformed at the time of failure. A nonlinear fracture-mechanics analysis would be needed here.

mechanics. How large could the crack get before it fails? Would heat-treating the part compensate for the loss of strength due to the crack?

Given   The material is steel with S_y = 540 MPa and K_c = 66 MPa-m$^{0.5}$. The length l = 6 m, width b = 80 mm, and thickness t = 3 mm. The crack-width a = 10 mm. The crack is completely through the thickness at one edge of the 80-mm width, similar to Figure 5-14c.

Assumptions   The load is static and the assembly is at room temperature. The ratio a / b is < 0.13 ,which allows use of equation 5.14d.

Solution   See the *TKSolver* file EX05-03.

1   First calculate the nominal stress in the uncracked part based on total cross section.

$$\sigma_{nom} = \frac{P}{A} = \frac{60\,000}{3(80)} = 250 \text{ MPa} \tag{a}$$

2   This is a uniaxial stress so is both the principal and the von Mises stress as well. The safety factor against yielding using the distortion-energy theory is (equation 5.8):

$$N_{vm} = \frac{S_y}{\sigma'} = \frac{540}{250} = 2.16 \tag{b}$$

3   The stress intensity K at the crack tip can be found for this case from equation 5.14d if the ratio a/b < 0.13:

$$\frac{a}{b} = \frac{10}{80} = 0.125$$

and   (c)

$$K = 1.12\sigma_{nom}\sqrt{\pi a} = 1.12(250)\sqrt{10\pi} = 49.63 \text{ MPa}\sqrt{m}$$

4   The safety factor against sudden crack propagation is found from equation 5.15.

$$N_{FM} = \frac{K_C}{K} = \frac{66}{49.63} = 1.33 \tag{d}$$

Note that failure is now predicted to be sudden at a 33% overload at which point the nominal stress in the part is still below the yield strength. This is too small a safety factor to allow the part to be used in the face of possible sudden fracture.

5   The crack size necessary for failure can be found approximately by substituting K_c for K in equation 5.14b and solving for a. The result is a crack of about 18-mm width. Note however that the a/b ratio would now exceed that recommended for 10% accuracy with this equation. A more accurate equation for this case could be obtained from one of the references if desired.

6   Assuming the steel has enough carbon to allow heat treatment, through hardening will increase the yield strength but the **ductility and the fracture toughness K_c will decrease** making the part **less safe** against a fracture-mechanics failure.

5.4 USING THE STATIC-LOADING FAILURE THEORIES

It is neither practical nor possible to test all engineering materials under all combinations of applied stresses. The failure theories for static loading presented here provide a means to relate the stress states present in parts subjected to combined stresses to the stress state of the simple, uniaxial tensile test. The concept of an effective stress that "converts" the combination of applied stresses to an equivalent value that can be compared to a tensile-test result is extremely useful. However, the designer must be aware of its limitations in order to properly apply the effective-stress concept.

A fundamental assumption in this chapter is that the materials in question are macroscopically homogeneous and isotropic. Most engineering metals and many engineering polymers are in this category. The assumed presence of microscopic cracks does not preclude the use of conventional failure theories as long as detectable, macroscopic cracks are not in evidence. If so, fracture-mechanics theory should be used.

Composite materials are finding increased use in applications requiring high strength-to-weight ratios. These materials are typically nonhomogeneous and anisotropic (or orthotropic) and thus require different and more complicated failure theories than those presented here. For more information, the reader is directed to the literature on composite materials, some of which is noted in this chapter's bibliography.

Another fundamental assumption of these static failure theories is that the loads are slowly applied and remain essentially constant with time. That is, they are static loads. **When the loads (and thus the stresses) vary with time or are suddenly applied, the failure theories of this chapter may not be the limiting factor.** The next chapter will discuss other failure theories suited to the dynamic loading case and extend the consideration of fracture mechanics to dynamic loading. When fracture mechanics is used in dynamic loading situations a dynamic fracture toughness value K_d (K_{Id}, K_{IId}, or K_{IIId}) is used instead of the static fracture toughness K_c discussed above.

STRESS CONCENTRATION due to geometric discontinuities or sharp contours needs to be taken into account in some cases of static loading before applying the appropriate failure theory. The concept of stress concentration was explained and discussed in Section 4.15 (p. 230). It was pointed out that, **under static loading**, stress concentration can be ignored if the material is ductile because the high stress at the discontinuity will cause local yielding that reduces its effect. However, it is worth repeating that for brittle materials under static loading, the effects of stress concentration should be applied to the calculated stresses before converting them to effective stresses for comparison to the failure theories described here. The one exception to this is with some cast materials (such as gray cast iron) in which the number of inherent stress-raisers within the cast material is so large that the addition of a few more geometric stress-raisers is considered to have little additional effect.

TEMPERATURE AND MOISTURE are also factors in failure. Most of the available test data for materials are generated at room temperature and low humidity. Virtually all material properties are a function of temperature. Metals typically become less strong and more ductile at elevated temperatures. A ductile material can become brittle at low temperatures. Polymers exhibit similar trends over much smaller ranges of temperature than metals. Boiling water is hot enough to soften some polymers and a cold winter day can make them extremely brittle. If your application involves either high or low temperatures or aqueous/corrosive environments, you need to obtain strength data for these conditions from the material manufacturer before applying any failure theories.

CRACKS   If macroscopic cracks are present, or are anticipated in service, then the fracture mechanics (FM) theory should be applied. Once actual cracks are discovered in the field, FM should be used to predict failure and determine the safety of the particular part. If previous experience with similar equipment indicates that service cracking is a problem, then FM should be used in the design of future assemblies and regular field inspections should be done to detect cracks as they occur.

5.5   CASE STUDIES IN STATIC FAILURE ANALYSIS

We will now continue some case studies whose forces were analyzed in Chapter 3 and whose stresses were analyzed in Chapter 4. The same case-study number is retained for a given design throughout the text and successive installments are designated by a letter suffix. For example, Chapter 4 presented four case studies labeled 1B, 2B, 3B, and 4B. This chapter will continue these case studies as 1C, 2C, 3C, and 4C. The reader can review the earlier installments of any case study by referring to its common case number. See the list of case studies in the table of contents to locate each part. Since stresses vary continuously over a part, we made an engineering judgment in Chapter 4 as to where the stresses would be the largest and calculated them for those locations. We wish now to determine their safety factors using the appropriate failure theories.

CASE STUDY 1C

Bicycle Brake Lever Failure Analysis

Problem	Determine the factors of safety at critical points in the brake lever shown in Figures 3-1 and 5-18.
Given	The stresses are known from the previous Case Study 1B (p. 252). The material of the brake lever is die-cast aluminum alloy ASTM G8A with S_{ut} = 310 MPa (45 kpsi) and S_y = 186 MPa (27 kpsi). The elongation to fracture is 8% making it a marginally ductile material.
Assumptions	The most likely failure points are the two holes where the pins insert and at the root of the cantilever-beam lever handle.
Solution	See Figure 5-18 and the TKSolver file CASE1C.

1 The bending stress at point P in Figure 5-18 at the root of the cantilever was found from equation 4.11b in Case Study 1B and is

$$\sigma_x = \frac{Mc}{I} = \frac{54.6 \text{ N-m}\left(\dfrac{0.0143}{2}\right) \text{m}}{\dfrac{\pi(0.0143)^4}{64} \text{ m}^4} = 190 \text{ MPa} \qquad (a)$$

2 This is the only applied stress at this point so it is also the principal stress. The von Mises effective stress $\sigma' = \sigma_1$ in this case (see equation 5.8c.) The safety factor against yielding at point P is then (from Eq. 5.8a),

$$N_{yield} = \frac{S_y}{\sigma'} = \frac{186}{190} = 0.98 \qquad (b)$$

This design clearly has a problem here. With the anticipated maximum load of a strong hand force (712 N), we are at the yield point. Note that in this simple stress situation, the distortion-energy theory gives identical results to the maximum shear theory because the ellipse and hexagon are coincident at the point $x = \sigma_1$, $y = 0$ in Figure 5-4.

3 Since this is a cast material with limited ductility, it would be interesting to also compute the modified-Mohr safety factor against brittle fracture from equation 5.12a. It could be argued that the brake handle might still be usable despite slight yielding having occurred:

$$N_{fracture} = \frac{S_{ut}}{\sigma_1} = \frac{310}{190} = 1.6 \qquad (c)$$

This shows that there is not much margin for overloads that might occur from a fall of the bicycle, etc., and we have not accounted for stress concentrations at the root of the cantilever which could reduce the fracture safety factor to less than 1.

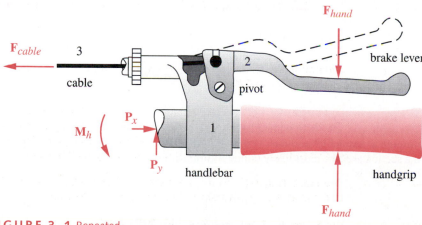

FIGURE 3-1 Repeated

Bicycle Brake Lever Assembly

4   The shear stress due to bending at point Q in Figure 5-18 was calculated from equation 4.15c as

$$\tau_{xy} = \frac{4V}{3A} = \frac{4(712) \text{ N}}{\dfrac{3\pi(14.3)^2}{4} \text{ mm}^2} = 6 \text{ MPa} \qquad (d)$$

This shear stress is also the maximum since no other stresses act at that point. The safety factor using the distortion-energy theory for pure shear at point Q is

$$N = \frac{S_{ys}}{\tau_{max}} = \frac{0.577S_y}{\tau_{max}} = \frac{0.577(186)}{6} = 18 \qquad (e)$$

Clearly, there is no danger of transverse-shear failure at point Q.

5   The compressive bearing stress in the hole at point A in Figure 5-18 is

$$A_{bearing} = \text{dia} \cdot \text{thickness} = 8(2)(6.4) = 102 \text{ mm}^2$$

$$\sigma_{bearing} = \frac{F_{12}}{A_{bearing}} = \frac{2\,993}{102} = 30 \text{ MPa} \qquad (f)$$

and this stress, acting alone, is also the principal and the von Mises stress. Assuming that the compressive strength of this material is equal to its tensile strength (an even material), the safety factor against bearing failure in the hole is

$$N = \frac{S_{yc}}{\sigma'} = \frac{186}{30} = 6.2 \qquad (g)$$

6   Tearout in this case requires that (4) 6.4-mm-thick sections fail in shear through the material between hole A and the edge. (See also Figure 4-13 on p. 192.)

$$A_{tearout} = \text{length} \cdot \text{thickness} = 7.1(4)(6.4) = 181 \text{ mm}^2$$

$$\tau_{tearout} = \frac{F_{12}}{A_{tearout}} = \frac{2\,993}{181} = 17 \text{ MPa} \qquad (h)$$

This is a pure shear case and the safety factor is found from

$$N = \frac{S_{ys}}{\tau_{max}} = \frac{0.577S_y}{\tau_{max}} = \frac{0.577(186)}{17} = 6.3 \qquad (i)$$

7   The cable-end inserts in a blind hole which is half-slotted to allow the cable to pass through at assembly as shown in Figure 5-18. This slot weakens the part and makes the section at R the most likely failure location at this joint. The bending stress at the outer fiber is

$$\sigma_x = \frac{Mc}{I} = \frac{\dfrac{2\,858}{2}\left(\dfrac{5}{2}\right)(4)}{\dfrac{10(5)^3}{12}} = 137 \text{ MPa} \qquad (j)$$

As the only applied stress at the outer fiber of this section, this is also the principal stress and the von Mises stress. The bending safety factor at point R is

$$N = \frac{S_{yc}}{\sigma'} = \frac{186}{137} = 1.4 \qquad (k)$$

8   The shear due to transverse loading at the neutral axis in the section at R is: (Eq. 4.14b)

$$\tau_{xy} = \frac{3V}{2A} = \frac{3(2\,858)}{2(10)(5)} = 76 \text{ MPa} \qquad (l)$$

This is the maximum shear stress at the neutral axis and the transverse-shear safety factor at point R is

$$N = \frac{S_{ys}}{\tau_{max}} = \frac{0.577S_y}{\tau_{max}} = \frac{0.577(186)}{76} = 1.4 \qquad (m)$$

It is interesting to note that the transverse-shear safety factor is the same as the bending safety factor at point R because the beam is so short. Compare this result with that of point P.

9   Some redesign of this part appears to be in order. Failure is predicted at point P and the safety factor for point R is low. The die-cast aluminum alloy chosen is one of the strongest available. Either a change of geometry to increase the section size and/or reduce stress concentrations or a change in material or manufacturing method or both are indicated for points P and R. A forged-aluminum part would be stronger but would increase the cost. Thicker sections would increase the weight slightly, but probably not prohibitively. Increasing the diameter of the handle around point P by 26% to 18 mm (with perhaps a more generous transition radius also) would double the safety factor there since the section modulus is a function of d^3. A similar increase of section at R would make its safety factor acceptable.

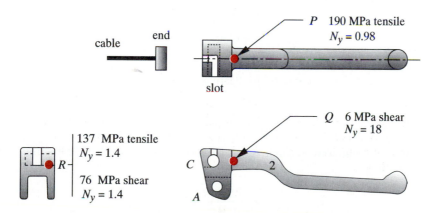

FIGURE 5-18

Stresses at Selected Points on the Bicycle Brake Lever

Even though some of the other safety factors may seem excessive, it may be impractical to reduce those sections due to difficulties in casting thin sections. Other considerations may take into account the appearance of a part intended for a consumer
application such as a bicycle. If the dimensions do not look "right" to the customer,
it may give an undesired impression of cheapness. Sometimes it is better economics
to provide more thickness than is necessary for a suitable safety factor in order to provide a quality appearance.

CASE STUDY 2C

Crimping-Tool Failure Analysis

Problem Determine the factors of safety at critical points in the crimping-tool
shown in Figures 3-3 and 5-19.

Given The stresses are known from the previous Case Study 2B on p. 255.
All material is 1095 ductile steel as-rolled with $S_y = 83$ kpsi. It is an
even material.

Assumptions The most likely failure points are link 3 as a column, the holes where
the pins insert, the connecting pins in shear, and link 4 in bending.

Solution See Figures 3-3 and 5-19 and the TKSolver file CASE2C.

1 The previous case study found the critical column load in link 3 to be 3.65 times larger
than the applied load. This is the safety factor against buckling, which is expressed
in terms of load rather than stress.

2 Any link can fail in bearing in the 0.25-dia holes. The bearing stress (Eqs. 4.7 and
4.10) is:

$$\sigma_b = \frac{P}{A_{bearing}} = \frac{1\,560}{0.125(0.25)} = 49\,920 \text{ psi} \qquad (a)$$

As the only applied stress at this element, this is the principal stress and also the von
Mises stress. The safety factor for bearing stress on either hole or pin is then

$$N = \frac{S_y}{\sigma'} = \frac{83\,000}{49\,920} = 1.7 \qquad (b)$$

4 The 0.25 dia pins are in single shear. The worst-case direct-shear stress from equation 4.9 is

$$\tau = \frac{P}{A_{shear}} = \frac{1\,560}{\frac{\pi(0.25)^2}{4}} = 31\,780 \text{ psi} \qquad (c)$$

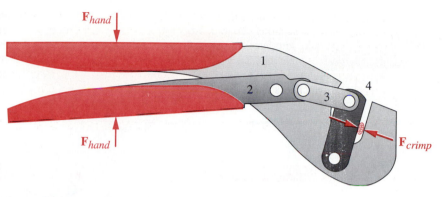

FIGURE 3-3 Repeated

Wire Connector Crimping Tool

As the only stress on this section, this is also the maximum shear stress. The safety factor for the pins in single shear from equation 5.9a is

$$N = \frac{0.577 S_y}{\tau_{max}} = \frac{(0.577)83\,000}{31\,780} = 1.5 \qquad (d)$$

5 Link 4 is a 1.55-in-long beam, simply supported at the pins and loaded with the 2 000-lb crimp force at 0.35 in from point C. The beam depth at the point of maximum moment is 0.75 in and the thickness is 0.187. The bending stress is then

$$\sigma = \frac{Mc}{I} = \frac{541.8\left(\dfrac{0.75}{2}\right)}{\dfrac{0.187(0.75)^3}{12}} = 30\,905 \text{ psi} \qquad (e)$$

As the only applied stress on this element at the outer fiber of the beam, this is the principal stress and also the von Mises stress. The safety factor for link 4 in bending is then

$$N = \frac{S_y}{\sigma'} = \frac{83\,000}{30\,905} = 2.7 \qquad (f)$$

6 Link 1 has a tensile stress due to bending in the inner fiber at point P on the curved beam superposed on an axial tensile stress at the same point. Their sum is the maximum principal stress:

$$\sigma_i = \frac{M}{eA}\left(\frac{c_i}{r_i}\right) = \frac{2\,240}{0.08(0.31)}\left(\frac{0.42}{0.60}\right) = 64\,200 \text{ psi}$$

$$\sigma_a = \frac{F}{A} = \frac{2\,000}{(1.0 - 0.25)(0.31)} = 8\,533 \text{ psi} \qquad (g)$$

$$\sigma_1 = \sigma_a + \sigma_i = 72\,733 \text{ psi}$$

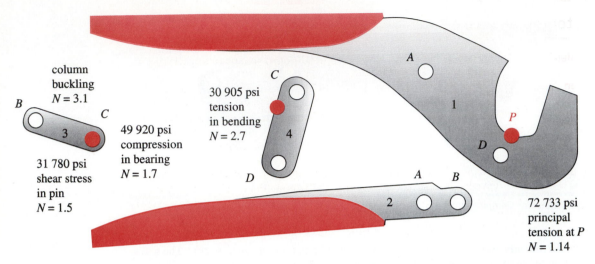

FIGURE 5-19

Significant Stresses and Safety Factors (N) at Critical Points in a Crimping Tool

There is no applied shear stress so this is the principal stress and also the von Mises stress. The safety factor for bending at the inner fiber of the curved beam at point P from equation 5.8a is

$$N = \frac{S_y}{\sigma'} = \frac{83\,000}{72\,733} = 1.14 \qquad (h)$$

7 At the hole in link 1, there is only the axial tensile stress σ_a from equation (g) and the safety factor is found from equation 5.8a:

$$N = \frac{S_y}{\sigma'} = \frac{83\,000}{8\,533} = 9.7 \qquad (i)$$

8 Some of these safety factors, such as the $N = 1.14$ for bending in link 1 at point P, are too low. The safety factors for the pins in shear could also be increased. Either a stronger steel such as SAE 4140 could be selected or the section sizes of the parts could be increased slightly. A small change in link thickness would achieve acceptable safety factors in the existing material. Note that the geometry of this tool has been simplified for this example from that of the actual device. The stresses and safety factors calculated here are not necessarily the same as those in the actual tool, which is a well-tested and safe design. See the *TKSolver* file CASE3B for further details on the solution.

CASE STUDY 3C

Automobile Scissors-Jack Failure Analysis

Problem   Determine safety factors at critical points in a scissors jack.

Given   The stresses are known from the previous Case Study 3B on p. 261. The design load is 2 000 lb total or 1 000 lb per side. The width of the links is 1.032 in and their thickness is 0.15 in. The screw is a 1/2-13 UNC thread with root dia = 0.406 in. The material of all parts is ductile steel with E = 30 Mpsi and S_y = 60 kpsi.

Assumptions The most likely failure points are the links as columns, the holes in bearing where the pins insert, the connecting pins in shear, the gear teeth in bending, and the screw in tension. There are two sets of links, one set on each side, Assume the two sides share the load equally. The jack is typically used for very few cycles over its lifetime so a static analysis is appropriate.

Solution   See Figures 3-5 and 5-20 and the TKSolver files CASE3C-1 and CASE3C-2.

1   The stresses on this jack assembly for the position shown were calculated in the previous installment of this case study in Chapter 4 (p. 261). Please see that case also.

2   The jack screw is in axial tension. The tensile stress was found from equation 4.7.

$$\sigma_x = \frac{P}{A} = \frac{878}{0.129} = 6\ 782 \text{ psi} \tag{a}$$

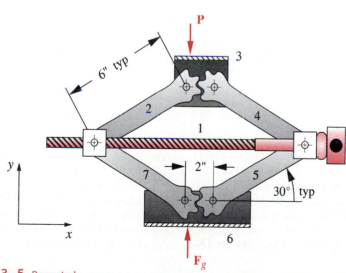

FIGURE 3-5 Repeated

An Automobile Scissors Jack

This is a uniaxial tension stress, so is also the principal and the von Mises stress. The safety factor is

$$N = \frac{S_y}{\sigma'} = \frac{40\ 000}{6\ 782} = 5.9 \tag{b}$$

3   Link 2 is loaded as a beam-column. Its safety factor against buckling was calculated in the last installment of this case study and is $N = 2.3$.

4   The bearing stress in the most heavily loaded hole at C is

$$\sigma_{bearing} = \frac{P}{A_{bearing}} = \frac{1\ 026}{0.15(0.437)} = 15\ 652 \text{ psi} \tag{c}$$

This is a uniaxial compression stress, so is also the principal and the von Mises stress. The safety factor is

$$N = \frac{S_y}{\sigma'} = \frac{40\ 000}{15\ 652} = 2.6 \tag{d}$$

The pins' shear stress is

$$\tau = \frac{P}{A_{shear}} = \frac{1\ 026}{\dfrac{\pi(0.437)^2}{4}} = 6\ 841 \text{ psi} \tag{e}$$

This is a pure shear stress, so is also the maximum shear stress. The safety factor is

$$N = \frac{0.577 S_y}{\tau_{max}} = \frac{0.577(40\ 000)}{6\ 841} = 3.4 \tag{f}$$

5   The bending stress at the root of the gear tooth on link 2 is

$$\sigma = \frac{Mc}{I} = \frac{91(0.22)}{\dfrac{0.15(0.44)^3}{12}} = 18\ 727 \text{ psi} \tag{g}$$

This is a uniaxial bending stress, so is also the principal and the von Mises stress. The safety factor is

$$N = \frac{S_y}{\sigma'} = \frac{40\ 000}{18\ 727} = 2.1 \tag{h}$$

6   This analysis should be continued, looking at other points in the assembly and, more importantly, at stresses and safety factors when the jack is in different positions. We have used an arbitrary position for this case study but, as the jack moves to the lowered position, the link and pin forces will increase due to poorer transmission angles. A complete stress and safety factor analysis should be done for multiple positions. See the *TKSolver* files CASE3C-1 and CASE3C-2 for a more complete analysis and to see all the equations and data used.

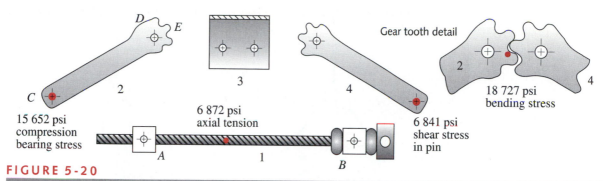

FIGURE 5-20

Stresses in the Scissors Jack Assembly

CASE STUDY 4C

Bicycle Brake Arm Factors of Safety

Problem   Choose appropriate material alloys to obtain a safety factor of at least 2 at critical points in the bicycle brake arm in Figures 3-9 and 5-21.

Given   The stresses are known from the previous Case Study 4B on p. 265. The arm is cast-aluminum and the pivot pin is steel.

Assumptions   Since the arm is a brittle material the modified-Mohr theory will be used for it. The pin is ductile, so the maximum distortion-energy theory will be used for it.

Solution   See Figures 3-9 and 5-21 and the TKSolver file CASE4C.

1   The bending stresses at the inner and outer fibers of the arm were found to be

$$\sigma_i = +\frac{M}{eA}\left(\frac{c_i}{r_i}\right) = \frac{44\,572(6.062)}{(0.187)(225)(58)} = 110.6 \text{ MPa}$$

$$\sigma_o = -\frac{M}{eA}\left(\frac{c_o}{r_o}\right) = \frac{44\,572(8.937)}{(0.187)(225)(73)} = -129.5 \text{ MPa}$$

(a)

For a safety factor of 2 at this point we need a material with an ultimate tensile strength of at least 221 MPa and a compressive strength of at least 260 MPa.

2   The stress on the left half of Section B-B is the sum of the bending stress and the axial tension stress:

$$\sigma_{hub} = \frac{Mc}{I_{hub}} + \frac{F_{32_y}}{A_{hub}} = \frac{44\,572(12.5)}{33\,948} + \frac{356}{399} = 17.3 \text{ MPa}$$

(g)

This only needs a material tensile strength of 35 MPa for a safety factor of 2.

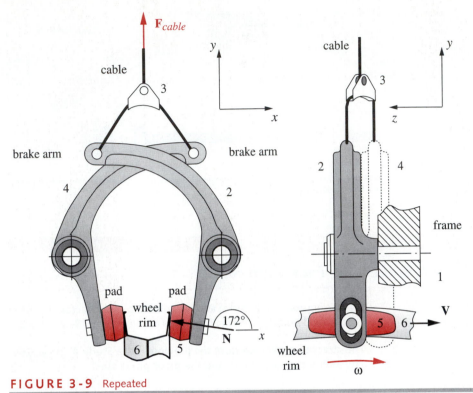

FIGURE 3-9 Repeated

Center-Pull Bicycle Brake Arm Assembly

3   The bending stress at the outer fiber of the 23-mm side of the straight portion of the brake arm is

$$\sigma_y = \sigma_{y_1} + \sigma_{y_2} = 80.75 + 16.05 = 96.8 \text{ MPa} \qquad (b)$$

For a safety factor of 2 at this point we need a material with an S_{ut} of at least 194 MPa.

4   Another possible failure point is the slot in the arm.   The tearout shear stress is

$$\tau = \frac{F_{52_z}}{A_{tearout}} = \frac{399}{32} = 12.5 \text{ MPa} \qquad (c)$$

For a uniaxial applied shear stress, the stresses are all in the first quadrant of Figure 5-9 and the modified-Mohr theory is identical to the maximum normal-stress theory. The equivalent tensile stress is then twice the maximum shear stress which requires an ultimate tensile strength greater than 50 MPa for a safety factor of 2 in this case.

5   The worst case among the points calculated is the curved beam root in (1) above. From the material properties data for cast-aluminum in Appendix C, we find that the A390-T6 sand-casting alloy has an S_{ut} of 280 MPa, which satisfies the requirement for a safety factor > 2 in this case. This is a brittle material with an elongation of <1%.

6   The pivot pin is subjected to a maximum bending stress of

$$\sigma_{pin} = \frac{M_{pin}c_{pin}}{I_{pin}} = \frac{71\,270\left(\dfrac{11}{2}\right)}{\dfrac{\pi(11)^4}{64}} = 545 \text{ MPa} \qquad (d)$$

For a safety factor of 2, this requires a steel with a yield strength of at least 1090 MPa. One of the strongest alloy steels is AISI 4340, which has $S_y = 862$ MPa in the normalized condition. When this alloy is hardened, quenched, and tempered at 1000°F, it will reach a yield strength of about 1100 MPa and provide the desired safety factor.

Perhaps a better approach in this case would be to increase the pin diameter to reduce the stress and allow the use of a less expensive steel without the heat treatment. Increasing the pin diameter from 11 mm to 14 mm the pin stress becomes

$$\sigma_{pin} = \frac{M_{pin}c_{pin}}{I_{pin}} = \frac{71\,270\left(\dfrac{14}{2}\right)}{\dfrac{\pi(14)^4}{64}} = 265 \text{ MPa} \qquad (e)$$

and an AISI 1095 cold-rolled steel with a yield strength of 572 MPa in the as-rolled condition will give a safety factor of 2.2. The loss of material in the hub from this increase in pin-hole diameter only raises its stress to 19 MPa, which still has a large safety factor compared to the pin.

7   See the *TKSolver* file CASE4C for more detailed information on the equations and calculations.

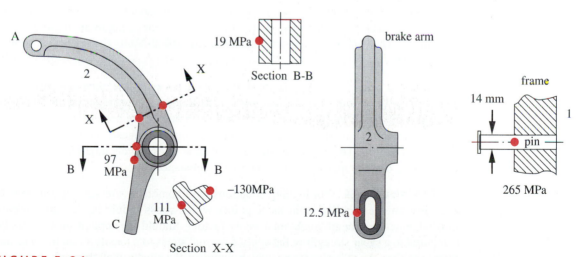

FIGURE 5-21

Brake Arm Stresses at Selected Points

Note that in most of these case studies, some redesign was necessary after (and only after) getting to the "bottom line" of determining safety factors for the geometry and loading assumed at the outset. Some of the safety factors were found to be low. This is typical of design problems and demonstrates their iterative nature. We cannot know whether our assumptions are valid until we spend the considerable time and energy to thoroughly analyze the proposed design. We should not be too disappointed when we find that our first design doesn't work. That is not necessarily a reflection on the designer's abilities. It is rather just the "nature of the beast." The value of setting up the analytical model in a computer tool such as a spreadsheet or equation solver should by now be obvious. The analysis of redesigned geometry as proposed in these latest case studies above can literally be accomplished in minutes if the time was taken initially to computerize the model. If not, we are faced with a much larger amount of work to reanalyze the modified design.

5.6   SUMMARY

This chapter has presented several theories of failure for materials under static loading. Two of these theories seem to fit the experimental data best. Both assume that the material is reasonably homogeneous and isotropic in the macro. In addition, the mechanism of crack propagation based on fracture mechanics theory is presented.

The **distortion-energy theory**, also called the **von Mises theory**, is best for ductile, even materials whose compressive and tensile strengths are approximately the same and whose shear strengths are smaller than their tensile strengths. These materials are considered to fail from shear stress and the distortion-energy theory best predicts their failure.

Uneven, brittle materials such as cast iron typically have tensile strengths that are much lower than their compressive strengths, and their shear strengths are between those two values. They are weakest in tension and the **modified-Mohr theory** best describes their failure.

Note that when the loading is not static, but varies with time, *then neither of these theories is appropriate to describe failure*. A different set of criteria for time-varying loading is discussed in the next chapter. If known cracks are present, then the possibility of sudden failure due to crack propagation must be investigated using fracture-mechanics theory. The crack can suddenly open at stress levels much lower than the yield strength of the material under certain circumstances.

EFFECTIVE STRESS   For loading situations with combined stresses (such as tensile and shear stresses applied at the same point), which stress should be used to compare to what material strength to obtain a safety factor? Should the applied shear stress be compared to a shear strength or the applied normal stress to a tensile strength? The answer is neither. An **effective stress**, which combines the effects of all the applied stresses combined at the point must be calculated that can be compared to the "pure" normal-stress state of the tensile-test specimen. These effective stresses are a useful means to create a stress-loading criterion for comparison with published material strength data on an "apples for apples" basis even when the applied stress situation is quite different from the test specimen's loading. The effective-stress approach is also valid when only one stress is applied at the point and so can be universally used. The calculation for the effective stress differs with the type of material, either ductile or brittle, however.

For ductile, even materials, the **von Mises effective stress** is calculated either directly from the applied stresses (Eq. 5.7b, 5.7d) or from the principal stresses that result from those applied stresses (Eq. 5.7a, 5.7c). Note that this effective stress calculation converts **any combination** of 2-D or 3-D applied stresses at a point into a **single stress value** σ' that can be compared with a suitable strength criterion in order to obtain a safety factor. For ductile materials under static loads, the desired strength criterion is the **tensile-yield strength**. (See Section 5.1.)

For brittle, uneven materials, a **modified-Mohr effective stress** can be calculated using the principal stresses that result from the particular combination of applied stresses at the point in question (Eq. 5.12a–d). The resulting effective stress is compared to the ultimate tensile strength of the material (not the yield strength) to obtain a safety factor. (See Section 5.2.)

FRACTURE MECHANICS   In addition to possible failure by yielding or breaking, a part can fail at much lower stresses by crack propagation if a crack of sufficient size is present. The theory of fracture mechanics provides a means to predict such sudden failure based on a calculated **stress intensity factor** compared to a tested **fracture toughness** criterion for the material. (See Section 5.3.)

The failure analysis process for static loading can be summarized in a series of steps as shown in the flow chart of Figure 5-22. Note that the first five steps are the same as those in the chart of Figure 4-57.

Important Equations Used in this Chapter

See the referenced sections for information on the proper use of these equations.

Von Mises Effective Stress for 3-Dimensions (Section 5.1):

$$\sigma' = \sqrt{\sigma_1^2 + \sigma_2^2 + \sigma_3^2 - \sigma_1\sigma_2 - \sigma_2\sigma_3 - \sigma_1\sigma_3} \tag{5.7a}$$

$$\sigma' = \sqrt{\frac{\left(\sigma_x - \sigma_y\right)^2 + \left(\sigma_y - \sigma_z\right)^2 + \left(\sigma_z - \sigma_x\right)^2 + 6\left(\tau_{xy}^2 + \tau_{yz}^2 + \tau_{zx}^2\right)}{2}} \tag{5.7b}$$

Von Mises Effective Stress for 2-Dimensions (Section 5.1):

$$\sigma' = \sqrt{\sigma_1^2 - \sigma_1\sigma_3 + \sigma_3^2} \tag{5.7c}$$

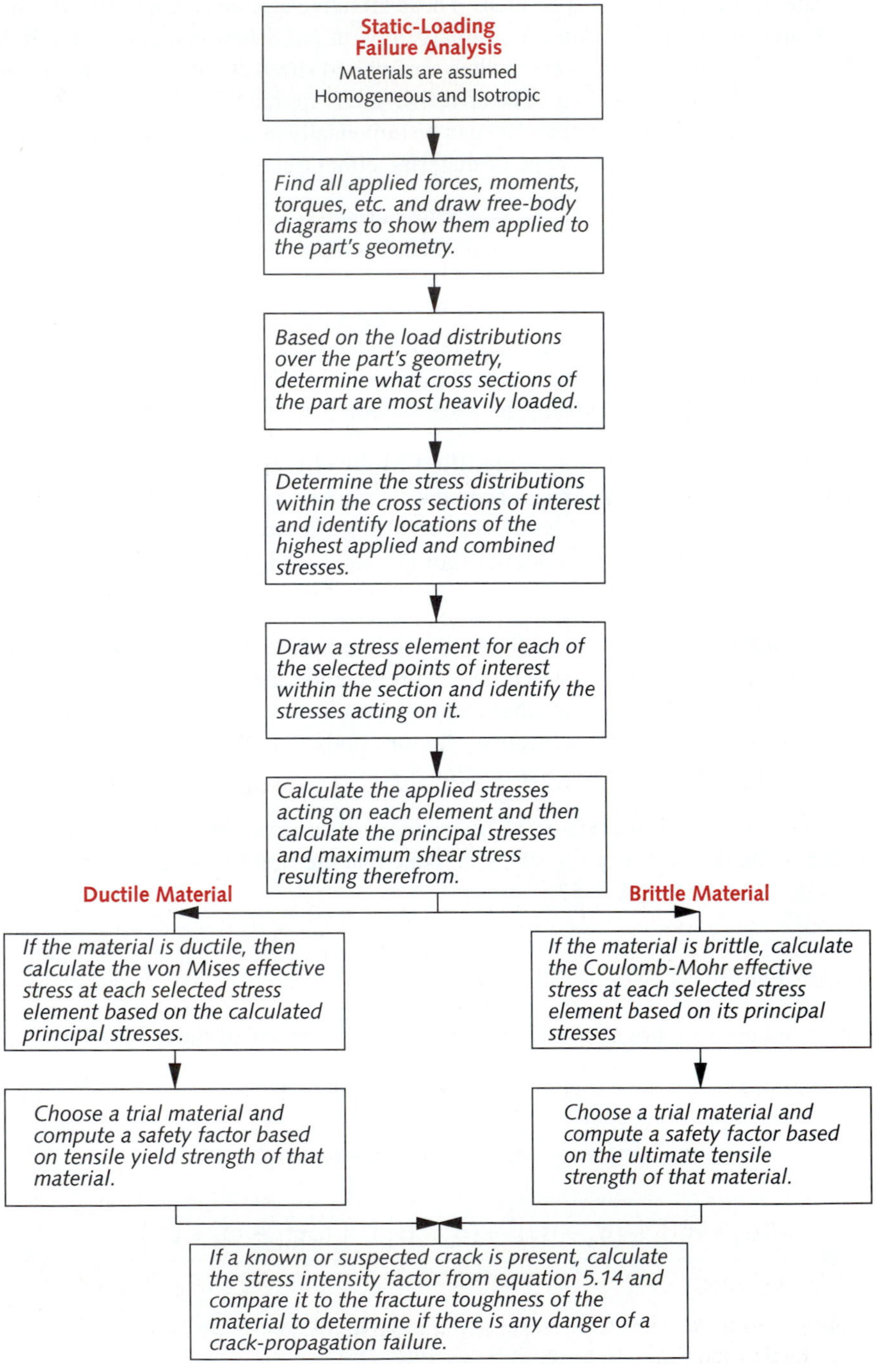

5

**Static-Loading
Failure Analysis**
Materials are assumed
Homogeneous and Isotropic

*Find all applied forces, moments,
torques, etc. and draw free-body
diagrams to show them applied to
the part's geometry.*

*Based on the load distributions
over the part's geometry,
determine what cross sections of
the part are most heavily loaded.*

*Determine the stress distributions
within the cross sections of interest
and identify locations of the
highest applied and combined
stresses.*

*Draw a stress element for each of
the selected points of interest
within the section and identify the
stresses acting on it.*

*Calculate the applied stresses
acting on each element and then
calculate the principal stresses
and maximum shear stress
resulting therefrom.*

Ductile Material

*If the material is ductile, then
calculate the von Mises effective
stress at each selected stress
element based on the calculated
principal stresses.*

*Choose a trial material and
compute a safety factor based
on tensile yield strength of that
material.*

Brittle Material

*If the material is brittle, calculate
the Coulomb-Mohr effective
stress at each selected stress
element based on its principal
stresses*

*Choose a trial material and
compute a safety factor based
on the ultimate tensile
strength of that material.*

*If a known or suspected crack is present, calculate
the stress intensity factor from equation 5.14 and
compare it to the fracture toughness of the
material to determine if there is any danger of a
crack-propagation failure.*

FIGURE 5-22

Flow Chart for Static Failure Analysis

$$\sigma' = \sqrt{\sigma_x^2 + \sigma_y^2 - \sigma_x\sigma_y + 3\tau_{xy}^2} \qquad (5.7d)$$

Safety Factor for Ductile Materials Under Static Loading (Section 5.1):

$$N = \frac{S_y}{\sigma'} \qquad (5.8a)$$

Shear Yield Strength as a Function of Tensile Yield Strength (Section 5.1):

$$S_{ys} = 0.577 S_y \qquad (5.9b)$$

Modified-Mohr Effective stress for 3-Dimensions (Section 5.2):

$$C_1 = \frac{1}{2}\left[|\sigma_1 - \sigma_2| + \frac{S_{uc} + 2S_{ut}}{S_{uc}}(\sigma_1 + \sigma_2)\right]$$

$$C_2 = \frac{1}{2}\left[|\sigma_2 - \sigma_3| + \frac{S_{uc} + 2S_{ut}}{S_{uc}}(\sigma_2 + \sigma_3)\right] \qquad (5.12c)$$

$$C_3 = \frac{1}{2}\left[|\sigma_3 - \sigma_1| + \frac{S_{uc} + 2S_{ut}}{S_{uc}}(\sigma_3 + \sigma_1)\right]$$

$$\tilde{\sigma} = \text{MAX}(C_1, C_2, C_3, \sigma_1, \sigma_2, \sigma_3)$$

$$\tilde{\sigma} = 0 \quad if \text{ MAX} < 0 \qquad (5.12d)$$

Safety Factor for Brittle Materials Under Static Loading (Section 5.2):

$$N = \frac{S_{ut}}{\tilde{\sigma}} \qquad (5.12e)$$

Stress Intensity Factor (Section 5.3):

$$K = \beta \sigma_{nom} \sqrt{\pi a} \qquad (5.14b)$$

Safety Factor for Crack Propagation (Section 5.3):

$$N_{FM} = \frac{K_c}{K} \qquad (5.15)$$

5.7   REFERENCES

1 **J. P. D. Hartog**, *Strength of Materials*. Dover Press: New York, p. 222, 1961.

2 **J. Marin**, *Mechanical Behavior of Engineering Materials*. Prentice-Hall: Englewood Cliffs, N. J., pp. 117-122, 1962.

3 **S. P. Timoshenko**, *History of Strength of Materials*. McGraw-Hill: New York, 1953.

4   **N. E. Dowling**, *Mechanical Behavior of Materials*. Prentice-Hall: Englewood Cliffs, N. J., p. 252, 1993.

5   *Ibid.*, pp. 262-264.

6   **T. J. Dolan**, *Preclude Failure: A Philosophy for Material Selection and Simulated Service Testing*. SESA J. Exp. Mech., Jan. 1970.

7   **N. E. Dowling**, *Mechanical Behavior of Materials*. Prentice-Hall: Englewood Cliffs N. J., p. 280, 1993.

8   *Interim Report of a Board of Investigation to Inquire into the Design and Methods of Construction of Welded Steel Merchant Vessels*, USCG Ship Structures Committee, Wash., D.C. 20593-0001, June 3, 1944.

9   **C. F. Tipper**, *The Brittle Fracture Story*. Cambridge University Press: New York, 1962.

10   **R. C. Juvinall**, *Engineeering Considerations of Stress, Strain, and Strength*. McGraw-Hill: New York, p. 71, 1967.

11   **J. M. Barsom and S. T. Rolfe**, *Fracture and Fatigue Control in Structures*. 2nd ed. Prentice-Hall: Englewood Cliffs N. J., p. 203, 1987.

12   **J. A. Bannantine, J. J. Comer, and J. L. Handrock**, *Fundamentals of Metal Fatigue Analysis*. Prentice-Hall: Englewood Cliffs N. J., p. 94, 1990.

13   **D. Broek**, *The Practical Use of Fracture Mechanics*. Kluwer Academic Publishers: Dordrecht, The Netherlands, pp. 8-10, 1988.

14   *Ibid*, p. 11.

15   **N. E. Dowling**, *Mechanical Behavior of Materials*. Prentice-Hall: Englewood Cliffs N. J., pp. 306-307, 1993.

5.8   BIBLIOGRAPHY

For further information on fracture mechanics see:

J. M. Barsom and S. T. Rolfe, *Fracture and Fatigue Control in Structures*. 2nd ed. Prentice-Hall: Englewood Cliffs N.J., 1987.

D. Broek, *The Practical Use of Fracture Mechanics*. Kluwer: Dordrecht, The Netherlands, 1988.

N. E. Dowling, *Mechanical Behavior of Materials*. Prentice-Hall: Englewood Cliffs, N.J., 1993.

R. C. Rice, ed. *Fatigue Design Handbook*. 2nd ed. SAE: Warrendale Pa. 1988.

For information on stress intensity factors see:

Y. Murakami, ed.. *Stress Intensity Factors Handbook*. Pergamon Press: Oxford, U.K. 1987.

D. P. Rooke and D. J. Cartwright, *Compendium of Stress Intensity Factors*. Her Majesty's Stationery Office: London, 1976.

H. Tada, P. C. Paris, and G. R. Irwin, *The Stress Analysis of Cracks Handbook.* 2nd ed. Paris Productions Inc.: 226 Woodbourne Dr., St. Louis Mo., 1985.

For information on fracture toughness of materials see:

Battelle, *Aerospace Structural Metals Handbook.* Metals and Ceramics Information Center, Battelle Columbus Labs.: Columbus Ohio, 1991.

J. P. Gallagher, ed. *Damage Tolerant Design Handbook.* Metals and Ceramics Information Center, Battelle Columbus Labs.: Columbus Ohio. 1983.

C. M. Hudson and S. K. Seward, *A Compendium of Sources of Fracture Toughness and Fatigue Crack Growth Data for Metallic Alloys.* Int. J. of Fracture, **14**(4): R151-R184, 1978.

B. Marandet and G. Sanz, *Evaluation of the Toughness of Thick Medium Strength Steels . . .,* in *Flaw Growth and Fracture.* Am. Soc. for Testing and Materials: Philadelphia, Pa. p. 72-95, 1977.

For information on failure of composite materials see:

R. Juran, ed., *Modern Plastics Encyclopedia.* McGraw-Hill: New York. 1992.

A. Kelly, ed., *Concise Encyclopedia of Composite Materials.* Pergamon Press: Oxford, U.K. 1989.

M. M. Schwartz, *Composite Materials Handbook.* McGraw-Hill: New York, 1984.

5.9   PROBLEMS

*†5-1   A differential stress element has a set of applied stresses on it as indicated in each row of Table P5-1. For the row(s) assigned, draw the stress element showing the applied stresses, Find the principal stresses and von Mises stresses.

†5-2   A 400-lb chandelier is to be hung from two 10-ft-long solid, low-carbon steel cables in tension. Size the cables for a safety factor of 4. State all assumptions.

†5-3   For the bicycle pedal arm assembly in Figure P5-1 with a rider-applied force of 1 500 N at the pedal, determine the von Mises stress in the 15-mm-dia pedal arm. The pedal attaches to the arm with a 12 mm thread. Find the von Mises stress in the screw? Find the safety factor against static failure if the material has $S_y = 350$ MPa.

*†5-4   The trailer hitch shown in Figure P5-2 and Figure 1-1 (p. 12) has loads applied as shown. The tongue weight of 100 kg acts downward and the pull force of 4 905 N acts horizontally. Using the dimensions of the ball bracket shown in Figure 1-5 (p. 15) and $S_y = 300$ MPa ductile steel, determine static safety factors for

    (a) The shank of the ball where it joins the ball bracket.
    (b) Bearing failure in the ball bracket hole.
    (c) Tearout failure in the ball bracket.
    (d) Tensile failure in the attachment bolts if they are 19-mm dia.
    (e) Bending failure in the ball bracket as a cantilever.

†5-5   Repeat Problem 5-4 for the loading conditions of Problem 3-5 on p. 169.

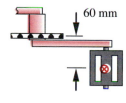

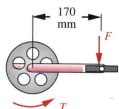

FIGURE P5-1

Problem 5-3

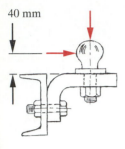

40 mm

FIGURE P5-2

Problems 5-4, 5-5, 5-6

Table P5-1   Data for Problem 5-1

Rows *a-g* are two-dimensional, others are 3-D problems

Row	σ_x	σ_y	σ_z	τ_{xy}	τ_{yz}	τ_{zx}
a	1 000	0	0	500	0	0
b	−1 000	0	0	750	0	0
c	500	−500	0	1 000	0	0
d	0	−1 500	0	750	0	0
e	750	250	0	500	0	0
f	−500	1 000	0	750	0	0
g	1 000	0	−750	0	0	250
h	750	500	250	500	0	0
i	1 000	−250	−750	250	500	750
j	−500	750	250	100	250	1 000

*†5-6   Repeat Problem 5-4 for the loading conditions of Problem 3-6 on p. 169.

*†5-7   Design the wrist pin of Problem 3-7 (p. 169) for safety factor = 3.0 if S_y = 100 kpsi.

*†5-8   A paper mill processes rolls of paper having a density of 984 kg/m^3. The paper roll is 1.50-m outside dia. (OD) x 0.22-m inside diameter (ID) x 3.23-m long and is on a simply supported, hollow, steel shaft with S_y = 300 MPa. Find the shaft ID needed to obtain a static safety factor of 5 if the shaft OD is 22 cm.

†5-9   For the ViseGrip® plier-wrench drawn to scale in Figure P5-3, and for which the forces were analyzed in Problem 3-9 and stresses in Problem 4-9, find the safety factors for each pin for an assumed clamping force of P = 4 000 N in the position shown. The pins are 8-mm dia, S_y = 400 MPa, and are all in double shear.

*†5-10   An overhung diving board is shown in Figure P5-4a. Assume cross-sectional dimensions of 305 mm x 32 mm. Find the largest principal stress in the board when

* Answers to these problems are provided in Appendix G.

† These problems are based on similar problems in previous chapters with the same –number, e.g., Problem 5-4 is based on Problem 4-4, etc. Problems in succeeding chapters may also continue and extend these problems.

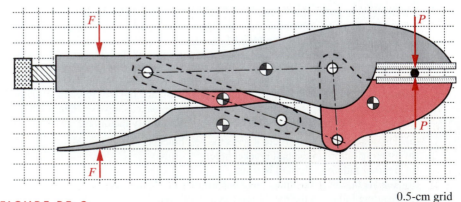

0.5-cm grid

FIGURE P5-3

Problem 5-9

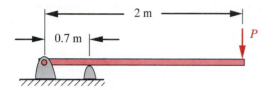

(a) Overhung diving board

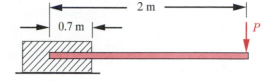

(b) Cantilevered diving board

FIGURE P5-4

Problems 5-10 through 5-13

a 100-kg person is standing at the free end. What is the static safety factor if the material is brittle fiberglass with S_{ut} = 130 MPa in the longitudinal direction?

*†5-11   Repeat Problem 5-10 assuming the 100-kg person in Problem 5-10 jumps up 25 cm and lands back on the board. Assume the board weighs 29 kg and deflects 13.1 cm statically when the person stands on it. What is the static safety factor if the material is brittle fiberglass with S_{ut} = 130 MPa in the longitudinal direction?

†5-12   Repeat Problem 5-10 using the cantilevered diving board design in Figure P5-4b.

†5-13   Repeat Problem 5-11 using the diving board design shown in Figure P5-4b. Assume the board weighs 19 kg and deflects 8.5 cm statically when the person stands on it.

†5-14   Figure P5-5 shows a child's toy called a *pogo stick*. The child stands on the pads, applying half her weight on each side. She jumps up off the ground, holding the pads up against her feet, and bounces along with the spring cushioning the impact and storing energy to help each rebound. Assume a 60-lb child and a spring constant of 100 lb/in. The pogo stick weighs 5 lb. Design the aluminum cantilever beam sections on which she stands to survive jumping 2 in off the ground with a safety factor of 2. Use 1100 series aluminum. Define the beam shape and size.

*†5-15   What is the safety factor for the shear pin as defined in Problem 4-15?

†5-16   A track to guide bowling balls is designed with two round rods as shown in Figure P-5-6. The rods have a small angle between them. The balls roll on the rods until they fall between them and drop onto another track. Each rod's unsupported length is 30 in and the angle between them is 3.2°. The balls are 4.5-in dia and weigh 2.5 lb. The center distance between the 1-in dia rods is 4.2 in at the narrow end. Find the static safety factor for the 1 in dia SAE 1045 normalized steel rods.

(a) Assume rods are simply supported at each end.
(b) Assume rods are fixed at each end.

*†5-17   A pair of ice tongs is shown in Figure P5-7. The ice weighs 50 lb and is 10 in wide across the tongs. The distance between the handles is 4 in, and the mean radius r of a tong is 6 in. The rectangular cross-sectional dimensions are 0.750 in x 0.312 in. Find the safety factor for the tongs if their S_y = 30 kpsi.

5-18   Repeat Problem 5-17 with the tongs made of Class 20 gray cast iron.

*†5-19   Determine the size of the clevis pin shown in Figure P5-8 needed to withstand an applied force of 130 000 lb. Also determine the required outside radius of the clevis

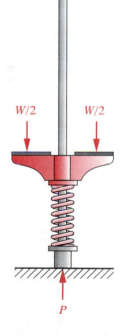

$W/2$ $W/2$

P

FIGURE P5-5

Problem 5-14

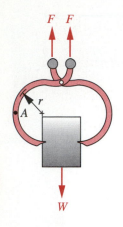

F F

r

A

W

FIGURE P5-7

Problem 5-17

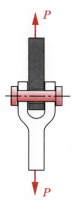

P

P

FIGURE P5-8

Problems 5-19

* Answers to these problems are provided in Appendix G.

† These problems are based on similar problems in previous chapters with the same –number, e.g., Problem 5-4 is based on Problem 4-4, etc. Problems in succeeding chapters may also continue and extend these problems.

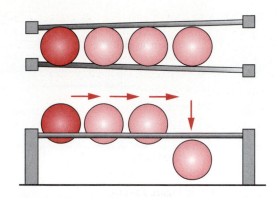

FIGURE P5-6

Problem 5-16

end to not fail in either tearout or bearing if the clevis flanges each are each 2.5 in thick. Use a safety factor of 3 for all modes of failure. Assume S_y = 89.3 kpsi for the pin and S_y = 35.5 kpsi for the clevis.

5-20   A 100 N-m torque is applied to a 1-m-long, solid, round shaft. Design it to limit its angular deflection to 2° and select a steel alloy to have a yielding safety factor of 2.

†5-21   Figure P5-9 shows an automobile wheel with two styles of lug wrench, a single ended wrench in (*a*), and a double-ended wrench in (*b*). The distance between points *A* and *B* is 1 ft in both cases and the handle diameter is 0.625 in. What is the maximum force possible before yielding the handle if the material S_y = 45 kpsi?

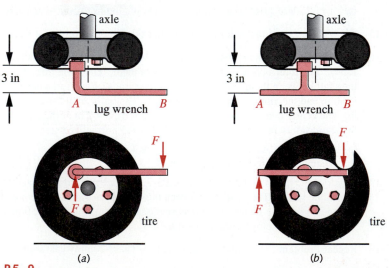

(a)   (b)

FIGURE P5-9

Problem 5-21

*†5-22   An inline "roller-blade" skate is shown in Figure P5-10. The polyurethane wheels
         are 72-mm dia and spaced on 104-mm centers. The skate-boot-foot combination
         weighs 2 kg. The effective "spring rate" of the person-skate system is 6 000 N/m.
         The axles are 10-mm-dia steel pins in double shear with S_y = 400 MPa. Find the
         safety factor for the pins when a 100 kg person lands a 0.5 m jump on one foot.

         (a) Assume that all four wheels land simultaneously.
         (b) Assume that one wheel absorbs all the landing force.

*†5-23   A beam is supported and loaded as shown in Figure P5-11a For the data given in
         the assigned row(s) in Table P5-2, find the static safety factor:

         (a) If the beam is a ductile material with S_y = 300 MPa,
         (b) If the beam is a cast-brittle material with S_{ut} = 150 MPa.

*†5-24   A beam is supported and loaded as shown in Figure P5-11b. For the data given in
         the assigned row(s) in Table -P5-2, find the static safety factor:

         (a) If the beam is a ductile material with S_y = 300 MPa,
         (b) If the beam is a cast-brittle material with S_{ut} = 150 MPa.

*†5-25   A beam is supported and loaded as shown in Figure P5-11c. For the data given in
         the assigned row(s) in Table P5-2, find the static safety factor:

         (a) If the beam is a ductile material with S_y = 300 MPa,
         (b) If the beam is a cast-brittle material with S_{ut} = 150 MPa.

*†5-26   A beam is supported and loaded as shown in Figure P5-11d. For the data given in
         the assigned row(s) in Table P5-2, find the static safety factor:

FIGURE P5-10

Problem 5-22

5

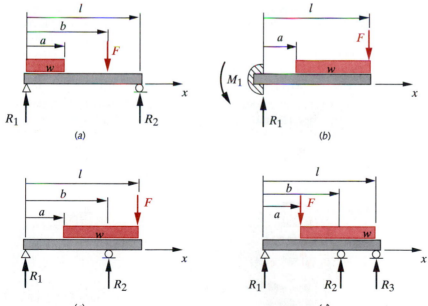
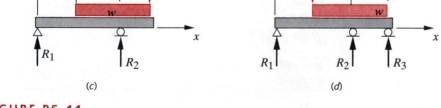

(a)

(b)

(c)

(d)

FIGURE P5-11

Beams and Beam Loadings for Problems 5-23 - See Table P5-2 for Data

* Answers to these
problems are provided in
Appendix G.

† These problems are based
on similar problems in
previous chapters with the
same –number, e.g.,
Problem 5-4 is based on
Problem 4-4, etc. Problems
in succeeding chapters may
also continue and extend
these problems.

Table P5-2   Data for Problems 5-23 through 5-26

Use only data relevant to the particular problem. Lengths in m, forces in N, I in m^4.

Row	l	a	b	w	F	I	c	E
a	1.00	0.40	0.60	200	500	2.85E–08	2.00E–02	steel
b	0.70	0.20	0.40	80	850	1.70E–08	1.00E–02	steel
c	0.30	0.10	0.20	500	450	4.70E–09	1.25E–02	steel
d	0.80	0.50	0.60	65	250	4.90E–09	1.10E–02	steel
e	0.85	0.35	0.50	96	750	1.80E–08	9.00E–03	steel
f	0.50	0.18	0.40	450	950	1.17E–08	1.00E–02	steel
g	0.60	0.28	0.50	250	250	3.20E–09	7.50E–03	steel
h	0.20	0.10	0.13	400	500	4.00E–09	5.00E–03	alum
i	0.40	0.15	0.30	50	200	2.75E–09	5.00E–03	alum
j	0.20	0.10	0.15	150	80	6.50E–10	5.50E–03	alum
k	0.40	0.16	0.30	70	880	4.30E–08	1.45E–02	alum
l	0.90	0.25	0.80	90	600	4.20E–08	7.50E–03	alum
m	0.70	0.10	0.60	80	500	2.10E–08	6.50E–03	alum
n	0.85	0.15	0.70	60	120	7.90E–09	1.00E–02	alum

    (a) If the beam is a ductile material with $S_y = 300$ MPa,

    (b) If the beam is a cast-brittle material with $S_{ut} = 150$ MPa.

*†5-27   A storage rack is to be designed to hold the paper roll of Problem 5-8 as shown in Figure P5-12. Determine suitable values for dimensions a and b in the figure. Make the static safety factor at least 1.5. The mandrel is solid and inserts halfway into the paper roll.

    (a) The beam is a ductile material with $S_y = 300$ MPa,

    (b) The beam is a cast-brittle material with $S_{ut} = 150$ MPa.

†5-28   Figure P5-13 shows a forklift truck negotiating a 15° ramp to drive onto a 4-ft-high loading platform. The truck weighs 5 000 lb and has a 42-in wheelbase. Design two (one for each side) 1-ft-wide ramps of steel to have a safety factor of 3 in the worst case of loading as the truck travels up them. Minimize the weight of the ramps by

* Answers to these problems are provided in Appendix G.

† These problems are based on similar problems in previous chapters with the same –number, e.g., Problem 5-4 is based on Problem 4-4, etc. Problems in succeeding chapters may also continue and extend these problems.

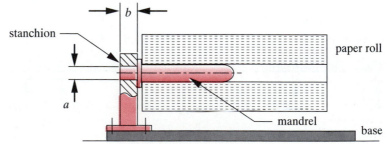

FIGURE P5-12

Problem 5-27

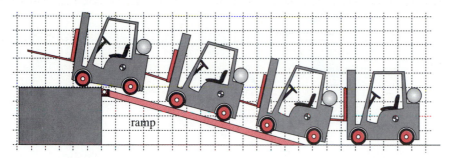

using a sensible cross-section geometry. Choose an appropriate steel or aluminum alloy.

5-29   A differential element is subjected to the stresses (in kpsi): $\sigma_1 = 10$, $\sigma_2 = 0$, $\sigma_3 = -20$. A ductile material has the strengths (in kpsi) $S_{ut} = 50$, $S_y = 40$, $S_{uc} = 50$. Calculate the safety factor and draw $\sigma_1 - \sigma_3$ diagrams of each theory showing the stress state using:

(a) Maximum shear-stress theory
(b) Distortion-energy theory

5-30   A differential element is subjected to the stresses (in kpsi): $\sigma_1 = 10$, $\sigma_2 = 0$, $\sigma_3 = -20$. A brittle material has the strengths (in kpsi) $S_{ut} = 50$, $S_y = 40$, $S_{uc} = 90$. Calculate the safety factor and draw $\sigma_1 - \sigma_3$ diagrams of each theory showing the stress state using:

(a) Coulomb-Mohr theory
(b) Modified-Mohr theory

5-31   Design a jack-stand in a tripod configuration that will support 2 tons of load with a safety factor of 3. Use SAE 1020 steel and minimize its weight.

*5-32   A part has a combined stress state and strengths (in kpsi) of: $\sigma_x = 10$, $\sigma_y = 5$, $\tau_{xy} = 4.5$, $S_{ut} = 20$, $S_{uc} = 80$, $S_y = 18$. Choose an appropriate failure theory based on the given data, find the effective stress and factor of safety against static failure.

*†5-33   For the bracket shown in Figure P5-14 and the data in the row(s) assigned from Table P5-3, determine the von Mises stresses at points A and B.

*†5-34   Calculate the safety factor for the bracket in Problem 5-33 using the distortion energy, the maximum shear-stress and the maximum normal-stress theories. Comment on their appropriateness. Assume a ductile material strength of $S_y = 400$ MPa (60 kpsi).

*†5-35   Calculate the safety factor for the bracket in Problem 5-33 using the Coulomb-Mohr and the modified-Mohr effective stress theories. Comment on their appropriateness. Assume a brittle material strength of $S_{ut} = 350$ MPa (50 kpsi) and $S_{uc} = 1\,000$ MPa (150 kpsi).

* Answers to these problems are provided in Appendix G.

† These problems are based on similar problems in previous chapters with the same –number, e.g., Problem 5-4 is based on Problem 4-4, etc. Problems in succeeding chapters may also continue and extend these problems.

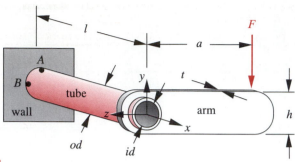

FIGURE P5-14

Problems 5-33 to 5-36

†5-36 For the bracket shown in Figure P5-14 and the data in the row(s) assigned from Table P5-3, redo Problem 5-33 considering the stress concentration at points A and B. Assume a stress concentration factor of 2.5 in both bending and torsion.

*†5-37 A semicircular curved beam as shown in Figure P5-15 has od = 150 mm, id = 100 mm, and t = 25 mm. For a load pair F = 14 kN applied along the diameter, find the safety factor at the inner and outer fibers: a) If the beam is a ductile material with S_y = 700 MPa, b) If the beam is a cast-brittle material with S_{ut} = 420 MPa, S_{uc} = 1200 MPa.

*5-38 Assume that the curved beam of Problem 5-37 has a crack on its inside surface of half-width a = 2 mm and a fracture toughness of 50 MPa-m$^{0.5}$. What is its safety factor against sudden fracture?

*5-39 Consider the failed 260-in dia by 0.73-in wall rocket case of Figure 5-14. The steel had S_y = 240 kpsi and a fracture toughness K_c = 79.6 kpsi-in$^{0.5}$. It was designed for an internal pressure of 960 psi but failed at 542 psi. Failure was attributed to a small crack that precipitated a sudden, brittle, fracture-mechanics failure. Find the nominal stresses in the wall and the yielding safety factor at the failure conditions and estimate the size of the crack that caused it to explode. Assume β = 1.0.

†5-40 Redesign the roll support of Problem 5-8 to be like Figure P5-16. The mandrels insert to 10% of the roll length. Design dimensions a and b for a safety factor of 2.

(a) If the beam is a ductile material with S_y = 300 MPa,
(b) If the beam is a cast-brittle material with S_{ut} = 150 MPa, S_{uc} = 1 200 MPa.

*†5-41 A 10-mm-ID steel tube carries liquid at 7 MPa. The steel has S_y = 400 MPa. Determine the safety factor for the wall if its thickness is: (a) 1 mm, (b) 5 mm.

†5-42 A cylindrical tank with hemispherical ends is required to hold 150 psi of pressurized air at room temperature. The steel has S_y = 400 MPa. Determine the safety factor if the tank diameter is 0.5 m with a 1 mm wall thickness, and its length is 1 m.

†5-43 The paper rolls in Figure P5-17 are 0.9-m OD x 0.22-m ID x 3.23-m long and have a density of 984 kg/m^3. The rolls are transferred from the machine conveyor (not shown) to the forklift truck by the V-linkage of the off-load station which is rotated through 90° by an air cylinder. The paper then rolls onto the waiting forks of the truck. The forks are 38-mm thick by 100-mm wide by 1.2-m long and are tipped at

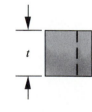

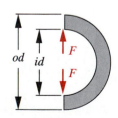

FIGURE P5-15

Problem 5-37

Table P5-3   Data for Problems 5-33 through 5-36

Use only data that are relevant to the particular problem. Lengths in mm, forces in N.

Row	l	a	t	h	F	od	id	E
a	100	400	10	20	50	20	14	steel
b	70	200	6	80	85	20	6	steel
c	300	100	4	50	95	25	17	steel
d	800	500	6	65	160	46	22	alum
e	85	350	5	96	900	55	24	alum
f	50	180	4	45	950	50	30	alum
g	160	280	5	25	850	45	19	steel
h	200	100	2	10	800	40	24	steel
i	400	150	3	50	950	65	37	steel
j	200	100	3	10	600	45	32	alum
k	120	180	3	70	880	60	47	alum
l	150	250	8	90	750	52	28	alum
m	70	100	6	80	500	36	30	steel
n	85	150	7	60	820	40	15	steel

a 3° angle from the horizontal and have $S_y = 600$ MPa. Find the safety factor for the two forks on the truck when the paper rolls onto it under two different conditions (state all assumptions):

(a) The two forks are unsupported at their free end.

(b) The two forks are contacting the table at point A.

[†]5-44   Determine a suitable thickness for the V-links of the off-loading station of Figure P5-17 to limit their deflections at the tips to 10 mm in any position during their rotation. Two V-links support the roll, at the 1/4 and 3/4 points along the roll's length and that each of the V arms is 10-cm wide by 1-m long.   What is their safety factor against yielding when designed to limit deflection as above? $S_y = 400$ MPa. See Problem 5-43 for more information.

[†]5-45   Determine the safety factor based on the critical load on the air-cylinder rod in Figure P5-17. The crank arm that it rotates is 0.3 m long and the rod has a maximum

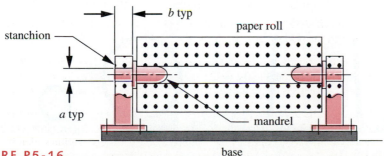

FIGURE P5-16

Problem 5-40

[*] Answers to these problems are provided in Appendix G.

[†] These problems are based on similar problems in previous chapters with the same –number, e.g., Problem 5-4 is based on Problem 4-4, etc. Problems in succeeding chapters may also continue and extend these problems.

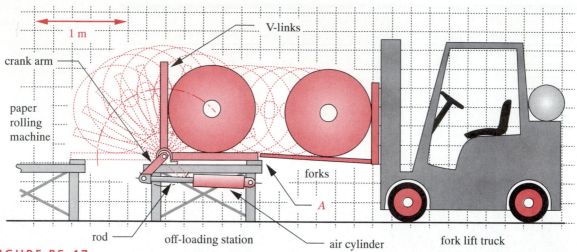

FIGURE P5-17

Problems 5-43 to 5-47

extension of 0.5 m. The 25 mm dia rod is solid steel with a yield strength of 400 MPa. State all assumptions.

†5-46   The V-links of Figure P5-17 are rotated by the crank arm through a shaft that is 60-mm dia by 3.23-m long. Determine the maximum torque applied to this shaft during the motion of the V-linkage and find the static safety factor against yielding for the shaft if its S_y = 400 MPa. See Problem 5-43 for more information.

†5-47   Determine the maximum forces on the pins at each end of the air cylinder of Figure P5-17. Determine the safety factor for these pins if they are 30-mm dia and in single shear. S_y = 400 MPa.

†5-48   Figure P5-18 shows an exerciser for a 100-kg wheelchair racer. The wheelchair has 65 cm dia drive wheels separated by a 70-cm track width. Two free-turning rollers on bearings support the rear wheels. The lateral movement of the chair is limited by the flanges. Design the 1-m-long rollers as hollow tubes of aluminum (select alloy) to minimize the height of the platform and also limit the roller deflections to 1 mm in the worst case. Specify suitable sized steel axles to support the tubes on bearings. Calculate all significant safety factors.

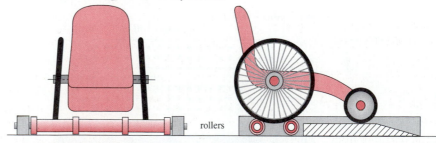

FIGURE P5-18

Problem 5-48

6

FATIGUE FAILURE
THEORIES

*Science is a first-rate piece of furniture for a
man's upper chamber, if he has common
sense on the ground floor.*
OLIVER WENDELL HOLMES

6.0   INTRODUCTION

Most failures in machinery are due to time-varying loads rather than to static loads. These failures typically occur at stress levels significantly lower than the yield strengths of the materials. Thus, using only the static failure theories of the last chapter can lead to unsafe designs when loads are dynamic.

Table 6-0 shows the variables used in this chapter and references the equations, tables, or sections in which they are used. At the end of the chapter, a summary section is provided that also groups the significant equations from this chapter for easy reference and identifies the chapter section in which their discussion can be found.

History of Fatigue Failure

This phenomenon was first noticed in the 1800's when railroad-car axles began failing after only limited time in service. They were made of ductile steel but exhibited sudden, brittle-like failures. Rankine published a paper in 1843, *On the Causes of Unexpected Breakage of Journals of Railway Axles* in which he postulated that the material had "crystallized" and become brittle due to the fluctuating stress. The axles had been designed with all the engineering expertise available at the time, which was based on experience with statically loaded structures. Dynamic loads were then a new phenomenon resulting from the introduction of steam-powered machinery. These axles were fixed to the wheels and turned with them. Thus the bending stress at any point on the surface of the axle varied cyclically from positive to negative as shown in Figure 6-1a.

Table 6-0   Variables Used in this Chapter

Symbol	Variable	ips units	SI units	See
A	amplitude ratio	none	none	Eq. 6.1d
A_{95}	95% stressed area	in^2	m^2	Eq. 6.7c
C_{load}	loading factor	none	none	Eq. 6.7a
C_{reliab}	reliability factor	none	none	Table 6-4
C_{size}	size factor	none	none	Eq. 6.7b
C_{surf}	surface factor	none	none	Eq. 6.7e
C_{temp}	temperature factor	none	none	Eq. 6.7f
d_{equiv}	equivalent diameter test specimen	in	m	Eq. 6.7d
K	stress intensity	kpsi-in$^{0.5}$	MPa-m$^{0.5}$	Sect. 6.1
K_c	fracture toughness	kpsi-in$^{0.5}$	MPa-m$^{0.5}$	Sect. 6.1
ΔK	stress-intensity factor range	kpsi-in$^{0.5}$	MPa-m$^{0.5}$	Eq. 6.3
ΔK_{th}	threshold stress-intensity factor	kpsi-in$^{0.5}$	MPa-m$^{0.5}$	Eq. 6.5
K_f	fatigue-stress-concentration factor	none	none	Eq. 6.11
K_{fm}	mean-stress fatigue-concentration factor	none	none	Eq. 6.15
N	number of cycles	none	none	Fig. 6-2, Sect. 6.2
N_f	safety factor in fatigue	none	none	Eq. 6.14, 6.17
q	material notch sensitivity	none	none	Eq. 6.13, Fig. 6-35
R	stress ratio	none	none	Eq. 6.1d
S_d	standard deviation	any	any	Sect. 6.6
S_e	corrected endurance limit	psi	Pa	Eq. 6.8
$S_{e'}$	uncorrected endurance limit	psi	Pa	Eq. 6.8
S_{el}	strength at elastic limit	psi	Pa	Eq. 6.2a–b, 6.6a–b
S_f	corrected endurance strength	psi	Pa	Eq. 6.8
$S_{f'}$	uncorrected endurance strength	psi	Pa	Eq. 6.8
S_m	mean strength at 10^3 cycles	psi	Pa	Eq. 6.2c–d, 6.6c–d
S_n	fatigue strength at any N	psi	Pa	Eq. 6.10
S_{us}	ultimate shear strength	psi	Pa	Eq. 6.9b
S_{yc}	yield strength in compression	psi	Pa	Fig. 6-41, Eq. 6.16a
β	stress-intensity geometry factor	none	none	Sect. 6.2
σ	normal stress	psi	Pa	Sect. 6.10
σ'	von Mises effective stress	psi	Pa	Sect. 6.11
σ_a', σ_m'	alternating and mean von Mises stress	psi	Pa	Sect. 6.11
$\sigma_{1,2,3}$	principal stresses	psi	Pa	Sect. 6.11
σ_a, σ_m	alternating and mean normal stress	psi	MPa	Sect. 6.4
σ_{max}	maximum applied normal stress	psi	MPa	Sect. 6.4
σ_{min}	minimum applied normal stress	psi	MPa	Sect. 6.4

The title-page photograph is of the fractured Liberty ship USS Schenectady, courtesy of the Ship Structures Committee, U. S. Government.

This loading is termed fully reversed. A German engineer, August Wohler, made the first scientific investigation (over a 12-year period) into what was known as *fatigue failure* by testing axles to failure in the laboratory under fully reversed loading. He published his findings in 1870, which identified the number of cycles of time-varying stress as the culprit and found the existence of an *endurance limit* for steels, i.e., a stress level that would be tolerable for millions of cycles of fully reversed stress. The *S-N*, or Wohler diagram shown in Figure 6-2 became the standard way to characterize the behavior of materials under completely reversed loading and is still in use, though other measures of material strength under dynamic loading are now also available.

The term "fatigue" was first applied to this situation by Poncelet in 1839. The mechanism of failure was not yet understood, and the brittle appearance of the failure surface in a ductile material caused speculation that the material had somehow become "tired" and embrittled from the load oscillations. Wohler later showed that the broken axle-halves were still as strong and ductile in tensile tests as was the original material. Nonetheless, the term *fatigue failure* stuck and is still used to describe any failure due to time-varying loads.

Fatigue failure constitutes a significant cost to the economy. Dowling suggests, based on data in a U.S. Government report by Reed et al.[1] that:

> "The annual cost of fatigue of materials to the U.S. economy in 1982 dollars is around $100 billion, corresponding to about 3% of the gross national product (GNP). These costs arise from the occurrence or prevention of fatigue failure for ground vehicles, rail vehicles, aircraft of all types, bridges, cranes, power plant equipment, offshore oil well structures, and a wide variety of miscellaneous machinery and equipment including everyday household items, toys and sports equipment."[2]

The cost can also involve human life. The first commercial passenger jet aircraft, the British *Comet,* suffered two fatal crashes in 1954 due to fatigue failure of the fuselage from the pressurization/depressurization cycles of the cabin.* More recently (1988), a Hawaiian Airlines Boeing 737 lost about one-third of its cabin top while in flight at 25 000 ft. It was landed safely with minimum loss of life. Many other recent examples of catastrophic fatigue failures exist. A great deal of work has been done over the past 150 years to determine the actual mechanism of fatigue failure. The demands placed on materials in aircraft and spacecraft applications since World War II have motivated increased investment in scientific research on this topic and it is now fairly well understood, though researchers continue to seek answers to questions about the fatigue mechanism. Table 6-1 shows a chronology of significant events in the history of fatigue failure research.

6.1   MECHANISM OF FATIGUE FAILURE

Fatigue failures always begin at a crack. The crack may have been present in the material since its manufacture, or the crack may have developed over time due to cyclic straining around stress concentrations. Fisher and Yen[3] have shown that virtually all

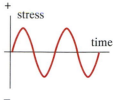

(a) Fully reversed

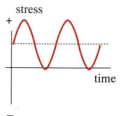

(b) Repeated

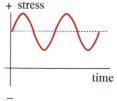

(c) Fluctuating

FIGURE 6-1

Time-Varying Stresses

---

* It is generally believed that these *Comet* failures also cost the U.K. its commercial airline industry. Britain led the field at the time, but the momentum lost in grounding and redesigning their planes gave the U.S. aircraft industry a chance to take the lead, which it holds to this day. Britain has only recently begun to achieve some market share with the European consortium-developed Airbus.

Table 6-1   Chronology of Fatigue Failure Research Events and Accomplishments
Source: "Fracture Mechanics & Fatigue," Union College, 1992, with permission

Year	Researcher	Event or Accomplishment
1829	Albert	First to document failure due to repeated loads.
1839	Poncelet	First to use the term fatigue.
1837	Rankine	Discusses the crystallization theory of fatigue.
1849	Stephenson	Discusses the product liability associated with railway axle fatigue failures.
1850	Braithwaite	First uses the term fatigue in an English publication and discusses the crystallization theory.
1864	Fairbairn	Reports the first experiments with repeated loads.
1871	Wohler	Publishes results of 20 years of investigation into axle failures, develops the rotating bending test, the S-N diagram and defines the endurance limit.
1871	Bauschinger	Develops a mirror extensometer with 10^{-6} sensitivity and studies inelastic stress-strain.
1886	Bauschinger	Proposes a cyclic "natural elastic limit" below which fatigue would not occur.
1903	Ewing/Humfrey	Discover slip lines, fatigue cracks, and crack growth to failure, disproving the crystallization theory.
1910	Bairstow	Verifies Bauschinger's theory of a natural elastic limit and Wohler's endurance limit.
1910	Basquin	Develops the exponential law of endurance tests (the Basquin equation).
1915	Smith/ Wedgewood	Separate cyclic plastic strain from total plastic strain.
1921	Griffith	Develops fracture criteria and relates fatigue to crack growth.
1927	Moore/Kommers	Quantify high-cycle-fatigue data for many materials in "The Fatigue of Metals."
1930	Goodman/Soderberg	Independently determine the influence of mean stresses on fatigue.
1937	Neuber	Publishes the Neuber equation for strain concentration in notches (English translation in 1946.)
1953	Peterson	Publishes "Stress Concentration Design Factors" providing an approach to account for notches.
1955	Coffin/Manson	Independently publish the strain-based low-cycle-fatigue law (Coffin-Manson law).
1961	Paris	Publishes the fracture mechanics Paris law for fatigue-crack growth.

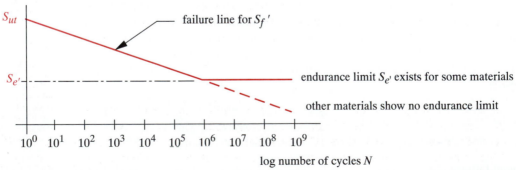

FIGURE 6-2

Woehler Strength-Life or S-N Diagram

structural members contain discontinuities ranging from microscopic (< 0.010 in) to macroscopic, introduced in the manufacturing or fabricating process. Fatigue cracks generally start at a notch or other stress concentration. (We will use the general term *notch* to represent any geometric contour that increases local stress.) The brittle failures of some of the World War II tankers (see Figure 5-11) were traced to cracks that began at an arc-strike left by a careless welder. The *Comet* airplane failures started at cracks smaller than 0.07 in long near the corners of windows that were nearly square, thus providing high stress concentrations. **Thus it is critical that dynamically loaded parts be designed to minimize stress concentrations as described in Section 4.15.**

There are three stages of fatigue failure, *crack initiation, crack propagation,* and *sudden fracture due to unstable crack growth.* The first stage can be of short duration, the second stage involves most of the life of the part, and the third stage is instantaneous.

Crack-Initiation Stage

Assume that the material is a ductile metal and as manufactured has no cracks present but has the usual collection of particles, inclusions, etc., that are common to engineering materials. At a microscopic scale, metals are not homogeneous and isotropic.[*] Assume further that there are some regions of geometric stress concentration (notches) in locations of significant time-varying stress that contains a tensile (positive) component as shown in Figure 6-1. As the stresses at the notch oscillate, local yielding may occur due to the stress concentration, even though the nominal stress in the section is well below the yield strength of the material. The localized plastic yielding causes distortion and creates slip bands (regions of intense deformation due to shear motion) along the crystal boundaries of the material. As the stress cycles, additional slip bands occur and coalesce into microscopic cracks. Even in the absence of a notch (as in smooth test specimens) this mechanism still operates as long as the yield strength is exceeded somewhere in the material. Preexisting voids or inclusions will serve as stress-raisers to start the crack.

Less ductile materials do not have the same ability to yield as ductile ones and will tend to develop cracks more rapidly. They are more *notch sensitive.* Brittle (especially cast) materials which do not yield may skip this initiation stage and proceed directly to crack propagation at sites of existing voids or inclusions that serve as microcracks.

Crack Propagation Stage

Once a microcrack is established, (or if present from inception) the mechanisms of fracture mechanics as described in Section 5.3 become operable. The sharp crack creates stress concentrations larger than those of the original notch, and a plastic zone develops at the crack tip each time a tensile stress opens the crack, blunting its tip and reducing the effective stress concentration. The crack grows a small amount. When the stress cycles to a compressive-stress regime, to zero, or to a sufficiently lower tensile stress as shown in Figure 6-1*a* through 6-1*c*, respectively, the crack closes, the yield-

[*] "When viewed at a sufficiently small size scale, all materials are anisotropic and inhomogeneous. For example, engineering metals are composed of an aggregate of small crystal grains. Within each grain the behavior is anisotropic due to the crystal planes, and if a grain boundary is crossed, the orientation of these planes changes. Inhomogeneities exist not only due to the grain structure but also because of the presence of tiny voids or particles of a different chemical composition than the bulk of the material, such as hard silicate or alumina inclusions in steel."[3]

ing momentarily ceases and the crack again becomes sharp, but now at its longer dimension. This process continues as long as the local stress is cycling from below the tensile yield to above the tensile yield at the crack tip. Thus, **crack growth is due to tensile stress** and the crack grows along planes normal to the maximum tensile stress. It is for this reason that fatigue failures are considered to be due to tensile stress even though shear stress starts the process in ductile materials as described above. Cyclic stresses that are always compressive will not cause crack growth as they tend to close the crack.

The crack-propagation growth rate is very small, on the order of 10^{-8} to 10^{-4} in per cycle,[5] but this adds up over a large number of cycles. If the failed surface is viewed at high magnification, the *striations* due to each stress cycle can be seen as in Figure 6-3 which shows the crack surface of a failed aluminum specimen at 12 000x magnification along with a representation of the stress-cycle pattern that failed it. The occasional large-amplitude stress cycles show up as larger striations than the more frequent small-amplitude ones, indicating that higher stress amplitudes cause larger crack growth per cycle.

CORROSION   Another mechanism for crack propagation is corrosion. **If a part containing a crack is in a corrosive environment, the crack will grow under static stress**. The combination of stress and corrosive environment has a synergistic effect and the material corrodes more rapidly than if unstressed. This combined condition is sometimes termed **stress-corrosion or environmentally assisted cracking**.

If the part is *cyclically stressed in a corrosive environment*, the crack will grow more rapidly than from either factor alone. This is also called **corrosion-fatigue**. While the

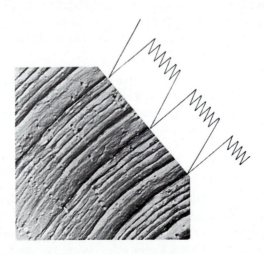

FIGURE 6-3

Fatigue Striations on the Crack Surface of an Aluminum Alloy. Spacing of the Striations Corresponds to the Cyclic Loading Pattern (From Fig. 1.5, p. 10, in Broek, D., 1988, *The Practical Use of Fracture Mechanics*, Kluwer Publishers, Dordrecht)

6

frequency of stress cycling (as opposed to the number of cycles) appears to have no detrimental effect on crack growth in a noncorrosive environment, in the presence of corrosive environments, it does. Lower cyclic frequencies allow the environment more time to act on the stressed crack tip while it is held open under tensile stress and this substantially increases the rate of crack growth per cycle.

Fracture

The crack will continue to grow as long as cyclical tensile stress and/or corrosion factors of sufficient severity are present. At some point, the crack size becomes large enough to raise the stress intensity factor K at the crack tip (Eq. 5.14) to the level of the material's fracture toughness K_c and sudden failure (as described in Section 5.3 on fracture mechanics) occurs instantaneously on the next tensile stress-cycle. This failure mechanism is the same whether the condition of $K = K_c$ was reached by reason of the crack propagating to a sufficient size (increasing a in Eq. 5.14) or by the nominal stress being raised sufficiently (increasing σ_{nom} in Eq. 5.14). The former is commonly the case in dynamic loading while the latter is more common in static loading. The result is the same, sudden and catastrophic failure with no warning.

Examination with the naked eye of parts failed by fatigue loading show a typical pattern as seen in Figure 6-4. There is a region emanating from the site of the original microcrack that appears burnished and a separate region that appears dull and rough,

origin

final rupture

rotation

(a)

(b)

FIGURE 6-4

Two Parts that Failed in Fatigue. Note the Beachmarks: (a) 1040 Steel Keyed Shaft Failed in Rotating Bending. Crack Started at Keyway. (b) Diesel-Engine Crankshaft Failed in Combined Bending and Torsion. Crack Started at Arrow. Source: D. J. Wulpi, *Understanding How Components Fail*. Am. Soc. for Metals: Metals Park, Ohio, 1990, Fig. 22, p. 149, and Fig. 25, p. 152.

looking like a brittle fracture. The burnished region was the crack and often shows *beachmarks*, so-called because they resemble ripples left on the sand by retreating waves. The beachmarks (not to be confused with the striations seen in Figure 6-3, which are much smaller and not discernible to the naked eye) are due to a starting and stopping of the crack growth and they surround the origin of the crack, usually at a notch or internal stress-raiser. Sometimes, if there was a lot of rubbing of the crack surfaces, the beach marks will be obscured. The brittle-failure zone is the portion that failed suddenly when the crack reached its size limit. Figure 6-5 shows drawings of failure surfaces of a variety of part geometries loaded in different ways at different levels of stress. The beachmarks can be seen in the crack zones. The brittle fracture zone can be a small remnant of the original cross section.

6.2   FATIGUE-FAILURE MODELS

There are three fatigue failure models in current use and each has a place and a purpose. They are the *stress-life (S-N)* approach, the *strain-life (ε-N)* approach and the *linear-elastic fracture-mechanics* (LEFM) approach. We will first discuss their application, advantages and disadvantages, compare them in a general way and then analyze some of them in more detail.

Fatigue Regimes

Based on the number of stress or strain cycles that the part is expected to undergo in its lifetime, it is relegated to either a **low-cycle fatigue (LCF)** regime or a **high-cycle fatigue (HCF)** regime. There is no sharp dividing line between the two regimes and various investigators suggest slightly different divisions. Dowling[6] defines HCF as starting at around 10^2 to 10^4 cycles of stress/strain variation with the number varying with the material. Juvinall[7] and Shigley[8] suggest 10^3 cycles and Yen[9] defines 10^3 to 10^4 cycles as the cutoff. In this text, we will assume that $N = 10^3$ cycles is a reasonable approximation of the divide between LCF and HCF.

The Stress-Life Approach

This is the oldest of the three models and is the most often used for high-cycle fatigue (HCF) applications where the assembly is expected to last for more than about 10^3 cycles of stress. It works best when the load amplitudes are predictable and consistent over the life of the part. It is a **stress-based model,** which seeks to determine a **fatigue strength** and/or an **endurance limit** for the material so that the cyclic stresses can be kept below that level and avoid failure for the required number of cycles. The part is then designed based on the material's fatigue strength (or endurance limit) and a safety factor. In effect, this approach attempts to keep local stresses in notches so low that the crack-initiation stage *never begins*. The assumption (and design goal) is that stresses

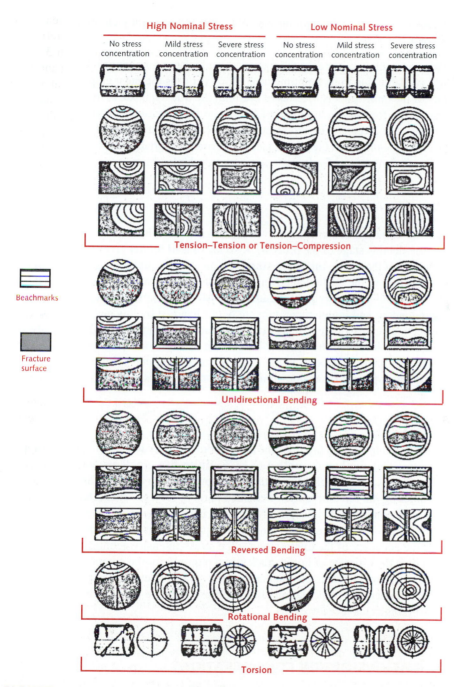

High Nominal Stress   Low Nominal Stress

No stress concentration | Mild stress concentration | Severe stress concentration | No stress concentration | Mild stress concentration | Severe stress concentration

Tension–Tension or Tension–Compression

Beachmarks

Fracture surface

Unidirectional Bending

Reversed Bending

Rotational Bending

Torsion

FIGURE 6-5

Schematic Representations of Fatigue-Fracture Surfaces of Various Smooth and Notched Cross Sections Under Various Loading Conditions and Stress Levels (From Metals Handbook, 1975, Am. Soc. for Metals, Metals Park Ohio, Vol. 10, 8th ed., p. 102, with permission)

and strains everywhere remain in the elastic region and no local yielding occurs to initiate a crack.

This approach is fairly easy to implement and a large amount of relevant strength data are available due to its longtime use. However, it is the most empirical and least accurate of the three models in terms of defining the true local stress/strain states in the part, especially for low-cycle fatigue (LCF) finite-life situations where the total number of cycles is expected to be less than about 10^3 and the stresses will be high enough to cause local yielding. On the other hand, with certain materials, the stress-life approach allows the design of parts for **infinite life** under cyclic loading.

The Strain-Life Approach

Because the initiation of a crack involves yielding, a stress-based approach cannot adequately model this stage of the process. A **strain-based model** gives a reasonably accurate picture of the **crack-initiation stage**. It can also account for cumulative damage due to variations in the cyclic load over the life of the part, such as overloads that may introduce favorable or unfavorable residual stresses to the failure zone. Combinations of fatigue loading and high temperature are better handled by this method because the creep effects can be included. This method is most often applied to **LCF, finite-life** problems where the cyclic stresses are high enough to cause local yielding. It is the most complicated of the three models to use and requires a computer solution. Test data are still being developed on the cyclic-strain behavior of various engineering materials.

The LEFM Approach

Fracture-mechanics theory provides the best model of the **crack-propagation stage** of the process. This method is applied to **LCF, finite-life** problems where the cyclic stresses are known to be high enough to cause the formation of cracks and is most useful in predicting the remaining life of cracked parts in service. It is often used in conjunction with nondestructive testing (NDT) in a periodic service-inspection program, especially in the aircraft/aerospace industry. Its application is fairly straightforward but relies on the accuracy of the expression for the stress-intensity geometry factor β (Eq. 5.14b) and on the estimate of initial crack size a required for the computation. One approach is to assume that a crack smaller than the smallest detectable crack already exists in order to begin the calculation. It gives more accurate results when a detectable and measurable crack already exists.

6.3   MACHINE-DESIGN CONSIDERATIONS

The choice of fatigue-failure models for machine-design purposes depends on the type of machinery being designed and on its intended use. The large class of **rotating machinery** (stationary or mobile) is well served by the stress-life *(S-N)* model because the required lives are usually in the HCF range. For example, consider the number of load cycles (revolutions) required of an automobile-engine crankshaft over its useful life. Assume a desired 100 000-mi life with no failure of the crankshaft. The average rolling radius of a car tire is about 1 ft and its circumference is then 6.28 ft. The axle will then rotate 5 280/6.28 = 841 rev/mile or 84E6 rev/100 000 mi. A typical final-drive ratio for a passenger car is about 3:1, meaning that the output shaft of the transmission is turning 3x the axle speed. If we assume that most of the car's life is spent in top gear (1:1) then the engine speed also averages 3x axle speed. This means that the crankshaft and most other rotating and oscillating components in the engine will see about 2.5E8 cycles in 100 000 miles (the valve train will see half that many). This is clearly in the HCF regime and doesn't even account for idling time. Also, the cyclic loads are reasonably predictable and consistent, so the *stress-life* approach is appropriate here.

As another example consider a typical, automated production machine as used in U.S. industry. Perhaps it is making batteries, or paper diapers, or filling soft-drink cans. Assume that its fundamental driveshaft speed is 100 rpm (a conservative estimate). Assume only one-shift operation (also conservative since many such machines run 2 or 3 shifts). How many cycles (revolutions) will the driveshaft and all the gears, cams, etc., driven by it see in a year? In one 8-hr day, it turns 100(60)(8) = 480 000 rev/shift-day. In a 260-day work-year it turns 125E6 rev/shift-year. Again we are in the HCF regime and the loads are usually quite predictable and consistent in amplitude.

One class of machinery that typically sees low cycle fatigue (LCF) is that of transportation (service) machinery. The airframe of an airplane, the hull of a ship, and the chassis of a land vehicle see a load-time history that can be quite variable due to storms, gusts/waves, hard landings/dockings, etc. (for the aircraft/ship) and overloads, potholes, etc. (for the land craft). The total number of load cycles seen in its life is also less predictable due to the vagaries of its use. Even though the number of low-magnitude stress cycles may be potentially large (and in the HCF regime) over its potential lifetime, the chance of higher-than-design loads causing local yielding is always present. A series of high-stress cycles, even if less than 10^3 in number, can cause significant crack growth due to local yielding.

Manufacturers of this kind of equipment develop extensive load-time or strain-time data by instrumenting actual vehicles while operating them in regular service or under controlled-test conditions. (Look ahead to Figure 6-7 for examples.) Computer-simulations are also developed and refined by comparison to experimental data. The simulated and experimental load-time histories are used, usually in conjunction with either the strain-life or LEFM models (or both), to more accurately predict failure and thus improve the design. Another example of the use of ε-N and LEFM models is in the design and analysis of gas-turbine rotor blades, which operate under high stresses at high temperatures and go through LCF thermal cycles at start-up and shutdown.

We will concentrate on the **stress-life model** in this text and also discuss the application of the **LEFM model** to cyclically loaded machine design problems. The **strain-life model** is best at depicting the conditions of crack-initiation and provides the most complete theoretical model but is less well-suited to the design of parts for HCF. A complete description of the **strain-life model** would require more space than is available in this introductory design text. The reader is directed to the references cited in the bibliography of this chapter, which provide thorough discussions of the strain-life approach (as well as of the other two approaches). The fracture-mechanics approach allows the determination of remaining life for service-cracked parts. The stress-life model is the most appropriate choice for the majority of rotating-machinery design problems due to the need for high-cycle (or infinite) life in most cases.

6.4　FATIGUE LOADS

Any loads that vary with time can potentially cause fatigue failure. The character of these loads may vary substantially from one application to another. In rotating machinery, the loads tend to be consistent in amplitude over time and repeat with some frequency. In service equipment (vehicles of all types), the loads tend to be quite variable in amplitude and frequency over time and may even be random in nature. The shape of the waveform of the load-time function seems not to have any significant effect on fatigue failure in the absence of corrosion, so we usually depict the function schematically as a sinusoidal- or sawtooth wave. Also, the presence or absence of periods of quiescence in the load history is not significant as long as the environment is noncorrosive. (Corrosion will cause continuous crack growth even in the absence of any load fluctuations.) The stress-time or strain-time waveform will have the same general shape and frequency as the load-time waveform. The significant factors are the amplitude and the average value of the stress-time (or strain-time) waveform and the total number of stress/strain cycles that the part has seen.

Rotating Machinery Loading

The typical stress-time functions experienced by rotating machinery can be modeled as shown in Figure 6-6, which shows them schematically as sine waves. Figure 6-6a shows

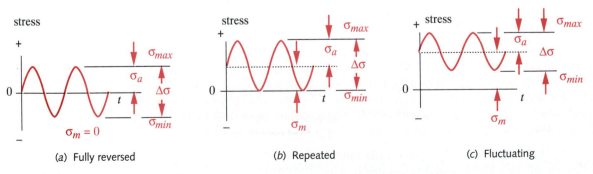

(a) Fully reversed　　　(b) Repeated　　　(c) Fluctuating

FIGURE 6-6

Alternating, Mean, and Range Values for Fully Reversed, Repeated, and Fluctuating Cyclic Stresses

the **fully reversed** case for which *the mean value is zero.* Figure 6-6b shows a **repeated stress** case in which the waveform ranges from *zero to a maximum with a mean value equal to the alternating component,* and Figure 6-6c shows one version of the more general case (called **fluctuating stress**) with *all component values nonzero.* (Note that any portion of the waveform could be in the compressive stress regime as well.) Any of these waveforms can be characterized by two parameters, their mean and alternating components, their maximum and minimum values, or ratios of these values.

The **stress range** $\Delta\sigma$ is defined as

$$\Delta\sigma = \sigma_{max} - \sigma_{min} \qquad (6.1a)$$

The **alternating component** σ_a is found from

$$\sigma_a = \frac{\sigma_{max} - \sigma_{min}}{2} \qquad (6.1b)$$

and the **mean component** σ_m is

$$\sigma_m = \frac{\sigma_{max} + \sigma_{min}}{2} \qquad (6.1c)$$

Two ratios can be formed:

$$R = \frac{\sigma_{min}}{\sigma_{max}} \qquad A = \frac{\sigma_a}{\sigma_m} \qquad (6.1d)$$

where R is the **stress ratio** and A is the **amplitude ratio**.

When the stress is fully reversed (Figure 6-6a), $R = -1$ and $A = \infty$. When the stress is repeated (Figure 6-6b), $R = 0$ and $A = 1$. When the maximum and minimum stresses have the same sign as in Figure 6-6c, both R and A are positive and $0 \leq R \leq 1$. These load patterns may result from bending, axial, torsional, or a combination of these types of stresses. We will see that the presence of a mean-stress component can have a significant effect on the fatigue life.

Service Equipment Loading

The character of the load-time function for service equipment is not so easily defined as for rotating machinery. The best data come from actual measurements made on equipment in service or operated under simulated service conditions. The automotive industry subjects prototype vehicles to test-track conditions that simulate various road surfaces and curves. The test vehicles are extensively instrumented with accelerometers, force transducers, strain gages, and other instruments that feed voluminous amounts of data to on board computers or telemeter it to stationary computers where it is digitized and stored for later analysis. The aircraft industry also instruments test-aircraft and records in-flight force, acceleration, and strain data. The same is done with ships and offshore oil platforms, etc.

Some examples of these in-service stress-time waveforms are shown in Figure 6-7, which depicts a simulated general loading case in (*a*), a typical pattern for a ship or off-shore platform in (*b*), and a pattern typical of a commercial aircraft in (*c*). These patterns are semi-random in nature as the events do not repeat with any particular period. Data such as these are used in computer simulation programs that calculate the cumulative fatigue damage based on either a strain-based model, a fracture-mechanics model, or a combination of both. The stress-life model is not able to deal as effectively with this type of loading history.

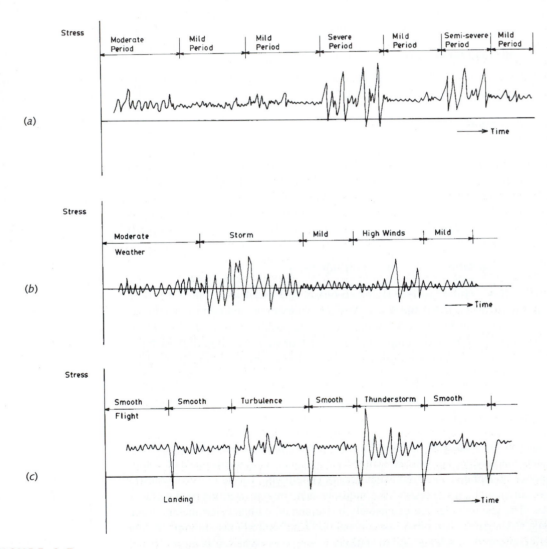

FIGURE 6-7

Semirandom Loading in Different Periods (voyages, months, flights) for (a) General Case, (b) Ship or Offshore Structure, (c) Commercial Aircraft (From Fig. 6.10, p. 186, in Broek, D., 1988, *The Practical Use of Fracture Mechanics*, Kluwer Publishers, Dordrecht)

6.5   MEASURING FATIGUE FAILURE CRITERIA

Several different testing techniques now exist for the purpose of measuring the response of materials to time-varying stresses and strains. The oldest approach is that of Wohler who loaded a rotating cantilever beam in bending to achieve variations in stress with time. R. R. Moore later adapted this technique to a simply supported rotating beam in fully reversed, pure bending. In the last 40 years the advent of the servohydraulically driven axial-testing machine has allowed much more flexibility in the patterns of either stress or strain that can be applied to a test specimen. Strain-based and fracture-mechanics data as well as stress-based data are obtained by this method. Most of the available fatigue-strength information is for a rotating beam in fully reversed bending, with less available for axial loading and less still for torsion, though this is changing as more axial fatigue data are developed. In some cases, no fatigue strength information for the desired material is available at all and we then need a means to estimate a value from available static strength data. This will be discussed in the next section.

Fully Reversed Stresses

This loading situation can be accomplished with the rotating bending, axial fatigue, cantilever bending, or torsional fatigue tests depending on the type of loading desired. The rotating bending test is a fully reversed, stress-based, HCF test that seeks to find the fatigue strength of the material under those conditions. The axial fatigue test can be used to generate similar fully reversed data to that of the rotating beam test in a given material and it can also be used to do strain-controlled tests. The principal advantage of the axial test is its ability to apply any combination of mean and alternating stresses. The cantilever bending test subjects a nonrotating beam to oscillations in bending stress. It can provide a mean stress as well as a fully reversed stress. The torsion test alternately twists a bar in opposite directions applying pure shear stresses.

ROTATING-BEAM TEST   The bulk of available fully reversed, fatigue strength data comes from the R. R. Moore rotating-beam test in which a highly polished specimen of about 0.3-in dia is mounted in a fixture which allows a constant magnitude, pure-bending moment to be applied while the specimen is rotated at 1725 rpm. This creates a fully reversed bending stress at any point on the circumference of the specimen as shown in Figure 6-6a. It is run at one particular stress level until it fails and the number of cycles to failure and the applied stress level is recorded. It takes about one-half day to reach 10^6 cycles and about 40 days to reach 10^8 cycles on one specimen. This test is repeated with multiple specimens of the same material loaded at different stress levels. The collected data are then plotted as normalized failure strength, S/S_{ut} against number of cycles, N (typically on log-log coordinates) to obtain an $S-N$ diagram.

Figure 6-8 shows the results of a number of rotating-beam tests on wrought steels of up to about 200 kpsi tensile strength. The data show that samples run at higher reversed-stress levels fail after fewer cycles. At lower stress levels, some (in the circle labeled *not broken*) do not fail at all before their tests are stopped at some number of cycles (here 10^7). Note the large amount of scatter in the data. This is typical of fa-

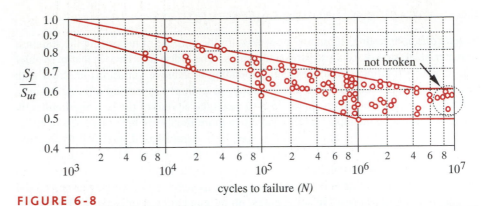

FIGURE 6-8

Log-Log Plot of Composite S-N Curves for Wrought Steels of S_{ut} < 200 ksi (From Fig. 11.7, p. 210, Juvinall, R. C., 1967, Stress, Strain, and Strength, McGraw-Hill, New York, with permission)

tigue strength tests. Differences among the many samples of material required to generate the entire curve may account for the scatter. Some samples may have contained more, or larger, defects to serve as localized stress-raisers. (The samples are all un-notched and are polished to a fine finish to minimize the possibility of surface defects starting a crack.) The solid lines are drawn to bracket the data.

ENDURANCE LIMIT   Note that the fatigue strength S falls steadily and linearly (on log-log coordinates) as a function of N until reaching a knee at about 10^6 to 10^7 cycles. This knee defines an **endurance limit** S_e' for the material which is a stress level below which it can be cycled infinitely without failure. At the lower bound of the scatter band beyond the knee, an approximate endurance limit can be defined

$$\text{for steels:} \qquad S_e' \cong 0.5\,S_{ut} \qquad\qquad S_{ut} < 200 \text{ ksi} \qquad (6.2a)$$

Not all materials exhibit this knee. *"Many low-strength carbon and alloy steels, some stainless steels, irons, molybdenum alloys, titanium alloys and some polymers"*[10] do. Other materials, such as *"aluminum, magnesium, copper, nickel alloys, some stainless steels, and high-strength carbon and alloy steels"*[10] show S-N curves that continue to fall with increasing N, though the slope may become smaller beyond about 10^7 cycles. For applications requiring $< 10^6$ cycles of operation, a **fatigue strength** S_f (sometimes also called **endurance strength**) can be defined at any N from these data. The term *endurance limit is used only to represent the infinite-life strength for those materials having one.*

The data in Figure 6-8 are for steels of S_{ut} < 200 kpsi. Steels with higher tensile strengths do not exhibit the relationship shown in equation 6.2a. Figure 6-9 shows the endurance limit S_e, plotted as a function of S_{ut}. There is a large scatter-band but the average behavior is a line of slope 0.5 up to 200 kpsi. Beyond that level the higher-strength steels' endurance limit falls off. The usual approach is to assume that the endurance strength for steels never exceeds 50% of 200 kpsi.

for steels : $\qquad S_{e'} \cong 100$ ksi $\qquad S_{ut} \geq 200$ ksi $\qquad$ (6.2b)

Figure 6-9 also shows scatter bands of fatigue strengths for severely notched specimens and for specimens in corrosive environments. Both of these factors have a severe effect on the fatigue strength of any material. An endurance limit only exists in the absence of corrosion. Materials in corrosive environments have *S-N* curves that continue to fall with increasing *N*. We will shortly consider these factors in determining useful, corrected fatigue strengths for materials.

Figure 6-10 shows the scatter-band results of rotating beam tests on aluminum alloys of various types including wrought alloys (with $S_{ut} < 48$ kpsi), die-cast, and sand-cast specimens. These are all unnotched and polished. Note the lack of a distinct knee though the slope becomes smaller at around 10^7 cycles. Aluminums do not have an

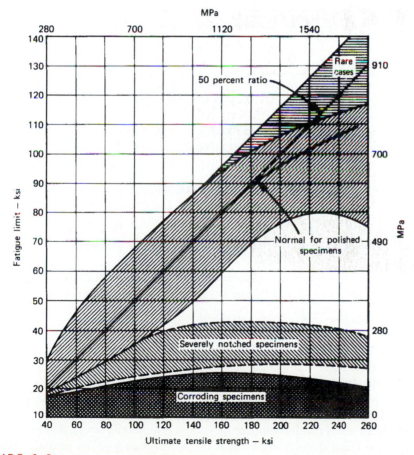

FIGURE 6-9

Relationship Between Fatigue Strengths and Ultimate Strength for Steel Specimens (From *Steel and its Heat Treatment*, by D. K. Bullens, John Wiley & Sons, New York, 1948, with permission of the publisher)

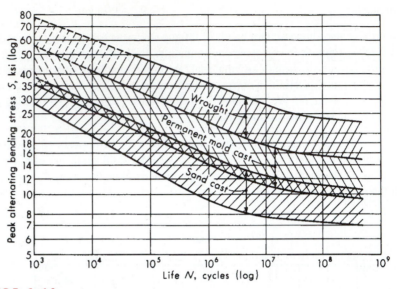

FIGURE 6-10

S-N Bands for Representative Aluminum Alloys, Excluding Wrought Alloys with $S_{ut} > 38$ kpsi (From Fig. 11.13, p. 216, Juvinall, R. C., 1967, *Stress, Strain, and Strength*, McGraw-Hill, New York, with permission)

endurance limit, thus their **fatigue strength** $S_{f'}$ is usually taken as the average failure stress at N = 5E8 cycles or some other value of N (which must be stated as part of the data).

Figure 6-11 shows the trend of fatigue strengths (at N = 5E8) for a number of aluminum alloys of varying static tensile strengths. The fatigue strength tracks the alloys' static tensile strengths at a ratio of

$$\text{for aluminums:} \quad S_{f'_{@5x10^8}} \cong 0.4\,S_{ut} \qquad\qquad S_{ut} < 48 \text{ kpsi} \qquad\qquad (6.2c)$$

up to a plateau at around $S_f' = 19$ kpsi indicating that aluminum alloys with $S_{ut} >$ about 48 kpsi "top out" at 19 kpsi of fatigue strength. ($S_{n'}$ in the figure is the same as $S_{f'}$.)

$$\text{for aluminums:} \quad S_{f'_{@5x10^8}} \cong 19 \text{ ksi} \qquad\qquad S_{ut} \geq 48 \text{ kpsi} \qquad\qquad (6.2d)$$

AXIAL FATIGUE TESTS   The *S-N* diagram can also be developed for a material by using an axial fatigue test wherein a specimen similar to that shown in Figure 2-1 is loaded cyclically in a servohydraulic test machine. The programmability of these machines allows any combination of mean and alternating components of stress to be applied including fully reversed loading (σ_m = 0). A principal difference versus the rotating beam test is that the entire cross section is uniformly stressed in axial tension/compression rather than having a linear stress distribution across its diameter that is maximum at the outer fiber and zero at the center. One result is that the fatigue strengths exhibited in the axial tests are typically lower than those seen in the rotating-beam test. This is thought to be due to the higher likelihood of a microcrack being present in the

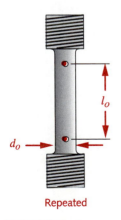

Repeated

FIGURE 2-1

A Tensile-Test Specimen

Alloys represented:
1100–0, H12, H14, H16, H18   2014–0, T4, and T6   6063–0, T42,T5,T6
3003–0, H12, H14, H16, H18   2024–T3,T36, and T4   7075–T6
5052–0, H32, H34, H36, H38   6061–0, T4, and T6

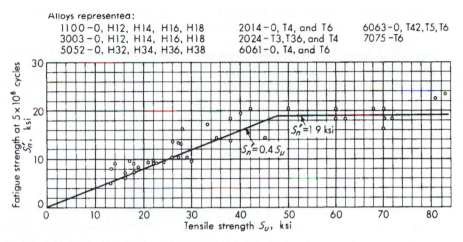

FIGURE 6-11

Fatigue Strength at 5×10^8 Cycles for Common Wrought-Aluminum Alloys (From Fig. 11.12, p. 215, Juvinall, R. C., 1967, *Stress, Strain, and Strength*, McGraw-Hill, New York, with permission)

much larger high-stress field of the axial specimen than in the smaller-volume outer regions of the rotating specimen that are highly stressed. The fact that it is difficult to create exact axial loading with no eccentricity may also be a factor in the lower strength values since eccentric loads will superpose bending moments on the axial loads.

Figure 6-12 shows two *S-N* curves for the same material (C10 steel) generated by a fully reversed axial test (labeled *push-pull*) and a rotating bending test. The axial data are seen to be at lower values than the rotating-beam data. Various authors report that the fatigue strength in reversed axial loading may be from 10%[11] to 30%[12] lower than

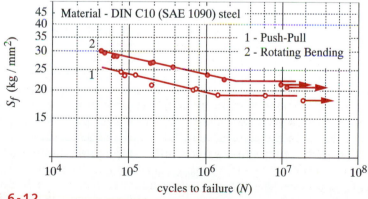

FIGURE 6-12

Fully Reversed Axial and Rotating-Beam *S-N* Curves Compared (From A. Esin, " A Method of Correlating Different Types of Fatigue Curves" *International Journal of Fatigue*, vol. 2, no. 4, pp.153-158, 1980)

the rotating beam data for the same material. If bending is known to be present in addition to the axial loading, then the reduction may be as large as 40%.[11]

Figure 6-13 shows the data for a fully reversed axial loading test on AISI 4130 steel, plotted on log-log coordinates. Note the slope change at around 10^3 cycles, which corresponds to the approximate transition from the LCF to the HCF region, and the change to essentially zero slope at about 10^6 cycles corresponding to the endurance limit for infinite life. The fatigue strength is approximately 80% of the material's static strength at about 10^3 cycles and about 40% of its static strength beyond 10^6 cycles, which are both 10% lower than the rotating-beam data of Figure 6-8.

CANTILEVER BENDING TESTS   If a cantilever beam is oscillated at its tip by a linkage mechanism, any combination of mean and alternating stresses as shown in Figure 6-6 can be achieved. This test is not used as often as the rotating-bending or axial test but is a less expensive alternative to the latter. Some examples of data from a cantilever test are shown in Figure 6-14 for various polymers. This is a semilog plot but still shows the presence of an endurance limit for some of these nonmetallic materials.

TORSIONAL FATIGUE TESTS   are done on a cylindrical specimen subjected to fully reversed, torsional loading. The failure points for reversed bending and reversed torsion in biaxial-stress tests are plotted on σ_1-σ_3 axes in Figure 6-15. Note the similarity to the distortion-energy ellipse of Figures 5-3 and 5-6, which are for static-loading failures. Thus the relationship between torsional strength and bending strength in cy-

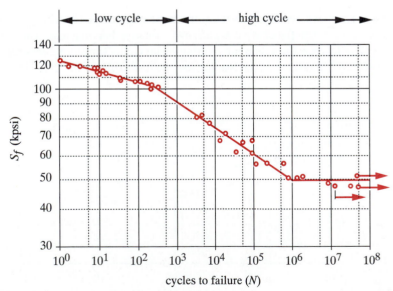

FIGURE 6-13

Fully Reversed Axial S-N Curve for AISI 4130 Steel, Showing Break at LCF/HCF Transition and an Endurance Limit (From Fig. 7-3, p. 273, in *Mechanical Engineering Design*, by Shigley and Mitchell, 4th ed.,1983, McGraw-Hill, New York, with permission. Data from NACA Technical Note #3866, Dec., 1966)

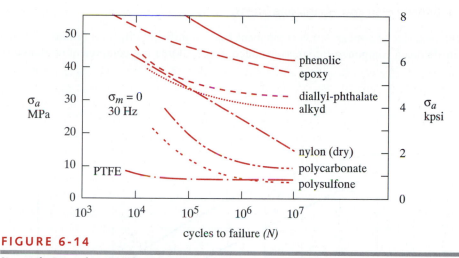

FIGURE 6-14

Stress-Life Curves from Cantilever Bending of Mineral and Glass-Filled Thermosets (Solid Lines) and Unfilled Thermoplastics (Dotted Lines) (From Fig. 9-22, p. 362, in *Mechanical Behavior of Materials*, by N. E. Dowling, Prentice-Hall, Englewood Cliffs, N.J., and based on data from M. N. Riddell, "A Guide to Better Testing of Plastics", *Plastics Engineering*, vol. 30, no. 4, pp. 71-78, with permissions)

clic loading is the same as in the static-loading case. The torsional fatigue strength (or the torsional endurance limit) for a ductile material can then be expected to be about 0.577 (58%) of the bending fatigue strength (or bending endurance limit).

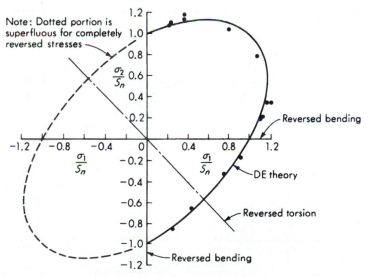

FIGURE 6-15

Fully Reversed Combined Torsional and Bending Biaxial Stress Failures Plotted on σ_1–σ_2 Axes (From *Behavior of Metals under Complex Static and Alternating Stresses* by G. Sines in *Metal Fatigue*, by G. Sines and J. Waisman (eds.), McGraw-Hill, New York, 1959, with data from W. Savert, Germany, 1943, for annealed mild steel)

Combined Mean and Alternating Stress

The presence of a mean-stress component has a significant effect on failure. When a tensile mean component of stress is added to the alternating component as shown in

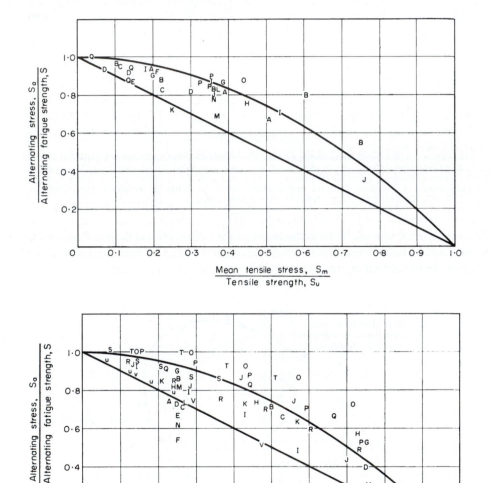

(a)

(b)

FIGURE 6-16

Effects of Mean Stress on Alternating Fatigue Strength at Long Life (a) Steels based on 10^7 to 10^8 Cycles (b) Aluminum Alloys Based on 5×10^8 Cycles. (From *Fatigue of Metals*, by P. G. Forrest, Pergamon Press, London)

Figures 6-6*b* and 6-6*c*, the material fails at lower alternating stresses than it does under fully reversed loading. Figure 6-16 shows the results of tests made on steels at $\approx 10^7$ cycles (*a*) and aluminum alloys at $\approx 5 \times 10^7$ cycles (*b*) for various levels of mean and alternating stresses in combination. The plots are normalized by dividing the alternating stress σ_a by the fatigue strength S_f of the material under fully reversed stress (at the same number of cycles), and dividing the mean stress σ_m by the ultimate tensile strength S_{ut} of the material. There is a great deal of scatter in the data but a parabola that intercepts 1 on each axis, called the **Gerber line**, can be fitted to the data with reasonable accuracy. A straight line connecting the fatigue strength (1 on the *y* axis) with the ultimate strength (1 on the *x* axis), called the **Goodman line**, is a reasonable fit to the lower envelope of the data. The Gerber line is a measure of the average behavior of these parameters (for ductile materials) and the Goodman line is a measure of their minimum behavior. The Goodman line is often used as a design criterion since it is safer than the Gerber line.

Figure 6-17 shows the effects of mean stresses (ranging from the compressive regime to the tensile regime) on failure when combined with alternating tensile stress for both aluminum and steel. It is clear from these data that compressive mean stresses have a beneficial effect and tensile mean stresses are detrimental. This fact provides an opportunity to mitigate the effects of alternating tensile stresses by the deliberate introduction of mean compressive stresses. One way to do this is to create **residual compressive stress** in the material in regions where large alternating components are expected. We will investigate ways to do this in later sections.

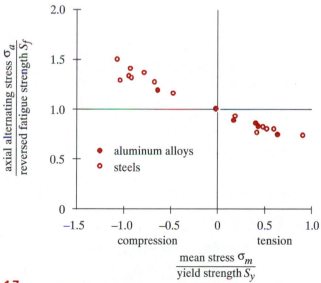

FIGURE 6-17

Compressive and Tensile Mean Stress Effect (From G. Sines, "Failure of Materials Under Combined Repeated Stresses with Superimposed Static Stresses," NACA Technical Note #3495, 1955)

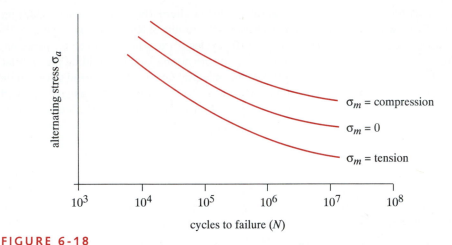

FIGURE 6-18

Effect of Mean Stress on Fatigue Life (From *Metal Fatigue in Engineering*, by Fuchs and Stephens, New York, 1980, reprinted by permission of John Wiley & Sons, Inc.)

Figure 6-18 shows another view of this phenomenon by plotting the *S-N* curve (on semilog axes) for a hypothetical material with compressive mean stress, no mean stress, and tensile mean stress added. The fatigue strength or endurance limit of the material is effectively increased by the introduction of a compressive mean stress, whether applied or residual.

Fracture-Mechanics Criteria

The static fracture-toughness test was described in Section 5.3. To develop fatigue strength data in terms of fracture-mechanics theory, a number of specimens of the same material are tested to failure at various levels of cyclical stress range $\Delta\sigma$. The test is done in an axial-fatigue machine and the load pattern is usually either repeated or fluctuating tensile stresses as shown in Figure 6-6*b* and 6-6*c*. Reversed-stress tests are seldom done for these data since compressive stress does not promote crack growth. The crack growth is continuously measured during the test. The applied stresses range from σ_{min} to σ_{max}. **A stress intensity factor range** ΔK can be calculated for each fluctuating-stress condition from

$$\Delta K = K_{max} - K_{min} : \qquad \text{if } K_{min} < 0 \quad \text{then } \Delta K = K_{max} \qquad (6.3a)$$

substituting the appropriate equation 5.14 gives:

$$\Delta K = \beta\sigma_{max}\sqrt{\pi a} - \beta\sigma_{min}\sqrt{\pi a}$$
$$= \beta\sqrt{\pi a}\left(\sigma_{max} - \sigma_{min}\right) \qquad (6.3b)$$

The log of the rate of crack growth as a function of cycles *da/dN* is calculated and plotted versus the log of the stress intensity factor range ΔK as shown in Figure 6-19.

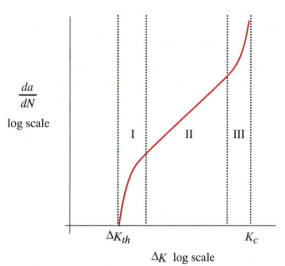

$\dfrac{da}{dN}$

log scale

I II III

ΔK_{th} K_c

ΔK log scale

FIGURE 6-19

Three Regions of the Crack-Growth-Rate Curve (From Fig. 3-12, p. 102, in *Fundamentals of Metal Fatigue Analysis*, by Bannantine et al., Prentice-Hall, Englewood Cliffs N.J.,1990, with permission)

The sigmoidal curve of Figure 6-19 is divided into three regions labeled I, II, and III. Region I corresponds to the crack initiation stage, region II to the crack-growth (crack-propagation) stage, and region III to unstable fracture. Region II is of interest in predicting fatigue life and that part of the curve is a straight line on log coordinates. Paris[13] defined the relationship in region II as

$$\frac{da}{dN} = A(\Delta K)^n \tag{6.4}$$

Barsom[14] tested a number of steels and developed empirical values for the coefficient A and the exponent n in equation 6.4. These are shown in Table 6-2.

Table 6-2 **Paris-Equation Parameters for Various Steels**
Source: J. M. Barsom[14], with permission

Steel	MKS units		ips units	
	A	n	A	n
Ferritic-Pearlitic	6.90 E–12	3.00	3.60 E–10	3.00
Martensitic	1.35 E–10	2.25	6.60 E–09	2.25
Austentitic Stainless	5.60 E–12	3.25	3.00 E–10	3.25

The fatigue-crack-growth life is found by integrating equation 6.4 between a known or assumed initial crack length and a maximum acceptable final crack length based on the particular load, geometry, and material parameters for the application.

Region I in Figure 6-19 is also of interest since it shows the existence of a minimum threshold ΔK_{th} below which no crack growth will occur. This *"threshold stress intensity factor ΔK_{th} has often been considered analogous to the unnotched fatigue limit S_e, since an applied stress intensity factor range ΔK below ΔK_{th} does not cause fatigue crack growth."*[15]

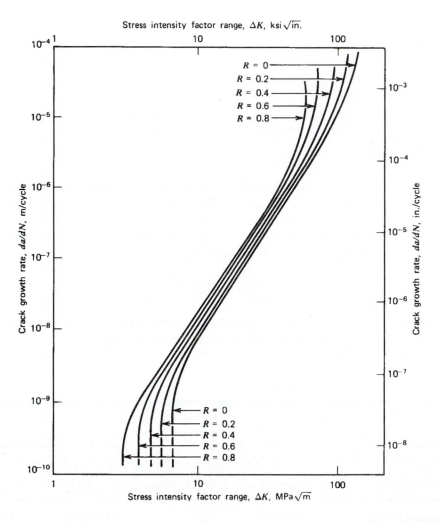

FIGURE 6-20

Schematic of the Effects of Mean Stress on the Crack-Growth-Rate Curve (From *Metal Fatigue in Engineering*, by Fuchs and Stephens, New York, 1980, reprinted by permission of John Wiley & Sons Inc.)

These axial fatigue tests have a mean-stress component present and the level of mean stress has an effect on the rate of crack propagation. Figure 6-20 shows a schematic set of *da/dN* curves for different levels of mean stress as defined by the stress ratio *R*. When *R* = 0, the stress is repeated as in Figure 6-6*b*. As *R* approaches 1, the minimum stress approaches the maximum stress (see Eq. 6.1*d*). Figure 6-20 shows little variation of the curves in Region II (crack growth), but shows significant changes with *R* in regions I and III. Crack initiation is thus affected by mean stress level. The threshold stress intensity factor ΔK_{th} may be reduced by a factor of 1.5 to 2.5 when *R* increases from 0 to 0.8[18]. This is consistent with the effects of mean stress on the *S-N* data in the stress-life model discussed above (see Figure 6-16). Figure 6-21 from reference 19

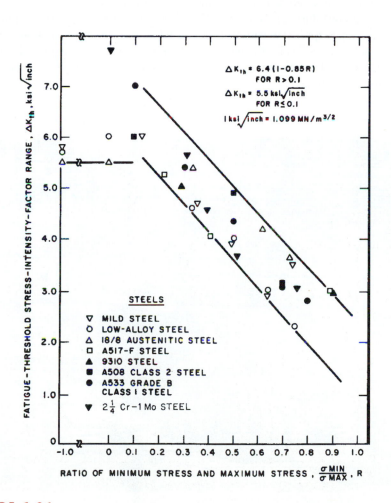

FIGURE 6-21

Effect of Mean Stress on Fatigue Threshold Stress Intensity Factor Range (From Fig 9.6, p. 285, in *Fracture and Fatigue Control in Structures*, by Barsom and Rolfe, Prentice-Hall, Englewood Cliffs, N.J.,1987, with permission)

provides test data showing the effect of stress-ratio R on the threshold stress intensity factor range ΔK_{th} for several steels.

Testing Actual Assemblies

While much strength data are available from test specimens, and the engineer can use these data as a starting point to estimate the strength of a particular part, the best data is obtained by testing the actual design under realistic load, temperature, and environmental conditions. This is an expensive proposition and is usually only done when the design's cost, quantity or threat to human safety demand it. Figure 6-22 shows an elaborate test fixture built to allow fatigue testing of the wing and fuselage assemblies of the Boeing 757 aircraft during production. The entire aircraft is placed in the fixture and time-varying loads applied to various elements while measurements of strains, deflections, etc., are made. This is obviously a costly process but serves to provide the most realistic data possible, applying the test to actual shapes, sizes, and materials rather than to laboratory specimens.

FIGURE 6-22

Boeing 757 Fatigue-Test Fixture for Wing and Fuselage Assemblies. (*Courtesy of Boeing Commercial Airplane Co., Seattle, Wash.*)

6.6   ESTIMATING FATIGUE FAILURE CRITERIA

The best information on a material's fatigue strength at some finite life, or its endurance limit at infinite life, comes from the testing of actual or prototype assemblies of the design as described above. If this is not practical or possible, the next best information comes from fatigue tests of specimens taken from the particular material as it is manufactured for the part (i.e., as cast, forged, machined etc.). Failing this, published fatigue strength data may be available in the literature or from the material manufacturers, but these data will be for small, polished specimens tested in controlled environments. In the absence of even these data, it will be necessary to make some estimation of the endurance limit or fatigue strength of the material based on data available from monotonic tests. This may be limited to information on the material's ultimate strength S_{ut} and yield strength S_y.

Estimating the Theoretical Fatigue Strength $S_{f'}$ or Endurance Limit $S_{e'}$

If published data are available for the fatigue strength $S_{f'}$ or endurance limit $S_{e'}$ of the material, they should be used and the correction factors discussed in the next section then applied to them. Published fatigue strength data are typically from fully reversed bending or axial loading tests on small, polished specimens. If no fatigue strength data are available, an approximate $S_{f'}$ or $S_{e'}$ can be crudely estimated from the published ultimate tensile strength of the material. Figure 6-23 shows the relationships between S_{ut} and $S_{f'}$ for wrought steels (a), wrought and cast irons (b), aluminum alloys (c), and wrought copper alloys (d). There is considerable scatter and the lines are fitted approximately to the upper and lower bounds. At high tensile strengths, the fatigue strengths tend to "top out" as described above. From these data, approximate relationships can be stated between S_{ut} and $S_{f'}$ or $S_{e'}$. These relationships for steels and aluminum alloys were stated in the previous section as equations 6.2 and are repeated here for convenience.

steels:
$$\begin{cases} S_{e'} \cong 0.5\,S_{ut} & \text{for } S_{ut} < 200 \text{ ksi } (1400 \text{ MPa}) \\ S_{e'} \cong 100 \text{ ksi } (700 \text{ MPa}) & \text{for } S_{ut} \geq 200 \text{ ksi } (1400 \text{ MPa}) \end{cases} \qquad (6.5a)$$

irons:
$$\begin{cases} S_{e'} \cong 0.4\,S_{ut} & \text{for } S_{ut} < 60 \text{ ksi } (400 \text{ MPa}) \\ S_{e'} \cong 24 \text{ ksi } (160 \text{ MPa}) & \text{for } S_{ut} \geq 60 \text{ ksi } (400 \text{ MPa}) \end{cases} \qquad (6.5b)$$

aluminums:
$$\begin{cases} S_{f'_{@5E8}} \cong 0.4\,S_{ut} & \text{for } S_{ut} < 48 \text{ ksi } (330 \text{ MPa}) \\ S_{f'_{@5E8}} \cong 19 \text{ ksi } (130 \text{ MPa}) & \text{for } S_{ut} \geq 48 \text{ ksi } (330 \text{ MPa}) \end{cases} \qquad (6.5c)$$

copper alloys:
$$\begin{cases} S_{f'_{@5E8}} \cong 0.4\,S_{ut} & \text{for } S_{ut} < 40 \text{ ksi } (280 \text{ MPa}) \\ S_{f'_{@5E8}} \cong 14 \text{ ksi } (100 \text{ MPa}) & \text{for } S_{ut} \geq 40 \text{ ksi } (280 \text{ MPa}) \end{cases} \qquad (6.5d)$$

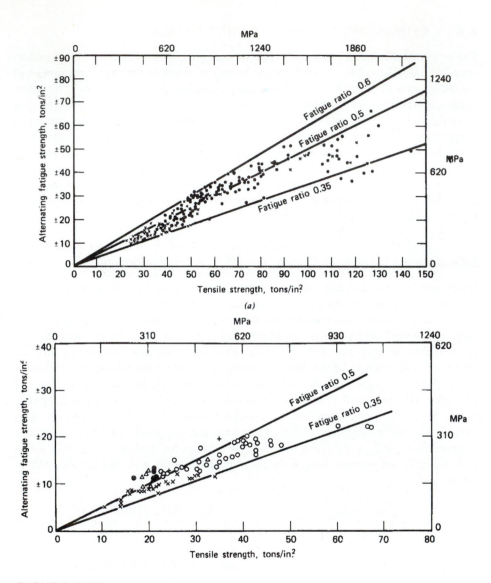

FIGURE 6-23

Relation Between Unnotched Rotating-Bending Fatigue Strength and Ultimate Strength. Part I —
(a) Steels, (b) Wrought and Cast Irons (From P. G. Forrest, Fatigue of Metals, Pergamon Press, London, 1962)

Correction Factors to the Theoretical Fatigue Strength or Endurance Limit

The fatigue strengths or endurance limits obtained from standard fatigue-test specimens or from estimates based on static tests must be modified to account for physical differences between the test specimen and the actual part being designed. Environmental and

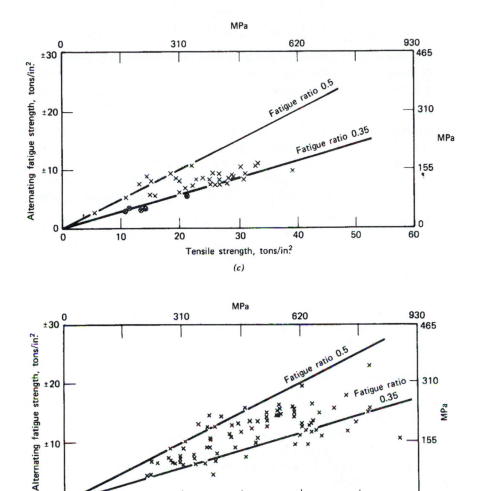

FIGURE 6-23 continued

Relation Between Unnotched Rotating-Bending Fatigue Strength and Ultimate Strength. Part II —
(c) Aluminum Alloys, (d) Copper Alloys (From P. G. Forrest, Fatigue of Metals, Pergamon Press, 1962)

temperature differences between the test conditions and the actual conditions must be
taken into account. Differences in the manner of loading need to be accounted for.
These and other factors are incorporated into a set of *strength-reduction factors* that are
then multiplied by the theoretical estimate to obtain a corrected fatigue strength or endurance limit for the particular application.

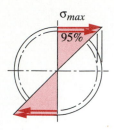

(a) Stress distribution

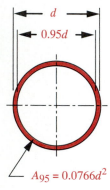

(b) Area above 95%

FIGURE 6-24

The Area in a Rotating Beam Specimen that is Stressed Above 95% of Maximum Stress

$$S_e = C_{load}\,C_{size}\,C_{surf}\,C_{temp}\,C_{reliab}\,S_{e'}$$
$$S_f = C_{load}\,C_{size}\,C_{surf}\,C_{temp}\,C_{reliab}\,S_{f'} \qquad (6.6)$$

where S_e represents a corrected endurance limit for a material that exhibits a knee in its S-N curve and S_f represents a corrected fatigue strength at a particular number of cycles N for a material that does not exhibit a knee. The strength reduction factors in equation 6.6 will now be defined.

LOADING EFFECTS Since the ratios described above and most published fatigue strength data are for rotating bending tests, a strength-reduction factor for axial loading must be applied. The differences between axial- and rotating-bending-test fatigue strengths were described in the previous section. Based on that discussion of axial and bending fatigue tests, we now define a strength-reduction **load factor** C_{load} as

$$\text{bending}: \qquad C_{load} = 1$$
$$\text{axial loading}: \qquad C_{load} = 0.70 \qquad (6.7a)$$

Note that a torsional-fatigue test shows a strength that is 0.577 times the rotating-bending fatigue strength as shown in Figure 6-15. For a pure-torsion fatigue case, we could compare the applied, alternating, torsional shear stress directly to the torsional fatigue strength. However, we will generally deal with the pure-torsional case (and all other cases as well) by calculating the von Mises effective stress from the applied stresses.* This gives an *effective alternating tensile stress* value that can be compared directly to a *bending fatigue strength*. So, for pure torsion cases, use $C_{load} = 1$ with this method.

SIZE EFFECTS The rotating-beam and static-test specimens are small (about 0.3-in dia). If the part is larger than that dimension, a strength-reduction **size factor** needs to be applied to account for the fact that larger parts fail at lower stresses due to the higher probability of a flaw being present in the larger stressed volume. Various authors have suggested different values for the size factor. Shigley and Mitchell[21] present a simple expression that is fairly conservative.

$$\text{for } d \le 0.3 \text{ in (8 mm)}: \qquad C_{size} = 1$$
$$\text{for } 0.3\,\text{in} \le d \le 10 \text{ in}: \qquad C_{size} = 0.869 d^{-0.097}$$
$$\text{for } 8\text{ mm} \le d \le 250\text{ mm}: \qquad C_{size} = 1.189 d^{-0.097} \qquad (6.7b)$$

For larger sizes use $C_{size} = 0.6$. (The test data on which these equations are based are for steel parts. The accuracy of equation 6.7b for nonferrous metals is questionable.)

Equation 6.7b is valid for cylindrical parts. For parts of other shapes, Kuguel[22] suggested that equating the nonround part's cross-sectional area stressed above 95% of its maximum stress with the similarly stressed area of a rotating beam specimen would provide an *equivalent diameter* to use in equation 6.7b. Since the stress is linearly distributed across the diameter d of a beam in rotating-bending, the area A_{95} stressed above 95% of the outer-fiber stress is that which lies between 0.95 d and 1.0 d as shown in Figure 6-24.

---

* One exception to this will be the analysis and design of coil springs in Chapter 13 for which most of the available strength data are torsional shear strength values. It then makes more sense to compare torsional stresses to torsional strengths directly without conversion of the stresses to von Mises equivalents.

$$A_{95} = \pi \left[\frac{d^2 - (0.95d)^2}{4} \right] = 0.0766\, d^2 \qquad (6.7c)$$

The equivalent-diameter rotating-beam specimen for any cross section is then

$$d_{equiv} = \sqrt{\frac{A_{95}}{0.0766}} \qquad (6.7d)$$

where A_{95} is the portion of the cross-sectional area of the nonround part that is stressed between 95% and 100% of its maximum stress. It is a simple task to compute the value of A_{95} for any cross section for which the loading is known. Shigley and Mitchell[21] have done so for several common sections and their results are shown in Figure 6-25.

SURFACE EFFECTS The rotating-beam specimen is polished to a mirror finish to preclude surface imperfections serving as stress raisers. It is usually impractical to provide such an expensive finish on a real part. Rougher finishes will lower the fatigue strength by the introduction of stress concentrations and/or by altering the physical properties of the surface layer. A forged surface is both rough and decarburized and the reduced carbon levels weaken the surface where stresses are often highest.[23] A strength-reduction **surface factor** C_{surf} is needed to account for these differences. Juvinall[24] provides a chart (Figure 6-26) that gives some guidance in selecting a surface factor for a number of common finishes on steel. Note that tensile strength is also a factor since higher-strength materials are more sensitive to the stress concentrations introduced by surface irregularities. In Figure 6-26 corrosive environments are seen to

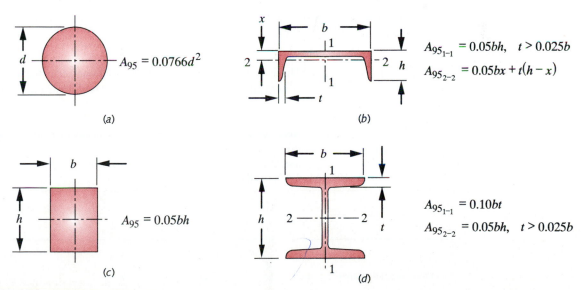

(a)   (b)

(c)   (d)

FIGURE 6-25

Formulas for 95% Stressed Areas of Various Sections (Source: *Mechanical Engineering Design* by Shigley and Mitchell, 1983, 4th ed., McGraw-Hill, New York, with permission)

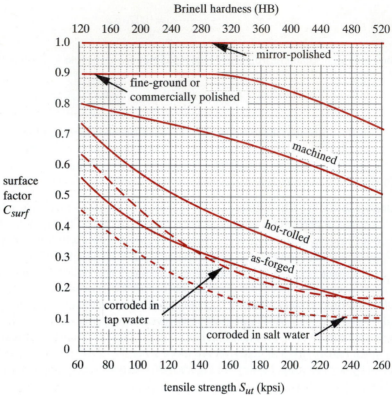

FIGURE 6-26

Surface Factors for Various Finishes on Steel (From Fig. 12.6, p. 234, R. C. Juvinall, 1967, *Stress, Strain, and Strength*, McGraw-Hill, New York, with permission)

drastically reduce strength. These surface factors have been developed for steels, and should only be applied to aluminum alloys and other ductile metals with the caution that

Table 6-3   Coefficients for the Surface-Factor Equation

Source: Shigley and Mischke, *Mechanical Engineering Design*, 5th ed., McGraw-Hill, New York, 1989, p. 283 with permission

Surface Finish	MPa		kpsi	
	A	b	A	b
Ground	1.58	−0.085	1.34	−0.085
Machined or cold-drawn	4.51	−0.265	2.7	−0.265
Hot-rolled	57.7	−0.718	14.4	−0.718
As-forged	272	−0.995	39.9	−0.995

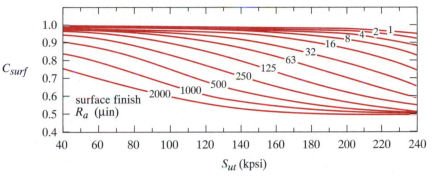

C_{surf}

surface finish
R_a (μin)

S_{ut} (kpsi)

FIGURE 6-27

Surface Factor as a Function of Surface Roughness and Ultimate Tensile Strength (From R. C. Johnson, *Machine Design, vol. 45, no. 11, 1967, p. 108*, Penton Publishing, Cleveland, Ohio., with permission)

testing of actual parts under realistic loading conditions be done in critical applications. Cast irons can be assigned a $C_{surf} = 1$ since their internal discontinuities dwarf the effects of a rough surface.

R. C. Johnson[25] provides the chart shown in Figure 6-27 that gives more detail for **machined and ground surfaces** by relating C_{surf} to tensile strength based on the measured surface-average-roughness R_a in microinches.* If R_a is known for a machined or ground part, Figure 6-27 can be used to determine a suitable surface factor C_{surf}. The surface-factor curves in Figure 6-26 for hot-rolled, forged and corroded surfaces should still be used as they account for decarburization and pitting effects as well as surface roughness.

Shigley and Mischke[39] suggest using an exponential equation of the form

$$C_{surf} \cong A\left(S_{ut}\right)^b \qquad \text{if } C_{surf} > 1.0, \text{ set } C_{surf} = 1.0 \qquad (6.7e)$$

to represent the surface factor with S_{ut} in either kpsi or MPa. The coefficients A and exponents b for various finishes were determined from data similar to that in Figure 6-26 and are shown in Table 6-3. This approach has the advantage of being computer programmable and eliminates the need to refer to charts such as Figure 6-26 and 6-27.

Surface treatments such as electroplating with certain metals can severely reduce the fatigue strength as shown in Figure 6-28 for chrome plating. Plating with soft metals such as cadmium, copper, zinc, lead, and tin appears not to severely compromise the fatigue strength. Electroplating with chrome and nickel is generally not recommended for parts stressed in fatigue unless additional surface treatments such as shot peening (see below) are also applied. An exception may be when the part is in a corrosive environment and the corrosion protection afforded by the plating outweighs its strength reduction. Most of the strength lost to plating can be recovered by shot-peening the surface to introduce beneficial compressive stresses as shown in Figure 6-29. Shot peening and other means of creating residual stresses are addressed in Section 6.8.

* There are many parameters used to characterize surface roughness, all of which are typically measured by passing a sharp, conical diamond stylus over the surface with controlled force and velocity. The stylus follows and encodes the microscopic contours and stores the surface profile in a computer. A number of statistical analyses are then done on the profile such as finding the largest peak-to-peak distance (R_t), the average of the 5 highest peaks (R_{pm}), etc. The most commonly used parameter is R_a (or A_a), which is the arithmetic average of the absolute values of the peak heights and valley depths. It is this parameter that is used in Figure 6-27. See Section 7.1 for more information on surface roughness.

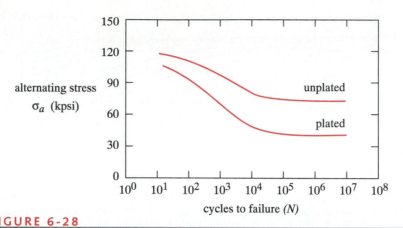

FIGURE 6-28

The Effect of Chrome Plating on Fatigue Strength of Steel (From *Fatigue Design*, by C. C. Osgood, Pergamon Press, London,1982, with permission)

TEMPERATURE   Fatigue tests are most commonly done at room temperature. The fracture toughness decreases at low temperatures and increases at moderately high temperatures (up to about 350°C). But, the endurance-limit-knee in the *S-N* diagram disappears at high temperatures making the fatigue strength continue to decline with number of cycles, *N*. Also, the yield strength declines continuously with temperatures above room ambient and, in some cases, this can cause yielding before fatigue failure. At temperatures above about 50% of the material's absolute melting temperature, creep becomes a significant factor and the stress-life approach is no longer valid. The strain-life approach can account for the combination of creep and fatigue under high-temperature conditions and should be used in those situations.

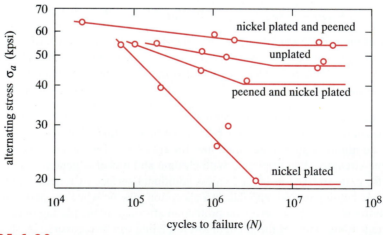

FIGURE 6-29

The Effect of Nickel Plating and Shot-Peening on Fatigue Strength of Steel (From *Residual Stresses and Fatigue in Metals*, by Almen & Black, McGraw-Hill, New York,1963)

Several approximate formulas have been proposed to account for the reduction in endurance limit at moderately high temperatures. A **temperature factor** C_{temp} can be defined. Shigley and Mitchell[26] suggest the following:

for $T \leq 450°C$ (840°F): $\qquad C_{temp} = 1$

for $450°C < T \leq 550°C$: $\qquad C_{temp} = 1 - 0.0058(T - 450)$ $\qquad$ (6.7f)

for $840°F < T \leq 1020°F$: $\qquad C_{temp} = 1 - 0.0032(T - 840)$

Note that these criteria are based on data for steels and should not be used for other metals such as Al, Mg, and Cu alloys.

RELIABILITY Many of the reported strength data are mean values. There is considerable scatter in multiple tests of the same material under the same test conditions. Haugen and Wirsching[27] report that the standard deviations of endurance strengths of steels seldom exceed 8% of their mean values. Table 6-4 shows reliability factors for an assumed 8% standard deviation. Note that a 50% reliability has a factor of 1 and the factor reduces as you choose higher reliability. For example, if you wish to have 99.99% probability that your samples meet or exceed the assumed strength, multiply the mean strength value by 0.702. The values in Table 6-4 provide strength-reduction factors C_{reliab} for chosen reliability levels.

ENVIRONMENT The environment can have significant effects on fatigue strength as is evidenced by the curves for corroded surfaces in Figure 6-26. Figure 6-30 shows schematically the relative effects of various environments on fatigue strength. Note that even room air reduces strength compared to vacuum. The higher the relative humidity and temperature, the larger will be the reduction of strength in air. The *presoak* line represents parts soaked in a corrosive environment (water or seawater) and then tested

Table 6-4
Reliability Factors
for $S_d = 0.08\ \mu$

Reliability %	C_{reliab}
50	1.000
90	0.897
99	0.814
99.9	0.753
99.99	0.702
99.999	0.659

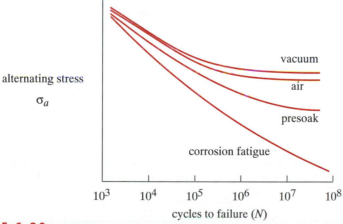

FIGURE 6-30

The Effect of Environment on Fatigue Strength of Steel (From *Metal Fatigue in Engineering*, by Fuchs and Stephens, New York, 1980, reprinted by permission of John Wiley & Sons, Inc.)

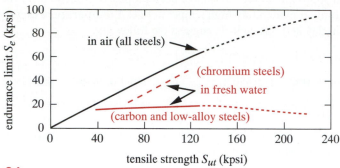

FIGURE 6-31

The Effect of Fresh Water on Fatigue Strength of Steel (From *Fatigue of Metals*, by P. G. Forrest Pergamon Press, Oxford, 1962)

in room air. The increased roughness of the corroded surface is thought to be the reason for the loss of strength. The *corrosion fatigue* line shows a drastic reduction of strength and elimination of the endurance-limit knee.

The **corrosion-fatigue** phenomenon is not yet fully understood but empirical data such as those in Figures 6-30 and 6-31 depict its severity. Figure 6-31 shows the effect of operation in fresh water on the *S-N* curves of carbon and low-alloy steels. The relationship between S_e, and S_{ut} becomes constant at about 15 kpsi. So low-strength carbon steel is as good as high-strength carbon steel in this environment. The only steels that retain some strength in water are chromium steels (including stainless steels) since that alloying element provides some corrosion protection. Figure 6-32 shows the ef-

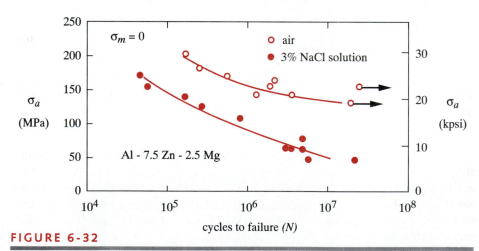

FIGURE 6-32

The Effect of Saltwater on the Fatigue Strength of Aluminum (From Stubbington and Forsyth, "Some Corrosion-Fatigue Observations on a High-Purity Aluminum-Zinc-Magnesium Alloy and Commercial D.T.D. 683 Alloy", *J. of the Inst. of Metals*, London, U.K., vol. 90, 1961-1962, pp. 347-354, with permission)

fects of saltwater on the fatigue strength of one alloy of aluminum. Only limited data are available for material strengths in severe environments. Thus it is difficult to define any universal strength reduction factors for environmental conditions. The best approach is to extensively test all designs and materials in the environment that they will experience. (This is difficult to do for situations in which the long-term effects of low-frequency loading are desired because it will take an unreasonable time to obtain the data.) Based on Figure 6-31, for carbon and low alloy steels in fresh water, the relationship between $S_{e'}$ and S_{ut} in equation 6.06a should be modified to

$$S_{e'} \cong 15 \text{ kpsi (100 MPa)} \quad \text{for carbon steel in fresh water} \quad (6.8)$$

Presumably, a saltwater environment would be even worse.

Calculating the Corrected Fatigue Strength S_f or Corrected Endurance Limit S_e

The strength reduction factors can now be applied to the uncorrected endurance limit $S_{e'}$ or to the uncorrected fatigue strength $S_{f'}$ using equation 6.6 to obtain corrected values for design purposes.

Creating Estimated S-N Diagrams

Equations 6.6 provide information about the material's strength in the high cycle region of the *S-N* diagram. With similar information for the low cycle region, we can construct an *S-N* diagram for the particular material and application. The bandwidth of interest is the HCF regime from 10^3 to 10^6 cycles and beyond. Let the material strength at 10^3 cycles be called S_m. Test data indicate that the following estimates of S_m are reasonable:[28]

$$\begin{aligned} \text{bending:} &\qquad S_m = 0.9 S_{ut} \\ \text{axial loading:} &\qquad S_m = 0.75 S_{ut} \end{aligned} \qquad (6.9)$$

The estimated *S-N* diagram can now be drawn on log-log axes as shown in Figure 6-33. The *x* axis runs from $N = 10^3$ to $N = 10^9$ cycles or beyond. The appropriate S_m from equation 6.9 for the type of loading is plotted at $N = 10^3$. Note that the correction factors from equation 6.6 are **not** applied to S_m.

If the material exhibits a knee, then the corrected S_e from equation 6.6 is plotted at $N_e = 10^6$ cycles and a straight line is drawn between S_m and S_e. The curve is continued horizontally beyond that point. If the material does not exhibit a knee, then the corrected S_f from equation 6.6 is plotted at the number of cycles for which it was generated (shown at $N_f = 5 \times 10^8$) and a straight line is drawn between S_m and S_f. This line may be extrapolated beyond that point, but its accuracy is questionable, though probably conservative (see Figure 6-10).

The equation of the line from S_m to S_e or S_f can be written as

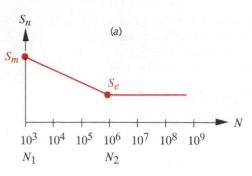
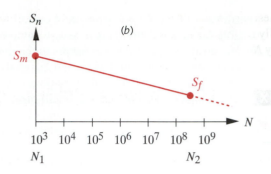

FIGURE 6-33

Estimated S-N Curves for (a) Materials with Knee, (b) Materials Without Knee

$$S_n = a N^b \qquad (6.10a)$$

or

$$\log S_n = \log a + b \log N \qquad (6.10b)$$

where S_n is the fatigue strength at any N and a, b are constants defined by the boundary conditions. For all cases, the y intercept is $S_n = S_m$ at $N = N_1 = 10^3$. For the endurance-limit case (Figure 6-33), $S_n = S_e$ at $N = N_2 = 10^6$. For a material that does not exhibit an endurance limit knee, the fatigue strength is taken at some number of cycles: $S_n = S_f$ at $N = N_2$ (Figure 6-33b). Substitute the boundary conditions in Eq. 6.10b and solve simultaneously for a and b:

$$b = \frac{1}{z} \log\left(\frac{S_m}{S_e}\right) \qquad \text{where} \qquad z = \log N_1 - \log N_2$$

$$(6.10c)$$

$$\log(a) = \log(S_m) - b \log(N_1) = \log(S_m) - 3b$$

Note that N_1 is always 1 000 cycles and its $\log_{10}$ is 3. For a material with a knee at $N_2 = 10^6$, $z = (6 - 3) = -3$ as shown in Table 6-5. This curve is valid only to the knee, beyond which $S_n = S_e$ as shown in Figure 6-33a.

For a material with no knee and $S_n = S_f$ at $N = N_2$ (Figure 6-33b), the values of z corresponding to various values of N_2 are easily calculated. For example, the curve in Figure 6-33b shows S_f at $N_2 = 5E8$ cycles. The value for z is then

$$z = \log(1000) - \log(5E8) = 3 - 8.699 = -5.699$$

$$(6.10d)$$

$$b = -\frac{1}{5.699} \log\left(\frac{S_m}{S_f}\right) \qquad \text{for } S_f @ N_2 = 5E8 \text{ cycles}$$

The constants for any other boundary conditions are determined in the same way. Some values of z for a range of values of N_2 with N_1 set to 10^3 are shown in Table 6-5.

Table 6-5

z Factors for Eq. 6-10c

N_2	z
1.E6	-3.000
5.E6	-3.699
1.E7	-4.000
5.E7	-4.699
1.E8	-5.000
5.E8	-5.699
1.E9	-6.000
5.E9	-6.699

These equations for the *S-N* curve allow the estimated finite life *N* to be found for any fully reversed fatigue strength S_n, or the estimated fatigue strength S_n can be found for any *N*.

EXAMPLE 6-1

Determining Estimated S-N Diagrams for Ferrous Materials

Problem   Create an estimated *S-N* diagram for a steel, axially loaded bar and define its equations. How many cycles of life can be expected if the alternating stress is 100 MPa?

Given   The S_{ut} has been tested at 600 MPa. The bar is 150 mm square and has a hot-rolled finish. The operating temperature is 500°C maximum. The loading will be pure axial, fully reversed.

Assumptions   Infinite life is required and is obtainable since this ductile steel will have an endurance limit. A reliability factor of 99.9% will be used.

Solution   See the *TKSolver* file EX06-01.

1   Since no endurance-limit or fatigue strength information is given, we will estimate S_e' based on the ultimate strength using equation 6.6*a*.

$$S_{e'} \cong 0.5 S_{ut} = 0.5(600) = 300 \text{ MPa} \qquad (a)$$

2   The loading is axial so the load factor from equation 6.7*a* is

$$C_{load} = 0.70 \qquad (b)$$

3   The part size is greater than the test specimen and the part is not round, so an equivalent diameter based on its 95% stressed area must be determined and used to find the size factor. From Figure 6-25 and equation 6.7*d*:

$$A_{95} = 0.05 bh = 0.05(150)(150) = 1125 \text{ mm}^2$$

$$d_{equiv} = \sqrt{\frac{A_{95}}{0.0766}} = 121 \text{ mm} \qquad (c)$$

and the size factor is found for this equivalent diameter from equation 6.7*b*:

$$C_{size} = 1.189 \left(d_{equiv} \right)^{-0.097} = 1.189(121)^{-0.097} = 0.747 \qquad (d)$$

4   The surface factor is found from equation 6.7*e* and the data in Table 6-3 for the specified hot-rolled finish.

$$C_{surf} = A S_{ut}^{b} = 57.7(600)^{-0.718} = 0.584 \qquad (e)$$

6

5   The temperature factor is found from equation 6.7f:

$$C_{temp} = 1 - 0.0058(T - 450) = 1 - 0.0058(500 - 450) = 0.71 \qquad (f)$$

6   The reliability factor is taken from Table 6-4 for the desired 99.9% and is

$$C_{reliab} = 0.753 \qquad (g)$$

7   The corrected endurance limit S_e can now be calculated from equation 6.6:

$$S_e = C_{load} C_{size} C_{surf} C_{temp} C_{reliab} S_{e'}$$
$$= 0.70(0.747)(0.584)(0.71)(0.753)(300) \qquad (h)$$
$$S_e = 49 \text{ MPa}$$

8   To create the S-N diagram, we also need a number for the estimated strength S_m at 10^3 cycles based on equation 6.9 for axial loading.

$$S_m = 0.75 S_{ut} = 0.75(600) = 450 \text{ MPa} \qquad (i)$$

9   The estimated S-N diagram is shown in Figure 6-34 with the above values of S_m and S_e. The expressions of the two lines are found from equations 6.10a through 6.10c assuming that S_e begins at 10^6 cycles.

$$b = -\frac{1}{3}\log\left(\frac{S_m}{S_e}\right) = -\frac{1}{3}\log\left(\frac{450}{49}\right) = -0.3222$$

$$\log(a) = \log(S_m) - 3b = \log[450] - 3(-0.3222): \quad a = 4\,168.24 \qquad (j)$$

$$S_n = aN^b = 4\,168.24\,N^{-0.3222} \text{ MPa} \qquad 10^3 \le N \le 10^6$$
$$\qquad (k)$$
$$S_n = S_e = 49 \text{ MPa} \qquad\qquad N > 10^6$$

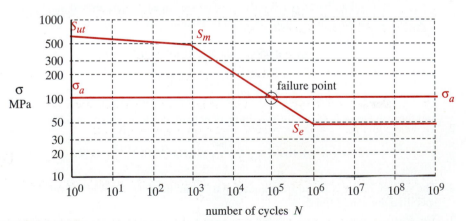

FIGURE 6-34

S-N Diagram and Alternating Stress Line Showing Failure Point

10 The number of cycles of life for any alternating stress level can now be found from equations (k). For the stated stress level of 100 MPa, we get

$$100 = 4\,168.24\,N^{-0.3222} \qquad\qquad 10^3 \leq N \leq 10^6$$
$$\log 100 = \log 4\,168.24 - 0.322\,246\,772 \log N$$
$$2 = 3.619\,952\,828 - 0.322\,246\,772 \log N \qquad\qquad (l)$$
$$\log N = 5.027\,056\,836$$
$$N = 106\,428 \text{ cycles}$$

Figure 6-34 shows the intersection of the alternating stress line with the failure line at $N = 106\,428$ cycles.

EXAMPLE 6-2

Determining Estimated S-N Diagrams for Nonferrous Materials

Problem   Create an estimated *S-N* diagram for an aluminum bar and define its equations. What is the corrected fatigue strength at 2E7 cycles?

Given   The S_{ut} for this 6061-T6 aluminum has been tested at 45 000 psi. The forged bar is 1.5 in round. The maximum operating temperature is 300°F. The loading is fully reversed torsion.

Assumptions A reliability factor of 99.0% will be used. The uncorrected fatigue strength will be taken at 5E8 cycles.

Solution   See the *TKSolver* file EX06-02.

1
Since no fatigue-strength information is given, we will estimate S_f' based on the ultimate strength using equation 6.5c.

$$S_f' \cong 0.4 S_{ut} \qquad \text{for } S_{ut} < 48 \text{ ksi}$$
$$S_f' \cong 0.4(45\,000) = 18\,000 \text{ psi} \qquad\qquad (a)$$

This value is at $N = 5E8$ cycles. There is no knee in an aluminum *S-N* curve.

2 The loading is *pure torsion* so the load factor from equation 6.7a is

$$C_{load} = 1.0 \qquad\qquad (b)$$

because the applied torsional stress will be converted to an equivalent von Mises normal stress for comparison to the *S-N* strength.

3 The part size is greater than the test specimen and it is round, so the size factor can be estimated with equation 6.7b, noting that this relationship is based on steel data:

$$C_{size} = 0.869(d_{equiv})^{-0.097} = 0.869(1.5)^{-0.097} = 0.835 \qquad (c)$$

4 The surface factor is found from equation 6.7e using the data in Table 6-3 for the specified forged finish, again with the caveat that these relationships were developed for steels and may be less accurate for aluminum.

$$C_{surf} = A S_{ut}^{b} = 39.9(45)^{-0.995} = 0.904 \qquad (d)$$

5 The temperature factor is found from equation 6.7f:

$$C_{temp} = 1 \qquad (f)$$

6 The reliability factor is taken from Table 6-4 for the desired 99.0% and is

$$C_{reliab} = 0.814 \qquad (f)$$

7 The corrected fatigue strength S_f at N=5E8 can now be calculated from equation 6.6:

$$S_f = C_{load} C_{size} C_{surf} C_{temp} C_{reliab} S_{f'}$$
$$= 1.0(0.835)(0.904)(1.0)(0.814)(18\ 000) = 11\ 063 \text{ psi} \qquad (g)$$

8 To create the S-N diagram, we also need a number for the estimated strength S_m at 10^3 cycles based on equation 6.9. Note that the bending value is used for torsion.

$$S_m = 0.90 S_{ut} = 0.90(45\ 000) = 40\ 500 \text{ psi} \qquad (h)$$

9 The coefficient and exponent of the corrected S-N line and its equation are found using equations 6.10a through 6.10c. The value of z is taken from Table 6-5 for S_f at 5E8 cycles.

$$b = -\frac{1}{5.699}\log\left(\frac{S_m}{S_f}\right) = -\frac{1}{5.699}\log\left(\frac{40\ 500}{11\ 063}\right) = -0.0989$$

$$(i)$$

$$\log(a) = \log(S_m) - 3b = \log[40\ 500] - 3(-0.0989): \quad a = 80\ 193$$

10 The fatigue strength at the desired life of $N = 2E7$ cycles can now be found from the equation for the corrected S-N line:

$$S_n = aN^b = 80\ 193 N^{-0.0989} = 80\ 193(2e7)^{-0.0989} = 15\ 209 \text{ psi} \qquad (j)$$

S_n is larger than S_f because it is at a shorter life than the published fatigue strength.

11 Note the order of operations. We first found an uncorrected fatigue strength $S_{f'}$ at some "standard" cycle life ($N = 5E8$), then corrected it for the appropriate factors from equations 6.7. Only then did we create equation 6.10a for the S-N line so that it passes through the **corrected** S_f at $N = 5E8$. If we had created Eq. 6.10a using the uncorrected $S_{f'}$, solved **it** for the desired cycle life ($N = 2E7$), and **then** applied the correction factors, we would get a different and incorrect result. Because these are exponential functions, superposition does not hold.

6.7   NOTCHES AND STRESS CONCENTRATIONS

Notch is a generic term in this context and refers to any geometric contour that disrupts the "force flow" through the part as described in Section 4.15. A **notch** can be *a hole, a groove, a fillet, an abrupt change in cross section, or any disruption to the smooth contours of a part.* The notches of concern here are those that are deliberately introduced to obtain engineering features such as O-ring grooves, fillets on shaft steps, fastener holes, etc. It is assumed that the engineer will follow good design practice and keep the radii of these notches as large as possible and reduce stress concentrations as described in Section 4.15. Notches of extremely small radii are poor design practice and if present, should be treated as cracks and the principles of fracture mechanics used to predict failure (see Sections 5.3 and 6.5).

A notch creates a stress concentration that raises the stresses locally and may even cause local yielding. In the discussion of stress concentration in Chapters 4 and 5 where only static loads were being considered, the effects of stress concentrations were only of concern for brittle materials. It was assumed that ductile materials would yield at the local stress concentration and lower the stress to acceptable levels. With dynamic loads, the situation is different since ductile materials behave as if brittle in fatigue failures.

The geometric (theoretical) stress-concentration factors K_t for normal stress and K_{ts} for shear stress were defined and discussed in Section 4.15. (We will refer to them both here as K_t.) These factors give an indication of the degree of stress concentration at a notch having a particular contour and are used as a multiplier on the nominal stress present in the cross section containing the notch (see equation 4.31). Many of these *geometric or theoretical stress-concentration factors* have been determined for various loadings and part geometries and are published in various references.[30], [31] For dynamic loading, we need to modify the theoretical stress-concentration factor based on the notch sensitivity of the material to obtain a **fatigue stress-concentration factor**, K_f, which can be applied to the nominal dynamic stresses.

Notch Sensitivity

Materials have different sensitivity to stress concentrations, which is referred to as the **notch sensitivity** of the material. In general, the more ductile the material, the less notch-sensitive it is. Brittle materials are more notch-sensitive. Since ductility and brittleness in metals are roughly related to strength and hardness, low-strength, soft materials tend to be less notch sensitive than high-strength, hard ones. Notch sensitivity is also dependent on the notch radius (which is a measure of notch sharpness). As notch radii approach zero, the notch sensitivity of materials **decreases** and also approaches zero. This is quite serendipitous since you will recall from Section 4.15 that the theoretical stress-concentration factor K_t approaches infinity as the notch radius goes to zero. If not for the reduced notch sensitivity of materials at radii approaching zero (i.e., cracks), engineers would be at a loss to design parts able to withstand *any nominal stress level* when notches are present.

Neuber[32] made the first thorough study of notch effects and published an equation for the fatigue stress-concentration factor in 1937. Kuhn[33] later revised Neuber's equation and experimentally developed data for the Neuber constant (a material property) needed in his equation. Peterson[30] subsequently refined the approach and developed the concept of notch sensitivity q defined as

$$q = \frac{K_f - 1}{K_t - 1} \qquad (6.11a)$$

where K_t is the theoretical (static) stress-concentration factor for the particular geometry and K_f is the fatigue (dynamic) stress-concentration factor. The notch sensitivity q varies between 0 and 1. This equation can be rewritten to solve for K_f.

$$K_f = 1 + q(K_t - 1) \qquad (6.11b)$$

The procedure is to first determine the theoretical stress concentration K_t for the particular geometry and loading, then establish the appropriate notch sensitivity for the chosen material and use them in equation 6.11b to find the dynamic stress-concentration factor K_f. The nominal dynamic stress for any situation is then increased by the factor K_f in the same manner as was done for the static case:

$$\sigma = K_f \, \sigma_{nom}$$
$$\tau = K_{fs} \, \tau_{nom} \qquad (6.12)$$

Note in equation 6.11 that when $q = 0$, $K_f = 1$ and it does not increase the nominal stress in equation 6.12. When $q = 1$, $K_f = K_t$ and the entire effect of the geometric stress-concentration factor is felt in equation 6.12.

The notch sensitivity q can also be defined from the Kunn-Hardrath formula[33] in terms of Neuber's constant a and the notch radius r.

$$q = \frac{1}{1 + \dfrac{\sqrt{a}}{\sqrt{r}}} \qquad (6.13)$$

Note that the Neuber constant is defined as the square root of a, not as a, so it is directly substituted in equation 6.13, while the value of r must have its square root taken. A plot of the Neuber constant $\sqrt{a}$ for three types of materials is shown in Figure 6-35, and Tables 6-6 to 6-8 show data taken from the figure. These data are incorporated into *TKSolver list functions* to allow their direct use in computations. Note in Figure 6-35 that, for torsional loads on steel, the value of $\sqrt{a}$ should be read for S_{ut} 20 kpsi higher than that of the material. Figures 6-36, part 1 and 2 show sets of notch sensitivity curves for steels and aluminums (respectively) generated with equation 6.13 using the data in Figure 6-35. (See also the *TKSolver* file NTCHSENS, which generates the plots.) These curves are for notches whose depth is less than four times the root radius and should be used with caution for deeper notches.

Table 6-6

Neuber's Constant
for Steels

S_{ut} (ksi)	$\sqrt{a}$
50	0.130
55	0.118
60	0.108
70	0.093
80	0.080
90	0.070
100	0.062
110	0.055
120	0.049
130	0.044
140	0.039
160	0.031
180	0.024
200	0.018
220	0.013
240	0.009

The full value of the fatigue stress-concentration factor K_f only applies to the high end of the HCF regime ($N = 10^6$ - 10^9 cycles). Some authors [10], [30], [34] recommend applying a reduced portion of K_f, designated $K_{f'}$ to the alternating stress at $N = 10^3$ cycles. For high-strength or brittle materials $K_{f'}$ is nearly equal to K_f but for low-strength, ductile materials, $K_{f'}$ approaches 1. Others [35] recommend applying the full value of K_f even at 10^3 cycles. The latter approach is more conservative since data indicate that the effects of stress concentration are less pronounced at lower N. We will adopt the conservative approach and apply K_f uniformly across the HCF range since the uncertainties surrounding the estimates of fatigue strength and its collection of modifying factors encourage conservatism.

Table 6-7

Neuber's Constant
for Annealed Aluminum

S_{ut} (kpsi)	$\sqrt{a}$
10	0.500
15	0.341
20	0.264
25	0.217
30	0.180
35	0.152
40	0.126
45	0.111

EXAMPLE 6-3

Determining Fatigue Stress-Concentration Factors

Problem A rectangular, stepped bar similar to that shown in Figure 4-36 is to be loaded in bending. Determine the fatigue stress-concentration factor for the given dimensions.

Given Using the nomenclature in Figure 4-36, $D = 2$, $d = 1.8$, and $r = 0.25$. The material has $S_{ut} = 100$ kpsi.

Solution See the *TKSolver* file EX06-03.

1 The geometric stress-concentration factor K_t is found from the equation in Figure 4-36:

$$K_t = A\left(\frac{r}{d}\right)^b \qquad (a)$$

where A and b are given in the same figure as a function of the D/d ratio, which is $2 / 1.8 = 1.111$. For this ratio, $A = 1.014\,7$ and $b = -0.217\,9$ giving

$$K_t = 1.0147\left(\frac{0.25}{1.8}\right)^{-0.2179} = 1.56 \qquad (b)$$

2 The notch sensitivity q of the material can be found by using the Neuber factor $\sqrt{a}$ from Figure 6-35 and Tables 6-6 to 6-8 in combination with equation 6.13, or by reading q directly from Figure 6-36. We will do the former. The Neuber factor from Table 6-6 for $S_{ut} = 100$ kpsi is 0.062. Note that this is the square root of a:

$$q = \frac{1}{1+\dfrac{\sqrt{a}}{\sqrt{r}}} = \frac{1}{1+\dfrac{0.062}{\sqrt{0.25}}} = 0.89 \qquad (c)$$

3 The fatigue stress-concentration factor can now be found from equation 6.11b:

$$K_f = 1 + q(K_t - 1) = 1 + 0.89(1.56 - 1) = 1.50 \qquad (d)$$

Table 6-8

Neuber's Constant
for Hardened Aluminum

S_{ut} (kpsi)	$\sqrt{a}$
15	0.475
20	0.380
30	0.278
40	0.219
50	0.186
60	0.162
70	0.144
80	0.131
90	0.122

6

6

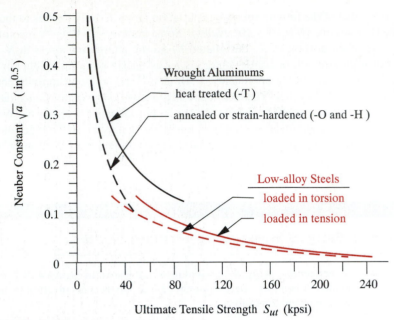

FIGURE 6-35

Neuber Constants for Steel and Aluminum. (From *ASME Paper* 843c, "The Prediction of Notch and Crack Strength under Static or Fatigue Loading," by P. Kuhn, April 1964)

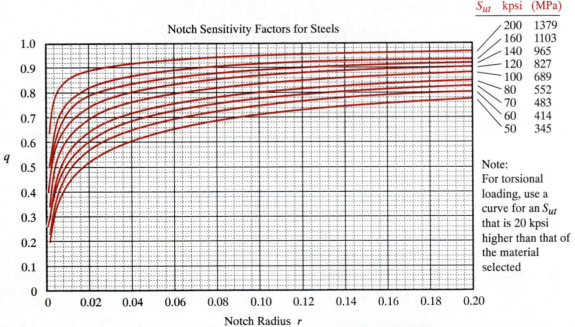

FIGURE 6-36   Part 1

Notch-Sensitivity Curves for Steels Calculated from Equation 6.13 using Data from Figure 6-35 as Originally Proposed by R. E. Peterson in "Notch Sensitivity," Chapter 13, in *Metal Fatigue* by G. Sines and J. Waisman, McGraw-Hill, New York, 1959.

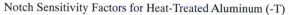

Notch Sensitivity Factors for Heat-Treated Aluminum (-T)

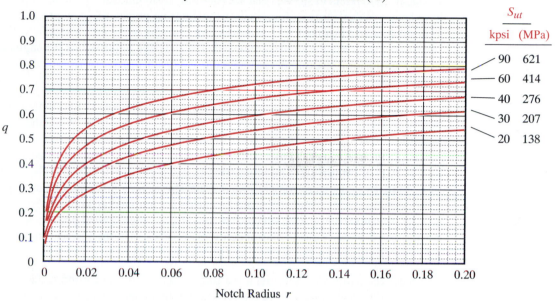

Notch Sensitivity Factors for Annealed and Strain-Hardened Aluminum (-O & -H)

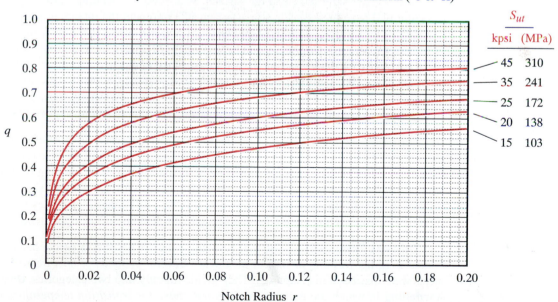

FIGURE 6-36 Part 2

Notch-Sensitivity Curves for Aluminums Calculated from Equation 6.13 using Data from Figure 6-35 as Originally Proposed by R. E. Peterson in "Notch Sensitivity," Chapter 13, in *Metal Fatigue* by G. Sines and J. Waisman, McGraw-Hill, New York, 1959.

6.8   RESIDUAL STRESSES

Residual stress refers to stresses that are "built-in" to an unloaded part. Most parts will have some residual stresses from their manufacturing processes. Any procedure such as forming or heat treatment that creates localized strains above the yield point will leave stresses behind when the strain is removed. Good design requires that the engineer try to tailor the residual stresses to, at a minimum, not create negative effects on the strength and preferably to create positive effects.

Fatigue failure is a tensile-stress phenomenon. Figures 6-17 and 6-18 show the beneficial effects of mean compressive stresses on fatigue strength. While the designer has little or no control over the presence or absence of a mean compressive stress in the loading pattern to which the part will be subjected, there are techniques that allow the introduction of **compressive residual stresses** in a part prior to its being placed in service. Properly done, these compressive residual stresses can make significant improvements in fatigue life. There are several methods for introducing compressive residual stresses, **thermal treatments**, **surface treatments,** and **mechanical prestressing treatments**, Most of them create biaxial compressive stresses at the surface, triaxial compressive stresses below the surface, and triaxial tensile stresses in the core.

Since the part is in equilibrium, the compressive stresses near the surface have to be balanced by tensile stresses in the core. If the treatment is overdone, the increased tensile core stresses can cause failure, so a balance must be struck. These treatments have the greatest value when the applied stress distribution due to loading is nonuniform and is maximally tensile at the surface, as in reversed bending. Bending in one direction will have peak tensile stress on one side only and the treatment can then be applied just to that side. Axial-tension loading is uniform across the section and so will not benefit from a nonuniform residual stress pattern unless there are notches at the surface to cause local increases in tensile stress. Then compressive residual stresses at the surface are very helpful. In fact, regardless of loading, the net tensile stresses at notches will be reduced by adding residual compressive stresses in those locations. Since designed notches are usually at the surface, a treatment can be applied to them.

The deliberate introduction of residual compressive stresses is most effective on parts made of high-yield-strength materials. If the yield strength of the material is low, then the residual stresses may not stay long in the part due to later yielding from high applied stresses in service. Faires [36] found that steels with $S_y < 80$ kpsi showed initial increases but little long-term improvement in fatigue strength. But, Heywood [37] reports 50% improvement in the fatigue strength of rolled threads in high-strength steel.

THERMAL TREATMENTS   Thermal stressing occurs whenever a part is heated and cooled as in hot-forming or in heat treatment. Several methods of heat treatment for steels were discussed in Chapter 2. They divide roughly into two categories, **through hardening** in which *the entire part is heated above the transition temperature then quenched*, and **case hardening** in which only *a relatively thin surface layer is heated above the transition temperature and quenched* or *the part is heated to a lower temperature in a special atmosphere that adds hardening elements to the surface.*

Through hardening causes *tensile residual stresses* in the surface. If the loading on the part creates high tensile stresses at the surface as in bending or torsion, or if notches at the surface of an axially loaded part cause high local tensile stress, then additional residual tensile stresses will worsen the situation. This makes through hardening a less desirable approach in these cases.

Case hardening by *carburizing, nitriding, flame-,* or *induction hardening* creates *compressive residual stresses* in the surface because the volume increase associated with the material's phase change (or element additions) is localized near the surface and the unchanged core pulls the case into compression. This surface compressive stress can have a significant beneficial effect on fatigue life. A particle of the material doesn't know or care whether the stress it feels is caused by some external force or by an internal, residual one. It feels a reduced net stress that is now the algebraic sum of the positive (applied-alternating) tensile stress and the negative (residual-mean) compressive stress. If a fatigue-loaded part is to be heat treated, case hardening offers distinct advantages over through hardening. Figure 6-37 shows the effects of nitriding and carburizing on the residual-stress state near the surface, and shows the distribution of compressive and tensile residual stresses across the thickness of a carburized part.

SURFACE TREATMENTS   The most common methods for introducing surface compressive stresses are **shot peening** and **cold forming**. Both involve a tensile yielding of the surface layer to some depth. Selective yielding of a portion of the material causes residual stresses of the opposite sign to be developed in that portion as the underlying, unstressed bulk of the material tries to force the yielded material back to its original size. The rule is, *to protect against later stresses in a particular direction, overstress the material (i.e., yield it) in the same direction as the applied stresses will.* Since we are attempting to protect against tensile stresses in fatigue loading, we want to yield the

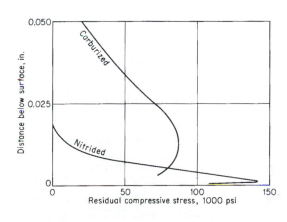

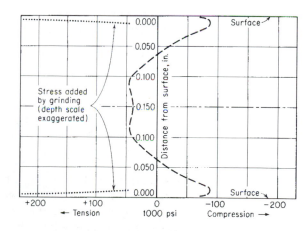

FIGURE 6-37

Distribution of Residual Stresses Due to Case Hardening (From Fig. 5.1, p. 51, and Fig. A-13, p. 202 in Almen and Black., 1963, *Residual Stresses and Fatigue in Metals*, McGraw-Hill, New York, with permission)

material in tension to develop compressive residual stresses. This technique of cold working the surface was known to ancient blacksmiths who hammered the surface of the sword or carriage-spring when cold as a final step to increase its strength.

Shot peening is relatively easy to do and can be applied to parts of almost any shape. The surface of the part is impacted with a stream of shot (like buckshot) made of steel, chilled cast iron, glass, walnut shells, or other material. The harder shot is used on steel parts and the softer shot on soft, nonferrous metals. The shot is fired at the part with high velocity either from a rotating wheel or air-blasted through a nozzle. The impacts of the shot dent the surface, yielding the material and creating a dimpled appearance. The surface is essentially stretched to a larger area and the underlying material pulls it back into a state of compressive residual stress. There can also be some cold working of the surface material which increases its hardness and yield strength.

Substantial levels of compressive stress can be achieved, up to about one-half the yield strength of the material. The depth of penetration of the compressive stress can be as much as 1 mm. It is difficult to accurately determine the level of residual stress in a shot-peened part as it must be destroyed to do so. If a slice is cut out to a depth below the peened layer, the cut will spring closed and the amount of closure is a measure of the residual stresses present. Figure 6-38 shows the distribution of residual stresses resulting from the shot peening of two steels of different yield strengths. The peak compressive stress occurs just below the surface and it decays rapidly with depth.

The degree of shot peening can be measured during the treatment by including a standard *Almen* test strip in the shot blast. The thin test strip is held in a fixture so that only one of its sides is peened. When removed from the fixture, the strip curls up because of the compressive stresses on one side. The height of its curve is converted to an Almen number that indicates the degree of peening that the part (and it) received. If no specific data are available for the level of residual stress present after shot peen-

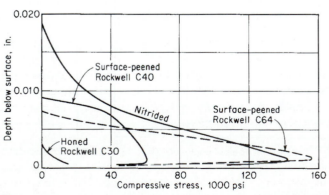

FIGURE 6-38

Distribution of Residual Stresses Due to Shot Peening (From Fig. 5.11, p 58, in Almen and Black., 1963, *Residual Stresses and Fatigue in Metals*, McGraw-Hill, N.Y., with permission)

ing, a conservative way to account for its benefit is to set the surface factor $C_{surf} = 1$ in calculating the corrected fatigue strength with equation 6.8.

Shot peening is widely used on parts such as chain-saw blades , crankshafts, connecting rods, gears, and springs.[38] On very large parts, **hammer peening** is sometimes done in which a hand-held air hammer is used to impact highly stressed portions of the surface (such as the roots of gear teeth) with a hardened ball. Parts of high-strength steel benefit most from peening. It is particularly beneficial on forgings and hot-rolled surfaces which are both rough and weak from decarburization. Chrome- and nickel-plated parts can be restored to their unplated levels of fatigue strength by peening. If the peening is done after plating, not only is the negative effect of the plating undone but the result is a higher fatigue strength than the original unplated part as shown in Figure 6-29. Properly peened helical coil springs can have their fatigue strengths increased to the point that they will fail by yielding before they will fail in fatigue.[38] So, shot peening is obviously a useful technique for improving the fatigue life of highly stressed parts and it does not add excessively to production costs.

Cold forming can be done to surfaces of revolution such as shafts, to flat surfaces that can be passed between rollers, and to the insides of holes. For example, a hardened roller can be impressed against a shaft as it is turned in a lathe. The high forces cause local yielding under the roller that results in compressive residual stresses at the surface which will protect it from the tensile effects of rotating-bending or reversed-torsional loading in service. Cold forming is particularly useful at fillets, grooves, or other stress-raisers.

Holes and bores can be cold-formed by forcing a mandrel of slightly larger diameter through the hole to expand the inside diameter by yielding and create compressive residual stresses. This is sometimes done to gun barrels (cannon) in a process called **autofrettage**. Autofrettage is also done by filling the gun barrel with a steel mandrel that leaves a small annulus (doughnut) of space, sealing the ends, filling the annulus with gasoline[*], and pressurizing it to over 200 000 psi. The hydrostatic pressure yields the inner surface in tension creating residual compressive stresses that protect it against fatigue failure from the cyclic tension stresses experienced when the cannon is fired.

The ends of holes in any part can be **coined** by yielding their edges with a conical tool to put compressive stresses around their stress concentration region at the surface. Reducing the dimensions of flat stock by cold rolling introduces residual compressive stress in the surface and tension in the core. Excessive rolling can cause tensile cracking by exceeding the static tensile strength at the center. The material can be annealed between successive rollings to prevent this.

MECHANICAL PRESTRESSING   For parts that are dynamically loaded in service in only one direction, such as support springs for vehicles, prestressing is a useful way to create residual stresses. Prestressing refers to the deliberate overloading of the part in the same direction as its service loading, prior to its being placed in service. The yielding that occurs during prestressing creates beneficial residual stresses.

* The ability of some liquids to transmit pressure rapidly is limited by their increase in viscosity at high pressures. Unleaded gasoline and some other liquids will withstand pressures to about 200 000 psi without serious degradation in their pressure transmissibility. Some fluids become solid at about 100 000 psi. Water will form ice-VI at about 155 000 psi and plug the tube or annulus.
Source: D. H. Newhall, Harwood Engineering Inc, Walpole, Mass., personal communication, 1994.

COLD PROCESSING OPERATIONS AND RESULTING STATE OF STRESS

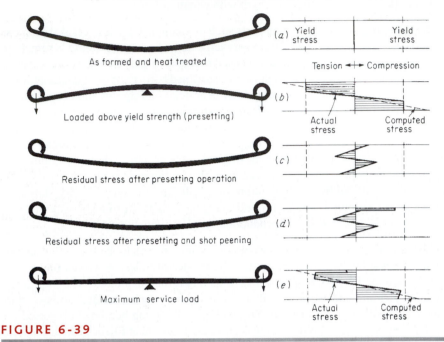

FIGURE 6-39

Residual Stresses from Prestressing and Shot Peening a Leaf Spring (From Fig. 6.2, p. 61, in Almen and Black, 1963, *Residual Stresses and Fatigue in Metals*, McGraw-Hill, New York, with permission)

Figure 6-39 shows an example of prestressing applied to a truck spring. The spring is initially formed with more contour than that needed at assembly. It is then placed in a fixture that loads it exactly as it will be loaded in service but at a level above its (tensile) yield strength to preset it. When the load is released, it springs back to a new shape, which is that desired for assembly. But, the elastic recovery has now placed the material that yielded into a residual-stress state, which will be in the opposite (compressive) direction than that of the applied load. Therefore this residual stress will act to protect the part against its tensile service loads. The residual stress patterns are shown in the figure and also indicate the result of shot peening the upper surface after presetting. The two treatments are additive on the upper surface in this case affording greater protection against fluctuating tensile stresses in service. Note that if the part were reverse-loaded in service to the point of yielding the upper surface in compression, it would relieve the beneficial compressive stress and compromise the part's life. Thus, this approach is most useful for parts whose service stresses are unidirectional.

SUMMARY   Residual stresses can be the "fatigue-designer's best friend." Properly configured, beneficial residual stresses can make an otherwise unworkable design safe. The designer should become thoroughly familiar with the means available for their creation. This brief description is intended to serve only as an introduction to a complicated topic and the reader is urged to consult the literature on residual stresses, some of which is noted in the bibliography of this chapter.

If quantitative data on the levels of residual stresses developed in a particular part can be obtained (usually by destructive testing) then these data can be used in the determination of safe applied-stress levels. Absent such quantitative information the designer is limited to considering the use of these treatments as providing some additional safety factor that is poorly quantified but is in the right direction.

6.9   DESIGNING FOR HIGH-CYCLE FATIGUE

We are now ready to consider how to apply all the information presented on fatigue failure *in order to avoid it* in the design of dynamically loaded parts. There are four basic categories that can be treated separately although three of them are just special cases of the fourth, general case. As we will see, the same general approach to the solution of all four categories is both possible and desirable. However, it may aid in understanding their solutions if we deal with them separately before presenting the general solution method.

Figure 6-40 shows the four categories in a matrix. The columns define the presence or absence of a mean stress. The fully reversed case has a zero mean stress and the fluctuating stress case has a nonzero mean value. Both have alternating components. The rows define the presence of applied-stress components on only one, or more than one axis. The uniaxial case represents simple loading cases such as pure axial loading, or pure bending. The multiaxial case is general and allows applied normal stress components on all axes in combination with applied shear stresses on any face of the stress cube. In reality, the pure loading cases are rare in practice. More often there will be some combination of multiaxial stresses on machine parts. Both the fully reversed and fluctuating stress cases are commonly encountered in practice, however.

We will first consider the simplest category (I), fully reversed uniaxial stresses. Many texts will further subdivide this category into bending loading, axial loading, and

	Fully reversed stresses ($\sigma_m = 0$)	Fluctuating stresses ($\sigma_m \neq 0$)
Uniaxial Stresses	Category I	Category II
Multiaxial Stresses	Category III	Category IV

FIGURE 6-40

Four Categories of Fatigue Design Situations

torsional loading and present separate approaches to each. We will combine them all into one category by calculating the von Mises effective stress and comparing it with the corrected bending-fatigue strength of the chosen material. This eliminates the need to consider pure torsion as a special case.

We will next consider fluctuating uniaxial stresses (Category II). This adds the complication of mean stresses and will employ the modified-Goodman diagram in addition to the (simpler) *S-N* diagram. We will use the von Mises effective stress to convert pure torsional loading into an equivalent form of tensile stress.

Finally we will investigate the general categories of multiaxial stresses in both fully reversed (III) and fluctuating (IV) load cases and present a recommended "universal approach" that will work in all categories for most common loading situations. The hope is that this approach will simplify a complicated topic and provide the student with one method that can be used for HCF design in a majority of situations.

6.10   DESIGNING FOR FULLY REVERSED UNIAXIAL STRESSES

The simplest example of fatigue loading is that of category I, fully reversed uniaxial stress with a mean stress of zero (see Figure 6-6a). Some common applications of this category are rotating bending of a shaft that supports a static load, or reversed torque on a shaft with large, oscillating inertia loads and a mean torque that is effectively zero compared to those oscillations. The process can be described in a set of general steps:

Design Steps for Fully Reversed Stresses with Uniaxial Loading:

1   Determine the number of cycles of loading N that the part will experience over its expected service life.

2   Determine the amplitude of the applied alternating loads from zero to peak. (Eq. 6.1) Note that a static load on a rotating shaft causes alternating stresses.

3   Create a tentative part-geometry design to withstand the applied loads based on good engineering practice. (See Chapters 3 and 4.)

4   Determine any appropriate geometric stress-concentration factors K_t (or K_{ts} for shear) at notches in your part's geometry. Try, of course, to minimize these through good design. (See Section 4.15.)

5   Choose a tentative material for the part and determine its properties of interest such as S_{ut}, S_y, S_e, (or S_f, at the life required), and q, from your own test data, from the literature, or from estimates as described in this chapter.

6   Convert the geometric stress-concentration factors K_t (or K_{ts} for shear) to fatigue concentration factors K_f using the material's notch sensitivity, q.

7   Calculate the nominal, alternating stress amplitudes σ_a (or τ_a if the load is pure shear) at critical locations in the part due to the alternating service loads based on

standard stress-analysis techniques (Chapter 4) and increase them as necessary with the appropriate fatigue stress-concentration factors (Sections 4.14 and 6.6).

8   Calculate the principal-stress amplitudes for the critical locations based on their states of applied stress (Chapter 4). Note that these contain the effects of stress concentrations. Calculate the von Mises effective stress for each location of interest.

9   Determine appropriate fatigue strength modification factors for the type of loading, size of part, surface, etc., as described in Section 6.6. Note that the loading factor C_{load} will differ based on whether there are axial or bending loads (see Eq. 6.7a). If the loading is pure torsion, then the von Mises effective stress calculation will convert it to a pseudo-tensile stress and C_{load} should then be set to 1.

10 Define the corrected fatigue strength S_f at the requisite cycle life N (or the corrected endurance limit S_e for infinite life if appropriate) and a "static" strength S_m @ $N = 10^3$ cycles from Eq. 6.9. Create an S-N diagram as shown in Figure 6-33, and/or write equation 6.10 for this tentative material choice.

11 Compare the alternating von Mises effective stress at the most highly stressed location with the material's corrected fatigue strength S_n taken from the S-N curve at the desired number of life cycles N. (Note that for infinite life situations in which the material has an S-N knee, $S_n = S_e$)

12 Calculate a safety factor for the design from the relationship,

$$N_f = \frac{S_n}{\sigma'} \tag{6.14}$$

where N_f is the safety factor in fatigue, S_n is the corrected fatigue strength at the required number of cycles of life taken from the S-N curve or Eq. 6.10, and σ' is the largest von Mises alternating stress at any location in the part, calculated to include all stress concentration effects.

13 Given the fact that the material was only tentatively chosen and that the design may not yet be as refined as possible, the result of the first pass through these steps will most likely be a failed design whose safety factor is either too large or too small. Iteration will be required (as it always is) to refine the design. Any subset of steps can be repeated as many times as necessary to obtain an acceptable design. The most common tactic is to return to step 3 and improve the geometry of the part to reduce stresses and stress concentrations and/or revisit step 4 to choose a more suitable material. Sometimes it will be possible to return to step 1 and redefine a shorter acceptable part life.

The design loads in step 2 may or may not be in the control of the designer. Usually they are not, unless the loading on the part is due to inertial forces; then increasing its mass to "add strength" will worsen the situation as this will proportionately increase the loads (see Section 3.6). Instead, the designer may want to lighten the part without excessively compromising its strength to reduce the forces. Whatever the particular circumstances, the designer must expect to cycle through these steps several times before converging on a usable solution. Equation solvers that allow rapid recalculation of the equations are a great help in this situation.

The best way to demonstrate the use of these steps for fatigue design is with an example.

EXAMPLE 6-4

Design of a Cantilever Bracket for Fully Reversed Bending

Problem   A feed-roll assembly is to be mounted at each end on support brackets cantilevered from the machine frame as shown in Figure 6-41. The feed rolls experience a fully reversed load of 1 000-lb amplitude, split equally between the two support brackets. Design a cantilever bracket to support a fully reversed bending load of 500-lb amplitude for 10^9 cycles with no failure. Its dynamic deflection cannot exceed 0.01 in.

Given   The load-time function shape is shown in Figure 6-41a. The operating environment is room air at a maximum temperature of 120°F. The space available allows a maximum cantilever length of 6 in. Only ten of these parts are required.

Assumptions   The bracket can be clamped between essentially rigid plates or bolted at its root. The normal load will be applied at the effective tip of the cantilever beam from a rod attached through a small hole in the beam. Since the bending moment is effectively zero at the beam tip, the stress concentration from this hole can be ignored. Given the small quantity required, machining of stock mill-shapes is the preferred manufacturing method.

Solution   See Figure 6-41, Tables 6-9 and 6-10, and *TKSolver* file EX06-04.

1 This is a typical design problem. Very little data are given except for the required performance of the device, some limitations on size, and the required cycle life. We will have to make some basic assumptions about part geometry, materials, and other factors as we go. Some iteration should be expected.

2 The first two steps of the process suggested above, finding the load amplitude and the number of cycles, are defined in the problem statement. We will begin at the third step, creating a tentative part-geometry design.

3 Figure 6-41a shows a tentative design configuration. A rectangular cross section is chosen to provide ease of mounting and clamping. A piece of cold-rolled bar stock from the mill could simply be cut to length and drilled to provide the needed holes, then clamped into the frame structure. This approach appears attractive in its simplicity because very little machining is required. The mill-finish on the sides could be adequate for this application. This design has some disadvantages however. The mill tolerances on the thickness are not tight enough to give the required accuracy on thickness, so the top and bottom would have to be machined or ground flat to dimension. Also, the sharp corners at the frame where it is clamped provide stress concentrations of about $K_t = 2$ and also create a condition called **fretting fatigue** due to the slight motions that will occur between the two parts as the bracket deflects. This motion continuously breaks down the protective oxide coating, exposing new metal to oxidation and speeding up the fatigue-failure process. The fretting could be a problem even if the edges of the frame pieces were radiused.

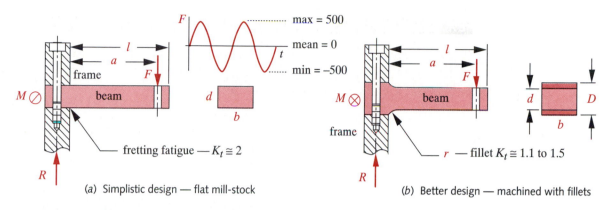

FIGURE 6-41

Design of a Cantilever Bracket for Fully Reversed Bending Loading

4 Figure 6-41b shows a better design in which the mill stock is purchased thicker than the desired final dimension and machined top and bottom to dimension D, then machined to thickness d over the length l. A fillet radius r is provided at the clamp point to reduce fretting fatigue and achieve a lower K_t. (See Figure 4-33.) Figure 4-31 shows that with suitable control of the r/d and D/d ratios for a stepped flat bar in bending, the geometric stress-concentration factor K_t can be kept under about 1.5.

5 Some trial dimensions must be assumed for b, d, D, r, a, and l. We will assume (guess) values of $b = 1$ in, $d = 0.75$ in, $D = 0.94$ in, $r = 0.25$, $a = 5.0$, and $l = 6.0$ in to the applied load for our first calculation. This length will leave some material around the hole and still fit within the 6-in-length constraint. These dimensions were put into the *TKSolver* model EX06-04 shown in Tables 6-9 and 6-10. This model incorporates all the relevant equations and performs steps 3-13 of the process suggested above for these problems. An equation solver makes it a simple task to adjust dimensions and recalculate the results once the equations are defined.

6 A material must also be chosen. For infinite life, low cost, and ease of fabrication, it is desirable to use a low-carbon alloy steel if possible and if environmental conditions permit. Since this is used in a controlled, indoor environment, carbon steel is acceptable on the latter point. The fact that the deflection is of concern is also a good reason to choose a material with a large E. Low-carbon, ductile steels have the requisite endurance-limit knee for the infinite life required in this case and also have low notch sensitivities. A low-carbon steel with $S_{ut} = 80\,000$ psi is selected for the first trial.

7 Table 6-9 shows the rule sheet for the *TKSolver* model. First the reaction force and reaction moment at the support are found using equations h from Example 4-5. Next the area moment of inertia of the cross section, the distance to the outer fiber, and the nominal alternating bending stress at the root are found using the alternating load's 500-lb amplitude (lines 2-4).

$$R = F = 500 \text{ lb} \tag{a}$$

$$M = Rl - F(l - a) = 500(6) - 500(6 - 5) = 2\,500 \text{ lb - in}$$

$$I = \frac{bd^3}{12} = \frac{1(0.75)^3}{12} = 0.035\,2 \text{ in}^4 \tag{b}$$

$$c = \frac{d}{2} = \frac{0.75}{2} = 0.375 \text{ in}$$

$$\sigma_{a_{nom}} = \frac{Mc}{I} = \frac{2\,500(0.375)}{0.035\,2} = 26\,667 \text{ psi} \tag{c}$$

8   At line 5, two ratios are calculated for use in Figure 4-36 in order to find the geometric stress-concentration factor for the assumed part dimensions. This is accomplished at line 6 by calling the list functions *coeff* and *exponent,* that were created in *TKSolver* to solve the equations in Figure 4-36 and which do a cubic interpolation between the listed values. This computation gives a value for K_t without the need for any reading of curves in Figure 4-36.

$$\frac{D}{d} = \frac{0.938}{0.75} = 1.25 \qquad\qquad \frac{r}{d} = \frac{0.25}{0.75} = 0.333 \tag{d}$$

from Figure 4 - 36 or Figure E - 10 : $A = 0.965\,8$ $\qquad$ $b = -0.266$ $\qquad$ (e)

$$K_t = A\left(\frac{r}{d}\right)^b = 0.965\,8(0.333)^{-0.266} = 1.29 \tag{f}$$

9   At line 7, the procedure *notch* is called to calculate the notch sensitivity q of the chosen material based on its ultimate strength and notch radius using equation 6.13 and the data for Neuber's constant in Table 6-6. The Neuber curves' data have been converted to cubic-interpolated *list functions* that are called from the procedure *notch,* which returns q for any combination of S_{ut} and notch radius r.

from Table 6 - 6 for S_{ut} = 80 kpsi : $\qquad$ $\sqrt{a} = 0.80$ $\qquad\qquad$ (g)

$$q = \frac{1}{1 + \dfrac{\sqrt{a}}{\sqrt{r}}} = \frac{1}{1 + \dfrac{0.80}{\sqrt{0.25}}} = 0.862 \tag{h}$$

10   The values of q and K_t are used at line 8 to find the fatigue stress-concentration factor K_f, which is in turn used to find the local alternating stress σ_a in the notch at line 9. Because we have the simplest case of a uniaxial tensile stress, the largest alternating principal stress σ_{1a} for this case is equal to the alternating tensile stress, as is the von Mises alternating stress σ'_a. See equations 4.6 and 5.7c.

$$K_f = 1 + q(K_t - 1) = 1 + 0.862(1.29 - 1) = 1.25 \tag{i}$$

$$\sigma_a = K_f \sigma_{a_{nom}} = 1.25(26\,667) = 33\,343 \text{ psi} \tag{j}$$

$$\tau_{ab} = \pm\sqrt{\left(\frac{\sigma_x - \sigma_y}{2}\right)^2 + \tau_{xy}^2} = \sqrt{\left(\frac{33\,343 - 0}{2}\right)^2 + 0} = 16\,672 \text{ psi}$$

$$\sigma_a, \sigma_b = \frac{\sigma_x + \sigma_y}{2} \pm \sqrt{\left(\frac{\sigma_x - \sigma_y}{2}\right)^2 + \tau_{ab}^2} = 33\,343 \text{ psi, 0 psi} \qquad (k)$$

$$\sigma' = \sqrt{\sigma_1^2 - \sigma_1\sigma_2 + \sigma_2^2} = \sqrt{33\,343^2 - 33\,343(0) + 0} = 33\,343 \text{ psi}$$

11 At line 10, the uncorrected endurance limit $S_{e'}$ is found from equation 6.6a. At line 11, the size factor for this rectangular part is determined by calculating the cross-sectional area stressed above 95% of its maximum stress (see Figure 6-25) and using that value in equation 6.7d to find an equivalent diameter test specimen for use in equation 6.7b to find C_{size} at line 12.

$$S_{e'} = 0.5 S_{ut} = 0.5(80\,000) = 40\,000 \text{ psi} \qquad (l)$$

$$A_{95} = 0.05db = 0.05(0.75)(1) = 0.04 \text{ in}^2$$

$$d_{equiv} = \sqrt{\frac{A_{95}}{0.0766}} = 0.700 \text{ in} \qquad (m)$$

$$C_{size} = 0.869\left(d_{equiv}\right)^{-0.097} = 0.900$$

12 The corrected endurance limit S_e is calculated at line 13. C_{load} is found from equation 6.7a and input manually as data on the variable sheet. C_{surf} for a machined finish is read from Figure 6-26 and also input as data. C_{temp} is found from equation 6.7e and C_{reliab} is chosen from Table 6-3 for a 99.99% reliability level and both are input as data. Note that the corrected S_e is only about 25% of S_{ut}.

$$S_e = C_{load}\, C_{size}\, C_{surf}\, C_{temp}\, C_{reliab}\, S_{e'}$$
$$= 1(0.900)(0.78)(1)(0.753)40\,000 = 21\,136 \text{ psi} \qquad (n)$$

13 The safety factor is calculated at line 14 using equation 6.14 and the beam deflection y is computed at line 15 using equation i from Example 4-5.

$$N_f = \frac{S_n}{\sigma'} = \frac{21\,136}{33\,343} = 0.63$$

$$y_{@x=l} = \frac{F}{6EI}\left[x^3 - 3ax^2 - \langle x - a\rangle^3\right] \qquad (o)$$
$$= \frac{500}{6(3E7)(0.0352)}\left[6^3 - 3(5)(6)^2 - (6-5)^3\right] = -0.026 \text{ in}$$

14 The results of all these computations for the first assumed design are seen in part 1 of Table 6-10. The deflection is not within the stated specification, and the design fails with a safety factor of less than one. So, more iterations are needed as was expected. Any of the dimensions can be changed, as can the material. The material was left unchanged but the beam cross-sectional dimensions and the notch radius were

Table 6-9   Design of a Cantilever Bracket for Fully Reversed Bending
TKSolver Rule Sheet (Partial) for Example 6-4 from file EX6-04

; 1- find the alternating reaction force and moment at support

$$R = F$$
$$M = R * l - F * (l - a)$$

; 2- find area moment of inertia

$$I = b * d^3 / 12$$

; 3- find distance to outer fiber

$$c = d / 2$$

; 4- find the nominal alternating bending stress at root

$$signom = M * c / I$$

; 5- create the ratios for Figure 4-36

$$Doverd = D / d$$
$$roverd = r / d$$

; 6- call list functions for Fig. 4-36 and interpolate coefficient and exponent to calculate Kt

$$Kt = coeff(Doverd) * roverd \wedge exponent(Doverd)$$

; 7- create Peterson's notch sensitivity factor

$$CALL\ notch\ (Sut,\ r\ ;\ q)$$

; 8- find fatigue stress-concentration factor from eq 6.11*b*

$$Kf = 1 + q * (Kt - 1)$$

; 9- calculate alternating stress including stress concentration

$$sigax = Kf * signom$$
$$CALL\ Stress2D\ (sigax,\ 0,\ 0\ ;\ sig1,\ sig2,\ sig3)$$; *Equation 4.6 for principal stress*
$$CALL\ Distort\ (sig1,\ sig2,\ sig3\ ;\ sigavm)$$; *Equation 6.18 for von Mises alternating stress*

; 10- estimate the endurance limit based on *Sut*

$$Seprime = Sut / 2$$

; 11- find the 95% area and the equivalent test-bar dia.

$$A95 = .05 * d * b$$
$$dequiv = SQRT(A95/.0766)$$

; 12- calculate the size factor from equation 6.7*b*

$$Csize = .869 * dequiv \wedge -.097$$

; 13- modify the endurance limit

$$Se = Cload * Csize * Csurf * Ctemp * Creliab * Seprime$$

; 14- compute the safety factor for infinite life

$$SF = Se / sigprime$$; *Equation 6.14*

; 15- calculate the deflection at point *lx* (see Eq. (*i*) in Example 4-5)

$$y = (\ -1/6 * CUBIC\ (lx,\ a,\ F) + 1/6 * R * lx^3 - .5 * M * lx^2\ ) / (E*I)$$

increased and the model rerun (this took only a few minutes) until the results shown in part 2 of Table 6-10 were achieved.

15   The final dimensions are $b = 2$ in, $d = 1$ in, $D = 1.125$ in, $r = 0.5$, $a = 5.0$, and $l = 6.0$ in. The safety factor is now 2.5 and the deflection is 0.005 in. These are both

Table 6-10 **Example 6-4 - Design of a Cantilever Bracket for Reversed Bending**

Part 1 of 2   First Iteration: An Unsuccessful Design – *TKSolver* Variable Sheet from EX06-04A

Input	Variable	Output	Unit	Comments
500	F		lb	applied load amplitude at point a
1	b		in	beam width
0.75	d		in	beam depth over length
0.94	D		in	beam depth in wall
0.25	r		in	fillet radius
6	l		in	beam length
5	a		in	distance to load F
6	lx		in	distance for deflection calculation
3E7	E		psi	modulus of elasticity
80 000	Sut		psi	ultimate tensile strength
1	$Cload$			load factor for bending
	$Csurf$	0.85		machined finish
1	$Ctemp$			room temperature
0.753	$Creliab$			99.9% reliability factor
	R	500	lb	reaction force at support
	M	2 500	in-lb	reaction moment at support
	I	0.035 2	in^4	area moment of inertia
	c	0.38	in	dist to outer fiber
	signom	26 667	psi	bending stress at root
	Doverd	1.25		bar width ratio $1.01 < D/d < 2$
	roverd	0.33		ratio radius to small dim
	Kt	1.29		geometric stress-concentration factor
	q	0.86		Peterson's notch sensitivity factor
	Kf	1.25		fatigue stress-concentration factor
	sigx	33 343	psi	concentrated stress at root
	sig1	33 343	psi	largest principal alternating stress
	sigvm	33 343	psi	von Mises alternating stress
	Seprime	40 000	psi	uncorrected endurance limit
	A95	0.04	in^2	95% stress area
	dequiv	0.7	in	equivalent diameter test specimen
	Csize	0.9		size factor based on 95% area
	Se	22 907	psi	corrected endurance limit
	N_{sf}	0.69		predicted safety factor
	y	−0.026	in	deflection at end of beam

Table 6-10   **Example 6-4 – Design of a Cantilever Bracket for Reversed Bending**

Part 2 of 2   Final Iteration: A Successful Design – *TKSolver* Variable Sheet from File EX06-04B

Input	Variable	Output	Unit	Comments
500	F		lb	applied load amplitude at point *a*
2	b		in	beam width
1	d		in	beam depth over length
1.125	D		in	beam depth in wall
0.5	r		in	fillet radius
6	l		in	beam length
5	a		in	distance to load *F*
6	lx		in	distance for deflection calculation
3E7	E		psi	modulus of elasticity
80 000	Sut		psi	ultimate tensile strength
1	Cload			load factor for bending
	Csurf	0.85		machined finish
1	Ctemp			room temperature
0.753	Creliab			99.9% reliability factor
	R	500	lb	reaction force at support
	M	2 500	in-lb	reaction moment at support
	I	0.166 7	in^4	area moment of inertia
	c	0.5	in	dist to outer fiber
	signom	7 500	psi	bending stress at root
	Doverd	1.13		bar width ratio $1.01 < D/d < 2$
	roverd	0.50		ratio radius to small dim
	Kt	1.18		geometric stress-concentration factor
	q	0.90		Peterson's notch sensitivity factor
	Kf	1.16		fatigue stress-concentration factor
	sigx	8 688	psi	concentrated stress at root
	sig1	8 688	psi	largest principal alternating stress
	sigvm	8 688	psi	von Mises alternating stress
	Seprime	40 000	psi	uncorrected endurance limit
	A95	0.10	in^2	95% stress area
	dequiv	1.14	in	equivalent diameter test specimen
	Csize	0.86		size factor based on 95% area
	Se	21 843	psi	corrected endurance limit
	N_{sf}	2.5		predicted safety factor
	y	−0.005	in	deflection at end of beam

satisfactory. Note how low the stress-concentration factor is at 1.16. The dimension D was deliberately chosen to be slightly less than a stock mill size so that material would be available for the cleanup and truing of the mounting surfaces. Also, with this design, hot-rolled steel (HRS) could be used, rather than the cold-rolled steel (CRS) initially assumed (Figure 6-41a). Hot-rolled steel is less expensive than CRS and, if normalized, has less residual stress, but its rough, decarburized surface needs to be removed by machining all over, or be treated with shot peening to strengthen it.

16 See the provided *TKSolver* file EX06-04 for more detailed information. The procedure and list functions used in lines 6 and 7 can be examined within the file as well. Only portions of its rule and variable sheets are reproduced in Tables 6-9 and 6-10 due to space limitations.

The above example should demonstrate that designing for fully reversed HCF loading is straightforward once the principles are understood. If the design called for fully reversed-torsional, -rotating-bending or -axial loading, the design procedure would be the same as in this example. The only differences would be in the choices of stress equations and strength modification factors as described in the previous sections. Note that calculation of the principal and von Mises stresses is somewhat redundant in this simple example since they are both identical to the applied stress. However it is done for the sake of consistency since these stresses will not be identical in more complicated applied stress situations. The value of using a computer and equation solver in this or any design problem cannot be overstated as it allows rapid iteration from initial guesses to final dimensions with minimum effort.

6.11 DESIGNING FOR FLUCTUATING UNIAXIAL STRESSES

Repeating or fluctuating stresses as shown in Figure 6-6b and 6-6c, have nonzero mean components and these must be taken into account when determining the safety factor. Figures 6-16, 6-17, 6-18, and 6-21 all show experimental evidence of the effect of mean-stress components on failure when present in combination with alternating stresses. This situation is quite common in machinery of all types.

Figure 6-42 shows the **modified-Goodman line**, **Gerber parabola**, **Soderberg line** ,and the **yield line** plotted on σ_m-σ_a axes. The Gerber parabola best fits the experimental failure data and the modified-Goodman line fits beneath the scatter in the data as shown in Figure 6-16, which superimposes these lines on the experimental failure points. Both of these lines intersect the corrected endurance limit S_e or fatigue strength S_f on the σ_a axis with S_{ut} on the σ_m axis. A yield line connecting S_y on both axes is also shown to serve as a limit on the first cycle of stress. (If the part yields, it has failed, regardless of its safety in fatigue.) The Soderberg line connects S_e or S_f to the yield strength S_y and is thus a more conservative failure criterion but does eliminate the need to invoke the yield line. It also eliminates otherwise safe σ_m-σ_a combinations as can be seen in Figure 6-16. Whichever lines are chosen to represent failure, safe combinations of σ_m and σ_a lie to the left and below their envelope. These failure lines are defined by

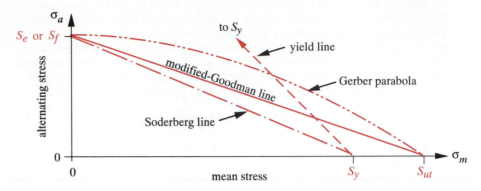

FIGURE 6-42

Various Failure Lines for Fluctuating Stresses

$$\text{Gerber parabola:} \qquad \sigma_a = S_e\left(1 - \frac{\sigma_m^2}{S_{ut}^2}\right) \qquad (6.15a)$$

$$\text{Modified - Goodman line:} \qquad \sigma_a = S_e\left(1 - \frac{\sigma_m}{S_{ut}}\right) \qquad (6.15b)$$

$$\text{Soderberg line:} \qquad \sigma_a = S_e\left(1 - \frac{\sigma_m}{S_y}\right) \qquad (6.15c)$$

While the Gerber line is a good fit to experimental data, making it useful for the analysis of failed parts, the modified-Goodman line is a more conservative and commonly used failure criterion when designing parts subjected to mean plus alternating stresses. The Soderberg line is less often used as it is overly conservative. We will now explore the use of the modified-Goodman line in more detail.

Creating the Modified-Goodman Diagram

Figure 6-43a shows a schematic plot of the three-dimensional surface formed by the alternating-stress component σ_a, the mean-stress component σ_m and the number of cycles N for a material possessing an endurance-limit knee at 10^6 cycles. If we look in at the σ_a-N plane as shown in Figure 6-43b, we see projections of lines on the surface that are S-N diagrams for various levels of mean stress. When $\sigma_m = 0$, the S-N diagram is the topmost line, connecting S_{ut} to S_e, as also shown in Figures 6-2 and 6-8. As σ_m increases, the σ_a intercept at $N = 1$ cycle reduces, becoming zero when $\sigma_m = S_{ut}$.

Figure 6-43c shows projections on the σ_a–σ_m plane for various values of N. This is called a **constant-life diagram** as each line on it shows the relationship between mean and alternating stress at a particular cycle life. When $N = 1$, the plot is a 45° line connecting S_{ut} on both axes. This is a static-failure line. The σ_a-intercept decreases as N

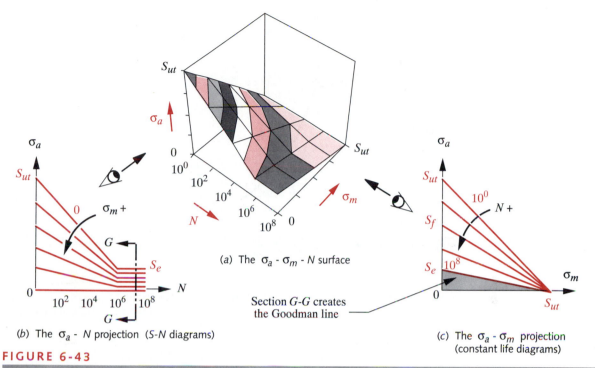

(a) The σ_a - σ_m - N surface

Section G-G creates
the Goodman line

(b) The σ_a - N projection (S-N diagrams)

(c) The σ_a - σ_m projection
(constant life diagrams)

FIGURE 6-43

Effect of a Combination of Mean and Alternating Stresses

increases, becoming equal to the endurance limit S_e beyond about 10^6 cycles. The line connecting S_e on the σ_a axis and S_{ut} on the σ_m axis in Figure 6-43c is the **modified-Goodman line**, taken at section G-G as shown in Figure 6-43a.

Figure 6-44 shows a plot of alternating stress σ_a versus mean stress σ_m, which we refer to as an "augmented" **modified-Goodman diagram**.[*] This is an embellishment of the modified-Goodman line shown in Figures 6-16 and 6-42. The yield lines and the compressive-stress region are included. Various failure points are noted. On the mean stress (σ_m) axis, the yield strength S_y and the ultimate tensile strength S_{ut} of the particular material are defined at points A, E, and F. On the alternating stress (σ_a) axis, the corrected fatigue strength S_f at some number of cycles (or the corrected endurance limit S_e) and the yield strength S_y of the particular material are defined at points C and G. Note that this diagram usually represents a section such as G-G from the three-dimensional surface in Figure 6-43. That is, the modified-Goodman diagram is usually drawn for the infinite-life or very high-cycle case ($N > 10^6$). But, it can be drawn for any section along the N axis in Figure 6-43, representing a shorter finite-life situation.

Lines defining failure can be drawn connecting various points on the diagram. The line CF is the Goodman line and can be extended into the compressive region (shown dotted) based on empirical data such as that shown in Figure 6-17. However, it is conventional to draw the more conservative horizontal line CB to represent a failure line

* Goodman's original diagram plotted the relationship between mean and alternating stresses on a different set of axes than shown here and included an assumption that the fatigue limit was 1/3 S_{ut}. Goodman's original approach is now seldom used. J. O. Smith[46] suggested the representation of the Goodman line shown in Figure 6-42, which has become known as the modified-Goodman diagram. Smith's version did not show the yield line or the compressive region as depicted in Figure 6-44. Thus the use of the term "augmented" here to note the addition of that information to the diagram. We will nevertheless refer to it as the modified-Goodman diagram or just MGD for simplicity. Also, references here to the "Goodman line" should be understood as shorthand for "modified-Goodman line," and not as a reference to Goodman's original representation.

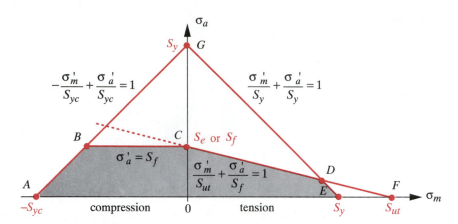

FIGURE 6-44

An "Augmented" Modified-Goodman Diagram

in the compressive region. This in effect ignores the beneficial effects of compressive mean stress and considers that situation to be identical to the fully reversed case of the previous section.

In the tensile region, the line *GE* defines static yielding and the failure envelope is defined as the lines *CD* and *DE* to account for the possibility of either fatigue or yield failure. If the mean component of stress were very large, and the alternating component very small, their combination could define a point in the region *DEF* that would be safely within the Goodman line but would yield on the first cycle. The entire failure envelope is then the shaded area labeled *ABCDEA*. Any combination of alternating and mean stress that falls within that envelope will be safe. Combinations landing on those lines are at failure and if outside the envelope will have failed.

In order to determine the safety factor of any fluctuating stress state, we will need expressions for the lines that form the failure envelope shown in Figure 6-44. The line *AG* defines yielding in compression and is

$$-\frac{\sigma'_m}{S_{yc}} + \frac{\sigma'_a}{S_{yc}} = 1 \qquad (6.16a)$$

Line *BC* defines fatigue failure with compressive mean stress and is

$$\sigma'_a = S_f \qquad (6.16b)$$

Line *CF* defines fatigue failure with tensile mean stress and is:

$$\frac{\sigma'_m}{S_{ut}} + \frac{\sigma'_a}{S_f} = 1 \qquad (6.16c)$$

Line *GE* defines yielding in tension and is

$$\frac{\sigma'_m}{S_y} + \frac{\sigma'_a}{S_y} = 1 \qquad (6.16d)$$

These equations are shown on Figure 6-44.

Applying Stress-Concentration Effects with Fluctuating Stresses

The alternating component of stress σ_a is treated the same way as it was for the case of fully reversed stress (see Example 6-3). That is, the geometric stress-concentration factor K_t is found, the material's notch sensitivity q is determined, and used in equation 6.11b to find a fatigue stress-concentration factor K_f. The local value of σ_a is then found from equation 6.12 for use in the modified-Goodman diagram.

The mean component of stress σ_m is treated differently depending on the ductility or brittleness of the material and, if ductile, on the amount of yielding possible at the notch. If the material is brittle, then the full value of the geometric stress concentration K_t is usually applied to the nominal mean stress to obtain the local mean stress at the notch using equation 4.31. If the material is ductile, Dowling [40] suggests one of three approaches based on Juvinall [41] depending on the relationship of the maximum local stresses to the yield strength of the ductile material.

A mean-stress fatigue-concentration factor K_{fm} is defined based on the level of local mean stress at the stress concentration versus the yield strength. Figure 6-45a shows a generalized fluctuating stress situation. Figure 6-45b depicts localized yielding that may occur around a stress concentration. For this analysis an *elastic-perfectly plastic* stress-strain relationship is assumed as shown in part (c). Three possibilities exist based on the relationship between σ_{max} and the material's yield strength S_y. If $\sigma_{max} < S_y$ no yielding occurs (see Figure 6-45d) and the full value of K_f is used for K_{fm}.

If $\sigma_{max} > S_y$ but $|\sigma_{min}| < S_y$, local yielding occurs on the first cycle (Figure 6-45e) after which the maximum stress cannot exceed S_y. The local stress at the concentration is relieved and a lower value of K_{fm} can be used as defined in Figure 6-45g, which plots the relationship between K_{fm} and σ_{max}.

The third possibility is that the stress range $\Delta\sigma$ exceeds $2S_y$ causing reversed yielding as shown in Figure 6-45f. The maximum and minimum stresses now equal $\pm S_y$ and the mean stress becomes zero (see equation 6.1c), making $K_{fm} = 0$.

These relationships can be summarized as follows:

$$\text{if } K_f|\sigma_{max}| < S_y \text{ then}: \qquad K_{fm} = K_f$$

$$\text{if } K_f|\sigma_{max}| > S_y \text{ then}: \qquad K_{fm} = \frac{S_y - K_f\sigma_a}{|\sigma_m|} \qquad (6.17)$$

$$\text{if } K_f|\sigma_{max} - \sigma_{min}| > 2S_y \text{ then}: \qquad K_{fm} = 0$$

The absolute values are used to account for either compressive or tensile situations. The value of the local mean stress σ_m for use in the modified-Goodman diagram is then found from equation 6.12 with K_{fm} substituted for K_f. Note that the stress-concentration factors should be applied to the nominal applied stresses, be they normal or shear stress.

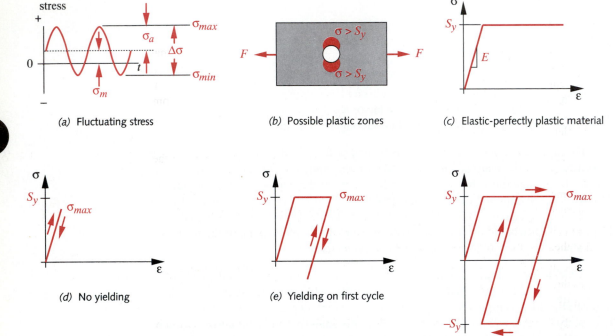

(a) Fluctuating stress

(b) Possible plastic zones

(c) Elastic-perfectly plastic material

(d) No yielding

(e) Yielding on first cycle

(f) Reversed yielding

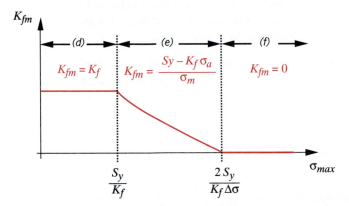

(g) K_{fm} as a function of maximum stress σ_{max}

FIGURE 6-45

Variation of Mean Stress-Concentration Factor with Maximum Stress in Ductile Materials with Possibility of Local Yielding
(Adapted from Fig. 10-14, p.415, N. E. Dowling, *Mechanical Behavior of Materials*, Prentice-Hall, Englewood Cliffs, N.J., 1993, with permission.)

The local applied stresses (with their fatigue stress concentration effects included) are used to calculate the alternating and mean von Mises stresses. This calculation is done separately for the alternating and mean components σ'_a and σ'_m. (See equations 6.22a and 6.22b.) We will use these von Mises components to find the safety factor.

Determining the Safety Factor with Fluctuating Stresses

Figure 6-46 shows four views of the tension side of the augmented modified-Goodman diagram and also shows a combination of mean and alternating von Mises stresses at point Z representing a part subjected to fluctuating stresses. The safety factor for any fluctuating-stress state depends on the manner in which the mean and alternating components can vary with respect to one another in service. There are four possible cases to consider as shown in Figure 6-46.

1   The alternating stress will remain essentially constant over the life of the part but the mean stress can increase under service conditions. (Line YQ in Figure 6-46a.)

2   The mean stress will remain essentially constant over the life of the part but the alternating stress can increase under service conditions. (Line XP in Figure 6-46b.)

3   Both alternating and mean stress components can increase under service conditions but their ratio will remain constant. (Line OR in Figure 6-46c.)

4   Both alternating and mean stress components can increase under service conditions but there is no known relationship between their amounts of increase. (Line ZS in Figure 6-46d.)

The safety factor for each of these cases is calculated differently. Note that S_f will be used in the following expressions to represent either the corrected fatigue strength at some defined number of cycles or the corrected endurance limit. So, S_e can be substituted for S_f in any of these expressions if appropriate to the material used.

FOR CASE 1   failure occurs at point Q and the safety factor is the ratio of the lines YQ/YZ. To express this mathematically, we can solve equation 6.16d for the value of $\sigma'_{m@Q}$ and divide that by $\sigma'_{m@Z}$.

$$\sigma'_{m@Q} = \left(1 - \frac{\sigma'_a}{S_y}\right)S_y$$

$$N_f = \frac{\sigma'_{m@Q}}{\sigma'_{m@Z}} = \frac{S_y}{\sigma'_m}\left(1 - \frac{\sigma'_a}{S_y}\right) \qquad (6.18a)$$

If σ'_a were so large and σ'_m so small that point Q was on line CD instead of DE, then equation 6.16c should be used instead to determine the value of $\sigma'_{m@Q}$.

FOR CASE 2   failure occurs at point P and the safety factor is the ratio of the lines XP/XZ. To express this mathematically, we can solve equation 6.16c for the value of $\sigma'_{a@P}$ and divide that by $\sigma'_{a@Z}$.

$$\sigma'_{a@P} = \left(1 - \frac{\sigma'_m}{S_{ut}}\right) S_f$$

$$N_f = \frac{\sigma'_{a@P}}{\sigma'_{a@Z}} = \frac{S_f}{\sigma'_a}\left(1 - \frac{\sigma'_m}{S_{ut}}\right) \qquad (6.18b)$$

If σ'_m were so large and σ'_a so small that point P was on line DE instead of CD, then equation 6.16d should be used instead to determine the value of $\sigma'_{a@P}$.

FOR CASE 3   failure occurs at point R and the safety factor is the ratio of the lines OR/OZ or by similar triangles, either of the ratios $\sigma'_{m@R} / \sigma'_{m@Z}$ or $\sigma'_{a@R} / \sigma'_{a@Z}$. To express this mathematically, we can solve equations 6.16c and the equation of line OR simultaneously for the value of $\sigma'_{m@R}$ and divide that by $\sigma'_{m@Z}$.

from eq. 6.16c : $\sigma'_{a@R} = \left(1 - \frac{\sigma'_{m@R}}{S_{ut}}\right) S_f$

$$\qquad (6.18c)$$

from line OR : $\sigma'_{a@R} = \left(\frac{\sigma'_{a@Z}}{\sigma'_{m@Z}}\right)\sigma'_{m@R} = \left(\frac{\sigma'_a}{\sigma'_m}\right)\sigma'_{m@R}$

The simultaneous solution of these gives

$$\sigma'_{m@R} = \frac{S_f}{\dfrac{\sigma'_a}{\sigma'_m} + \dfrac{S_f}{S_{ut}}} \qquad (6.18d)$$

which after substitution and some manipulation yields

$$N_f = \frac{\sigma'_{m@R}}{\sigma'_{m@Z}} = \frac{S_f S_{ut}}{\sigma'_a S_{ut} + \sigma'_m S_f} \qquad (6.18e)$$

There is also the possibility that point R may lie on line DE instead of CD in which case equation 6.16d should be substituted for 6.16c in the above solution.

FOR CASE 4   in which the future relationship between the mean and alternating stress components is either random or unknown, the point S on the failure line closest to the stress state at Z can be taken as a conservative estimate of the failure point. Line ZS is normal to CD, so its equation can be written and solved simultaneously with that of the line CD to find the coordinates of point S and the length ZS, which are

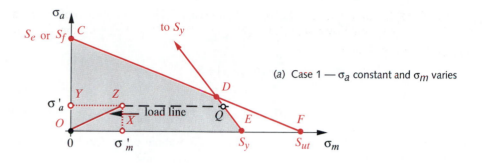

(a) Case 1 — σ_a constant and σ_m varies

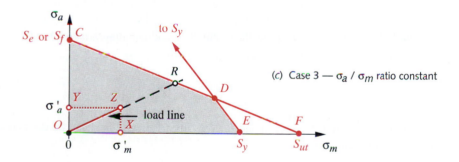

(b) Case 2 — σ_a varies and σ_m constant

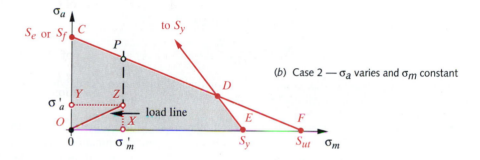

(c) Case 3 — σ_a / σ_m ratio constant

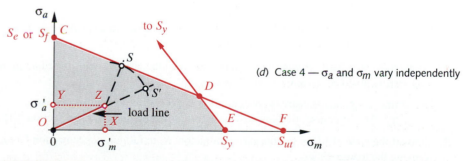

(d) Case 4 — σ_a and σ_m vary independently

FIGURE 6-46

Safety Factors from the Modified-Goodman Diagram for Four Possible Load-Variation Scenarios

$$\sigma'_{m @ S} = \frac{S_{ut}\left(S_f^2 - S_f\sigma'_a + S_{ut}\sigma'_m\right)}{S_f^2 + S_{ut}^2}$$

$$\sigma'_{a @ S} = -\frac{S_f}{S_{ut}}\left(\sigma'_{m @ S}\right) + S_f$$

$$ZS = \sqrt{\left(\sigma'_m - \sigma'_{m @ S}\right)^2 + \left(\sigma'_a - \sigma'_{a @ S}\right)^2} \qquad (6.18f)$$

To establish a ratio for the safety factor, swing point S about point Z to be coincident with line OZR at point S'. The safety factor is the ratio OS'/OZ.

$$OZ = \sqrt{\left(\sigma'_a\right)^2 + \left(\sigma'_m\right)^2}$$

$$N_f = \frac{OZ + ZS}{OZ} \qquad (6.18g)$$

There is also the possibility that point S may lie on line DE instead of CD in which case equation 6.16d should be substituted for 6.16c in the above solution.

Case 4 gives a more conservative safety factor than case 3. The same approach can be used to obtain safety factor expressions for stress-component combinations in the left half-plane of the modified-Goodman diagram. Also, if the diagram is drawn to scale, rough estimates of the safety factors can be scaled from it. The *TKSolver* file GOODMAN supplied with this text calculates all the safety factors defined in equations 6.18 for any supplied values of σ'_a and σ'_m, and plots the modified-Goodman diagram and the stress line OZ extended so that the failure intercept can be seen.

Design Steps for Fluctuating Stresses

A set of design steps similar to those listed for the fully reversed case can be defined for the case of fluctuating stresses:

1 Determine the number of cycles of loading N that the part will experience over its expected service life.

2 Determine the amplitude of the applied alternating loads (mean to peak) and of the mean load. (See Chapter 3 and equations 6.1.)

3 Create a tentative part-geometry design to withstand the applied loads based on good engineering practice. (See Chapters 3 and 4.)

4 Determine any geometric stress-concentration factors K_t at notches in the part's geometry. Try, of course, to minimize these through good design. (See Section 4.15.)

5 Convert the geometric stress-concentration factors K_t to fatigue-concentration factors K_f using the material's q.

6 Calculate the nominal, alternating tensile-stress amplitudes σ_a (see Figure 6.6c) at critical locations in the part due to the alternating service loads based on standard

stress analysis techniques (Chapter 4) and increase them as necessary with the appropriate fatigue stress-concentration factors from equation 6.11. (See Sections 4.14 and 6.6.) Calculate the nominal mean-stress amplitudes at the same critical locations and increase them as necessary with the appropriate mean fatigue stress-concentrations factors K_{fm} from equation 6.17.

7   Calculate the principal-stress and von Mises-stress amplitudes for the critical locations based on their states of applied stress. Do this separately for the mean and alternating components. (See Chapter 4 and equations 6.22.)

8   Choose a tentative material for the part and determine its properties of interest such as S_{ut}, S_y, S_e, (or S_f, at the life required), and notch sensitivity q, from your own test data, from the literature, or from estimates as described in this chapter.

9   Determine appropriate fatigue-strength-modification factors for the type of loading, size of part, surface, etc., as described in Section 6.6. Note that the loading factor C_{load} will differ based on whether there are axial or bending loads (see Eq. 6.7a). If the loading is pure torsion, then the von Mises effective stress calculation will convert it to a pseudo-tensile stress and C_{load} should then be set to 1.

10 Define the corrected fatigue strength S_f at the requisite cycle life N (or the corrected endurance limit S_e for infinite life if appropriate). Create a modified-Goodman diagram as shown in Figure 6-44 using the material's corrected fatigue strength S_f taken from the S-N curve at the desired number of cycles N. (Note that for infinite life situations in which the material has an S-N knee, $S_f = S_e$). Write equations 6.16 for the Goodman and yield lines.

11 Plot the mean and alternating von Mises stresses (for the most highly stressed location) on the modified-Goodman diagram and calculate a safety factor for the design from one of the relationships shown in equations 6.18.

12 Given the fact that the material was only tentatively chosen and that the design may not yet be as refined as possible, the result of the first pass through these steps will most likely be a failed design whose safety factor is either too large or too small. Iteration will be required (as it always is) to refine the design. Any subset of steps can be repeated as many times as necessary to obtain an acceptable design. The most common tactic is to return to step 3 and improve the geometry of the part to reduce stresses and stress concentrations and/or revisit step 4 to choose a more suitable material. Sometimes it will be possible to return to step 1 and redefine a shorter acceptable part life. The design loads in step 2 may or may not be in the control of the designer. Usually they are not, unless the loading on the part is due to inertial forces, then increasing its mass to "add strength" will worsen the situation as this will proportionately increase the loads (see Section 3.6) and the designer may want to lighten the part without compromising its strength excessively to reduce the forces. Whatever the particular circumstances, the designer must expect to cycle through these steps several times before converging on a usable solution. Equation solvers that allow rapid recalculation of the equations are a great help in this situation. If the equation solver can also "backsolve," allowing variables to be exchanged from input to output, the geometry needed to achieve a desired safety factor can by directly calculated by making the safety factor an input and the geometry variable an output.

The best way to demonstrate the use of these steps for fluctuating stress fatigue design is with an example. We will repeat the previous example, modifying its load pattern.

EXAMPLE 6-5

Design of a Cantilever Bracket for Fluctuating Bending

Problem   A feed-roll assembly is to be mounted at its ends on support brackets cantilevered from the machine frame as shown in Figure 6-47. The feed rolls experience a total fluctuating load that varies from a minimum of 200 lb to a maximum of 2 200 lb, split equally between the two support brackets. Design a cantilever bracket to support a fluctuating bending load of 100- to 1 100-lb amplitude for 10^9 cycles with no failure. Its dynamic deflection cannot exceed 0.02 in.

Given   The load-time function shape is shown in Figure 6-47. The operating environment is room air at a maximum temperature of 120°F. The space available allows a maximum cantilever length of 6 in. Only ten of these parts are required.

Assumptions The bracket can be clamped between essentially rigid plates bolted at its root. The normal load will be applied at the effective tip of the cantilever beam from a rod attached through a small hole in the beam. Since the bending moment is effectively zero at the beam tip, the stress concentration from this hole can be ignored. Given the small quantity required, machining of stock mill-shapes is the preferred manufacturing method.

Solution   See Figure 6-47, Tables 6-11 and 6-12, and *TKSolver* file EX06-05.

1 This is a typical design problem. Very little data are given except for the loading on the device, some limitations on size, and the required cycle life. We will have to make some basic assumptions about part geometry, materials and other factors as we go. Some iteration should be expected.

2 Figure 6-47 shows the same tentative design configuration as in Figure 6-41*b*. The mill stock is purchased thicker than the desired final dimension and machined top and bottom to dimension *D*, then machined to thickness *d* over the length *l*. A fillet radius *r* is provided at the clamp point to reduce fretting fatigue and achieve a lower K_t. (See Figure 4-37.) Figure 4-36 shows that with suitable control of the *r/d* and *D/d* ratios for a stepped flat bar in bending, the geometric stress-concentration factor K_t can be kept under about 1.5.

3 A material must be chosen. For infinite life, low cost, and ease of fabrication, it is desirable to use a low-carbon alloy steel if environmental conditions permit. Since this is used in a controlled, indoor environment, carbon steel is acceptable on the latter point. The fact that the deflection is of concern is also a good reason to choose a material with a large *E*. Low-carbon, ductile steels have the requisite endurance limit knee for the infinite life required in this case and also have low notch sensitivi-

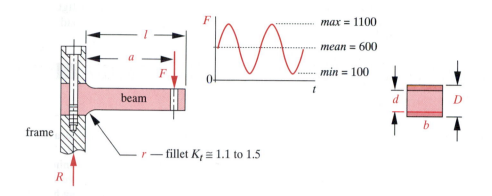

FIGURE 6-47

Design of a Cantilever Bracket for Fluctuating-Bending Loading

ties. A low-carbon steel with $S_{ut} = 80$ kpsi and $S_y = 60$ kpsi is selected for the first trial.

4 We will assume the trial dimensions to be the same as those of the successful solution to the fully reversed case from Example 6-4. These are $b = 2$ in, $d = 1$ in, $D = 1.125$ in, $r = 0.5$ in, $a = 5$ in, and $l = 6.0$ in. This a will leave some material around the hole and still fit within the 6-in-length constraint.

5 These dimensions were put into the *TKSolver* model EX06-05. This model incorporates all the relevant equations and performs steps 3 to 11 of the process suggested above for these problems. An equation solver makes it a simple task to adjust dimensions and recalculate the results once the equations are defined.

6 Table 6-11 shows a portion of the rule sheet for the *TKSolver* model. The number of required cycles is defined in the problem statement. The mean and alternating components of the load and their reaction forces can be calculated from the given maximum and minimum load.

$$F_m = \frac{F_{max} + F_{min}}{2} = \frac{1100 + 100}{2} = 600 \text{ lb}$$

$$F_a = \frac{F_{max} - F_{min}}{2} = \frac{1100 - 100}{2} = 500 \text{ lb} \tag{a}$$

$$R_a = F_a = 500 \text{ lb} \qquad R_m = F_m = 600 \text{ lb} \qquad R_{max} = F_{max} = 1100 \text{ lb} \tag{b}$$

From these, the mean and alternating moments, and the maximum moment acting at the root of the cantilever beam can be calculated.

$$M_a = R_a l - F_a(l - a) = 500(6) - 500(6 - 5) = 2\,500 \text{ lb - in}$$
$$M_m = R_m l - F_m(l - a) = 600(6) - 600(6 - 5) = 3\,000 \text{ lb - in} \tag{c}$$
$$M_{max} = R_{max} l - F_{max}(l - a) = 1\,100(6) - 1\,100(6 - 5) = 5\,500 \text{ lb - in}$$

Line 2 in Table 6-11 contains equations for area moment of inertia and the distance to the outer fiber.

$$I = \frac{bd^3}{12} = \frac{2.0(1.0)^3}{12} = 0.166\ 7\ \text{in}^4 \tag{d}$$

$$c = \frac{d}{2} = \frac{1.0}{2} = 0.5\ \text{in}$$

The nominal bending stresses at the root are found for both the alternating load and the mean load at line 4.

$$\sigma_{a_{nom}} = \frac{Mc}{I} = \frac{2\ 500(0.5)}{0.166\ 7} = 7\ 500\ \text{psi} \tag{e}$$

$$\sigma_{m_{nom}} = \frac{Mc}{I} = \frac{3\ 000(0.5)}{0.166\ 7} = 9\ 000\ \text{psi}$$

7   Line 5 calls the user functions *coeff* and *exponent* to solve the equations in Figure 4-36 and interpolate for the geometric stress-concentration factor K_t without the need for any reading of curves in Figure 4-36 (See also Appendix E, Figure E-10.)

$$\frac{D}{d} = \frac{1.125}{1.0} = 1.125 \qquad\qquad \frac{r}{d} = \frac{0.5}{1.0} = 0.5 \tag{f}$$

from the table in Figure 4 - 35 : $A = 1.013\ 7$ $b = -0.213\ 6$ $\hfill (g)$

$$K_t = A\left(\frac{r}{d}\right)^b = 1.013\ 7(0.5)^{-0.213\ 6} = 1.175 \tag{h}$$

8   At line 6, the procedure *notch* is called to calculate the notch sensitivity q of the chosen material based on its ultimate strength using equation 6.13 and the data for Neuber's constant in Table 6-6. The Neuber curves have been converted to cubic-interpolated *list functions* that are called from the procedure *notch*, which returns q for any combination of S_{ut} and notch radius r using equation 6.13. The values of q and K_t are used at line 7 to find the fatigue stress-concentration factor K_f using equation 6.11b. The function *Kfmean* is called and computes K_{fm} based on equation 6.17.

from Table 6 - 6 for S_{ut} = 80 ksi : $\sqrt{a} = 0.08$ $\hfill (i)$

$$q = \frac{1}{1 + \dfrac{\sqrt{a}}{\sqrt{r}}} = \frac{1}{1 + \dfrac{0.08}{\sqrt{0.5}}} = 0.903 \tag{j}$$

$$K_f = 1 + q(K_t - 1) = 1 + 0.903(1.175 - 1) = 1.158 \tag{k}$$

if $K_f |\sigma_{max}| < S_y$ then

$$K_{fm} = K_f$$

(l)

$$K_f \left| \frac{M_{max}c}{I} \right| = 1.158 \left| \frac{5\,500(0.5)}{0.166\,7} \right| = 19\,113 < 60\,000 : \quad K_{fm} = 1.158$$

9   Use these factors at line 8 to find the local mean and alternating notch stresses.

$$\sigma_a = K_f \sigma_{a_{nom}} = 1.158(7\,500) = 8\,688 \text{ psi}$$

(m)

$$\sigma_m = K_{fm} \sigma_{m_{nom}} = 1.158(9\,000) = 10\,425 \text{ psi}$$

The local stresses are used to compute the von Mises alternating and mean stresses from equations 6.22b using the rule function *Distort*.

$$\sigma'_a = \sqrt{\sigma_{x_a}^2 + \sigma_{y_a}^2 - \sigma_{x_a}\sigma_{y_a} + 3\tau_{xy_a}^2} = \sqrt{8\,688^2 + 0 - 8\,688(0) + 3(0)} = 8\,688$$

(n)

$$\sigma'_m = \sqrt{\sigma_{x_m}^2 + \sigma_{y_m}^2 - \sigma_{x_m}\sigma_{y_m} + 3\tau_{xy_m}^2} = \sqrt{10\,425^2 + 0 - 10\,425(0) + 3(0)} = 10\,425$$

10 At line 9, the uncorrected endurance limit $S_{e'}$ is determined either from equation 6.6a or from an input value. The *GIVEN* function will compute the value according to the formula listed as its third argument only if the input field for the variable is blank. This allows a value of $S_{e'}$ to be typed into the variable sheet and it will override the calculated estimate of $0.5\,S_{ut}$. At line 10, the *size factor* for this rectangular part is determined by calculating the cross-sectional area stressed above 95% of its maximum stress (see Figure 6-25) and using that value in equation 6.7d to find an equivalent-diameter test specimen. Line 11 calls the function *SizeUS* that executes equation 6.7b to find C_{size}.

$$S_{e'} = 0.5S_{ut} = 0.5(80\,000) = 40\,000 \text{ psi}$$

(o)

$$A_{95} = 0.05db = 0.05(1.0)(2.0) = 0.1 \text{ in}^2$$

$$d_{equiv} = \sqrt{\frac{A_{95}}{0.076\,6}} = \sqrt{\frac{0.1}{0.076\,6}} = 1.143 \text{ in}$$

(p)

$$C_{size} = 0.869 (d_{equiv})^{-0.097} = 0.859$$

11 The other four strength-modification factors are found in lines 11 and 12. Custom *user functions* have been created in *TKSolver* to compute these factors based either on curves fitted to empirical data or on the raw data placed in *list functions* for automatically-interpolated table lookup. Table 6-12 shows the variable sheet where the *temperature*, desired *reliability*, type of *finish*, and type of *loading* are input or calculated. These data are used by the user functions *Load*, *SurfUS*, *Tempfactor*, and *Reliab* to create C_{load}, C_{surf}, C_{temp}, and C_{reliab} using the relationships shown in Section 6.6. C_{surf} is computed from the data in Table 6-3. The endurance limit is modified at line 13.

Table 6-11   Design of a Cantilever Bracket for Fluctuating Bending
TKSolver Rule Sheet (Partial) for Example 6-5 from File EX06-05B

; 1- find the forces and moments

$Fm = (Fmax + Fmin) / 2$
$Fa = (Fmax - Fmin) / 2$
$Mmean = Fm * l - Fm * (l - a)$
$Malt = Fa * l - Fa * (l - a)$

; 2- find area moment of inertia and distance to outer fiber

$I = b * d^3 / 12$
$c = d / 2$

; 3- find the nominal bending stresses at root

$siganom = Malt * c / I$
$sigmnom = Mmean * c / I$

; 4- create the ratios for Figure 4-36

$Doverd = D / d$
$roverd = r / d$

; 5- call list functions for Fig. 4-36 and interpolate coefficient and exponent to calculate Kt

$Kt = coeff(Doverd) * roverd \char`\^ exponent(Doverd)$

; 6- create Peterson's notch sensitivity factor

$CALL \ notch (Sut, r \ ; q)$

; 7- find fatigue stress-concentration factors from equations 6.11*b* and 6.17

$Kf = 1 + q * (Kt - 1)$
$CALL \ Kfmean(sigmnom, siganom, Sy, Kf; Kfm)$

; 8- calculate stress including stress concentration

$sigax = Kf * siganom$; *note that sigax = the max principal stress in this uniaxial case*
$CALL \ Distort (sigax, 0, 0 ; sigavm)$; *Equation 6.21 for von Mises alternating stress*
$sigmx = Kfm * sigmnom$
$CALL \ Distort (sigmx, 0, 0 ; sigmvm)$; *Equation 6.21 for von Mises mean stress*

; 9- estimate the endurance limit based on *Sut*

$Seprime = GIVEN \ ('Seprime, Seprime, Sut / 2)$

; 10- find the 95% area and the equivalent test-bar dia.

$A95 = .05 * d * b$
$dequiv = SQRT(A95/.0766)$

; 11- calculate the size factor from eq. 6.7*b* and surface factor from eq. 7.7*e* and Table 6-4

$Csize = SizeUS(dequiv)$
$Csurf = SurfUS(finish)$

; 12- calculate the load, temperature, and reliability factors

$Cload = Load(loading)$; *Load is a List Function containing the load factors*
$Ctemp = Tempfactor(temper)$; *Tempfactor is a Rule Function containing equation 6.7f*
$Creliab = Reliab(percent)$; *Reliab is a List Function containing data from Table 6-5*

; 13- modify the endurance limit

$Se = Cload * Csize * Csurf * Ctemp * Creliab * Seprime$

; 14- compute the safety factor from Goodman diagram using all of equations 6.17

$CALL \ Safety(sigmvm, sigavm, Se, Sy, Sut; n1, n2, n3, n4)$

Table 6-12   Design of a Cantilever Bracket for Fluctuating Bending

TKSolver Variable Sheet (Partial) for Example 6-5 from File EX06-05B

Input	Variable	Output	Unit	Comments
2	*b*		in	beam width
1.2	*d*		in	beam depth over length
1.4	*D*		in	beam depth in wall
0.5	*r*		in	fillet radius
5	*a*		in	beam length to load *F*
80 000	*Sut*		psi	ultimate tensile strength
60 000	*Sy*		psi	yield strength
'machined	*finish*			'ground, 'machined, 'hotroll 'forged
'bending	*loading*			'bending, 'axial, 'shear
99.9	*percent*			Reliability % desired
1 100	*Fmax*		lb	maximum applied load
100	*Fmin*		lb	minimum applied load
	Fa	500	lb	alternating applied force
	Fm	600	lb	mean applied force
	Kt	1.22		geometric stress-conc. factor
	q	0.9		Peterson's notch sensitivity factor
	Kf	1.2		fatigue stress-conc. factor — alter.
	Kfm	1.2		fatigue stress-conc. factor — mean
	siganom	5 208	psi	alternating nominal stress
	siga	6 235	psi	alternating stress with concentratn
	sigavm	6 235	psi	von Mises alternating stress
	sigmnom	6 250	psi	mean nominal stress
	sigm	7 482	psi	mean stress with concentration
	sigmvm	7 482	psi	von Mises mean stress
	Seprime	40 000	psi	uncorrected endurance limit
	Cload	1		load factor for bending
	Csurf	0.85		machined finish
	Csize	0.85		size factor based on 95% area
	Ctemp	1		room temperature
	Creliab	0.753		99.9% reliability factor
	Se	21 658	psi	corrected endurance limit
	Nsf_1	8.6		FS for sigalt = constant
	Nsf_2	3.1		FS for sigmean = constant
	Nsf_3	2.6		FS for sigalt/sigmean = constant
	Nsf_4	2.3		FS for closest failure line

$$S_e = C_{load} \, C_{size} \, C_{surf} \, C_{temp} \, C_{reliab} \, S_e{'}$$

$$= 1(0.859)(0.85)(1)(0.753)40\,000 = 21\,883 \text{ psi} \qquad (q)$$

12   The four possible safety factors are calculated at line 14 using equations 6.18. The smallest or most appropriate one can be selected from those calculated. Equation (r) shows the Case 3 safety factor, which assumes that the alternating and mean components will have a constant ratio if they vary in maximum amplitude over the life of the part.

$$N_{f_3} = \frac{S_e S_{ut}}{\sigma'_a \, S_{ut} + \sigma'_m \, S_e} = \frac{21\,883(80\,000)}{8\,688(80\,000) + 10\,425(21\,883)} = 1.9 \qquad (r)$$

The *TKSolver* program also plots the modified-Goodman diagram for the solution which is shown in Figure 6-48. The maximum deflection is calculated using the maximum applied force F_{max}.

$$y_{@x=l} = \frac{F_{max}}{6EI}\left[x^3 - 3ax^2 - \langle x-a \rangle^3\right]$$

$$= \frac{1\,100}{6(3E7)(0.1667)}\left[6^3 - 3(5)(6)^2 - (6-5)^3\right] = -0.012 \text{ in} \qquad (s)$$

13   The dimensions of the successful design from Example 6-4, which were assumed in step 4 give a Case 4, minimum safety factor N_{f4} of 1.7 in this case. It should be expected that the addition of a mean stress to the same level of alternating stress as before would require a stronger part.

14   A few iterations of the dimensions gave the better results shown in Table 6-12. The final dimensions are $b = 2$ in, $d = 1.2$ in, $D = 1.4$ in, $r = 0.5$ in, $a = 5$ in, and $l = 6.0$ in. The smallest safety factor is now 2.3 and the maximum deflection is 0.007 in, These are both acceptable. The dimension D was deliberately chosen to be slightly less than a stock mill size so that material would be available for the cleanup and truing of the mounting surfaces.

15   See the provided *TKSolver* files EX06-05A and EX06-05B for more detailed information. The rule, procedure, and list functions can be examined within the file as well. Only portions of its rule and variable sheets are reproduced in Tables 6-11 and 6-12 due to space limitations.

The above example should demonstrate that designing for fluctuating HCF loads is straightforward once the principles are understood. If the design called for fluctuating torsional, bending or axial loading, the design procedure would be the same as in this example. The only differences would be in the choices of stress equations and strength modification factors as described in the previous sections. The value of using a computer and equation solver in this or any design problem cannot be overstated as it allows rapid iteration from initial guesses to final dimensions with minimum effort.

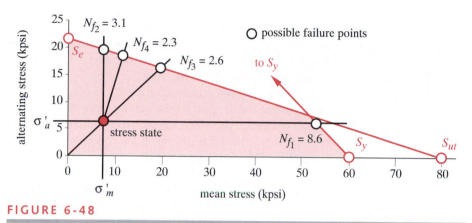

FIGURE 6-48

Modified-Goodman Diagram for Example 6-5

6.12 DESIGNING FOR MULTIAXIAL STRESSES IN FATIGUE

The previous discussions were limited to cases in which the loading produced uniaxial stresses in the part. It is quite usual in machinery to have combined loads that create simultaneous time-varying biaxial or triaxial stresses at the same point. A common example is a rotating shaft subjected to both a static bending moment and a torque. Because the shaft is turning, the static moment creates fully reversed normal stresses that are maximum at the shaft's outer fiber and the torque creates shear stresses that are also maximum at the outer fiber. There are many possible loading combinations. The torque might be constant, fully reversed, or fluctuating. If the torque is not constant, it could be synchronous, asynchronous, in- or out-of-phase with the bending moment. These factors complicate the stress calculation. We explored the case of combined stresses under static loading in Chapter 5 and used the *von Mises stress* to convert them to an equivalent tensile stress that could be used to predict failure in the static loading case. Similar techniques exist for handling combined stresses in dynamic loading.

Frequency and Phase Relationships

When multiple time-varying loads are present they may be periodic, random, or some combination of the two. If periodic, they can be mutually synchronous or asynchronous. If synchronous, they may have phase relationships from in-phase to 180° out-of-phase or anything in between. The possible combinations are quite varied and only a few such combinations have been studied to determine their effects on fatigue failure. Collins[49] suggests that the assumption that loads are synchronous and in-phase is usually accurate and usually (but not always) conservative.

The most-studied cases are those of periodic, synchronous, in-phase loads that cause combined stresses whose principal directions do not change with time. This is referred to as **simple multiaxial stress**. Sines [42] developed a model for this case in 1955. Pressure vessels or pipes that are subjected to time-varying internal pressures may see synchronous, in-phase multiaxial tensile stresses from a single load source. The case of a rotating shaft in combined bending and torsion can also be in this category if the torque is constant with time since the alternating component of principle stress, due only to bending, is then in a constant direction. If the torque is time-varying, then the alternating principle stress directions are not constant. Also when stress concentrations are present, such as a transverse hole through a shaft, the local stresses at the concentration will be biaxial. These situations in which the directions of the principal stresses vary with time or which are asynchronous or out-of-phase, are called **complex multiaxial stress**, and are still being studied. According to the *SAE Fatigue Design Handbook*,[51] *"Analysis of this situation is, in general, beyond the current state of the technology. The design process must proceed by very approximate analyses supported by extensive experimental studies simulating the material and geometry as well as the loading."* Analysis methods have been developed for some of these cases by Kelly ,[43] Garud [44], Brown [45], Langer,[48] and others. Some of these approaches are fairly complicated to use and reference 51 also *"caution(s) against direct use of these data unless the conditions examined match those being analyzed."* We will limit our discussion to a few approaches that are useful for design purposes and that should give approximate but conservative results in most machine design situations.

Fully Reversed Simple Multiaxial Stresses

Experimental data developed for simple biaxial stresses, such as that shown in Figure 6-15, indicate that for fully reversed, simple multiaxial stresses in ductile materials, the distortion energy theory is applicable if the von Mises stress is calculated for the alternating components using equation 5.7. For the three-dimensional case:

$$\sigma'_a = \sqrt{\sigma_{1_a}^2 + \sigma_{2_a}^2 + \sigma_{3_a}^2 - \sigma_{1_a}\sigma_{2_a} - \sigma_{2_a}\sigma_{3_a} - \sigma_{1_a}\sigma_{3_a}} \qquad (6.19a)$$

and for the two-dimensional case:

$$\sigma'_a = \sqrt{\sigma_{1_a}^2 - \sigma_{1_a}\sigma_{2_a} + \sigma_{2_a}^2} \qquad (6.19b)$$

Note that this form of the von Mises equation contains the alternating principal stresses which are computed from the alternating applied-stress components of the multiaxial stress state using equation 4.4c (for 3-D) or 4.6 (for 2-D) after those alternating components have been increased by all applicable fatigue stress-concentration factors. This effective alternating stress σ'_a can then be used to enter an *S-N* diagram to determine a safety factor using

$$N_f = \frac{S_n}{\sigma'_a} \qquad (6.20)$$

where S_n is the fatigue strength of the material at the desired life N and σ'_a is the von Mises alternating stress.

Fluctuating Simple Multiaxial Stresses

SINES METHOD Sines [42] developed a model for fluctuating simple multiaxial stresses which creates an equivalent mean stress as well as an equivalent alternating stress from the applied stress components. His equivalent alternating stress is, in fact, the von Mises alternating stress as defined in equation 6.19a above. However, we will present it in an alternate form that uses the applied stresses directly instead of the principal stresses. For a triaxial stress state:

$$\sigma'_a = \sqrt{\frac{\left(\sigma_{x_a} - \sigma_{y_a}\right)^2 + \left(\sigma_{y_a} - \sigma_{z_a}\right)^2 + \left(\sigma_{z_a} - \sigma_{x_a}\right)^2 + 6\left(\tau^2_{xy_a} + \tau^2_{yz_a} + \tau^2_{zx_a}\right)}{2}}$$

$$\sigma'_m = \sigma_{x_m} + \sigma_{y_m} + \sigma_{z_m}$$

(6.21a)

and, for a biaxial stress state:

$$\sigma'_a = \sqrt{\sigma^2_{x_a} + \sigma^2_{y_a} - \sigma_{x_a}\sigma_{y_a} + 3\tau^2_{xy_a}}$$

$$\sigma'_m = \sigma_{x_m} + \sigma_{y_m}$$

(6.21b)

The applied-stress components in equations 6.21 are the local stresses, increased by all applicable stress-concentration factors. The two equivalent stresses σ'_a and σ'_m are then used in a modified-Goodman diagram as described in the previous section, and the appropriate safety factor calculated from equations 6.18.

While the individual local stresses in equations 6.19 and 6.21 each can be increased by a different stress-concentration factor, there may be some conflicts when the corrected fatigue strength or endurance limit is calculated for a combined stress state. For example, a combination of bending and axial loading would give two choices for the load factors from equations 6.7a and 6.9. Use the axial factor if axial loads are present, with or without bending loads.

Note that the Sines equivalent mean stress σ'_m of equation 6.21 contains only normal stress components (which are the hydrostatic stress) while the von Mises equivalent alternating stress σ'_a of equation 6.21 contains both normal and shear stresses. The mean components of shear stress thus do not contribute to Sines model. This is consistent with experimental data for smooth, polished, unnotched, round bars tested in combined bending and torsion.[46] But, notched specimens under the same loading do show dependence on the value of mean torsional stress,[46] so equations 6.21 may be nonconservative in such cases.

VON MISES METHOD   Others [47], [49] recommend using the von Mises effective stress for both alternating and mean components of applied stress in simple multiaxial stress loading. Appropriate (and possibly different) stress-concentration factors can be applied to the alternating and mean components of the applied stresses as described in section 6.10. Then the von Mises effective stresses for the alternating and mean components are calculated for a triaxial stress state using

$$\sigma'_a = \sqrt{\frac{\left(\sigma_{x_a} - \sigma_{y_a}\right)^2 + \left(\sigma_{y_a} - \sigma_{z_a}\right)^2 + \left(\sigma_{z_a} - \sigma_{x_a}\right)^2 + 6\left(\tau^2_{xy_a} + \tau^2_{yz_a} + \tau^2_{zx_a}\right)}{2}}$$

$$\sigma'_m = \sqrt{\frac{\left(\sigma_{x_m} - \sigma_{y_m}\right)^2 + \left(\sigma_{y_m} - \sigma_{z_m}\right)^2 + \left(\sigma_{z_m} - \sigma_{x_m}\right)^2 + 6\left(\tau^2_{xy_m} + \tau^2_{yz_m} + \tau^2_{zx_m}\right)}{2}}$$

(6.22a)

or for a biaxial stress state using:

$$\sigma'_a = \sqrt{\sigma^2_{x_a} + \sigma^2_{y_a} - \sigma_{x_a}\sigma_{y_a} + 3\tau^2_{xy_a}}$$

$$\sigma'_m = \sqrt{\sigma^2_{x_m} + \sigma^2_{y_m} - \sigma_{x_m}\sigma_{y_m} + 3\tau^2_{xy_m}}$$

(6.22b)

These alternating and mean von Mises effective stresses are then used in a modified-Goodman diagram to determine a safety factor using the appropriate version of equations 6.18. This approach is more conservative than the Sines method and is thus more appropriate for situations involving stress concentrations due to notches.

Complex Multiaxial Stresses

This topic is still under investigation by numerous researchers. Many specific cases of complex multiaxial stresses have been analyzed but no overall design approach applicable to all situations has yet been developed.[50] Nishihara and Kawamoto[52] found that the fatigue strengths of two steels, a cast iron, and an aluminum alloy tested under complex multiaxial stress were not less than their in-phase fatigue strengths at any phase angle. For the common biaxial stress case of combined bending and torsion, such as occurs in shafts, several approaches have been proposed.[50] One of these, called *SEQA*, which is based on the ASME Boiler Code* will be discussed briefly. *SEQA* is an equivalent or effective stress (similar in concept to the von Mises effective stress), which combines the effects of normal and shear stresses and the phase relationship between them into an effective-stress value that can be compared to a ductile material's fatigue and static strengths on a modified-Goodman diagram. It is calculated from

* ASME Boiler and Pressure Vessel Code, Section III, Code Case N-47-12, American Society of Mechanical Engineers, New York, 1980.

$$SEQA = \frac{\sigma}{\sqrt{2}}\left[1 + \frac{3}{4}Q^2 + \sqrt{1 + \frac{3}{2}Q^2 \cos 2\phi + \frac{9}{16}Q^4}\right]^{\frac{1}{2}}$$

(6.23)

where σ = bending - stress amplitude including any stress - concentration effects

$$Q = 2\frac{\tau}{\sigma}$$

τ = torsional - stress amplitude including any stress - concentration effects

ϕ = phase angle between bending and torsion

The *SEQA* can be computed for both mean and alternating components of stress.

Figure 6-49 shows the variation of the *SEQA* effective stress, expressed as a ratio to the von Mises stress for the same bending-torsion combination as a function of two variables: the phase angle ϕ and the ratio τ/σ. Note in Figure 6-49*a* that when the bending and torsion are either in-phase, or 180° out-of-phase, the *SEQA* equals the von Mises stress σ'. At $\phi = 90°$, the *SEQA* is about 73% of σ'. The *SEQA* stress also varies with the relative values of τ and σ as shown in Figure 6-49*b*. When $\tau/\sigma = 0.575$, ($Q = 1.15$)

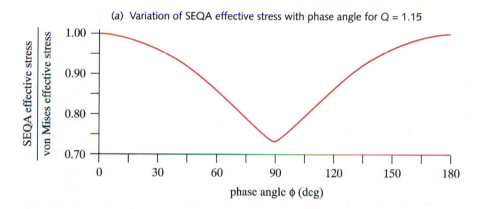

(a) Variation of SEQA effective stress with phase angle for Q = 1.15

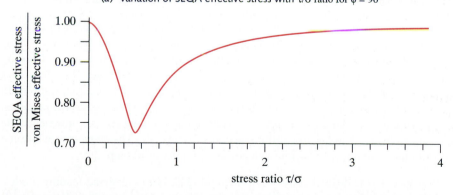

(a) Variation of SEQA effective stress with τ/σ ratio for $\phi = 90°$

FIGURE 6-49

Variation of SEQA Effective Stress with the τ/σ Ratio and the Phase Angle Between τ and σ

the reduction in *SEQA* stress at $\phi = 90°$ is maximal and it approaches σ' for large and small τ/σ ratios. This figure indicates that using the von Mises stress for combined bending and torsion fatigue gives a conservative result at any phase angle or τ/σ ratio. However, Garud has shown that this approach will be nonconservative for out-of-phase loading if the local strain is above about 0.13%.[44] Thus, this approach is not recommended for low-cycle fatigue situations. Tipton and Nelson[50] show that the SEQA approach is conservative for out-of-phase high-cycle fatigue (low-strain) applications. In fact, when the stress-concentration factors K_f and K_{fs} for the notch were set to 1, the SEQA and similar approaches* gave reasonably accurate predictions of HCF failure.[50]

The complex multiaxial fatigue analysis method presented above assumes that the applied loads are synchronous with a predictable phase relationship. If the sources of the multiple loads are decoupled and have a random or unknown time-phase relationship, then this method may not be sufficient to solve the problem. The reader should consult the literature as described in the bibliography to this chapter for more information on complex multiaxial loading cases. The best approach for unusual situations is a testing program of your own.

6.13   A GENERAL APPROACH TO HIGH-CYCLE FATIGUE DESIGN

The previous sections and examples have used a consistent approach regardless of the category of fatigue loading involved (See Figure 6-40 on p. 399). Even in the uniaxial stress cases, the von Mises stress was calculated for the alternating and mean stresses. It could be argued that this step is unnecessary with uniaxial stress since the von Mises stress will be identical to the applied stress. Nevertheless, for a slight additional computational burden (which is irrelevant if using a computer), we gain the advantage of consistency. In addition, the appropriate individual stress-concentration factors can be applied to the various stress components before incorporating them into the von Mises stress calculation. Frequently, the geometric stress-concentration factors for the same contour on a part will vary for different loadings (axial versus bending, etc.).

Whether the loading is uniaxial or multiaxial, bending or torsion, or any combination thereof, the safety factor with this approach is found in the same manner, comparing some combination of alternating and mean von Mises stresses to a line defined by the fatigue tensile strength and a static tensile strength of the material. This eliminates the need for computing separate torsional fatigue strengths. If you accept the approach outlined in the previous section for multiaxial loadings with stress concentrations, namely using the von Mises stress for both mean and alternating stress components, the difference between uniaxial and multiaxial cases disappears. The same computational algorithm then can apply to all four categories of Figure 6-40 on p. 399.

In respect to the difference between fluctuating and fully reversed loading modes, recall that the latter is just a special case of the former. We can treat all fatigue-loading cases as fluctuating and consistently apply the modified-Goodman diagram (MGD) criteria of failure with good results. Note in Figure 6-43 that the MGD and the *S-N* dia-

* A similar method based on the maximum shear-stress theory is also defined in reference [50]. This method, called SALT, gives similar but even more conservative results for HCF than those shown for the SEQA method in Figure 6-49. The caveats regarding its application only to HCF loading apply, though it gives better correlation to experimental results of low-cycle, strain-based multiaxial fatigue tests than does the SEQA method.[44]

gram are simply different views of the same three-dimensional relationship between mean stress σ'_m, alternating stress σ'_a, and number of cycles N. Figure 6-43c shows the Goodman section taken through the 3-D surface that relates the variables. A fully reversed stress state ($\sigma'_a \neq 0$, $\sigma'_m = 0$) can be plotted on a Goodman diagram and its safety factor calculated easily when you realize that the resulting data point will be on the σ'_a axis. Equation 6.18b gives its safety factor and is the same as equation 6.14 when $\sigma'_m = 0$. For that matter, a static loading problem ($\sigma'_m \neq 0$, $\sigma'_a = 0$) can also be plotted on the MGD and its data point will fall on the σ'_m axis. Its safety factor can be calculated from equation 6.18a which is identical to equation 5.8a when $\sigma'_a = 0$. Thus, the modified-Goodman diagram provides a universal tool to determine a safety factor for any stress problem, whether static, fully reversed fatigue, or fluctuating fatigue.

The recommended general approach to HCF design with uniaxial or synchronous multiaxial stresses is then:

1   Generate a suitable modified-Goodman diagram from tensile strength information for the particular material. This can be done for any desired finite life or for infinite life by taking the Goodman section at a different point N_2 along the N axis in Figure 6-42. This is automatically accomplished by the choice of S_f, at some number of cycles N_2 as shown in Figure 6-33 and equation 6.10. Apply the appropriate strength reduction factors from equations 6.7 to obtain a corrected fatigue strength.

2   Calculate the alternating and mean components of the applied stresses at all points of interest in the part and apply the appropriate stress-concentration factor to each applied stress component. (See Example 4-9 and the summary section of Chapter 4.)

3   Convert the alternating and mean components of the applied stresses at any point of interest in the loaded part to alternating and mean von Mises effective stresses using equations 6.22.

4   Plot the alternating and mean von Mises stresses on the modified-Goodman diagram and determine the appropriate safety factor from equations 6.18.

Recall from the discussion of static failure theories in Chapter 5 that the von Mises approach was recommended only for use with ductile materials as it accurately predicts yielding in the static-loading case where shear is the failure mechanism. Here we are using it for a slightly different purpose, namely to combine the multiaxial mean and alternating applied stresses into effective (pseudo-uniaxial) mean and alternating tensile stresses that can be compared with tensile fatigue and static strengths on a modified-Goodman diagram. As such, the von Mises approach can be used for both ductile and brittle materials in HCF fatigue loading since the (correct) assumption is that fatigue failures are tensile failures regardless of the ductility or brittleness of the material. In fact, it was long thought that ductile materials had somehow become embrittled under prolonged fatigue loading because their failure surfaces look like those of a statically-failed brittle material. However, this is now known to be untrue.

The designer should nevertheless be cautioned about using cast-brittle materials in fatigue-loaded situations because their tensile strengths tend to be lower than equivalent-density wrought materials, and the higher probability that they will contain stress raisers within the material from the casting process. Many successful applications of castings under fatigue loading can be cited such as IC engine crankshafts, camshafts, and connecting rods. These applications tend to be in smaller-size, low-power engines for lawn mowers, etc. Higher-powered automotive and truck engines will usually use (ductile) forged steel or nodular (ductile) cast iron rather than (brittle) gray cast iron for crankshafts and connecting rods, for example.

We now present an example of simple multiaxial fatigue using the same bracket that was investigated in Examples 4-9 and 5-1. This time the load is fluctuating with time.

EXAMPLE 6-6

Multiaxial Fluctuating Stresses

Problem	**Determine the safety factors for the bracket tube shown in Figure 5-7.**
Given	**The material is 2024-T4 aluminum with S_y = 47 000 psi, and S_{ut} = 68 000 psi. The tube length l = 6 in and arm a = 8 in. The tube outside diameter od = 2 in and inside diameter id = 1.5 in. The applied load varies sinusoidally from F = 340 to –200 lb.**
Assumptions	**The load is dynamic and the assembly is at room temperature. Consider shear due to transverse loading as well as other stresses. A finite-life design will be sought with a life of 6E7 cycles. The notch radius at the wall is 0.25 in and stress-concentration factors are for bending, K_t = 1.7, and for shear, K_{ts} = 1.35.**
Solution	**See Figure 5-7, Table 6-13, and file EX06-06 Also see Example 4-9 for a more complete explanation of the stress analysis for this problem.**

1   Aluminum does not have an endurance limit. Its endurance strength at 5E8 cycles can be estimated from equation 6.6c. Since the S_{ut} is larger than 48 kpsi, the uncorrected $S_{f'@5E8}$ = 19 kpsi.

2   The correction factors are calculated from equations 6.7 and used to find a corrected endurance strength at the standard 5E8 cycles.

$$C_{load} = 1: \quad \text{for bending}$$

$$C_{size} = 0.869\left(d_{equiv}\right)^{-0.097} = 0.869(2)^{-0.097} = 0.807$$

$$C_{surf} = 2.7\left(S_{ut}\right)^{-0.265} = 2.7(68)^{-0.265} = 0.883 \qquad (a)$$

$$C_{temp} = 1$$

$$C_{reliab} = 0.753: \quad \text{for 99.9\%}$$

$$S_{f_{@5e8}} = C_{load}\, C_{size}\, C_{surf}\, C_{temp}\, C_{reliab}\, S_{f'}$$

$$= (1)(0.807)(0.883)(1)(0.753)19\,000 = 10\,185 \text{ psi} \qquad (b)$$

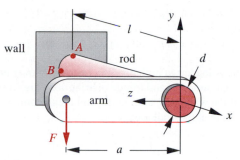

FIGURE 5-7   Repeated

Bracket for Examples 4-9, 5-1, and 6-6

Note that the bending value of C_{load} is used despite the fact that there is both bending and torsion present. The torsional shear stress will be converted to an equivalent tensile stress with the von Mises calculation. C_{surf} is calculated from equation 6.7e using data from Table 6-3. This corrected fatigue strength is still at the tested number of cycles, $N = 5E8$.

3 This problem calls for a life of 6E7 cycles, so a strength value at that life must be estimated from the S-N line of Figure 6-33b using the corrected fatigue strength at that life. Equation 6.10a for this line can be solved for the desired strength after we compute the values of its coefficient a and exponent b from equation 6.10c.

$$b = \frac{1}{z}\log\left(\frac{S_m}{S_f}\right) = \frac{1}{-5.699}\log\left[\frac{0.9(68\,000)}{10\,185}\right] = -0.1367$$

$$\log(a) = \log(S_m) - 3b = \log[0.9(68\,000)] - 3(-0.1367): \qquad a = 157\,297 \qquad (c)$$

$$S_n = aN^b = 157\,297N^{-0.1367} = 157\,297(6e7)^{-0.1367} = 13\,608 \text{ psi}$$

Note that S_m is calculated as 90% of S_{ut} because the loading is bending rather than axial (see Eq. 6.9). The value of z is taken from Table 6-5 for $N = 5E8$ cycles. This is a corrected fatigue strength for the shorter life required in this case and so is larger than the corrected test value, which was calculated at a longer life.

4 The notch sensitivity of the material must be found to calculate the fatigue stress-concentration factors. Table 6-8 shows the Neuber factors for hardened aluminum. Interpolation gives a value of 0.147 for $\sqrt{a}$ at the material's S_{ut}. Equation 6.13 gives the resulting notch sensitivity for an assumed notch radius of 0.25 in.

$$q = \frac{1}{1 + \dfrac{\sqrt{a}}{\sqrt{r}}} = \frac{1}{1 + \dfrac{0.147}{\sqrt{0.25}}} = 0.773 \qquad (d)$$

5 The fatigue stress-concentration factors are found from equation 6.11b using the given geometric stress-concentration factors for bending and torsion, respectively.

$$K_f = 1 + q(K_t - 1) = 1 + 0.773(1.7 - 1) = 1.541 \tag{e}$$

$$K_{fs} = 1 + q(K_{ts} - 1) = 1 + 0.773(1.35 - 1) = 1.270 \tag{f}$$

6 The bracket tube is loaded in both bending (as a cantilever beam) and in torsion. The shapes of the shear, moment and torque distributions are shown in Figure 4-30. All are maximum at the wall. The alternating and mean components of the applied force, moment, and torque at the wall are

$$F_a = \frac{F_{max} - F_{min}}{2} = \frac{340 - (-200)}{2} = 270 \text{ lb}$$
$$\tag{g}$$
$$F_m = \frac{F_{max} + F_{min}}{2} = \frac{340 + (-200)}{2} = 70 \text{ lb}$$

$$M_a = F_a l = 270(6) = 1\,620 \text{ lb} - \text{in}$$
$$\tag{h}$$
$$M_m = F_m l = 70(6) = 420 \text{ lb} - \text{in}$$

$$T_a = F_a a = 270(8) = 2\,160 \text{ lb} - \text{in}$$
$$\tag{i}$$
$$T_m = F_m a = 70(8) = 560 \text{ lb} - \text{in}$$

7 The fatigue stress-concentration factor for the mean stresses depends on the relationship between the maximum local stress in the notch and the yield strength as defined in equation 6.17, a portion of which is shown here.

$$\text{if } K_f |\sigma_{max}| < S_y \text{ then} \qquad K_{fm} = K_f, \qquad K_{fsm} = K_{fs}$$

$$K_f \left| \frac{M_{max} c}{I} \right| = K_f \left| \frac{F_{max} l c}{I} \right| = 1.541 \left| \frac{340(6)(1)}{0.5369} \right| = 5\,855 < 47\,000 \tag{j}$$

$$K_{fm} = 1.541 \qquad\qquad K_{fsm} = 1.270$$

In this case, there is no reduction in stress-concentration factors for the mean stress because there is no yielding at the notch to relieve the stress concentration.

8 The largest tensile bending stress will be in the top or bottom outer fiber at points A or A'. The largest torsional shear stress will be all around the outer circumference of the tube. (See Example 4-9 for more details.) First take a differential element at point A or A' where both of these stresses combine. (See Figure 4-32.) Find the alternating and mean components of the normal bending stress and of the torsional shear stress on point A using equations 4.11b and 4.24b, respectively.

$$\sigma_a = K_f \frac{M_a c}{I} = 1.541 \frac{1\,620(1)}{0.5369} = 4\,649 \text{ psi}$$
$$\tag{k}$$
$$\tau_a = K_{fs} \frac{T_a r}{J} = 1.270 \frac{2\,160(1)}{1.074} = 2\,556 \text{ psi}$$

$$\sigma_m = K_{fm} \frac{M_m c}{I} = 1.541 \frac{420(1)}{0.5369} = 1\,205 \text{ psi}$$

$$\tau_m = K_{fsm} \frac{T_m r}{J} = 1.270 \frac{560(1)}{1.074} = 663 \text{ psi}$$

(l)

9 Find the alternating and mean von Mises effective stresses at point A from Eq. 6.22b.

$$\sigma'_a = \sqrt{\sigma_{x_a}^2 + \sigma_{y_a}^2 - \sigma_{x_a}\sigma_{y_a} + 3\tau_{xy_a}^2}$$

$$= \sqrt{4\,649^2 + 0^2 - 4\,649(0) + 3(2\,556^2)} = 6\,419 \text{ psi}$$

$$\sigma'_m = \sqrt{\sigma_{x_m}^2 + \sigma_{y_m}^2 - \sigma_{x_m}\sigma_{y_m} + 3\tau_{xy_m}^2}$$

$$= \sqrt{1\,205^2 + 0^2 - 1\,205(0) + 3(663^2)} = 1\,664 \text{ psi}$$

(m)

10 Because the moment and torque are both caused by the same applied force, they are synchronous and in-phase and any changes in them will be in a constant ratio. This is a Case 3 situation and the safety factor is found using equation 6.18e.

$$N_f = \frac{S_f S_{ut}}{\sigma'_a S_{ut} + \sigma'_m S_f} = \frac{13\,608(68\,000)}{6\,419(68\,000) + 1\,664(13\,608)} = 2.0$$

(n)

11 Since the tube is a short beam, we need to check the shear due to transverse loading at point B on the neutral axis where the torsional shear is also maximal. The maximum transverse shear stress at the neutral axis of a hollow, thin-walled, round tube was given as equation 4.15d.

$$\tau_{a_{bending}} = K_f \frac{2V_a}{A} = 1.541 \frac{2(270)}{1.374} = 605 \text{ psi}$$

$$\tau_{m_{bending}} = K_{fm} \frac{2V_m}{A} = 1.541 \frac{2(70)}{1.374} = 157 \text{ psi}$$

(o)

Point B is in pure shear. The total shear stress at point B is the sum of the transverse shear stress and torsional shear stress which act on the same planes of the element.

$$\tau_{a_{total}} = \tau_{a_{bending}} + \tau_{a_{torsion}} = 605 + 2\,556 = 3\,161 \text{ psi}$$

$$\tau_{m_{total}} = \tau_{m_{bending}} + \tau_{m_{torsion}} = 157 + 663 = 819 \text{ psi}$$

(p)

12 Find the alternating and mean von Mises effective stresses for point B from Eq. 6.22b.

$$\sigma'_a = \sqrt{\sigma_{x_a}^2 + \sigma_{y_a}^2 - \sigma_{x_a}\sigma_{y_a} + 3\tau_{xy_a}^2} = \sqrt{0 + 0 - 0 + 3(3\,161)} = 5\,475 \text{ psi}$$

$$\sigma'_m = \sqrt{\sigma_{x_m}^2 + \sigma_{y_m}^2 - \sigma_{x_m}\sigma_{y_m} + 3\tau_{xy_m}^2} = \sqrt{0 + 0 - 0 + 3(819)} = 1\,419 \text{ psi}$$

(q)

13   The safety factor for point B is found using equation 6.18e.

$$N_f = \frac{S_f S_{ut}}{\sigma'_a S_{ut} + \sigma'_m S_f} = \frac{13\,608(68\,000)}{5\,475(68\,000) + 1\,419(13\,608)} = 2.4 \qquad (r)$$

Both points A and B are safe against fatigue failure.

14   This example can be examined in more detail in the *TKSolver* file EX06-06 on the accompanying disk.

6.14   A CASE STUDY IN FATIGUE DESIGN

The following case study contains all the elements of a HCF fatigue design problem. It is an actual design problem from the author's consulting experience and serves to illustrate many of the points of this chapter. While it is long and fairly complicated, its careful study will prove to be worth the time invested. Several *TKSolver* files are provided that solve the various designs that were analyzed.

CASE STUDY 6

Redesign of a Failed Laybar for a Water-Jet Power Loom

Problem   The laybars in a number of water-jet looms have begun to fail in fatigue. The owner of the weaving works had increased the speed of the looms to boost production. The original design of a painted steel laybar had lasted with no failures for 5 years of 3-shift operation at the lower speed but began failing within months of the speed increase. The owner had a local machine shop make painted steel replacements similar to the original and these failed in six months of use. The owner substituted an aluminum replacement laybar of his own design, which lasted 3 months. He then sought engineering assistance. Analyze the failures of the three existing designs and redesign the part to last for an additional 5 years at the higher speed.

Given   The laybar is 54 in long and is carried between the rockers of two identical Grashof crank-rocker fourbar linkages that are driven synchronous and in-phase by gear trains connected through a 54 in-long transmission shaft. The loom arrangement is shown in Figure 6-50 and the linkage is shown in Figure 6-51. Details of its operation are discussed below. Cross sections of the failed designs are shown in Figure 6-53 and photographs in Figure 6-54. The new design cannot be any wider than the widest existing one (2.5 in). The original loom speed was 400 rpm and the new speed is 500 rpm. The cost of a new design should be competitive with the cost of current (failed) designs (about $300 each in lots of 50).

Assumptions   The major fluctuating loading on the part is inertial due to its own mass plus that of the reed carried on it being accelerated and decelerated by the linkage motion. There is also a "beat-up" force on the reed when it strikes the cloth to push the latest weave thread into place. This force causes a repeated torque on the laybar that may or may not be significant in the failure. The magnitude of the beat-up force is not accurately known and will vary with the weight of the cloth being woven. It is estimated to be 10 lb/in of cloth width (540-lb total). The environment is wet with fresh water and all failed specimens show evidence of corrosion.

Solution   See Figures 6-50 to 6-56 and TKSolver files CASE6-0 through CASE6-7

1   Some additional background information is needed to understand the problem before pursuing its solution. Weaving looms for the making of cloth are very old devices and were originally human-powered. The power loom was invented during the industrial revolution and presently exists in many forms. Figure 6-50 shows parts of the water-jet power loom of concern here. Perhaps the best way to understand a loom's fundamental operation is to consider a hand-powered one, which the reader may have seen in a museum, a custom-weaving shop, or in a hobbyist's workroom. Its basic elements are similar to some of those in the figure.

A set of threads called the *warp* is strung across the loom. Each thread is grabbed by a device (not shown) that can pull it up or down. These devices are activated by a mechanism that, in a hand-loom, is typically operated by foot pedals. When one pedal is pushed, every other warp thread is raised up and the alternate ones pulled down to create a "tunnel" if observed from the edge of the cloth. This tunnel is called the *shed*. A *shuttle*, which looks like a miniature canoe, and contains a bobbin of thread within it, is next "thrown" through the shed by the weaver's hand. The shuttle trails a single

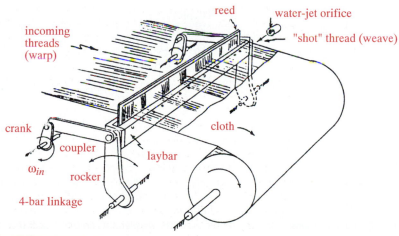

FIGURE 6-50

Warp, Weave, Laybar, Reed, and Laybar Drive for a Water-Jet Loom thread called the *weave* through the warp shed. The weaver then pulls on the *laybar*, which carries a comb-like device called a *reed*. The warp threads are strung through the teeth of this reed-comb. The reed pushes the new weave-thread sideways into the previous ones to "beat-up" the cloth and create a tight weave. The weaver next switches pedals on the shed and the "up" threads of the warp become the "down" threads and vice versa creating a new tunnel (shed) of crossed threads. The shuttle is again thrown through the warp (from the other side), weaving another thread to be beat-up by the reed.

The original power looms simply mechanized the manual process, replacing the weavers hands and feet with linkages and gears. The throwing of the wooden shuttle was accomplished by literally hitting it with a stick, flying it through the shed and catching it on the other side. The dynamics of this (pre-NASA) "shuttle flight" became the limiting factor on the loom's speed. Shuttle looms can only go at about 100 picks (threads) per minute (ppm). Much effort was expended to develop faster looms and these usually involved eliminating the shuttle whose mass limited the speed. Both air-jet and water-jet looms were developed in this century that shoot the weave thread across the shed on a jet of air or water. Figure 6-50 shows the orifice through which the thread is fed. At the right time in the cycle, a small piston pump shoots a jet of water through the orifice and surface tension pulls the thread across the shed. The water-jet loom can operate at up to about 500 ppm. The looms in question were designed to operate at 400 ppm but the owner changed their gearing to increase the speed to 500 ppm. Failures soon ensued because the dynamic loads increased with the square of the speed and exceeded the loads for which the machine was designed.

2   Figures 6-50 and 6-51 show the laybar, which is carried between two identical fourbar linkages that move it in an arc to push the reed into the cloth at the right point in the cycle. The linkage pivots are in self-aligning ball-bearings, which allow us to model the laybar as a simply supported beam that carries a uniformly distributed load

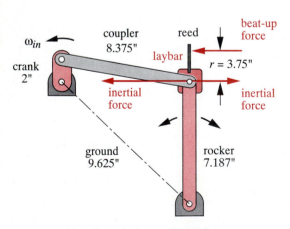

(a) Linkage, laybar, reed, and dimensions   (b) Acceleration on laybar and force on reed

FIGURE 6-51

Fourbar Linkage for Laybar Drive, Showing Forces and Accelerations on Laybar that is equal to its total mass times its acceleration plus the beat-up force. The total mass is the sum of the laybar mass and the mass of its 10-lb payload, the reed. Figure 6-51 shows the laybar-linkage geometry, its dimensions and a polar plot of the acceleration vectors at the mass-center of the laybar. The tangential components of acceleration are the largest and create bending moments in the directions of the inertial forces shown in the same figure. Figure 6-52 plots the tangential acceleration component of the laybar mass-center for 1 cycle and shows the beat-up force in its phase relationship with acceleration. The acceleration creates a fluctuating bending moment and the beat-up force, being offset 3.75 in from the mass center of the laybar, creates a repeated torque on the laybar. Depending on the laybar's cross-sectional geometry, this combination of loads can create a case of synchronous, in-phase, *simple multiaxial stress* at locations of maximum stress (see Section 6.12). Because the loading is largely inertial, the design of the laybar should minimize its mass (to reduce the inertial loading) while simultaneously maximizing its stiffness and strength. These are conflicting constraints, making the design task more challenging.

3 Since this is a case of fluctuating loading, we will follow the set of design steps recommended in Section 6.11, the first of which is to determine the number of loading cycles expected over the service life. The owner has requested that the new design last for 5 years of 3-shift operation. Assuming 2080 hours per shift in a standard work-year, this amounts to

$$N = 500 \ \frac{\text{cycles}}{\text{min}} \left(\frac{60 \ \text{min}}{\text{hr}} \right) \left(\frac{2 \ 080 \ \text{hr}}{\text{shift - yr}} \right) (3 \ \text{shifts})(5 \ \text{yr}) = 9.4\text{E8 cycles} \qquad (a)$$

This is clearly in the HCF regime and could benefit from the use of a material with an endurance limit.

The owner reports that his steel replacement laybar lasted about 6 months and his aluminum design lasted 3 months. (See Figures 6-53 and 6-54). The cycle lives are:

$$6 \ \text{mos}: \ N = 500 \ \frac{\text{cycles}}{\text{min}} \left(\frac{60 \ \text{min}}{\text{hr}} \right) \left(\frac{2 \ 080 \ \text{hr}}{\text{yr}} \right) (3 \ \text{shifts})(0.5 \ \text{yr}) = 9.4\text{E7 cycles}$$

$$(b)$$

$$3 \ \text{mos}: \ N = 500 \ \frac{\text{cycles}}{\text{min}} \left(\frac{60 \ \text{min}}{\text{hr}} \right) \left(\frac{2 \ 080 \ \text{hr}}{\text{yr}} \right) (3 \ \text{shifts})(0.25 \ \text{yr}) = 4.7\text{E7 cycles}$$

4 Since the amplitudes of the applied bending loads are a function of acceleration (which is determined) and the mass of the part (which will vary with the design), it is best to express the bending loads in terms of $F = ma$. The applied torque is assumed to be the same for any design based on the owner's estimate of a typical beat-up force. These data are shown in Figure 6-52 and the mean and alternating terms are

bending :

$$F_{mean} = ma_{mean} = m \frac{a_{max} + a_{min}}{2} = m \frac{8 \ 129 + (-4 \ 356)}{2} = 1 \ 886.5m$$

$$(c)$$

$$F_{alt} = ma_{alt} = m \frac{a_{max} - a_{min}}{2} = m \frac{8 \ 129 - (-4 \ 356)}{2} = 6 \ 242.5m$$

tangential acceleration (in/sec²)

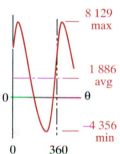

8 129 max

1 886 avg

0 —— θ

−4 356 min

0   360

beat-up force (lb)

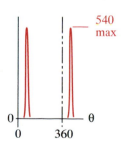

540 max

0 —— θ

0   360

FIGURE 6-52

Time-Varying Acceleration and Load on the Laybar at 500 rpm Showing their Phase Relationship

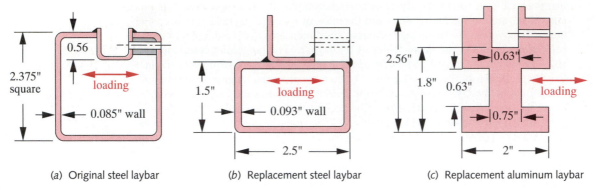

(a) Original steel laybar   (b) Replacement steel laybar   (c) Replacement aluminum laybar

FIGURE 6-53

6

Existing Laybar Designs, All of Which Failed in Fatigue

torque :

$$T_{mean} = \left(\frac{F_{beat\,max} + F_{beat\,min}}{2} \right) r = \left(\frac{540 + 0}{2} \right) 3.75 = 1\,012.5 \text{ lb - in}$$

$$T_{alt} = \left(\frac{F_{beat\,max} - F_{beat\,min}}{2} \right) r = \left(\frac{540 - 0}{2} \right) 3.75 = 1\,012.5 \text{ lb - in}$$

(d)

5   We are in the enviable position of having test data for typical parts run under actual service conditions available in the form of the failed specimens. In effect, the owner has inadvertently run a test program (to his chagrin) and has determined stress levels that cause failure in this application. Thus the first step will be to analyze the existing failed designs in order to learn more about the problem. We know that the original design (Figure 6-53a) survived for 5 years at the lower stress levels associated with 400-rpm operation. They only began to fail when he increased the speed, which increased the inertial loading.

There are many factors involved in this application that are difficult to quantify. Corrosion is evident on the failed parts. The steel laybars are rusted with pitted surfaces. The unanodized aluminum part is also pitted. The designers did not take great care to minimize stress concentrations and the fatigue fractures can be seen to have started (as is typical) at stress raisers. The failed aluminum part in Figure 6-54b shows that the crack started at a tapped hole, which is a very sharp notch. The cracks in the steel part (Figure 6-54b) appear to have started at a weld bead used to attach the reed supports. Welds are notorious stress-raisers and often leave tensile residual stresses behind. These lessons should be taken into account in our redesign and attempts made to reduce these negative factors. **By definition, a failed part has a safety factor of 1**. Knowing this, a model of the part's loading, stresses, and safety factor can be created and then *back-calculated with the safety factor set to 1* in order to determine various of the above factors that are difficult to quantify for a particular application.

(a)   (b)

FIGURE 6-54

Photographs of Failed Laybars (a) Replacement steel laybar — after 6 mos, (b) Replacement aluminum laybar — after 3 mos.

6   A *TKSolver* model was created to solve the equations for this case. Data specific to each of the three failed designs were input and the model modified as necessary to account for differences in geometry and material among the three designs. The same model was further modified to accommodate the proposed new designs shown in Figure 6-55. Eight versions of the *TKSolver* model resulted, and are included with this text. They are labeled CASE6-0 through CASE6-7. Space does not permit discussion of the contents of all eight models, so only two will be discussed in detail and the results of the others will be compared in summary. The failed original design and the final new design will be presented. The reader is encouraged to investigate all the models within *TKSolver* using the program and data files provided.

7   The analysis of the original laybar design is contained in CASE6-1. The section geometry and the beam mass must be calculated to determine the bending stresses.

$$area = 2.375^2 - 2.205^2 + 2(0.56)(0.085) = 0.874 \text{ in}^2$$

$$weight = area(length)(\gamma) = 0.874(54)(0.286) = 13.5 \text{ lb}$$

$$m = \frac{weight + reed}{386.4} = \frac{13.5 + 10}{386.4} = 0.061 \text{ blobs}$$

$$I = \frac{b_{out}h_{out}^3 - b_{in}h_{in}^3}{12} = \frac{2.375^4 - 2.205^4}{12} = 0.68 \text{ in}^4$$

(e)

Note that the area calculation includes the reed-trough sides since they add mass but the calculation for I ignores them since they add a negligible amount to that quantity. The specific weight γ is for steel and the mass unit is blobs or lb-sec^2/in.

8   The nominal mean and alternating components of the inertial force and bending moment can now be calculated.

$$F_{mean} = 1\,886.5m = 1\,886.5(0.061) = 115 \text{ lb}$$

$$F_{alt} = 6\,242.5m = 6\,242.5(0.061) = 380 \text{ lb} \tag{f}$$

$$M_{mean} = \frac{wl^2}{8} = \frac{(wl)l}{8} = \frac{Fl}{8} = \frac{115(54)}{8} = 775 \text{ lb - in}$$

$$M_{alt} = \frac{wl^2}{8} = \frac{(wl)l}{8} = \frac{Fl}{8} = \frac{380(54)}{8} = 2\,565 \text{ lb - in}$$

The moment equations are for the maximum moment in the center of a simply-supported beam with uniformly distributed load (see Figure D-2 in Appendix D). The nominal bending stresses (not including any stress concentration) are then

$$\sigma_{m_{nom}} = \frac{M_{mean}(c)}{I} = \frac{775(1.188)}{0.68} = 1\,351 \text{ psi}$$

$$\sigma_{a_{nom}} = \frac{M_{alt}(c)}{I} = \frac{2\,565(1.188)}{0.68} = 4\,470 \text{ psi} \tag{g}$$

9   The nominal shear stresses due to torsion in a hollow-square section are maximum in the centers of the four sides and so occur at points of maximum bending stress. The shear stress is found from $\tau_{max} = T/Q$ (Eq. 4.26a) where Q for the particular geometry is found in Table 4-7:

$$Q = 2t(a-t)^2 = 2(0.085)\left(\frac{2.375}{2} - 0.085\right)^2 = 0.207 \text{ in}^3 \tag{h}$$

where t is the wall thickness and a is the half-width of the cross section. The nominal mean and alternating shear stresses are then

$$\tau_{m_{nom}} = \frac{T_{mean}}{Q} = \frac{1\,012.5}{0.207} = 4\,900 \text{ psi}$$

$$\tau_{a_{nom}} = \frac{T_{alt}}{Q} = \frac{1\,012.5}{0.207} = 4\,900 \text{ psi} \tag{i}$$

10   The stress-concentration factors for bending and shear need to be found or estimated. Peterson[30] provides a chart for the case of a hollow, rectangular section in torsion, and from that a $K_{ts} = 1.08$ is found. No suitable data were found for the bending stress-concentration factor for this case. The corrosion and pitting in combination with rough welds would predict a large K_t. The approach taken here was to back-solve for K_t with the safety factor set to 1 and all other material factors and the nominal stresses specified. The result was $K_t = 4.56$ in this failed part. This result is presented at this point to provide continuity of narrative but it must be understood that the value of K_t was found by back-solving the *TK* model using iteration after all other factors were defined. The local alternating and mean stresses and K_t were then solved for simultaneously with $N_f = 1$.

11   The material's notch sensitivity and the fatigue stress-concentration factors for alternating bending and shear are found from equations 6.11b and 6.13 following the procedure used in Example 6-3. The results are: $q = 0.8$, $K_f = 3.86$, and $K_{fs} = 1.06$. The corresponding fatigue stress-concentration factors for the mean stress are found from

equation 6.17 and since the local stress is below the yield point for both bending and torsion in this case, they are identical to the factors for the alternating stress: $K_{fm} = K_f$, and $K_{fms} = K_{fm}$.

12 The local-stress components can now be found using the fatigue stress-concentration factors:

$$\sigma_m = K_{fm}\sigma_{m_{nom}} = 3.86(1\,351) = 5\,212 \text{ psi}$$

$$\sigma_a = K_f\sigma_{a_{nom}} = 3.86(4\,470) = 17\,247 \text{ psi}$$

(j)

$$\tau_m = K_{fsm}\tau_{m_{nom}} = 1.06(4\,900) = 5\,214 \text{ psi}$$

$$\tau_a = K_{fsm}\tau_{a_{nom}} = 1.06(4\,900) = 5\,214 \text{ psi}$$

(k)

13 Since we have a case of combined, fluctuating, biaxial stresses that are synchronous and in-phase, and that have stress concentration, the general method using von Mises effective stresses for both mean and alternating components is appropriate (equation 6.22b). These calculate equivalent alternating and mean stresses for the biaxial case.

$$\sigma'_a = \sqrt{\sigma_{x_a}^2 + \sigma_{y_a}^2 - \sigma_{x_a}\sigma_{y_a} + 3\tau_{xy_a}^2}$$

$$= \sqrt{17\,247 + 0 - 0 + 3(5\,214)} = 19\,468 \text{ psi}$$

(l)

$$\sigma'_m = \sqrt{\sigma_{x_m}^2 + \sigma_{y_m}^2 - \sigma_{x_m}\sigma_{y_m} + 3\tau_{xy_m}^2}$$

$$= \sqrt{5\,212 + 0 - 0 + 3(5\,214)} = 10\,428 \text{ psi}$$

14 The material properties must now be determined. A laboratory test was done on a sample from the failed part and its chemical composition matched that of an AISI 1018 cold rolled steel. Strength values for this material were obtained from published data (see Appendix C) and are: $S_{ut} = 64\,000$ psi and $S_y = 50\,000$ psi. A shear yield strength was calculated from $S_{ys} = 0.577\,S_y = 28\,850$ psi. An uncorrected endurance limit was taken as $S_{e'} = 0.5\,S_{ut} = 32\,000$ psi.

15 The strength modification factors were found from the equations and data in Section 6.6. The loading is a combination of bending and torsion. However, we have incorporated the torsion stresses into the von Mises equivalent stress, which is a normal stress, so

$$C_{load} = 1$$

(m)

The equivalent diameter test specimen is found from the 95% stress area using equations 6.6c. and 6.6d. The size factor is then found from equation 6.7b:

$$A_{95} = 0.05bh = 0.05(2.375)^2 = 0.282 \text{ in}^2$$

$$d_{equiv} = \sqrt{\frac{A_{95}}{0.0766}} = 1.92 \text{ in}$$

(n)

$$C_{size} = 1.189(d_{equiv})^{-0.097} = 1.189(1.92)^{-0.097} = 0.81$$

The surface factor is found from equation 6.7e and the data in Table 6-3 for machined or cold-drawn surfaces. The material of the laybar appeared to have been originally cold-drawn but was corroded. Corrosion could dictate the use of a lower-valued surface factor, but it was decided to allow the geometric stress-concentration factor K_t to account for the effects of pitting in this case as described above and the machined-surface factor was applied.

$$C_{surf} = A(S_{ut})^b = 2.7(64)^{-0.265} = 0.897 \tag{o}$$

The temperature factor and the reliability factor were both set to 1. The reliability was taken as 50% for this back-calculation in order to place all the uncertainty in the highly-variable stress-concentration factor.

16   A corrected endurance limit can now be calculated from

$$S_e = C_{load} C_{size} C_{surf} C_{temp} C_{reliab} S_e'$$
$$S_e = (1)(0.81)(0.90)(1)(1)(32\ 000) = 23\ 258 \text{ psi} \tag{p}$$

17   The safety factor is calculated from equation 6.18e. Safety factor case 3 is applicable here since, with inertial loading, the mean and alternating components of bending stress will maintain a constant ratio with changes in speed. Because the minimum beat-up force is always zero, that ratio is also constant regardless of the maximum force.

$$\frac{1}{N_{sf}} = \frac{1}{1} = \frac{\sigma'_a}{S_e} + \frac{\sigma'_m}{S_{ut}}$$
$$N_{sf} = 1 = \frac{19\ 468}{23\ 258} + \frac{10\ 428}{64\ 000} \tag{q}$$

The alternating and mean stresses were back-solved and can now be used to plot a modified-Goodman diagram. Since we forced the safety factor to 1 to represent the known failure of this part, the applied stress point σ'_a, σ'_m falls on the Goodman line.

18   The above analysis was repeated, changing the operating speed to the original design value of 400 rpm. Using the same stress-concentration factor of 4.56 that was backed-out of the failed-part analysis, the safety factor at 400 rpm is 1.3 indicating why the original design survived at the design rpm. (See file Case6-0)

19   The analysis of this and the other failed parts provides some insight into the constraints of the problem and allows a better design to be created. Some of the factors that influenced the new design were the corrosive environment, which makes steel a less desirable material despite its endurance limit. The painted finishes had not protected the now rusty steel parts. The one failed aluminum sample examined also showed significant pitting in only 3 months of use. If aluminum is to be used, an anodized finish should be applied to protect it from oxidation.

Another obvious factor is the role of stress concentration, which appears to be quite high in this part. The presence of welds and tapped holes in regions of high stress clearly contributed to the failures. Any new design should reduce stress concentrations by moving the required screw holes for reed attachment to locations of lower stress. Welds in high-stress regions should be avoided if possible. Surface treatments such as shot-peening should be considered in order to provide beneficial compressive residual stresses.

The poorly defined but potentially significant levels of torsional stresses are a concern. The weaving of heavier cloth will provide larger levels of torsional stress. Thus the geometry of any new design should be resistant to torsional stress as well as to bending stress. Finally, a new design should not be much heavier than the existing design since additional mass will cause the higher inertial forces to be transmitted to all other parts of the machine, possibly engendering failures of other parts.

20   Because the loading on the beam is primarily inertial in nature and because it carries a fixed-weight payload, there should be an optimum cross section for any design. A beam's resistance to bending is a function of its area moment of inertia I. The inertial loading is an inverse function of its area A. If the cross section were solid, its I would be the largest possible for a given outside dimension but so would its area, mass and inertial load. If the wall were made paper-thin, its mass would be minimal but so would its I. Both A and I are nonlinear functions of its dimensions. Thus, there must be some particular wall thickness that maximizes the safety factor, all else constant.

(a) Square design

With all the above factors in mind, two designs were considered as shown in Figure 6-55, a square and a round cross section with integral external ears to support the reed. They both share some common features. The contours have generous radii to minimize K_t. (The round section is the ultimate in this regard). The reed-support ears, which must contain threaded holes, are close to the neutral axis where the bending stress is lower and they are external to the basic geometric structure. There may not need to be any welds if the shape can be extruded as shown. Both are basically closed sections that can resist torques and the round section is the optimum shape for torsional loading. The square section will have a larger I and will thus resist bending better than a round shape of the same overall dimension.

Two materials were considered, mild steel and aluminum. (Titanium would be ideal in terms of its strength-to-weight ratio (SWR) and endurance limit, but its high cost precludes consideration.) Aluminum (if anodized) has the advantage of better corrosion resistance in water, but steel has the advantage of an endurance limit if protected from corrosion. The overall weight of the new design is of concern. High-strength aluminum has a better SWR than mild steel. (High-strength steel does not show an endurance limit and is notch sensitive as well as expensive.) Aluminum can be custom extruded with integral ears for low tooling cost thus making a short production run economically feasible. Tooling for a custom cross section in steel would require very large quantities be purchased to amortize the tooling cost. So, a steel design will be limited to stock mill-shapes and will require welding-on of the ears.

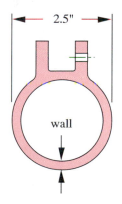

(b) Round design

21   Each of the geometries in Figure 6-55 was designed separately in both steel and aluminum. The wall thickness was varied within each *TKSolver* model as a list of values from very thin to nearly solid in order to determine the optimum dimension. The details of each design can be explored by the reader within the *TKSolver* files supplied. The final design chosen was a round section in 6061-T6 extruded aluminum with a wall thickness of 0.5 in. Its design will now be discussed though it must be understood that a great deal of iteration was required to arrive at the result presented here. Space does not permit discussion of all the iterations.

FIGURE 6-55

Two New Designs for the Water-Jet Laybar

22　The previous calculation of *cycle life* (equation *a*, step 3), *acceleration* and *beat-up force* (equations *c* and *d*, step 4) are still applicable. The section properties are

$$area = \pi\left(\frac{2.5^2 - 2.0^2}{4}\right) + 2(0.5)(0.75) = 3.142 \text{ in}^2$$

$$weight = area(length)(\gamma) = 3.142(54)(0.10) = 21.1 \text{ lb}$$

$$m = \frac{weight + reed}{386.4} = \frac{21.1 + 10}{386.4} = 0.080 \text{ blobs} \tag{r}$$

$$I = \pi\left(\frac{2.5^4 - 2.0^4}{64}\right) = 1.669 \text{ in}^4$$

Note that the area calculation includes the ears since they add mass but the calculation for *I* ignores them since they add a negligible amount to that quantity. The weight density γ for aluminum is in lb/in^3 and the mass unit is blobs or $\text{lb-sec}^2/\text{in}$.

23　The mean and alternating components of the inertial force and bending moment are

$$F_{mean} = 1\,886.5m = 1\,886.5(0.08) = 152 \text{ lb}$$

$$F_{alt} = 6\,242.5m = 6\,242.5(0.08) = 502 \text{ lb}$$

$$M_{mean} = \frac{Fl}{8} = \frac{152(54)}{8} = 1\,023 \text{ lb-in} \tag{s}$$

$$M_{alt} = \frac{Fl}{8} = \frac{502(54)}{8} = 3\,386 \text{ lb-in}$$

The moment equations are for the maximum moment in the center of a simply-supported beam with uniformly distributed load. The nominal bending stresses (not including stress concentration) are then

$$\sigma_{m_{nom}} = \frac{M_{mean}(c)}{I} = \frac{1\,023(1.25)}{1.669} \cong 766 \text{ psi}$$

$$\sigma_{a_{nom}} = \frac{M_{alt}(c)}{I} = \frac{3\,386(1.25)}{1.669} \cong 2\,536 \text{ psi} \tag{t}$$

If we compare these results to those of the original design (see step 8), the forces and moments are now greater due to the heavier part but the stresses are less due to the larger *I* of the cross section.

24　The torsional shear stresses in a hollow-round section are maximum at the outer fiber and so occur at points of maximum bending stress. The nominal shear stress is found from $\tau_{max} = Tr/J$ (Eq. 4.24*b*) where *J* for its geometry is found from equation 4.25*b*:

$$J = \pi\left(\frac{d_{out}^4 - d_{in}^4}{32}\right) = \pi\left(\frac{2.5^4 - 2.0^4}{32}\right) = 3.338 \text{ in}^4 \tag{u}$$

The nominal mean and alternating shear stresses are then

$$\tau_{m_{nom}} = \frac{T_{mean}(r)}{J} = \frac{1\,012.5}{3.338} = 379 \text{ psi}$$

$$\tau_{a_{nom}} = \frac{T_{alt}(r)}{J} = \frac{1\,012.5}{3.338} = 379 \text{ psi}$$

(v)

25 Because of the large radius and smooth contours of this round part, K_t and K_{ts} were taken as 1. There will be larger stress concentrations at the roots of the ears but the bending stress is much lower there and the shear stress in this torsionally optimum shape is very low. The material's notch sensitivity is irrelevant when $K_t = 1$ making both $K_f = 1$ and $K_{fs} = 1$. The fatigue stress-concentration factors for the mean stress are also 1 with the above assumptions. The local-stress components are then the same as the nominal stress components found in equations t and v.

26 Since we have a case of combined, fluctuating, biaxial stresses that are synchronous and in-phase, and the notches have been designed out, the method of Sines is now appropriate (equation 6.21b). This calculates equivalent alternating and mean stresses for the biaxial, unnotched case.

$$\sigma'_a = \sqrt{\sigma_{x_a}^2 + \sigma_{y_a}^2 - \sigma_{x_a}\sigma_{y_a} + 3\tau_{xy_a}^2}$$

$$= \sqrt{2\,536 + 0 - 0 + 3(379)} = 2\,619 \text{ psi}$$

$$\sigma'_m = \sigma_{x_m} + \sigma_{y_m} = 766 + 0 = 766 \text{ psi}$$

(w)

27 The material properties must now be determined. Aluminum does not have an endurance limit but fatigue strengths at particular cycle lives are published. A search of the literature showed that of all the aluminum alloys, 7075 and 5052 offered the largest fatigue strengths S_f. However, no aluminum extruders were found locally that could extrude either of those alloys. The strongest extruded alloy available was 6061-T6 with a published $S_{f'} = 13\,500$ psi at $N = 5E7$, $S_{ut} = 45\,000$, and $S_y = 40\,000$ psi. Since we need a life of about $N = 9.4E8$ cycles, the equation for this material's S-N curve must be written and solved for $S_{n'}$ at $N = 9.4E8$ cycles. To do so we need the material's strength S_m at 10^3 cycles from equation 6.9a:

$$S_m = 0.9 S_{ut} = 0.9(45\,000) = 40\,500 \text{ psi}$$

(x)

Using equations 6.10d and 6.10a to find the coefficient and exponent of the S-N line:

$$a = \frac{(S_m)^2}{S_f} = \frac{40\,500^2}{13\,500} = 121\,500$$

$$b = -\frac{1}{4.699} \log\left(\frac{S_m}{S_f}\right) = -0.1015$$

(y)

$$S_{n'} = a N^b = 121\,500 N^{-0.1015} = 121\,500(9.4E8)^{-0.1015}$$

$$S_{n'}{}_{@9.4E8} \cong 14\,920 \text{ psi} = S_{f'}$$

This value will be used as an uncorrected $S_{f'}$ at the desired life.

6

28 The strength modification factors were found from the equations and data in Section 6.6. The loading is a combination of bending and torsion, which appears to create a conflict in the selection of a loading factor from equation 6.7*a*. However, we have incorporated the torsion stresses into the Sines equivalent stress, which is a normal stress, so $C_{load} = 1$. The equivalent diameter for a round part is its outside diameter. The size factor is found from equation 6.7*b*:

$$C_{size} = 1.189(d_{equiv})^{-0.097} = 1.189(2.5)^{-0.097} = 0.79 \qquad (z)$$

The surface factor is found from equation 6.7*e* and the data in Table 6-3 for machined or cold-drawn surfaces. Corrosion could dictate the use of a lower-valued surface factor, but since the part will be anodized for corrosion resistance, the machined-surface factor was applied.

$$C_{surf} = A S_{ut}^b = 2.7(45)^{-0.265} = 0.98 \qquad (aa)$$

Note that S_{ut} is in kpsi for equation 6.7*e*. The temperature factor was set to 1 since it operates in a room-temperature environment. C_{reliab} was set to 0.702 from Table 6-4 to represent a desired 99.99% reliability for the new design.

29 A corrected fatigue strength can now be calculated from

$$S_f = C_{load} C_{size} C_{surf} C_{temp} C_{reliab} S_{f'}$$
$$S_f = 1(0.79)(0.98)(1)(0.702)(14\,920) \cong 8\,100 \text{ psi} \qquad (ab)$$

30 The equivalent alternating and mean stresses can now be plotted on a modified-Goodman diagram or the safety factor can be calculated from equation 6.18*e* for a case 3 situation as described in step 17.

$$\frac{1}{N_f} = \frac{\sigma_a}{S_f} + \frac{\sigma_m}{S_{ut}} = \frac{2\,536}{8\,100} + \frac{766}{45\,000}$$
$$N_f = 2.9 \qquad (ac)$$

N_f

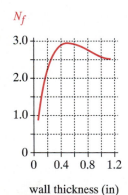

wall thickness (in)

FIGURE 6-56

Safety Factor *N* as a Function of Wall Thickness for a Round Aluminum Laybar

This safety factor is quite acceptable, but for an additional measure of safety, the finished parts were shot peened before anodizing. The variation of safety factor with wall thickness is shown in Figure 6-56. The peak occurs at a wall thickness of about 0.5 in, which was the value used in the design. Curves of safety factor versus wall thickness are plotted for all designs within their *TKSolver* files. They are similar in shape to Figure 6-56 and show an optimum wall thickness to maximize the safety factor.

31 The *TKSolver* file names and pertinent data for the seven designs are shown in Table 6-13. The original design is shown for both the machine's design speed of 400 rpm at which it performed successfully and for the increased speed of 500 rpm, at which it failed. The only difference is the safety factor, which went from 1.3 to 1. The K_t factors are back-calculated for the failed designs (safety factor = 1) as described above but are estimated for the new designs. The steel designs use the back-calculated K_t from the failed original laybar to account for possible corrosion and weld concentration effects. The square aluminum design has an elevated K_t for its internal corners.

Table 6-13 **Data for Various Laybar Designs**

Case Study 6: Dimensions in inches and pounds

Design	See Figure	Rpm	Material	TKSolver File Name	Wall Thickness	Beam Depth	K_t	Weight Factor	Safety Factor	Comment
Original Laybar	6-53a	400	1018 Steel	CASE6-0	0.085	2.38	4.6	1	1.4	safe at design speed
Original Laybar	6-53a	500	1018 Steel	CASE6-1	0.085	2.38	4.6	1	1	failed at higher speed
Steel Replacement	6-53b	500	1020 Steel	CASE6-2	0.093	2.50	3.2	1	1	failed in 6 mos.
Aluminum Replacement	6-53c	500	6061-T6 Aluminum	CASE6-3	solid	2.00	7.2	1.4	1	failed in 3 mos.
Square Steel	6-55a	500	1020 Steel	CASE6-4	0.062	2.50	4.6	1.4	0.5	rejected design
Round Steel	6-55b	500	1020 Steel	CASE6-5	0.10	2.50	4.6	1.4	0.5	rejected design
Square Aluminum	6-55a	500	6061-T6 Aluminum	CASE6-6	0.35	2.50	2.0	1.4	1.6	rejected design
Round Aluminum	6-55b	500	6061-T6 Aluminum	CASE6-7	0.50	2.50	1.0	1.4	2.9	selected design

The weight factor is calculated as the ratio of total weight (with reed) of the new design over the total weight (with reed) of the original laybar. The values for safety factor in the table are the largest attainable without exceeding the weight factor (1.4) of the heaviest replacement laybar that operated without inflicting damage on the rest of the machine (Figure 6-53c). Larger safety factors than those shown in Table 6-13 are achievable with some of the rejected designs, but only at a weight penalty. Thus the selected round design is seen to be the best one on a safety factor to weight basis.

Other factors that entered into the decision to use the round-aluminum design were the corrosion resistance of anodized aluminum, its availability at reasonable cost in custom-extruded shapes, which eliminated all welding except at the ends, and the superior torsional resistance and lack of stress concentration of a round cross section. Approximately 100 laybars of this design were made and installed in 1971-1972. They operated without a single failure for over 7 years. The machinery was later sold and shipped overseas and the author lost track of them.

6.15 SUMMARY

Time-varying loads are more the rule than the exception in machinery. Designing to avoid failure under these conditions is more challenging than is design for static loading. The fatigue failure mechanism is now reasonably well understood, though research continues on its many details. Two loading regimes are considered, low-cycle fatigue (LCF) where the total number of stress oscillations over the life of the part is less than about 1000, and high-cycle fatigue (HCF), which sees cycles into the millions or more.

A strain-based analysis is the most accurate method for determining fatigue strengths and is preferred for LCF situations where the local stresses may occasionally exceed the yield strength of the material on particular cycles. An example is an airframe, which sees occasionally severe overloads within a series of lower-level stress oscillations during its life as shown in Figure 6-7. Fracture mechanics (FM) is an increasingly useful tool for predicting incipient failure in assemblies that can be inspected for cracks. The crack growth is monitored and FM theory used to calculate a projected time to failure. The part is then replaced on a maintenance schedule that precludes its failure in service. This is regularly done in the aircraft industry. Most factory-based machinery and some land-transport vehicles see more uniform-magnitude oscillations in stress and are also expected to endure them for many millions of cycles. For these cases the more approximate, but easily applied, methods of stress-based HCF analysis are appropriate.

Rules of thumb and approximations are used to estimate material strengths under dynamic loading conditions especially for the high-cycle fatigue case. Many of these err on the side of conservatism. If specific test data are available for the fatigue strength of the chosen material, those data should always be used in preference to a calculated estimate. Lacking specific test data, the uncorrected fatigue strength can be estimated as a percentage of the ultimate tensile strength. In either event, the uncorrected fatigue strength is then reduced with a collection of factors as defined in Section 6.6 and equations 6.7 to account for differences between the actual part and the test specimen from which the ultimate strength was measured. A modified-Goodman diagram is then constructed using estimates of the material's "static" strength at 1 000 cycles and its corrected fatigue strength at some larger number of cycles appropriate to the part's expected life. (See Section 6.11.)

A general approach to designing for HCF cases is presented in Section 6.13. The von Mises effective stress equation is used to create effective alternating and mean components of stress for the most highly loaded points within the part. In some cases the mean stress component may be zero. All appropriate stress-concentration effects should be included in these stress calculations. The mean and alternating von Mises components are then plotted on the modified-Goodman diagram and a safety factor calculated based on an assumption about the way in which the mean and alternating stresses may vary in service. See Section 6.11 and equations 6.18.

Important Equations Used in this Chapter

Fluctuating Stress Components (Section 6.4):

$$\Delta\sigma = \sigma_{max} - \sigma_{min} \tag{6.1a}$$

$$\sigma_a = \frac{\sigma_{max} - \sigma_{min}}{2} \tag{6.1b}$$

$$\sigma_m = \frac{\sigma_{max} + \sigma_{min}}{2} \tag{6.1c}$$

$$R = \frac{\sigma_{min}}{\sigma_{max}} \qquad\qquad A = \frac{\sigma_a}{\sigma_m} \qquad\qquad (6.1d)$$

Uncorrected Fatigue Strength Estimates (Section 6.5):

steels:
$$\begin{cases} S_{e'} \cong 0.5\,S_{ut} & \text{for } S_{ut} < 200 \text{ ksi } (1400 \text{ MPa}) \\ S_{e'} \cong 100 \text{ ksi } (700 \text{ MPa}) & \text{for } S_{ut} \geq 200 \text{ ksi } (1400 \text{ MPa}) \end{cases} \qquad (6.5a)$$

irons:
$$\begin{cases} S_{e'} \cong 0.4\,S_{ut} & \text{for } S_{ut} < 60 \text{ ksi } (400 \text{ MPa}) \\ S_{e'} \cong 24 \text{ ksi } (160 \text{ MPa}) & \text{for } S_{ut} \geq 60 \text{ ksi } (400 \text{ MPa}) \end{cases} \qquad (6.5b)$$

aluminums:
$$\begin{cases} S_{f'_{@5E8}} \cong 0.4\,S_{ut} & \text{for } S_{ut} < 48 \text{ ksi } (330 \text{ MPa}) \\ S_{f'_{@5E8}} \cong 19 \text{ ksi } (130 \text{ MPa}) & \text{for } S_{ut} \geq 48 \text{ ksi } (330 \text{ MPa}) \end{cases} \qquad (6.5c)$$

copper alloys:
$$\begin{cases} S_{f'_{@5E8}} \cong 0.4\,S_{ut} & \text{for } S_{ut} < 40 \text{ ksi } (280 \text{ MPa}) \\ S_{f'_{@5E8}} \cong 14 \text{ ksi } (100 \text{ MPa}) & \text{for } S_{ut} \geq 40 \text{ ksi } (280 \text{ MPa}) \end{cases} \qquad (6.5d)$$

Correction Factors for Fatigue Strength (Section 6.6):

bending:	$C_{load} = 1$	
axial loading:	$C_{load} = 0.70$	$(6.7a)$

$$\begin{array}{lll} \text{for } d \leq 0.3 \text{ in } (8 \text{ mm}): & C_{size} = 1 & \\ \text{for } 0.3 \text{ in} \leq d \leq 10 \text{ in}: & C_{size} = 0.869 d^{-0.097} & (6.7b) \\ \text{for } 8 \text{ mm} \leq d \leq 250 \text{ mm}: & C_{size} = 1.189 d^{-0.097} & \end{array}$$

$$C_{surf} \cong A\left(S_{ut}\right)^b \qquad \text{if } C_{surf} > 1.0, \text{ set } C_{surf} = 1.0 \qquad (6.7e)$$

$$\begin{array}{lll} \text{for } T \leq 450°C\ (840°F): & C_{temp} = 1 & \\ \text{for } 450°C < T \leq 550°C: & C_{temp} = 1 - 0.0058(T - 450) & (6.7f) \\ \text{for } 840°F < T \leq 1020°F: & C_{temp} = 1 - 0.0032(T - 840) & \end{array}$$

Corrected Fatigue Strength Estimates (Section 6.6):

$$\begin{aligned} S_e &= C_{load}\,C_{size}\,C_{surf}\,C_{temp}\,C_{reliab}\,S_{e'} \\ S_f &= C_{load}\,C_{size}\,C_{surf}\,C_{temp}\,C_{reliab}\,S_{f'} \end{aligned} \qquad (6.6)$$

Approximate Strength at 1 000 Cycles (Section 6.6):

bending:	$S_m = 0.9 S_{ut}$	
axial loading:	$S_m = 0.75 S_{ut}$	(6.9)

S-N Diagram (Section 6.6):

$$\log S_n = \log a + b \log N \tag{6.10b}$$

$$b = \frac{1}{z} \log\left(\frac{S_m}{S_e}\right) \qquad \text{where} \qquad z = \log N_1 - \log N_2$$

$$\log(a) = \log(S_m) - b \log(N_1) = \log(S_m) - 3b \tag{6.10c}$$

Notch Sensitivity (Section 6.7):

$$q = \frac{1}{1 + \dfrac{\sqrt{a}}{\sqrt{r}}} \tag{6.13}$$

Fatigue Stress-concentration factors (Sections 6.7 and 6.10):

$$K_f = 1 + q(K_t - 1) \tag{6.11b}$$

$$\text{if } K_f |\sigma_{max}| < S_y \text{ then:} \qquad K_{fm} = K_f$$

$$\text{if } K_f |\sigma_{max}| > S_y \text{ then:} \qquad K_{fm} = \frac{S_y - K_f \sigma_a}{|\sigma_m|} \tag{6.17}$$

$$\text{if } K_f |\sigma_{max} - \sigma_{min}| > 2S_y \text{ then:} \qquad K_{fm} = 0$$

Safety Factor - Fully Reversed Stresses (Section 6.9):

$$N_f = \frac{S_n}{\sigma'} \tag{6.14}$$

Modified-Goodman Diagram (Section 6.10):

$$-\frac{\sigma'_m}{S_{yc}} + \frac{\sigma'_a}{S_{yc}} = 1 \tag{6.16a}$$

$$\sigma'_a = S_f \tag{6.16b}$$

$$\frac{\sigma'_m}{S_{ut}} + \frac{\sigma'_a}{S_f} = 1 \tag{6.16c}$$

$$\frac{\sigma'_m}{S_y} + \frac{\sigma'_a}{S_y} = 1 \tag{6.16d}$$

Safety Factor - Fluctuating Stresses (Section 6.10):

Case 1:

$$N_f = \frac{\sigma'_{m@Q}}{\sigma'_{m@Z}} = \frac{S_y}{\sigma'_m}\left(1 - \frac{\sigma'_a}{S_y}\right) \tag{6.18a}$$

Case 2:

$$N_f = \frac{\sigma'_{a@P}}{\sigma'_{a@Z}} = \frac{S_f}{\sigma'_a}\left(1 - \frac{\sigma'_m}{S_{ut}}\right) \qquad (6.18b)$$

Case 3:

$$N_f = \frac{\sigma'_{m@R}}{\sigma'_{m@Z}} = \frac{S_f S_{ut}}{\sigma'_a S_{ut} + \sigma'_m S_f} \qquad (6.18e)$$

$$ZS = \sqrt{\left(\sigma'_m - \sigma'_{m@S}\right)^2 + \left(\sigma'_a - \sigma'_{a@S}\right)^2} \qquad (6.18f)$$

$$OZ = \sqrt{\left(\sigma'_a\right)^2 + \left(\sigma'_m\right)^2}$$

Case 4:

$$N_f = \frac{OZ + ZS}{OZ} \qquad (6.18g)$$

General Multiaxial Stresses in Fatigue - 2-D (Section 6.11):

$$\sigma'_a = \sqrt{\sigma_{x_a}^2 + \sigma_{y_a}^2 - \sigma_{x_a}\sigma_{y_a} + 3\tau_{xy_a}^2}$$

$$\sigma'_m = \sqrt{\sigma_{x_m}^2 + \sigma_{y_m}^2 - \sigma_{x_m}\sigma_{y_m} + 3\tau_{xy_m}^2} \qquad (6.22b)$$

General Multiaxial Stresses in Fatigue - 3-D (Section 6.11):

$$\sigma'_a = \sqrt{\frac{\left(\sigma_{x_a} - \sigma_{y_a}\right)^2 + \left(\sigma_{y_a} - \sigma_{z_a}\right)^2 + \left(\sigma_{z_a} - \sigma_{x_a}\right)^2 + 6\left(\tau_{xy_a}^2 + \tau_{yz_a}^2 + \tau_{zx_a}^2\right)}{2}}$$

$$\sigma'_m = \sqrt{\frac{\left(\sigma_{x_m} - \sigma_{y_m}\right)^2 + \left(\sigma_{y_m} - \sigma_{z_m}\right)^2 + \left(\sigma_{z_m} - \sigma_{x_m}\right)^2 + 6\left(\tau_{xy_m}^2 + \tau_{yz_m}^2 + \tau_{zx_m}^2\right)}{2}} \qquad (6.22a)$$

Sines Method for Multiaxial Stresses in Fatigue - 2-D (Section 6.11):

$$\sigma'_a = \sqrt{\sigma_{x_a}^2 + \sigma_{y_a}^2 - \sigma_{x_a}\sigma_{y_a} + 3\tau_{xy_a}^2}$$

$$\sigma'_m = \sigma_{x_m} + \sigma_{y_m} \qquad (6.21b)$$

Sines Method for Multiaxial Stresses in Fatigue - 3-D (Section 6.11):

$$\sigma'_a = \sqrt{\frac{\left(\sigma_{x_a} - \sigma_{y_a}\right)^2 + \left(\sigma_{y_a} - \sigma_{z_a}\right)^2 + \left(\sigma_{z_a} - \sigma_{x_a}\right)^2 + 6\left(\tau_{xy_a}^2 + \tau_{yz_a}^2 + \tau_{zx_a}^2\right)}{2}}$$

$$\sigma'_m = \sigma_{x_m} + \sigma_{y_m} + \sigma_{z_m} \qquad (6.21a)$$

6

SEQA Method for Complex Multiaxial Stresses in Fatigue (Section 6.11):

$$SEQA = \frac{\sigma}{\sqrt{2}} \left[1 + \frac{3}{4}Q^2 + \sqrt{1 + \frac{3}{2}Q^2 \cos 2\phi + \frac{9}{16}Q^4} \right]^{\frac{1}{2}} \qquad (6.23)$$

Fracture Mechanics in Fatigue (Section 6.5):

$$\Delta K = \beta \sigma_{max} \sqrt{\pi a} - \beta \sigma_{min} \sqrt{\pi a}$$
$$= \beta \sqrt{\pi a} \left(\sigma_{max} - \sigma_{min} \right) \qquad (6.3b)$$

$$\frac{da}{dN} = A(\Delta K)^n \qquad (6.4)$$

6.16   REFERENCES

1   **R. P. Reed, J. H. Smith, and B. W. Christ**, *The Economic Effects of Fracture in the United States: Part I*, Special Pub. 647-1, U. S. Dept. of Commerce, National Bureau of Standards, Washington, D.C., 1983.

2   **N. E. Dowling**, *Mechanical Behavior of Materials*. Prentice-Hall: Englewood Cliffs, N. J., p. 340, 1993.

3   **J. W. Fischer and B. T. Yen**, *Design, Structural Details, and Discontinuities in Steel, Safety and Reliability of Metal Structures*, ASCE, Nov. 2, 1972.

4   **N. E. Dowling**, *Mechanical Behavior of Materials*. Prentice-Hall: Englewood Cliffs, N. J., p. 355, 1993.

5   **D. Broek**, *The Practical Use of Fracture Mechanics*. Kluwer Academic Publishers: Dordrecht, The Netherlands, p. 10, 1988.

6   **N. E. Dowling**, *Mechanical Behavior of Materials*. Prentice-Hall: Englewood Cliffs, N. J., p. 347, 1993.

7   **R. C. Juvinall**, *Engineering Considerations of Stress, Strain and Strength*. McGraw-Hill: New York, p. 280, 1967.

8   **J. E. Shigley and C. R. Mischke**, *Mechanical Engineering Design*. 5th ed. McGraw-Hill: New York, p. 278, 1989.

9   **A. F. Madayag**, *Metal Fatigue: Theory and Design*. John Wiley & Sons: New York, p. 117, 1969.

10   **N. E. Dowling**, *Mechanical Behavior of Materials*. Prentice-Hall: Englewood Cliffs, N. J., p. 418, 1993.

11   **R. C. Juvinall**, *Engineering Considerations of Stress, Strain and Strength*. McGraw-Hill: New York, p. 231, 1967.

12   **J. A. Bannantine, J. J. Comer, and J. L. Handrock**, *Fundamentals of Metal Fatigue*. Prentice-Hall: Englewood Cliffs, N. J., p. 13, 1990.

13 **P. C. Paris and F. Erdogan**, *A Critical Analysis of Crack Propagation Laws. Trans. ASME, J. Basic Eng.*, **85**(4): p. 528, 1963.

14 **J. M. Barsom**, *Fatigue-Crack Propagation in Steels of Various Yield Strengths. Trans. ASME, J. Eng. Ind.*, **Series B**(4): p. 1190, 1971.

15 **H. O. Fuchs and R. I. Stephens**, *Metal Fatigue in Engineering*. John Wiley & Sons: New York, p. 88, 1980.

16 **J. A. Bannantine, J. J. Comer, and J. L. Handrock**, *Fundamentals of Metal Fatigue*. Prentice-Hall: Englewood Cliffs, N. J., p. 106, 1990.

17 **J. M. Barsom and S. T. Rolfe**, *Fracture and Fatigue Control in Structures*, 2nd ed. Prentice-Hall: Englewood Cliffs, N. J., p. 256, 1987.

18 **H. O. Fuchs and R. I. Stephens**, *Metal Fatigue in Engineering*. John Wiley & Sons: New York, p. 89, 1980.

19 **J. M. Barsom and S. T. Rolfe**, *Fracture and Fatigue Control in Structures*. 2nd ed. Prentice-Hall: Englewood Cliffs, N. J., p. 285, 1987.

20 **P. G. Forrest**, *Fatigue of Metals*. Pergamon Press: London, 1962.

21 **J. E. Shigley and L. D. Mitchell**, *Mechanical Engineering Design*. 4th ed. McGraw-Hill: New York, p. 293, 1983.

22 **R. Kuguel**, *A Relation Between Theoretical Stress-concentration factor and Fatigue Notch Factor Deduced from the Concept of Highly Stressed Volume*. Proc. ASTM, **61**: p. 732-748, 1961.

23 **R. C. Juvinall**, *Engineering Considerations of Stress, Strain and Strength*. McGraw-Hill: New York, p. 233, 1967.

24 *Ibid.*, p. 234

25 **R. C. Johnson**, *Machine Design*, vol. 45, p. 108, 1973.

26 **J. E. Shigley and L. D. Mitchell**, *Mechanical Engineering Design*. 4th ed. McGraw-Hill: New York, p. 300, 1983.

27 **E. B. Haugen and P. H. Wirsching**, "Probabilistic Design." *Machine Design*, vol. 47, p. 10-14,

28 **R. C. Juvinall and K. M. Marshek**, *Fundamentals of Machine Component Design*. 2nd ed. John Wiley & Sons: New York, p. 270, 1967.

29 *Ibid.*, p. 267.

30 **R. E. Peterson**, *Stress-concentration factors*. John Wiley & Sons: New York, 1974.

31 **R. J. Roark and W. C. Young**, *Formulas for Stress and Strain*. 5th ed. McGraw-Hill: New York, 1975.

32 **H. Neuber**, *Theory of Notch Stresses*. J. W. Edwards Publisher Inc.: Ann Arbor Mich., 1946.

33 **P. Kuhn and H. F. Hardrath**, *An Engineering Method for Estimating Notch-size Effect in Fatigue Tests on Steel*. Technical Note 2805, NACA, Washington, D.C., Oct. 1952.

34   **R. B. Heywood**, *Designing Against Fatigue*. Chapman & Hall Ltd.: London, 1962.

35   **R. C. Juvinall**, *Engineering Considerations of Stress, Strain and Strength*. McGraw-Hill: New York, p. 280, 1967.

36   **V. M. Faires**, *Design of Machine Elements*. 4th ed. Macmillan: London, p. 162, 1965.

37   **R. B. Heywood**, *Designing Against Fatigue of Metals*. Reinhold: New York, p. 272, 1962.

38   **H. O. Fuchs and R. I. Stephens**, *Metal Fatigue in Engineering*. John Wiley & Sons: New York, p. 130, 1980.

39   **J. E. Shigley and C. R. Mischke**, *Mechanical Engineering Design*. 5th ed. McGraw-Hill: New York, p. 283, 1989.

40   **N. E. Dowling**, *Mechanical Behavior of Materials*. Prentice-Hall: Englewood Cliffs, N. J., p. 416, 1993.

41   **R. C. Juvinall**, *Engineering Considerations of Stress, Strain and Strength*. McGraw-Hill: New York, p. 280, 1967.

42   **G. Sines**, *Failure of Materials under Combined Repeated Stresses Superimposed with Static Stresses*, Technical Note 3495, NACA, 1955.

43   **F. S. Kelly**, *A General Fatigue Evaluation Method*, Paper 79-PVP-77, ASME, New York, 1979.

44   **Y. S. Garud**, *A New Approach to the Evaluation of Fatigue Under Multiaxial Loadings,* in *Methods for Predicting Material Life in Fatigue,* W. J. Ostergren and J. R. Whitehead, ed. ASME: New York. pp. 249-263, 1979.

45   **M. W. Brown and K. J. Miller**, *A Theory for Fatigue Failure under Multiaxial Stress-Strain Conditions. Proc. Inst. Mech. Eng.*, **187**(65): pp. 745-755, 1973.

46   **J. O. Smith**, *The Effect of Range of Stress on the Fatigue Strength of Metals.* Univ. of Ill., *Eng. Exp. Sta. Bull.*, (334), 1942.

47   **J. E. Shigley and L. D. Mitchell**, *Mechanical Engineering Design*. 4th ed. McGraw-Hill: New York, p. 333, 1983.

48   **B. F. Langer**, *Design of Vessels Involving Fatigue,* in *Pressure Vessel Engineering,* R. W. Nichols, ed. Elsevier: Amsterdam. pp. 59-100, 1971.

49   **J. A. Collins**, *Failure of Materials in Mechanical Design*. 2nd ed. J. Wiley & Sons: New York, pp. 238-254, 1993.

50   **S. M. Tipton and D. V. Nelson**, *Fatigue Life Predictions for a Notched Shaft in Combined Bending and Torsion,* in *Multiaxial Fatigue,* K. J. Miller and M. W. Brown, Editors. ASTM: Philadelphia, Pa. p. 514-550, 1985.

51   **R. C. Rice**, ed. *Fatigue Design Handbook*. 2nd ed. Soc. of Automotive Engineers: Warrendale, Pa. 260, 1988.

52   **T. Nishihara and M. Kawamoto**, *The Strength of Metals under Combined Alternating Bending and Torsion with Phase Difference.* Memoirs College of Engineering, Kyoto Univ., Japan, **11**(85), 1945.

6.17   BIBLIOGRAPHY

For more information on fatigue design, see:

J. A. Bannantine, J. J. Comer, and J. L. Handrock, *Fundamentals of Metal Fatigue*. Prentice-Hall: Englewood Cliffs, N. J., 1990.

H. E. Boyer, ed., *Atlas of Fatigue Curves*. Amer. Soc. for Metals: Metals Park, Ohio. 1986.

H. O. Fuchs and R. I. Stephens, *Metal Fatigue in Engineering*. John Wiley & Sons: New York, 1980.

R. C. Juvinall, *Engineering Considerations of Stress, Strain and Strength*.

A. J. McEvily, ed. *Atlas of Stress-Corrosion and Corrosion Fatigue Curves*. Amer. Soc. for Metals: Metals Park, Ohio. 1990. McGraw-Hill: New York, 1967.

For more information on the strain-life approach to low-cycle fatigue, see:

N. E. Dowling, *Mechanical Behavior of Materials*. Prentice-Hall: Englewood Cliffs, N. J., 1993.

R. C. Rice, ed. *Fatigue Design Handbook*. 2nd ed. Soc. of Automotive Engineers: Warrendale, Pa. 1988.

For more information on the fracture mechanics approach to fatigue, see:

J. M. Barsom and S. T. Rolfe, *Fracture and Fatigue Control in Structures*. 2nd ed. Prentice-Hall: Englewood Cliffs, N. J., 1987.

D. Broek, *The Practical Use of Fracture Mechanics*. Kluwer Academic Publishers: Dordrecht, The Netherlands, 1988.

For more information on residual stresses, see:

J. O. Almen and P. H. Black, *Residual Stresses and Fatigue in Metals*. McGraw-Hill: New York, 1963.

For more information on multiaxial stresses in fatigue, see:

A. Fatemi and D. F. Socie, *A Critical Plane Approach to Multiaxial Fatigue Damage Including Out-of-Phase Loading*. Fatigue and Fracture of Engineering Materials and Structures, **11**(3): pp. 149-165, 1988.

Y. S. Garud, *Multiaxial Fatigue: A Survey of the State of the Art*. J. Test. Eval., **9**(3), 1981.

K. F. Kussmaul, D. L. McDiarmid and D. F. Socie, ed. *Fatigue Under Biaxial and Multiaxial Loading*. Mechanical Engineeering Publications Ltd.: London. 1991.

G. E. Leese and D. Socie, ed. *Multiaxial Fatigue: Analysis and Experiments*. Soc. of Automotive Engineers: Warrendale, Pa., 1989.

K. J. Miller and M. W. Brown, ed. *Multiaxial Fatigue*. Vol. STP 853. ASTM: Philadelphia, Pa., 1985.

6

G. Sines, *Behavior of Metals Under Complex Static and Alternating Stresses,* in *Metal Fatigue,* G. Sines and J. L. Waisman, ed. McGraw-Hill: New York. 1959.

R. M. Wetzel, ed., *Fatigue Under Complex Loading: Analyses and Experiments.* SAE Pub. No. AE-6, Soc. of Automotive Engineers: Warrendale, Pa., 1977.

Table P6-1

Data for Problem 6-1

Row	σ_{max}	σ_{min}
a	1 000	0
b	1 000	–1 000
c	1 500	500
d	1 500	–500
e	500	–1 000
f	2 500	–1 200
g	0	–4 500
h	2 500	1 500

Table P6-2

Data for Probs. 6-2, 6-4

Row	S_{ut} (psi)	mat'l
a	90 000	steel
b	250 000	steel
c	120 000	steel
d	150 000	steel
e	25 000	alum.
f	70 000	alum.
g	40 000	alum.
h	35 000	alum.

* Answers to these problems are provided in Appendix G.

† These problems are based on similar problems in previous chapters with the same –number, e.g., Problem 6-9 is based on Problem 5-9 etc. Problems in succeeding chapters may also continue and extend these problems.

6.18   PROBLEMS

*6-1   For the data in the row(s) assigned in Table P6-1, find the stress range, alternating stress component, mean stress component, stress ratio, and amplitude ratio.

6-2   For the strength data in the row(s) assigned in Table P6-2, calculate the uncorrected endurance limit and draw a strength-life (*S-N*) diagram for the material, assuming it to be a steel.

*6-3   For the bicycle pedal arm assembly in Figure P6-1 assume a rider-applied force that ranges from 0 to 1 500 N at the pedal each cycle. Determine the fluctuating stresses in the 15-mm-dia pedal arm. Find the fatigue safety factor if $S_{ut} = 500$ MPa.

6-4   For the strength data in the row(s) assigned in Table P6-2, calculate the uncorrected fatigue strength at 5E8 cycles and draw a strength-life (*S-N*) diagram for the material, assuming it to be an aluminum alloy.

6-5   For the data in the row(s) assigned in Table P6-3, find the corrected endurance strength (or limit), create equations for the *S-N* line and draw the *S-N* diagram.

*†6-6   Design the trailer hitch from Problem 3-6 on p. 169 (also see Figures P6-2 and 1-5) assuming that the horizontal impact force of the trailer on the ball is fully reversed. Use steel with $S_{ut} = 600$ MPa and $S_y = 450$ MPa. Find the infinite-life fatigue safety factors for all modes of failure.

*†6-7   Design the wrist pin of Problem 3-7 (p. 169) for infinite life with a safety factor of 1.5 if the 2 500-g acceleration is fully reversed and $S_{ut} = 130$ kpsi.

*†6-8   A paper machine processes rolls of paper having a density of 984 kg/m^3. The paper roll is 1.50-m outside dia (OD) x 0.22-m inside dia (ID) x 3.23 m long and is on a simply-supported, hollow, steel shaft with $S_{ut} = 400$ MPa. Find the shaft ID needed to obtain a dynamic safety factor of 2 for a 10-year life if the shaft OD is 22 cm and the roll turns at 50 rpm.

†6-9   For the ViseGrip® plier-wrench drawn to scale in Figure P6-3, and for which the forces were analyzed in Problem 3-9 and stresses in Problem 4-9, find the safety factors for each pin for an assumed clamping force of $P = 4\,000$ N in the position shown. The steel pins are 8-mm dia with $S_y = 400$ MPa, $S_{ut} = 520$ MPa, and are all in double shear. Assume a desired finite life of 5E4 cycles.

*†6-10   An overhung diving board is shown in Figure P6-4a. A 100-kg person is standing on the free end. Assume cross-sectional dimensions of 305 mm x 32 mm. What is the fatigue safety factor for finite life if the material is brittle fiberglass with $S_f = 39$ MPa @ N = 5E8 cycles and $S_{ut} = 130$ MPa in the longitudinal direction?

*†6-11   Repeat Problem 6-10 assuming the 100-kg person in Problem 6-10 jumps up 25 cm and lands back on the board. Assume the board weighs 29 kg and deflects 13.1 cm

Table P6-3   Data For Problem 6-5

Row	Material	Sut kpsi	Shape	Size inches	Surface Finish	Loading	Temp °F	Reliability
a	steel	110	round	2	ground	torsion	room	99.9
b	steel	90	square	4	mach.	axial	600	99.0
c	steel	80	I-beam	16 x 18[*]	hot rolled	bending	room	99.99
d	steel	200	round	5	forged	torsion	–50	99.999
e	steel	150	square	7	cold rolled	axial	900	50
f	aluminum	70	round	9	mach.	bending	room	90
g	aluminum	50	square	9	ground	torsion	room	99.9
h	aluminum	85	I-beam	24 x 36[*]	cold rolled	axial	room	99.0
i	aluminum	60	round	4	ground	bending	room	99.99
j	aluminum	40	square	6	forged	torsion	room	99.999
k	ductile iron	70	round	5	cast	axial	room	50
l	ductile iron	90	square	7	cast	bending	room	90
m	bronze	60	round	9	forged	torsion	50	90
n	bronze	80	square	6	cast	axial	212	99.999

[*] width x height

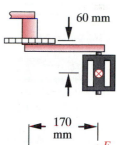

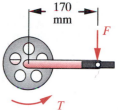

FIGURE P6-1

Problem 6-3

statically when the person stands on it. What is the fatigue safety factor for finite life if the material is brittle fiberglass with S_f = 39 MPa @ N = 5E8 cycles and S_{ut} = 100 MPa in the longitudinal direction?

[†]6-12   Repeat Problem 6-10 using the cantilevered diving board design in Figure P6-4b.

[†]6-13   Repeat Problem 6-11 using the diving board design shown in Figure P6-4b. Assume the board weighs 19 kg and deflects 8.5 cm statically when the person stands on it.

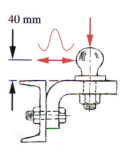

FIGURE P6-2

Problem 6-6

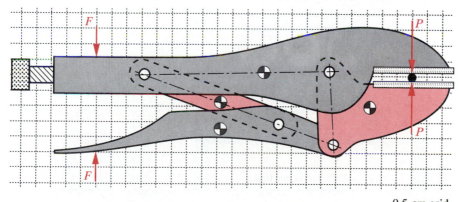

FIGURE P6-3

Problem 6-9

0.5-cm grid

---

[*] Answers to these problems are provided in Appendix G.

[†] These problems are based on similar problems in previous chapters with the same –number, e.g., Problem 6-9 is based on Problem 5-9 etc. Problems in succeeding chapters may also continue and extend these problems.

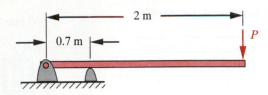

(a) Overhung diving board   (b) Cantilevered diving board

FIGURE P6-4

Problems 6-10 through 6-13

†6-14   Figure P6-5 shows a child's toy called a *pogo stick*. The child stands on the pads, applying half her weight on each side. She jumps up off the ground, holding the pads up against her feet, and bounces along with the spring cushioning the impact and storing energy to help each rebound. Assume a 60-lb child and a spring constant of 100 lb/in. The pogo stick weighs 5 lb. Design the aluminum cantilever beam sections on which she stands to survive jumping 2 in off the ground with a dynamic safety factor of 2 for a finite life of 5E4 cycles. Use 2000 series aluminum. Define the beam shape and size.

*†6-15   For a notched part having notch dimension r, geometric stress-concentration factor K_t, and material strength S_{ut} as shown in the assigned row(s) of Table P6-4, find the Neuber factor a, the material's notch sensitivity q, and the fatigue stress-concentration factor K_f.

†6-16   A track to guide bowling balls is designed with two round rods as shown in Figure P-6-6. The rods have a small angle between them. The balls roll on the rods until they fall between them and drop onto another track. Each rod's unsupported length is 30 in and the angle between them is 3.2°. The balls are 4.5-in-dia and weigh 2.5 lb. The center distance between the 1-in-dia rods is 4.2 in at the narrow end. Find the infinite-life safety factor for the 1-in-dia SAE 1045 normalized steel rods.

(a) Assume rods are simply supported at each end.
(b) Assume rods are fixed at each end.

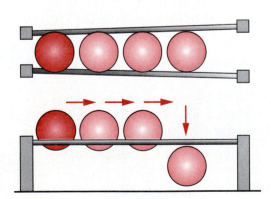

FIGURE P6-5

Problem 6-14

FIGURE P6-6

Problem 6-16

Table P6-4   Data For Problem 6-15

Row	S_{ut} (kpsi)	K_t	r (in)	mat'l
a	100	3.3	0.25	steel
b	90	2.5	0.55	steel
c	150	1.8	0.75	steel
d	200	2.8	1.22	steel
e	20	3.1	0.25	soft alum.
f	35	2.5	0.28	soft alum.
g	45	1.8	0.50	soft alum.
h	50	2.8	0.75	hard alum.
i	30	3.5	1.25	hard alum.
j	90	2.5	0.25	hard alum.

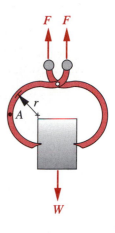

FIGURE P6-7

Problem 6-17

*†6-17   A pair of ice tongs is shown in Figure P6-7. The ice weighs 50 lb and is 10-in wide across the tongs. The distance between the handles is 4 in, and the mean radius r of a steel tong is 6 in. The rectangular cross-sectional dimensions are 0.750 in x 0.312 in. Find the safety factor for the tongs for 5E5 cycles if their S_{ut} = 50 kpsi.

  6-18   Repeat Problem 6-17 with the tongs made of Class 40 gray cast iron.

*†6-19   Determine the size of the clevis pin shown in Figure P6-8 needed to withstand an applied repeated force of 0 to 130 000 lb for infinite life. Also determine the required outside radius of the clevis end to not fail in either tearout or bearing if the clevis flanges each are 2.5 in thick. Use a safety factor of 3. Assume S_{ut} = 140 kpsi for the pin and S_{ut} = 80 kpsi for the clevis.

  6-20   A ±100 N-m torque is applied to a 1-m-long, solid, round steel shaft. Design it to limit its angular deflection to 2° and select a steel alloy to have a fatigue safety factor of 2 for infinite life.

†6-21   Figure P6-9 shows an automobile wheel with two styles of lug wrench, a single ended wrench in (a), and a double-ended wrench in (b). The distance between points A and B is 1 ft in both cases and the handle diameter is 0.625 in. How many cycles of tightening can be expected before a fatigue failure if the average tightening torque is 100 ft-lb and the material S_{ut} = 60 kpsi?

*†6-22   An inline "roller-blade" skate is shown in Figure P6-10. The polyurethane wheels are 72-mm dia and spaced on 10-mm centers. The skate-boot-foot combination weighs 2 kg. The effective "spring rate" of the person-skate system is 6 000 N/m. The axles are 10-mm-dia steel pins in double shear with S_y = 550 MPa. Find the fatigue safety factor for the pins when a 100-kg person lands a 0.5-m jump on one foot assuming infinite life.

  (a) Assume that all four wheels land simultaneously.
  (b) Assume that one wheel absorbs all the landing force.

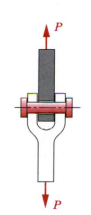

FIGURE P6-8

Problems 6-19

* Answers to these problems are provided in Appendix G.

† These problems are based on similar problems in previous chapters with the same –number, e.g., Problem 6-9 is based on Problem 5-9 etc. Problems in succeeding chapters may also continue and extend these problems.

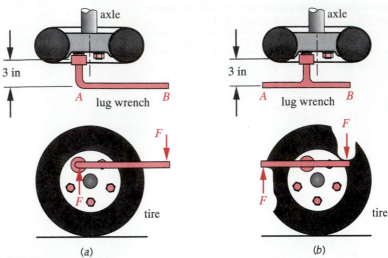

FIGURE P6-9

Problem 6-21

*6-23   The beam in Figure P6-11a is subjected to a sinusoidal force time function with $F_{max} = F$ and $F_{min} = -F/2$, where F and the beam's other data are given in the row(s) assigned from Table P6-5. Find the stress state in the beam due to this loading and choose a material specification that will give a safety factor of 3 for $N = 5E8$ cycles.

*6-24   The beam in Figure P6-11b is subjected to a sinusoidal force time function with $F_{max} = F$ N and $F_{min} = F/2$, where F and the beam's other data are given in the row(s) assigned from Table P6-5. Find the stress state in the beam due to this loading and choose a material specification that will give a safety factor of 1.5 for $N = 5E8$ cycles.

*6-25   The beam in Figure P6-11c is subjected to a sinusoidal force time function with $F_{max} = F$ and $F_{min} = 0$, where F and the beam's other data are given in the row(s) assigned from Table P6-5. Find the stress state in the beam due to this loading and choose a material specification that will give a safety factor of 2.5 for $N = 5E8$ cycles.

*6-26   The beam in Figure P6-11d is subjected to a sinusoidal force time function with $F_{max} = F$ lb and $F_{min} = -F$, where F and the beam's other data are given in the row(s) assigned from Table P6-5. Find the stress state in the beam due to this loading and choose a material specification that will give a safety factor of 6 for $N = 5E8$ cycles.

*†6-27   A storage rack is to be designed to hold the paper roll of Problem 6-8 as shown in Figure P6-12. Determine suitable values for dimensions a and b in the figure for an infinite-life fatigue safety factor of 2. The mandrel is solid and inserts halfway into the paper roll:

    a)   If the beam is a ductile material with $S_{ut} = 600$ MPa,

    b)   If the beam is a cast-brittle material with $S_{ut} = 300$ MPa.

FIGURE P6-10

Problem 6-22

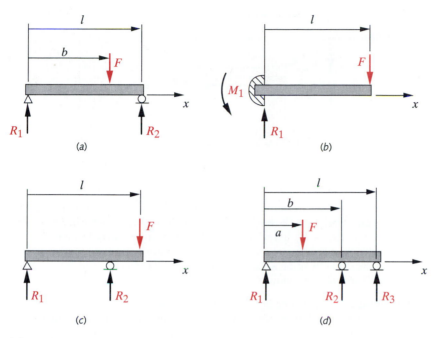

FIGURE P6-11

Beams and Beam Loadings for Problems 6-23 to 6-26: See Table P6-5 for Data

Table P6-5   Data for Problems 6-23 through 6-26

Use Only Data Relevant to a Particular Problem. Lengths in m, Forces in N, I in m^4

Row	l	a	b	w	F	I	c	E
a	1.00	0.40	0.60	200	500	2.85E–08	2.00E–02	steel
b	0.70	0.20	0.40	80	850	1.70E–08	1.00E–02	steel
c	0.30	0.10	0.20	500	450	4.70E–09	1.25E–02	steel
d	0.80	0.50	0.60	65	250	4.90E–09	1.10E–02	steel
e	0.85	0.35	0.50	96	750	1.80E–08	9.00E–03	steel
f	0.50	0.18	0.40	450	950	1.17E–08	1.00E–02	steel
g	0.60	0.28	0.50	250	250	3.20E–09	7.50E–03	steel
h	0.20	0.10	0.13	400	500	4.00E–09	5.00E–03	alum.
i	0.40	0.15	0.30	50	200	2.75E–09	5.00E–03	alum.
j	0.20	0.10	0.15	150	80	6.50E–10	5.50E–03	alum.
k	0.40	0.16	0.30	70	880	4.30E–08	1.45E–02	alum.
l	0.90	0.25	0.80	90	600	4.20E–08	7.50E–03	alum.
m	0.70	0.10	0.60	80	500	2.10E–08	6.50E–03	alum.
n	0.85	0.15	0.70	60	120	7.90E–09	1.00E–02	alum.

* Answers to these problems are provided in Appendix G.

† These problems are based on similar problems in previous chapters with the same –number, e.g., Problem 6-9 is based on Problem 5-9 etc. Problems in succeeding chapters may also continue and extend these problems.

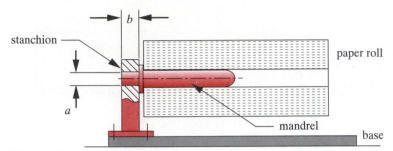

FIGURE P6-12

Problem 6-27

†6-28 Figure P6-13 shows a forklift truck negotiating a 15° ramp to drive onto a 4-ft high loading platform. The truck weighs 5 000 lb and has a 42-in wheelbase. Design two (one for each side) 1-ft-wide ramps of steel to have a safety factor of 2 for infinite life in the worst case of loading as the truck travels up them. Minimize the weight of the ramps by using a sensible cross-sectional geometry. Choose an appropriate steel or aluminum alloy.

*6-29 A bar 22 mm x 30 mm in cross section is loaded axially in tension with $F(t) = \pm 8$ kN. A 10-mm hole passes through the center of the 30-mm side. Find the safety factor for infinite life if the material has $S_{ut} = 500$ MPa.

6-30 Repeat Problem 6.29 with $F_{min} = 0$, $F_{max} = 16$ kN.

*6-31 Repeat Problem 6.29 with $F_{min} = 8$ kN, $F_{max} = 24$ kN.

6-32 Repeat Problem 6.29 with $F_{min} = -4$ kN, $F_{max} = 12$ kN.

*†6-33 The bracket in Figure P6-14 is subjected to a sinusoidal force time function with $F_{max} = F$ and $F_{min} = -F$, where F and the beam's other data are given in the row(s) assigned from Table P6-6. Find the stress states at points A and B due to this fully reversed loading and choose a ductile steel or aluminum material specification that will give a safety factor of 2 for infinite life if steel or $N = 5E8$ cycles if aluminum. Assume a geometric stress-concentration factor of 2.5 in bending and 2.8 in torsion.

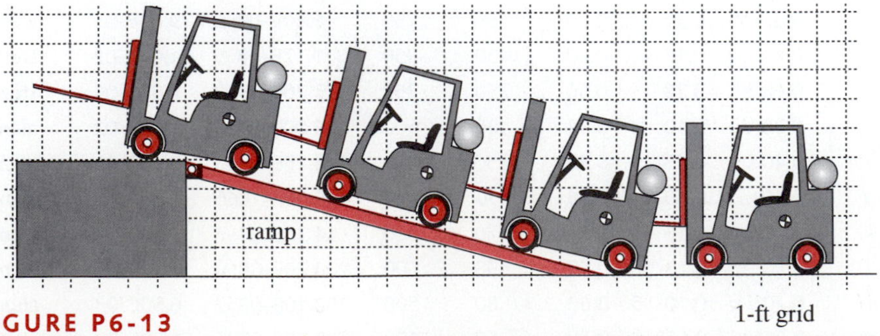

ramp

FIGURE P6-13

1-ft grid

Problem 6-28

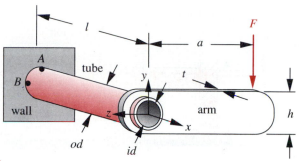

FIGURE P6-14

Problems 6-33 to 6-36

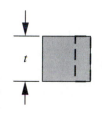

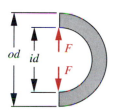

FIGURE P6-15

Problem 6-37

*†6-34   The bracket in Figure P6-14 is subjected to a sinusoidal force time function with $F_{max} = F$ and $F_{min} = 0$, where F and the beam's other data are given in the row(s) assigned from Table P6-6. Find the stress states at points A and B due to this repeated loading and choose a ductile steel or aluminum material specification that will give a safety factor of 2 for infinite life if steel or $N = 5E8$ cycles if aluminum. Assume a geometric stress-concentration factor of 2.8 in bending and 3.2 in torsion.

6-35   Repeat Problem 6-33 using a cast-iron material.

6-36   Repeat Problem 6-34 using a cast-iron material.

Table P6-6   Data for Problems 6-33 through 6-36

Use only data that are relevant to the particular problem. Lengths in mm, forces in N.

Row	l	a	t	h	F	od	id	E
a	100	400	10	20	50	20	14	steel
b	70	200	6	80	85	20	6	steel
c	300	100	4	50	95	25	17	steel
d	800	500	6	65	160	46	22	alum.
e	85	350	5	96	900	55	24	alum.
f	50	180	4	45	950	50	30	alum.
g	160	280	5	25	850	45	19	steel
h	200	100	2	10	800	40	24	steel
i	400	150	3	50	950	65	37	steel
j	200	100	3	10	600	45	32	alum.
k	120	180	3	70	880	60	47	alum.
l	150	250	8	90	750	52	28	alum.
m	70	100	6	80	500	36	30	steel
n	85	150	7	60	820	40	15	steel

* Answers to these problems are provided in Appendix G.

† These problems are based on similar problems in previous chapters with the same –number, e.g., Problem 6-9 is based on Problem 5-9 etc. Problems in succeeding chapters may also continue and extend these problems.

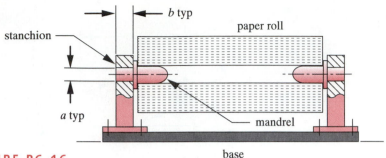

FIGURE P6-16

Problem 6-40

6

*†6-37 A semicircular curved beam as shown in Figure P6-15 has od = 150 mm, id = 100 mm and t = 25 mm. For a load pair F = ±3 kN applied along the diameter, find the fatigue safety factor at the inner and outer fibers:

(a) if the beam is steel with S_{ut} = 700 MPa,
(b) if the beam is a cast-iron with S_{ut} = 420 MPa.

6-38 A 42-mm-dia steel shaft with a 19-mm transverse hole is subjected to a sinusoidal combined loading of σ = ±400 MPa bending stress and a steady torsion of 210 MPa. Find its safety factor for infinite life if S_{ut} = 1 GPa.

*6-39 A 42-mm-dia steel shaft with a 19-mm transverse hole is subjected to a combined loading of σ = ±400 MPa bending stress and an alternating torsion of ±210 MPa which are 90° out-of-phase. Find its safety factor for infinite life if S_{ut} = 1 GPa.

6-40 Redesign the roll support of Problem 5-8 to be like Figure P6-16. The mandrels insert to 10% of the roll length. Design dimensions a and b for an infinite-life safety factor of 2.

(a) If the beam is steel with S_{ut} = 600 MPa,
(b) If the beam is a cast iron with S_{ut} = 300 MPa.

*†6-41 A 10-mm-ID steel tube carries liquid at 7 MPa. The pressure varies periodically from zero to maximum. The steel has S_{ut} = 400 MPa. Determine the infinite-life fatigue safety factor for the wall if its thickness is:

(a) 1 mm,
(b) 5 mm.

†6-42 A cylindrical tank with hemispherical ends is required to hold 150 psi of pressurized air at room temperature. The pressure cycles from zero to maximum. The steel has S_{ut} = 500 MPa. Determine the infinite-life fatigue safety factor if the tank diameter is 0.5 m, the wall thickness is 1 mm, and its length is 1 m.

†6-43 The paper rolls in Figure P6-17 are 0.9-m-OD x 0.22-m-ID x 3.23 m long and have a density of 984 kg/m³. The rolls are transferred from the machine conveyor (not shown) to the forklift truck by the V-linkage of the off-load station, which is rotated through 90° by an air cylinder. The paper then rolls onto the waiting forks of the truck. The forks are 38-mm-thick by 100-mm-wide by 1.2 m long and are tipped at a 3° angle from the horizontal and have S_{ut} = 600 MPa. Find the infinite-life fatigue

* Answers to these problems are provided in Appendix G.

† These problems are based on similar problems in previous chapters with the same –number, e.g., Problem 6-9 is based on Problem 5-9 etc. Problems in succeeding chapters may also continue and extend these problems.

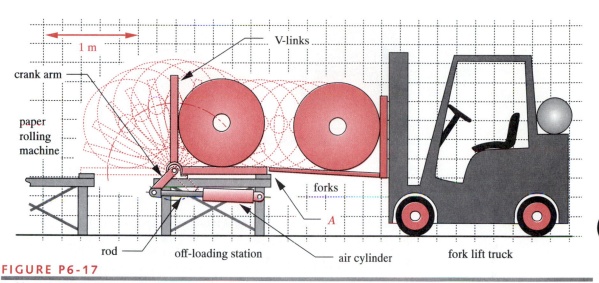

FIGURE P6-17

Problems 6-43 to 6-47

safety factor for the two forks on the truck when the paper rolls onto it under 2 different conditions (state all assumptions):

(a) The two forks are unsupported at their free end.
(b) The two forks are contacting the table at point A.

†6-44 Determine a suitable thickness for the V-links of the off-loading station of Figure P6-17 to limit their deflections at the tips to 10 mm in any position during their rotation. Two V-links support the roll, at the 1/4 and 3/4 points along the roll's length and that each of the V arms is 10 cm wide by 1 m long. What is their infinite-life fatigue safety factor when designed to limit deflection as above? S_{ut} = 600 MPa. See Problem 5-42 for more information.

†6-45 Determine the infinite-life fatigue safety factor based on the tension load on the air-cylinder rod in Figure P6-17. The tension load cycles from zero to maximum (compression loads below the critical buckling load will not affect fatigue life). The crank arm that it rotates is 0.3 m long and the rod has a maximum extension of 0.5 m. The 25-mm-dia rod is solid steel with S_{ut} = 600 MPa. State all assumptions.

†6-46 The V-links of Figure P6-17 are rotated by the crank arm through a shaft that is 60-mm dia by 3.23-m long. Determine the maximum torque applied to this shaft during the motion of the V-linkage and find the infinite-life fatigue safety factor for the shaft if its S_{ut} = 600 MPa. See Problem 6.43 for more information.

*†6-47 Determine the maximum forces on the pins at each end of the air cylinder of Figure P6-17. Determine the infinite-life fatigue safety factor for these pins if they are 30-mm dia and in single shear. S_{ut} = 600 MPa.

* Answers to these problems are provided in Appendix G.

† These problems are based on similar problems in previous chapters with the same –number, e.g., Problem 6-9 is based on Problem 5-9 etc. Problems in succeeding chapters may also continue and extend these problems.

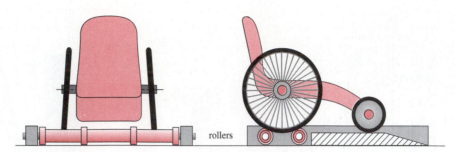

rollers

FIGURE P6-18

Problem 6-48

6

†6-48   Figure P6-18 shows an exerciser for a 100-kg wheelchair racer. The wheelchair has 65-cm-dia drive wheels separated by a 70-cm track width. Two free-turning rollers on bearings support the rear wheels. The lateral movement of the chair is limited by the flanges. Design the 1-m-long rollers as hollow tubes of aluminum (select alloy) to minimize the height of the platform and also limit the roller deflections to 1 mm in the worst case. Specify suitable sized steel axles to support the tubes on bearings. Calculate the fatigue safety factors at a life of N = 5E8 cycles.

* Answers to these problems are provided in Appendix G.

† These problems are based on similar problems in previous chapters with the same –number, e.g., Problem 6-9 is based on Problem 5-9 etc. Problems in succeeding chapters may also continue and extend these problems.

7

SURFACE FAILURE

Use it up, wear it out;
Make it do, or do without.
NEW ENGLAND MAXIM

7.0   INTRODUCTION

There are only three ways in which parts or systems can "fail": *obsolescence, breakage,* or *wearing out*. My old Apple II computer still works well but is obsolete and no longer of any use to me. My wife's favorite vase is now in pieces since I dropped it on the floor, and it is irrecoverable. However, my 120 000 mile automobile is still quite serviceable and useful despite showing some signs of wear. Most systems are subject to all three types of possible failure. Failure by obsolescence is somewhat arbitrary. (My niece is now getting good use of the Apple II.) Failure by breakage is often sudden and may be permanent. Failure by "wearing out" is generally a gradual process and is sometimes repairable. Ultimately, any system that does not fall victim to one of the other two modes of failure will inevitably wear out if kept in service long enough. Wear is the final mode of failure, which nothing escapes. Thus, we should realize that we cannot design to avoid all types of wear completely, only to postpone them.

The previous chapters have dealt with failure of parts by distortion (yielding) and breakage (fracture). **Wear** is a broad term that encompasses many types of failures, all of which involve changes to the **surface** of the part. Some of these so-called *wear mechanisms* are still not completely understood and rival theories exist in some cases. Most experts describe five general categories of wear: **adhesive wear**, **abrasive wear**, **erosion, corrosion wear**, and **surface fatigue**. The following sections discuss these topics in detail. In addition, there are other types of surface failure that do not fit neatly into one of the four categories or that can fit into more than one. **Corrosion fatigue** has aspects of the last two categories as does **fretting corrosion**. For simplicity, we will discuss these hybrids in concert with one of the four main categories listed above.

Table 7-0   Variables Used in this Chapter

Symbol	Variable	ips units	SI units	See
a	half-width of contact patch	in	m	Sect. 7.8–7.10
A_a	apparent area of contact	in^2	m^2	Sect. 7.2
A_r	real area of contact	in^2	m^2	Eq. 7.1
B	geometry factor	1/in	1/m	Eq. 7.9b
b	half-length of contact patch	in	m	Sect. 7.8–7.10
d	depth of wear	in	m	Eq. 7.7
E	Young's modulus	psi	Pa	all
F, P	force or load	lb	N	all
f	friction force	lb	N	Eq. 7.2
f_{max}	maximum tangential force	lb	N	Eq. 7.22f
K	wear coefficient	none	none	Eq. 7.7
l	length of linear contact	in	m	Eq. 7.7
L	length of cylindrical contact	in	m	Eq. 7.14
m_1, m_2	material constants	1/psi	m^2/N	Eq. 7.9a
H	penetration hardness	psi	kg/mm^2	Eq. 7.7
N	number of cycles	none	none	Eq. 7.25
N_f	safety factor in surface fatigue	none	none	Example 7-5
p	pressure in contact patch	psi	N/m^2	Sect. 7.8–7.10
p_{avg}	avg pressure in contact patch	psi	N/m^2	Sect. 7.8–7.10
p_{max}	max pressure in contact patch	psi	N/m^2	Sect. 7.8–7.10
R_1, R_2	radii of curvature	in	m	Eq. 7.9b
S_{us}	ultimate shear strength	psi	Pa	Sect. 7.3
S_{ut}	ultimate tensile strength	psi	Pa	Sect. 7.3
S_y	yield strength	psi	Pa	Sect. 7.3
S_{yc}	yield strength in compression	psi	Pa	Sect. 7.3
V	volume	in^3	m^3	Eq. 7.7
x, y, z	generalized length variables	in	m	all
μ	coefficient of friction	none	none	Eq. 7.2–7.6
ν	Poisson's ratio	none	none	all
σ	normal stress	psi	Pa	all
σ_1	principal stress	psi	Pa	Sect. 7.11
σ_2	principal stress	psi	Pa	Sect. 7.11
σ_3	principal stress	psi	Pa	Sect. 7.11
τ	shear stress	psi	Pa	all
τ_{13}	maximum shear stress	psi	Pa	Sect. 7.11
τ_{21}	principal shear stress	psi	Pa	Sect. 7.11
τ_{32}	principal shear stress	psi	Pa	Sect. 7.11

7

Failure from wear usually involves the loss of some material from the surfaces of solid parts in the system. The wear motions of interest are sliding, rolling, or some combination of both. Wear is a serious cost to the national economy.* It only requires the loss of a very small volume of material to render the entire system nonfunctional. Rabinowicz[1] estimates that a 4 000-lb automobile, when completely "worn out" will have lost only a few ounces of metal from its working surfaces. Moreover, these damaged surfaces will not be visible without extensive disassembly, so it is often difficult to monitor and anticipate the effects of wear before failure occurs.

Table 7-0 shows the variables used in this chapter and references the equations, tables, or sections in which they are used. At the end of the chapter, a summary section is provided that also groups all the significant equations from this chapter for easy reference and identifies the chapter section in which their discussion can be found.

7.1   SURFACE GEOMETRY

Before discussing the types of wear mechanisms in detail it will be useful to define the characteristics of an engineering surface that are relevant to these processes. (Material strength and hardness will also be factors in wear.) Most solid surfaces that are subject to wear in machinery will be either machined or ground, though some will be as-cast or as-forged. In any case, the surface will have some degree of roughness that is concomitant with its finishing process. Its degree of roughness or smoothness will have an effect on both the type and degree of wear that it will experience.

Even an apparently smooth surface will have microscopic irregularities. These can be measured by any of several methods. A profilometer passes a lightly loaded, hard (e.g., diamond) stylus over the surface at controlled (low) velocity and records its undulations. The stylus has a very small (about 0.5μm) radius tip that acts, in effect, as a low-pass filter since contours smaller than its radius are not sensed. Nevertheless, it gives a reasonably accurate profile of the surface with a resolution of 0.125 μm or better. Figure 7-1 shows the profiles and SEM photographs (100x) of both ground (*a*) and machined (*b*) surfaces of hardened steel cams. The profiles were measured with a Hommel T-20 profilometer that digitizes 8000 data points over the sample length (here 2.5 mm). The microscopic "mountain peaks" on the surfaces are called **asperities**.

From these profiles a number of statistical measures may be calculated. ISO defines at least 19 such parameters. Some of them are shown in Figure 7-2 along with their mathematical definitions. Perhaps the most commonly used parameters are R_a, which is the average of the absolute values of the measured points or R_q, which is their rms average. These are very similar in both value and meaning. Unfortunately many engineers specify only one of these two parameters, neither of which tells enough about the surface. For example, the two surfaces shown in Figures 7-3*a* and *b* have the same R_a and R_q values but are clearly different in nature. One has predominantly positive, and the other predominantly negative, features. These two surfaces will react quite differently to sliding or rolling against another surface.

* A 1977 study sponsored by the ASME estimated that the energy cost to the U.S. economy associated with the replacement of equipment that failed from wear accounted for 1.3% of total U. S. energy consumption. This was then equivalent to about 160 million barrels of oil per year. See O. Pinkus and D. F. Wilcock, *Strategy for Energy Conservation through Tribology*, ASME, New York, 1977, p. 93.

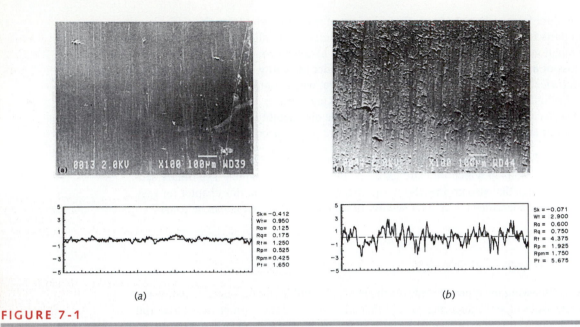

(a)   (b)

FIGURE 7-1

Scanning Electron Microscope Surface-Replica Photographs (100x) and Profiles of Ground (a) and Milled (b) Cam Surfaces

In order to differentiate these surfaces that have identical R_a or R_q values, other parameters should be calculated. Skewness S_k is a measure of the average of the first derivative of the surface contour. A negative value of S_k indicates that the surface has a predominance of valleys (Figure 7-3a) and a positive S_k defines a predominance of peaks (Figure 7-3b). Many other parameters can be computed (see Figure 7-2). For example, R_t defines the largest peak-to-valley dimension in the sample length, R_p the largest peak height above the mean line, and R_{pm} the average of the 5 largest peak heights. All the roughness measurements are calculated from an electronically filtered measurement that zeros out any slow-changing waves in the surface. An average line is computed from which all peak/valley measurements are then made. In addition to these roughness measurements (denoted by R), the waviness W_t of the surface can also be computed. The W_t computation filters out the high-frequency contours and preserves the long-period undulations of the raw surface measurement. If you want to completely characterize the surface-finish condition, note that using only R_a or R_q is not sufficient.

7.2   MATING SURFACES

When two surfaces are pressed together under load, their apparent area of contact A_a is easily calculated from geometry but their real area of contact A_r is affected by the asperities present on their surfaces and is more difficult to accurately determine. Figure 7-4 shows two parts in contact. The tops of the asperities will initially contact the mating

R_a	Arithmetic	The arithmetic average value of filtered rough-
(CLA)	mean rough-	ness profile determined from deviations about
(AA)	ness value	the centre line within the evaluation length l_m.

DIN 4768
DIN 4762
ISO 4287/1

$$R_a = \frac{1}{l_m} \int_{x=0}^{x=l_m} |y|\,dx$$

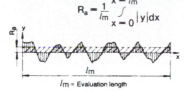

l_m = Evaluation length

R_q	Root mean	The RMS value obtained from the deviations
(RMS)	square rough-	of the filtered roughness profile over the
	ness value	evaluation length l_m.

DIN 4762
ISO 4287/1

$$R_q = \sqrt{\frac{1}{l_m} \int_0^{l_m} y^2(x)\,dx}$$

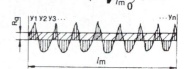

R_{pm}	Mean peak
	height value
	above the
	mean line

DIN 4762

The arithmetic average value of the five single highest peaks above the mean line $R_{p1} - R_{p5}$, similar to the R_z (DIN) definition specified in DIN 4768.
The five highest peaks are determined from the "centre line" of the filtered roughness profile each from a single sampling length l_e.

$$R_{pm} = \frac{1}{5} \cdot (R_{p1} + R_{p2} + \ldots + R_{p5})$$

R_p	Single highest
	peak above
	mean line

DIN 4762
ISO 4287/1

The value of the highest single peak above the centre line of the filtered profile as obtained from R_{pm}.

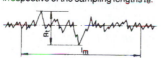

S_k	Skewness of	A measure of the shape or symmetry of the
	the profile	amplitude distribution curve obtained from the
		filtered roughness profile. A negative skew-
		ness would represent good bearing properties.

Amplitude-distribution curve

DIN 4762
ISO 4287/1

$$S_k = \frac{1}{R_q^3} \cdot \frac{1}{n} \sum_{i=1}^{i=n} (y_i - \bar{y})^3$$

A graph of the frequency in % of profile amplitudes.

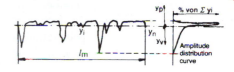

R_t	Maximum peak-	The maximum peak-to-valley height of the
(R_h)	to-valley height	filtered profile over the evaluation length l_m
(R_d)		irrespective of the sampling lengths l_e.

DIN 4762
(1960)
since 1978
it is R_{max}.

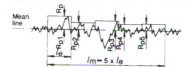

| W_t | Waviness depth | The maximum peak-to-valley height of level- |
| | | led waviness profile (roughness eliminated) |

DIN 4774

within the evaluation length l_m.

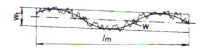

7

FIGURE 7-2

DIN and ISO Surface Roughness, Waviness, and Skewness Parameter Definitions (*Courtesy of Hommel America Inc., New Britain, CT*)

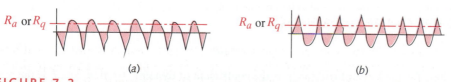

(a)   (b)

FIGURE 7-3

Different Surface Contours Can Have the Same R_a or R_q Values

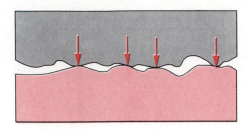

FIGURE 7-4

The Actual Contact Between Two Surfaces is Only at the Asperity Tips

part and the initial area of contact will be extremely small. The resulting stresses in the asperities will be very high and can easily exceed the compressive yield strength of the material. As the mating force is increased, the asperity tips will yield and spread until their combined area is sufficient to reduce the average stress to a sustainable level, i.e., some *compressive penetration strength* of the weaker material.

We can get a measure of a material's compressive penetration strength from conventional hardness tests (Brinell, Rockwell, etc.), which force a very smooth stylus into the material and deform (yield) the material to the stylus' shape. (See Section 2.4). The penetration strength S_p is easily calculated from these test data and tends to be of the order of 3 times the compressive yield strength S_{yc} of most materials.[2]

The real area of contact can then be estimated from

$$A_r \cong \frac{F}{S_p} \cong \frac{F}{3S_{yc}} \tag{7.1}$$

where F is the force applied normal to the surface and the strengths are as defined in the above paragraph, taken for the weaker of the two materials. *Note that the contact area for a material of particular strength under a given load will be the same regardless of the apparent area of the mating surfaces.*

7.3   FRICTION

Note that the real area of contact A_r (Eq. 7.1) is independent of the apparent area A_a that is defined by the geometry of the mating parts. This is the reason that Coulomb friction between two solids is also independent of the apparent area of contact A_a. The equation for Coulomb sliding friction is

$$f = \mu F \tag{7.2a}$$

where f is the force of friction, μ is the coefficient of dynamic friction, and F is the normal force.

The normal force presses the two surfaces together and creates elastic deformations and adhesions (see next section) at the asperities' tips. We can define the dynamic Coulomb friction force f as being the force necessary to shear the adhered and elastically interlocked asperities in order to allow a sliding motion. This shearing force is equal to the product of the shear strength of the weaker material and the actual contact area A_r plus a "plough force" P.

$$f = S_{us} A_r + P \qquad (7.2b)$$

The plough force P is due to loose particles digging into the surfaces and is negligible compared to the shear force[*], so can be ignored. Recalling equation 7.1 gives

$$A_r \cong \frac{F}{3S_{yc}} \qquad (7.2c)$$

Substituting equation 7.2c in 7.2b (ignoring P) gives

$$f \cong F \frac{S_{us}}{3S_{yc}} \qquad (7.2d)$$

Combining equations 7.2a and 7.2d gives

$$\mu = \frac{f}{F} \cong \frac{S_{us}}{3S_{yc}} \qquad (7.3)$$

which indicates that the coefficient of friction μ is a function only of a ratio of material strengths of the weaker of the two materials in contact.

The ultimate shear strength can be estimated based on the ultimate tensile strength of the material.

$$\begin{array}{lll} \text{steels:} & S_{us} \cong 0.8 S_{ut} & \\ & & (7.4) \\ \text{other ductile metals:} & S_{us} \cong 0.75 S_{ut} & \end{array}$$

The yield strength in compression as a fraction of the ultimate tensile strength varies with the material and alloy over a fairly broad range, perhaps

$$0.5 S_{ut} < S_{yc} < 0.9 S_{ut} \qquad (7.5)$$

Substituting equations 7.4 and 7.5 in equation 7.3 gives

$$\frac{0.75 S_{ut}}{3(0.9 S_{ut})} < \mu < \frac{0.8 S_{ut}}{3(0.5 S_{ut})}$$

$$0.28 < \mu < 0.53 \qquad (7.6)$$

which is approximately the range of values of μ common for dry metals in air. Note that if the metals are thoroughly cleaned, their μ will be as much as twice these values.

[*] This will be true only if the two surfaces have about the same hardness. If one surface is harder and rougher than the other, there could be a significant plough force.

In vacuum, the μ for clean surfaces can approach infinity due to cold-welding. There is so much variation in coefficient of friction with contaminant levels and other factors that the engineer should develop test data for the actual materials used under realistic service conditions. It is a simple test to perform.

Effect of Roughness on Friction

One might expect the surface roughness to have a strong influence on the friction coefficient. Tests show only a weak relationship, however. At extremely smooth finishes, below about 10 μin R_a, the coefficient of friction μ does increase by as much as a factor of 2 due to an increase in the real contact area. At very rough finishes, above about 50 μin R_a, μ also increases slightly due to the energy needed to overcome asperity interferences (plowing) in addition to shearing their adhesion bonds.

Effect of Velocity on Friction

Kinetic Coulomb friction is usually modeled as being independent of sliding velocity V except for a discontinuity at $V = 0$ where a larger, static coefficient is measured. In reality, there is a continuous, nonlinear drop in μ with increasing V. This function is approximately a straight line when plotted against the log of V and its negative slope is a few percent per decade.[7] It is believed that some of this is due to the increased interface temperatures resulting from the higher velocities reducing the material's shear yield strength in equation 7.3.

Rolling Friction

When a part rolls on another without any sliding, the coefficient of friction is much lower with μ in the range of 5E–3 to 5E–5. The friction force will vary as some power of the load (from 1.2 to 2.4) and inversely with the radius of curvature of the rolling elements. Surface roughness does have an effect on rolling friction and most such joints are finished by grinding to minimize their roughness. High hardness materials are usually used to obtain the needed strengths and promote smooth ground finishes. There is little variation of rolling friction with velocity.[7]

Effect of Lubricant on Friction

Introduction of a lubricant to a sliding interface has several beneficial effects on the friction coefficient. Lubricants may be liquid or solid, but will share the properties of low shear strength and high compressive strength. A liquid lubricant such as petroleum oil is essentially incompressible at the levels of compressive stress encountered in bearings but it readily shears. Thus, it becomes the weakest material in the interface and its low shear strength in equation 7.3 reduces the coefficient of friction. Lubricants also act as contaminants to the metal surfaces and coat them with monolayers of molecules

Table 7-1   Coefficients of Friction for Some Material Combinations

Material 1	Material 2	Static Dry	Static Lubricated	Dynamic Dry	Dynamic Lubricated
mild steel	mild steel	0.74		0.57	0.09
mild steel	cast iron		0.183	0.23	0.133
mild steel	aluminum	0.61		0.47	
mild steel	brass	0.51		0.44	
hard steel	hard steel	0.78	0.11-0.23	0.42	0.03-0.19
hard steel	babbitt	0.42-0.70	0.08-0.25	0.34	0.06-0.16
teflon	Teflon	0.04			0.04
steel	Teflon	0.04			0.04
cast iron	cast iron	1.10		0.15	0.07
cast iron	bronze			0.22	0.077
aluminum	aluminum	1.05		1.4	

Source: *Mark's Mechanical Engineers' Handbook*, T. Baumeister, ed., McGraw-Hill, New York.

that inhibit adhesion even between compatible metals (see next section). Many commercial lubricant oils are mixed with various additives that react with the metals to form monolayer contaminants. So-called EP (*Extreme Pressure*) lubricants add fatty acids or other compounds to the oil that attack the metal chemically and form a contaminant layer that protects and reduces friction even when the oil film is squeezed out of the interface by high contact loads. Lubricants, especially liquids, also serve to remove heat from the interface. Lower temperatures reduce surface interactions and wear. Lubricants and lubrication phenomena will be discussed in more detail in Chapter 10. Table 7-1 shows some typical values of friction coefficients for commonly encountered pairs.

7.4   ADHESIVE WEAR

When (clean) surfaces such as those shown in Figure 7-1 are pressed against one another under load, some of the asperities in contact will tend to adhere to one another due to the attractive forces between the surface atoms of the two materials.[3] As sliding between the surfaces is introduced, these adhesions are broken, either along the original interface or along a new plane through the material of the asperity peak. In the latter case, a piece of part *A* is transferred to part *B* causing surface disruption and damage. Sometimes, a particle of one material will be broken free and become debris in the interface, which can then scratch the surface and plough furrows in both parts. This damage is sometimes called **scoring** or **scuffing*** of the surface. Figure 7-5 shows an example of a shaft failed by adhesive wear in the absence of adequate lubricant.[6]

* Note that scuffing is often associated with gear teeth, which typically experience a combination of rolling and sliding. See Chapter 11 for further discussion.

7

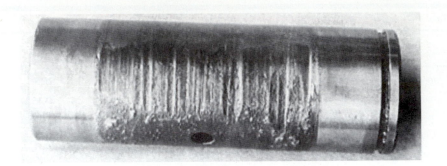

FIGURE 7-5

Adhesive Wear on a Shaft. Source: D. J. Wulpi, *Understanding How Components Fail.* Amer. Soc. for Metals: Metals Park, Ohio, 1990, with permission.

The original adhesion theory postulated that all asperity contacts would result in yielding and adhesion due to the high stresses present. It is now believed that in most cases of contact, especially with repeated rubbing, only a small fraction of the asperity contacts actually result in yielding and adhesion; elastic deformations of the asperities also play a significant role in the tractive forces (friction) developed at the interface.[32]

COMPATIBILITY   An important factor affecting adhesion is the metallurgical compatibility of the mating materials. **Metallurgical compatibility** between two metals is defined as *high mutual solubility* or *the formation of intermetallic compounds*.[4] Davies defines two conditions for metallurgical **incompatibility**, meaning that the *metals can then slide on one another with relatively little scoring*.[5]

1   *The metals must be insoluble in each other, with neither material dissolving in the other nor forming an alloy with it.*

2   *At least one of the materials must be from the B-subgroup, i.e., the elements to the right of the Ni-Pd-Pt column in the Periodic Table.**

Unfortunately this terminology can be confusing because the word *compatibility* usually means an ability to work together, whereas in this context it means that they do not work (slide) together well. Their metallurgical "compatibility" in this case is one of *adhering together*, which acts to prevent sliding, making them **frictionally incompatible**.

Rabinowicz[33] groups material pairs into (metallurgically) identical, compatible, partially compatible, partially incompatible, and incompatible categories based on the above criteria. The identical and compatible combinations should not be run together in unlubricated sliding contact. The incompatible and the partial categories can be run together. Figure 7-6 shows a compatibility chart for commonly used metals based on his categories. The dotted circles indicate **metallurgically compatible** metals (i.e., not acceptable for sliding contact). A dark quarter circle indicates **partially compatible**, and a dark half circle **partially incompatible** combinations. The latter are better in slid-

---

* Some metals in this B-subgroup that are of possible interest for affordable bearing alloys are (in alphabetical order): Aluminum (Al), Antimony (Sb), Bismuth (Bi), Cadmium (Cd), Carbon (C), Copper (Cu), Lead(Pb), Silicon (Si), Tin (Sn), Zinc (Zn).

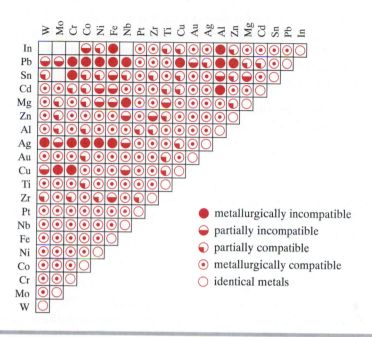

FIGURE 7-6

Compatibility Chart for Metal Pairs Based on Binary Phase Diagrams (Adapted from Figure 7, p. 491, E. Rabinowicz, *Wear Coefficients—Metals*, in *Wear Control Handbook*, M. B. Peterson and W. O. Winer, eds., ASME, New York, 1980, with permission)

ing contact than the former. The solid-color circles indicate **metallurgically incompatible** pairs that can be expected to resist adhesive wear best of any combinations shown.

CONTAMINANTS Adhesive bonding at the asperities can only occur if the material is clean and free of contaminants. Contaminants can take the form of oxides, skin oils from human handling, atmospheric moisture, etc. Contaminants in this context also include materials deliberately introduced to the interface such as coatings or lubricants. In fact, one of the chief functions of a lubricant is to prevent these adhesions and thus reduce friction and surface damage. A lubricant film effectively isolates the two materials and can prevent adhesion even between identical materials.

SURFACE FINISH It is not necessary for the surfaces to be "rough" for this adhesive-wear mechanism to operate. The fine-ground finish of the part in Figure 7-1*a* is seen to have as many asperities available for this process as the rougher milled surface of Figure 7-1*b*.

COLD-WELDING If the mating materials are metals, are compatible, and are extremely clean, the adhesive forces will be high and the sliding friction can generate enough localized heat to weld the asperities together. If the clean metal surfaces are also finished to a low roughness value (i.e., polished), and then rubbed together (with sufficient force) they can cold-weld (seize) with a bond virtually as strong as the parent metal.

This process is enhanced if done in a vacuum as the absence of air eliminates contamination from surface oxidation. The **roll-bonding** process in which two compatible metals are cold-welded (in air) by rolling or coining them together under high normal loads is used commercially to make bimetallic strips for thermostats, and dimes and quarters for your pocket.

GALLING   describes a situation of incomplete cold-welding where, for whatever reason (usually contamination), the parts do not completely weld together. But, portions of the surfaces do adhere, causing material to be transferred from one part to the other in large streaks visible to the naked eye. Galling generally ruins the surface in one pass.

These factors explain the reasons for what is common knowledge among machinists and experienced engineers; *the same material should generally not be run against itself.* There are some exceptions to this rule, notably for hardened steel on hardened steel, but other combinations such as aluminum on aluminum *must* be avoided.

The Adhesive Wear Coefficient

In general, wear is inversely proportional to hardness. The rate of wear can be determined by running a pin against a rotating disk under controlled loading and lubrication conditions over a known sliding distance and measuring the loss in volume. The volume of wear is independent of the velocity of sliding and can be expressed as

$$V = K\frac{Fl}{H} \tag{7.7a}$$

where V = volume of wear from the softer of the two materials, F = normal force, l = length of sliding, and H is the penetration hardness in kg_f/mm^2 or psi. H may be expressed as Brinell (HB), Vickers (HV), or other absolute hardness units. A Rockwell hardness reading can be used if it is first converted to one of the other scales having true units. The factor K is the **wear coefficient** and is a dimensionless property of the sliding system. K will be a function of the materials used and also of the lubrication situation. (See Table 2-3.)

Since the depth of wear d may be of more engineering interest than the volume, equation 7.7a can be written in those terms as

$$d = K\frac{Fl}{HA_a} \tag{7.7b}$$

where A_a is the apparent area of contact of the interface.

Values of K obtained for the same materials tested under the same conditions vary by about a factor of 2 from test to test. This kind of variability is also seen in tests of friction coefficients, which typically have standard deviations of ±20%. The reasons for these variations are not fully understood but are generally attributed to the difficulty in accurately reproducing the same surface conditions from test to test.[33] Despite this

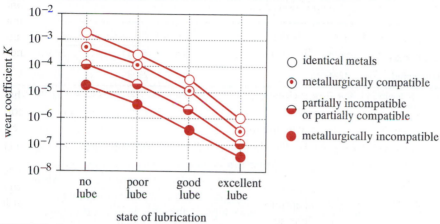

FIGURE 7-7

Adhesion Wear Coefficient as a Function of Compatibility and Lubrication (Adapted from Figure 11, p. 495, E. Rabinowicz, *Wear Coefficients—Metals*, in *Wear Control Handbook*, M. B. Peterson and W. O. Winer, eds., ASME, New York, 1980, with permission)

large variation, the data obtained are nevertheless better than no data and are useful to make estimates of wear rates for situations in which testing of the actual design is not feasible.

Tables of empirical values for various combinations of material pairs and lubrication conditions have been published in the literature.[33] While it is possible to find wear coefficient data for combinations that approximate your particular situation, the sheer number of possible permutations dictates that some design situations will not fit the available data. Figure 7-7 shows a generalized plot of the wear coefficient K as a function of both the lubrication condition and Rabinowicz's compatibility categories. An approximate value of K can be obtained from this figure for any design situation. Recognize that only a testing program can give reasonably accurate wear life data for a design. Note that the data in Figure 7-7 are based on the wear coefficient associated with material loss. Material is also transferred from one of the sliding materials to the other by adhesion. The wear coefficient for adhesive transfer is about three times that for material loss from the system.[33]

7.5   ABRASIVE WEAR

Abrasion occurs in two modes, referred to as the *two-body* and *three-body* abrasive wear processes.[9] **Two-body abrasion** refers to a *hard rough material sliding against a softer one*. The hard surface digs into and removes material from the softer one. An example is a file used to contour a metal part. **Three-body abrasion** refers to the *introduction of hard particles between two sliding surfaces, at least one of which is softer than the*

Table 7-2	Wear Coefficient K for Abrasion			
	File	**Abrasive Paper, New**	**Loose Abrasive Grains**	**Coarse Polishing**
Dry Surfaces	5E–2	1E–2	1E–3	1E–4
Lubricated	1E–1	2E–2	2E–3	2E–4

Source: E. Rabinowitz, *Wear Coefficients—Metals*, in *Wear Control Handbook*, M.B. Peterson and W. O. Winer, eds., ASME, New York, 1980.

particles. The hard particles abrade material from one or both surfaces. Lapping and polishing are in this category. **Abrasion** is then *a material removal process in which the affected surfaces lose mass at some controlled or uncontrolled rate*. Abrasive wear also obeys the wear equation 7.7. See Figure 7-8 for an indication of wear coefficients for abrasive wear. Table 7-2 also shows typical abrasive wear coefficients.

UNCONTROLLED ABRASION   Earth-moving equipment such as backhoes, bulldozers, and mining equipment operate in a relatively uncontrolled three-body abrasion mode since the dug earth or minerals often contain materials harder than the steel surfaces of the equipment. Silica (sand) is the most abundant solid material on the earth's surface and it is harder than most metals (absolute hardness of 800 kg/mm^2). Soft steel's absolute hardness is only about 200 kg/mm^2, but hardened tool steels can be as high as 1 000 kg/mm^2 and so can survive in these applications. Hard-steel files can thus be used

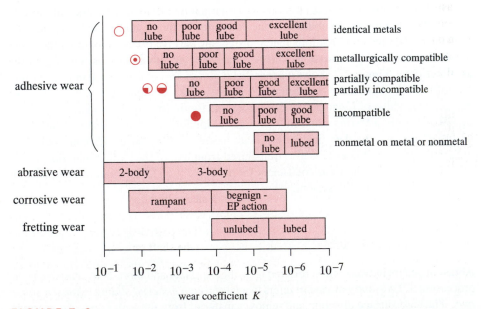

FIGURE 7-8

Wear Coefficients for Various Sliding Situations Source: E. Rabinowicz, *Wear Coefficients–Metals*. in *Wear Control Handbook*, M.B. Peterson and W.O. Winer, eds., ASME, New York, 1980

FIGURE 7-9

A Backhoe Tooth: (a) New, (b) Abrasive Wear on Soft Back Face, (c) Abrasive Wear on Hard Front
Face   Source: D. J. Wulpi, *Understanding How Components Fail*, Amer. Soc. for Metals: Metals Park, Ohio, 1990

to abrade softer metals, nonmetals, and even glass (which is a form of silica). Many
machine design applications involve the handling of materials of production that are
abrasive. Pumping of wet concrete, rock crushing, and transport of ceramic parts all
are obvious cases of abrasive media. Figure 7-9 shows the effects of uncontrolled abrasion on the replaceable tooth of a backhoe. A new part (a) is contrasted with the worn
back side (b) and front side (c) of a used part. The front side is made of 8640 medium-
hard steel while the back side is soft 1010 steel.[6]

Machine parts that operate in cleaner environments can be designed to minimize
or eliminate abrasive wear through proper selection of materials and finishes. Smooth,
hard materials will not abrade soft ones in two-body contact. Sleeve bearings and shafts
are typically finished to very low roughness and made of suitable pairs of materials as
indicated in Section 7.4. The smooth finishes minimize abrasion at the outset and, unless hard particulate contaminants are later introduced to the interface in service, that
situation should continue. One reason for making sleeve bearings of soft materials (running against hard shafts) is to promote the embedding in the soft bearing material of any
hard particles that find their way into the bearing. The particles are then trapped (bur-
ied) in the soft material and their potential damage to the shaft minimized. Particles may
enter the bearing either as foreign contaminants in the lubricant or as oxidation products generated within the bearing. Iron oxides are harder than the steel that spawns them
and will abrade the shaft. If hydrostatic lubrication is used (in which the lubricant is
actively circulated—see Chapter 10), then filtration of the lubricant should be done to
remove any particulates that enter the system. A properly designed, hydrostatically lubricated bearing should experience no abrasive wear if it has sufficient, clean lubricant.

CONTROLLED ABRASION   In addition to designing systems to avoid abrasion, engineers also design them to *create* controlled abrasive wear. Controlled abrasion is widely used in manufacturing processes. Two-body **grinding** is perhaps the most common example, in which abrasive media such as silicon carbide (Carborundum) are forced against the part under high sliding velocities to remove material and control size and finish. A coolant is often used both to protect the material from an unwanted heat treatment and to enhance the abrasion process. Moisture increases the abrasion rate by about 15% over dry abrasion.[10] Abrasive paper and cloth provide a means of applying an abrasive medium to a compound- curved surface as well.   Sandblasting is an example of erosion, where one body is the sand and the other is the surface to be abraded or eroded.

A common example of controlled three-body abrasion in manufacturing is the **tumbling process** in which parts are placed in a drum along with some abrasive particles and then tumbled together. The parts rub and bounce off one another in the abrasive mix. The result is removal of burrs and sharp edges and a general polishing of all exposed surfaces. Another example is the **surface polishing** process, which involves the use of very fine, hard particulates trapped between a relatively soft and conforming material (e.g., cloth) and the surface to be polished. Relative velocities are high and moisture is often added.

PARTICLE SIZE   has an effect on the efficiency of an abrasion process. There is a threshold particle size for any situation above which abrasion will continue rapidly but below which the wear rate slows. It is believed that the wear rate is compromised when the size of the particles abraded from the workpiece is the same or larger than the abrasive particles. If so, they prevent the abrasives from digging into the workpiece.

Abrasive Materials

The two requirements for an abrasive are hardness and sharpness. The abrasive must be harder than the material to be abraded. Excessive hardness beyond about 150% of the workpiece hardness does not increase the wear rate but does prolong the useful "sharpness life" of the abrasive, which itself loses its cutting ability over time.[11] Sharpness is achieved by using brittle materials that break into sharp-edged particles. The classes of materials that best meet these two criteria are ceramics and hard nonmetals. Most commercial abrasives are of these types. Table 7-3 shows some common abrasive materials and their hardness. Aluminum oxide (corundum) and silicon carbide (Carborundum) are the most used due to their favorable combination of relatively high hardness and low cost. Boron carbide and diamond are used for applications requiring the hardest materials, but both are expensive.

Abrasion-Resistant Materials

Some engineering materials are better suited to abrasive wear applications than others based largely on their hardness. However with hardness usually comes brittleness and

---

Table 7-3   Materials for Use as Abrasives

Material	Composition	Hardness (kg/mm^2)
Diamond	C	8 000
Boron Carbide	B_4C	2 750
Carborundum (Silicon Carbide)	SiC	2 500
Titanium Carbide	TiC	2 450
Corundum (Alumina)	Al_2O_3	2 100
Zirconium Carbide	ZrC	2 100
Tungsten Carbide	WC	1 900
Garnet	$Al_2O_3.3FeO.3SiO_2$	1 350
Zirconia	ZrO_2	150
Quartz, Silica, Sand	SiO_2	800
Glass	Silicate	$\cong 500$

Source: *Friction and Wear of Materials,* E. Rabinowitz, 1965, reprinted by permission of John Wiley & Sons, Inc., New York

---

thus their resistance to impact or fatigue loads can be less than optimum. Table 7-4 shows the hardness of some materials that are suitable for abrasive wear applications.

COATINGS   Some ceramic materials can be plasma-sprayed onto metal substrates to provide a hard facing that also has high corrosion and chemical resistance. These plasma-sprayed coatings are quite rough upon application (like severely orange-peeled paint) and thus must be diamond ground to obtain a finish suitable for a sliding joint. These coatings are also very brittle and may chip from the substrate if overstressed either mechanically or thermally.

Note that aluminum oxide can be created in controlled fashion on aluminum by anodizing and will have a finish as good as the substrate. So-called hard-anodizing is merely a thicker anodized coating than used for corrosion protection and is commonly used to protect aluminum parts in abrasive wear conditions. (See Section 2.5.)

7.6   CORROSION WEAR

CORROSION   occurs in normal environments with virtually all materials except those termed noble, i.e., gold, platinum, etc. The most common form of corrosion is oxidation. Most metals react with the oxygen in air or water to form oxides. In some materials, such as aluminum, the oxidation is self-limiting as long as the surface is undisturbed. Aluminum in air forms an oxide layer that gradually builds to a thickness of 0.02 μm at which point the reaction ceases because the nonporous aluminum-oxide

Table 7-4   Materials Resistant to Abrasion

Material	Hardness (kg/mm^2)	Relative Wear
Tungsten carbide (sintered)	1 400-1 800	0.5-5
High-chromium white cast iron		5-10
Tool steel	700-1 000	20-30
Bearing steel	700-950	
Chromium (electroplated)	900	
Carburized steel	900	20-30
Nitrided steel	900-1 250	20-30
Pearlitic white iron		25-50
Austenitic manganese steel		30-50
Pearlitic low-alloy (0.7%C) steel	480	30-60
Pearlitic unalloyed (0.7%C) steel	300	50-70
As-rolled or normalized low-carbon (0.2%C) steel		100

Sources: *Friction and Wear of Materials*, E. Rabinowitz, 1965, reprinted by permission of *John Wiley & Sons, Inc., New York*
T. E. Norman, *Abrasive Wear of Metals*, in *Handbook of Mechanical Wear*, C. Lipson, ed.,U. Mich. Press, 1961.

film seals the substrate from further contact with the oxygen in the air. (This is the principle of anodizing, which creates a uniform, controlled-thickness layer of aluminum oxide on the part before it is placed in service.) Iron alloys, on the other hand, form a discontinuous and porous film of oxide that readily flakes off by itself to expose new substrate material. Oxidation will continue until all the iron is converted to oxides. Elevated temperatures greatly increase the rate of all chemical reactions.

CORROSION WEAR   adds to the chemically corrosive environment a mechanical disruption of the surface layer due to a sliding or rolling contact of two bodies. This surface contact can act to break up the oxide (or other) film and expose new substrate to the reactive elements thus increasing the rate of corrosion. If the products of the chemical reaction are hard and brittle (as with oxides) flakes of this layer can become loose particles in the interface and contribute to other forms of wear such as abrasion. See Figure 7-8 for an indication of wear coefficients for corrosion wear.

Some reaction products of metals such as metallic chlorides, phosphates, and sulfides are softer than the metal substrate and are also not brittle. These corrosion products can act as beneficial contaminants to reduce adhesive wear by blocking the adhesion of the metal asperities. This is the reason for adding compounds containing chlorine, sulphur, and other reactive agents to EP (extreme pressure) oils. The strategy is to trade a slow rate of corrosive wear for a more rapid and damaging rate of adhesive wear on metal surfaces such as gear teeth and cams, which can have poor lubrication due to their nonconforming geometry.

Corrosion Fatigue

Chapter 6 discussed the mechanisms of fracture mechanics and fatigue failure in detail and made brief mention of the phenomenon variously called corrosion fatigue or stress-corrosion. This mechanism is not yet fully understood, but the empirical evidence of its result is strong and unequivocal. When a part is stressed in the presence of a corrosive environment, the corrosion process is accelerated and failure occurs more rapidly than would be expected from either the stress state alone or the corrosion process alone.

Static stresses are sufficient to accelerate the corrosion process. The combination of stress and corrosive environment has a synergistic effect and the material corrodes more rapidly than if unstressed. This combined condition of static stress and corrosion is termed **stress corrosion**. If the part is *cyclically stressed in a corrosive environment*, the crack will grow more rapidly than from either factor alone. This is called **corrosion fatigue**. While the frequency of stress cycling (as opposed to the number of cycles) appears to have no detrimental effect on crack growth in a noncorrosive environment, in the presence of corrosive environments, it does. Lower cyclic frequencies allow the environment more time to act on the stressed crack tip while it is held open under tensile stress and this substantially increases the rate of crack growth. See Figures 6-30 to 6-32 and their discussion in Chapter 6 for more information on this phenomenon.

Fretting Corrosion

When two metal surfaces are in intimate contact, such as press-fit or clamped, one would expect no severe corrosion to occur at the interface especially if in air. However, these kinds of contacts are subject to a phenomenon called **fretting corrosion** (or **fretting**) that can cause significant loss of material from the interface. Even though no gross sliding motions are possible in these situations, even small deflections (of the order of thousandths of an inch) are enough to cause fretting. Vibrations are another possible source of small fretting motions.

The fretting mechanism is believed to be some combination of abrasion, adhesion, and corrosion.[12] Free surfaces will oxidize in air but the rate will slow as the oxides formed on the surface gradually block the substrate from the atmosphere. As discussed above, some metals actually self-limit their oxidation if left undisturbed. The presence of vibrations or repeated mechanical deflections tends to disturb the oxide layer, scraping it loose and exposing new base metal to oxygen. This promotes adhesion of the "cleaned" metal asperities between the parts and also provides abrasive media in the form of hard oxide particles in the interface for three-body abrasion. All of these mechanisms tend to slowly reduce the solid volume of the materials and produce a "dust" or "powder" of abraded/oxidized material. Over time, significant dimensional loss can occur at the interface. In other cases, the result can be only a minor discoloration of the surfaces or adhesion similar to galling. All this from a joint that has no designed-in relative motion and was probably thought of by the designer as rigid! Of course, nothing is truly rigid and fretting is evidence that microscopic motions are enough to

FIGURE 7-10

Fretting Wear on a Shaft Beneath a Press-Fit Hub Source: D. J. Wulpi, *Understanding How Components Fail*, Amer. Soc. for Metals: Metals Park, Ohio, 1990

cause wear. Figure 7-10 shows fretting on a shaft where a hub was press-fitted.[6] See Figure 7-8 for an indication of wear coefficients for fretting wear.

Some techniques that have proven to reduce fretting are the reduction of deflections (i.e., stiffer designs or tighter clamping) and the introduction of dry or fluid lubricants to the joint to act as an oxygen barrier and friction reducer. The introduction of a gasket, especially one with substantial elasticity (such as rubber) to absorb the vibrations has been shown to help. Harder and smoother surfaces on the metal parts are more resistant to abrasion and will reduce fretting damage. Corrosion-resistant platings such as chromium are sometimes used. The best method (impractical in most instances) is to eliminate the oxygen by operating in a vacuum or inert-gas atmosphere.

7.7   SURFACE FATIGUE

All the surface-failure modes discussed above apply to situations in which the relative motions between the surfaces are essentially pure sliding. When two surfaces are in **pure-rolling** contact, or are primarily rolling in combination with a small percentage of sliding, a different surface failure mechanism comes into play, called **surface fatigue**. Many applications of this condition exist such as ball and roller bearings, cams with roller followers, nip rolls, and spur or helical gear-tooth contact. All except the gear teeth and nip rolls typically have essentially pure rolling with only about 1% sliding. Gear teeth have significant sliding at portions of their tooth interface and this will change the stress state significantly compared to the pure rolling cases, as we shall see. Other types of gears such as spiral bevel, hypoid, and worm sets have essentially pure sliding at their interfaces and one or more of the wear mechanisms discussed above will apply. Nip rolls (such as those used to roll sheet steel) can be run with or without sliding depending on their purpose.

The stresses introduced in two materials contacting at a rolling interface are highly dependent on the geometry of the surfaces in contact as well as on the loading and material properties. The general case allows any three-dimensional geometry on each contacting member and, as would be expected, its calculation is the most complex. Two special-geometry cases are of practical interest and are also somewhat simpler to analyze. These are *sphere-on-sphere*, and *cylinder-on-cylinder*. In all cases, the radii of curvature of the mating surfaces will be significant factors. By varying the radii of curvature of one mating surface, these special cases can be extended to include the subcases of *sphere-on-plane*, *sphere-in-cup*, *cylinder-on-plane*, and *cylinder-in-trough*. It is only necessary to make the radii of curvature of one element infinite to obtain a plane, and negative radii of curvature define a concave cup, or concave trough surface. For example, some ball bearings can be modeled as *sphere-on-plane* and some roller bearings as *cylinder-in-trough*.

As a ball passes over another surface, the theoretical contact patch is a point of zero dimension. A roller against a cylindrical or flat surface theoretically contacts along a line of zero width. Since the area of each of these theoretical contact geometries is zero, any applied force will then create an infinite stress. We know that this cannot be true as the materials would instantly fail. In fact, the materials must deflect to create sufficient contact area to support the load at some finite stress. This deflection creates a semi-ellipsoidal pressure distribution over the contact patch. In the general case, the contact patch is elliptical as shown in Figure 7-11a. Spheres will have a circular contact patch and cylinders create a rectangular contact patch as shown in Figure 7-11b.

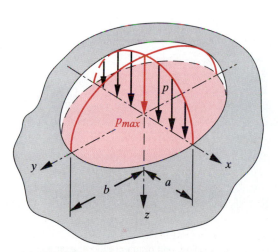

(a) Ellipsoidal pressure distribution in general contact— for spherical contact $a = b$

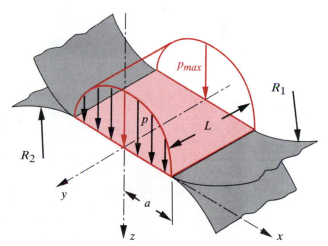

(b) Ellipsoidal-prism pressure distribution in cylindrical contact

FIGURE 7-11

Pressure Distributions and Contact Zones of Spherical, Cylindrical, and General Hertzian Contact

Consider the case of a spherical ball rolling in a straight line against a flat surface with no slip, and under a constant normal load. If the load is such as to stress the material only below its yield point, the deflection in the contact patch will be elastic and the surface will return to its original curved geometry after passing through contact. The same spot on the ball will contact the surface again on each succeeding revolution. The resulting stresses in the contact patch are called **contact stresses or Hertzian stresses**. The contact stresses in this small volume of the ball are **repeated** at the ball's rotation frequency. This creates a fatigue-loading situation that eventually leads to a **surface fatigue failure**.

This repeated loading is similar to the tensile fatigue-loading case shown in Figure 6-1*b*. The significant difference in this case is that the principal contact stresses at the center of the contact patch are all compressive, not tensile. Recall from Chapter 6 that fatigue failures are considered to be initiated by shear stress and continued to failure by tensile stress. There is also a shear stress associated with these compressive contact stresses, and it is believed to be the cause of crack formation after many stress-cycles. Crack growth then eventually results in failure by **pitting**—*the fracture and dislodgment of small pieces of material from the surface*. Once the surface begins to pit, its surface finish is compromised and it rapidly proceeds to failure by **spalling**—*the loss of large pieces of the surface*. Figure 7-12 shows some examples of pitted and spalled surfaces.

If the load is large enough to raise the contact stress above the material's compressive yield strength, then the contact-patch deflection will create a permanent flat on the ball. This condition is sometimes called **false brinelling** because it has a similar ap-

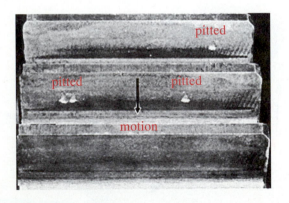

(a) Mild pitting on gear teeth

(b) Severe pitting, spalling, and disintegration of gear teeth

FIGURE 7-12

Examples of Surfaces Failed by Pitting and Spalling due to Surface Fatigue Source: J. D. Graham, *Pitting of Gear Teeth*, in *Handbook of Mechanical Wear*, C. Lipson, ed., U. Mich. Press, 1961, pp. 138, 143, with permission

pearance to the indentation made to test a material's Brinell hardness. Such a flat on even one of its balls (or rollers) makes a ball (or roller) bearing useless.

We will now investigate the contact-patch geometries, pressure distributions, stresses and deformations in rolling contacts starting with the relatively simple geometry of *sphere-on-sphere*, next dealing with the *cylinder-on-cylinder* case, and finally discussing the *general* case. Derivation of the equations for these cases are among the more complex sets of examples from the theory of elasticity. The equations for the area of contact, deformation, pressure distribution and contact stress on the centerline of two bodies with static loading were originally derived by Hertz in 1881[13] an English translation of which can be found in reference 14. Many others have since added to the understanding of this problem.[15],[16],[17],[18]

7.8   SPHERICAL CONTACT

Cross-sections of two spheres in contact are shown in Figure 7-13. The dotted lines indicate the possibilities of one being a flat plane or a concave cup. The difference is only in the magnitude or the sign of its radius of curvature (convex +, concave −). Figure 7-11*a* shows the general semi-ellipsoidal pressure distribution over the contact patch. For a sphere-on-sphere, it will be a hemisphere with a circular contact patch ($a = b$).

Contact Pressure and Contact Patch in Spherical Contact

The contact pressure is a maximum p_{max} at the center and zero at the edge. The total applied load F on the contact patch is equal to the volume of the hemisphere:

$$F = \frac{2}{3} \pi a^2 p_{max} \tag{7.8a}$$

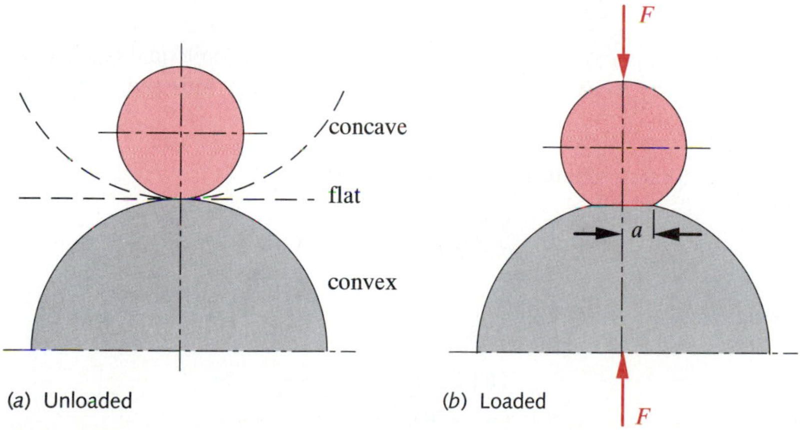

(a) Unloaded   (b) Loaded

FIGURE 7-13

Contact Zone of Two Spheres or Cylinders

where a is the half-width (radius) of the contact patch. This can be solved for the maximum pressure:

$$p_{max} = \frac{3}{2}\frac{F}{\pi a^2} \tag{7.8b}$$

The average pressure on the contact patch is the applied force divided by its area:

$$p_{avg} = \frac{F}{area} = \frac{F}{\pi a^2} \tag{7.8c}$$

and substituting equation 7.8c in 7.8b gives:

$$p_{max} = \frac{3}{2}P_{avg} \tag{7.8d}$$

We now define *material constants* for the two spheres

$$m_1 = \frac{1-v_1^2}{E_1} \qquad\qquad m_2 = \frac{1-v_2^2}{E_2} \tag{7.9a}$$

where E_1, E_2 and v_1, v_2 are the Young's moduli and Poisson's ratios for the materials of sphere 1 and sphere 2, respectively.

The dimensions of the contact area are typically very small compared to the radii of curvature of the bodies, which allows the radii to be considered constant over the contact area despite the small deformations occurring there. We can define a *geometry constant* that depends only on the radii R_1 and R_2 of the two spheres,

$$B = \frac{1}{2}\left(\frac{1}{R_1}+\frac{1}{R_2}\right) \tag{7.9b}$$

To account for the case of a sphere-on-plane, R_2 becomes infinite making $1/R_2$ zero. For a sphere-in-cup, R_2 becomes negative. (See Figure 7-13.) Otherwise R_2 is finite and positive, as is R_1.

The contact patch radius a is then found from

$$a = \frac{\pi}{4}P_{max}\frac{m_1+m_2}{B} \tag{7.9c}$$

Substitute equation 7.8b in 7.9c:

$$a = \frac{\pi}{4}\left(\frac{3}{2}\frac{F}{\pi a^2}\right)\frac{m_1+m_2}{B}$$

$$a = \sqrt[3]{0.375\frac{m_1+m_2}{B}F} \tag{7.9d}$$

The pressure distribution within the hemisphere is

$$p = p_{max} \sqrt{1 - \frac{x^2}{a^2} - \frac{y^2}{a^2}} \qquad (7.10)$$

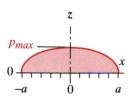

We can normalize the pressure p to the magnitude of p_{avg} and the patch dimension x or y to the patch radius a and then plot the normalized pressure distribution across the patch, which will be an ellipse as shown in Figure 7-14.

FIGURE 7-14

Static Stress Distributions in Spherical Contact

Pressure Distribution
Across Contact Patch

The pressure on the contact patch creates a three-dimensional stress state in the material. The three applied stresses σ_x, σ_y, and σ_z are compressive and are maximal at the sphere's surface in the center of the patch. They diminish rapidly and nonlinearly with depth and with distance from the axis of contact. They are called **Hertzian stresses** in honor of their original discoverer. A complete derivation of these equations can be found in reference 19. Note that these applied stresses in the x, y, and z directions are also the principal stresses in this case. If we look at these stresses as they vary along the z-axis (with z increasing into the material) we find

$$\sigma_z = p_{max}\left[-1 + \frac{z^3}{\left(a^2 + z^2\right)^{3/2}}\right] \qquad (7.11a)$$

$$\sigma_x = \sigma_y = \frac{p_{max}}{2}\left[-(1+2\nu) + 2(1+\nu)\left(\frac{z}{\sqrt{a^2 + z^2}}\right) - \left(\frac{z}{\sqrt{a^2 + z^2}}\right)^3\right] \qquad (7.11b)$$

Poisson's ratio is taken for the sphere of interest in this calculation. These normal, (and principal) stresses are maximal at the surface, where $z = 0$:

$$\sigma_{z_{max}} = -p_{max} \qquad (7.11c)$$

$$\sigma_{x_{max}} = \sigma_{y_{max}} = -\frac{1+2\nu}{2}p_{max} \qquad (7.11d)$$

There is also a shear stress induced from these normal stresses:

$$\tau_{13} = \frac{p_{max}}{2}\left[\frac{(1-2\nu)}{2} + (1+\nu)\frac{z}{\sqrt{a^2 + z^2}} - \frac{3}{2}\left(\frac{z}{\sqrt{a^2 + z^2}}\right)^3\right] \qquad (7.12a)$$

which is not maximum at the surface but rather at a small distance $z_{@\tau_{max}}$ below the surface.

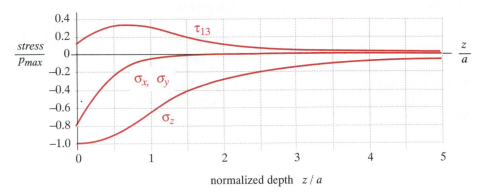

7

FIGURE 7-15

Normalized Stress Distribution Along the z axis in Static Spherical Contact - *xyz* Stresses are Principal

$$\tau_{13_{max}} = \frac{p_{max}}{2}\left[\frac{(1-2\nu)}{2} + \frac{2}{9}(1+\nu)\sqrt{2(1+\nu)}\right] \qquad (7.12b)$$

$$z_{@\tau_{max}} = a\sqrt{\frac{2+2\nu}{7-2\nu}} \qquad (7.12c)$$

Figure 7-15 shows a plot of the principal normal and maximum shear stresses as a function of depth z along a radius of the sphere. The stresses are normalized to the maximum pressure p_{max}, and the depth is normalized to the half-width a of the contact patch. This plot provides a dimensionless picture of the stress distribution on the centerline under a spherical contact. Note that all the stresses have diminished to <10% of p_{max} within $z = 5a$. The subsurface location of the maximum shear stress can also be seen. If both materials are steel, it occurs at a depth of about $0.63a$ and its magnitude is about $0.34\,p_{max}$. The shear stress is about $0.11\,p_{max}$ at the surface on the z axis.

The subsurface location of the maximum shear stress is believed by some to be a significant factor in surface-fatigue failure. The theory says cracks that begin below the surface eventually grow to the point that the material above the crack breaks out to form a pit as shown in Figure 7-12.

Figure 7-16 shows a photoelastic model of the contact stresses in a cam immediately beneath a loaded roller follower.[20] Experimental photoelastic stress analysis uses a physical model of the part to be analyzed made from a transparent plastic material (Lexan in this example) that shows fringes of constant stress magnitude when loaded and viewed in polarized light. The maximum shear stress can be clearly seen a small distance into the cam directly under the follower. While this is a cylindrical rather than a spherical contact, their stress distributions along the centerline are similar, as will be seen in the next section.

follower

cam

FIGURE 7-16

Photoelastic Analysis of Contact Stresses Under a Cam Follower Source: V. S. Mahkijani, *Study of Contact Stresses as Developed on a Radial Cam using Photoelastic Model and Finite Element Analysis.* M.S. Thesis, Worcester Polytechnic Institute, 1984

When we move off the centerline of the contact patch on the surface of the sphere, the stresses diminish. At the edge of the patch the radial stress σ_z is zero but there is a condition of pure shear stress with the magnitude:

$$\tau_{xy} = \frac{1-2\nu}{3} p_{max} \qquad (7.13a)$$

Picture the Mohr's circle for a pure shear case. The two nonzero principal stresses will be $\pm \tau_{xy}$, which means that there is also a tensile stress at that point of

$$\sigma_{1_{edge}} = \frac{1-2\nu}{3} p_{max} \qquad (7.13b)$$

EXAMPLE 7-1

Stresses in a Ball Thrust Bearing

Problem   A ball thrust bearing with 7 balls is loaded axially across its races through the balls. What is the size of the contact patch on a race and what are the stresses developed in balls and races? What is the depth of the maximum shear stress in a ball?

Given   The 7 spherical balls are 10-mm (0.394-in) dia and the races are flat. All parts are hardened steel. The axial load is 151 lb or 21.5 lb per ball.

Assumptions The 7 balls share the load equally. The rotational speed is sufficiently slow that this can be considered a static loading problem.

Solution   See the *TKSolver* file EX07-01.

1   We need to first determine the size of the contact patch, for which we need to find the geometry constant and material constants from equation 7.9a and b.

$$B = \frac{1}{2}\left(\frac{1}{R_1} + \frac{1}{R_2}\right) = \frac{1}{2}\left(\frac{1}{0.197} + \frac{1}{\infty}\right) = 2.54 \qquad (a)$$

Note the infinite radius of curvature for R_2.

$$m_1 = m_2 = \frac{1 - v_1^2}{E_1} = \frac{1 - 0.28^2}{3E7} = 3.072E - 8 \qquad (b)$$

Note both materials are the same in this example. The material and geometry constants can now be used in equation 7.9d.

$$a = \sqrt[3]{\frac{3}{8}\frac{m_1 + m_2}{B}F} = \sqrt[3]{0.375\frac{2(3.072E - 8)}{2.54}21.5} = 0.005\ 8\ \text{in} \qquad (c)$$

where a is the half-width (radius) of the contact patch. The circular contact patch area is then

$$area = \pi a^2 = \pi\left(0.0058^2\right) = 1.057E - 4\ \text{in}^2 \qquad (d)$$

2   The average and maximum contact pressure can now be found from equations 7.8c and d.

$$p_{avg} = \frac{F}{area} = \frac{21.5}{1.057E - 4} = 203\ 587\ \text{psi} \qquad (e)$$

$$p_{max} = \frac{3}{2}p_{avg} = \frac{3}{2}(203\ 587) = 305\ 381\ \text{psi} \qquad (f)$$

3   The maximum normal stresses in the center of the contact patch at the surface are then found using equations 7.11c and d.

$$\sigma_{z_{max}} = -p_{max} = -305\ 381\ \text{psi} \qquad (g)$$

$$\sigma_{x_{max}} = \sigma_{y_{max}} = -\frac{1 + 2v}{2}p_{max} = -\frac{1 + 2(0.28)}{2}(305\ 381) = -238\ 197\ \text{psi} \qquad (h)$$

4   The maximum shear stress and its location under the surface are found from equations 7.12b and c.

$$\tau_{yz_{max}} = \frac{p_{max}}{2}\left[\frac{(1 - 2v)}{2} + \frac{2}{9}(1 + v)\sqrt{2(1 + v)}\right]$$

$$= \frac{305\ 381}{2}\left[\frac{(1 - 2(0.28))}{2} + \frac{2}{9}(1 + 0.28)\sqrt{2(1 + 0.28)}\right] = 103\ 083\ \text{psi} \qquad (i)$$

$$z_{@\tau_{max}} = a\sqrt{\frac{2+2v}{7-2v}} = 0.005\,8\sqrt{\frac{2+2(0.28)}{7-2(0.28)}} = 0.003\,7 \text{ in} \qquad (j)$$

5   All the stresses found so far exist on the centerline of the patch. At the edge of the patch, at the surface, there will be a shear stress of

$$\tau_{xy} = \frac{1-2v}{3}p_{max} = \frac{1-2(0.28)}{3}(305\,381) = 44\,789 \text{ psi} \qquad (k)$$

and a tensile stress of the same magnitude.

5   Since both parts are the same material, all of these stresses apply to both.

7.9   CYLINDRICAL CONTACT

Cylindrical contact is common in machinery. Contacting rollers are often used to pull web material such as paper through machinery or to change the thickness of a material in the rolling or calendering process. Roller bearings are another application. The cylinders can both be convex, one convex and one concave (cylinder-in-trough), or in the limit, a cylinder-on-plane. In all such contacts there is the possibility of sliding as well as rolling at the interface. The presence of tangential sliding forces has a significant effect on the stresses compared to pure rolling. We will first consider the case of two cylinders in pure rolling and later introduce a sliding component.

Contact Pressure and Contact Patch in Parallel Cylindrical Contact

When two cylinders roll together, their contact patch will be rectangular as shown in Figure 7-11b. The pressure distribution will be a semi-elliptical prism of half-width a. The contact zone will look as shown in Figure 7-13. The contact pressure is a maximum p_{max} at the center and zero at the edges as shown in Figure 7-14. The applied load F on the contact patch is equal to the volume of the half-prism:

$$F = \frac{1}{2}\pi\,aL\,p_{max} \qquad (7.14a)$$

where F is the total applied load and L is the length of contact along the cylinder axis. This can be solved for the maximum pressure:

$$p_{max} = \frac{2F}{\pi\,aL} \qquad (7.14b)$$

The average pressure is the applied force divided by the contact patch area:

$$p_{avg} = \frac{F}{area} = \frac{F}{2aL} \qquad (7.14c)$$

Substituting equation 7.8*c* in 7.8*b* gives

$$p_{max} = \frac{4}{\pi} P_{avg} \cong 1.273 P_{avg} \qquad (7.14d)$$

We now define a cylindrical geometry constant that depends on the radii R_1 and R_2 of the two cylinders, (Note that it is the same as equation 7.9*b* for spheres.)

$$B = \frac{1}{2}\left(\frac{1}{R_1} + \frac{1}{R_2} \right) \qquad (7.15a)$$

To account for the case of a cylinder-on-plane, R_2 becomes infinite making $1/R_2$ zero. For a cylinder-in-trough, R_2 becomes negative. Otherwise R_2 is finite and positive, as is R_1. The contact-patch half-width a is then found from

$$a = \sqrt{\frac{2}{\pi} \frac{m_1 + m_2}{B} \frac{F}{L}} \qquad (7.15b)$$

where m_1 and m_2 are material constants as defined in equation 7.9*a*.

The pressure distribution within the semi-elliptical prism is

$$p = p_{max} \sqrt{1 - \frac{x^2}{a^2}} \qquad (7.16)$$

which is an ellipse as shown in Figure 7-11.

Static Stress Distributions in Parallel Cylindrical Contact

Hertzian stress analysis is for static loading but is also applied to pure-rolling contact. The stress distributions within the material are similar to those shown in Figure 7-15 for the sphere-on-sphere case. Two cases are possible: *plane stress,* where the cylinders are very short axially as in some cam roller-followers, and *plane strain,* where the cylinders are long axially such as in squeeze-rollers. In the plane-stress case, one of the principal stresses is zero. In plane-strain, all three principal stresses may be nonzero.

Figure 7-17 shows the principal stress, maximum shear and von Mises stress distributions across the patch width at the surface and along the *z* axis (where they are largest) for two cylinders in static or pure-rolling contact. The normal stresses are all compressive and are maximal at the surface. They diminish rapidly with depth into the material and also diminish away from the centerline, as shown in the figure.

At the surface on the centerline, the maximum applied normal stresses are

$$\sigma_x = \sigma_z = -p_{max}$$

$$\sigma_y = -2\nu p_{max} \qquad (7.17a)$$

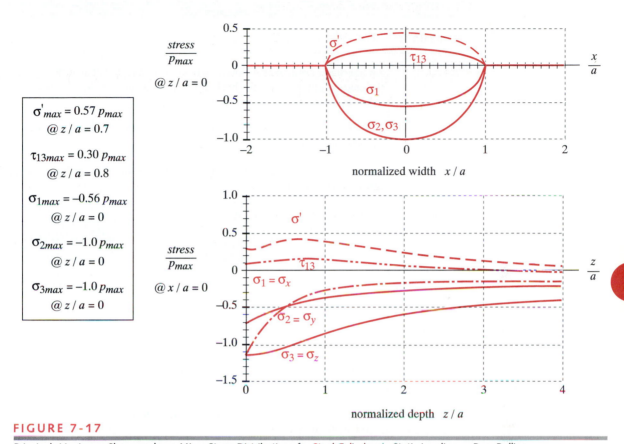

$\sigma'_{max} = 0.57\,p_{max}$
$@\,z\,/\,a = 0.7$

$\tau_{13max} = 0.30\,p_{max}$
$@\,z\,/\,a = 0.8$

$\sigma_{1max} = -0.56\,p_{max}$
$@\,z\,/\,a = 0$

$\sigma_{2max} = -1.0\,p_{max}$
$@\,z\,/\,a = 0$

$\sigma_{3max} = -1.0\,p_{max}$
$@\,z\,/\,a = 0$

FIGURE 7-17

Principal, Maximum Shear, and von Mises Stress Distributions for Steel Cylinders in Static Loading or Pure Rolling

These stresses are principal since there is no applied shear stress. The maximum shear stress τ_{13} on the z axis that results from the combination of stresses on the Mohr's-circle plane is beneath the surface as it was in the spherical-contact case. For two steel cylinders in static contact, the peak value and location of the maximum shear stress are[19]

$$\tau_{13_{max}} = 0.304\,p_{max}$$

$$z_{@\tau_{max}} = 0.786a \tag{7.17b}$$

However, note in Figure 7-17 that, on the z-axis, the shear stress is not zero, but is $0.22\,p_{max}$ at the surface and does not vary greatly over the depth $0 < z < 2a$.

EXAMPLE 7-2

Stresses in Cylindrical Contact

Problem   An overhead crane wheel runs slowly on a steel rail. What is the size of the contact patch between wheel and rail and what are the stresses? What is the depth of the maximum shear stress?

Given   The wheel is 12-in dia and the rail is flat. Both parts are steel. The radial load is 5 000 lb.

Assumptions   The rotational speed is sufficiently slow that this can be considered a static loading problem.

Solution   See the *TKSolver* file EX07-02.

1   First determine the size of the contact patch, for which the geometry constant and material constants are found from equations 7.15a and 7.9a.

$$B = \frac{1}{2}\left(\frac{1}{R_1} + \frac{1}{R_2}\right) = \frac{1}{2}\left(\frac{1}{6} + 0\right) = 0.083 \qquad (a)$$

Note the infinite radius of curvature for R_2.

$$m_1 = m_2 = \frac{1-v_1^2}{E_1} = \frac{1-0.28^2}{3E7} = 3.072E-8 \qquad (b)$$

Note both materials are the same in this example. The material and geometry constants can now be used in equation 7.15b.

$$a = \sqrt{\frac{2}{\pi}\frac{m_1 + m_2}{B}\frac{F}{L}} = \sqrt{\left(\frac{2}{\pi}\right)\frac{2(3.072E-8)}{0.083}\left(\frac{5\,000}{0.875}\right)} = 0.051\,8 \text{ in} \qquad (c)$$

where a is the half-width of the contact patch. The rectangular contact patch area is

$$area = 2aL = 2(0.051\,8)(0.875) = 0.091 \text{ in}^2 \qquad (d)$$

2   The average and maximum contact pressure are found from equations 7.14b and c.

$$p_{avg} = \frac{F}{area} = \frac{5\,000}{0.168} = 29\,795 \text{ psi} \qquad (e)$$

$$p_{max} = \frac{2F}{\pi a L} = \frac{2(5\,000)}{\pi(0.051\,8)(1.5)} = 70\,243 \text{ psi} \qquad (f)$$

3   The maximum normal stresses in the center of the contact patch at the surface are then found using equations 7.17a.

$$\sigma_{z_{max}} = \sigma_{x_{max}} = -p_{max} = -70\,243 \text{ psi} \qquad (g)$$

$$\sigma_{y_{max}} = -2\nu p_{max} = -2(0.28)(70\,243) = -39\,336 \text{ psi} \qquad (h)$$

4   The maximum shear stress and its location (depth) are found from equations 7.17b.

$$\tau_{13_{max}} = 0.304 p_{max} = 0.304(70\,243) = 21\,354 \text{ psi}$$

$$z_{@\tau_{max}} = 0.786a = 0.786(0.083) = 0.004 \text{ in}$$

$$\qquad (i)$$

5   All the stresses found exist on the z axis and the normal stresses are principal. These stresses apply to the wheel and rail as both are steel.

7.10   GENERAL CONTACT

When the geometry of the two contacting bodies is allowed to have any general curvature, the contact patch is an ellipse and the pressure distribution is a semi-ellipsoid, as shown in Figure 7-11a. Even the most general curvature can be represented as a radius of curvature over a small angle with minimal error. The size of the contact patch for most practical materials in these applications is so small that this approximation is reasonable. Thus the compound curvature of each body is represented by two mutually orthogonal radii of curvature at the contact point.

Contact Pressure and Contact Patch in General Contact

The contact pressure is a maximum p_{max} at the center and zero at the edge. The total applied load F on the contact patch is equal to the volume of the semi-ellipsoid:

$$F = \frac{2}{3}\pi\, ab\, p_{max} \qquad (7.18a)$$

where a is the half-width of the major axis and b the half-width of the minor axis of the contact patch ellipse. This can be solved for the maximum pressure:

$$p_{max} = \frac{3}{2}\frac{F}{\pi\, ab} \qquad (7.18b)$$

The average pressure on the contact patch is the applied force divided by its area:

$$p_{avg} = \frac{F}{area} = \frac{F}{\pi\, ab} \qquad (7.18c)$$

and substituting equation 7.8c in 7.8b gives

$$p_{max} = \frac{3}{2} p_{avg} \qquad (7.18d)$$

We must define two geometry constants that depend on the radii of curvature of the two bodies,

$$A = \frac{1}{2}\left(\frac{1}{R_1} + \frac{1}{R_1'} + \frac{1}{R_2} + \frac{1}{R_2'} \right) \qquad (7.19a)$$

$$B = \frac{1}{2}\left[\left(\frac{1}{R_1} - \frac{1}{R_1'} \right)^2 + \left(\frac{1}{R_2} - \frac{1}{R_2'} \right)^2 + 2\left(\frac{1}{R_1} - \frac{1}{R_1'} \right)\left(\frac{1}{R_2} - \frac{1}{R_2'} \right)\cos 2\theta \right]^{\frac{1}{2}} \qquad (7.19b)$$

$$\phi = \cos^{-1}\left(\frac{B}{A} \right) \qquad (7.19c)$$

where R_1 and R_1' are the two radii of curvature* of body 1, R_2 and R_2' are the radii* of body 2, and θ is the angle between the planes containing R_1 and R_2.

The contact patch dimensions a and b are then found from

$$a = k_a \sqrt[3]{\frac{3F(m_1 + m_2)}{4A}} \qquad b = k_b \sqrt[3]{\frac{3F(m_1 + m_2)}{4A}} \qquad (7.19d)$$

where m_1 and m_2 are material constants as defined in equation 7.9a and the values of k_a and k_b are found from Table 7-5 corresponding to the value of ϕ from equation 7.19c.

The pressure distribution within the semi-ellipsoid is

$$p = p_{max}\sqrt{1 - \frac{x^2}{a^2} - \frac{y^2}{b^2}} \qquad (7.20)$$

which is an ellipse as shown in Figure 7-11.

Stress Distributions in General Contact

The stress distributions within the material are similar to those shown in Figure 7-17 for the cylinder-on-cylinder case. The normal stresses are all compressive and are maximal at the surface. They diminish rapidly with depth into the material and away from the centerline. At the surface on the centerline, the maximum normal stresses are[19]

* Measured in mutually perpendicular planes.

Table 7-5		Factors For Use in Equation 7.19d																
ϕ	0	10	20	30	35	40	45	50	55	60	65	70	75	80	85	90		
k_a	∞	6.612	3.778	2.731	2.397	2.136	1.926	1.754	1.611	1.486	1.378	1.284	1.202	1.128	1.061	1		
k_b	0	0.319	0.408	0.493	0.530	0.567	0.604	0.641	0.678	0.717	0.759	0.802	0.846	0.893	0.944	1		

Sources: H. Hertz, *Contact of Elastic Solids*, in *Miscellaneous Papers*, P. Lenard, ed. Macmillan & Co. Ltd.: London. pp. 146-162, 1896.
H. L. Whittemore and S. N. Petrenko, *Natl. Bur. Std. Tech. Paper* 201, 1921.

$$\sigma_x = -\left[2v + (1-2v)\frac{b}{a+b}\right]p_{max}$$

$$\sigma_y = -\left[2v + (1-2v)\frac{a}{a+b}\right]p_{max} \qquad (7.21a)$$

$$\sigma_z = -p_{max}$$

$$k_3 = \frac{b}{a} \qquad\qquad k_4 = \frac{1}{a}\sqrt{a^2 - b^2} \qquad (7.21b)$$

These applied stresses are also the principal stresses. The maximum shear stress at the surface associated with these stresses can be found from equation 4.5. The largest shear stress occurs slightly below the surface with that distance dependent on the ratio of the semiaxes of the contact ellipse. For $b/a = 1.0$, the largest shear stress occurs at $z = 0.63a$, and for $b/a = 0.34$ at $z = 0.24a$. Its peak magnitude is approximately 0.34 p_{max}.[19]

At the ends of the major axis of the contact ellipse the shear stress at the surface is[19]

$$\tau_{xz} = (1-2v)\frac{k_3}{k_4^2}\left(\frac{1}{k_4}\tanh^{-1}k_4 - 1\right)p_{max} \qquad (7.21c)$$

At the ends of the minor axis of the contact ellipse the shear stress at the surface is

$$\tau_{xz} = (1-2v)\frac{k_3}{k_4^2}\left[1 - \frac{k_3}{k_4}\tan^{-1}\left(\frac{k_4}{k_3}\right)\right]p_{max} \qquad (7.21d)$$

The location of the largest surface shear stress will vary with the ellipse ratio k_3. For some cases it is as shown in equation 7.21c, but in others it moves to the center of the ellipse and is found from the principal stresses in equation 7.21a using equation 4.5.

EXAMPLE 7-3

Stresses in a Crowned Cam Follower

Problem   A crowned cam roller-follower has a gentle radius transverse to its rolling direction to eliminate the need for critical alignment of its axis with that of the cam. The cam's radius of curvature and dynamic load vary around its circumference. What is the size of the contact patch between cam and follower and what are the worst-case stresses?

Given   The roller radius is 1 in with a 20-in crown radius at 90° to the roller radius. The cam's radius of curvature at the point of maximum load is 3.46 in and it is flat axially. The rotational axes of the cam and roller are parallel, which makes the angle between the two bodies zero. The force is 250 lb, normal to the contact plane.

Assumptions   **Materials are steel. The relative motion is rolling with <1% sliding.**

Solution   See the *TKSolver* file EX07-03.

1 Find the material constants from equation 7.9b.

$$m_1 = m_2 = \frac{1-v_1^2}{E_1} = \frac{1-0.28^2}{3E7} = 3.072E-8 \tag{a}$$

2 Two geometry constants are needed from equations 7.19a.

$$A = \frac{1}{2}\left(\frac{1}{R_1} + \frac{1}{R_1'} + \frac{1}{R_2} + \frac{1}{R_2'}\right) = \frac{1}{2}\left(\frac{1}{1} + \frac{1}{20} + \frac{1}{3.46} + \frac{1}{\infty}\right) = 0.669\,5 \tag{b}$$

$$B = \frac{1}{2}\left[\left(\frac{1}{R_1} - \frac{1}{R_1'}\right)^2 + \left(\frac{1}{R_2} - \frac{1}{R_2'}\right)^2 + 2\left(\frac{1}{R_1} - \frac{1}{R_1'}\right)\left(\frac{1}{R_2} - \frac{1}{R_2'}\right)\cos 2\theta\right]^{\frac{1}{2}} \tag{c}$$

$$B = \frac{1}{2}\left[\left(\frac{1}{1} - \frac{1}{20}\right)^2 + \left(\frac{1}{3.46} - \frac{1}{\infty}\right)^2 + 2\left(\frac{1}{1} - \frac{1}{20}\right)\left(\frac{1}{3.46} - \frac{1}{\infty}\right)\cos 2(0)\right]^{\frac{1}{2}} = 0.619\,5$$

The angle ϕ is found from their ratio,

$$\phi = \cos^{-1}\left(\frac{B}{A}\right) = \cos^{-1}\left(\frac{0.6695}{0.6195}\right) = 22.3° \tag{d}$$

and used in Table 7-6 to find the factors k_a and k_b. Cubic interpolation[*] for k_a and linear interpolation[*] for k_b gives

$$k_a = 3.444 \qquad\qquad k_b = 0.427 \tag{e}$$

3 The material and geometry constants can now be used in equation 7.19d.

$$a = k_a \sqrt[3]{\frac{3F(m_1 + m_2)}{4A}} = 3.444 \sqrt[3]{\frac{3(250)2(3.072E-8)}{4(0.6695)}} = 0.088\,9 \tag{f}$$

$$b = k_b \sqrt[3]{\frac{3F(m_1 + m_2)}{4A}} = 0.427 \sqrt[3]{\frac{3(250)2(3.072E-8)}{4(0.6695)}} = 0.011\,0$$

where a is the half-width of the major axis, and b is the half-width of the minor axis of the contact patch. The contact patch area is then

$$area = \pi ab = \pi(0.0889)(0.011) = 0.0031 \text{ in}^2 \tag{g}$$

4 The average and maximum contact pressure can be found from equations 7.18b and c.

[*] The different interpolation methods are used to best fit the functions, one of which is linear and the other nonlinear. Plot the values in Table 7-6 to see this.

$$P_{avg} = \frac{F}{area} = \frac{250}{0.003\,1} = 81\,119 \text{ psi} \qquad (h)$$

$$P_{max} = \frac{3}{2}P_{avg} = \frac{3}{2}(81\,119) = 121\,679 \text{ psi} \qquad (i)$$

5 The maximum normal stresses in the center of the contact patch at the surface are then found using equations 7.21a.

$$\sigma_x = -\left[2v + (1-2v)\frac{b}{a+b}\right]P_{max}$$

$$= -\left[2(.28) + \left(1 - 2(.28)\right)\frac{0.011}{0.0889 + 0.011}\right]121\,679 = -74\,051 \text{ psi}$$

$$\sigma_y = -\left[2v + (1-2v)\frac{a}{a+b}\right]P_{max} \qquad (j)$$

$$= -\left[2(.28) + \left(1 - 2(.28)\right)\frac{0.0889}{0.0889 + 0.011}\right]121\,679 = -115\,678 \text{ psi}$$

$$\sigma_z = -P_{max} = -121\,679 \text{ psi}$$

These stresses are principal: $\sigma_1 = \sigma_x$, $\sigma_2 = \sigma_y$, $\sigma_3 = \sigma_z$. The maximum shear stress associated with them at the surface will be (from Eq. 4.5)

$$\tau_{13} = \left|\frac{\sigma_1 - \sigma_3}{2}\right| = \left|\frac{-74\,051 + 121\,679}{2}\right| = 23\,814 \text{ psi (at surface)} \qquad (k)$$

6 The largest shear stress under the surface on the z axis is approximately

$$\tau_{13} \cong 0.32\,p_{max} = 0.32(121\,679) \cong 39\,000 \text{ psi (below surface)} \qquad (l)$$

7 All the stresses found so far exist on the centerline of the patch. At the edge of the patch, at the surface, there will also be a shear stress. Two constants are found from equation 7.21b for this calculation.

$$k_3 = \frac{b}{a} = \frac{0.0110}{0.0889} = 0.124$$

$$\qquad (m)$$

$$k_4 = \frac{1}{a}\sqrt{a^2 - b^2} = \frac{1}{0.0889}\sqrt{0.0889^2 - 0.0110^2} = 0.992$$

These constants are used in equations 7.21c and d to find the shear stresses on the surface at the ends of the major and minor axes.

$$\tau_{xz} = (1-2v)\frac{k_3}{k_4^2}\left(\frac{1}{k_4}\tanh^{-1}k_4 - 1\right)P_{max}$$

$$\qquad (n)$$

$$\tau_{xz} = (1-0.56)\frac{0.124}{(0.992)^2}\left(\frac{1}{0.992}\tanh^{-1}0.992 - 1\right)121\,679 = 12\,131 \text{ psi}$$

$$\tau_{xz} = (1 - 2\nu) \frac{k_3}{k_4^2} \left[1 - \frac{k_3}{k_4} \tan^{-1}\left(\frac{k_4}{k_3} \right) \right] p_{max}$$

(o)

$$\tau_{xz} = (1 - 0.56) \frac{0.124}{(0.992)^2} \left[1 - \frac{0.124}{0.992} \tan^{-1}\left(\frac{0.992}{0.124} \right) \right] 121\,679 = 5\,528 \text{ psi}$$

7.11   DYNAMIC CONTACT STRESSES

The equations presented above for contact stresses assume that the load is pure rolling. When rolling and sliding are both present, the stress field is distorted by the tangential loading. Figure 7-18 shows a photoelastic study of a cam-follower pair[20] loaded statically (a) and dynamically with sliding (b). The distortion of the stress field from the sliding motion can be seen in part b. This is a combination of rolling contact with relatively low velocity sliding. Increased sliding causes more distortion of the stress field.

Effect of a Sliding Component on Contact Stresses

Smith and Lui[18] analyzed the case of parallel rollers in combined rolling and sliding and developed the equations for the stress distribution beneath the contact point. The sliding (frictional) load has a significant effect on the stress field. The stresses can be expressed as separate components, one set due to the normal load on the rolls (denoted by a subscript n) and the other set due to the tangential friction force (denoted by a subscript t). These are then combined to obtain the complete stress situation. The stress field can be two-dimensional in a very short roll, such as a thin plate cam or thin gear, assumed to be in plane stress. If the rolls are long axially, then a plane strain condition will exist in regions away from the ends giving a three-dimensional stress state.

The contact geometry is as shown in Figure 7-11b with the x axis aligned to the direction of motion, the z axis radial to the rollers and the y axis axial to the rollers. The stresses due to the normal loading p_{max} are

$$\sigma_{x_n} = -\frac{z}{\pi} \left[\frac{a^2 + 2x^2 + 2z^2}{a} \alpha - \frac{2\pi}{a} - 3x\beta \right] p_{max}$$

$$\sigma_{z_n} = -\frac{z}{\pi} [a\beta - x\alpha] p_{max}$$

(7.22a)

$$\tau_{xz_n} = -\frac{1}{\pi} z^2 \alpha p_{max}$$

and those due to the frictional force f_{max} are

follower

cam

(a) Static loading

(b) Dynamic loading

FIGURE 7-18

Photoelastic Study of Stresses for Two Cylinders in Contact in Static (a) and Dynamic Pure Rolling Loading (b) Source: V. S. Mahkijani, *Study of Contact Stresses as Developed on a Radial Cam using Photoelastic Model and Finite Element Analysis*, M.S. Thesis, Worcester Polytechnic Institute, 1984

$$\sigma_{x_t} = -\frac{1}{\pi}\left[\left(2x^2 - 2a^2 - 3z^2\right)\alpha + 2\pi\frac{x}{a} + 2\left(a^2 - x^2 - z^2\right)\frac{x}{a}\beta\right]f_{max}$$

$$\sigma_{z_t} = -\frac{1}{\pi}z^2\alpha\,f_{max} \qquad\qquad (7.22b)$$

$$\tau_{xz_t} = -\frac{1}{\pi}\left[\left(a^2 + 2x^2 + 2z^2\right)\frac{z}{a}\beta - 2\pi\frac{z}{a} - 3xz\alpha\right]f_{max}$$

where the factors α and β are given by

$$\alpha = \frac{\pi}{k_1}\frac{1 - \sqrt{\dfrac{k_1}{k_2}}}{\sqrt{\dfrac{k_1}{k_2}}\sqrt{2\sqrt{\dfrac{k_1}{k_2}} + \left(\dfrac{k_1 + k_2 - 4a^2}{k_1}\right)}} \qquad\qquad (7.22c)$$

$$\beta = \frac{\pi}{k_1}\frac{1 + \sqrt{\dfrac{k_1}{k_2}}}{\sqrt{\dfrac{k_1}{k_2}}\sqrt{2\sqrt{\dfrac{k_1}{k_2}} + \left(\dfrac{k_1 + k_2 - 4a^2}{k_1}\right)}} \qquad\qquad (7.22d)$$

$$k_1 = (a+x)^2 + z^2 \qquad\qquad k_2 = (a-x)^2 + z^2 \qquad (7.22e)$$

The tangential unit force f_{max} is found from the normal load and a coefficient of friction μ.

$$f_{max} = \mu p_{max} \qquad (7.22f)$$

The independent variables in these equations are then the coordinates x, z in the cross section of the roller, referenced to the contact point, the half-width a of the contact patch, and the maximum normal load p_{max} at the contact point.

Equations 7.22 define the behavior of the stress functions below the surface but when $z = 0$, the factors α and β become infinite and these equations fail. Other forms are needed to account for the stresses on the surface of the contact patch.

when $z = 0$

$$\text{if } |x| \le a \text{ then } \sigma_{x_n} = -p_{max}\sqrt{1 - \frac{x^2}{a^2}} \text{ else } \sigma_{x_n} = 0$$

$$\sigma_{z_n} = \sigma_{x_n} \qquad (7.23a)$$

$$\tau_{xz_n} = 0$$

$$\text{if } x > a \text{ then } \sigma_{x_t} = -2f_{max}\left(\frac{x}{a} - \sqrt{\frac{x^2}{a^2} - 1}\right)$$

$$\text{if } x < a \text{ then } \sigma_{x_t} = -2f_{max}\left(\frac{x}{a} + \sqrt{\frac{x^2}{a^2} - 1}\right) \qquad (7.23b)$$

$$\text{if } |x| \le a \text{ then } \sigma_{x_t} = -2f_{max}\frac{x}{a}$$

$$\sigma_{z_t} = 0 \qquad (7.23c)$$

$$\text{if } |x| \le a \text{ then } \tau_{xz_t} = -f_{max}\sqrt{1 - \frac{x^2}{a^2}} \text{ else } \tau_{xz_t} = 0$$

The total stress on each Cartesian plane is found by superposing the components due to the normal and tangential loads:

$$\sigma_x = \sigma_{x_n} + \sigma_{x_t}$$

$$\sigma_z = \sigma_{z_n} + \sigma_{z_t} \qquad (7.24a)$$

$$\tau_{xz} = \tau_{xz_n} + \tau_{xz_t}$$

For short rollers in plane stress, σ_y is zero, but if the rollers are long axially then a plane strain condition will exist away from the ends and the stress in the y direction will be:

$$\sigma_y = \nu(\sigma_x + \sigma_z) \qquad\qquad (7.24b)$$

where ν is Poisson's ratio.

These stresses are maximum at the surface and decrease with depth. Except at very low ratios of tangential force to normal force, (< about 1/9)[18],[21] the maximum shear stress occurs at the surface as well, unlike the pure rolling case. A computer program was written to evaluate equations 7.22 and 7.23 for the conditions at the surface and plot them. (See the file CONTACT.EXE) The stresses are all normalized to the maximum normal load p_{max} and the locations normalized to the patch half-width a. A coefficient of friction of 0.33 and steel rollers with $\nu = 0.28$ were assumed for the examples. The magnitudes and shapes of the stress distributions will be a function of these factors.

Figure 7-19a shows the x direction stresses at the surface, which are due to the normal and tangential loads, and also shows their sum from the first of equations 7.24a. Note that the stress component σ_{x_t} due to the tangential force is tensile from the contact point to and beyond the trailing edge of the contact patch. This should not be surprising as one can picture that the tangential force is attempting to pile up material in front of the contact point and stretch it behind that point, just as a carpet bunches up in front of anything you try to slide across it. The stress component σ_{x_n} due to the normal force is compressive everywhere. However, the sum of the two σ_x components has a significant normalized tensile value of twice the coefficient of friction (here 0.66 p_{max}) and a compressive peak of about $-1.2\ p_{max}$. Figure 7-19b shows all of the applied stresses in x, y, and z directions across the surface of the contact zone. Note that the stress fields on the surface extend beyond the contact zone when a tangential force is present, unlike the situation in pure rolling where they extend beyond the contact zone only under the surface. (See Figure 7-17 and program CONTACT.EXE).

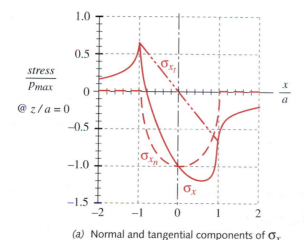

(a) Normal and tangential components of σ_x

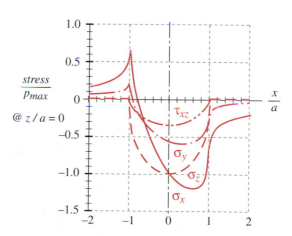

(b) All applied stresses at surface of contact patch

FIGURE 7-19

Applied Tangential and Normal and Shear Stresses at Surface for Cylinders in Combined Rolling and Sliding with $\mu = 0.33$

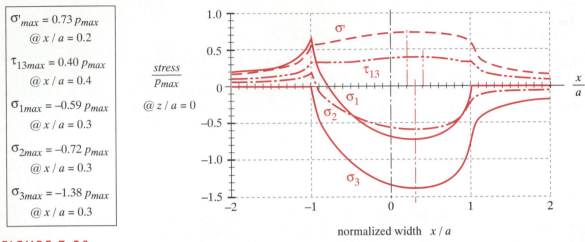

$\sigma'_{max} = 0.73\,p_{max}$
   @ $x / a = 0.2$

$\tau_{13max} = 0.40\,p_{max}$
   @ $x / a = 0.4$

$\sigma_{1max} = -0.59\,p_{max}$
   @ $x / a = 0.3$

$\sigma_{2max} = -0.72\,p_{max}$
   @ $x / a = 0.3$

$\sigma_{3max} = -1.38\,p_{max}$
   @ $x / a = 0.3$

$\dfrac{stress}{p_{max}}$

@ $z / a = 0$

normalized width x / a

FIGURE 7-20

7

Principal and von Mises Stresses Across Contact Zone at Surface for Cylinders in Combined Rolling and Sliding with $\mu = 0.33$

Figure 7-20 shows the principal stresses, maximum shear stress, and von Mises stress for the plane-strain, applied-stress state in Figure 7-19. Note that the magnitude of the largest compressive principal stress is about 1.38 p_{max} and the largest tensile principal stress is 0.66 p_{max} at the trailing edge of the contact patch. The presence of an applied tangential shear stress in this example increases the peak compressive stress by 40% over a pure rolling case and introduces a tensile stress in the material. The principal shear stress reaches a peak value of 0.40 p_{max} at $x / a = 0.4$. All the stresses shown in Figures 7-19 and 7-20 are at the surface of the rollers.

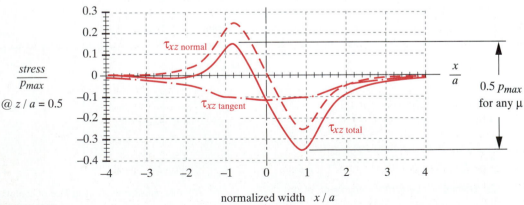

$\dfrac{stress}{p_{max}}$

@ $z / a = 0.5$

τ_{xz} normal

τ_{xz} tangent

τ_{xz} total

$0.5\,p_{max}$
for any μ

normalized width x / a

FIGURE 7-21

Shear Stresses Below Surface at $z / a = 0.5$ for Cylinders in Combined Rolling and Sliding—Plotted with $\mu = 0.33$

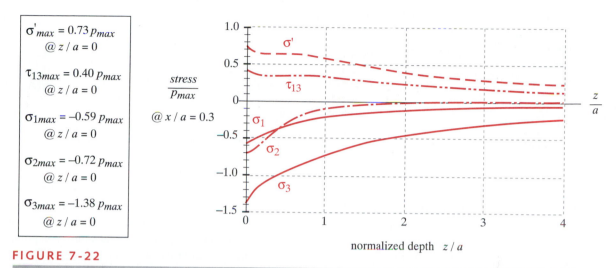

FIGURE 7-22

Principal and von Mises Stresses Below Surface at $x / a = 0.3$ for Cylinders in Combined Rolling and Sliding with $\mu = 0.33$

Beneath the surface, the magnitudes of the compressive stresses due to the normal load reduce. However, the shear stress τ_{xz_n} due to the normal loading increases with depth becoming a maximum beneath the surface at $z = 0.5a$, as shown in Figure 7-21. Note the sign reversal at the midpoint of the contact zone. There are fully reversed shear stress components acting on each differential element of material as it passes through the contact zone. The peak-to-peak range of this fully reversed shear stress in the xz plane is greater in magnitude than the range of the maximum shear stress and is considered by some to be responsible for subsurface pitting failures.[17]

Figure 7-22 plots the principal stresses, maximum shear stress, and von Mises stress (calculated for $\mu = 0.33$ and a plane-strain condition) versus the normalized depth z / a taken at the $x / a = 0.3$ plane (where they are maximum as shown in Figure 7-20). All the stresses are maximum at the surface. The principal stresses diminish rapidly with depth but the shear and von Mises stresses remain nearly constant over the first $1a$ of depth.

At the surface, the maximum shear stress is relatively uniform across the patch width with a peak of 0.4 at $x / a = 0.4$ when $\mu = 0.33$, as shown in Figure 7-20. This τ_{max} peak location moves versus the patch centerline with increasing depth but its magnitude varies only slightly with depth. Figure 7-23 plots the largest peak value of the shear stress τ_{13} occurring at any value of x across the patch zone, and so is a composite plot of the peak shear stress value in each z plane. For $0 < \mu < 0.5$ the peak value remains within 60-80% of its largest value over the first a of depth and is still 58-70% of its peak value at $z / a = 2.0$. As the coefficient of friction is increased to 0.5 or greater, the normalized maximum shear stress value becomes equal to μ and is constant across the contact-patch surface.

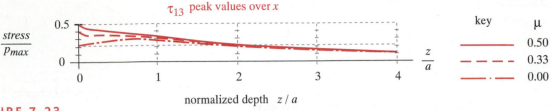

FIGURE 7-23

Peak Values of Maximum Shear Stress at all Values of x / a for Cylinders in Combined Rolling and Sliding with $0 \le \mu \le 0.5$

The limited variation of τ_{max} over small z depths may explain why some pitting failures appear to start at the surface and some below it. With a relatively uniform-magnitude maximum shear stress over the entire near-surface region, any inclusion in that region of the material creates a stress concentration and serves as a crack initiation point. The fact that the peak value of the maximum shear stress occurs at slightly different transverse locations at different depths within the contact zone is irrelevant since an inclusion at any particular depth will pass through that location once per revolution and be exposed to the peak stress value.

The stress functions can be calculated and plotted at any depth using the computer program CONTACT.EXE or the *TKSolver* file SUBSURF provided on disk. These programs solve and plot the general equations 7.22, 7.23, and 7.24 for any user-specified roller geometry, materials, load, and coefficient of friction, at any specified range of locations below and along the surface.

EXAMPLE 7-4

Stresses in Combined Rolling and Sliding of Cylinders

Problem   A pair of calendering rolls are run together with a combination of rolling and sliding. Find the maximum tensile, compressive, and shear stresses in the rollers.

Given   The roller radii are 1.25 and 2.5 in and are each 24 in long. The force is 5000 lb, normal to the contact plane.

Assumptions   Both materials are steel. The coefficient of friction is 0.33.

Solution   See the *TKSolver* file EX07-04.

1   The contact-patch geometry is found in the same way as was done in Example 7-2. Find the material constants from equation 7.9a.

$$m_1 = m_2 = \frac{1 - v_1^2}{E_1} = \frac{1 - 0.28^2}{3E7} = 3.072E - 8 \qquad (a)$$

The geometry constant is found from equation 7.15a

$$B = \frac{1}{2}\left(\frac{1}{R_1} + \frac{1}{R_2}\right) = \frac{1}{2}\left(\frac{1}{1.25} + \frac{1}{2.5}\right) = 0.600 \qquad (b)$$

and the patch half-width from equation 7.15b.

$$a = \sqrt{\frac{2}{\pi}\frac{m_1 + m_2}{B}\frac{F}{L}} = \sqrt{\left(\frac{2}{\pi}\right)\frac{2(3.072E-8)}{0.600}\left(\frac{5000}{24}\right)} = 0.0037 \text{ in} \qquad (c)$$

where a is the half-width of the contact patch. The rectangular contact-patch area is then

$$area = 2aL = 2(0.002\ 3)(24) = 0.176\ 9 \text{ in}^2 \qquad (d)$$

2   The average and maximum contact pressure can now be found from equations 7.14b and c.

$$p_{avg} = \frac{F}{area} = \frac{5000}{0.1769} = 28\ 266 \text{ psi} \qquad (e)$$

$$p_{max} = \frac{2F}{\pi aL} = \frac{2(5000)}{\pi(0.003\ 7)24} = 35\ 989 \text{ psi} \qquad (f)$$

The tangential pressure is found from equation 7.22f:

$$f_{max} = \mu p_{max} = 0.25(22\ 761) = 5\ 690 \text{ psi} \qquad (g)$$

3   With $\mu = 0.33$, the principal stresses in the contact zone will be maximal on the surface $(z = 0)$ at $x = 0.3a$ from the centerline as shown in Figures 7-20 and 7-22. The applied-stress components are found from equation 7.23a for the normal force and equation 7.23b for the tangential force.

$$\sigma_{x_n} = -p_{max}\sqrt{1 - \frac{x^2}{a^2}} = -35\ 989\sqrt{1 - 0.3^2} = -34\ 331 \text{ psi}$$
$$\qquad (h)$$

$$\sigma_{x_t} = -2f_{max}\frac{x}{a} = -2(11\ 876)(0.3) = -7\ 126 \text{ psi}$$

$$\sigma_{z_n} = -p_{max}\sqrt{1 - \frac{x^2}{a^2}} = -35\ 989\sqrt{1 - 0.3^2} = -34\ 331 \text{ psi}$$
$$\qquad (i)$$

$$\sigma_{z_t} = 0 \qquad\qquad \tau_{xz_n} = 0$$

$$\tau_{xz_t} = -f_{max}\sqrt{1 - \frac{x^2}{a^2}} = -11\ 876\sqrt{1 - 0.3^2} = -11\ 329 \text{ psi} \qquad (j)$$

4   Equations 7.24a and 7.24b can now be solved for the total applied stresses along the x, y, and z axes.

$$\sigma_x = \sigma_{x_n} + \sigma_{x_t} = -34\,331 - 7\,126 = -41\,457 \text{ psi} \qquad (k)$$

$$\sigma_z = \sigma_{z_n} + \sigma_{z_t} = -34\,331 + 0 = -34\,331 \text{ psi} \qquad (l)$$

$$\tau_{xz} = \tau_{xz_n} + \tau_{xz_t} = 0 - 11\,329 = -11\,329 \text{ psi} \qquad (m)$$

5   Since the rollers are long, we expect a plane-strain condition to exist. The stress in the third dimension is found from equation 7.23b:

$$\sigma_y = v(\sigma_x + \sigma_z) = 0.28(-41\,457 + 34\,331) = -21\,221 \text{ psi} \qquad (n)$$

6   Unlike the pure-rolling case, these stresses are not principal because of the applied shear stress. The principal stresses are found from equation 4.4 using a cubic root-finding solution (*TKSolver* file STRESS3D or program MOHR.EXE).

$$\sigma_1 = -21\,221 \text{ psi}$$
$$\sigma_2 = -26\,018 \text{ psi} \qquad (o)$$
$$\sigma_3 = -49\,771 \text{ psi}$$

The maximum shear stress is found from the principal stresses using equation 4.5.

$$\tau_{13} = \frac{|\sigma_1 - \sigma_3|}{2} = \frac{|-21\,221 + 49\,771|}{2} = 14\,275 \text{ psi} \qquad (p)$$

7   The principal stresses are maximum at the surface as seen in Figures 7-20 and 7-22.

7.12   SURFACE FATIGUE FAILURE MODELS—DYNAMIC CONTACT

There is still some disagreement among experts as to the actual mechanism of failure that results in pitting and spalling of surfaces. The possibility of having a maximum shear stress at a subsurface location (in pure rolling) has led some to conclude that pits begin at or near that location. Others have concluded that pitting begins at the surface. It is possible that both mechanisms are at work in these cases since failure initiation usually begins at an imperfection, which may be on or below the surface. Figure 7-24 shows both surface and subsurface cracks in a case-hardened steel roll subjected to heavy rolling loads.[22]

An extensive experimental study of pitting under rolling contact was done by Way[23] in 1935. Over 80 tests were made with contacting, pure-rolling, parallel rollers of different materials, lubricants, and loads, run for up to 18 million cycles, though most samples failed between 0.5E6 and 1.5E6 cycles. The samples were monitored for the appearance of minute surface cracks, which inevitably presaged a pitting failure within less than about 100 000 additional cycles in the presence of a lubricant.

surface crack   surface crack

surface

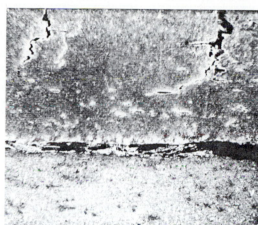

subsurface crack

FIGURE 7-24

Photomicrograph (100x) of Surface and Subsurface Cracks in a Carburized and Hardened Roll (HRC 52-58) Subjected to a Heavy Rolling Load   Source: J. D. Graham, *Pitting of Gear Teeth*, in C. Lipson, *Handbook of Mechanical Wear*, U. Mich. Press, 1961, p.137, with permission.

7

Harder and smoother surfaces better resisted pitting failure. Highly polished samples did not fail in over 12E6 cycles. Nitrided rolls with very hard cases on a soft core were longer-lived than other materials tested. **No pitting occurred on any samples in the absence of a lubricant** even though dry-running produced surface cracks. The cracked parts would continue to run dry with no failure for as many as 5E6 cycles until some lubricant was added. Then the surface cracks would rapidly enlarge and turn to pits of a characteristic arrowhead shape within 100 000 additional cycles.

The suggested explanation for the deleterious effect of the lubricant was that once suitably oriented surface cracks form, they are pumped full of oil on approaching the roll-nip, and then are pressed closed within the roll-nip, pressurizing the fluid trapped in the crack. The fluid pressure creates tensile stress at the crack tip causing rapid crack growth and subsequent break-out of a pit. Higher-viscosity lubricants did not eliminate metal-to-metal contact but did delay the pitting failure indicating that the fluid must be able to readily enter the crack to do the damage.

Way reached a number of conclusions regarding how to design rollers to delay surface fatigue failure.[23]

1   Use no oil (though he was quick to point out that this is not a practical solution as it promotes other types of wear as discussed in previous sections).

2   Increase the viscosity of the lubricant.

3   Polish the surfaces (though this is expensive to do).

4   Increase the surface hardness (preferably on a softer, tough core).

Table 7-6   Modes of Surface Failure and Their Causes

Mode of Failure	Factors That Promote Occurrance
Inclusion origin	Frequency and severity of oxide or other hard inclusions
Geometric stress concentration	End-of-contact geometry. Misalignment and deflections. Possible lubricant-film thickness effects.
Point-surface origin (PSO)	Low lubricant viscosity. Thin elastohydrodynamic film compared to asperities on contact surfaces. Tangential forces and/or gross sliding.
Peeling (superficial pitting)	Low lubricant viscosity. Frerquent asperities in surface finish exceeding elastohydrodynamic-film thickness. Loss of elastohydrodynamic pressure due to side leakage or scratches in contact surface.
Subcase Fatigue (on carburized components)	Low core hardness. Thin case depth relative to radius of curvature for elements in contact.

Source: W. E. Littmann and R. L. Widner, Propagation of Contact Fatigue from Surface and Subsurface Origins. *J. Basic Eng. Trans.* ASME, vol. 88: pp. 624-636, 1966.

7

No conclusions were drawn with respect to the reasons for the initiation of the initial cracks on the surface. Though, with pure rolling, the shear stresses are not maximal at the surface, they **are** nonzero there at some locations (see Figures 7-12 and 7-17).

Littman and Widner[24] performed an extensive analytical and experimental study on contact fatigue in 1966 and describe five different modes of failure in rolling contact. These are listed in Table 7-6 along with some factors that promote their occurrence. Some of these modes address the crack initiation issue and others the crack propagation issue. We will briefly discuss each in the order listed.

INCLUSION ORIGIN   describes a mechanism for crack initiation and is similar to the one discussed in Section 6.1 on fatigue failure. It is assumed that the crack originates in a shear-stress field at a subsurface or surface location containing a small inclusion of "foreign" matter. The most commonly identified inclusions are oxides of the material that formed during processing and were captured within it. These are typically hard and irregular in shape and create stress concentration.   Several researchers[25],[26],[27] have published photomicrographs of (or otherwise identified) subsurface cracks starting at oxide inclusions. "These oxide inclusions are often present as stringers or elongated aggregates of particles . . ., which provide a much greater chance for a point of high stress concentration to be in an unfavorable position with respect to the applied stress."[28] The propagation of a crack from the inclusion may remain subsurface, or break out to the surface. In the latter case it provides a site for hydraulic pressure propagation, as described above. In either case, it ultimately results in pitting or spalling.

GEOMETRIC STRESS CONCENTRATION (GSC)   was discussed in Chapter 4. This mechanism can act on a surface when, for example, one contacting part is shorter axially than the other (common with cam-follower joints and roller bearings). The ends of the shorter roller create line-contact stress concentration in the mating roller as shown in Figure 7-25a and pitting or spalling will likely occur at that location. This is one reason for using crowned rollers, which have a large crown-radius of curvature in the yz plane in addition to their roller radius in the xz plane. If the contact load is predictable, the crown radius can be sized to provide a more uniform stress distribution across the axial length of the contact area due to the deflections of the rollers as shown in Figure 7-25b. However, at lighter loads, there will be reduced contact area and thus higher stresses at the center and at higher than design loads the stress concentrations at the ends will return. A partial crown can be used as shown in Figure 7-25c but may cause some stress concentration at the transition from straight to crown. Reusner[29] has shown that a logarithmic curve on the crown as shown in Figure 7-25d will give a more uniform stress distribution under varied load levels.

POINT SURFACE ORIGIN (PSO) is the phenomenon described by Way and discussed above. Littman et al.[24] consider PSO to be more a manner of crack propagation than crack initiation and suggest that an inclusion at or near the surface may be responsible for starting the crack. Handling nicks or dents can also provide a crack nucleus on the surface. Once present, and if pointing in the right direction to capture oil, the crack rapidly propagates to failure. Once spalling starts, the debris can create new nicks to serve as additional crack sites.

PEELING refers to a situation in which the fatigue cracks are at shallow depth and extend over a large area such that the surface "peels" away from the substrate. Rough surfaces exacerbate peeling if the surface asperities are larger than the lubricant film thickness.

SUBCASE FATIGUE also called *case crushing*, occurs only on case-hardened parts and is more likely if the case is so thin that the subsurface stresses extend into the softer, weaker core material. The fatigue crack starts below the case and eventually causes the case to either collapse into the failed subsurface material or break out in pits or spalls.

Whatever the detailed cause of the start of a crack, once started the outcome is predictable. So, the designer needs to take all possible precautions to improve the part's resistance to pitting as well as to all other wear modes. The summary section to this chapter will attempt to set some guidelines to this end.

7.13 SURFACE FATIGUE STRENGTH

Repeated, time-varying loads tend to fail parts at lower stress levels than the material can stand in static-load applications. Bending and axial fatigue strength were discussed extensively in Chapter 6. The concept of **surface fatigue strength** is similar except for one main difference. While steels and a few other materials loaded in bending or axial fatigue show an endurance limit, *no materials* show an equivalent property when loaded in surface fatigue. Thus, we should expect that our machine, though carefully designed to be safe against all other forms of failure will eventually succumb to surface fatigue if so loaded for enough cycles.

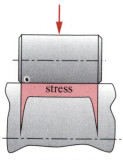

(a) Straight roller

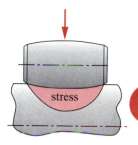

(b) Crowned roller

(c) Part-crown roller

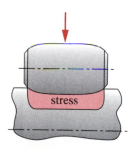

(d) Logarithmic roller

FIGURE 7-25

Stress Concentrations Beneath Variously Shaped Rollers

Morrison[30] and Cram[31] report separately on an experimental study of the surface-fatigue strength of materials done at USM Corp. from 1932 to 1956. Four wear-testing machines were operated **24 hours per day for 24 years** to gather surface fatigue strength data on cast iron, steel, bronze, aluminum, and nonmetallic materials. Their tests included rollers in pure rolling as well as rolling plus varying percentages of sliding. Most of their roll/slide data are done at 9% sliding since that simulates the average conditions on spur and helical gear teeth. The percent sliding figure is defined as the relative sliding velocity between the rollers or gear teeth divided by the pitch-line velocity of the interface.

Previous sections have shown the complexity of the stress state that exists in the surface and subsurface regions of the contact zones of mating cylinders, spheres, or other bodies. The discussion of crack initiation mechanisms above indicates that the location of an incipient crack is quite unpredictable given the random distribution of inclusions in the material. Therefore it is more difficult to accurately predict the condition of stress at an expected point of failure in a contact zone than was the case in designing a cantilever beam, for example. This dilemma is resolved by using one, easily calculated contact-zone stress as a *reference value* to compare to material strengths. The one chosen is the largest negative (compressive) principal contact stress. In a pure-rolling case, its magnitude will be equal to the applied maximum contact pressure P_{max}. But it will be greater than that value if sliding is present.

To develop allowable surface fatigue strengths, the material is typically run under controlled loading conditions (i.e., controlled p_{max}) and the number of cycles to failure recorded and reported along with other loading factors such as percent sliding, lubrication, body geometry, etc. This "virtual strength" can be compared to the peak magnitude of compressive stress in other applications having similar loading factors. Thus the reported surface fatigue strength has only an indirect relationship to the actual stresses that may have been present in the test piece and in your similarly loaded part since the Hertzian stress equations are only valid for static loading.

The expression for the normal, compressive Hertzian static stress in cylindrical contact is found by combining equations 7.14b and 7.17a:

$$\sigma_z = -p_{max} = \frac{2F}{\pi a L} \tag{7.25a}$$

Substitute the expression for a from equation 7.15b, square both sides, and simplify:

$$\sigma_z^2 = \frac{2}{\pi} \frac{F}{L} \frac{B}{(m_1 + m_2)} \tag{7.25b}$$

Rearrange to solve for the load F,

$$F = \sigma_z^2 \frac{\pi L (m_1 + m_2)}{2B} \tag{7.25c}$$

and collect terms in a constant K,

$$F = \frac{KL}{2B} \qquad (7.25d)$$

where

$$K = \pi(m_1 + m_2)\sigma_z^2 \qquad (7.25e)$$

This factor K is termed an *experimental load factor* and is used to determine the safe endurance load F at a specified number of cycles or the number of cycles that can be expected before failure occurs at a given load.

Table 7-7 shows experimentally determined load factors, K, fatigue strengths, S_c, and strength factors for a number of materials either running against themselves or against hardened tool steel.[30] See the original reference for a complete listing as some materials were omitted here due to lack of space. Two different loading modes are also addressed in separate sections of the table, pure rolling, and rolling with 9% sliding. The first column of the table defines the material. In each section, the next two columns give the K value and the surface fatigue strength at 1E8 cycles as tested. The next two columns contain strength factors λ and ζ, which represent the slope and intercept of the S-N diagram (on log-log coordinates) for the surface fatigue strength of the material as determined by regression on large amounts of test data. These factors can be used in the equation of the statistically fitted S-N line to find the expected cycle life N for the applied stress level.

$$\log_{10} K = \frac{\zeta - \log_{10} N}{\lambda} \qquad (7.26)$$

The K values in Table 7-7 can be used directly in equation 7.25d to calculate an allowable load F for the selected material at 1E8 cycles of stress. For other desired design cycle lives, first calculate the largest-negative (compressive) radial stress for your design from the appropriate equations as defined in the preceding sections. Then calculate K from equation 7.25e and use it and the values of λ and ζ from Table 7-7 to find the value of N for the application from equation 7.26. Since there is no endurance limit for surface fatigue loading, we can expect pitting to begin after approximately N stress cycles at the level of nominal stress contained in your calculated K factor.

Alternatively, a desired number of cycles N can be chosen and an allowable design-stress level σ_z for a chosen material computed from equations 7.25e and 7.26. A safety factor can be applied either by selecting a material with a longer cycle life than required for the application or by sizing the parts to have a stress level below the calculated allowable stress level for a necessary number of cycles.

The strength values in Table 7-7 were obtained using rollers in contact, lubricated with a light mineral oil of 280-320 SSU at 100°F. The researchers report that "an orderly transition occurs from pitting fatigue to abrasive wear as percent sliding is increased." Pitting failures were observed under as high as 300% sliding on some cast irons and abrasive wear was seen at as low as 9% sliding on hardened steels under high

Table 7-7   Surface Fatigue Strength Data for Various Materials

Part 1: Materials Running Against an HRC 60-62 Tool-Steel Roller

#	Material	Pure Rolling				Rolling & 9% Sliding			
		K	Sc @ 1E8 cycles, psi	λ	ζ	K	Sc @ 1E8 cycles, psi	λ	ζ
1	1020 steel, carburized, 0.045 in min depth HRC 50-60	12 700	256 000	7.39	38.33	10 400	99 000	13.20	61.06
2	1020 steel, HB 130-150	—	—	—	—	1 720	94 000	4.78	23.45
3	1117 steel, HB 130-150	1 500	89 000	4.21	21.41	1 150	77 000	3.63	19.12
4	X1340 steel, induction hardened, 0.045 in. min depth HRC 45-58	10 000	227 000	6.56	34.24	8 200	206 000	8.51	41.31
5	4150 steel, h-t, HB 270-300, flash-chrome plated	6 060	177 000	11.18	50.29	—	—	—	—
6	4150 steel, h-t, HB 270-300, phosphate coated	9 000	216 000	8.80	42.81	6 260	180 000	11.56	51.92
7	4150 cast steel, h-t, HB 270-300	—	—	—	—	2 850	121 000	17.86	69.72
8	4340 steel, induction hardened, 0.045 in. min depth HRC 50-58	13 000	259 000	14.15	66.22	9 000	216 000	14.02	63.44
9	4340 steel, h-t, HB 270-300	—	—	—	—	5 500	169 000	18.05	75.55
10	6150 steel, HB 300-320	1 170	78 000	3.10	17.51	—	—	—	—
11	6150 steel, HB 270-300	—	—	—	—	1 820	97 000	8.30	35.06
12	18% Ni maraging tool steel, air hardened, HRC 48-50					4 300	146 000	3.90	22.18
13	Gray iron, Cl. 20, HB 140-160	790	49 000	3.83	19.09	740	47 000	4.09	19.72
14	Gray iron, Cl. 30, HB 200-220	1 120	63 000	4.24	20.92	—	—	—	—
15	Gray iron, Cl. 30, h-t (austempered) HB 255-300, phosphate-coated	2 920	102 000	5.52	27.11	2 510	94 000	6.01	28.44
16	Gray iron, Cl. 35, HB 225-255	2 000	86 000	11.62	46.35	1 900	84 000	8.39	35.51
17	Gray iron, Cl. 45, HB 220-240	—	—	—	—	1 070	65 000	3.77	19.41
18	Nodular iron, Gr. 80-60-03, h-t HB 207-241	2 100	96 000	10.09	41.53	1 960	93 000	5.56	26.31
19	Nodular iron, Gr. 100-70-03, h-t HB 240-260	—	—	—	—	3 570	122 000	13.04	54.33
20	Nickel bronze, HB 80-90	1 390	73 000	6.01	26.89	—	—	—	—
21	SAE 65 phosphor-bronze sand casting, HB 65-75	730	52 000	2.84	16.13	350	36 000	2.39	14.08
22	SAE 660 cont-cast bronze, HB 75-80	—	—	—	—	320	33 000	1.94	12.87
23	Aluminum bronze	2 500	98 000	5.87	27.97	—	—	—	—
24	Zinc die-casting, HB 70	250	28 000	3.07	15.35	220	26 000	3.11	15.29
25	Acetal resin	620	—	—	—	580	—	—	—
26	Polyurethane rubber	240	—	—	—	—	—	—	—

Table 7-7   Surface Fatigue Strength Data for Various Materials
Part 2: Materials Running Against the Same Material

#	Material	Pure Rolling				Rolling & 9% Sliding			
		K	Sc @ 1E8 cycles, psi	λ	ζ	K	Sc @ 1E8 cycles, psi	λ	ζ
27	1020 steel, HB 130-170, and same but phosphate coated	2 900	122 000	7.84	35.17	1 450	87 000	6.38	28.23
28	1144 steel CD steel, HB 260-290, (stress-proof)	—	—	—	—	2 290	109 000	4.10	21.79
29	4150 steel, h-t, HB 270-300, and same but phosphate coated	6 770	187 000	10.46	48.09	2 320	110 000	9.58	40.24
30	4150 leaded steel, phosphate coated, h-t, HB 270-300	—	—	—	—	3 050	125 000	6.63	31.1
31	4340 steel, h-t, HB 320-340, and same but phosphate coated	10 300	230 000	18.13	80.74	5 200	164 000	26.19	105.31
32	Gray iron, Cl. 20, HB 130-180	960	45 000	3.05	17.10	920	43 900	3.55	18.52
33	Gray iron, Cl. 30, h-t (austempered) HB 270-290	3 800	102 000	7.25	33.97	3 500	97 000	7.87	35.90
34	Nodular iron, Gr. 80-60-03, h-t HB 207-241	3 500	117 000	4.69	24.65	1 750	82 000	4.18	21.56
35	Meehanite, HB 190-240	1 600	80 000	4.77	23.27	1 450	76 500	4.94	23.64
36	6061-T6 aluminum, hard anodized coating	350	—	10.27	34.15	260	—	5.02	20.12
37	HK31XA-T6 magnesium, HAE coating	175	—	6.46	22.53	275	—	11.07	35.02

Source: R. A. Morrison, "Load/Life Curves for Gear and Cam Materials," *Machine Design*, vol. 40, pp. 102-108, Aug. 1, 1968, A Penton Publication, Cleveland, Ohio, with the publisher's permission.

7

stress. They also note that the addition of oxide coatings, fortified (EP) lubricants, or lead as an alloying element all reduced tangential stress levels and increased fatigue life or allowable % sliding. The addition of phosphate coatings to the surfaces reduced sparking and flashing of lubricant, reduced the friction coefficient, and also increased fatigue life. They saw evidence of pitting starting both at the surface under high % sliding and below the surface in pure-rolling or low-percent-sliding situations.[30] Increased sliding percentages reduce fatigue life but not linearly. Figure 7-26 shows some *S-N* curves (from reference 30) for three materials with various percentages of sliding.

The speed of stress cycling only affected nonmetallic materials, wherein friction heat blistered or yielded the material. A material's stiffness is a factor, however. Lower-modulus materials reduce the contact stress because their larger deflections increase the contact-patch area. Cast iron on cast iron had longer life than cast iron on hardened steel. The free graphite in cast iron also makes it a good choice in contact situations as it acts to retard adhesion as well as being a dry lubricant, though the lower grades of CI have strengths too low to be useful in this situation. Nodular iron in its harder forms may be a better choice. Hardness of a material was not found to correlate closely with its surface endurance. Some softer steels performed better than some harder ones.[30]

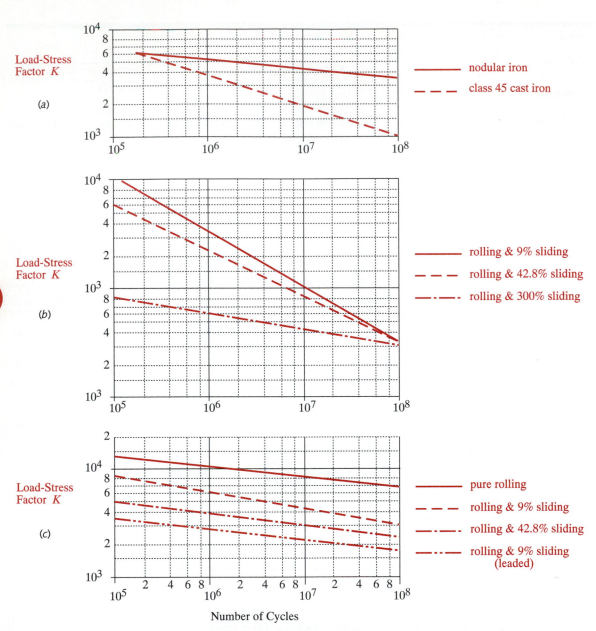

Typical curves showing load-life relationships for common gear and cam materials. Curves in (*a*) are for 100-70-30 nodular iron (HB 240-260) and class 45 gray cast iron (HB 220-240), both materials running on carbon tool steels (HRC 60-62). Curves in (*b*) are for continuous-cast bronze running on hardened steel. Curves in (*c*) are for heat-treated 4150 steel running against the same material, but phosphate coated. In all charts, 9% sliding velocity is 54 fpm; 42.8% sliding velocity is 221 fpm.

FIGURE 7-26

Load-Life Curves for Some Combinations of Materials in Combined Rolling and Sliding Source: R. A. Morrison, "Load/Life Curves for Gear and Cam Materials," *Machine Design*, vol. 40, pp. 102-108, Aug. 1, 1968, A Penton Publication, Cleveland, Ohio, with permission

---

EXAMPLE 7-5

Finding the Safety Factor in Surface Fatigue

Problem   Choose a material to provide 10 years of life for the rollers in Example 7-4.

Given   The roller radii are 1.25 and 2.5 in and are each 24 in long. The principal stresses are as shown in Example 7-4. The smaller roll is at 4 000 rpm.

Assumptions There is 9% sliding combined with rolling. Both materials will be of the same steel. The machine will operate 3 shifts/day for 345 days/year.

Solution   See the *TKSolver* file EX07-05.

1   Calculate the required cycle life from the given data:

$$cycles = 4\ 000 \frac{rev}{min} \cdot 60 \frac{min}{hr} \cdot 24 \frac{hr}{day} \cdot 345 \frac{day}{yr} \cdot 10\ yr = 6.0E10 \qquad (a)$$

2   The maximum normal stress calculated in Example 7-4 is 49 771 psi compressive. Its K factor can be calculated from equation 7.25d. The previously calculated material constants m_1 and m_2 are needed:

$$m_1 = m_2 = \frac{1 - v_1^2}{E_1} = \frac{1 - 0.28^2}{3E7} = 3.072E - 8 \qquad (b)$$

$$K = \pi(m_1 + m_2)\sigma_z^2 = 2\pi(3.072E - 8)(49\ 771)^2 = 478 \qquad (c)$$

3   A trial material must be selected from Table 7-8. With a K this low virtually any of the steels can probably be used. We will try the HB 130-170 SAE 1020, phosphate-coated steel (#27 in part 2 of the table), since the same materials are running together. The slope and intercept factors of this steel for rolling with 9% sliding are

$$\lambda = 6.38 \qquad \qquad \zeta = 28.23 \qquad (d)$$

4   These are used in equation 7.26 along with the value of K from equation c above to find the number of cycles that can be expected at this load before pitting begins.

$$\log_{10} K = \frac{\zeta - \log_{10} N_{life}}{\lambda}$$

$$\log_{10} N_{life} = \zeta - \lambda \log_{10} K = 28.23 - 6.38 \log_{10}(478) \qquad (e)$$

$$N_{life} = 10^{(28.23 - 6.38 \log_{10}(478))} = 1.4E11$$

5   A safety factor against pitting can now be calculated from the ratio of the projected cycle life and the desired number of cycles.

$$N_f = \frac{N_{life}}{cycles} = \frac{1.4E11}{6.0E10} = 2.3 \qquad (f)$$

7.14   SUMMARY

This chapter has presented a brief introduction to the very broad topic of surface wear. Wear is generally considered to be divisible into four general categories, *adhesive wear, abrasive wear, corrosive wear*, and *surface fatigue*. Other mechanisms such as *corrosion fatigue* and *fretting corrosion* combine elements from more than one category.

Wear usually requires some relative motion to be present between two surfaces. **Adhesive wear** occurs when the asperities of two mating surfaces adhere to one another and then break when sliding occurs, transferring material from one part to the other, or out of the system. **Abrasive wear** involves a hard, rough surface-abrading material from a softer one, or loose, hard particles trapped between two surfaces and abrading both.

Corrosion wear occurs when a corrosive atmosphere (such as oxygen) is present to attack the surface of the material in combination with sliding that breaks the oxides or other contaminants free from the surfaces. This exposes new material to the corrosive elements and also turns the often hard corrosion products into abradants. Corrosion fatigue refers to the combination of a corrosive environment with cyclic stresses. This combination is particularly deadly and greatly shortens the fatigue life of materials. Fretting corrosion occurs in tight joints (such as press fits) where no gross motion is present. Tiny vibratory motions are sufficient to set up a corrosive wear process called fretting that can remove significant volumes of material over time.

Surface fatigue occurs in pure-rolling or roll-sliding contact, but not in pure-sliding situations. The very high contact stresses engendered by the small areas of contact act to cause fatigue failure of the material after many thousands of cycles of repeated stress. **Pitting** is the loss of small pieces of material from the surface, leaving behind pits. Pits will grow into larger areas of flaked-off surface material, which is then called **spalling**. An audible warning is usually noticeable when the pitting process begins. If unattended, it will proceed to gross damage of the part.

High-strength, smooth materials are required in contact-stress applications. **No materials show an endurance limit against surface fatigue** and all will eventually fail by this mechanism if subjected to a sufficient number of contact-stress reversals.

Designing to Avoid Surface Failure

There are a number of precautions that a designer may take to reduce the chances of a wear failure by any of the mechanisms described in this chapter.

1   **Proper choice of materials:** The issues of material compatibility must be observed. Careful attention to surface finish and hardness as well as to strength is necessary to reduce abrasion and increase surface fatigue life. Corrosive environments require special materials. Coatings should be considered in some situations. Homogeneity of materials in contact-stress situations is desirable. More expensive steels processed to create more uniform and inclusion-free microstructures can give superior service in highly stressed surface fatigue cases and may be cheaper in the long run. In

general, higher surface hardness reduces both adehisive and abrasive wear as well as surface fatigue.

2 **Proper lubricants:** It is rare that a heavily loaded joint is run dry (and then only if some overriding concern is present such as fear of contamination of the product with escaped lubricant). Hydrodynamic or hydrostatic lubrication should be used where feasible. Boundary lubrication is less desirable though is often unavoidable. If boundary lubrication is used, an EP lubricant may significantly reduce adhesive wear at the expense of some corrosive wear. See Chapter 10 for more on lubrication.

3 **Cleanliness:** Reasonable measures should be taken to ensure that no external or environmental contaminants can enter bearings or joints. Seals or other means to protect them should be provided. If particulate contamination cannot be avoided (as in dirty environments) soft materials should be chosen for bearings to allow embedment of trapped particles.

4 **Stress:** Avoid or minimize stress concentrations especially in fatigue-loaded applications. Consider using a less-stiff material to increase the contact-patch area and reduce stresses in surface-fatigue cases. Be extremely wary of situations in which any kind of fatigue loading (not just surface fatigue) is present in combination with a corrosive environment since corrosion fatigue is then a problem. A test program is probably necessary in such situations as little data are available on this phenomenon.

5 **Fretting:** Consider the possibility of fretting failure if vibration or repeated deflections are present in combination with press fits or tight joints.

Important Equations Used in this Chapter

See the referenced sections for information on the proper use of these equations.

Real Area of Contact (Section 7.3):

$$A_r \cong \frac{F}{S_p} \cong \frac{F}{3S_{yc}}$$ (7.1)

Coefficient of Friction (Section 7.3):

$$\mu = \frac{f}{F} \cong \frac{S_{us}}{3S_{yc}}$$ (7.3)

Volume of Wear (Section 7.4):

$$V = K\frac{Fl}{H}$$ (7.7a)

Maximum Pressure—Spherical Contact (Section 7.8):

$$p_{max} = \frac{3}{2}\frac{F}{\pi a^2}$$ (7.8b)

Material Constants (Section 7.8):

$$m_1 = \frac{1 - v_1^2}{E_1} \qquad\qquad m_2 = \frac{1 - v_2^2}{E_2} \qquad (7.9a)$$

Geometry Constant for Spherical and Cylindrical Contact (Section 7.8):

$$B = \frac{1}{2}\left(\frac{1}{R_1} + \frac{1}{R_2}\right) \qquad (7.9b)$$

Radius of Contact Patch—Spherical Contact (Section 7.8):

$$a = \sqrt[3]{0.375 \frac{m_1 + m_2}{B} F} \qquad (7.9d)$$

Maximum Stresses—Spherical Contact (Section 7.8):

$$\sigma_{z_{max}} = -p_{max} \qquad (7.11c)$$

$$\sigma_{x_{max}} = \sigma_{y_{max}} = -\frac{1 + 2v}{2} p_{max} \qquad (7.11d)$$

$$\tau_{13_{max}} = \frac{p_{max}}{2}\left[\frac{(1 - 2v)}{2} + \frac{2}{9}(1 + v)\sqrt{2(1 + v)}\right] \qquad (7.12b)$$

$$z_{@\tau_{max}} = a\sqrt{\frac{2 + 2v}{7 - 2v}} \qquad (7.12c)$$

Maximum Pressure—Cylindrical Contact (Section 7.9):

$$p_{max} = \frac{2F}{\pi a L} \qquad (7.14b)$$

Half-Width of Cylindrical Contact Patch (Section 7.9):

$$a = \sqrt{\frac{2}{\pi}\frac{m_1 + m_2}{B}\frac{F}{L}} \qquad (7.15b)$$

Maximum Stresses—Cylindrical Contact (Section 7.9):

$$\sigma_x = \sigma_z = -p_{max}$$
$$\sigma_y = -2v p_{max} \qquad (7.17a)$$

$$\tau_{13_{max}} = 0.304 p_{max}$$
$$z_{@\tau_{max}} = 0.786a \qquad (7.17b)$$

Maximum Pressure—General Contact (Section 7.10):

$$p_{max} = \frac{3}{2}\frac{F}{\pi\, ab} \tag{7.18b}$$

Half-Dimensions of Elliptical Contact Patch (Section 7.10—See Table 7-5 for k_a and k_b):

$$a = k_a \sqrt[3]{\frac{3F(m_1 + m_2)}{4A}} \qquad\qquad b = k_b \sqrt[3]{\frac{3F(m_1 + m_2)}{4A}} \tag{7.19d}$$

$$A = \frac{1}{2}\left(\frac{1}{R_1} + \frac{1}{R_1{}'} + \frac{1}{R_2} + \frac{1}{R_2{}'}\right) \tag{7.19a}$$

Maximum Stresses—General Contact (Section 7.10):

$$\sigma_x = -\left[2\nu + (1 - 2\nu)\frac{b}{a+b}\right]p_{max}$$

$$\sigma_y = -\left[2\nu + (1 - 2\nu)\frac{a}{a+b}\right]p_{max} \tag{7.21a}$$

$$\sigma_z = -p_{max}$$

Friction Unit Force—Parallel Cylinders Rolling and Sliding (Section 7.11):

$$f_{max} = \mu\, p_{max} \tag{7.22f}$$

Maximum Stresses—Parallel Cylinders Rolling and Sliding (Section 7.11):

when $z = 0$

$$\text{if } |x| \le a \quad \text{then} \quad \sigma_{x_n} = -p_{max}\sqrt{1 - \frac{x^2}{a^2}} \quad \text{else } \sigma_{x_n} = 0$$

$$\sigma_{z_n} = \sigma_{x_n} \tag{7.23a}$$

$$\tau_{xz_n} = 0$$

$$\text{if } x > a \text{ then } \sigma_{x_t} = -2f_{max}\left(\frac{x}{a} - \sqrt{\frac{x^2}{a^2} - 1}\right)$$

$$\text{if } x < a \text{ then } \sigma_{x_t} = -2f_{max}\left(\frac{x}{a} + \sqrt{\frac{x^2}{a^2} - 1}\right) \tag{7.23b}$$

$$\text{if } |x| \le a \text{ then } \sigma_{x_t} = -2f_{max}\frac{x}{a}$$

$$\sigma_{z_t} = 0 \tag{7.23c}$$

$$\text{if } |x| \le a \text{ then } \tau_{xz_t} = -f_{max}\sqrt{1 - \frac{x^2}{a^2}} \quad \text{else } \tau_{xz_t} = 0$$

$$\sigma_x = \sigma_{x_n} + \sigma_{x_t}$$

$$\sigma_z = \sigma_{z_n} + \sigma_{z_t} \tag{7.24a}$$

$$\tau_{xz} = \tau_{xz_n} + \tau_{xz_t}$$

$$\sigma_y = \nu\left(\sigma_x + \sigma_z\right) \tag{7.24b}$$

Material Surface Fatigue Strength Factor (Section 7.12):

$$K = \pi\left(m_1 + m_2\right)\sigma_z^2 \tag{7.25e}$$

S-N line Equation for Surface Fatigue (Section 7.12—See Table 7-7 for λ and ζ):

$$\log_{10} K = \frac{\zeta - \log_{10} N}{\lambda} \tag{7.26}$$

7.15   REFERENCES

1   **E. Rabinowicz**, *Friction and Wear of Materials*. John Wiley & Sons: New York, pp. 110, 1965.

2   *Ibid.*, pp. 21, 33.

3   *Ibid.*, p. 125.

4   *Ibid.*, p. 30.

5   **R. Davies**, Compatibility of Metal Pairs, in *Handbook of Mechanical Wear*, C. Lipson, ed. Univ. of Mich. Press: Ann Arbor. p. 7, 1961.

6   **D. J. Wulpi**, *Understanding How Components Fail*. American Society for Metals: Metals Park, OH, 1990.

7   **E. Rabinowicz**, *Friction and Wear of Materials*. John Wiley & Sons: New York, pp. 60, 1965.

8   *Ibid.*, p. 85.

9   **J. T. Burwell**, Survey of Possible Wear Mechanisms. *Wear*, **1**: pp. 119-141, 1957.

10 **E. Rabinowicz**, *Friction and Wear of Materials*. John Wiley & Sons: New York, pp. 179, 1965.

11 *Ibid.*, p. 180.

12 **J. R. McDowell**, Fretting and Fretting Corrosion, in *Handbook of Mechanical Wear*, C. Lipson and L. V. Colwell, ed. Univ. of Mich. Press: Ann Arbor. pp. 236-251, 1961.

13 **H. Hertz**, On the Contact of Elastic Solids. *J. Math.*, **92**: pp. 156-171, 1881 (in German).

14   **H. Hertz**, Contact of Elastic Solids, in *Miscellaneous Papers,* P. Lenard, ed. Macmillan & Co. Ltd.: London. pp. 146-162, 1896.

15   **H. L. Whittemore and S. N. Petrenko**, *Friction and Carrying Capacity of Ball and Roller Bearings*, Technical Paper 201, National Bureau of Standards, Washington, D.C., 1921.

16   **H. R. Thomas and V. A. Hoersch**, *Stresses Due to the Pressure of One Elastic Solid upon Another*, Bulletin 212, U. Illinois Engineering Experiment Station, Champaign, Ill., July 15,1930.

17   **E. I. Radzimovsky**, *Stress Distribution and Strength Condition of Two Rolling Cylinders*, Bulletin 408, U. Illinois Engineering Experiment Station, Champaign, Ill., Feb 1953.

18   **J. O. Smith and C. K. Lui**, Stresses Due to Tangential and Normal Loads on an Elastic Solid with Application to Some Contact Stress Problems. *J. Appl. Mech. Trans. ASME*, **75**: pp. 157-166, 1953.

19   **S. P. Timoshenko and J. N. Goodier**, *Theory of Elasticity*, 3rd ed., McGraw-Hill: New York, pp. 403-419, 1970.

20   **V. S. Mahkijani,** *Study of Contact Stresses as Developed on a Radial Cam using Photoelastic Model and Finite Element Analysis.* M.S. Thesis, Worcester Polytechnic Institute, 1984.

21   **J. Poritsky**, Stress and Deformations due to Tangential and Normal Loads on an Elastic Solid with Applications to Contact of Gears and Locomotive Wheels, *J. Appl. Mech*. Trans ASME, **72**: p. 191, 1950.

22   **E. Buckingham and G. J. Talbourdet**, *Recent Roll Tests on Endurance Limits of Materials*. in Mechanical Wear Symposium. ASM: 1950.

23   **S. Way**, Pitting Due to Rolling Contact. *J. Appl. Mech. Trans. ASME*, **57**: p. A49-58, 1935.

24   **W. E. Littmann and R. L. Widner**, Propagation of Contact Fatigue from Surface and Subsurface Origins, *J. Basic Eng. Trans. ASME*, **88**: pp. 624-636, 1966.

25   **H. Styri**, *Fatigue Strength of Ball Bearing Races and Heat-Treated 52100 Steel Specimens*. Proceedings ASTM, **51**: p. 682, 1951.

26   **T. L. Carter, et al.**, Investigation of Factors Governing Fatigue Life with the Rolling Contact Fatigue Spin Rig, *Trans. ASLE,* **1**: p. 23, 1958.

27   **H. Hubbell and P. K. Pearson**, *Nonmetallic Inclusions and Fatigue under Very High Stress Conditions, in Quality Requirements of Super Duty Steels.* AIME Interscience Publishers: p. 143, 1959.

28   **W. E. Littmann and R. L. Widner**, Propagation of Contact Fatigue from Surface and Subsurface Origins, *J. Basic Eng. Trans. ASME*, **88**: p. 626, 1966.

29   **H. Reusner**, The Logarithmic Roller Profile - the Key to Superior Performance of Cylindrical and Tapered Roller Bearings, *Ball Bearing Journal*, SKF, **230** , June 1987.

30   **R. A. Morrison**, "Load/Life Curves for Gear and Cam Materials." *Machine Design*, v. 40, pp. 102-108, Aug. 1, 1968.

31   **W. D. Cram**, Experimental Load-Stress Factors, in *Handbook of Mechanical Wear*, C. Lipson and L. V. Colwell, eds., Univ. of Mich. Press: Ann Arbor. pp. 56-91, 1961.

32   **J. F. Archard**, "Wear Theory and Mechanisms," in *Wear Control Handbook*, M. B. Peterson and W. O. Winer, eds., McGraw-Hill: New York, pp. 35-80, 1980.

33   **E. Rabinowicz**, "Wear Coefficients—Metals," in *Wear Control Handbook*, M. B. Peterson and W. O. Wincr, eds., McGraw-Hill: New York, pp. 475-506, 1980.

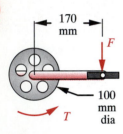

FIGURE P7-1

Problem 7-3

7.16   PROBLEMS

*7-1   Two 3 x 5 cm blocks of steel with machined finish R_a = 0.6 μm are rubbed together with a normal force of 400 N. Estimate the true area of contact between them if their S_y = 400 MPa.

*7-2   Estimate the dry coefficient of friction between the two pieces in Problem 7-1 if their S_{ut} = 600 MPa.

*†7-3   For the bicycle pedal-arm assembly in Figure P7-1 assume a rider-applied force that ranges from 0 to 1 500 N at the pedal each cycle. Determine the maximum contact stresses at one sprocket tooth-chain roller interface. Assume that the one tooth takes all the applied torque, that the chain roller is 8-mm dia, the sprocket has a nominal (pitch) dia of 100 mm, and that the sprocket tooth is essentially flat at the point of contact. The roller is X1340 steel at HRC45-58 and the sprocket is 4340 steel at HB270-300. The roller and sprocket contact over a length of 8 mm. Assuming rolling plus 9% sliding, estimate the number of cycles to failure for this particular tooth-roller combination.

*†7-4   For the trailer hitch from Problem 3-4 on p. 169 (also see Figures P7-2 and 1-5), determine the contact stresses in the ball and the ball cup (not shown). Assume that the ball is 2-in dia and the ill-fitting ball cup that surrounds it is an internal spherical surface 10% larger in diameter than the ball.

7-5   For the trailer hitch from Problem 3-5 on p. 169 (also see Figures P7-2 and 1-5) determine the contact stresses in the ball and the ball cup (not shown). Assume that the ball is 2-in dia and the ill-fitting ball cup that surrounds it is an internal spherical surface 10% larger in diameter than the ball.

†7-6   For the trailer hitch from Problem 3-6 on p. 169 (also see Figures P7-2 and 1-5) determine the contact stresses in the ball and the ball cup (not shown). Assume that the ball is 2-in dia and the ill-fitting ball cup that surrounds it is an internal spherical surface 10% larger in diameter than the ball.

†7-7   For the 12-mm dia steel wrist pin of Problem 3-7 (p. 169) find the maximum contact stress if the 2 500 g acceleration is fully reversed. The aluminum piston has a hole for the wrist pin that is 2% larger than the pin.

*7-8   A paper machine processes rolls of paper having a density of 984 kg/m^3. The paper roll is 1.50-m OD x 0.22-m ID x 3.23-m long and has an effective modulus of

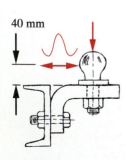

FIGURE P7-2

Problem 7-6

8

DESIGN CASE STUDIES

*No man's knowledge here
can go beyond his experience.*

JOHN LOCKE

8.0 INTRODUCTION

This chapter will introduce and set up several larger-scale design case studies than those presented in the earlier chapters. These case studies will then be used throughout the remainder of the book as common examples to illustrate the application of the design process for various aspects of each design problem. The following chapters will each explore a different type of design element such as shafts, gears, springs, etc., which are commonly found in machinery. This collection of elements cannot be exhaustive but is rather intended to be illustrative of the way that the principles of the first part of the book are applied to practical design problems. The particular machine elements selected for study are chosen in part because of their common usage and partly because of their ability to exemplify some of the design and failure criteria discussed in Part I. Table 8-0 shows the variables used in this chapter and references the case studies in which they are used.

Design is, by its nature, an iterative process. When presented with a design-problem statement, some simplifying assumptions are always necessary in order to get started. As the design takes shape, the results of later design choices will inevitably force the designer to revisit earlier assumptions about parts already designed and change them to suit the new conditions. A simple example of this could be the design of a set of gears mounted on shafts. Whether one starts by designing the shafts or the gears (say the shafts), when it is time to address the second of the two elements (say the gears), their requirements may force a change to some of the assumptions made regarding the shaft design done earlier. Eventually, one arrives at a compromise that satisfies all the constraints, but only after some iteration that inevitably involves the redesign of parts done earlier.

Table 8-0   Variables Used in this Chapter

Symbol	Variable	ips units	SI units	See
A	area	in^2	m^2	Case 7a
a	acceleration	in/sec^2	m/sec^2	Case 8a
c	damping constant	lb-sec/in	N-sec/m	Case 8a, 9a
C_f	coefficient of fluctuation	none	none	Case 9a
d	diameter	in	m	Case 7a
E	energy	in-lb	Joules	Case 9a
F	force or load	lb	N	all
g	gravitational acceleration	in/sec^2	m/sec^2	Case 8a
k	gas law exponent	none	none	Case 7a
k	spring rate or spring constant	lb/in	N/m	Case 8a, 9a
l	length	in	m	Case 7a
m	mass	lb-sec^2/in	Kg	all
P	power	hp	watts	Case 8a
p	pressure	psi	Pa	Case 7a
r	radius	in	m	Case 7a
T	torque	lb-in	N-m	all
v	volume	in^3	m^3	Case 7a
v	linear velocity	in/sec	m/sec	Case 8a
W	weight	lb	N	Case 8a
y	displacement	in	m	all
ω	rotational or angular velocity	rad/sec	rad/sec	all
ω_n	natural frequency	rad/sec	rad/sec	Case 9a
ζ	damping ratio	none	none	Case 9a

8

* The student may not have yet been exposed to all aspects of these broad scale problems in his or her studies to date, but should nevertheless not be dismayed if some details of these case studies seem obscure. One will undoubtedly encounter more detailed explanation of these topics in other courses, in later experience, or in self-study. One of the more interesting aspects of design engineering is its breadth. One must continually learn new things in order to be able to solve real engineering problems. An engineering education only begins in college and is far from complete at graduation. One should welcome the challenge of exploring new topics throughout one's career.

Because of the need for iteration, any time spent to set up the problem solution in a computerized tool such as a spreadsheet or equation solver will be well rewarded when you have to redesign each part several times. Lacking a computer-based model, you will be faced with redoing your calculations from scratch for each iteration, which is not a pleasant prospect. We will make extensive use of computer-aided design tools in these case studies.

8.1   A PORTABLE AIR COMPRESSOR[*]

A building contractor needs a small, gasoline-engine powered air compressor to use for driving air hammers on remote job sites. A preliminary design concept is shown in Figure 8-1. A single-cylinder, two-stroke engine with flywheel is coupled to a gearset to

MACHINE DESIGN

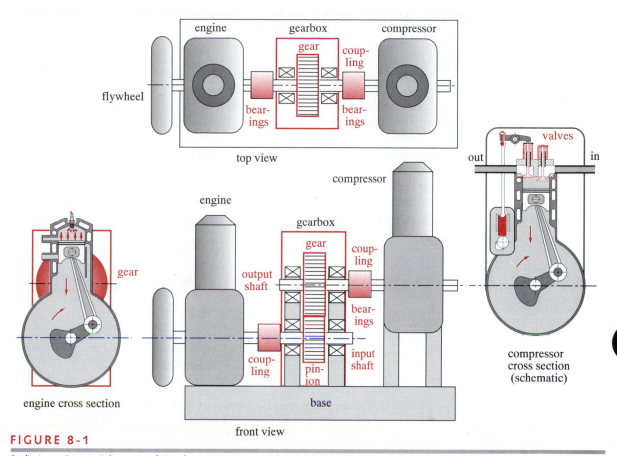

FIGURE 8-1

Preliminary Design Schematic of Gasoline-Engine-Powered Portable Air Compressor, Gearbox, Couplings, Shafts, and Bearings

reduce the engine speed and boost its torque appropriately. The ratio for this gearset is to be determined. The 2.5 hp gasoline engine is governed at 3 800 rpm. The gearset's output shaft drives the crankshaft of a single-cylinder Schramm (poppet valve) piston compressor through a keyed coupling. Some preliminary thermodynamic calculations (see CASE7A) have indicated that the desired flow rate of 9 cfm at a mean effective pressure of 26 psig can be obtained from a 25-in^3 stroke-volume compressor running at 1 500 rpm.

Figure 8-1 shows the engine mounted on a base (which could be on wheels) with its output shaft connected via a coupling to the input shaft of a gearbox. The gearbox contains a single gearset to reduce the high engine speed to a lower one suitable for the compressor. The required gear ratio is 1 500:3 800 or 0.39:1. The output shaft from the gearbox is connected via a coupling to the crankshaft of the compressor. The shafts in the gearbox housing are carried in suitable bearings. The cross section of the compressor shows the active exhaust valve driven by a cam-operated pushrod and rocker-arm train. The intake valve is passive, i.e., opened and closed by the pressure differentials and its light spring. The valve spring on the exhaust valve must be strong enough to keep the follower in contact with the cam.

There are several aspects of this device that we will investigate. We will assume that the gasoline engine will be purchased as a unit. The compressor will dictate the loads on the elements between the engine and itself, so some information is needed on the load-time characteristics of the compressor. The shafts, couplings, bearings, and gears that deliver the power from engine to compressor will be the principal elements to be designed in this case study. We will also look at a few elements within the compressor such as the headbolts and valve spring since they provide excellent examples of fatigue design. Because of their complicated geometries, the design of other parts of the compressor such as the piston, connecting rod, and crankshaft is more suited to the application of *Finite Element Analysis* (FEA) and will not be addressed in this case study.

CASE STUDY 7A

Preliminary Design of a Compressor Drive Train

Problem	Determine the force-time function within the compressor's cylinder and the torque-time function acting on the input shaft of the compressor during any one cycle.
Given	Compressor speed is 1 500 rpm. The compressor has a bore of 3.125 in, a stroke of 3.26 in, and a connecting rod to crank ratio of 3.5. Inlet pressure is atmospheric (14.7 psia), peak cylinder pressure is 132 psig and the mean effective pressure (mep) is 26 psig. Flow is 8.9 cfm at mep giving 1.6 hp.
Assumptions	The piston weight is 1 lb, the connecting rod weighs 2 lb with its center of mass at the 1/3 point from the big end. The crankshaft weighs 5.4 lb including a counterbalance that optimally overbalances it to minimize the shaking force. The exponent for the gas law equation is $k = 1.13$.
Solution	See Figures 8-1 to 8-3 and TKSolver file CASE7-A.

1   The force-time function in the cylinder depends on the compressed-gas pressure, which in turn depends on the slider-crank mechanism's geometry and the gas-law:

$$p_1 v_1^k = p_2 v_2^k \qquad (a)$$

where p_1 is atmospheric pressure (in psia), v_1 is the expanded cylinder volume at bottom dead center (BDC), and p_2, v_2 are the pressure and volume of the compressed gas at top dead center (TDC) or at any other position. The gas-law exponent k is assumed to be 1.13 since the process is neither isothermal ($k = 1$) nor adiabatic ($k = 1.4$). A compression ratio of 10.9:1 is also assumed as typical. An expression for piston displacement y referenced to BDC (assuming constant crankshaft ω) is

$$y = \left(r\cos\theta + l\sqrt{1 - \left(\frac{r}{l}\sin\theta\right)^2} \right) - l + r \qquad (b)$$

where r = crank radius, l = connecting rod length, and θ = crank angle. See reference 1 for a derivation of the expression in parentheses.

2   Combining these functions with the assumed pressure ranges and gas law constant gives an approximate function for the cylinder pressure p as a function of crank angle for this particular problem's data:

$$\text{If } \pi \le \theta \le 2\pi \text{ then } p \cong 924\left(\frac{\theta}{\pi}\right)^6 - 792\left(\frac{\theta}{\pi}\right)^7 \text{ else } p \cong 0 \qquad (c)$$

This function is shown in Figure 8-2.* The force F_g on the piston and cylinder head due to the gas pressure is then

$$F_g = pA_p = \frac{\pi}{4}pd_p^2 \qquad (d)$$

where A_p is piston area and d_p is piston diameter. This is the same function as shown in Figure 8-2 multiplied by a constant. A second scale is shown on the ordinate giving gas force F_g in addition to the gas pressure p for this problem.

3   The torque required to drive the compressor crankshaft has two components, one from the gas force F_g and another from the inertial forces F_i due to the accelerations. [1]

$$T = T_g + T_i$$

where

$$T_g \cong F_g r\sin\theta\left(1 + \frac{r}{l}\cos\theta\right) \qquad (e)$$

and

$$T_i \cong \frac{1}{2}mr^2\omega^2\left(\frac{r}{2l}\sin\theta - \sin 2\theta - \frac{3r}{2l}\sin 3\theta\right)$$

The mass m is taken as that of the piston and wrist pin plus the portion of the connecting rod (about 1/3) considered to be acting at the piston. [1] When the data for this problem are substituted into (e), the torque time function is as shown in Figure 8-3.*

The force-time and torque-time functions shown in the Figures 8-2 and 8-3 assume that the shaft speed is essentially constant. This is a reasonable assumption for the steady-state condition since the engine driving it is governed and has a flywheel to smooth its own speed oscillations. These force and torque functions define the time-varying loading that the shafts, couplings, and gears will feel and are thus a starting point for their design. Because of the time-variation of the loads, all the parts will be subjected to fatigue loading and must be designed accordingly using the theories outlined in Chapters 6 and 7.

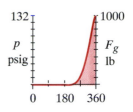

FIGURE 8-2

Pressure and Force Within Cylinder During One Cycle

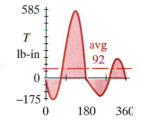

FIGURE 8-3

Total Torque-Time Function at Crankshaft with Constant ω

* The functions in Figures 8-2 and 8-3 were generated with program ENGINE from reference 1.

8.2   A HAY-BALE LIFTER

A dairy farmer in Bellows Falls, Vermont, needs a small winch-hoist to use for the lifting of hay bales into the barn loft. A preliminary design concept is shown in Figure 8-4. An electric motor is coupled to a worm-wheel gearset to reduce its speed and boost the torque appropriately. The best ratio for this gearset is to be determined. The gearbox output shaft is coupled to the winch-drum shaft and both turn in bearings to be selected. The drum serves as a capstan around which is wound a rope that has a forged hook at its end. The entire winch assembly will ultimately be suspended from the rafters in the hayloft above a central floor-hatch. Hay bales will be manually attached below and manually removed above. The electric motor is reversible and the worm-wheel set must be designed to be self-locking in order to hold the load when the motor is unpowered.

The above problem statement is very unstructured as it gives no information about the size and weight of a hay bale nor what number of bales should be lifted at a time for the best efficiency. These considerations in combination with the choice of winch-drum diameter will determine the torque requirements that the drive train will have to meet. The start-up load can be significantly higher than the steady-state lifting load due to shock loading when the slack is first taken out of the line and the load lifted. The dynamic loading at start up will be modeled using a differential-equation solver.

CASE STUDY 8A

Preliminary Design of a Winch Lift

Problem   Determine the force-time function in the lifting cable, the necessary drum diameter, and the torque-time function acting on the shaft of the winch drum during any 1 cycle. Define the gear ratio and the power and torque requirements for the motor.

Given   A hay bale's weight varies depending on its moisture content but can be assumed to average about 60 lb. The hay truck holds 100 bales and the farmer would like to unload it in 30 min. The lift height is 24 ft.

Assumptions   Nylon rope of 3/4 in dia has a minimum break strength of about 8000 lb and a spring constant of about 50 000 lb/in per ft of length in axial tension.

Solution   See Figures 8-4 to 8-6 and TKSolver file CASE8-A.

1   The nominal load depends on the number of bales to be lifted at one time and the weight of any structure used to support the bales. To unload 100 bales from the truck, one at a time in 30 min, requires that the average bale rate be 100 / 30 = 3.3 bales/min, or an 18-sec average period per bale. Since some of this time must be used to return the empty lift to the ground, we cannot use the entire 18 sec to lift the load. We must also allow some time for manual loading and unloading of the bales at top and bot-

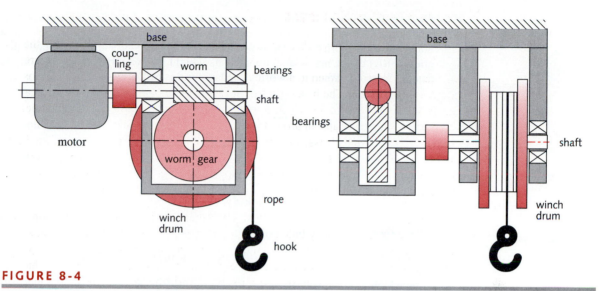

FIGURE 8-4

Motor-Driven Winch with Gear Train, Shafts, Bearings, and Couplings

tom. The portion of the total period during which the mechanism is working is called its *duty cycle*. Let's assume that 1/3 of the period is used to load/unload, 1/3 to lift, and 1/3 to lower. This allows 6 sec per bale if we lift only one bale at a time. The average velocity of the lift would then have to be 24 ft. / 6 sec = 4 ft/sec. A better arrangement would seem to be to load two bales at a time on the lift. This doubles the period to 36 sec, doubles the time available for lifting to 12 sec and halves the average velocity to 2 ft/sec, still keeping the same duty cycle.

2   The payload with two bales on the lift is 120 lb. The dead-load will be the weight of the rope, hook, and any platform or structure used to support the bales. Since this structure has yet to be designed, its weight is unknown. We will assume that we can keep this deadweight under 50 lb. The total nominal load will then be 170 lb for the lift phase, and 50 lb for the lowering phase.

3   At steady state, the load on the rope should be the above number. However, at start-up, the load can be significantly higher due to the need to accelerate the load to its steady-state velocity and also due to the fact that there are both spring and mass in the system. A combination of spring and mass in a dynamic system allows oscillations to occur as the kinetic energy of the moving mass is transferred to potential energy in the elastic spring and vice versa. The rope is a spring. When the slack in the rope is suddenly taken up against the mass of the load, the rope will stretch, storing potential energy. When the force in the stretched rope becomes sufficient to move the load, it will accelerate the mass upward, increasing its velocity and transferring the spring's potential energy to kinetic energy in the mass. If the mass accelerates sufficiently it will take the rope slack again. When the mass falls to take up the slack, the cycle repeats. Thus, as it starts up, the force in the rope can oscillate from zero to some value significantly greater than the steady-state nominal load. To calculate the dynamic loading requires writing and solving the differential equations of motion for the system.

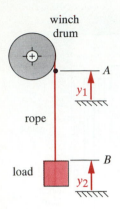

winch
drum

A

y_1

rope

B

load

y_2

(a) Dynamic system

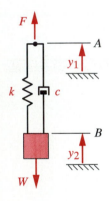

F

A

y_1

k ▨ *c*

B

y_2

W

(b) Lumped model

F_s F_d

W

(c) Free-body diagram

FIGURE 8-5

Dynamic System,
Lumped Model, and
Free-Body Diagram of a
Rope Hoist

4   Figure 8-5*a* shows a simplified schematic of the portion of the dynamic system containing the lift mass and the rope spring. Figure 8-5*b* shows the system modeled as a lumped mass supported by a spring and a damper. Figure 8-5*c* shows a free-body diagram (FBD) of the mass acted upon by its weight W, the spring force F_s, and the damper force F_d. Writing Newton's 2nd law for this FBD gives

$$\sum F = ma$$

$$F_s + F_d - W = \frac{W}{g}\ddot{y}_2 \tag{a}$$

where

$$F_s = k(y_1 - y_2)$$

$$F_d = c(\dot{y}_1 - \dot{y}_2) \tag{b}$$

Substitute the initial conditions:

when $t = 0$ $y_1(0) = 0,$ $\dot{y}_1(0) = v_0,$ $y_2(0) = 0,$ $\dot{y}_2(0) = 0$ (c)

from which $F_s(0) = 0,$ $F_d(0) = 0$ (d)

$$m\ddot{y}_2 = k(y_1 - y_2) + c(\dot{y}_1 - \dot{y}_2) - W$$

$$\ddot{y}_2 = \frac{1}{m}\Big[k(y_1 - y_2) + c(\dot{y}_1 - \dot{y}_2) - W\Big]$$

let $\dot{y}_1 = v, \quad y_1 = vt$

then

$$\ddot{y}_2 = \Big[\frac{k}{m}(vt - y_2) + \frac{c}{m}(v - \dot{y}_2) - g\Big] \tag{e}$$

5   The constants for this equation are defined as follows:

$$v = 24\frac{in}{sec}$$

$$W = 170 \text{ lb,} \qquad m = \frac{W}{g} = \frac{170}{386} = 0.44\ \frac{lb\text{-}sec^2}{in} \tag{f}$$

$$k = 50\,000\ \frac{lb/in}{ft}\Big/24\,ft = 2\,083\ \frac{lb}{in}$$

The critical damping c_c is easily calculated from the known mass and spring constant values. This system is only lightly damped by the rope's internal friction. We will assume that its ratio of actual damping to critical damping, z, is about 10% (0.1) and use this to calculate a damping value for equation (e).

$$c_c = 2m\sqrt{\frac{k}{m}} = 2(0.44)\sqrt{\frac{2\,083}{0.44}} = 61\ \frac{lb\text{-}sec}{in}$$

$$c = \zeta c_c = 0.10(61) = 6.1\ \frac{lb\text{-}sec}{in} \tag{g}$$

$$\frac{c}{m} = \frac{6.1}{0.44} = 14$$

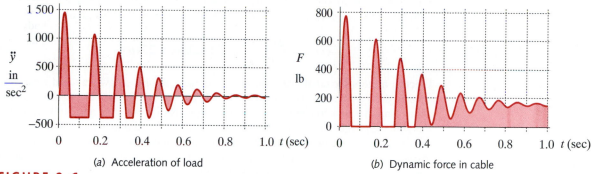

(a) Acceleration of load   (b) Dynamic force in cable

FIGURE 8-6

Acceleration and Cable Force at Startup of Load-Lift

6   Equation (e) was solved with the ASDEQ* simulation package. The acceleration of
the load over the first second of operation is shown in Figure 8-6a. Downward acceleration (gravity) is taken as negative. Note the periods of negative acceleration (at
a limiting value of $-g$) during which the load is in free-fall and the rope is slack with
no tension. The force in the rope over the first second of operation is shown in Figure 8-6b. Note that the tension force rises to over four times the nominal load on the
first oscillation and then drops to zero as the rope goes slack since it cannot support a
compressive force. This pattern repeats for 3 cycles, at which point the damping has
reduced the oscillations to the point that the rope is always in tension. After about 10
cycles, it has settled down to the value of the nominal load.

7   The torque required to drive the drum shaft will depend on the dynamic loads just calculated and on the diameter of drum selected. Too small a diameter will cause high
stresses and wear on the rope. A large drum diameter will increase the required torque
and increase the package size. A 3/4-in-dia rope can wrap around a 20-in-dia sheave.
Since $T = Fr$, the torque required on the shaft will then be 10 times the tension in the
rope (using in-lb) and will have the same time variations as shown in Figure 8-6b.

8   The average power required can be easily found from the change in potential energy
over the time desired. To raise a 170 lb load 24 ft in 12 sec requires

$$P = \frac{170 \text{ lb } (24 \text{ ft})}{12 \text{ sec}} = 340 \frac{\text{ft - lb}}{\text{sec}} = 0.62 \text{ hp} \qquad (h)$$

Since there will be losses in the gear train and winch, we will need an input power
larger than this, say 1 hp for a first trial. It is desirable to keep it at or below this level
since larger horsepower motors will require higher voltage than 110V.

This average power is based on the nominal load. The peak load at startup requires
more power. Rather than size the motor to accommodate the transient start-up load,
another approach is to provide sufficient flywheel in the system to supply the transient
pulse of energy to get it past the start-up phase. It is possible that the rotational inertia of the drum and worm gear will supply enough flywheel effect assuming that the
winch is up to speed before the slack is first taken out of the rope.

* Automatic Solution of
Differential EQuations, by
Maj. Abram Jack and Maj.
James D. Brown, U. S.
Military Academy, West
Point, N.Y. Note that this
package is available from the
U.S.M.A. and is included
with this text.

8   The average drum angular velocity is determined from the required average linear velocity of the rope which, is 2 ft/sec. At a 10-in drum radius this gives

$$\omega = \frac{v}{r} = \frac{24 \text{ in/sec}}{10 \text{ in}} = 2.4 \text{ rad/sec} \cong 23 \text{ rpm} \qquad\qquad (i)$$

9   Electric motors for 60-Hz AC operation are made in only a few standard rotational speeds, the most common of which are 1 725 rpm and 3 450 rpm. These speeds come about from the line-frequency-synchronous speeds of 1 800 rpm and 3 600 rpm minus some slippage in these nonsynchronous motors. To minimize the gear ratio in the wormset, we should choose the slower of the two standard speeds, or 1 725 rpm. This gives a desired gear ratio of 23:1 725 or 1:75. This ratio is obtainable in one stage of a worm-wheel combination, and so is feasible.

10   To summarize the parameters determined from this preliminary design study, we are looking to design a system that has a 1 hp, 1725 rpm, 110-V AC electric motor, driving a 1:75 reduction worm-wheel set, that, in turn, drives a 20-in-dia winch drum at 23 rpm. A 3/4-in rope is capstan-wound around the drum and its forged hook attached to a platform that weighs no more than 50 lb and stably supports two hay bales of up to 60 lb each. These constitute a set of task specifications for our design.

This formerly unstructured problem now has some structure that can be used as a starting point for more detailed design of the various components. Some of the components of this case study will be addressed in the ensuing chapters as the relevant topics such as shafts, gears, bearings, etc., are presented. Note that though the load is relatively steady with time in this device, the oscillations at startup and the repeated cycles of use make this a fatigue design problem, as virtually any machine will be. The parts are subjected to fatigue loading and must be designed accordingly using the theories outlined in Chapters 6 and 7.

8.3   A CAM-TESTING MACHINE

A machine is needed that will allow the dynamic characteristics of cams to be measured. This machine must itself be dynamically quiet, have minimal deflections, and provide a virtually constant but adjustable rotational speed in the face of variations in torque loading from the cams. Instrumentation will be provided to measure the dynamic forces and accelerations of the cam-follower. The mounting of the 1-in rise test cams can be custom designed to fit the test machine. The cam profiles are defined. The rotational speed is to be as high as possible without causing any follower jump. The cams must be easily and quickly replaceable on the machine. The cams will run in an oil bath that must be contained within the machine.

This is also an unstructured problem statement that allows the designer a great deal of latitude in respect to the solution. We will now attempt to further bound the problem with assumptions and preliminary calculations in order to allow a more detailed design to take place.

CASE STUDY 9A

Preliminary Design of a Cam Dynamic Test Fixture (CDTF)

Problem Define a preliminary design concept to satisfy the general constraints of the problem listed above. Determine the force-time function acting on the follower and the torque-time function on the camshaft during any one cycle. Define the drive ratio and the power and torque requirements for the motor.

Given The four-dwell cams have a minimum diameter of 6 in and a maximum diameter of 8 in. The rise is 1 in. The roller follower is 2-in dia. The cams are run at 180 rpm. The cam shape is shown in Figure 8-7.

Assumptions Plain bearings must be used throughout since rolling-element bearings introduce too much noise. A speed-controlled DC motor will be used.

Solution See Figures 8-7 to 8-13.

1   A preliminary design is shown in Figure 8-8. The camshaft is tapered to receive a matching taper in the cam. This avoids the use of keyways, which can introduce vibration and noise on torque reversal. The cam will be axially clamped to the shaft for concentric location. A dowel pin at a large radius keys the cam to the hub to establish a zero position. This arrangement allows quick removal and installation when changing cams.

2   A flywheel is attached to the camshaft to provide modulation of the speed oscillations during torque variations. The flywheel also serves as a sheave for a flat belt from the motor's smaller driving sheave in order to reduce the camshaft speed appropriately.

3   The follower arm is pivoted 12 in from the camshaft and carries a commercial roller follower running in a plain bearing. A helical-coil tension spring loads the follower-arm roller against the cam. This spring must be de-tensioned and removed to replace a cam and then re-tensioned for the new cam. The cover, which is pivoted at the follower-arm pivot, applies tension to the spring when closed and releases it when opened.

4   Accelerometers and force transducers are fitted on and between the roller pivot and follower arm to measure the desired parameters.

5   The whole is mounted on a box-structure base that supports it, provides rigidity, and also contains an openable oil chamber around the cam. The base can be supported either on casters for mobility, or on jacking legs for stability. The motor is mounted to the base on rubber vibration isolators. The electronics for motor control and instrumentation are contained within the base.

6   The design of the cam itself along with its rotational velocity determines the magnitudes and shapes of the acceleration of the follower arm. This follower-acceleration function, generated from program DYNACAM, [2] is shown in Figure 8-9. The acceleration function multiplied by the effective mass of the follower is one component of the dynamic force needed for stress calculations. The dynamic system of cam and

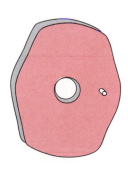

FIGURE 8-7

Four-Dwell Cam

8

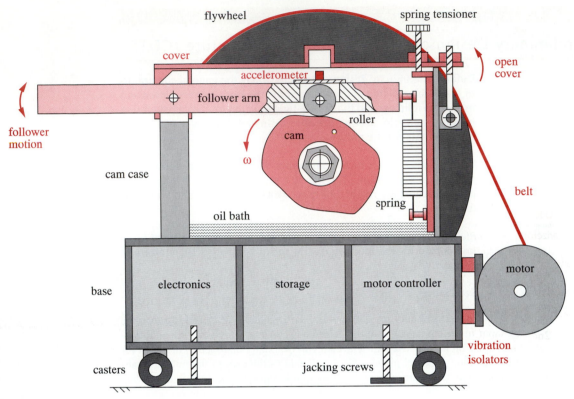

FIGURE 8-8

Cam Dynamic Test Fixture—General Design Scheme

follower can be modeled as a linearized, lumped parameter, single-degree-of-freedom system as shown in Figure 8-10. The motion of the roller centerline on the rotating

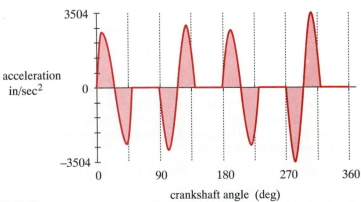

FIGURE 8-9

Acceleration Function for Cam Over 1 Cycle

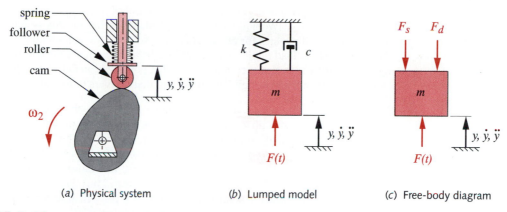

(a) Physical system   (b) Lumped model   (c) Free-body diagram

FIGURE 8-10

Linearized Cam-Follower System (a), Lumped Parameter Model (b), and Free-Body Diagram (c)

follower arm is actually along an arc, but that arc is quite flat in this design due to the length of the arm's radius. The error in assuming the roller's motion to be linear over its short excursion is minimal. A portion of the moving follower mass is then considered to be lumped at the roller so as to be dynamically equivalent. From Newton's 2nd law, the differential equation for the lumped system in Figure 8-10 is [3]

$$\sum F = ma$$

$$F(t) - F_d - F_s = m\ddot{y}$$

where $F_d = c\dot{y}, \quad F_s = ky + F_{pl}$

and

$$F(t) = m\ddot{y} + c\dot{y} + \left(ky + F_{pl}\right) \qquad (a)$$

The spring force F_s has two components. The spring constant k times the deflection y is added to whatever initial force F_{pl} was created by preloading the spring at installation. The damping force F_d is proportional to velocity by a damping coefficient c.

7   Equation (a) can be solved kinetostatically in this case since we are maintaining a constant angular velocity and the displacement (y), velocity $(\dot{y})$, and acceleration $(\ddot{y})$ are defined functions of time. The value of m will depend on our design of the follower arm and anything attached to it as moving mass such as the roller. The value of the damping factor c is sometimes difficult to predict and the usual way to estimate its value is to define an expected damping ratio ζ for the type of system and calculate damping from

$$c = 2\zeta m \omega_n$$

where

$$\omega_n = \sqrt{\frac{k}{m}} \qquad (b)$$

Koster [4] found that a typical value of ζ for cam follower systems is about 0.06. The term ω_n is the undamped natural frequency of the system.

The value of the spring constant k is under the control of the designer as is the amount of spring preload F_{pl}. We will later design a suitable spring for this system that will provide appropriate values for these variables. Note that we cannot numerically calculate the dynamic load on the system until we have preliminary designs of the moving parts in order to define their masses. At that point we can define a desired spring k and F_{pl} and then try to design a viable spring that delivers those values.

8   For a given mass, damping ratio, and acceleration function, the choice of k and F_{pl} determine whether the follower will jump off the cam during the fall. Figure 8-11a shows the effect on the dynamic force $F(t)$ of a too-small combination of k and F_{pl}. The shaded areas highlight portions where the dynamic force is negative. A cam-follower joint cannot deliver a negative (tension) force any more than the rope of the previous case study could support a compressive force. Thus the spring constant and preload must be increased in some combination until the dynamic-force function remains positive throughout the cycle, as shown in Figure 8-11b.

9   For this case study we will define the follower-arm geometry as shown in Figure 8-12. It is a solid 2 in x 2.5 in rectangular cross section of aluminum, relieved internally around the follower for clearance. The distance from pivot to roller follower is 12 in with a 10 in extension beyond the pivot for balance. It extends 6 in beyond the roller follower to attach the spring. The effective mass of the follower arm reflected to the roller centerline, plus the mass of roller and its pivot is 0.02 lb-sec²/in. Using this value of effective mass and applying a spring constant at the end of the arm of $k = 25$ lb/in and a spring preload $F_{pl} = 25$ lb, (which translates to effective values of $k = 56.25$ and $F_{pl} = 37.5$ lb at the follower) to achieve the dynamic force function shown in Figure 8-11b. The peak dynamic force is 110 lb and the minimum force is 13 lb **at the cam follower**. The deflection at the spring is 1.5 in.

10   The camshaft torque can be found from [5]

$$T(t) = \frac{F(t)\dot{y}}{\omega} \tag{c}$$

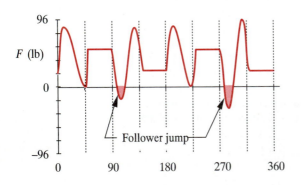

(a) Insufficient spring force allows follower jump

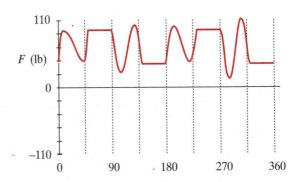

(b) Sufficient spring force keeps the dynamic force positive

FIGURE 8-11

Dynamic Force Between Cam and Cam Follower

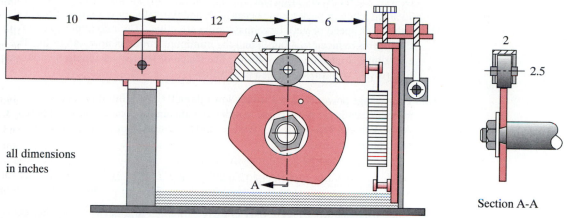

FIGURE 8-12

Dimensions of Follower Arm

This torque function for the values assumed above is shown in Figure 8-13. The maximum torque is 176 in-lb and the minimum torque is −204 in-lb. The average torque is 7 lb-in.

11   The flywheel is 24-in-dia and 1.88-in-thick solid steel. Its mass moment of inertia is $I = 44$ blob-in^2. The coefficient of fluctuation C_f for this flywheel is found by integrating the torque-time function of Figure 8-13, pulse by pulse, to find the maximum energy oscillation E over one cycle. That integration was done numerically with program DYNACAM [2] and gives $E = 3\,980$ in-lb of energy over one cycle. The coefficient of fluctuation is then [6]

$$C_f = \frac{E}{I\,\omega^2} = \frac{3\,980}{44(18.85)^2} = 0.25 \qquad\qquad (d)$$

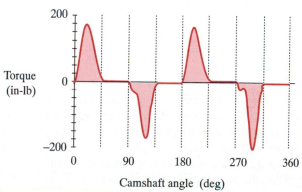

FIGURE 8-13

Camshaft Torque Without Flywheel

Despite the relatively large size and weight (227 lb) of this flywheel, it gives only a 75% reduction in peak torque because its angular velocity is so low. Flywheels need either high speed or very large mass to be effective. The maximum torque with this flywheel is reduced to 47 in-lb and the minimum torque is now –48 in-lb. The average torque is unchanged at 7 lb-in. The shape of the torque function is the same as Figure 8-13 with these reduced peak values.

12   The average power required is very low (about 0.02 hp), but it is necessary to size the motor to handle the peak torque in order to maintain the constant speed required. Using the peak torque value with the flywheel and the camshaft operating speed gives a minimum power level of

$$P_{peak} = T_{peak}\,\omega = 47 \text{ lb - in}\left(18.9\frac{\text{rad}}{\text{sec}}\right) = 888\frac{\text{in - lb}}{\text{sec}} = 74\frac{\text{ft - lb}}{\text{sec}} = 0.14 \text{ hp} \qquad (e)$$

Since frictional losses were only crudely estimated in this case and other cams will operate at higher speeds, a 1/2-hp speed-controlled DC motor was selected to drive the camshaft. This uses 110-V AC power to the motor's rectifier/speed controller.

13   We need to select a speed ratio for the belt drive from motor to flywheel. Since the motor is speed-controlled over an operating range of 0-1 800 rpm, we can afford to allow a broader range of operating speeds than dictated by this particular cam, which needs 180 rpm. A range of 0-400 rpm is reasonable since that will put the motor in the middle of its speed range for this cam and allow other cams to be run either faster or slower. The drive pulley diameter is then

$$d_{in} = d_{out}\frac{\omega_{out}}{\omega_{in}} = 24\frac{400}{1\,800} = 5.33 \text{ in} \qquad (f)$$

Many more details remain to be worked out to create a finished design, but these preliminary calculations seem to indicate that the proposed design is feasible. For more detailed information on the modeling of the cam dynamics in this case study, see Chapters 9 and 17 of reference 1. Later chapters of this text will continue the study of various aspects of this case as they relate to subsequent topics such as bearing and spring design.

8.4   SUMMARY

This chapter has presented preliminary design calculations for some case studies of relatively simple machines. The intent is to incorporate further study of these cases in subsequent chapters that will deal with the design of elements common to a wide variety of machinery. Space will not permit a complete treatment of all the design details involved in any one of these case studies but it is hoped that their presentation will provide some insight into the way design must integrate a wide variety of often conflicting requirements to obtain a working product.

A number of open-ended design projects are also suggested at the end of this chapter. These can be used as term-long project assignments for individual or group effort. Alternatively, subsets of the suggested design projects can serve as multi-week design assignments.

8.5   REFERENCES

1   **R. L. Norton**, *Design of Machinery*. McGraw-Hill: New York, pp. 529-550, 1992.

2   *Ibid.*, pp. 382-389.

3   *Ibid.*, p. 650.

4   **M. P. Koster**, *Vibrations of Cam Mechanisms*. Macmillan: London, 1974.

5   **R. L. Norton**, *Design of Machinery*. McGraw-Hill: New York, p. 656, 1992.

6   *Ibid.*, p. 489.

8.6   DESIGN PROJECTS

These large-scale problems are deliberately unstructured and are typical of real engineering design problems. In fact, most of them are real problems. They then have many valid solutions. While some of these project problems were "invented" for this chapter, most come either from the author's consulting experience or from senior projects assigned to and completed by his students at Worcester Polytechnic Institute. In the latter cases, the projects were typically done by a team of 2-4 students over a three- or four-term period (21-28 weeks) and often resulted in a working prototype of the solution. The versions of such senior projects stated here have been simplified or truncated with the intention that they be tractable for solution by a team of students over a one-term course. The consulting problems have also been abridged to suit the structure and time available in a typical junior or senior design course. Some of the projects listed here have been used successfully by the author as term-long projects in the course for which this text is intended.

8-1   Complete the design of the portable air compressor in Case Study 7A. Note that some parts of this design are addressed in later chapters.

8-2   Complete the design of the winch lift in Case Study 8A. Note that some parts of this design are addressed in later chapters.

8-3   Complete the design of the cam-testing machine in Case Study 9A. Note that some parts of this design are addressed in later chapters.

8-4   Case Study 5-A and 5-B in Chapter 3 describe the design of a fourbar linkage demonstrator machine. Complete its detailed design based on the information given in those case studies. Some aspects not addressed therein are the size of the gear train to reduce the motor speed, the bearings, torque coupling, flywheel, and stresses.

8-5   Design a safer log splitter with the following characteristics:
- Able to be towed at highway speeds behind a full-size pickup truck.
- An 8-hp gasoline engine drives a two-stage hydraulic pump that in turn pressurizes a hydraulic cylinder to split the log.
- Accommodates a 2-ft-long log.
- Generates 15 tons of force on the log against a stationary splitting wedge.
- Has a safety cage that covers the log/wedge/cylinder area during splitting to prevent injury to the operator. This cage slides (manually) out of the way to load/unload logs and is interlocked such that it must be in place before the hydraulic cylinder will move.

8-6   Design an inspection capsule to be lowered into an oil well to a depth of 5 000 ft. The capsule must fit in a 6-in-dia hole with at least 10% diametral clearance, have a suitable attachment for the lowering cable, and a 1.5-in-dia quartz lens port in the sidewall. A 0.5-in-dia power and communications cable passes through the top end of the cylinder. Some design concerns are the hydrostatic pressure at that depth, the abrasive nature of the rock walls of the well hole, and the hydraulic integrity of the seals around the lens port and power cable. The capsule has a dry nitrogen atmosphere inside at 800 psig. Design for a finite life of at least 1E4 insertions/removals from a well.

8-7   Design a battery-powered, motorized shopping cart capable of carrying a 200-lb person plus 50 lb of groceries around the aisles of a supermarket. It should hold at least half the volume of foodstuffs of a conventional, manual shopping cart, be speed limited, safe against tipover, and require constant pressure on its control to run (i.e., a "dead-man" switch). When the power is cut, an automatic brake should stop it within 1 ft. Intended users are elderly or infirm shoppers. It should run for 1 hour between recharges.

8-7   Figure P8-1 shows the geometry and dimensions of a popular off-road motorcycle rear suspension system. The wheel is carried at the end of link 4, which is part of the fourbar linkage 1-2-3-4, where 1 is the frame of the cycle, 2 is the triangular rocker, and 3 is a binary coupler connecting 2 to 4. The shock strut 5 is pivoted to link 2 and slides into the shock cylinder 6. The shock cylinder 6 is pivoted to the frame 1. The total vertical travel of the rear axle is about 12 in.

Figure P8-2 shows the result of a dynamic simulation of a 250-lb cycle, with a 200-lb rider, traveling at 18 mph, jumping 3 ft vertically and landing on the rear wheel.[*] The graphs in Figure P8-2 show the resulting dynamic force at the rear axle and at the pivot of link 4. Design the rear suspension system based on the given loading and geometry data. Some design concerns are the shock strut as a column, the pivot pins in shear, and the links in bending plus tension or compression in some instances. It will be of value to inspect a similar motorcycle suspension system to obtain additional information on its general design.

8-8   Off-road motorcycles typically have chain and sprocket drives from the transmission output shaft to the rear wheel. Some road bikes use enclosed transmission shafts and gear drives instead of a chain and sprockets. The advantage of chain drives is light weight but exposure to the dust and mud of off-road riding reduces their reliability. Enclosed shaft drives are protected from the elements. Design a

---

[*] This simulation was done using the software package *Working Model* by Knowledge Revolution.

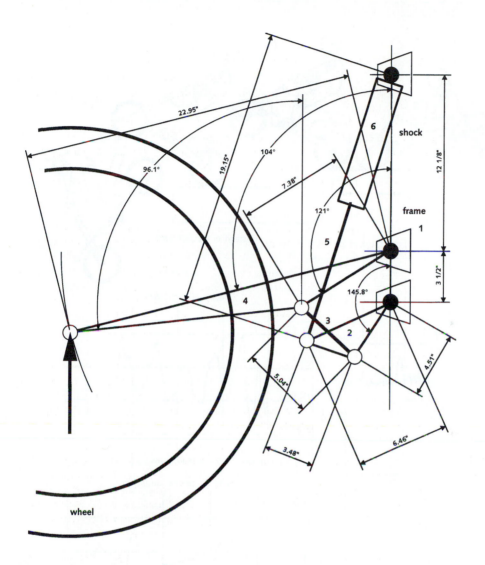

FIGURE P8-1

Geometry of an Off-Road Motorcycle Rear-Wheel Suspension System

lightweight shaft-drive system for the off-road motorcycle of Figures P8-1 and P8-2. Assume an engine of 60 hp at 9 000 rpm. Low gear in the transmission has a ratio of 1:4 and the final drive ratio from transmission output shaft to rear wheel should be approximately 1:3.5. At least one universal joint will be needed in the driveshaft to accommodate the suspension motion. Some combination of spur (or helical) and bevel gears will be needed. Suitable bearings, housings, and seals should be specified. It will be of value to inspect a similar motorcycle shaft-drive system to obtain additional information on its general design.

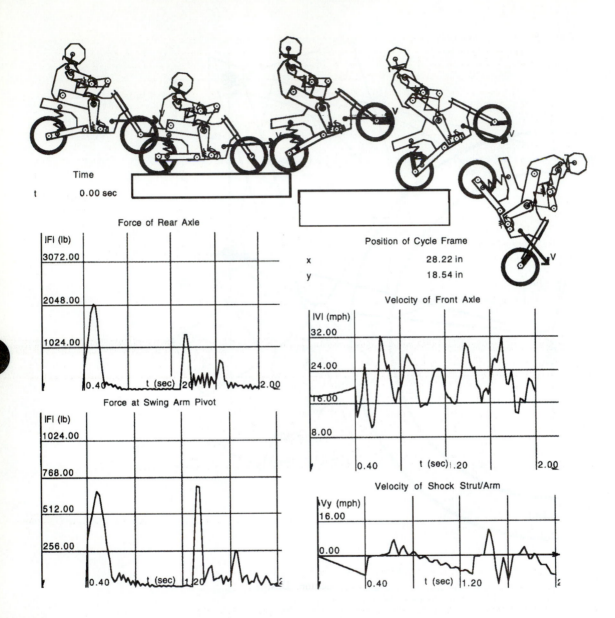

FIGURE P8-2

Simulation of Forces Generated by a Motorcycle Jumping and Landing on its Rear Wheel

8-9   The Army wants a machine to test army boots for durability. This machine should mimic, as closely as possible, the geometry and forces of a typical soldier walking in combat boots as shown in Figure P8-3. It should repeat this motion for an unlimited

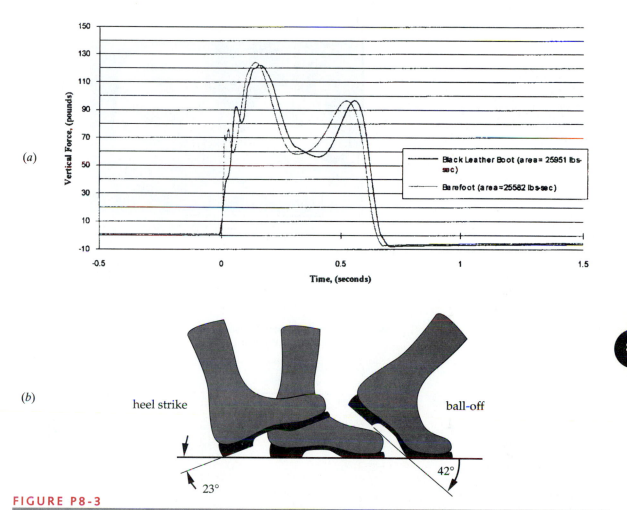

FIGURE P8-3

Typical Walking Forces (a) and Geometry (b)

number of cycles until the boot leather breaks down. The single boot will be fitted to a prosthetic foot that is attached to the test machine. Design it for infinite life.

8-11   A company that manufactures molded urethane jounce bumpers as shown in Figure P8-4, wants a test machine that will repeatedly impact these bumpers to determine their durability. The static force-deflection characteristic of one size bumper is shown in Figure P8-5. The suggested concept is a drop weight that will impact the bumper. A resetting mechanism to lift and repeatedly drop the weight is required. A dynamic analysis should be done to determine the size and height of weight needed to fully compress the specified bumper.

8-12   The crimping tool shown in Figure P8-6 is used to crimp (yield) metal connectors onto wire. The wire is inserted into the connector (not shown), the pair then put into the crimping tool's jaws and the handles squeezed. Case Study 2 analyzed the

FIGURE P8-4

Two Sizes of Urethane Jounce Bumpers for Use in Automotive Suspension Systems

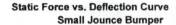

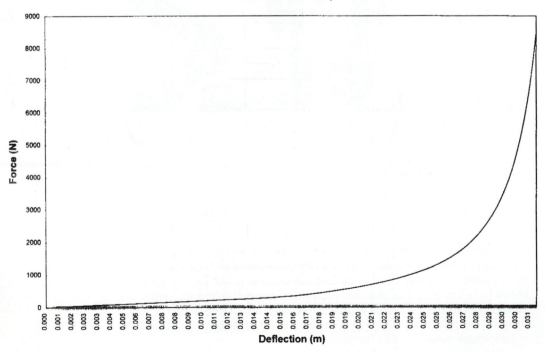

FIGURE P8-5

Static Force-Deflection Curve for One Size of Urethane Jounce Bumper

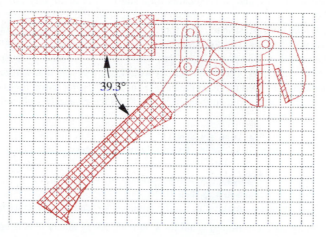

1-cm grid

FIGURE P8-6

Crimping Tool Geometry

Angle Between Handles, Deg.	Torque, lb$_f$-in	Distance Between Handles, in	Force, lb$_f$
39.3	0	2.89	0
24.4	49.17	1.89	11.84
22.3	73.27	1.74	17.36
20.9	97.53	1.63	22.89
19.7	122.01	1.54	28.42
18.7	146.62	1.46	33.95
17.8	171.38	1.40	39.47
16.5	220.91	1.30	50.52
15.1	271.10	1.19	61.57
13.7	321.77	1.08	72.63
12.6	372.41	1.00	83.68
11.2	423.77	0.89	94.73
9.5	475.78	0.75	105.79
8.0	502.66	0.64	111.32
7.6	528.12	0.60	116.84
5.5	580.52	0.43	127.89
4.8	606.27	0.39	133.42
0.8	653.34	0.06	143.29

FIGURE P8-7

Crimping Tool Force and Torque-Displacement data

forces, stresses, and safety factors for a similar tool of slightly different design. Figure P8-7 shows a table of data measured from the tool in Figure P8-6 while crimping the largest wire. The forces required at the handle are quite large. The distances between handles are measured at a 4.56-in radius. Repeated manual use of such a tool at these force levels frequently leads to physical problems such as carpal-tunnel syndrome. The manufacturer would like a device that will mechanically create the required force on the handles of this existing crimp tool to eliminate manual use and avoid such injuries. Some of the constraints placed on this device are as follows:

(a) No larger than 12 in x 4 in x 6 in.
(b) Self-contained, including a portable energy supply.
(c) Weigh less than 10 lb.
(d) Require only one or two hand operation by one person.
(e) Perform a crimp in less than 10 sec.
(f) Accommodate maximum handle angle of 40° and handle splay of 7 in.

8-13 A motorized, roll-around stand for a portable X-ray machine is needed as shown in Figure 8-8. The X-ray head weighs 65 lb and is 10 in x 8 in x 16 in high. The head must be motor-adjusted over a vertical range from 41 in to 82 in measured from the floor to the 18 in long cantilevered arm from which the X-ray head is suspended. The head must traverse its maximum vertical travel in 20 sec or less and stop within 0.5 in of a desired location. Limit switches must automatically shut off the motor at each end of the head's vertical travel. The stand must be able to pass through a standard 3-ft x 7-ft interior hospital door and operate from a 15-A, 110-V AC circuit.. Certain Underwriters Laboratories (UL) specifications also must be met such as:

- UL 27.4-A — must not tip on a 10° incline with the head in the lowest position.
- UL 27.4-B — must not tip on a 5° incline with the head in the highest position.
- UL 27.4-C — must not tip on a 0° incline when a horizontal force of 25% of its weight is applied at the transport handle.

8-14 Design a dumping-bed attachment to retrofit an existing full-size pickup truck. This device should require minimal modification of the truck and be capable of remote activation from the cab. It should lift and dump up to 3/4 ton of cargo. It will take its power from the truck engine and can be electrically, mechanically, or hydraulically driven, or any combination thereof.

8-15 Design a wheelchair lift device to operate in a residential garage. The garage floor is 1 m below the main floor of the house. The device should safely raise or lower the wheelchair and its 100-kg occupant through that vertical distance.

8-16 Design a device to transfer a 100-kg paraplegic patient safely from bed to wheelchair and vice versa. The patient has good upper body strength but no control of the lower extremities. Your design should be operable by the patient with minimal assistance. Safety is of paramount concern.

8-17 Design an indoor bicycle exerciser similar in concept to the wheelchair exerciser shown in Problem 6-48. The concept is to provide twin rollers to support the rear wheel and a single roller for the front wheel. The rear rollers will be attached in some kinematic fashion (to be designed) to a DC generator whose output is shunted through an electrical load that can be varied by the rider to provide a dynamometric resistance. Design all parts with suitable geometries and materials for infinite life.

8

FIGURE P8-8

Motorized, Portable, Roll-Around X-ray Stand

8-18   Design a service stand for an off-road motorcycle that will allow the bike to be
suspended at a convenient working height and rotated to allow access to all
serviceable systems. Stability in all rotated positions, and with various states of
partial disassembly, is of paramount importance.

8-19   Fluorescent light bulbs must be coated inside with tin oxide while hot. The 46-in-
long, straight glass tubes are carried through a baking oven on a metal chain
conveyor that travels at a constant velocity of 5 500 bulbs per hour. The bulbs are
spaced 2 in apart on the chain. As they exit the oven at 550° C, a mechanism
carrying two spray heads chases a pair of bulbs, accelerates to match their constant
velocity, travels with them for a short distance, and sprays the tin oxide into the hot
bulbs within 3/4 sec. The two spray heads mount on a 6 in x 10 in rectangular table
that is carried on linear bearings. The spray equipment bolted to the table weighs 10
lb. All elements exposed to the tin-oxide spray must be stainless steel to resist
chemical attack from the hydrochloric acid by-product in the spray.

A cam drives the table to match the conveyor velocity and returns it in time to accelerate and catch the next two bulbs. A plate cam has been designed to accomplish this action and it is defined in the file SPRAY.CAM supplied on disk with this text. This diskfile can be input to the program DYNACAM (also supplied on disk) to obtain the necessary dynamic data for the required design. The cam is driven from the conveyor sprocket that has an 8.931-in pitch dia.

What is required is a detailed design of the spray table, its bearings and mounting hardware for infinite life. The dynamic loads on the cam and follower will be highly dependent on the mass of this moving assembly. Once an estimate of the moving mass is available from your preliminary design, program DYNACAM can be used to quickly calculate the dynamic forces at the cam-follower interface. The stresses in the various parts of the assembly can then be estimated based on the level of dynamic forces present. See reference 2 for information on using program DYNACAM.

8

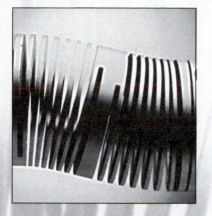

9

SHAFTS, KEYS, AND COUPLINGS

The greater our knowledge increases, the greater our ignorance unfolds.

JOHN F. KENNEDY

9.0 INTRODUCTION

Transmission shafts, or just shafts, are used in virtually every piece of rotating machinery to transmit rotary motion and torque from one location to another. Thus, the machine designer is often faced with the task of designing shafts. This chapter will explore some of the common problems encountered in that endeavor. Table 9-0 shows the variables used in this chapter and notes the equations or sections in which they occur.

At a minimum, a shaft typically transmits torque from the driving device (motor, or engine) through the machine. Sometimes shafts will carry gears, sheaves (pulleys), or sprockets, which transmit the rotary motion via mating gears, belts, or chains from shaft to shaft. The shaft may be an integral part of the driver such as a motor shaft or engine crankshaft, or it may be a freestanding shaft connected to its neighbor by a coupling of some design. Automated production machinery often has line shafts that extend the length of the machine (as much as 100 ft) and carry the power to all the workstations. Shafts are carried in bearings, in a simply-supported (straddle-mounted) configuration, cantilevered, or overhung, depending on the machine configuration. The pros and cons of these mounting and coupling arrangements will also be discussed.

9.1 SHAFT LOADS

The loading on rotating transmission shafts is principally one of two types: torsion due to the transmitted torque or bending from transverse loads at gears, sheaves, and sprock

Table 9-0   Variables Used in this Chapter

Symbol	Variable	ips units	SI units	See
A	area	in^2	m^2	various
c	distance to outer fiber	in	m	Sect. 9.7
C_f	coefficient of fluctuation	none	none	Eq. 9.19
d	diameter	in	m	various
e	eccentricity of a disk	in	m	Eq. 9.26
E	Young's modulus	psi	Pa	various
E_k, E_p	kinetic energy, potential energy	in-lb	Joule	Eq. 9.25
F	force or load	lb	N	various
F_l	fluctuation (in speed)	rad/sec	rad/sec	Eq. 9.19
f_n	natural frequency	Hz	Hz	Eq. 9.24
g	gravitational acceleration	in/sec^2	m/sec^2	Sect. 9.12
G	shear modulus, modulus of rigidity	psi	Pa	various
I, J	2nd moment, polar 2nd moment of area	in^4	m^4	Sects. 9.7, 9.12
I_m, I_s	mass moment of inertia about axis	$lb\text{-}in\text{-}sec^2$	$N\text{-}m\text{-}sec^2$	Sect. 9.11
k	spring rate or spring constant	lb/in	N/m	Sect. 9.12
K_f, K_{fm}	fatigue stress concentration factors	none	none	Sects. 9.7, 9.12
K_t, K_{ts}	geometric stress concentration factors	none	none	Sect. 9.7
l	length	in	m	various
m	mass	$lb\text{-}sec^2/in$	kg	Sect. 9.12
M	moment, moment function	lb-in	N-m	various
N_f	safety factor in fatigue	none	none	Eq. 9.6-9.8
N_y	safety factor in yielding	none	none	Ex. 9-7
P	power	hp	watts	Eq. 9.1
p	pressure	psi	N/m^2	Sect. 9.11
r	radius	in	m	various
S_e, S_f	corrected endurance limit, fatigue strength	psi	Pa	Eqs. 9.6-9.8
S_{ut}, S_y	ultimate tensile strength, yield strength	psi	Pa	Eqs. 9.6-9.8
T	torque	lb-in	N-m	Eq. 9.1
W	weight	lb	N	various
α	angular acceleration	rad/sec^2	rad/sec^2	Eq. 9.18
δ	deflection	in	m	various
ν	Poisson's ratio	none	none	various
θ	angular deflection or beam slope	rad	rad	various
γ	weight density	lb/in^3	N/m^3	Eq. 9.23
σ	normal stress (with various subscripts)	psi	Pa	various
σ'	von Mises stress (with various subscripts)	psi	Pa	various
τ	shear stress (with various subscripts)	psi	Pa	various
ω	angular velocity	rad/sec	rad/sec	Eq. 9.1
ω_n	natural frequency	rad/sec	rad/sec	Sect. 9.12
ζ	damping ratio	none	none	Sect. 9.12

9

Title page photograph courtesy of Helical Products Co. Inc., Santa Maria, Calif. 93456.

ets. These loads often occur in combination since, for example, the transmitted torque may be associated with forces at the teeth of gears or sprockets attached to the shafts. The character of both the torque and bending loads may either be steady (constant) or may vary with time. Steady and time-varying torque and bending loads can also occur in any combination on the same shaft.

If the shaft is stationary (nonrotating) and the sheaves or gears rotate with respect to it (on bearings), then it becomes a statically loaded member as long as the applied loads are steady with time. However, such a nonrotating shaft is not a transmission shaft since it is not transmitting any torque. It is merely an axle, or round beam, and can be designed as such. This chapter is concerned with rotating, transmission shafts and their design for fatigue loading.

Note that a rotating shaft subjected to a steady, transverse-bending load will experience a fully reversed stress state as shown in Figure 9-1a. Any one stress element on the shaft surface goes from tension to compression each cycle as the shaft turns. Thus, even for steady bending loads, a rotating shaft must be designed against fatigue failure. If either or both the torque and transverse loads vary with time, the fatigue loading becomes more complex, but the fatigue design principles remain the same, as outlined in Chapter 6. The torque, for example, could be repeated or fluctuating as shown in Figures 9-1b and c, as could the bending loads.

We will deal primarily with the general case, which allows for the possibility of both steady and time-varying components in both bending and torsion loads. If either load lacks a steady or time-varying component in a given case, it will merely force a term in the general equations to zero and simplify the calculation.

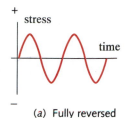

(a) Fully reversed

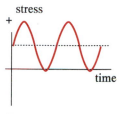

(b) Repeated

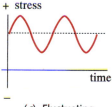

(c) Fluctuating

FIGURE 9-1

Time-Varying Stresses

9.2   ATTACHMENTS AND STRESS CONCENTRATIONS

While it is sometimes possible to design useful transmission shafts that have no changes in section diameter over their length, it is most common for shafts to have a number of steps or shoulders where the diameter changes to accommodate attached elements such as bearings, sprockets, gears, etc., as shown in Figure 9-2, which also shows a collection of features commonly used to attach or locate elements on a shaft. Steps or shoulders are necessary to provide accurate and consistent axial location of the attached elements as well as to create the proper diameter to fit standard parts such as bearings.

Keys, snap rings, or cross-pins are often used to secure attached elements to the shaft in order to transmit the required torque or to capture the part axially. Keys require a groove in both shaft and part and may need a setscrew to prevent axial motion. Snap rings groove the shaft, and cross-pins create a hole through the shaft. Each of these changes in contour will contribute some stress concentration and this must be accounted for in the fatigue-stress calculations for the shaft. Use generous radii where possible, and techniques such as those shown in Figures 4-37, 4-38, and 9-2 (at the sheave and snap ring) to reduce the effects of these stress concentrations.

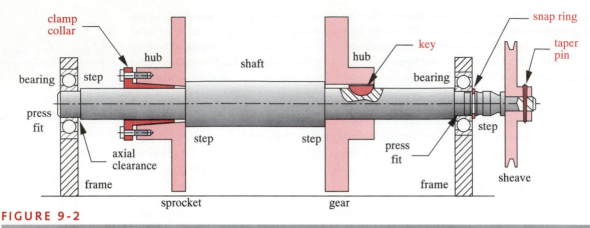

FIGURE 9-2

Various Methods to Attach Elements to Shafts

Keys and pins can be avoided by using friction to attach elements (gears, sprockets) to a shaft. Many designs of **clamp collars** (keyless fits*) are available, which squeeze the outside diameter (OD) of the shaft with high compressive force to clamp something to it as shown on the sprocket hub in Figure 9-2 and in Figure 9-34. The hub has a gently tapered bore and a matching taper on this type of clamp collar is forced into the space between hub and shaft by tightening the bolts. Axial slits in the tapered portion of the collar allow it to change diameter and squeeze the shaft, creating sufficient friction to transmit the torque. Another type of clamp collar, called a **split collar** uses a screw to close a radial slit and clamp the collar to the shaft. **Press and shrink fits** are also used for this purpose and will be discussed in a later section of this chapter. However, as we will see, these friction couplings also create stress concentrations in the shaft and can cause fretting corrosion as described in Section 7-6.

A standard taper pin is sometimes used to couple elements to shafts as seen in the sheave of Figure 9-2. The hole is reamed to match the standardized pin-taper and the purchased pin is driven into place. The shallow taper locks it by friction. It must be driven out for disassembly. This technique should be used with caution in locations of large bending moment as it weakens the shaft as well as creating stress concentration.

Rolling-element bearings as shown in Figure 9-2 are intended to have their inner and outer races be press-fitted to shaft and housing, respectively. This requires close-tolerance machining of the shaft diameter and requires a step shoulder to provide a stop for the press fit and for axial location. Thus, one must start with a larger stock shaft diameter than the bearing inside diameter (ID) and machine the shaft to fit the selected bearing whose sizes are standardized (and are metric). A snap ring is sometimes used to guarantee no axial movement of the shaft versus the bearing as shown at the sheave end of the shaft in Figure 9-2. Snap rings are commercially available in a variety of styles and require a small, close-tolerance groove of specific dimension be machined in the shaft. Note in Figure 9-2 how the axial location of the shaft is achieved by capturing only one of the bearings (the right one) axially. The other bearing at the left-hand

* See the ANSI/AGMA
Standard 9003-A91, *Flexible
Couplings—Keyless Fits*

end has axial clearance between it and the step. This is to prevent axial stresses being generated by thermal expansion of the shaft between the two bearings.

So, it appears that we cannot escape the problems of stress concentration in practical machinery. In the case of shafts, we need to provide steps, snap rings, or other means to accurately locate components axially on the shaft, and we have to key, pin, or squeeze the shaft to transmit the torque.

Each of these methods of attachment has its own advantages and disadvantages. A key is simple to install and sizes are standardized to the shaft diameter. It provides accurate phasing[*] and is easily disassembled and repaired. It may have no resistance to axial movement and it does not always provide a truly tight torque coupling due to the slight clearance between key and keyway. Torque reversals can cause slight backlash.

A taper pin creates a truly tight torque coupling and locates axially as well as radially with phasing but weakens the shaft. It can be disassembled with slightly more difficulty than a key. A clamp collar is easy to install but has no repeatable phasing. This is only a disadvantage if timing of the shaft rotation to other shafts in the system is required. It allows easy (though inaccurate) adjustment of phasing if desired. Press fits are semipermanent connections that require special equipment to disassemble. They do not provide repeatable phasing.

9.3   SHAFT MATERIALS

In order to minimize deflections, steel is the logical choice for a shaft material because of its high modulus of elasticity, though cast or nodular iron is sometimes also used, especially if gears or other attachments are integrally cast with the shaft. Bronze or stainless steel is sometimes used for marine or other corrosive environments. Where the shaft also serves as the journal, running against a sleeve bearing, hardness can become an issue. Through- or case-hardened steel may be the material of choice for the shaft in these cases. See Chapter 10 for a discussion of desired relative hardness and material combinations for shafts and bearings. Rolling element bearings do not need hardened shafts.

Most machine shafts are made from low- to medium-carbon steel, either cold-rolled or hot-rolled, though alloy steels are also used where their higher strengths are needed. Cold-rolled steel is more often used for smaller diameter shafts (< about 3-in dia) and hot-rolled used for larger sizes. The same alloy when cold-rolled has higher mechanical properties than if hot-rolled due to the cold-working, but this comes at the cost of residual tensile stresses in the surface. Machining for keyways, grooves, or steps relieves these residual stresses locally and can cause warping. Hot-rolled bars must be machined all over to remove the carburized outer layer, whereas portions of a cold-rolled surface can be left as-rolled except where machining to size is needed for bearings, etc. Pre-hardened (30HRC) and ground precision (straight) steel shafting can be purchased in small sizes and can be machined with carbide tools. Full-hard, ground, precision shafting (60HRC) is also available but cannot be machined.

* Phasing means the relative angular locations of the various elements attached to the shaft.

9.4   SHAFT POWER

The power transmitted through a shaft can be found from first principles. In any rotating system, instantaneous power is the product of torque and angular velocity,

$$P = T\omega \qquad (9.1a)$$

where ω must be expressed in radians per unit time. Whatever the base units used for calculations, power is usually converted to units of horsepower (hp) in any English system or to kilowatts (kW) in any metric system. (See Table 1-2 for conversion factors.) Both torque and angular velocity can be time-varying, though much of rotating machinery is designed to operate at constant or near-constant speeds for large blocks of time. In such cases, the torque will often vary with time. The average power is found from

$$P_{avg} = T_{avg}\omega_{avg} \qquad (9.1b)$$

9.5   SHAFT LOADS

The most general shaft-loading case is that of a fluctuating torque and a fluctuating moment in combination. There can be axial loads as well, if the shaft axis is vertical or if fitted with helical or worm gears having an axial force component. (A shaft should be designed to minimize the portion of its length subjected to axial loads by taking them to ground through suitable thrust bearings as close to the source of the load as possible.) Both torque and moment can be time-varying, as shown in Figure 9-1, and can have both mean and alternating components.

The combination of a bending moment and a torque on a rotating shaft creates multiaxial stresses. The issues discussed in Section 6.12 on multiaxial stresses in fatigue are then germane. If the loadings are asynchronous, random, or misphased, then it is will be a *complex multiaxial stress* case. But, even if the moment and torque are in-phase (or 180° out-of-phase), it may still be a complex multiaxial stress case. The critical factor in determining whether it has simple or complex multiaxial stresses is the direction of the principal alternating stress on a given shaft element. If its direction is constant with time, then it is considered a simple multiaxial stress case. If it varies with time, then it is a complex multiaxial stress case. Most rotating shafts loaded in both bending and torsion will be in the complex category. While the direction of the alternating bending stress component will tend to be constant, the torsional component's direction varies as the element rotates around the shaft. Combining them on the Mohr's circle will show that the result is an alternating principal stress of varying direction. One exception to this is the case of a constant torque superposed on a time-varying moment. Since the constant torque has no alternating component to change the direction of the principal alternating stress, this becomes a simple multiaxial stress case. However even this exception cannot be taken if there are stress concentrations present, such as holes or keyways in the shaft, since they will introduce local biaxial stresses and require a complex multiaxial fatigue analysis.

Assume that the bending moment function over the length of the shaft is known or is calculable from the given data and that it has both a mean component M_m and an alternating component M_a. Likewise, assume the torque on the shaft is known or calculable from given data and has both mean and alternating components, T_m and T_a. Then the general approach follows that outlined in the list labeled **Design Steps for Fluctuating Stresses** in Section 6.11 in combination with the multiaxial-stress issues addressed in Section 6.12. Any locations along the length of the shaft that appear to have large moments and/or torques (especially if in combination with stress concentrations) need to be examined for possible stress failure and the cross-sectional dimensions or material properties adjusted accordingly.

9.6   SHAFT STRESSES

With the understanding that the following equations will have to be calculated for a multiplicity of points on the shaft and their combined multiaxial effects also considered, we must first find the applied stresses at all points of interest. The largest alternating and mean bending stresses are at the outside surface and are found from

$$\sigma_a = k_f \frac{M_a c}{I} \qquad\qquad \sigma_m = k_{fm} \frac{M_m c}{I} \qquad (9.2a)$$

where k_f and k_{fm} are the bending fatigue stress-concentration factors for the alternating and mean components, respectively (see equations 6.11 and 6.15). Since the typical shaft is a solid-round cross-section,* we can substitute for c and I:

$$c = r = \frac{d}{2} \qquad\qquad I = \frac{\pi d^4}{64} \qquad (9.2b)$$

giving

$$\sigma_a = k_f \frac{32 M_a}{\pi d^3} \qquad\qquad \sigma_m = k_{fm} \frac{32 M_m}{\pi d^3} \qquad (9.2c)$$

where d is the local shaft diameter at the section of interest.

The alternating and mean torsional shear stresses are found from

$$\tau_a = k_{fs} \frac{T_a r}{J} \qquad\qquad \tau_m = k_{fsm} \frac{T_m r}{J} \qquad (9.3a)$$

where k_{fs} and k_{fsm} are the torsional fatigue stress-concentration factors for the alternating and mean components, respectively (see equation 6.11 for k_{fs} and use the applied shear stresses and shear yield strength in equation 6.17 to get k_{fsm}). For a solid-round cross section,* we can substitute for r and J:

---

* For a hollow shaft, substitute the appropriate expressions for I and J.

$$r = \frac{d}{2} \qquad\qquad J = \frac{\pi d^4}{32} \qquad\qquad (9.3b)$$

giving

$$\tau_a = k_{fs} \frac{16 T_a}{\pi d^3} \qquad\qquad \tau_m = k_{fsm} \frac{16 T_m}{\pi d^3} \qquad\qquad (9.3c)$$

A tensile axial load F_z, if any is present, will typically have only a mean component (such as the weight of the components) and can be found from

$$\sigma_{m_{axial}} = k_{fm} \frac{F_z}{A} = k_{fm} \frac{4 F_z}{\pi d^2} \qquad\qquad (9.4)$$

9.7   SHAFT FAILURE IN COMBINED LOADING

Extensive studies of fatigue failure of both ductile steels and brittle cast irons in combined bending and torsion were done originally in England in the 1930s by Davies[3] and Gough and Pollard.[5] These early results are shown in Figure 9-3, which is taken from the ANSI/ASME Standard B106.1M-1985 on the *Design of Transmission Shafting*. Data from later research is also included on these plots.[2],[4] The combination of torsion and bending on ductile materials in fatigue was found to generally follow the elliptical relationship as defined by the equations in the figure. Cast brittle materials (not shown) were found to fail based on the maximum principal stress. These findings are similar to those for combined torsional and bending stresses in fully reversed loading shown in Figure 6-15.

9.8   SHAFT DESIGN

Both stresses and deflections need to be considered in shaft design. Often, deflection can be the critical factor since excessive deflections will cause rapid wear of shaft bearings. Gears, belts, or chains driven from the shaft can also suffer from misalignment introduced by shaft deflections. Note that the stresses in a shaft can be calculated locally for various points along the shaft based on known loads and assumed cross sections. But, the deflection calculations require that the entire shaft geometry be defined. So, a shaft is typically first designed using stress considerations and then the deflections calculated once the geometry is completely defined. The relationship between the shaft's natural frequencies (in both torsion and bending) and the frequency content of the force- and torque-time functions can also be critical. If the forcing functions are close in frequency to the shaft's natural frequencies, resonance can create vibrations, high stresses, and large deflections.

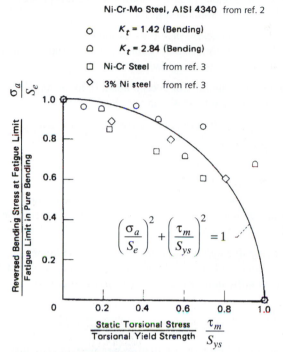

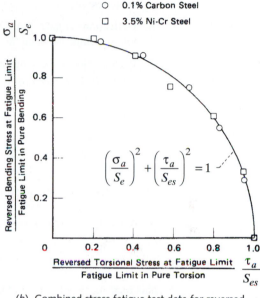

(a) Combined stress fatigue test data for reversed bending combined with static torsion (from ref. 4)

(b) Combined stress fatigue test data for reversed bending combined with reversed torsion (from ref. 5)

FIGURE 9-3

Results of Fatigue Tests of Steel Specimens Subjected to Combined Bending and Torsion (From *Design of Transmission Shafting*, American Society of Mechanical Engineers, New York, ANSI/ASME Standard B106.1M-1985, with permission)

General Considerations

Some general rules of thumb for shaft design can be stated as follows:

1. To minimize both deflections and stresses, the shaft length should be kept as short as possible and overhangs minimized.

2. A cantilever beam will have a larger deflection than a simply supported (straddle mounted) one for the same length, load, and cross section, so straddle mounting should be used unless a cantilever shaft is dictated by design constraints. (*Figure 9-2 shows a situation in which an overhung or cantilevered section of shaft is required for serviceability. The sheave on the right-hand end of the shaft carries an endless V-belt. If the sheave were mounted between the bearings, then the shaft assembly would have to be disassembled to change a belt, which is undesirable. In such cases, the cantilevered shaft can be the lesser of the evils.*)

3. A hollow shaft has a better stiffness/mass ratio (specific stiffness) and higher natural frequencies than a comparably stiff or strong solid shaft, but will be more expensive and larger in diameter.

4   Try to locate stress raisers away from regions of large bending moment if possible and minimize their effects with generous radii and reliefs.

5   If minimizing deflection is the primary concern, then low-carbon steel may be the preferred material since its stiffness is as high as that of more expensive steels and a shaft designed for low deflection will tend to have low stresses.

6   Deflections at gears carried on the shaft should not exceed about 0.005 in and the relative slope between the gear axes should be less than about 0.03°.[1]

7   If plain (sleeve) bearings are used, the shaft deflection across the bearing length should be less than the oil-film thickness in the bearing.[1]

8   If non-self-aligning rolling element bearings are used, the shaft angular deflection should be kept less than about 0.04° at the bearing.[1]

9   If axial thrust loads are present, they should be taken to ground through a single thrust bearing per load direction. Do not split axial loads between thrust bearings as thermal expansion of the shaft can overload the bearings.

10 The first natural frequency of the shaft should be at least three times the highest forcing frequency expected in service, and preferably much more. (A factor of 10x or more is preferred, but this is often difficult to achieve in mechanical systems).

Design for Fully Reversed Bending and Steady Torsion

This loading case is a subset of the general case of fluctuating bending and fluctuating torsion, and because of the absence of an alternating component of torsional stress, is considered to be a simple multiaxial fatigue case. (The presence of local stress concentrations can cause complex multiaxial stresses, however.) This simple loading case has been experimentally investigated and data exist for failure of parts so loaded as shown in Figure 9-3. The ASME has defined an approach for the design of shafts loaded in this manner.

THE ASME METHOD   An ANSI/ASME Standard for the *Design of Transmission Shafting* is published as B106.1M-1985. This standard presents a simplified approach to the design of shafts. The ASME approach *assumes that the loading is fully reversed bending (zero mean bending component) and steady torque (zero alternating torque component)* at a level that creates stresses below the torsional yield strength of the material. The standard makes the case that many machine shafts are in this category. It uses the elliptical curve of Figure 9-3 fitted through the bending endurance strength on the σ_a axis and the **tensile yield strength** on the σ_m axis as the failure envelope. The tensile yield strength is substituted for the torsional yield strength by using the von Mises relationship of equation 5.9. The derivation of the ASME shaft equation is as follows.

Starting with the relationship for the failure envelope shown in Figure 9-3a:

$$\left(\frac{\sigma_a}{S_e}\right)^2 + \left(\frac{\tau_m}{S_{ys}}\right)^2 = 1 \qquad (9.5a)$$

introduce a safety factor N_f

$$\left(N_f \frac{\sigma_a}{S_e}\right)^2 + \left(N_f \frac{\tau_m}{S_{ys}}\right)^2 = 1 \qquad (9.5b)$$

Recall the von Mises relationship for S_{ys} from equation 5.9:

$$S_{ys} = S_y / \sqrt{3} \qquad (9.5c)$$

and substitute it in equation 9.5b.

$$\left(N_f \frac{\sigma_a}{S_e}\right)^2 + \left(N_f \sqrt{3} \frac{\tau_m}{S_y}\right)^2 = 1 \qquad (9.5d)$$

Substitute the expressions for σ_a and τ_m from equations 9.2c and 9.3c, respectively:

$$\left[\left(k_f \frac{32M_a}{\pi d^3}\right)\left(\frac{N_f}{S_e}\right)\right]^2 + \left[\left(k_{fsm} \frac{16T_m}{\pi d^3}\right)\left(\frac{N_f \sqrt{3}}{S_y}\right)\right]^2 = 1 \qquad (9.5e)$$

which can be rearranged to solve for the shaft diameter d as

$$d = \left\{\frac{32N_f}{\pi}\left[\left(k_f \frac{M_a}{S_f}\right)^2 + \frac{3}{4}\left(k_{fsm} \frac{T_m}{S_y}\right)^2\right]^{\frac{1}{2}}\right\}^{\frac{1}{3}} \qquad (9.6a)$$

The notation used in equation 9.6 is slightly different than that of the ANSI/ASME standard in order to remain consistent with the notation used in this text. The standard uses the approach of reducing the fatigue strength S_f by the fatigue-stress-concentration factor k_f rather than using k_f as a stress increaser as is done consistently in this text. In most cases (including this one) the result is the same. Also, the ASME standard assumes the stress concentration for mean stress k_{fsm} to be 1 in all cases, which gives

$$d \doteq \left\{\frac{32N_f}{\pi}\left[\left(k_f \frac{M_a}{S_f}\right)^2 + \frac{3}{4}\left(\frac{T_m}{S_y}\right)^2\right]^{\frac{1}{2}}\right\}^{\frac{1}{3}} \qquad (9.6b)$$

Be careful to apply equation 9.6 only to situations where the loads are as it assumes them to be, namely constant torque and fully reversed moment. The ASME standard gives nonconservative results if either of the loading components that it assumes to be zero are in fact nonzero in a given case. In such situations, the more general approach of equation 9.8 should be used.

Figure 9-4 shows the Gough elliptical failure line of Figure 9-3 superposed on the Gerber, Soderberg, and modified-Goodman lines. Note that the elliptical line closely

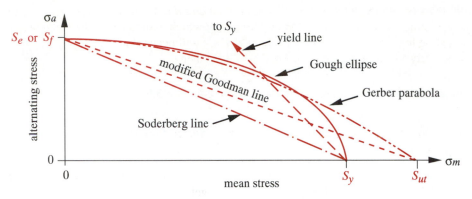

FIGURE 9-4

Elliptical Failure Line Using Yield Strength Shown with Other Failure Lines for Fluctuating Stresses

matches the Gerber line at the left-hand end but diverges to intersect the yield strength on the mean stress axis. The elliptical line has the advantage of accounting for possible yielding without needing to introduce an additional constraint involving the yield line. However, the Gough elliptical line, while a good fit to the failure data, is less conservative than the combination of Goodman line and yield line used as a failure envelope.

Design for Fluctuating Bending and Fluctuating Torsion

When the torque is not constant, its alternating component will create a complex multiaxial stress state in the shaft. Then the approach described in Section 6.12, which computes the von Mises components of the alternating and mean stresses using equations 6.22, can be used. A rotating shaft in combined bending and torsion has a biaxial stress state, which allows the two-dimensional version of equation 6.22b to be used.

$$\sigma'_a = \sqrt{\sigma_a^2 + 3\tau_a^2}$$

$$\sigma'_m = \sqrt{\left(\sigma_m + \sigma_{m_{axial}}\right)^2 + 3\tau_m^2}$$

$$(9.7a)$$

These von Mises stresses can now be entered into a modified-Goodman diagram (MGD) for a chosen material to find a safety factor, or equations 6.17 can be applied without drawing the MGD.

For design purposes, where the diameter of the shaft is the desired quantity to be found, equations 9.2, 9.3, and 9.7 as presented require iteration to find a value for d given some known loads and assumed material properties. This is not a great difficulty if an equation solver with iterative ability such as *TKSolver* is used. However, a hand calculator solution is cumbersome with the equations in this form. If a particular failure case is assumed for the MGD, the equations can be manipulated to provide a design equation (similar to equation 9.6) for shaft diameter d at the section of interest. If the failure model used is Case 3 from Section 6.11, which assumes that the mean and alternating loads maintain a constant ratio,[*] failure will occur at point R in Figure 6-46c. The safety factor as defined in equation 6.18e is then

$$\frac{1}{N_f} = \frac{\sigma'_a}{S_f} + \frac{\sigma'_m}{S_{ut}} \tag{9.7b}$$

where N_f is the desired safety factor, S_f is the corrected fatigue strength at the selected cycle life (from equation 6.10), and S_{ut} is the ultimate tensile strength of the material.

If we now also assume that the axial load on the shaft is zero and substitute equations 9.2c, 9.3c, and 9.7a into equation 9.7b we get

$$d = \left\{ \frac{32N_f}{\pi} \left[\frac{\sqrt{\left(k_f M_a\right)^2 + \frac{3}{4}\left(k_{fs} T_a\right)^2}}{S_f} + \frac{\sqrt{\left(k_{fm} M_m\right)^2 + \frac{3}{4}\left(k_{fsm} T_m\right)^2}}{S_{ut}} \right] \right\}^{\frac{1}{3}} \tag{9.8}$$

which can be used as a design equation to find a shaft diameter for any combination of bending and torsional loading with the assumptions noted above of zero axial load and a constant ratio between alternating and mean values of load over time.

EXAMPLE 9-1

Shaft Design for Steady Torsion and Fully Reversed Bending

Problem   Design a shaft to support the attachments shown in Figure 9-5 with a minimum design safety factor of 2.5.

Given   A preliminary design of the shaft configuration is shown in Figure 9-3. It must transmit 2 hp at 1 725 rpm. The torque and the force on the gear are both constant with time.

Assumptions   There are no applied axial loads. Steel will be used for infinite life. Assume a stress-concentration factor of 3.5 for the step radii in bending, 2 for step radii in torsion, and 4 at the keyways.[†] Since the torque is steady and the bending moment fully reversed, the ASME method of equation 9.6 can be used.

Solution   See the *TKSolver* files EX09-01a, EX09-01b, EX09-01c, and EX09-01d.

1   First determine the transmitted torque from the given power and angular velocity using equation 9.1.

[*] Note that this assumption is also implicit in equation 9.6 for the ASME method.

[†] See R. E. Peterson, *Stress Concentration Factors*, J. Wiley, 1974, Figures 72, 79, and 183 which show these numbers as approximate maxima for these contours and loadings. Since we are creating a preliminary design at this stage and have not yet defined the shaft geometry in detail, it is not fruitful to try to define these factors any more accurately. This can be done later and the design refined accordingly.

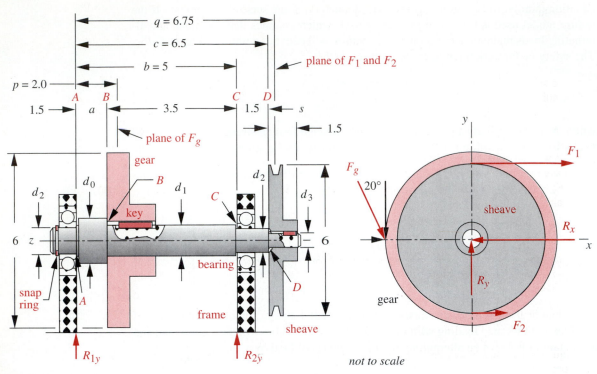

FIGURE 9-5

Geometry of a Preliminary Design for Examples 9-1 through 9-3

$$T = \frac{P}{\omega} = \frac{2 \text{ hp} \left(6\,600 \, \dfrac{\text{in - lb/sec}}{\text{hp}} \right)}{1\,725 \text{ rpm} \left(\dfrac{2\pi}{60} \, \dfrac{\text{rad/sec}}{\text{rpm}} \right)} = 73.1 \text{ lb - in} \qquad (a)$$

This torque exists only over the portion of shaft between the sheave and the gear and is uniform in magnitude over that length as shown in Figure 9-6.

2   The tangential forces on sheave and gear are found from the torque and their respective radii. A V-belt has tension on both sides and the ratio between the force F_1 on the tight side and F_2 on the "slack" side is usually assumed to be about 5. The net force associated with the driving torque is $F_n = F_1 - F_2$, but the force that bends the shaft is $F_s = F_1 + F_2$. Combining these relationships gives $F_s = 1.5F_n$. Looking from the sheave end:

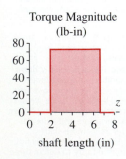

Torque Magnitude
(lb-in)

FIGURE 9-6

Torque in Example 9-1

$$F_n = \frac{T}{r} = \frac{73.1 \text{ lb - in}}{3 \text{ in}} = 24.36 \; \hat{i} \; \text{lb}$$

$$F_s = 1.5 F_n = 36.54 \; \hat{i} \; \text{lb} \qquad (b)$$

3   The tangential force at the spur-gear tooth is

$$F_{g_{tangential}} = \frac{T}{r} = \frac{73.1 \text{ lb - in}}{3 \text{ in}} = -24.36 \,\hat{j} \text{ lb} \qquad (c)$$

The spur gear has a 20° pressure angle as shown, which means that there will also be a radial component of force at the gear tooth of

$$F_{g_{radial}} = F_{g_{tangential}} \tan(20°) = 8.87 \,\hat{i} \text{ lb} \qquad (d)$$

4   We will consider the gear and sheave forces to be concentrated at their centers. Solve for the reaction forces in the xz and yz planes using $\Sigma F_x = 0$, $\Sigma M_x = 0$ and $\Sigma F_y = 0$, $\Sigma M_y = 0$ with the trial beam dimensions, $a = 1.5$, $b = 5$, and $c = 6.5$, which make $p = 2$ and $q = 6.75$.

$$\sum M_A = R_2 b + F_g p + F_s q = 0$$
$$R_2 = -\frac{1}{b}\left(F_g p + F_s q\right) = -\frac{1}{5}\left(2F_g + 6.75 F_s\right) = -0.40 F_g - 1.35 F_s \qquad (e)$$

$$\sum F = R_1 + F_g + F_s + R_2 = 0$$
$$R_1 = -F_g - F_s - R_2 = -F_g - F_s - \left(-0.40 F_g - 1.35 F_s\right) = -0.60 F_g + 0.35 F_s \qquad (f)$$

Equations (e) and (f) can be solved for R_1 and R_2 in each plane, using the appropriate components of the applied loads F_g and F_s.

$$R_{1_x} = -0.60 F_{g_x} + 0.35 F_{s_x} = -0.60(8.87) + 0.35(36.54) = 7.47 \text{ lb}$$
$$R_{1_y} = -0.60 F_{g_y} + 0.35 F_{s_y} = -0.60(-24.36) + 0.35(0) = 14.61 \text{ lb}$$
$$\qquad (g)$$
$$R_{2_x} = -0.40 F_{g_x} - 1.35 F_{s_x} = -0.40(8.87) - 1.35(36.54) = -52.87 \text{ lb}$$
$$R_{2_y} = -0.40 F_{g_y} - 1.35 F_{s_y} = -0.40(-24.36) - 1.35(0) = 9.74 \text{ lb}$$

5   The shear load and bending moment acting on the shaft can now be found. Write an equation for the loading function q using singularity functions, integrate it to get the shear function V, and integrate again for the moment function M.

$$q = R_1 \langle z - 0 \rangle^{-1} + F_g \langle z - 2 \rangle^{-1} + R_2 \langle z - 5 \rangle^{-1} + F_s \langle z - 6.75 \rangle^{-1} \qquad (h)$$

$$V = R_1 \langle z - 0 \rangle^0 + F_g \langle z - 2 \rangle^0 + R_2 \langle z - 5 \rangle^0 + F_s \langle z - 6.75 \rangle^0 \qquad (i)$$

$$M = R_1 \langle z - 0 \rangle^1 + F_g \langle z - 2 \rangle^1 + R_2 \langle z - 5 \rangle^1 + F_s \langle z - 6.75 \rangle^1 \qquad (j)$$

Recall that the integration constants C_1 and C_2 are zero when we include the reaction forces in the equation.

6   Substitute the values of the loads and reaction forces for each coordinate direction into equations (h), (i), and (j) and evaluate them for all values of z along the shaft

Shear in *xz* Plane

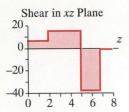

Shear in *yz* Plane

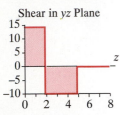

Shear Magnitude

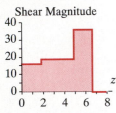

Moment in *xz* Plane

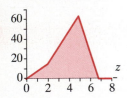

Moment in *yz* Plane

Moment Magnitude

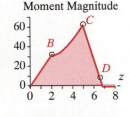

FIGURE 9-7

Loading in Example 9-1

axis. Then combine the moment-function components in the *xz* and *yz* planes (using the Pythagorean theorem) to find the maximum magnitude of the moment function.

The shear and moment distributions over the shaft length are shown in Figure 9-7. The applied torque is uniform over the portion of shaft between points *B* and *D* as shown in Figure 9-6. Within that length, there are three locations of concern where a moment occurs in combination with a stress concentration, point *B* at the step and keyway under the gear (M_b = ±33 lb-in), point *C* at the right bearing where there is a step with a small radius to fit the bearing (M_c = ±63 lb-in), and point *D* at the sheave step (M_d = ±9 lb-in). Note that because of its high stress concentration, the snap-ring groove used for axial location has been placed at the end of the shaft where the moment and torque are both zero.

7   A trial material needs to be selected for the computations. We will first try an inexpensive, low-carbon, cold-rolled steel such as SAE 1020 with S_{ut} = 65 kpsi and S_y = 38 kpsi. Though not exceptionally strong, this material has low notch sensitivity, which will be an advantage given the large stress concentrations. Calculate the uncorrected endurance strength using equation 6.5:

$$S_{e'} = 0.5 S_{ut} = 0.5(65\ 000) = 32\ 500 \text{ psi} \qquad (k)$$

This must be reduced by various factors to account for differences between the part and the test specimen.

$$S_e = C_{load}\,C_{size}\,C_{surf}\,C_{temp}\,C_{reliab}\,S_{e'}$$
$$S_e = (1)(1)(0.84)(1)(1)(32\ 500) = 27\ 300 \text{ psi} \qquad (l)$$

The loading is bending and torsion, so C_{load} is 1. Since we don't yet know the part size, we will temporarily assume C_{size} =1 and adjust it later. C_{surf} is chosen for a machined finish from either Figure 6-26 or equation 6.7e. The temperature is not elevated so C_{temp} = 1 and we assume 50% reliability at this preliminary design stage with C_{reliab} = 1.

8   The notch sensitivity of the material is found from either equation 6.13 or Figure 6-35 and is *q* = 0.5 for an assumed notch radius of 0.01 in.

9   The fatigue stress-concentration factor is found from equation 6.11b using the assumed geometric stress-concentration factor noted above. For the bending stress in the step at point *C*:

$$K_f = 1 + q(K_t - 1) = 1 + 0.5(3.5 - 1) = 2.25 \qquad (m)$$

The stress concentration for a step loaded in torsion is less than for the same geometry loaded in bending:

$$K_{fs} = 1 + q(K_{ts} - 1) = 1 + 0.5(2 - 1) = 1.5 \qquad (n)$$

From equation 6.17 we find that in this case, the same factor should be used on the mean torsional stress component:

$$K_{fsm} = K_{fs} = 1.5 \qquad (o)$$

10 The shaft diameter at point C can now be found from equation 9.6 using the moment magnitude at that point of 63.9 in-lb.

$$d_2 = \left\{ \frac{32 N_f}{\pi} \left[\left(k_f \frac{M_a}{S_f} \right)^2 + \frac{3}{4} \left(k_{fsm} \frac{T_m}{S_y} \right)^2 \right] \right\}^{\frac{1}{3}}$$

$$= \left\{ \frac{32(2.5)}{\pi} \left[\left(2.25 \frac{63.9}{27\,300} \right)^2 + \frac{3}{4} \left(1.5 \frac{73.1}{38\,000} \right)^2 \right] \right\}^{\frac{1}{3}} = 0.528 \text{ in} \qquad (p)$$

If k_{fsm} is set to 1 as ASME recommends, then equation 9.6 gives $d = 0.520$ in. If the more general equation 9.8 is used, the result is $d = 0.554$ in. Note that the ASME method is less conservative than equation 9.8 as it gives smaller shaft diameters for the same safety factor. A modified-Goodman diagram for this stress element is shown in Figure 9-8b. It predicts failure from fatigue.

11 At point B, under the gear, the moment is less, but the stress-concentration factor is greater so it should also be checked. The bending-fatigue stress-concentration factor at B is

$$K_f = 1 + q(K_t - 1) = 1 + 0.5(4 - 1) = 2.5 \qquad (q)$$

The torsion fatigue stress-concentration factor is the same as the bending factor in this case.

12 The minimum recommended diameter at point B from equation 9.6 is

$$d_1 = \left\{ \frac{32 N_f}{\pi} \left[\left(k_f \frac{M_a}{S_f} \right)^2 + \frac{3}{4} \left(k_{fsm} \frac{T_m}{S_y} \right)^2 \right] \right\}^{\frac{1}{3}}$$

$$= \left\{ \frac{32(2.5)}{\pi} \left[\left(2.5 \frac{32.8}{27\,300} \right)^2 + \frac{3}{4} \left(2.5 \frac{73.1}{38\,000} \right)^2 \right] \right\}^{\frac{1}{3}} = 0.502 \text{ in} \qquad (r)$$

If k_{fsm} is set to 1 as ASME recommends, then equation 9.6 gives $d = 0.444$ in. If the more general equation 9.8 is used the result is $d = 0.513$ in. A modified-Goodman diagram for this stress element is shown in Figure 9-8a. It predicts failure from fatigue.

13 Another location of possible failure is the step against which the sheave seats, at point D. The moment is lower than at C, being about 9.1 lb-in. (See Figure 9-7.) However, the shaft will be stepped smaller there and will have the same order of stress concentration as at C. (The keyway for the sheave is in a region of zero moment and so will be ignored.) Using those data in equation 9.6 for point D:

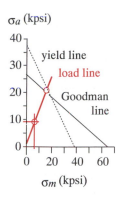

(a) Stresses at point B

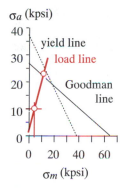

(b) Stresses at point C

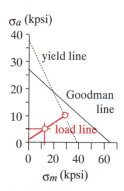

(c) Stresses at point D

FIGURE 9-8

Modified-Goodman Diagrams for Three Points on Shaft in Example 9-1

$$d_3 = \left\{ \frac{32 N_f}{\pi} \left[\left(k_f \frac{M_a}{S_f} \right)^2 + \frac{3}{4} \left(k_{fsm} \frac{T_m}{S_y} \right)^2 \right] \right\}^{\frac{1}{3}}$$

$$= \left\{ \frac{32(2.5)}{\pi} \left[\left(2.25 \frac{9.1}{27\,300} \right)^2 + \frac{3}{4} \left(1.5 \frac{73.1}{38\,000} \right)^2 \right] \right\}^{\frac{1}{3}} = 0.402 \text{ in} \qquad (s)$$

If k_{fsm} is set to 1 as ASME recommends, then equation 9.6 gives $d = 0.360$ in. If the more general equation 9.8 is used the result is $d = 0.381$ in. A modified-Goodman diagram for this stress element is shown in Figure 9-8c. It predicts failure from yielding.

14   From these preliminary calculations, we can determine reasonable sizes for the three step diameters, d_0, d_1, d_2, d_3 of Figure 9-3. The next largest standard ball-bearing diameter to the $d_2 = 0.554$ in calculated for point C is 15 mm or 0.591 in. Selecting this value for d_2, we set $d_3 = 0.50$ in, and $d_1 = 0.625$ in. The stock size d_0 is then 0.75 in, left as-rolled for the outside diameter at the gear flange. These dimensions will give safety factors that meet or exceed the specification. The stresses and safety factors at all three points should now be recalculated using more accurate strength-reduction (e.g., C_{size}) and stress-concentration factors based on the final dimensions.

EXAMPLE 9-2

Shaft Design for Repeated Torsion with Repeated Bending

Problem   Design a shaft to support the attachments shown in Figure 9-5 with a minimum design safety factor of 2.5.

Given   The torque and the moment on the shaft are both varying with time in repeated fashion, i.e., their alternating and mean components are of equal magnitude. The mean and alternating components of torque are both 73 lb-in, making the peak torque twice the mean value of Example 9-1. The mean and alternating components of moment are equal in magnitude. Figure 9-9 shows the peak moment and peak torque which are each twice the value of their fully reversed counterparts of Figure 9-5 and Example 9-1 due to the presence of the mean moment.

Assumptions   There are no applied axial loads. Steel will be used for infinite life. Assume a stress-concentration factor of 3.5 for the step radii in bending, 2 for step radii in torsion, and 4 at the keyways. Since the torsional load is not steady and the bending moment is not fully reversed, the ASME method of equation 9.6 should not be used.

Solution   See Figures 9-9, 9-10, Table 9-1, and the *TKSolver* files EX09-01a, EX09-02b, EX09-02c, and EX09-02d.

1   For comparison purposes, we will keep all factors except the loading configuration the same as in the previous example. The same, low-carbon, cold-rolled steel SAE 1020, which has S_{ut} = 65 kpsi, S_y = 38 kpsi, and a corrected S_e = 27.3 kpsi, is used. Its notch sensitivity is 0.5.

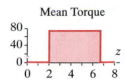

Mean Torque

2   There are three points of interest, labeled B, C, and D in Figure 9-3. The fatigue stress-concentration factors are assumed to be the same at C and D and are larger at B. See Example 9-1 for their calculation.

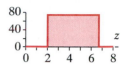

Alternating Torque

3   The required shaft diameter at point C can be found from equation 9.8.

$$d_2 = \left\{ \frac{32N_f}{\pi} \left[\frac{\sqrt{\left(k_f M_a\right)^2 + \frac{3}{4}\left(k_{fs} T_a\right)^2}}{S_f} + \frac{\sqrt{\left(k_{fm} M_m\right)^2 + \frac{3}{4}\left(k_{fsm} T_m\right)^2}}{S_{ut}} \right] \right\}^{\frac{1}{3}}$$

$$= \left\{ \frac{32(2.5)}{\pi} \left[\frac{\sqrt{[2.25(64)]^2 + \frac{3}{4}[1.5(73.1)]^2}}{27\,300} + \frac{\sqrt{[2.25(64)]^2 + \frac{3}{4}[1.5(73.1)]^2}}{65\,000} \right] \right\}^{\frac{1}{3}}$$

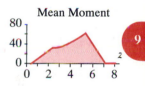

Peak Torque

$d_2 = 0.611$ (a)

Compare this to the value of 0.554 from the same equation in the prior example where the loads were steady.

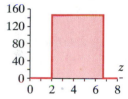

Mean Moment

4   At point B the required diameter from equation 9.8 is

$$d_1 = \left\{ \frac{32(2.5)}{\pi} \left[\frac{\sqrt{[2.5(32.8)]^2 + \frac{3}{4}[2.5(73.1)]^2}}{27\,300} + \frac{\sqrt{[2.5(32.8)]^2 + \frac{3}{4}[2.5(73.1)]^2}}{65\,000} \right] \right\}^{\frac{1}{3}}$$

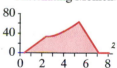

Alternating Moment

$d_1 = 0.618$ (b)

Compare this to the value of 0.513 from the same equation in the prior example where the loads were steady.

5   And at point D:

$$d_3 = \left\{ \frac{32(2.5)}{\pi} \left[\frac{\sqrt{[2.25(9.1)]^2 + \frac{3}{4}[1.5(73.1)]^2}}{27\,300} + \frac{\sqrt{[2.25(9.1)]^2 + \frac{3}{4}[1.5(73.1)]^2}}{65\,000} \right] \right\}^{\frac{1}{3}}$$

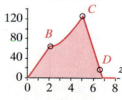

Peak Moment

shaft length (in)

$d_3 = 0.505$ (c)

FIGURE 9-9

Torque and Moments in Example 9-2 (lb-in)

σa (kpsi)

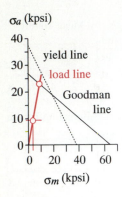

(a) Stresses at point B

σa (kpsi)

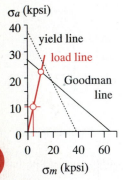

(b) Stresses at point C

σa (kpsi)

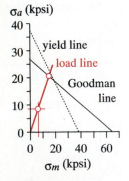

(c) Stresses at point D

FIGURE 9-10

Modified-Goodman
Diagrams for Three
Points on Shaft in
Example 9-2

Table 9-1	Comparison of Shaft Design Results from Examples 9-1 and 9-2								
	Minimum Diameters Give $N_f = 2.5$ at Each Point								
Design	Max Alt Torque	Max Mean Torque	Max Alt Moment	Max Mean Moment	d_0 (in) nom	d_1 (in) min / nom	d_2 (in) min / nom	d_3 (in) min / nom	S_f at C
Ex. 9-1	0	73.1	63.9	0	0.750	0.513 / 0.625	0.554 / 0.591	0.381 / 0.500	2.5 / 3.0
Ex. 9-2	73.1	73.1	63.9	63.9	0.875	0.618 / 0.750	0.611 / 0.669	0.505 / 0.531	2.5 / 3.8

Compare this to the value of $d_3 = 0.381$ from the same equation in the prior example where the loads were steady.

6   The presence of repeated stresses requires a larger shaft to maintain the same safety factor. We need the next larger standard bearing at C, which is a 17 mm (0.669 in) inside diameter. Selecting this value for d_2, we try values of $d_3 = 0.531$ in, which is a smaller standard inch size, and $d_1 = 0.750$ in, which is a larger standard inch size. The stock size is now 0.875 in, and left as-rolled for the outside diameter at the gear flange. These dimensions will give safety factors that meet or exceed the specification as shown in the modified-Goodman diagrams of Figure 9-10. The stresses and safety factors at all three points should now be recalculated using more accurate strength-reduction and stress-concentration factors based on the final dimensions.

7   Table 9-1 compares the results from Examples 9-1 and 9-2 to show the differences in shaft diameter needed for steady loading or fluctuating loading. Note that the peak load in Example 9-1 is half that of Example 9-2. The final safety factors are larger than the design minimum due to the need to size-up the shaft to fit an available stock bearing size.

9.9   SHAFT DEFLECTION

A shaft is a beam that deflects transversely and is also a torsion bar that deflects torsionally. Both modes of deflection need to be analyzed. The principles of deflection analysis were reviewed in Chapter 4 and will not be detailed again here. Section 4.10 developed an approach for calculating beam deflections using singularity functions and Section 4.11 investigated torsional deflection.

Shafts as Beams

The methods of Section 4.10 are directly applicable. The only complicating factor is the usual presence of steps in a shaft that change the cross-sectional properties along its length. The integration of the M/EI function becomes much more complicated due to the fact that both I and M are now functions of the dimension along the shaft-beam. Rather than do an analytical integration as was done in Section 4.10 for the case of constant I, we will use a numerical integration technique such as Simpson's rule or the trapezoidal rule to form the slope and deflection functions from the M/EI function. This will be demonstrated in an example. If the transverse loads and moment are time-varying, then the absolute maximum magnitudes should be used to calculate the deflections. The deflection function will depend on the loading and the beam boundary conditions, i.e., whether simply supported, cantilevered, or overhung.

Shafts as Torsion Bars

The methods of Section 4.12 are directly applicable, particularly equation 4.24, since the only practical shaft cross-section is circular. The angular deflection θ (in radians) for a shaft of length l, shear modulus G, polar moment of inertia J, with torque T is

$$\theta = \frac{Tl}{GJ} \qquad (9.9a)$$

from which we can form the expression for the torsional spring constant:

$$k_t = \frac{T}{\theta} = \frac{GJ}{l} \qquad (9.9b)$$

If the shaft is stepped, the changing cross sections complicate the torsional deflection and spring constant calculation due to the changing polar moment of inertia J.

Any collection of adjacent, different-diameter sections of shaft can be considered as a set of springs in series since their deflections add and the torque passes through unchanged. An effective spring constant or an effective J can be computed for any segment of shaft in order to find the relative deflection between its ends. For a segment of a shaft containing three sections of differing cross sections J_1, J_2, and J_3 with corresponding lengths l_1, l_2, l_3, the total deflection is merely the sum of the deflections of each section subjected to the same torque. We assume that the material is consistent throughout,

$$\theta = \theta_1 + \theta_2 + \theta_3 = \frac{T}{G}\left(\frac{l_1}{J_1} + \frac{l_2}{J_2} + \frac{l_3}{J_3}\right) \qquad (9.9c)$$

The effective spring constant k_{eff} of a three-segment stepped shaft is

$$\frac{1}{k_{t_{eff}}} = \frac{1}{k_{t_1}} + \frac{1}{k_{t_2}} + \frac{1}{k_{t_3}} \qquad (9.9d)$$

These expressions can be extended to any number of segments of a stepped shaft.

EXAMPLE 9-3

Designing a Stepped Shaft to Minimize Deflection

Problem   Design the same shaft as in Example 9-2 to have a maximum bending deflection of 0.002 in and a maximum angular deflection of 0.5° between sheave and gear.

Given   The loading is the same as in Example 9-2. The peak torque is 146 lb-in. Figure 9-9 shows the distribution of the peak moment over the shaft length. The values are 65.6 lb-in at point *B*, 127.9 lb-in at point *C*, and 18.3 lb-in at point *D*.

Assumptions   The lengths will remain the same as in previous example, but diameters can be changed to stiffen the shaft if necessary. The material is the same as in Example 9-2.

Solution   See Figures 9-11 to 9-13 and the *TKSolver* files EX09-03a and EX09-03b.

1   The torsional deflection is found from equations 9.9. The lengths of each segment are (from Figure 9-3): *AB* = 1.5 in, *BC* = 3.5 in, and *CD* = 1.5 in. The polar area moments of inertia are first calculated for each segment of different diameter.

$$\text{from } A \text{ to } B: \qquad J = \frac{\pi d_{AB}^4}{32} = \frac{\pi (0.875)^4}{32} = 0.057\,5 \text{ in}^4$$

$$\text{from } B \text{ to } C: \qquad J = \frac{\pi d_{BC}^4}{32} = \frac{\pi (0.750)^4}{32} = 0.031\,1 \text{ in}^4 \qquad (a)$$

$$\text{from } C \text{ to } D: \qquad J = \frac{\pi d_{CD}^4}{32} = \frac{\pi (0.669)^4}{32} = 0.019\,7 \text{ in}^4$$

and used in equation 9.9c.

$$\theta = \frac{T}{G}\left(\frac{l_1}{J_1} + \frac{l_2}{J_2} + \frac{l_3}{J_3}\right)$$

$$= \frac{146}{1.2E7}\left(\frac{1.5}{0.057\,5} + \frac{3.5}{0.031\,1} + \frac{1.5}{0.019\,7}\right) = 0.385 \text{ deg} \qquad (b)$$

This deflection is within the requested specification.

2   The moment function for this shaft was derived using singularity functions as equation (*j*) in Example 9-1. It must now be divided by the product of *E* and the area moment of inertia *I* at each point along the shaft axis. While *E* is constant, the value of *I* changes with each diametral change in the stepped shaft.

$$\frac{M}{EI} = \frac{1}{EI}\left[R_1\langle z - 0\rangle^1 + F_g\langle z - 1.5\rangle^1 + R_2\langle z - 5\rangle^1 + F_s\langle z - 6.5\rangle^1\right] \qquad (c)$$

Figure 9-11a shows the moment function for this shaft as derived in the previous examples and Figure 9-11b shows the M/EI function for the section diameters defined in Example 9-2.

3   The bending deflection is found by integrating the moment function twice.

$$\theta = \int \frac{M}{EI}dz + C_3 \qquad\qquad (d)$$

$$\delta = \int\int \frac{M}{EI}dz + C_3 z + C_4 \qquad\qquad (e)$$

4   The first integration of the M/EI function from equation (c) gives the beam slope and the second integration gives the deflection function. In previous discussions of beam deflection, (see Section 4.10 and examples 4-4 to 4-7) the cross section I of the beam was constant across its length. In a stepped shaft, I is a function of the shaft length. This makes the analytical integration of the M/EI function much more complicated. A simpler approach is to numerically integrate the function twice using a trapezoidal or Simpson's rule. This numerical integration must be done for each coordinate direction to obtain the x and y components of deflection. These are then combined vectorially to get the deflection magnitude and phase angle functions over the shaft length.

5   Since the shaft deflection is zero at $z = 0$, $C_4 = 0$. The other constant of integration C_3 can be determined numerically. Figure 9-12a shows the beam slope in the y direction as integrated by a trapezoidal rule, and also shows the corrected slope function. The integrated result is shifted up by the integration constant C_3. However, we do not know where the proper zero crossover is for this function, so cannot determine C_3 from the beam-slope function.

6   The as-integrated deflection function in Figure 9-12b does not equal zero at the second support. Since the deflection is really zero there, the error in this integrated

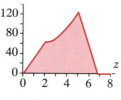

(a) Moment magnitude

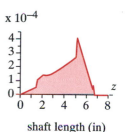

(b) M/EI

FIGURE 9-11

Moment and M/EI Functions in Example 9-3

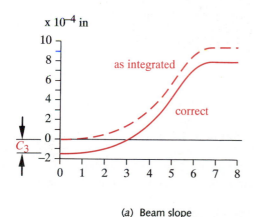

(a) Beam slope

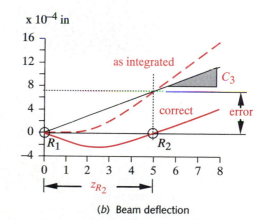

(b) Beam deflection

FIGURE 9-12

Numerical Integration of Moment Function and Finding the Integration Constant C_3

x deflection

x 10⁻³ in

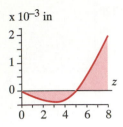

y deflection

x 10⁻³ in

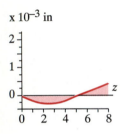

deflection
magnitude

x 10⁻³ in

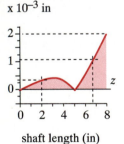

shaft length (in)

FIGURE 9-13

Deflection Functions for
Example 9-3

* ANSI standard B17.1-
1967, *Keys and Keyseats,*
and B17.2-1967, *Woodruff
Keys and Keyseats,* available
from the American Society
of Mechanical Engineers,
345 East 47th St., New York,
N.Y. 10017.

function can be used to determine the constant of integration C_3. A line is drawn in Figure 9-12*b* from the origin to the point on the curve at $z = 5$ where the function should be zero. The slope of this straight line is the constant C_3 for the y direction, which is found from

$$C_{3_y} = \frac{error_y}{z_{R2}} = \frac{0.0007}{5.0} = 0.00014 \text{ in} \qquad (f)$$

The constant for the x direction is found in similar manner. The functions are then recalculated using the correct values of C_3.

7   These deflection functions are plotted in Figure 9-13 for the shaft diameters of $d_0 = 0.875$, $d_1 = 0.750$, $d_2 = 0.669$, $d_3 = 0.531$ from Example 9-2. The magnitude of the deflection at the gear is 0.000 3 in, which is well within the requested specification. At the sheave the deflection is 0.001 in, also within the specification. The deflection at the right-hand end of the shaft is 0.002 in.

9.10   KEYS AND KEYWAYS

The ASME defines a **key** as "*a demountable machinery part which, when assembled into keyseats, provides a positive means for transmitting torque between the shaft and hub.*" Keys are standardized as to size and shape in several styles.* A **parallel key** is square or rectangular in cross section and of constant height and width over its length. (See Figure 9-14*a*.) A **tapered key** is of constant width but its height varies with a linear taper of 1/8 in per foot and is driven into a tapered slot in the hub until it locks. It may either have no head or have a **gib head** to facilitate removal. (See Figure 9-14*b*.) A **Woodruff key** is semicircular in plan and of constant width. It fits in a semicircular keyseat milled in the shaft with a standard circular cutter. (See Figure 9-14*c*.) The tapered key serves to lock the hub axially on the shaft, but the parallel or Woodruff keys require some other means for axial fixation. Retaining rings or clamp collars are sometimes used for this purpose.

Parallel Keys

Parallel keys are the most commonly used. The ANSI Standard defines particular key cross-sectional sizes and keyseat depths as a function of shaft diameter at the keyseat. A partial reproduction of this information is provided in Table 9-2 for the lower range of shaft diameters. Consult the standard for larger shaft sizes. Square keys are recommended for shafts up to 6.5-in dia and rectangular keys for larger diameters. The parallel key is placed with half of its height in the shaft and half in the hub, as shown in Figure 9-14*a*.

Parallel keys are typically made from standard cold-rolled bar stock, which is conventionally "negatively toleranced" meaning it will never be larger than its nominal

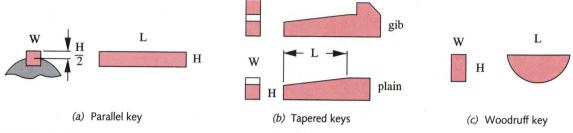

(a) Parallel key   (b) Tapered keys   (c) Woodruff key

FIGURE 9-14

Various Styles of Keys

dimension, only smaller. For example, a nominal 1/4-in square bar will have a tolerance on width and height of +0.000, –0.002 in. Thus, the keyseat can be cut with a standard 1/4-in milling cutter and the bar stock key will fit with slight clearance. A special
keystock is also available, which is positively toleranced (e.g., 0.250 +0.002, –0.000),
which is used when a closer fit between key and keyseat is desired, and may require machining of the keystock to final dimension.

The key fit can be of concern when the torque loading is alternating from positive
to negative each cycle. When the torque changes sign, any clearance between key and
keyway will be suddenly taken up with resulting impact and high stresses. This is
termed **backlash**. A setscrew in the hub, placed at 90° from the key can both hold the
hub axially and stabilize the key against backlash. The ANSI Standard also defines the
size of setscrew to be used with each size key as shown in Table 9-2. Key length should

Table 9-2   USA Standard Key and Setscrew Sizes for Inch-Sized Shafts

Shaft Diameters (in)	Nominal Key Width (in)	Setscrew Diameter (in)
$0.312 < d \le 0.437$	0.093	#10
$0.437 < d \le 0.562$	0.125	#10
$0.562 < d \le 0.875$	0.187	0.250
$0.875 < d \le 1.250$	0.250	0.312
$1.250 < d \le 1.375$	0.312	0.375
$1.375 < d \le 1.750$	0.375	0.375
$1.750 < d \le 2.250$	0.500	0.500
$2.250 < d \le 2.750$	0.625	0.500
$2.750 < d \le 3.250$	0.750	0.625
$3.250 < d \le 3.750$	0.875	0.750
$3.750 < d \le 4.500$	1.000	0.750
$4.500 < d \le 5.500$	1.250	0.875
$5.500 < d \le 6.500$	1.500	1.000

be less than about 1.5 times the shaft diameter to avoid excessive twisting with shaft deflection. If more strength is needed, two keys can be used, oriented at 90° and 180° from the key.

Tapered Keys

The width of a tapered key for a given shaft diameter is the same as for a parallel key as shown in Table 9-2. The taper and gib-head size are defined in the standard. The taper is a locking one, which means that the friction force between the surfaces holds the key in place axially. The gib-head is optional and provides a surface for prying the key out when the small end is not accessible. Tapered keys tend to create eccentricity between hub and shaft as they drive all the radial clearance to one side.

Woodruff Keys

Woodruff keys are used on smaller shafts. They are self-aligning, so are preferred for tapered shafts. The penetration of a Woodruff key into the hub is the same as that of a square key, i.e., half the key width. The semicircular shape creates a deeper keyseat in the shaft, which resists key-rolling, but weakens the shaft compared to a square or tapered keyseat. Woodruff key widths as a function of shaft diameter are essentially the same as for square keys shown in Table 9-2. The other dimensions of the Woodruff key are defined in the ANSI Standard and keyseat cutters are readily available to match these dimensions. Table 9-3 reproduces a sample of the key size specifications from the standard. Each key size is given a key number, which encodes its dimensions. The ANSI Standard states: "The last two digits give the nominal key diameter in eighths of an inch and the digits preceding the last two give the nominal width in thirty-seconds of an inch." For example, the key number 808 defines a key size of 8/32 x 8/8 or 1/4 wide x 1-in dia. See reference 6 for complete dimensional information on keys.

Stresses in Keys

There are two modes of failure in keys, shear and bearing. A shear failure occurs when the key is sheared across its width at the interface between the shaft and hub. Bearing failure occurs by crushing either side in compression.

SHEAR FAILURE   The average stress due to direct shear was defined in equation 4.9, repeated here:

$$\tau_{xy} = \frac{F}{A_{shear}} \qquad (9.10)$$

where F is the applied force and A_{shear} is the shear area being cut. In this case A_{shear} is the product of the key's width and length. The force on the key can be found from the quotient of the shaft torque and the shaft radius. If the shaft torque is constant with time, the force will be also and the safety factor can be found by comparing the shear

Table 9-3   ANSI Standard Sizes of Woodruff Keys
A Partial List: See the Standard for Complete Information and Figure 9-12c for Labels

Key Number	Nominal Key Size W x L	Height H
202	0.062 x 0.250	0.106
303	0.093 x 0.375	0.170
404	0.125 x 0.500	0.200
605	0.187 x 0.625	0.250
806	0.250 x 0.750	0.312
707	0.218 x 0.875	0.375
608	0.187 x 1.000	0.437
808	0.250 x 1.000	0.437
1208	0.375 x 1.000	0.437
610	0.187 x 1.250	0.545
810	0.250 x 1.250	0.545
1210	0.187 x 1.250	0.545
812	0.250 x 1.500	0.592
1212	0.375 x 1.500	0.592

stress to the shear yield strength of the material. If the shaft torque is time-varying, then a fatigue failure of the key in shear is possible. The approach then is to compute the mean and alternating shear stress components and use them to compute the mean and alternating von Mises stresses. These can then be used in a modified-Goodman diagram to find the safety factor as described in Section 6.13.

BEARING FAILURE   The average bearing stress is defined as

$$\sigma_x = \frac{F}{A_{bearing}}$$ (9.11)

where F is the applied force and the bearing area is the area of contact between the key side and the shaft or the hub. For a square key this will be its half-height times its length. A Woodruff key has a different bearing area in the hub than in the shaft. The hub's Woodruff bearing area is much smaller and will fail first. The bearing stress should be calculated using the maximum applied force, whether constant or time-varying. Since compressive stresses do not cause fatigue failures, bearing stresses can be considered static. The safety factor is found by comparing the maximum bearing stress to the material yield strength in compression.

Key Materials

Because keys are loaded in shear, ductile materials are used. Soft, low-carbon steel is the most common choice unless a corrosive environment requires a brass or stainless steel key. Square or rectangular keys are often made from cold-rolled bar stock and merely cut to length. The special keystock mentioned above is used when a closer fit is required between key and keyway. Tapered and Woodruff keys are also usually made from soft, cold-rolled steel.

Key Design

Only a few design variables are available when sizing a key. The shaft diameter at the keyseat determines the key width. The key height (or its penetration into the hub) is also determined by the key width. This leaves only the length of the key and the number of keys used per hub as design variables. A straight or tapered key can be as long as the hub allows. A Woodruff key can be had in a range of diameters for a given width, which effectively determines its length of engagement in the hub. Of course, as the Woodruff key diameter is increased, it further weakens the shaft with its deeper keyseat. If a single key cannot handle the torque at reasonable stresses, an additional key can be added, rotated 90° from the first.

It is common to size the key so that it will fail before the keyseat or other location in the shaft fails in the event of an overload. The key then acts like a shear pin in an outboard motor to protect the more expensive elements from damage. A key is inexpensive and relatively easy to replace if the keyseat is undamaged. This is one reason to use only soft, ductile materials for the key, having lower strength than that of the shaft so that a bearing failure will selectively affect the key rather than the keyway if the system sees an overload beyond its design range.

Stress Concentrations in Keyways

Since keys have relatively sharp corners, (< 0.02-in radius) keyseats must also. This causes significant stress concentrations. The keyway is broached in the hub and runs through its length but the keyway must be milled into the shaft and has one or two ends.

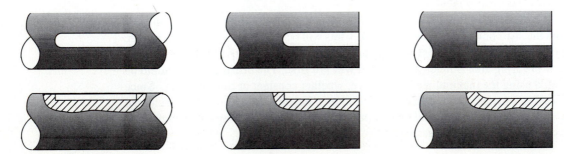

(a) End-milled keyway—double-ended   (b) End-milled keyway—single-ended   (c) Sled-runner keyway—single-ended

FIGURE 9-15

Various Styles of Keyways in Shafts

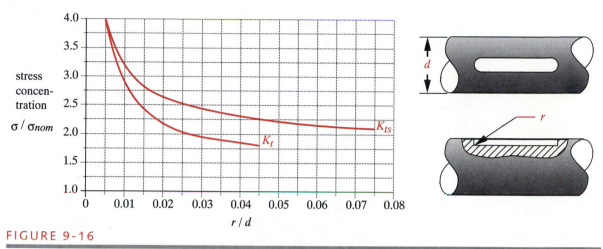

FIGURE 9-16

Stress-Concentration Factors for an End-Milled Keyseat in Bending (K_t) and Torsion (K_{ts}) Source: *Stress Concentration Factors*, R. C. Peterson, 1974, reprinted by permision of John Wiley & Sons, Inc.

If an end-mill is used, the keyway will look like Figure 9-15a and will have sharp corners in the side view at one or both ends as well as along each side. If instead, a sled-runner keyway is cut as shown in Figure 9-14b.) the sharp corner at the end is eliminated and the stress concentration reduced. A Woodruff keyseat in the shaft also has a large radius in the side view but it (and all) keyseats suffer from sharp corners on the sides.

Peterson[7] shows experimentally derived stress-concentration curves for end-milled keyseats in shafts under either bending or torsional loading. These are reproduced in Figure 9-16. These factors range from about 2 to about 4 depending on the ratio of the corner radius to the shaft diameter. Curve-fits to Figure 9-16 have been done and *TKSolver* functions created for these curves so that the stress-concentration factor can be determined "on the fly" during a shaft-design computation See the file SHFTDES. These factors should be applied to the bending and shear stresses in the shaft at the keyway location as was done in Examples 9-1 and 9-2.

EXAMPLE 9-4

Designing Shaft Keys

Problem   Design the keys for the shaft in Examples 9-2 and 9-3 and refine the estimate of the shaft's safety factors based on the preliminary design dimensions from that earlier example in conjunction with refined stress-concentration factors.

Given   The loading is the same as in Example 9-2. The peak torque is 146 lb-in. Figure 9-9 shows the distribution of the peak moment over the shaft length. The values are 65.6 lb-in at point *B*, 127.9 lb-in at point *C*, and

18.3 lb-in at point D. The preliminary shaft diameters are $d_0 = 0.875$ in, $d_1 = 0.750$ in, $d_2 = 0.669$ in, and $d_3 = 0.531$ in. See Figure 9-3 for labels.

Assumptions   Use square, parallel keys with end-milled keyways. The shaft material is the same as in Example 9-3. A lower-carbon steel, SAE 1010, will be used for the keys. Its $S_{ut} = 53$ kpsi and $S_y = 44$ kpsi. S_e is calculated to be 22 990 psi. See Figure 9-16 for stress-concentration factors.

Solution   See the *TKSolver* files EX09-04a and EX09-04b.

1 There are two locations with keys on this shaft, at points B and D. The design diameters chosen for these sections in Example 9-3 were $d_1 = 0.750$, $d_2 = 0.669$ in, and $d_3 = 0.531$ in, respectively. Table 9-2 shows that the standard key width for d_1 is 0.187 in and for d_3 is 0.125 in. The key length can be adjusted for each location.

2 At point B, the mean and alternating components of force on the key are found from the torque component divided by the shaft radius at that point.

$$F_a = \frac{T_a}{r} = \frac{73.1}{0.375} = 194.67 \text{ lb}$$

$$F_m = \frac{T_m}{r} = \frac{73.1}{0.375} = 194.67 \text{ lb} \tag{a}$$

3 Assume a key length of 0.5 in and calculate the alternating and mean shear stress components from

$$\tau_a = \frac{F_a}{A_{shear}} = \frac{194.67}{0.187(0.500)} = 2\,082 \text{ psi}$$

$$\tau_m = \frac{F_m}{A_{shear}} = \frac{194.67}{0.187(0.500)} = 2\,082 \text{ psi} \tag{b}$$

4 To find a safety factor for shear fatigue of the key, compute the von Mises equivalent stresses for each of these components from equation 5.7d,

$$\sigma'_a = \sqrt{\sigma_x^2 + \sigma_y^2 - \sigma_x\sigma_y + 3\tau_{xy}^2} = \sqrt{3(2\,082)^2} = 3\,606 \text{ psi}$$

$$\sigma'_m = \sqrt{\sigma_x^2 + \sigma_y^2 - \sigma_x\sigma_y + 3\tau_{xy}^2} = \sqrt{3(2\,082)^2} = 3\,606 \text{ psi} \tag{c}$$

and use them in equation 6.18e to determine the fatigue safety factor:

$$N_f = \frac{1}{\dfrac{\sigma'_a}{S_e} + \dfrac{\sigma'_m}{S_{ut}}} = \frac{1}{\dfrac{3\,606}{22\,990} + \dfrac{3\,606}{53\,000}} = 4.4 \tag{d}$$

5 The bearing stress on the key is compression and thus can be considered a static load. It is calculated using the maximum force on the key:

$$\sigma_{max} = \frac{F_m + F_a}{A_{bearing}} = \frac{194.67 + 194.67}{0.093\,5(0.500)} = 8\,328 \text{ psi} \tag{e}$$

6 Calculate the safety factor for bearing failure from:

$$N_s = \frac{S_y}{\sigma_{max}} = \frac{44\,000}{4\,164} = 5.3 \qquad (f)$$

7 At point D, the force on the key is

$$F_a = \frac{T_a}{r} = \frac{73.1}{0.266} = 275 \text{ lb}$$

$$F_m = \frac{T_m}{r} = \frac{73.1}{0.266} = 275 \text{ lb} \qquad (g)$$

8 Assume a key length of 0.50 in and calculate the alternating and mean shear stress components from

$$\tau_a = \frac{F_a}{A_{shear}} = \frac{275}{0.125(0.50)} = 4\,400 \text{ psi}$$

$$\tau_m = \frac{F_m}{A_{shear}} = \frac{275}{0.125(0.50)} = 4\,400 \text{ psi} \qquad (h)$$

9 Compute the von Mises equivalent stresses for each of these components from equation 5.7d,

$$\sigma'_a = \sqrt{\sigma_x^2 + \sigma_y^2 - \sigma_x\sigma_y + 3\tau_{xy}^2} = \sqrt{3(4\,399)^2} = 7\,620 \text{ psi}$$

$$\sigma'_m = \sqrt{\sigma_x^2 + \sigma_y^2 - \sigma_x\sigma_y + 3\tau_{xy}^2} = \sqrt{3(4\,399)^2} = 7\,620 \text{ psi} \qquad (i)$$

and use them in equation 6.18e:

$$N_f = \frac{1}{\dfrac{\sigma'_a}{S_e} + \dfrac{\sigma'_m}{S_{ut}}} = \frac{1}{\dfrac{7\,620}{22\,990} + \dfrac{7\,620}{53\,000}} = 2.1 \qquad (j)$$

10 The bearing stress on the key is calculated using the maximum force on the key:

$$\sigma_{max} = \frac{F_m + F_a}{A_{bearing}} = \frac{275 + 275}{0.0625(0.50)} = 17\,600 \text{ psi} \qquad (k)$$

11 Calculate the safety factor for bearing failure from

$$N_s = \frac{S_y}{\sigma_{max}} = \frac{44\,000}{17\,600} = 2.5 \qquad (l)$$

12 The safety factors for the shaft at these locations can now be recalculated using a stress-concentration factor at the keyways that takes into account the actual shaft diameter. Our previous design calculation in Example 9-2 used a worst-case assumption for these values. Figure 9-16 shows the stress-concentration functions for end-milled keyways in both bending and torsion. To use these charts we must

calculate the r/d ratio of the end-mill radius versus the shaft diameter. Assume a radius on the end mill of 0.010 in. The r/d ratios for the two points are then

for point B:
$$\frac{r}{d} = \frac{0.010}{0.750} = 0.0133$$

for point D:
$$\frac{r}{d} = \frac{0.010}{0.531} = 0.0188 \qquad (m)$$

The corresponding stress-concentration factors are read from Figure 9-16 as

for point B: $K_t = 2.5$ $K_{ts} = 2.9$

for point D: $K_t = 2.2$ $K_{ts} = 2.7$ (n)

13 These are used in equations (m), (n), and (o) of Example 9-1 to obtain the fatigue stress-concentration factors, which for a material notch sensitivity $q = 0.4$, are

for point B: $K_f = 1.7$ $K_{fs} = 1.9$

for point D: $K_f = 1.6$ $K_{fs} = 1.8$ (o)

for both points: $K_{fm} = K_f$ $K_{fsm} = K_{fs}$

14 The new safety factors are then calculated using equation 9.8 with the data from equations (b) and (c) from Example 9-2 with the design values for shaft diameter and the above stress-concentration values inserted:

for point B:

$$0.75 = \left\{ \frac{32(N_f)}{\pi} \left[\frac{\sqrt{[1.7(32.8)]^2 + \frac{3}{4}[1.9(73.1)]^2}}{27\,300} + \frac{\sqrt{[1.7(32.8)]^2 + \frac{3}{4}[1.9(73.1)]^2}}{38\,000} \right] \right\}^{\frac{1}{3}}$$

$N_f = 5.1$ $\qquad (p)$

for point D:

$$0.531 = \left\{ \frac{32(N_f)}{\pi} \left[\frac{\sqrt{[1.6(9.1)]^2 + \frac{3}{4}[1.8(73.1)]^2}}{27\,300} + \frac{\sqrt{[1.6(9.1)]^2 + \frac{3}{4}[1.8(73.1)]^2}}{38\,000} \right] \right\}^{\frac{1}{3}}$$

$N_f = 2.1$ $\qquad (q)$

At point B the safety factor is greater than the specified value of 2.5. At point D it is lower. Increasing the diameter at D to 0.562 in gives the desired safety factor of 2.5. Then the safety factors for key failure (4.4 at B and 2.1 at D) are lower than those for shaft failure, which is desirable since the keys will then fail before the shafts in an overload situation. This is now a viable and acceptable design.

9.11 SPLINES

When more torque must be transmitted than can be handled by keys, splines can be used instead. Splines are essentially "built-in keys" formed by contouring the outside of the shaft and inside of the hub with toothlike forms. Early splines had teeth of square cross section but these have been supplanted by involute spline teeth, as shown in Figure 9-17. The involute tooth form is universally used on gears and the same cutting techniques are used to manufacture splines. In addition to its manufacturing advantage, the involute tooth has less stress concentration than a square tooth and is stronger. The SAE defines standards for both square and involute spline tooth forms and ANSI publishes involute spline standards.* The standard involute spline has a pressure angle of 30° and half the depth of a standard gear tooth. The tooth size is defined by a fraction whose numerator is the diametral pitch (which defines tooth width—see Chapter 11 for more information on these terms) and the denominator controls tooth depth (and is always double the numerator). Standard diametral pitches are 2.5, 3, 4, 5, 6, 8, 10, 12, 16, 20, 24, 32, 40, and 48. Standard splines can have from 6 to 50 teeth. Splines may have either a flat root or a filleted root, both shapes being shown in Figure 9-17. See reference 8 for complete dimensional information on standard splines.

Some advantages of splines are maximum strength at the root of the tooth, accuracy of tooth form due to the use of standard cutters, and good machined surface finish from the standard gear-cutting (hobbing) process, which eliminates the need for grinding. A major advantage of splines over keys is their ability (with proper clearances) to accommodate large axial movements between shaft and hub while simultaneously transmitting torque. They are used to connect the transmission output shaft to the driveshaft in automobiles and trucks where the suspension movement causes axial motion between the members. They are also used within nonautomatic, nonsynchromesh truck transmissions to couple the axially shiftable gears to their shafts. In addition, engine torque is usually passed into the transmission through a spline that connects the engine clutch

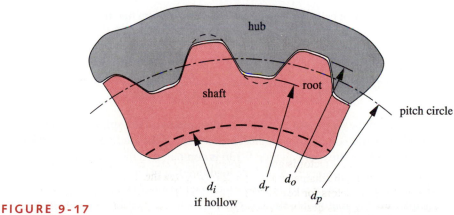

FIGURE 9-17

Involute Spline Geometry

* ANSI Standard B92.1 and B92.2M, American National Standards Institute, 11 West 42nd St., New York, N.Y. 10036

to the transmission input shaft and allows the axial motion necessary to disengage the clutch from the flywheel.

The loading on a spline is typically pure torsion, either steady or fluctuating. While it is possible to have bending loads superposed, good design practice will minimize the bending moments by proper placement of bearings and by keeping a cantilevered spline as short as possible. As with keys, two failure modes are possible, bearing or shear. Shear failure is usually the limiting mode. Unlike keys, many teeth are available to share the load to some degree. Ideally, the spline length l needs to be only as long as is required to develop a combined tooth shear strength equal to the torsional shear strength of the shaft itself. If the spline were made perfectly with no variation in either tooth thickness or spacing, all teeth would share the load equally. However, the realities of manufacturing tolerances prohibit this ideal condition. The SAE states that "actual practice has shown that due to inaccuracies in spacing and tooth form, the equivalent of about 25% of teeth are in contact, so that a good approximate formula for a splined shaft (length) is"

$$l \cong \frac{d_r^3\left(1 - d_i^4/d_r^4\right)}{d_p^2} \qquad (9.12)$$

where d_r is the root diameter of the external spline, d_i is the internal diameter (if any) of a hollow shaft, and d_p is the pitch diameter of the spline, which is approximately at mid-tooth. The variable l represents the actual engaged length of the spline teeth with one another and should be considered as the minimum value needed to develop the strength in the teeth of an equivalent diameter shaft.

The shear stress is calculated at the pitch diameter of the splines where the shear area is

$$A_{shear} = \frac{\pi d_p l}{2} \qquad (9.13a)$$

The shear stress can be calculated using the SAE assumption that only 25% of the teeth are actually sharing the load at any one time by considering only 1/4 of the shear area to be stressed:

$$\tau \cong \frac{4F}{A_{shear}} = \frac{4T}{r_p A_{shear}} = \frac{8T}{d_p A_{shear}} = \frac{16T}{\pi d_p^2 l} \qquad (9.13b)$$

where T is the torque on the shaft. Any bending stresses on the spline must also be calculated and properly combined with this shear stress. If the loading is pure torsion and static, then the shear stress of equation 9.13b is compared to the shear yield strength of the material to obtain a safety factor. If the loads are fluctuating or if bending is present, then the applied stresses should be converted to equivalent von Mises tensile stresses and compared to the appropriate strength criteria using the modified-Goodman diagram.

9.12   INTERFERENCE FITS

Another common means of coupling a hub to a shaft is to use a **press** or **shrink fit**, also called an **interference fit**. A press fit is obtained by machining the hole in the hub to a slightly smaller diameter than that of the shaft as shown in Figure 9-18. The two parts are then forced together slowly in a press, preferably with oil lubricant applied to the joint. The elastic deflection of both shaft and hub act to create large normal and frictional forces between the parts. The friction force transmits the shaft torque to the hub and resists axial motion as well. The American Gear Manufacturers Association (AGMA) publishes a standard AGMA 9003-A91, *Flexible Couplings—Keyless Fits* which defines formulas for the calculation of interference fits.

Only relatively small parts can be press-fitted without exceeding the force capacity of a typical shop press. For larger parts, a **shrink fit** can be made by heating the hub to expand its inside diameter and/or an **expansion fit** can be made by cooling the shaft to reduce its diameter. The hot and cold parts can be slipped together with little axial force, and when they equilibrate to room temperature, their dimensional change creates the desired interference for frictional contact. Another method is to hydraulically expand the hub with pressurized oil delivered through passageways in the shaft or hub. This technique can be used to remove a hub as well.

The amount of interference needed to create a tight joint varies with the diameter of the shaft. Approximately 0.001 to 0.002 units of diametral interference per unit of shaft diameter is typical, (the rule of thousandths) the smaller amount being used with larger shaft diameters. For example, the interference for a diameter of 2 in would be about 0.004 in, but a diameter of 8 in would receive only about 0.009 to 0.010 in of interference. Another (and simpler) machinist's rule of thumb is to use 0.001 in of interference for diameters up to 1 in, and 0.002 in for diameters from 1 to 4 in.

Stresses in Interference Fits

An interference fit creates the same stress state in the shaft as would a uniform external pressure on its surface. The hub experiences the same stresses as a thick-walled cylinder subjected to internal pressure. The equations for stresses in a thick-walled cylinder were presented in Section 4.16 and depend on the applied pressures and the radii of the elements. The pressure p created by the press fit can be found from the deformation of the materials caused by the interference.

$$p = \frac{0.5\delta}{\dfrac{r}{E_o}\left(\dfrac{r_o^2 + r^2}{r_o^2 - r^2} + \nu_o\right) + \dfrac{r}{E_i}\left(\dfrac{r^2 + r_i^2}{r^2 - r_i^2} - \nu_i\right)} \qquad (9.14a)$$

where $\delta = 2\Delta r$ is the total diametral interference between the two parts, r is the nominal radius of the interface between the parts, r_i is the inside radius (if any) of a hollow shaft, and r_o is the outside radius of the hub as shown in Figure 9-18. E and ν are the Young's moduli and Poisson's ratios of the two parts, respectively.

The torque that can be transmitted by the interference fit can be defined in terms of the pressure p at the interface, which creates a friction force at the shaft radius.

$$T = 2\pi r^2 \mu\, pl \qquad\qquad (9.14b)$$

where l is the length of the hub engagement, r is the shaft radius, and μ is the coefficient of friction between shaft and hub. The AGMA standard suggests a value of $0.12 \le \mu \le 0.15$ for hydraulically expanded hubs and $0.15 \le \mu \le 0.20$ for shrink or press-fit hubs. AGMA assumes (and recommends) a surface finish of 32 μin rms (1.6 μm R_a) which requires a ground finish on both diameters. Equations 9.14a and 9.14b can be combined to give an expression that defines the torque obtainable from a particular deformation, coefficient of friction, and geometry.

$$T = \frac{\pi l r \mu\, \delta}{\dfrac{1}{E_o}\left(\dfrac{r_o^2 + r^2}{r_o^2 - r^2} + \nu_o\right) + \dfrac{1}{E_i}\left(\dfrac{r^2 + r_i^2}{r^2 - r_i^2} - \nu_i\right)} \qquad\qquad (9.14c)$$

The pressure p is used in equations 4.47 to find the radial and tangential stresses in each part. For the shaft:

$$\sigma_{t_{shaft}} = -p\frac{r^2 + r_i^2}{r^2 - r_i^2} \qquad\qquad (9.15a)$$

$$\sigma_{r_{shaft}} = -p \qquad\qquad (9.15b)$$

where r_i is the inside radius of a hollow shaft. If the shaft is solid, r_i will be zero.

For the hub:

$$\sigma_{t_{hub}} = p\frac{r_o^2 + r^2}{r_o^2 - r^2} \qquad\qquad (9.16a)$$

$$\sigma_{r_{hub}} = -p \qquad\qquad (9.16b)$$

These stresses need to be kept below the yield strengths of the materials to maintain the fit. If the materials yield, the hub will become loose on the shaft.

Stress Concentration in Interference Fits

Even though there may be no disruption of the smooth surface of the press-fit shaft by shoulders or keyways, an interference fit nevertheless creates stress concentrations in the shaft and hub at the ends of the hub due to the abrupt transition from uncompressed

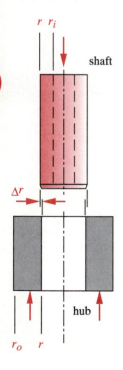

FIGURE 9-18

An Interference Fit

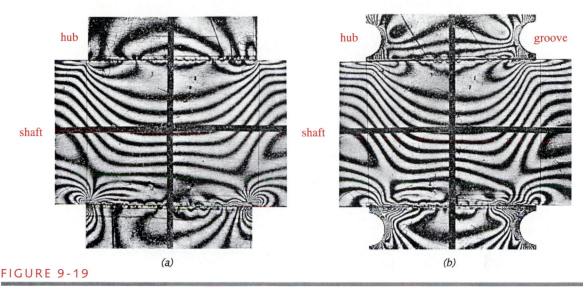

hub   hub   groove

shaft   shaft

(a)   (b)

FIGURE 9-19

Photoelastic Stress Analysis of (a) A Plain Press-fit Assembly and (b) A Grooved-Hub Press-fit Assembly   Source: R. E. Peterson and A. M. Wahl, Fatigue of Shafts at Fitted Members, with a Related Photoelastic Analysis, *ASME J. App., Mech.*, vol. 57, p. A1, 1935.

to compressed material. Figure 9-19*a* shows a photoelastic study of a press-fit hub on a shaft. The fringes show stress concentration at the corners. Figure 9-19*b* shows how the stress concentration can be reduced by providing circumferential relief grooves in the faces of the hub close to the shaft diameter. These grooves make the material at the edge of the hub more compliant enabling it to deflect away from the shaft and reduce the stress locally. This approach is similar to the techniques for stress-concentration reduction shown in Figure 4-37.

Figure 9-20 shows curves of stress-concentration factors for interference fits between hubs and shafts developed from the photoelastic study of Figure 9-19*a*. The values on the abscissa are ratios of hub length to shaft diameter. These geometric stress-concentration factors are applied in the same manner as before. For static loading they need to be used to determine whether local yielding will compromise the interference fit. For dynamic loading, they are modified by the material's notch sensitivity to get a fatigue stress-concentration factor to use in equation 9.8 for shaft design.

Fretting Corrosion

This problem was discussed in Chapter 7. Interference fits are the primary victims of fretting problems. Though the fretting mechanism is not yet fully understood, certain precautions are known to help reduce its severity. See Section 7.6 for more details.

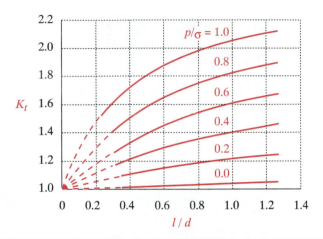

$$p/\sigma = \frac{\text{nominal press-fit pressure}}{\text{nominal bending stress}}$$

$$l/d = \frac{\text{length of hub}}{\text{diameter of shaft}}$$

FIGURE 9-20

Stress Concentration in a Press-Fit or Shrink-Fit Hub on a Shaft   Source: R. E. Peterson and A. M. Wahl, Fatigue of Shafts at Fitted Members, with a Related Photoelastic Analysis, *ASME J. App., Mech.*, vol. 57, p. A73, 1935.

EXAMPLE 9-5

Designing an Interference Fit

Problem   Redesign the attachment of the gear to the shaft in Figure 9-3 to make it an interference fit rather than a keyed connection. Define the shaft and gear hole dimensions and their tolerances for a press fit.

Given   The loading is the same as in Example 9-2. The peak torque at the gear (point *B*) is 146 lb-in. Use the nominal shaft diameters of $d_0 = 0.875$ in, $d_1 = 0.750$ in. See Figure 9-3 for labels. The gear hub diameter is 3 in and its length is 1.5 in.

Assumptions   The shaft material is the same as in Example 9-2. Class 30 gray cast iron is used for the gear with $S_{ut} = 32$ kpsi and $E = 14$ Mpsi. See Figure 9-20 for stress-concentration factors. The 0.750 shaft diameter will be increased slightly to 0.780 nominal, where it press fits into the gear hub to allow the gear to slip over the rest of the shaft at assembly.

Solution   See the *TKSolver* file EX09-05.

1   The nominal diameter of the shaft at the gear hub is 0.780 in. Based on the rule of thousandths, a reasonable diametral interference would be 0.0015 in. From this assumption, the bearing pressure after pressing can be found from equation 9.14.

$$p = \frac{0.5\delta}{\frac{r}{E_o}\left(\frac{r_o^2 + r^2}{r_o^2 - r^2} + v_o\right) + \frac{r}{E_i}\left(\frac{r^2 + r_i^2}{r^2 - r_i^2} - v_i\right)}$$

$$= \frac{0.5(0.001\,5)}{\frac{0.390}{1.4E7}\left(\frac{1.5^2 + 0.390^2}{1.5^2 - 0.390^2} + 0.28\right) + \frac{0.390}{3.0E7}\left(\frac{0.390^2 + 0}{0.390^2 - 0} - 0.28\right)} \quad (a)$$

$$p = 15\,288 \text{ psi}$$

2 The stresses in the shaft after pressing are found from equations 9.15.

$$\sigma_{t_{shaft}} = -p\frac{r^2 + r_i^2}{r^2 - r_i^2} = -15\,288\frac{0.390^2 + 0}{0.390^2 - 0} = -15\,288 \text{ psi} \quad (b)$$

$$\sigma_{r_{shaft}} = -p = -15\,288 \text{ psi} \quad (c)$$

3 The stresses in the hub after pressing are found from equations 9.16.

$$\sigma_{t_{hub}} = p\frac{r_o^2 + r^2}{r_o^2 - r^2} = 15\,288\frac{1.5^2 + 0.390^2}{1.5^2 - 0.390^2} = 17\,505 \text{ psi} \quad (d)$$

$$\sigma_{r_{hub}} = -p = -15\,288 \text{ psi} \quad (e)$$

4 To find the stress-concentration factor we need the hub length to shaft diameter ratio, l/d:

$$\frac{l}{d} = \frac{1.500}{0.780} = 1.923 \quad (f)$$

and the ratio of press-fit pressure to nominal bending stress:

$$\sigma = \frac{Mc}{I} = \frac{65.6(0.390)64}{\pi(0.780^4)} = 22\,529 \text{ psi} \quad (g)$$

$$\frac{p}{\sigma} = \frac{15\,288}{22\,529} = 0.68 \quad (h)$$

5 Taking these values to Figure 9-20, we find that the l/d ratio is off the chart. Extrapolation gives an approximate value for the stress-concentration factor of:

$$k_t \cong 1.82 \quad (i)$$

6 The safety factors against failure during press fitting can now be found:

$$N_{s_{shaft}} = \frac{S_y}{k_t \sigma_{t_{shaft}}} = \frac{-38\,000}{1.82(-15\,288)} = 1.37 \quad (j)$$

$$N_{s_{hub}} = \frac{S_{ut}}{k_t \sigma_{t_{hub}}} = \frac{32\,000}{1.82(17\,505)} = 1.01 \quad (k)$$

Table 9-4   Safety Factors for Various Interferences in Example 9-5

Interference (in)	p (psi)	p/σ	K_t	$N_{s_{shaft}}$	$N_{s_{hub}}$
0.000 8	8 154	0.36	1.46	3.2	2.3
0.000 9	9 173	0.41	1.51	2.7	2.0
0.001 0	10 192	0.45	1.55	2.4	1.8
0.001 1	11 212	0.50	1.60	2.1	1.6
0.001 2	12 231	0.54	1.64	1.9	1.4
0.001 3	13 250	0.59	1.69	1.7	1.2
0.001 4	14 269	0.63	1.75	1.5	1.1
0.001 5	15 288	0.68	1.82	1.4	1.0

7   The shaft is safe, but the hub is at failure with this much interference. The computations were repeated for a range of interference values from 0.0008 to 0.0015 in and those results are shown in Table 9-4. The hub safety factors range from 2.3 to 1.0.

8   We would like a tolerance range of at least 0.0002 in on each part or a total variation in interference of at least 0.0004 in for a set of mass-produced parts. So we choose to set the range of interference from 0.0008 to 0.0013 in, a total range of 0.0005 in.

9   The part dimensions are then established as

$$hub\ dia = 0.7800 + 0.0000/-0.0003 = \frac{0.7800}{0.7797}\ \text{in}$$

$$shaft\ dia = 0.7808 + 0.0002/-0.0000 = \frac{0.7810}{0.7808}\ \text{in}$$

(l)

giving a range of interference of

$$\text{min interference} = 0.7808 - 0.7800 = 0.0008\ \text{in}$$

$$\text{max interference} = 0.7810 - 0.7797 = 0.0013\ \text{in}$$

(m)

10 What torque will this pressed joint transmit at its minimum interference assuming $\mu = 0.15$? From equation 9.14c:

$$T = \frac{\pi l r \mu \delta}{\dfrac{1}{E_o}\left(\dfrac{r_o^2 + r^2}{r_o^2 - r^2} + \nu_o\right) + \dfrac{1}{E_i}\left(\dfrac{r^2 + r_i^2}{r^2 - r_i^2} - \nu_i\right)}$$

$$T = \frac{\pi(1.5)(0.375)(0.15)(0.000\ 8)}{\dfrac{0.375}{1.4E7}\left(\dfrac{1.5^2 + 0.390^2}{1.5^2 - 0.390^2} + 0.28\right) + \dfrac{0.375}{3.0E7}\left(\dfrac{0.390^2 + 0}{0.390^2 - 0} - 0.28\right)}$$

(n)

$$T = 1\ 753\ \text{in - lb}$$

This is well in excess of the peak operating torque of 146 in-lb, so will work.

9.13   FLYWHEEL DESIGN

A flywheel is used to smooth out variations in the speed of a shaft caused by torque fluctuations. Many machines have load patterns that cause the torque time function to vary over the cycle. Piston compressors, punch presses, rock crushers, etc., all have time-varying loads. The prime mover can also introduce torque oscillations to the transmission shaft. Internal combustion engines with one or two cylinders are an example. Other systems may have both smooth torque sources and smooth loads such as an electrical generator driven by a steam turbine. These smooth-acting devices have no need for a flywheel. If the source of the driving torque or the load torque have a fluctuating nature, then a flywheel is usually called for.

A flywheel is an energy storage device. It absorbs and stores kinetic energy when speeded up and returns energy to the system when needed by slowing its rotational speed. The kinetic energy E_k in a rotating system is

$$E_k = \frac{1}{2} I_m \omega^2 \tag{9.17a}$$

where I_m is the mass moment of inertia of all rotating mass on the shaft about the axis of rotation and ω is the rotational velocity. This includes the I_m of the motor rotor and anything else rotating with the shaft plus that of the flywheel.

Flywheels may be as simple as a cylindrical disk of solid material, or be of spoked construction with a hub and rim. The latter arrangement is more efficient of material, especially for large flywheels, as it concentrates the bulk of its mass in the rim, which is at the largest radius. Since the mass moment of inertia I_m of a flywheel is proportional to mr^2, mass at larger radius contributes much more. If we assume a solid-disk geometry with inside radius r_i and outside radius r_o, the mass moment of inertia is

$$I_m = \frac{m}{2} \left(r_o^2 + r_i^2 \right) \tag{9.17b}$$

The mass of a hollow circular disk of constant thickness t is:

$$m = \frac{W}{g} = \pi \frac{\gamma}{g} \left(r_o^2 - r_i^2 \right) t \tag{9.17c}$$

Substituting in equation 9.17b gives an expression for I_m in terms of the disk geometry:

$$I_m = \frac{\pi}{2} \frac{\gamma}{g} \left(r_o^4 - r_i^4 \right) t \tag{9.17d}$$

where γ is the material's weight density and g is the gravitational constant.

There are two stages to the design of a flywheel. First, the amount of energy required for the desired degree of smoothing must be found and the moment of inertia needed to absorb that energy determined. Then a flywheel geometry must be defined

*   Portions of this section are adapted from R. L. Norton, *Design of Machinery,* McGraw-Hill, 1992, pp. 482-489, with the publisher's permission.

that both supplies that mass moment of inertia in a reasonably sized package and that is safe against failure at design speeds.

Energy Variation in a Rotating System

Figure 9-21 shows a flywheel, designed as a flat circular disk, attached to a motor shaft. The motor supplies a torque magnitude T_m, which we would like to be as constant as possible, i.e., to be equal to the average torque T_{avg}. Assume that the load on the other side of the flywheel, demands a torque T_l, which is time-varying, as shown in Figure 9-22. This torque variation can cause the shaft speed to vary depending on the torque-speed characteristic of the driving motor. We need to determine how much I_m to add in the form of a flywheel to reduce the speed variation of the shaft to an acceptable level. Write Newton's law for the free-body diagram in Figure 9-21.

$$\sum T = I_m \, \alpha$$

$$T_l - T_m = I_m \, \alpha \qquad (9.18a)$$

but we want

$$T_m = T_{avg}$$

so

$$T_l - T_{avg} = I_m \, \alpha \qquad (9.18b)$$

substituting

$$\alpha = \frac{d\omega}{dt} = \frac{d\omega}{dt}\left(\frac{d\theta}{d\theta}\right) = \omega \, \frac{d\omega}{d\theta}$$

gives

$$T_l - T_{avg} = I_m \, \omega \, \frac{d\omega}{d\theta}$$

$$\left(T_l - T_{avg}\right) d\theta = I_m \, \omega \, d\omega \qquad (9.18c)$$

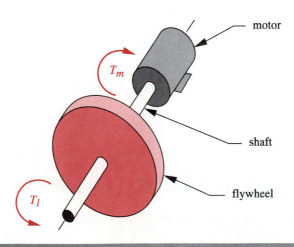

motor

T_m

shaft

flywheel

T_l

FIGURE 9-21

A Flywheel on a Transmission Shaft

integrating

$$\int_{\theta\,@\,\omega_{min}}^{\theta\,@\,\omega_{max}} \left(T_l - T_{avg}\right) d\theta = \int_{\omega_{min}}^{\omega_{max}} I_m\,\omega\,d\omega$$

$$\int_{\theta\,@\,\omega_{min}}^{\theta\,@\,\omega_{max}} \left(T_l - T_{avg}\right) d\theta = \frac{1}{2} I_m \left(\omega_{max}^2 - \omega_{min}^2\right) \qquad (9.18d)$$

The left-hand side of this expression represents the change in kinetic energy E_k between the maximum and minimum shaft ω's and is equal to the area under the torque-time diagram of Figure 9-22 between those extreme values of ω. The right-hand side of equation 9.18c is the change in kinetic energy stored in the flywheel. The only way to extract kinetic energy from the flywheel is to slow it down as shown in equation 9.17a. Adding kinetic energy speeds it up. It is impossible to obtain exactly constant shaft velocity in the face of changing energy demands by the load. The best we can do is to minimize the speed variation $(\omega_{max} - \omega_{min})$ by providing a flywheel with sufficiently large I_m.

EXAMPLE 9-6

Determining the Energy Variation in a Torque-Time Function

Problem   Find the energy variation per cycle in a torque-time function that needs to be absorbed by a flywheel for smooth operation.

Given   A torque-time function that varies over its cycle as shown in Figure 9-22. The torque varies during the 360° cycle about its average value.

Assumptions   The one cycle of torque variation shown is representative of the steady-state condition. Energy delivered from the source to the load will be considered positive and energy returned from load to source, negative.

Solution

1   Calculate the average value of the torque-time function over one cycle using numerical integration. In this case it is 7 020 lb-in. (Note that in some cases the average value may be zero.)

2   Note that the integration on the left-hand side of equation 9.18c is done with respect to the average line of the torque function, not with respect to the θ axis. (From the definition of the average, the sum of positive area above an average line is equal to the sum of negative area below that line.) The integration limits in equation 9.18 are from the shaft angle θ, at which the shaft ω is a minimum to the shaft angle θ, at which ω is a maximum.

3   The minimum ω will occur after the maximum positive energy has been delivered from the motor to the load, i.e., at a point θ where the summation of positive energy (area) in the torque pulses is at its largest positive value.

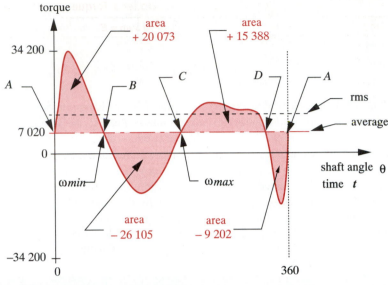

FIGURE 9-22

Integrating the Pulses Above and Below the Average Value in the Torque-Time Function

9

4   The maximum ω will occur after the maximum negative energy has been returned to the load, i.e., at a point θ where the summation of energy (area) in the torque pulses is at its largest negative value.

5   To find these locations in θ corresponding to the maximum and minimum ω's and thus find the amount of energy needed to be stored in the flywheel, we need to numerically integrate each pulse of this function from crossover to crossover with the average line. The crossover points have been labeled *A, B, C,* and *D* and the areas between them are shown in Figure 9-22.

6   The remaining task is to accumulate these pulse areas beginning at an arbitrary crossover (in this case point *A*) and proceeding pulse by pulse across the cycle. Table 9-5 shows this process and the result.

8   Note in Table 9-5 that the minimum shaft speed occurs after the largest accumulated positive energy pulse (+20 073 in-lb) has been delivered from the driveshaft to the system. This delivery of energy slows the motor down. The maximum shaft speed occurs after the largest accumulated negative energy pulse (–6 032 in-lb) has been received back from the load by the shaft. This return of stored energy will tend to speed up the motor. The total energy variation is the algebraic difference between these two extreme values, which in this example is –26 105 in-lb. This energy coming back from the load needs to be absorbed by the flywheel and later returned to the system within each cycle to smooth the variations in shaft speed.

From	ΔArea = ΔE	Accumulated Sum = E	Min & Max
A to B	+20 073	+20 073	ω_{min} @ B
B to C	−26 105	−6 032	ω_{max} @ C
C to D	+15 388	+9 356	
D to A	−9 202	+154	

Table 9-5   Accumulation of Energy Pulses Under a Torque-Time Curve

Total ΔEnergy = E @ ω_{max} − E @ ω_{min}

= (−6 032) − (+20 073) = −26 105 in-lb

Determining the Flywheel Inertia

We must now determine how large a flywheel is needed to absorb this energy with an acceptable change in speed. The change in shaft speed during a cycle is called its fluctuation *Fl* and is equal to

$$Fl = \omega_{max} - \omega_{min} \qquad (9.19a)$$

We can normalize this to a dimensionless ratio by dividing it by the average shaft speed. This ratio is called the coefficient of fluctuation C_f.

$$C_f = \frac{\left(\omega_{max} - \omega_{min}\right)}{\omega_{avg}} \qquad (9.19b)$$

This coefficient of fluctuation is a design parameter to be chosen by the designer. It is typically set to a value between 0.01 and 0.05 for precision machinery, and as high as 0.20 for crushing or hammering machinery, which correspond to a 1 to 5% fluctuation in shaft speed. The smaller this chosen value, the larger the flywheel will have to be. This presents a design trade-off. A larger flywheel will add more cost and weight to the system, which are factors that have to be considered against the smoothness of operation desired.

We found the required change in kinetic energy E_k by integrating the torque curve,

$$\int_{\theta\,@\,\omega_{min}}^{\theta\,@\,\omega_{max}} \left(T_l - T_{avg}\right) d\theta = E_k \qquad (9.20a)$$

and can now set it equal to the right-hand side of equation 9.18c:

$$E_k = \frac{1}{2} I_m \left(\omega_{max}^2 - \omega_{min}^2\right) \qquad (9.20b)$$

Factoring this expression:

$$E_k = \frac{1}{2} I_m \left(\omega_{max} + \omega_{min} \right) \left(\omega_{max} - \omega_{min} \right) \qquad (9.20c)$$

If the torque-time function were a pure harmonic then its average value could be expressed exactly as

$$\omega_{avg} = \frac{\left(\omega_{max} + \omega_{min} \right)}{2} \qquad (9.21)$$

Our torque functions will seldom be pure harmonics, but the error introduced by using this expression as an approximation of the average is acceptable. We can now substitute equations 9.19b and 9.21 into equation 9.20c to get an expression for the mass moment of inertia I_s needed in the entire rotating system in order to obtain the selected coefficient of fluctuation.

$$E_k = \frac{1}{2} I_s \left(2\omega_{avg} \right) \left(C_f \omega_{avg} \right)$$

$$I_s = \frac{E_k}{C_f \omega_{avg}^2} \qquad (9.22)$$

Equation 9.22 can be used to design the physical flywheel by choosing a desired coefficient of fluctuation C_f, and using the value of E_k from the numerical integration of the torque curve (see Table 9-5 for an example) and the average shaft ω to compute the needed system I_s. The physical flywheel's mass moment of inertia I_m is then set equal to the required system I_s. But, if the moments of inertia of the other rotating elements on the same shaft (such as the motor) are known, the physical flywheel's required I_m can be reduced by those amounts.

The most efficient flywheel design in terms of maximizing I_m for minimum material used is one in which the mass is concentrated in its rim and its hub is supported on spokes, like a carriage wheel. This puts the majority of the mass at the largest radius possible and minimizes the weight for a given I_m. Even if a flat, solid circular disk flywheel design is chosen, either for simplicity of manufacture or to obtain a flat surface for other functions (such as an automobile clutch), the design should be done with an eye to reducing weight and thus cost. Since in general, $I_m = mr^2$, a thin disk of large diameter will need fewer pounds of material to obtain a given I_m than will a thicker disk of smaller diameter. Dense materials such as cast iron and steel are the obvious choices for a flywheel. Aluminum is seldom used. Though many metals (lead, gold, silver, platinum) are more dense than iron and steel, one can seldom get the accounting department's approval to use them in a flywheel.

Figure 9-23 shows the change in the torque of Figure 9-22 after the addition of a flywheel sized to provide a coefficient of fluctuation of 0.05. The oscillation in torque about the unchanged average value is now 5%, much less than what it was without the flywheel. Note that the peak value is now 87 instead of 372 lb-in. A much smaller-horsepower motor can now be used because the flywheel is available to absorb the energy returned from the load during the cycle.

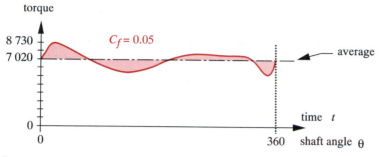

FIGURE 9-23

Torque-Time Function of Figure 9-22 After Adding a Flywheel with $C_f = 0.05$

Stresses in Flywheels

As a flywheel spins, the centrifugal force acts upon its distributed mass and attempts to pull it apart. These centrifugal forces are similar to those caused by an internal pressure in a cylinder. Thus, the stress state in a spinning flywheel is analogous to a thick-walled cylinder under internal pressure (see Section 4.16). The tangential stress of a solid disk flywheel as a function of its radius r is

$$\sigma_t = \frac{\gamma}{g}\omega^2\left(\frac{3+\nu}{8}\right)\left(r_i^2 + r_o^2 + \frac{r_i^2 r_o^2}{r^2} - \frac{1+3\nu}{3+\nu}r^2\right) \qquad (9.23a)$$

and the radial stress is

$$\sigma_r = \frac{\gamma}{g}\omega^2\left(\frac{3+\nu}{8}\right)\left(r_i^2 + r_o^2 - \frac{r_i^2 r_o^2}{r^2} - r^2\right) \qquad (9.23b)$$

where γ = material weight density, ω = angular velocity in rad/sec, ν = Poisson's ratio, r is the radius to a point of interest, and r_i, r_o are inside and outside radii of the solid disk flywheel.

Figure 9-24 shows how these stresses vary over the radius of the flywheel. The tangential stress is a maximum at the inner radius. The radial stress is zero at inside and outside radii and peaks at an interior point but is everywhere less than the tangential stress. The point of most interest is then at the inside radius. The tangential tensile stress at that point is what fails a flywheel, and when it fractures at that point, it typically fragments and explodes with extremely dangerous results. Since the forces causing the stress are a function of rotational speed, there will always be some speed that will fail the flywheel. A maximum safe operating speed should be calculated for a flywheel and some means taken to preclude its operation at higher speeds, such as a speed control or governor. A safety factor against overspeeding can be determined as the quotient of the speed that will cause yielding over the operating speed, $N_{os} = \omega / \omega_{yield}$.

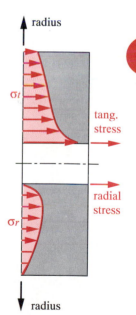

FIGURE 9-24

Stress Disrtibutions Along the Radius of a Spinning Flywheel

Failure Criteria

If the flywheel spends most of its life operating at essentially constant speed, then it can be considered to be statically loaded and the yield strength used as a failure criterion. The number of start-stop cycles in its operating regime will determine whether a fatigue-loading situation needs to be considered. Each runup to operational speed and rundown to zero constitutes a fluctuating stress cycle. If the number of these start-stop cycles is large enough over the projected life of the system, then fatigue-failure criteria should be applied. A low-cycle fatigue regime may require a strain-based fatigue failure analysis rather than a stress-based one, particularly if there exists the possibility of any transient overloads that may cause the local stresses to exceed the yield stress at stress concentrations.

EXAMPLE 9-7

Designing a Solid-Disk Flywheel

Problem	Design a suitable flywheel for the system of Example 9-6. An overspeed safety factor of at least 2 is desired.
Given	An input torque-time function which varies over its cycle as shown in Figure 9-20. The torque is varying during the 360° cycle about its average value and the energy variation per cycle is 26 105 in-lb as shown in Table 9-5. The shaft $\omega = 800$ rad/sec.
Assumptions	The one cycle of torque variation shown is representative of the steady-state condition. The desired coefficient of fluctuation is 0.05. The system is in continuous operation with minimal start-stop cycles. A steel of 62 kpsi yield strength will be used. No keyway will be used in order to reduce stress concentrations. Instead, a tapered locking hub will frictionally couple it to the shaft and the hub will be axially bolted to the flywheel.
Solution	See Figure 9-25, Table 9-6 and the *TKSolver* files EX09-07a and EX09-07b.

1 We know the amount of energy needed from the torque-time diagram from Example 9-6 and have defined the shaft ω and a desired coefficient of fluctuation. From these data we can determine a required mass moment of inertia for the system I_s using equation 9.22.

$$I_s = \frac{E}{C_f \omega_{avg}^2} = \frac{26\,105}{0.05(800)^2} = 0.816 \text{ lb - in sec}^2 \qquad (a)$$

The flywheel need only supply a portion of this amount if other rotating masses such as the motor armature are present. However, we will assume in this example that the flywheel will provide all of the required inertia making $I_m = I_s$.

2   The flywheel dimensions for this moment of inertia can be defined from equation 9.17d. We will assume a steel material with $\gamma = 0.28$ lb/in^3, and an inside radius $r_i = 1$ in.

$$I_m = \frac{\pi}{2}\frac{\gamma}{g}\left(r_o^4 - r_i^4\right)t = 0.816$$

$$0.816 = \frac{\pi}{2}\left(\frac{0.28}{386}\right)\left(r_o^4 - 1^4\right)t$$

$$716.14 = \left(r_o^4 - 1\right)t \qquad (b)$$

So, the required I_m can be obtained with an infinity of combinations of flywheel radius r_o and thickness t for the assumed data.

3   The best solution to equation (b) will be one that balances the conflicting factors of flywheel size, weight, stresses, and safety factor. Consider two possible designs, one with a small thickness t and the other with a large t. The thin flywheel will be larger in diameter but considerably lighter than the thick one due to the nonlinearity of the terms involving r_o. But, as r_o increases, so will the stresses because the mass at larger radius exerts more centrifugal force on the material.

4   To get a value of r_o that is consistent with any desired safety factor, the tangential stress equation 9.23a can be back-solved with assumed values of $\sigma_t = S_y / N_y$, r_i, and material parameters v and γ,

$$\sigma_t = \frac{\lambda}{g}\omega^2\left(\frac{3+v}{8}\right)\left(r_i^2 + r_o^2 + \frac{r_i^2 r_o^2}{r^2} - \frac{1+3v}{3+v}r^2\right) = \frac{S_y}{N_y}$$

$$\frac{62\,000}{N_y} = \frac{0.28}{386}(800)^2\left(\frac{3+0.28}{8}\right)\left(1 + r_o^2 + \frac{r_o^2}{1} - \frac{1+3(0.28)}{3+0.28}(1)\right) \qquad (c)$$

$$r_o^2 = \frac{162.9}{N_y} - 0.535$$

That value of r_o can then be used in (b) to find the flywheel thickness. For a design safety factor against yielding of 2.5 and the values assumed in (c), we get $r_o = 8.06$ in and $t = 0.172$ in.

5   With the flywheel geometry now defined, the rotational speed at which yielding will begin can be computed from equation 9.17d using the yield strength for the stress value.

$$\sigma_t = S_y = \frac{\gamma}{g}\omega^2\left(\frac{3+v}{8}\right)\left(r_i^2 + r_o^2 + \frac{r_i^2 r_o^2}{r^2} - \frac{1+3v}{3+v}r^2\right)$$

$$62\,000 = \frac{0.28}{g}\omega_{yield}^2\left(\frac{3+0.28}{8}\right)\left(1 + 8.06^2 + \frac{8.06^2}{1} - \frac{1+3(0.28)}{3+0.28}(1)\right) \qquad (d)$$

$$\omega_{yield} = 1\,265 \text{ rad/sec}$$

Table 9-6	Data for Example 9-7 with $I_m = 0.816$ lb-in-sec^2						
Thickness (in)	Diameter (in)	r/t ratio	Weight (lb)	Stress (psi)	Safety Factor Yielding	Safety Factor Overspeed
0.125	17.40	69.6	8.20	28 896	2.1	1.5
0.250	14.63	29.3	11.60	20 459	3.0	1.7
0.375	13.22	17.6	14.10	16 722	3.7	1.9
0.500	12.31	12.3	16.20	14 494	4.3	2.1
0.625	11.64	9.3	18.10	12 974	4.8	2.2
0.750	11.12	7.4	19.70	11 852	5.2	2.3
0.875	10.70	6.1	21.30	10 980	5.6	2.4
1.000	10.35	5.2	22.70	10 277	6.0	2.5
1.125	10.05	4.5	24.00	9 695	6.4	2.5
1.250	9.79	3.9	25.20	9 202	6.7	2.6

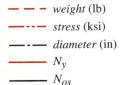

- – – – weight (lb)
- –·–· stress (ksi)
- –·– diameter (in)
- ——— N_y
- ——— N_{os}

9

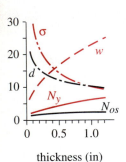

thickness (in)

FIGURE 9-25

Variation of Weight, Stress, Safety Factors, and Diameter with Flywheel Thickness in Example 9-7

Since this operating speed causes failure, a safety factor against overspeeding can be computed from

$$N_{os} = \frac{\omega_{yield}}{\omega} = \frac{1\,265\ \text{rad/sec}}{800\ \text{rad/sec}} = 1.6 \qquad (e)$$

6 To show the variation of parameters with flywheel geometry, this set of equations was solved for a list of possible thicknesses t chosen over a reasonable design range of 0.125 to 1.25 in. Table 9-6 shows the resulting data and Figure 9-25 plots the trends. Note that the weight *increases* as the outside diameter *decreases* and the thickness increases. The maximum tangential stress at the inside radius *decreases* with *decreasing* r_o, the safety factor against yielding increases from 2.1 to 6.7, and the safety factor for overspeed varies from 1.5 to 2.6 over this range of thicknesses.

7 The final design chosen is $t = 0.438$ in and $r_o = 6.36$ in because it has a reasonable mix of parameter values (size, weight) and provides a safety factor of 2 against overspeeding. In other words, the flywheel could be run at up to twice its design speed before it would yield. The safety factor against yielding at the lower design speed will then always be higher and is now 4. Choosing larger safety factors will exact a weight penalty, as can be seen in Figure 9-25.

9.14   CRITICAL SPEEDS OF SHAFTS

All systems containing energy-storage elements will possess a set of natural frequencies at which the system will vibrate with potentially large amplitudes. Any moving mass stores kinetic energy and any spring stores potential energy. All machine elements are made of elastic materials and thus can act as springs. All elements have mass, and if they also have a velocity, they will store kinetic energy. When a dynamic system vibrates, a transfer of energy from potential to kinetic to potential, etc., repeatedly takes place within the system. Shafts meet these criteria, rotating with some velocity and deflecting both in torsion and in bending.

If a shaft, or any element for that matter, is subjected to a time-varying load, it will vibrate. Even if it receives only a transient load, such as a hammer blow, it will vibrate at its natural frequencies, just as a bell rings when struck. This is called a **free vibration**. Such a transient or free vibration will eventually die out due to the damping present in the system. If the time-varying loading is sustained, as, for example, in a sinusoidal manner, the shaft or other element will continue to vibrate at the driving function's **forcing frequency**. If the forcing frequency happens to coincide with one of the element's natural frequencies, then the amplitude of the vibratory response will be much larger than the amplitude of the driving function. The element is then said to be in **resonance**.

Figure 9-26a shows the amplitude response of a forced-vibration, and Figure 9-26b a self-excited vibration, as a function of the ratio of the forcing frequency to the system's natural frequency ω_f / ω_n. When this ratio is 1, the system is in resonance and the amplitude of the response approaches infinity in the absence of damping. The amplitude responses in Figure 9-26 are shown as a dimensionless ratio of the output to input amplitudes. Any damping, shown as a damping ratio ζ, reduces the amplitude ratio at resonance. A natural frequency is also called a **critical frequency** or **critical speed**. *Exciting a system at or near its critical (resonant) frequencies must be avoided since the resulting deflections will often cause stresses large enough to quickly fail the part.*

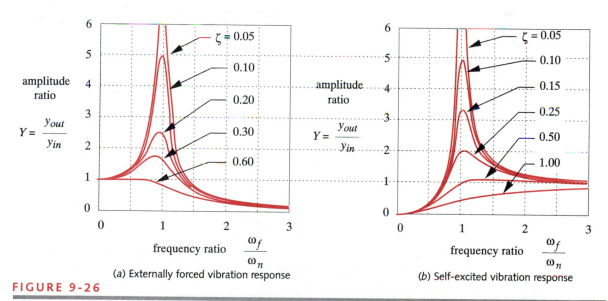

FIGURE 9-26

Response of a Single-Degree-of-Freedom System to Varying Forcing or Self-Excitation Frequencies

A system consisting of discrete lumps of mass connected with discrete spring elements can be considered to have a finite number of natural frequencies equivalent to its number of kinematic degrees of freedom. But a continuous system such as a beam or shaft has an infinite number of particles, each of which is capable of elastic motion versus its neighboring particles. Thus a continuous system has an infinity of natural frequencies. In either case, the lowest, or fundamental, natural frequency is usually of most interest.

The natural frequencies of vibration of a system can be expressed either as a circular frequency ω_n with units of rad/sec or rpm, or as a linear frequency f_n with units of hertz (Hz). They are the same frequencies expressed in different units. The general expression for the fundamental natural frequency is

$$\omega_n = \sqrt{\frac{k}{m}} \quad \text{rad/sec} \tag{9.24a}$$

$$f_n = \frac{1}{2\pi}\sqrt{\frac{k}{m}} \quad \text{Hz} \tag{9.24b}$$

where k is the spring constant of the system and m is its mass. The natural frequencies are a physical property of the system and once built, it retains them essentially unchanged unless it loses or gains mass or stiffness during its useful life. Equations 9.24 define the undamped natural frequency. Damping reduces the natural frequency slightly. Shafts, beams, and most machine parts tend to be lightly damped, so the undamped value can be used with little error.

The usual design strategy is to keep all forcing, or self-exciting, frequencies below the first critical frequency by some comfort margin. The larger this margin, the better, but a factor of at least 3 or 4 is desirable. This keeps the amplitude-response ratio close to one or zero, as shown in Figures 9-26a and 9-26b. In some cases, the fundamental frequency of a shaft system cannot be made higher than the required rotating frequency. If the system can be accelerated rapidly enough through resonance, before the vibrations have a chance to build up amplitude, then the system can be run at a speed higher than resonance. Stationary power plants are in this category. The massiveness of the turbines and generators gives a low fundamental frequency (see equation 9.24) but they must be run at a high speed to generate the proper line frequency of AC current. Thus they operate to the right of the peak in Figure 9-14b.) where the amplitude ratio approaches one at high ratios of ω_f/ω_n. Their start-ups and shutdowns may be infrequent but must always be accomplished rapidly to get through the resonance peak before any damage is caused by excessive deflections. Also, sufficient driving power must be available to provide the energy absorbed at resonance by the oscillations in addition to accelerating the rotating mass. If the driver lacks sufficient power, then the system may stall in resonance, unable to increase its speed in the face of the potentially destructive vibrations. This is called the Sommerfeld effect.[9]

There are three types of shaft vibration of concern:

1   Lateral vibration

2   Shaft whirl

3   Torsional vibration

The first two involve bending deflections and the last torsional deflection of the shaft.

Lateral Vibration of Shafts and Beams—Rayleigh's Method

A complete analysis of the natural frequencies of a shaft or beam is a complicated prob-lem, especially if the geometry is complex, and is best solved with the aid of *Finite-Element Analysis* software. A so-called **modal analysis** can be done on a finite-element model of even complex geometries and will yield a large number of natural frequen-cies (in three dimensions) from the fundamental up. This is the preferred and frequently used approach when analyzing a completed or mature design in detail. However, in the early stages of design, when the part geometries are still not fully defined, a quick and easily applied method for finding at least an approximate fundamental frequency for a proposed design is very useful. **Rayleigh's method** serves that purpose. It is an en-ergy method that gives results within a few percent of the true ω_n. It can be applied to a continuous system or to a lumped-parameter model of the system. The latter approach is usually preferred for simplicity.

RAYLEIGH'S METHOD equates the potential and kinetic energies in the system. The potential energy is in the form of strain energy in the deflected shaft and is maximum at the largest deflection. The kinetic energy is a maximum when the vibrating shaft passes through the undeflected position with maximum velocity. This method assumes that the lateral vibrating motion of the shaft is sinusoidal and that some external exci-tation is present to force the lateral vibration (Figure 9-26a).

To illustrate the application of this method, consider a shaft with three disks (gears, sheaves, etc.) on it as shown in Figure 9-27. We will model this as three discrete lumps of known mass on a massless shaft. The shaft's geometry will define the bending spring constant, thus lumping all the "spring" into the shaft. The total potential energy stored at maximum deflection is the sum of the potential energies of each lumped mass:

$$E_p = \frac{g}{2}\left(m_1\delta_1 + m_2\delta_2 + m_3\delta_3\right) \qquad (9.25a)$$

where the deflections are all taken as positive regardless of the local shape of the de-flection curve because the strain energy is not affected by the external coordinate sys-tem. The energy of the deflected shaft is ignored as small compared to the disk energy.

The total kinetic energy is the sum of the individual kinetic energies:

$$E_k = \frac{\omega_n^2}{2}\left(m_1\delta_1^2 + m_2\delta_2^2 + m_3\delta_3^2\right) \qquad (9.25b)$$

where the velocities are taken as positive.

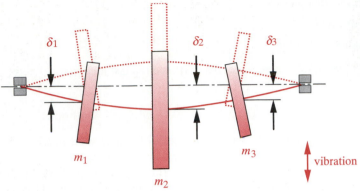

FIGURE 9-27

A Shaft in Lateral Vibration (amplitude greatly exaggerated)

Equating these gives

$$\omega_n = \sqrt{g \frac{\sum_{i=1}^{n} m_i \delta_i}{\sum_{i=1}^{n} m_i \delta_i^2}} = \sqrt{g \frac{\sum_{i=1}^{n} (W_i/g) \delta_i}{\sum_{i=1}^{n} (W_i/g) \delta_i^2}} = \sqrt{g \frac{\sum_{i=1}^{n} W_i \delta_i}{\sum_{i=1}^{n} W_i \delta_i^2}} \qquad (9.25c)$$

the second version resulting from substituting $m = W/g$, where the W_i are the weight forces of the discrete lumps into which we divided the system and the δ_i are the dynamic deflections at the locations of the weights due to their vibration. The weight forces and their deflections are all taken as positive to represent the maximum stored energies.

The problem is that we typically do not know the dynamic deflections of the system *a priori*. Rayleigh showed that virtually any estimate of the deflection curve, provided that it reasonably represents the maximum deflection and the boundary conditions of the actual dynamic curve will suffice. The static deflection curve due to the weights of the assumed lumps (including the weight of the shaft or not as desired) is a very suitable estimate. Note that any applied external loads are not included in this deflection calculation, only the gravitational forces. The resulting approximate ω_n will always be higher than the true fundamental frequency by a few percent regardless of the assumed deflection curve shape. If more than one estimated deflection curve is tried, the one yielding the lowest value of ω_n should be used as it will be the closest approximation.

Equation 9.25c can be applied to a system of any complexity by breaking it into a large number of lumps. If gears, pulleys, etc., are on the shaft, they make logical lumped masses. If the shaft mass is significant or dominant, it can be broken into discrete elements along its length with each piece providing a term in the summation.

Rayleigh's method can theoretically be used to find higher frequencies than the fundamental but to do so is difficult without a good estimate of the shape of the higher-order deflection curve. More accurate methods for approximating both the fundamental

frequency and the higher frequencies exist but are somewhat more complicated to implement. Ritz modified Rayleigh's method (Rayleigh-Ritz) to allow iteration to the higher
frequencies. Holzer's method is more accurate and can find multiple frequencies. See
references 10 and 11 for more information.

Shaft Whirl

Shaft whirl is a self-excited vibratory phenomenon to which all shafts are potentially
subject. While it is common and recommended practice to dynamically balance all rotating elements in machinery (especially if operated at high speeds), it is not possible
to achieve exact dynamic balance except by chance. (See reference 12 for a discussion
of dynamic balancing.) Any residual unbalance of a rotating element causes its true mass
center to be eccentric from the axis of shaft rotation. This eccentricity creates a centrifugal force that tends to deflect the shaft in the direction of the eccentricity, increasing it and thus further increasing the centrifugal force. The only resistance to this force
comes from the elastic stiffness of the shaft as shown in Figure 9-28. The initial shaft
eccentricity is labeled e and the dynamic deflection is δ. A free-body diagram shows
the forces acting to be

$$k\delta = m(\delta + e)\omega^2 \qquad (9.26a)$$

$$\delta = \frac{e\omega^2}{(k/m) - \omega^2} \qquad (9.26b)$$

The dynamic deflection of the shaft from this centrifugal force causes it to whirl
about its axis of rotation with points at the center of the deflected shaft describing circles
about the axis. Note in equation 9.26b that the deflection becomes infinite when
$\omega^2 = k/m$. As the rotational speed of the shaft approaches the fundamental natural (or
critical) speed of lateral vibration, a similar resonance phenomenon to that of lateral vibration occurs. Note in equation 9.26b that when $\omega^2 = k/m$, $\delta = \infty$. Equation 9.26b
can be normalized to a nondimensional form, which clearly shows the relationship:

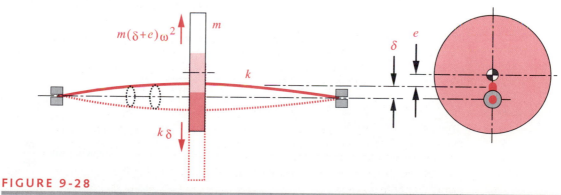

FIGURE 9-28
Shaft Whirl (amplitude greatly exaggerated)

$$\frac{\delta}{e} = \frac{(\omega/\omega_n)^2}{1-(\omega/\omega_n)^2} \qquad\qquad (9.26c)$$

Equation 9.26c and Figure 9-29 show the amplitude of shaft deflection normalized to the original eccentricity (δ / e) as a function of the ratio between the rotational frequency and critical frequency ω / ω_n. Note that at $\omega / \omega_n = 0$, there is no response, unlike the forced vibration of the previous section. This is because no centrifugal force exists unless the shaft is rotating. As the shaft speed increases, the deflection rapidly increases. If no damping is present ($\zeta = 0$), at $\omega / \omega_n = 0.707$, the deflection of the shaft is equal to the eccentricity and it becomes theoretically infinite at resonance ($\omega / \omega_n = 1$). Of course, there will always be some damping present, but if ζ is small, the deflections will be very large at resonance and can cause stresses large enough to fail the shaft.

Note what happens when the shaft speed passes through ω_n. The phase shifts 180°, which means that the deflection switches sides abruptly at resonance. At higher ratios of ω / ω_n, the deflection approaches $-e$, which means that the system is then rotating about the mass center of the eccentric mass, and the shaft centerline is eccentric. Conservation of energy makes the system want to rotate about its true mass center. In any system where the rotating elements are eccentric and large compared to the shaft this will occur. Perhaps you have observed a pivoted ceiling fan rotating with its motor center orbiting about the axis of rotation. The fan blades are usually not in perfect balance and the assembly rotates about the mass center of the blade assembly rather than about the motor/shaft centerline.

It should be clear that rotation of a system at or near its critical frequency is to be strictly avoided. The critical frequency for shaft whirl is the same as for lateral vibration and can be found using Rayleigh's method or any other suitable technique. Because this shaft-whirl vibration amplitude ratio starts at zero rather than at one (as with forced vibrations) the forcing frequency can be closer to the critical frequency than for the lateral vibration case. Keeping the operating speed below about half the shaft-whirl critical

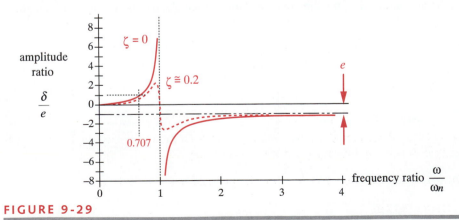

FIGURE 9-29

Amplitude Response of a Self-Excited Shaft-Whirl System as a Function of Frequency Ratio

frequency will usually provide good results unless the initial eccentricity is excessive (which should not be allowed anyway).

Note the difference between shaft *lateral vibration* and *shaft whirl*. **Lateral vibration** is a *forced vibration*, requiring some outside source of energy such as vibrations from other parts of the machine to precipitate it and the shaft then vibrates in one or more lateral planes whether or not it is rotating. **Shaft whirl** is a *self-excited vibration* caused by the shaft rotation acting on an eccentric mass. It will *always* occur when both rotation and eccentricity are present. The shaft assumes a deflected shape, which then rotates or whirls about the axis much like a jump-rope being swung by children.

Torsional Vibration

Just as a shaft can vibrate laterally, it can also vibrate torsionally and will have one or more torsional natural frequencies. The same equations that describe lateral vibrations can be used for torsional ones. The systems are analogous. Force becomes torque, mass becomes mass moment of inertia, and linear spring constant becomes torsional spring constant. Equation 9.24 for the circular natural frequency becomes, for a single degree of freedom rotating system:

$$\omega_n = \sqrt{\frac{k_t}{I_m}} \quad \text{rad/sec} \tag{9.27a}$$

The torsional spring constant k_t for a solid circular shaft is

$$k_t = \frac{GJ}{l} \quad \text{lb - in/rad or N - m/rad} \tag{9.27b}$$

where G is the material's modulus of rigidity, and l is the shaft length. The polar second moment of area J of a solid circular shaft is

$$J = \frac{\pi d^4}{32} \quad \text{in}^4 \text{ or } \text{m}^4 \tag{9.27c}$$

If the shaft is stepped, then an equivalent polar second moment of area J_{eff} is found from

$$J_{eff} = \frac{l}{\sum_{i=1}^{n} \frac{l_i}{J_i}} \tag{9.27d}$$

where l is total shaft length, and J_i, and l_i are the polar moments and lengths of the subsections of shaft of differing diameters, respectively.

The mass moment of inertia of a solid circular shaft about its axis is:

$$I_m = \frac{mr^2}{2} \quad \text{in - lb - sec}^2 \text{ or } \text{kg - m}^2 \tag{9.27e}$$

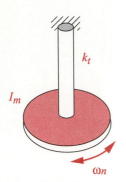

FIGURE 9-30

A Disk on Axle in
Torsional Vibration

where r is the shaft radius and m is its mass.

These equations are sufficient to find the critical frequency of a single disk mounted on a fixed axle, as shown in Figure 9-30.

Two Disks on a Common Shaft

A more interesting problem is that of two (or more) disks displaced on a common shaft as shown in Figure 9-31. The two disks shown will oscillate torsionally at the same natural frequency, 180° out-of-phase. There will be a point, called a node, somewhere on the shaft, at which there will be no angular deflection. On either side of the node, points on the shaft rotate in opposite angular directions when vibrating. The system can be modeled as two separate, single-mass systems coupled at this stationary node. One has mass moment and spring constant I_1, k_1 and the other I_2, k_2. Their common natural frequency is then

$$\omega_n = \sqrt{\frac{k_1}{I_1}} = \sqrt{\frac{k_2}{I_2}} \qquad (9.28a)$$

The spring constants of the shaft segments are each found from $k_t = JG/l$ assuming that the J is constant across the node.

$$\sqrt{\frac{JG}{l_1 I_1}} = \sqrt{\frac{JG}{l_2 I_2}}$$

and

$$l_1 I_1 = l_2 I_2 = I_2(l - l_1)$$

so

$$l_1 = \frac{I_2 l}{I_1 + I_2} \qquad (9.28b)$$

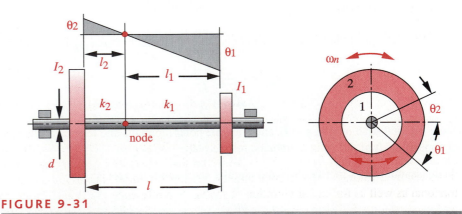

FIGURE 9-31

Torsional Vibration of Two Disks on a Common Shaft

Equation 9.28b allows the location of the node to be found. Substituting this expression into equation 9.28a gives

$$\omega_n = \sqrt{\frac{k_1}{I_1}} = \sqrt{\frac{JG}{l_1 I_1}} = \sqrt{\frac{JG}{l}\frac{I_1 + I_2}{I_1 I_2}}$$

$$\omega_n = \sqrt{k_t \frac{I_1 + I_2}{I_1 I_2}} \quad \text{or} \quad \omega_n^2 = k_t \frac{I_1 + I_2}{I_1 I_2} \qquad (9.28c)$$

which defines the critical speed for torsional vibration in terms of the known mass properties of the two disks and the overall spring constant of the shaft.

The critical frequency needs to be avoided in any forcing functions applied to the shaft in order to avoid torsional resonance that will overload it. Devices attached to the shaft, such as piston engines or piston pumps, will have frequencies in their torque-time functions that correspond to the pulses of their operation multiplied by their rotational frequency. For example, a four-cylinder engine will have a strong forcing frequency component at four times its rpm. If this fourth harmonic coincides with the shaft's critical frequency, there could be a problem. When designing a shaft, the frequency characteristics of the driving and driven rotating devices attached to it must be taken into account along with their primary rotational frequency.

Multiple Disks on a Common Shaft

Two disks on a common shaft have one node and one torsional natural frequency. Three disks will have two nodes and two natural frequencies. This pattern will hold for any number of disks assuming in all cases that the masses of the disks dominate the shaft mass allowing it to be ignored. N disks will have $N-1$ nodes and natural frequencies. The degree of the equation for the natural frequencies will also be $N-1$ if we consider the variable to be ω_n^2 rather than ω_n. Note that equation 9.28c for two masses is linear with this assumption. The equation for three masses is quadratic in ω_n^2 and for four masses is a cubic in ω_n^2.

For the three-mass case, the squares of the natural frequencies are the two roots of

$$I_1 I_2 I_3 \left(\omega_n^2\right)^2 - \left[k_2(I_1 I_2 + I_1 I_3) + k_1(I_2 I_3 + I_1 I_3)\right]\omega_n^2 + k_1 k_2(I_1 + I_2 + I_3) = 0 \quad (9.29)$$

Higher-order polynomials can be derived for additional masses and an iterative root-finding method can be used to solve them.

Approximate methods are also available to solve for the torsional natural frequencies with any number of masses. These allow the shaft mass to be easily accounted for if desired by breaking it into discrete masses. Holzer's method is commonly used for torsional as well as for lateral vibration of shafts. See references 10 and 11 or any vibrations text for a derivation and discussion of these methods. Space does not allow a complete treatment of them here.

Controlling Torsional Vibrations

When shafts are long and/or have a number of masses distributed along their length, torsional vibrations can be a serious design problem. Internal-combustion engine crankshafts are an example. The crank-throw geometry severely reduces the torsional stiffness, which lowers their natural frequency. This, in combination with the presence of strong higher harmonics from the cylinder explosions in the torque can lead to early failure from torsional fatigue. The straight-eight engine popular in the 1930s through 1940s was less successful than its straight-six cousin due in part to the problems of torsional vibrations in the long, eight-throw crankshaft. The vee-eight engine with its shorter, stiffer, four-throw crankshaft has completely supplanted the straight-eight. Even in these shorter engines, torsional crankshaft vibrations can be a problem.

Several methods can be used to counter the effects of an unwanted correspondence between the forcing frequencies and the system natural frequencies. The first line of defense is to redesign the mass and stiffness properties of the system to get the critical frequencies as far above the highest forcing frequency as possible. This usually involves increasing stiffness while removing mass, something not always easy to do. Effective use of geometry to obtain the maximum stiffness with a minimum of material is required. The term **specific stiffness** refers to the *stiffness to mass ratio* of an object. We want to maximize the specific stiffness to increase the natural frequencies. Finite element analysis can be very useful in refining a design's geometry to alter its natural frequencies because of the detailed information obtainable from that analysis.

Another approach is the addition of a **tuned absorber** to the system. A tuned absorber is a mass-spring combination added to the system whose presence alters the set of natural frequencies away from any dominant forcing frequencies. The system is effectively tuned away from the undesired frequencies. This approach can be quite effective in some cases and is used in linear-motion as well as in torsional systems.

A **torsional damper** is usually added to the end of an engine crankshaft to reduce its oscillations. This device, also called a Lanchester damper after its inventor, is a disk coupled to the shaft through an energy absorbent medium such as rubber or oil. The oil-coupling provides viscous damping and the rubber has significant internal hysteresis damping. Its effect is to reduce the peak amplitude at resonance, as can be seen in Figure 9-26 for larger values of ζ. The reader is referred to references [10] and [11] for more information on all of these methods.

EXAMPLE 9-8

Determining the Critical Frequencies of a Shaft

Problem   Find the shaft whirl and torsional critical frequencies of the shaft in Example 9-2 and compare them to its forcing frequency.

Given   The steel shaft dimensions are 0.875-in dia for 1.5 in, 0.750-in dia for 3.5 in, 0.669-in dia for 1.5 in, and 0.531-in dia for 1.5 in. Its rotational speed is 1 725 rpm. Shaft supports are at 0 and 5 in on an 8-in-long shaft. The steel gear weighs 10 lb and acts at $z = 2$ in. It has a mass moment of inertia of 0.23 lb-in-sec^2. The aluminum sheave weighs 3 lb and acts at $z = 6.75$ in. It has a mass moment of inertia of 0.07 lb-in-sec^2.

Assumptions   The static deflection of the shaft due to the weights of gear and sheave will be used as an estimate for Rayleigh's method, but the gear and sheave weights will be applied in the directions that give the largest static deflection. The shaft weight will be ignored.

Solution   See Figures 9-3 and 9-32, and the *TKSolver* file EX09-08.

1   The deflection of the stepped shaft is found by the same technique used in Example 9-3. In this instance, the loads are taken as just the weights of the two disks. But, we will consider the gear weight-force to act downward and the sheave weight-force to act upward since that arrangement better represents the dynamic situation where the inertia forces act outward from the axis in whatever direction increases the deflection. If we directed both weight forces downward in this case, we would get a smaller maximum deflection and a different curve shape than that of the dynamic deflection. Figure 9-32 shows the applied weight forces and the deflection curve for this shaft. The magnitude of the deflection at the gear is 6.0E–5 in and at the sheave is 1.25E–4 in. These values are needed in equation 9.25c.

2   Calculate the critical frequency for shaft whirl from equation 9.25c:

$$\omega_n = \sqrt{g \frac{\sum_{i=1}^{n} W_i \delta_i}{\sum_{i=1}^{n} W_i \delta_i^2}} = \sqrt{386.4 \frac{10(6.0E-5)+3(1.25E-4)}{10(6.0E-5)^2+3(1.25E-4)^2}} = 2\,241 \text{ rad/sec} \quad (a)$$

Note that the magnitudes of the weight forces and their corresponding deflections are all taken as positive regardless of their vector directions in the static deflection case.

3   Compare the critical whirl frequency to the forcing frequency.

$$\frac{\omega_n}{\omega_f} = \frac{30/\pi(2\,241 \text{ rad/sec})}{1\,725 \text{ rpm}} = \frac{21\,405 \text{ rpm}}{1\,725 \text{ rpm}} = 12.4 \quad (b)$$

This is a very comfortable margin. If, in this example, the shaft weight is included in both the deflection calculation and in the critical frequency calculation, the critical frequency becomes 20 860 rpm, which is 12.1 times the forcing frequency. Even with the relatively light disks on this shaft, ignoring the shaft weight does not introduce a large error. Both of these values from Rayleigh's method are higher than the actual natural frequency. The reader can examine their differences in the *TKSolver* model supplied by simply changing the density of the shaft material on the variable sheet from 0.28 lb/in^3 (for steel) to zero to eliminate the shaft's effect from the computation. The *TKSolver* model includes the shaft weight in the computation by dividing it into 50 increments over its length.

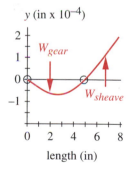

y (in x 10^{-4})

FIGURE 9-32

Static Deflection Due to Weight Forces of Disks on Shaft Oriented to Give the Maximum Deflection of a Shape Similar to the Dynamic Deflection

4   To find the torsional critical frequency of a stepped shaft requires that an effective spring constant for the combined stepped sections be found. The shaft portion of interest is between the sheave and gear. The spring constant of any one section is

$$k_t = \frac{GJ}{l} = \frac{G\pi d^4}{32l} \text{ lb - in/rad}$$

$$k_{t_1} = \pi \frac{11.5E6(0.75)^4}{32(3.5)} = 104\,006 \text{ lb - in/rad} \qquad (c)$$

$$k_{t_2} = \pi \frac{11.5E6(0.669)^4}{32(1.5)} = 153\,636 \text{ lb - in/rad}$$

Since the shaft sections all have the same torque but different deflections (which sum to the total deflection as shown in equation 9.9b), they act as springs in series, The effective spring constant k_{teff} of the portion of stepped shaft between the two torque loads is found from equation 9.9d:

$$\frac{1}{k_{t_{eff}}} = \frac{1}{k_{t_1}} + \frac{1}{k_{t_2}} = \frac{1}{104\,006} + \frac{1}{153\,636}$$

$$k_{t_{eff}} = 62\,020 \text{ lb - in/rad} \qquad (d)$$

5   The torsional critical frequency is found from equation 9.28c.

$$\omega_n = \sqrt{k_{t_{eff}} \frac{I_1 + I_2}{I_1 I_2}} = \sqrt{62\,020 \frac{0.23 + 0.07}{(0.23)(0.07)}} = 1\,074 \text{ rad/sec} \qquad (e)$$

6   Compare the critical torsional frequency to the forcing frequency.

$$\frac{\omega_n}{\omega_f} = \frac{30/\pi(1\,024 \text{ rad/sec})}{1\,725 \text{ rpm}} = \frac{10\,258 \text{ rpm}}{1\,725 \text{ rpm}} = 5.9 \qquad (f)$$

This is an acceptable margin.

9.15   COUPLINGS

A wide variety of commercial shaft couplings are available, ranging from simple keyed, rigid couplings to elaborate designs that utilize gears, elastomers, or fluids to transmit the torque from one shaft to another or to other devices in the presence of various types of misalignment. Couplings can be roughly divided into two categories, rigid and compliant. Compliant in this context means that the coupling can absorb some misalignment between the two shafts and rigid implies that no misalignment is allowed between the connected shafts.

Rigid Couplings

Rigid couplings lock the two shafts together allowing no relative motion between them, though some axial adjustment is possible at assembly. These are used when accuracy and fidelity of torque transmission are of paramount importance, as, for example, when the phase relationship between driver and driven device must be accurately maintained. Automated production machinery driven by long line-shafts often uses rigid couplings between shaft sections for this reason. Servomechanisms also need zero backlash connections in the drive train. The trade-off is that the alignment of the coupled shaft axes must be adjusted with precision to avoid introducing large side forces and moments when the coupling is clamped in place.

Figure 9-33 shows some examples of commercial rigid couplings. There are three general types, setscrew couplings, keyed couplings, and clamp couplings.

SETSCREW COUPLINGS   use a hard setscrew that digs into the shaft to transmit both torque and axial loads. These are not recommended for any but light-load applications and can loosen with vibration.

KEYED COUPLINGS   use standard keys as discussed in an earlier section and can transmit substantial torque. Setscrews are often used in combination with a key, the screw being located 90° from the key. For proper holding against vibration, a cup-point setscrew is used to dig into the shaft. For better security, the shaft should be dimpled with a shallow drilled hole under the setscrew to provide a mechanical interference against axial slip rather than relying on friction.

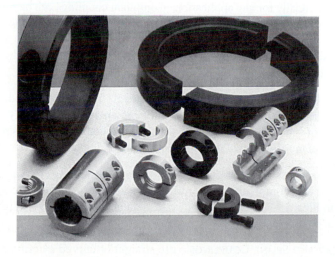

FIGURE 9-33

Various Types and Sizes of Rigid Shaft Couplings *Courtesy of Ruland Manufacturing Inc., Watertown, Mass.*

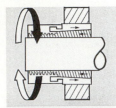

FIGURE 9-34

A *Trantorque* Taper-Lock Coupling - *Courtesy of Fenner Manheim, Manheim, PA 17545*

9

CLAMP COUPLINGS   are made in several designs, the most common being one- or two-piece split couplings that clamp around both shafts and transmit torque through friction as shown in Figure 9-33. A taper-lock coupling uses a split-tapered collet which is squeezed between shaft and the tapered coupling housing to clamp the shaft as shown in Figure 9-34.

Compliant Couplings

A shaft as a rigid body has six potential degrees of freedom (DOF) with respect to a second shaft. However, due to symmetry only four of these DOF are of concern. They are **axial**, **angular**, **parallel**, and **torsional** misalignment, as shown in Figure 9-35. These can occur singly or in combination and may be present at assembly due to manufacturing tolerances or may occur during operation due to the relative motions of the two shafts. The final driveline of an automobile has relative motion between the ends of the driveshaft. The drive end is affixed to the frame and the driven end is on the road. The frame and road are separated by the car's suspension so the driveshaft couplings must absorb both angular and axial misalignment as the car traverses bumps.

Unless care is taken to align two adjacent shafts there can be axial, angular, and parallel misalignment in any machinery. Torsional misalignment occurs dynamically when a driven load attempts to lead or lag the driver. If the coupling allows any torsional clearance there will be backlash when the torque reverses sign. This is undesirable if accurate phasing is needed, as in servomechanisms. Torsional compliance in a coupling may be desirable if large shock loads or torsional vibrations must be isolated from the driver.

Numerous designs of compliant couplings are manufactured and each offers a different combination of features. The designer can usually find a suitable coupling available commercially for any application. Compliant couplings can be roughly divided into several subcategories, which are listed in Table 9-7 along with some of their characteristics. The torque ratings are not shown as these vary widely with size and materials. Various size couplings can handle power levels from subfractional horsepower to thousands of horsepower.

JAW COUPLINGS   have two (often identical) hubs with protruding jaws as shown in Figure 9-36. These jaws overlap axially and interlock torsionally through a compliant insert of rubber or soft-metal material. The clearances allow some axial, angular, and parallel misalignment, but can also allow some undesirable backlash.

FLEXIBLE-DISK COUPLINGS   are similar to jaw couplings in that their two hubs are connected by a compliant member (disk) of elastomeric or metallic-spring material, as shown in Figure 9-37. These allow axial, angular, and parallel misalignment, with some torsional compliance but little or no backlash.

GEAR AND SPLINE COUPLINGS   use straight or curved external gear teeth in mesh with internal teeth, as shown in Figure 9-38. These can allow substantial axial movement between shafts and depending on the tooth shapes and clearances can absorb small angular and parallel misalignment as well. They have high torque capacity due to the number of teeth in mesh.

axial misalignment

HELICAL AND BELLOWS COUPLINGS are one-piece designs that use their elastic deflections to allow axial, angular, and parallel misalignment with little or no backlash. Helical couplings (Figure 9-39 and Chapter 9 title page photograph) are made from a solid metal cylinder cut with a helical slit to increase its compliance. Metal-bellows couplings (Figure 9-40) are made of thin sheet metal by welding a series of cupped washers together, by hydraulically forming a tube into the shape, or by electroplating a thick coating on a mandrel. These couplings have limited torque capability compared to other designs but offer zero backlash and high torsional stiffness in combination with axial, angular, and parallel misalignment.

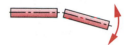

angular misalignment

LINKAGE COUPLINGS or Schmidt couplings (Figure 9-41) connect two shafts through a network of links that allow significant parallel misalignment with no side loads or torque losses and no backlash. Some designs allow small amounts of angular and axial misalignment as well. These couplings are often used where large parallel adjustments or dynamic motions are needed between shafts.

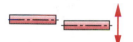

parallel misalignment

UNIVERSAL JOINTS are of two common types, the Hooke coupling (Figure 9-42), which does not have constant velocity (CV) and the Rzeppa coupling, which does. Hooke couplings are generally used in pairs to cancel their velocity error. Both types can handle very large angular misalignment, and in pairs provide large parallel offsets as well. These are used in automobile driveshafts, a pair of Hooke couplings in a rear-drive driveshaft and Rzeppas (called CV joints) in a front-drive automobile.

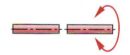

torsional misalignment

The variety of couplings available makes it necessary for the designer to seek more detailed information on their capabilities from their manufacturers, who are always willing to help in the selection of the proper type of coupling for any application. Manufacturers can often supply test data on the load and alignment capabilities of their particular couplings.

FIGURE 9-35

Types of Shaft
Misalignment

FIGURE 9-36

Exploded View of a Jaw Coupling Showing Jaws and Elastomer Insert   *Courtesy of Magnaloy Coupling Company, Alpena, Mich. 49707*

FIGURE 9-37

A Flexible Disk Coupling
*Courtesy of Zero-Max,
Minneapolis, Minn. 55441*

FIGURE 9-38

A Flexible Gear Coupling
*Courtesy of Amerigear/Zurn
Industries, Inc., Erie, Pa.
16514*

FIGURE 9-39

A Helical Coupling
*Courtesy of Helical Products
Co., Inc., Santa Maria, Calif.
93456*

Table 9-7   Characteristics of Various Types of Couplings

Class	Misalignment Tolerated				Comments
	Axial	Angular	Parallel	Torsional	
Rigid	large	none	none	none	requires accurate alignment
Jaw	slight	slight (<2°)	slight (3% d)	moderate	shock absorption— significant backlash
Gear	large	slight (<5°)	slight (<1/2% d)	none	slight backlash—large torque capacity
Spline	large	none	none	none	slight backlash—large torque capacity
Helical	slight	large (20°)	slight (<1% d)	none	one piece - compact—no backlash
Bellows	slight	large (17°)	moderate (20% d)	none	subject to fatigue failure
Flexible disc	slight	slight (3°)	slight (2% d)	slight to none	shock absorption—no backlash
Linkage (Schmidt)	none	slight (5°)	large (200% d)	none	no backlash—no sideloads on shaft
Hooke	none	large	large (in pairs)	none	slight backlash—speed variation unless used in pairs
Rzeppa	none	large	none	none	constant velocity

9.16   CASE STUDY

We will now address the design of shafts in one of the Case Study assemblies that were defined in Chapter 8.

Designing Driveshafts for a Portable Air Compressor

The preliminary design of this device is shown in Figure 8-1 (repeated here). There are two shafts, input and output. The output shaft will have the larger torque in this case since it is at the slower speed of 1500 rpm. The torque on that shaft was defined in the earlier Case Study 7A in Chapter 8 and is shown in Figure 8-3 (repeated here). Since this is a time-varying torque, the shaft must be designed for fatigue loading. In addition to the torque on the shafts, there will be side loads from the gears that apply bending moments making a combined-loading situation. Note in Figure 8-1 that the shafts are shown as short, only being long enough to accommodate the gear and bearings. This is done to minimize the bending moments from the gear forces.

Since the gears are yet to be designed, we will have to make some assumptions about their diameters and thicknesses to do a preliminary design of the shafts. Later selection of bearings may also dictate some changes to our shaft design. This is typi-

cal of design problems since all of their elements interact. Iteration is necessary to refine all the elements' designs.

CASE STUDY 7B

Preliminary Design of Shafts for a Compressor Drive Train

Problem   Determine reasonable sizes for the input and output shafts of the
gearbox in Figure 8-1 based on the loadings defined in Case Study
7A and specify a suitable type of coupling.

Given   The torque-time function on the output shaft is as shown in Figure
8-3. The required gear ratio is a 2.5:1 reduction in velocity from the
input to the output shaft.

Assumptions   Try an input gear (pinion) diameter of 4 in and output gear diameter
of 10 in, both of 2 in thickness and 20° pressure angle. Ball bear-
ings of standard diameters will be used on all shafts.

Solution   See Figures 8-1, 8-3, 9-43 and TKSolver file CASE7-B.

1   The time-varying torque on the output shaft is defined in Figure 8-3 as varying between
−175 and +585 lb-in. From these data and the assumed gear diameters, we can determine the forces at the gear mesh that are felt by the shaft. Figure 9-43 shows a free-
body diagram of a gearset. Because of the pressure angle ϕ between the gears, there
are both radial and tangential components of force at the gear mesh. The tangential
component is found from the known torque and the assumed gear radius:

$$F_{t_{max}} = \frac{T_{max}}{r_g} = \frac{585 \text{ lb - in}}{5 \text{ in}} = 117 \text{ lb}$$

$$F_{t_{min}} = \frac{T_{min}}{r_g} = \frac{-175 \text{ lb - in}}{5 \text{ in}} = -35 \text{ lb} \qquad (a)$$

2   The maximum and minimum resultant forces are found from:

$$F_{max} = \frac{F_{t_{max}}}{\cos\phi} = \frac{117 \text{ lb}}{\cos 20°} = 124.5 \text{ lb}$$

$$F_{min} = \frac{F_{t_{min}}}{\cos\phi} = \frac{-35 \text{ lb}}{\cos 20°} = -37.25 \text{ lb} \qquad (b)$$

Note that these forces will be the same on the input shaft, whose torque is 0.4 times
that of the output shaft because of the 1:2.5 gear ratio.

3   The maximum and minimum moments on the shaft can now be found. We will assume that the gears will be centered between the simply supported bearings that are
set 4 in apart. The bearing reaction forces are then half of the gear forces and the bending moments peak in the center with a magnitude of

FIGURE 9-40

Metal-Bellows Coupling
*Courtesy of Senior Flexonics
Inc., Metal Bellows Division,
Sharon, Mass. 02067*

FIGURE 9-41

Schmidt Offset Coupling
*Courtesy of
Zero-Max/Helland Co.,
Minneapolis, Minn. 55441*

FIGURE 9-42

A Hooke's Coupling
*Courtesy of Lovejoy, Inc.,
Downwers Grove, Ill. 60515*

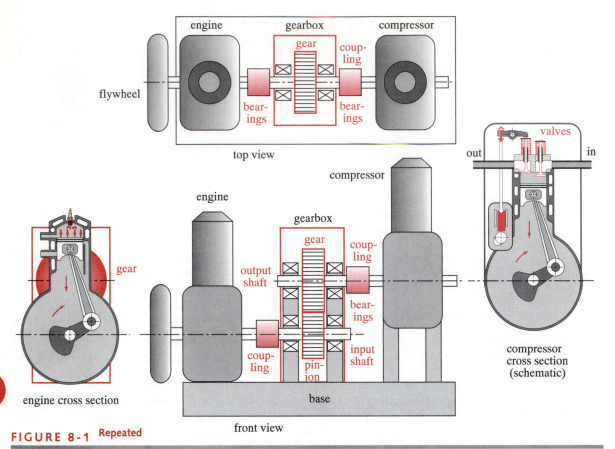

FIGURE 8-1 Repeated

Preliminary Design Schematic of Gasoline-Engine-Powered Portable Air Compressor, Gearbox, Couplings, Shafts, and Bearings

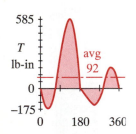

Repeated
FIGURE 8-3

Total Torque-Time
Function at Crankshaft
with Constant ω

$$M_{max} = F_{max} \frac{l}{4} = 124.5 \frac{4}{4} = 124.5 \text{ lb - in}$$

$$M_{min} = F_{min} \frac{l}{4} = -37.25 \frac{4}{4} = -37.25 \text{ lb - in} \qquad (c)$$

4   The shafts will ultimately need to be stepped for the bearings at each end and either stepped or snap-ringed to provide axial location for the gears. At this stage of the design, we will assume a constant diameter shaft in order to get an approximate size for the torque and moment loading.

Since a keyway will probably be needed at the gear, assume a stress-concentration factor of 3 for both bending and torsion at that critical location where both moment and torque components are largest. (See Figure 9-16.) After the gears are designed and the bearings selected, we can refine the design including stepped shoulders and using more accurate stress-concentration factors.

5   The loading is a combination of a fluctuating moment and a fluctuating torque that are synchronous. The mean and alternating components of both moment and torque are needed for the stress calculations.

$$M_m = \frac{M_{max} + M_{min}}{2} = \frac{124.5 - 37.25}{2} = 43.6 \text{ lb-in}$$

$$M_a = \frac{M_{max} - M_{min}}{2} = \frac{124.5 + 37.25}{2} = 80.9 \text{ lb-in} \tag{d}$$

$$T_m = \frac{T_{max} + T_{min}}{2} = \frac{585 - 175}{2} = 205 \text{ lb-in}$$

$$T_a = \frac{T_{max} - T_{min}}{2} = \frac{585 + 175}{2} = 380 \text{ lb-in} \tag{e}$$

6   A trial material needs to be selected for the computations. We will first try an inexpensive, low-carbon, cold-rolled steel such as SAE 1018 with S_{ut} = 64 kpsi and S_y = 54 kpsi. Though not exceptionally strong, this material has low notch sensitivity, which will be an advantage given the stress concentrations. Calculate the uncorrected endurance strength using equation 6.5:

$$S_{e'} = 0.5 S_{ut} = 0.5(64\,000) = 32\,000 \text{ psi} \tag{f}$$

This must be reduced by various factors to account for differences between the part and the test specimen.

$$S_e = C_{load}\, C_{size}\, C_{surf}\, C_{temp}\, C_{reliab}\, S_{e'}$$

$$S_e = (1)(1)(0.84)(1)(1)(32\,000) \cong 27\,000 \text{ psi} \tag{g}$$

The loading is bending and torsion, so C_{load} is 1. Since we don't yet know the part size, we will temporarily assume C_{size} =1 and adjust it later. C_{surf} is chosen for a machined finish from either Figure 6-26 or equation 6.7e. The temperature is not elevated so C_{temp} = 1 and we assume 50% reliability with C_{reliab} = 1.

7   The notch sensitivity of the material is found from either equation 6.13 or Figure 6-36 and is q = 0.5 for an assumed notch radius of 0.01 in.

8   The fatigue stress-concentration factor is found from equation 6.11b using the assumed geometric stress-concentration factor. For bending stress in the keyway:

$$K_f = 1 + q(K_t - 1) = 1 + 0.5(3.0 - 1) = 2.0 \tag{h}$$

The stress concentration for the keyway in torsion is assumed to be the same.

9   From equation 6.17 we find that in this case, the same factor should be used on the mean stress components:

$$K_{fm} = K_{fsm} = K_{fs} = 2.0 \tag{i}$$

10 The shaft diameter can now be found from equation 9.8 using an assumed safety factor of 3 to account for the uncertainties in this preliminary design. Note that the

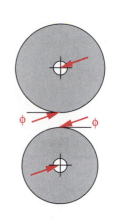

FIGURE 9-43

Forces on a Gearset

9

ASME equation (9.6) cannot be safely used in this case since it assumes constant torque. The more general, modified-Goodman line approach of equation 9.8 can be used.

$$d_2 = \left\{ \frac{32N_{sf}}{\pi} \left[\frac{\sqrt{\left(k_f M_a\right)^2 + \frac{3}{4}\left(k_{fs} T_a\right)^2}}{S_f} + \frac{\sqrt{\left(k_{fm} M_m\right)^2 + \frac{3}{4}\left(k_{fsm} T_m\right)^2}}{S_{ut}} \right] \right\}^{\frac{1}{3}}$$

$$= \left\{ \frac{32(3)}{\pi} \left[\frac{\sqrt{[2.0(81)]^2 + \frac{3}{4}[2.0(380)]^2}}{27\,000} + \frac{\sqrt{[2.0(44)]^2 + \frac{3}{4}[2.0(205)]^2}}{64\,000} \right] \right\}^{\frac{1}{3}}$$

$$d_2 = 0.991 \text{ in} \tag{j}$$

So a 1-in nominal shaft diameter seems acceptable for the output shaft.

11   The input shaft has the same mean and alternating bending moments as the output shaft, but its torque is only 40% of the output shaft's. The mean and alternating torques on it are 82 and 152 lb-in. When these are put in equation 9.8 with all other factors the same, a smaller shaft diameter results.

$$d_2 = \left\{ \frac{32(3)}{\pi} \left[\frac{\sqrt{[2.0(81)]^2 + \frac{3}{4}[2.0(152)]^2}}{27\,000} + \frac{\sqrt{[2.0(44)]^2 + \frac{3}{4}[2.0(82)]^2}}{64\,000} \right] \right\}^{\frac{1}{3}}$$

$$d_2 = 0.751 \text{ in} \tag{k}$$

The input shaft can then be a nominal 0.75 in diameter.

12   The couplings between the engine and input shaft and between the output shaft and compressor must be able to accommodate some angular and parallel misalignment due to tolerances in the mounting of the three subassemblies of engine, gearbox, and compressor. Torsional compliance in the couplings would also serve to absorb some of the shock associated with the torque reversal seen in Figure 8-3.

With these constraints, a jaw-type coupling with an elastomer insert might be a good choice for both couplings. The backlash inherent in these couplings would not be a problem since the phasing of the three subassemblies is not critical in this application. A flexible-disc coupling could also be used. A rigid coupling would require more accurate mounting of the subassemblies than is otherwise warranted. See Table 9-7 for information on coupling characteristics.

9.17   SUMMARY

Shafts are used in all rotating machinery. Steel is the usual choice of material to obtain high stiffness for low deflections. Shafts may be of soft, low-carbon steel or of medium to high-carbon steel for higher strength or if a hard surface finish is needed for wear resistance. Machine shafts usually have stepped shoulders for axial location of attached elements such as bearings, gears, or sprockets (sheaves). These shoulders create stress concentrations that must be considered in the stress analysis. Keyways or interference fits also create stress concentrations.

The loading on shafts is usually a combination of torsion and bending, either or both of which can be time-varying. The general loading case of fluctuating torque combined with fluctuating bending requires a modified-Goodman diagram approach to its failure analysis. For the common case of the torque and bending moment being related through common forces, the modified-Goodman approach is captured in equation 9.8 which provides a design tool to determine a shaft diameter for known fluctuating loads, stress concentrations, material strengths, and a chosen safety factor. The ASME shaft design (Eq. 9.6) addresses a subset of the general loading case in which the torque is considered steady and the steady bending moment is fully reversed due to the shaft rotation. The ASME equation should be used only in situations that match this loading limitation. A better strategy may be to use the general design equation 9.8, which encompasses the case addressed in the ASME equation, and is somewhat more conservative.

Several techniques or devices such as keys, splines, and interference fits are commonly used to attach elements to shafts. Keys are standardized to the shaft diameter. Consult the ANSI standard or reference 3 for data on size ranges not reproduced in this chapter. Splines provide greater torque capacity than keys. Interference fits may be by direct press fit or by thermally expanding or shrinking one or both members. Very high stresses can be created by these techniques, possibly failing the part during assembly.

Flywheels are used when some torque or velocity smoothing is needed. The flywheel must be sized to give the desired coefficient of speed fluctuation and then checked for stress at the operating speed. The maximum stresses in a flywheel occur at the inside diameter. A maximum safe speed must be determined as the stresses increase with the square of the rotational speed. When a flywheel fails while spinning it typically flies apart and can cause serious injury.

All spinning shafts will have critical frequencies at which they will resonate with large deflections, causing failure. The fundamental lateral and torsional frequencies will be different and both must be avoided in operation by keeping the rotational speed well below the lowest critical frequency of the shaft.

A wide variety of shaft couplings are commercially available. Some types and their characteristics are briefly discussed in this chapter. The manufacturers should be consulted for more complete and definitive information.

Important Equations Used in this Chapter

See the referenced sections for information on the proper use of these equations.

Power-Torque Relationship (Section 9.4):

$$P = T\omega \tag{9.1a}$$

ASME Shaft Design Equation (Section 9.8):

$$d = \left\{ \frac{32N_f}{\pi} \left[\left(k_f \frac{M_a}{S_f} \right)^2 + \frac{3}{4} \left(\frac{T_m}{S_y} \right)^2 \right] \right\}^{\frac{1}{3}} \tag{9.6b}$$

General Shaft Design Equation (Section 9.8):

$$d = \left\{ \frac{32N_f}{\pi} \left[\frac{\sqrt{\left(k_f M_a \right)^2 + \frac{3}{4}\left(k_{fs} T_a \right)^2}}{S_f} + \frac{\sqrt{\left(k_{fm} M_m \right)^2 + \frac{3}{4}\left(k_{fsm} T_m \right)^2}}{S_{ut}} \right] \right\}^{\frac{1}{3}} \tag{9.8}$$

Shaft Torsional Deflection (Section 9.9):

$$\theta = \frac{Tl}{GJ} \tag{9.9a}$$

Pressure Generated by an Interference Fit (Section 9.11):

$$p = \frac{0.5\delta}{\dfrac{r}{E_o}\left(\dfrac{r_o^2 + r^2}{r_o^2 - r^2} + \nu_o \right) + \dfrac{r}{E_i}\left(\dfrac{r^2 + r_i^2}{r^2 - r_i^2} - \nu_i \right)} \tag{9.14a}$$

Tangential Stresses in Shaft and Hub of an Interference Fit (Section 9.11):

$$\sigma_{t_{shaft}} = -p\frac{r^2 + r_i^2}{r^2 - r_i^2} \tag{9.15a}$$

$$\sigma_{t_{hub}} = p\frac{r_o^2 + r^2}{r_o^2 - r^2} \tag{9.16a}$$

Energy Stored in a Spinning Flywheel (Section 9.13):

$$E_k = \frac{1}{2}I_m\omega^2 \tag{9.17a}$$

Mass Moment of Inertia of a Solid Disk Flywheel (Section 9.13):

$$I_m = \frac{\pi}{2}\frac{\gamma}{g}\left(r_o^4 - r_i^4\right)t \tag{9.17d}$$

Flywheel Inertia Needed for a Chosen Coefficient of Fluctuation (Section 9.13):

$$E_k = \frac{1}{2}I_s\left(2\omega_{avg}\right)\left(C_f\omega_{avg}\right)$$

$$I_s = \frac{E_k}{C_f\omega_{avg}^2} \tag{9.22}$$

Tangential Stress in a Spinning Flywheel (Section 9.13):

$$\sigma_t = \frac{\gamma}{g}\omega^2\left(\frac{3+\nu}{8}\right)\left(r_i^2 + r_o^2 + \frac{r_i^2 r_o^2}{r^2} - \frac{1+3\nu}{3+\nu}r^2\right) \tag{9.23a}$$

Natural Frequency of Single-Degree-of-Freedom System (Section 9.14):

$$\omega_n = \sqrt{\frac{k}{m}} \quad \text{rad/sec} \tag{9.24a}$$

$$f_n = \frac{1}{2\pi}\sqrt{\frac{k}{m}} \quad \text{Hz} \tag{9.24b}$$

First Lateral Critical Frequency (Approximate) (Section 9.14):

$$\omega_n = \sqrt{g\frac{\sum_{i=1}^{n}m_i\delta_i}{\sum_{i=1}^{n}m_i\delta_i^2}} = \sqrt{g\frac{\sum_{i=1}^{n}(W_i/g)\delta_i}{\sum_{i=1}^{n}(W_i/g)\delta_i^2}} = \sqrt{g\frac{\sum_{i=1}^{n}W_i\delta_i}{\sum_{i=1}^{n}W_i\delta_i^2}} \tag{9.25c}$$

First Torsional Critical Frequency for Two Masses on Weightless Shaft (Section 9.14):

$$\omega_n = \sqrt{k_t\frac{I_1 + I_2}{I_1 I_2}} \quad \text{or} \quad \omega_n^2 = k_t\frac{I_1 + I_2}{I_1 I_2} \tag{9.28c}$$

9.18 REFERENCES

1 **R. C. Juvinall and K. M. Marshek**, *Fundamentals of Machine Component Design*. John Wiley & Sons: New York, pp. 656, 1991.

2 **D. B. Kececioglu and V. R. Lalli**, Reliability Approach to Rotating Component Design, *Technical Note TN D-7846*, NASA, 1975.

3 **V. C. Davies, H. J. Gough, and H. V. Pollard**, Discussion to The Strength of Metals Under Combined Alternating Stresses, *Proc. of the Inst. Mech Eng.*, **131**(3): pp. 66-69, 1935.

4   **S. H. Loewenthal**, Proposed Design Procedure for Transmission Shafting Under Fatigue Loading, *Technical Note TM-78927*, NASA, 1978.

5   **H. J. Gough and H. V. Pollard**, The Strength of Metals Under Combined Alternating Stresses. *Proc. of the Inst. Mech Eng.*, **131**(3): p. 3-103, 1935.

6   **E. Oberg and F. D. Jones**, eds. *Machinery's Handbook*, 17th ed., Industrial Press Inc.: New York. pp. 867-883, 1966.

7   **R. C. Peterson**, *Stress-concentration factors*. John Wiley & Sons: New York, pp. 266-267, 1974.

8   **E. Oberg and F. D. Jones**, eds. *Machinery's Handbook*, 17th ed., Industrial Press Inc.: New York. pp. 884-913, 1966.

9   **M. Dimentberg et al.**, "Passage Through Critical Speed with Limited Power by Switching System Stiffness," AMD-Vol. 192 / DE-Vol. 78, Nonlinear and Stochastic Dynamics, ASME 1994.

10 **R. M. Phelan**, *Dynamics of Machinery*. McGraw-Hill: New York, 1967.

11 **C. R. Mischke**, *Engineering Analysis*. Addison Wesley: Reading Mass., 1963.

12 **R. L. Norton**, *Design of Machinery*. McGraw-Hill: New York, p. 500, 1992.

9.19   PROBLEMS

*†9-1   A simply supported shaft is shown in Figure P9-1. A constant magnitude transverse load P is applied as the shaft rotates subject to a time-varying torque that varies from T_{min} to T_{max}. For the data in the row(s) assigned from Table P9-1, find the diameter of shaft required to obtain a safety factor of 2 in fatigue loading if the shaft is steel of S_{ut} = 108 kpsi and S_y = 62 kpsi. The dimensions are in inches, the force in pounds, and the torque is in lb-in. Assume no stress concentrations are present.

*†9-2   A simply supported shaft is shown in Figure P9-2. A constant magnitude distributed unit load p is applied as the shaft rotates subject to a time-varying torque that varies

* Answers to these problems are provided in Appendix G.
_____

† These problems are based on similar problems in previous chapters with the same –number, e.g., Problem 9-3 is based on Problem 6-3, etc. Problems in succeeding chapters may also continue and extend these problems.

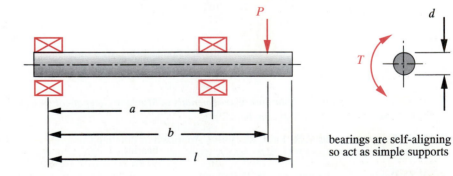

bearings are self-aligning so act as simple supports

FIGURE P9-1

Shaft Design for Problems 9-1, 9-4, and 9-15

Table P9-1   Data for Problems

Row	l	a	b	P or p	T_{min}	T_{max}
a	20	16	18	1 000	0	2 000
b	12	2	7	500	−100	600
c	14	4	12	750	−200	400
d	8	4	8	1 000	0	2 000
e	17	6	12	1 500	−200	500
f	24	16	22	750	1 000	2 000

from T_{min} to T_{max}. For the data in the row(s) assigned from Table P9-1, find the diameter of shaft required to obtain a safety factor of 2 in fatigue loading if the shaft is steel of $S_{ut} = 745$ MPa and $S_y = 427$ MPa. The dimensions are in cm, the distributed force in N/cm, and the torque in N-m. Assume no stress concentrations are present.

†9-3   For the bicycle pedal-arm assembly in Figure P6-1 assume a rider-applied force that ranges from 0 to 1500 N at each pedal each cycle. Design a suitable shaft to connect the two pedal arms and carry the sprocket against a step. Use the fatigue safety factor of 2 and a material with $S_{ut} = 500$ MPa. The shaft has a square detail on each end where it inserts into the pedal arms.

*9-4   Determine the maximum deflections in torsion and in bending of the shaft shown in Figure P9-1 for the data in the row(s) assigned in Table P9-1 if the steel shaft diameter is 1.75 in.

*9-5   Determine the maximum deflections in torsion and in bending of the shaft shown in Figure P9-2 for the data in the row(s) assigned in Table P9-1 if the steel shaft diameter is 4 cm.

*9-6   Determine the size of key necessary to give a safety factor of at least 2 against both shear and bearing failure for the design shown in Figure P9-3 using the data from the row(s) assigned in Table P9-1. Assume a shaft diameter of 1.75 in. The shaft is

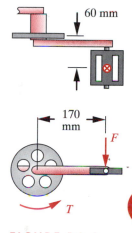

FIGURE P6-1

Problem 9-3

9

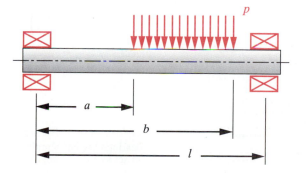

FIGURE P9-2

Shaft Design for Problems 9-2, 9-5, and 9-16

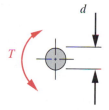

bearings are self-aligning so act as simple supports

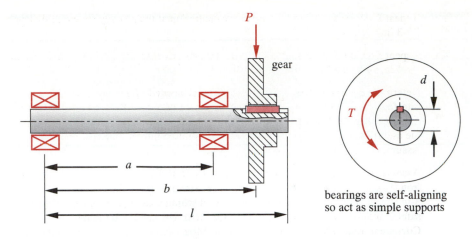

FIGURE P9-3

Shaft Design for Problems 9-6, 9-9. 9-11, and 9-12

steel of S_{ut} = 108 kpsi and S_y = 62 kpsi. The key is steel of S_{ut} = 88 kpsi and S_y = 52 kpsi.

9-7 Determine the size of key necessary to give a safety factor of at least 2 against both shear and bearing failure for the design shown in Figure P9-4 using the data from the row(s) assigned in Table P9-1. Assume a shaft diameter of 4 cm. The shaft is steel of S_{ut} = 745 MPa and S_y = 427 MPa. The key is steel of S_{ut} = 600 MPa and S_y = 360 MPa.

*†9-8 A paper machine processes rolls of paper having a density of 984 kg/m³. The paper roll is 1.50-m outside dia. (OD) x 22-cm inside diameter (ID) x 3.23-m long and is on a simply supported, hollow, steel shaft with S_{ut} = 400 MPa. Find the shaft ID needed to obtain a dynamic safety factor of 2 for a 10-year life if the shaft OD is 22 cm and the roll turns at 50 rpm with 1.2 hp absorbed.

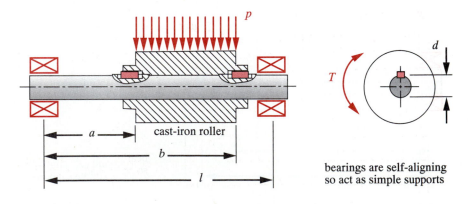

FIGURE P9-4

Shaft Design for Problems 9-7, 9-10, and 9-14

*9-9   Repeat Problem 9-1 taking the stress concentration at the keyway shown in Figure P9-3 into account.

9-10   Repeat Problem 9-2 taking the stress concentrations at the keyways shown in Figure P9-4 into account.

*9-11   Determine the amount of diametral interference needed to provide a suitable interference fit for the 6-in-dia by 1-in-thick gear of Figure P9-3 using a shaft dia of 1.75 in, such that the stresses in the hub and shaft will be safe and the torque from the assigned row(s) in Table P9-1 can be transmitted through the interference fit. Assume that $S_{ut} = 108$ kpsi and $S_y = 62$ kpsi.

9-12   Assume that the device shown keyed to the shaft of Figure P9-3 is a Class 50, cast-iron flywheel of 20-in outside diameter and 1-in thickness. The hub is 4-in dia and 3-in thick. Determine the maximum speed at which it can safely be run using a safety factor of 2. Use the dimensions and other appropriate data from Problem 9-6. Consider the transverse force P to be zero in this case.

*9-13   Determine the critical frequency of shaft whirl for the assembly shown in Figure P9-3 using the dimensions from the assigned row(s) of Table P9-1 and a steel shaft diameter of 2 in. Use the flywheel dimensions of Problem 9-12.

9-14   Determine the critical frequency of shaft whirl for the assembly shown in Figure P9-4 using the dimensions from the assigned row(s) of Table P9-1 and a steel shaft diameter of 4 cm. The cast-iron roller diameter is 3 times the shaft diameter.

*9-15   What are the maximum, minimum, and average power values for the shaft shown in Figure P9-1 for the data in the row(s) assigned in Table P9-1 if the shaft speed is 750 rpm?

*9-16   What are the maximum, minimum, and average power values for the shaft shown in Figure P9-2 for the data in the row(s) assigned in Table P9-1 if the shaft speed is 50 rpm?

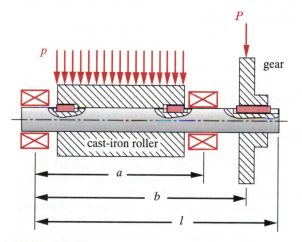
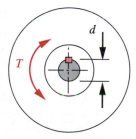

bearings are self-aligning
so act as simple supports

FIGURE P9-5

Shaft Design for Problems 9-17 and 9-18

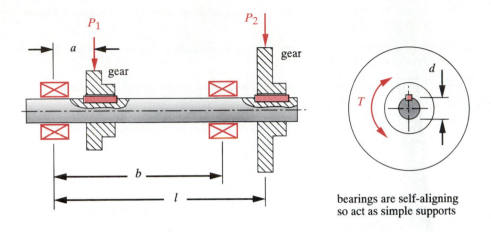

FIGURE P9-6

Shaft Design for Problem 9-19

*9-17   Figure P9-5 shows a roller assembly driven by a gear. The roller extends over 80% of length a and is centered in that dimension. The roller occupies 95% of the exposed shaft length between the bearing faces. The shaft is steel with $S_y = 427$ MPa and $S_{ut} = 745$ MPa. For the data in the assigned row(s) of Table P9-1, find:

(a) The safety factor against fatigue failure for a shaft diameter of 40 mm.

(b) The maximum torsional deflection between gear and roller.

(c) The torsional natural frequency of the shaft.

*9-18   Figure P9-5 shows a roller assembly driven by a gear. The roller extends over 80% of length a and is centered in that dimension. The roller occupies 95% of the exposed shaft length between the bearing faces. For the data in the assigned row(s) of Table P9-1, find the maximum bending deflection of the 40-mm-dia. shaft.

*9-19   Figure P9-6 shows two gears on a common shaft. Assume that the constant radial force P_1 is 40% of P_2. For the data in the row(s) assigned from Table P9-1, find the diameter of shaft required to obtain a safety factor of 2 in fatigue loading if the shaft is steel of $S_{ut} = 108$ kpsi and $S_y = 62$ kpsi. The dimensions are inches, the force lb, and the torque lb-in.

9-20   A 12-in-long, solid, straight shaft is supported in self-aligning bearings at each end. A gear is attached at the middle of the shaft with a 3/8-in square steel key in a slot. The geometric stress-concentration factor in the keyslot is 2.5 and its corner radius is 0.02 in. The gear drives a fluctuating load which creates a bending moment that varies from +100 lb-in to + 900 lb-in and a torque that varies from −300 lb-in to +1 500 lb-in each cycle. The material chosen is cold-drawn 4140 steel, hardened and tempered to Rockwell C45 ($S_{ut} = 180$ kpsi). Design the shaft for infinite life and determine the diameter needed for a safety factor of 1.5.

* Answers to these problems are provided in Appendix G.

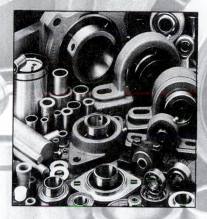

BEARINGS AND LUBRICATION

10

The knowledge of man is as the waters,
some descending from above,
and some springing from beneath.

FRANCIS BACON

10.0 INTRODUCTION

We use the term **bearing** here in its most general sense. Whenever two parts have relative motion, they constitute a bearing by definition, regardless of their shape or configuration. Usually, lubrication is needed in any bearing to reduce friction and remove heat. Bearings may roll or slide or do both simultaneously.

A *plain bearing* is formed by any two materials rubbing on one another whether a sleeve around a shaft or a flat surface under a slider. In a plain bearing, one of the moving parts usually will be steel or cast iron or some other structural material in order to achieve the required strength and hardness. For example, transmission shafts, links, and pins are in this category. The parts that they move against will usually be made of a "bearing" material such as bronze, babbitt, or a nonmetallic polymer. A radial plain bearing may be split axially to assemble it to the shaft, or may be a complete circle called a **bushing**. A **thrust bearing** supports axial loads.

Alternatively, a rolling-element bearing, which has hardened steel balls or rollers captured between hardened steel raceways, may be used to provide very low friction. Plain bearings are typically custom designed for the application while rolling-element bearings are typically selected from manufacturers' catalogs to suit the loads, speeds, and desired life of the particular application. Rolling-element bearings can support radial, thrust, or combinations of those loads depending on their design.

This chapter will discuss bearings in general, both the sliding (or plain) variety and also rolling-element bearings. Lubrication theory also will be discussed as it applies

Table 10-0   Variables Used in this Chapter

Symbol	Variable	ips units	SI units	See
A	area	in^2	m^2	Sect. 10.5
a, b	half-width, length of contact patch	in	m	Eq. 10.18
C	basic dynamic load rating	lb	N	Sect. 10.10
C_0	basic static load rating	lb	N	Sect. 10.10
c_d, c_r	diametral and radial clearance	in	m	Sect. 10.5
d	diameter	in	m	various
E'	effective Young's modulus	psi	Pa	Eq. 10.16
F	force (with various subscripts)	lb	N	Eq. 10.20
f	friction force	lb	N	Eq. 10.11
h	lubricant film thickness	in	m	Sect. 10.5
K_ε	dimensionless parameter	none	none	Sect. 10.5
l	length	in	m	Sect. 10.5
L	fatigue life of rolling bearings	10^6 revs	10^6 revs	Eq. 10.19
n'	angular velocity	rps	rps	Sect. 10.5
O_N	Ocvirk number	none	none	Eq. 10.12
P	force or load	lb	N	Sect. 10.5
p	pressure	psi	N/m^2	Sect. 10.5
r	radius	in	m	Sect. 10.5
R'	effective radius	in	m	Eq. 10.16
S	Sommerfeld number	none	none	Sect. 10.5
T	torque	lb-in	N-m	Sect. 10.5
U	linear velocity	in/sec	m/sec	Sect. 10.5
X, Y	radial and axial force factors	none	none	Eq. 10.20
α	pressure-viscosity exponent	in^2/lb	m^2/N	Eq. 10.15
ε	eccentricity ratio	none	none	Eq. 10.3
ε_x	empirical eccentricity ratio	none	none	Eq. 10.3
ϕ	angle to resultant force	rad	rad	Sect. 10.5
Φ	power	hp	watts	Eq. 10.10
η	absolute viscosity	reyn	cP	Eq. 10.1
Λ	specific film thickness	none	none	Eq. 10.14
μ	coefficient of friction	none	none	Eq. 10.11
ν	Poisson's ratio	none	none	Eq. 10.16
θ_{max}	angle to maximum pressure	rad	rad	Sect. 10.5
ρ	mass density	$blob/in^3$	kg/mm^3	Eq. 10.1
τ	shear stress (with various subscripts)	psi	Pa	Sect. 10.5
υ	kinematic viscosity	in^2/sec	cS	Eq. 10.1
ω	angular velocity	rad/sec	rad/sec	Eq. 10.10

10

to these types of bearings. Table 10-0 shows the variables used in this chapter and notes the equations or sections in which they first occur.

A. G. M. Michell, a pioneer in bearing theory and design, and one of the inventors of the tilted-pad bearing, once defined what is desired in a bearing as follows:

> To the machine designer all bearings are of course only necessary evils, contributing nothing to the product or function of the machine; and any virtues they can have are only of a negative order. Their merits consist in absorbing as little power as possible, wearing out as slowly as possible, occupying as little space as possible, and costing as little as possible.[1]

A Caveat

Lubrication theory for surfaces in relative motion is extremely complex mathematically. Solutions to the partial differential equations that govern the behavior are based on simplifying assumptions that yield only approximate solutions. This chapter does not attempt to present a complete discussion or explanation of all the complicated phenomena of dynamic lubrication as that is far beyond the scope of this text. Rather an introductory discussion of a few of the common cases encountered in machine design is presented. Boundary, hydrostatic, hydrodynamic, and elastohydrodynamic lubrication are introduced and described, and the theory for the last two conditions is discussed without presentation of complete derivations of the governing equations due to space limitations.

Such topics as squeeze-film theory and oil whirl are not addressed at all, nor are the issues of lubricant supply to, and heat transfer from, the bearing. Entire books have been written on these topics and the reader is directed to those sources for more complete information. Derivations of the governing equations are presented in most of the referenced works. Reference 2 provides an excellent introduction to lubrication theory with minimal mathematics, and reference 3 is a very complete, up-to-date, and mathematically rigorous treatment of the subject.

In this chapter, we present a simple and reasonably accurate approach to the design of short journal bearings that will allow them to be designed for the loads and speeds required in most common machinery. We also address the lubrication of nonconforming contacts such as gear teeth and cam-follower joints. Finally, a discussion of rolling-element bearing selection from manufacturers' information is provided. Rolling-element bearings are a topic that is again as complicated as that of journal bearings and books have been written on this subject as well. The reader is directed to references 3 and 4 for up-to-date and complete treatments of rolling-contact bearing theory and lubrication. The references to this chapter identify additional readings on the complex subject of lubrication and bearing design. We will barely "wet the surface" of this complicated subject here. We hope that it "whets your appetite" to learn more about the subject.

Table 10-1   Types of Liquid Lubricants

Type	Properties	Typical Uses
Petroleum oils (mineral oils)	Basic lubrication ability fair, but additives produce great improvement. Poor lubrication action at high temperatures	Very wide and general
Ployglycols	Quite good lubricants, do not form sludge on oxidizing	Brake fluid
Silicones	Poor lubrication ability, especially against steel. Good thermal stability	Rubber seals. Mechanical dampers
Chlorofluorocarbons	Good lubricants, good thermal stability	Oxygen compressors. Chemical processing equipment
Polyphenyl ethers	Very wide liquid range. Excellent thermal stability. Fair lubricating ability	High-temperature sliding systems
Phosphate esters	Good lubricants—EP action	Hydraulic fluid + lubricant
Dibasic esters	Good lubricating properties. Can stand higher temperatures than mineral oils.	Jet engines

Source: *Friction and Wear of Materials*, E. Rabinowitz, 1965, reprinted by permission of John Wiley & Sons, Inc.

10.1   LUBRICANTS

Introduction of a lubricant to a sliding interface has several beneficial effects on the friction coefficient. Lubricants may be gaseous, liquid, or solid. Liquid or solid lubricants share the properties of low shear strength and high compressive strength. A liquid lubricant such as petroleum oil is essentially incompressible at the levels of compressive stress encountered in bearings, but it readily shears. Thus, it becomes the weakest material in the interface and its low shear strength reduces the coefficient of friction (see equation 7.4). Lubricants can also act as contaminants to the metal surfaces and coat them with monolayers of molecules that inhibit adhesion even between compatible metals.

Liquid lubricants are the most commonly used and mineral oils the most common liquid. Greases are oils mixed with soaps to form a thicker, stickier lubricant used where liquids cannot be supplied to or retained on the surfaces. Solid lubricants are used in situations where liquids either cannot be kept on the surfaces or lack some required property such as high-temperature resistance. Gaseous lubricants are used in special situations such as air bearings to obtain extremely low friction. Lubricants, especially liquids, also remove heat from the interface. Lower bearing temperatures reduce surface interactions and wear.

Table 10-2   **Types of Solid Film Lubricants**

Type	Properties	Typical Uses
Graphite and/or MoS_2 + binder	Best general purpose lubricants. Low friction (0.12-0.06) reasonably long life ($\cong 10^4$-10^6 cycles)	Locks and other intermittent mechanisms
Teflon + binder	Life not quite as long as previous type, but resistance to some liquids better.	As above
Rubbed graphite or MoS_2 film	Friction very low (0.10-0.04), but life quite short (10^2-10^4 cycles)	Deep drawing and other metalworking
Soft metal (lead, indium, cadmium)	Friction higher (0.30-0.15) and life not as long as resin-bonded types	Running-in protection (temporary)
Phosphate, anodized film. Other chemical coatings	Friction high ($\cong 0.20$). Galling preventatives leave "spongy" surface layer.	Undercoating for resin-bonded film

Source: *Friction and Wear of Materials*, E. Rabinowitz, 1965, reprinted by permission of John Wiley & Sons, Inc.

LIQUID LUBRICANTS are largely petroleum-based or synthetic oils, though water is sometimes used as a lubricant in aqueous environments. Many commercial lubricant oils are mixed with various additives that react with the metals to form monolayer contaminants. So-called EP (*Extreme Pressure*) lubricants add fatty acids or other compounds to the oil that attack the metal chemically and form a contaminant layer that protects and reduces friction even when the oil film is squeezed out of the interface by high contact loads. Oils are classified by their viscosity as well as by the presence of additives for EP applications. Table 10-1 shows some common liquid lubricants, their properties, and typical uses. Lubricant manufacturers should be consulted for particular applications.

SOLID-FILM LUBRICANTS are of two types: materials that exhibit low shear stress such as graphite and molybdenum disulfide, which are added to the interface, and coatings such as phosphates, oxides, or sulfides that are caused to form on the material surfaces. The graphite and MoS_2 materials are typically supplied in powder form and can be carried to the interface in a binder of petroleum grease or other material. These dry lubricants have the advantage of low friction and high-temperature resistance though the latter may be limited by the choice of binder. Coatings such as phosphates or oxides can be chemically or electrochemically deposited. These coatings are thin and tend to wear through in a short time. The EP additives in some oils provide a continuous renewal of sulfide or other chemically induced coatings. Table 10-2 shows some common solid-film lubricants, their properties, and typical uses.

10.2   VISCOSITY

Viscosity is a measure of a fluid's resistance to shear. Viscosity varies inversely with temperature and directly with pressure, both in a nonlinear fashion. It can be expressed either as an **absolute viscosity** η or as **kinematic viscosity** υ. They are related as

$$\eta = \upsilon\rho \qquad\qquad (10.1)$$

where ρ is the mass density of the fluid. The units of absolute viscosity η are either lb-sec/in^2 (reyn) in the English system or Pa-s in SI units. These are often expressed as μreyn and mPa-s to better suit their typical magnitudes. A centipoise (cP) is 1 mPa-s. Typical absolute viscosity values at 20°C (68°F) are 0.0179 cP (0.0026 μreyn) for air, 1.0 cP (0.145 μreyn) for water, and 393 cP (57μreyn) for SAE 30 engine oil. Oils operating in hot bearings typically have viscosities in the 1 to 5 μreyn range. The term viscosity used without modifiers implies absolute viscosity.

KINEMATIC VISCOSITY   is measured in a *viscometer,* which may be either rotational or capillary. A capillary viscometer measures the rate of flow of the fluid through a capillary tube at a particular temperature, typically 40 or 100°C. A rotational viscometer measures the torque and speed of rotation of a vertical shaft or cone running inside a bearing with its concentric annulus filled with the test fluid at the test temperature. The SI units of kinematic viscosity are cm^2/sec (stoke) and the English units are in^2/sec. Stokes are quite large, so centistokes (cS) are often used.

ABSOLUTE VISCOSITY   is needed for calculation of lubricant pressures and flows within bearings. It is determined from the measured kinematic viscosity and the density of the fluid at the test temperature. Figure 10-1 shows a plot of the variation of absolute viscosity with temperature for a number of common petroleum oils, designated by their ISO numbers and by SAE numbers on both the engine oil and gear oil scales.

10.3   TYPES OF LUBRICATION

There are three general types of lubrication that can occur in bearings, **full-film, mixed film**, and **boundary lubrication**. Full-film lubrication describes a situation in which the bearing surfaces are fully separated by a film of lubricant, eliminating any contact. Full-film lubrication can be **hydrostatic**, **hydrodynamic**, or **elastohydrodynamic**, each discussed below. Boundary lubrication describes a situation where for reasons of geometry, surface roughness, excessive load, or lack of sufficient lubricant, the bearing surfaces physically contact and adhesive or abrasive wear may occur. Mixed-film lubrication describes a combination of partial lubricant film plus some asperity contact between the surfaces.

Figure 10-2 shows a curve depicting the relationship between friction and the relative sliding speed in a bearing. At slow speeds, boundary lubrication occurs with concomitant high friction. As the sliding speed is increased beyond point *A*, a hydrodynamic fluid film begins to form, reducing asperity contact and friction in the mixed-film re-

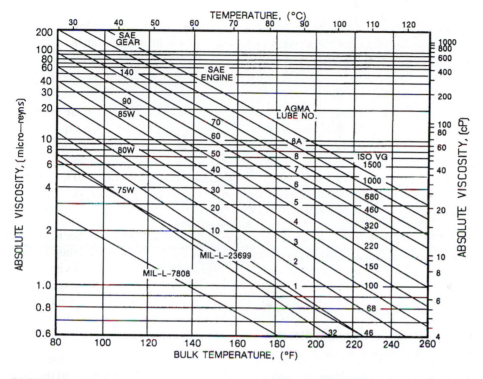

gime. At higher speeds, a full film is formed at point *B*, separating the surfaces completely with reduced friction. (This is the same phenomenon that causes automobile tires to *aquaplane* on wet roads. If the relative velocity of the tire versus the wet road exceeds a certain value, the tire motion pumps a film of water into the interface, lifting the tire off the road. The tire's coefficient of friction is drastically reduced, and the sudden loss of traction can cause a dangerous skid.) At still higher speeds the viscous losses in the sheared lubricant increase friction.

In rotating journal (sleeve) bearings, all three of these regimes will be experienced during start-up and shutdown. As the shaft begins to turn it will be in boundary lubrication. If its top speed is sufficient, it will pass through the mixed regime and reach the desired full-film regime where wear is reduced virtually to zero if the lubricant is kept clean and not overheated. We will briefly discuss the conditions that determine these lubrication states, and then explore a few of them in somewhat greater detail.

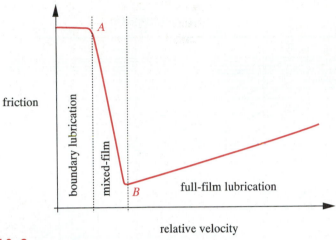

FIGURE 10-2

Change in Friction with Relative Velocity in a Sliding Bearing

Full-Film Lubrication

Three mechanisms can create full-film lubrication, **hydrostatic**, **hydrodynamic**, and **elastohydrodynamic** lubrication.

HYDROSTATIC LUBRICATION   refers to the continuous supply of a flow of lubricant (typically an oil) to the sliding interface at some elevated hydrostatic pressure ($\approx 10^2$– 10^4 psi). This requires a reservoir (sump) to store, a pump to pressurize, and plumbing to distribute the lubricant. When properly done, with appropriate bearing clearances, this approach can eliminate all metal-to-metal contact at the interface during sliding. The surfaces are separated by a film of lubricant, which if kept clean and free of contaminants, reduces wear rates to virtually zero. At zero relative velocity, the friction is essentially zero. With relative velocity, the coefficient of friction in a hydrostatically lubricated interface is about 0.002 to 0.010. This is also the principle of a so-called air bearing, used on "air pallets" to lift (thrust) a load from a surface, allowing it to be moved sideways with very little effort. Hovercraft operate on a similar principle. Water is sometimes used in hydrostatic bearings. Denver's Mile High Stadium has a 21 000 seat grandstand that slides back on hydrostatic water films to convert the stadium from baseball to football.[5] Hydrostatic thrust bearings are more common than radial ones.

HYDRODYNAMIC LUBRICATION   refers to the supply of sufficient lubricant (typically an oil) to the sliding interface to allow the relative velocity of the mating surfaces to pump the lubricant within the gap and separate the surfaces on a dynamic film of liquid. This technique is most effective in journal bearings, where the shaft and bearing create a thin annulus within their clearance that can trap the lubricant and allow the shaft to pump it around the annulus. A leakage path exists at the ends, so a continuous supply of oil must be provided to replace the losses. This supply may be either gravity fed,

ply of oil must be provided to replace the losses. This supply may be either gravity fed, or pressure fed. This is the system used to lubricate the crankshaft and camshaft bearings in an internal combustion engine. Filtered oil is pumped to the bearings under relatively low pressure to replenish the oil lost through the bearing ends, but the condition within the bearing is hydrodynamic, creating much higher pressures to support the bearing loads.

In a hydrodynamic sleeve bearing at rest, the shaft or journal sits in contact with the bottom of the bearing, as shown in Figure 10-3a. As it begins to rotate, the shaft centerline shifts eccentrically within the bearing and the shaft acts as a pump to pull the film of oil clinging to its surface around with it as shown in Figure 10-3b. (The "outer side" of the oil film is stuck to the stationary bearing.) A flow is set up within the small thickness of the oil film. With sufficient relative velocity, the shaft "climbs up" on a wedge of pumped oil and ceases to have metal-to-metal contact with the bearing as shown in Figure 10-3c.

Thus, a hydrodynamically lubricated bearing only touches its surfaces together when stopped or when rotating below its "aquaplane speed." This means that adhesive wear can only occur during the transients of start-up and shutdown. As long as sufficient lubricant and velocity are present to allow hydrodynamic lifting of the shaft off the bearing at its operating speed, there is essentially no adhesive wear. This greatly increases wear life over that of a continuous-contact situation. As with hydrostatic lubrication, the oil must be kept free of contaminants to preclude other forms of wear such as abrasion. The coefficient of friction in a hydrodynamically lubricated interface is about 0.002 to 0.010.

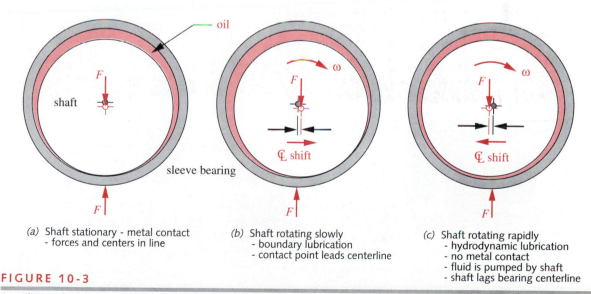

(a) Shaft stationary - metal contact
    - forces and centers in line

(b) Shaft rotating slowly
    - boundary lubrication
    - contact point leads centerline

(c) Shaft rotating rapidly
    - hydrodynamic lubrication
    - no metal contact
    - fluid is pumped by shaft
    - shaft lags bearing centerline

FIGURE 10-3

Boundary and Hydrodynamic Lubrication Conditions in a Sleeve Bearing—Clearance and Motions Exaggerated

ELASTOHYDRODYNAMIC LUBRICATION   When the contacting surfaces are nonconforming, as with the gear teeth or cam and follower shown in Figure 10-4, then it is more difficult to form a full film of lubricant since the nonconforming surfaces tend to expel rather than entrap the fluid. At low speeds these joints will be in boundary lubrication and high wear rates can result with possible scuffing and scoring. The load creates a contact patch from the elastic deflections of the surfaces, as discussed in Chapter 7. This small contact patch can provide enough of a flat surface to allow a full hydrodynamic film to form if the relative sliding velocity is high enough (see Figure 10-2). This condition is termed **elastohydrodynamic lubrication** (EHD) as it depends on the elastic deflections of the surfaces and the fact that the high pressures (100 to 500 kpsi) within the contact zone greatly increase the viscosity of the fluid. (In contrast, the film pressure in conforming bearings is only several 1 000 psi and the change in viscosity due to this pressure is small enough to ignore.)

Gear teeth can operate in any of the three conditions depicted in Figure 10-2. Boundary lubrication occurs in start-stop operation and if prolonged will cause severe wear. Cam-follower joints can also experience any of the regimes in Figure 10-2 but are more likely to be in a boundary-lubricated mode at locations of small radius of curvature on the cam. Rolling-element bearings can see any of the three regimes as well.

The most important parameter that determines which situation occurs in nonconforming contacts is the ratio of the oil-film thickness to the surface roughness. To get full-film lubrication and avoid asperity contact, the rms average surface roughness (R_q) needs to be no more than about 1/2 to 1/3 of the oil-film thickness. An EHD full-film thickness is normally of the order of 1 μm. At very high loads, or low speeds, the EHD film thickness may become too small to separate the surface asperities and mixed-film or boundary lubrication conditions may recur. Factors that have the most effect in creating EHD conditions are increased relative velocity, increased lubricant viscosity, and increased radius of curvature at the contact. Reduction in unit load and reduced stiffness of the material have less effect.[6]

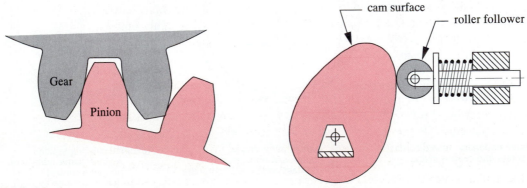

FIGURE 10-4

Open Joints that can have EHD, Mixed, or Boundary Lubrication

Boundary Lubrication

Boundary lubrication refers to situations in which some combination of the geometry of the interface, high load levels, low velocity, or insufficient lubricant quantity preclude the initiation of a hydrodynamic condition. The properties of the contacting surfaces and lubricant properties other than its bulk viscosity determine friction and wear in this situation. Viscosity of the lubricant is not a factor. Note in Figure 10-2 that friction is independent of velocity in boundary lubrication. This is consistent with the definition of Coulomb friction in Section 7.3. See Table 7-1 on p. 479.

Boundary lubrication implies that there is always some metal-to-metal contact in the interface. If the lubricant film is not thick enough to "bury" the asperities on the surfaces this will be true. Rough surfaces could cause this condition. If the relative velocity or the supply of lubricant to a hydrodynamic interface is reduced, it will revert to a boundary-lubrication condition. Surfaces such as gear teeth and cam/follower interfaces (see Figure 10-4) that do not envelop each other can be in a boundary lubrication mode if EHD conditions do not prevail. Ball and roller bearings can also operate in boundary-lubrication mode if the combination of speeds and loads do not allow EHD to occur.

Boundary lubrication is a less desirable condition than the other types described above because it allows the surface asperities to contact and wear rapidly. It is sometimes unavoidable as in the examples of cams, gears, and rolling-element bearings cited. The EP lubricants mentioned above were created for these boundary lubrication applications, especially for hypoid gears, which experience both high sliding velocities and high loads. The coefficient of friction in a boundary-lubricated sliding interface depends on the materials used as well as on the lubricant but ranges from about 0.05 to 0.15 with most being about 0.10.

10.4 MATERIAL COMBINATIONS IN SLIDING BEARINGS

Figure 7-6 (p. 481) shows material combinations and their predicted sliding ability based on their mutual insolubility and other factors. This section will discuss some combinations of materials that have either proven successful or unsuccessful in engineering applications of bearings and sliders.

Some properties sought in a bearing material are relative softness (to absorb foreign particles), reasonable strength, machinability (to maintain tolerances), lubricity, temperature and corrosion resistance, and in some cases, porosity (to absorb lubricant). A bearing material should be less than one-third as hard as the material running against it in order to provide embedability of abrasive particles.[7] In addition, the compatibility issues addressed in Section 7.4 on adhesive wear are of concern and these also depend on the mating material. Several different classes of materials can be useful as bearings, typically those based on lead, tin, or copper. Aluminum, alone, is not a good bearing material although it is used as an alloying element in some bearing materials.

BABBITTS   A whole family of alloys based on lead and tin in combination with other elements are very effective especially when electroplated in thin films on a stronger substrate such as steel. Babbitt is probably the most common example of this family and is used for crankshaft and camshaft bearings in internal combustion engines. Its softness allows particulate embedment and it can be finished to low roughness. A thin electroplated babbitt layer has better fatigue resistance than a thick babbitt bushing, but cannot embed particles as well. Good hydrodynamic or hydrostatic lubrication is required as babbitt has a low melt temperature and will quickly fail under boundary-lubricated conditions. Shafts for babbitt bearings should have a minimum hardness of 150-200HB and a ground surface finish of R_a = 0.25 to 0.30 µm (10 to 12 µin).[8]

BRONZES   The copper-alloy family, principally bronzes, are an excellent choice for running against steel or cast iron. Bronze is softer than ferrous materials but has good strength, machinability, and corrosion resistance and runs well against ferrous alloys when lubricated. There are five common copper alloys used in bearings, copper-lead, leaded bronze, tin bronze, aluminum bronze, and beryllium copper. They have a range of hardness from that of babbitts to close to that of steel.[8] Bronze bushings can withstand boundary lubrication and can support high loads and high temperatures. Bronze bushings and flat stock are available commercially in a variety of sizes, both solid and sintered (see below).

GRAY CAST IRON AND STEEL   are reasonable bearing materials when run against each other at low velocities. The free graphite in the cast iron adds lubricity but liquid lubricant is needed as well. Steel can also be run against steel if both parts are hardened and lubricated. This is a common choice in rolling contact as in rolling-element bearings. In fact, hardened steel will run against almost any material with proper lubrication. Hardness seems to protect against adhesion in general.

SINTERED MATERIALS   are formed from powder and remain microscopically porous after heat treatment. (See Section 2.8.) Their porosity allows them to take up significant amounts of lubricant and hold it by capillary action, releasing it into the bearing when hot. Sintered bronze is widely used for running against steel or cast iron.

NONMETALLIC MATERIALS   of some types offer the possibility of dry running if they have sufficient lubricity. Graphite is one example. Some thermoplastics such as nylon, acetal, and filled Teflon offer a low coefficient of friction µ against any metal but have low strengths and low melt temperatures, which when combined with their poor heat conduction severely limits the loads and speeds of operation that they can sustain. Teflon has a very low µ (approaching rolling values) but requires fillers to raise its strength to usable levels. Inorganic fillers such as talc or glass fiber add significant strength and stiffness to any of the thermoplastics but at the cost of a higher µ and increased abrasiveness. Graphite and MoS_2 powder are also used as fillers and these add lubricity as well as strength and temperature resistance. Some mixtures of polymers such as acetal-teflon are also offered. Thermoplastic bearings are usually only practical where loads and temperatures are low. The practical combinations of shaft and bearing materials are really quite limited. Table 10-3 shows some usable combinations of metallic bearing materials and indicates their hardness ratios versus typical shaft-steel.[9]

10

Table 10-3   Recommended Bearing Materials for Sliding Against Steel or Cast Iron

Bearing Material	Hardness kg/mm^2	Minimum Shaft Hardness kg/mm^2	Hardness Ratio
Lead-base babbitt	15-20	150	8
Tin-base babbitt	20-30	150	6
Alkali-hardened lead	22-26	200-250	9
Copper-lead	20-23	300	14
Silver (overplated)	25-50	300	8
Cadmium base	30-40	200-250	6
Aluminum alloy	45-50	300	6
Lead bronze	40-80	300	5
Tin bronze	60-80	300-400	5

Source: Wilcock and Booser, *Bearing Design and Application*, McGraw-Hill, 1957.

10.5   HYDRODYNAMIC LUBRICATION THEORY

Consider the sleeve bearing shown in Figure 10-3. Figure 10-5a shows a similar journal and bearing, but concentric and with the axis vertical. The diametral clearance c_d between journal and bearing is very small, typically about one-thousandth of the diameter. We can model this as two flat plates because the gap h is so small compared to the radius of curvature. Figure 10-5b shows two such plates separated by an oil film with a gap of dimension h. If the plates are parallel, the oil film will not support a transverse load. This is also true of a concentric journal and bearing. A concentric horizontal journal will become eccentric from the weight of the shaft, as in Figure 10-3. If the axis is vertical as in Figure 10-5a, the journal can spin concentric with the bearing since there is no transverse gravity force.

Petroff's Equation for No-Load Torque

If we hold the lower plate of Figure 10-5b stationary and move the upper plate to the right with a velocity U, the fluid between the plates will be sheared in the same manner as in the concentric gap of Figure 10-5a. The fluid wets and adheres to both plates making its velocity zero at the stationary plate and U at the moving plate. Figure 10-5c shows a differential element of fluid in the gap. The velocity gradient causes the angular distortion β. In the limit, $\beta = dx / dy$. The shear stress τ_x acting on a differential element of fluid in the gap is proportional to the shear rate:

$$\tau_x = \eta \frac{d\beta}{dt} = \eta \frac{d}{dt}\frac{dx}{dy} = \eta \frac{d}{dy}\frac{dx}{dt} = \eta \frac{du}{dy} \qquad (10.2a)$$

and the constant of proportionality is the viscosity η. In a film of constant thickness h, the velocity gradient $du / dy = U / h$ and is constant. The force to shear the entire film is

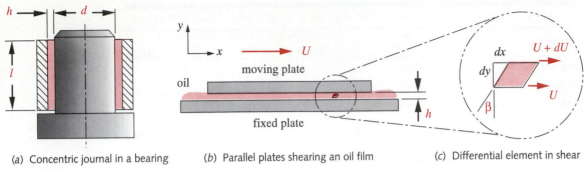

(a) Concentric journal in a bearing   (b) Parallel plates shearing an oil film   (c) Differential element in shear

FIGURE 10-5

* The amount of angle needed to create a supporting force is surprisingly small. For example, in a bearing of about 32-mm dia, the circumference is 100 mm. A typical entrance gap h_{max} might be 25 μm (0.001 in), and the exit gap h_{min}, 12.5 μm (0.0005 in). The slope is then 0.0000125 / 100 or about 7 / 1000 of a degree (26 seconds of arc). This is equivalent to about a 3-cm rise over a 300-yard-long football field.

$$F = A\tau_x = \eta A \frac{U}{h} \qquad (10.2b)$$

where A is the area of the plate.

For the concentric journal and bearing of Figure 10-5a, let the gap $h = c_d / 2$ where c_d is the diametral clearance. The velocity is $U = \pi dn'$ where n' is revolutions per second, and the shear area is $A = \pi dl$. The torque T_0 required to shear the film is then

$$T_0 = \frac{d}{2}F = \frac{d}{2}\eta A \frac{U}{h} = \frac{d}{2}\eta \pi \, dl \frac{\pi dn'}{c_d/2}$$

$$T_0 = \eta \frac{\pi^2 d^3 l n'}{c_d} \qquad (10.2c)$$

This is *Petroff's equation* for the no-load torque in a fluid film.

Reynolds' Equation for Eccentric Journal Bearings

† In England, in the 1880's, Beauchamp Tower was experimentally investigating the friction in hydrodynamically lubricated bearings for the railroad industry (though the term hydrodynamic and its theory was only then about to be discovered). His results showed much lower friction coefficients than expected. He drilled a radial hole through the bearing in order to add oil while running but was surprised to find that oil flowed out of the hole when the shaft turned. He corked the hole but the cork was expelled. He plugged the hole with wood but that also popped out. When he put a pressure gage in the hole he measured pressures well

To support a transverse load, the plates of Figure 10-5 must be nonparallel. If we rotate the lower plate of Figure 10-5a slightly counterclockwise and move the upper plate to the right with a velocity U, the fluid between the plates will be carried into the decreasing gap as shown in Figure 10-6a, developing a pressure that will support a transverse load P. The angle between the plates is analogous to the varying clearance due to the eccentricity e of the journal and bearing in Figure 10-6b.* When a transverse load is applied to a journal, it must assume an eccentricity with respect to the bearing in order to form a changing gap to support the load by developing pressure in the film.†

Figure 10-6b shows a greatly exaggerated eccentricity e and gap h for a journal bearing. The eccentricity e is measured from the center of the bearing O_b to the center of the journal O_j. The zero to π axis for the independent variable θ is established along the line O_bO_j as shown in Figure 10-6b. The maximum possible value of e is $c_r = c_d / 2$, where c_r is the radial clearance. The eccentricity can be converted to a dimensionless eccentricity ratio ε:

$$\varepsilon = \frac{e}{c_r} \tag{10.3}$$

which varies from 0 at no load to 1 at maximum load when the journal contacts the bearing. An approximate expression for the film thickness h as a function of θ is

$$h = c_r (1 + \varepsilon \cos \theta) \tag{10.4a}$$

The film thickness h is maximum at $\theta = 0$ and minimum at $\theta = \pi$, found from

$$h_{min} = c_r (1 - \varepsilon) \qquad\qquad h_{max} = c_r (1 + \varepsilon) \tag{10.4b}$$

Consider the journal bearing shown in Figure 10-7. In the analysis that follows, the gap is given by equation 10.4a. We can take the origin of an xy coordinate system at any point on the circumference of the bearing such as O. The x axis is then tangent to the bearing, the y axis is through the bearing center O_b, and the z axis (not shown) is parallel to the axis of the bearing. Generally, the bearing is stationary and only the journal rotates, but in some cases the reverse may be true, or both may rotate as in the planet shaft of an epicyclic gear train. Thus we show a tangential velocity U_1 for the bearing as well as a tangential velocity T_2 for the journal. Note that their directions (angles) are not the same due to the eccentricity. The tangential velocity T_2 of the journal can be resolved into components in the x and y directions as U_2 and V_2, respectively. The angle between T_2 and U_2 is so small that its cosine is essentially 1 and we can set $U_2 \cong T_2$. The component V_2 in the y direction is due to the closing (or opening) of the gap h as it rotates and is $V_2 = \partial h / \partial x$.

above the average pressure expected from a calculation of load / area. He then mapped the pressure distribution over 180° of the bearing and discovered the now familiar pressure distribution (see Figure 10-8) whose average value is load / area. On learning of this discovery, Osborne Reynolds set out to develop the mathematical theory to explain it, publishing the results in 1886.[12]

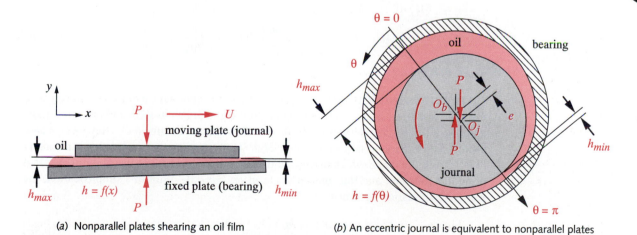

(a) Nonparallel plates shearing an oil film

(b) An eccentric journal is equivalent to nonparallel plates

FIGURE 10-6

An Oil Film Sheared Between Nonparallel Surfaces Can Support a Transverse Load

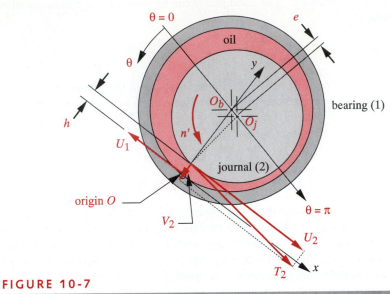

FIGURE 10-7

Velocity Components in an Eccentric Journal Bearing

Using the above assumptions, we can write Reynolds' equation[*] relating the changing gap thickness h, the relative velocities between the journal and bearing V_2 and $U_1 - U_2$, and the pressure in the fluid p as a function of the two dimensions x and z, assuming that the journal and bearing are parallel in the z direction and the viscosity η is constant,

$$\frac{1}{6\eta}\left[\frac{\partial}{\partial x}\left(h^3\frac{\partial p}{\partial x}\right) + \frac{\partial}{\partial z}\left(h^3\frac{\partial p}{\partial z}\right)\right] = (U_1 - U_2)\frac{\partial h}{\partial x} + 2V_2$$

$$= (U_1 - U_2)\frac{\partial h}{\partial x} + 2U_2\frac{\partial h}{\partial x} = (U_1 + U_2)\frac{\partial h}{\partial x} = U\frac{\partial h}{\partial x} \qquad (10.5a)$$

where $U = U_1 + U_2$.

LONG BEARING SOLUTION   Equation 10.5a does not have a closed-form solution, but can be solved numerically. Raymondi and Boyd did so in 1958 and provide a large number of design charts for its application to finite-length bearings[11] . Reynolds solved a simplified version in series form (in 1886)[12] by assuming that the bearing is infinitely long in the z direction, which makes the flow zero and the pressure distribution over that direction constant, and thus makes the term $\partial p / \partial z = 0$. With this simplification, the Reynolds' equation becomes

$$\frac{\partial}{\partial x}\left(h^3\frac{\partial p}{\partial x}\right) = 6\eta U\frac{\partial h}{\partial x} \qquad (10.5b)$$

---

[*] For a derivation of Reynolds' equation, see reference 2, 3, 4, or 10.

In 1904, A. Sommerfeld found a closed-form solution for the infinitely long-bearing equation 10.5b as

$$p = \frac{\eta U r}{c_r^2}\left[\frac{6\varepsilon(\sin\theta)(2+\varepsilon\cos\theta)}{(2+\varepsilon^2)(1+\varepsilon\cos\theta)^2}\right] + p_0 \qquad (10.6a)$$

which gives the pressure p in the lubricant film as a function of angular position θ around the bearing for particular dimensions of journal radius r, radial clearance c_r, eccentricity ratio ε, surface velocity U, and viscosity η. The term p_0 accounts for any supply pressure at the otherwise zero pressure position at $\theta = 0$. Equation 10.6a is referred to as the *Sommerfeld solution* or the *long-bearing solution*.

If p is computed from this equation over $\theta = 0$ to 2π, it will predict negative pressures from $\theta = \pi$ to 2π with absolute magnitudes equal to the positive pressures from 0 to π. Since a fluid cannot withstand large negative pressure without cavitation, the equation is typically evaluated only from 0 to π and the pressure assumed to be p_0 over the other half of the circumference. This is referred to as the half-Sommerfeld solution.

Sommerfeld also determined an equation for the total load P on a long bearing as

$$P = \frac{\eta U l r^2}{c_r^2}\frac{12\pi\varepsilon}{(2+\varepsilon^2)(1+\varepsilon^2)^{1/2}} \qquad (10.6b)$$

This equation can be rearranged in nondimensional form to provide a characteristic bearing number called the **Sommerfeld number** S. First rearrange terms:

$$\frac{(2+\varepsilon^2)(1+\varepsilon^2)^{1/2}}{12\pi\varepsilon} = \eta\frac{Ul}{P}\left(\frac{r}{c_r}\right)^2 \qquad (10.6c)$$

The average pressure p_{avg} on the bearing is

$$p_{avg} = \frac{P}{A} = \frac{P}{ld} \qquad (10.6d)$$

The velocity $U = \pi d n'$ where n' is revolutions per second, and $c_r = c_d/2$. Substituting gives

$$\frac{(2+\varepsilon^2)(1+\varepsilon^2)^{1/2}}{12\pi\varepsilon} = \eta\frac{(\pi d n')l}{dl\,p_{avg}}\left(\frac{d}{c_d}\right)^2 = \eta\left(\frac{\pi n'}{p_{avg}}\right)\left(\frac{d}{c_d}\right)^2 = S \qquad (10.6e)$$

Note that S is a function only of the eccentricity ratio ε, but can also be expressed in terms of geometry, pressure, velocity, and viscosity.

SHORT-BEARING SOLUTION   Long bearings are not often used in modern machinery for several reasons. Small shaft deflections or misalignment can reduce the radial clearance to zero in a long bearing, and packaging considerations often require short bearings. Typical l/d ratios of modern bearings are in the range of 1/4 to 1. The long-bearing (Sommerfeld) solution assumes no end leakage of oil from the bearing, but at these small l/d ratios, end leakage can be a significant factor. Ocvirk and DuBois[13], [14], [15], [16] solved a form of Reynolds' equation that includes the end leakage term.

$$\frac{\partial}{\partial z}\left(h^3\frac{\partial p}{\partial z}\right) = 6\eta U\frac{\partial h}{\partial x} \tag{10.7a}$$

This form neglects the term that accounts for the circumferential flow of oil around the bearing on the premise that it will be small in comparison to the flow in the z direction (leakage) in a short bearing. Equation 10.7a can be integrated to give an expression for pressure in the oil film as a function of both θ and z:

$$p = \frac{\eta U}{rc_r^2}\left(\frac{l^2}{4} - z^2\right)\frac{3\varepsilon\sin\theta}{(1+\varepsilon\cos\theta)^3} \tag{10.7b}$$

Equation 10.7b is known as the *Ocvirk solution* or the *short-bearing solution*. It is typically evaluated for $\theta = 0$ to π, with the pressure assumed to be zero over the other half of the circumference. Figure 10-8 typical pressure distributions over θ and z. The $\theta = 0$

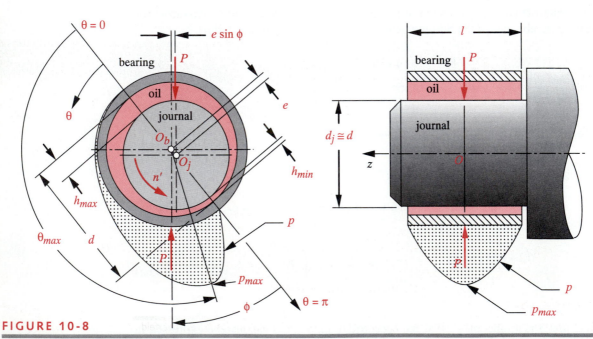

FIGURE 10-8

Pressure Distribution in a Short Journal Bearing—Film Thickness Greatly Exaggerated

position is taken at $h = h_{max}$ and the θ axis goes through O_b and O_j. The pressure distribution p with respect to z is parabolic and peaks at the center of the bearing length l and is zero at $z = \pm l / 2$. Pressure p varies nonlinearly over θ and peaks in its second quadrant. The value of θ_{max} at p_{max} can be found from

$$\theta_{max} = \cos^{-1}\left(\frac{1 - \sqrt{1 + 24\varepsilon^2}}{4\varepsilon}\right) \qquad (10.7c)$$

and the value of p_{max} can be found by substituting $z = 0$ and $\theta = \theta_{max}$ in equation 10.7b.

Figure 10-9 compares the variation of pressure p in the film over $\theta = 0$ to π for the Sommerfeld long-bearing solution (taken as the reference at 100%) and the Ocvirk short-bearing solution at several l / d ratios from 1/4 to 1. Note the large error that would occur if the long bearing solution were used for ratios < 1. At $l / d = 1$, the two solutions give similar results with the Ocvirk solution predicting slightly higher peak pressure than the Sommerfeld solution. DuBois and Ocvirk found in tests[13], [14] that the short-bearing solution gave results that closely matched experimental measurements for l / d ratios from 1/4 to 1 and also matched experimental data up to $l / d = 2$ if that ratio was kept at 1 for the computation of bearings with actual ratios between 1 and 2. Because most modern bearings tend to have l / d ratios between 1/4 and 2, the Ocvirk solution provides a convenient and reasonably accurate method to use. The Sommerfeld solution gives accurate results for l / d ratios above about 4. Boyd and Raimondi's method[11] gives more accurate results for intermediate l / d ratios but is more cumbersome to use.

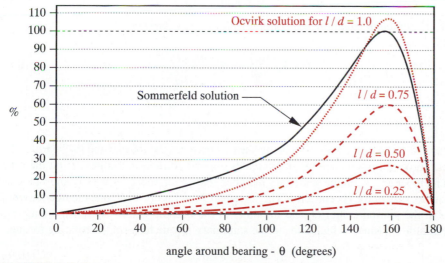

FIGURE 10-9

Comparison of the Ocvirk Short-Bearing Approximation for Various l / d ratios with the Sommerfeld Long-Bearing Approximation of Pressure in the Oil Film from 0 to 180°

Note in Figure 10-8 that the peak pressure occurs at an angle θ_{max} as defined in equation 10.7c. This angle is measured from the zero θ axis, which lies along the line of centers of the bearing and journal. But what determines the angle of this eccentricity line between the centers O_b and O_j? Typically, the line of action of the force P applied to the journal is defined by external factors. This force P is shown vertical in the figure and the angle between this force and the $\theta = \pi$ axis is shown as ϕ. (Angle ϕ is used rather than the angle θ_P measured from $\theta = 0$ because ϕ will always be an acute angle.) Angle ϕ can be found from

$$\phi = \tan^{-1}\left(\frac{\pi\sqrt{1-\varepsilon^2}}{4\varepsilon}\right) \tag{10.8a}$$

and the magnitude of the resultant force P is related to the bearing parameters as

$$P = K_\varepsilon \frac{\eta U l^3}{c_r^2} \tag{10.8b}$$

where K_ε is a dimensionless parameter that is a function of the eccentricity ratio ε:

$$K_\varepsilon = \frac{\varepsilon\left[\pi^2\left(1-\varepsilon^2\right)+16\varepsilon^2\right]^{\frac{1}{2}}}{4\left(1-\varepsilon^2\right)^2} \tag{10.8c}$$

The linear velocity U can be expressed as

$$U = \pi d n' \tag{10.8d}$$

and substituted in equation 10.8b along with $c_r = c_d / 2$ to get

$$P = K_\varepsilon \frac{\eta U l^3}{c_r^2} = K_\varepsilon \frac{4\pi\eta d n' l^3}{c_d^2} \tag{10.8e}$$

Torque and Power Losses in Journal Bearings

Figure 10-8 shows the fluid film being sheared between the journal and the bearing. The shear force acting on each member creates opposite direction torques, T_r on the rotating member and T_s on the stationary member. However, these torques T_r and T_s are not equal because of the eccentricity. The force pair P, in Figure 10-8, one member of which acts at the journal center O_j and the other at the bearing center O_b, form a couple of magnitude $P e \sin \phi$, which adds to the stationary torque to form the rotating torque.

$$T_r = T_s + P e \sin \phi \tag{10.9a}$$

The stationary torque T_s can be found from

$$T_s = \eta \frac{d^2 l (U_2 - U_1)}{c_d} \frac{\pi}{\left(1 - \varepsilon^2\right)^{1/2}} \tag{10.9b}$$

Substitute equation 10.8d in 10.9b to put it in terms of the rotational velocities of journal and bearing:

$$T_s = \eta \frac{d^3 l (n_2' - n_1')}{c_d} \frac{\pi^2}{\left(1 - \varepsilon^2\right)^{1/2}} \tag{10.9c}$$

Note the similarity of equation 10.9c to the Petroff equation 10.2c for the concentric-journal, no-load torque T_0. We can form a ratio of the stationary torque in an eccentric bearing to the no-load torque as

$$\frac{T_s}{T_0} = \frac{1}{\left(1 - \varepsilon^2\right)^{1/2}} \tag{10.9d}$$

which, not surprisingly, is a function of only the eccentricity ratio ε. A similar ratio between rotating torque T_r and the Petroff no-load torque can also be formed.

The power Φ lost in the bearing can be found from the rotating torque T_r and the rotational velocity n'.

$$\Phi = T_r \omega = 2\pi T_r (n_2' - n_1') \qquad \text{N-m/s} \quad \text{or} \quad \text{in-lb/s} \tag{10.10}$$

This can be converted to watts or horsepower as appropriate to the units system used.

COEFFICIENT OF FRICTION   The coefficient of friction in the bearing can be determined as the ratio between the tangential shear force and the applied normal force P.

$$\mu = \frac{f}{P} = \frac{T_r/r}{P} = \frac{2T_r}{Pd} \tag{10.11}$$

10.6   DESIGN OF HYDRODYNAMIC BEARINGS

Usually the applied force P that the bearing is expected to support and the speed of rotation n' are known. The bearing diameter may or may not be known, but often will have been defined by stress, deflection, or other considerations. The design of the bearing requires finding a suitable combination of bearing diameter and/or length that will operate with a suitable viscosity of fluid, have reasonable and manufacturable clearance and have an eccentricity ratio that will not allow metal to metal contact under load or any expected overload conditions.

Design Load Factor—The Ocvirk Number

A convenient way to approach this problem is to define a dimensionless load factor against which various bearing parameters can be computed, plotted, and compared. Equation 10.8e can be rearranged to provide such a factor. Solve Eq. 10.8e for K_ε:

$$K_\varepsilon = \frac{Pc_d^2}{4\eta\pi dn'l^3} \qquad (10.12a)$$

Substitute equation 10.6d for the load P to introduce the average film pressure p_{avg}.

$$K_\varepsilon = \frac{p_{avg}ldc_d^2}{4\eta\pi dn'l^3}\frac{d}{d} = \frac{1}{4\pi}\left[\left(\frac{p_{avg}}{\eta n'}\right)\left(\frac{d}{l}\right)^2\left(\frac{c_d}{d}\right)^2\right] = \frac{1}{4\pi}O_N \qquad (10.12b)$$

The term in brackets is the desired dimensionless **load factor** or **Ocvirk number** O_N.

$$O_N = \left(\frac{p_{avg}}{\eta n'}\right)\left(\frac{d}{l}\right)^2\left(\frac{c_d}{d}\right)^2 = 4\pi K_\varepsilon = \frac{\pi\varepsilon\left[\pi^2\left(1-\varepsilon^2\right)+16\varepsilon^2\right]^{\frac{1}{2}}}{\left(1-\varepsilon^2\right)^2} \qquad (10.12c)$$

This expression contains the parameters over which the designer has control and shows that any combination of those parameters that yields the same Ocvirk number will have the same eccentricity ratio ε. The eccentricity ratio gives an indication of how close to failure the oil film is since $h_{min} = c_r(1 - \varepsilon)$. Compare the Ocvirk number to the Sommerfeld number of equation 10.6e. The concept is the same.

Figure 10-10 shows a plot of eccentricity ratio ε as a function of Ocvirk number O_N and also shows experimental data from reference 10 for the same parameters. An empirical curve is fitted through the data which shows that the theory understates the magnitude of the eccentricity ratio. The empirical curve can be approximated by

$$\varepsilon_x \cong 0.21394 + 0.38517\log O_N - 0.0008(O_N - 60) \qquad (10.13)$$

The calculation of load, torque, average and maximum pressures in the oil film, and other bearing parameters can be done using this empirical value of ε in equations 10.7 to 10.11 and the minimum film thickness calculated from equation 10.4b.

Other dimensionless ratios can be formed from equations 10.7 to 10.11 for use as design aids. Figure 10-11 shows ratios of p_{max}/p_{avg}, and T_s/T_0 as a function of Ocvirk number for both theoretical and experimental values of ε. Figure 10-12 shows the theoretical and experimental variation in the angles θ_{max} and ϕ with the Ocvirk number.

Design Procedures

Load and speed are typically known. If the shaft has been designed for stress or deflection its diameter will be known, A bearing length or l/d ratio should be chosen based

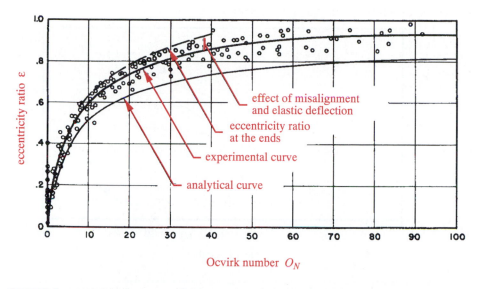

FIGURE 10-10

Analytical and Experimental Relationship Between Eccentricity Ratio ε and Ocvirk Number O_N
Source: G. B. DuBois and F. W. Ockvirk, The Short Bearing Approximation for Plain Journal Bearings, *Trans. ASME*, vol. 77, pp. 1173-1178, 1955.

on packaging considerations. Larger l/d ratios will give lower film pressures, all else equal. The **clearance ratio** is defined as C_d/d. Clearance ratios are typically in the range of 0.001 to 0.002 and sometimes as high as 0.003. Larger clearance ratios will rapidly increase the load number O_N as C_d/d is squared in equation 10.12c. Higher Ocvirk numbers give larger eccentricity, pressure, and torque as can be seen in Figures 10-10 and 10-11, but these factors increase more slowly at higher O_N. An advantage of larger clearance ratios is higher lubricant flow, which promotes cooler running. Large l/d ratios may require greater clearance ratios to accommodate shaft deflection.* An Ocvirk number can be chosen and the required viscosity of the lubricant found from equations 10.7 to 10.11. Some iteration will usually be required to obtain a balanced design.

If the dimensions of the shaft are not yet determined, a diameter and length of bearing can be found from iteration of the bearing equations with an assumed Ocvirk number. A trial lubricant must be chosen and its viscosity found for the assumed operating temperatures from charts such as Figure 10-1. After the bearing is designed, a fluid flow and heat transfer analysis can be done to determine its required oil flow rates and predicted operating temperatures. These aspects are not addressed here due to lack of space, but can be found in many references, such as 3 and 10.

The choice of Ocvirk number has a significant effect on the design. G. B Dubois has offered some guidance by suggesting that a load number of $O_N = 30$ (ε = 0.82) be

---

* Note that if the bearing is short enough to prevent any metal contact at its ends due to shaft slope or deflection, then the bearing can be considered to give simple support to the shaft.

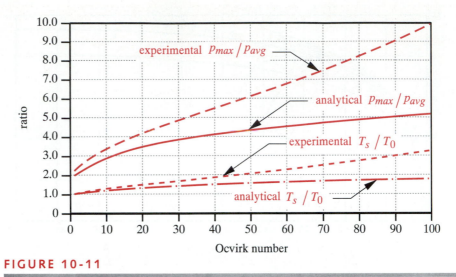

FIGURE 10-11

Pressure Ratios and Torque Ratios for Short Bearings as a Function of Ocvirk Number

considered an upper limit for "moderate" loading, $O_N = 60$ ($\varepsilon = 0.90$) an upper limit for "heavy" loading, and $O_N = 90$ ($\varepsilon = 0.93$) a limit for "severe" loading. At load numbers above about 30, care should be taken to carefully control manufacturing tolerances, surface finishes, and deflections. For general bearing applications it is probably better to stay below an O_N of about 30. The design procedure is best shown with an example.

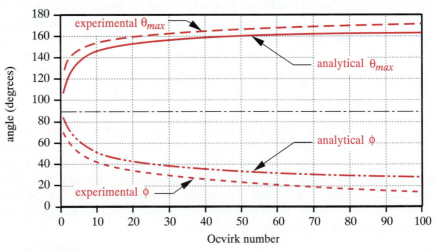

FIGURE 10-12

Angles θ_{max} and ϕ as a Function of the Ocvirk Number

EXAMPLE 10-1

Sleeve Bearing Design for a Defined Shaft Diameter

Problem   Design sleeve bearings to replace the rolling-element bearings on the shaft shown in Figure 9-5 (repeated here). The shaft was designed in Example 9-1.

Given   The maximum transverse loads on the shaft at the bearings are 16 lb at R_1 and 54 lb at R_2. Since the load at R_2 is 4x that at R_1, one design can be created for R_2 and used also at R_1. Shaft diameters at R_1 and R_2 are 0.591 in. The shaft speed is 1725 rpm. The bearings are stationary.

Assumptions   Use a clearance ratio of 0.0017 and an l/d ratio of 0.75. Keep the Ocvirk number at 30 or below, preferably about 20.

Find   The bearing eccentricity ratio, maximum pressure and its location, minimum film thickness, coefficient of friction, torque, and power lost in bearing. Choose a suitable lubricant to operate at 190°F.

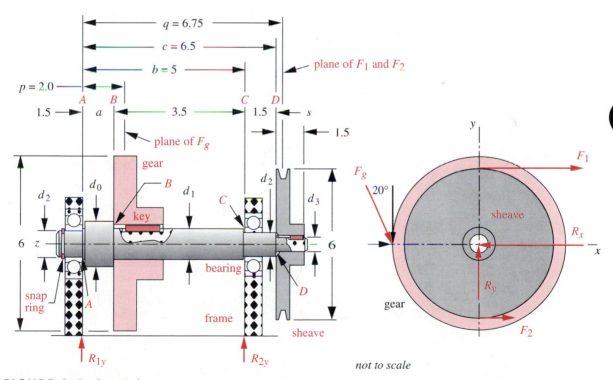

FIGURE 9-5   Repeated

Geometry of a Preliminary Design for Examples 9-1 through 9-3

Solution   See Figure 9-5, and the *TKSolver* files EX10-01a and EX10-01b.

1 Convert the speed given in rpm to rps and find the tangential velocity U.

$$n' = 1\,725 \,\frac{\text{rev}}{\text{min}}\left(\frac{1\,\text{min}}{60\,\text{sec}}\right) = 28.75 \text{ rps}$$

$$U = \pi d n' = \pi(0.591)(28.75) = 53.38 \text{ in/sec} \qquad (a)$$

2 The diametral and radial clearances are found from the given diameter and the assumed clearance ratio:

$$c_d = 0.0017(0.591) = 0.0010 \text{ in}$$

$$c_r = c_d/2 = 0.0005 \text{ in} \qquad (b)$$

3 The bearing length is found from the assumed $l\,/\,d$ ratio of 0.75.

$$l = 0.75(0.591) = 0.443 \text{ in} \qquad (c)$$

4 Find the experimental eccentricity ratio from equation 10.13 or from Figure 10-10 using the suggested value of $O_N = 20$.

$$\varepsilon_x \cong 0.21394 + 0.38517 \log O_N - 0.0008(O_N - 60)$$

$$\cong 0.21394 + 0.38517 \log 20 - 0.0008(20 - 60) = 0.747 \qquad (d)$$

5 Find the dimensionless parameter K_ε from equation 10.8c.

$$K_\varepsilon = \frac{\varepsilon\left[\pi^2\left(1-\varepsilon^2\right) + 16\varepsilon^2\right]^{\frac{1}{2}}}{4\left(1-\varepsilon^2\right)^2}$$

$$= \frac{0.747\left[\pi^2\left(1-0.747^2\right) + 16(0.747)^2\right]^{\frac{1}{2}}}{4\left(1-0.747^2\right)^2} = 3.487 \qquad (e)$$

6 The viscosity η of lubricant required to support the design load P can now be found by rearranging equation 10.8b:

$$\eta = \frac{Pc_r^2}{K_\varepsilon U l^3} = \frac{54(0.0005)^2}{3.487(53.38)(0.443)^3} = 0.833 \text{ μreyn} \qquad (f)$$

Enter Figure 10-1 to find that an oil of about ISO VG 32 will provide this value at 190°F. This oil is equivalent to an SAE 10W engine oil.

7 The average pressure in the oil film is found from equation 10.6d.

$$p_{avg} = \frac{P}{ld} = \frac{54}{0.443(0.591)} = 206 \text{ psi} \qquad (g)$$

8   The angle θ_{max} at which the pressure is maximum can be found either from equation
    10.7c using the experimental value of $\varepsilon = 0.747$,

$$\theta_{max} = \cos^{-1}\left(\frac{1 - \sqrt{1 + 24\varepsilon^2}}{4\varepsilon}\right) = \cos^{-1}\left(\frac{1 - \sqrt{1 + 24(0.747)^2}}{4(0.747)}\right) = 159.2° \qquad (h)$$

or it can be read from the experimental curve in Figure 10-12 for $O_N = 20$ as 159°.

9   The maximum pressure can be found by substituting θ_{max} in equation 10.7b with
    $z = 0$ since it is maximum at the center of the bearing length l.

$$p = \frac{\eta U}{rc_r^2}\left(\frac{l^2}{4} - z^2\right)\frac{3\varepsilon\sin\theta}{(1 + \varepsilon\cos\theta)^3}$$

$$= \frac{(8.33E - 7)(53.38)}{0.296(0.0005)^2}\left(\frac{(0.443)^2}{4} - 0^2\right)\frac{3(0.747)\sin(159.2°)}{(1 + 0.747\cos(159.2°))^3} = 857 \text{ psi} \qquad (i)$$

or the ratio of p_{max}/p_{avg} can be read from the experimental curve in Figure 10-11
for $O_N = 20$ as 4.16 and multiplied by p_{avg} from step (g) above to get the same result.

10 Find the angle ϕ, which locates the $\theta = 0$ to π axis with respect to the applied load P
    from equation 10.8a.

$$\phi = \tan^{-1}\left(\frac{\pi\sqrt{1 - \varepsilon^2}}{4\varepsilon}\right) = \tan^{-1}\left(\frac{\pi\sqrt{1 - (0.747)^2}}{4(0.747)}\right) = 34.95° \qquad (j)$$

11 The stationary and rotating torques can now be found from equations 10.9b and
    10.9a using the angle ϕ.

$$T_s = \eta\frac{d^3 l(n_2' - n_1')}{c_d}\frac{\pi^2}{(1 - \varepsilon^2)^{1/2}}$$

$$= (8.33E - 7)\frac{(0.591)^3(0.443)(28.75 - 0)}{0.001}\frac{\pi^2}{\left(1 - (0.747)^2\right)^{1/2}} = 0.032\ 5 \text{ lb - in} \qquad (k)$$

$$T_r = T_s + Pe\sin\phi = 0.0325 + 54(0.000\ 37)\sin 34.95° = 0.044\ 1 \text{ lb - in} \qquad (l)$$

12 The power loss in the bearing is found from equation 10.10.

$$\Phi = 2\pi T_r(n_2' - n_1') = 2\pi(0.044\ 1)(28.75 - 0) = 7.963\ \frac{\text{in - lb}}{\text{sec}} = 0.001 \text{ hp} \qquad (m)$$

13 The coefficient of friction in the bearing can be found from the ratio of the shear
    force to the normal force using equation 10.11.

$$\mu = \frac{2T_r}{Pd} = \frac{2(0.044\ 1)}{54(0.591)} = 0.003 \qquad (n)$$

14　The minimum film thickness is found from equation 10.4b.

$$h_{min} = c_r(1 - \varepsilon) = 0.0005(1 - 0.747) = 0.000\,126 \text{ in } (126 \text{ µin})　\qquad (o)$$

This is a reasonable value since the composite rms surface roughness (see equation 10.14*a* in next section) needs to be no more than about a third to a fourth of the minimum film thickness to avoid asperity contact (see Figure 10-13) and a 30-40 µin R_q finish or better is easily obtainable by precision milling, grinding, or honing.

15　All of the computations shown above were done with the *TKSolver* model EX10-01a. The reader may run that model to see more detail of the solution than is presented here and to see plots of the variables over a range of Ocvirk numbers.

16.　A safety factor against asperity contact can be estimated by back-solving the model using a minimum film thickness equal to the assumed average surface finish of say 40 µin, and determining what Ockvirk number and load *P* would be required to reduce the minimum oil film thickness to that value. This was easily done in the *TKSolver* model EX10-01b by switching h_{min} to input status and both *P* and O_N to output status, providing a guess value for O_N and iterating to a solution. The result is:

when　$h_{min} = 40$ µin,　　　$O_N = 72.3$,　　$\varepsilon = 0.92$,　　$P = 586$ lb

and　　　　　　　　　　$N = \dfrac{586}{54} = 10.9$　　　　　　　　　　(p)

which is an ample reserve for overloads.

17　If this safety factor calculation had indicated that a small overload could put the bearing in trouble, redesigning the bearing for a lower Ocvirk number would give more margin against failure under overloads. Equation 10.11*c*, repeated here as (q) shows what could be changed to reduce O_N :

$$O_N = \left(\frac{p_{avg}}{\eta n'}\right)\left(\frac{d}{l}\right)^2\left(\frac{c_d}{d}\right)^2　\qquad (q)$$

It would require some combination of: decreasing the clearance ratio, decreasing the *d / l* ratio, or using a higher viscosity oil, Assuming the rotational speed, load, and shaft diameter remain unchanged, the bearing length could be increased or the diametral clearance reduced as well as η increased to improve the design.

10.7　NONCONFORMING CONTACTS

Nonconforming contacts such as gear teeth, cam-follower joints, and rolling-element bearings (balls, rollers) can operate in boundary, mixed, or elastohydrodynamic (EHD) modes of lubrication. The principal factor that determines which of these situations will occur is the specific film thickness Λ, which is defined as the minimum film thickness at the patch center divided by the composite rms surface roughness of the two surfaces.

$$\Lambda = h_c \Big/ \sqrt{R_{q_1}^2 + R_{q_2}^2} \qquad (10.14a)$$

where h_c is the film thickness of the lubricant at the center of the contact patch and R_{q_1} and R_{q_2} are the rms average roughnesses of the two contacting surfaces. The denominator of Eq. 10.14a is termed the **composite surface roughness**. (See Section 7.2 for a discussion of surface roughness.) The film thickness at the center of the contact patch can be related to the minimum film thickness h_{min} at the trailing edge of contact by

$$h_c \cong \frac{4}{3} h_{min} \qquad (10.14b)$$

Figure 10-13a shows the experimentally measured frequency of asperity contact within an EHD gap as a function of specific film thickness.[28] When $\Lambda < 1$, the surfaces are in continuous metal-to-metal contact, i.e, in boundary lubrication. When $\Lambda > 3$ to 4, there is essentially no asperity contact. Between these values there is some combination of partial EHD and boundary-lubrication conditions. A majority of Hertzian contacts in gears, cams and rolling-element bearings operate in this partial EHD (mixed lubrication) region of Figure 10-2.[17] From Figure 10-13a we can conclude that Λ needs to be > 1 for partial EHD to begin[17] and > 3 to 4 for full film EHD.[6], [17] Effective partial EHD conditions begin at about $\Lambda = 2$ and if $\Lambda < 1.5$, it indicates an effective boundary lubrication condition in which significant asperity contact occurs.[17]

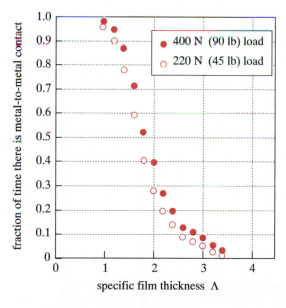

(a) Penetration of EHL film by surface asperities (ref 28)

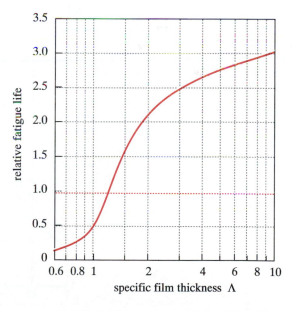

(b) Effect of film thickness on fatigue life (ref 29)

FIGURE 10-13

Effect of Specific Film Thickness Λ on the Asperity Contacts and Fatigue Life

Figure 10-13*b* shows the effect of specific film thickness on fatigue life of a rolling bearing.[29] The ordinate defines a ratio of expected life over the predicted catalog life for a bearing. This plot also shows the desirability of maintaining $\Lambda > 1.5$ in order to obtain the catalog life. A small increase in Λ from 1.5 to about 2 can double the fatigue life. Further increases in Λ have less dramatic effect on life and may cause higher friction due to viscous drag losses if a heavier oil is used to obtain the greater Λ.

Surface roughness is fairly easy to measure and control. The lubricant film thickness is more difficult to predict. Chapter 7 discusses calculation of the Hertzian pressure in surface contact and shows that the pressures in the contact zone between stiff materials in nonconforming (theoretical point or line) contact are extremely high, commonly as much as 80 to 500 kpsi (0.5 to 3 GPa) if both materials are steel, It was once believed that lubricants could not withstand these pressures and thus could not separate the metal surfaces. It is now known that viscosity is an exponential function of pressure, and at typical contact pressures, oil can become essentially as stiff as the metals it separates. Figure 10-14 shows the viscosity-pressure relationship for several common lubricants on a semilog plot. The curve for mineral oils can be approximated by

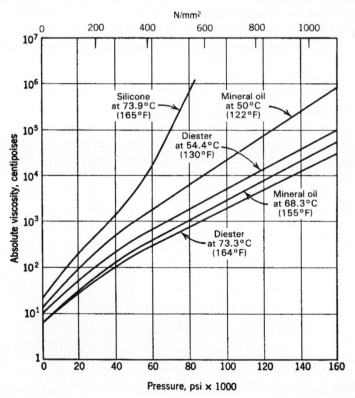

FIGURE 10-14

Absolute Viscosity Versus Pressure of Various Lubricating Oils   Source: ASME Research Committee on Lubrication "Pressure Viscosity Report–Vol. 11," 1953.

$$\eta = \eta_0 e^{\alpha p} \tag{10.15a}$$

where η_0 is the absolute viscosity at atmospheric pressure and p is the pressure in kpsi. An approximate expression for the pressure-viscosity exponent α for mineral oils is:[27]

$$\alpha \cong 7.74E - 4\left(\frac{\upsilon_0}{10^4}\right)^{0.163} \cong 7.74E - 4\left(\frac{\eta_0}{\rho(10^4)}\right)^{0.163} \tag{10.15b}$$

CYLINDRICAL CONTACT   Dowson and Higginson[18], [19] determined a formula for the minimum film thickness in an EHD contact between cylindrical rollers as

$$h_{min} = 2.65R'(\alpha E')^{0.54}\left(\frac{\eta_0 U}{E'R'}\right)^{0.7}\left(\frac{P}{lE'R'}\right)^{-0.13} \tag{10.16}$$

where P is the transverse load, l is the length of axial contact, U = average velocity $(U_1 + U_2)/2$, η_0 is the lubricant viscosity at atmospheric pressure and operating temperature, and α is the pressure-viscosity exponent for the particular lubricant from equation 10.15b. The effective radius R' is defined as

$$\frac{1}{R'} = \frac{1}{R_{1_x}} + \frac{1}{R_{2_x}} \tag{10.17a}$$

where R_{1_x} and R_{2_x} are the radii of the contacting surfaces in the direction of rolling. The effective modulus is defined as

$$E' = \frac{2}{m_1 + m_2} = \frac{2}{\dfrac{1-v_1^2}{E_1} + \dfrac{1-v_2^2}{E_2}} \tag{10.17b}$$

where E_1, E_2 are Young's moduli, and v_1, v_2 are Poisson's ratio for each material.

GENERAL CONTACT   In general point contact, the contact patch is an ellipse as discussed in Chapter 7. The contact ellipse is defined by its major and minor half-axis dimensions, a and b, respectively. Contact between two spheres, or between a sphere and a flat plate, will have a circular contact patch, which is a special case of elliptical contact wherein $a = b$. Hamrock and Dowson[21] developed an equation for the minimum film thickness in generalized point contact as

$$h_{min} = 3.63R'(\alpha E')^{0.49}\left(\frac{\eta_0 U}{E'R'}\right)^{0.68}\left(1-e^{-0.68\psi}\right)\left[\frac{P}{E'(R')^2}\right]^{-0.073} \tag{10.18}$$

where ψ is the ellipticity ratio of the contact patch a/b (see Section 7.10).

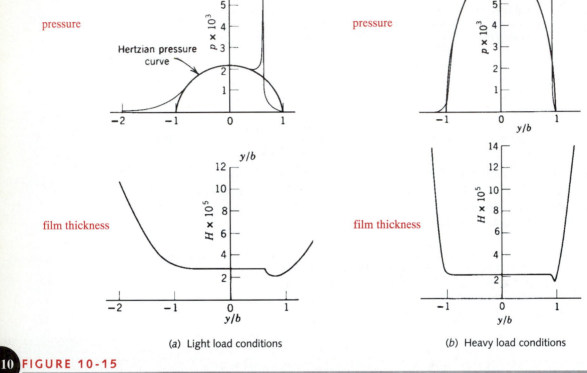

(a) Light load conditions (b) Heavy load conditions

10 **FIGURE 10-15**

Pressure Distribution and Film Thickness in an EHD Joint. Source: D. Dowson and G. Higginson, "The Effect of Material Properties on the Lubrication of Elastic Rollers", *J. Mech. Eng, Sci*. vol. 2, no. 3, 1960, with permission

In all of these equations, the film thickness is most dependent on speed and lubricant viscosity, but is relatively insensitive to load. Figure 10-15 shows pressure distribution and film thickness plots for light and heavy load conditions in an EHD contact between steel rollers lubricated with mineral oil.[22] Note that the fluid pressure is the same as the dry Hertzian contact pressure except for the pressure spike that occurs as the film thickness contracts near the exit. Except for that local contraction, the film thickness is essentially constant throughout the contact patch.

Equations 10.16, 10.17, and 10.18 allow a minimum film thickness to be calculated for a nonconforming contact joint such as a pair of gear teeth, cam-follower, or rolling-element bearing. The specific film thickness from equation 10.14 will indicate whether EHD or boundary lubrication can be expected in the contact. An oil with EP additives is needed if EHD is not present.

EXAMPLE 10-2

Lubrication in a Crowned Cam-Follower Interface

Problem   A cam-follower system was analyzed for contact-patch geometry and contact stresses in Example 7-3 (p. 505). Determine the film thickness parameter and the lubrication condition for a ground roller running against both a ground cam and a milled cam.

Given   The roller follower radius is 1 in with a 20-in crown radius at 90° to the roller radius with an rms surface roughness of $R_q = 7$ µin. The cam's minimum radius of curvature is 1.72 inches, in the direction of rolling. It is flat axially. It forms an elliptical contact patch with the cam. The half-dimensions of this ellipse are $a = 0.0889$ in and $b = 0.0110$ in. The cam angular velocity is 18.85 rad/sec and the radius to its surface at the point of minimum radius of curvature is 3.92 in. The bulk oil temperature is 200°F. The ground cam has an rms surface roughness of $R_q = 7$ µin and the milled cam has $R_q = 30$ µin.

Assumptions Try an ISO VG 68 oil with an assumed specific gravity of 0.9. The roller has 1% slip versus the cam.

Find   The specific film thickness and lubrication condition for the assumed lubricant and the viscosity of lubricant required to obtain effective partial or full EHD conditions for each cam, if possible.

Solution   See the *TKSolver* file EX10-02.

1   Figure 10-1 gives the viscosity of an ISO VG 68 oil as about 1.2 µreyn at 200°F.

2   Find the mass density ρ of the oil from the given specific gravity SG of the oil and the weight density of water.

$$\rho = SG \frac{\gamma}{g} = 0.9 \left(0.036\,11 \frac{lb}{in^3} \bigg/ 386 \frac{in}{sec^2} \right) = 84.2E - 6 \frac{lb - sec^2}{in^4} \text{ or } \frac{blob}{in^3} \qquad (a)$$

3   Find the approximate pressure-viscosity exponent α from equation 10.15b.

$$\alpha \cong 7.74E - 4 \left(\frac{\eta_0}{\rho(10^4)} \right)^{0.163} \cong 7.74E - 4 \left(\frac{1.2E - 6}{84.2E - 6(10^4)} \right)^{0.163} = 1.737E - 3 \qquad (b)$$

4   Find the effective radius from equation 10.16b.

$$\frac{1}{R'} = \frac{1}{R_{1_x}} + \frac{1}{R_{2_x}} = \frac{1}{1} + \frac{1}{1.720} \qquad R' = 0.632 \text{ in} \qquad (c)$$

5   Find the effective modulus of elasticity from equation 10.16c.

$$E' = \frac{2}{\dfrac{1-v_1^2}{E_1} + \dfrac{1-v_2^2}{E_2}} = \frac{2}{\dfrac{1-0.28^2}{3E7} + \dfrac{1-0.28^2}{3E7}} = 3.255E7 \qquad (d)$$

6   Find the average velocity U. The roller has 99% of the velocity of the cam.

$$U_2 = r\omega = 3.92 \text{ in } (18.85 \text{ rad/sec}) = 73.892 \text{ in/sec}$$

$$U_1 = 0.99U_2 = 0.99(73.892) = 73.153 \text{ in/sec}$$

$$U = (U_1 + U_2)/2 = (73.892 + 73.153)/2 = 73.523 \text{ in/sec} \qquad (e)$$

7   Find the ellipticity ratio = major / minor axis. The minor axis is in the direction of rolling in this case.

$$\psi = a/b = 0.0889/0.0110 = 8.082 \qquad (f)$$

8   Find the minimum film thickness from equation 10.16a.

$$h_{min} = 3.63R'(\alpha E')^{0.49}\left(\frac{\eta_0 U}{2E'R'}\right)^{0.68}\left(1 - e^{-0.68\psi}\right)\left[\frac{P}{E'(R')^2}\right]^{-0.073}$$

$$= 3.63(0.632)\left[1.737E - 3(3.255E7)\right]^{0.49}\left[\frac{(1.2E-6)(73.523)}{2(3.255E7)(0.632)}\right]^{0.68}$$

$$\times \left[1 - e^{-0.68(8.082)}\right]\left[\frac{250}{3.255E7(0.632)^2}\right]^{-0.073} = 20.0 \text{ }\mu\text{in} \qquad (g)$$

9   Convert this minimum value at the exit to an approximate thickness at the center of the contact patch with equation 10.14b.

$$h_c \cong \frac{4}{3}h_{min} = \frac{4}{3}(20.0) = 26.7 \qquad (h)$$

10   The specific film thickness values for each cam can now be found from equation 10.14a.

ground cam : $\Lambda = h_c\big/\sqrt{R_{q_1}^2 + R_{q_2}^2} = 26.7\big/\sqrt{7^2 + 7^2} = 2.7$

$\qquad\qquad\qquad\qquad\qquad\qquad\qquad\qquad\qquad\qquad\qquad\qquad (i)$

milled cam : $\Lambda = h_c\big/\sqrt{R_{q_1}^2 + R_{q_2}^2} = 26.7\big/\sqrt{7^2 + 30^2} = 0.87$

which indicates that the milled cam is in boundary lubrication and the ground cam is in partial EHD with the specified oil. These are common conditions for ground or milled cams, respectively, running against a ground roller-follower.

11   To determine what viscosity of oil would be needed to put each system into partial or full EHD condition, the *TKSolver* model was list-solved for a range of possible η_0

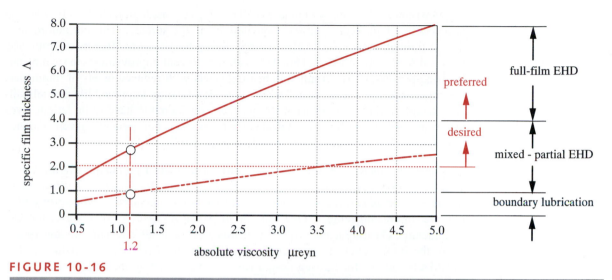

FIGURE 10-16

Variation of Specific Film Thickness Λ with Lubricant Viscosity η_0 in Example 10-2

values from 5E–7 to 5E–6 taken from Figure 10-1 at 200°F. A plot of the results is shown in Figure 10-16. It shows that an oil with $\eta_0 \geq 2$ μreyn is needed to put this ground cam into full EHD and that a $\eta_0 > 3.5$ μreyn oil will get this milled cam to $\Lambda > 2$ for an acceptable level of partial EHD. It would require an oil of ISO VG 1500 at 200°F to achieve $\Lambda > 4$ full-film EHD with the milled cam. See the *TKsolver* file EX10-02 for a plot of the functions in Figure 10-16 over a larger range of η.

10.8 ROLLING-ELEMENT BEARINGS

Rollers have been known as a means to move heavy objects since ancient times and there is evidence of the use of ball-thrust bearings in the first century B.C.; but it was only in the 20th century that improved materials and manufacturing technology allowed precision rolling-element bearings to be made. The need for higher-speed, higher-temperature-resistant low-friction bearings was engendered by the development of aircraft gas turbines. Considerable research effort since World War II has resulted in high-quality, high precision, rolling-element bearings (REB) being available at quite reasonable cost.

It is interesting to note that from their earliest designs around 1900, ball and roller bearings have been standardized worldwide in metric sizes. It is possible to remove a REB from the wheel assembly of an antique automobile made in most any country in the 1920's, for example, and find a replacement in a current bearing manufacturer's catalog which will fit. The new bearing will be much improved over the original in terms of design, quality, and reliability, but it will have the same external dimensions.

MATERIALS   A majority of modern ball bearings are made from AISI 5210 steel and hardened to a high degree, either throughout or on the surface. This chromium-steel alloy is through-hardenable to HRC 61-65. Roller bearings are often made from case hardened AISI 3310, 4620, and 8620 steel alloys. Recent improvements in steel manufacturing processes have resulted in bearing steels with reduced levels of impurities. Bearings made with these "clean" steels show significantly improved life and reliability. Though rolling bearings have always been considered to have finite fatigue lives and "standard" ones still do, REB made of "clean" steels have recently given evidence of an infinite-life endurance limit in surface fatigue.[23]

MANUFACTURING   Rolling-element bearings are made by all major bearing manufacturers worldwide to standard dimensions defined by the *Anti-Friction Bearing Manufacturers Association* (AFBMA) and/or the *International Standards Organization* (ISO), and are interchangeable. One can be reasonably assured that selecting any manufacturer's bearing made to these standards will not result in an unrepairable assembly in the future, even if that manufacturer goes out of the bearing business. The AFBMA Standards for bearing design have been adopted by the American National Standards Institute (ANSI). Some of the information in this section is taken from ANSI/AFBMA Standard 9-1990 for ball bearings[24] and —Standard 11-1990 for rolling bearings[25]. The standards also define tolerance classes for bearings. Radial bearings are classed by ANSI into ABEC-1 through -9 tolerance classes, precision increasing with class number. ISO defines Class 6 through Class 2 with precision varying inversely with class number. Cost increases with increased precision.

Comparison of Rolling and Sliding Bearings

Rolling-element bearings have a number of advantages over sliding contact bearings and vice versa. Hamrock[26] lists the following advantages of rolling over sliding bearings:

1   low starting and good operating friction, $\mu_{static} \cong \mu_{dynamic}$

2   can support combined radial and thrust loads

3   less sensitive to interruptions of lubrication

4   no self-excited instabilities

5   good low-temperature starting

6   can seal lubricant within bearing and "lifetime-lubricate"

7   typically require less space in axial direction

The following are disadvantages of rolling bearings compared to hydrodynamic conformal sliding bearings[26]:

1   rolling bearings may eventually fail from fatigue

2   require more space in radial direction

3   poor damping ability

4   higher noise level

5   more severe alignment requirements

6   higher cost

7   higher friction

Types of Rolling-Element Bearings

Rolling-element bearings can be grouped into two broad categories, **ball** and **roller** bearings, both of which have many variants within these divisions.

BALL BEARINGS   capture a number of hardened and ground steel spheres between two raceways, an inner and outer race for radial bearings, or top and bottom races for thrust bearings. A retainer (also called a cage or separator) is used to keep the balls properly spaced around the raceways, as shown in Figure 10-17. Ball bearings can support combined radial and thrust loads to varying degrees depending on their design and construction. Figure 10-17a shows a deep-groove or Conrad-type ball bearing that will support both radial and moderate thrust loads. Figure 10-17b shows an angular contact ball bearing designed to handle larger thrust loads in one direction as well as radial loads. Some ball bearings are available with shields to keep out foreign matter and seals to retain factory-applied lubricant. Ball bearings are less expensive for smaller sizes and lighter loads.

ROLLER BEARINGS   use straight, tapered, or contoured rollers running between raceways as shown in Figure 10-18. In general, roller bearings can support larger static and dynamic (shock) loads than ball bearings because of their line contact and are less expensive for larger sizes and heavier loads. Unless the rollers are tapered or contoured, they can only support a load in one direction, either radial or thrust according to the bearing design. Figure 10-18a shows a straight, cylindrical roller bearing designed to support only radial loads. It has very low friction and floats axially, which can be an advantage on long shafts where thermal expansion can load-up a ball-bearing pair in the axial direction if not properly mounted. Figure 10-18b shows a needle bearing that uses

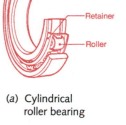

(a) Cylindrical
roller bearing

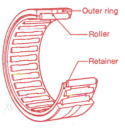

(b) Needle roller
bearing

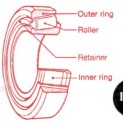

(c) Tapered roller
bearing

10

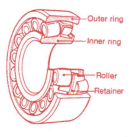

(d) Spherical
roller bearing

FIGURE 10-18

Roller-Type Bearings
Courtesy of NTN Corporation

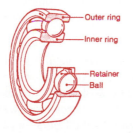

(a) Deep-groove (Conrad) ball bearing

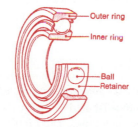

(b) Angular-contact ball bearing

FIGURE 10-17

Ball Bearings *Courtesy of NTN Corporation*

small diameter rollers and may or may not have an inner race. Its advantages are higher load capacity due to the full complement of rollers and its compact radial dimension, especially if used without an inner race. In such cases the shaft against which the rollers run must be hardened and ground. Figure 10-18*c* shows a tapered roller bearing designed to support large thrust and radial loads. These are often used as wheel bearings in automobiles and trucks. Tapered (and other) roller bearings can be split apart axially which makes assembly easier than with ball bearings that are usually permanently assembled. Figure 10-18*d* shows a spherical roller bearing that is self aligning, allowing no moment to be supported at the bearing.

THRUST BEARINGS   Ball and roller bearings are also made for pure thrust loads as shown in Figure 10-19. Cylindrical-roller-thrust bearings have higher friction than ball-thrust bearings due to the sliding that occurs between roller and raceways (because only one point on the roller can match the varying linear velocity over the raceways' radii), and should not be used in high-speed applications.

BEARING CLASSIFICATIONS   Figure 10-20 shows a classification of REB types. Each of the main categories of ball and roller divides into radial and thrust subcategories. Within these divisions, much variety is possible. Single- or double-row configurations are offered with the latter giving higher load capacity. Unidirectional or angular contact is another choice, the former accepting "pure" radial or thrust loading, and the latter accepting a combination of both. Deep-groove ball bearings are capable of handling both large radial loads and limited thrust loads in both directions, and are the most commonly used.

The angular-contact ball bearing can stand larger thrust loads than the deep-groove ball bearing, but only in one direction. They are often used in pairs to absorb axial loads in both directions. The maximum-capacity ball bearings have a filling slot to allow more balls to be inserted than can be accommodated by eccentric displacement of the races at assembly as is done with the deep-groove (Conrad-type) ball bearing, but the filling slot limits its axial load capacity.

Self-aligning designs have the advantage of accommodating some shaft misalignment and also create simple support for the shaft. They also have very low friction. If non-self-aligning bearings are used on a shaft, the bearing mounts must be carefully aligned for colinearity and angularity to avoid creating residual loads on the bearings at assembly, which will severely shorten their life.

Figure 10-21 shows size ranges and one bearing manufacturer's ratings and recommendations regarding the use of various types of bearings as an example. Note that a few types are available in inch sizes, but most are available only in metric dimensions. The columns labeled *Capacity* indicate relative ability to accommodate radial and thrust loads for each type. The *Limiting Speed* column uses the Conrad-type bearing as the standard of comparison since it has one of the best high-speed capabilities. Consult the bearing manufacturers' catalogs for additional information on other types and series of bearings. Many more are available than are shown in these few figures.

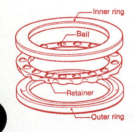

(a) Ball thrust bearing

10

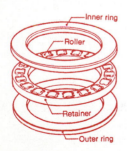

(b) Roller thrust bearing

FIGURE 10-19

Thrust Bearings *Courtesy of NTN Corporation*

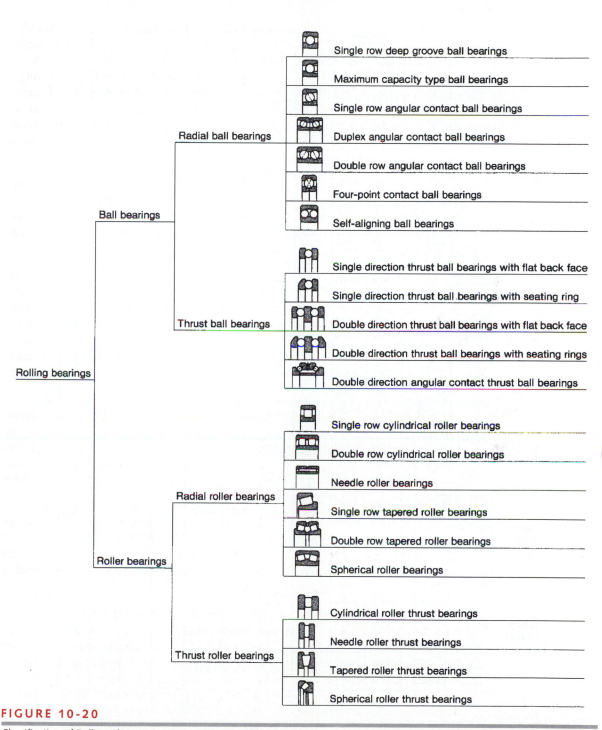

FIGURE 10-20

Classification of Rolling-Element Bearings   *Courtesy of NTN Corporation*

	TYPE	SIZE RANGE IN INCHES		AVERAGE RELATIVE RATINGS				AVAILABLE WITH			DIMENSIONS	
		Bore	O.D.	Capacity: Radial	Capacity: Thrust	Limiting Speed	Permissible Misalignment	Shields	Seals	Snap Rings	Metric	Inch
BALL BEARINGS	CONRAD TYPE 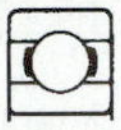	.1181 to 41.7323	.3750 to 55.1181	Good	Fair ↔	Conrad is basis for comparison 1.00	± 0° 8′ Std. Radial Clearance. ± 0° 12′ C3 Clear	X	X	X	X	X
BALL BEARINGS	MAXIMUM TYPE 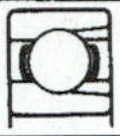	.6693 to 4.3307	1.5748 to 8.4646	Excellent	Poor ↔	1.00	± 0° 3′	X		X	X	
BALL BEARINGS	ANGULAR CONTACT 15°/40° 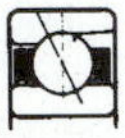 15° / 40°	.3937 to 7.4803	1.0236 to 15.7480	Good	Good (15°) Excellent (40°) ←	1.00 / 0.70	± 0° 2′				X	
BALL BEARINGS	ANGULAR CONTACT 35° 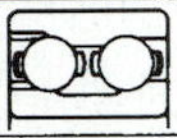	.3937 to 4.3307	1.1811 to 9.4488	Excellent	Good ←	0.70	0°				X	
BALL BEARINGS	SELF-ALIGNING 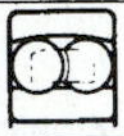	.1969 to 4.7244	.7480 to 9.4488	Fair	Fair ↔	1.00	± 4°				X	
CYLINDRICAL ROLLER BEARINGS	SEPARABLE INNER RING NON-LOCATING 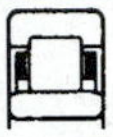	.4724 to 19.6850	1.2598 to 28.3465	Excellent	0	1.00	± 0° 4′				X	
CYLINDRICAL ROLLER BEARINGS	SEPARABLE INNER RING ONE DIR. LOCATING 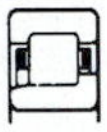	.4724 to 12.5984	1.2598 to 22.8346	Excellent	Poor ←	1.00	± 0° 4′				X	
CYLINDRICAL ROLLER BEARINGS	SELF-CONTAINED TWO DIR. LOCATING 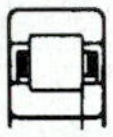	.4724 to 3.9370	1.4567 to 8.4646	Excellent	Poor ↔	1.00	± 0° 4′				X	
TAPERED ROLLER BEARINGS	SEPARABLE 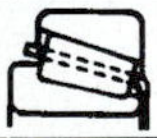	.6205 to 6.0000	1.5700 to 10.0000	Good	Good →	0.60	± 0° 2′				X	X
SPHERICAL ROLLER BEARINGS	SELF-ALIGNING 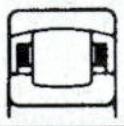	.9843 to 12.5984	2.0472 to 22.8346	Good	Fair ↔	0.50	± 4°				X	
SPHERICAL ROLLER BEARINGS	SELF-ALIGNING 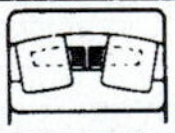	.9843 to 35.4331	2.0472 to 46.4567	Excellent	Good ↔	0.75	± 1°				X	
NEEDLE BEARINGS	COMPLETE BEARINGS 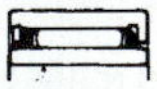 with or without locating rings & lubricating groove	.2362 to 14.1732	.6299 to 17.3228	Good	0	0.60	± 0° 2′		X		X	X
NEEDLE BEARINGS	DRAWN CUP	.1575 to 2.3622	.3150 to 2.6772	Good	0	0.30	± 0° 2′				X	X
THRUST BEARINGS	SINGLE DIRECTION BALL Grooved Race 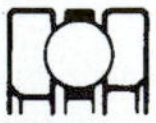	.2540 to 46.4567	.8130 to 57.0866	Poor	Excellent →	0.30	0°				X	X
THRUST BEARINGS	SINGLE DIRECTION CYL. ROLLER 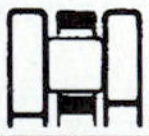	1.1811 to 23.6220	1.8504 to 31.4960	0	Excellent →	0.20	0°				X	
THRUST BEARINGS	SELF-ALIGNING SPHERICAL ROLLER 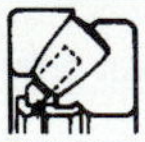	3.3622 to 14.1732	4.3307 to 22.0472	Poor	Excellent →	0.50	± 3°				X	

FIGURE 10-21

Relative Performance, Size, and Availability Information for Rolling-Element Bearings *Courtesy of FAG Bearings Corp., Stamford, Conn.*

10.9 FAILURE OF ROLLING-ELEMENT BEARINGS

If sufficient, clean lubricant is provided, failure in rolling bearings will be by surface fatigue, as described in Chapter 7. Failure is considered to occur when either raceway or balls (rollers) exhibit the first pit. Typically the raceway will fail first. The bearing will give an audible indication that pitting has begun by emitting noise and vibration. It can be run beyond this point, but as the surface continues to deteriorate, the noise and vibration will increase, eventually resulting in spalling or fracture of the rolling elements and possible jamming and damage to other connected elements. If you have ever had a wheel bearing fail on your automobile, you know the growling sound of a pitted or spalled rolling-element bearing *in extremis*.

Any large sample of bearings will exhibit wide variations in life among its members. The failures do not distribute statistically in a symmetrical Gaussian manner, but rather according to a Weibull distribution, which is skewed. Bearings are typically rated based on the life, stated in revolutions (or in hours of operation at the design speed), that 90% of a random sample of bearings of that size can be expected to reach or exceed at their design load. In other words, 10% of the batch can be expected to fail at that load before the design life is reached. This is called the L_{10} life.[*] For critical applications, a smaller failure percentage can be designed for, but most manufacturers have standardized on the L_{10} life as a means of defining the load-life characteristic of a bearing. The rolling-bearing selection process largely involves using this parameter to obtain whatever life is desired under the anticipated loading or overloading conditions expected in service.

Figure 10-22 shows a curve of bearing failure and survival percentages as a function of relative fatigue life. The L_{10} life is taken as the reference. The curve is relatively linear to 50% failure, which occurs at a life 5 times that of the reference. In other words, it should take 5 times as long for 50% of the bearings to fail as it does for 10% to do so. After that point the curve becomes quite nonlinear, showing that it will take about 10 times as long to fail 80% of the bearings as to fail 10%, and at 20 times the L_{10} life there are still a few percent of the original bearings running.

10.10 SELECTION OF ROLLING-ELEMENT BEARINGS

Once a bearing type suited to the application is chosen based on considerations discussed above and outlined in Figure 10-21, selection of an appropriate-size bearing depends on the magnitudes of applied static and dynamic loads and the desired fatigue life.

Basic Dynamic Load Rating C

Extensive testing by bearing manufacturers, based on well-established theory, has shown that the fatigue life L of rolling bearings is inversely proportional to the third power of the load for ball bearings, and to the 10/3 power for roller bearings. These relationships can be expressed as

[*] Some bearing manufacturers refer to this as the B_{90} or C_{90} life, referring to the survival of 90% of the bearings rather than the failure of 10%.

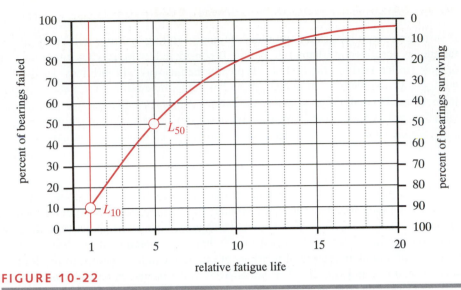

FIGURE 10-22

Typical Life Distribution in Rolling-Element Bearings *Adapted from SKF USA Inc.*

ball bearings :
$$L = \left(\frac{C}{P}\right)^{3}$$
(10.19a)

roller bearings :
$$L = \left(\frac{C}{P}\right)^{10/3}$$
(10.19b)

where L is fatigue life expressed in millions of revolutions, P is the constant applied load,[*] and C is the *basic dynamic load rating* for the particular bearing that is defined by the manufacturer and published for each bearing in the bearing catalogs. The **basic dynamic load rating** C is defined as *the load that will give a life of 1 million revolutions of the inner race.* This load C is typically larger than any practical load that one would subject the particular bearing to because the desired life is usually much higher than 1 million revolutions. In fact some bearings will fail statically if actually subjected to a load equal to C. It is simply a reference value that allows bearing life to be predicted at any level of actual applied load. Figure 10-23 shows a page from a bearing manufacturer's catalog that specifies the value of C for each bearing. Each bearing's maximum limiting speed is also defined.

Basic Static Load Rating C_0

Permanent deformations on rollers or balls can occur at even light loads because of the very high stresses within the small contact area. The limit on static loading in a bearing is defined as the load that will produce a total permanent deformation in the raceway and rolling element at any contact point of 0.0001 times the diameter d of the rolling element. Larger deformations will cause increased vibration and noise, and can lead to premature fatigue failure. The stresses required to cause this $0.0001d$ static deformation in bearing steel are quite high, ranging from about 4 GPa (580 kpsi) in roller bearings to 4.6 GPa (667 kpsi) in ball bearings. Bearing manufacturers publish a basic static load rating C_0 for each bearing, calculated according to AFBMA standards. This loading can sometimes be exceeded without failure, especially if rotating speeds are low, which avoids vibration problems. It usually takes a load of $8C_0$ or larger to fracture a bearing. Figure 10-23 shows a page from a bearing manufacturer's catalog that specifies the value of C_0 for each bearing.

Combined Radial and Thrust Loads

If both radial and thrust loads are applied to a bearing, an equivalent load must be calculated for use in equation 10.19. The AFBMA recommends the following equation:

$$P = XVF_r + YF_a \qquad (10.20)$$

where P = equivalent load
   F_r = applied constant radial load
   F_a = applied constant thrust load
   V = a rotation factor (see Figure 10-24)
   X = a radial factor (see Figure 10-24)
   Y = a thrust factor (see Figure 10-24)

The rotation factor V is 1 for a bearing with a rotating inner ring. If the outer ring rotates, V is increased to 1.2 for certain types of bearings. The factors X and Y vary with bearing type and relate to that type's ability to accommodate thrust as well as radial loads. Values of $V, X,$ and Y are defined by bearing manufacturers in tables such as the one reproduced in Figure 10-24. Bearing types, such as cylindrical roller, that cannot support thrust loads are not included in this table. A factor e is also specified for the bearing types in Figure 10-24 and defines a minimum ratio between the axial and radial forces below which the axial force can be ignored (set to zero) in equation 10.20.

$$\text{if} \quad \frac{F_a}{VF_r} \le e \quad \text{then} \ X = 1 \text{ and } Y = 0 \qquad (10.21)$$

Calculation Procedures

Equations 10.19 and 10.20 can be solved together for any situation in which either the applied load or a desired fatigue life is known. Usually, the radial and thrust loads acting on each bearing location will be known from a load analysis of the design. Often an approximate shaft size will be known from stress or deflection calculations. A bearing catalog should then be consulted, a trial bearing (or bearings) selected, and the values of $C, C_0, V, X,$ and Y extracted. The effective load P can be found from equation 10.20 and used in equation 10.19 with C to find the predicted fatigue life L.

OPEN	ONE SHIELD	TWO SHIELDS	ONE SEAL	TWO SEALS	SEAL & SHIELD	OPEN, WITH SNAP RING†	RADIAL SEAL & SHIELD
Suffix	.Z	.2Z	.RS	.2RS	.RSZ	.NR	.RSRZR

This configuration only shown to illustrate new standard enclosures. Some bearings are now being converted.

BEARING NUMBER*	BOUNDARY DIMENSIONS						SNAP RING DIMENSIONS inches			MAX. FILLET RADIUS Shaft & Hsg. inch	APPROX. WEIGHT lb.	S_L LIMITING SPEED ‡ rpm	C DYNAMIC LOAD RATING lb.	C_o STATIC LOAD RATING lb.
	BORE		O. DIAM.		WIDTH		H	S	t					
	mm	inch	mm	inch	mm	inch								
6300	10	.3937	35	1.3780	11	.4331	.125	1.562	.044	.025	.13	22000	1400	850
6301	12	.4724	37	1.4567	12	.4724	.125	1.625	.044	.040	.15	20000	1700	1040
6302	15	.5906	42	1.6535	13	.5118	.125	1.821	.044	.040	.20	18000	1930	1200
6303	17	.6693	47	1.8504	14	.5512	.141	2.074	.044	.040	.25	16000	2320	1460
6304	20	.7874	52	2.0472	15	.5906	.141	2.276	.044	.040	.34	14000	3000	1930
6305	25	.9843	62	2.4409	17	.6693	.195	2.665	.067	.040	.58	11000	3800	2550
6306	30	1.1811	72	2.8346	19	.7480	.195	3.091	.067	.040	.83	9500	5000	3400
6307	35	1.3780	80	3.1496	21	.8268	.195	3.406	.067	.060	1.07	8500	5700	4000
6308	40	1.5748	90	3.5433	23	.9055	.226	3.799	.097	.060	1.41	7500	7350	5300
6309	45	1.7717	100	3.9370	25	.9843	.226	4.193	.097	.060	1.95	6700	9150	6700
6310	50	1.9685	110	4.3307	27	1.0630	.226	4.587	.097	.080	2.50	6000	10600	8150
6311	55	2.1654	120	4.7244	29	1.1417	.271	5.104	.111	.080	3.30	5300	12900	10000
6312	60	2.3622	130	5.1181	31	1.2205	.271	5.498	.111	.080	3.81	5000	14000	10800
6313	65	2.5591	140	5.5118	33	1.2992	.304	5.892	.111	.080	4.64	4500	16000	12500
6314	70	2.7559	150	5.9055	35	1.3780	.304	6.286	.111	.080	5.68	4300	18000	14000
6315	75	2.9528	160	6.2992	37	1.4567	.304	6.679	.111	.080	6.60	4000	19300	16300
6316	80	3.1496	170	6.6929	39	1.5354	.346	7.198	.122	.080	9.53	3800	21200	18000
6317	85	3.3465	180	7.0866	41	1.6142	.346	7.593	.122	.100	11.00	3400	21600	18600
6318	90	3.5433	190	7.4803	43	1.6929	.346	7.986	.122	.100	11.60	3400	23200	20000
6319	95	3.7402	200	7.8740	45	1.7717	.346	8.380	.122	.100	13.38	3200	24500	22400
6320	100	3.9370	215	8.4646	47	1.8504	—	—	—	.100	16.34	3000	28500	27000
6321	105	4.1338	225	8.8582	49	1.9291	—	—	—	.100	17.8	2800	30500	30000
6322	110	4.3307	240	9.4488	50	1.9685	—	—	—	.100	21.0	2600	32500	32500
6324	120	4.7244	260	10.2362	55	2.1654	—	—	—	.100	32.3	2400	36000	38000
6326	130	5.1181	280	11.0236	58	2.2835	—	—	—	.12	40.1	2200	39000	43000
6328	140	5.5118	300	11.8110	62	2.4409	—	—	—	.12	48.1	2000	44000	50000
6330	150	5.9055	320	12.5984	65	2.5590	—	—	—	.12	57.8	1900	49000	60000

·Bearing numbers listed are for open bearings only. For shields, seals and snap rings, add suffix or prefix indicated below bearing diagram. Eg. 6300.Z, 6300.RS, 6300.NR, etc. Check availability of closures for larger sizes.
·Snap ring bearings available with shields or seals. Add both suffixes. Eg. 6300.ZNR, etc.
:For grease lubricated bearings without seals. For other conditions, see Page 114.
For mounting data, shaft and housing fits and shoulder diameters, see Pages 124-132.

FIGURE 10-23

Dimensions and Load Ratings for 6300 Series, Medium, Metric, Deep-Groove (Conrad-type) Ball Bearings *Courtesy of FAG Bearings Corporation, Stamford Conn.*

Factors V, X, and Y for Radial Bearings

Bearing Type			In Relation to the Load the Inner Ring is		Single Row Bearings 1) $\frac{F_a}{VF_r} > e$		Double Row Bearings 2) $\frac{F_a}{VF_r} \leqq e$		$\frac{F_a}{VF_r} > e$		e
			Rotating V	Stationary V	X	Y	X	Y	X	Y	
3)	4) $\frac{F_a}{C_0}$	5) $\frac{F_a}{i Z D_w^2}$	V	V							
Radial Contact Groove Ball Bearings	0.014	25								2.30	0.19
	0.028	50								1.99	0.22
	0.056	100								1.71	0.26
	0.084	150				1.55				1.55	0.28
	0.11	200	1	1.2	0.56	1.45	1	0	0.56	1.45	0.30
	0.17	300				1.31				1.31	0.34
	0.28	500				1.15				1.15	0.38
	0.42	750				1.04				1.04	0.42
	0.56	1000				1.00				1.00	0.44
20°					0.43	1.00		1.09	0.70	1.63	0.57
25°					0.41	0.87		0.92	0.67	1.44	0.68
30°			1	1.2	0.39	0.76	1	0.78	0.63	1.24	0.80
35°					0.37	0.66		0.66	0.60	1.07	0.95
40°					0.35	0.57		0.55	0.57	0.93	1.14
Self-Aligning Ball Bearings			1	1	0.40	0.4 cot α	1	0.42 cot α	0.65	0.65 cot α	1.5 tan α
Self-Aligning and Tapered Roller Bearings			1	1.2	0.40	0.4 cot α	1	0.45 cot α	0.67	0.67 cot α	1.5 tan α

1) For single row bearings, when $\frac{F_a}{VF_r} \leqq e$ use $X = 1$ and $Y = 0$.

For two single row angular contact ball or roller bearings mounted "face-to-face" or "back-to-back" the values of X and Y which apply to double row bearings. For two or more single row bearings mounted "in tandem" use the values of X and Y which apply to single row bearings.

2) Double row bearings are presumed to be symmetrical.

3) Permissible maximum value of $\frac{F_a}{C_0}$ depends on the bearing design.

4) C_0 is the basic static load rating.
5) Units are pounds and inches.
Values of X, Y and e for a load or contact angle other than shown in the table are obtained by linear interpolation.

FIGURE 10-24

V, X, and Y Factors for Radial Bearings *Courtesy of SKF USA Inc.*

Alternatively, since V, X, and Y depend only on the type but not the size of a bearing, they can be determined first and equations 10.19 and 10.20 solved simultaneously for the value of dynamic load factor C required to achieve a desired life L. The bearing catalogs can then be consulted to find a suitably sized bearing with the necessary C value. In either case, the static load should also be compared to the static load factor C_0 for the chosen bearing to guard against excessive deformations.

EXAMPLE 10-3

Selection of Ball Bearings for a Designed Shaft

Problem	Select radial ball bearings for the shaft shown in Figure 9-5 (repeated on p. 665). The shaft was designed in Example 9-1.
Given	The maximum transverse loads on the shaft at the bearings are 16 lb at R_1 and 54 lb at R_2. Since the load at R_2 is 4x that at R_1, one design can be created for R_2 and used also at R_1. Shaft diameter at both R_1 and R_2 is 0.591 in, based on the tentative choice of a 15 mm ID bearing in Example 9-1. The shaft speed is 1 725 rpm.
Assumptions	Thrust loads are negligible.
Find	The bearing fatigue lives at both shaft locations.

Solution

1   From Figure 10-23, choose a #6302 bearing with 15-mm inside diameter. Its dynamic load rating factor is $C = 1\,930$ lb. The static load rating $C_0 = 1\,200$ lb. The static applied load of 54 lb is obviously well below the bearing's static rating.

2   Calculate the projected life with equation 10.19. Note that the equivalent load in this case is simply the applied radial load due to the absence of any thrust load. For the larger reaction load of 54 lb at R_2:

$$L = \left(\frac{C}{P}\right)^3 = \left(\frac{1\,930}{54}\right)^3 = 45E3 \text{ millions of revs } = 45E9 \text{ revs} \qquad (a)$$

3   For the smaller reaction load of 16 lb at R_1:

$$L = \left(\frac{C}{P}\right)^3 = \left(\frac{1\,930}{16}\right)^3 = 1.75E6 \text{ millions of revs } = 1.75E12 \text{ revs} \qquad (b)$$

This shows the nonlinear relationship between load and life. A 3.5x reduction in load results in a 38x increase in fatigue life. These bearings are obviously very lightly loaded, but their size was dictated by considerations of stresses in the shaft.

4   From Figure 10-23, this bearing's limiting speed is 18 000 rpm, well above the operating speed of 1 725 rpm.

EXAMPLE 10-4

Selection of Ball Bearings for Combined Radial and Thrust Loads

Problem   Select a deep-groove ball bearing for specified loading and desired life.

Given   The radial load $F_r = 1\,686$ lb (7 500 N) and the axial load $F_a = 1\,012$ lb (4 500 N). The shaft speed is 2 000 rpm.

Assumptions   A Conrad-type deep-groove ball bearing will be used. The inner ring rotates.

Find   A suitable size bearing to give an L_{10} life of 5E8 revolutions.

Solution   See the *TKSolver* file EX10-04.

1 Try a #6316 bearing from Figure 10-23 and extract its data as: $C = 21\,200$ lb (94 300 N), $C_0 = 18\,000$ lb (80 000 N), and maximum rpm = 3 800.

2 Calculate the ratio F_a / C_0:

$$\frac{F_a}{C_0} = \frac{1\,012}{18\,000} = 0.056 \qquad (a)$$

and take this value to Figure 10-24 to find the corresponding value of $e = 0.26$ for radial-contact groove ball bearings.

3 Form the ratio $F_a / (V F_r)$ and compare it to the value of e.

$$\frac{F_a}{VF_r} = \frac{1\,012}{1(1\,686)} = 0.6 > e = 0.056 \qquad (b)$$

Note that $V = 1$ because the inner ring is rotating.

4 Since the ratio in step 3 is $> e$, extract the X and Y factors from Figure 10-24 as $X = 0.56$ and $Y = 1.71$, and use them to calculate the equivalent load from equation 10.20.

$$P = XVF_r + YF_a = 0.56(1)(1\,686) + 1.71(1\,012) = 2\,675 \text{ lb} \qquad (c)$$

5 Use the equivalent load in equation 10.19a to find the L_{10} life for this bearing.

$$L = \left(\frac{C}{P}\right)^3 = \left(\frac{21\,200}{2\,675}\right)^3 = 5.0E2 \text{ millions of revs } = 5.0E8 \text{ revs} \qquad (d)$$

This result actually required some iteration, trying several bearing numbers before finding that this one would give the desired life. See the *TKSolver* file EX10-04, which has the bearing data from Figures 10-23 and 10-24 built in as list functions for automatic table look-up, which makes iterating to a solution a simple matter.

10.11 BEARING MOUNTING DETAILS

Rolling bearings are made with close tolerances on their inside and outside diameters to allow press-fitting on the shaft or in the housing. The bearing races (rings) should be tightly coupled to the shaft and housing to guarantee that motion only occurs inside the low-friction bearing. Press-fitting both rings can make for a difficult assembly or disassembly in some cases. Various clamping arrangements are commonly used to capture either the inner or outer ring without a press fit, the other being secured by pressing. The inner ring is usually located against a shoulder on the shaft. Bearing catalog tables provide recommended shaft shoulder diameters, which should be observed to avoid interference with seals or shields. Maximum allowable fillet radii to clear the rings' corners are also defined by the manufacturers.

Figure 10-25a shows a nut and lock washer arrangement used to clamp the inner ring to the shaft to avoid a press fit. Bearing manufacturers supply special nuts and washers standardized to fit their bearings, Figure 10-25b shows a snap ring used to axially locate the inner ring, which would be pressed to the shaft. Figure 10-25c shows the outer ring clamped axially to the housing and the inner ring located by a sleeve spacer between the inner ring and an external accessory flange on the same shaft.

Pairs of bearings on the same shaft are commonly needed to provide moment support. Figure 10-26 shows one possible arrangement to axially capture the assembly without risking the introduction of axial forces to the bearings from thermal expansion of the parts. The inner races of both bearings are clamped axially with a nut on the left and a spacer between them. The outer race of the bearing on the right is captured (clamped) axially to the housing, but the outer race of the one on the left is "floating"

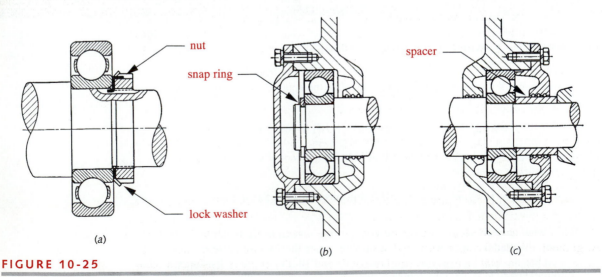

(a)   (b)   (c)

FIGURE 10-25

Bearing Mounting Methods *Source: SKF Engineering Data, SKF USA Inc., 1968*

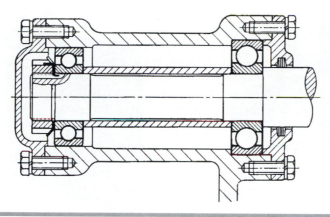

FIGURE 10-26

Bearings on Shaft, One Fixed Axially and One Floating Axially   *Source: SKF Engineering Data, SKF USA Inc., 1968*

axially in the housing to allow for thermal expansion. It might be tempting to capture both right and left bearings axially but this would be unwise. It is considered good practice to capture a long assembly axially at only one location to avoid expansion-induced axial forces on the bearings, which would seriously shorten their life. Another way to accomplish this is to use only one bearing that can support an axial load (e.g., a ball bearing) and use a cylindrical roller or other type bearing that cannot support axial load across its rolling elements at the other end of the shaft.

10.12 SPECIAL BEARINGS

Many other types and arrangements of rolling-element bearings are available. Pillow blocks and flange-units package standard ball or roller bearings in cast-iron housings that make it easy to attach bearings to horizontal or vertical surfaces. Figure 10-27 shows a pillow block and a flange-unit bearing.

Cam followers, as shown in Figure 10-28, are made with ball or roller bearings and a special outer race that can be run directly against a cam surface. They are available with integral mounting studs (as shown) or with a hole to mount on a rod or stud. Rod ends are typically a single spherical ball in a socket designed to attach to rods and provide a self-aligning, low-friction connection between links in a mechanism as shown in Figure 10-29.

Linear motion is easily guided on plain bushings, but will have moderate friction levels. For lower friction in linear motion, ball bushings are available as shown in Figure 10-30. These require special hardened and ground shafts made to close tolerances. Alignment of parallel shafts must be done very precisely to obtain the low friction advantages of linear ball bearings. They are not as able to absorb shock loads as a plain bushing, however.

(a)   (b)

FIGURE 10-27

Pillow Block (a) and Flange-Mount Bearing Units (b)   *Courtesy of McGill Manufacturing Co. Inc., Bearing Division, Valparaiso, Ind.*

10.13 CASE STUDY

Case Study 9, which was set up in Chapter 8, describes the design of a test fixture for the dynamic measurement of cam-follower accelerations and forces. The sensitive nature of these measurements requires that only sliding bearings be used throughout because the vibrations and noise from rolling-element bearings would contaminate the measurements. We will now continue that case study for the design of its main camshaft bearings.

FIGURE 10-28

Roller-Bearing Cam Follower   *Courtesy of Roller Bearing Company of America, Newtown Pa.*

FIGURE 10-29

Spherical Rod End   *Courtesy of Morse Chain, Division of Borg-Warner Corp., Aurora, Ill.*

FIGURE 10-30

Linear Ball Bushing
Courtesy of Thompson Industries, Inc., Manhasset, N.Y.

CASE STUDY 9B

Design of Hydrodynamic Bearings for a Cam Test Fixture.

Problem   **Determine the hydrodynamic conditions in the proposed bearings for the camshaft in the CDTF.**

Given   **The cam generates a peak dynamic force of 110 lb at a maximum speed of 180 rpm (3 rps), as defined in Case Study 9a (Chapter 8). The flywheel weighs 220 lb and is located midway between the two bearings. Bulk oil temperature is controlled to 200°F. The camshaft is 2-in dia and the preliminary design of the bearings allows up to 2 in of length each.**

Assumptions   **Plain bearings must be used throughout since rolling-element bearings introduce too much noise. Porous bronze bushings are proposed. Use a clearance ratio of 0.001. Try a mineral oil of SAE 30W (ISO VG 100). Gravity oilers will be provided at each bearing.**

Solution   **See Figures 8-8 and 10-31 and the TKSolver file CASE9B.**

1   Find the reaction forces acting on each bearing from the applied forces and dimensions defined in Figure 10-31. Sum moments about R_1 and assume upward forces positive.

$$\sum M = 0 = 110(4.5) - 120(3.125) + 6.25R_2$$

$$R_2 = -19.2 \text{ lb} \qquad\qquad (a)$$

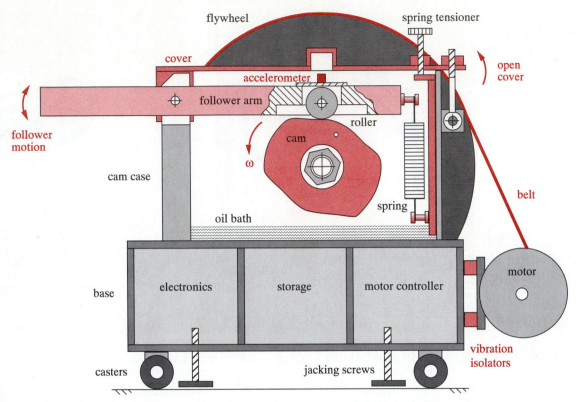

flywheel   spring tensioner

cover

accelerometer

open
cover

follower arm

follower
motion

roller

cam

ω

cam case

belt

oil bath

spring

base   electronics   storage   motor controller

motor

vibration
isolators

casters   jacking screws

FIGURE 8-8 Repeated

Cam Dynamic Test Fixture—General Design Scheme

$$\sum F = 0 = -110 - 120 - 19.2 + R_1$$

$$R_1 = 210.8 \text{ lb} \tag{b}$$

The bearing at R_1 takes most of the load so we will design for that force.

2   An ISO VG100 oil was suggested. Figure 10-1 gives a viscosity for this oil at 200°F
    of about 1.5 µreyn.

3   Find the average pressure in the bearing for an assumed length of 2 in.

$$p_{avg} = \frac{P}{ld} = \frac{210.8}{2(2)} = 52.7 \text{ psi} \tag{c}$$

4   Find the diametral clearance in the bearing for the assumed clearance ratio.

$$c_d = 0.001(2) = 0.002 \text{ in}$$
$$c_r = c_d/2 = 0.001 \text{ in} \tag{d}$$

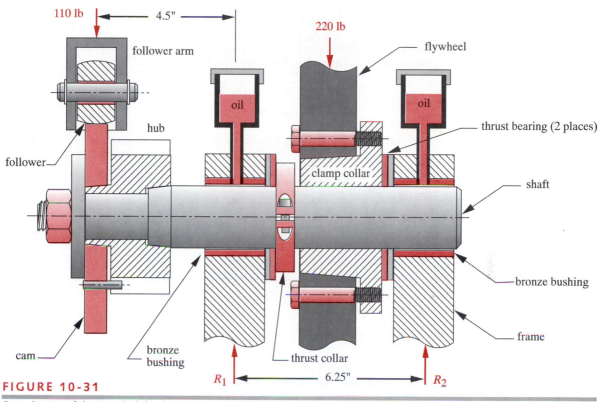

110 lb

4.5"

follower arm

220 lb

flywheel

oil

oil

thrust bearing (2 places)

hub

follower

clamp collar

shaft

bronze bushing

frame

cam

bronze bushing

thrust collar

R_1 6.25" R_2

FIGURE 10-31

Cross-Section of the Camshaft for the Cam Dynamic Test Fixture of Case Study 9

5   Find the surface velocity of the shaft in the bearing.

$$U = \pi dn' = \pi(2)(3) = 18.85 \text{ in/sec} \qquad (e)$$

6   Since the load and speed are known and the bearing dimensions are assumed, as is the viscosity, equation 10.8b can be solved for the dimensionless parameter K_ε.

$$K_\varepsilon = \frac{Pc_r^2}{\eta U l^3} = \frac{210.8(0.001)^2}{1.5(18.85)(2)^3} = 0.932 \qquad (f)$$

7   The eccentricity ratio can now be found from equation 10.8c.

$$K_\varepsilon = \frac{\varepsilon\left[\pi^2\left(1-\varepsilon^2\right)+16\varepsilon^2\right]^{\frac{1}{2}}}{4\left(1-\varepsilon^2\right)^2} = 0.932$$

$$\varepsilon = 0.543 \qquad (g)$$

This equation must be solved for ε with a root-finding algorithm as it is implicit. The value was obtained with *TKSolver*.

8   The Ocvirk number can now be found from the approximate expression for the experimental curve of Figure 10-10, equation 10.12.

$$\varepsilon_x \cong 0.21394 + 0.38517 \log O_N - 0.0008(O_N - 60) = 0.543$$

$$O_N \cong 5.5 \tag{h}$$

This must also be solved with a root-finding algorithm and was done in *TKSolver*. We conclude that the bearing design is viable based on this acceptable Ocvirk number.

9   The minimum film thickness is found from equation 10.4*b*.

$$h_{min} = c_r(1 - \varepsilon) = 0.001(1 - 0.543) = 0.000\ 457 \text{ in } (457\ \mu\text{in}) \tag{i}$$

This is an ample film to protect even a poorly finished bearing, which this is not.

10   We conclude that the design is acceptable as proposed.

10.14 SUMMARY

Low friction in sliding or rotating joints can be obtained either with hydrostatically or hydrodynamically lubricated plain bearings or with rolling-element bearings. Each has its own set of advantages and disadvantages.

Hydrostatic bearings use a high-pressure source of fluid to separate the surfaces even when no relative motion is present. Air, water, or oil can be used as the fluid. Air bearings have essentially zero friction and wear. A Hovercraft is supported on an "air bearing," for example.

Hydrodynamic bearings use the relative motion of the surfaces to pump the entrained lubricant (usually an oil) around the annulus between shaft and bearing. A properly designed hydrodynamic bearing separates the two parts on a film of oil when in motion and has no metal-to-metal contact except at start-up and shutdown. If the oil is kept clean and plentiful, essentially zero wear and very low friction are possible. Two surfaces that "conform" geometrically such as a shaft in a hole, entrap the lubricant and thus readily form the supporting oil film.

Geometrically nonconforming joints such as cam-follower contacts, gear teeth, and rolling bearings, tend to expel the fluid rather than entrap it making it more difficult to achieve full-film separation of the surfaces. **Elastohydrodynamic lubrication** (EHD) refers to the combination of elastic deflection of a contact patch between two nonconforming surfaces (analogous to the contact patch between your tire and the road) and the pumping of fluid between the "flattened" surfaces to create at least a partial hydrodynamic film. These joints often have some combination of fluid film and metal-to-metal contact at the surface asperities. Thus wear can be higher than in a conforming hydrodynamic joint. The minimum fluid-film thickness between the surfaces in comparison to their composite surface roughness determines how much asperity contact occurs. In the absence of sufficient lubricant, speed, or geometry to form a separating fluid film, a bearing will revert to **boundary lubrication** in which significant metal contact and wear occurs.

Rolling bearings are commercially available in a variety of configurations that use either balls or rollers of hardened steel captured between hardened steel raceways or rings. Since the contact is rolling, with little or no sliding, the friction is low both statically and dynamically. Start-up torque is significantly lower for rolling bearings than for hydrodynamic ones[*] (which require a relative velocity to establish the low-friction fluid film.) Rolling bearings are available that can accommodate radial, thrust, or a combination of both types of loads. The lubrication state in rolling bearings will be either elastohydrodynamic, boundary, or some combination of the two referred to as partial EHD. Designing with rolling bearings largely involves the proper selection of a bearing from those commercially available. The manufacturers define a load-life parameter based on the load at which 90% of a batch of bearings can be expected to survive for 1 million revolutions of the inner race. This and other manufacturer-supplied data are used to calculate a projected life for a particular bearing under the given load and speed conditions of the application.

Important Equations Used in this Chapter

Absolute Viscosity Versus Kinematic Viscosity (Section 10.2):

$$\eta = \upsilon\rho \tag{10.1}$$

Petroff's Equation for No-Load Torque (Section 10.5):

$$T_0 = \eta\frac{\pi^2 d^3 l n'}{c_d} \tag{10.2c}$$

Eccentricity Ratio (Section 10.5):

$$\varepsilon = \frac{e}{c_r} \tag{10.3}$$

Lubricant Film Thickness in a Hydrodynamic Bearing (Section 10.5):

$$h = c_r(1 + \varepsilon\cos\theta) \tag{10.4a}$$

$$h_{min} = c_r(1-\varepsilon) \qquad\qquad h_{max} = c_r(1+\varepsilon) \tag{10.4b}$$

Average Pressure in a Hydrodynamic Bearing (Section 10.5):

$$P_{avg} = \frac{P}{A} = \frac{P}{ld} \tag{10.6d}$$

[*] When railroad cars were converted from hydrodynamic plain bearings to rolling-element bearings many years ago, long freight trains that had required two engines to start them moving (but only one to keep them moving) could be started moving with only one engine.

Sommerfeld's Equations for Pressure and Load in an Infinite Bearing (Section 10.5):

$$p = \frac{\eta U r}{c_r^2} \left[\frac{6\varepsilon (\sin\theta)(2 + \varepsilon\cos\theta)}{\left(2 + \varepsilon^2\right)(1 + \varepsilon\cos\theta)^2} \right] + p_0 \qquad (10.6a)$$

$$P = \frac{\eta U l r^2}{c_r^2} \frac{12\pi\varepsilon}{\left(2 + \varepsilon^2\right)\left(1 + \varepsilon^2\right)^{1/2}} \qquad (10.6b)$$

Ocvirk's Equations for Pressure and Load in a Short Bearing (Section 10.5):

$$p = \frac{\eta U}{r c_r^2} \left(\frac{l^2}{4} - z^2 \right) \frac{3\varepsilon\sin\theta}{(1 + \varepsilon\cos\theta)^3} \qquad (10.7b)$$

$$P = K_\varepsilon \frac{\eta U l^3}{c_r^2} \qquad (10.8b)$$

$$K_\varepsilon = \frac{\varepsilon\left[\pi^2\left(1 - \varepsilon^2\right) + 16\varepsilon^2\right]^{\frac{1}{2}}}{4\left(1 - \varepsilon^2\right)^2} \qquad (10.8c)$$

Location of Maximum Pressure in a Short Bearing (Section 10.5):

$$\theta_{max} = \cos^{-1}\left(\frac{1 - \sqrt{1 + 24\varepsilon^2}}{4\varepsilon} \right) \qquad (10.7c)$$

Location of Resultant Load in a Short Bearing (Section 10.5):

$$\phi = \tan^{-1}\left(\frac{\pi\sqrt{1 - \varepsilon^2}}{4\varepsilon} \right) \qquad (10.8a)$$

Torque in a Hydrodynamic Bearing (Section 10.5):

$$T_s = \eta \frac{d^3 l \left(n_2' - n_1'\right)}{c_d} \frac{\pi^2}{\left(1 - \varepsilon^2\right)^{1/2}} \qquad (10.9c)$$

$$T_r = T_s + Pe\sin\phi \qquad (10.9a)$$

Power Loss in a Hydrodynamic Bearing (Section 10.5):

$$\Phi = T_r\omega = 2\pi T_r\left(n_2' - n_1'\right) \qquad \text{N - m/s or in - lb/s} \qquad (10.10)$$

Coefficient of Friction in a Hydrodynamic Bearing (Section 10.5):

$$\mu = \frac{f}{P} = \frac{T_r/r}{P} = \frac{2T_r}{Pd} \tag{10.11}$$

The Ocvirk Number for a Short Hydrodynamic Bearing (Section 10.6):

$$O_N = \left(\frac{p_{avg}}{\eta n'}\right)\left(\frac{d}{l}\right)^2\left(\frac{c_d}{d}\right)^2 = 4\pi K_\varepsilon = \frac{\pi\varepsilon\left[\pi^2\left(1-\varepsilon^2\right)+16\varepsilon^2\right]^{\frac{1}{2}}}{\left(1-\varepsilon^2\right)^2} \tag{10.12c}$$

Empirical Relationship Between Ocvirk Number and Eccentricity Ratio (Section 10.6):

$$\varepsilon_x \cong 0.21394 + 0.38517\log O_N - 0.0008(O_N - 60) \tag{10.13}$$

Specific Film Thickness (Section 10.7):

$$\Lambda = h_c \Big/ \sqrt{R_{q_1}^2 + R_{q_2}^2} \tag{10.14a}$$

$$h_c \cong \frac{4}{3}h_{min} \tag{10.14b}$$

Minimum Film Thickness for EHD Cylindrical Contact (Section 10.7):

$$h_{min} = 2.65R'(\alpha E')^{0.54}\left(\frac{\eta_0 U}{E'R'}\right)^{0.7}\left(\frac{P}{lE'R'}\right)^{-0.13} \tag{10.16}$$

$$\frac{1}{R'} = \frac{1}{R_{1_x}} + \frac{1}{R_{2_x}} \qquad\qquad E' = \frac{2}{m_1 + m_2} \tag{10.17}$$

Minimum Film Thickness for EHD General (Elliptical) Contact (Section 10.7):

$$h_{min} = 3.63R'(\alpha E')^{0.49}\left(\frac{\eta_0 U}{E'R'}\right)^{0.68}\left(1 - e^{-0.68\psi}\right)\left[\frac{P}{E'(R')^2}\right]^{-0.073} \tag{10.18}$$

Load-Life Relationship for Rolling-Element Bearings (Section 10.10):

ball bearings :
$$L = \left(\frac{C}{P}\right)^3 \tag{10.19a}$$

roller bearings :
$$L = \left(\frac{C}{P}\right)^{10/3} \tag{10.19b}$$

Equivalent Load for Rolling-Element Bearings (Section 10.10):

$$P = XVF_r + YF_a \tag{10.20}$$

10.15 REFERENCES

1   **A. G. M. Michell**, Progress of Fluid-Film Lubrication, *Trans. ASME*, **51**: pp. 153-163, 1929.

2   **A. Cameron**, *Basic Lubrication Theory*. John Wiley & Sons: New York, 1976.

3   **B. J. Hamrock**, *Fundamentals of Fluid Film Lubrication*. McGraw-Hill: New York, 1994.

4   **T. A. Harris**, *Rolling Bearing Analysis*. John Wiley & Sons: New York, 1991.

5   **R. C. Elwell**, "Hydrostatic Lubrication," in *Handbook of Lubrication,* E. R. Booser, ed., CRC Press: Boca Raton Fla., p. 105, 1983.

6   **J. L. Radovich**, "Gears," in *Handbook of Lubrication,* E. R. Booser, ed., CRC Press: Boca Raton, Fla., p. 544, 1983.

7   **E. Rabinowitz**, *Friction and Wear of Materials*. John Wiley & Sons: New York, p. 182, 1965.

8   **W. Glaeser**, "Bushings," in *Wear Control Handbook,* M. B. Peterson and W. O. Winer, Editor. ASME: Wear Control Handbook, p. 598, 1980.

9   **D. F. Wilcock and E. R. Booser**, *Bearing Design and Application*. McGraw-Hill: New York, 1957.

10   **A. H. Burr and J. B. Cheatham**, *Mechanical Analysis and Design*, 2nd ed. Prentice-Hall: Englewood Cliffs, N.J., pp. 31-51, 1995.

11   **A. A. Raimondi and J. Boyd**, A Solution for the Finite Journal Bearing and its Application to Analysis and Design–Parts I, II, and III, *Trans. Am. Soc. Lubrication Engineers*, **1**(1): pp. 159-209, 1958.

12   **O. Reynolds**, On the Theory of Lubrication and its Application to Mr. Beauchamp Tower's Experiments. *Phil. Trans. Roy. Soc. (London)*, **177**: pp. 157-234, 1886.

13   **G. B. Dubois and F. W. Ocvirk**, The Short Bearing Approximation for Full Journal Bearings. *Trans. ASME*, **77**: p. 1173-1178, 1955.

14   **G. B. Dubois, F. W. Ocvirk, and R. L. Wehe**, *Experimental Investigation of Eccentricity Ratio, Friction, and Oil Flow of Long and Short Journal Bearings–With Load Number Charts,* TN3491, NACA, 1955.

15   **F. W. Ocvirk**, *Short Bearing Approximation for Full Journal Bearings*, TN2808, NACA, 1952.

16   **G. B. Dubois and F. W. Ocvirk**, *Analytical Derivation and Experimental Evaluation of Short Bearing Approximation for Full Journal Bearings*, TN1157, NACA, 1953.

17   **H. S. Cheng**, "Elastohydrodynamic Lubrication," in *Handbook of Lubrication,* E. R. Booser, ed., CRC Press: Boca Raton Fla., pp. 155-160, 1983.

18   **D. Dowson and G. Higginson**, A Numerical Solution to the Elastohydrodynamic Problem, *J. Mech. Eng. Sci.*, **1**(1): p. 6, 1959.

10

19   **D. Dowson and G. Higginson**, New Roller Bearing Lubrication Formula, *Engineering*, **192**: p. 158-9, 1961.

20   **G. Archard and M. Kirk**, Lubrication at Point Contacts, *Proc. Roy. Soc. (London) Ser.* **A261**, pp. 532-550, 1961.

21   **B. Hamrock and D. Dowson**, Isothermal Elastohydrodynamic Lubrication of Point Contacts—Part III—Fully Flooded Results, *ASME J. Lubr. Technol.*, **99**: pp. 264-276, 1977.

22   **D. Dowson and G. Higginson**, *Proceedings of Institution of Mechanical Engineers*, **182** (Part 3A): pp. 151-167, 1968.

23   **T. A. Harris**, *Rolling Bearing Analysis.* John Wiley & Sons: New York, pp. 872-888, 1991.

24   *Load Ratings and Fatigue Life for Ball Bearings*, ANSI/AFBMA Standard 9-1990, American National Standards Institute, New York, 1990.

25   *Load Ratings and Fatigue Life for Roller Bearings*, ANSI/AFBMA Standard 11-1990, American National Standards Institute, New York, 1990

26   **B. J. Hamrock**, *Fundamentals of Fluid Film Lubrication.* McGraw-Hill: New York, p. 16, 1994.

27   ASME Research Committee on Lubrication "Pressure-Viscosity Report—Vol. 11," ASME, 1953.

28   **T. E. Tallian**, Lubricant Films in Rolling Contact of Rough Surfaces, *ASLE Trans.*, **7**(2): pp. 109-126, 1964.

29   **E. N. Bamberger, et al.,** Life Adjustment factors for Ball and Roller Bearings, *ASME Engineering Design Guide*, 1971.

10.16 PROBLEMS

[*][†]10-1   The shaft shown in Figure P10-1 was designed in Problem 9-1. For the data in the row(s) assigned from Table P10-1, and the corresponding diameter of shaft found in Problem 9-1, design suitable bearings to support the load for at least 7E7 cycles at 1 500 rpm. State all assumptions.

   (a)   Using hydrodynamically lubricated bronze sleeve bearings with $O_N = 20$, $l/d = 1.25$, and a clearance ratio of 0.001 5.

   (b)   Using deep-groove ball bearings.

[†]10-2   The shaft shown in Figure P10-2 was designed in Problem 9-2. For the data in the row(s) assigned from Table P10-1, and the corresponding diameter of shaft found in Problem 9-2, design suitable bearings to support the load for at least 3E8 cycles at 2 500 rpm. State all assumptions.

   (a)   Using hydrodynamically lubricated bronze sleeve bearings with $O_N = 30$, $l/d = 1.0$, and a clearance ratio of 0.002.

   (b)   Using deep-groove ball bearings.

[*] Answers to these problems are provided in Appendix G.

[†] These problems are based on similar problems in previous chapters (particularly Chapters 7 and 9) with the same −number, e.g., Problem 10-1 is based on Problem 9-1, etc. Problems in succeeding chapters may also continue and extend these problems.

Table P10-1   Data for Problems

Row	l	a	b	P or p	T_{min}	T_{max}
a	20	16	18	1000	0	2000
b	12	2	7	500	−100	600
c	14	4	12	750	−200	400
d	8	4	8	1000	0	2000
e	17	6	12	1500	−200	500
f	24	16	22	750	1000	2000

*10-3   An oil has a kinematic viscosity of 300 centistokes. Find its absolute viscosity in centipoise (cP). Assume a specific gravity of 0.89.

10-4   An oil has an absolute viscosity of 2 μreyn. Find its kinematic viscosity in in²/sec. Assume a specific gravity of 0.87.

*10-5   Find the Petroff no-load torque for the journal bearing designed in Case Study 9b.

*10-6   Find the minimum film thickness for a long bearing with: 45-mm dia, 200 mm long, $\varepsilon = 0.55$, clearance ratio = 0.001, 2 500 rpm, ISO VG 46 oil at 150°F.

*10-7   Find the torques and power lost in the bearing of Problem 10-6.

*†10-8   A paper machine processes rolls of paper having a density of 984 kg/m³. The paper roll is 1.50-m OD x 22-cm ID x 3.23-m long and is on a simply supported, 22-cm OD, steel shaft. The roll turns at 50 rpm. Design suitable hydrodynamically-lubricated full-film bronze short bearings of $l / d = 0.75$ to support the shaft at each end. Specify the viscosity of lubricant needed at 180°F. State all assumptions.

10-9   Find the minimum film thickness for a long bearing with the following data: 30-mm dia, 130 mm long, 0.0015 clearance ratio, 1 500 rpm, ISO VG 100 oil at 200°F, and supporting a load of 7 kN.

*10-10   Find the minimum film thickness for a bearing with these data: 45-mm dia, 30 mm long, 0.001 clearance ratio, 2 500 rpm, $O_N = 25$, ISO VG 46 oil at 150°F.

10-11   Find the minimum film thickness for a bearing with these data: 30-mm dia, 25 mm long, 0.0015 clearance ratio, 1 500 rpm, $O_N = 30$, ISO VG 220 oil at 200°F.

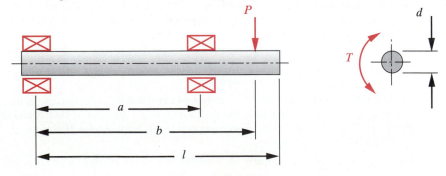

FIGURE P10-1

Shaft Design for Problems 10-1

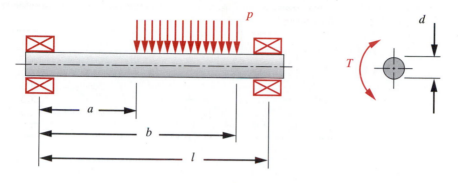

FIGURE P10-2

Shaft Design for Problems 10-2

†10-12   Problem 7-12 estimated the volume of adhesive wear to expect from a steel shaft of 40 mm dia rotating at 250 rpm for 10 years in a plain bronze bushing with a transverse load of 1 000 N for conditions of both poor and good lubrication. If the bushing has $l/d = 0.5$ and a clearance ratio of 0.001, define the lubricant viscosity in µreyn needed to obtain good lubrication.

10-13   Find the torques and power lost in the bearing of Problem 10-9.

*10-14   Find the torques and power lost in the bearing of Problem 10-10.

10-15   Find the torques and power lost in the bearing of Problem 10-11.

†10-16   Problem 7-16 determined the size of the contact patch and the maximum contact stresses for a 20-mm-dia steel ball rolled against a flat aluminum plate with 1 kN of force. Assuming the ball rolls at 1 200 rpm, determine its lubrication condition with ISO VG 68 oil at 150°F. Assume $R_q = 16$ µin (ball), $R_q = 64$ µin (plate).

*†10-17   The shaft shown in Figure P10-3 was designed in Problem 9-17. For the data in the row(s) assigned from Table P10-1, and the corresponding diameter of shaft found in Problem 9-17, design suitable bearings to support the load for at least 1E8 cycles at 1800 rpm. State all assumptions.

   (a)   Using hydrodynamically lubricated bronze sleeve bearings with $O_N = 15$, $l/d = 0.75$, and a clearance ratio of 0.001.
   (b)   Using deep-groove ball bearings.

†10-18   Problem 7-18 determined the size of the contact patch and the maximum contact stresses for a 40-mm-dia steel cylinder, 25 cm long, rolled against a flat aluminum plate with 4 kN of force. If the ball rolls at 800 rpm, determine its lubrication condition with ISO VG 22 oil at 200°F. $R_q = 64$ µin (cylinder); $R_q = 32$ µin (plate).

†10-19   The shaft shown in Figure P10-4 was designed in Problem 9-19. For the data in the row(s) assigned from Table P10-1, and the corresponding diameter of shaft found in Problem 9-19, design suitable bearings to support the load for at least 5E8 cycles at 1200 rpm. State all assumptions.

   (a)   Using hydrodynamically-lubricated bronze sleeve bearings with $O_N = 40$, $l/d = 0.80$, and a clearance ratio of 0.002 5.
   (b)   Using deep-groove ball bearings.

* Answers to these problems are provided in Appendix G.

† These problems are based on similar problems in previous chapters (particularly Chapters 7 and 9) with the same –number, e.g., Problem 10-1 is based on Problem 9-1, etc. Problems in succeeding chapters may also continue and extend these problems.

10

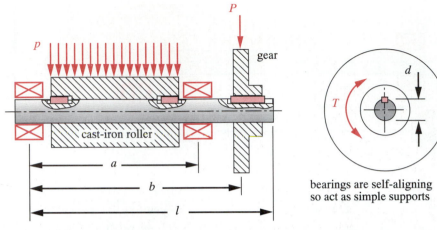

FIGURE P10-3

Shaft Design for Problems 10-17

*†10-20 Problem 7-20 determined the size of the contact patch and the maximum contact stresses for a 20-mm-dia steel ball rolled against a 40-mm-dia steel cylinder, 25 cm long with 10 kN of force. Assuming the ball rolls at 1 800 rpm, determine its lubrication condition with ISO VG 46 oil at 120°F. Assume R_q = 32 µin (for both).

†10-21 Problem 7-21 stated: "A cam-follower system has a dynamic load of 0 to 20 kN. The cam is cylindrical with a minimum radius of curvature of 20 mm. The roller follower is crowned with radii of 15 mm in one direction and 150 mm in the other. Find the contact stresses and estimate the cycle life if the follower is 4150 steel at 300HB and the cam is nodular iron at 207HB." For this problem, find the lubrication condition between cam and follower if lubricated with an ISO VG 68 oil at 200°F. Assume R_q = 8 µin (follower), R_q = 32 µin (cam). Follower rpm = 300.

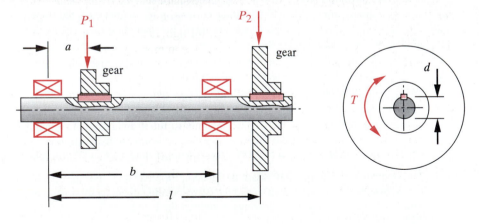

FIGURE P10-4

Shaft Design for Problem 10-19

SPUR GEARS 11

*The beginning of wisdom is to call things by
their right names.*

CHINESE PROVERB

*Tim was so learned that he could name a
horse in nine languages: so ignorant that he
bought a cow to ride on.*

BENJAMIN FRANKLIN

11.0 INTRODUCTION

Gears are used to transmit torque and angular velocity in a wide variety of applications. There is also a wide variety of gear types to choose from. This chapter will deal with the simplest type of gear, the spur gear, designed to operate on parallel shafts and having teeth parallel to the shaft axis. Other gear types such as helical, bevel, and worm can accommodate nonparallel shafts. These will be dealt with in the next chapter.

Gears are now highly standardized as to tooth shape and size. The *American Gear Manufacturers Association* (AGMA) supports research in gear design, materials, and manufacturing and publishes standards for their design, manufacture, and assembly.[1], [2],[3] We will follow the AGMA methods and recommendations as defined in those standards.

Gears have a long history. The ancient Chinese *South-Pointing Chariot*, supposedly used to navigate across the Gobi desert in pre-Biblical times, contained gears. Leonardo DaVinci shows many gear arrangements in his drawings. Early gears were most likely made crudely of wood and other easily worked materials, their teeth merely being pegs inserted in a disk or wheel. It was not until the industrial revolution that machines demanded, and manufacturing techniques allowed, the creation of gears as we now know them with specially shaped teeth formed or cut into a metal disk.

There is a great deal of specialized terminology for gears and it is necessary that the reader become familiar with these terms. As indicated in the above epigraphs, calling things by their correct names is important but is not sufficient to ensure complete understanding of the topic. The variables used in this chapter are listed in Table 11-0.

Title page photograph
courtesy of Boston Gear,
Division of IMO Industries,
Quincy, Mass.

Table 11-0   Variables Used in this Chapter

Part 1 of 2

Symbol	Variable	ips units	SI units	See
a	addendum	in	m	Fig. 11-8
b	dedendum	in	m	Fig. 11-8
C	center distance	none	none	Eq. 11.22b
C_f	surface finish factor	none	none	Sect. 11.8
C_H	hardness factor	none	none	Eq. 11.26, -27
C_p	elastic coefficient	none	none	Eq. 11.23
d	pitch diameter	in	m	various
F	face width	in	m	Eq. 11.14
HB	Brinell hardness	none	none	Sect. 11.8
I	AGMA surface geometry factor	none	none	Eq. 11.22a
J	AGMA bending geometry factor	none	none	Sect. 11.8
K_a, C_a	application factor	none	none	Sect. 11.8
K_B	rim bending factor	none	none	Sect. 11.8
K_I	idler factor	none	none	Eq. 11.15
K_L, C_L	life factor	none	none	Fig. 11-24, -26
K_m, C_m	load distribution factor	none	none	Sect. 11.8
K_R, C_R	reliability factor	none	none	Table 11-19
K_s, C_s	size factor	none	none	Sect. 11.8
K_T, C_T	temperature factor	none	none	Eq. 11.24a
K_v, C_v	dynamic factor	none	none	Sect. 11.8
m	module	–	mm	Eq. 11.4c
M	moment, moment function	lb-in	N-m	Fig. 11-21
m_A	mechanical advantage	none	none	Eq. 11.1b
m_G	gear ratio	none	none	Eq. 11.1c
m_p	contact ratio	none	none	Eq. 11.7a
m_V	angular velocity ratio	none	none	Eq. 11.1a
N	number of cycles or number of teeth	none	none	Fig. 11.24
N_b, N_c	factors of safety—bending and contact	none	none	various
p_b	base pitch	in	m	Eq. 11.3b
p_c	circular pitch	in	mm	Eq. 11.3a
p_d	diametral pitch	1/in	–	Eq. 11.4a
Q_v	gear quality index	none	none	Fig. 11-22
r	pitch radius	in	m	various
S_{fb}	corrected bending-endurance strength	psi	Pa	Eq. 11.24
S'_{fb}	uncorrected bending-endurance strength	psi	Pa	Eq. 11.24

11

* Portions of this chapter
including Figures 11-1
through 11-8, 11-11, 11-13
through 11-16 and their
discussions are adapted from
R. L. Norton, *Design of
Machinery*, McGraw-Hill,
1992, Chapter 10, with the
publisher's permission.

Table 11-0   Variables Used in this Chapter
Part 2 of 2

Symbol	Variable	ips units	SI units	See
S_{fc}	corrected surface-endurance strength	psi	Pa	Eq. 11.25
S'_{fc}	uncorrected surface-endurance strength	psi	Pa	Eq. 11.25
T	torque	lb-in	N-m	Eq. 11.13a
V_t	pitch line velocity	in/sec	m/sec	Eq. 11.16
W	total force on gear teeth	lb	N	Eq. 11.13c
W_r	radial force on gear teeth	lb	N	Eq. 11.13b
W_t	tangential force on gear teeth	lb	N	Eq. 11.13a
x_1, x_2	addendum modification coefficients	none	none	Sect. 11.3
Y	Lewis form factor	none	none	Eq. 11.14
Z	length of action	in	m	Eq. 11.2
ϕ	pressure angle	deg	deg	various
ρ	radius of curvature	in	m	Eq. 11.22b
σ_b	bending stress	psi	Pa	Eq. 11.15
σ_c	surface stress	psi	Pa	Eq. 11.21
ω	angular velocity	rad/sec	rad/sec	Eq. 11.1a

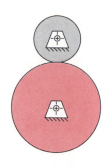

(a) External set

(b) Internal set

FIGURE 11-1

Rolling Cylinders

11.1   GEAR TOOTH THEORY[*]

The simplest means of transferring rotary motion from one shaft to another is a pair of rolling cylinders. They may be an external set of rolling cylinders, as shown in Figure 11-1a or an internal set, as in Figure 11-1b. If sufficient friction is available at the rolling interface, this mechanism will work quite well. There will be no slip between the cylinders until the maximum available frictional force at the joint is exceeded by the demands of torque transfer.

The principal drawbacks to the rolling-cylinder drive mechanism are its relatively low torque capability and the possibility of slip. Some drives require absolute phasing of the input and output shafts for timing purposes. This requires adding some meshing teeth to the rolling cylinders. They then become gears, as shown in Figure 11-2, and are together called a *gearset*. When two gears are placed in mesh to form a gearset such as this one, it is conventional to refer to the *smaller of the two gears as the* **pinion** and to the other as the **gear**.

The Fundamental Law of Gearing

Conceptually, teeth of any shape will prevent gross slip. Old water-powered mills and windmills used wooden gears whose teeth were merely round wooden pegs stuck into

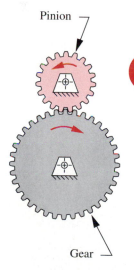

Pinion

Gear

FIGURE 11-2

An External Gearset

the rims of the cylinders. Even ignoring the crudity of construction of these early examples of gearsets, there was no possibility of smooth velocity transmission because the geometry of the tooth "pegs" violated the **fundamental law of gearing**, which states that *the angular velocity ratio between the gears of a gearset must remain constant throughout the mesh*. The angular velocity ratio m_V is equal to the ratio of the pitch radius of the input gear to that of the output gear.

$$m_V = \frac{\omega_{out}}{\omega_{in}} = \pm \frac{r_{in}}{r_{out}} \qquad (11.1a)$$

The pitch radii in equation 11.1a are those of the rolling cylinders to which we are adding the teeth. The positive or negative sign accounts for internal or external cylinder sets as shown in Figure 11-1. An external set reverses the direction of rotation between the cylinders and requires the negative sign. An internal gearset (and a belt or chain drive) will have the same direction of rotation on input and output shafts and require the positive sign in equation 11.1a. The surfaces of the rolling cylinders will become the **pitch circles**, and their diameters the **pitch diameters** of the gears. The contact point between the cylinders lies on the line of centers, as shown in Figure 11-4, and this point is called the **pitch point**.

The **torque ratio** or **mechanical advantage** m_A is the reciprocal of the velocity ratio m_V;

$$m_A = \frac{1}{m_V} = \frac{\omega_{in}}{\omega_{out}} = \pm \frac{r_{out}}{r_{in}} \qquad (11.1b)$$

Thus a gearset is essentially a device to exchange torque for velocity or vice versa. A common gearset application reduces velocity and increases torque to drive heavy loads, as in your automobile transmission. Other applications require an increase in velocity, for which a reduction in torque must be accepted. In either case, it is usually desirable to maintain a constant ratio between the gears as they rotate. Any variation in ratio will show up as oscillation in the output velocity and torque even if the input is constant with time.

For calculation purposes, the **gear ratio** m_G is taken as the magnitude of either the velocity ratio or the torque ratio, whichever is > 1.

$$m_G = |m_V| \quad \text{or} \quad m_G = |m_A|, \quad \text{for} \quad m_G \geq 1 \qquad (11.1c)$$

In other words, the gear ratio is always a positive number > 1 regardless of which direction the power flows through the gearset.

In order for the fundamental law of gearing to be true, the gear tooth contours on mating teeth must be conjugates of one another. There is an infinite number of possible conjugate pairs that could be used, but only a few curves have seen practical application as gear teeth. The **cycloid** is still used as a tooth form in some watches and clocks, but most gears use the **involute** of a circle for their shape.

The Involute Tooth Form

The involute of a circle is a curve that can be generated by unwrapping a taut string from a cylinder, as shown in Figure 11-3. Note the following about this involute curve:

1 The string is always tangent to the base circle.

2 The center of curvature of the involute is always at the point of tangency of the string with the base circle.

3 A tangent to the involute is always normal to the string, which is the instantaneous radius of curvature of the involute curve.

Figure 11-4 shows two involutes on separate cylinders in contact or "in mesh." These represent gear teeth. The cylinders from which the strings are unwrapped are called the **base circles** of the respective gears. Note that the base circles are necessarily smaller than the pitch circles, which are at the radii of the original rolling cylinders, r_p and r_g. The gear tooth must project both below and above the rolling cylinder surface (pitch circle) and the *involute only exists outside of the base circle*. The amount of tooth that sticks out above the pitch circle is the **addendum**, shown as a_p and a_g for pinion and gear, respectively. These are equal for standard, full-depth gear teeth.

There is a **common tangent** to both involute tooth curves at the contact point, and a common normal, perpendicular to the common tangent. Note that the common normal is, in fact, the "strings" of both involutes, which are colinear. Thus the **common**

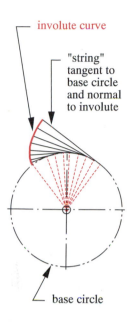

FIGURE 11-3

Development of the Involute of a Circle

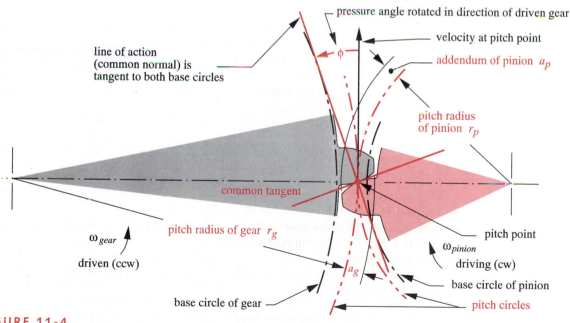

FIGURE 11-4

Contact Geometry and Pressure Angle of Involute Gear Teeth

normal, which is also the **line of action**, always passes through the **pitch point** regardless of where in the mesh the two teeth are contacting. The pitch point has the same linear velocity in both pinion and gear, called the **pitch line velocity**. The angle between the line of action and the velocity vector is the **pressure angle** φ.

Pressure Angle

The **pressure angle** φ in a gearset is defined as the angle between the line of action (common normal) and the direction of velocity at the pitch point such that the line of action is rotated φ degrees in the direction of rotation of the driven gear, as shown in Figures 11-4 and 11-5. Pressure angles of gearsets are standardized at a few values by the gear manufacturers. These are defined at the nominal center distance for the gearset as cut. The standard values are 14.5, 20, and 25° with 20° being the most commonly used and 14.5° now being obsolete. Any custom pressure angle can be made, but its expense over the available stock gears with standard pressure angles would be hard to justify. Special cutters would have to be made. Gears to be run together must be cut to the same nominal pressure angle.

Mesh Geometry

Figure 11-5 shows a pair of involute tooth forms in two positions, just beginning contact and about to leave contact. The common normals of both these contact points still pass through the same pitch point. It is this property of the involute that causes it to obey the fundamental law of gearing. The ratio of the driving gear radius to the driven gear radius remains constant as the teeth move into and out of mesh.

From this observation of the behavior of the involute we can restate the **fundamental law of gearing** in a more kinematically formal way as: *the common normal of the tooth profiles, at all contact points within the mesh, must always pass through a fixed point on the line of centers, called the pitch point.* The gearset's velocity ratio will then be a constant defined by the ratio of the respective radii of the gears to the pitch point.

The points of beginning and leaving contact define the **mesh** of the pinion and gear. The distance along the line of action between these points within the mesh is called the **length of action** Z, defined by the intersections of the respective addendum circles with the line of action, as shown in Figure 11-5. The distance along the pitch circle within the mesh is the **arc of action**, and the angles subtended by these points and the line of centers are the **angle of approach** and **angle of recess**. These are shown only on the gear in Figure 11-5 for clarity, but similar angles exist for the pinion. The arc of action on both pinion and gear pitch circles must be the same length for zero slip between the theoretical rolling cylinders. The length of action Z can be calculated from the gear and pinion geometry:

$$Z = \sqrt{\left(r_p + a_p\right)^2 - \left(r_p \cos\phi\right)^2} + \sqrt{\left(r_g + a_g\right)^2 - \left(r_g \cos\phi\right)^2} - C\sin\phi \qquad (11.2)$$

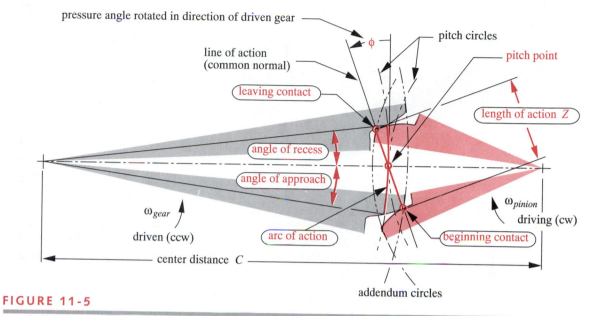

pressure angle rotated in direction of driven gear

ϕ

pitch circles

line of action
(common normal)

pitch point

leaving contact

length of action Z

angle of recess

angle of approach

ω_{gear}

ω_{pinion}
driving (cw)

driven (ccw)

arc of action

beginning contact

center distance C

addendum circles

FIGURE 11-5

Length of Action, Arc of Action, and Angles of Approach and Recess During the Meshing of a Gear and Pinion

where r_p and r_g are the pitch circle radii, and a_p and a_g the addenda of pinion and gear, respectively. C is the center distance and ϕ is the pressure angle.

Rack and Pinion

If the diameter of the base circle of a gear is increased without limit, the base circle will become a straight line. If the "string" wrapped around this base circle to generate the involute were still in place after the base circle's enlargement to an infinite radius, the string would be pivoted at infinity and would generate an involute that is a straight line. This linear gear is called a **rack**. Figure 11-6 shows a rack and pinion and the geometry of a standard, full-depth rack. Its teeth are trapezoids, yet are true involutes. This fact makes it easy to create a cutting tool to generate involute teeth on circular gears, by accurately machining a rack and hardening it to cut teeth in other gears. This is another advantage of the involute tooth form. Rotating the gear blank with respect to the rack cutter while moving the cutter axially back and forth across the gear blank will shape, or develop, a true involute tooth on the circular gear.

The most common application of the rack and pinion is in rotary to linear motion conversion or vice versa. It can be backdriven, so it requires a brake if used to hold a load. An example of its use is in rack-and-pinion steering in automobiles. The pinion is attached to the bottom end of the steering column and turns with the steering wheel. The rack meshes with the pinion and is free to move left and right in response to your angular input at the steering wheel. The rack is also one link in a multibar linkage, which

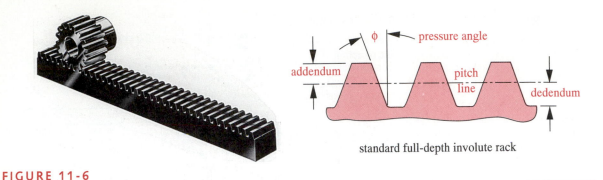

standard full-depth involute rack

FIGURE 11-6

A Rack and Pinion   *Photo Courtesy of Martin Sprocket and Gear Co., Austin, Texas*

converts the linear translation of the rack to the proper amount of angular motion of a rocker link attached to the front wheel assembly in order to steer the car.

Changing Center Distance

When involute teeth (or any teeth) have been cut into a cylinder with respect to a particular base circle to create a single gear, we do not yet have a pitch circle. The pitch circle only comes into being when we mate this gear with another to create a *pair of gears*, or **gearset**. There will be some range of center-to-center distances over which we can achieve a mesh between the gears. There will also be an ideal center distance that will give us the nominal pitch diameters for which the gears were designed. However, the limitations of the manufacturing process give a low probability that we will be able to exactly achieve this ideal center distance in every case. More likely, there will be some error in the center distance, even if small.

If the gear tooth form is **not** an involute, then an error in center distance will cause variation, or "ripple," in the output velocity. The output angular velocity will then not be constant for a constant input velocity, violating the fundamental law of gearing. However, **with an involute tooth form**, *center distance errors do not affect the velocity ratio*. This is the principal advantage of the involute over all other possible tooth forms and is the reason why it is nearly universally used for gear teeth. Figure 11-7 shows what happens when the center distance is varied on an involute gearset. Note that the common normal still goes through the pitch point, and also through all contact points within the mesh. Only the pressure angle is affected by the change in center distance.

Figure 11-7 also shows the pressure angles for two different center distances. As the center distance increases, so will the pressure angle and vice versa. This is one result of a change, or error, in center distance when using involute teeth. Note that the fundamental law of gearing still holds in the modified center distance case. The common normal is still tangent to the two base circles and still goes through the pitch point.

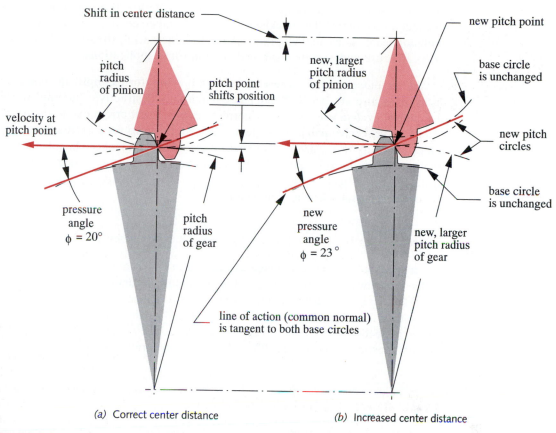

Shift in center distance

new pitch point

pitch
radius
of pinion

new, larger
pitch radius
of pinion

base circle
is unchanged

pitch point
shifts position

velocity at
pitch point

new pitch
circles

base circle
is unchanged

pressure
angle
$\phi = 20°$

pitch
radius
of gear

new
pressure
angle
$\phi = 23°$

new, larger
pitch radius
of gear

line of action (common normal)
is tangent to both base circles

(a) Correct center distance   (b) Increased center distance

FIGURE 11-7

Changing Center Distance of Involute Gears Changes Only the Pressure Angle and Pitch Diameters

The pitch point has moved in proportion to the shifts of the center distance and the pitch radii. The velocity ratio is unchanged despite the shift in center distance. In fact, the velocity ratio of involute gears is fixed by the ratio of their base circle diameters, which are unchanging once the gear is cut.

Backlash

Another factor affected by changing center distance C is backlash. Increasing C will increase the backlash and vice versa. **Backlash** is defined as *the gap between mating teeth measured along the circumference of the pitch circle*. Manufacturing tolerances preclude a zero backlash, as all teeth cannot be exactly the same dimensions, and all must mesh without jamming. So, there must be some small difference between the tooth thickness and the space width, (see Figure 11-8). As long as the gearset is run with a nonreversing torque, the backlash should not be a problem.

However, whenever the torque changes sign, the teeth will move from contact on one side to the other. The backlash gap will be traversed and the teeth will impact with noticeable noise and vibration. As well as increasing stresses and wear, backlash can cause undesirable positional error in some applications.

In servomechanisms, where motors are driving, for example, the control surfaces on an aircraft, backlash can cause potentially destructive "hunting" in which the control system tries in vain to correct the positional errors due to the backlash "slop" in the mechanical drive system. Such applications need **antibacklash gears,** which are really two gears back-to-back on the same shaft that can be rotated slightly at assembly (or by springs) with respect to one another, so as to take up the backlash. In less critical applications, such as the propeller drive on a boat, backlash on torque reversal will not even be noticed.

Relative Tooth Motion

The relative motion between involute teeth is pure rolling at the pitch point. At points on the tooth away from the pitch point, some sliding occurs in combination with rolling. The average amount of sliding in an involute-tooth mesh is about 9%, as discussed in Section 7.13. The surface stresses are increased by the sliding component, as discussed in Section 7.11. Table 7-7 shows surface fatigue-strength data developed from extensive testing of rolling plus 9% sliding between various material combinations.

11.2 GEAR TOOTH NOMENCLATURE

Figure 11-8 shows two teeth of a gear with the standard nomenclature defined. **Pitch circle** and **base circle** have been defined above. The tooth height is defined by the **addendum** (*added on*) and the **dedendum** (*subtracted from*), which are referenced to the nominal pitch circle. The dedendum is slightly larger than the addendum to provide a small amount of **clearance** between the tip of one mating tooth (**addendum circle**) and the bottom of the tooth space of the other (**dedendum circle**). The **tooth thickness** is measured at the pitch circle, and the tooth **space width** is slightly larger than the tooth thickness. The difference between these two dimensions is the **backlash.** The **face width** of the tooth is measured along the axis of the gear. The **circular pitch** is the arc length along the pitch circle circumference measured from a point on one tooth to the same point on the next. The circular pitch defines the tooth size. The definition of **circular pitch** p_c is

$$p_c = \frac{\pi d}{N} \tag{11.3a}$$

where d = pitch diameter and N = number of teeth. The tooth pitch can also be measured along the base circle circumference and then is called the **base pitch** p_b.

$$p_b = p_c \cos \phi \tag{11.3b}$$

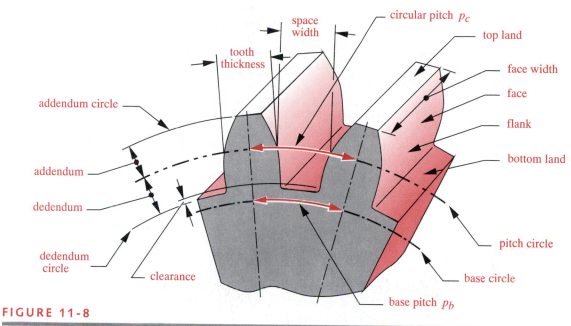

FIGURE 11-8

Gear Tooth Nomenclature

The units of p_c are inches or millimeters. A more convenient way to define tooth size is to relate it directly to the diameter d of the pitch circle rather than to its circumference. The **diametral pitch** p_d is

$$p_d = \frac{N}{d} \qquad (11.4a)$$

The units of p_d are reciprocal inches, or number of teeth per inch. This measure is only used in U.S. specification gears. Combining equations 11.3a and 11.4a gives the relationship between circular pitch and diametral pitch.

$$p_d = \frac{\pi}{p_c} \qquad (11.4b)$$

The SI system, used for metric gears, defines a parameter called the **module**, which is *the reciprocal of diametral pitch* with pitch diameter d measured in millimeters.

$$m = \frac{d}{N} \qquad (11.4c)$$

The units of the module are millimeters. Unfortunately, metric gears are not interchangeable with U.S. gears, despite both being involute tooth forms, as their standards for tooth sizes are different (see Table 11-5). In the United States, gear tooth sizes are specified by diametral pitch. The conversion from one standard to the other is

$$m = \frac{25.4}{p_d} \qquad (11.4d)$$

The velocity ratio m_V of the gearset can be put into a more convenient form by substituting equation 11.4a into equation 11.1, noting that the diametral pitch of meshing gears must be the same.

$$m_V = \pm \frac{r_{in}}{r_{out}} = \pm \frac{d_{in}}{d_{out}} = \pm \frac{N_{in}}{N_{out}} \qquad (11.5a)$$

Thus the **velocity ratio** can be computed from the number of teeth on the meshing gears, which are integers. Note that a minus sign implies an external gearset and a positive sign an internal gearset, as shown in Figure 11-1. The gear ratio m_G, can be expressed as the number of teeth on the gear N_g over the number of teeth on the pinion N_p.

$$m_G = \frac{N_g}{N_p} \qquad (11.5b)$$

STANDARD GEAR TEETH   Standard, full-depth gear teeth have equal addenda on pinion and gear, with the dedendum being slightly larger for clearance. The standard tooth dimensions are defined in terms of the diametral pitch. Table 11-1 shows the dimensions of standard, full-depth gear teeth as defined by the AGMA and Figure 11-9 shows their shapes for three standard pressure angles, Figure 11-10 shows the actual sizes of 20° pressure angle, standard, full-depth teeth from $p_d = 4$ to $p_d = 80$. Note the inverse relationship between p_d and tooth size.

While there are no theoretical restrictions on the possible values of diametral pitch, a set of standard values is defined based on available gear cutting tools. These standard tooth sizes are shown in Table 11-2 in terms of diametral pitch and in Table 11-3 in terms of metric module.

11.3   INTERFERENCE AND UNDERCUTTING

The involute tooth form is only defined outside of the base circle. In some cases, the dedendum will be large enough to extend below the base circle. If so, then the portion of tooth below the base circle will not be an involute and will interfere with the tip of the tooth on the mating gear, which is an involute. If the gear is cut with a standard gear shaper or a "hob," the cutting tool will also interfere with the portion of tooth below the base circle and will cut away the interfering material. This results in an undercut tooth, as shown in Figure 11-11. Undercutting weakens the tooth by removing material at its root. The maximum moment and maximum shear from the tooth loaded as a cantilever beam both occur in this region. Severe undercutting will cause early tooth failure.

Interference and its attendant undercutting can be prevented simply by avoiding gears with too few teeth. If a pinion has a large number of teeth, they will be small com-

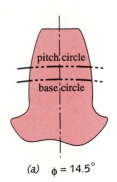

(a) $\phi = 14.5°$

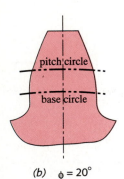

(b) $\phi = 20°$

11

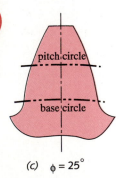

(c) $\phi = 25°$

FIGURE 11-9

AGMA Full-Depth Tooth Profiles for Three Pressure Angles

Table 11-1   AGMA Full-Depth Gear Tooth Specifications

Parameter	Coarse Pitch ($p_d < 20$)	Fine Pitch ($p_d \geq 20$)
Pressure angle ϕ	20° or 25°	20°
Addendum a	1.000 / p_d	1.000 / p_d
Dedendum b	1.250 / p_d	1.250 / p_d
Working depth	2.000 / p_d	2.000 / p_d
Whole depth	2.250 / p_d	2.200 / p_d + 0.002 in
Circular tooth thickness	1.571 / p_d	1.571 / p_d
Fillet radius—basic rack	0.300 / p_d	not standardized
minimum basic clearance	0.250 / p_d	0.200 / p_d + 0.002 in
min width of top land	0.250 / p_d	not standardized
Clearance (shaved or ground teeth)	0.350 / p_d	0.350 / p_d + 0.002 in

pared to its diameter. As the number of teeth is reduced for a fixed diameter pinion, the teeth must become larger. At some point, the dedendum will exceed the radial distance

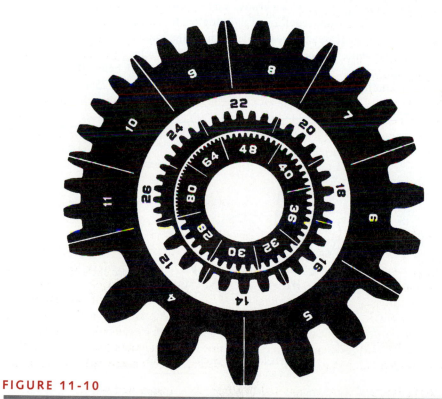

FIGURE 11-10

Actual Gear Tooth Sizes for Various Diametral Pitches *Courtesy of Barber-Colman Co., Loves Park, Ill.*

Table 11-2

Standard Diametral Pitches

Coarse ($p_d < 20$)	Fine ($p_d \geq 20$)
1	20
1.25	24
1.5	32
1.75	48
2	64
2.5	72
3	80
4	96
5	120
6	
8	
10	
12	
14	
16	
18	

Table 11-3
Standard Metric Modules

Metric Module (mm)	Equivalent p_d (in^{-1})
0.3	84.67
0.4	63.50
0.5	50.80
0.8	31.75
1	25.40
1.25	20.32
1.5	16.93
2	12.70
3	8.47
4	6.35
5	5.08
6	4.23
8	3.18
10	2.54
12	2.12
16	1.59
20	1.27
25	1.02

between the base circle and the pitch circle, and interference will occur. The minimum number of full-depth teeth required to avoid interference on a pinion running against a standard rack can be calculated from:

$$N_{min} = \frac{2}{\sin^2 \phi} \qquad (11.6)$$

Table 11-4 shows the minimum number of teeth required to avoid undercutting against a standard rack as a function of pressure angle. Table 11-5 shows the minimum number of full-depth pinion teeth that can be used against a selection of full-depth gears of various sizes (for $\phi = 20°$). As the mating gear gets smaller, the pinion can have fewer teeth and still avoid interference.

Unequal-Addendum Tooth Forms

In order to avoid interference on small pinions, the tooth form can be changed from the standard, full-depth shapes of Figure 11-9 that have equal addenda on both pinion and gear to an involute shape with a longer addendum on the pinion and a shorter one on the gear. These are called **profile-shifted gears**. The AGMA defines addendum modification coefficients, x_1 and x_2, which always sum to zero, being equal in magnitude and opposite in sign. The positive coefficient x_1 is applied to increase the pinion addendum and the negative x_2 decreases the gear addendum by the same amount. The total tooth depth remains the same. The net effect is to shift the pitch circles away from the pinion's base circle and eliminate that noninvolute portion of pinion tooth below the base circle. The standard coefficients are ±0.25 and ±0.50, which add/subtract 25% or 50% of the

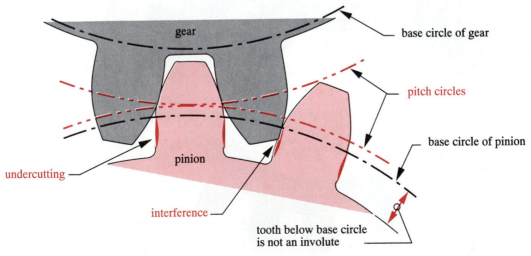

FIGURE 11-11

Interference and Undercutting of Teeth Below the Base Circle

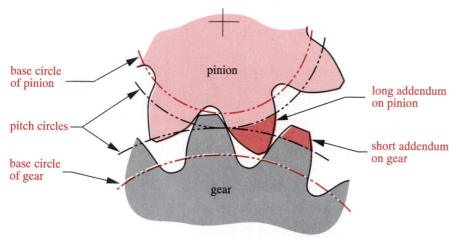

FIGURE 11-12

Profile-Shifted Gear Teeth with Long and Short Addenda to Avoid Interference and Undercutting

standard addendum, respectively. The limit of this approach occurs when the pinion tooth becomes pointed.

There are some secondary benefits to this technique. The pinion tooth becomes thicker at its base and thus stronger. The gear tooth is correspondingly weakened but since a full-depth gear tooth is stronger than a full-depth pinion tooth, this shift brings them to nearly equal strength. A disadvantage of unequal-addendum tooth forms is an increase in the sliding velocity at the tooth tip. The percent of sliding between the teeth is then greater than with equal addendum teeth. This increases the tooth-surface stresses, as discussed in Section 7.11. The friction losses in the gear mesh are also increased by higher sliding velocities. Dudley[10] recommends avoiding more than 25% long addendum pinion teeth in spur or helical gears because of the disadvantages associated with their high sliding velocities. Figure 11-12 shows the contours of profile-shifted involute teeth. Compare these to the standard tooth shapes in Figure 11-9.

11.4   CONTACT RATIO

The contact ratio m_p defines the average number of teeth in contact at any one time. It is calculated from

$$m_p = \frac{Z}{p_b} \qquad (11.7a)$$

where Z is the length of action from equation 11.2 and p_b is the base pitch from equation 11.3b. Substituting equations 11.3b and 11.4b into 11.7a defines m_p in terms of diametral pitch:

Table 11-4

Minimum Number of Pinion Teeth to Avoid Interference Between a Full-Depth Pinion and a Full-Depth Rack

Pressure Angle (deg)	Minimum Number of Teeth
14.5	32
20	18
25	12

Table 11-5

Minimum Number of Pinion Teeth to Avoid Interference Between a 20° Full-Depth Pinion and Full-Depth Gears of Various Sizes

Minimum Pinion Teeth	Maximum Gear Teeth
17	1 309
16	101
15	45
14	26
13	16

$$m_p = \frac{p_d Z}{\pi \cos \phi} \qquad\qquad (11.7b)$$

If the contact ratio is 1, then one tooth is leaving contact just as the next is beginning contact. This is undesirable because slight errors in the tooth spacing will cause oscillations in the velocity, vibration, and noise. In addition, the load will be applied at the tip of the tooth, creating the largest possible bending moment. At larger contact ratios than 1, there is the possibility of load sharing among the teeth. For contact ratios between 1 and 2, which are common for spur gears, there will still be times during the mesh when one pair of teeth will be taking the entire load. However, these will occur toward the center of the mesh region where the load is applied at a lower position on the tooth, rather than at its tip. This point is called the **highest point of single-tooth contact** or HPSTC. The minimum acceptable contact ratio for smooth operation is 1.2. A minimum contact ratio of 1.4 is preferred and larger is better. Most spur gearsets will have contact ratios between 1.4 and 2. Equation 11.7b shows that for smaller teeth (larger p_d) and larger pressure angle, the contact ratio will be larger.

EXAMPLE 11-1

Determining Gear Tooth and Gear Mesh Parameters

Problem   Find the gear ratio, circular pitch, base pitch, pitch diameters, pitch radii, center distance, addendum, dedendum, whole depth, clearance, outside diameters, and contact ratio of a gearset with the given parameters. If the center distance is increased 2% what is the new pressure angle?

Given   A 6 p_d, 20° pressure angle, 19-tooth pinion is meshed with a 37 tooth gear.

Assumptions The tooth forms are standard AGMA full-depth involute profiles.

Solution   See the *TKSolver* file EX11-01.

1   The gear ratio is easily found from the given tooth numbers on pinion and gear using equation 11.5b.

$$m_B = \frac{N_g}{N_p} = \frac{37}{19} = 1.947 \qquad\qquad (a)$$

2   The circular pitch can be found either from equation 11.3a or 11.4b.

$$p_c = \frac{\pi}{p_d} = \frac{\pi}{6} = 0.524 \ \text{in} \qquad\qquad (b)$$

3   The base pitch measured on the base circle is (from equation 11.3b):

$$p_b = p_c \cos\phi = 0.524\cos(20°) = 0.492 \text{ in} \tag{c}$$

4   The pitch diameters and pitch radii of pinion and gear are found from equation 11.4a.

$$d_p = \frac{N_p}{P_d} = \frac{19}{6} = 3.167 \text{ in}, \qquad r_p = \frac{d_p}{2} = 1.583 \text{ in} \tag{d}$$

$$d_g = \frac{N_g}{P_d} = \frac{37}{6} = 6.167 \text{ in}, \qquad r_p = \frac{d_p}{2} = 3.083 \text{ in} \tag{e}$$

5   The nominal center distance C is the sum of the pitch radii:

$$C = r_p + r_g = 4.667 \text{ in} \tag{f}$$

6   The addendum and dedendum are found from the equations in Table 11-1:

$$a = \frac{1.0}{P_d} = 0.167 \text{ in}, \qquad b = \frac{1.25}{P_d} = 0.208 \text{ in} \tag{g}$$

7   The whole depth h_t is the sum of the addendum and dedendum.

$$h_t = a + b = 0.167 + 0.208 = 0.375 \text{ in} \tag{h}$$

8   The clearance is the difference between dedendum and addendum.

$$c = b - a = 0.208 - 0.167 = 0.042 \text{ in} \tag{i}$$

9   The outside diameter of each gear is the pitch diameter plus two addenda:

$$D_{o_p} = d_p + 2a = 3.500 \text{ in}, \qquad D_{o_g} = d_g + 2a = 6.500 \text{ in} \tag{j}$$

10   The contact ratio is found from equations 11.2 and 11.7.

$$Z = \sqrt{(r_p + a_p)^2 - (r_p \cos\phi)^2} + \sqrt{(r_g + a_g)^2 - (r_g \cos\phi)^2} - C\sin\phi$$
$$= \sqrt{(1.583 + 0.167)^2 - (1.583\cos 20)^2}$$
$$+ \sqrt{(3.083 + 0.167)^2 - (3.083\cos 20)^2} - 4.667\sin 20 = 0.798 \text{ in}$$

$$m_p = \frac{Z}{p_b} = \frac{0.798}{0.492} = 1.62 \tag{k}$$

11   If the center distance is increased from the nominal value due to assembly errors or other factors, the effective pitch radii will change by the same percentage. The gears' base radii will remain the same. The new pressure angle can be found from the changed geometry. For a 2% increase in center distance (1.02x):

$$\phi_{new} = \cos^{-1}\left(\frac{r_{\text{base circle}_p}}{1.02 r_p}\right) = \cos^{-1}\left(\frac{r_p \cos\phi}{1.02 r_p}\right) = \cos^{-1}\left(\frac{\cos 20°}{1.02}\right) = 22.89° \tag{l}$$

11.5   GEAR TRAINS

A **gear train** is any collection of two or more meshing gears. A pair of gears, or gear-set, is then the simplest form of gear train and *is usually limited to a ratio of about* 10:1. The gearset will become large and hard to package beyond that ratio if the pinion is kept above the minimum numbers of teeth shown in Tables 11-4 and 11-5. Gear trains may be **simple**, **compound**, or **epicyclic**. What follows is a brief review of the kinematic design of gear trains. For more complete information, see reference 4.

Simple Gear Trains

A simple gear train is one in which each shaft carries only one gear, the most basic two-gear example of which is shown in Figure 11-2. The *velocity ratio* (also called *train ratio*) of a gearset is given by equation 11.5*a*. Figure 11-13 shows a simple gear train with five gears in series. Equation 11.8 shows the expression for this train's velocity ratio:

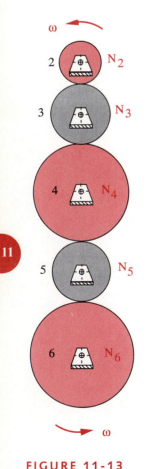

$$m_V = \left(-\frac{N_2}{N_3}\right)\left(-\frac{N_3}{N_4}\right)\left(-\frac{N_4}{N_5}\right)\left(-\frac{N_5}{N_6}\right) = +\frac{N_2}{N_6} \tag{11.8}$$

Each gearset potentially contributes to the overall train ratio, but in the case of a simple (series) train, the numerical effects of all gears except the first and last cancel out. The train ratio of a simple train is always just the ratio of the first gear over the last. Only the sign of the overall train ratio is affected by the intermediate gears, which are called *idlers*, because no power is typically taken from their shafts. If all gears in the train are external and there is an even number of gears in the train, the output direction will be opposite to that of the input. If there is an odd number of external gears in the train, the output will be in the same direction as the input. Thus a single, external, idler gear *of any diameter* can be used to change the direction of the output gear without affecting its velocity magnitude.

It is common practice to insert a single idler gear to change direction, but more than one idler is superfluous. There is little justification for designing a gear train as is shown in Figure 11-13. If the need is to connect two shafts that are far apart, a simple train of many gears could be used but will be much more expensive than a chain or belt drive for the same application. If the need is to get a larger train ratio than can be obtained with a single gearset, it is clear from equation 11.8 that the simple gear train will be of no help.

Compound Gear Trains

To get a train ratio of greater than about 10:1 with gears it is necessary to **compound the train** (unless an epicyclic train is used—see next section). A **compound train** *is one in which at least one shaft carries more than one gear*. This will be a parallel or series-parallel arrangement, rather than the pure series connections of the simple gear train. Figure 11-14*a* shows a compound train of four gears, two of which, gears 3 and

FIGURE 11-13

A Simple Gear Train

4, are attached to the same shaft and thus have the same angular velocity. The train ratio is now

$$m_V = \left(-\frac{N_2}{N_3} \right)\left(-\frac{N_4}{N_5} \right)$$ (11.9a)

This can be generalized for any number of gears in the train as

$$m_V = \pm \frac{\text{product of number of teeth on driver gears}}{\text{product of number of teeth on driven gears}}$$ (11.9b)

Note that these intermediate ratios do not cancel and the overall train ratio is the product of the ratios of parallel gearsets. Thus a larger ratio can be obtained in a compound gear train despite the approximately 10:1 limitation on individual gearset ratios. The plus or minus sign in equation 11.9b depends on the number and type of meshes in the train, whether external or internal. Writing the expression in the form of equation 11.9a and carefully noting the sign of each mesh ratio in the expression will result in the correct algebraic sign for the overall train ratio.

Reverted Compound Trains

In Figure 11-14a the input and output shaft are in different locations. This may well be acceptable or even desirable in some cases, depending on other packaging constraints in the overall machine design. Such a gear train, whose *input and output shafts are not*

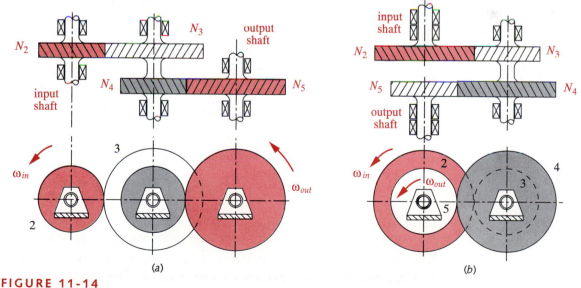

(a)   (b)

FIGURE 11-14

Two-Stage Compound Gear Trains: (a) Nonreverted, (b) Reverted

coincident, is called a **nonreverted compound train**. In some cases, such as in automobile transmissions, it is desirable or even necessary to have the *output shaft concentric with the input shaft,* as shown in Figure 11-14*b*. This is referred to as "reverting the train" or "bringing it back onto itself." The design of a **reverted compound train** is more complicated because of the additional constraint that the center distances of the stages must be equal. See reference 4 for more information.

EXAMPLE 11-2

Designing a Compound Gear Train

Problem   Design a compound, spur-gear train for an overall train ratio of 29:1.

Given   Use 25° pressure angle gears with a module of 3 mm in all stages.

Assumptions The largest ratio in any one gearset should be limited to about 10:1.
                 The minimum number of teeth on any pinion is 12 (from Table 11-4).

Solution

1   The required ratio is too large for one stage (one gearset), but two will each be within the 10:1 limitation. We can get an idea of the gearset ratios needed by taking the square root of the desired train ratio: $(29)^{0.5} = 5.385$. Thus, two gearsets with this ratio will do it.

2   Since the number of teeth on each gear must be an integer, see how close we can come to the 5.385:1 gearset ratio with integer combinations of teeth, starting with the smallest possible pinion:

$$12(5.385) = 64.622$$
$$13(5.385) = 70.007 \hspace{3em} (a)$$
$$14(5.385) = 75.392$$

The second of these will be very close to the correct ratio when rounded to an integer.

3   Try two gearsets of 13t and 70t. What will be the train ratio?

$$\left(\frac{70}{13}\right)\left(\frac{70}{13}\right) = 28.994 \hspace{3em} (b)$$

4   If this is close enough for the application, the problem is solved. The only situation in which it might not be acceptable would be where an exact ratio was required to provide a timing function.

5   Note that using two identical gearsets in a compound train automatically reverts it, allowing the input and output shafts to be concentric.

Epicyclic or Planetary Gear Trains

The conventional gear trains described in the previous sections are all one degree of freedom (DOF) devices. Another class of gear train, the **epicyclic or planetary train**, has wide application. This is a 2-DOF device. Two inputs are needed to obtain a predictable output. In some cases, such as the automotive differential, one input is provided (the driveshaft) and two frictionally coupled outputs are obtained (the two driving wheels).

Epicyclic or planetary trains have several advantages over conventional trains, among which are higher train ratios obtainable in smaller packages, reversion by default, and simultaneous, concentric, bidirectional outputs available from a single unidirectional input. These features make planetary trains popular as automatic transmissions in automobiles and trucks, etc.

Figure 11-15a shows a conventional, 1-DOF gearset in which link 1 is immobilized as the ground link. Figure 11-15b shows the same gearset with link 1 now free to rotate as an **arm** that connects the two gears. Now only the pinion pivot is grounded and the system DOF = 2. This has become an **epicyclic** train with a **sun gear** and a **planet gear** orbiting around the sun, held in orbit by the **arm**. Two inputs are required. Typically, the arm and the sun gear will each be driven in some direction at some velocities. In many cases, one of these inputs will be zero velocity, i.e., a brake applied to either the arm or the sun gear. Note that a zero-velocity input to the arm merely makes a conventional gear train out of the epicyclic train, as shown in Figure 11-15a. Thus the conventional gear train is simply a special case of the more complex epicyclic train, in which the arm is held stationary.

In the simple example of an epicyclic train in Figure 11-15, the only gear left to act as output, after putting inputs to sun and arm, is the planet. It is a bit difficult to

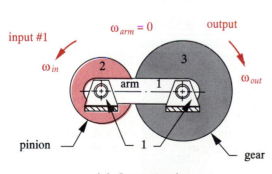

(a) Conventional gearset

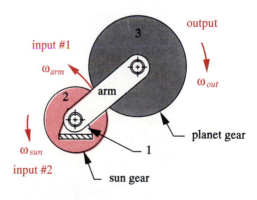

(b) Planetary or epicyclic gearset

FIGURE 11-15

Conventional Gearsets are Special Cases of Planetary or Epicyclic Gearsets get a usable output from this orbiting planet gear because its pivot is moving. A more useful configuration is shown in Figure 11-16, to which a ring gear has been added. This **ring gear** meshes with the planet and pivots concentric with the pinion, so it can be easily tapped as the output member. Most planetary trains will be arranged with ring gears to bring the planetary motion back to a grounded pivot. Note how the sun gear, ring gear, and arm are all brought out as concentric hollow shafts so that each can be accessed to tap its angular velocity and torque either as an input or an output.

While it is relatively easy to visualize the power flow through a conventional gear train and observe the directions of motion for its member gears, it is very difficult to determine the behavior of a planetary train by observation. We must do the necessary calculations to determine its behavior and may be surprised at the often counterintuitive results. Since the gears are rotating with respect to the arm and the arm itself has motion, the velocity-difference equation must be used:

$$\omega_{gear} = \omega_{arm} + \omega_{gear \, / \, arm} \qquad\qquad (11.10)$$

Equations 11.10 and 11.5a are all that are needed to solve for the velocities in an epicyclic train, provided that the tooth numbers and two input conditions are known. Rearrange equation 11.10 to solve for the velocity difference term. Then, let ω_F represent the angular velocity of the first gear in the train (chosen at either end), and ω_L represent the angular velocity of the last gear in the train (at the other end).

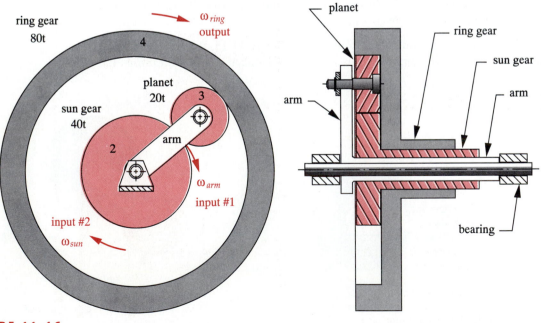

FIGURE 11-16

Planetary Gear Train with Ring Gear Used as Output

For the first gear in the system:

$$\omega_{F/arm} = \omega_F - \omega_{arm} \qquad (11.11a)$$

For the last gear in the system:

$$\omega_{L/arm} = \omega_L - \omega_{arm} \qquad (11.11b)$$

Dividing the last by the first:

$$\frac{\omega_{L/arm}}{\omega_{F/arm}} = \frac{\omega_L - \omega_{arm}}{\omega_F - \omega_{arm}} = m_V \qquad (11.11c)$$

This gives an expression for the overall train ratio m_V. The left-most side of equation 11.11c involves only the velocity-difference terms, which are relative to the arm. This fraction is equal to the ratio of the products of tooth numbers of the gears from first to last in the train (as defined in equation 11.9b), which can be substituted for the left-most side of equation 11.11c.

$$\pm\frac{\text{product of number of teeth on driver gears}}{\text{product of number of teeth on driven gears}} = \frac{\omega_L - \omega_{arm}}{\omega_F - \omega_{arm}} \qquad (11.12)$$

Equation 11.12 can be solved for any one of the three variables on the right-hand side provided that the other two are defined as inputs to this 2-DOF gear train. Either the velocities of the arm and one gear must be known or the velocities of two gears, the first and last (as so designated), must be known. Another limitation of this method is that both the first and last gears chosen must be pivoted to ground (not orbiting), and there must be a path of meshes connecting them, which may include orbiting planet gears.

EXAMPLE 11-3

Analyzing an Epicyclic Gear Train

Problem   Determine the train ratio for the epicyclic train shown in Figure 11-16.

Given   The sun has 40 teeth, the planet 20 teeth, and the ring gear 80 teeth. The arm is the input and the sun is the output. The ring gear is held stationary.

Assumptions   The sun gear is the first gear in the train and the ring gear is the last. Let the arm have an ω of 1 rpm. The train ratio desired is sun/arm.

Solution   See the *TKSolver* file EX11-03.

1   Equation 11.12 defines the kinematics of an epicyclic train:

$$\left(-\frac{N_2}{N_3}\right)\left(+\frac{N_3}{N_4}\right) = \frac{\omega_L - \omega_{arm}}{\omega_F - \omega_{arm}}$$

$$\left(-\frac{40}{20}\right)\left(+\frac{20}{80}\right) = \frac{0-1}{\omega_F - 1} \qquad\qquad (a)$$

$$\omega_F = 3$$

which defines the train ratio as +3. The sun gear rotates three times as fast and in the same direction as the arm. Note the signs on the gearset ratios. One is an external set (−) and one an internal set (+).

11.6   GEAR MANUFACTURING

There are several methods used to manufacture gears. They can be divided into two categories, forming and machining. Machining further divides into roughing and finishing operations. Forming refers to the direct casting, molding, drawing, or extrusion of tooth forms in molten, powdered, or heat-softened materials. Roughing and finishing are material removal techniques used to cut or grind the tooth shape into a solid blank at room temperature. Roughing methods are often used alone without any subsequent finishing operation for nonprecision gears. Despite its name, the roughing processes actually create a smooth and accurate gear tooth. Only when high precision and quiet running are required is the added cost of secondary finishing operations justified.

Forming Gear Teeth

In all tooth-forming operations, the teeth on the gear are formed all at once from a mold or die into which the tooth shapes have been machined. The accuracy of the teeth is entirely dependent on the quality of the die or mold and in general is much less than can be obtained from roughing or finishing methods. Most of these methods have high tooling costs, making them suitable only for high production quantities. See Chapter 2 for a more general discussion of these manufacturing processes.

CASTING   Teeth can be sand cast or die cast in various metals. The advantage is low cost as the tooth shape is built into the mold. No finishing operations on the teeth are typically done after casting, though they could be. The resulting teeth are of low precision and only used in noncritical applications such as toys and small appliances or cement-mixer barrels where noise and excessive backlash are not detrimental to operation. Sand casting is an economical way to obtain low-quality gear teeth in small quantities since the tooling costs are reasonable, but surface finish and dimensional accuracy are very poor. Die casting provides a better surface finish and accuracy than sand casting but has high tooling costs, requiring large production volume to justify its use.

11

INVESTMENT CASTING also known as lost-wax casting, can provide reasonably accurate gears in a wide variety of materials. The mold is made of a refractory material that allows high melt-temperature materials to be cast. Accuracy is a function of the original master pattern used to make the mold.

SINTERING Powdered metals (PM) are pressed into a gear-shaped metal mold cavity, removed and heat-treated (sintered) to increase their strength. These PM gears have similar accuracy to die-cast gears but their material properties can be controlled by mixing various metal powders. This technique is typically used for small-sized gears.

INJECTION MOLDING is used to make nonmetallic gears in various thermoplastics such as nylon and acetal. These are low-precision gears in small sizes but have the advantages of low cost and the ability to be run without lubricant at light loads.

EXTRUDING is used to form teeth on long rods, which are then cut into usable lengths and machined for bores and keyways etc. Nonferrous materials such as aluminum and copper alloys are commonly extruded rather than steels.

COLD DRAWING forms teeth on steel rods by drawing them through hardened dies. The cold-working increases strength and reduces ductility. The rods are then cut into usable lengths and machined for bores and keyways, etc.

STAMPING Sheet metal can be stamped with tooth shapes to form low-precision gears at low cost in high quantities. Surface finish and accuracy are poor.

Machining

The bulk of metal gears used to transmit power in machinery are made by a machining process from cast, forged, or hot-rolled blanks. Roughing processes include milling the tooth shape with formed cutters or generating the shape with a rack cutter, a shaper cutter, or a hob. Finishing processes include shaving, burnishing, lapping, honing, or grinding. Each of these methods will be briefly described.

Roughing Processes

FORM MILLING requires a shaped milling cutter, as shown in Figure 11-17 (labeled 1). The cutter must be made to the shape of the gear tooth space for the tooth geometry and number of teeth of each particular gear. The rotating cutter is plunged into the blank to cut one tooth at a time. The gear blank is then rotated through one circular pitch and the next tooth cut. Because a different shape cutter is needed for each gear size to be made, the tooling cost becomes high. To reduce costs, the same cutter is often used for multiple-sized gears resulting in profile errors for all but one number of teeth. This method is the least accurate of the roughing methods.

RACK GENERATION A rack cutter for any involute pitch can be easily made since its tooth shape is a trapezoid (see Figure 11-6). The hardened and sharpened rack (see

FIGURE 11-17

A Collection of Gear-Cutting Tools: 1—Milling Cutter, 2—Rack Cutter, 3—Shaper Cutter,
4—Hob *Courtesy of Pfauter-Maag Cuting Tools.Limited Partnership, Loves Park, Ill..*

2 in Figure 11-17) is then reciprocated along the axis of the gear blank and fed into it
while being rotated around the gear blank so as to generate the involute tooth on the gear.
The rack and gear blank must be periodically repositioned to complete the circumference. This repositioning can introduce errors in the tooth geometry, making this method
less accurate than others still to be discussed.

GEAR SHAPING   uses a cutting tool in the shape of a gear (see 3 in Figure 11-17),
which is reciprocated axially across the gear blank to cut the teeth while the blank rotates around the shaper tool as shown in Figure 11-18. It is a true shape-generation pro-

11

FIGURE 11-18

A Gear Shaper Cutting a Helical Gear *Courtesy of Pfauter-Maag Cutting Tools, Ltd Prtship, Loves Park, Ill.*

cess in that the gear-shaped tool cuts itself into mesh with the gear blank. The accuracy is good, but any errors in even one tooth of the shaper cutter will be directly transferred to the gear. Internal gears can be cut with this method as well.

HOBBING   A hob, labeled 4 in Figure 11-17, is analogous to a thread tap. Its teeth are shaped to match the tooth space and are interrupted with grooves to provide cutting surfaces. It rotates about an axis perpendicular to that of the gear blank, cutting into the rotating blank to generate the teeth. It is the most accurate of the roughing processes since no repositioning of tool or blank is required and each tooth is cut by multiple hob-teeth, averaging out any tool errors. Excellent surface finish can be achieved by this method and it is one of the most widely used for production gears.

Finishing Processes

When high precision is required, secondary operations can be done to gears made by any of the above roughing methods. Finishing operations typically remove little or no material but improve dimensional accuracy, surface finish, and/or hardness.

SHAVING   is similar to gear shaping, but uses accurate shaving tools to remove small amounts of material from a roughed gear to correct profile errors and improve finish.

GRINDING   uses a contoured grinding wheel that is passed over the machined surface of the gear teeth, typically computer-controlled, to remove small amounts of material and improve surface finish. It can be used on gears that have been hardened after roughing to correct the heat-treatment distortion, as well as to achieve the other advantages listed above.

BURNISHING   runs the rough-machined gear against a specially hardened gear. The high forces at the tooth interface cause plastic yielding of the gear tooth surface, which both improves finish and work hardens the surface, creating beneficial compressive residual stresses.

LAPPING AND HONING   both employ an abrasive-impregnated gear or gear-shaped tool that is run against the gear to abrade the surface. In both cases, the abrasive tool drives the gear in what amounts to an accelerated and controlled run-in to improve surface finish and accuracy.

Gear Quality

The AGMA Standard 2000-A88 defines dimensional tolerances for gear teeth and a quality index Q_v that ranges from the lowest quality (3) to the highest precision (16). The manufacturing method essentially determines the quality index Q_v of the gear.

Formed gears will typically have quality indices of 3-4. Gears made by the roughing methods listed above generally fall within a Q_v range of 5-7. If the gears are fin-

Application	Qv
Cement mixer drum drive	3-5
Cement Kiln	5-6
Steel mill drives	5-6
Corn picker	5-7
Cranes	5-7
Punch press	5-7
Mining conveyor	5-7
Paper-box making machine	6-8
Gas meter mechanism	7-9
Small power drill	7-9
Clothes washing machine	8-10
Printing press	9-11
Computing mechanism	10-11
Automotive transmission	10-11
Radar antenna drive	10-12
Marine propusion drive	10-12
Aircraft engine drive	10-13
Gyroscope	12-14

11

Table 11-7
Recommended Gear
Quality Numbers for
Pitch Line Velocity

Pitch Velocity	Qv
0-800 fpm	6-8
800-2000 fpm	8-10
2000-4000 fpm	10-12
Over 4000 fpm	12-14

ished by shaving or grinding, Q_v can be in the 8-11 range. Lapping and honing can achieve higher-quality indices. Obviously, the cost of the gear will be a function of Q_v.

Table 11-6 shows the quality indices recommended by AGMA for a number of common applications of gears. Another way to select a suitable quality index is based on the linear velocity of the gear teeth at the pitch point, called the pitch-line velocity. Inaccuracies in tooth spacing will cause impacts between teeth and impact forces are increased at higher velocities. Table 11-7 shows recommended gear-quality indices Q_v as a function of the pitch-line velocity of the gear mesh. Spur gears are seldom used with pitch-line velocities greater than 10 000 ft/min (50 m/s) due to excessive noise and vibration. Helical gears (discussed in the next chapter) are preferred in such applications.

Gear quality can have a significant effect on the load sharing between teeth. If the tooth spacings are not accurate and uniform, the teeth in the mesh will not all be in simultaneous contact. This will obviate the advantage of a large contact ratio. Figure 11-19 shows two gears with large contact ratio but low accuracy. Only one pair of teeth are in contact and taking load in the same direction. The others in the mesh are taking no load. Despite an apparent contact ratio of about 5, the actual contact ratio at this point in the mesh is only 1.

11.7 LOADING ON SPUR GEARS

The analysis of loading on meshing gear teeth can be done by the standard methods of load analysis, as discussed in Chapter 3 using equations 3.2 or 3.3 as appropriate. We will only briefly discuss their application to gear teeth. Figure 11-20 shows a pair of gear teeth. The teeth are actually meshed (in contact) at the pitch point, but are shown separated for clarity. A torque T_p is being delivered by the pinion to the gear. Both are shown as free-body diagrams. At the pitch point, the only force that can be transmitted from one tooth to the other, neglecting friction, is the force W acting along the line of action at the pressure angle. This force can be resolved into two components, W_r acting in the radial direction and W_t in the tangential direction. The magnitude of the tangential component W_t can be found from

$$W_t = \frac{T_p}{r_p} = \frac{2T_p}{d_p} = \frac{2p_d T_p}{N_p} \qquad (11.13a)$$

where T_p is the torque on the pinion shaft, r_p is the pitch radius, d_p the pitch diameter, N_p the number of teeth and p_d the diametral pitch of the pinion.

The radial component W_r is

$$W_r = W_t \tan \phi \qquad (11.13b)$$

and the resultant force W is

$$W = \frac{W_t}{\cos \phi} \qquad (11.13c)$$

Note that equation 11.13a could as well have been written for the gear rather than for the pinion since the force W is equal and opposite on gear and pinion.

The reaction force R, and its components R_t, and R_r at the pivots are equal and opposite to the corresponding forces acting at the pitch point of the respective gear or pinion. The pinion forces are equal and opposite to those acting on the gear.

Depending on the contact ratio, the teeth can take all or part of the load W at any location from the tip of the tooth to a point near the dedendum circle as it rotates through the mesh. Obviously, the worst loading condition is when W acts at the tooth tips. Then, its tangential component W_t has the largest possible moment arm acting on the tooth as a cantilever beam. The bending moment and the transverse shear force due to bending will both be maximum at the root of the tooth. For contact ratios > 1, there will be a highest point of single-tooth contact (HPSTC) somewhere below the tip and this will create the maximum bending moment on any one tooth provided that the gears' accuracies are good enough to allow load sharing. If the teeth are of low quality, as shown in Figure 11-19, then tip loading with the full value of W will occur regardless of the contact ratio.

Even if the torque T_p is constant with time, each tooth will experience repeated loading as it comes through the mesh, creating a fatigue-loading situation. The bending moment-time function for a gearset will be as shown in Figure 11-21a. There are equal mean (M_m) and alternating (M_a) components of bending moment. There is some advantage in avoiding integer values of gear ratio m_G for gearsets in order to prevent the same teeth from contacting one another every m_G revolutions. Noninteger ratios will distribute the contact more evenly among all the teeth.

FIGURE 11-19

Actual Spur-Gear Teeth in Mesh Showing Poor Load Sharing Due to Tooth Inaccuracy Source: R. L. Thoen, Minimizing Backlash in Spur Gears, *Gear Technology*, May/June 1994, p. 27, with permission.

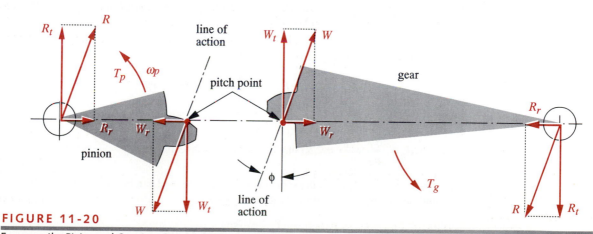

FIGURE 11-20

Forces on the Pinion and Gear in a Gearset (Gears Separated for Illustration—Pitch Points are Actually in Contact)

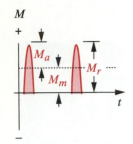

(a) Repeated moment on nonidler tooth

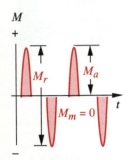

(b) Reversed moment on idler tooth

FIGURE 11-21

Time-Varying Bending Moments on Gear Teeth

11

If an idler gear is inserted between the pinion and gear to change the output direction, each of the idler's teeth will experience a fully reversed moment, as shown in Figure 11-21b, because the force W acts on opposite sides of each idler tooth in alternate meshes. Note that the range M_r of the moment magnitude on the idler is twice that of the nonidler gears, making it a more heavily loaded gear even though its mean moment is zero. The same is true for planet gears, as shown in Figure 11-16.

EXAMPLE 11-4

Load Analysis of a Spur-Gear Train

Problem   Determine the torques and transmitted loads on the gear teeth in a 3-gear train containing a pinion, an idler gear, and a gear. Find the gear diameters and the mean and alternating components of transmitted load on each gear.

Given   The pinion shaft passes 20 hp at 2 500 rpm. The train ratio is 3.5:1. The pinion has 14 teeth, a 25° pressure angle, and $p_d = 6$. The idler has 17 teeth. *units = 1/inch*

Assumptions   The pinion meshes with the idler and the idler meshes with the gear.

Solution   See the *TKSolver* file EX11-04.

1 Find the number of gear teeth from the given information:

$$N_g = m_G N_p = 3.5(14) = 49 \text{ teeth} \qquad (a)$$

2 The torque on the pinion shaft is found from equation 9.1:

$$T_p = \frac{P}{\omega_p} = \frac{20 \text{ hp} \left(6\,600 \; \frac{\text{in - lb}}{\text{sec}} \Big/ \text{hp} \right)}{2\,500 \text{ rpm} \; (2\pi/60) \; \frac{\text{rad}}{\text{sec}} \Big/ \text{rpm}} = 504 \text{ lb - in} \qquad (b)$$

3 The output torque is

$$T_g = m_G T_p = 3.5(504) = 1\,765 \text{ lb - in} \qquad (c)$$

4 The pitch diameters are

$$d_p = \frac{N_p}{p_d} = \frac{14}{6} = 2.33 \text{ in}, \qquad d_i = \frac{17}{6} = 2.83 \text{ in} \qquad d_g = \frac{49}{6} = 8.17 \text{ in} \qquad (d)$$

5 The transmitted load is the same on all three gears and can be found from the torque and radius of any one of the gears:

$$W_t = \frac{T_p}{d_p/2} = \frac{504}{2.33/2} = 432 \text{ lb} \qquad (e)$$

6 The radial component of load is

$$W_r = W_t \tan\phi = 432 \tan 25° = 202 \text{ lb} \tag{f}$$

7 The total load is

$$W = \frac{W_t}{\cos\phi} = \frac{432}{\cos 25°} = 477 \text{ lb} \tag{g}$$

8 The repeated loads on any pinion or gear tooth are

$$W_{t_{alternating}} = \frac{W_t}{2} = 216 \text{ lb} \qquad W_{t_{mean}} = \frac{W_t}{2} = 216 \text{ lb} \tag{h}$$

9 The fully reversed loads on the idler are

$$W_{t_{alternating}} = W_t = 432 \text{ lb} \qquad W_{t_{mean}} = 0 \text{ lb} \tag{i}$$

11.8 STRESSES IN SPUR GEARS

There are two modes of failure that affect gear teeth, **fatigue fracture** due to fluctuating bending stresses at the root of the tooth and **surface fatigue** (pitting) of the tooth surfaces. Both failure modes must be checked when designing gears. Fatigue fracture due to bending can be prevented with proper design by keeping the stress state within the modified-Goodman line for the material as discussed in Chapter 6. Since most heavily loaded gears are made of ferrous materials that do have a bending endurance limit, infinite life can be obtained for the bending loads. However, as was discussed in Chapter 7, materials do not exhibit an endurance limit for repeated surface-contact stresses. Therefore, it is not possible to design gears for infinite life against surface failure. Properly designed gearsets should never fracture a tooth in normal service (barring overloads greater than they were designed for) but must be expected to eventually fail by one of the wear mechanisms discussed in Chapter 7. Pitting is the most common mode of failure, though abrasive or adhesive wear (scuffing or scoring) may occur, especially if the gears are not properly lubricated in service. We will address each of the two principal failure modes in turn, using the AGMA recommended procedures.

Bending Stresses

THE LEWIS EQUATION The first useful equation for the bending stress in a gear tooth was developed by W. Lewis in 1892. He recognized that the tooth is a cantilever beam with its critical section at the root. Starting with the equation for bending stress in a cantilever beam, he derived what is now known as the Lewis equation:

$$\sigma_b = \frac{W_t \, p_d}{FY} \tag{11.14}$$

11

where W_t is the tangential force on the tooth, p_d the diametral pitch, F the face width, and Y is a dimensionless geometry factor, defined by him, and now called the Lewis form factor. His form factor took the tooth geometry into account to determine its effective strength at the root fillet. He published a table of Y values for gears of different pressure angles and numbers of teeth.[5] Note that the radial component W_r is ignored as it puts the tooth in compression and acts to reduce the dangerous tensile bending stress. Ignoring the radial stress is thus conservative and also simplifies the analysis.

The Lewis equation is no longer used in its original form, but it serves as the basis for a more modern version as defined by the AGMA and based on the work of Lewis and many others. The principles of Lewis' equation are still valid, but it has been augmented with additional factors to account for failure mechanisms only later understood. His form factor Y has been supplanted by a new geometry factor J, which includes the effects of stress concentration at the root fillet[3]. Stress concentration was still waiting to be discovered in Lewis's day.

THE AGMA BENDING STRESS EQUATION   as defined in AGMA Standard 2001-B88 is valid only for certain assumptions about the tooth and gear-mesh geometry:

1   The contact ratio is between 1 and 2.

2   There is no interference between the tips and root fillets of mating teeth and no undercutting of teeth above the theoretical start of the active profile.

3   No teeth are pointed.

4   There is nonzero backlash.

5   The root fillets are standard, assumed smooth, and produced by a generating process.

6   The friction forces are neglected.

The first assumption comes about in spite of the theoretical desirability of high contact ratios because the actual load-sharing between teeth in such situations is subject to factors of tooth accuracy and stiffness that are difficult to predict, making the problem indeterminate. Assumption 1 is then conservative with larger contact ratios. Assumption 2 limits the analysis to pinion-gear combinations that obey the minimum-tooth limitations described in Tables 11-4 and 11-5. If smaller numbers of teeth are needed for packaging purposes, then unequal-addendum teeth should be used and the AGMA method applied with the appropriate geometry factor J used in the equation. Assumption 3 deals with the limits of unequal-addendum pinions. Assumption 4 recognizes that gears with zero backlash will not run freely together due to excessive friction. Assumption 5 accounts for the use of stress concentration factors for root fillets based on work by Dolan and Broghammer.[6] Assumption 6 is self-explanatory. Also, this method is only valid for external gear teeth. The geometry of internal gear teeth is sufficiently different to require another approach to the calculation of bending stresses. See the AGMA standard for more information.

The AGMA bending stress equation differs slightly for U.S. and SI specification gears due to the reciprocal relationship between diametral pitch and module. We will list both versions with suffixes of *us* or *si* on the equation numbers where applicable.

$$\sigma_b = \frac{W_t\, p_d}{FJ}\, \frac{K_a K_m}{K_v}\, K_s K_B K_I \qquad (11.15us)$$

$$\sigma_b = \frac{W_t}{FmJ}\, \frac{K_a K_m}{K_v}\, K_s K_B K_I \qquad (11.15si)$$

The core of this equation is Lewis's formula with the updated geometry factor J substituted for his form factor Y. W_t, F, and p_d have the same meanings as in equation 11.14 and m is the metric module. The K factors are modifiers to account for various conditions. We will now discuss each of the empirical terms in equation 11.15.

BENDING STRENGTH GEOMETRY FACTOR J   The geometry factor J can be calculated from a complicated algorithm defined in AGMA Standard 908-B89. The same standard also provides tables of J factors for standard, full-depth teeth and for 25% and 50% unequal addendum teeth, all with 14.5, 20, and 25° pressure angles. These J factors vary with the numbers of teeth on the pinion and gear and are given only for a range of combinations which obey assumption 2 above. The AGMA recommends that tooth-number combinations that create interference be avoided.

Tables 11-8 through 11-15[*] replicate the AGMA geometry factors J for a subset of the gear-tooth combinations covered in the standard. In these eight tables, two gear-tooth designs are covered (the full-depth tooth, and the 25% long-addendum tooth), each for two pressure angles (20 and 25°), and for both tip loading and for loading at the highest point of single-tooth contact (HPSTC). See the standard for other combinations.

Note in these tables that the J factors are different for the pinion and gear (labeled P and G) in each mesh combination. This results in different bending stress levels in the pinion teeth than in the gear teeth. The letter U in the tables indicates that undercutting occurs with that combination due to interference between the tip of the gear tooth and the root-flank of the pinion. The choice between tip-loaded or HPSTC J factors should be based on the manufacturing precision of the gearset. If the manufacturing tolerances are small (high-precision gears) then load-sharing between the teeth can be assumed and the HPSTC tables used. If not, then it is likely that only one pair of teeth will take all the load at the tip in the worst case, as seen in Figure 11-19. See the AGMA Standard 908-B89 for more information on acceptable manufacturing variations in base pitch to ensure HPSTC.

DYNAMIC FACTOR K_v   The dynamic factor K_v attempts to account for internally generated vibration loads from tooth-tooth impacts induced by nonconjugate meshing of the gear teeth. These vibration loads are called **transmission error** and will be worse with low accuracy gears. Precision gears will more closely approach the ideal of smooth, constant-velocity-ratio torque transmission. In the absence of test data that define the

* Extracted from AGMA Standard 908-B89, INFORMATION SHEET, *Geometry Factors for Determining the Pitting Resistance and Bending Strength of Spur, Helical, and Herringbone Gear Teeth*, with the permission of the publisher, American Gear Manufacturers Association, 1500 King St., Suite 201, Alexandria, Va., 22134.

Table 11-8   AGMA Bending Geometry Factor J for 20°, Full-Depth Teeth with Tip Loading

Gear teeth	Pinion teeth															
	12		14		17		21		26		35		55		135	
	P	G	P	G	P	G	P	G	P	G	P	G	P	G	P	G
12	U	U														
14	U	U	U	U												
17	U	U	U	U	U	U										
21	U	U	U	U	U	U	0.24	0.24								
26	U	U	U	U	U	U	0.24	0.25	0.25	0.25						
35	U	U	U	U	U	U	0.24	0.26	0.25	0.26	0.26	0.26				
55	U	U	U	U	U	U	0.24	0.28	0.25	0.28	0.26	0.28	0.28	0.28		
135	U	U	U	U	U	U	0.24	0.29	0.25	0.29	0.26	0.29	0.28	0.29	0.29	0.29

Table 11-9   AGMA Bending Geometry Factor J for 20°, Full-Depth Teeth with HPSTC Loading

Gear teeth	Pinion teeth															
	12		14		17		21		26		35		55		135	
	P	G	P	G	P	G	P	G	P	G	P	G	P	G	P	G
12	U	U														
14	U	U	U	U												
17	U	U	U	U	U	U										
21	U	U	U	U	U	U	0.33	0.33								
26	U	U	U	U	U	U	0.33	0.35	0.35	0.35						
35	U	U	U	U	U	U	0.34	0.37	0.36	0.38	0.39	0.39				
55	U	U	U	U	U	U	0.34	0.40	0.37	0.41	0.40	0.42	0.43	0.43		
135	U	U	U	U	U	U	0.35	0.43	0.38	0.44	0.41	0.45	0.45	0.47	0.49	0.49

Table 11-10   AGMA Bending Geometry Factor J for 20°, 25% Long-Addendum Teeth with Tip Loading

Gear teeth	Pinion teeth															
	12		14		17		21		26		35		55		135	
	P	G	P	G	P	G	P	G	P	G	P	G	P	G	P	G
12	U	U														
14	U	U	U	U												
17	U	U	U	U	0.27	0.19										
21	U	U	U	U	0.27	0.21	0.27	0.21								
26	U	U	U	U	0.27	0.22	0.27	0.22	0.28	0.22						
35	U	U	U	U	0.27	0.24	0.27	0.24	0.28	0.24	0.28	0.24				
55	U	U	U	U	0.27	0.26	0.27	0.26	0.28	0.26	0.28	0.26	0.29	0.26		
135	U	U	U	U	0.27	0.28	0.27	0.28	0.28	0.28	0.28	0.28	0.29	0.28	0.30	0.28

Table 11-11   AGMA Bending Geometry Factor J for 20°, 25% Long-Addendum Teeth with HPSTC Loading

Gear teeth	Pinion teeth															
	12		14		17		21		26		35		55		135	
	P	G	P	G	P	G	P	G	P	G	P	G	P	G	P	G
12	U	U														
14	U	U	U	U												
17	U	U	U	U	0.36	0.24										
21	U	U	U	U	0.37	0.26	0.39	0.27								
26	U	U	U	U	0.37	0.29	0.39	0.29	0.41	0.30						
35	U	U	U	U	0.37	0.32	0.40	0.32	0.41	0.33	0.43	0.34				
55	U	U	U	U	0.38	0.35	0.40	0.36	0.42	0.36	0.44	0.37	0.47	0.39		
135	U	U	U	U	0.39	0.39	0.41	0.40	0.43	0.41	0.45	0.42	0.48	0.44	0.51	0.46

Table 11-12   AGMA Bending Geometry Factor J for 25°, Full-Depth Teeth with Tip Loading

Gear teeth	Pinion teeth															
	12		14		17		21		26		35		55		135	
	P	G	P	G	P	G	P	G	P	G	P	G	P	G	P	G
12	U	U														
14	U	U	0.28	0.28												
17	U	U	0.28	0.30	0.30	0.30										
21	U	U	0.28	0.31	0.30	0.31	0.31	0.31								
26	U	U	0.28	0.33	0.30	0.33	0.31	0.33	0.33	0.33						
35	U	U	0.28	0.34	0.30	0.34	0.31	0.34	0.33	0.34	0.34	0.34				
55	U	U	0.28	0.36	0.30	0.36	0.31	0.36	0.33	0.36	0.34	0.36	0.36	0.36		
135	U	U	0.28	0.38	0.30	0.38	0.31	0.38	0.33	0.38	0.34	0.38	0.36	0.38	0.38	0.38

Table 11-13   AGMA Bending Geometry Factor J for 25°, Full-Depth Teeth with HPSTC Loading

Gear teeth	Pinion teeth															
	12		14		17		21		26		35		55		135	
	P	G	P	G	P	G	P	G	P	G	P	G	P	G	P	G
12	U	U														
14	U	U	0.33	0.33												
17	U	U	0.33	0.36	0.36	0.36										
21	U	U	0.33	0.39	0.36	0.39	0.39	0.39								
26	U	U	0.33	0.41	0.37	0.42	0.40	0.42	0.43	0.43						
35	U	U	0.34	0.44	0.37	0.45	0.40	0.45	0.43	0.46	0.46	0.46				
55	U	U	0.34	0.47	0.38	0.48	0.41	0.49	0.44	0.49	0.47	0.50	0.51	0.51		
135	U	U	0.35	0.51	0.38	0.52	0.42	0.53	0.45	0.53	0.48	0.54	0.53	0.56	0.57	0.57

11

Table 11-14 AGMA Bending Geometry Factor J for 25°, 25% Long-Addendum Teeth with Tip Loading

Gear teeth	Pinion teeth															
	12		14		17		21		26		35		55		135	
	P	G	P	G	P	G	P	G	P	G	P	G	P	G	P	G
12	0.32	0.20														
14	0.32	0.22	0.33	0.22												
17	0.32	0.25	0.33	0.25	0.34	0.25										
21	0.32	0.27	0.33	0.27	0.34	0.27	0.36	0.27								
26	0.32	0.29	0.33	0.29	0.34	0.29	0.36	0.29	0.36	0.29						
35	0.32	0.31	0.33	0.31	0.34	0.31	0.36	0.31	0.36	0.31	0.37	0.31				
55	0.32	0.34	0.33	0.34	0.34	0.34	0.36	0.34	0.36	0.34	0.37	0.34	0.38	0.34		
135	0.32	0.37	0.33	0.37	0.34	0.37	0.36	0.37	0.36	0.37	0.37	0.37	0.38	0.37	0.39	0.37

Table 11-15 AGMA Bending Geometry Factor J for 25°, 25% Long-Addendum Teeth with HPSTC Loading

Gear teeth	Pinion teeth															
	12		14		17		21		26		35		55		135	
	P	G	P	G	P	G	P	G	P	G	P	G	P	G	P	G
12	0.38	0.22														
14	0.38	0.25	0.40	0.25												
17	0.38	0.29	0.40	0.29	0.43	0.29										
21	0.38	0.32	0.41	0.32	0.43	0.33	0.46	0.33								
26	0.39	0.35	0.41	0.35	0.44	0.36	0.46	0.36	0.48	0.37						
35	0.39	0.38	0.41	0.39	0.44	0.39	0.47	0.40	0.49	0.41	0.51	0.41				
55	0.39	0.42	0.42	0.43	0.44	0.44	0.47	0.44	0.49	0.45	0.52	0.46	0.55	0.47		
135	0.40	0.47	0.42	0.48	0.45	0.49	0.48	0.49	0.50	0.50	0.53	0.51	0.56	0.53	0.59	0.55

11

level of transmission error to be expected in a particular gear design, the designer must estimate the dynamic factor. The AGMA provides empirical curves for K_v as a function of pitch-line velocity V_t. Figure 11-22[*] shows a family of these curves, which vary with the quality index Q_v of the gearset. Empirical equations for the curves numbered 6 through 11 in Figure 11-22 are

* Extracted from AGMA Standard 2001-B88, *Fundamental Rating Factors and Calculation Methods for Involute Spur and Helical Gear Teeth*, with the permission of the publisher, American Gear Manufacturers Association, 1500 King St., Suite 201, Alexandria, Va., 22134.

$$K_v = \left(\frac{A}{A + \sqrt{V_t}} \right)^B \qquad (11.16us)$$

$$K_v = \left(\frac{A}{A + \sqrt{200 V_t}} \right)^B \qquad (11.16si)$$

where V_t is the pitch-line velocity of the gear mesh in units of ft/min (U.S.) or m/s (SI). The factors A and B are defined as

$$A = 50 + 56(1 - B) \tag{11.17a}$$

$$B = \frac{(12 - Q_v)^{2/3}}{4} \quad \text{for} \quad 6 \le Q_v \le 11 \tag{11.17b}$$

in which Q_v is the quality index of the lower-quality gear in the mesh.

Note in Figure 11-22 that these empirical curves each end abruptly at particular values of V_t. They can be extrapolated beyond those points but the experimental data from which they were generated did not extend beyond those limits. The terminal values of V_t for each curve can be calculated from

$$V_{t_{max}} = \left[A + (Q_v - 3)\right]^2 \quad \text{ft/min} \tag{11.18us}$$

$$V_{t_{max}} = \frac{\left[A + (Q_v - 3)\right]^2}{200} \quad \text{m/s} \tag{11.18si}$$

For gears with $Q_v \le 5$, a different equation is used:

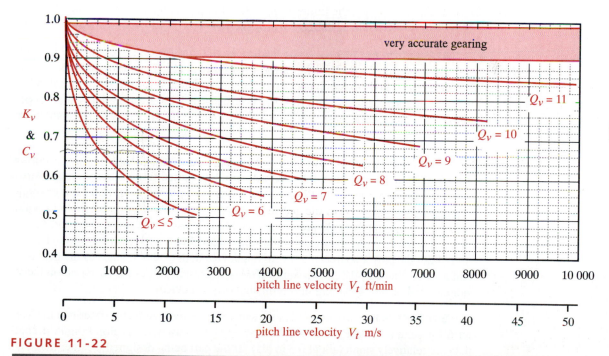

FIGURE 11-22

AGMA Dynamic Factors K_v and C_v

Table 11-16
Load Distribution
Factors K_m

Face Width		K_m
in	(mm)	
<2	(50)	1.6
6	(150)	1.7
9	(250)	1.8
≥20	(500)	2.0

$$K_v = \frac{50}{50 + \sqrt{V_t}} \qquad (11.19us)$$

$$K_v = \frac{50}{50 + \sqrt{200V_t}} \qquad (11.19si)$$

This relationship is only valid for $V_t \leq 2\,500$ ft/min (13 m/s) as can be seen from, the $Q_v = 5$ line in Figure 11-22. Above that velocity gears of higher Q_v should be used. See Table 11-7.

Mark[7] derives a method for computing the transmission error in parallel-axis gears that takes into account bearing misalignment, dynamic shaft misalignment, variations in tooth spacing, tooth profile modifications, and the stiffness of the structures supporting the bearings. If the actual dynamic loading due to transmission errors is known and is taken into account by increasing the applied load W_t, then the dynamic factor K_v can be set to 1.

LOAD DISTRIBUTION FACTOR K_M Any axial misalignment or axial deviation in tooth form will cause the transmitted load W_t to be unevenly distributed over the face width of the gear teeth. This problem becomes more pronounced with larger face widths. An approximate and conservative way to account for less than uniform load distribution is to apply the factor K_m to increase the stresses for larger face widths. Some suggested values are shown in Table 11-16. A useful rule of thumb is to keep the face width F of a spur gear within the limits $8 / p_d < F < 16 / p_d$, with a nominal value of $12 / p_d$. This ratio is referred to as the **face width factor**.

APPLICATION FACTOR K_A The loading model discussed in Section 11-7 assumed that the transmitted load W_t was uniform with time. The fluctuating moments on the teeth described in that section are due to the teeth coming into and out of mesh under a uniform or average load. If either the driving or driven machine has time-varying torques or forces, then these will increase the loading felt by the gear teeth above the average values.

In the absence of definitive information about the dynamic loads in the driving and driven machines, an application factor K_a can be applied to increase the tooth stress based on the "shockiness" of the machinery connected to the gear train. For example, if the gear train connects an electric motor to a centrifugal water pump (both of which are smooth-running devices) there is no need to increase the average loads and $K_a = 1$. But, if a single-cylinder, internal-combustion engine drives a rock crusher through a gear train, both the power source and the driven device deliver shock loads to the gear teeth and $K_a > 1$. Table 11-17 shows some AGMA-suggested values for K_a based on the assumed level of shock loading in driving and driven devices.

SIZE FACTOR K_S can be used in the same way as the size factor described in Chapter 6 for general fatigue loading. The test specimens used to develop fatigue strength data are relatively small (about 0.3 in dia). If the part being designed is larger than that,

* Extracted from AGMA Standard 2001-B88, *Fundamental Rating Factors and Calculation Methods for Involute Spur and Helical Gear Teeth*, with the permission of the publisher, American Gear Manufacturers Association, 1500 King St., Suite 201, Alexandria, Va., 22134.

11

Table 11-17 Application Factors K_a

Driving Machine	Driven Machine		
	Uniform	Moderate Shock	Heavy Shock
Uniform (Electric motor, turbine)	1.00	1.25	1.75 or higher
Light Shock (Multicylinder engine)	1.25	1.50	2.00 or higher
Medium Shock (Single-cylinder engine)	1.50	1.75	2.25 or higher

it may be weaker than indicated by the test data. The K_s factor allows a modification of the tooth stress to account for such situations. However, much of the available gear strength data have been developed from tests of actual gear teeth and thus better represent reality than the general strength data of Chapter 6. The AGMA has not yet established standards for size factors and recommends that K_s be set to 1 unless the designer wishes to raise its value to account for particular situations, such as very large teeth. A value of 1.25 to 1.5 would be a conservative assumption in such cases.

RIM THICKNESS FACTOR K_B   This factor was recently introduced by the AGMA to account for situations in which a large diameter gear, made with a rim and spokes rather than as a solid disk, has a thin rim depth in comparison to the tooth depth. Such designs can fail with a radial fracture across the rim rather than through a tooth root. The AGMA defines a **backup ratio m_B** as

$$m_B = \frac{t_R}{h_t} \qquad (11.20a)$$

where t_R is the rim thickness from the tooth root diameter to the rim's inside diameter and h_t is the whole depth of the tooth (the sum of addendum and dedendum) as shown in Figure 11-23.[*] This ratio is used to define the rim thickness factor from

$$K_B = -2m_B + 3.4 \qquad 0.5 \le m_B \le 1.2$$
$$K_B = 1.0 \qquad\qquad m_B > 1.2 \qquad (11.20b)$$

Backup ratios < 0.5 are not recommended. Solid-disk gears will always have $K_B = 1$.

IDLER FACTOR K_I   An idler gear is subjected both to more cycles of stress per unit time and to larger-magnitude alternating loads than its nonidler cousins. To account for this situation, the factor K_I is set to 1.42 for an idler gear or 1.0 for a nonidler gear. The AGMA uses the reciprocal of this factor to reduce the apparent strength of the material for an idler gear, but that is not consistent with the approach used in this text of applying factors that affect the stress state of a part to the stress equation, not to the material's strength.

FIGURE 11-23

Parameters for AGMA Rim Thickness Factor K_B

11

* Extracted from AGMA Standard 2001-B88, *Fundamental Rating Factors and Calculation Methods for Involute Spur and Helical Gear Teeth*, with the permission of the publisher, American Gear Manufacturers Association, 1500 King St., Suite 201, Alexandria, Va., 22134.

EXAMPLE 11-5

Bending Stress Analysis of a Spur-Gear Train

Problem   Determine a suitable face width and the bending stresses in the gear teeth of the 3-gear train from Example 11-4.

Given   The transmitted load on the teeth is 432 lb. The pinion has 14 teeth, a 25° pressure angle, and $p_d = 6$. The idler has 17 teeth and the gear 49 teeth. Pinion speed is 2 500 rpm. See Example 11-3 for other dimensional information.

Assumptions   The teeth are standard AGMA full-depth profiles. The load and source are both uniform in nature. A gear quality index of 6 will be used.

Solution   See the *TKSolver* file EX11-05.

1   Even though the transmitted load is the same, the bending stress in the teeth of each size gear will be different because of its slightly different tooth geometry. The general formula for tooth-bending stress is equation 11.15:

$$\sigma_b = \frac{W_t\,P_d}{FJ}\frac{K_a K_m}{K_v}K_s K_B K_I \qquad (a)$$

W_t, p_d, F, K_a, K_m, K_v, and K_s are common to all gears in the set and J, K_B and K_I are potentially different for each gear.

2   A first approximation of the face width can be estimated as a function of the diametral pitch. Take the middle of the recommended face width factor range $8 / p_d < F < 16 / p_d$ for a first calculation:

$$F \cong \frac{12}{p_d} = \frac{12}{6} = 2 \text{ in} \qquad (b)$$

3   Based on the assumption of uniform load and source, the application factor K_a can be set to 1.

4   The load distribution factor can be estimated from Table 11-16 based on the assumed face width: $K_m = 1.6$.

5   The velocity factor K_v can be calculated from equations 11.16 and 11.17 based on the assumed gear-quality index Q_v and the pitch-line velocity V_t.

$$V_t = \frac{d_p}{2}\omega_p = \frac{2.33 \text{ in}}{2(12)}(2\,500 \text{ rpm})(2\pi) = 1\,527\,\frac{\text{ft}}{\text{min}} \qquad (c)$$

$$B = \frac{(12 - Q_v)^{2/3}}{4} = \frac{(12 - 6)^{2/3}}{4} = 0.826 \qquad (d)$$

$$A = 50 + 56(1 - B) = 50 + 56(1 - 0.826) = 59.745 \qquad (e)$$

$$K_v = \left(\frac{A}{A + \sqrt{V_t}}\right)^B = \left(\frac{59.745}{59.745 + \sqrt{1\,527}}\right)^{0.826} = 0.660 \qquad (f)$$

6.   The size factor $K_s = 1$ for all three gears.

7.   These gears are all too small to have a rim and spokes, so $K_B = 1$.

8.   The idler factor $K_I = 1$ for the pinion and gear and $K_I = 1.42$ for the idler gear.

9.   The bending geometry factor J for the 25°, 14-tooth pinion in mesh with the 17-tooth idler is found from Table 11-13 to be $J_{pinion} = 0.33$. The pinion-tooth bending stress is then

$$\sigma_{b_p} = \frac{W_t \, P_d}{FJ} \frac{K_a K_m}{K_v} K_s K_B K_I = \frac{432(6)}{2(0.33)} \frac{1(1.6)}{0.66}(1)(1)(1) = 9\,526 \text{ psi} \qquad (g)$$

10.   The bending geometry factor J for the 25°, 17-tooth idler in mesh with the 14-tooth pinion is found from Table 11-13 to be $J_{idler} = 0.36$. The idler-tooth bending stress is then

$$\sigma_{b_i} = \frac{W_t \, P_d}{FJ} \frac{K_a K_m}{K_v} K_s K_B K_I = \frac{432(6)}{2(0.36)} \frac{1(1.6)}{0.66}(1)(1)(1.42) = 12\,400 \text{ psi} \qquad (h)$$

Note in Table 11-13 that the idler has a different J factor when considered to be the "gear" in mesh with the smaller pinion (0.36) than when considered to be the "pinion" in mesh with the larger gear (0.37). The smaller value of the two is used because it gives the higher stress.

11.   The bending geometry factor J for the 25°, 49-tooth gear in mesh with the 17-tooth idler is found (by interpolation) from Table 11-13 to be $J_{gear} = 0.46$. The gear tooth bending stress is then

$$\sigma_{b_g} = \frac{W_t \, P_d}{FJ} \frac{K_a K_m}{K_v} K_s K_B K_I \cong \frac{432(6)}{2(0.46)} \frac{1(1.6)}{0.66}(1)(1)(1) \cong 6\,834 \text{ psi} \qquad (i)$$

12.   If this is an acceptable stress level then the assumed face width can be used. This issue will be revisited for this design in a later example.

Surface Stresses

Mating gear teeth have a combination of rolling and sliding at their interface. At the pitch point, their relative motion is pure rolling. The percentage of sliding increases with distance from the pitch point. An average value of 9% sliding is sometimes taken to represent the combined roll-slide motion between the teeth.[8] The stresses at the tooth surface are dynamic Hertzian contact stresses in combined rolling and sliding, as defined in Section 7.11. These stresses are three-dimensional and have peak values either at the surface or slightly below it depending on the amount of sliding present in combination with rolling. Depending on the surface velocity, tooth radii of curvature, and lubricant viscosity, a condition of full or partial elastohydrodynamic (EHD) lubrication or of boundary lubrication may exist in the interface as described in Chapter 10. If sufficient, clean lubricant of an appropriate type is provided in order to create at least partial EHD lubrication (specific film thickness $\Lambda > 2$) and prevent surface failure by the adhesive, abrasive, or corrosive mechanisms described in Chapter 7, the ultimate failure mode will be pitting and spalling due to surface fatigue. See Section 7.7 for a discussion of this mechanism and Figure 7-12 for examples of gear tooth surface failure.

The surface stresses in gear teeth were first investigated in a systematic way by Buckingham[9] who recognized that two cylinders having the same radius of curvature as the gear teeth at the pitch point, and radially loaded in rolling contact, could be used to simulate gear tooth contact while controlling the necessary variables. His work led to the development of an equation for surface stresses in gear teeth that is now known as the Buckingham equation. It serves as the basis for the AGMA pitting resistance formula[2], which is

$$\sigma_c = C_p \sqrt{\frac{W_t}{FId} \frac{C_a C_m}{C_v} C_s C_f} \qquad (11.21)$$

where W_t is the tangential force on the tooth, d the pitch diameter, F the face width, and I is a dimensionless **surface geometry factor** for pitting resistance. C_p is an **elastic coefficient** that accounts for differences in the gear and pinion material constants. The factors C_a, C_m, C_v, and C_s are equal, respectively, to K_a, K_m, K_v, and K_s as defined for the bending stress equation 11.15. The new factors I, C_p, and C_f will now be defined.

SURFACE GEOMETRY FACTOR I   This factor takes into account the radii of curvature of the gear teeth and the pressure angle. AGMA defines an equation for I:

$$I = \frac{\cos \phi}{\left(\dfrac{1}{\rho_p} \pm \dfrac{1}{\rho_g} \right) d_p} \qquad (11.22a)$$

where ρ_p and ρ_g are the radii of curvature of the pinion and gear teeth, respectively, ϕ is the pressure angle, and d_p is the pitch diameter of the pinion. The $\pm$ sign accounts for external and internal gearsets. Use the upper sign for external gearsets in all related expressions. The radii of curvature of the teeth are calculated from the mesh geometry:

$$\rho_p = \sqrt{\left(r_p + \frac{1+x_p}{P_d} \right)^2 - \left(r_p \cos \phi \right)^2} - \frac{\pi}{P_d} \cos \phi$$

$$\rho_g = C \sin \phi \mp \rho_p \qquad (11.22b)$$

where p_d is the diametral pitch, r_p is the pitch radius of the pinion, ϕ is the pressure angle, C is the center distance between pinion and gear, and x_p is the pinion addendum coefficient, which is equal to the decimal percentage of addendum elongation for unequal addendum teeth. For standard, full-depth teeth, $x_p = 0$. For 25% long-addendum teeth, $x_p = 0.25$, etc. Note the choice of sign in the second equation of 11.22b. Use the upper sign for an external gearset and the lower one for an internal gearset.

ELASTIC COEFFICIENT C_p   The elastic coefficient accounts for differences in tooth materials and is found from

$$C_p = \sqrt{\frac{1}{\pi\left[\left(\dfrac{1-v_p^2}{E_p}\right)+\left(\dfrac{1-v_g^2}{E_g}\right)\right]}} \qquad (11.23)$$

where E_p and E_g are, respectively, the moduli of elasticity for pinion and gear, and v_p and v_g are their respective Poisson's ratios. The units of C_p are either $(psi)^{0.5}$ or $(MPa)^{0.5}$. Table 11-18[*] shows values of C_p for various combinations of common gear and pinion materials based on an assumed $v = 0.3$ for all the materials.

SURFACE FINISH FACTOR C_F   is used to account for unusually rough surface finishes on gear teeth. The AGMA has not yet established standards for surface-finish factors and recommends that C_f be set to 1 for gears made by conventional methods. Its value can be increased to account for unusually rough surface finishes or for the known presence of detrimental residual stresses.

Table 11-18   AGMA Elastic Coefficient C_p in Units of $[psi]^{0.5}$ ($[MPa]^{0.5}$)[*†]

Pinion Material	E_p psi (MPa)	Gear Material					
		Steel	Malleable Iron	Nodular Iron	Cast Iron	Aluminum Bronze	Tin Bronze
Steel	30E6 (2E5)	2 300 (191)	2 180 (181)	2 160 (179)	2 100 (174)	1 950 (162)	1 900 (158)
Malleable Iron	25E6 (1.7E5)	2 180 (181)	2 090 (174)	2 070 (172)	2 020 (168)	1 900 (158)	1 850 (154)
Nodular Iron	24E6 (1.7E5)	2 160 (179)	2 070 (172)	2 050 (170)	2 000 (166)	1 880 (156)	1 830 (152)
Cast Iron	22E6 (1.5E5)	2 100 (174)	2 020 (168)	2 000 (166)	1 960 (163)	1 850 (154)	1 800 (149)
Aluminum Bronze	17.5E6 (1.2E5)	1 950 (162)	1 900 (158)	1 880 (156)	1 850 (154)	1 750 (145)	1 700 (141)
Tin Bronze	16E6 (1.1E5)	1 900 (158)	1 850 (154)	1 830 (152)	1 800 (149)	1 700 (141)	1 650 (137)

[†]The values of E_p in this table are approximate and $v = 0.3$ was used as an approximation of Poisson's ratio for all materials. If more accurate numbers are available for E_p and v, they should be used in equation 11.23 to obtain C_p.

[*] Extracted from AGMA Standard 2001-B88, *Fundamental Rating Factors and Calculation Methods for Involute Spur and Helical Gear Teeth*, with the permission of the publisher, American Gear Manufacturers Association, 1500 King St., Suite 201, Alexandria, Va., 22134.

EXAMPLE 11-6

Surface-Stress Analysis of a Spur-Gear Train

Problem　　Determine the surface stresses in the gear teeth of the 3-gear train from Examples 11-4 and 11-5.

Given　　The transmitted load on the teeth is 432 lb. The pinion has 14 teeth, a 25° pressure angle, and $p_d = 6$. The idler has 17 teeth and the gear 49 teeth. Pinion speed is 2 500 rpm. Face width is 2 in. See Example 11-3 for other dimensional information.

Assumptions　　The teeth are standard AGMA full-depth profiles. The load and source are both uniform in nature. A gear-quality index of 6 will be used. All gears are steel with $\nu = 0.28$.

Solution　　See the *TKSolver* file EX11-06.

1　The general formula for tooth-surface stress is equation 11.21:

$$\sigma_c = C_p \sqrt{\frac{W_t}{FId} \frac{C_a C_m}{C_v} C_s C_f} \qquad (a)$$

W_t, F, C_a, C_m, C_v, and C_s are common to all gears in the set and d and C_f are potentially different for each gear. C_p and I are potentially different for each pair in mesh.

2　The face width can be estimated as a function of the diametral pitch. Take the middle of the recommended range $8/p_d < F < 16/p_d$ for a first calculation:

$$F \cong \frac{12}{p_d} = \frac{12}{6} = 2 \text{ in} \qquad (b)$$

3　Based on the assumption of uniform load and source, the application factor C_a can be set to 1.

4　The load distribution factor can be estimated from Table 11-16 based on the assumed face width: $C_m = K_m = 1.6$. So $C_v = K_v = 1.66$

5　The velocity factor C_v can be calculated from equations 11.16 and 11.17 based on the assumed gear-quality index Q_v and the pitch-line velocity V_t.

$$V_t = \frac{d_p}{2}\omega_p = \frac{2.33 \text{ in}}{2(12)}(2\,500 \text{ rpm})(2\pi) = 1\,527 \frac{\text{ft}}{\text{min}} \qquad (c)$$

$$B = \frac{(12 - Q_v)^{2/3}}{4} = \frac{(12 - 6)^{2/3}}{4} = 0.826 \qquad (d)$$

$$A = 50 + 56(1 - B) = 50 + 56(1 - 0.826) = 59.745 \qquad (e)$$

11

$$C_v = \left(\frac{A}{A + \sqrt{V_t}}\right)^B = \left(\frac{59.745}{59.745 + \sqrt{1\,527}}\right)^{0.826} = 0.660 \qquad (f)$$

6   The size factor $C_s = 1$ for all three gears.

7   The surface factor $C_f = 1$ for well-finished gears made by conventional methods.

8   The elastic coefficient C_p is found from equation 11.23.

$$C_p = \sqrt{\frac{1}{\pi\left[\left(\dfrac{1-v_p^2}{E_p}\right) + \left(\dfrac{1-v_g^2}{E_g}\right)\right]}} = \sqrt{\frac{1}{\pi\left[\left(\dfrac{1-0.28^2}{30E6}\right) + \left(\dfrac{1-0.28^2}{30E6}\right)\right]}} = 2\,276 \qquad (g)$$

9   The pitting geometry factor I is calculated for a pair of gears in mesh. Since we have two meshes (pinion/idler and idler/gear) there will be two different values of I to be calculated using equations 11.22. We will need the pitch diameter and pitch radius of each gear for this calculation. From the data in Example 11-4:

$$\begin{array}{ccc} d_p = 2.333 & d_i = 2.833 & d_g = 8.167 \\[2mm] r_p = 1.167 & r_i = 1.417 & r_g = 4.083 \end{array} \qquad (h)$$

10   For the pinion/idler pair, let $I_{pi} = I$, $d_1 = d_p$, $r_1 = r_p$, and $r_2 = r_i$, then

$$\begin{aligned} \rho_1 &= \sqrt{\left(r_1 + \frac{1}{P_d}\right)^2 - (r_1 \cos\phi)^2} - \frac{\pi}{P_d}\cos\phi \\ &= \sqrt{\left(1.167 + \frac{1}{6}\right)^2 - (1.167\cos 25°)^2} - \frac{\pi}{6}\cos 25° = 0.338 \text{ in} \end{aligned} \qquad (i)$$

$$\begin{aligned} \rho_2 &= C\sin\phi - \rho_1 = (r_1 + r_2)\sin\phi - \rho_1 \\ &= (1.167 + 1.417)\sin 25° - 0.338 = 0.754 \text{ in} \end{aligned} \qquad (j)$$

$$I_{pi} = \frac{\cos\phi}{\left(\dfrac{1}{\rho_1} \pm \dfrac{1}{\rho_2}\right)d_1} = \frac{\cos 25°}{\left(\dfrac{1}{0.338} + \dfrac{1}{0.754}\right)2.33} = 0.091 \qquad (k)$$

11   For the idler/gear pair, let $I_{ig} = I$, $d_1 = d_i$, $r_1 = r_i$, and $r_2 = r_g$, then

$$\begin{aligned} \rho_1 &= \sqrt{\left(r_1 + \frac{1}{P_d}\right)^2 - (r_1 \cos\phi)^2} - \frac{\pi}{P_d}\cos\phi \\ &= \sqrt{\left(1.417 + \frac{1}{6}\right)^2 - (1.417\cos 25°)^2} - \frac{\pi}{6}\cos 25° = 0.452 \text{ in} \end{aligned} \qquad (l)$$

11

$$\rho_2 = C\sin\phi - \rho_1 = (r_1 + r_2)\sin\phi - \rho_1$$
$$= (1.417 + 4.083)\sin 25° - 0.452 = 1.872 \text{ in} \qquad (m)$$

$$I_{ig} = \frac{\cos\phi}{\left(\dfrac{1}{\rho_1} \pm \dfrac{1}{\rho_2}\right)d_1} = \frac{\cos 25°}{\left(\dfrac{1}{0.452} + \dfrac{1}{1.872}\right)2.83} = 0.116 \qquad (n)$$

12 The surface stress for the pinion teeth is then

$$\sigma_{c_p} = C_p\sqrt{\frac{W_t}{FI_{pi}d_p}\frac{C_aC_m}{C_v}C_sC_f}$$

$$= 2\,276\sqrt{\frac{432}{2(0.091)(2.33)}\frac{1(1.6)}{0.66}(1)(1)} = 113\,315 \text{ psi} \qquad (o)$$

13 The surface stress for the idler teeth is then

$$\sigma_{c_i} = C_p\sqrt{\frac{W_t}{FI_{pi}d_p}\frac{C_aC_m}{C_v}C_sC_f}$$

$$= 2\,276\sqrt{\frac{432}{2(0.091)(2.83)}\frac{1(1.6)}{0.66}(1)(1)} = 102\,831 \text{ psi} \qquad (p)$$

14 The surface stress for the gear teeth is then

$$\sigma_{c_g} = C_p\sqrt{\frac{W_t}{FI_{ig}d_p}\frac{C_aC_m}{C_v}C_sC_f}$$

$$= 2\,276\sqrt{\frac{432}{2(0.116)(8.17)}\frac{1(1.6)}{0.66}(1)(1)} = 53\,422 \text{ psi} \qquad (q)$$

11

11.9   GEAR MATERIALS

Only a limited number of metals and alloys are suitable for gears that transmit significant power. Table 11-18 shows some of them. Steels, cast irons, and malleable and
nodular irons are the most common choices for gears. Surface- or through-hardening
is recommended (on those alloys that allow it) to obtain sufficient strength and wear resistance. Where high corrosion resistance is needed, such as in marine environments,
bronzes are often used. The combination of a bronze gear and a steel pinion has advantages in terms of material compatibility and conformity, as discussed in Chapter 7
and this combination is often used in nonmarine applications as well.

CAST IRONS   are commonly used for gears. The gray cast irons (CI) have advantages of low cost, ease of machining, high wear resistance, and internal damping (due to the graphite inclusions), which makes them acoustically quieter than steel gears. However, they have low tensile strength, which requires larger teeth than steel gears to obtain sufficient bending strength. Nodular irons have higher tensile strength than gray CI and retain the other advantages of machinability, wear resistance, and internal damping, but are more costly. The combination of a steel pinion (for strength in the higher-stressed member) and a cast-iron gear is often used.

STEELS   are also commonly used for gears. They have superior tensile strength to cast iron and are cost-competitive in their low-alloy forms. They need heat treatment to get a surface hardness that will resist wear, but soft steel gears are sometimes used in low-load, low-speed applications or where long life may not be a prime concern. For heat treatment, either a medium-to-high carbon (0.35 to 0.60% C) plain or alloy steel is needed. Small gears are typically through-hardened and larger gears flame or induction hardened to minimize distortion. Lower-carbon steels can be case hardened by carburizing or nitriding. A case-hardened gear has the advantage of a tough core and a hard surface, but if the case is not deep enough, the teeth may fail in bending fatigue beneath the case in the soft, weaker core material. It is often necessary to use secondary finishing methods such as grinding, lapping, and honing to remove the heat-treatment distortion from hardened gears if high accuracy is needed.

BRONZES   are the most common nonferrous metals used for gears. The lower modulus of elasticity of these copper alloys provides greater tooth deflection and improves load sharing between the teeth. Since bronze and steel run well together, the combination of a steel pinion and a bronze gear is often used.

NONMETALLIC GEARS   are often made of injection-molded thermoplastics such as nylon and acetal, sometimes filled with inorganics such as glass or talc. Teflon is sometimes added to nylon or acetal to lower the coefficient of friction. Dry lubricants such as graphite and molybdenum disulphide (MoS_2) can be added to the plastic to allow dry running. Composite gears of cloth-reinforced thermosetting phenolic have long been used for applications such as the camshaft-drive (timing) gear driven by a steel pinion in some gasoline engines. Nonmetallic gears have very low noise but are limited in torque capacity by their low material strengths.

Material Strengths

Since both of the gear-failure modes involve fatigue loading, material fatigue strength data is needed, both for bending stresses and for surface contact stresses. The methods for estimating fatigue strength outlined in Chapter 6 could be used for gear applications since the principles involved are the same. However, better data are available on the fatigue strengths of gear alloys because of the extensive testing programs that have been done for this application over the past century. Test data for fatigue strengths of most gear materials have been compiled by the AGMA. As stated in Section 6.6 of this text:

> The best information on a material's fatigue strength at some finite life, or its endurance limit at infinite life, comes from the testing of actual or prototype assemblies of the design . . .. If published data are available for the fatigue strength S_f, or endurance limit S_e, of the material, they should be used . . ..

So, it would not make sense to start by assuming an uncorrected fatigue strength as a fraction of the static ultimate tensile strength and then reducing it by the collection of correction factors outlined in Section 6.6 if we have more nearly correct fatigue-strength data available.

AGMA Bending-Fatigue Strengths for Gear Materials

The published AGMA data for both bending- and surface-fatigue strengths are, in effect, partially-corrected fatigue strengths since they are generated with appropriately sized parts having the same geometry, surface finish, etc., as the gears to be designed. The AGMA refers to material strengths as *allowable stresses,* which is not consistent with our procedure of reserving the term *stress* for the results of applied loading and using the term *strength* to refer to material properties. For internal consistency within this text we will designate the published AGMA bending-fatigue strength data as S_{fb}', to differentiate it from the completely uncorrected fatigue strength S_f' of Chapter 6. There are still three correction factors that need to be applied to the published AGMA bending-fatigue strength data in order to obtain what we will designate as the corrected bending-fatigue strength for gears S_{fb}.

The AGMA bending-fatigue strength data are all stated at 1E7 cycles of repeated stress (rather than the 1E6 or 1E8 cycles sometimes used for other materials), and for a 99% reliability level (rather than the 50% reliability common for general fatigue- and static-strength data). These strengths are compared to the peak stress σ_b calculated from equation 11.15 using the load W_t. The Goodman-line analysis is encapsulated in this direct comparison because the strength data are obtained from a test that provides a fluctuating stress state identical to that of the actual gear loading.

The correction formula for the bending-fatigue strength of gears is

$$S_{fb} = \frac{K_L}{K_T K_R} S_{fb}'$$ (11.24)

where S_{fb}' is the published AGMA bending-fatigue strength as defined above, S_{fb} is the corrected strength and the K factors are modifiers to account for various conditions. These modifiers will now be defined and briefly discussed.

LIFE FACTOR K_L   Since the test data are for a life of 1E7 cycles, a shorter or longer cycle life will require modification of the bending-fatigue strength based on the *S-N* relationship for the material. The number of load cycles in this case is defined as the number of mesh contacts, under load, of the gear tooth being analyzed. Figure 11-24[*] shows *S-N* curves for the bending-fatigue strength of steels having several different tensile strengths as defined by their Brinell hardness numbers. Curve-fitted equations are also

11

* Extracted from AGMA Standard 2001-B88, *Fundamental Rating Factors and Calculation Methods for Involute Spur and Helical Gear Teeth*, with the permission of the publisher, American Gear Manufacturers Association, 1500 King St., Suite 201, Alexandria, Va., 22134.

used

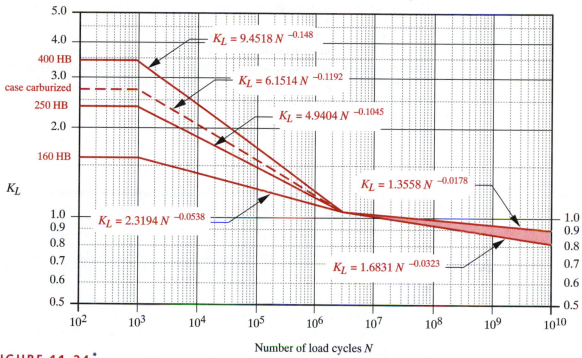

FIGURE 11-24 [*]

AGMA Bending Strength Life Factor K_L

shown in the figure for each S-N line. These equations can be used to compute the appropriate K_L factor for a required number of load cycles N. The AGMA suggests that:

> The upper portion of the shaded zone can be used for commercial applications. The lower portion of the shaded zone is typically used for critical service applications where little pitting and tooth wear is permissible and where smoothness of operation and low vibration levels are required.

Unfortunately, similar data have not yet been developed for gear materials other than these steels.

TEMPERATURE FACTOR K_T   The lubricant temperature is a reasonable measure of gear temperature. For steel materials in oil temperatures up to about 250°F, K_T can be set to 1. For higher temperatures, K_T can be estimated from

$$K_T = \frac{460 + T_F}{620} \tag{11.24a}$$

where T_F is the oil temperature in °F. Do not use this relationship for materials other than steel.

RELIABILITY FACTOR K_R   The AGMA strength data are based on a statistical probability of 1 failure in 100 samples, or a 99% reliability. If this is satisfactory, set $K_R = 1$.

[*] Extracted from AGMA Standard 2001-B88, *Fundamental Rating Factors and Calculation Methods for Involute Spur and Helical Gear Teeth*, with the permission of the publisher, American Gear Manufacturers Association, 1500 King St., Suite 201, Alexandria, Va., 22134.

11

$$S = k_L \frac{S_b}{k_T k_R}$$

Table 11-19
AGMA Factor K_R

Reliability %	K_R
90	0.85
99	1.00
99.9	1.25
99.99	1.50

If either a higher or lower reliability factor is desired, K_R can be set to one of the values in Table 11-19.[*]

BENDING-FATIGUE STRENGTH DATA　Table 11-20 shows AGMA bending-fatigue strengths for a number of commonly-used gear materials. The AGMA standard also defines heat treatment specifications where applicable. A plot showing ranges of AGMA bending-fatigue strengths for steels as a function of their Brinell hardness is shown in Figure 11-25.[*] See the referenced standard for the metallurgical properties required for AGMA grades of steels. To achieve the strength values in Table 11-20 and Figure 11-25, the material should be specified to comply with that standard.

Table 11-20　AGMA Bending-Fatigue Strengths S_{fb}' for a Selection of Gear Materials[*]

Material	AGMA Class	Material Designation	Heat Treatment	Minimum Surface Hardness	Bending-Fatigue Strength psi x 10^3	MPa
Steel	A1-A5		Through-hardened	≤ 180 HB	25-33	170-230
			Through-hardened	240 HB	31-41	210-280
			Through-hardened	300 HB	36-47	250-325
			Through-hardened	360 HB	40-52	280-360
			Through-hardened	400 HB	42-56	290-390
			Flame- or induction-hardened	Type A pattern 50-54 HRC	45-55	310-380
			Flame- or induction-hardened	Type B pattern	22	150
			Carburized and case hardened	55-64HRC	55-75	380-520
		AISI 4140	Nitrided	84.6 15N	34-45	230-310
		AISI 4340	Nitrided	83.5 15N	36-47	250-325
		Nitralloy 135M	Nitrided	90.0 15N	38-48	260-330
		Nitralloy N	Nitrided	90.0 15N	40-50	280-345
		2.5% Chrome	Nitrided	87.5-90.0 15N	55-65	380-450
Cast iron	20	Class 20	As cast		5	35
	30	Class 30	As cast	175 HB	8	69
	40	Class 40	As cast	200 HB	13	90
Nodular (ductile) iron	A-7-a	60-40-18	Annealed	140 HB	22-33	150-230
	A-7-c	80-55-06	Quenched and tempered	180 HB	22-33	150-230
	A-7-d	100-70-03	Quenched and tempered	230 HB	27-40	180-280
	A-7-e	120-90-02	Quenched and tempered	230 HB	27-40	180-280
Malleable iron (pearlitic)	A-8-c	45007		165 HB	10	70
	A-8-e	50005		180 HB	13	90
	A-8-f	53007		195 HB	16	110
	A-8-i	80002		240 HB	21	145
Bronze	Bronze 2	AGMA 2C	Sand cast	40 ksi min tensile strength	5.7	40
	Al/Br 3	ASTM B-148 78 alloy 954	Heat-treated	90 ksi min tensile strength	23.6	160

used

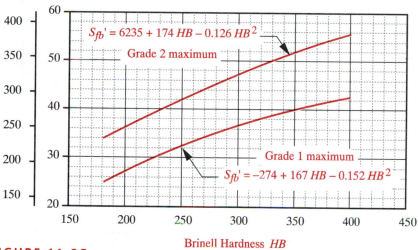

FIGURE 11-25

AGMA Bending-Fatigue Strengths S_{fb}' for Steels[*]

AGMA Surface-Fatigue Strengths for Gear Materials

We will designate the published AGMA surface-fatigue strength data as S_{fc}'. There are four correction factors that need to be applied to the published AGMA data in order to obtain what we will designate as the corrected surface-fatigue strength for gears S_{fc},

$$S_{fc} = \frac{C_L C_H}{C_T C_R} S_{fc}' \tag{11.25}$$

where S_{fc}' is the published surface-fatigue strength as defined above, S_{fc} is the corrected strength and the C factors are modifiers to account for various conditions. The factors C_T and C_R are identical, respectively, to K_T and K_R and can be chosen as described in the previous section. The life factor C_L has the same purpose as K_L in equation 11.24 but references a different S-N diagram. C_H is a hardness-ratio factor for pitting resistance. These two different factors will now be defined.

SURFACE-LIFE FACTOR C_L   Since the published surface-fatigue test data are for a life of 1E7 cycles, a shorter or longer cycle life will require modification of the surface-fatigue strength based on the S-N relationship for the material. The number of load cycles is defined as the number of mesh contacts, under load, of the gear tooth being analyzed. Figure 11-26[*] shows S-N curves for the surface-fatigue strength of steels. Curve-fitted equations are also shown in the figure for the S-N lines. These equations can be used to compute the appropriate C_L factor for a required number of load cycles N. The AGMA suggests that: "The upper portion of the shaded zone can be used for commercial applications. The lower portion of the shaded zone is typically used for criti-

[*] Extracted from AGMA Standard 2001-B88, *Fundamental Rating Factors and Calculation Methods for Involute Spur and Helical Gear Teeth*, with the permission of the publisher, American Gear Manufacturers Association, 1500 King St., Suite 201, Alexandria, VA 22134.

11

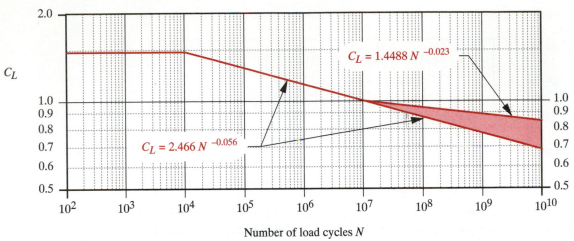

FIGURE 11-26 [*]

AGMA Surface-Fatigue Strength Life Factor C_L

cal service applications where little pitting and tooth wear is permissible and where smoothness of operation and low vibration levels are required." Unfortunately, similar data have not yet been developed for gear materials other than these steels.

HARDNESS RATIO FACTOR C_H   This factor is a function of the gear ratio and the relative hardness of pinion and gear. C_H is in the numerator of equation 11.25 and is always ≥ 1.0, so it acts to increase the apparent strength of the gear. It accounts for situations in which the pinion teeth are harder than the gear teeth and thus act to work harden the gear-tooth surfaces when run-in. C_H is only applied to the gear-tooth strength, not to the pinion. Two formulas for its calculation are suggested in the standard. The choice of one versus the other depends on the relative hardness of pinion and gear teeth.

For through-hardened pinions running against through-hardened gears:

$$C_H = 1 + A(m_G - 1) \tag{11.26a}$$

where m_G is the gear ratio and A is found from

$$\text{if} \quad \frac{HB_p}{HB_g} < 1.2 \quad \text{then } A = 0 \tag{11.26b}$$

$$\text{if} \quad 1.2 \leq \frac{HB_p}{HB_g} \leq 1.7 \quad \text{then } A = 0.008\,98\,\frac{HB_p}{HB_g} - 0.008\,29 \tag{11.26c}$$

$$\text{if} \quad \frac{HB_p}{HB_g} > 1.7 \quad \text{then } A = 0.006\,98 \tag{11.26d}$$

where HB_p and HB_g are the Brinell harnesses of pinion and gear, respectively.

[*] Extracted from AGMA Standard 2001-B88, *Fundamental Rating Factors and Calculation Methods for Involute Spur and Helical Gear Teeth*, with the permission of the publisher, American Gear Manufacturers Association, 1500 King St., Suite 201, Alexandria, VA 22134.

For surface-hardened pinions (>48 HRC) run against through-hardened gears, C_H is found from

$$C_H = 1 + B(450 - HB_g)$$ (11.27)

$$B = 0.000\ 75\ e^{-0.0112 R_q}$$ (11.28us)

$$B = 0.000\ 75\ e^{-0.052 R_q}$$ (11.28si)

where R_q is the rms surface roughness of the pinion teeth (see Section 7.1).

Table 11-21[*] shows AGMA surface-fatigue strengths for a number of commonly used gear materials. The AGMA standard defines the heat-treatment specifications for the case-hardened steels. A plot showing ranges of AGMA surface-fatigue strengths for steels as a function of their Brinell hardness is shown in Figure 11-27.[*] See the referenced standard for the metallurgical properties required for AGMA Grades 1, 2, and 3 steels. To achieve the strength values in Table 11-21 and Figure 11-27, the material should be specified to comply with that standard.

EXAMPLE 11-7

Material Selection and Safety Factor for Spur Gears

Problem   Select suitable materials and calculate the safety factors for both bending and surface stresses in the 3-gear train from Examples 11-4, 11-5 and 11-6.

Given   The stresses are as calculated in Examples 11-5 and 11-6.

Assumptions   The service life required is 5 years of one-shift operation. All gears are steel. Operating temperature is 200 °F.

Solution   See the *TKSolver* file EX11-07.

1   An estimate of the uncorrected bending-fatigue strength can be made from the curves of Figure 11-25. We will try an AGMA Grade 2 steel, through-hardened to 250HB. The uncorrected fatigue strength in bending is found from the upper curve of the figure to be

$$S_{fb}' = 6\ 235 + 174 HB - 0.126 Hb^2$$

$$= 6\ 235 + 174(250) - 0.126(250)^2 = 41\ 860 \text{ psi}$$ (a)

2   This value needs to be corrected for certain factors using equation 11.24.

3   The life factor K_L is found from the appropriate equation in Figure 11-24 based on the required number of cycles in the life of the gears. The pinion sees the largest

[*] Extracted from AGMA Standard 2001-B88, *Fundamental Rating Factors and Calculation Methods for Involute Spur and Helical Gear Teeth*, with the permission of the publisher, American Gear Manufacturers Association, 1500 King St., Suite 201, Alexandria, Va., 22134.

11

Table 11-21 AGMA Surface-Fatigue Strengths S_{fc}' for a Selection of Gear Materials[*]

Material	AGMA Class	Material Designation	Heat Treatment	Minimum Surface Hardness	Surface-Fatigue Strength psi x 10^3	Surface-Fatigue Strength MPa
Steel	A1-A5		Through-hardened	≤ 180 HB	85-95	590-660
			Through-hardened	240 HB	105-115	720-790
			Through-hardened	300 HB	120-135	830-930
			Through-hardened	360 HB	145-160	1000-1100
			Through-hardened	400 HB	155-170	1100-1200
			Flame- or induction-hardened	50 HRC	170-190	1200-1300
			Flame- or induction-hardened	54 HRC	175-195	1200-1300
			Carburized and case hardened	55-64HRC	180-225	1250-1300
		AISI 4140	Nitrided	84.6 15N	155-180	1100-1250
		AISI 4340	Nitrided	83.5 15N	150-175	1050-1200
		Nitralloy 135M	Nitrided	90.0 15N	170-195	1170-1350
		Nitralloy N	Nitrided	90.0 15N	195-205	1340-1410
		2.5% Chrome	Nitrided	87.5 15N	155-172	1100-1200
		2.5% Chrome	Nitrided	90.0 15N	192-216	1300-1500
Cast iron	20	Class 20	As cast		50-60	340-410
	30	Class 30	As cast	175 HB	65-70	450-520
	40	Class 40	As cast	200 HB	75-85	520-590
Nodular (ductile) iron	A-7-a	60-40-18	Annealed	140 HB	77-92	530-630
	A-7-c	80-55-06	Quenched and tempered	180 HB	77-92	530-630
	A-7-d	100-70-03	Quenched and tempered	230 HB	92-112	630-770
	A-7-e	120-90-02	Quenched and tempered	230 HB	103-126	710-870
Malleable iron (pearlitic)	A-8-c	45007		165 HB	72	500
	A-8-e	50005		180 HB	78	540
	A-8-f	53007		195 HB	83	570
	A-8-i	80002		240 HB	94	650
Bronze	Bronze 2	AGMA 2C	Sand cast	40 ksi min tensile strength	30	450
	Al/Br 3	ASTM B-148 78 alloy 954	Heat-treated	90 ksi min tensile strength	65	450

number of repeated tooth-loadings, so we calculate the life based on it. First, calculate the number of cycles N for the required life of 5 years, one shift.

$$N = 2\,500 \text{ rpm} \left(\frac{60 \text{ min}}{\text{hr}} \right)\left(\frac{2\,080 \text{ hr}}{\text{shift - yr}} \right)(5 \text{ yr})(1 \text{ shift}) = 1.56E9 \text{ cycles} \qquad (c)$$

The value of K_L is found from

$$K_L = 1.3558 N^{-0.0178} = 1.3558(1.56E9)^{-0.0178} = 0.9302 \qquad (d)$$

4 At the specified operating temperature, $K_T = 1$.

[*] Extracted from AGMA Standard 2001-B88, *Fundamental Rating Factors and Calculation Methods for Involute Spur and Helical Gear Teeth*, with the permission of the publisher, American Gear Manufacturers Association, 1500 King St., Suite 201, Alexandria, Va., 22134.

11

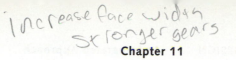
Increase face width
stronger gears

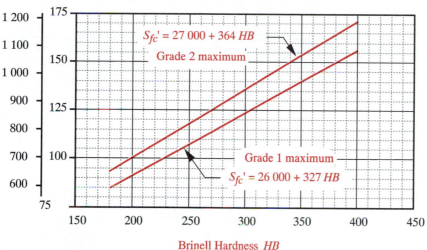

FIGURE 11-27

AGMA Surface-Fatigue Strengths S_{fc}' for Steels[*]

5 The gear-material data are all taken at a reliability level of 99%. This is satisfactory in this case, making $K_R = 1$.

6 The corrected bending-fatigue strength is then

$$S_{fb} = \frac{K_L}{K_T K_R} S_{fb}' = \frac{0.9302}{1(1)} 41\,860 = 38\,937 \text{ psi} \qquad (e)$$

7 An estimate of the uncorrected surface fatigue strength can be made from the curves of Figure 11-27. For an AGMA Grade 2 steel, through-hardened to 250HB, the strength is found from the upper curve of the figure to be

$$S_{fc}' = 27\,000 + 364HB = 27\,000 + 364(250) = 118\,000 \text{ psi} \qquad (f)$$

8 This value needs to be corrected for certain factors using equation 11.25:

$$S_{fc} = \frac{C_L C_H}{C_T C_R} S_{fc}' \qquad (g)$$

9 The life factor C_L is found from the appropriate equation in Figure 11-26 based on the required number of cycles N found above.

$$C_L = 1.4488 N^{-0.023} = 1.4488(1.56E9)^{-0.023} = 0.8904 \qquad (h)$$

10 $C_T = K_T = 1$ and $C_R = K_R = 1$.

11 Since the gears and pinion are of the same hardness material in this case, $C_H = 1$.

12 The corrected surface-fatigue strength is then

$$S_{fc} = \frac{C_L C_H}{C_T C_R} S_{fc}{}' = \frac{0.8904(1)}{1(1)} 118\,000 = 105\,063 \text{ psi} \qquad (i)$$

13 The safety factor against bending failure is found by comparing the corrected bending strength to the bending stress for each gear in the mesh:

$$N_{b_{pinion}} = \frac{S_{fb}}{\sigma_{b_{pinion}}} = \frac{38\,937}{9\,526} = 4.1 \qquad (j)$$

$$N_{b_{idler}} = \frac{S_{fb}}{\sigma_{b_{idler}}} = \frac{38\,937}{12\,400} = 3.1 \qquad (k)$$

$$N_{b_{gear}} = \frac{S_{fb}}{\sigma_{b_{gear}}} = \frac{38\,937}{6\,834} = 5.7 \qquad (l)$$

which are acceptable.

13 The safety factor against surface failure is found by comparing the corrected surface strength to the surface stress for each gear in the mesh:

$$N_{c_{pinion}} = \frac{S_{fc}}{\sigma_{c_{pinion}}} = \frac{105\,063}{113\,315} = 0.93 \qquad (m)$$

$$N_{c_{idler}} = \frac{S_{fc}}{\sigma_{c_{idler}}} = \frac{105\,063}{102\,831} = 1.0 \qquad (n)$$

$$N_{c_{gear}} = \frac{S_{fc}}{\sigma_{c_{gear}}} = \frac{105\,063}{53\,422} = 2.0 \qquad (o)$$

which are low for the pinion and idler.

14 A small change to the design will improve these. Increasing the face width of the gears from the current 2.0 to 2.5 in ($15 / p_d$) reduces all stresses and gives new safety factors of

$$N_{b_{pinion}} = 5.1 \qquad N_{b_{idler}} = 3.9 \qquad N_{b_{gear}} = 7.1$$
$$\qquad (p)$$
$$N_{c_{pinion}} = 1.0 \qquad N_{c_{idler}} = 1.1 \qquad N_{c_{gear}} = 2.2$$

15 These gears are very safe against tooth breakage and based on the assumptions and calculations, should have a 99% probability of lasting the required 5 years before pitting of the pinion or idler begins.

11.10 LUBRICATION OF GEARING

With the exception of lightly loaded plastic gears, all gearsets must be lubricated to avoid premature failure from one of the surface-failure modes discussed in Chapter 7, such as adhesive or abrasive wear. Controlling temperature at the mesh interface is important in reducing scuffing and scoring of the teeth. Lubricants remove heat as well as separate the metal surfaces to reduce friction and wear. Sufficient lubricant must be provided to transfer the heat of friction to the environment without allowing excessive local temperatures in the mesh.

The usual and preferred approach is to provide an oil bath by housing the gears in an oil-tight box, called a gearbox. The gearbox is partially filled with an appropriate lubricant such that at least one member of each gearset is partially submerged. (The box is never completely filled with oil.) Gear rotation will carry the lubricant to the meshes and keep the unsubmerged gears oiled. The oil must be kept clean and free of contaminants and should be changed periodically. A much less desirable arrangement, sometimes used for situations in which a gearbox is not practical, is to periodically apply grease lubricant to the gears when they are stopped for servicing. Grease is merely petroleum oil suspended in a soap emulsion. This topical, grease lubrication does little for heat removal and is only recommended for low-velocity, lightly loaded gears.

Gear lubricants are typically petroleum-based oils of differing viscosity depending on the application. Light oils (10-30W) are sometimes used for gears with velocities high enough and/or loads low enough to promote elastohydrodynamic lubrication (see Chapter 10). In highly loaded and/or low-velocity gearsets, or ones with large sliding components, extreme pressure (EP) lubricants are often used. These are typically 80-90W gear oils with fatty-acid type additives that provide some protection against scuffing under boundary-lubricated conditions. See Section 7.5 and Chapter 10 for more information on lubrication and lubricants. The AGMA provides extensive data in its standards on the proper selection of gear lubricants. The reader is referred to that source and to other sources such as lubricant vendors for more detailed information on lubricants.

11

11.11 DESIGN OF SPUR GEARS

The design of gears usually requires some iteration. Not enough information typically exists in the problem statement to directly solve for the unknowns. The values of some parameters must be assumed and a trial solution done. Many approaches are possible.

Usually, the gear ratio and either the power and speed, or the torque and speed, of one shaft are defined. The parameters to be determined are the pinion and gear pitch diameters, the diametral pitch, the face width, the material(s), and the safety factors. Some design decisions regarding the mesh-accuracy required, the number of cycles, the pressure angle, the tooth form (standard or long-addendum), the gear manufacturing method (for surface finish considerations), operating temperature range, and desired reliability must be made. With at least preliminary information on these factors, the design process can begin.

We ultimately will need to calculate safety factors for both bending-fatigue and surface-fatigue failures. These can be investigated in either order, but the better strategy is to calculate the bending stresses first because increasing the surface hardness of the material has a greater effect on wear life than on bending strength. Thus, if the chosen material will survive the bending stresses, its hardness can be adjusted to improve its wear life with no other design change. Also, increasing tooth size has a greater effect on bending strength than on wear life, and tooth size is the primary variable in the calculations.

Before any stress calculations can be done, the loads must be determined. The tangential load on the gear teeth can be found from the known torque on the shaft and an assumed pitch radius for its pinion or gear (see equation 11.13a). Note that a larger pitch radius reduces the tooth load, but increases the pitch-line velocity. A reasonable compromise between these factors must be determined. Also, too small a pitch radius may result in a pinion with too few teeth to avoid interference, depending on the diametral pitch or module selected. Once a trial diametral pitch has been selected, its smallest acceptable pinion diameter can be used as the first choice in order to keep the package size small. The first design attempt should use a standard tooth form to keep costs low. If the design needs to be smaller than the standard tooth form allows, a long-addendum form can be investigated.

Since the bending strength of the gear tooth is directly related to tooth size as defined by its diametral pitch or module, a common starting point for the stress calculation is to assume values for diametral pitch or module and also for face width, then solve for the bending stress using equation 11.15. (Note that face width can also be roughly expressed as a range-function of diametral pitch $(8 / p_d < F < 16 / p_d)$. See the discussion of the factor K_m above.

A trial material is then chosen and its corrected bending-fatigue strength calculated from equation 11.24. If the resulting safety factor is either too large or too small, the assumed values are adjusted and the calculation repeated until it converges to an acceptable solution.

The surface stress and surface-fatigue strength are then calculated from equations 11.21 and 11.25 and a safety factor against wear determined. Material hardness can be adjusted at this point if necessary or the whole process can be repeated with adjusted values of either pitch or face width or both.

One useful strategy is to tailor the safety factors for bending failure to be higher than those for surface failure. Bending failure is sudden and catastrophic, resulting in tooth breakage and a disabling of the machine. Surface failure gives audible warning and the gears can be run for some time after the noise begins before having to be replaced. Thus, surface failure is the more desirable design limit on gear life.

11.12 CASE STUDY

We will now address the design of spur gears in one of the case study assemblies that were defined in Chapter 8.

CASE STUDY 7C

Design of Spur Gears for a Compressor Drive Train

Problem   Design a spur gearset for the compressor gearbox in Figure 8-1 based on the loadings defined in Case Study 7A and specify suitable materials and heat treatments.

Given   The torque-time function on the output shaft is as shown in Figure 8-3. The required gear ratio is a 2.5:1 reduction in velocity from the input to the output shaft. Output shaft velocity is 1 500 rpm.

Assumptions   A 10-year life of 1-shift operation is desired. AGMA standard full-depth teeth will be used. Based on the data in Tables 11-6 and 11-7, set $Q_v = 10$ Both pinion and gear will be through-hardened steel.

Solution   See Figures 8-1, 8-3, and TKSolver files CASE7-C1 and CASE7-C2.

1 The time-varying torque on the output shaft is defined in Figure 8-3 as varying between -175 and $+585$ lb-in. In Case Study 7B, in which the shafts for this same machine were designed, we assumed a 4-in pitch diameter, 20° pinion and a 10-in gear. For a first attempt at the gearset design we will retain those assumptions. From these data we can determine the forces at the gear mesh. The tangential component is found from the known output torque and the assumed gear radius:

$$W_{t_{max}} = \frac{T_{max}}{r_g} = \frac{585 \text{ lb - in}}{5 \text{ in}} = 117 \text{ lb}$$

$$W_{t_{min}} = \frac{T_{min}}{r_g} = \frac{-175 \text{ lb - in}}{5 \text{ in}} = -35 \text{ lb} \qquad (a)$$

2 We will take the positive peak value as the transmitted load, $W_t = 117$ lb. The -35-lb peak force acts on the opposite sides of the teeth, loading both gear and pinion similar to an idler gear. We will take that aspect of the loading into account with the application factor K_a.

3 Assume a pinion with $N_p = 20$ teeth. The gear then has 2.5 $N_p = 50$ teeth. The diametral pitch for that combination is

$$p_d = \frac{N}{d} = \frac{20}{4} = 5 \qquad (b)$$

which is a standard pitch (Table 11-2).

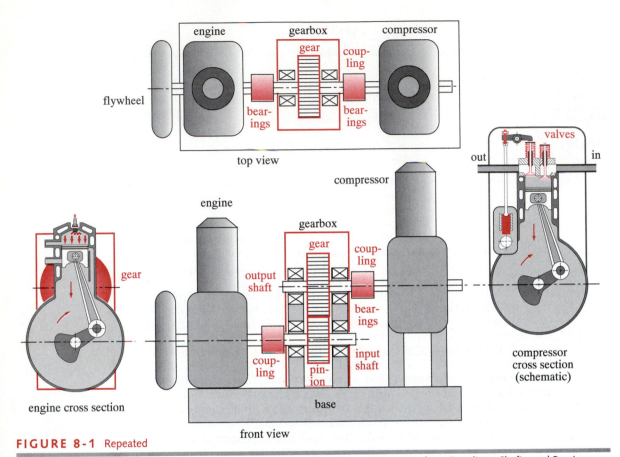

FIGURE 8-1 Repeated

Preliminary Design Schematic of Gasoline-Engine-Powered Portable Air Compressor, Gearbox, Couplings, Shafts, and Bearings

11

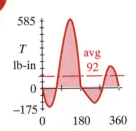

crank angle (deg)

FIGURE 8-3

Total Torque-Time
Function at Crankshaft
with Constant ω

4   The bending geometry factors J for this combination are found in Table 11-9 for loading at the highest point of single-tooth contact (HPSTC) and are approximately:

$$J_p = 0.34 \qquad\qquad J_g = 0.40 \qquad\qquad (c)$$

5   The velocity factor K_v (C_v) is calculated from equations 11.16 and 11.17 based on the assumed gear-quality index Q_v and the pitch-line velocity V_t.

$$V_t = \frac{d_p}{2}\omega_p = \frac{4.0 \text{ in}}{2(12)}(1\,500 \text{ rpm})(2\pi) = 3\,927\ \frac{\text{ft}}{\text{min}} \qquad (d)$$

$$B = \frac{(12 - Q_v)^{2/3}}{4} = \frac{(12 - 10)^{2/3}}{4} = 0.397 \qquad (e)$$

$$A = 50 + 56(1 - B) = 50 + 56(1 - 0.397) = 83.77 \qquad (f)$$

$$K_v = C_v = \left(\frac{A}{A + \sqrt{V_t}}\right)^B = \left(\frac{83.77}{83.77 + \sqrt{3\,927}}\right)^{0.397} = 0.801 \qquad (g)$$

6   V_t should be checked against the maximum allowable pitch-line velocity for this quality gear using equation 11.18:

$$V_{t_{max}} = \left[A + (Q_v - 3)\right]^2 = \left[83.77 + (10 - 3)\right]^2 = 8\,239 \ \text{ft/min} \qquad (h)$$

which is larger than V_t, so is acceptable.

7   Assuming a face width factor of 12 the face width can be estimated as

$$F \cong \frac{12}{P_d} = \frac{12}{5} = 2.4 \ \text{in} \qquad (i)$$

8   This value is used to interpolate in Table 11-16 for K_m (C_m).

$$K_m = C_m \cong 1.61 \qquad (j)$$

9   The application factor K_a is intended to account for shockiness in the driving and driven machinery. This machine has both, since it is driven by a single-cylinder engine and drives a single-cylinder compressor. In many such cases, only an average value of the transmitted torque is known, based on the average transmitted power. In this case, we have computed (in Case Study 7A) a fairly accurate torque-time function for the compressor, which in effect defines the "overloads" in the driven part of the system. We used the peak torque rather than the average torque to define the transmitted load. So, the full value of the application factor recommended in Table 11-17 may not be needed here. We will use it to account for the partially reversed loading on the gear teeth (Figure 8-3) as well as the shock loading associated with the driver (engine) and estimate it at $K_a = C_a = 2$.

10   The size factor K_s (C_s), and the rim bending factor K_B are all 1 for these small gears.

11   The bending stresses in pinion and gear can now be computed.

$$\sigma_{b_p} = \frac{W_t\, P_d}{FJ} \frac{K_a K_m}{K_v} K_s K_B K_I = \frac{117(5)}{2.4(0.34)} \frac{1(1.61)}{0.801}(1)(1)(1) = 2\,881 \ \text{psi} \qquad (k)$$

$$\sigma_{b_g} = \frac{W_t\, P_d}{FJ} \frac{K_a K_m}{K_v} K_s K_B K_I = \frac{117(5)}{2.4(0.40)} \frac{1(1.61)}{0.801}(1)(1)(1) = 2\,449 \ \text{psi} \qquad (l)$$

12   Additional factors are needed for the surface stress calculation. Table 11-18 shows an approximate elastic coefficient of 2 300 for steel on steel. In Example 11-6, we computed a more accurate value of $C_p = 2\,276$. The surface finish factor C_f is 1.

13   The surface geometry factor I is calculated from equations 11.22:

11

$$\rho_1 = \sqrt{\left(r_p + \frac{1}{p_d}\right)^2 - \left(r_p \cos\phi\right)^2} - \frac{\pi}{p_d}\cos\phi$$

$$= \sqrt{\left(2.0 + \frac{1}{5}\right)^2 - \left(2.0\cos 20°\right)^2} - \frac{\pi}{5}\cos 20° = 0.553 \text{ in} \qquad (m)$$

$$\rho_2 = C\sin\phi - \rho_1 = \left(r_p + r_g\right)\sin\phi - \rho_1$$

$$= (2.0 + 5.0)\sin 20° - 0.553 = 1.841 \text{ in} \qquad (n)$$

$$I = \frac{\cos\phi}{\left(\dfrac{1}{\rho_p} \pm \dfrac{1}{\rho_g}\right)d_p} = \frac{\cos 20°}{\left(\dfrac{1}{0.553} + \dfrac{1}{1.841}\right)2.0} = 0.100 \qquad (o)$$

14   The surface stresses in pinion and gear can now be computed.

$$\sigma_{c_p} = C_p\sqrt{\frac{W_t}{FId_p}\frac{C_aC_m}{C_v}C_sC_f}$$

$$= 2\,276\sqrt{\frac{117}{2.4(0.100)(4.0)}\frac{1(1.61)}{0.801}(1)(1)} = 50\,393 \text{ psi} \qquad (p)$$

$$\sigma_{c_g} = C_p\sqrt{\frac{W_t}{FId_p}\frac{C_aC_m}{C_v}C_sC_f}$$

$$= 2\,276\sqrt{\frac{117}{2.4(0.100)(10.0)}\frac{1(1.61)}{0.801}(1)(1)} = 31\,871 \text{ psi} \qquad (q)$$

15   An estimate of the uncorrected bending-fatigue strength can be made from the curves of Figure 11-25. We will try an AGMA Grade 1 steel, through-hardened to 250HB. The uncorrected bending-fatigue strength is found from the lower curve of the figure:

$$S_{fb}' = -274 + 167HB - 0.152Hb^2$$

$$= -274 + 167(250) - 0.152(250)^2 = 31\,976 \text{ psi} \qquad (r)$$

16   This value needs to be corrected for certain factors using equation 11.24. The life factor K_L is found from the appropriate equation in Figure 11-24 based on the required number of cycles in the life of the gears. The pinion sees the largest number of repeated tooth-loadings, so we calculate the life based on it. First, calculate the number of cycles N for the required life of 10 years, one shift.

$$N = 1\,500 \text{ rpm}\left(\frac{60 \text{ min}}{\text{hr}}\right)\left(\frac{2080 \text{ hr}}{\text{shift - yr}}\right)(10 \text{ yr})(1 \text{ shift}) = 4.7E9 \text{ cycles} \qquad (s)$$

The value of K_L is found from

$$K_L = 1.3558N^{-0.0178} = 1.3558(4.7E9)^{-0.0178} = 0.9121 \qquad (t)$$

17 At the specified operating temperature, $K_T = 1$.

18 The gear-material data are all taken at a reliability level of 99%. This is satisfactory in this case, making $K_R = 1$.

19 The corrected bending-fatigue strength is then

$$S_{fb} = \frac{K_L}{K_T K_R} S_{fb'} = \frac{0.9121}{1(1)} 31\,9760 = 29\,167 \text{ psi} \qquad (u)$$

20 An estimate of the uncorrected surface fatigue strength can be made from the curves of Figure 11-27. For an AGMA Grade 1 steel, through-hardened to 250HB, the strength is found from the lower curve of the figure:

$$S_{fc'} = 26\,000 + 327HB = 26\,000 + 327(250) = 107\,750 \text{ psi} \qquad (v)$$

21 This value needs to be corrected for certain factors using equation 11.25. The life factor C_L is found from the appropriate equation in Figure 11-26 based on the required number of cycles N found above.

$$C_L = 1.4488N^{-0.023} = 1.4488(4.7E9)^{-0.023} = 0.8681 \qquad (w)$$

22 $C_T = K_T = 1$ and $C_R = K_R = 1$.

23 Since the gears and pinion are of the same hardness material in this case, $C_H = 1$.

24 The corrected surface-fatigue strength is then

$$S_{fc} = \frac{C_L C_H}{C_T C_R} S_{fc'} = \frac{0.8681(1)}{1(1)} 107\,750 = 93\,543 \text{ psi} \qquad (x)$$

25 The safety factors against bending failure are found by comparing the corrected bending strength to the bending stress for each gear in the mesh:

$$N_{b_{pinion}} = \frac{S_{fb}}{\sigma_{b_{pinion}}} = \frac{29\,167}{2\,881} = 10.1$$

$$\qquad (y)$$

$$N_{b_{gear}} = \frac{S_{fb}}{\sigma_{b_{gear}}} = \frac{29\,167}{2\,449} = 11.9$$

which are too high, causing the package to be larger than necessary.

26 The safety factors against surface failure are found by comparing the corrected surface strength to the surface stress for each gear in the mesh:

11

Table 11-22   Case Study 7C – Spur Gear Train Design

Partial list—See the *TKSolver* file CASE7C-2 for complete data

Input	Variable	Output	Unit	Comments
2.50	*ratio*			gear ratio
8	p_d		1/in	diametral pitch
20	*phi*		deg	pressure angle
170	W_t		lb	tangential force
22	N_{pinion}			no. of teeth on pinion
	N_{gear}	55		no. of teeth on gear
	d_{pinion}	2.75	in	pitch dia of pinion
	d_{gear}	6.88	in	pitch dia of gear
1.50	*face*		in	face width
0.34	J_{pinion}			geometry factor—pinion
0.40	J_{gear}			geometry factor— gear
	I	0.10		I factor for pinion/gear mesh
2.0	K_a			application factor
1.6	K_m			load distribution factor
2 276	C_p			elastic coefficient
10	Q_v			gear-quality index
	V_t	2 700	ft/min	pitch-line velocity
	V_{tmax}	8 239	ft/min	max allowable pitch-line velocity
	K_v	0.826		dynamic factor
	$\sigma_{bpinion}$	10 346	psi	bending stress—pinion tooth
	σ_{bgear}	8 794	psi	bending stress—gear tooth
	$\sigma_{cpinion}$	90 026	psi	surface stress in pinion
	σ_{cgear}	56 937	psi	surface stress in gear
	Sf_{bprime}	31 976	psi	uncorrected bending strength
	Sf_b	29 167	psi	corrected bending strength
	Sf_{cprime}	107 750	psi	uncorrected surface strength
	Sf_c	93 543	psi	corrected surface strength
	K_R	1.00		reliability factor
	cycles	4.7E+9		number of repeated pinion cycles
	K_L	0.91		life factor—bending fatigue
	C_L	0.87		life factor—surface fatigue
	N_{bp}	2.8		bending safety factor for pinion
	N_{bg}	3.3		bending safety factor for gear
	N_{cp}	1.0		surface safety factor for pinion
	N_{cg}	1.6		surface safety factor for gear

$$N_{c_{pinion}} = \frac{S_{fc}}{\sigma_{c_{pinion}}} = \frac{107\,750}{50\,393} = 1.9$$

$$N_{c_{gear}} = \frac{S_{fc}}{\sigma_{c_{gear}}} = \frac{107\,750}{31\,871} = 2.9$$

(z)

which are also higher than they need to be.

27 The diametral pitch was increased from the current 5 to 8 (decreasing the tooth size) in order to reduce the pitch diameters, increase the stresses, and lower the safety factors. The face width becomes 1.5 in for the same face width factor of 12. The pinion was increased to 22 teeth giving 55 teeth on the gear with the new p_d. The computations were redone with the results shown in Table 11-22, taken from the *TKSolver* variable sheet. The new safety factors are

$$N_{b_{pinion}} = 2.8 \qquad\qquad N_{b_{gear}} = 3.3$$

$$N_{c_{pinion}} = 1.0 \qquad\qquad N_{c_{gear}} = 1.6$$

(aa)

28 These gears are still very safe against tooth breakage and based on the assumptions and calculations and, if properly lubricated, should have a 99% probability of lasting the required 10 years before pitting of the pinion begins.

29 Note that the change of gear pitch diameters over those assumed in Case Study 7B for the shaft design also increases the transverse gear loads on the shaft by 45%. This will require another iteration of that shaft design.

11.13 SUMMARY

There are two principal types of gear failure: tooth breakage from bending stresses and pitting from surface (Hertzian) stresses. Of the two, bending failure is more catastrophic, since tooth breakage usually disables the machine. Pitting failure comes on gradually and gives audible and visible warning (if the teeth can be inspected). Gears can run for some time after pitting begins before having to be replaced.

Both modes of failure are fatigue failures due to the repeated stressing of individual teeth as they come in and out of mesh. The principles of fatigue analysis (Chapter 6) apply and a modified-Goodman analysis is needed. However, the similar nature of the loading on all gear teeth allows the Goodman analysis to be captured in a standardized approach as defined by the AGMA.

Proper involute tooth geometry is crucial to the operation and life of gears. The AGMA defines a standard tooth profile plus several modifications to that standard for special situations. Geometry factors necessary for proper stress calculations are defined for these geometries. Extensive testing of gear materials under realistic loading conditions in combination with years of experience by gear manufacturers have resulted in a set of proven equations for the calculation of both stresses and corrected endurance strengths in bending and surface fatigue for gears.

This chapter summarizes the AGMA approach to spur-gear design and presents a number of charts and empirical formulas for calculation. The reader is directed to the AGMA standards for more complete information.

Important Equations Used in this Chapter

Circular Pitch (Section 11.2):

$$p_c = \frac{\pi d}{N} \tag{11.3a}$$

Diametral Pitch (Section 11.2):

$$p_d = \frac{N}{d} \tag{11.4a}$$

Metric Module (Section 11.2):

$$m = \frac{d}{N} \tag{11.4c}$$

Gear Ratio (Section 11.2):

$$m_G = \frac{N_g}{N_p} \tag{11.5b}$$

Contact Ratio (Section 11.4):

$$m_p = \frac{p_d Z}{\pi \cos \phi} \tag{11.7b}$$

$$Z = \sqrt{\left(r_p + a_p\right)^2 - \left(r_p \cos \phi\right)^2} + \sqrt{\left(r_g + a_g\right)^2 - \left(r_g \cos \phi\right)^2} - C \sin \phi \tag{11.2}$$

Tangential Load on Gear Teeth (Section 11.5):

$$W_t = \frac{T_p}{r_p} = \frac{2T_p}{d_p} = \frac{2p_d T_p}{N_p} \tag{11.13a}$$

AGMA Bending Stress Equations (Section 11.8):

$$\sigma_b = \frac{W_t\, p_d}{FJ}\frac{K_a K_m}{K_v} K_s K_B K_I \tag{11.15us}$$

$$\sigma_b = \frac{W_t}{FmJ}\frac{K_a K_m}{K_v} K_s K_B K_I \tag{11.15si}$$

AGMA Surface Stress Equation (Section 11.8):

$$\sigma_c = C_p \sqrt{\frac{W_t}{FId} \frac{C_a C_m}{C_v} C_s C_f} \qquad (11.21)$$

AGMA Bending-Fatigue Strength Equation (Section 11.9):

$$S_{fb} = \frac{K_L}{K_T K_R} S_{fb'} \qquad (11.24)$$

AGMA Surface-Fatigue Strength Equation (Section 11.9):

$$S_{fc} = \frac{C_L C_H}{C_T C_R} S_{fc'} \qquad (11.25)$$

11.14 REFERENCES

1   **AGMA**, *Gear Nomenclature, Definitions of Terms with Symbols*. ANSI/AGMA Standard 1012-F90. American Gear Manufacturers Association, 1500 King St., Suite 201, Alexandria, Va., 22314, 1990.

2   **AGMA**, *Fundamental Rating Factors and Calculation Methods for Involute Spur and Helical Gear Teeth*. ANSI/AGMA Standard 2001-B88. American Gear Manufacturers Association, 1500 King St., Suite 201, Alexandria, Va., 22314, 1988.

3   **AGMA**, *Geometry Factors for Determining the Pitting Resistance and Bending Strength of Spur, Helical, and Herringbone Gear Teeth*. ANSI/AGMA Standard 908-B89. American Gear Manufacturers Association, 1500 King St., Suite 201, Alexandria, Va., 22314, 1989.

4   **R. L. Norton**, *Design of Machinery: An Introduction to the Synthesis and Analysis of Mechanisms and Machines*. McGraw-Hill: New York, pp. 394-428, 1992.

5   **W. Lewis**, "Investigation of the Strength of Gear Teeth, an address to the Engineer's Club of Philadelphia, October 15, 1892." reprinted in *Gear Technology*, vol. 9, no. 6, p. 19, Nov./Dec. 1992.

6   **T. J. Dolan and E. L. Broghammer**, *A Photoelastic Study of the Stresses in Gear Tooth Fillets*. Bulletin 335, U. Illinois Engineering Experiment Station: , 1942.

7   **W. D. Mark**, The Generalized Transmission Error of Parallel-Axis Gears. *Journal of Mechanisms, Transmissions, and Automation in Design*. **111**: pp. 414-423, 1989.

8   **R. A. Morrison** "Load/Life Curves for Gear and Cam Materials." *Machine Design*, pp. 102-108, Aug. 1, 1968.

9   **E. Buckingham**, *Analytical Mechanics of Gears*, McGraw-Hill: New York, 1949.

10   **D. W. Dudley**, Gear Wear, in *Wear Control Handbook*, M. B. Peterson and W. O. Winer, Editor. ASME: New York, p. 764, 1980.

11

11.15 PROBLEMS

*11-1   A 20° pressure angle, 27-tooth spur gear has a diametral pitch $p_d = 5$. Find the pitch dia., addendum, dedendum, outside dia., and circular pitch.

11-2   A 25° pressure angle, 43-tooth spur gear has a diametral pitch $p_d = 8$. Find the pitch dia., addendum, dedendum, outside dia., and circular pitch.

*11-3   A 57-tooth spur gear is in mesh with a 23-tooth pinion. The $p_d = 6$ and $\phi = 25°$. Find the contact ratio.

11-4   A 78-tooth spur gear is in mesh with a 27-tooth pinion. The $p_d = 6$ and $\phi = 20°$. Find the contact ratio.

*11-5   What will the pressure angle be if the center distance of the spur gearset in Problem 11-3 is increased by 5%?

11-6   What will the pressure angle be if the center distance of the spur gearset in Problem 11-4 is increased by 7%?

*11-7   If the spur gearsets in Problems 11-3 and 11-4 are compounded as shown in Figure 11-14, what will the overall train ratio be?

†11-8   A paper machine processes rolls of paper having a density of 984 kg/m³. The paper roll is 1.50-m outside dia. (OD) x 0.22-m inside dia (ID) x 3.23 m long and is on a simply-supported, hollow, steel shaft with $S_{ut} = 400$ MPa. Design a 2.5:1 reduction spur gearset to drive this roll shaft to obtain a minimum dynamic safety factor of 2 for a 10-year life if the shaft OD is 22 cm and the roll turns at 50 rpm with 1.2 hp absorbed.

*11-9   Design a two-stage compound spur gear train for an overall ratio of approximately 47:1. Specify tooth numbers for each gear in the train.

11-10   Design a three-stage compound spur gear train for an overall ratio of approximately 656:1. Specify tooth numbers for each gear in the train.

*11-11   An epicyclic spur gear train as shown in Figure 11-16 has a sun gear of 33 teeth and a planet gear of 21 teeth. Find the required number of teeth in the ring gear and determine the ratio between the arm and sun gear if the planet is held stationary. Hint: Consider the arm to rotate at 1 rpm.

11-12   An epicyclic spur gear train as shown in Figure 11-16 has a sun gear of 23 teeth and a planet gear of 31 teeth. Find the required number of teeth in the ring gear and determine the ratio between the arm and ring gear if the sun is held stationary. Hint: Consider the arm to rotate at 1 rpm.

11-13   An epicyclic spur gear train as shown in Figure 11-16 has a sun gear of 23 teeth and a planet gear of 31 teeth. Find the required number of teeth in the ring gear and determine the ratio between the sun and ring gear if the arm is held stationary. Hint: Consider the sun to rotate at 1 rpm.

*11-14   If the gearset in Problem 11-3 transmits 125 HP at 1 000 pinion rpm, find the torque on each shaft.

* Answers to these problems are provided in Appendix G.

† These problems are based on similar problems in previous chapters with the same –number or the number noted in the problem statement, e.g., Problem 11-8 is based on Problem 9-8 etc. Problems in succeeding chapters may also continue and extend these problems.

11-15   If the gearset in Problem 11-4 transmits 33 kW at 1 600 pinion rpm, find the torque on each shaft.

*11-16   Size the spur gears in Problem 11-14 for a bending safety factor of at least 2 assuming a steady torque, 25° pressure angle, full-depth teeth, a face width factor of 8, $Q_v = 9$, an AISI 4140 steel pinion and a class 40 cast-iron gear.

11-17   Size the spur gears in Problem 11-15 for a bending safety factor of 2.5 assuming a steady torque, 20° pressure angle, full-depth teeth, a face width factor of 12, $Q_v = 11$, an AISI 4340 steel pinion and an A-7-d nodular iron gear.

*11-18   Size the spur gears in Problem 11-14 for a surface safety factor of at least 2 assuming a steady torque, 25° pressure angle, full-depth teeth, a face width factor of 8, $Q_v = 9$, an AISI 4140 steel pinion and a class 40 cast-iron gear.

11-19   Size the spur gears in Problem 11-15 for a surface safety factor of 1.2 assuming a steady torque, 20° pressure angle, full-depth teeth, a face width factor of 12, $Q_v = 11$, an AISI 4340 steel pinion and an A-7-d nodular-iron gear.

*11-20   If the gearset in Problem 11-11 transmits 83 kW at 1 200 arm rpm, find the torque on each shaft.

11-21   If the gearset in Problem 11-12 transmits 39 HP at 2 600 arm rpm, find the torque on each shaft.

11-22   If the gearset in Problem 11-13 transmits 23 kW at 4 800 sun rpm, find the torque on each shaft.

*11-23   Size the spur gears in Problem 11-20 for a bending safety factor of at least 2.8 assuming a steady torque, 25° pressure angle, full-depth teeth, a face width factor of 10, Qv = 9, an AISI 4140 steel pinion and a class 40 cast-iron gear.

11-24   Size the spur gears in Problem 11-21 for a bending safety factor of at least 2.4 assuming a steady torque, 20° pressure angle, full-depth teeth, a face width factor of 12, Qv = 11, an AISI 4340 steel pinion and an A-7-d nodular-iron gear.

*11-25   Size the spur gears in Problem 11-20 for a surface safety factor of at least 1.7 assuming a steady torque, 25° pressure angle, full-depth teeth, a face width factor of 10, Qv = 9, an AISI 4140 steel pinion and a class 40 cast-iron gear.

11-26   Size the spur gears in Problem 11-21 for a surface safety factor of at least 1.3 assuming a steady torque, 20° pressure angle, full-depth teeth, a face width factor of 12, Qv = 11, an AISI 4340 steel pinion and an A-7-d nodular-iron gear.

*11-27   If the gearset in Problem 11-10 transmits 190 kW at 1800 input pinion rpm, find the torque on each of the four shafts.

*11-28   Size the spur gears in Problem 11-27 for a bending safety factor of at least 3.2 and a surface safety factor of at least 1.7 assuming a steady torque, 25° pressure angle, full-depth teeth, a face width factor of 11, $Q_v = 8$, and AISI 4140 steel for all gears.

*11-29   Design a two-stage compound spur gear train for an overall ratio of approximately 78:1. Specify tooth numbers for each spur gear in the train.

11

* Answers to these problems are provided in Appendix G.

† These problems are based on similar problems in previous chapters with the same –number or the number noted in the problem statement, e.g., Problem 11-8 is based on Problem 9-8 etc. Problems in succeeding chapters may also continue and extend these problems.

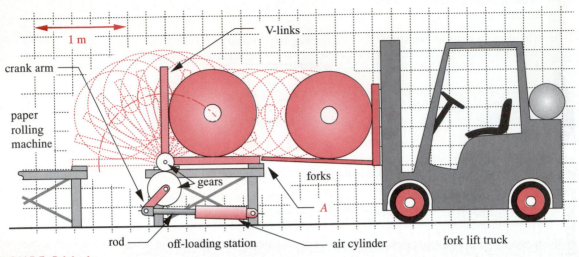

FIGURE P11-1

Problems 11-32

11-30   Design a nonreverted compound transmission based on the arrangement shown in Figure 11-14*a* for an overall train ratio of approximately 90:1. It should be capable of transmitting 50 hp at 1 000 rpm input shaft speed. State all assumptions.

11-31   Design a reverted compound transmission based on the arrangement shown in Figure 11-14*b* for an overall train ratio of approximately 80:1. It should be capable of transmitting 30 hp at 1 500 rpm input shaft speed. State all assumptions.

†11-32   Figure P11-1 shows the same paper machine that was analyzed in Problem 6-46 and in other problems from previous chapters. The paper rolls in Figure P11-1 are 0.9-m OD x 0.22-m ID x 3.23 m long and have a density of 984 kg/m^3. The rolls are transferred from the machine conveyor (not shown) to the forklift truck by the V-linkage of the off-load station, which is rotated through 90° by an air cylinder. The paper then rolls onto the waiting forks of the truck. The machine makes 30 rolls per hour and runs 2 shifts. The V-links are rotated by the crank arm through a shaft that is 60-mm dia by 3.23 m long. A redesign of the V-link rotating mechanism is desired in order to introduce a gearset between the crank arm and the V-link shaft with a 2:1 ratio. This will reduce the required stroke of the air cylinder by 50% and improve its geometry. Design a suitable spur-gear set for this application for a 10-year life against surface failure. State all assumptions.

11

12

HELICAL, BEVEL, AND WORM GEARS

Cycle and Epicycle, Orb in Orb , , ,
JOHN MILTON, PARADISE LOST

12.0 INTRODUCTION

Chapter 11 explored the topic of straight-toothed or spur gears in some detail. Gears are available in many other tooth configurations for particular applications. This chapter will present a brief introduction to designing with helical, bevel, and worm gears. The complexity of the design problem increases significantly when these more complicated gear tooth shapes are used. The *American Gear Manufacturers Association* (AGMA) presents detailed data and algorithms for their calculation. We will base this presentation on the AGMA recommendations, but cannot give a complete treatment of this complex subject, in the space available here. The reader is encouraged to consult the AGMA standards for more information when faced with a real design problem involving gearing. Table 12-0 lists the variables used in this chapter and indicates the section or equation in which they appear. A summary and list of important equations appears at the end of the chapter.

12.1 HELICAL GEARS

Helical gears are very similar to spur gears. Their teeth are involutes. The difference is that their teeth are angled with respect to the axis of rotation at a helix angle ψ, as shown in Figure 12-1. The helix angle may typically range from about 10 to 45°. If the gear were long enough axially, any one tooth would wrap around the circumference 360°. The teeth form a helix, which may be either right- or left-handed. A pair of op

Table 12-0   Variables Used in this Chapter

Part 1 of 2

Symbol	Variable	ips units	SI units	See
a	addendum	in	m	Eq. 12.18
b	dedendum	in	m	Eq. 12.18
C	center distance	none	none	Eq. 12.16
C_f	surface finish factor	none	none	Eq. 12.10
C_H	hardness factor	none	none	Eq. 12.11
C_{md}	mounting factor	none	none	Eq. 12.11
C_p	elastic coefficient	none	none	Eq. 12.10
C_R	reliability factor	none	none	Eq. 12.11
C_s	materials factor	none	none	Eq. 12.24
C_T	temperature factor	none	none	Eq. 12.11
C_{xc}	crowning factor	none	none	Eq. 12.10
d	pitch diameter (with various subscripts)	in	m	various
e	efficiency	none	none	Eq. 12.30
F	face width	in	m	Eq. 12.9, 12.19
I	AGMA surface geometry factor	none	none	Eq. 12.10
J	AGMA bending geometry factor	none	none	Eq. 12.9
K_a, C_a	application factor	none	none	Eq. 12.9
K_B	rim bending factor	none	none	Eq. 11.15
K_I	idler factor	none	none	Eq. 11.15
K_m, C_m	load distribution factor	none	none	Eq. 12.9, 12.10
K_s, C_s	size factor	none	none	Eq. 12.9, 12.10
K_v, C_v	dynamic factor, velocity factor	none	none	Eq. 12.9, 12.26
K_x	curvature factor	none	none	Eq. 12.9
L	length, lead	in	m	Eq. 12.6, 12.7
m	module	none	mm	Eq. 12.9
m_F	axial contact ratio	none	none	Eq. 12.5
m_G	gear ratio	none	none	Eq. 12.15
m_N	load sharing ratio	none	none	Eq. 12.6
m_p	transverse contact ratio	none	none	Eq. 12.6
N	number of teeth (with various subscripts)	none	none	various
N_b, N_c	factors of safety—bending and contact	none	none	various
p_c	circular pitch	in	mm	Eq. 12.1c
p_d	diametral pitch	1/in	none	Eq. 12.1c
p_t	transverse pitch	in	m	Eq. 12.1a
p_x	axial pitch	in	mm	Eq. 12.1b
S_{fb}	corrected bending-endurance strength	psi	Pa	Ex. 12-2
S_{fc}	corrected surface-endurance strength	psi	Pa	Ex. 12-2
S'_{fc}	uncorrected surface-endurance strength	psi	Pa	Eq. 12.11

12

Table 12-0   Variables Used in this Chapter
Part 2 of 2

Symbol	Variable	ips units	SI units	See
T	torque (with various subscripts)	lb-in	N-m	various
V_t	pitch line velocity	in/sec	m/sec	Eq. 12.27
W	total force on gear teeth	lb	N	Eq. 12.3
W_a	axial force on gear teeth	lb	N	Eq. 12.3
W_f	friction force on gear teeth	lb	N	Eq. 12.28
W_r	radial force on gear teeth	lb	N	Eq. 12.3
W_t	tangential force on gear teeth	lb	N	Eq. 12.3
α	pitch cone angle	deg	deg	Eq. 12.7
ϕ	pressure angle	deg	deg	various
ψ	helix angle or spiral angle	deg	deg	various
λ	lead angle	deg	deg	Eq. 12.12
μ	coefficient of friction	none	none	Eq. 12.28
ω	angular velocity	rad/sec	rad/sec	Ex. 12-2
ρ	radius of curvature	in	m	Eq. 12.6
Φ	power	hp	W	Eq. 12.20
σ_b	bending stress	psi	Pa	Eq. 12.9
σ_c	surface stress	psi	Pa	Eq. 12.10

(a) Opposite-hand pair meshed on parallel axes

posite-hand helical gears mesh with their axes parallel as shown in Figure 12-1a. Same-hand helical gears can be meshed with their axes skewed and are then called crossed-axis or just crossed helical gears as shown in Figure 12-1b.

PARALLEL HELICAL GEARS   mesh with a combination of rolling and sliding with contact starting at one end of the tooth and "wiping" across its face width. This is quite different than spur gear tooth contact, which occurs all at once along a line across the tooth face at the instant of tooth contact. One result of this difference is that helical gears run quieter and with less vibration than spur gears because of the gradual tooth contact. Automotive transmissions use helical gears almost exclusively in order to obtain quiet operation. One common exception is in the reverse-gear mesh of a nonautomatic transmission, which often uses spur gears to enable shifting them in and out of engagement. In such a transmission, a noticeable "gear whine" can be heard when backing up the vehicle, due to the resonances of the spur-gear teeth being exited by the sudden impacts of tooth-tooth line contact. The helical forward-gear meshes are essentially silent. Parallel helical gears are also capable of transmitting high power levels.

CROSSED-HELICAL GEARS   mesh differently than parallel-helical gears; their teeth slide without rolling and are in theoretical point contact rather than the line contact of parallel gears. This severely reduces their load carrying capacity. Crossed-helical gears are not recommended for applications that must transmit large torque or power. They are nevertheless frequently used in light-load applications, such as distributor and speedometer drives of automobiles.

(b) Same-hand pair meshed on crossed axes

FIGURE 12-1

Helical Gears *Courtesy of Boston Gear, Division of IMO Industries, Quincy, MA*

12

Helical Gear Geometry

Figure 12-2 shows the geometry of a basic helical rack. The teeth form the helix angle ψ with the "axis" of the rack. The teeth are cut at this angle and the tooth form is then in the **normal plane**. The **normal pitch** p_n and the **normal pressure angle** ϕ_n are measured in this plane. The **transverse pitch** p_t and the **transverse pressure angle** ϕ_t are measured in the **transverse plane**. These dimensions are related to one another by the helix angle. The transverse pitch is the hypotenuse of the right triangle ABC.

$$p_t = p_n/\cos \psi \qquad (12.1a)$$

An axial pitch p_x can also be defined as the hypotenuse of the right triangle BCD.

$$p_x = p_n/\sin \psi \qquad (12.1b)$$

P_t corresponds to circular pitch P_c, measured in the pitch plane of a circular gear. Diametral pitch is more commonly used to define tooth size and is related to circular pitch by

$$p_d = \frac{N}{d} = \frac{\pi}{P_c} = \frac{\pi}{P_t} \qquad (12.1c)$$

where N is number of teeth and d is pitch diameter.

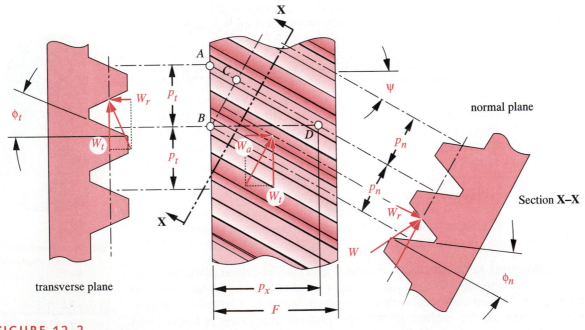

FIGURE 12-2

Basic Helical Rack Showing Normal and Transverse Planes and Resolution of Forces

The diametral pitch in the normal plane is

$$P_{nd} = P_t / \cos \psi \qquad (12.1d)$$

The pressure angles in the two planes are related by

$$\tan \phi_t = \tan \phi = \tan \phi_n / \cos \psi \qquad (12.2)$$

Helical Gear Forces

A set of forces acting on a tooth is shown schematically in Figure 12-2. The resultant force W is at a compound angle defined by the pressure angle and the helix angle in combination. The tangential force component W_t at the mesh can be found from the torque applied either to gear or pinion, as defined in equation 11.13a for the pinion.

$$W_t = \frac{T_p}{r_p} = \frac{2T_p}{d_p} = \frac{2p_d T_p}{N_p} \qquad (11.13a)$$

In addition to the radial component W_r due to the pressure angle, there is now also a component of force W_a, which tends to separate the gears axially. Bearings with axial thrust capability must be used with helical gears to resist this force component, unless helical gears are mounted in opposite-hand pairs on the same shaft to cancel the axial force component. Sometimes for this purpose, both left- and right-hand sets of teeth are cut side by side on the same gear blank with a groove between them to clear the cutter. These are called **double helical gears**. If the clearance groove is eliminated and the opposite-hand teeth cut to run into one another, it is called a **herringbone gear**.

The components of force in a helical gearmesh are

$$W_r = W_t \tan \phi \qquad (12.3a)$$

$$W_a = W_t \tan \psi \qquad (12.3b)$$

$$W = \frac{W_t}{\cos \psi \cos \phi_n} \qquad (12.3c)$$

Virtual Number of Teeth

Another advantage of helical gears over spur gears besides quiet operation is their relatively stronger teeth for a gear with the same normal pitch, pitch diameter and number of teeth. The reason for this can be seen in Figure 12-2. The component of force that transmits the torque is W_t, which lies in the transverse plane. The tooth size (normal pitch) is defined in the normal plane. The thickness of the tooth in the transverse plane is $1 / \cos \psi$ times that of a spur gear of equal normal pitch. Another way to visualize this is to consider the fact that the intersection of the normal plane and the pitch cylinder of diameter d is an ellipse whose radius is $r_e = (d / 2) / \cos^2 \psi$. We can then define a virtual number of teeth N_e as the quotient of the circumference of a virtual pitch circle of radius r_e and the normal pitch p_c:

$$N_e = \frac{2\pi r_e}{p_n} = \frac{\pi d}{p_n \cos^2 \psi} \tag{12.4a}$$

Substitute equation 12.1a for p_n:

$$N_e = \frac{\pi d}{p_t \cos^3 \psi} \tag{12.4b}$$

and substitute $p_t = \pi_d / N$ from equation 12.1c to get

$$N_e = \frac{N}{\cos^3 \psi} \tag{12.4c}$$

This defines a **virtual gear** that is equivalent to a spur gear with N_e teeth thus giving a stronger tooth in both bending and surface fatigue than a spur gear with the same physical number of teeth as the helical gear. The larger number of virtual teeth also reduces undercutting in small pinions allowing a lower minimum number of teeth for helical gears than for spur gears.

Contact Ratios

The **transverse contact ratio** m_p was defined for spur gears in equation 11.7 and is the same for helical gears. The helix angle introduces another ratio called the **axial contact ratio** m_F, which is defined as the quotient of face width F and axial pitch p_x:

$$m_F = \frac{F}{p_x} = \frac{F p_d \tan \psi}{\pi} \tag{12.5}$$

This ratio should be at least 1.15 and indicates the degree of **helical overlap** in the mesh.

Just as a larger transverse contact ratio allows multiple teeth to share the load, a wider face width for a given helix angle will increase the overlapping of teeth and also promote load sharing. However, effective load sharing will still be limited by the accuracy with which the gears are made (see Figure 11-19). Note that larger helix angles will increase the axial contact ratio allowing narrower-width gears to be used, but this will be at the expense of larger axial force components.

If m_F is kept above 1 as desired, the gears are considered conventional helicals. If $m_F < 1$, then they are called **low axial contact ratio** (LACR) gears and their calculation involves additional steps. Consult the AGMA standards[1], [2], [3] for more information on LACR gears. We will only consider conventional helical gears here.

12

Stresses in Helical Gears

The AGMA equations for bending stress and surface stress in spur gears are also used for helical gears. These equations were presented in Chapter 11 with extensive explanation and definition of terms that will not be repeated here. The equations from that chapter are, for bending stress:

$$\sigma_b = \frac{W_t\, p_d}{FJ} \frac{K_a K_m}{K_v} K_s K_B K_I \qquad (11.15us)$$

$$\sigma_b = \frac{W_t}{FmJ} \frac{K_a K_m}{K_v} K_s K_B K_I \qquad (11.15si)$$

and for surface stress:

$$\sigma_c = C_p \sqrt{\frac{W_t}{FId} \frac{C_a C_m}{C_v} C_s C_f} \qquad (11.21)$$

The only significant differences in their application to helical gears involves the geometry factors I and J. The values of J for various combinations of helix angle (10, 15, 20, 25, 30°), pressure angle (14.5, 20, 25°), and addendum ratio (0, 0.25, 0.5) are presented in reference 3. A few examples are reproduced here as Tables 12-1 to 12-6.[*] Consult the AGMA standard for more complete information.

The calculation of I for conventional helical gear pairs requires the inclusion of one additional term in equation 11.22a, which becomes[†]:

$$I = \frac{\cos \phi}{\left(\dfrac{1}{\rho_p} \pm \dfrac{1}{\rho_g}\right) d_p m_N} \qquad (12.6a)$$

The new term m_N is a **load-sharing ratio** defined as

$$m_N = \frac{F}{L_{min}} \qquad (12.6b)$$

where F is the face width. Calculation of the **minimum length of the lines of contact** L_{min} requires several steps. First, two factors must be formed from the residuals of the transverse contact ratio m_p and the axial contact ratio m_F.

$$n_r = \text{fractional part of } m_p$$

$$n_a = \text{fractional part of } m_F \qquad (12.6c)$$

and

$$\text{if } n_a \leq 1 - n_r \text{ then} \qquad L_{min} = \frac{m_p F - n_a n_r p_x}{\cos \psi_b} \qquad (12.6d)$$

* Extracted from AGMA 908-B89, Geometry Factors for Determining the Pitting Resistance and Bending Strength of Spur, Helical, and Herringbone Gear Teeth, with the permission of the American Gear Manufacturers Association, 1500 King St., Suite 201, Alexandria, VA 22314.

† A second additional term is required for LACR helical gears but these will not be addressed here as previously noted.

12

Table 12-1   AGMA Bending Geometry Factor J for $\phi = 20°$, $\psi = 10°$ Full-Depth Teeth with Tip Loading

Gear teeth	Pinion teeth															
	12		14		17		21		26		35		55		135	
	P	G	P	G	P	G	P	G	P	G	P	G	P	G	P	G
12	U	U														
14	U	U	U	U												
17	U	U	U	U	U	U										
21	U	U	U	U	U	U	0.46	0.46								
26	U	U	U	U	U	U	0.47	0.49	0.49	0.49						
35	U	U	U	U	U	U	0.48	0.52	0.50	0.53	0.54	0.54				
55	U	U	U	U	U	U	0.49	0.55	0.52	0.56	0.55	0.57	0.59	0.59		
135	U	U	U	U	U	U	0.50	0.60	0.53	0.61	0.57	0.62	0.60	0.63	0.65	0.65

Table 12-2   AGMA Bending Geometry Factor J for $\phi = 20°$, $\psi = 20°$ Full-Depth Teeth with Tip Loading

Gear teeth	Pinion teeth															
	12		14		17		21		26		35		55		135	
	P	G	P	G	P	G	P	G	P	G	P	G	P	G	P	G
12	U	U														
14	U	U	U	U												
17	U	U	U	U	0.44	0.44										
21	U	U	U	U	0.45	0.46	0.47	0.47								
26	U	U	U	U	0.45	0.49	0.27	0.22	0.50	0.50						
35	U	U	U	U	0.46	0.51	0.27	0.24	0.51	0.53	0.54	0.54				
55	U	U	U	U	0.47	0.54	0.27	0.26	0.52	0.56	0.55	0.57	0.58	0.58		
135	U	U	U	U	0.48	0.58	0.27	0.28	0.54	0.60	0.57	0.61	0.60	0.62	0.64	0.64

Table 12-3   AGMA Bending Geometry Factor J for $\phi = 20°$, $\psi = 30°$ Full-Depth Teeth with Tip Loading

Gear teeth	Pinion teeth															
	12		14		17		21		26		35		55		135	
	P	G	P	G	P	G	P	G	P	G	P	G	P	G	P	G
12	U	U														
14	U	U	0.39	0.39												
17	U	U	0.39	0.41	0.41	0.41										
21	U	U	0.40	0.43	0.42	0.43	0.44	0.44								
26	U	U	0.41	0.44	0.43	0.45	0.45	0.46	0.46	0.46						
35	U	U	0.41	0.46	0.43	0.47	0.45	0.48	0.47	0.48	0.49	0.49				
55	U	U	0.42	0.49	0.44	0.49	0.46	0.50	0.48	0.50	0.50	0.51	0.52	0.52		
135	U	U	0.43	0.51	0.45	0.52	0.47	0.53	0.49	0.53	0.51	0.54	0.53	0.55	0.56	0.56

12

Table 12-4   AGMA Bending Geometry Factor J for $\phi = 25°$, $\psi = 10°$ Full-Depth Teeth with Tip Loading

Gear teeth	Pinion teeth															
	12		14		17		21		26		35		55		135	
	P	G	P	G	P	G	P	G	P	G	P	G	P	G	P	G
12	U	U														
14	U	U	0.47	0.47												
17	U	U	0.48	0.51	0.52	0.52										
21	U	U	0.48	0.55	0.52	0.55	0.56	0.56								
26	U	U	0.49	0.58	0.53	0.58	0.57	0.59	0.60	0.60						
35	U	U	0.50	0.61	0.54	0.62	0.57	0.63	0.61	0.64	0.64	0.64				
55	U	U	0.51	0.65	0.55	0.66	0.58	0.67	0.62	0.68	0.65	0.69	0.70	0.70		
135	U	U	0.52	0.70	0.56	0.71	0.60	0.72	0.63	0.73	0.67	0.74	0.71	0.75	0.76	0.76

Table 12-5   AGMA Bending Geometry Factor J for $\phi = 25°$, $\psi = 20°$ Full-Depth Teeth with Tip Loading

Gear teeth	Pinion teeth															
	12		14		17		21		26		35		55		135	
	P	G	P	G	P	G	P	G	P	G	P	G	P	G	P	G
12	0.47	0.47														
14	0.47	0.50	0.50	0.50												
17	0.48	0.53	0.51	0.54	0.54	0.54										
21	0.48	0.56	0.51	0.57	0.55	0.58	0.58	0.58								
26	0.49	0.59	0.52	0.60	0.55	0.60	0.69	0.61	0.62	0.62						
35	0.49	0.62	0.53	0.63	0.56	0.64	0.60	0.64	0.62	0.65	0.66	0.66				
55	0.50	0.66	0.53	0.67	0.57	0.67	0.60	0.68	0.63	0.69	0.67	0.10	0.71	0.71		
135	0.51	0.70	0.54	0.71	0.58	0.72	0.62	0.72	0.65	0.73	0.68	0.74	0.72	0.75	0.76	0.76

Table 12-6   AGMA Bending Geometry Factor J for $\phi = 25°$, $\psi = 30°$ Full-Depth Teeth with Tip Loading

Gear teeth	Pinion teeth															
	12		14		17		21		26		35		55		135	
	P	G	P	G	P	G	P	G	P	G	P	G	P	G	P	G
12	0.46	0.46														
14	0.47	0.49	0.49	0.49												
17	0.47	0.51	0.50	0.52	0.52	0.52										
21	0.48	0.54	0.50	0.54	0.53	0.55	0.55	0.55								
26	0.48	0.56	0.51	0.56	0.53	0.57	0.56	0.57	0.58	0.58						
35	0.49	0.58	0.51	0.59	0.54	0.59	0.56	0.60	0.58	0.60	0.61	0.61				
55	0.49	0.61	0.52	0.61	0.54	0.62	0.57	0.62	0.59	0.63	0.62	0.64	0.64	0.64		
135	0.50	0.64	0.53	0.64	0.55	0.65	0.58	0.66	0.60	0.66	0.62	0.67	0.65	0.68	0.68	0.68

12

if $n_a > 1 - n_r$ then $L_{min} = \dfrac{m_p F - (1 - n_a)(1 - n_r) p_x}{\cos \psi_b}$ (12.6e)

All the factors in these equations are defined either in this section or in Chapter 11 except for ψ_b, the base helix angle, which is

$$\psi_b = \cos^{-1}\left(\cos \psi \, \frac{\cos \phi_n}{\cos \phi} \right) \qquad (12.6f)$$

Also, the radius of curvature of a helical pinion for equation 12.6a is calculated with a different formula than that used for spur gears. Instead of equation 11.22b, use

$$\rho_p = \sqrt{\left\{ 0.5\left[(r_p + a_p) \pm (C - r_g + a_g) \right] \right\}^2 - (r_p \cos \phi)^2}$$

$$\rho_g = C \sin \phi \mp \rho_p \qquad (12.6g)$$

where (r_p, a_p) and (r_g, a_g) are the (pitch radius, addendum) of the pinion and gear, respectively, and C is the actual (operating) center distance.

The bending and surface stresses can be calculated from the above equations using the data in Tables 12-1 to 12-6. The material strengths can be found in Chapter 11 and safety factors calculated in the same manner as described there for spur gears.

EXAMPLE 12-1

Stress Analysis of a Helical-Gear Train

Problem
Redesign the spur-gear train of Examples 11-4 through 11-7 using helical gears and compare their safety factors.

Given
The referenced examples address, respectively, the kinematics, bending stresses, surface stresses, and safety factors for a 3-gear train with the following data: $W_t = 432.17$ lb, $N_p = 14$, $N_{idler} = 17$, $N_g = 49$, $\phi = 25°$, $p_d = 6$, $F = 2.667$ in, pinion speed = 2 500 rpm, and 20 hp. The velocity factor $K_v = 0.66$ from previous calculations.

Assumptions
The teeth are standard AGMA full-depth profiles. The load and source are both uniform in nature. A gear quality index of 6 will be used. All gears are steel with $v = 0.28$. The service life required is 5 years of one-shift operation. Operating temperature is 200 °F. Based on the assumption of uniform load and source, the application factor K_a can be set to 1. The load distribution factor can be estimated from Table 11-16 based on the assumed face width: $K_m = 1.6$. The idler factor $K_I = 1$ for the pinion and gear and $K_I = 1.42$ for the idler gear. The size factor $K_s = 1$ for all three gears. $C_f = 1$. $K_B = 1$. Keep the same ϕ and p_d as the previous examples' spur gears and try a 20° helix angle.

Solution   See the *TKSolver* file EX12-01.

1 The bending geometry factor J for a 25° pressure angle, 20° helix angle, 14-tooth pinion in mesh with the 17-tooth idler is found from Table 12-5 to be $J_{pinion} = 0.51$. The pinion-tooth bending stress is then

$$\sigma_{b_p} = \frac{W_t \, P_d}{FJ} \frac{K_a K_m}{K_v} K_s K_B K_I \cong \frac{432.17(6)}{2.667(0.51)} \frac{1(1.6)}{0.66}(1)(1)(1) \cong 4\,620 \text{ psi} \qquad (a)$$

2 The bending geometry factor J for the 25°, 17-tooth idler in mesh with the 14-tooth pinion from Table 11-13 is $J_{idler} = 0.54$. The idler-tooth bending stress is then

$$\sigma_{b_i} = \frac{W_t \, P_d}{FJ} \frac{K_a K_m}{K_v} K_s K_B K_I \cong \frac{432.17(6)}{2.667(0.54)} \frac{1(1.6)}{0.659}(1)(1)(1.42) \cong 6\,200 \text{ psi} \qquad (b)$$

Note in Table 11-13 that the idler has a different J factor when considered to be the "gear" in mesh with the smaller pinion than when considered to be the "pinion" in mesh with the larger gear. The smaller value of the two is used because it gives the higher stress.

3 The bending geometry factor J for the 25°, 49-tooth gear in mesh with the 17-tooth idler is found from Table 12-5 as $J_{gear} = 0.57$. The gear-tooth bending stress is then

$$\sigma_{b_g} = \frac{W_t \, P_d}{FJ} \frac{K_a K_m}{K_v} K_s K_B K_I \cong \frac{432.17(6)}{2.667(0.57)} \frac{1(1.6)}{0.66}(1)(1)(1) \cong 4\,130 \text{ psi} \qquad (c)$$

4 The pitting geometry factor I is calculated for a pair of gears in mesh. Since we have two meshes (pinion/idler and idler/gear) there will be two different values of I to be calculated using equations 12.6.

$$I_{pi} = 0.144 \qquad\qquad I_{gi} = 0.208 \qquad (d)$$

5 The elastic coefficient C_p is found from equation 11.23 and, as before, is 2 276.

6 The surface stress for the pinion teeth is then

$$\sigma_{c_p} = C_p \sqrt{\frac{W_t}{FI_{pi} d_p} \frac{C_a C_m}{C_v} C_s C_f}$$

$$\cong 2\,276 \sqrt{\frac{432.17}{2.667(0.144)(2.33)} \frac{1(1.6)}{0.66}(1)(1)} \cong 77\,700 \text{ psi} \qquad (e)$$

7 The surface stress for the idler teeth is then

$$\sigma_{c_i} = C_p \sqrt{\frac{W_t}{FI_{pi} d_p} \frac{C_a C_m}{C_v} C_s C_f}$$

$$\cong 2\,276 \sqrt{\frac{432.17}{2.667(0.144)(2.83)} \frac{1(1.6)}{0.66}(1)(1)} \cong 70\,500 \text{ psi} \qquad (f)$$

12

8   The surface stress for the gear teeth is then

$$\sigma_{c_g} = C_p \sqrt{\frac{W_t}{F I_{ig} d_p} \frac{C_a C_m}{C_v} C_s C_f}$$

$$\cong 2\,276 \sqrt{\frac{432.17}{2.667(0.208)(8.17)} \frac{1(1.6)}{0.66}} (1)(1) \cong 34\,600 \text{ psi} \qquad (g)$$

9   The corrected bending-fatigue strength of the steel from Example 11-7 is 38 937 psi.

10   The corrected surface-fatigue strength from Example 11-7 is 105 063 psi.

11   The safety factors against bending failure are found by comparing the corrected bending strength to the bending stress for each gear in the mesh:

$$N_{b_{pinion}} = \frac{S_{fb}}{\sigma_{b_{pinion}}} = \frac{38\,937}{4\,620} \cong 8.4 \qquad (h)$$

$$N_{b_{idler}} = \frac{S_{fb}}{\sigma_{b_{idler}}} = \frac{38\,937}{6\,200} = 6.3 \qquad (i)$$

$$N_{b_{gear}} = \frac{S_{fb}}{\sigma_{b_{gear}}} = \frac{38\,937}{4\,130} = 9.4 \qquad (j)$$

12   The safety factors against surface failure are found by comparing the corrected surface strength to the surface stress for each gear in the mesh:

$$N_{c_{pinion}} = \frac{S_{fc}}{\sigma_{c_{pinion}}} = \frac{105\,063}{77\,700} \cong 1.4 \qquad (k)$$

$$N_{c_{idler}} = \frac{S_{fc}}{\sigma_{c_{idler}}} = \frac{105\,063}{70\,500} \cong 1.5 \qquad (l)$$

$$N_{c_{gear}} = \frac{S_{fc}}{\sigma_{c_{gear}}} = \frac{105\,063}{34\,600} = 3.0 \qquad (m)$$

13   Compare these results to the safety factors for the spur-gear train in Example 11-7:

$$N_{b_{pinion}} = 5.1 \qquad N_{b_{idler}} = 3.9 \qquad N_{b_{gear}} = 7.1$$

$$N_{c_{pinion}} = 1.0 \qquad N_{c_{idler}} = 1.1 \qquad N_{c_{gear}} = 2.2 \qquad (n)$$

14   The helical gears have significantly larger safety factors than same-pitch spur gears.

12.2   BEVEL GEARS

Bevel gears are cut on mating cones rather than the mating cylinders of spur or helical gears. Their axes are nonparallel and intersect at the apices of the mating cones. The angle between their axes can be any value and is often 90°. If the teeth are cut parallel to the cone axis, they are **straight bevel gears**, analogous to spur gears. If the teeth are cut at a spiral angle ψ to the cone axis they are **spiral bevel gears**, analogous to helical gears. Contact between the teeth of straight or spiral bevel gears has the same attributes as their analogous cylindrical counterparts, with the result that spiral bevels run quieter and smoother than straight bevels and spirals can be smaller in diameter for the same load capacity.

Figure 12-3*a* shows a pair of straight bevel gears and Figure 12-3*b* a pair of spiral bevels. Another form is the *ZEROL*® gear (not shown), which has curved teeth like a spiral gear but has a zero spiral angle like a straight bevel gear. Zerol gears have some of the quietness and smooth-running characteristics of spiral gears. Spirals are the ultimate in smoothness and quiet running and are recommended for speeds up to 8 000 fpm (40 m/sec). Higher speeds require precision-finished gears. Straight helicals are limited to speeds of about 1 000 fpm (10 m/sec). Zerols can run as fast as spirals. As with spur and helical gears, a maximum reduction of 10:1 is recommended for any one bevel or spiral gear set. A 5:1 limit is recommended when used as a speed increaser. The torque on the pinion is used as a rating parameter. The most common pressure angle for bevels or spirals is $\phi = 20°$. Spirals most often have a 35° spiral angle ψ. Bevel gears in general are not interchangeable. They are made and replaced as matched sets of pinion and gear.

(a)

(b)

FIGURE 12-3

(a) Straight Bevel Gears *Courtesy of Martin Sprocket and Gear Co. Arlington Tex.* and (b) Spiral Bevel Gears *Courtesy of the Boston Gear Division of IMO Industries, Quincy Mass.*

Bevel Gear Geometry and Nomenclature

Figure 12-4 shows a cross section of two bevel gears in mesh. Their pitch-cone angles are denoted by α_p and α_g for pinion and gear, respectively. The pitch diameters are defined at the large end, on the back cones. The tooth size and shape are defined on the back cone and are similar to a spur-gear tooth with a long addendum pinion to minimize interference and undercutting. The addendum ratio varies with the gear ratio from equal addenda (full-depth teeth) at a 1:1 gear ratio to about a 50% longer pinion addendum at gear ratios above 6:1. The face width F is generally limited to $L / 3$ with L as defined in Figure 12-4. From geometry:

$$L = \frac{r_p}{\sin \alpha_p} = \frac{d_p}{2 \sin \alpha_p} = \frac{d_g}{2 \sin \alpha_g} \tag{12.7a}$$

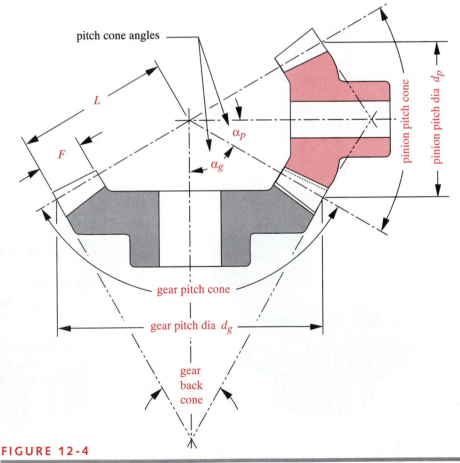

FIGURE 12-4

Bevel Gear Geometry and Nomenclature Source: Extracted from AGMA Standard 2005-B88, *Design Manual for Bevel Gears*, with the permission of the publisher, the American Gear Manufacturers Association, 1500 King St., Suite 201, Alexandria, Va., 22314

The gear ratio m_G for a 90° bevel set can be defined in terms of the pitch cone angles as

$$m_G = \frac{\omega_p}{\omega_g} = \frac{N_g}{N_p} = \frac{d_g}{d_p} = \tan \alpha_g = \cot \alpha_p \tag{12.7b}$$

See also equations 11.1*a* through 11.1*c*.

Bevel Gear Mounting

Straddle mounting (bearings on both sides of the tooth plane) is preferred for best support, but is difficult to achieve on both pinion and gear with intersecting shafts. The gear is usually straddle mounted and the pinion cantilevered unless there is room enough to provide a bearing on the inside of the pinion to straddle mount it as well.

Forces on Bevel Gears

As in helical gears, there are tangential, radial, and axial force components acting on a bevel or spiral gear. For a straight bevel gear:

$$W_a = W_t \tan \phi \sin \alpha$$
$$W_r = W_t \tan \phi \cos \alpha \tag{12.8a}$$
$$W = W_t / \cos \phi$$

For a spiral bevel gear:

$$W_a = \frac{W_t}{\cos \psi} (\tan \phi_n \sin \alpha \mp \sin \psi \cos \alpha)$$

$$W_r = \frac{W_t}{\cos \psi} (\tan \phi_n \cos \alpha \pm \sin \psi \sin \alpha) \tag{12.8b}$$

where the upper signs of the $\pm$ and $\mp$ are used for a driving pinion with a RH spiral rotating CW viewed from its large end, or a driving pinion with a LH spiral rotating CCW viewed from its large end, and the lower signs used for the opposite conditions.

In equations 12.8*a* and 12.8*b*, use the appropriate pitch cone angle α_p for the pinion or α_g for the gear in place of α to obtain the forces on each member. The tangential load W_t can be found from the torque applied to either member in combination with its pitch diameter d.

$$W_t = \frac{2T}{d} \tag{12.8c}$$

where the torque acting on, and the diameter of, the same element (gear or pinion) are used to find the common transmitted force.

Stresses in Bevel Gears

The calculation of stresses and life estimates for bevel gears is more complicated than for spur or helical gears. The AGMA standards[4], [5] provide more complete information than can be presented here and should be consulted for any actual design applications. We will present only a brief summary of the approach to bevel gear design as an introduction to the subject, suitable for a basic understanding of the factors involved and for the execution of some exercises.[*]

BENDING STRESS IN BEVEL GEARS   The bending stress in straight or spiral bevel gears is found with essentially the same equation as used for spur or helical gears. The principal difference is accounted for with the value of the J factor.

$$\sigma_b = \frac{2T_p}{d}\frac{P_d}{FJ}\frac{K_aK_mK_s}{K_vK_x} \qquad \text{psi} \qquad (12.9us)$$

$$\sigma_b = \frac{2000T_p}{d}\frac{1}{FmJ}\frac{K_aK_mK_s}{K_vK_x} \qquad \text{MPa} \qquad (12.9si)$$

Note that the applied load is expressed in terms of the pinion torque T_p by substituting equation 12.8c rather than using W_t as in equation 11.15. The pitch diameter d in equation 12.9 is that of the pinion.   (The SI formula has lengths expressed in mm.) For our purposes, the factors K_a, K_m, K_s, and K_v can be taken to be the same as were defined in Chapter 11 for spur gears. However, some of these factors have slightly different definitions for bevel gears in the AGMA standard[4], [5] and it should be consulted for the most accurate formulas in any actual design application. The factor $K_x = 1$ for straight bevel gears and is a function of the cutter radius for spiral or *Zerol* gears. Use $K_x = 1.15$ as an approximation in the latter two cases.

SURFACE STRESS IN BEVEL GEARS   The surface stress in straight or spiral bevel gears is calculated in similar fashion to that of spur or helical gears, but with some additional adjustment factors included. As with the bending stress for bevel gears, the applied load is expressed as pinion torque rather than as a tangential load.

$$\sigma_c = C_pC_b\sqrt{\frac{2T_D}{FId^2}\left(\frac{T_p}{T_D}\right)^z\frac{C_aC_m}{C_v}C_sC_fC_{xc}} \qquad (12.10)$$

For our purposes, the factors C_p, C_a, C_m, C_v, C_s, and C_f can be taken to be the same as those defined in Chapter 11. However, some of these factors have slightly different definitions for bevel gears in the AGMA standard[5] and it should be consulted for the most accurate formulas in any actual design application. The adjustment factors new to this version of the surface-stress equation versus equation 11.21 are C_b, which is a **stress adjustment constant**, defined as 0.634 by the current AGMA standard,[5] and C_{xc}, a **crowning factor** defined as 1.0 for uncrowned teeth and 1.5 for crowned teeth. The exponent z is 0.667 when $T_p < T_D$ and 1.0 otherwise.

* Extracted from AGMA Standard 2005-B88, *Design Manual for Bevel Gears*, and/or AGMA 2003-A86, *Rating the Pitting Resistance and Bending Strength of Generated Straight Bevel, ZEROL® Bevel, and Spiral Bevel Gear Teeth*, with the permission of the American Gear Manufacturers Association, 1500 King St., Suite 201, Alexandria, Va., 22314.

12

The two torque terms T_D and T_p require some explanation. T_p is the **operating pinion torque**, defined by the applied loads, the applied torque, or the power and speed and may be time-varying. T_D is the **design pinion torque**, which is the minimum value needed to produce the full (optimum) contact patch on the gear teeth. In most cases T_D is the torque necessary to create a contact stress equal to the allowable contact stress for the material as defined in Table 11-21. T_D can be estimated from

$$T_D = \frac{F}{2} \frac{IC_v}{C_s C_{md} C_f C_a C_{xc}} \left(\frac{S'_{fc} d}{C_p C_b} \frac{0.774 C_H}{C_T C_R} \right)^2 \quad \text{psi} \qquad (12.11us)$$

$$T_D = \frac{F}{2000} \frac{IC_v}{C_s C_{md} C_f C_a C_{xc}} \left(\frac{S'_{fc} d}{C_p C_b} \frac{0.774 C_H}{C_T C_R} \right)^2 \quad \text{MPa} \qquad (12.11si)$$

where S'_{fc} is the material's surface fatigue strength from Table 11-21 and the C-factors are as defined above or in Chapter 11. (See equation 11.25 for C_H, C_T, and C_R.) C_{md} is a **mounting factor** to account for cantilever or straddle mounting of one or both gears. If the gear teeth are crowned, C_{md} varies from 1.2 for both members straddle mounted to 1.8 if both are cantilevered. Use a value between these two numbers if one member is cantilevered and the other straddle mounted. For uncrowned teeth, double these numbers. See the AGMA standard[5] for more detailed information.

GEOMETRY FACTORS I AND J   The geometry factors for straight and spiral bevel gears are different than those for either spur or helical gears. The AGMA standard provides charts of these factors for straight, Zerol, and spiral gears. A few of these charts are reproduced here as Figures 12-5 through 12-8.[*]

SAFETY FACTORS   The safety factors against bending or pitting failure are calculated in the same manner as outlined for spur gears in Chapter 11.

EXAMPLE 12-2

Stress Analysis of a Bevel-Gear Train

Problem   Determine the bending and surface stresses and safety factors in a straight bevel gearset made of the same steel materials, and operating under the same conditions for the same 5-year life as in Example 11-7.

Given $N_p = 20$, $N_g = 35$, $\phi = 25°$, and $p_d = 8$, passing 10 hp at 2 500 rpm. From Example 11-7: the corrected bending strength is 38 937 psi, and the surface strength is 118 000 psi uncorrected and 105 063 psi corrected.

Assumptions   From Example 11-7: $K_a = C_a = K_s = C_s = C_f = C_H = C_R = C_T = 1$, $K_m = C_m = 1.6$, $K_v = C_v = 0.652$, $C_L = 0.890$, and $C_p = 2\ 276$. From this section: $C_{xc} = K_x = 1$, $C_b = 0.634$, and $C_{md} = 1.5$.

12

* Extracted from AGMA Standard 2005-B88, *Design Manual for Bevel Gears,* and/or AGMA 2003-A86, *Rating the Pitting Resistance and Bending Strength of Generated Straight Bevel, ZEROL® Bevel, and Spiral Bevel Gear Teeth*, with the permission of the American Gear Manufacturers Association, 1500 King St., Suite 201, Alexandria, Va., 22314.

Solution   See the *TKSolver* file EX12-02

1   Determine the pinion torque from the given power and speed.

$$T_p = \frac{P}{\omega_p} = \frac{10 \text{ hp} \left(6\,600 \frac{\text{in - lb}}{\text{sec}} \big/ \text{hp} \right)}{2\,500 \text{ rpm} \ (2\pi/60) \frac{\text{rad}}{\text{sec}} \big/ \text{rpm}} = 252.1 \text{ lb - in} \qquad (a)$$

2   Find the pitch diameters of pinion and gear.

$$d_p = \frac{N_p}{p_d} = \frac{20}{8} = 2.50 \text{ in}, \qquad\qquad d_g = \frac{35}{8} = 4.375 \text{ in} \qquad (b)$$

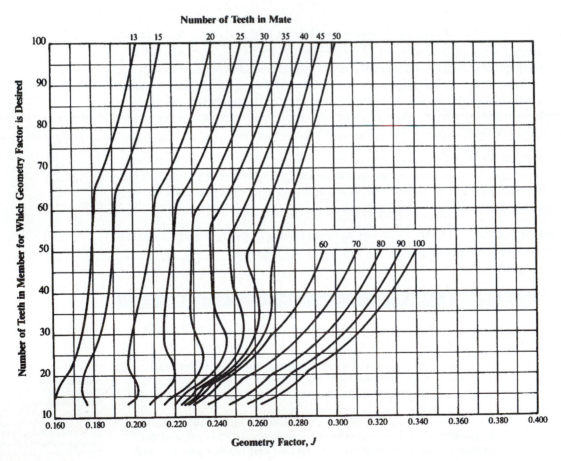

FIGURE 12-5

Geometry Factor *J* for Straight Bevel Gears with $\phi = 20°$ and $0.120/p_d$ Tool-Edge Radius Source: Extracted from AGMA Standard 2003-A86, *Rating the Pitting Resistance and Bending Strength of Generated Straight Bevel, ZEROL*® *Bevel, and Spiral Bevel Gear Teeth*, with the permission of the publisher, the American Gear Manufacturers Association, 1500 King St., Suite 201, Alexandria, Va., 22314

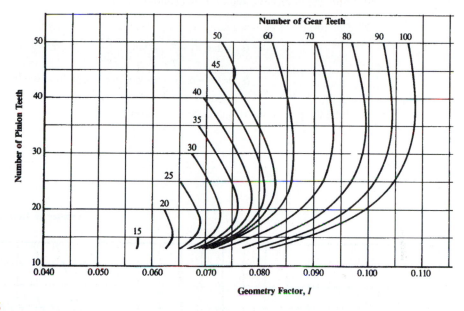

FIGURE 12-6

Geometry Factor *I* for Straight Bevel Gears with ϕ = 20° and 0.120/p_d Tool-Edge Radius Source: Extracted from AGMA Standard 2003-A86, *Rating the Pitting Resistance and Bending Strength of Generated Straight Bevel, ZEROL® Bevel, and Spiral Bevel Gear Teeth*, with the permission of the publisher, the American Gear Manufacturers Association, 1500 King St., Suite 201, Alexandria, Va., 22314

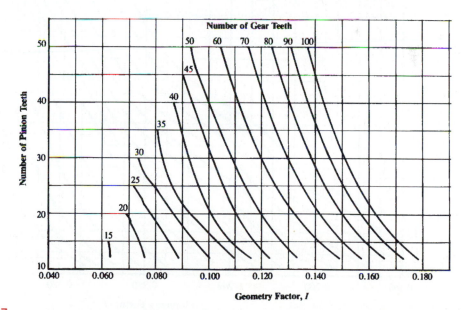

FIGURE 12-7

Geometry Factor *I* for Spiral Bevel Gears with ϕ = 20°, Spiral Angle ψ = 35°, and 0.240/p_d Tool-Edge Radius Source: Extracted from AGMA Standard 2003-A86, *Rating the Pitting Resistance and Bending Strength of Generated Straight Bevel, ZEROL® Bevel, and Spiral Bevel Gear Teeth*, with the permission of the publisher, the American Gear Manufacturers Association, 1500 King St., Suite 201, Alexandria, Va., 22314

3   Find the pitch cone angles from equation 12.7*b*:

$$\alpha_g = \tan^{-1}\left(\frac{N_g}{N_p}\right) = \tan^{-1}\left(\frac{35}{20}\right) = 60.26°$$

$$\alpha_p = 90 - \alpha_g = 90 - 60.26 = 29.74° \qquad (c)$$

4   Find the pitch cone length *L* from equation 12.7*a*:

$$L = \frac{d_p}{2\sin\alpha_p} = \frac{2.50}{2\sin 29.74} = 2.519 \text{ in} \qquad (d)$$

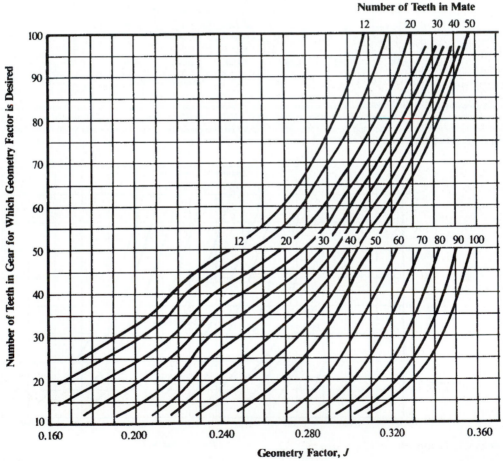

FIGURE 12-8

Geometry Factor *J* for Spiral Bevel Gears with $\phi = 20°$, Spiral Angle $\psi = 35°$, and $0.240/p_d$ Tool-Edge Radius. Source: Extracted from AGMA 2003-A86, *Rating the Pitting Resistance and Bending Strength of generated Straight Bevel, ZEROL® Bevel, and Spiral Bevel Gear Teeth*, with the permission of the publisher, the American Gear Manufacturers Association, 1500 King St., Suite 201, Alexandria, VA, 22314

5 Use the pitch cone length L to find a suitable face width, set to the maximum recommended value.

$$F = \frac{L}{3} = \frac{2.519}{3} = 0.840 \text{ in} \qquad (e)$$

6 Look up the bending geometry factors for pinion and gear in Figure 12-5 to find $J_p = 0.237$ and $J_g = 0.201$.

7 Find the bending stress in the pinion from equation 12.9 using J_p.

$$\sigma_{b_{pinion}} = \frac{2T_p}{d} \frac{P_d}{FJ} \frac{K_a K_m K_s}{K_v K_x} = \frac{2(252.1)}{2.5} \frac{8}{0.840(0.237)} \frac{1(1.6)(1)}{0.652(1)} \cong 19\,880 \text{ psi} \qquad (f)$$

8 Find the bending stress in the gear from equation 12.9 using J_g.

$$\sigma_{b_{gear}} = \frac{2T_p}{d} \frac{P_d}{FJ} \frac{K_a K_m K_s}{K_v K_x} = \frac{2(252.1)}{2.5} \frac{8}{0.840(0.201)} \frac{1(1.6)(1)}{0.652(1)} \cong 23\,440 \text{ psi} \qquad (g)$$

Note that the gear tooth is more highly stressed than the pinion tooth because the long addendum on the pinion makes it stronger at the expense of the short-addendum gear tooth.

9 Look up the surface geometry factor for this combination of pinion and gear in Figure 12-6 to find $I = 0.076$. Use this in equation 12.11 to find T_D.

$$T_D = \frac{F}{2} \frac{IC_v}{C_s C_{md} C_f C_a C_{xc}} \left(\frac{S'_{fc}d}{C_p C_b} \frac{0.774 C_H}{C_T C_R} \right)^2$$

$$= \frac{0.840}{2} \frac{0.076(0.652)}{1(1.5)(1)(1)(1)} \left(\frac{118\,000(2.5)}{2\,276(0.634)} \frac{0.774(1)}{1(1)} \right)^2 \cong 275.48 \text{ lb-in} \qquad (h)$$

10 Since $T_D > T_p$, $z = 0.667$. Use these data to find the surface stress with equation 12.10.

$$\sigma_c = C_p C_b \sqrt{ \frac{2T_D}{FId^2} \left(\frac{T_p}{T_D} \right)^z \frac{C_a C_m}{C_v} C_s C_f C_{xc} }$$

$$= 2\,276(0.634) \sqrt{ \frac{2(275.48)}{0.840(0.076)(2.5)^2} \left(\frac{252.1}{275.48} \right)^{0.667} \frac{1(1.6)}{0.652} (1)(1)(1) }$$

$$\cong 81\,538 \text{ psi} \qquad (i)$$

11 The safety factors can now be found as

$$N_{b_{pinion}} = \frac{S_{fb}}{\sigma_{b_{pinion}}} = \frac{38\,937}{19\,880} \cong 2.0 \qquad (j)$$

12

$$N_{b_{gear}} = \frac{S_{fb}}{\sigma_{b_{gear}}} = \frac{38\,937}{23\,440} \cong 1.7 \qquad (k)$$

$$N_c = \frac{S_{fc}}{\sigma_c} = \frac{105\,063}{81\,538} \cong 1.3 \qquad (l)$$

12   These are acceptable safety factors. See the *TKSolver* file EX12-02 for more information.

worm gear

worm

FIGURE 12-9

A Single-Enveloping Wormset, Consisting of a Worm and an Enveloping Worm Gear *Courtesy of Martin Sprocket and Gear Co., Arlington, Tex.*

12

---

* Extracted from AGMA Standard 6022-C93, *Design Manual for Cylindrical Wormgearing,* and/or AGMA Standard 6034-B92, *Practice for Enclosed Cylindrical Wormgear Speed Reducers and Gearmotors,* with the permission of the American Gear Manufacturers Association, 1500 King St., Suite 201, Alexandria, Va., 22314.

12.3   WORMSETS

Worm gearing is more complicated to design than conventional gearing. We present only a brief look at the process here as an introduction to the topic. The AGMA standards contain much more information. For any real applications, the reader is encouraged to consult the AGMA documents.[6], [7] They contain many tables of data needed for a complete design. Most of the relevant equations have been excerpted from the standard but its tabular data are not presented here. Instead, empirical equations from the AGMA standard's appendix[7] are included for calculation of the tabular data.*

A wormset consists of a worm and a worm gear (also called a worm wheel) as shown in Figure 12-9. They connect nonparallel, nonintersecting shafts, typically at right angles to one another. The worm is, in effect, a helical gear with a helix angle so large that a single tooth wraps continuously around its circumference. The worm is analogous to a screw thread and the worm gear is analogous to its nut. The distance that a point on the mating gear (nut) moves axially in one revolution of the worm is called the **lead** L and the lead divided by the pitch circumference πd of the worm is the tangent of its **lead angle** λ.

$$\tan\lambda = \frac{L}{\pi d} \qquad (12.12)$$

Worms commonly have only one tooth (or thread) and thus can create ratios as large as the number of teeth on the worm gear. This ability to provide high ratios in a compact package is one of the principle advantages of a wormset over other possible gearing configurations, most of which are limited to about a 10:1 ratio per pair of gears. Wormsets can be produced with ratios ranging from 1:1 to 360:1, though the usual range available from catalogs is 3:1 to 100:1. Ratios above 30:1 usually have a single-thread worm, and ratios below that value often use multiple-thread worms. The number of threads on the worm is also referred to as its number of *starts*. A 2- or 3-start worm might be used for a low ratio wormset, for example. The axial pitch p_x of the worm equals the circular pitch p_c of the worm gear and is related to the lead L by the chosen number of starts or number of teeth on the worm N_w.

$$p_x = \frac{L}{N_w} = p_c = \frac{\pi d_g}{N_g} \qquad\qquad (12.13)$$

where d_g is the pitch diameter and N_g is the number of teeth on the worm gear. The number of starts N_w is typically between 1 and 10 for commercial wormsets, though more starts may be used on large wormsets.

Another advantage of wormsets over other types of gearsets is their ability to self-lock. If the wormset is self-locking, it will not backdrive, i.e., a torque applied to the worm gear will not rotate the worm. A self-locking wormset can only be driven "forward" from worm to worm gear. Thus it can be used to hold a load as, for example, in jacking up a car. Whether or not a particular wormset will be self-locking depends on a number of factors, including the ratio of tan λ to the coefficient of friction μ, the surface finish, lubrication, and vibration. Generally, self-locking occurs at lead angles below 6° and may occur at lead angles as high as 10°.[8] (See Section 14.2 for a complete discussion of self-locking as it applies to power screws, the principles of which are equally applicable to wormsets.)

Standard pressure angles for wormsets are 14.5, 17.5, 20, 22.5, 25, 27.5, or 30°. Higher pressure angles give higher tooth strengths at the expense of higher friction, larger bearing loads, and higher bending stresses in the worm. For high-power applications at high speed, a relatively fine-pitch worm gear should be used. High torques at low speeds need coarse pitch and larger worm diameters.

The tooth forms for worms and worm gears are not involutes and there are large sliding-velocity components in the mesh. Worms and worm gears are not interchangeable, but are made and replaced as matched sets. To increase contact area between the teeth, either single-enveloping or double-enveloping tooth forms are used. A single-enveloping set (as shown in Figure 12-9) wraps the worm gear teeth partially around the worm. A double-enveloping set also wraps the worm partially around the gear, making the worm an hourglass shape instead of a cylinder. These configurations increase manufacturing complexity and cost, but also increase load capacity. Both types are available commercially.

Materials for Wormsets

Only a few materials are suitable for wormsets. The worm is highly stressed and requires a hardened steel. Low-carbon steels such as AISI 1020, 1117, 8620, or 4320 are used and case-hardened to HRC 58-62. Medium-carbon steels such as AISI 4140 or 4150 are also used, induction- or flame-hardened to a case of HRC 58-62. They need to be ground or polished to a finish of 16 μin (0.4 μm) R_a or better. The wormgear needs to be made from a material soft and compliant enough to run-in and conform to the hard worm under the high-sliding conditions. Sand cast, chill-cast, centrifugal-cast, or forged bronze is typically used for the worm gear. Phosphor or tin bronze is used for high-power applications and manganese bronze for small, slower-speed worms. Cast iron, soft steel, and plastics are sometimes used for lightly loaded, low-speed applications.

Lubrication in Wormsets

The lubrication condition in a wormset can range from boundary lubrication to partial or full EHD depending on loads, velocities, temperatures, and lubricant viscosity as discussed in Chapter 10. The lubrication situation is more like that of sliding bearings than rolling bearings in this instance because of the dominant sliding velocities. Their high percentage of sliding makes wormsets less efficient than conventional gearsets. Lubricants containing extreme pressure (EP) additives are sometimes used in wormsets.

Forces in Wormsets

A three-dimensional loading condition exists at the mesh of a wormset. Tangential, radial, and axial components act on each member. With a (typical) 90° angle between the axes of worm and worm gear, the magnitude of the tangential component on the worm gear W_{tg} equals the axial component on the worm W_{aw} and vice versa. These components can be defined as

$$W_{tg} = W_{aw} = \frac{2T_g}{d_g} \tag{12.14a}$$

where T_g and d_g are the torque on, and the pitch diameter of, the worm gear. The axial force W_{ag} on the worm gear and the tangential force on the worm W_{tw} are

$$W_{ag} = W_{tw} = \frac{2T_w}{d} \tag{12.14b}$$

where T_w is the torque on, and d is the pitch diameter of, the worm. The radial force W_r separating the two elements is

$$W_r = \frac{W_{tg} \tan \phi}{\cos \lambda} \tag{12.14c}$$

where ϕ is the pressure angle and λ is the lead angle.

Wormset Geometry

The pitch diameters and numbers of teeth of nonworm gearsets have a unique relationship but this is not so in wormsets. Once the decision is made regarding the number of starts or teeth N_w desired on the worm, the number of teeth on the worm gear N_g is defined by the required gear ratio m_G:

$$N_g = m_G N_w \tag{12.15}$$

However, the pitch diameter of the worm is not tied to these tooth numbers as with other gearsets. The worm can theoretically be any diameter as long as its tooth cross section (axial pitch) matches the circular pitch of the worm gear. (This is analogous to different diameter machine screws having the same thread pitch as with #6-32, 8-32, and 10-32 sizes). Thus the worm's pitch diameter d can be selected independent of the worm gear diameter d_w and, for any given d_w, changes in d will just vary the center distance C between worm and worm gear but will not affect the gear ratio. The AGMA recommends minimum and maximum values for the worm pitch diameter as

$$\frac{C^{0.875}}{3} \le d \le \frac{C^{0.875}}{1.6} \qquad (12.16a)$$

and Dudley[9] recommends using

$$d \cong \frac{C^{0.875}}{2.2} \qquad (12.16b)$$

which is about midway between the AGMA limits.

The pitch diameter of the worm gear d_w can be related to that of the worm through the center distance C.

$$d_g = 2C - d \qquad (12.17)$$

The addendum a and dedendum b of the teeth are found from

$$a = 0.3183 p_x \qquad\qquad b = 0.3683 p_x \qquad (12.18)$$

The face width of the worm gear is limited by the diameter of the worm. The AGMA recommends a maximum value for face width F as

$$F_{max} \le 0.67d \qquad (12.19)$$

Table 12-7 shows AGMA recommended minimum numbers of wormgear teeth as a function of pressure angle.

Rating Methods

Unlike helical and bevel gearsets in which calculations are made separately for the bending and surface stresses in the gear teeth and compared to material properties, wormsets are rated by their ability to handle a level of input power. The **AGMA power rating** is based on their pitting and wear resistance, because experience has shown that to be the usual failure mode. Because of the high sliding velocities in wormsets, the temperature of the oil film separating the gear teeth becomes an important factor and this is taken into account in the AGMA standard.[6], [7] These standards are based on a duty cycle of 10 continuous hours per day of service under uniform load, defined as a service factor of 1.0. Materials for worm and worm gear are assumed to be as defined above.

The rating of a wormset can be expressed as the allowable input power Φ, output power Φ_o, or as the allowable torque T at a given speed at either the input or output shaft, these being related by the general power-torque-speed relationship (Eq. 9.1a). The AGMA defines an input power rating formula as

Table 12-7
AGMA Suggested Minimum Numbers of Teeth for Worm Gears
Source: Reference 6.

ϕ	N_{min}
14.5	40
17.5	27
20	21
22.5	17
25	14
27.5	12
30	10

$$\Phi = \Phi_o + \Phi_l \tag{12.20}$$

where Φ_l is the power lost to friction in the mesh. Output power Φ_o is defined as

$$\Phi_o = \frac{nW_{tg}d_g}{126\,000\,m_G} \quad \text{hp} \tag{12.21us}$$

$$\Phi_o = \frac{nW_{tg}d_g}{1.91E7\,m_G} \quad \text{kW} \tag{12.21si}$$

and the power lost Φ_l is defined as

$$\Phi_l = \frac{V_t W_f}{33\,000} \quad \text{hp} \tag{12.22us}$$

$$\Phi_l = \frac{V_t W_f}{1\,000} \quad \text{kW} \tag{12.22si}$$

These are mixed-unit equations. The rotational speed n is in rpm. The tangential sliding velocity V_t is in fpm (m/s) and is taken at the worm diameter d in inches (mm). The loads W_{tg} and W_f are in lb (N). The power is in hp (kW).

The tangential load W_{tg} on the worm gear in lb (N) is found from

$$W_{tg} = C_s C_m C_v d_g^{0.8} F \tag{12.23us}$$

$$W_{tg} = C_s C_m C_v d_g^{0.8} F / 75.948 \tag{12.23si}$$

where C_s is a materials factor defined by AGMA for chill-cast bronze[*] as

if	$C < 8$ in	$C_s = 1000$
if	$C \ge 8$ in	$C_s = 1\,411.651\,8 - 455.825\,9\log_{10} d_g$

$$\tag{12.24}$$

and C_m is a ratio correction factor defined by AGMA as

if	$3 < m_G \le 20$	$C_m = 0.0200\sqrt{-m_G^2 + 40m_G - 76} + 0.46$
if	$20 < m_G \le 76$	$C_m = 0.0107\sqrt{-m_G^2 + 56m_G - 5\,145}$
if	$76 < m_G$	$C_m = 1.1483 - 0.00658m_G$

$$\tag{12.25}$$

C_v is a velocity factor defined by AGMA as

if	$0 < V_t \le 700$ fpm	$C_v = 0.0659e^{-0.0011V_t}$
if	$700 < V_t \le 3\,000$ fpm	$C_v = 13.31V_t^{-0.571}$
if	$3\,000 < V_t$ fpm	$C_v = 65.52V_t^{-0.774}$

$$\tag{12.26}$$

[*] Note that AGMA defines material factors for other bronzes as well. Consult the standard for more information (see reference 8).

The tangential velocity at the worm pitch diameter is

$$V_t = \frac{\pi n d}{12 \cos \lambda} \text{ fpm} \qquad (12.27)$$

The friction force W_f on the gear is

$$W_f = \frac{\mu W_{tg}}{\cos \lambda \cos \phi} \qquad (12.28)$$

The coefficient of friction in a wormgear mesh is not constant. It is a function of velocity. The AGMA suggests the following relationships:

if $V_t = 0$ fpm $\mu = 0.15$

if $0 < V_t \le 10$ fpm $\mu = 0.124 e^{\left(-0.074 V_t^{0.645}\right)}$ (12.29)

if $10 < V_t$ fpm $\mu = 0.103 e^{\left(-0.110 V_t^{0.450}\right)} + 0.012$

The efficiency of the wormset alone (exclusive of bearings, oil churn, etc.) is

$$e = \frac{\Phi_o}{\Phi} \qquad (12.30)$$

The rated output torque can be found from equations 12.14 and 12.23:

$$T_g = W_{tg} \frac{d_g}{2} \qquad (12.31)$$

A Design Procedure for Wormsets

A common design specification for a wormset will define the desired input (or output) speed and gear ratio. Some information on the output loading, either in terms of force or torque, or the required output power, will usually be known. There may also be some package size limits specified. One approach (of many possible) is to assume a number of starts for the worm and calculate the kinematic data for the worm and worm gear. Then assume a trial center distance C and use it to find a trial pitch diameter d for the worm from equation 12.16. Find a suitable face width F for the gear that obeys equation 12.19. The pitch diameter of the gear can then be found from equation 12.17 and used in equations 12.23 and 12.28 to find the tangential forces in the mesh. From these data, the rated (allowable) power and torque levels for a wormset of the assumed size can be found from equations 12.20 to 12.22 and 12.31. If these power and torque values are large enough to satisfy the design requirement with suitable safety margins, the design is done. If not (which is likely) the original assumptions regarding the number of starts, worm diameter, center distance, etc., must be revised and the calculation repeated until an acceptable combination is found. The center distance can be adjusted further to obtain a diametral pitch or module that matches available hobs. An equation solver can make this task much easier by allowing rapid iteration of the equations.

12.4   CASE STUDY

Case Study 8A in Chapter 8 set up a design problem involving a winch to hoist hay bales into a barn. The proposed device is to be powered by an electric motor connected to a winch with a 75:1 reduction gearset that needs to be self-locking to hold the load. A wormset is a reasonable solution in this application. We will now address the design of that gear train.

CASE STUDY 8B

Design of a Wormset Speed Reducer for a Winch Lift

Problem   Size the worm and worm gear for the winch lift defined in Case Study 8A (p. 542) as shown in Figure 8-4 repeated here.

Given   The force-time function was estimated in the previous study to be as shown in Figure 8-6b (repeated here). For an assumed winch-drum radius of 10 in, the peak torque will be about 7 800 lb-in. The average output power required was calculated to be about 0.6 hp. A 75:1 reduction is required. Input speed to the worm is 1 725 rpm. Output speed is 23 rpm.

Assumptions A single start worm with a 20° pressure angle will be tried. The worm will be of steel case hardened to 58HRC and the worm gear of chill-cast phosphor bronze. A self-locking wormset is needed.

Solution   See Figures 8-4, 8-6, and TKSolver file CASE8-B.

1 A single-start worm will require a 75-tooth worm gear for the desired 75:1 ratio. This number of worm-gear teeth is well above the minimum recommended in Table 12-7.

2 Assume a center distance of 5.5 in for a trial calculation and find a suitable worm diameter based on that assumption from equation 12.16b.

$$d \cong \frac{C^{0.875}}{2.2} \cong \frac{5.5^{0.875}}{2.2} = 2.02 \text{ in} \qquad (a)$$

3 Find a suitable worm gear diameter from equation 12.17.

$$d_g = 2C - d = 2(5.5) - 2.02 = 8.98 \text{ in} \qquad (b)$$

4 Find the lead from equation 12.13.

$$L = \pi d_g \frac{N_w}{N_g} = \pi(8.98)\frac{1}{75} = 0.376 \text{ in} \qquad (c)$$

5 Find the lead angle from equation 12.12.

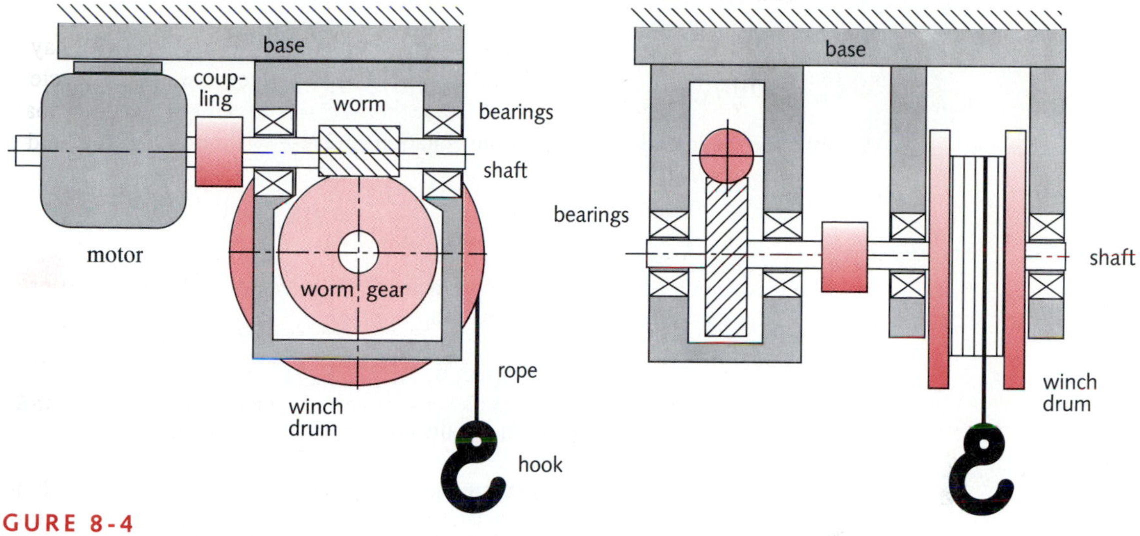

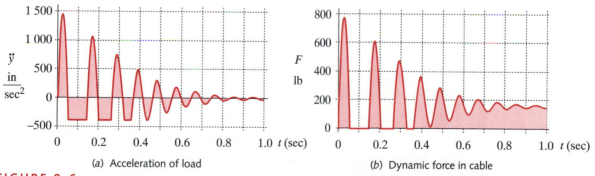

FIGURE 8-4

Motor-Driven Winch with Gear Train, Shafts, Bearings, and Couplings

$$\lambda = \tan^{-1}\frac{L}{\pi d} = \tan^{-1}\frac{0.376}{\pi(2.02)} = 3.39° \qquad (d)$$

This is less than 6°, so will be self-locking, as required.

6   Find the maximum recommended face width from equation 12.19.

$$F_{max} \cong 0.67d = 0.67(2.02) = 1.354 \text{ in} \qquad (e)$$

7   Find the materials factor C_s from equation 12.24. Since $C < 8$ in, $C_s = 1\,000$.

FIGURE 8-6

Acceleration and Cable Force at Startup of Load-Lift

8   Find the ratio correction factor C_m from equations 12.25. Based on $m_G = 75$, the second of the expressions in that equation set will be used.

$$C_m = 0.0107\sqrt{-m_G^2 + 56m_G - 5\,145} = 0.0107\sqrt{-75^2 + 56(75) - 5\,145} = 0.653 \qquad (f)$$

9   Find the tangential velocity V_t from equation 12.27.

$$V_t = \frac{\pi n d}{12\cos\lambda} = \frac{\pi(1\,725)(2.02)}{12\cos(3.392°)} = 913.9 \text{ fpm} \qquad (g)$$

10   Use this velocity to find the velocity factor C_v from equations 12.26. For this value of V_t, the second of these equations is appropriate.

$$C_v = 13.31(913.9)^{-0.571} = 0.271 \qquad (h)$$

11   Find the tangential load W_t from equation 12.23.

$$W_{tg} = C_s C_m C_v d_g^{0.8} F = 1000(0.653)(0.271)(8.98)^{0.8}(1.354) = 1\,388 \text{ lb} \qquad (i)$$

12   Find the coefficient of friction from the third expression in equation 12.29.

$$\mu = 0.103e^{\left(-0.110V_t^{0.450}\right)} + 0.012 = 0.103e^{\left(-0.110[913.9]^{0.450}\right)} + 0.012 = 0.022 \qquad (j)$$

13   Find the friction force W_f from equation 12.28.

$$W_f = \frac{\mu W_{tg}}{\cos\lambda\cos\phi} = \frac{0.022(1\,388)}{\cos 3.392°\cos 20°} = 32 \text{ lb} \qquad (k)$$

14   Find the rated output power from equation 12.21.

$$\Phi_o = \frac{n W_{tg} d_g}{126\,000\,m_G} = \frac{1\,725(1\,388)(8.98)}{126\,000(75)} = 2.274 \text{ hp} \qquad (l)$$

15   Find the power lost in the mesh from equation 12.22.

$$\Phi_l = \frac{V_t W_f}{33\,000} = \frac{913.9(32)}{33\,000} = 0.888 \text{ hp} \qquad (m)$$

16   Find the rated input power from equation 12.20.

$$\Phi = \Phi_o + \Phi_l = 2.274 + 0.888 = 3.162 \text{ hp} \qquad (n)$$

17   The efficiency of the gearset is

$$e = \frac{\Phi_o}{\Phi} = \frac{2.274}{3.162} = 71.9\% \qquad (o)$$

18   Find the rated output torque from equation 12.31.

$$T_g = W_{tg}\frac{d_g}{2} = 1\,388\frac{8.98}{2} = 6\,230 \text{ lb - in} \qquad (p)$$

19   While the power rating appears to be adequate for this application, the output torque rating falls short of the projected peak torque of 7 800 lb-in modeled in Case Study 8A; thus some redesign is in order.

20   The original assumption for center distance was increased to 6.531 in and the model recalculated. The center distance was also adjusted slightly to give an integer diametral pitch of 7 in^{-1}. This increased the worm gear diameter to 10.714 in and the output torque rating to 9 131 lb-in. The new input power rating is 4.52 hp and the power loss is 1.18 hp for an efficiency of 73.8 %. The output power rating is 3.33 hp. The new lead angle is 3.48°, so is still self-locking.

21   While this new design appears feasible based on the loading calculations done in the previous case study, one of the original assumptions regarding the electric motor size will need to be revised. The average net power required was estimated to be 0.62 hp. It was hoped that a 1- to 1.25-hp motor would be adequate which would allow 110-V operation. This now appears impossible due to the 1.18-hp loss in the wormset which would leave too little power available to lift the load even if a 1.25-hp motor were used. The flywheel effect of the rotating winch drum can supply bursts of energy to get past the peaks of load oscillation shown in Figure 8-6, but cannot provide a sustained increase in power above the average available. So a 220-V motor of about 2 or 2.25 hp appears to be necessary for this design. The input power rating of the gearset should easily accommodate that level of power with no overheating problems.

22   The reader is encouraged to look at the provided *TKSolver* files CASE8B-1 and CASE8B-2 which, respectively, contain the first (unsuccessful) and second (successful) solutions to this problem to see more detail than has been presented here.

12.5   SUMMARY

Several specialized forms of gears exist. This chapter has presented a brief introduction to the application and design of three types: helical, bevel, and worm gearsets.

HELICAL GEARS   are formed from rolling cylinders and can perform essentially the same function as spur gears, connecting parallel shafts for speed reduction/increase and torque multiplication/division. The teeth of helical gears are angled with respect to the axis at a helix angle, which can be a few degrees or as much as 45°. Their helix is either right or left-handed. Opposite-hand helical gears with the same helix angle mesh with their axes parallel. Same-hand helical gears mesh with their (nonintersecting) axes skewed or crossed and have theoretical point contact between the teeth. This limits their load-carrying capacity compared to parallel-axis helicals, which mate with a roll-slide motion similar to spur gears, but engage their teeth in a smooth, wiping action across the face width.

The principle advantages of helical over spur gearsets is their quieter running and greater strength for the same size gear. The disadvantage is greater cost than spur gears and the introduction of an axial component of force that requires thrust bearings on the shaft. Parallel-axis helical gears are used extensively in vehicle transmissions, both manual and automatic-shifting, principally because of their quiet running.

The design of helical gears is very similar to that of spur gears. The same bending and surface stress equations apply, but with different values for the geometry factors *I* and *J*. Some additional factors are also introduced to the equations. These factors are obtained from the AGMA standards, which contain tables of data for helical gears of various pressure angles, helix angles, and addendum ratios. A few of these tables are reproduced here. Consult the AGMA standards for more complete information. The materials used are the same as for spur gears. A helical gear will have lower stresses and higher safety factors than a spur gear of the same pitch and diameter because its angled tooth is thicker in the direction of the applied load.

BEVEL GEARS   are formed from rolling cones and thus connect intersecting axes. They are used principally to take motion and torque "around a corner." Their teeth taper with the changing cone diameter and they are specified by the diameter and tooth size at their large end. Their teeth may be straight and parallel to the axis (analogous to spur gears), in which case they are called **straight bevel gears**, or the teeth may be angled with respect to the axis at a spiral angle (analogous to the helix angle of helical gears), in which case they are called **spiral bevel** or just spiral gears. Spiral bevel gears have advantages over straight bevels similar to those of helicals over spur gears. Because their angled teeth engage gradually, spirals run smoother and quieter than straight bevels and have less vibration. Because a spiral's tooth is thicker in the direction of loading, it is stronger than a straight bevel gear of the same diameter and pitch. A spiral-type gear with curved teeth but zero spiral angle, called a ZEROL® gear, is also made to obtain the smooth running of a spiral without the additional tooth load introduced by the spiral angle.

Bevel gears are seldom made with equal addenda on pinion and gear. A long-addendum pinion is used with the percent increase varying from zero at a 1:1 ratio to over 50% at large gear ratios. This makes the pinion tooth stronger and the gear tooth weaker to balance the design as discussed in Chapter 11 for spur gears.

The design of bevel gears is very similar to that of spur or helical gears. The same bending and surface stress equations apply, but use different values for the geometry factors *I* and *J*. Some additional factors are also introduced to the equations. These factors are obtained from the AGMA standards, which contain charts plotting *I* and *J* for bevel and spiral gears of various pressure angles, spiral angles, and addendum ratios. A few of these charts are reproduced here. Consult the AGMA standards for more complete information. The materials used for bevel gears are the same as for spur or helical gears.

WORMS AND WORM GEARS   connect nonparallel, nonintersecting axes. The worm is similar to a screw thread, having one or a few teeth wrapped around it at what is, in effect, a very large helix angle. The worm mates with a special gear called a worm gear or worm wheel that is analogous to a nut being advanced by the thread of the worm.

12

Their axes are typically at 90° to one another. The wormset can give very large gear ratios (up to about 360:1) in a compact package because of the small number of teeth on the worm. If the lead angle of the worm is small enough, (< about 6°) the wormset can be self-locking, meaning that it cannot be backdriven from the worm gear, i.e., it will hold a load. Its main disadvantage is its relatively low efficiency compared to other gearsets. The relative motion at the teeth is sliding rather than rolling, which generates significant heat. Heat transfer from the gearbox, rather than the stresses in the teeth, can limit a wormset's life. The oil temperature in the mesh should be kept below about 200°F for long tooth life.

The design of wormsets is different than for other gearsets. The AGMA defines an **input power rating** equation for wormsets. This equation, in combination with a number of AGMA defined empirical factors, allows the wormset to be sized for a given power or torque-speed combination. Consult the AGMA standards for more complete information. The materials used for wormsets are quite limited. The worm is usually a steel, case-hardened to 58HRC and the worm gear is a bronze alloy. The soft gear runs in against the hard worm in the first few hours of operation and conforms to its particular contours. If properly run-in without overloading or overheating, a properly sized (rated) wormset can be expected to give a very long cycle life before succumbing to its ultimate demise by pitting in surface fatigue. Bending failure of the teeth in wormsets is rare. A wormset's load capacity can be increased by designing it as a single- or double-enveloping configuration. A single-enveloping set wraps the gear partially around the worm to gain surface-contact area. A double-enveloping set does the above and also wraps the worm partially around the gear in an hourglass shape to obtain even more contact area.

Important Equations Used in this Chapter

Helical Gear Geometry (Section 12.1):

$$p_t = p_n/\cos \psi \qquad (12.1a)$$

$$p_x = p_n/\sin \psi \qquad (12.1b)$$

$$P_d = \frac{N}{d} = \frac{\pi}{p_c} = \frac{\pi}{p_t} \qquad (12.1c)$$

Helical Gear Forces (Section 12.1):

$$W_r = W_t \tan \phi \qquad (12.3a)$$

$$W_a = W_t \tan \psi \qquad (12.3b)$$

$$W = \frac{W_t}{\cos \psi \cos \phi_n} \qquad (12.3c)$$

Stresses in Helical Gears (Section 12.1):

$$\sigma_b = \frac{W_t\, P_d}{FJ}\, \frac{K_a K_m}{K_v}\, K_s K_B K_I \qquad\qquad (11.15us)$$

$$\sigma_b = \frac{W_t}{FmJ}\, \frac{K_a K_m}{K_v}\, K_s K_B K_I \qquad\qquad (11.15si)$$

$$\sigma_c = C_p \sqrt{\frac{W_t}{FId}\, \frac{C_a C_m}{C_v}\, C_s C_f} \qquad\qquad (11.21)$$

Surface Geometry Factor for Helical Gears (Section 12.1):

$$I = \frac{\cos\phi}{\left(\dfrac{1}{\rho_p} \pm \dfrac{1}{\rho_g}\right) d_p m_N} \qquad\qquad (12.6a)$$

Gear Ratio for a Bevel Gear Set (Section 12.2):

$$m_G = \frac{\omega_p}{\omega_g} = \frac{N_g}{N_p} = \frac{d_g}{d_p} = \tan\alpha_g = \cot\alpha_p \qquad\qquad (12.7b)$$

Forces on Straight Bevel Gears (Section 12.2):

$$W_a = W_t \tan\phi \sin\alpha$$
$$W_r = W_t \tan\phi \cos\alpha \qquad\qquad (12.8a)$$
$$W = W_t / \cos\phi$$

Stresses in Bevel Gears (Section 12.2):

$$\sigma_b = \frac{2T_p}{d}\, \frac{P_d}{FJ}\, \frac{K_a K_m K_s}{K_v K_x} \qquad \text{psi} \qquad (12.9us)$$

$$\sigma_b = \frac{2000 T_p}{d}\, \frac{1}{FmJ}\, \frac{K_a K_m K_s}{K_v K_x} \quad \text{MPa} \qquad (12.9si)$$

$$\sigma_c = C_p C_b \sqrt{\frac{2T_D}{FId^2}\left(\frac{T_p}{T_D}\right)^z \frac{C_a C_m}{C_v} C_s C_f C_{xc}} \qquad (12.10)$$

Design Torque for Bevel Gears (Section 12.2):

$$T_D = \frac{F}{2}\, \frac{IC_v}{C_s C_{md} C_f C_a C_{xc}} \left(\frac{S'_{fc} d}{C_p C_b}\, \frac{0.774 C_H}{C_T C_R}\right)^2 \quad \text{psi} \qquad (12.11us)$$

$$T_D = \frac{F}{2000}\, \frac{IC_v}{C_s C_{md} C_f C_a C_{xc}} \left(\frac{S'_{fc} d}{C_p C_b}\, \frac{0.774 C_H}{C_T C_R}\right)^2 \quad \text{MPa} \qquad (12.11si)$$

Lead and Lead angle of a Worm (Section 12.3):

$$\tan \lambda = \frac{L}{\pi d} \qquad (12.12)$$

Forces in Wormsets (Section 12.3):

$$W_{tg} = W_{aw} = \frac{2T_g}{d_g} \qquad (12.14a)$$

$$W_{ag} = W_{tw} = \frac{2T_w}{d} \qquad (12.14b)$$

$$W_r = \frac{W_{tg} \tan \phi}{\cos \lambda} \qquad (12.14c)$$

Recommended Worm Pitch Diameter (Section 12.3):

$$d \cong \frac{C^{0.875}}{2.2} \qquad (12.16b)$$

Worm Gear Pitch Diameter (Section 12.3):

$$d_g = 2C - d \qquad (12.17)$$

Recommended Maximum Face Width of Worm Gear   (Section 12.3):

$$F_{max} \leq 0.67d \qquad (12.19)$$

Rated Power for a Wormset (Section 12.3):

$$\Phi_o = \frac{nW_{tg}d_g}{126\,000\,m_G}\ \text{hp} \qquad (12.21us)$$

$$\Phi_o = \frac{nW_{tg}d_g}{1.91E7\,m_G}\ \text{kW} \qquad (12.21si)$$

$$\Phi_l = \frac{V_t W_f}{33\,000}\ \text{hp} \qquad (12.22us)$$

$$\Phi_l = \frac{V_t W_f}{1\,000}\ \text{kW} \qquad (12.22si)$$

$$\Phi = \Phi_o + \Phi_l \qquad (12.20)$$

Tangential Force on a Worm Gear (Section 12.3):

$$W_{tg} = C_s C_m C_v d_g^{0.8} F \qquad (12.23us)$$

$$W_{tg} = C_s C_m C_v d_g^{0.8} F / 75.948 \qquad (12.23si)$$

Friction Force on a Worm Gear (Section 12.3):

$$W_f = \frac{\mu W_{tg}}{\cos \lambda \cos \phi}$$

(12.28)

Rated Output Torque of a Worm Gear (Section 12.3):

$$T_g = W_{tg} \frac{d_g}{2}$$

(12.31)

Efficiency of a Wormset (Section 12.3):

$$e = \frac{\Phi_o}{\Phi}$$

(12.30)

12.6   REFERENCES

1   **AGMA**, *Gear Nomenclature, Definitions of Terms with Symbols*. ANSI/AGMA 1012-F90. American Gear Manufacturers Association, 1500 King St., Suite 201, Alexandria, Va., 22314, 1990.

2   **AGMA**, *Fundamental Rating Factors and Calculation Methods for Involute Spur and Helical Gear Teeth*. ANSI/AGMA Standard 2001-B88. American Gear Manufacturers Association, 1500 King St., Suite 201, Alexandria, Va., 22314, 1988.

3   **AGMA**, *Geometry Factors for Determining the Pitting Resistance and Bending Strength of Spur, Helical, and Herringbone Gear Teeth*. ANSI/AGMA Standard 908-B89. American Gear Manufacturers Association, 1500 King St., Suite 201, Alexandria, Va., 22314, 1989.

4   **AGMA**, *Design Manual for Bevel Gears*. ANSI/AGMA Standard 2005-B88. American Gear Manufacturers Association, 1500 King St., Suite 201, Alexandria, Va., 22314, 1988.

5   **AGMA**, *Rating the Pitting Resistance and Bending Strength of Generated Straight Bevel, ZEROL® Bevel, and Spiral Bevel Gear Teeth*. ANSI/AGMA Standard 2003-A86. American Gear Manufacturers Association, 1500 King St., Suite 201, Alexandria, Va., 22314, 1986.

6   **AGMA**, *Design Manual for Cylindrical Wormgearing*. ANSI/AGMA Standard 6022-C93. American Gear Manufacturers Association, 1500 King St., Suite 201, Alexandria, Va., 22314, 1993.

7   **AGMA**, *Practice for Enclosed Cylindrical Wormgear Speed Reducers and Gearmotors*. ANSI/AGMA Standard 6034-B92. American Gear Manufacturers Association, 1500 King St., Suite 201, Alexandria, Va., 22314, 1992.

8   **D. W. Dudley**, *Handbook of Practical Gear Design*. McGraw-Hill: New York, p. 3.66, 1984.

9   *Ibid.*, p. 3.67.

12

12.7   PROBLEMS

*12-1   A 20° pressure angle, 30° helix angle, 27-tooth helical gear has a diametral pitch p_d = 5. Find the pitch diameter, addendum, dedendum, outside diameter, normal, transverse, and axial pitch.

12-2   A 25° pressure angle, 20° helix angle, 43-tooth helical gear has a diametral pitch p_d = 8. Find the pitch diameter, addendum, dedendum, outside diameter, normal, transverse, and axial pitch.

*12-3   A 57-tooth, 10° helix angle, helical gear is in mesh with a 23-tooth pinion. The p_d = 6 and ϕ = 25°. Find the transverse and axial contact ratios.

12-4   A 78-tooth, 30° helix angle helical gear is in mesh with a 27-tooth pinion. The p_d = 6 and ϕ = 20°. Find the transverse and axial contact ratios.

*12-5   A 90° straight bevel gear set is needed to give a 9:1 reduction. Determine the pitch cone angles, pitch diameters, and gear forces if the 25° pressure angle pinion has 14 teeth of p_d = 6, and the transmitted power is 746 W at 1000 pinion rpm.

12-6   A 90° straight bevel gear set is needed to give a 4.5:1 reduction. Determine the pitch cone angles, pitch diameters, and gear forces if the 20° pressure angle pinion has 18 teeth of p_d = 5, and the transmitted power is 7 460 W at 800 pinion rpm.

*12-7   A 90° spiral bevel gear set is needed to give a 5:1 reduction. Determine the pitch cone angles, pitch diameters, and gear forces if the 20° pressure angle pinion has 16 teeth of p_d = 7, and the transmitted power is 3 hp at 600 pinion rpm

†12-8   A paper machine processes rolls of paper having a density of 984 kg/m³. The paper roll is 1.50-m outside dia. (OD) x 0.22-m inside dia. (ID) x 3.23 m long and is on a simply-supported, hollow, steel shaft with S_{ut} = 400 MPa. Design a 2.5:1 reduction helical gearset to drive this roll shaft to obtain a minimum dynamic safety factor of 2 for a 10-year life if the shaft OD is 22 mm and the roll turns at 50 rpm with 1.2 hp absorbed.

*12-9   A 2-start wormset has d = 50 mm, p_x = 10 mm, m_G = 22:1. Find the lead, lead angle, worm gear diameter, and center distance. Will it self-lock? The input speed is 2 200 rpm.

12-10   A 3-start wormset has d = 1.75 in, p_x = 0.2 in, m_G = 17:1. Find the lead, lead angle, worm gear diameter, and center distance. Will it self-lock? The input speed is 1 400 rpm.

*12-11   A 1-start wormset has d = 40 mm, p_x = 5 mm, m_G = 82:1. Find the lead, lead angle, worm gear diameter, and center distance. Will it self-lock? The input speed is 4 500 rpm.

*12-12   Determine the power transmitted and the torques and forces in the mesh for the wormset in Problem 12-9 if it runs at 1 000 worm rpm.

12-13   Determine the power transmitted and the torques and forces in the mesh for the wormset in Problem 12-10 if it runs at 500 worm rpm.

*12-14   If the gearset in Problem 12-3 transmits 125 HP at 1 000 pinion rpm, find the torque on each shaft.

12

†   These problems are based on similar problems in previous chapters with the same –number or the number noted in the problem statement, e.g., Problem 12-8 is based on Problem 11-8 etc. Problems in succeeding chapters may also continue and extend these problems.

12-15   If the gearset in Problem 12-4 transmits 33 kW at 1 600 pinion rpm, find the torque on each shaft.

*12-16   Size the helical gears in Problem 12-14 for a bending safety factor of at least 2 assuming a steady torque, 25° pressure angle, full-depth teeth, a face width factor of 10, $Q_v = 9$, an AISI 4140 steel pinion and a class 40 cast-iron gear.

12-17   Size the helical gears in Problem 12-15 for a bending safety factor of 2.5 assuming a steady torque, 20° pressure angle, full-depth teeth, a face width factor of 12, $Q_v = 11$, an AISI 4340 steel pinion and an A-7-d nodular iron gear.

*12-18   Size the helical gears in Problem 12-14 for a surface safety factor of at least 1.6 assuming a steady torque, 25° pressure angle, full-depth teeth, a face width factor of 10, $Q_v = 9$, an AISI 4140 steel pinion and a class 40 cast-iron gear.

12-19   Size the helical gears in Problem 12-15 for a surface safety factor of 1.2 assuming a steady torque, 20° pressure angle, full-depth teeth, a face width factor of 12, $Q_v = 11$, an AISI 4340 steel pinion and an A-7-d nodular iron gear.

*12-20   Size the bevel gears in Problem 12-5 for a bending safety factor of 2 assuming a 5-year, 1-shift life, steady torque, $Q_v = 9$, an AISI 4140 steel pinion and gear.

12-21   Size the bevel gears in Problem 12-6 for a bending safety factor of 2.5 assuming a 15-year, 3-shift life, steady torque, $Q_v = 11$, an AISI 4340 steel pinion and gear.

12-22   Size the bevel gears in Problem 12-7 for a bending safety factor of 2.2 assuming a 10-year, 3-shift life, steady torque, $Q_v = 8$, an AISI 4340 steel pinion and gear.

*12-23   Size the bevel gears in Problem 12-5 for a minimum safety factor of 1.4 for any mode of failure of pinion or gear assuming a 5-year, 1-shift life, steady torque, $Q_v = 9$, an AISI 4140 steel pinion and gear.

12-24   Size the bevel gears in Problem 12-6 for a surface safety factor of 1.3 assuming a 15-year, 3-shift life, steady torque, $Q_v = 11$, an AISI 4340 steel pinion and gear.

12-25   Size the bevel gears in Problem 12-7 for a surface safety factor of 1.4 assuming a 10-year, 3-shift life, steady torque, $Q_v = 8$, an AISI 4340 steel pinion and gear.

12-26   Find the rated power and rated output torque of the wormset in Problem 12-9 with an input speed of 2 200 rpm.

*12-27   Find the rated power and rated output torque of the wormset in Problem 12-10 with an input speed of 1 400 rpm.

12-28   Find the rated power and rated output torque of the wormset in Problem 12-11 with an input speed of 4 500 rpm.

* Answers to these problems are provided in Appendix G.

† These problems are based on similar problems in previous chapters with the same –number or the number noted in the problem statement, e.g., Problem 12-8 is based on Problem 11-8 etc. Problems in succeeding chapters may also continue and extend these problems.

13

SPRING DESIGN

Not to know is bad;
not to wish to know is worse.

NIGERIAN PROVERB

13.0 INTRODUCTION

Virtually any part made from an elastic material has some "spring" to it. The term *spring* in the context of this chapter refers to parts made in particular configurations to provide a range of force over a significant deflection and/or to store potential energy. Springs are designed to provide a *push*, a *pull*, or a *twist* force (torque), or to primarily store energy, and can be divided into those four general categories. Within each category, many configurations of springs are possible. Springs may be made of round or rectangular wire bent into a suitable form such as a coil, or made of flat stock loaded as a beam. This chapter's opening photograph shows a few such spring configurations. Many standard spring configurations are available as stock catalog items from spring manufacturers. It is usually more economical for the designer to use a stock spring if possible. Sometimes the task requires a custom spring design. Custom springs may perform secondary functions, such as the location or mounting of other parts. In any case, the designer must understand, and properly use, spring design theory in order to specify or design the part. Table 13-0 defines the variables used in this chapter and references the section or equation(s) in which they are used.

13.1 SPRING RATE

Regardless of the spring configuration, it will possess a **spring rate** k, defined as the slope of its force-deflection curve. If the slope is constant, it can be defined as

Title page photograph
courtesy of Associated
Spring, Barnes Group Inc.

Table 13-0   Variables Used in this Chapter

Part 1 of 2

Symbol	Variable	ips units	SI units	See
A	area	in^2	m^2	Eq. 13.8a
C	spring index	none	none	Eq. 13.5
d	wire diameter	in	m	various
D	mean coil diameter	in	m	various
D_i	inside diameter	in	m	various
D_o	outside diameter	in	m	various
E	Young's modulus	psi	Pa	various
F	force or load	lb	N	various
F_a	alternating load	lb	N	Eq. 13.15
F_i	initial tension—extension spring	lb	N	Eq. 13.20
F_m	mean load	lb	N	Eq. 13.15
F_{max}	maximum fluctuating load	lb	N	Eq. 13.15
F_{min}	minimum fluctuating load	lb	N	Eq. 13.15
f_n	natural frequency	Hz	Hz	Eq. 13.11
g	gravitational acceleration	in/sec^2	m/sec^2	various
G	shear modulus, modulus of rigidity	psi	Pa	various
h	height of cone	in	m	Eq. 13.35
k	spring rate or spring constant	lb/in	N/m	Eq. 13.1
K_b	Wahl factor - bending	none	none	Eq. 13.23b
K_c	curvature factor	none	none	Eq. 13.10
K_s	direct shear factor	none	none	Eq. 13.8b
K_w	Wahl factor—torsion	none	none	Eq. 13.9b
L_b	body length—extension spring	in	m	Eq. 13.19
L_f	free length—compression spring	in	m	various
L_{max}	coil length—torsion spring	in	m	Eq. 13.3
L_s	shut height—compression spring	in	m	Ex. 13-3
M	moment	lb-in	N-m	Eq. 13.27
N	number of cycles	none	none	various
N_a	number of active coils	none	none	various
N_{f_s}	safety factor in fatigue—torsion	none	none	Eq. 13.16a
N_{f_b}	safety factor in fatigue—bending	none	none	Eq. 13.34a
N_t	number of total coils	none	none	various
N_s	safety factor—static yielding	none	none	Eq. 13.14
r	radius	in	m	various
R	stress ratio	none	none	various
R_d	diameter ratio	none	none	Eq. 13.35b

The author wishes to express
his appreciation to
**Associated Spring, Barnes
Group Inc.**, 10 Main St.,
Bristol Conn., for permission
to use material from their
*Design Handbook:
Engineering Guide to Spring
Design*, 1987 edition.

13

Table 13-0   Variables Used in this Chapter

Part 2 of 2

Symbol	Variable	ips units	SI units	See
R_F	force ratio	none	none	Eq. 13.15b
S_f, S_e	bending fatigue, endurance strengths—$R = -1$	psi	Pa	Eq. 13.34
S_{fs}, S_{es}	torsional fatigue, endurance strengths—$R = -1$	psi	Pa	Eq. 13.16b
S_{fw}, S_{ew}	wire torsional fatigue strengths—$R = 0$	psi	Pa	Eq 13.12
S_{fwb}, S_{ewb}	wire bending fatigue strengths— $R = 0$	psi	Pa	Eq 13.33
S_{ys}, S_y	shear, tensile yield strengths	psi	Pa	Tables 13-6, -13
S_{ms}	mean torsional strength at 10^3 cycles	psi	Pa	Eq. 13.13
S_{us}	ultimate shear strength	psi	Pa	Eq. 13.4
S_{ut}	ultimate tensile strength	psi	Pa	Eq. 13.3
t	thickness	in	m	Eq. 13.35
T	torque	lb-in	N-m	Eq. 13.8a
W	weight	lb	N	Eq. 13.11b
y	deflection	in	m	various
ν	Poisson's ratio	none	none	Eq. 13.35
θ	angular deflection—torsion	rad	rad	Eq. 13.27
γ	weight density	lb/in^3	N/m^3	Eq. 13.11
σ	normal (bending) stress	psi	Pa	Eq. 13.23a
τ	shear stress	psi	Pa	various
ω_n	natural frequency	rad/sec	rad/sec	Eq. 13.11a

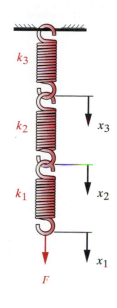

(a) Series

$$k = \frac{F}{y} \qquad (13.1)$$

where F is the applied force and y the deflection. Since the deflection function can always be determined for any known geometry and loading, and because the deflection function expresses a relationship between the applied load and the deflection, it can simply be rearranged algebraically to express k as equation 13.1.

The spring rate may be a constant value (linear spring) or may vary with deflection (nonlinear spring). Both have their applications, but we often want a linear spring to control loading. Many spring configurations have constant spring rates and a few have zero rate (constant force).

When multiple springs are combined, the resulting spring rate depends on whether they are combined in series or parallel. Series combinations have the same force passing through all springs and each contributes a part of the total deflection, as shown in Figure 13-1a. Parallel springs all have the same deflection and the total force splits among the individual springs, as shown in Figure 13-1b. For springs in parallel, the individual spring rates add directly:

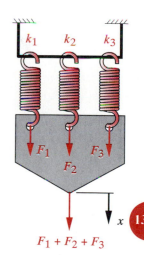

(b) Parallel

FIGURE 13-1

Springs in Series and Springs in Parallel

13

$$k_{total} = k_1 + k_2 + k_3 \ldots + k_n \tag{13.2a}$$

For springs in series the spring rates add reciprocally:

$$\frac{1}{k_{total}} = \frac{1}{k_1} + \frac{1}{k_2} + \frac{1}{k_3} \ldots + \frac{1}{k_n} \tag{13.2b}$$

13.2   SPRING CONFIGURATIONS

Springs can be categorized in different ways. The four load types mentioned in Section 13.0 are one way. Another is by the spring's physical configuration. We will use the latter approach. Figure 13-2 shows a selection of spring configurations. Additional examples can be found in reference 1. Wire-form springs come in **helical compression**, **helical tension**, **helical torsion**, and custom forms. Flat springs are typically **cantilever** or **simply supported beams** and can have many shapes. Spring washers come in a variety of styles: **curved**, **wave**, **finger**, and **Belleville**. Flat-wound springs can be **motor** (clock) springs, **volute**, or **constant-force** springs. We will discuss all of these configurations briefly and the design of some of them in detail.

Figure 13-2a shows five forms of **helical compression spring**. All provide a push force and are capable of large deflections. Common applications are valve-return springs in engines, die springs, etc. The standard form has a constant coil diameter, constant pitch (axial distance between coils), and constant spring rate. It is the most common spring configuration and many sizes are available off-the-shelf. Most are made of round wire, but they can be made from rectangular wire as well. The pitch can be varied to create a **variable rate spring**. The low-rate coils will shut first, increasing the effective rate when they touch one another or "bottom."

Conical springs can be made with either a constant rate or an increasing one. Their spring rate is usually nonlinear, increasing with deflection because the smaller diameter coils have greater resistance to deflection and the larger coils deflect first. By varying the coil pitch, a nearly constant spring rate can be obtained. Their main advantage is the ability to close to a height as small as one wire diameter if the coils nest. Barrel and hourglass springs can be thought of as two conical springs back to back, also having a nonlinear spring rate. The **barrel** and **hourglass** shapes are used primarily to change the natural frequency of the spring from that of a standard form.

Figure 13-2b shows a **helical extension spring** with hooks on each end. It provides a pull force and is capable of large deflections. These springs are commonly used in door-closers and counterbalancers. The hook is more highly stressed than the coils and usually fails first. Anything suspended from the hook will fall when the extension spring breaks making it a potentially unsafe design. Figure 13-2c shows a **drawbar spring**, which overcomes this problem by using a helical compression spring in a pull mode. The drawbars compress the spring, and if it breaks, will still support the load safely. Figure 13-2d shows a helical torsion spring, which is wound similar to the helical ex-

13

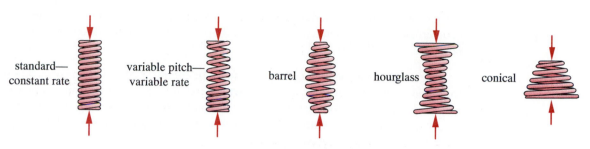

(a) Helical compression springs. *Push*—wide load and deflection range—round or rectangular wire. Standard has constant coil diameter, pitch, and rate. Barrel, hourglass, and variable-pitch springs are used to minimize resonant surging and vibration. Conical springs can be made with minimum solid height and with constant or increasing rate.

(b) Helical extension springs. *Pull*—wide load and deflection range—round or rectangular wire, constant rate.

(c) Drawbar springs. *Pull*—uses compression spring and drawbars to provide extension pull with fail-safe, positive stop.

(d) Torsion springs. *Twist*—round or rectangular wire—constant rate.

Belleville   wave   slotted   finger   curved

(e) Spring washers. *Push*—Belleville has high loads and low deflections—choice of rates (constant, increasing, or decreasing). Wave has light loads, low deflection, uses limited radial space. Slotted has higher deflections than Belleville. Finger is used for axial loading of bearings. Curved is used to absorb axial end play.

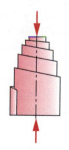
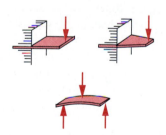

(f) Volute spring. *Push*—may have an inherently high friction damping.

(g) Beam springs. *Push* or *Pull*—wide load but low deflection range—rectangular or shaped cantilever, or simply supported.

(h) Power or motor springs. *Twist*—exerts torque over many turns. Shown in, and removed from, retainer.

(i) Constant Force. *Pull*—long deflection at low or zero rate.

FIGURE 13-2

Spring Configurations (Adapted from: *Design Handbook: Engineering Guide to Spring Design*, 1987, Associated Spring, Barnes Group Inc., 10 Main St., Bristol Conn., with permission)

13

tension spring but is loaded in twist (torque). Common applications are garage-door counterbalancers, mousetraps, etc. Many different shapes and details on its "legs" are possible.

Figure 13-2e shows five common varieties of **spring washer**. All provide a push force and are commonly used to load something axially, such as to take up end play on a bearing. They typically have small deflections and, except for the Belleville, can only supply light loads. The **volute spring** in Figure 13-2f provides a push force but has significant friction and hysteresis.

Figure 13-2g shows three varieties of **beam springs**. Any beam type can serve as a spring. Cantilever and simply supported beams are the most common. A beam spring can be constant width or shaped as in the trapezoidal example shown. The spring rate and stress distribution can be controlled with changes in beam width or depth along its length. Loads can be high but deflections are limited.

Figure 13-2h shows one type of **power spring**, also called a **motor** or **clock spring**. It is used primarily to store energy and provide twist. Windup clocks and toys make use of this type of spring. Figure 13-2i shows a constant force (Neg'ator) spring used for counterbalancing loads, returning typewriter carriages, and to make constant torque spring motors. They provide very large deflection strokes at a nearly constant pull force (zero spring rate).

We will discuss the design of some of these spring types. For more information on the others, see reference 1.

13.3 SPRING MATERIALS

There is a limited number of materials and alloys suitable for use as springs. The ideal spring material would have a high ultimate strength, high yield point, and a low modulus of elasticity in order to provide maximum energy storage (area under the elastic portion of the stress-strain curve). For dynamically loaded springs, the fatigue strength properties of the material are of primary importance. High strengths and yield points are attainable from medium- to high-carbon and alloy steels and these are the most common spring materials, despite their high modulus of elasticity. A few stainless-steel alloys are suitable for springs, as are beryllium copper and phosphor bronze, among the copper alloys.

Most light-duty springs are made of cold-drawn, round or rectangular wire or of thin, cold-rolled, flat-strip stock. Heavy-duty springs, such as vehicle-suspension parts, are typically made from hot-rolled or forged forms. Spring materials are typically hardened in order to obtain the required strength. Small cross sections are work-hardened in the cold-drawing process. Large sections are typically heat treated. Low-temperature heat treatments (175 to 510°C) are used after forming to relieve residual stresses and stabilize dimensions [1], even in small-section parts. High-temperature quenching and tempering is used to harden larger springs that must be formed in the annealed condition.

Spring Wire

Round wire is by far the most common spring material. It is available in a selection of alloys and a wide range of sizes. Rectangular wire is available only in limited sizes. Some common wire alloys and their descriptions are shown in Table 13-1, identified both by ASTM and SAE designations. Commonly available stock wire sizes are shown in Table 13-2, along with an indication of the size ranges available for the common steel alloys, identified by ASTM number. The designer should try to use these sizes for best

Table 13-1 Common Spring Wire Materials
Source: Reference 2

ASTM #	Material	SAE #	Description
A227	Cold-drawn wire ("hard-drawn")	1066	Least expensive general-purpose spring wire. Suitable for static loading but not good for fatigue or impact. Temperature range 0°C to 120°C (250°F).
A228	Music wire	1085	Toughest, most widely used material for small coil springs. Highest tensile and fatigue strength of all spring wire. Temperature range 0°C to 120°C (250°F).
A229	Oil-tempered wire	1065	General-purpose spring steel. Less expensive and available in larger sizes than music wire. Suitable for static loading but not good for fatigue or impact. Temperature range 0°C to 180°C (350°F).
A230	Oil-tempered wire	1070	Valve-spring quality—suitable for fatigue loading.
A232	Chrome vanadium	6150	Most popular alloy spring steel. Valve-spring quality—suitable for fatigue loading. Also good for shock and impact loads. For temperatures to 220°C (425°F). Available annealed or pretempered.
A313 (302)	Stainless steel	30302	Suitable for fatigue applications.
A401	Chrome silicon	9254	Valve-spring quality—suitable for fatigue loading. Second highest strength to music wire and has higher temperature resistance to 220°C (425°F).
B134, #260	Spring brass	CA-260	Low strength—good corrosion resistance.
B159	Phosphor bronze	CA-510	Higher strength than brass—better fatigue resistance—good corrosion resistance. Cannot be heat-treated or bent along the grain.
B197	Beryllium copper	CA-172	Higher strength than brass—better fatigue resistance—good corrosion resistance. Can be heat-treated and bent along the grain.
-	Inconel X-750	-	Corrosion resistance

Table 13-2
Preferred Wire Diameters

U.S. (in)	SI (mm)
0.004	0.10
0.005	0.12
0.006	0.16
0.008	0.20
0.010	0.25
0.012	0.30
0.014	0.35
0.016	0.40
0.018	0.45
0.020	0.50
0.022	0.55
0.024	0.60
0.026	0.65
0.028	0.70
0.030	0.80
0.035	0.90
0.038	1.00
0.042	1.10
0.045	
0.048	1.20
0.051	
0.055	1.40
0.059	
0.063	1.60
0.067	
0.072	1.80
0.076	
0.081	2.00
0.085	2.20
0.092	
0.098	2.50
0.105	
0.112	2.80
0.125	3.00
0.135	3.50
0.148	
0.162	4.00
0.177	4.50
0.192	5.00
0.207	5.50
0.225	6.00
0.250	6.50
0.281	7.00
0.312	8.00
0.343	9.00
0.362	
0.375	
0.406	10.0
0.437	11.0
0.469	12.0
0.500	13.0
0.531	14.0
0.562	15.0
0.625	16.0

13

Table 13-3   Relative Costs of Common Spring Wires
Source: Reference 1

| ASTM # | Material | Relative Cost of 2 mm (0.08 in) Dia Wire | |
		Mill Quantities	Warehouse Lots
A227	Cold-drawn wire	1.0	1.0
A229	Oil-tempered wire	1.3	1.3
A228	Music wire	2.6	1.4
A230	Oil-tempered wire	3.1	1.9
A401	Chrome silicon	4.0	3.9
A313 (302)	302 Stainless steel	7.6	4.7
B159	Phosphor bronze	8.0	6.7
A313 (631)	17-7ph Stainless steel	11.0	8.7
B197	Beryllium copper	27.0	17.0
-	Inconel X-750	44.0	31.0

cost and availability, though other sizes not shown are also made. Table 13-3 shows the relative costs of a selection of common round steel spring-wire materials.

TENSILE STRENGTH   The relationship between wire size and tensile strength shown in Figure 13-3 is a serendipitous situation. As discussed in Section 2.7 and Table 2-8, when materials are made very small in cross section, they begin to approach the very high theoretical strengths of their atomic bonds. Thus, the tensile strengths of fine steel wire become quite high. The same steel that may break at 200 000 psi in a 0.3 in (7.4 mm) diameter test-specimen can have nearly twice that strength after being cold-drawn down to 0.010 in (0.25 mm). The cold-drawing process is responsible for hardening and strengthening the material at the expense of much of its ductility.

Figure 13-3 is a semilog plot of wire strength versus diameter based on extensive testing by Associated Spring, Barnes Group Inc. The data for five of the materials shown in the figure can be fitted quite closely with an exponential function of the form

$$S_{ut} \cong Ad^b \qquad (13.3)$$

where A and b are defined in Table 13-4 for these wire materials over the specified ranges of diameters. These empirical functions provide a convenient means to calculate steel wire tensile strengths within a spring-design computer program and allow rapid iterating to a suitable design solution. Figure 13-4 plots these empirical strength functions to show, on linear axes, the change in strength with reduction of diameter.

SHEAR STRENGTH   Extensive testing has determined that a reasonable estimate of the ultimate strength in torsion[*] of common spring materials is 67% of the ultimate tensile strength.[1]

$$S_{us} \cong 0.67 S_{ut} \qquad (13.4)$$

13

* Note that this relates the ultimate strength in shear to the ultimate strength in tension. It is a different relationship than the distortion energy (Von Mises) criterion, which relates the material's yield strengths in shear and tension as $S_{ys} = 0.577 S_y$.

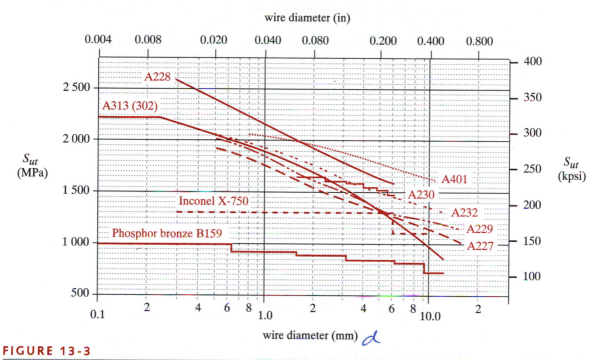

FIGURE 13-3

Minimum Tensile Strengths of Spring Wire—Identified by ASTM Number—See Table 13-1 Source: *Design Handbook: Engineering Guide to Spring Design*, 1987, Associated Spring, Barnes Group Inc., Bristol Conn.

Flat Spring Stock

Medium- to high-carbon steel strip stock is the most commonly used material for flat (beam) springs, volutes, clock and power springs, spring washers, etc. When corrosion resistance is needed, stainless-steel alloys 301, 302, and 17-7ph, beryllium copper, or phosphor bronze are also used for flat springs.

Table 13-4   Coefficients and Exponents for Equation 13.3
Source: Reference 1

ASTM #	Material	Range mm	Range in	Exponent b	Coefficient A MPa	Coefficient A psi	Correlation Factor
A227	Cold-drawn	0.5-16	0.020-0.625	−0.182 2	1 753.3	141 040	0.998
A228	Music wire	0.3-6	0.010-0.250	−0.1625	2 153.5	184 649	0.9997
A229	Oil-tempered	0.5-16	0.020-0.625	−0.183 3	1 831.2	146 780	0.999
A232	Chrome-v.	0.5-12	0.020-0.500	−0.145 3	1 909.9	173 128	0.998
A401	Chrome-s.	0.8-11	0.031-0.437	−0.093 4	2 059.2	220 779	0.991

13

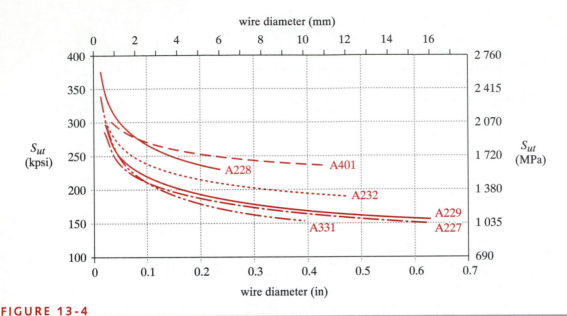

FIGURE 13-4

Minimum Tensile Strengths of Spring Steel Wire as Generated from Equation 13.3 and Table 13-4

Cold-rolled AISI 1050, 1065, 1074, and 1095 steel are the usual flat-stock alloys used. These are available in annealed or pretempered conditions referred to as 1/4 hard, 1/2 hard, 3/4 hard, and full-hard. The two softer tempers can be readily formed (bent) into shapes, but the harder tempers have less formability. Full-hard steel can be formed with gentle contours, but not bent with small radii. The advantage of forming prehardened steel strip is to avoid heat-treatment distortion of the formed part. If sharp bends are required, annealed material will have to be used and then hardened after forming.

The cold-rolling process creates a "grain" in the material analogous to (but much less pronounced than) the grain in wood. Just as wood will readily crack if bent along the grain, metal will not allow bends of small radius along its "grain" without cracking. The grain runs in the direction of rolling, which for strip stock is along the long axis. Thus, formed sheet-metal parts with sharp contours should be bent across the grain. If orthogonal bends are required, the grain should be oriented at 45° to the bends. A dimensionless bend factor $2r/t$ (where r is the bend radius and t the stock thickness) is defined to indicate the relative formability of strip stock. Low values of $2r/t$ indicate greater formability. Full- and 3/4-hard steel strip will fracture if bent along the grain.

Spring-steel strip is produced to a specified hardness that relates to its tensile strength. Figure 13-5 shows a plot of tensile strength versus hardness for quenched and tempered steels. Any of the carbon levels noted in the above AISI spring steels can be hardened to values in the range shown in Figure 13-5, meaning that final hardness rather than carbon content is the defining factor for tensile strength. Table 13-5 shows the strengths, hardness, and bend factors of some common flat-spring materials.

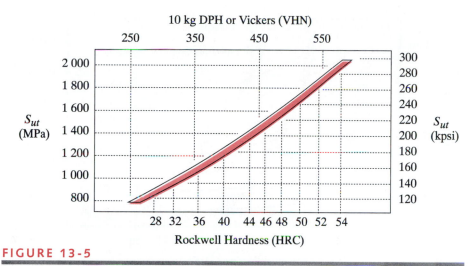

FIGURE 13-5

Tensile Strength Versus Hardness of Quenched and Tempered Spring Steel Strip   Source: Reference 1

Figure 13-6 shows the minimum bending radii that flat spring steel can withstand transverse to the grain. Three ranges of steel strengths are shown as bands that depend on material thickness and hardness. These correspond to three points taken from the upper range of Figure 13-5. The lines represent minimum bend radii for the strength steel that they intersect. Interpolation may be done between either lines or bands.

Table 13-5   Typical Properties of Spring Temper Alloy Strip
Source: Reference 1

Material	Sut MPa (ksi)	Rockwell Hardness	Elongation %	Bend Factor	E GPa (Mpsi)	Poisson's Ratio
Spring steel	1 700 (246)	C50	2	5	207 (30)	0.30
Stainless 301	1 300 (189)	C40	8	3	193 (28)	0.31
Stainless 302	1 300 (189)	C40	5	4	193 (28)	0.31
Monel 400	690 (100)	B95	2	5	179 (26)	0.32
Monel K500	1 200 (174)	C34	40	5	17.9 (26)	0.29
Inconel 600	1 040 (151)	C30	2	2	214 (31)	0.29
Inconel X-750	1 050 (152)	C35	20	3	214 (31)	0.29
Beryllium copper	1 300 (189)	C40	2	5	128 (18.5)	0.33
Ni-Span-C	1 400 (203)	C42	6	2	186 (27)	–
Brass CA 260	620 (90)	B90	3	3	11 (16)	0.33
Phosphor bronze	690 (100)	B90	3	2.5	103 (15	0.20
17-7PH RH950	1 450 (210)	C44	6	flat	203 (29.5)	0.34
17-7PH Cond. C	1 650 (239)	C46	1	2.5	203 (29.5)	0.34

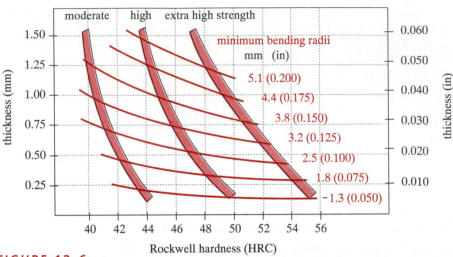

FIGURE 13-6

Minimum Transverse (across grain) Bending Radii for Various Tempers and Thicknesses of Tempered Spring Steel Source: Reference 1

(a)

number of coils = N_t

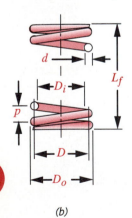

(b)

FIGURE 13-7

(a) Sample Springs and (b) Dimensional Parameters for Helical Compression Springs

13.4   HELICAL COMPRESSION SPRINGS

The most common helical compression spring is a constant coil diameter, constant pitch, round-wire spring, as shown in Figure 13-2a. We will refer to this as the standard helical compression spring (HCS). Other arrangements are possible, such as conical, barrel, hourglass, and variable pitch, also shown in Figure 13-2a. All provide a push force. A helical spring may be coiled either left handed or right handed.

Sample springs and dimensional parameters for a standard helical compression spring are shown in Figure 13-7. The **wire diameter** is d, the **mean coil diameter** is D, and these two dimensions along with the **free length** L_f and the **number of coils** N_t or the **coil pitch** p, are used to define the spring geometry for calculation and manufacturing purposes. The outside diameter D_o and the inside diameter D_i are of interest mainly to define the minimum hole in which it will fit or the maximum pin over which it can be placed. They are found by adding or subtracting the wire diameter d to or from the mean coil diameter D. The minimum recommended diametral clearances between the D_o and a hole or between D_i and a pin are $0.10D$ for $D < 0.5$ in (13 mm) or $0.05D$ for $D > 0.5$ in (13mm).[1]

Spring Lengths

Compression springs have several lengths and deflections of interest as shown in Figure 13-8. **Free length** L_f is the overall spring length in the unloaded condition, i.e., as manufactured. **Assembled length** L_a is the length after installation to its initial deflection $y_{initial}$. This initial deflection in combination with the spring rate k determines the amount of **preload** force at assembly. The **working load** is applied to further compress the spring through its **working deflection** $y_{working}$. The **minimum working length** L_m is the shortest dimension to which it is compressed in service. The **shut height** or **solid height** L_s is its length when compressed such that all coils are in contact. The **clash allowance** y_{clash} is the difference between the minimum working length and the shut height, expressed as a percentage of the working deflection. A minimum clash allowance of 10-15% is recommended to avoid reaching the shut height in service with out-of tolerance springs, or with excessive deflections.

End Details

There are four types of end details available on helical compression springs: *plain, plain-ground, squared,* and *squared-ground* as shown in Figure 13-9. Plain ends result from simply cutting the coils and leaving the ends with the same pitch as the rest of the spring. This is the least expensive end detail, but provides poor alignment to the surface against which the spring is pressed. The end coils can be ground flat and perpendicular to the spring axis to provide normal surfaces for load application. Squaring the ends involves yielding the end coils to flatten them and remove their pitch. This improves the alignment. A flat surface on the end coil of at least 270° is recommended for proper operation.[1] Squaring and grinding combined provides a 270-330° flat surface for load application. It is the most expensive end treatment, but is nevertheless recommended for machinery springs unless the wire diameter is very small (< 0.02 in or 0.5 mm), in which case they should be squared but not ground.[1]

Active Coils

The total number of coils N_t may or may not contribute actively to the spring's deflection depending on the end treatment. The number of active coils N_a is needed for cal-

$N_a = N_t$

(a) Plain ends

$N_a = N_t - 1$

(b) Plain-ground ends

$N_a = N_t - 2$

(c) Squared ends

$N_a = N_t - 2$

(d) Squared-ground ends

FIGURE 13-9

Four Styles of End-Coil Treatments for Helical Compression Springs

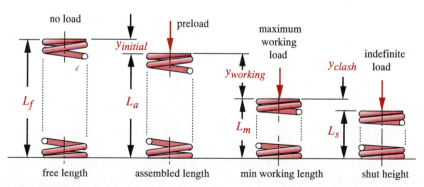

FIGURE 13-8

Various Lengths of a Helical Compression Spring in Use

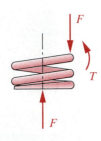

culation purposes. Squared ends effectively remove two coils from active deflection. Grinding by itself removes 1 active coil. Figure 13-9 shows the relationships between total coils N_t and active coils N_a for each of the four end-coil conditions. The calculated number of active coils is usually rounded to the nearest 1/4 coil as the manufacturing process cannot achieve better than that accuracy.

Spring Index

The spring index C is the ratio of coil diameter D to wire diameter d.

$$C = \frac{D}{d} \tag{13.5}$$

The preferred range of C is from 4 to 12.[1] At $C < 4$, the spring is difficult to manufacture and at $C > 12$ it is prone to buckling and also tangles easily when handled in bulk.

Spring Deflection

FIGURE 13-10

Forces and Torques on the Coils of a Helical Compression Spring

Figure 13-10 shows a portion of a helical coil spring with compressive axial loads applied. Note that even though the load on the spring is compression, the spring wire is in *torsion* as the load on any coil tends to twist the wire about its axis. A simplified model of this loading, neglecting the curvature of the wire, is a torsion bar as shown in Figure 4-28 in Section 4.12. A helical compression spring is, in fact, a torsion bar wrapped into a helical form, which packages better. The deflection of a round-wire helical compression spring is

$$y = \frac{8FD^3 N_a}{d^4 G} \tag{13.6}$$

where F is the applied axial load on the spring, D is mean coil diameter, d is wire diameter, N_a is number of active coils, and G is the shear modulus of the material.

Spring Rate

The equation for spring rate is found by rearranging the deflection equation:

$$k = \frac{F}{y} = \frac{d^4 G}{8D^3 N_a} \tag{13.7}$$

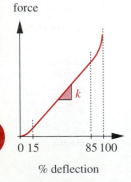

FIGURE 13-11

Force-Deflection Curve for a Standard Helical Compression Spring (Adapted from reference 1)

The standard helical compression spring has a spring rate k that is essentially linear over most of its operating range, as shown in Figure 13-11. The first and last few percent of its deflection have a nonlinear rate. When it reaches its **shut height** L_s, all the coils are in contact and the spring rate approaches the modulus of elasticity of the material.

13

The spring rate should be defined between about 15% and 85% of its total deflection[1] and its working deflection range $L_a - L_m$ kept in that region (see Figure 13-8).

Stresses in Helical Compression Spring Coils

The free-body diagram in Figure 13-10 shows that there will be two components of stress on any cross section of a coil, a torsional shear stress from the torque T and a direct shear stress due to the force F. These two shear stresses have the distributions across the section shown in Figure 13-12a and 13-12b. The two stresses add directly and the maximum shear stress τ_{max} occurs at the inner fiber of the wire's cross section, as shown in Figure 13-12c.

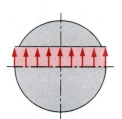

(a) Direct shear stress distribution across section

$$\tau_{max} = \frac{Tr}{J} + \frac{F}{A} = \frac{F(D/2)(d/2)}{\pi d^4/32} + \frac{F}{\pi d^2/4}$$

$$= \frac{8FD}{\pi d^3} + \frac{4F}{\pi d^2}$$

(13.8a)

We can substitute the expression for spring index C from equation 13.5 in Eq. 13.8a.

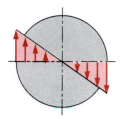

(b) Torsional shear stress distribution across section

$$\tau_{max} = \frac{8FC}{\pi d^2} + \frac{4F}{\pi d^2} = \frac{8FC + 4F}{\pi d^2}$$

$$= \frac{8FC}{\pi d^2}\left(1 + \frac{1}{2C}\right) = \frac{8FD}{\pi d^3}\left(1 + \frac{0.5}{C}\right)$$

$$\tau_{max} = K_s \frac{8FD}{\pi d^3} \qquad \text{where} \quad K_s = \left(1 + \frac{0.5}{C}\right)$$

(13.8b)

This manipulation has put the direct shear term of equation 13.8a into a **direct shear factor** K_s. The two equations are identical in value, but the second version (Eq. 13.8b) is preferred.

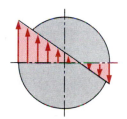

(c) Combined direct shear and torsional stress

If the wire were straight and were subjected to the combination of direct shear force F and torque T shown in Figure 13-10, equation 13.8 would be the exact solution. However, this wire is curved into a coil. We learned in Section 4.10 that curved beams have a stress concentration on the inner surface of curvature. While our spring is not loaded as a beam, the same principal applies and there is higher stress at the inner surface of the coil. Wahl[3] determined the stress-concentration factor for this application and defined a factor K_w, which includes both the direct shear effects and the stress concentration due to curvature.

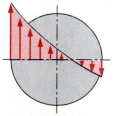

(d) Effects of stress concentration at inside edge

$$K_w = \frac{4C-1}{4C-4} + \frac{0.615}{C}$$

(13.9a)

$$\tau_{max} = K_w \frac{8FD}{\pi d^3}$$

(13.9b)

FIGURE 13-12

Stress Distributions Across Wire in a Helical Compression Spring

13

This combined stress is shown in Figure 13-12*d*.

Since Wahl's factor K_w includes both effects, we can separate them into a curvature factor K_c and the direct shear factor K_s using

$$K_w = K_s K_c \,; \qquad\qquad K_c = \frac{K_w}{K_s} \qquad\qquad (13.10)$$

If a spring is statically loaded, then yielding is the failure criterion. If the material yields, it will relieve the local stress concentration due to the curvature factor K_c and equation 13.8*b* can be used to account just for the direct shear. But, if the spring is dynamically loaded, then failure will be by fatigue at stresses well below the yield point and equation 13.9*b* should be used to incorporate both the direct shear and curvature effects. In a fatigue-loading case with both mean and alternating loads, equation 13.8*b* can be used to compute the mean stress component and equation 13.9*b* used for the alternating stress component.

Residual Stresses

When a wire is coiled into a helix, tensile residual stresses are developed at the inner surface and compressive residual stresses occur at the outer surface. Neither of these residual stresses is beneficial. They can be removed by stress relieving (annealing) the spring.

SETTING   Beneficial residual stresses can be introduced by a process confusingly called both "set removal" and "setting the spring" by the manufacturers. Setting can increase the static load capacity by 45-65% and double the spring's energy storage capacity per lb of material.[1] Setting is done by compressing the spring to its shut height and yielding the material to introduce beneficial residual stresses. Recall from Section 6.8 that the rule for introducing beneficial residual stresses is to *overstress (yield) the material in the same direction as the stresses applied in service will*. The "set" spring loses some free length but gains the benefits described above. In order to achieve the advantages of setting, the initial free length must be made longer than the desired (postset) length and should be designed to give a stress at shut height about 10 to 30% greater than the yield strength of the material. Less than that amount of overload will not create sufficient residual stress. More than 30% overstress adds little benefit and increases distortion.[1]

The allowable stress (i.e., the strength) of a spring that has been" set" is significantly higher than that of an as-wound spring. In addition, equation 13.8*b* with its smaller K_s factor can be used rather than equation 13.9*b* to calculate the stress in a "set" spring since, for static loading, the yielding during setting relieves the curvature stress concentration. Setting is of greatest value for statically loaded springs, but also has value in cyclic loading.

Not all commercial springs are set as this increases their cost. The designer should specify setting if it is desired. Do not assume it will be done as a matter of course. Sometimes a setting operation is specified as part of the assembly process rather than as part of the spring-manufacturing process. If convenient, a spring can be deliberately cycled once to its shut height prior to, or when assembled into, its final location in a machine.

LOAD REVERSAL Whether set or not, coil springs typically will have some residual stresses in them. For this reason it is not acceptable to apply reversed loads to them. Assuming that the residual stresses have been arranged to be beneficial against the expected direction of loading, reversed loading will obviously exacerbate the residual stresses and cause early failure. A compression spring should never be loaded in tension nor a tension spring in compression. Even torsion springs, as we shall see, need to have a unidirectional torque applied to avoid premature failure.

SHOT PEENING is another way to obtain beneficial residual stresses in springs and is most effective against cyclic loading in fatigue. It has little benefit for statically loaded springs. The shot-peening process was discussed in Section 6.8. For wire springs, shot diameters of 0.008 in (0.2 mm) to 0.055 in (1.4 mm) are typically used. Springs of very small wire diameter will not benefit from shot peening as much as will ones of larger-diameter wire. Also, if the coil pitch is small (i.e., a tightly wound spring), the shot cannot effectively impact the inner surfaces of the coils.

free to tip

fixed end

(a) Nonparallel ends

Buckling of Compression Springs

A compression spring is loaded as a column and can buckle if it is too slender. A slenderness ratio was developed for solid columns in Chapter 4. That measure is not directly applicable to springs due to their much different geometry. A similar slenderness factor is created as the aspect ratio of free length to mean coil diameter L_f / D. If this factor is > 4 the spring may buckle. Gross buckling can be prevented by placing the spring in a hole or over a rod. However, rubbing of the coils on these guides will take some of the spring force to ground through friction and reduce the load delivered at the spring end.

Just as with solid columns, the end constraints of the spring affect its tendency to buckle. If one end is free to tip as shown in Figure 13-13*a*, the spring will buckle with a smaller aspect ratio than if it is held against parallel plates at each end as shown in Figure 13-13*b*.

The ratio of the spring's deflection to its free length also affects its tendency to buckle. Figure 13-14 shows a plot of two lines that depict the stability of the two end-constraint cases of Figure 13-13. Springs with aspect ratio-deflection ratio combinations to the left of these lines are stable against buckling.

constrained parallel

fixed end

(b) Parallel ends

13

FIGURE 13-13

End Conditions Determine Critical Buckling Situation Adapted from Reference 1

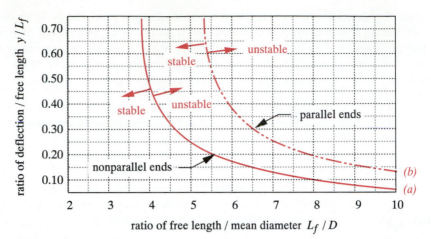

FIGURE 13-14

Critical Buckling Condition Curves Adapted from Reference 1

Compression-Spring Surge

Any device with both mass and elasticity will have one or more natural frequencies, as was discussed in Chapter 9 relative to shaft vibrations. Springs are no exception to this rule and can vibrate both laterally and longitudinally when dynamically exited near their natural frequencies. If allowed to go into resonance, the waves of longitudinal vibrations, called surging, cause the coils to impact one another. The large forces from both the excessive coil deflections and impacts will fail the spring. To avoid this condition, the spring should not be cycled at a frequency close to its natural frequency. Ideally, the natural frequency of the spring should be greater than about 13 times that of any applied forcing frequency.

The natural frequency ω_n or f_n of a helical compression spring depends on its boundary conditions. Fixing both ends is the more common and desirable arrangement as its f_n will be twice that of a spring with one end fixed and the other free. For the fixed-fixed case:

$$\omega_n = \pi \sqrt{\frac{kg}{W_a}} \ \ \text{rad/sec} \qquad f_n = \frac{1}{2} \sqrt{\frac{kg}{W_a}} \ \ \text{Hz} \qquad (13.11a)$$

where k is the spring rate, W_a is the weight of the spring's active coils and g is the gravitational constant. It can be expressed either as angular frequency ω_n or linear frequency f_n. The weight of the active coils can be found from

$$W_a = \frac{\pi^2 d^2 D N_a \gamma}{4} \qquad (13.11b)$$

where γ is the material's weight density. For total spring weight substitute N_t for N_a.

Substituting equations 13.7 and 13.11*a* into 13.11*b* gives

$$f_n = \frac{2}{\pi N_a} \frac{d}{D^2} \sqrt{\frac{Gg}{32\gamma}} \text{ Hz} \qquad (13.11c)$$

for the natural frequency of a fixed-fixed helical coil spring. If one end of the spring is fixed and the other free, it acts like a fixed-fixed spring of twice its length. Its natural frequency can be found by using a number for N_a in equation 13.11*c* that is twice the actual number of active coils present in the fixed-free spring.

Allowable Strengths for Compression Springs

Extensive test data are available on the failure strengths of round-wire helical compression springs, both statically and dynamically loaded. The relationships of ultimate tensile strength to wire diameter were discussed in Section 13.3. For spring design, additional strength data are needed for yield and fatigue strengths.

TORSIONAL YIELD STRENGTH   The torsional yield strengths of spring wire vary depending on the material and whether the spring has been set or not. Table 13-6 shows recommended torsional yield-strength factors for several common spring wires as a percentage of the wire's ultimate tensile strength. These factors should be used to determine an estimated strength for a helical compression spring in static loading.

TORSIONAL FATIGUE STRENGTH   over the $10^3 \le N \le 10^7$ cycles range varies with the material and whether it is shot peened or not. Table 13-7 shows recommended values for several wire materials in the peened and unpeened conditions at three points on their *S-N* diagrams, 10^5, 10^6, and 10^7 cycles. Note that these are torsional fatigue strengths and are determined from test springs loaded with equal mean and alternating stress components (stress ratio $R = \tau_{min}/\tau_{max} = 0$). So, they are not directly comparable to any of the fully reversed fatigue strengths generated from rotating-bending specimens discussed in Chapter 6 because of both the torsional loading and the presence of a mean stress component. We will use the designation S_{fw}, for these wire fatigue strengths to differentiate them from the fully reversed fatigue strengths of Chapter 6. These fatigue strengths S_{fw}, are nonetheless very useful in that they represent an actual (and typical) spring-fatigue-loading situation and are generated from spring samples, not test specimens, so the geometry and size are correct. Note that the fatigue strengths in Table 13-7 are declining with increasing numbers of cycles, even above 10^6 cycles, where steels usually display an endurance limit.

TORSIONAL ENDURANCE LIMIT   Steel can have an endurance limit for infinite life. High-strength materials tend to show a "topping out" of their endurance limits with increasing ultimate strength. Figures 6-9 and 6-11 show this trend and equation 6.5*a*[*] defines an uncorrected tensile endurance limit for fully reversed bending of steels with S_{ut} > 200 kpsi that remains constant with increasing tensile strength above that value. Note in Figure 13-3 that most spring wires smaller than about 10-mm diameter are in this ultimate-strength category. This would imply that these spring-wire materials should have

13

---

[*] See pages 361, 363, and 373, respectively.

Table 13-6   Maximum Torsional Yield Strength S_{ys} for Helical Compression Springs in Static Applications

Bending or Buckling Stresses Not Included   Source: Reference 1

Material	Maximum Percent of Ultimate Tensile Strength	
	Before Set Removed (Use Eq. 13.9*b*)	After Set Removed (Use Eq. 13.8*b*)
Patented and cold-drawn carbon steel	45%	60-70%
Hardened and tempered carbon and low-alloy steel	50	65-75
Austenitic stainless steel	35	55-65
Nonferrous alloys	35	55-65

a torsional endurance limit that is independent of size or of their particular alloy composition. Other research bears this out. Zimmerli[4] reports that all spring-steel wire of < 10-mm diameter exhibits a torsional endurance limit for infinite life with stress ratio $R = 0$ (which, to differentiate from the fully reversed endurance limit, we will call $S_{ew'}$).

$$S_{ew'} \cong 45.0 \text{ kpsi} \ (310 \text{ MPa}) \quad \text{for unpeened springs}$$

$$\text{(13.12)}$$

$$S_{ew'} \cong 67.5 \text{ kpsi} \ (465 \text{ MPa}) \quad \text{for peened springs}$$

There is no need in this case to apply surface-, size-, or loading-correction factors to either $S_{fw'}$ or $S_{ew'}$, since the test data were developed with actual conditions for those aspects of the wire materials. Table 13-7 notes that the fatigue strength data are taken at room temperature, in a noncorrosive environment, with no surging present. This is also true of Zimmerli's data. If the spring is to operate at elevated temperature or in a corrosive environment, the fatigue strength or endurance limit can be downrated accordingly. A temperature factor K_{temp}, and/or a reliability factor K_{reliab}, can be applied. See Chapter 6, equations 6.7*f* and 6.8. Figure 6-30 gives some information regarding cor-

Table 13-7   Maximum Torsional Fatigue Strength $S_{fw'}$ for Round Wire Helical Compression Springs in Cyclic Applications (Stress Ratio, R = 0)

No Surging, Room Temperature, and Noncorrosive Environment. Source: Ref. 1

Fatigue Life (cycles)	Percent of Ultimate Tensile Strength			
	ASTM 228, Austenitic Stainless Steel and Nonferrous		ASTM A230 and A232	
	Unpeened	Peened	Unpeened	Peened
10^5	36%	42%	42%	49%
10^6	33	39	40	47
10^7	30	36	38	46

rosive environments. We will use the uncorrected values of $S_{fw'}$ for S_{fw} and $S_{ew'}$ for S_{ew} in our discussion here, assuming room temperature, no corrosion, and 50% reliability.

The Torsional-Shear S-N Diagram for Spring Wire

A torsional-shear S-N diagram for a particular wire material and size can be constructed from the information in Tables 13-4 and 13-7 using the method described in Section 6.6 for creating estimated S-N diagrams. The region of interest for high-cycle fatigue is from $N = 1\,000$ cycles to $N = 1E7$ cycles and beyond. The endurance limit for infinite life of spring wire is defined in equation 13.12. The tensile strength S_m at 1 000 cycles is typically taken as 90% of the ultimate strength S_{ut} at 1 cycle (the static strength). Since this is a torsional-loading situation, the wire tensile strengths shown in Figure 13-3 and defined by equation 13.3 and Table 13-4 must be converted to torsional strengths with equation 13.4. This makes the torsional strength S_{ms} at 1 000 cycles equal to

$$S_{ms} \cong 0.9 S_{us} \cong 0.9 (0.67 S_{ut}) \cong 0.6 S_{ut} \tag{13.13}$$

EXAMPLE 13-1

Constructing the S-N Diagram for a Spring Wire Material

Problem　　Create the torsional-shear S-N diagrams for a range of spring-wire sizes.

Given　　ASTM A228 music wire, unpeened.

Assumptions　　Three diameters will be used: 0.010 in (0.25 mm), 0.042 in (1.1 mm), and 0.250 in (6.5 mm).

Solution　　See Figure 13-15.

1　The tensile strength of each wire size is found from equation 13.3 in combination with the coefficient and exponent from Table 13-4 for this material.

$$S_{ut} \cong 184\,649\,d^{-0.1625}$$

$$= 184\,649(0.010)^{-0.1625} = 390\,239 \text{ psi}$$

$$= 184\,649(0.042)^{-0.1625} = 309\,071 \text{ psi} \tag{a}$$

$$= 184\,649(0.250)^{-0.1625} = 231\,301 \text{ psi}$$

2　These values are converted to shear strengths at 1 000 cycles using equation 13.13:

$$S_{ms} \cong 0.6 S_{ut}$$

$$d = 0.010: \qquad S_{ms} \cong 0.6(390\,239) = 234\,143 \text{ psi}$$

$$d = 0.042: \qquad S_{ms} \cong 0.6(309\,071) = 185\,443 \text{ psi} \tag{b}$$

$$d = 0.250: \qquad S_{ms} \cong 0.6(231\,301) = 138\,781 \text{ psi}$$

13

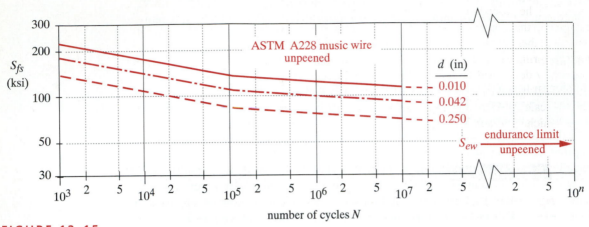

S_{fs} (ksi)

ASTM A228 music wire
unpeened

d (in)

--- 0.010
--- 0.042
--- 0.250

S_{ew} — endurance limit unpeened →

number of cycles N

FIGURE 13-15

Torsional-Fatigue S-N Diagrams for Music Wire of Various Diameters

3   The torsional fatigue strengths S_{fw} at three values of N are provided as percentages of the tensile strength in Table 13-7 for unpeened A228 music wire.

$$@N = 1E5: \qquad S_{fw} \cong 0.36(390\,239) = 140\,486 \text{ psi}$$

$$@N = 1E6: \qquad S_{fw} \cong 0.33(309\,071) = 101\,993 \text{ psi} \qquad (c)$$

$$@N = 1E7: \qquad S_{fw} \cong 0.30(231\,301) = 69\,390 \text{ psi}$$

These values are plotted in combination with the result from equation 13.13 to generate the S-N curves.

4   Figure 13-15 shows the S-N curves. There are two separate portions to each S-N curve: the $1E3 \le N < 1E5$ segment and the segment for $N \ge 1E5$. The unpeened wire endurance limit for infinite life S_{ew} is also shown at 45 000 psi (equation 13.12).

5   If desired, any of these S-N curves can be fitted to an exponential equation (Eq. 6.10) by the method shown in Section 6-6. Evaluating the coefficients and exponents separately for the two pieces of the S-N curve allows the estimated wire fatigue strength S_{fw} to be easily found for any number of cycles.

6   It is important to remember that the S_{fw} data in Table 13-7 are for a repeated-stress state, not a fully reversed stress condition, which means that this S-N diagram is taken at some point along the σ_m axis in Figure 6-42.

13

The Modified-Goodman Diagram for Spring Wire

A modified-Goodman diagram can be constructed for any spring-loading situation. In Section 6.13 of Chapter 6 we presented a general approach to fatigue design that involved finding the von Mises effective stresses for any combined loading case in order to simplify the procedure. It was noted there that a pure-torsional loading situation can be solved in that manner as well, converting the shear stresses to von Mises stresses and comparing them to the material's tensile strengths. However, in the case of helical compression spring design, it makes little sense to use the von Mises approach because the empirically developed fatigue strengths for wire are expressed as torsional strengths. Thus it will be easier to construct a Goodman diagram using torsional strengths and apply the calculated torsional stresses to it directly. The results will be the same regardless of which approach is used.

EXAMPLE 13-2

Creating the Modified-Goodman Diagram for a Helical Spring

Problem Construct the Goodman line for the spring wire of Example 13-1.

Given The required cycle life is $N = 1E6$ cycles. Wire is 0.042-in (1.1-mm) dia.

Assumptions The torsional strengths and torsional shear stresses will be used on the Goodman diagram.

Solution See Figure 13-16.

1 The material's ultimate tensile strength from Figure 13-3 or equation 13.3, converted to ultimate torsional strength with equation 13.4 using data from Table 13-4 allows one point on the Goodman line to be determined.

$$S_{ut} \cong 184\,649\,(0.042)^{-0.1625} = 309\,071 \text{ psi} \qquad (a)$$

$$S_{us} \cong 0.67 S_{ut}$$
$$= 0.67(309\,071) = 207\,078 \text{ psi} \qquad (b)$$

This value is plotted as point A on the diagram in Figure 13-16.

2 The S-N diagrams each provide one data point (S_{fw} or S_{ew} depending on whether for finite or infinite life) on the modified-Goodman line for a material/size combination in pure torsional loading. The fatigue strength S_{fw} for that wire material and condition is taken from the S-N line of Figure 13-15 or calculated from the data in Table 13-7 as

$@N = 1E6:$ $S_{fw} \cong 0.33(309\,071) = 101\,993 \text{ psi}$ $\qquad (c)$

The x and y intercepts are $0.707\,S_{fw} = 72\,110$ psi. This is plotted as point B on the diagram in Figure 13-16. Note that for infinite life the value of S_{ew} from equation 13.12 would be plotted at B instead of this value of S_{fw} for a finite life.

3 Note in Figure 13-16 that the wire fatigue strength S_{fw} is plotted on a 45° line from the origin to correspond to the test conditions of equal mean and alternating stress

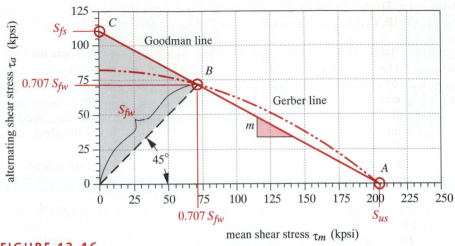

FIGURE 13-16

Torsional-Stress Modified-Goodman Diagram for 0.045-in Dia ASTM A228 Wire at N = 1E6 Cycles

components (stress ratio $R = \tau_{min}/\tau_{max} = 0$). Point B is then connected with the ultimate shear strength S_{us} on the mean-stress axis at point A to draw the Goodman line, which is extended to point C.

4　We can now find the value of the fully reversed fatigue strength ($R = -1$), which is point C on the diagram. This value can be found from the equation for the Goodman line, defined in terms of its two known points, A and B:

$$m = -\frac{0.707 S_{fw}}{S_{us} - 0.707 S_{fw}}$$

$$S_{fs} = -m S_{us}$$

$$S_{fs} = 0.707 \frac{S_{fw} S_{us}}{S_{us} - 0.707 S_{fw}} \qquad (d)$$

5　The Gerber line is also plotted through points A and B. Recall from Figure 6-16 (p. 366) that the Gerber line is a better fit to experimental failure data for combined mean and alternating stresses. Thus, the extrapolation of the Goodman line back to the τ_a axis in this case would seem to be nonconservative, and it is so for the shaded area in Figure 13-16. However, this use of the Goodman line is conservative for stress ratios $R \geq 0$ and its use is justified in this case because springs should always be loaded in the same direction. Helical compression springs tend to have stress ratios between 0 and 0.8, which puts their stress coordinates to the right of the 45° line in the figure, where the Goodman line is more conservative than the Gerber line.

6　Any other combination of mean and alternating stress with a stress ratio $R \geq 0$ for this material and number of cycles can now be plotted on this diagram to obtain a safety factor.

13.5   DESIGNING HELICAL COMPRESSION SPRINGS FOR STATIC LOADING

The functional requirements for a spring design can be quite varied. There may be a requirement for a particular force at some deflection or the spring rate may be defined for a range of deflection. In some cases there are limitations on the outside diameter, inside diameter, or working length. The approach to design will vary depending on these requirements. In any case, spring design is inherently an iterative problem. Some assumptions must be made to establish the values of enough variables to calculate the stresses, deflections and spring rate. Because wire size appears to the third or fourth power in the stress and deflection equations, and because material strength is dependent on wire size, the safety of the design is very sensitive to this parameter.

Many approaches may be taken to spring design and more than one combination of spring parameters can satisfy any set of functional requirements. It is possible to optimize parameters such as spring weight for a given set of performance specifications. To minimize weight and cost, the stress levels should be made as high as possible without causing static yielding in service.

A trial wire diameter d should be assumed and a reasonable spring index C chosen, from which the coil diameter D can be calculated using equation 13.5. A trial spring material is chosen and the relevant material strengths calculated for the trial wire diameter. It is convenient to calculate the stress before computing the deflection because, while both involve d and D, only the deflection depends on N_a. If a required force F is defined, the stress at that force can be computed with equation 13.8 or 13.9 as appropriate. If two operating forces are defined with a specified deflection between them, they will define the spring rate.

The stress state is compared to the yield strength for static loading. The safety factor for static loading is

$$N_s = \frac{S_{ys}}{\tau} \tag{13.14}$$

If the calculated stress is too high compared to the material strength, the wire diameter, spring index, or material can be changed to improve the result. When the calculated stress at the required operating force seems reasonable compared to the material strength, a trial number of coils and a clash allowance can be assumed and further calculations for spring rate, deflection, and free length done using equations 13.6 and 13.7. Unreasonable values of any of these parameters will require further iteration with changed assumptions.

After several iterations, a reasonable combination of parameters can usually be found. Some of the things that need to be checked before the design is complete will be the stress at shut height, the D_i, D_o, and free length of the coil with respect to packaging considerations. In addition, the possibility of buckling needs to be checked.

13

If the above process seems complicated, the reader should appreciate the value of making a computer do the "dirty work." Spring design, like any iterative design procedure, is an obvious task for a computer solution. Equation solvers that allow automatic iteration are extremely well suited to these kinds of tasks since they solve all aspects of the problem simultaneously. We will now present some examples of spring design problems and show how an equation solver can be used to expedite their solutions.

EXAMPLE 13-3

Design of a Helical Compression Spring for Static Loading

Problem   Design a compression spring for a static load over a known deflection.

Given   The spring must give a minimum force of 100 lb and a maximum force of 150 lb over an adjustment range of 0.75-in deflection.

Assumptions   Use the least expensive, unpeened, cold-drawn spring wire (ASTM A227) since the loads are static.

Solution   See Table 13-8 and *TKSolver* files EX13-03a and EX13-03b.

1 Assume a trial wire diameter of 0.162 in from the available sizes in Table 13-2.

2 Assume a spring index of 8, which is in the middle of the recommended range, and calculate the mean coil diameter D from equation 13.5.

$$D = Cd = 8(0.162) = 1.30 \text{ in} \qquad (a)$$

3 Find the direct shear factor K_s and use it to calculate the shear stress in the coil at the larger force.

$$K_s = 1 + \frac{0.5}{C} = 1 + \frac{0.5}{8} = 1.06 \qquad (b)$$

$$\tau = K_s \frac{8FD}{\pi d^3} = 1.06 \frac{8(150)(1.30)}{\pi(0.162)^3} = 123\,714 \text{ psi} \qquad (c)$$

4 Find the ultimate tensile strength of this wire material from equation 13.3 and Table 13-4 and use it to find the torsional yield strength from Table 13-6, assuming that the set has been removed and using the low end of the recommended range.

$$S_{ut} = Ad^b = 141\,040(0.162)^{-0.1822} = 196\,503 \text{ psi} \qquad (d)$$

$$S_{ys} = 0.60 S_{ut} = 0.60(196\,503) = 117\,902 \text{ psi} \qquad (e)$$

5 Find the safety factor against yielding at this working deflection from equation 13.14.

$$N_s = \frac{S_{ys}}{\tau} = \frac{117\ 902\ \text{psi}}{123\ 714\ \text{psi}} = 0.95 \qquad (f)$$

This is obviously not acceptable, so the design must be iterated with some parameters changed.

6 Try increasing the wire diameter slightly, perhaps to 0.192 in, keeping the same spring index. Recalculate the coil diameter, stress, strength, and safety factor.

$$D = Cd = 8(0.192) = 1.54\ \text{in} \qquad (g)$$

$$\tau = K_s \frac{8FD}{\pi d^3} = 1.06 \frac{8(150)(1.54)}{\pi(0.192)^3} = 88\ 074\ \text{psi} \qquad (h)$$

$$S_{ut} = Ad^b = 141\ 040(0.192)^{-0.1822} = 190\ 513\ \text{psi} \qquad (i)$$

$$S_{ys} = 0.60S_{ut} = 0.60(190\ 513) = 114\ 308\ \text{psi} \qquad (j)$$

$$N_s = \frac{S_{ys}}{\tau} = \frac{114\ 308\ \text{psi}}{88\ 074\ \text{psi}} = 1.30 \qquad (k)$$

This appears to be acceptable, so we can go on to design the other spring parameters.

7 The spring rate is defined in this problem because of the two specified forces at a particular relative deflection.

$$k = \frac{\Delta F}{y} = \frac{150 - 100}{0.75} = 66.67\ \text{lb/in} \qquad (l)$$

8 To achieve this spring rate, the number of active coils must satisfy equation 13.7:

$$k = \frac{d^4 G}{8D^3 N_a} \quad \text{or} \quad N_a = \frac{d^4 G}{8D^3 k} = \frac{(0.192)^4 11.5E6}{8(1.54)^3 66.67} = 8.09 \cong 8 \qquad (m)$$

Note that we round it to the nearest 1/4 coil as the manufacturing tolerance cannot achieve better than that accuracy.

9 Assume squared and ground ends making the total number of coils from Figure 13-9:

$$N_t = N_a + 2 = 8 + 2 = 10 \qquad (n)$$

10 The shut height can now be determined.

$$L_s = dN_t = 0.192(10) = 1.92\ \text{in} \qquad (o)$$

11 The initial deflection to reach the smaller of the two specified loads is

$$y_{initial} = \frac{F_{initial}}{k} = \frac{100}{66.67} = 1.5\ \text{in} \qquad (p)$$

12 Assume a clash allowance of 15% of the working deflection:

13

Table 13-8   Example 13-3—Program for Helical Compression Spring Design

Part 1 of 2   *TKSolver* Rule Sheet for Static Loading of Spring. See the File EX13-03b.

; select available wire size close to specified dia & calculate spring index C or coil mean dia D

$\quad d = Wiregage(dia)$

$\quad C = D / d$

; calculate initial and working deflections

$\quad yinit = Fmin/k$

$\quad y = (Fmax - Fmin) / k$

$\quad ymax = yinit + y$

; calculate clash allowance as % of deflection

$\quad yclash = clash * y$

; calculate spring constant or number of coils for assumed wire dia

$\quad k = (d^4 * G) / (8 * D^3 * N)$

; round no of coils to nearest 1/4 coil & calculate total coils for end condition

$\quad Na = INT(N * 4 + 0.5) / 4$

$\quad Ntot = Na + Nend(end)$ " - using LIST FUNCTION Nend

; calculate free length, shut ht. and other lengths

$\quad Lshut = Ntot * d$

$\quad Lf = yinit + y + yclash + Lshut$

$\quad Linstal = Lf - Fmin / k$

$\quad Lcomp = Lf - Fmax / k$

; calculate shut deflection and force at shut ht & calculate mean and inside coil diameter

$\quad yshut = Lf - Lshut$

$\quad Fshut = yshut * k$

$\quad Dout = D + d$

$\quad Din = D - d$

; calculate Wahl factor and static stress conc factor ks

$\quad Kw = ((4 * C - 1) / (4 * C - 4)) + (.615 / C)$

$\quad Ks = 1 + .5 / C$

;calculate shut height stress and stress at initial preload

$\quad taushut = Ks * (8 * Fshut * D) / (PI() * d^3)$

$\quad tauinit = Ks * (8 * Fmin * D) / (PI() * d^3)$

$\quad taustatic = Ks * (8 * Fmax * D) / (PI() * d^3)$

; calculate Sut based on material selection & calculate Sus from eq. 13.4

$\quad Sut = A(matl) * d^b(matl)$; - using LIST FUNCTION

$\quad Sus = .67 * Sut$

; calculate shear yield strength - Table 13-6 - in *LIST FUNCTIONS*

$\quad$ IF setflag = 'set THEN Sys = Syieldset(matl) * Sut ELSE Sys = Syieldunset(matl)

; calculate safety factors

$\quad Ns\_shut = Sys / taushut$

$\quad Ns\_static = Sys / taustatic$

; check for buckling using Figure 13-14

$\quad R1 = Lf / D$

$\quad R2 = ymax / Lf$

Table 13-8　Example 13-3 – Helical Compression Spring Design for Static Loading
Part 2 of 2　　See the *TKSolver* File EX13-03b for Complete Information

Input	Variable	Output	Unit	Comments
8	*C*			trial spring index
0.192	*dia*			trial wire diameter
0.750	*y*		in	deflection of spring
15	*clash*		%	% of deflect for clash allowance
'hdrawn	*matl*			'music, 'oiltemp, 'hdrawn, 'chromev, etc.
'sqgrnd	*end*			one of 'plain, 'pgrnd, 'square, 'sqgrnd
'unpeen	*surface*			one of 'unpeen or 'peen
'set	*setflag*			one of 'set or 'unset
150	*Fmax*		lb	maximum applied force
100	*Fmin*		lb	mimimum applied force
	Fshut	158	lb	force at shut height
	k	66.67	lb/in	spring rate
	Na	8		no of active coils—to nearest 1/4 coil
	Ntot	10		no of total coils
	D	1.54	in	mean coil diameter
	Dout	1.73	in	outside coil diameter
	Din	1.34	in	inside coil diameter
	Ks	1.06		static factor—direct shear - Eq. 13.8
	Kw	1.18		Wahl Factor—Eq. 13.9
	tauinit	58 716	psi	shear stress at installed length
	taustat	88 074	psi	shear stress at Fmax for static loading
	taushut	92 478	psi	stress at shut height
	Sut	190 513	psi	tensile strength—Eq. 13.3 & Table 13-4
	Sus	127 644	psi	ultimate shear strength—Eq. 13.4
	Sys	114 308	psi	shear yield based on Table 13-6
	Ns_static	1.30		safety factor—static loading at Fmax
	Ns_shut	1.24		safety factor—shut height (yielding)
	Lf	4.28	in	free length
	Linstal	2.78	in	installed length
	Lcomp	2.03	in	compressed length
	Lshut	1.92	in	shut height
	yinit	1.50	in	inital deflection at assembly
	ymax	2.25	in	max working deflection
	yclash	0.113	in	coil clash allowance
	yshut	2.36	in	deflection to shut height

13

$$y_{clash} = 0.15y = 0.15(0.75) = 0.113 \text{ in} \qquad (q)$$

13 The free length (see Figure 13-8) can now be found from

$$L_f = L_s + y_{clash} + y_{working} + y_{initial} = 1.92 + 0.113 + 0.75 + 1.5 = 4.28 \text{ in} \qquad (r)$$

14 The deflection to the shut height is

$$y_{shut} = L_f - L_s = 4.25 - 1.92 = 2.36 \text{ in} \qquad (s)$$

15 The force at this shut-height deflection is

$$F_{shut} = k y_{shut} = 66.67(2.36) = 158 \text{ lb} \qquad (t)$$

16 The shut-height stress and safety factor are

$$\tau_{shut} = K_s \frac{8FD}{\pi d^3} = 1.06 \frac{8(158)(1.54)}{\pi(0.192)^3} = 92\,478 \text{ psi} \qquad (u)$$

$$N_{s_{shut}} = \frac{S_{sy}}{\tau_{shut}} = \frac{114\,308 \text{ psi}}{92\,478 \text{ psi}} = 1.24 \qquad (v)$$

which is acceptable.

17 To check for buckling, two ratios need to be calculated, L_f / D and y_{max} / L_f.

$$\frac{L_f}{D} = \frac{4.28}{1.54} = 2.79$$

$$\frac{y_{max}}{L_f} = \frac{y_{initial} + y_{working}}{L_f} = \frac{1.5 + 0.75}{4.28} = 0.53 \qquad (w)$$

Take these two values to Figure 13-14 and find that their coordinates are safely within the zones that are stable against buckling for either end-condition case.

18 The inside and outside coil diameters are

$$D_o = D + d = 1.54 + 0.192 = 1.73 \text{ in}$$
$$D_i = D - d = 1.54 - 0.192 = 1.34 \text{ in} \qquad (x)$$

19 The smallest hole and largest pin that should be used with this spring are

$$hole_{min} = D_o + 0.05D = 1.73 + 0.05(1.54) = 1.81 \cong 1\frac{13}{16} \text{ in}$$

$$pin_{max} = D_i - 0.05D = 1.34 - 0.05(1.54) = 1.26 \cong 1\frac{1}{4} \text{ in} \qquad (y)$$

20 The total weight of the spring is

$$W_t = \frac{\pi^2 d^2 D N_t \rho}{4} = \frac{\pi^2(0.192)^2(1.54)(10)(0.28)}{4} = 0.40 \text{ lb} \qquad (z)$$

13

21   We now have a complete design specification for this A227 wire spring:

$$d = 0.192 \text{ in} \qquad OD = 1.73 \text{ in} \qquad N_t = 10, \text{ sq \& g} \qquad L_f = 4.28 \text{ in} \qquad (aa)$$

22   Other spring parameters are shown in Table 13-8, which reproduces portions of the *TKSolver* rule and variable sheets for this example. Note in the table and in the supplied *TKSolver* files, that selection of the available wire diameters and the various factors for material strengths taken from the tables of this chapter have been auto-mated using **list functions**. For example, a code such as '*hdrawn* for hard-drawn wire is input and used as the argument for the list functions *A* and *b* that return the appropriate coefficient and exponent for use in equation 13.3 to calculate the tensile strength as a function of the wire diameter.

The functions *Syieldset* and *Syieldunset* return the appropriate yield strength percent-ages from Table 13-6 for the input code that defines the material. The list function *Wiregage* contains the available wire sizes from Table 13-2 and returns the closest value to the trial diameter input. The list function *Nend* returns the proper factor to calculate the total number of coils based on an input code such as '*sqgrnd*.

Encoding these empirical factors in *list functions* allows their values to be accessed "on the fly" while the program computes all the other equations. This means that the user need only specify the particular input values needed for the problem and select the desired codes for material, etc. The program then computes the result automati-cally. Of course, any variable on the sheet can be switched from input to output or vice-versa by moving it from one column to the other. Thus, the program is flexible enough to design a spring for any compatible set of defined input parameters.

13.6 DESIGNING HELICAL COMPRESSION SPRINGS FOR FATIGUE LOADING

When the spring loads are dynamic (time-varying), a fatigue-stress situation exists in the spring. The design process for dynamic loading is similar to that for static loading with some significant differences. A dynamically loaded spring will operate between two force levels, F_{min} and F_{max}. From these values, the alternating and mean compo-nents of force can be calculated from

$$F_a = \frac{F_{max} - F_{min}}{2}$$

$$F_m = \frac{F_{max} + F_{min}}{2} \qquad (13.15a)$$

A **force ratio** R_F can also be defined as:

$$R_F = \frac{F_{min}}{F_{max}} \qquad (13.15b)$$

13

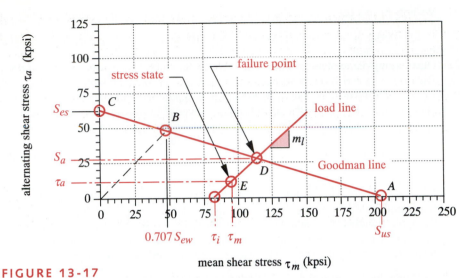

FIGURE 13-17

Modified-Goodman Diagram Showing Load Line and Data for Safety Factor Calculation

In the most common spring-loading cases F_{min} and F_{max} are both positive, with the force ratio about $0 < R_F < 0.8$. As described in the earlier discussion of residual stresses, bi-directional loading of coil springs is to be avoided as it causes early failure.

The fatigue-design procedure is essentially as outlined in the previous section for static loading. It is still an iterative problem. A trial wire diameter d should be assumed and a reasonable spring index C chosen, from which the coil diameter D can be calculated using equation 13.5. A trial spring material is chosen and the relevant material strengths calculated for the trial wire diameter. The ultimate shear strength, shear yield strength, and the endurance limit (or fatigue strength at some number of cycles) are all needed. The problem statement usually contains sufficient information to estimate the number of life-cycles required. For dynamic loading, the alternating and mean stresses are calculated separately (using F_{min} and F_{max} from equation 13.15a).

Unidirectional loading, also called fluctuating or repeated loading in Chapter 6, has nonzero mean stress and thus requires a Goodman diagram failure analysis. Since all significant stresses in this spring are torsional-shear stresses and most spring-wire material strength data are for torsional loading, we will use a **torsional Goodman diagram**, as discussed earlier. The modified-Goodman diagram is constructed as shown in Figures 13-16 and 13-17 with the torsional wire-fatigue strength S_{fw} or wire-endurance strength S_{ew} defined along a 45° line from the origin to represent the test data that was generated at $R_F = 0$. Figure 13-17 uses the value of torsional wire-endurance limit S_{ew} for infinite life of peened steel wire in combination with the torsional ultimate strength S_{us} to create the torsional Goodman line.

The load line, which represents the applied stress state, is not drawn from the origin in this case, but rather from a point on the τ_m axis representing the initial stress τ_i in the coils at assembly, as shown in Figure 13-17. This assumes that some spring preload is applied, which is usually the case. We do not want $F_{min} = 0$ in a dynamic-loading situation as that will create impact loads on the coils (see Section 3.8). If $F_{min} = 0$, the load line would start at the origin. The safety factor for torsional fatigue N_{f_s} can be expressed as the ratio of the alternating strength S_a at the intersection of the load line and the Goodman line (point D) to the applied alternating stress τ_a at point E.

$$N_{fs} = S_a / \tau_a \tag{13.16a}$$

This ratio can be derived from the geometry of the two lines. Let x represent the independent variable on the mean-stress axis, m represent the slope of a line and b its y intercept. Let the value on the load line for any x be y_{load}. The load line equation is

$$y_{load} = m_{load}x + b_{load}$$

from geometry $\qquad m_{load} = \dfrac{\tau_a}{\tau_m - \tau_i} \quad$ and $\quad b_{load} = -m_{load}\tau_i$

$$y_{load} = \frac{\tau_a}{\tau_m - \tau_i}(x - \tau_i) \tag{a}$$

Let the value on the Goodman line for any x be y_{Good}. Then its equation is:

$$y_{Good} = m_{Good}x + b_{Good}$$

from geometry $\qquad m_{Good} = -\dfrac{S_{es}}{S_{us}} \quad$ and $\quad b_{Good} = S_{es}$

$$y_{Good} = -\frac{S_{es}}{S_{us}}x + S_{es} = S_{es}\left(1 - \frac{x}{S_{us}}\right) \tag{b}$$

At the failure point, $y_{load} = y_{Good}$. Setting (a) and (b) equal and solving for x gives

$$S_{es}\left(1 - \frac{x}{S_{us}}\right) = \frac{\tau_a}{\tau_m - \tau_i}(x - \tau_i)$$

$$x = \frac{S_{us}\left[S_{se}(\tau_i - \tau_m) - \tau_a\tau_i\right]}{S_{se}(\tau_i - \tau_m) - S_{us}\tau_a} \tag{c}$$

Substitute (b) in equation 13.16a:

$$N_{fs} = \frac{S_a}{\tau_a} = \frac{y_{Good}}{\tau_a} = \frac{m_{Good}x + S_{se}}{\tau_a} \tag{d}$$

Substitute (c) in (d) and simplify to get the safety factor as

$$N_{f_s} = \frac{S_{es}(S_{us} - \tau_i)}{S_{es}(\tau_m - \tau_i) + S_{us}\tau_a} \tag{13.16b}$$

13

where the fully reversed endurance limit (point *C*) from Example 13-3 is

$$S_{es} = 0.707 \frac{S_{ew}S_{us}}{S_{us} - 0.707S_{ew}} \qquad (13.16c)$$

This approach assumes that the initial preload will not vary significantly over the life of the part and also that any increase in the loading will be such as to maintain a constant ratio between the alternating and mean components of stress. This corresponds to Case 3 of Section 6.11. If this is not the situation, then one of the other cases of Section 6.11 should be used to find the safety factor. The stress at the initial preload must still be accounted for, and itself may vary under service conditions.

If the safety factor is too low, the wire diameter, spring index, or the material can be changed to improve the result. When the fatigue safety factor is acceptable, a trial number of coils and a clash allowance can be assumed and further calculations for spring rate, deflection and free length done using equations 13.6 and 13.7. Unreasonable values of any of these parameters will require further iteration with changed assumptions.

After several iterations, a reasonable combination of parameters can usually be found. Some of the things that need to be checked before the design is complete will be the stress at shut height versus the yield stress, and the D_i, D_o, and free length of the coil with respect to packaging considerations. In addition, the possibility of buckling needs to be checked, and for dynamic loading, the natural frequency of the spring must be compared to any forcing frequencies in the system to guard against surging.

Spring design for fatigue loading benefits greatly from a computer solution. Equation solvers that allow automatic iteration are extremely well-suited to these kinds of tasks since they solve all aspects of the problem simultaneously. We will now present an example of spring design for fatigue loading.

13

EXAMPLE 13-4

Design of a Helical Compression Spring for Cyclic Loading

Problem	**Design a compression spring for a dynamic load over a given deflection.**
Given	**The spring must give a minimum force of 50 lb and a maximum force of 180 lb over a dynamic deflection of 1.00 in. The forcing frequency is 1000 rpm. A 10-year life of 1-shift operation is desired.**
Assumptions	**Music wire (ASTM A228) will be used since the loads are dynamic. Peening will be used to obtain a higher endurance strength.**
Solution	**See Table 13-9, Figures 13-18 and 13-19 , and *TKSolver* files EX13-04a and EX13-04b.**

1   Find the number of cycles that the spring will see over its design life.

$$N_{life} = 1000 \frac{rev}{min} \left(\frac{60 \ min}{hr} \right) \left(\frac{2080 \ hr}{shift \text{-} yr} \right) (10 \ yr) = 1.2E9 \ cycles \qquad (a)$$

This large a number requires that an endurance limit for infinite life be used.

2   Find the mean and alternating forces from equation 13.15a:

$$F_a = \frac{F_{max} - F_{min}}{2} = \frac{180 - 50}{2} = 65 \ lb$$

$$F_m = \frac{F_{max} + F_{min}}{2} = \frac{180 + 50}{2} = 115 \ lb \qquad (b)$$

3   Assume a 0.207-in wire diameter from the available sizes in Table 13-2. Assume a spring index of 9 and calculate the mean coil diameter D from equation 13.5.

$$D = Cd = 9(0.207) = 1.863 \ in \qquad (c)$$

4   Find the direct shear factor K_s and use it to calculate the stress τ_i at the initial deflection (lowest defined force), and the mean stress τ_m:

$$K_s = 1 + \frac{0.5}{C} = 1 + \frac{0.5}{9} = 1.056 \qquad (d)$$

$$\tau_i = K_s \frac{8F_i D}{\pi d^3} = 1.056 \frac{8(50)(1.863)}{\pi(0.207)^3} = 28 \ 229 \ psi \qquad (e)$$

$$\tau_m = K_s \frac{8F_m D}{\pi d^3} = 1.056 \frac{8(115)(1.863)}{\pi(0.207)^3} = 64 \ 926 \ psi \qquad (f)$$

5   Find the Wahl factor K_w and use it to calculate the alternating shear stress τ_a in the coil.

$$K_w = \frac{4C - 1}{4C - 4} + \frac{0.615}{C} = \frac{4(9) - 1}{4(9) - 4} + \frac{0.615}{9} = 1.162 \qquad (g)$$

$$\tau_a = K_w \frac{8F_a D}{\pi d^3} = 1.162 \frac{8(65)(1.863)}{\pi(0.207)^3} = 40 \ 401 \ psi \qquad (h)$$

6   Find the ultimate tensile strength of this music-wire material from equation 13.3 and Table 13-4 and use it to find the ultimate shear strength from equation 13.4 and the torsional yield strength from Table 13-6, assuming that the set has been removed and using the low end of the recommended range.

$$S_{ut} = Ad^b = 184 \ 649(0.207)^{-0.1625} = 238 \ 507 \ psi$$

$$S_{us} = 0.67 S_{ut} = 159 \ 800 \ psi \qquad (i)$$

$$S_{ys} = 0.60 S_{ut} = 0.60(238 \ 507) = 143 \ 104 \ psi \qquad (j)$$

13

7   Find the wire endurance limit for peened springs in repeated loading from equation
    13.12 and convert it to a fully reversed endurance strength with equation 13.16c.

$$S_{ew} = 67\ 500 \text{ psi} \qquad (k)$$

$$S_{es} = 0.707\frac{S_{ew}S_{us}}{S_{us} - 0.707S_{ew}} = 0.707\frac{67\ 500(159\ 800)}{159\ 800 - 0.707(67\ 500)} = 68\ 043 \text{ psi} \qquad (l)$$

8   The safety factor is calculated from equation 13.16b.

$$N_{fs} = \frac{S_{es}(S_{us} - \tau_i)}{S_{es}(\tau_m - \tau_i) + S_{us}\tau_a}$$

$$= \frac{68\ 043(159\ 800 - 28\ 229)}{68\ 043(64\ 926 - 28\ 229) + 159\ 800(40\ 401)} = 1.00 \qquad (m)$$

This is obviously not an acceptable design. To get some idea of what to change to
improve it, the *TKSolver* model was solved for a list of values of the spring index C
from 4 to 14, keeping all other parameters as defined above. The resulting values of
coil diameter, free length, spring weight, and torsional fatigue safety factor are
plotted in Figure 13-18. Note that the wire diameter was held constant to develop the
variation of parameters with spring index as shown in Figure 13-18. If another
parameter such as mean coil diameter D were held constant instead, a different set of
functions would result for the free length, weight, safety factor, etc.

The safety factor increases with decreasing spring index, so a reduction in our
assumed value for C will improve the safety factor even with no change in wire
diameter. Note however that the free length increases exponentially with decreasing
spring index. If package size is limited, we may not want to decrease the spring

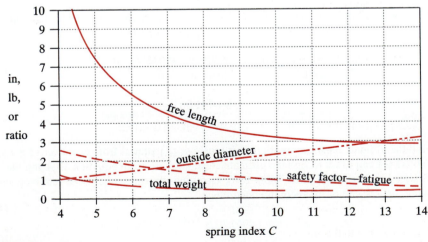

FIGURE 13-18

Variation of Helical Compression Spring Parameters with Spring Index—Constant Wire Diameter

index too much in order to avoid excessive spring length. The coil diameter increases linearly with the spring index for a constant wire diameter. Spring weight decreases slowly with increasing spring index.

If we decrease the spring index from 9 to 7, keeping all other parameters the same, we will obtain an acceptable design in this case with $N_f = 1.3$. Table 13-9 shows the results of this new calculation from the *TKSolver* variable sheet and also shows the results of the calculations outlined below to complete the design. Figure 13-19 shows the modified-Goodman diagram for the final design. A summary of the changed values is

$$C = 7 \qquad D = 1.45 \text{ in} \qquad K_w = 1.21 \qquad K_s = 1.07 \qquad (n)$$

$$\tau_i = 22\ 286 \text{ psi} \qquad \tau_a = 32\ 796 \text{ psi} \qquad \tau_m = 51\ 258 \text{ psi} \qquad N_{f_s} = 1.3$$

9   The spring rate is defined from the two specified forces at their relative deflection.

$$k = \frac{\Delta F}{y} = \frac{180 - 50}{1.0} = 130 \text{ lb/in} \qquad (o)$$

10 To get the defined spring rate, the number of active coils must satisfy equation 13.7:

$$k = \frac{d^4 G}{8D^3 N_a} \quad \text{or} \quad N_a = \frac{d^4 G}{8D^3 k} = \frac{(0.207)^4 11.5E6}{8(1.45)^3(130)} = 6.67 \cong 6\frac{3}{4} \qquad (p)$$

Note that we round it to the nearest 1/4 coil as the manufacturing tolerance cannot achieve better than that accuracy.

11 Assume squared and ground ends making the total number of coils from Figure 13-9:

$$N_t = N_a + 2 = 6.75 + 2 = 8.75 \qquad (q)$$

12 The shut height can now be determined.

$$L_s = dN_t = 0.207(8.75) = 1.81 \text{ in} \qquad (r)$$

13 The initial deflection to reach the smaller of the two specified loads is

$$y_{initial} = \frac{F_{initial}}{k} = \frac{50}{130} = 0.38 \text{ in} \qquad (s)$$

14 Assume a clash allowance of 15% of the working deflection:

$$y_{clash} = 0.15y = 0.15(1.0) = 0.15 \text{ in} \qquad (t)$$

15 The free length (see Figure 13-8) can now be found from

$$L_f = L_s + y_{clash} + y_{working} + y_{initial} = 1.81 + 0.15 + 1.0 + 0.38 = 3.35 \text{ in} \qquad (u)$$

16 The deflection to the shut height is

$$y_{shut} = L_f - L_s = 3.35 - 1.81 = 1.53 \text{ in} \qquad (v)$$

alternating stress (kpsi)

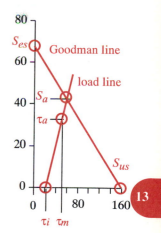

mean stress (kpsi)

FIGURE 13-19

Torsional Stress Goodman Diagram for Example 13-4

Table 13-9   Example 13-4 -Helical Compression Spring for Dynamic Loading
See the *TKSolver* File EX13-04b for Complete Information

Input	Variable	Output	Unit	Comments
1 000	rpm		rpm	excitation frequency
7	C			trial spring index
0.207	d		in	available wire diameter (List Function)
1	y		in	deflection of spring
'music	matl			one of 'music, 'oiltemp, 'hdrawn, etc.
'sqgrnd	end			one of 'plain,'pgrnd,'square,'sqgrnd
'peen	surface			one of 'unpeen or 'peen
'set	setflag			set for a set spring—'unset otherwise
180	Fmax		lb	maximum applied force
50	Fmin		lb	mimimum applied force
	Falt	65	lb	alternating force
	Fmean	115	lb	mean force
	Fshut	200	lb	force at shut height
	k	130	lb/in	spring rate
	Na	6.75		no. of active coils—rounded to 1/4 coil
	Nt	8.75		no. of total coils
	D	1.45	in	mean coil diameter
	Dout	1.66	in	outside coil diameter
	Din	1.24	in	inside coil diameter
	Ks	1.07		static factor—direct shear—Eq. 13.8
	Kw	1.21		Wahl Factor—Eq. 13.9
	tauinit	22 286	psi	shear stress at installed length
	taushut	88 921	psi	stress at shut height
	taualt	32 796	psi	alternating shear stress for fatigue
	taumean	51 258	psi	mean shear stress for fatigue
	Sut	238 507	psi	tensile strength—Eq. 13.3 & Table 13-4
	Sus	159 800	psi	ultimate shear strength—Eq. 13.4
	Sys	143 104	psi	shear yield based on Table 13-6
	Sew	67 500	psi	wire endurance limit—Eq. 13.12
	Ses	68 043	psi	fully reversed endurance limit Eq. 13.16*b*
	Nf	1.30		safety factor—fatigue—Eq. 13.14
	Nshut	1.61		safety factor—shut height (yielding)
	Lf	3.35	in	free length
	Lshut	1.81	in	shut height
	yinit	0.38	in	inital deflection at assembly
	yshut	1.53	in	deflection to shut height
	nf	206	Hz	natural frequency in Hz
	FreqFac	12.40		ratio of nat. freq to excitation freq.

13

17   The force at this shut-height deflection is

$$F_{shut} = k\,y_{shut} = 130(1.53) = 200 \text{ lb} \qquad (w)$$

18   The shut-height stress and safety factor are

$$\tau_{shut} = K_s\frac{8F_{shut}D}{\pi d^3} = 1.07\frac{8(200)(1.45)}{\pi(0.207)^3} = 88\,921 \text{ psi} \qquad (x)$$

$$N_{s_{shut}} = \frac{S_{ys}}{\tau_{shut}} = \frac{143\,104 \text{ psi}}{88\,921 \text{ psi}} = 1.6 \qquad (y)$$

which is acceptable.

19   To check for buckling, two ratios need to be calculated, L_f / D and y_{max} / L_f.

$$\frac{L_f}{D} = \frac{3.35}{1.45} = 2.31$$

$$(z)$$

$$\frac{y_{max}}{L_f} = \frac{y_{initial} + y_{working}}{L_f} = \frac{0.38 + 1.0}{3.35} = 0.41$$

Take these two values to Figure 13-14 and find that their coordinates are safely within the zones that are stable against buckling for either end-condition case.

20   The weight of the spring's active coils from equation 13.11b is

$$W_a = \frac{\pi^2 d^2 D N_a \gamma}{4} = \frac{\pi^2(0.207)^2(1.449)(6.75)(0.285)}{4} = 0.295 \text{ lb} \qquad (aa)$$

21   The natural frequency of this spring is found from equation 13.11a and is.

$$f_n = \frac{1}{2}\sqrt{\frac{kg}{W}} = \frac{1}{2}\sqrt{\frac{130(386)}{0.295}} = 206.32 \text{ Hz} = 12\,379 \; \frac{\text{cycles}}{\text{min}} \qquad (ab)$$

The ratio between the natural frequency and the forcing frequency is

$$\frac{12\,379}{1\,000} = 12.4 \qquad (ac)$$

which is sufficiently high.

22   The design specification for this A228 wire spring is

$$d = 0.207 \text{ in} \qquad D_o = 1.66 \text{ in} \qquad N_t = 8.75, \text{ sq \& g} \qquad L_f = 3.35 \qquad (ad)$$

23   Other spring parameters are shown in Table 13-9, which reproduces portions of the *TKSolver* variable sheet for this example. Note in the table and in the supplied *TKSolver* files, that selection of the various factors for material strengths taken from the tables of this chapter have been automated using **list functions**, as they were in Example 13-3. See that example for more information.

13.7   HELICAL EXTENSION SPRINGS

Helical extension springs are similar to helical compression springs, but are loaded in tension, as shown in Figure 13-2b. Figure 13-20 shows the significant dimensions of an extension spring. Hooks or loops are provided to allow a pull force to be applied. A standard loop and standard hook are shown in the figure, but many variations are possible. See reference 1 for descriptions of other possible hook and loop configurations. The standard ends are formed simply by bending the last coil at 90° to the coil body. The hooks and loops can have higher stresses than the coil body and these may limit the safety of the design. Setting of the coils is not done with extension springs and shot-peening is impractical since the tightly wound coils shield one another from the shot.

Active Coils in Extension Springs

All coils in the body are considered active coils, but one coil is typically added to the number of active coils to obtain the body length L_b.

$$N_t = N_a + 1 \qquad (13.18)$$

$$L_b = dN_t \qquad (13.19)$$

The free length is measured from the inside of one end loop (or hook) to the other and can be varied by changing the end configuration without changing the number of coils.

Spring Rate of Extension Springs

Extension-spring coils are wound tightly together, and the wire is twisted as it is wound, creating a preload in the coils that must be overcome to separate them. Figure 13-21 shows a typical load-deflection curve for a helical extension spring. The spring rate k is linear except for the initial portion. The preload F_i is measured by extrapolating the linear portion of the curve back to the force axis. The spring rate can be expressed as

$$k = \frac{F - F_i}{y} = \frac{d^4 G}{8 D^3 N_a} \qquad (13.20)$$

Note that no deflection occurs until the applied force exceeds the preload force F_i that is built into the spring.

Spring Index of Extension Springs

The spring index is found from equation 13.5 and should be kept in the same range of about 4 to 12, as is recommended for compression springs.

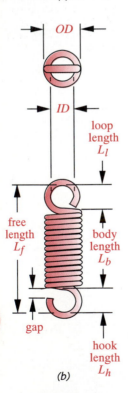

(a)

(b)

FIGURE 13-20

(a) Sample Springs and (b) Dimensions of an Extension Spring

13

Coil Preload in Extension Springs

The preload F_i can be controlled to some degree in the manufacturing process, and should be designed to keep the initial coil stress within the preferred range shown in Figure 13-22.[1] This shows desired ranges for initial coil stress as a function of spring index. Values outside the range are possible but are difficult to manufacture. Cubic functions have been fitted to the curves of Figure 13-22 to allow their use in a computer program. The approximate cubic expressions are shown on the figure and are

$$\tau_i \cong -4.231C^3 + 181.5C^2 - 3\,387C + 28\,640 \qquad (13.21a)$$

$$\tau_i \cong -2.987C^3 + 139.7C^2 - 3\,427C + 38\,404 \qquad (13.21b)$$

where τ_i is in psi. The average of the two values computed from these functions can be taken as a good starting value for initial coil stress.

Deflection of Extension Springs

Coil deflection is found from the same equation as used for the compression spring with a modification for the preload.

$$y = \frac{8(F - F_i)D^3 N_a}{d^4 G} \qquad (13.22)$$

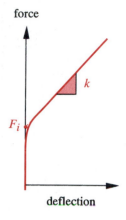

force

F_i

deflection

FIGURE 13-21

Force-Deflection Curve of a Helical Extension Spring Showing its Initial Tension

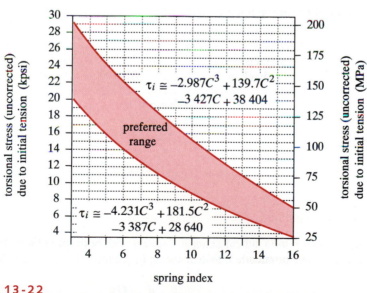

FIGURE 13-22

Preferred Range of Initial Stress in Extension Springs as a Function of Spring Index

Coil Stresses in Extension Springs

The stresses in the coils are found from the same formulas used for compression springs. See equations 13.8 and 13.9. The factors K_s and K_w are used as before.

End Stresses in Extension Springs

The standard hooks or loops have two locations of high stress, as shown in Figure 13-23. The maximum torsional stress occurs at point B, where the bend radius is smallest. There is also a bending stress in the hook or loop at point A, since the end is loaded as a curved beam. Wahl also defines a stress-concentration factor K_b for bending in a curved wire.

The bending stress at point A is found from

$$\sigma_A = K_b \frac{16DF}{\pi d^3} + \frac{4F}{\pi d^2} \qquad (13.23a)$$

where

$$K_b = \frac{4C_1^2 - C_1 - 1}{4C_1(C_1 - 1)} \qquad (13.23b)$$

and

$$C_1 = \frac{2R_1}{d} \qquad (13.23c)$$

R_1 is the mean loop radius, as shown in Figure 13-23. Note that for a standard end, the mean loop radius is the same as the mean coil radius.

The torsional stress at point B is found from

$$\tau_B = K_{w_2} \frac{8DF}{\pi d^3} \qquad (13.24a)$$

where

$$K_{W_2} = \frac{4C_2 - 1}{4C_2 - 4} \qquad (13.24b)$$

and

$$C_2 = \frac{2R_2}{d} \qquad (13.24c)$$

R_2 is the side-bend radius, as shown in Figure 13-23. C_2 should be greater than 4.[1]

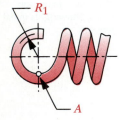

R_1

A

maximum
bending stress

R_2

B

maximum
torsional stress

FIGURE 13-23

Points of Maximum
Stress in Hook or Loop of
an Extension Spring

Surging in Extension Springs

The natural frequency of a helical extension spring with both ends fixed against axial deflection is the same as that for a helical spring in compression (see equation 13.11):

$$f_n = \frac{2}{\pi N_a} \frac{d}{D^2} \sqrt{\frac{Gg}{32\gamma}} \quad \text{Hz} \qquad (13.25)$$

Table 13-10   Maximum Torsional Yield Strengths S_y and S_{ys} for Helical Extension Springs in Static Applications

No Set Removal and Low-Temperature Heat Treatment Applied Source: Reference 1

| Material | Maximum Percent of Ultimate Tensile Strength | | |
| | S_{ys} in Torsion | | S_y in Bending |
	Body	End	End
Patented and cold-drawn carbon steel	45%	40%	75%
Hardened and tempered carbon and low-alloy steel	50	40	75
Austenitic stainless steel and nonferrous alloys	35	30	55

Material Strengths for Extension Springs

The same wire materials are used for both extension and compression springs. Some of the strength data developed for compression springs is applicable to extension springs as well. Table 13-10 shows recommended strengths for static yielding of the coil body and the ends in both torsion and bending. Note that the wire torsion strengths are the same as those for compression springs in Tables 13-6 and 13-7. Table 13-11 shows recommended fatigue strengths for two materials at several cycle lives, giving separate data for the body coils and the ends. The endurance limits of equation 13.12 are valid for extension springs and should be converted to fully reversed values with equation 13.16c to use in the Goodman-line safety factor expression of equation 13.16b.

Design of Helical Extension Springs

The design procedure for extension springs is essentially the same as for compression springs with the additional complication of the end details. Assumptions must be made

Table 13-11   Maximum Torsional Fatigue Strength S_{fw}' for ASTM A228 and Type 302 Stainless Steel Wire Helical Extension Springs in Cyclic Applications (Stress Ratio, R = 0)

No Surging, No Shot Peening, Room Temperature, and Low Temperature Heat Treatment Applied Source: Reference 1

| Fatigue Life (cycles) | Percent of Ultimate Tensile Strength | | |
| | In Torsion | | In Bending |
	Body	End	End
10^5	36%	34%	51%
10^6	33	30	47
10^7	30	28	45

13

for sufficient design parameters to allow a trial computation. The assumed values are adjusted based on the result and the design iterated to an acceptable solution.

It is often convenient in extension-spring design problems to assume a spring index and a wire diameter as was done for the compression springs. The mean coil diameter can then be found from equation 13.5. The assumed spring index can be used with equations 13.21 to obtain an approximate initial coil-winding stress. Using that initial-stress value, the coil preload F_i can be computed from the stress-equation 13.8. The coil stresses and end stresses can be found next and appropriate adjustments made to the assumed values to get acceptable safety factors.

The deflection or number of coils can be found from equation 13.22 with the other one assumed or specified. The spring rate can then be found using the maximum design force and the preload in combination with the assumed or calculated deflection using equation 13.20. Buckling is not an issue with extension springs, but the natural frequency should be compared to the forcing frequency in dynamic situations.

The safety factors are found from equations 13.14 and 13.16, being careful to use the appropriate material strength for torsion in the coils and for bending or shear in the ends. A Goodman-line analysis is needed for cyclically loaded springs, and is encapsulated in equation 13.16. A fatigue analysis is needed for the ends as well as the coils. Note that for bending stresses, the tensile endurance limit and tensile yield strengths are needed. The von Mises relationship between torsion and tension can be used to convert the available torsional fatigue data to tensile strength. Divide the torsional data by 0.577 to obtain tensile strengths.

EXAMPLE 13-5

Design of a Helical Extension Spring for Cyclic Loading

Problem   Design an extension spring for a dynamic load over a given deflection.

Given   The spring must give a minimum force of 50 lb and a maximum force of 100 lb over a dynamic deflection of 0.50 in. The forcing frequency is 500 rpm. An infinite life is desired.

Assumptions Standard hooks will be used at each end. Music wire (ASTM A228) will be used since the loads are dynamic. Setting and peening cannot be used to obtain a higher endurance strength in an extension spring.

Solution   See Table 13-12, Figure 13-24, and *TKSolver* files EX13-05a and EX13-05b.

1   Assume an 0.177 in trial wire diameter from the available sizes in Table 13-2. Assume a spring index of $C = 9$ and use it to calculate the mean coil diameter D from equation 13.5.

$$D = Cd = 9(0.177) = 1.59 \text{ in} \qquad (a)$$

2   Use the assumed value of C to find an appropriate value of initial coil stress τ_i from equations 13.21:

$$\tau_{i_1} \cong -4.231C^3 + 181.5C^2 - 3\,387C + 28\,640$$

$$= -4.231(9)^3 + 181.5(9)^2 - 3\,387(9) + 28\,640 = 9\,774 \text{ psi} \qquad (b)$$

$$\tau_{i_2} \cong -2.987C^3 + 139.7C^2 - 3\,427C + 38\,404$$

$$= -2.987(9)^3 + 139.7(9)^2 - 3\,427(9) + 38\,404 = 16\,699 \text{ psi} \qquad (c)$$

$$\tau_i \cong \frac{\tau_{i_1} + \tau_{i_2}}{2} = \frac{9\,774 + 16\,699}{2} = 13\,237 \text{ psi} \qquad (d)$$

3   Find the direct shear factor:

$$K_s = 1 + \frac{0.5}{C} = 1 + \frac{0.5}{9} = 1.06 \qquad (e)$$

4   Use the value of τ_i from (c) in equation 13.8 to find the corresponding initial coil-tension force F_i:

$$F_i = \frac{\pi d^3 \tau_i}{8K_s D} = \frac{\pi(0.177)^3(13\,237)}{8(1.06)(1.59)} = 17.1 \text{ lb} \qquad (f)$$

Check that this force is less than the required minimum applied force F_{min}, which in this case, it is. Any applied force smaller than F_i will not deflect the spring.

5   Find the mean and alternating forces from equation 13.15a:

$$F_a = \frac{F_{max} - F_{min}}{2} = \frac{100 - 50}{2} = 25 \text{ lb}$$

$$\qquad (g)$$

$$F_m = \frac{F_{max} + F_{min}}{2} = \frac{100 + 50}{2} = 75 \text{ lb}$$

6   Use the direct shear factor K_s and previously assumed values to find the mean stress τ_m:

$$\tau_m = K_s \frac{8F_m D}{\pi d^3} = 1.06 \frac{8(75)(1.59)}{\pi(0.177)^3} = 57\,913 \text{ psi} \qquad (h)$$

7   Find the Wahl factor K_w and use it to calculate the alternating shear stress τ_a in the coil.

$$K_w = \frac{4C - 1}{4C - 4} + \frac{0.615}{C} = \frac{4(9) - 1}{4(9) - 4} + \frac{0.615}{9} = 1.16 \qquad (i)$$

13

$$\tau_a = K_w \frac{8F_a D}{\pi d^3} = 1.16 \frac{8(25)(1.59)}{\pi(0.177)^3} = 21\,253 \text{ psi} \qquad (j)$$

8   Find the ultimate tensile strength of this music-wire material from equation 13.3 and Table 13-4. Use it to find the ultimate shear strength from equation 13.4 and the torsional yield strength for the coil body from Table 13-10, assuming no set removal.

$$S_{ut} = A d^b = 184\,649(0.177)^{-0.1625} = 244\,633 \text{ psi}$$
$$S_{us} = 0.667 S_{ut} = 163\,918 \text{ psi} \qquad (k)$$

$$S_{ys} = 0.45 S_{ut} = 0.45(244\,633) = 110\,094 \text{ psi} \qquad (l)$$

9   Find the wire endurance limit for unpeened springs from equation 13.13 and convert it to a fully reversed endurance strength with equation 13.16c.

$$S_{ew} = 45\,000 \text{ psi} \qquad (m)$$

$$S_{es} = 0.707 \frac{S_{ew} S_{us}}{S_{us} - 0.707 S_{ew}} = 0.707 \frac{45\,000(163\,918)}{163\,918 - 0.707(45\,000)} = 39\,477 \text{ psi} \qquad (n)$$

10   The fatigue safety factor for the coils in torsion is calculated from equation 13.16b.

$$N_{f_s} = \frac{S_{es}(S_{us} - \tau_{min})}{S_{es}(\tau_m - \tau_{min}) + S_{us}\tau_a}$$
$$= \frac{39\,477(163\,918 - 38\,609)}{39\,477(57\,913 - 38\,609) + 163\,918(21\,253)} = 1.17 \qquad (o)$$

Note that the minimum stress due to force F_{min} is used in this calculation, **not** the coil-winding stress from (d).

11   The stresses in the end hooks also need to be determined. The bending stresses in the hook are found from equation 13.23:

$$C_1 = \frac{2R_1}{d} = \frac{2D}{2d} = C = 9$$

$$K_b = \frac{4C_1^2 - C_1 - 1}{4C_1(C_1 - 1)} = \frac{4(9)^2 - (9) - 1}{4(9)(9-1)} = 1.09 \qquad (p)$$

$$\sigma_a = K_b \frac{16DF_a}{\pi d^3} + \frac{4F_a}{\pi d^2} = 1.09 \frac{16(1.59)(25)}{\pi(0.177)^3} + \frac{4(25)}{\pi(0.177)^2} = 40\,895 \text{ psi}$$
$$\qquad (q)$$

$$\sigma_m = K_b \frac{16DF_m}{\pi d^3} + \frac{4F_m}{\pi d^2} = 1.09 \frac{16(1.59)(75)}{\pi(0.177)^3} + \frac{4(75)}{\pi(0.177)^2} = 122\,685 \text{ psi}$$

$$\sigma_{min} = K_b \frac{16DF_{min}}{\pi d^3} + \frac{4F_{min}}{\pi d^2} = 1.09 \frac{16(1.59)(50)}{\pi(0.177)^3} + \frac{4(50)}{\pi(0.177)^2} = 81\,790 \text{ psi} \qquad (r)$$

13

12   Convert the torsional endurance strength to a tensile endurance strength with the von Mises relationship and use it and the ultimate tensile strength in equation 13.16 to find a fatigue safety factor for the hook in bending:

$$S_e = \frac{S_{es}}{0.577} = \frac{39\ 477}{0.577} = 68\ 418 \text{ psi}$$

$$N_{f_b} = \frac{S_e(S_{ut} - \sigma_{min})}{S_e(\sigma_{mean} - \sigma_{min}) + S_{ut}\sigma_{alt}}$$

$$= \frac{68\ 418(244\ 633 - 81\ 790)}{68\ 418(122\ 685 - 81\ 790) + 244\ 633(40\ 895)} = 0.87 \qquad (s)$$

13   The torsional stresses in the hook are found from equation 13.24 using an assumed value of $C_2 = 5$.

$$R_2 = \frac{C_2 d}{2} = \frac{5(0.177)}{2} = 0.44 \text{ in}$$

$$K_{w_2} = \frac{4C_2 - 1}{4C_2 - 4} = \frac{4(5) - 1}{4(5) - 4} = 1.19 \qquad (t)$$

$$\tau_{B_a} = K_{w_2}\frac{8DF_a}{\pi d^3} = 1.19\frac{8(1.59)25}{\pi(0.177)^3} = 21\ 717 \text{ psi}$$

$$\tau_{B_m} = K_{w_2}\frac{8DF_m}{\pi d^3} = 1.19\frac{8(1.59)75}{\pi(0.177)^3} = 65\ 152 \text{ psi} \qquad (u)$$

$$\tau_{B_{min}} = K_{w_2}\frac{8DF_{min}}{\pi d^3} = 1.19\frac{8(1.59)50}{\pi(0.177)^3} = 43\ 435 \text{ psi}$$

14   The fatigue safety factor for the hook in torsion is calculated from equation 13.16b.

$$N_{f_s} = \frac{S_{es}(S_{us} - \tau_{min})}{S_{es}(\tau_m - \tau_{min}) + S_{us}\tau_a}$$

$$= \frac{39\ 477(163\ 918 - 43\ 435)}{39\ 477(65\ 152 - 43\ 435) + 163\ 918(21\ 717)} = 1.08 \qquad (v)$$

All three of these safety factors are too low, making this an unacceptable design. To get some idea of what to change to improve it, the *TKSolver* model was solved for a list of values of the spring index from 4 to 14, keeping all other parameters as defined above. The resulting values of coil diameter, free length, spring weight, and fatigue safety factor are plotted in Figure 13-24.

The safety factor decreases with increasing spring index, so a reduction in our assumed value for C will improve it even with no change in wire diameter. Note however that the spring free length shows a minimum value at a spring index of about 7.5. The coil diameter increases linearly with the spring index for a constant wire diameter. Spring weight decreases with increasing spring index.

If we decrease the spring index from 9 to 7.5, and increase the wire diameter one size to 0.192 in, keeping all other parameters the same, we will obtain an acceptable design in this case with the smallest $N_f = 1.36$ for the hook in bending. Table 13-12 shows the results of this new calculation from the *TKSolver* variable sheet and also shows the results of the calculations outlined below to complete the design. A summary of the changed values is

$$C = 7.5 \qquad D = 1.44 \text{ in} \qquad K_w = 1.20 \qquad K_s = 1.07$$
$$\tau_i = 15\,481 \text{ psi} \qquad \tau_{min} = 27\,631 \text{ psi} \qquad \tau_a = 15\,509 \text{ psi} \qquad \tau_m = 41\,447 \text{ psi} \quad (w)$$
$$N_{s_{coil}} = 1.97 \qquad N_{f_{s_{coil}}} = 1.74 \qquad N_{f_{s_{hook}}} = 1.68 \qquad N_{f_{b_{hook}}} = 1.36$$

15 The spring rate is defined from the two specified forces at their relative deflection.

$$k = \frac{\Delta F}{y} = \frac{100 - 50}{0.5} = 100 \text{ lb/in} \qquad (x)$$

16 To get the defined spring rate, the number of active coils must satisfy equation 13.7:

$$k = \frac{d^4 G}{8D^3 N_a} \quad \text{or} \quad N_a = \frac{d^4 G}{8D^3 k} = \frac{(0.192)^4 11.5E6}{8(1.44)^3 (100)} = 6.54 \cong 6\frac{1}{2} \qquad (y)$$

Note that we round it to the nearest 1/4 coil as the manufacturing tolerance cannot achieve better than that accuracy.

17 The total number of coils in the body and the body length are

$$N_t = N_a + 1 = 6.5 + 1 = 7.5$$
$$L_b = N_t d = 7.5(0.192) = 1.44 \text{ in} \qquad (z)$$

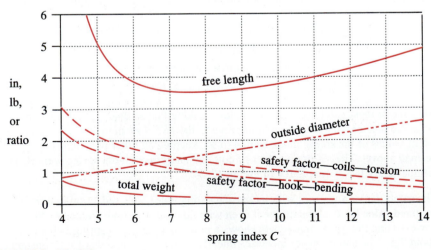

FIGURE 13-24

Variation of Helical Extension Spring Parameters with Spring Index—Constant Wire Diameter

Table 13-12   **Example 13-5—Design of a Helical Extension Spring for Cyclic Loads**

Part 1 of 2   See the *TKSolver* File EX13-05b for More Information

Input	Variable	Output	Unit	Comments
500	*rpm*		rpm	excitation frequency
7.50	*C*			trial spring index
0.192	*d*		in	trial wire diameter (List Function)
0.50	*y*		in	deflection bet Fmin & Fmax
	ymax	0.72	in	maximum deflection at Fmax
	ymin	0.22	in	minimum deflection at Fmin
'music	*matl*			'music, 'oiltemp, 'hdrawn, etc.
100	*Fmax*		lb	maximum applied force
50	*Fmin*		lb	mimimum applied force
	Finit	28.01	lb	force from initial tension
	Flow	50.00	lb	lowest applied force on spring
	Falt	25.00	lb	alternating force
	Fmean	75.00	lb	mean force
	k	100	lb/in	spring rate
	N	6.54		no of active coils—exact
	Na	6.50		no of active coils—nearest 1/4 coil
	Ntot	7.50		no of total coils
	D	1.44	in	mean coil diameter
	Dout	1.63	in	outside coil diameter
	Din	1.25	in	inside coil diameter
	Ks	1.07		static factor—Eq. 13.8
	Kw	1.20		Wahl Factor—Eq. 13.9
	tauinit	15 481	psi	shear stress from initial tension
	taumin	27 631	psi	shear stress at Fmin
	taumax	55 262	psi	shear stress at Fmax
	taualt	15 509	psi	alternating shear stress for fatigue
	taumean	41 447	psi	mean shear stress for fatigue
	Sut	241 441	psi	tensile strength—Eq 13.3 & Table13-4
	Sus	161 765	psi	ultimate shear strength—Eq. 13.4
	Ssy	108 648	psi	shear yield based on Table 13-10
	Ssyh	96 576	psi	shear yield in hook—Table 13-10
	Sew	45 000	psi	wire endurance limit—Eq. 13.12
	Ses	39 604	psi	fully reversed endur. limit Eq. 13.16*b*
	Sy	181 081	psi	yield strength in bending
	Se	68 638	psi	endurance limit in tension

13

Table 13-12   Example 13-5—Design of a Helical Extension Spring for Cyclic Loads
Part 2 of 2   See the *TKSolver* File EX13-05b for More Information

Input	Variable	Output	Unit	Comments
	N_f	1.74		SF coils—fatigue—Eq. 13.14
	N_s	1.97		SF coils—static loading at Fmax
	N_{fht}	1.68		SF hook—torsion fatigue
	N_{sht}	1.57		SF hook—torsional yielding
	N_{fhs}	1.36		SF hook—bending fatigue
	N_{shs}	1.53		SF hook—bending yielding
	L_{body}	1.44	in	length of coil body
	hook1	1.25	in	length of hook on one end
	hook2	1.25	in	length of hook on other end
	L_f	3.94	in	free length inside hooks
	W_{total}	0.28	lb	weight of total coils Eq. 13-11b
	n_f	98.80	Hz	natural frequency in Hz
	c_{rpm}	5 929	rpm	natural frequency in rpm
	FreqFac	11.90		ratio—nat. freq. to forcing freq.
5.00	C_2			should be set > 4
	R_2	0.48	in	side bend radius at hook root
	K_{hook}	1.19		K factor for hook in torsion
	$t_{maxhook}$	61 522	psi	maximum torsional stress in hook
	$t_{minhook}$	30 761	psi	minimum torsional stress in hook
	$t_{althook}$	15 381	psi	alternating torsional stress in hook
	t_{mnhook}	46 142	psi	mean torsional stress in hook
	C_1	7.50		spring index for hook

18   The free length can now be determined. The length of a standard hook is equal to the coil inside diameter:

$$L_f = L_b + 2L_{hook} = 1.44 + 2(1.25) = 3.94 \text{ in} \qquad (aa)$$

19   The deflection to reach the larger of the two specified loads is

$$y_{max} = \frac{F_{max} - F_{initial}}{k} = \frac{100 - 28}{100} = 0.72 \text{ in} \qquad (ab)$$

20   The natural frequency of this spring is found from equation 13.25 and is

$$f_n = \frac{2}{\pi N_a} \frac{d}{D^2} \sqrt{\frac{Gg}{32\gamma}} = \frac{2(0.192)}{\pi(6.5)(1.44)^2} \sqrt{\frac{11.5E6(386)}{32(0.285)}} = 200.2 \text{ Hz} = 12\ 011 \text{ rpm} \qquad (ac)$$

The ratio between the natural frequency and the forcing frequency is

$$\frac{12\ 011}{500} = 24 \qquad\qquad (ad)$$

which is sufficiently high.

21 The design specification for this A228 wire spring is

$$d = 0.192 \text{ in} \qquad OD = 1.63 \text{ in} \qquad N_t = 6.5 \qquad L_f = 3.94 \qquad (ae)$$

22 Other spring parameters are shown in Table 13-12, which reproduces portions of the *TKSolver* variable sheet for this example. Note in the table and in the supplied *TKSolver* files, that selection of the various factors for material strengths taken from the tables of this chapter have been automated using **list functions,** as they were in Example 13-3. See that example for more information.

13.8   HELICAL TORSION SPRINGS

A helical coil spring can be loaded in torsion instead of compression or tension and is then called a **torsion spring**. The ends of the coil are extended tangentially to provide lever arms on which to apply the moment load, as shown in Figure 13-25. These coil ends can have a variety of shapes to suit the application. The coils are usually close-wound like an extension spring but do not have any initial tension. The coils can also be wound with spacing like a compression spring and this will avoid friction between the coils. However, most torsion springs are close-wound.

The applied moment on the coils puts the wire in bending as a curved beam, as shown in Figure 13-26. The applied moment should always be arranged to close the coils rather than open them because the residual stresses from coil-winding are favorable against a closing moment. The applied moment should never be reversed in service. Dynamic loading should be repeated or fluctuating with the stress ratio $R \geq 0$.

Radial support must be provided at three or more points on the coil diameter to take the reaction forces. This support is usually accomplished by means of a rod placed inside the coil. The rod should be no larger in diameter than about 90% of the coils smallest inside diameter when "wound-up" under load in order to avoid binding.

The manufacturing specifications of a torsion spring should define the parameters indicated in Figure 13-26 as well as wire diameter, outside coil diameter, number of coils, and spring rate. The load should be defined at an angle α between the tangent ends in the loaded position rather than at a deflection from the free position.

Because the load is bending, rectangular wire is more efficient in terms of stiffness per unit volume (larger *I* for same dimension). Nevertheless, most helical torsion springs are made from round wire simply because of its lower cost and the greater variety of available sizes and materials.

13

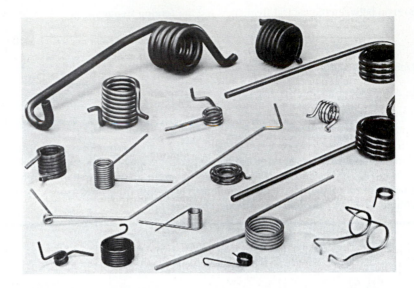

FIGURE 13-25

A Variety of End Details is Possible on Helical Torsion Springs   *Courtesy of Associated Spring, Barnes Group Inc., Bristol, Conn.*

Terminology for Torsion Springs

The following parameters have the same meaning for torsion springs as for helical compression springs: mean coil diameter D, wire diameter d, spring index C, outside diameter D_o, inside diameter D_i, and number of active coils N_a. The spring rate k is expressed as moment per unit of angular deflection.

Number of Coils in Torsion Springs

The active coils are equal to the number of body turns N_b plus some contribution from the ends, which also bend. For straight ends, the contribution can be expressed as an equivalent number of coils N_e:

$$N_e = \frac{L_1 + L_2}{3\pi D} \qquad (13.26a)$$

where L_1 and L_2 are the respective lengths of the tangent-ends of the coil. The number of active coils is then

$$N_a = N_b + N_e \qquad (13.26b)$$

where N_b is the number of coils in the spring body.

Deflection of Torsion Springs

The angular deflection of the coil-end is normally expressed in radians but is often converted to revolutions. We will use revolutions. Since it is essentially a beam in bending, the (angular) deflection can be expressed as

$$\theta_{rev} = \frac{1}{2\pi} \theta_{rad} = \frac{1}{2\pi} \frac{ML_w}{EI} \qquad (13.27a)$$

where M is the applied moment, L_w is the length of wire, E is Young's modulus for the material, and I is the second moment of area for the wire cross section about the neutral axis.

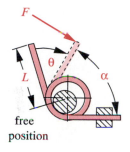

free
position

For torsion springs of round wire, we can substitute the appropriate geometry to get

$$\theta_{rev} = \frac{ML_w}{EI} = \frac{1}{2\pi} \frac{M(\pi DN_a)}{E\left(\pi d^4/64\right)}$$

$$= \frac{64}{2\pi} \frac{MDN_a}{d^4 E} \qquad (13.27b)$$

$$\theta_{rev} \cong 10.2 \frac{MDN_a}{d^4 E}$$

The factor 10.2 is usually increased to 10.8 to account for the friction between coils, based on experience, and the equation becomes[1]

$$\theta_{rev} \cong 10.8 \frac{MDN_a}{d^4 E} \qquad (13.27c)$$

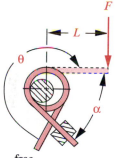

free
position

Spring Rate of Torsion Springs

The spring rate can always be obtained from the deflection formula:

$$k = \frac{M}{\theta_{rev}} \cong \frac{d^4 E}{10.8 D N_a} \qquad (13.28)$$

Coil Closure

When a torsion spring is loaded to close the coils (as it should be), the coil diameter decreases and its length increases as the coil is "wound up." The minimum inside coil diameter at full deflection is

$$D_{i_{min}} = \frac{DN_b}{N_b + \theta_{rev}} - d \qquad (13.29)$$

specify:
α—angle between ends
F—load on ends at α
L—moment arm
θ—angular deflection from free position

FIGURE 13-26

Specifying Load and Deflection Requirements for Torsion Springs.
Source: Reference 1

where D is the unloaded mean coil diameter. Any pin that the coil works over should be limited to about 90% of this minimum inside diameter.

The maximum coil-body length at full "windup" is:

$$L_{max} = d(N_b + 1 + \theta) \qquad (13.30)$$

Coil Stresses in Torsion Springs

The stress in the outer fiber of a straight beam is $M c / I$, but this is a curved beam, and we learned in Section 4.10 that stress is concentrated at the inside of a curved beam. Wahl[3] derived the stress-concentration factor for the inside of a coiled round wire in bending as

$$K_{b_i} = \frac{4C^2 - C - 1}{4C(C-1)} \qquad (13.31a)$$

and at the outside of the coil

$$K_{b_o} = \frac{4C^2 + C - 1}{4C(C+1)} \qquad (13.31b)$$

where C is the spring index.

The maximum compressive bending stress at the inside coil diameter of a round-wire helical torsion spring (loaded to close its coils) is then

$$\sigma_{i_{max}} = K_{b_i} \frac{M_{max} c}{I} = K_{b_i} \frac{M_{max}(d/2)}{\pi d^4/64} = K_{b_i} \frac{32 M_{max}}{\pi d^3} \qquad (13.32a)$$

and the tensile bending stress components at the outside coil diameter are

$$\sigma_{o_{min}} = K_{b_o} \frac{32 M_{min}}{\pi d^3}; \qquad \sigma_{o_{max}} = K_{b_o} \frac{32 M_{max}}{\pi d^3} \qquad (13.32b)$$

$$\sigma_{o_{mean}} = \frac{\sigma_{o_{max}} + \sigma_{o_{min}}}{2}; \qquad \sigma_{o_{alt}} = \frac{\sigma_{o_{max}} - \sigma_{o_{min}}}{2} \qquad (13.32c)$$

Note that for static failure (yielding) of a torsion spring loaded to close its coils, the higher-magnitude compressive stress $\sigma_{i_{max}}$ at the inside of the coil is of most concern, but for fatigue failure, which is a tensile-stress phenomenon, the slightly lower maximum tensile stress at the outside of the coils is the concern. Thus the alternating and mean stress components are calculated at the outside of the coil. If the spring is loaded to open the coils, (which is not recommended) it must be stress relieved to eliminate the residual stresses from coiling and then the inside coil stress should be used to calculate the components for the fatigue safety factor calculation.

Table 13-13   Maximum Recommended Bending Yield Strength S$_y$ for Helical Torsion Springs in Static Applications

Source: Reference 1

Material	Maximum Percent of Ultimate Tensile Strength	
	Stress Relieved	Favorable Residual Stress
Patented and cold-drawn carbon steel	80%	100%
Hardened and tempered carbon and low-alloy steel	85	100
Austenitic stainless steel and nonferrous alloys	60	80

Material Parameters for Torsion Springs

Yield and endurance strengths in bending are needed in this instance. Table 13-13 shows suggested yield strengths for several wire materials as a percentage of their ultimate tensile strength. Note that favorable residual stresses allow the material's ultimate strength to be used as a yield criterion in some cases. Table 13-14 shows bending-fatigue-strength percentages for several wires at 10^5 and 10^6 cycles in both peened and unpeened states. The same limitations on effective shot peening apply to close-wound torsion springs as to extension springs since the closely spaced coils prevent the shot from impacting the inside diameter of the coil. Shot-peening may not be effective in many torsion springs.

The torsional-endurance limit data for helical compression springs shown in equation 13.12 can be adapted for bending by using the von Mises relationship between torsion and tension loading.

$$S_{ew_b}' = \frac{S_{ew}'}{0.577} \qquad (13.33a)$$

Table 13-14   Maximum Recommended Bending Fatigue Strength S$_{fw}$' for Helical Torsion Springs in Cyclic Applications (Stress Ratio, R = 0)

Stress Relieved, No Surging—Shot Peening May Not be Possible in All Cases.
Source: Reference 1

Fatigue Life (cycles)	Percent of Ultimate Tensile Strength			
	ASTM A228 or 302 Stainless		ASTM A230 and A232	
	Unpeened	Shot peened	Unpeened	Shot peened
10^5	53%	62%	55%	64%
10^6	50	60	53	62

13

which gives

$$S_{ew_b}' \cong \frac{45.0}{0.577} = 78 \text{ kpsi} \quad (537 \text{ MPa}) \quad \text{for unpeened springs}$$

$$(13.33b)$$

$$S_{ew_b}' \cong \frac{67.5}{0.577} = 117 \text{ kpsi} \quad (806 \text{ MPa}) \quad \text{for peened springs}$$

Safety Factors for Torsion Springs

Failure in yielding is anticipated at the inside coil surface and the safety factor can be found from

$$N_y = \frac{S_y}{\sigma_{i_{max}}}$$

$$(13.34a)$$

Note that the available fatigue and endurance data are for a repeated-stress situation (equal mean and alternating components) and so must be converted to fully reversed values before being used to calculate the safety factor in fatigue with equations 13.16. Because the bending notation is slightly different, we repeat equations 13.16 here with the appropriate substitutions of variables for the torsion spring case.

$$N_{f_b} = \frac{S_e\left(S_{ut} - \sigma_{o_{min}}\right)}{S_e\left(\sigma_{o_{mean}} - \sigma_{o_{min}}\right) + S_{ut}\sigma_{o_{alt}}}$$

$$(13.34b)$$

$$S_e = 0.707 \frac{S_{ew_b}S_{ut}}{S_{ut} - 0.707S_{ew_b}}$$

$$(13.34c)$$

Designing Helical Torsion Springs

The process for designing helical torsion springs is very similar to that for helical compression springs. The best way to illustrate it is with an example.

EXAMPLE 13-6

Design of a Helical Torsion Spring for Cyclic Loading

Problem   Design a torsion spring for a dynamic load over a given deflection.

Given   The spring must give a minimum moment of 50 lb-in and a maximum moment of 80 lb-in over a dynamic deflection of 0.25 revolutions (90°). An infinite life is desired.

Assumptions Use unpeened music wire (ASTM A228). Use 2-in-long straight tangent ends. The coil is loaded to close it.

Solution See Figure 13-27 and *TKSolver* file EX13-06.

1 Assume a 0.192 in trial wire diameter from the available sizes in Table 13-2. Assume a spring index of $C = 9$ and use it to calculate the mean coil diameter D from equation 13.5.

$$D = Cd = 9(0.192) = 1.73 \text{ in} \qquad (a)$$

2 Find the mean and alternating moments:

$$M_m = \frac{M_{max} + M_{min}}{2} = \frac{80 + 50}{2} = 65 \text{ lb} \qquad (b)$$

$$M_a = \frac{M_{max} - M_{min}}{2} = \frac{80 - 50}{2} = 15 \text{ lb}$$

3 Find the Wahl bending factor for the inside surface K_{b_i} and use it to calculate the maximum compressive stress in the coil at the inner surface.

$$K_{b_i} = \frac{4C^2 - C - 1}{4C(C-1)} = \frac{4(9)^2 - 9 - 1}{4(9)(9-1)} = 1.090 \qquad (c)$$

$$\sigma_{i_{max}} = K_{b_i}\frac{32M_{max}}{\pi d^3} = 1.09\frac{32(80)}{\pi(0.192)^3} = 125\,523 \text{ psi} \qquad (d)$$

4 Find the Wahl bending factor K_{b_o} for the outside surface and calculate the maximum, minimum, alternating, and mean tensile stresses in the coil at the outer surface.

$$K_{b_o} = \frac{4C^2 + C - 1}{4C(C+1)} = \frac{4(9)^2 + 9 - 1}{4(9)(9+1)} = 0.9222 \qquad (e)$$

$$\sigma_{o_{min}} = K_{b_i}\frac{32M_{min}}{\pi d^3} = 0.9222\frac{32(50)}{\pi(0.192)^3} = 66\,359 \text{ psi} \qquad (f)$$

$$\sigma_{o_{max}} = K_{b_i}\frac{32M_{max}}{\pi d^3} = 0.9222\frac{32(80)}{\pi(0.192)^3} = 106\,175 \text{ psi}$$

$$\sigma_{o_{mean}} = \frac{\sigma_{o_{max}} + \sigma_{o_{min}}}{2} = \frac{66\,359 + 106\,175}{2} = 86\,267 \qquad (g)$$

$$\sigma_{o_{alt}} = \frac{\sigma_{o_{max}} - \sigma_{o_{min}}}{2} = \frac{66\,359 - 106\,175}{2} = 19\,908$$

4 Find the ultimate tensile strength of this music-wire material from equation 13.3 and Table 13-4 and use it to find the bending yield strength for the coil body from Table 13-13, assuming no stress relieving.

13

$$S_{ut} = Ad^b = 184\,649(0.192)^{-0.1625} = 241\,441 \text{ psi} \tag{h}$$

$$S_y = 1.0S_{ut} = 241\,441 \text{ psi} \tag{i}$$

5   Find the wire bending endurance limit for unpeened springs from equation 13.33 and convert it to a fully reversed bending endurance strength with equation 13.34b.

$$S_{ew_b}' \cong \frac{45\,000}{0.577} = 77\,990 \text{ psi} \tag{j}$$

$$S_e = 0.707\frac{S_{ew_b}S_{ut}}{S_{ut} - 0.707S_{ew_b}} = 0.707\frac{77\,990(241\,441)}{241\,441 - 0.707(77\,990)} = 71\,458 \text{ psi} \tag{k}$$

6   The fatigue safety factor for the coils in bending is calculated from equation 13.34a.

$$N_{f_b} = \frac{S_e\left(S_{ut} - \sigma_{o_{min}}\right)}{S_e\left(\sigma_{o_{mean}} - \sigma_{o_{min}}\right) + S_{ut}\sigma_{o_{alt}}}$$

$$= \frac{71\,458(241\,441 - 66\,359)}{71\,458(86\,267 - 66\,359) + 241\,441(19\,908)} = 2.0 \tag{l}$$

7   The static safety factor against yielding is

$$N_{y_b} = \frac{S_y}{\sigma_{i_{max}}} = \frac{241\,441}{125\,523} = 1.9 \tag{m}$$

These are both acceptable safety factors.

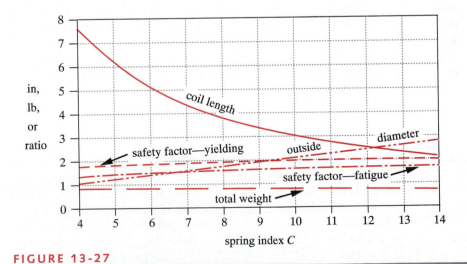

FIGURE 13-27

Variation of Helical Torsion Spring Parameters with Spring Index - Constant Wire Diameter

8   The spring rate is defined from the two specified moments at their relative deflection.

$$k = \frac{\Delta M}{\theta} = \frac{80 - 50}{0.25} = 120 \text{ lb - in/rev} \qquad (n)$$

9   To get the defined spring rate, the number of active coils must satisfy equation 13.28:

$$k = \frac{d^4 E}{10.8 D N_a} \quad \text{or} \quad N_a = \frac{d^4 E}{10.8 Dk} = \frac{(0.192)^4 30E6}{10.8(1.73)(120)} = 18.2 \qquad (o)$$

The ends contribute to the active coils as

$$N_e = \frac{L_1 + L_2}{3\pi D} = \frac{2 + 2}{3\pi(1.73)} = 2.2 \qquad (p)$$

and the number of body coils in the spring are

$$N_b = N_a - N_e = 18.2 - 2.2 = 16 \qquad (q)$$

10   The angular deflections at the specified loads from equation 13.27c are

$$\theta_{min} \cong 10.8 \frac{M_{min} D N_a}{d^4 E} = 10.8 \frac{50(1.73)(18.2)}{(0.192)^4 (30E6)} = 0.417 \text{ rev} = 150 \text{ deg} \qquad (r)$$

$$\theta_{max} \cong 10.8 \frac{M_{max} D N_a}{d^4 E} = 10.8 \frac{80(1.73)(18.2)}{(0.192)^4 (30E6)} = 0.667 \text{ rev} = 240 \text{ deg} \qquad (s)$$

In the interest of brevity, only a successful solution to the above example is shown. However, just as in the preceding examples, it was necessary to iterate through some unsuccessful ones to get to it. See the *TKsolver* file EX13-06 for more information. Change the variable values and recompute to see their effects. The trends in coil length, outside diameter, static and fatigue safety factors, and spring weight of a torsion spring as a function of its spring index are shown in Figure 13-27. Unlike compression and extension springs, safety factors for torsion springs increase with the spring index.

13.9   BELLEVILLE SPRING WASHERS

Belleville washers, patented in France by J. F. Belleville in 1867, have a nonlinear force-deflection characteristic which makes them very useful in certain applications. A selection of commercial Belleville washers is shown in Figure 13-28. Their cross-section is a coned shape with a material thickness t and inside height of cone h, as shown in Figure 13-29. They are extremely compact and are capable of large push forces, but their deflections are limited. If they are placed on a flat surface, their maximum deflection is h, which puts them in the "flat" condition and should be operated only between about 15% and 85% of the deflection to flat. We will later show how they can be de-

FIGURE 13-28

Commercially Available Belleville Washers *Courtesy of Associated Spring, Barnes Group Inc., Bristol, Conn.*

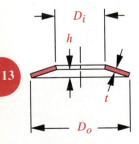

FIGURE 13-29

A Belleville Spring
Washer

flected beyond the flat position to achieve some interesting effects. These springs are used where high loads over small deflections are needed in compact spaces such as metal-forming die-stripper pins, gun recoil mechanisms, etc. In their zero-spring-rate (constant force) form, they are used to load clutches and seals, which need a uniform load over a small deflection.

The ratio of D_o to D_i, called R_d, affects their behavior. At about $R_d = 2$ the spring has maximum energy storage capacity. Depending on the h / t ratio, the spring rate can be essentially linear, can increase or decrease with increasing deflection, or can be essentially constant over a portion of the deflection.

Figure 13-30 shows force-deflection curves for Belleville washers with h / t ratios ranging from 0.4 to 2.8. These curves are normalized on both axes to the spring's condition when compressed flat. Zero deflection and force are taken at the free position as shown in Figure 13-29. One hundred percent deflection represents the flat condition, and 100% force represents the force of that spring at the flat condition. The absolute values of force and deflection will vary with the h / t ratio, thickness t, and material.

At $h / t = 0.4$, the spring rate is close to linear and resembles a helical spring's curve. As h / t is increased above 0.4, the rate becomes increasingly nonlinear and at $h / t = 1.414$, the curve has a portion of nearly constant value, centered around the flat position. Its force deviates less than ±1% of the force value at 100% deflection over the range of 80% to 120% of the deflection-to-flat and is within ±10% over 55% to 145% of deflection-to-flat, as shown in Figure 13-31.

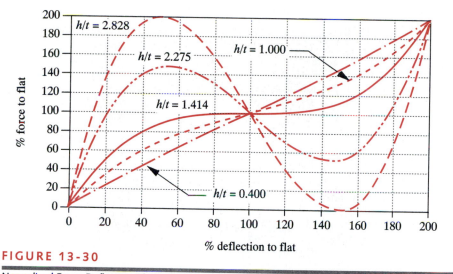

FIGURE 13-30

Normalized Force-Deflection Characteristics of Belleville Springs for Various h/t Ratios.

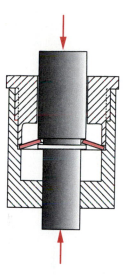

(a) Washer above center

At h/t ratios above 1.414, the curve becomes bimodal. A given applied force corresponds to more than one possible deflection. If such a spring is mounted to allow it to go beyond the flat condition as shown in Figure 13-32, it will be bistable, requiring a force in either direction to trip it past center. The mounting technique shown in Figure 13-32 is also useful for springs of smaller h/t ratios as it allows twice the potential deflection and can use the entire constant-force section of a 1.414 h/t ratio spring.

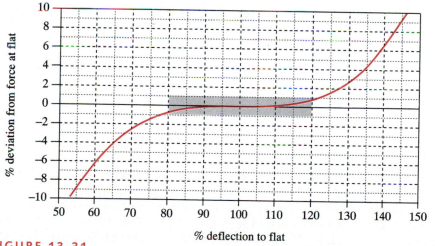

FIGURE 13-31

Percent Error in a Constant-Force Belleville Spring Around its Flat Position ($R_d = 2.0$, $h/t = 1.414$)

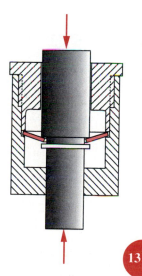

(b) Washer below center

FIGURE 13-32

Mounting a Belleville Spring Washer to Allow Deflection Past the Flat Position

Load-Deflection Function for Belleville Washers

The load-deflection relationship is nonlinear, so we cannot state it as a spring rate. It can be calculated from

$$F = \frac{4Ey}{K_1 D_o^2 (1 - v^2)} \left[(h - y) \left(h - \frac{y}{2} \right) t + t^3 \right] \qquad (13.35a)$$

where
$$K_1 = \frac{6}{\pi \ln R_d} \left[\frac{(R_d - 1)^2}{R_d^2} \right] \quad \text{and} \quad R_d = \frac{D_o}{D_i} \qquad (13.35b)$$

The load at the flat position ($y = h$) is

$$F_{flat} = \frac{4Eht^3}{K_1 D_o^2 (1 - v^2)} \qquad (13.35c)$$

The curves in Figure 13-30 were generated with these equations.

Stresses in Belleville Washers

The stresses are not uniformly distributed in the washer, but are concentrated at the edges of inside and outside diameters, as shown in Figure 13-33. The largest stress σ_c occurs at the inside radius on the convex side and is compressive. The edges on the concave side have tensile stresses with the outside edge stress σ_{t_o} usually larger than the inside edge stress σ_{t_i}. The expressions for stresses at the locations defined in Figure 13-33 are

$$\sigma_c = -\frac{4Ey}{K_1 D_o^2 (1 - v^2)} \left[K_2 \left(h - \frac{y}{2} \right) + K_3 t \right] \qquad (13.36a)$$

$$\sigma_{t_i} = \frac{4Ey}{K_1 D_o^2 (1 - v^2)} \left[-K_2 \left(h - \frac{y}{2} \right) + K_3 t \right] \qquad (13.36b)$$

$$\sigma_{t_o} = \frac{4Ey}{K_1 D_o^2 (1 - v^2)} \left[K_4 \left(h - \frac{y}{2} \right) + K_5 t \right] \qquad (13.36c)$$

where
$$K_2 = \frac{6}{\pi \ln R_d} \left(\frac{R_d - 1}{\ln R_d} - 1 \right) \quad \text{and} \quad R_d = \frac{D_o}{D_i} \qquad (13.36d)$$

$$K_3 = \frac{6}{\pi \ln R_d} \left(\frac{R_d - 1}{2} \right) \qquad (13.36e)$$

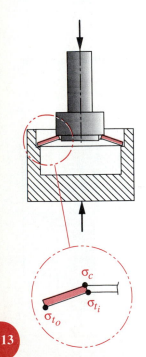

13

FIGURE 13-33

Points of Highest Stress in a Belleville Spring Washer

$$K_4 = \left[\frac{R_d \ln R_d - (R_d - 1)}{\ln R_d} \right] \left[\frac{R_d}{(R_d - 1)^2} \right] \quad (13.36f)$$

$$K_5 = \frac{R_d}{2(R_d - 1)} \quad (13.36g)$$

See equation 13.35b for K_1. A typical variation of these stresses with deflection is shown in Figure 13-34. The steel spring dimensions in this case are $t = 0.012$ in, $h = 0.017$ in, $h/t = 1.414$, $D_o = 1$ in, $D_i = 0.5$ in.

Static Loading of Belleville Washers

The compressive stress σ_c usually controls the design in static loading, but because the stress is highly concentrated at the edges, local yielding will occur to relieve it and the stress throughout the spring will be less. Because the local σ_c is higher than the average stress, it can be compared to a strength value larger than the ultimate compressive strength S_{uc}. Since even materials are typically used for springs, $S_{uc} = S_{ut}$. Table 13-15 shows some recommended percentages of S_{ut} for comparison to σ_c in static loading. Recognize that the material in general cannot withstand these stress levels. They are just a means to predict failure based on the localized stress σ_c. Setting (set removal) can be used to introduce favorable residual stresses and the allowable stress increases substantially, as shown in Table 13-15.

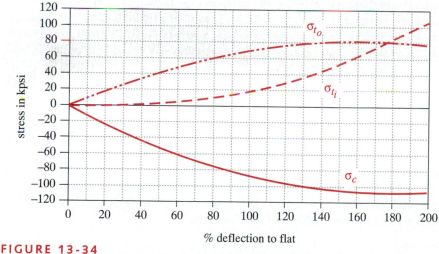

FIGURE 13-34

Stresses in a Carbon Steel Belleville Spring (with $R_d = 2.0$, $h/t = 1.414$, $t = 0.012$, $h = 0.017$ in)

(a) Parallel stack

Table 13-15   Maximum Recommended Compressive Stress Levels σ_c for Belleville Washers in Static Applications, Assuming $S_{uc} = S_{ut}$

Source: Reference 1

Material	Maximum Percent of Ultimate Tensile Strength	
	Before Set Removed	After Set Removed
Carbon or alloy steel	120%	275%
Nonferrous and austenitic stainless steel	95	160

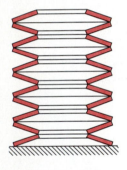

(b) Series stack

Dynamic Loading

If the spring is dynamically loaded, the maximum and minimum tensile stresses σ_{t_i} and σ_{t_o} at the extremes of its deflection range should be calculated from equations 13.36 and the alternating and mean components determined from them. A Goodman diagram analysis can then be done and the safety factor found from equation 13.34a. The endurance limit for the material can be found using the methods of Chapter 6. Shot peening can be used to increase fatigue life.

Stacking Springs

The maximum deflection of a Belleville spring tends to be small. To get more deflection, they can be stacked in series as shown in Figure 13-35b. The total force will be the same as for one spring but the deflections will add. They can also be stacked in parallel, as shown in Figure 13-35a, in which case the total deflection will be the same as for one spring and the forces will add. Series-parallel combinations are also possible. Note however, that series or series-parallel stacks are inherently unstable and must be guided over a pin or in a hole, in which case friction will reduce the available load. Interleaf friction can also be substantial in parallel stacks.

Designing Belleville Springs

The design of Belleville springs requires iteration. Trial values of the diameter ratio R_d and the h/t ratio must be chosen. The type of force deflection curve desired will suggest an appropriate h/t ratio (see Figure 13-30). If a force or range of forces is specified, the associated deflection can be calculated from equations 13.35 once an outside diameter and thickness are assumed. It is possible to estimate the thickness required to obtain a particular force at the flat position from

(c) Series-parallel stack

FIGURE 13-35

Belleville Washers Can Be Combined in Series, Parallel, or Series-Parallel

$$t = \sqrt[4]{\frac{F_{flat}}{19.2E7} \frac{D_o^2}{h/t}} \qquad\qquad (13.37us)$$

13

$$t = \frac{1}{10} \sqrt[4]{\frac{F_{flat}}{132.4} \frac{D_o^2}{h/t}}$$ (13.37si)

This value, in combination with other assumed values can be used in equations 13.35 and 13.36 to find deflections and stresses. The best way to illustrate this process is with an example.

EXAMPLE 13-7

Design of a Belleville Spring for Static Loading

Problem   Design a Belleville spring to give a constant load over a given deflection.

Given   An end seal on a shaft requires a nearly constant load over small motions associated with temperature change. The spring must apply a nominal force of 10 lb ±5% over a range of ±0.006 in at any convenient nominal deflection. The spring must fit in a 1.25-in-diameter hole.

Assumptions   Assume a diameter ratio $R_d = 2$. Use unset carbon spring steel 50HRC

Solution   See the *TKSolver* file EX13-07.

1   Assume an outside diameter D_o of 1.2 in to allow some clearance in the hole.

2   Since a constant-force spring is needed, the h/t ratio is 1.414 (see Figure 13.30).

3   The required force variation of not more than ±5% can be met by choosing an appropriate deflection range to operate in from Figure 13-31. If the deflection is kept between about 65% and 135% of the flat deflection, this tolerance will be achieved. The nominal force will then occur at the flat position and the spring must operate on both sides, so must be mounted in similar fashion to that shown in Figure 13-33.

4   Use the above assumptions and the specified nominal force in equation 13.37a to find an appropriate spring thickness t:

$$t = \sqrt[4]{\frac{F_{flat}}{19.2E7} \frac{D_o^2}{h/t}} = \sqrt[4]{\frac{10}{19.2E7} \frac{(1.2)^2}{1.414}} = 0.015 \text{ in}$$ (a)

5   The height h can now be found:

$$h = 1.414t = 1.414(0.015) = 0.021 \text{ in}$$ (b)

6   Based on the choices in (3) above, find the minimum and maximum deflections:

$$y_{min} = 0.65h = 0.65(0.021) = 0.014 \text{ in}$$

$$y_{max} = 1.35h = 1.35(0.021) = 0.029 \text{ in}$$ (c)

13

The difference between these distances is greater than the required deflection range, so the force tolerance can be met over that range.

7 Figure 13-34 shows that the worst stress state will occur at the largest deflection y_{max}, so solve equations 13.36 for stresses at that deflection:

$$K_1 = \frac{6}{\pi \ln R_d}\left[\frac{(R_d-1)^2}{R_d^2}\right] = \frac{6}{\pi \ln 2}\left[\frac{(2-1)^2}{2^2}\right] = 0.689 \qquad (d)$$

$$K_2 = \frac{6}{\pi \ln R_d}\left(\frac{R_d-1}{\ln R_d}-1\right) = \frac{6}{\pi \ln 2}\left(\frac{2-1}{\ln 2}-1\right) = 1.220 \qquad (e)$$

$$K_3 = \frac{6}{\pi \ln R_d}\left(\frac{R_d-1}{2}\right) = \frac{2}{\pi \ln 2}\left(\frac{2-1}{2}\right) = 1.378 \qquad (f)$$

$$K_4 = \left[\frac{R_d \ln R_d-(R_d-1)}{\ln R_d}\right]\left[\frac{R_d}{(R_d-1)^2}\right] = \left[\frac{2\ln 2-(2-1)}{\ln 2}\right]\left[\frac{2}{(2-1)^2}\right] = 1.115 \quad (g)$$

$$K_5 = \frac{R_d}{2(R_d-1)} = \frac{2}{2(2-1)} = 1 \qquad (h)$$

$$\sigma_c = -\frac{4Ey}{K_1 D_o^2(1-n^2)}\left[K_2\left(h-\frac{y}{2}\right)+K_3 t\right]$$

$$= -\frac{4(30E6)(0.029)}{0.689(1.2)^2(1-0.3^2)}\left[1.220\left(0.021-\frac{0.029}{2}\right)+1.378(0.015)\right] \qquad (i)$$

$$\sigma_c = -112\ 227 \text{ psi}$$

$$\sigma_{t_i} = \frac{4Ey}{K_1 D_o^2(1-v^2)}\left[-K_2\left(h-\frac{y}{2}\right)+K_3 t\right]$$

$$= \frac{4(30E6)(0.029)}{0.689(1.2)^2(1-0.3^2)}\left[-1.220\left(0.021-\frac{0.029}{2}\right)+1.378(0.015)\right] \qquad (j)$$

$$\sigma_{t_i} = 46\ 600 \text{ psi}$$

$$\sigma_{t_o} = \frac{4Ey}{K_1 D_o^2(1-v^2)}\left[K_4\left(h-\frac{y}{2}\right)+K_5 t\right]$$

$$= \frac{4(30E6)(0.029)}{0.689(1.2)^2(1-0.3^2)}\left[1.115\left(0.021-\frac{0.029}{2}\right)+(1)(0.015)\right] \qquad (k)$$

$$\sigma_{t_o} = 87\ 628 \text{ psi}$$

8 Table 13-5 gives S_{ut} = 246 kpsi for this material. Table 13-15 indicates that 120% of this value can be used for an unset spring. The safety factor for static loading is then

$$N_s = \frac{1.2 S_{ut}}{\sigma_c} = \frac{1.2(246\ 000)}{112\ 227} = 2.6 \qquad\qquad (l)$$

which is acceptable

9   A summary of this spring design is

$$D_o = 1.2 \qquad D_i = 0.6 \qquad t = 0.015 \qquad h = 0.021 \qquad (m)$$

13.10 CASE STUDIES

We will now address the design of a spring in one of the Case Study assemblies that were defined in Chapter 8.

Designing a Return Spring for a Cam-Testing Machine

The preliminary design of this device is shown in Figure 13-36. The follower arm is loaded against the cam by an extension spring with loops at each end. The calculations done in Case Study 9A indicate that a spring constant of 25 lb/in and a preload of 25 lb will keep the follower force positive between values of 13 and 110 lb at the design speed of 180 rpm. The length of the spring should be suitable to the package as shown in Figure 13-36, i.e., of the same order as the cam diameter, which is 8 in. The attachment point of spring to ground is adjustable.

CASE STUDY 9C

Design of a Return-Spring for a Cam-Follower Arm

Problem   Design an extension spring for the cam-follower arm in Figure 8-8 based on the loadings defined in Case Study 9A in Chapter 8.

Given   The spring rate is 25 lb/in with a preload of 25 lb. The spring's dynamic deflection is 1.5 in.

Assumptions   The spring operates in an oil bath whose temperature is below 250°F. Infinite life is required. Use ASTM A228 music wire and standard loops on each end.

Solution   See Figure 13-36 and TKSolver file CASE9-C.

1   Assume an 0.177 in trial wire diameter from the available sizes in Table 13-2. Assume a spring index of $C = 8$ and calculate the mean coil diameter D from equation 13.5.

$$D = Cd = 8(0.177) = 1.42 \text{ in} \qquad\qquad (a)$$

13

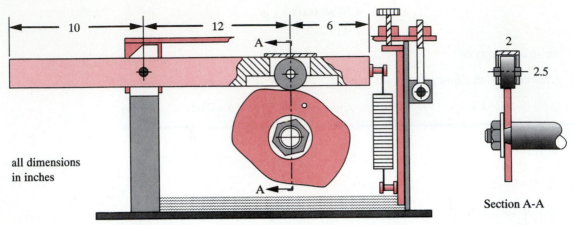

FIGURE 13-36

Cam-Follower Arm with Helical Extension Return Spring

2   Use the assumed value of C to find an appropriate value of initial coil stress τ_i from equations 13.21 as the average value of the functions that bracket the acceptable range of spring preloads in Figure 13-22:

$$\tau_{i_1} \cong -4.231C^3 + 181.5C^2 - 3\,387C + 28\,640$$

$$= -4.231(8)^3 + 181.5(8)^2 - 3\,387(8) + 28\,640 = 10\,994 \text{ psi} \qquad (b)$$

$$\tau_{i_2} \cong -2.987C^3 + 139.7C^2 - 3\,427C + 38\,404$$

$$= -2.987(8)^3 + 139.7(8)^2 - 3\,427(8) + 38\,404 = 18\,399 \text{ psi} \qquad (c)$$

$$\tau_i \cong \frac{\tau_{i_1} + \tau_{i_2}}{2} = \frac{10\,994 + 18\,399}{2} = 14\,697 \text{ psi} \qquad (d)$$

3   Find the direct shear factor:

$$K_s = 1 + \frac{0.5}{C} = 1 + \frac{0.5}{8} = 1.0625 \qquad (e)$$

4   Use the value of τ_i from (c) in equation 13.8 to find the corresponding initial coil-tension force F_i:

$$F_i = \frac{\pi d^3 \tau_i}{8K_s D} = \frac{\pi(0.177)^3(14\,697)}{8(1.0625)(1.416)} = 21.272 \text{ lb} \qquad (f)$$

Check that this force is less than the required 25-lb minimum applied force F_{min}, which in this case, it is. Any force smaller than F_i in the spring will not deflect it.

5   Find the maximum force from the given rate and deflection and use them to find the mean and alternating forces from equation 13.15a:

$$F_{max} = F_{min} + ky = 25 + 25(1.5) = 62.5 \text{ lb}$$

$$F_a = \frac{F_{max} - F_{min}}{2} = \frac{62.5 - 25}{2} = 18.75 \text{ lb} \qquad (g)$$

$$F_m = \frac{F_{max} + F_{min}}{2} = \frac{62.5 + 25}{2} = 43.75 \text{ lb}$$

6   Use the direct shear factor K_s and previously assumed values to find the minimum stress τ_{min} and the mean stress τ_m:

$$\tau_{min} = K_s \frac{8F_{min}D}{\pi d^3} = 1.0625 \frac{8(25)(1.416)}{\pi(0.177)^3} = 17\,272 \text{ psi}$$

$$(h)$$

$$\tau_m = K_s \frac{8F_m D}{\pi d^3} = 1.0625 \frac{8(43.75)(1.416)}{\pi(0.177)^3} = 30\,227 \text{ psi}$$

7   Find the Wahl factor K_w and use it to calculate the alternating shear stress in the coil.

$$K_w = \frac{4C-1}{4C-4} + \frac{0.615}{C} = \frac{4(8)-1}{4(8)-4} + \frac{0.615}{8} = 1.184 \qquad (i)$$

$$\tau_a = K_w \frac{8F_a D}{\pi d^3} = 1.184 \frac{8(18.75)(1.416)}{\pi(0.177)^3} = 14\,436 \text{ psi} \qquad (j)$$

8   Find the ultimate tensile strength of this music-wire material from equation 13.3 and Table 13-4 and use it to find the ultimate shear strength from equation 13.4 and the torsional yield strength for the coil body from Table 13-10, assuming no set removal.

$$S_{ut} = Ad^b = 184\,649(0.177)^{-0.1625} = 244\,653 \text{ psi}$$

$$S_{us} = 0.67S_{ut} = 163\,918 \text{ psi} \qquad (k)$$

$$S_{ys} = 0.45S_{ut} = 0.45(244\,653) = 110\,094 \text{ psi} \qquad (l)$$

9   Find the wire endurance limit for unpeened springs from equation 13.12 and convert it to a fully reversed endurance strength with equation 13.16c.

$$S_{ew} = 45\,000 \text{ psi} \qquad (m)$$

$$S_{es} = 0.707 \frac{S_{ew}S_{us}}{S_{us} - 0.707S_{ew}} = 0.707 \frac{45\,000(163\,918)}{163\,918 - 0.707(45\,000)} = 39\,477 \text{ psi} \qquad (n)$$

10   The fatigue safety factor for the coils in torsion is calculated from equation 13.16b.

$$N_{f_s} = \frac{S_{es}(S_{us} - \tau_{min})}{S_{es}(\tau_m - \tau_{min}) + S_{us}\tau_a}$$

$$= \frac{39\,477(163\,918 - 17\,272)}{39\,477(30\,227 - 17\,272) + 163\,918(14\,436)} = 2.0 \qquad (o)$$

13

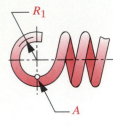

maximum
bending stress

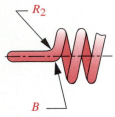

B

maximum
torsional stress

Repeated

FIGURE 13-23

Points of Maximum
Stress in Hook or Loop of
an Extension Spring

13

Note that the minimum stress due to force F_{min} is used in this calculation, **not** the coil-winding stress from (d).

11 The stresses in the end hooks also need to be determined. The bending stresses in the hook at point A in Figure 13-23 (repeated here) are found from equation 13.23:

$$C_1 = \frac{2R_1}{d} = \frac{2D}{2d} = C = 8$$

$$K_b = \frac{4C_1^2 - C_1 - 1}{4C_1(C_1 - 1)} = \frac{4(8)^2 - 8 - 1}{4(8)(8 - 1)} = 1.103 \qquad (p)$$

$$\sigma_a = K_b \frac{16DF_a}{\pi d^3} + \frac{4F_a}{\pi d^2} = 1.103\frac{16(1.416)(18.75)}{\pi(0.177)^3} + \frac{4(18.75)}{\pi(0.177)^2} = 27\,650 \text{ psi}$$
$$(q)$$

$$\sigma_m = K_b \frac{16DF_m}{\pi d^3} + \frac{4F_m}{\pi d^2} = 1.103\frac{16(1.416)(43.75)}{\pi(0.177)^3} + \frac{4(43.75)}{\pi(0.177)^2} = 64\,517 \text{ psi}$$

$$\sigma_{min} = K_b \frac{16DF_{min}}{\pi d^3} + \frac{4F_{min}}{\pi d^2} = 1.103\frac{16(1.416)(25)}{\pi(0.177)^3} + \frac{4(25)}{\pi(0.177)^2} = 38\,867 \text{ psi} \qquad (r)$$

12 Convert the torsional endurance strength to a tensile endurance strength with the von Mises relationship and use it and the ultimate tensile strength in equation 13.16 to find a fatigue safety factor for the hook in bending:

$$S_e = \frac{S_{es}}{0.577} = \frac{39\,477}{0.577} = 68\,418 \text{ psi}$$

$$N_{f_b} = \frac{S_e(S_{ut} - \sigma_{min})}{S_e(\sigma_m - \sigma_{min}) + S_{ut}\sigma_a}$$

$$= \frac{68\,418(244\,633 - 38\,867)}{68\,418(64\,517 - 38\,867) + 244\,633(27\,650)} = 1.6 \qquad (s)$$

13 Find the torsional stresses at point B of the hook in Figure 13-23 using equation 13.24 and an assumed value of $C_2 = 5$.

$$R_2 = \frac{C_2 d}{2} = \frac{5(0.177)}{2} = 0.443 \text{ in}$$

$$K_{w_2} = \frac{4C_2 - 1}{4C_2 - 4} = \frac{4(5) - 1}{4(5) - 4} = 1.188 \qquad (t)$$

$$\tau_{B_a} = K_{w_2} \frac{8DF_a}{\pi d^3} = 1.188\frac{8(1.416)18.75}{\pi(0.177)^3} = 14\,478 \text{ psi}$$

$$\tau_{B_m} = K_{w_2} \frac{8DF_m}{\pi d^3} = 1.188\frac{8(1.416)43.75}{\pi(0.177)^3} = 33\,783 \text{ psi} \qquad (u)$$

$$\tau_{B_{min}} = K_{w_2} \frac{8DF_{min}}{\pi d^3} = 1.188\frac{8(1.416)25}{\pi(0.177)^3} = 19\,304 \text{ psi}$$

14 The fatigue safety factor for the hook in torsion is calculated from equation 13.16*b*.

$$N_{f_s} = \frac{S_{es}(S_{us} - \tau_{min})}{S_{es}(\tau_m - \tau_{min}) + S_{us}\tau_a}$$

$$= \frac{39\ 477(163\ 918 - 19\ 304)}{39\ 477(33\ 783 - 19\ 304) + 163\ 918(14\ 478)} = 1.9 \qquad (v)$$

All three of these safety factors are acceptable.

15 To get the specified spring rate, the number of active coils must satisfy equation 13.7:

$$k = \frac{d^4 G}{8D^3 N_a} \quad \text{or} \quad N_a = \frac{d^4 G}{8D^3 k} = \frac{(0.177)^4 11.5E6}{8(1.416)^3(25)} = 19.88 \cong 20 \qquad (w)$$

Note that we round it to the nearest 1/4 coil, as the manufacturing tolerance cannot achieve better than that accuracy.

16 The total number of coils in the body and the body length are

$$N_t = N_a + 1 = 20 + 1 = 21$$

$$L_b = N_t d = 21(0.177) = 3.72 \text{ in} \qquad (x)$$

17 The length of a standard loop is equal to the coil inside diameter. The free length is

$$L_f = L_b + 2L_{hook} = 3.72 + 2(1.24) = 6.2 \text{ in} \qquad (y)$$

18 The maximum deflection and spring length at that deflection are

$$y_{max} = \frac{F_{max} - F_{initial}}{k} = \frac{62.5 - 21.27}{25} = 1.65 \text{ in}$$

$$L_{max} = L_f + y_{max} = 6.2 + 1.65 = 7.85 \text{ in} \qquad (z)$$

This length is within the maximum cam diameter so is acceptable.

19 The natural frequency of this spring is found from equation 13.25*a* and is

$$f_n = \frac{2}{\pi N_a} \frac{d}{D^2} \sqrt{\frac{Gg}{32\gamma}} = \frac{2(0.177)}{\pi(20)(1.416)^2} \sqrt{\frac{11.5E6(386)}{32(0.285)}} = 62 \text{ Hz} = 3\ 720 \text{ rpm} \qquad (aa)$$

The ratio between the natural frequency and the forcing frequency is

$$\frac{3\ 720}{180} = 20.7 \qquad (ab)$$

which is sufficiently high.

20 The design specification for this A228 wire spring is

$$d = 0.177 \text{ in} \qquad OD = 1.593 \text{ in} \qquad N_t = 21 \qquad L_f = 6.2 \text{ in} \qquad (ac)$$

13

13.11 SUMMARY

Springs are widely used in machinery of all types to provide push, pull, or twist forces, or to store potential energy. This chapter has discussed the uses of a variety of springs and the design of a few types of commonly used springs, the helical compression, helical extension, helical torsion, and Belleville. The names of the first three define the type of external load applied, not the type of stress present. It is easy to confuse these two aspects. A helical compression or extension spring has torsional stress in its coils, and a helical torsion spring has tensile and compressive stress in its coils. These three types of springs are made of coiled wire. The wire is usually round, though rectangular wire is sometimes used. Belleville spring washers are made of flat stock formed into a cone shape. The helical springs typically have an essentially linear force-deflection characteristic (constant spring rate). Belleville springs have a highly nonlinear characteristic, which can be used to advantage to obtain a near-zero rate or a bistable action.

Extensive data have been developed on the strength characteristics of spring wire and flat spring stock. Much of this literature is reproduced in this chapter. Material strength generally increases as the wire cross-section size is reduced with the result that fine wire has very high break strength under static loads. The endurance strength of high static-strength materials tends to saturate ("top-out") at reasonably high levels rather than be a function of static strength. Estimates of the fatigue strength of various spring materials are also quoted from the literature in this chapter.

The design of springs, whether for static or dynamic loading, is an inherently iterative exercise. Assumptions must be made for the values of several parameters in order to do the calculations. Usually, the first result is an unsuccessful design, requiring changes to the assumed values and recalculation. A computer is an indispensable aid in this process. Many fully worked-out examples of spring design are provided in the chapter and the reader is encouraged to study them along with their accompanying *TKSolver* files which, provide more information than can be presented in tables.

Important Equations Used in this Chapter

Spring Rate (Section 13.1):

$$k = \frac{F}{y} \tag{13.1}$$

Combining Springs in Series (Section 13.1):

$$k_{total} = k_1 + k_2 + k_3 \ldots + k_n \tag{13.2a}$$

Combining Springs in Parallel (Section 13.1):

$$\frac{1}{k_{total}} = \frac{1}{k_1} + \frac{1}{k_2} + \frac{1}{k_3} \ldots + \frac{1}{k_n} \tag{13.2b}$$

13

Spring Index (Section 13.4):

$$C = \frac{D}{d} \tag{13.5}$$

Deflection of Helical Compression Spring (Section 13.4):

$$y = \frac{8FD^3N_a}{d^4G} \tag{13.6}$$

Deflection of Helical Extension Spring (Section 13.7):

$$y = \frac{8(F - F_i)D^3N_a}{d^4G} \tag{13.22}$$

Deflection of Round-Wire Helical Torsion Spring (Section 13.8):

$$\theta_{rev} \cong 10.8\frac{MDN_a}{d^4E} \tag{13.27c}$$

Spring Rate of Helical Compression Spring (Section 13.4):

$$k = \frac{F}{y} = \frac{d^4G}{8D^3N_a} \tag{13.7}$$

Spring Rate of Helical Extension Spring (Section 13.7):

$$k = \frac{F - F_i}{y} = \frac{d^4G}{8D^3N_a} \tag{13.20}$$

Spring Rate of Round-Wire Helical Torsion Spring (Section 13.8):

$$k = \frac{M}{\theta_{rev}} \cong \frac{d^4E}{10.8DN_a} \tag{13.28}$$

Static Stress in Helical Compression or Extension Spring (Section 13.7):

$$\tau_{max} = K_s\frac{8FD}{\pi d^3} \qquad \text{where} \quad K_s = \left(1 + \frac{0.5}{C}\right) \tag{13.8b}$$

Dynamic Stress in Helical Compression or Extension Spring (Section 13.7):

$$K_w = \frac{4C - 1}{4C - 4} + \frac{0.615}{C} \tag{13.9a}$$

$$\tau_{max} = K_w\frac{8FD}{\pi d^3} \tag{13.9b}$$

13

Stress in Helical Torsion Spring at Inside Diameter (Section 13.8):

$$K_{b_i} = \frac{4C^2 - C - 1}{4C(C-1)} \tag{13.31a}$$

$$\sigma_{i_{max}} = K_{b_i} \frac{M_{max}c}{I} = K_{b_i} \frac{M_{max}(d/2)}{\pi d^4/64} = K_{b_i} \frac{32M_{max}}{\pi d^3} \tag{13.32a}$$

Stress in Helical Torsion Spring at Outside Diameter (Section 13.8):

$$K_{b_o} = \frac{4C^2 + C - 1}{4C(C+1)} \tag{13.31b}$$

$$\sigma_{o_{min}} = K_{b_o} \frac{32M_{min}}{\pi d^3}; \qquad \sigma_{o_{max}} = K_{b_o} \frac{32M_{max}}{\pi d^3} \tag{13.32b}$$

Ultimate Tensile Strength of Steel Wire—See Table 13-4 for Constants (Section 13.4):

$$S_{ut} \cong A d^b \tag{13.3}$$

Ultimate Shear Strength of Wire (Section 13.4):

$$S_{us} \cong 0.67 S_{ut} \tag{13.4}$$

Torsional Endurance Limits for Spring-Steel Wire for Stress Ratio R = 0 (Section 13.4):

$$S_{ew'} \cong 45.0 \text{ kpsi } (310 \text{ MPa}) \quad \text{for unpeened springs}$$

$$S_{ew'} \cong 67.5 \text{ kpsi } (465 \text{ MPa}) \quad \text{for peened springs} \tag{13.12}$$

Torsional Endurance Limits for Spring-Steel Wire for Stress Ratio R = –1 (Section 13.4):

$$S_{es} = 0.707 \frac{S_{ew}S_{us}}{S_{us} - 0.707 S_{ew}} \tag{13.16c}$$

Bending Endurance Limits for Spring-Steel Wire for Stress Ratio R = 0 (Section 13.4):

$$S_{ew_b}' = \frac{S_{ew'}}{0.577} \tag{13.33a}$$

Bending Endurance Limits for Spring-Steel Wire for Stress Ratio R = –1 (Section 13.4):

$$S_e = 0.707 \frac{S_{ew_b}S_{ut}}{S_{ut} - 0.707 S_{ew_b}} \tag{13.34c}$$

Static Safety Factor for Helical Compression or Extension Spring (Section 13.5):

$$N_s = \frac{S_{ys}}{\tau} \tag{13.14}$$

Dynamic Safety Factor for Helical Compression or Extension Spring (Section 13.4):

$$N_{f_s} = \frac{S_{es}(S_{us} - \tau_i)}{S_{es}(\tau_m - \tau_i) + S_{us}\tau_a} \qquad (13.16b)$$

Dynamic Safety Factor for Helical Torsion Spring (Section 13.8):

$$N_{f_b} = \frac{S_e\left(S_{ut} - \sigma_{o_{min}}\right)}{S_e\left(\sigma_{o_{mean}} - \sigma_{o_{min}}\right) + S_{ut}\sigma_{o_{alt}}} \qquad (13.34b)$$

Load-Deflection Function for a Belleville Washer (Section 13.9):

$$F = \frac{4Ey}{K_1 D_o^2 \left(1 - v^2\right)}\left[(h - y)\left(h - \frac{y}{2}\right)t + t^3\right] \qquad (13.35a)$$

where $\quad K_1 = \dfrac{6}{\pi \ln R_d}\left[\dfrac{(R_d - 1)^2}{R_d^2}\right] \quad$ and $\quad R_d = \dfrac{D_o}{D_i} \qquad (13.35b)$

Maximum Compressive Stress in a Belleville Washer (Section 13.9):

$$\sigma_c = -\frac{4Ey}{K_1 D_o^2 \left(1 - v^2\right)}\left[K_2\left(h - \frac{y}{2}\right) + K_3 t\right] \qquad (13.36a)$$

where $\quad K_2 = \dfrac{6}{\pi \ln R_d}\left(\dfrac{R_d - 1}{\ln R_d} - 1\right) \quad$ and $\quad R_d = \dfrac{D_o}{D_i} \qquad (13.36d)$

$$K_3 = \frac{6}{\pi \ln R_d}\left(\frac{R_d - 1}{2}\right) \qquad (13.36e)$$

Required Thickness of a Belleville Spring for Force at Flat (Section 13.9):

$$t = \sqrt[4]{\frac{F_{flat}}{19.2E7}\frac{D_o^2}{h/t}} \qquad (13.37us)$$

13.12   REFERENCES

1   **Associated Spring**, *Design Handbook: Engineering Guide to Spring Design.* Associated Spring, Barnes Group Inc., Bristol, Conn., 1987.

2   **H. C. R. Carlson**, Selection and Application of Spring Material. *Mechanical Engineering,* **78**: pp. 331-334, 1956.

3   **A. M. Wahl**, *Mechanical Springs.* McGraw-Hill: New York, 1963.

4   **F. P. Zimmerli**, "Human Failures in Spring Design." *The Mainspring, Associated Spring Corp.,* Aug.-Sept. 1957.

13

13.13 PROBLEMS

*13-1   A linear spring is to give 200 N at its maximum deflection of 150 mm and 40 N at its minimum deflection of 50 mm. What is its spring rate?

13-2   Find the ultimate tensile strength, the ultimate shear strength, and the torsional yield strength of a 1.8-mm-dia, A229 oil-tempered steel wire.

*13-3   Find the torsional yield and ultimate shear strength of an 0.105-in-dia, unset A230 wire to be used in a helical compression spring.

*13-4   What is the torsional fatigue strength of the wire in Problem 13-3 at $N = 5E6$ cycles?

13-5   Draw the modified-Goodman diagram for the wire of Problem 13-3.

*13-6   What are the spring rate and spring index of a squared and ground compression spring with $d = 1$ mm, $D = 10$ mm, and 12 total coils?

*13-7   Find the natural frequency of the spring in Problem 13-6.

†13-8   A paper machine processes rolls of paper having a density of 984 kg/m^3. The paper roll is 1.50-m outside dia (OD) x 0.22-m inside dia (ID) x 3.23 m long and is on a simply supported, hollow, steel shaft with 22-cm OD x 20-cm ID and as long as the paper roll. Find the spring rate of the shaft and the fundamental natural frequency of the shaft-roll assembly.

13-9   Determine the minimum allowable bending radius for an 50HRC strip-steel spring of 1-mm thickness.

*†13-10 An overhung diving board is shown in Figure P13-1a. A 100-kg person is standing on the free end. Assume cross-sectional dimensions of 305 mm x 32 mm and a material $E = 10.3$ GPa. What is the spring rate and fundamental natural frequency of the diver-board combination?

*13-11 Design a helical compression spring for a static load of 45 lb at a deflection of 1.25 in with a safety factor of 2.5. Use $C = 7.5$. Specify all parameters necessary to manufacture the spring.

†13-12 Repeat Problem 13-10 using the cantilevered diving board design in Figure P13-1b.

*13-13 Given $d = 0.312$ in, $y_{working} = 0.75$ in, 15% clash allowance, unpeened chrome-vanadium wire, squared ends, $F_{max} = 250$ lb, $F_{min} = 50$ lb, find N_a, D, L_f, L_{shut}, k,

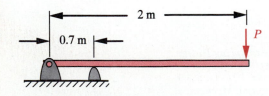

(a) Overhung diving board

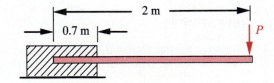

(b) Cantilevered diving board

FIGURE P13-1

Problems 13-10 and 13-12

$y_{initial}$, and the minimum hole diameter for the spring. Infinite life is desired with a safety factor of 2. Choose an acceptable spring index. Setting will be done.

†13-14   Figure P13-2 shows a child's toy called a *pogo stick*. The child stands on the pads, applying half her weight on each side. She jumps up off the ground, holding the pads up against her feet, and bounces along with the spring cushioning the impact and storing energy to help each rebound. Assume a 60-lb child and a spring constant of 100 lb/in. The pogo stick weighs 5 lb. Design the helical compression spring to survive jumping 2 in off the ground with a dynamic safety factor of 2 for a finite life of 5E4 cycles. Determine the fundamental natural frequency of the system.

13-15   Draw the modified-Goodman diagram for the following spring design and find its safety factor: S_{fw} = 40 kpsi, S_{us} = 200 kpsi, τ_a = 12 kpsi, τ_m = 95 kpsi, τ_i = 75 kpsi.

†13-16   Problem 6-16 describes a track for bowling balls that are 4.5-in dia and 2.5-lb weight. Design a spring-loaded launcher that will allow quadriplegic bowlers to launch the balls down the bowling alley from the point where the track of Problem 6-16 drops them with only a switch closure that releases the launcher. The launcher's plunger will be cocked by an assistant and the energy stored in the helical compression spring, which you will design will drive the plunger into the ball and roll it down the bowling alley. You will have to determine appropriate constraints and make many assumptions about such things as the size of the alley, the friction losses, and the energy needed to knock over the pins.

*13-17   Design a helical extension spring to handle a dynamic load that varies from 175 N to 225 N over a 0.85 cm working deflection. Use music wire and standard hooks. The forcing frequency is 1500 rpm. Infinite life is desired. Minimize the package size. Find safety factors against fatigue, yielding, and surging.

13-18   Design a helical extension spring with standard hooks to handle a dynamic load that varies from 300 lb to 500 lb over a 2-in working deflection. Use chrome-vanadium wire. The forcing frequency is 1 000 rpm. Infinite life is desired. Minimize the package size. Find safety factors against fatigue, yielding, and surging

*13-19   Design a helical compression spring to handle a dynamic load that varies from 175 lb to 225 lb over a 0.85-in working deflection. Use squared and ground, unpeened music wire and a 10% clash allowance. The forcing frequency is 500 rpm. Infinite life is desired. Minimize the package size. Find safety factors against fatigue, yielding, and surging.

13-20   Design a helical compression spring to handle a dynamic load that varies from 30 lb to 50 lb over a 1.25-in working deflection. Use squared, peened chrome-vanadium wire and a 15% clash allowance. The forcing frequency is 250 rpm. Infinite life is desired. Minimize the package size. Find safety factors against fatigue, yielding, and surging.

*13-21   Design a helical compression spring for a static load of 400 N at a deflection of 45 mm with a safety factor of 2.5. Use C = 8. Specify all parameters necessary to manufacture the spring. State all assumptions.

*13-22   Design a straight-ended helical torsion spring for a static load of 200 N-m at a deflection of 45° with a safety factor of 2. Specify all parameters necessary to manufacture the spring. State all assumptions.

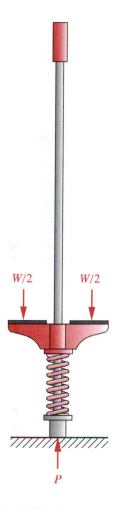

$W/2$ $W/2$

P

FIGURE P13-2

Problem 13-14

13

* Answers to these problems are provided in Appendix G.

† These problems are based on similar problems in previous chapters with the same –number, e.g., Problem 13-8 is based on Problem 6-8, etc. Problems in succeeding chapters may also continue and extend these problems.

13-23   Design a straight-ended helical torsion spring for a static load of 300 in-lb at a deflection of 75° with a safety factor of 2. Specify all parameters necessary to manufacture the spring. State all assumptions.

*13-24   Design a straight-ended helical torsion spring for a dynamic load of 50-150 N-m over a deflection of 80° with a safety factor of 2.5. Specify all parameters necessary to manufacture the spring. State all assumptions.

13-25   Design a straight-ended helical torsion spring for a dynamic load of 150-350 in-lb over a deflection of 50° with a safety factor of 2. Specify all parameters necessary to manufacture the spring. State all assumptions.

*13-26   Design a Belleville spring to give a constant 400 N ± 10% force over a 1-mm deflection.

13-27   Design a Belleville spring for bimodal operation between ± 50 N.

13-28   Given the following data for a helical compression spring loaded in fatigue, design the spring for infinite life. State all assumptions and sources of empirical data used. $C = 8.5$, $d = 0.312$ in, 625 rpm, working deflection = 0.75 in, 15% clash allowance, unpeened music wire, squared ends, preset, $F_{max} = 250$ lb, $F_{min} = 50$ lb.

13-29   A helical extension spring, loaded in fatigue, has been designed for infinite life with the following data. $C = 9$, $d = 0.312$ in, working deflection = 2 in, unpeened chrome-silicon wire, $F_{max} = 240$ lb, $F_{min} = 150$ lb, $F_{init} = 53$ lb, 13.75 active coils, $D = 2.81$ in.   Find the safety factors for failure in the standard hooks. State all assumptions and sources of empirical data used.

13-30   Given the following data for a helical torsion spring loaded in fatigue, find the spring index, unloaded coil dia, minimum loaded coil dia, and safety factor in fatigue. State all assumptions and sources of empirical data used. Deflection at assembly = 0.25 rev, working deflection = 0.5 rev, $k = 100$ lb-in/rev, $N_a = 20$, 0.192 in music wire, unpeened.

13-31   A helical compression spring is required to provide a minimum force of 150 lb at installation and have a working deflection of 1 in. The spring rate is 75 lb/in. The coil must fit in a 2.1-in-dia hole with 0.1 in clearance. Use 0.25-in diameter unpeened music wire and squared/ground ends. Using a 15% clash allowance, find:

(a) The stresses and safety factor for infinite life in fatigue
(b) The shut height
(c) The stress and safety factor at shut height
(d) The total number of coils
(e) The free length
(f) The natural frequency in Hz.
(g) Draw a Goodman diagram and show the safety factor from (a) on it.

13-32   A helical compression coil spring is needed to provide a time-varying force that ranges from a minimum of 100 lb to a maximum of 300 lb over a deflection of 1 in. It needs to work free over a shaft of 1.25-in dia. Use a cold-drawn carbon steel wire having an $S_{ut} = 250\ 000$ psi. A spring index of 6, a clash allowance of 15%, and squared and ground ends are desired.

13

SCREWS AND FASTENERS

For want of a nail the shoe is lost;
For want of a shoe the horse is lost;
And, for want of a horse the rider is lost.
GEORGE HERBERT

14.0   INTRODUCTION

The "nuts and bolts" of a design might seem to be one of its least interesting aspects, but in fact is one of the most fascinating. The success or failure of a design can hinge on proper selection and use of its fasteners. Moreover, the design and manufacture of fasteners is very big business and is a significant part of our economy. Literally thousands of different designs of fasteners are offered by vendors and thousands to millions of fasteners are used in a single complex assembly such as an automobile or aircraft. The Boeing 747 uses about 2.5 million fasteners, some of which cost several dollars each.[1]

There is a tremendous variety of fasteners available commercially, ranging from the mundane nuts and bolts to multipiece devices for quick-release of panels or for hidden-fastener applications. Figure 14-1 shows a small sample of the variety available. We cannot cover all of these varieties in one chapter. Whole books have been written on fasteners and some of them are noted in the bibliography of this chapter. We will limit our discussion to the design and selection of conventional fasteners such as bolts, screws, nuts, etc., used for machine-design applications in which significant loads and stresses are encountered.

Screws are used both to hold things together as fasteners and to move loads as so-called power screws or lead screws. We will investigate both of these applications. Screws as fasteners can be arranged to take tensile loads, shear loads, or both. We will explore the application of preloads to screw fasteners, which can have significant benefit to their load-carrying abilities. Table 14-0 shows the variables used in this chapter and indicates the equations or sections in which they can be found.

The opening epigraph on the previous page is often erroneously attributed to Benjamin Franklin, who popularized Herbert's maxim as a prefix to his *Poor Richard's Almanac* over a century later. In any event, the truth of this maxim is borne out by contemporary experience. The *Boston Sunday Globe* of October 16, 1994 reported that in the summer of 1994, three radioactive fuel assemblies at the Seabrook, N.H., nuclear power plant were damaged when a foot-long, 5-pound bolt was swept into the reactor by cooling water after it vibrated loose from the pump it was attached to. The shutdown for repairs would cost consumers millions of dollars for replacement electricity. The accident was attributed to poor fastener design. For want of a bolt ..., etc.

Table 14-0   Variables Used in this Chapter

Part 1 of 2

Symbol	Variable	ips units	SI units	See
A	area (with various subscripts)	in^2	m^2	various
A_b	total area of bolt	in^2	m^2	Sect. 14.2
A_m	effective area of material in clamp zone	in^2	m^2	Eq. 14.17
A_t	tensile-stress area of bolt	in^2	m^2	Sect. 14.2
C	joint stiffness constant	none	none	Eq. 14.13
C_{load}	loading factor	none	none	Ex. 14-3
C_{reliab}	reliability factor	none	none	Ex. 14-3
C_{size}	size factor	none	none	Ex. 14-3
C_{surf}	surface factor	none	none	Ex. 14-3
C_{temp}	temperature factor	none	none	Ex. 14-3
d	diameter (with various subscripts)	in	m	various
D	diameter (with various subscripts)	in	m	various
e	efficiency	none	none	Eq. 14.7
E	Young's modulus	psi	Pa	various
F	force (with various subscripts)	lb	N	various
f	friction force	lb	N	Eq. 14.4
F_b	maximum force in bolt	lb	N	Sect. 14.6
F_i	preload force	lb	N	Sect. 14.6
F_m	minimum force in material	lb	N	Sect. 14.6
HRC	Rockwell C hardness	none	none	various
J	polar second moment of area	in^4	m^4	Eq. 14.9
k	spring rate (with various subscripts)	lb/in	N/m	Sect. 14.6
k_b	bolt stiffness (spring rate)	lb/in	N/m	Sect. 14.6
k_m	material stiffness (spring rate)	lb/in	N/m	Sect. 14.6
K_f	fatigue stress concentration factor	none	none	Eq.14.15b
K_{fm}	mean-stress fatigue-concentration factor	none	none	Eq. 14.15b
l	length (with various subscripts)	in	m	various
L	lead of thread	in	mm	Sect. 14.2
n	number of fasteners	none	none	Sect. 14.8
N	number of threads per unit length	none	none	Sect. 14.2
N_f	safety factor in fatigue	none	none	Eq. 14.16
N_{leak}	safety factor—leakage	none	none	Case St. 7D
N_y	safety factor—static yielding	none	none	Ex. 14-2
N_{sep}	safety factor—separation	none	none	Eq. 14.14d
p	thread pitch	in	mm	various

14

Title page photograph courtesy of Fastbolt Inc., South Hackensack, N.J., 07606

Table 14-0   Variables Used in this Chapter
Part 2 of 2

Symbol	Variable	ips units	SI units	See
P	load (with various subscripts)	lb	N	various
P_b	portion of load felt by preloaded bolt	lb	N	Eq. 14.13
P_m	portion of load felt by preloaded material	lb	N	Eq. 14.13
r	radius	in	m	Sect. 14.8
S_e	corrected endurance limit	psi	Pa	various
$S_{e'}$	uncorrected endurance limit	psi	Pa	various
S_p	bolt proof strength	psi	Pa	Sect. 14.5
S_{us}	ultimate shear strength	psi	Pa	various
S_{ut}	ultimate tensile strength	psi	Pa	various
S_y	yield strength	psi	Pa	various
S_{ys}	shear yield strength	psi	Pa	Ex. 14-6
T	torque	lb-in	N-m	Eq. 14.5
w_i, w_o	thread-geometry factors	none	none	Table 14-5
W	work	in-lb	joules	Eq. 14.7
x, y	generalized length variables	in	m	Sect. 14.8
α	radial angle of thread	deg	deg	Eq. 14.5
δ	deflection	in	m	Sect. 14.6
λ	lead angle	deg	deg	Sect. 14.2
μ	coefficient of friction	none	none	Sect. 14.2
σ	normal stress (with various subscripts)	psi	Pa	various
τ	shear stress (with various subscripts)	psi	Pa	Sect. 14.3

14.1   STANDARD THREAD FORMS

The common element among screw fasteners is their thread. In general terms, the thread is a helix that causes the screw to advance into the workpiece or nut when rotated. Threads may be external (screw) or internal (nut or threaded hole). Thread forms originally differed in each major manufacturing country but after World War II were standardized in Great Britain, Canada, and the United States to what is now called the Unified National Standard (UNS) series, as shown in Figure 14-2. A European standard is also defined by ISO and has essentially the same thread cross-sectional shape, but uses metric dimensions, so is not interchangeable with UNS threads. Both UNS and ISO threads are in general use in the United States. Both use a 60° included angle and define thread size by the nominal outside (major) diameter d of an external thread. The thread pitch p is the distance between adjacent threads. The crests and roots are defined as flats to reduce the stress concentration from that of a sharp corner. The specifications allow for rounding of these flats due to tool wear. The pitch diameter d_p and the

FIGURE 14-1

A Sample of the Variety of Commercially Available Fasteners *Courtesy of Bolt Products Inc., City of Industry, Calif., 91745*

root diameter d_r are defined in terms of the thread pitch p with slightly different ratios used for UNS and ISO threads.

The lead L of the thread is the distance that a mating thread (nut) will advance axially with one revolution of the nut. If it is a **single thread,** as shown in Figure 14-2, the lead will equal the pitch. Screws can also be made with **multiple threads**, also called **multiple-start** threads. A **double thread** (2-start) has two parallel grooves wrapped around the diameter, like a pair of helical "railroad tracks." In this case the lead will be twice the pitch. A **triple thread** (3-start) will have a lead of three times the pitch, etc. The advantages of multiple threads are smaller thread height and the increased lead for fast advance of the nut. Some automotive power-steering screws use 5-start threads. However, most screws are made with only a single thread (1-start).

Three standard series of thread pitch families are defined in UNS threads, coarse-pitch (UNC), fine-pitch (UNF), and extra-fine-pitch (UNEF). ISO also defines coarse and fine series of threads. The **coarse series** is most common and is recommended for ordinary applications, especially where repeated insertions and removals of the screw are required or where the screw is threaded into a softer material. Coarse threads are less likely to cross or strip the soft material on insertion. **Fine threads** are more resistant to loosening from vibrations than coarse threads because of their smaller helix angle and so are used in automobiles, aircraft, and other applications that are subject to vibration. **Extra fine series** threads are used where wall thickness is limited and their short threads are an advantage.

14

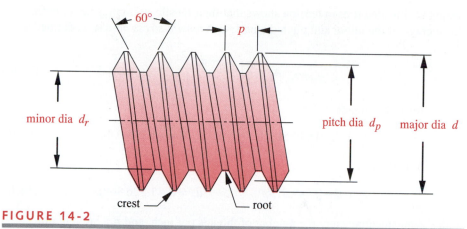

FIGURE 14-2

Unified National and ISO Standard Thread Form

The Unified National and ISO standards define tolerance ranges for both internal and external threads to control their fit. The UNS defines three fit classes, labeled class 1, 2, and 3. Class 1 has the broadest tolerances and is used for "hardware quality" (i.e., inexpensive) fasteners intended for casual use around the home, etc. Class 2 defines closer tolerances for a better quality fit between mating threads and is suitable for general machine design applications. Class 3 is the highest precision and can be specified where closer fits are needed. Cost increases with higher class of fit. A letter designator indicates an external (A) or internal (B) thread.

A thread is specified with a code that defines its series, diameter, pitch, and class of fit. The pitch of UNS threads is defined reciprocally as the number of threads per inch, while metric (ISO) thread pitch is specified by the pitch dimension in mm. An example of a UNS thread specification is

$$1/4\text{-}20 \text{ UNC-}2A$$

which defines a 0.250 in diameter by 20 threads per inch, coarse series, class 2 fit, external thread. An example of a metric thread specification is

$$M8 \times 1.25$$

which defines an 8-mm diameter by 1.25-mm pitch thread in the ISO coarse series. All standard threads are right-hand (RH) by default unless specified as left-hand by adding the letters LH to the specification.[*] A right-hand thread will advance the nut (or screw) away from you when either is turned clockwise.

Tensile Stress Area

If a threaded rod as shown in Figure 14-2 is subjected to pure tensile loading, one might expect its strength to be limited by the area of its minor (root) diameter, d_r. However,

14

[*] A left-hand-threaded nut often has a circumferential groove cut around its hex flats to identify it as a left-hand (LH) nut.

testing of threaded rods in tension shows that their tensile strength is better defined by the average of the minor and pitch diameters. A *tensile-stress area A_t* is defined as

$$A_t = \frac{\pi}{4}\left(\frac{d_p + d_r}{2}\right)^2 \qquad (14.1a)$$

where, for UNS threads:

$$d_p = d - 0.649\,519/N \qquad\qquad d_r = d - 1.299\,038/N \qquad (14.1b)$$

and for ISO threads:

$$d_p = d - 0.649\,519\,p \qquad\qquad d_r = d - 1.226\,869\,p \qquad (14.1c)$$

with d = outside diameter, N = number of threads per inch, and p = pitch in mm.

The stress in a threaded rod due to a pure axial tensile load F is then

$$\sigma_t = \frac{F}{A_t} \qquad (14.2)$$

Standard Thread Dimensions

Table 14-1 shows the principal dimensions of UNS threads and Table 14-2 shows the same for ISO threads. UNS threads smaller than 0.25-in diameter are designated by a gage number. A useful algorithm for determining the diameter of numbered threads is to multiply the gage number by 13 and add 60. The result is its approximate major diameter in thousandths of an inch. The minor diameter = major diameter – pitch. See references 2, 3, and 4 for more detailed dimensional information on standard threads, including tolerances for the different classes of fit.

14.2   POWER SCREWS

Power screws, also called lead screws, are used to convert rotation to linear motion in actuators, production machines, and jacks, among many other applications. They are capable of very large mechanical advantages and so can lift or move large loads. In such cases, a very strong thread form is needed. While the standard thread forms described above are well-suited to use in fasteners, they may not be strong enough for all power-screw applications. Other thread profiles have been standardized for such applications.

Square, Acme, and Buttress Threads

The square thread shown in Figure 14-3a provides the greatest strength and efficiency and also eliminates any radial component of force between the screw and nut. How-

Table 14-1   Principal Dimensions of Unified National Standard Screw Threads
Data Calculated from Equations 14.1—See Reference 3 for More Information

Size	Major Diameter d (in)	Coarse Threads—UNC			Fine Threads—UNF		
		Threads per inch	Minor Diameter d_r (in)	Tensile Stress Area A_t (in²)	Threads per inch	Minor Diameter d_r (in)	Tensile Stress Area A_t (in²)
0	0.0600	–	–	–	80	0.0438	0.0018
1	0.0730	64	0.0527	0.0026	72	0.0550	0.0028
2	0.0860	56	0.0628	0.0037	64	0.0657	0.0039
3	0.0990	48	0.0719	0.0049	56	0.0758	0.0052
4	0.1120	40	0.0795	0.0060	48	0.0849	0.0066
5	0.1250	40	0.0925	0.0080	44	0.0955	0.0083
6	0.1380	32	0.0974	0.0091	40	0.1055	0.0101
8	0.1640	32	0.1234	0.0140	36	0.1279	0.0147
10	0.1900	24	0.1359	0.0175	32	0.1494	0.0200
12	0.2160	24	0.1619	0.0242	28	0.1696	0.0258
1/4	0.2500	20	0.1850	0.0318	28	0.2036	0.0364
5/16	0.3125	18	0.2403	0.0524	24	0.2584	0.0581
3/8	0.3750	16	0.2938	0.0775	24	0.3209	0.0878
7/16	0.4375	14	0.3447	0.1063	20	0.3725	0.1187
1/2	0.5000	13	0.4001	0.1419	20	0.4350	0.1600
9/16	0.5625	12	0.4542	0.1819	18	0.4903	0.2030
5/8	0.6250	11	0.5069	0.2260	18	0.5528	0.2560
3/4	0.7500	10	0.6201	0.3345	16	0.6688	0.3730
7/8	0.8750	9	0.7307	0.4617	14	0.7822	0.5095
1	1.0000	8	0.8376	0.6057	12	0.8917	0.6630
1 1/8	1.1250	7	0.9394	0.7633	12	1.0167	0.8557
1 1/4	1.2500	7	1.0644	0.9691	12	1.1417	1.0729
1 3/8	1.3750	6	1.1585	1.1549	12	1.2667	1.3147
1 1/2	1.5000	6	1.2835	1.4053	12	1.3917	1.5810
1 3/4	1.7500	5	1.4902	1.8995			
2	2.0000	4.5	1.7113	2.4982			
2 1/4	2.2500	4.5	1.9613	3.2477			
2 1/2	2.5000	4	2.1752	3.9988			
2 3/4	2.7500	4	2.4252	4.9340			
3	3.0000	4	2.6752	5.9674			
3 1/4	3.2500	4	2.9252	7.0989			
3 1/2	3.5000	4	3.1752	8.3286			
3 3/4	3.7500	4	3.4252	9.6565			
4	4.0000	4	3.6752	11.0826			

14

Table 14-2   Principal Dimensions of ISO Metric Standard Screw Threads

Data Calculated from Equations 14.1—See Reference 4 for More Information

Major Diameter d (mm)	Coarse Threads			Fine Threads		
	Pitch p mm	Minor Diameter d_r (mm)	Tensile Stress Area A_t (mm^2)	Pitch p mm	Minor Diameter d_r (mm)	Tensile Stress Area A_t (mm^2)
3.0	0.50	2.39	5.03			
3.5	0.60	2.76	6.78			
4.0	0.70	3.14	8.78			
5.0	0.80	4.02	14.18			
6.0	1.00	4.77	20.12			
7.0	1.00	5.77	28.86			
8.0	1.25	6.47	36.61	1.00	6.77	39.17
10.0	1.50	8.16	57.99	1.25	8.47	61.20
12.0	1.75	9.85	84.27	1.25	10.47	92.07
14.0	2.00	11.55	115.44	1.50	12.16	124.55
16.0	2.00	13.55	156.67	1.50	14.16	167.25
18.0	2.50	14.93	192.47	1.50	16.16	216.23
20.0	2.50	16.93	244.79	1.50	18.16	271.50
22.0	2.50	18.93	303.40	1.50	20.16	333.06
24.0	3.00	20.32	352.50	2.00	21.55	384.42
27.0	3.00	23.32	459.41	2.00	24.55	495.74
30.0	3.50	25.71	560.59	2.00	27.55	621.20
33.0	3.50	28.71	693.55	2.00	30.55	760.80
36.0	4.00	31.09	816.72	3.00	32.32	864.94
39.0	4.00	34.09	975.75	3.00	35.32	1028.39

ever, it is more difficult to cut because of its perpendicular face. A modified square thread (not shown) is made with a 10° included angle to improve its manufacturability. The Acme thread of Figure 14-3*b* has a 29° included angle, making it easier to manufacture and also allowing the use of a split nut that can be squeezed radially against the screw to take up wear. An Acme stub thread (not shown) is also available, with teeth 0.3 p high instead of the standard 0.5 p. Its advantage is more uniform heat-treatability. The Acme thread is a common choice for power screws that must take loads in both directions. If the axial load on the screw is unidirectional, the buttress thread (Figure 14-3*c*) can be used to obtain greater strength at the root than either of the others shown. Table 14-3 shows some principal dimensions for standard Acme threads.

Power Screw Application

Figure 14-4 shows one possible arrangement of a power screw used as a jack to lift a load. The nut is turned by an applied torque T and the screw translates up to lift the load

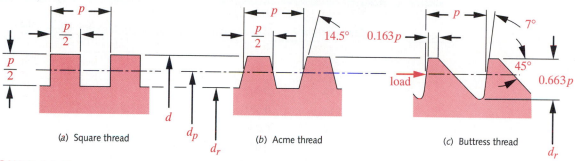

FIGURE 14-3

Square, Acme, and Buttress Threads

P or down to lower it. There needs to be some friction at the load surface to prevent the screw from turning with the nut. Once the load *P* is engaged, this is not a problem. Alternatively, the screw could be turned against a fixed nut to lift the load. In either case there will be significant friction between the screw and the nut as well as friction between the nut and base, requiring that a thrust bearing be provided as shown. If a plain (i.e., nonrolling) thrust bearing is used, it is possible for the bearing interface to generate larger friction torque than the threads. Ball-thrust bearings are often used in this application to reduce these losses.

Other applications for power screws are in linear actuators, which operate on the same principle as shown in Figure 14-4, but either motorize the nut rotation to translate the screw or motorize the screw rotation to translate the nut, as shown in Figure 14-5. These devices are used in machine tools to move the table and workpiece under the cutting tool, in assembly machines to position parts, and in aircraft to move the control surfaces, as well as in many other applications. If the rotating input is provided by a servomotor or stepping motor in combination with a precision lead screw, very accurate positioning can be obtained.

Power Screw Force and Torque Analysis

SQUARE THREADS   A screw thread is essentially an inclined plane that has been wrapped around a cylinder to create a helix. If we unwrapped one revolution of the helix, it would look like Figure 14-6a, which shows a block representing the nut being slid up the inclined plane of a square thread. The forces acting on the nut as a free-body diagram are also shown. Figure 14-6b shows the free-body diagram of the same nut as it slides down the plane. The friction force, of course, always opposes motion. The inclination of the plane is called the lead angle λ.

$$\tan \lambda = \frac{L}{\pi d_p} \qquad (14.3)$$

For the load-lifting case in Figure 14-6a, sum forces in *x* and *y* directions:

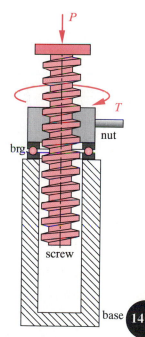

FIGURE 14-4

An Acme-Thread
Power-Screw Jack

14

Table 14-3   Principal Dimensions of American Standard Acme Threads

See Reference 2 for More Complete Dimensional and Tolerance Information

Major Diameter (in)	Threads per inch	Thread Pitch (in)	Pitch Diameter (in)	Minor Diameter (in)	Tensile Stress Area (in²)
0.250	16	0.063	0.219	0.188	0.032
0.313	14	0.071	0.277	0.241	0.053
0.375	12	0.083	0.333	0.292	0.077
0.438	12	0.083	0.396	0.354	0.110
0.500	10	0.100	0.450	0.400	0.142
0.625	8	0.125	0.563	0.500	0.222
0.750	6	0.167	0.667	0.583	0.307
0.875	6	0.167	0.792	0.708	0.442
1.000	5	0.200	0.900	0.800	0.568
1.125	5	0.200	1.025	0.925	0.747
1.250	5	0.200	1.150	1.050	0.950
1.375	4	0.250	1.250	1.125	1.108
1.500	4	0.250	1.375	1.250	1.353
1.750	4	0.250	1.625	1.500	1.918
2.000	4	0.250	1.875	1.750	2.580
2.250	3	0.333	2.083	1.917	3.142
2.500	3	0.333	2.333	2.167	3.976
2.750	3	0.333	2.583	2.417	4.909
3.000	2	0.500	2.750	2.500	5.412
3.500	2	0.500	3.250	3.000	7.670
4.000	2	0.500	3.750	3.500	10.321
4.500	2	0.500	4.250	4.000	13.364
5.000	2	0.500	4.750	4.500	16.800

$$\sum F_x = 0 = F - f\cos\lambda - N\sin\lambda = F - \mu N\cos\lambda - N\sin\lambda$$

$$F = N(\mu\cos\lambda + \sin\lambda) \tag{14.4a}$$

$$\sum F_y = 0 = N\cos\lambda - f\sin\lambda - P = N\cos\lambda - \mu N\sin\lambda - P$$

$$N = \frac{P}{(\cos\lambda - \mu\sin\lambda)} \tag{14.4b}$$

where μ is the coefficient of friction between screw and nut and the other variables are defined in Figure 14-6. Combine these equations to get an expression for the force F:

14

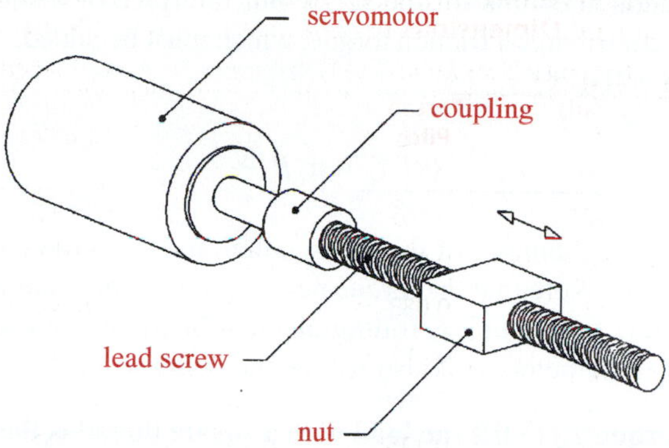

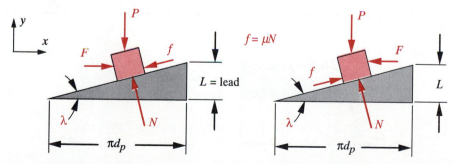

FIGURE 14-5

Servomotor-Driven Lead Screw for Use as a Positioning Device *Courtesy of J. Karsberg, Gillette Co. Inc.*

$$F = P\frac{(\mu \cos\lambda + \sin\lambda)}{(\cos\lambda - \mu \sin\lambda)} \tag{14.4c}$$

The screw torque T_{s_u} required to lift the load is

$$T_{s_u} = F\frac{d_p}{2} = \frac{Pd_p}{2}\frac{(\mu \cos\lambda + \sin\lambda)}{(\cos\lambda - \mu \sin\lambda)} \tag{14.4d}$$

It is sometimes more convenient to express this as a function of lead L rather than lead angle λ, so divide the numerator and denominator of in equation 14.4d by $\cos\lambda$ and substitute the right-hand side of equation 14.3 for $\tan\lambda$:

$$T_{s_u} = \frac{Pd_p}{2}\frac{(\mu\pi d_p + L)}{(\pi d_p - \mu L)} \tag{14.4e}$$

(a) Lifting the load up the plane   (b) Lowering the load down the plane

FIGURE 14-6

Force Analysis at the Screw-Nut Interface

This expression accounts for the screw-nut interface of a square thread, but the thrust collar also contributes a friction torque, which must be added. The torque required to turn the thrust collar is

$$T_c = \mu_c P \frac{d_c}{2} \tag{14.4f}$$

where d_c is the mean diameter of the thrust collar and μ_c is the coefficient of friction in the thrust bearing. Note that the torque needed to overcome collar friction can equal or exceed the screw torque unless rolling-element bearings are used in the thrust collar. Smaller collar diameters will also reduce the collar torque.

The **total torque** T_u to **lift the load** with a square thread is then

$$T_u = T_{s_u} + T_c = \frac{P d_p}{2} \frac{\left(\mu \pi d_p + L\right)}{\left(\pi d_p - \mu L\right)} + \mu_c P \frac{d_c}{2} \tag{14.4g}$$

The same analysis can be done for the case of lowering the load, as shown in Figure 14-6b. The applied- and friction-force signs change and the **torque** T_d to **lower the load** is

$$T_d = T_{s_d} + T_c = \frac{P d_p}{2} \frac{\left(\mu \pi d_p - L\right)}{\left(\pi d_p + \mu L\right)} + \mu_c P \frac{d_c}{2} \tag{14.4h}$$

ACME THREADS   The radial angle of an Acme (or other) thread introduces an additional factor in the torque equations. The normal force between screw and nut is angled in two planes, at the lead angle λ as shown in Figure 14-6, and also at the $\alpha = 14.5°$ angle of the Acme thread as shown in Figure 14-7. A similar derivation as done for the square thread will give expressions for lifting and lowering torques of

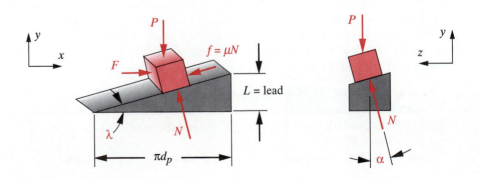

FIGURE 14-7

Force Analysis for an Acme-Thread Screw-Nut Interface

$$T_u = T_{S_u} + T_c = \frac{Pd_p}{2} \frac{\left(\mu \pi d_p + L \cos \alpha\right)}{\left(\pi d_p \cos \alpha - \mu L\right)} + \mu_c P \frac{d_c}{2} \qquad (14.5a)$$

$$T_d = T_{S_d} + T_c = \frac{Pd_p}{2} \frac{\left(\mu \pi d_p - L \cos \alpha\right)}{\left(\pi d_p \cos \alpha + \mu L\right)} + \mu_c P \frac{d_c}{2} \qquad (14.5b)$$

These equations reduce to those for the square thread when angle $\alpha = 0$.

Friction Coefficients

Experiments indicate that the coefficient of friction in an oil-lubricated thread-nut combination is about 0.15 ± 0.05.[5] The friction in a plain (nonrolling) thrust bearing is about the same as in the threads. Steel on bronze or steel on cast iron are common plain-bearing combinations. If a rolling-element bearing is used for the thrust washer, its coefficient of friction will be about 1/10 of the plain bearing values (i.e., 0.01 to 0.02).

Self-Locking and Back-Driving of Power Screws

Self-locking refers to a condition in which the screw cannot be turned by the application of any magnitude of force applied axially (not as a torque) to the nut. In other words, a self-locking screw will hold the load in place without any application of torque. It does not need a brake to hold the load. This is a very useful situation. For example, if you jacked up your car with a screw jack that was not self-locking, as soon as you let go of the jack handle the car would run the jack back down. You would have to be pretty fast with the lug wrench to change a tire in that case.

The opposite situation to self-locking is a screw that can be back driven, which means that pushing axially on the nut will cause the screw to turn. While of no value for a jack application, this is a useful feature in other situations. One example is a so-called *Yankee screwdriver*, which has a high-lead thread on its barrel that is attached to the blade. The handle is the nut. As you push down axially on the handle, the barrel turns, driving the wood screw into place. Any application in which you want to convert linear motion to rotary motion is a candidate for a back-drivable lead screw.

The condition of self-locking for a power or lead screw is easily predicted if the coefficient of friction in the screw-nut joint is known. The relationship between the friction coefficient and the screw's lead angle determines its self-locking condition. A screw will self-lock if

$$\mu \geq \frac{L}{\pi d_p} \cos \alpha, \qquad \text{or} \qquad \mu \geq \tan \lambda \cos \alpha \qquad (14.6a)$$

If it is a square thread $\cos \alpha = 1$, and this reduces to

14

$$\mu \geq \frac{L}{\pi d_p}, \quad \text{or} \quad \mu \geq \tan \lambda \qquad (14.6b)$$

Note that these relationships presume a static-loading situation. The presence of any vibration from dynamic loading or other sources can cause an otherwise self-locking screw to back down. Any vibrations that cause relative motion between screw and nut will inevitably cause slippage down the thread's incline.

Screw Efficiency

The efficiency of any system is defined as *work out / work in*. The work done on a power screw is the product of torque and angular displacement (in radians), which for one revolution of the screw is

$$W_{in} = 2\pi T \qquad (14.7a)$$

The work delivered over one revolution is the load-force times the lead:

$$W_{out} = PL \qquad (14.7b)$$

The efficiency is then

$$e = \frac{W_{out}}{W_{in}} = \frac{PL}{2\pi T} \qquad (14.7c)$$

Substituting equation 14.5a (neglecting the collar friction term) gives

$$e = \frac{L}{\pi d_p} \frac{\pi d_p \cos \alpha - \mu L}{\pi \mu d_p + L \cos \alpha} \qquad (14.7d)$$

This can be simplified by substituting equation 14.3:

$$e = \frac{\cos \alpha - \mu \tan \lambda}{\cos \alpha + \mu \cot \lambda} \qquad (14.7e)$$

Note that efficiency is a function only of the screw geometry and the coefficient of friction. For a square thread, $\alpha = 0$ and

$$e = \frac{1 - \mu \tan \lambda}{1 + \mu \cot \lambda} \qquad (14.7f)$$

Figure 14-8 shows plots of the efficiency function for an Acme thread over a range of coefficients of friction neglecting collar friction. Higher coefficients of friction reduce efficiency as should be expected. Note that the efficiency is zero when the lead angle $\lambda = 0$ because no useful work is being done to raise the load, but friction is still present. The efficiency also approaches zero at high lead angles because the torque is simply increasing the normal force (and thus friction) without any useful component to

14

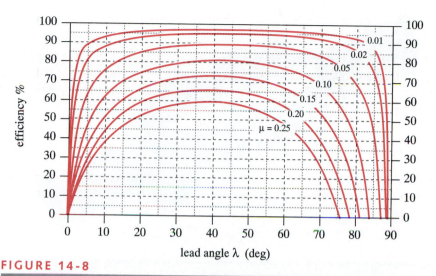

FIGURE 14-8

Efficiency for an Acme-Thread Power Screw (Neglecting Thrust-Collar Friction)

Table 14-4

Lead Angle and Efficiency for Standard Acme Threads with Coefficient of Friction $\mu = 0.15$

Size	Lead Angle (deg)	Effic- iency %
1/4 - 16	5.2	36
5/16 - 14	4.7	34
3/8 - 12	4.5	34
7/16 - 12	3.8	30
1/2 - 10	4.0	31
5/8 - 8	4.0	31
3/4 - 6	4.5	34
7/8 - 6	3.8	30
1 - 5	4.0	31
1 1/8 - 5	3.6	28
1 1/4 - 5	3.2	26
1 3/8 - 4	3.6	29
1 1/2 - 4	3.3	27
1 3/4 - 4	2.8	24
2 - 4	2.4	21
2 1/4 - 3	2.9	25
2 1/2 - 3	2.6	23
2 3/4 - 3	2.4	21
3 - 2	3.3	27
3 1/2 - 2	2.8	24
4 - 2	2.4	21
4 1/2 - 2	2.1	19
5 - 2	1.9	18

rotate the nut. The total efficiency including collar friction will be lower than is shown in Figure 14-8.

Figure 14-8 points out the major drawback of conventional power screws, their potentially low efficiency. Standard Acme screws have lead angles that vary between about 2 and 5°, as shown in Table 14-4. This puts them at the extreme left of the set of curves in Figure 14-8. The efficiencies of standard Acme screws for an assumed coefficient of friction of 0.15 are seen in Table 14-4 to vary between 18 and 36%. If the thread friction can be reduced, significant efficiency increases can be realized.

Ball Screws

A significant reduction in thread friction can be obtained with the use of ball screws, which use a train of ball bearings in the nut to create an approximate rolling contact with the screw threads, as shown in Figure 14-9. The thread form is shaped to fit the spherical balls and is usually hardened and ground for long life. The friction coefficient is similar to that of conventional ball bearings, putting them in the range of the top two curves in Figure 14-8, with correspondingly higher efficiency.

The low friction of ball screws makes them back-drivable and thus **not** self-locking. Thus a brake must be used to hold a load driven by a ball screw. So, ball screws can be used to convert linear to rotary motion. They have very high load capacity, typically larger than a conventional screw of the same diameter and they do not suffer from the stick-slip characteristics of sliding joints.

14

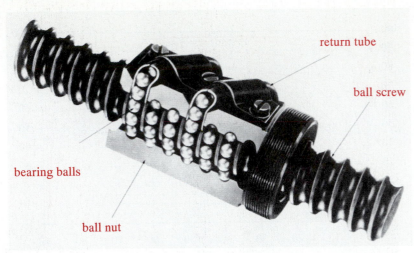

FIGURE 14-9

A Ball Screw and Ball Nut *Courtesy of Thompson-Saginaw Ball Screw Co., Saginaw, Mich.*

Ball screws are used in many applications from aircraft control-surface and landing-gear actuators to machine-tool controls, automotive steering mechanisms, and hospital bed mechanisms. Many manufacturers supply ball-screw assemblies and they should be consulted for technical information regarding their proper application.

EXAMPLE 14-1

Torque and Efficiency of a Power Screw

Problem	Determine the lifting and lowering torques and the efficiency of the power screw shown in Figure 14-4 using an Acme screw and nut. Is it self-locking? What is the contribution of collar friction versus screw friction if the collar has *a*) sliding friction, *b*) rolling friction?
Given	The screw is a single-start Acme 1.25–5. The axial load is 1 000 lb. The mean collar diameter is 1.75 in.
Assumptions	The screw and nut are lubricated with oil. Sliding friction μ = 0.15, Rolling friction μ = 0.02.
Solution	See *TKSolver* files EX14-01a and EX14-01b.

1   There are several aspects to this problem. We need to compute the lifting and lowering torques for two cases, one with a sliding-friction collar and one with a ball-bearing collar. In both cases we will calculate the screw and collar contributions to torque and efficiency separately for comparison and also combine them. First do the case of the sliding collar.

14

2   Since it is a single-start thread, the lead L equals the pitch p, which is $1/N = 0.2$. The pitch diameter of the thread d_p is found in Table 14-3. The torque to lift the load is found from equation 14.5a:

$$T_u = T_{s_u} + T_c = \frac{Pd_p}{2} \frac{\left(\mu \pi d_p + L\cos\alpha\right)}{\left(\pi d_p \cos\alpha - \mu L\right)} + \mu_c P \frac{d_c}{2}$$

$$= \frac{1\,000(1.15)}{2} \frac{(0.15\pi(1.15) + 0.2\cos14.5)}{\left(\pi(1.15)\cos14.5 - 0.15(0.2)\right)} + 0.15(1\,000)\frac{1.75}{2} \qquad (a)$$

$$T_u = 122.0 + 131.2 = 253.2 \ \text{in - lb}$$

Note that the collar friction exceeds the screw friction.

3   The torque to lower the load is found from equation 14.5b:

$$T_d = T_{s_d} + T_c = \frac{Pd_p}{2} \frac{\left(\mu \pi d_p - L\cos\alpha\right)}{\left(\pi d_p \cos\alpha + \mu L\right)} + \mu_c P \frac{d_c}{2}$$

$$= \frac{1\,000(1.15)}{2} \frac{(0.15\pi(1.15) - 0.2\cos14.5)}{\left(\pi(1.15)\cos14.5 - 0.15(0.2)\right)} + 0.15(1\,000)\frac{1.75}{2} \qquad (b)$$

$$T_d = 56.8 + 131.2 = 188.0 \ \text{in - lb}$$

4   The efficiency in the lifting mode will be less than for lowering and is found from equations 14.7. We choose the version shown as equation 14.7c in order to account for both screw and collar components.

$$e = \frac{PL}{2\pi T}$$

for the screw $$e_{screw} = \frac{1\,000(0.2)}{2\pi(122.0)} = 0.26 \qquad (c)$$

for both combined $$e = \frac{1\,000(0.2)}{2\pi(253.2)} = 0.13$$

5   Now recalculate the collar torque and total torque to lift the load with a ball-bearing thrust washer, using equation 14.4f.

$$T_c = \mu_c P \frac{d_c}{2} = 0.02(1\,000)\frac{1.75}{2} = 17.5 \ \text{in - lb} \qquad (d)$$

$$T_u = T_{s_u} + T_c = 122.0 + 17.5 = 139.5 \ \text{in - lb} \qquad (e)$$

6   The efficiency with the ball bearing thrust-washer is now:

$$e = \frac{PL}{2\pi T} = \frac{1\,000(0.2)}{2\pi(139.5)} = 0.23 \qquad (f)$$

The improvement in efficiency is significant and shows why it is poor practice to use anything but a rolling-element bearing as a thrust washer on a power screw.

14

7   The self-locking aspects of the screw are independent of the thrust-collar friction and can be found from equation 14.6a.

$$\mu \geq \frac{L}{\pi d_p} \cos \alpha$$

$$0.15 \geq \frac{0.2}{\pi(1.15)} \cos(14.5°) \qquad\qquad (g)$$

$$0.15 \geq 0.06 \qquad\qquad \text{so it self - locks}$$

Note that the positive lowering torque in step 3 also indicates that the screw is self-locking. A negative lowering torque means that a braking torque of opposite sense to the lift torque must be applied to hold the load.

14.3   STRESSES IN THREADS

A similar situation exists in mating threads as in mating gear teeth. Section 11.4 discusses the desirability of having multiple gear teeth in contact in a gearmesh (contact ratio > 1) and Figure 11-19 shows a gearmesh in which two teeth are taking all the load due to inaccuracies in tooth spacing despite an apparently large contact ratio. When a nut engages a thread, theoretically all the threads in engagement should share the load. In actuality, inaccuracies in thread spacing, cause virtually all the load to be taken by the first pair of threads. Thus, the conservative approach in calculating thread stresses is to assume the worst case of one thread-pair taking the entire load. The other extreme is to assume that all the engaged threads share the load equally. Both of these assumptions can be used to calculate estimated thread stresses. The true stress will be between these extremes, but most likely closer to the one-thread assumption. Power screws and fasteners for high-load applications are usually made of high-strength steel and are often hardened. Power-screw nuts may also be of hardened material for strength and wear resistance. Fastener nuts on the other hand are often made of soft material and are thus typically weaker than the screw. This promotes local yielding in the nut threads when the fastener is tightened, which can improve the thread fit and promote load-sharing between threads. Hardened nuts are used on high-strength, hardened bolts.

Axial Stress

A power screw can see axial loads of either tension or compression. A threaded fastener typically sees only axial tension. The tensile stress area of a screw thread was discussed earlier and is defined in equation 14.1 and Tables 14-1, 14-2, and 14-3 for various types of threads. Equation 14.2 can be used to compute the axial tension stress in a screw. For power screws loaded in compression, the possibility of column buckling must be investigated using the methods outlined in Section 4.15. Use the screw's minor diameter to compute the slenderness ratio.

14

Shear Stress

One possible shear-failure mode involves stripping of the threads either out of the nut or off the screw. Which, if either, of these scenarios occurs is dependent on the relative strengths of the nut and screw materials. If the nut material is weaker (as is often the case), it may strip its threads at its major diameter. If the screw is weaker, it may strip its threads at its minor diameter. If both materials are of equal strength, the assembly may strip along the pitch diameter. In any event, we must assume some degree of load sharing among the threads in order to calculate a stress. One approach is to consider that since complete failure requires all threads to strip, all can be considered to share the load equally. This is probably a good assumption as long as the nut or screw (or both) is ductile to allow each thread to yield as the assembly begins to fail. However, if both parts are brittle (e.g., high-hardness steels or cast iron) and the thread fit is poor, one can envision each thread taking the entire load in turn until it fractures and passes the job to the next thread. Reality is again somewhere between these extremes. If we express the shear area in terms of number of threads in engagement, a judgment can be made in each case as to what degree of load sharing is appropriate.

The stripping-shear area A_s for one screw thread is the area of the cylinder of its minor diameter d_r:

$$A_s = \pi d_r w_i p \qquad (14.8a)$$

where p is the thread pitch and w_i is a factor that defines the percentage of the pitch occupied by metal at the minor diameter. Values of w_i for several common thread forms are shown in Table 14-5. The area for one thread pitch from equation 14.8a can be multiplied by all, one, or some fraction of the total number of threads in engagement based on the designer's judgment of the factors discussed above for the particular case.

For the nut stripping at its major diameter, the shear area for one screw thread is

$$A_s = \pi d w_o p \qquad (14.8b)$$

where the value for w_o at the major diameter is found in Table 14-5.

The shear stress for thread stripping τ_s is then found from:

$$\tau_s = \frac{F}{A_s} \qquad (14.8c)$$

MINIMUM NUT LENGTH   If the nut is long enough, the load required to strip the threads will exceed the load needed to fail the screw in tension. The equations for both modes of failure can be combined and a minimum nut length computed for any particular screw size. For any UNS/ISO threads or Acme threads of $d \leq 1$ in, a nut length of at least $0.5\,d$ will have a strip strength in excess of the screw's tensile strength. For larger-diameter Acme threads, the strip-strength of a nut with length $\geq 0.6\,d$ will exceed the screw's tensile strength. These figures are valid only if the screw and nut are of the same material, which is usually the case.

Table 14-5

Area Factors for Thread Stripping Shear Area

Thread Type	w_i (minor)	w_o (major)
UNS/ISO	0.80	0.88
Square	0.50	0.50
Acme	0.77	0.63
Buttress	0.90	0.83

14

MINIMUM TAPPED HOLE ENGAGEMENT   When a screw is threaded into a tapped hole rather than a nut, a longer thread engagement is needed. For same-material combinations, a thread engagement length at least equal to the nominal thread diameter d is recommended. For a steel screw in cast iron, brass, or bronze, use $1.5d$. For steel screws in aluminum, use $2d$ of minimum thread-engagement length.

Torsional Stress

When a nut is tightened on a screw, or when a torque is transmitted through a power-screw nut, a torsional stress can be developed in the screw. The torque that twists the screw is dependent on the friction at the screw-nut interface. If the screw and nut are well lubricated, less of the applied torque is transmitted to the screw and more is absorbed between the nut and the clamped surface. If the nut is rusted to the screw, all the applied torque will twist the screw, which is why rusty bolts usually shear even when you attempt to loosen the nut. In a power screw, if the thrust collar has low friction, all the applied torque at the nut will create torsional stress in the screw (since little torque is taken to ground through the low-friction collar). Thus, to accommodate the worst case of high thread friction, use the total applied torque in the equation for torsional stress in a round section (see Section 4.11):

$$\tau = \frac{Tr}{J} = \frac{16T}{\pi d_r^3} \qquad (14.9)$$

The minor diameter d_r of the thread should be used in this calculation.

14.4   TYPES OF SCREW FASTENERS

There is a wide variety of screw styles available, many of which are for specialized applications. Conventional bolts and nuts typically use standard threads as defined in Section 14.1. Variations in the standard thread are possible among certain varieties of screws, especially those intended for self-tapping applications. Fasteners can be classified in different ways: by their intended use, by thread type, by head style, and by their strength. Fasteners of all types are available in a variety of materials including steel, stainless steel, aluminum, brass, bronze, and plastics.

Classification by Intended Use

BOLTS AND MACHINE SCREWS   The same fastener may take on a different name when used in a particular manner. For example, a **bolt** is a fastener with a head and a straight, threaded shank intended to be used with a **nut** to clamp an assembly together. However, the same fastener is called a **machine screw** or **cap screw** when it is threaded into a tapped hole rather than used with a nut. This is only a semantic distinction, but one in which some purists take great stock. The ANSI standards distinguish between a bolt

14

and a screw by noting that a bolt is intended to be held stationary while a nut is torqued onto it to make the joint, whereas a screw is intended to be turned into its receptacle, be that a tapped or untapped hole, by torquing its head. (Nevertheless, it is not yet illegal in most states to place a nut on a machine screw, but be forewarned—you will change it instantly into a bolt by doing so!)

STUDS   A **stud** is a headless fastener, threaded on both ends and intended to be semipermanently threaded into one-half of an assembly. A hole in the mating part then drops over the protruding stud and is secured with a nut. Each end of the stud can have either the same or different pitch thread. The permanent end sometimes has a higher-class thread fit to grip the tapped hole and resist loosening when the nut is removed from the top half.

Figure 14-10 shows a bolt (with nut and washer), a machine screw, and a stud. Another distinction between screws and bolts is that a bolt has only straight, uniform threads whereas a screw can have any thread form, including tapered or interrupted as shown in Figure 14-11. Thus, there are **wood screws** but not "wood bolts" (though *carriage bolts* are used to fasten wood assemblies).

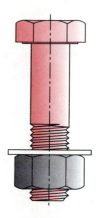

(a) Bolt, washer, and nut

Classification by Thread Type

TAPPING SCREWS   All fasteners intended to make their own hole or make their own threads are called **tapping screws** as in *self-tapping, thread-forming, thread-cutting,* and *self-drilling screws.* Figure 14-11 shows a selection of the many thread types available in tapping screws. Tapping-screw threads are similar to the standard form but are often spaced further apart (i.e., larger pitch) for use in sheet metal or plastic to allow the displaced material a place to go as the screw forces its way into a small pilot hole while forming the threads. Thread-cutting screws have a standard thread form but are relieved with axial grooves and hardened to provide a cutting edge to tap the part as the screw is inserted. Self-drilling screws (not shown) have a drill-bit shape at their tip to make the pilot hole. They also form the threads as they are driven in.

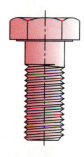

(b) Machine (cap) screw

Classification by Head Style

SLOTTED SCREWS   Many different head styles are made including straight-slot, cross-slot (Phillips), hexagonal, hexagonal socket, and others. Head shape can be round, flat (recessed), filister, pan, etc., as shown in Figure 14-12. These head styles in combination with slotted or Phillips grooves are typically used only on smaller machine or tapping screws as the maximum torque obtainable with these slots is limited. The torques needed with larger screws are more readily obtained with hexagonal heads or hexagonal-socket heads, as shown in Figures 14-10 and 14-13. Hex heads are the most popular style for larger bolts and machine screws that require substantial torque unless space is limited, in which case the socket-head cap screw is a better choice.

14

(c) Stud

FIGURE 14-10

Bolt and Nut, Machine Screw and Stud

(a) Socket head

(b) Socket flat head

(c) Socket button head

(d) Shoulder screw

(e) Socket set screw

FIGURE 14-13

Various Styles of
Socket-Head Cap Screws
*Courtesy of Cordova Bolt
Inc., Buena Park, Calif.*

14

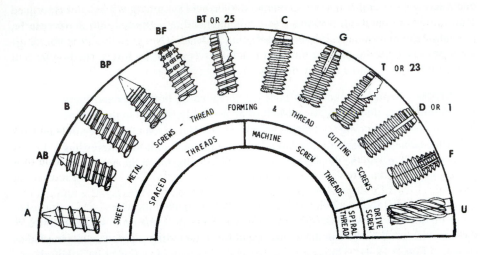

FIGURE 14-11

Various Styles of Threads Used on Tapping Screws *Courtesy of Cordova Bolt Inc., Buena Park, Calif., 90621*

SOCKET-HEAD CAP SCREWS as shown in Figure 14-13 are typically made of high-strength, hardened steel, stainless steel, or other metals, and are used extensively in machinery. The hex socket allows sufficient torque to be applied with special hexagonal Allen wrenches. The standard socket head style (Figure 14-13*a*) is designed to be placed in a counterbore so that the head is flush-to-below the surface. The flat-head cap screw (Figure 14-13*b*) is countersunk flush. The shoulder screw (Figure 14-13*d*) has a close-

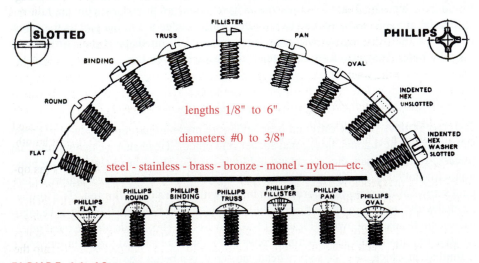

FIGURE 14-12

Various Styles of Heads Used on Small Machine Screws *Courtesy of Cordova Bolt Inc., Buena Park, Calif.*

tolerance, ground-finish shank that can be used as a pivot or to accurately locate a part. The screw is torqued tightly against its shoulder and yet a properly sized part can be free to pivot between its head and the surface into which it is screwed. Set screws (Figure 14-13e) are used to lock collars and hubs to shafts, as described in Chapter 9.

Nuts and Washers

NUTS   Figure 14-14 shows a sample of the large variety of nuts available. The jam nut is a thinner version of the standard hex nut and is used in combination with a standard nut to lock the standard nut to the bolt. The castle nut has grooves for insertion of a cotter pin through a cross-hole in the bolt to prevent the nut from working loose. An acorn nut is used for decorative purposes and a wing nut allows removal without tools.

LOCK NUTS   A universal concern is the prevention of spontaneous loosening of nuts due to vibration. Two nuts jammed together on the bolt or a castle nut with cotter pin both reliably achieve this goal. Many other proprietary designs of lock nuts are offered by manufacturers. A few are shown in Figure 14-15. The elliptical lock nut has its top few threads upset into an elliptical shape after the nut is made. These threads provide interference with those on the bolt and, when forced on, grip the thread and resist loosening. Nuts are also made with nylon inserts in the threads that deform when forced onto the bolt. The nylon flows into the thread clearances and grips the bolt. A pin lock nut has a steel pin that allows tightening, but digs into the bolt threads to prevent loosening. Nuts are also made with serrations on one face that act to dig into the clamped part and resist loosening.

WASHERS   A plain washer is simply a flat, doughnut-shaped part that serves to increase the area of contact between the bolt head or nut and the clamped part (see Figure 14-10). Hardened-steel washers are used where the bolt compression load on the clamped part needs to be distributed over a larger area than the bolt head or nut provides. A soft washer will yield in bending rather than effectively distribute the load. Any plain washer also prevents marring of the part surface by the nut when it is tightened. Nonmetallic washers are used when electrical insulation of the bolt from the part is required. Flat-washer sizes are standardized to bolt size (see reference 2). If washers larger than the standard diameter are needed, **fender washers** (which have larger outside diameter) can be used. **Belleville washers** (see Section 13.9) are sometimes used under nuts or screw heads to provide a controlled axial force over changes in bolt length.

LOCK WASHERS   To help prevent spontaneous loosening of standard nuts (as opposed to locknuts) lock washers can be used under the nut of a bolt or under the head of a machine screw. Figure 14-16 shows several of the many styles of lock washers available. The split washer is hardened steel and acts as a spring under the nut. Its sharp corners also tend to dig into the clamped surfaces. Various styles of toothed washers are also offered. Their turned-up teeth are compressed when clamped, and dig into the nut and part surfaces. Lock washers are generally considered to be less effective in preventing loosening than locknuts, which are preferred.

(a) Standard hex nut

(b) Hex jam nut

(c) Hex castle nut

(d) Hex acorn nut

(e) Wing nut

14

FIGURE 14-14

Various Styles of Standard Nuts
Courtesy of Cordova Bolt Inc., Buena Park, Calif., 90621

(a) Split lock washer

(b) Internal-tooth washer

(c) External-tooth washer

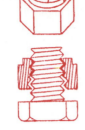

(d) Internal-external tooth

(e) Countersunk tooth

FIGURE 14-16

Types of Lock Washers
Courtesy of Cordova Bolt Inc., Buena Park, Calif.

(a) Elliptical lock nut   (b) Nylon insert lock nut   (c) Pin lock nut   (d) Flange lock nut

FIGURE 14-15

A Small Sample of the Types of Lock Nuts Available *Courtesy of Cordova Bolt Inc., Buena Park, Calif.*

SEMS   are combinations of nuts and captive lockwashers that remain with the nut. Many styles are made, one of which is shown in Figure 14-17. Their main advantage is to ensure that the lockwasher will not be left out at assembly or reassembly.

14.5   MANUFACTURING FASTENERS

THREAD CUTTING   Several techniques are available for making threads. Internal threads are usually cut with a special tool called a **tap** that has the desired thread form and looks like a screw. A tap is made of hardened tool steel and has axial grooves that interrupt its threads to provide cutting edges in the shape of the threads. A pilot hole is drilled with a proper-size tap drill and the lubricated tap is turned slowly into the hole while being advanced at a suitable rate. Nuts too large to tap are threaded in a lathe with a thread-shaped, single-point tool that is advanced axially through the hole by a lead screw to control its lead and pitch. External threads can also be cut with a single-point tool in a lathe or alternatively with a **die**, which is the external-thread equivalent of a tap. The rod to be threaded is the same size as the outside diameter of the thread. Specialized machines called *screw machines* are used to produce screws, bolts, and nuts (as well as other turned parts) in high quantity at low cost. All threads made by the methods described above are classified as **cut threads**.

THREAD ROLLING   Another, and superior, method for making external threads is by **thread-rolling**, also called **thread-forming**. Hardened-steel dies in the form of threads are forced into the surface of the rod being threaded. The dies cold-flow the material into the thread shape. The final outside diameter of the thread is larger than the initial diameter of the rod because material is forced out of the roots and into the crests of the threads.

There are several advantages of rolling versus cutting threads. The cold forming work-hardens and strengthens the material, creates radii at root and crest, and introduces favorable compressive residual stresses at the thread roots. The disruption of the material's shape into the thread form causes a reorientation of the material's grain to the thread shape. In contrast, thread cutting interrupts the grain. All of these factors contribute to a significant increase in the strength of rolled threads compared to cut threads. In addition to their better strength, rolled threads have less waste than cut threads as no material is removed, and the blank is consequently smaller in volume. High-strength fasteners are usually hardened steel. Thread-rolling should be done after hardening the bolt, if possible, as the thermal hardening process will relieve the desirable residual stresses introduced by rolling.

FIGURE 14-17

Nut with Captive Lock Washer (SEM) *Courtesy of Cordova Bolt Inc., Buena Park, Calif., 90621*

Figure 14-18 shows the profiles and grain structure of cut and rolled threads. In any application where the loads on fasteners are high or where fatigue loads are present, rolled threads should always be used. In noncritical or lightly loaded applications, the weaker and less-expensive cut threads may be used.

HEAD FORMING   The heads of bolts and screws are typically cold-formed in an *up-setting* process. To picture this process, imagine taking a rod of modeling clay in your hand, leaving a short length sticking up above your fist. Smack the top of the clay rod axially with your other hand while tightly gripping the rod in your fist and you will mushroom the rod-end into a shorter but larger-diameter head. In similar fashion, the shank of the *bolt-to-be* is gripped tightly in the cold-heading machine with the appropriate length sticking out. A die of the desired head diameter surrounds this exposed end. When the hammer comes down, it cold-flows the material into a round head. Similar improvements in grain orientation in the head are achieved as were described above for thread-rolling. Bolts over about 3/4-in diameter must be heated before being headed. Hexagonal sockets and Phillips slots are formed in the cold (or hot) heading process. Hexagonal flats or screw slots are later machined into the head.

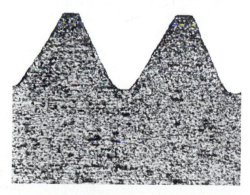

(a) Cut threads

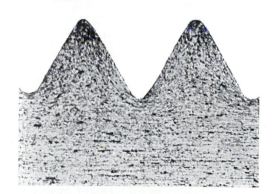

(b) Rolled threads

FIGURE 14-18

Grain Structure of Cut and Rolled Threads   Source: R. D. Barer and B. F. Peters, *Why Metals Fail*, p. 23, Gordon and Breach, New York, 1970

14

14.6 STRENGTHS OF STANDARD BOLTS AND MACHINE SCREWS

Bolts and screws for structural or heavily loaded applications should be selected based on their proof strength S_p as defined in SAE, ASTM, or ISO specifications. These organizations define bolt grades or classes that specify material, heat treatment, and a **minimum proof strength** for the bolt or screw. **Proof strength S_p** is *the stress at which the bolt begins to take a permanent set*, and is close to, but lower than, the yield strength of the material. The grade or class of each bolt is indicated by marks (or their absence) on the head. Table 14-6 shows strength information for several SAE grades of bolts and Table 14-7 shows similar information for metric bolts. The head markings for each grade or class are shown in Figures 14-19 and 14-20.

14.7 PRELOADED FASTENERS IN TENSION

One of the primary applications of bolts and nuts is clamping parts together in situations where the applied loads put the bolt(s) in tension, as shown in Figure 14-21. It is common practice to preload the joint by tightening the bolt(s) with sufficient torque to create tensile loads that approach their proof strength. For statically loaded assemblies, a preload that generates bolt stress as high as 90% of the proof strength is sometimes used. For dynamically loaded assemblies (fatigue loading) a preload of 75% or more of proof strength is commonly used. Assuming that the bolts are suitably sized for the applied loads, these high preloads make it very unlikely that the bolts will break in service if they do not break while being tensioned (tightened). The reasons for this are subtle and require an understanding of how the elasticities of the bolt and the clamped members interact when the bolt is tightened and when the external load is later applied.

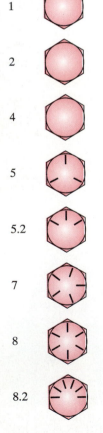

SAE
Grade

1

2

4

5

5.2

7

8

8.2

FIGURE 14-19

Head Marks for SAE Bolts

Table 14-6   SAE Specifications and Strengths for Steel Bolts

SAE Grade Number	Size Range Outside Diameter (in)	Minimum Proof Strength (kpsi)	Minimum Yield Strength (kpsi)	Minimum Tensile Strength (kpsi)	Material
1	0.25-1.5	33	36	60	low or medium carbon
2	0.25-0.75	55	57	74	low or medium carbon
2	0.875-1.5	33	36	60	low or medium carbon
4	0.25-1.5	65	100	115	medium carbon, cold drawn
5	0.25-1.0	85	92	120	medium carbon, Q&T*
5	1.125-1.5	74	81	105	medium carbon, Q&T
5.2	0.25-1.0	85	92	120	low carbon martensite, Q&T
7	0.25-1.5	105	115	133	medium carbon alloy, Q&T
8	0.25-1.5	120	130	150	medium carbon alloy, Q&T
8.2	0.25-1.0	120	130	150	low carbon martensite, Q&T

* Quenched and Tempered.

S_p σ_{yp} σ_{ult}

Table 14-7 Metric Specifications and Strengths for Steel Bolts

Class Number	Size Range Outside Diameter (mm)	Minimum Proof Strength (MPa)	Minimum Yield Strength (MPa)	Minimum Tensile Strength (MPa)	Material
4.6	M5-M36	225	240	400	low or medium carbon
4.8	M1.6-M16	310	340	420	low or medium carbon
5.8	M5-M24	380	420	520	low or medium carbon
8.8	M16-M36	600	660	830	medium carbon, Q&T
9.8	M1.6-M16	650	720	900	medium carbon, Q&T
10.9	M5-M36	830	940	1 040	low carbon martensite, Q&T
12.9	M1.6-M36	970	1 100	1 220	alloy, quenched & tempered

Metric Class

 4.6

 4.8

 5.8

 8.8

 9.8

 10.9

 12.9

FIGURE 14-20

Head Marks—Metric Bolts

Figure 14-22 shows a bolt clamping a spring, which is the analogue of the clamped material in Figure 14-21. Whatever material is clamped, it will have a spring constant and will compress when the bolt is tightened. (The bolt, also being elastic, will stretch when tightened.) We show a spring as the clamped material in Figure 14-22a in order to exaggerate its compression for illustration purposes. For the same reason, we also postulate an unusual method for tensioning this particular bolt. It seems that we misplaced our wrench, and so had to ask Crusher Casey to grab that nut and pull it down with 100 lb of force while we stuck a scrap of steel between the nut and the ground plane to serve as a stop as shown in (b). The bolt now has 100 lb of tensile preload in it and the spring (i.e., the material) has 100 lb of compressive preload. This preload remains in the assembly even after Crusher has let go (c). The situation depicted in (c) is identical to that which would result if the nut had been tightened conventionally to compress the spring the same amount.

Figure 14-22d shows a new load of 90 lb applied to the bolt. Note that the tension in the bolt is still 100 lb and will be so regardless of the external load applied until that

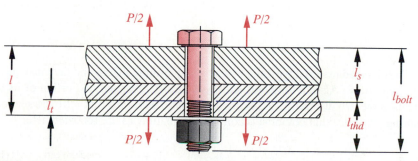

FIGURE 14-21

A Bolted Assembly in Tension

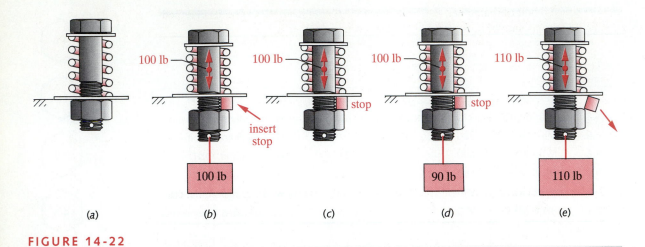

FIGURE 14-22

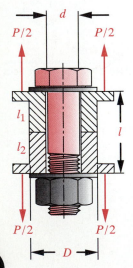

FIGURE 14-23

A Preloaded Bolt
Compressing a Cylinder
to Which External Loads
are Applied

load exceeds the preload of 100 lb in this case. Figure 14-22*e* shows that a load larger than the preload further compresses the spring, breaking the contact between nut and ground plane, and the bolt tension is now equal to the new applied load of 110 lb. When the bolt and the material separate as in (*e*) the bolt takes the full amount of the applied load. This diagram hints at why the presence of a preload is advantageous, especially when the applied loads are varying with time. To fully understand why requires further examination of the joint's elastic behavior under loading.

Figure 14-23 shows a bolt clamping a cylinder of known cross section and length. We wish to examine the loads, deflections, and stresses in both bolt and cylinder under preload and after an external load is applied. The spring constant of a bar in tension is found from the equation for the deflection of a tension bar:

$$\delta = \frac{Fl}{AE} \tag{14.10a}$$

and

$$k = \frac{F}{\delta} = \frac{AE}{l} \tag{14.10b}$$

The clamped material typically contains two or more pieces and they may be of different materials. Also, a long bolt will have threads over only a portion of its length and thus have two different cross-sectional areas. These different-stiffness sections act as springs in series that combine according to equation 13.2*b*, repeated here:

$$\frac{1}{k_{total}} = \frac{1}{k_1} + \frac{1}{k_2} + \frac{1}{k_3} \cdots + \frac{1}{k_n} \tag{13.2b}$$

For a round bolt of diameter d and axially loaded thread length l_t within its clamped zone of length l as shown in Figure 14-21 , the spring constant is

$$\frac{1}{k_b} = \frac{l_t}{A_t E_b} + \frac{l - l_t}{A_b E_b} = \frac{l_t}{A_t E_b} + \frac{l_s}{A_b E_b} \tag{14.11a}$$

where A_b is the total cross-sectional area and A_t is the tensile stress area of the bolt, and $l_s = (l - l_t)$ is the length of the unthreaded shank. The length of the threaded portion is standardized as twice the bolt diameter plus 1/4 in for bolts up to 6 in long. An additional 1/4 in of thread is provided on longer bolts. Bolts shorter than the standard thread length are threaded as close to the head as possible.[2]

For the cylindrical material geometry in Figure 14-23 (ignoring the flanges), the material spring constant becomes

$$\frac{1}{k_m} = \frac{l_1}{A_{m_1} E_1} + \frac{l_2}{A_{m_2} E_2} = \frac{4 l_1}{\pi D_{eff_1}^2 E_1} + \frac{4 l_2}{\pi D_{eff_1}^2 E_2} \tag{14.11b}$$

where the A_m are the effective areas of the clamped materials and the D_{eff} are the effective diameters of those areas.

If both clamped materials are the same

$$k_m = \frac{A_m E_m}{l} \tag{14.11c}$$

where A_m is the effective area of the clamped material (also see Section 14.8). If A_m can be defined as a solid cylinder with an effective diameter D_{eff}, equation 14.11c becomes

$$k_m = \frac{\pi D_{eff}^2}{4} \frac{E_m}{l} \tag{14.11d}$$

Preloaded Bolts Under Static Loading

Figure 14-24a plots the load-deflection behavior of both bolt and material on common axes with their initial length taken as zero deflection δ. Note that the slope of the bolt line is positive because its length increases with increased force. The slope of the material line is negative as its length decreases with increasing force. The material is shown stiffer than the bolt since its area is typically larger and we are assuming the same material for both. The force in both bolt and material is the same as long as they remain in contact. As a preload force F_i is introduced by tightening the bolt, the deflections of bolt δ_b and material δ_m are controlled by their spring rates and reach points A and B on their respective load-deflection curves as shown in Figure 14-24a. With our assumptions for the relative magnitudes of k_b and k_m, the bolt stretches more (δ_b) than the material compresses (δ_m).

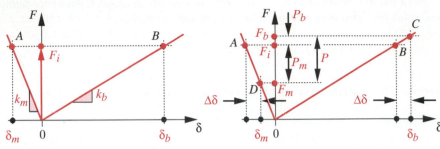

(a) Preload force and initial deflections   (b) Load deflection and resulting forces

FIGURE 14-24

Effects on Bolt and Material from Preload (a) Preload and (b) Applied Load

When an external load P is applied to the joint of Figure 14-23, there is an additional deflection $\Delta\delta$ introduced to both bolt and material as shown in Figure 14-24b. This deflection must be the same in both the bolt and the material unless the applied load is large enough to separate the joint (i.e., $P_m > F_i$ as shown in Figure 14-22e). The additional deflection $\Delta\delta$ creates a new load situation in both the bolt and material as shown in Figure 14-24b. The load in the material is reduced by P_m and moves down the material stiffness line to point D with a new value F_m. The load in the bolt is increased by P_b and moves up the bolt stiffness line to point C with a new value F_b. Note that the applied load P is split into two components, one (P_m) taken by the material and one (P_b) taken by the bolt.

$$P = P_m + P_b \tag{14.12a}$$

The load F_m in the material is now

$$F_m = F_i - P_m \tag{14.12b}$$

and the load F_b in the bolt becomes

$$F_b = F_i + P_b \tag{14.12c}$$

Note what has happened as a result of the preload force F_i. The "spring" of the material was "wound up" under preloading. Any applied loads are partially supported by the "unwinding" of this spring. If the relative stiffness of bolt and material are as shown in Figure 14-24 (i.e., material stiffer than bolt), the material supports the majority of the applied load and the bolt feels little additional load above that of the initial preload. This is one aspect of the justification for the earlier statement that "if the bolt doesn't fail when preloaded, it probably won't fail in service." There is also another reason for this being true and that will be discussed in a later section.

Note however, that if the applied load P is large enough to cause the component P_m to exceed the preload force F_i, then the joint will separate and the bolt will feel the full value of the applied load P. The material can no longer contribute to supporting the load if the joint is separated. This is one reason for the very large recommended preloads as a percentage of bolt proof strength. In order to get the full benefit of material load sharing, the preload should be high.

We can summarize the information in Figure 14-24 in the following way. The common change in deflection $\Delta\delta$ due to the applied load P is

$$\Delta\delta = \frac{P_b}{k_b} = \frac{P_m}{k_m} \tag{14.13a}$$

or :
$$P_b = \frac{k_b}{k_m} P_m \tag{14.13b}$$

Substitute equation 14.12a to get

$$P_b = \frac{k_b}{k_m + k_b} P$$

or
$$\tag{14.13c}$$

$$P_b = CP \qquad \text{where} \quad C = \frac{k_b}{k_m + k_b}$$

The term C is called the joint's *stiffness constant* or just the **joint constant**. Note that C is typically < 1 and if k_b is small compared to k_m, C will be a small fraction. This confirms that the bolt will only see a portion of the applied load P.

In like fashion,

$$P_m = \frac{k_m}{k_b + k_m} P = (1 - C)P \tag{14.13d}$$

These expressions for P_b and P_m can be substituted into equations 14.12b and 14.12c to get expressions for the bolt and material loads in terms of the applied load P:

$$F_m = F_i - (1 - C)P \tag{14.14a}$$

$$F_b = F_i + CP \tag{14.14b}$$

Equation 14.14b can be solved for the preload F_i needed for any given combination of applied load P and maximum allowable bolt (proof) load F_b, provided that the joint constant C is known.

The load P_0 required to separate the joint can be found from equation 14.14a by setting F_m to zero.

$$P_0 = \frac{F_i}{(1 - C)} \tag{14.14c}$$

14

A safety factor against joint separation can be found from

$$N_{sep} = \frac{P_0}{P} = \frac{F_i}{P(1-C)} \qquad (14.14d)$$

EXAMPLE 14-2

Preloaded Fasteners in Static Loading

Problem   Determine a suitable bolt size and preload for the joint shown in Figure 14-23. Find its safety factor against yielding and separation. Determine the optimum preload as a percentage of proof strength to maximize the safety factors.

Given   The joint dimensions are $D = 1$ in and $l = 2$ in. The applied load $P = 2\,000$ lb.

Assumptions   Both of the clamped parts are steel. The effects of the flanges on the joint stiffness will be ignored. A preload of 90% of the bolt's proof strength will be applied as a first trial.

Solution   See Figure 14-25 and *TKSolver* files EX14-02a and EX14-02b.

1 As with most design problems, there are too many unknown variables to solve the necessary equations in one pass. Trial values must be chosen for various parameters and iteration used to find a good solution. We actually made several iterations to solve this problem, but will only present two in the interest of brevity. Thus the trial values used here have already been massaged to reasonable values.

2 The bolt diameter is the principal trial value to be chosen along with a thread series and a bolt class to define the proof strength. We choose a 5/16-18 UNC-2A steel bolt of SAE class 5.2. (This was actually our third trial choice.) For a clamp length of 2 in, assume a bolt length of 2.5 in to allow sufficient protrusion for the nut. The preload is taken at 90% of proof strength as assumed above.

3 Table 14-6 shows the proof strength of this bolt to be 85 kpsi. The tensile stress area from Table 14-1 is 0.052 in^2. The preload is then

$$F_i = 0.9 S_p A_t = 0.9(85\,000)(0.052\,43) = 4\,011 \text{ lb} \qquad (a)$$

4 Find the lengths of thread l_{thd} and shank l_s of the bolt as shown in Figure 14-21:

$$l_{thd} = 2d + 0.25 = 2(0.3125) + 0.25 = 0.875 \text{ in}$$
$$l_s = l - l_{thd} = 2.5 - 0.875 = 1.625 \text{ in} \qquad (b)$$

from which we can find the length of thread l_t that is in the clamp zone:

$$l_t = l - l_s = 2.0 - 1.625 = 0.375 \text{ in} \qquad (c)$$

14

5   Find the stiffness of the bolt from equation 14.11a.

$$\frac{1}{k_b} = \frac{l_t}{A_t E} + \frac{l_s}{A_b E} = \frac{0.375}{0.052\,43(30E6)} + \frac{1.625(4)}{\pi(0.3125)^2(30E6)}$$

$$k_b = 1.059E6 \text{ lb/in} \qquad\qquad (d)$$

6   The calculation for the stiffness of the clamped material is simplified in this example by its relatively small diameter. We can assume in this case that the entire cylinder of material is compressed by the bolt force. (We will soon address the problem of finding the clamped area in a continuum of material.) The material stiffness from equation 14.11d is

$$k_m = \frac{\pi\left(D^2 - d^2\right)}{4}\frac{E_m}{l} = \frac{\pi\left(1.0^2 - 0.312^2\right)}{4}\frac{(30E6)}{2.0} = 1.063E7 \text{ lb/in} \qquad (e)$$

7   The joint stiffness factor from equation 14.13c is

$$C = \frac{k_b}{k_m + k_b} = \frac{1.059E6}{1.063E7 + 1.059E6} = 0.090\,56 \qquad\qquad (f)$$

8   The portions of the applied load P felt by the bolt and the material can now be found from equations 14.13.

$$P_b = CP = 0.090\,56(2\,000) = 181 \text{ lb}$$

$$(g)$$

$$P_m = (1 - C)P = (1 - 0.090\,56)(2\,000) = 1\,819 \text{ lb}$$

9   Find the resulting loads in bolt and material after the load P is applied.

$$F_b = F_i + P_b = 4\,011 + 181 = 4\,192 \text{ lb}$$

$$(h)$$

$$F_m = F_i - P_m = 4\,011 - 1\,819 = 2\,192 \text{ lb}$$

Note how little the applied load adds to the preload in the bolt.

10   The maximum tensile stress in the bolt is

$$\sigma_b = \frac{F_b}{A_t} = \frac{4\,192}{0.052\,43} = 79\,955 \text{ psi} \qquad\qquad (i)$$

Note that no stress-concentration factor is used because it is static loading.

11   This is a uniaxial stress situation, so the principal stress and von Mises stress are identical to the applied tensile stress. The safety factor against yielding is then

$$N_y = \frac{S_y}{\sigma_b} = \frac{92\,000}{79\,657} = 1.15 \qquad\qquad (j)$$

The yield strength is found from Table 14-1.

14

12 The load required to separate the joint and the safety factor against joint separation are found from equations 14.14c and 14.14d.

$$P_0 = \frac{F_i}{(1-C)} = \frac{4\,011}{(1-0.09056)} = 4\,410 \text{ lb} \qquad (k)$$

$$N_{sep} = \frac{P_0}{P} = \frac{4\,410}{2\,000} = 2.2 \qquad (l)$$

13 The safety factor against separation is acceptable, but the yielding safety factor is too low. The *TKSolver* model was list-solved for the entire range of possible preloads from zero to 100 percent of proof strength and these two safety factors plotted versus preload percentage. The results are shown in Figure 14-25.

The separation safety factor rises linearly with increased preload, but is < 1 until the preload exceeds about 42% of the proof strength. At least that much preload is needed to keep the joint together under the applied load.

The yielding safety factor is high at low preloads and decreases nonlinearly with increasing preload. The two lines cross at a preload of about 65% of proof strength at point A. This preload gives a better solution to this particular problem as it balances the safety factors against both modes of failure at a value of 1.6.

14 The recommended design is then a 5/16-18 UNC-2A, grade 5.2 bolt, 2.5 in long, preloaded to 65% of proof strength with a preload force of

$$F_i = 0.65 S_p A_t = 0.65(85\,000)(0.052) \cong 2\,900 \text{ lb} \qquad (m)$$

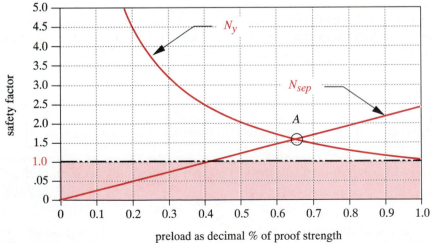

FIGURE 14-25

Safety Factors Versus Preload for the Statically Loaded Bolt in Example 14-2

Preloaded Bolts Under Dynamic Loading

The value of preloading is even greater for dynamically loaded joints than for statically loaded ones. Consider again the joint shown in Figure 14-23 but let the applied force P be a function of time, varying between some minimum and maximum values P_{min} and P_{max}, both positive. A very common situation is that of a fluctuating load ($P_{min} = 0$) such as in a bolted pressure vessel that is cycled from zero to maximum pressure.

Figure 14-26 shows the load deflection diagram of a bolted assembly subjected to a fluctuating load. When the fluctuating load drops to zero, the diagram looks like Figure 14-26a, i.e., with only the static preload F_i present. When the load rises to a maximum, the diagram looks like Figure 14-26b. P_{max} is split between the bolt and the material in the same manner as the static loading case of Figure 14-24. The bolt feels only a portion of the fluctuating load due to the presence of the preload, *which causes the material to absorb the bulk of the load's oscillations*. This drastically reduces the dangerous alternating tensile stress in the bolt from what it would be with no preload. The compressive stress oscillations in the material are of no concern with respect to fatigue failure, which is always due to tensile stress.

The mean and alternating forces felt by the bolt are

$$F_{alt} = \frac{F_b - F_i}{2}, \qquad F_{mean} = \frac{F_b + F_i}{2} \qquad (14.15a)$$

where F_b is found from equation 14.14b with $P = P_{max}$.

The mean and alternating stresses in the bolt are

$$\sigma_{alt} = K_f \frac{F_{alt}}{A_t}, \qquad \sigma_{mean} = K_{fm} \frac{F_{mean}}{A_t} \qquad (14.15b)$$

where A_t is the bolt's tensile stress area from Table 14-1 or 14-2, K_f is the fatigue stress-concentration factor for the bolt and K_{fm} is the mean-stress-concentration factor from equation 6.15. Note that K_{fm} will typically be 1.0 for preloaded bolts.

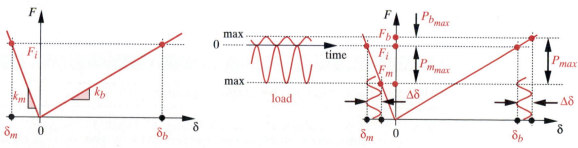

(a) Load condition when $P = 0$ (b) Load condition when $P = P_{max}$

FIGURE 14-26

Effects on Bolt and Material of a Load Fluctuating from Zero to P_{max}

14

Table 14-8   Fatigue Stress-Concentration Factors for Bolts

Brinell Hardness	SAE Grade (UNS)	SAE Class (ISO)	K_f Rolled Threads	K_f Cut Threads	K_f Fillet
< 200 (annealed)	≤ 2	≤ 5.8	2.2	2.8	2.1
> 200 (hardened)	≥ 4	≥ 6.6	3.0	3.8	2.3

The stress due to the preload force F_i is

$$\sigma_i = K_{fm}\frac{F_i}{A_t} \tag{14.15c}$$

Peterson[6] reports that about 15% of bolt failures occur at the fillet under the head, 20% at the end of the threads on the shank, and about 65% in the thread at the nut face. Table 14-8 shows some suggested fatigue stress-concentration factors for these locations on bolts with cut or rolled threads. Rolled threads have lower stress-concentration factors due to their favorable grain orientation. High-strength bolts usually have rolled threads.

The stresses calculated from equations 14.15 need to be compared to a suitable set of material strength parameters on a modified-Goodman diagram, as discussed in Section 6.11. The endurance strength can be calculated by the methods of Section 6.6 using a machined-finish factor for either rolled or cut threads. This will be demonstrated in an example. The fatigue safety factor can be calculated without drawing the Goodman diagram by employing equation 13.34b, repeated here with a new equation number.

$$N_f = \frac{S_e(S_{ut} - \sigma_i)}{S_e(\sigma_m - \sigma_i) + S_{ut}\sigma_a} \tag{14.16}$$

The value of high preloads in reducing the effects of fatigue loading should be clear from the previous discussion. If no preload were applied to the joint, the mean and alternating loads and stresses felt by the bolt would increase by the factor $1/C$, which is potentially a large number as C is typically small.

EXAMPLE 14-3

Preloaded Fasteners in Dynamic Loading

Problem   Repeat Example 14-2 with a fluctuating load applied to the joint. Determine a suitable bolt size and preload for the joint shown in Figure 14-23. Find its safety factors against fatigue, yielding, and separation. Determine the optimum preload as a percentage of proof strength to maximize the fatigue, yielding, and separation safety factors.

14

Given   The joint dimensions are $D = 1$ in and $l = 2$ in. The applied load fluctuates between $P = 0$ and $P = 2\,000$ lb.

Assumptions   The bolt has rolled threads. Both of the clamped parts are steel. The effects of the flanges on the joint stiffness will be ignored. A preload of 90% of the bolt's proof strength will be applied as a first trial. Use 99% reliability and an operating temperature of 300°F.

Solution   See Figures 14-27 to 14-29 and *TKSolver* files EX14-03a and EX14-03b.

1   Again, there are too many unknown variables to solve the necessary equations in one pass. Trial values must be chosen for various parameters and iteration used to find a good solution. We actually made several iterations to solve this problem, but will only present one in the interest of brevity. The trial values used here have already been massaged to reasonable values.

2   The bolt diameter is the principal trial value to be chosen along with a thread series and a bolt class to define the proof strength. We choose a 5/16-18 UNC-2A steel bolt of SAE class 5.2 based on its successful use in the static-loading problem of Example 14-2 on which this example is based. For a clamp length of 2 in, assume a bolt length of 2.5 in to allow sufficient protrusion for the nut. The preload is taken at 90% of proof strength as assumed above.

3   The proof strength, tensile-stress area, preload force, bolt stiffness, material stiffness, and joint constant are all the same as were found in Example 14-2 for the 90% preload factor. See that example for details. In summary they are

$$S_p = 85 \text{ kpsi} \qquad A_t = 0.052\,43 \text{ in}^2 \qquad F_i = 4\,010.95 \text{ lb}$$

$$k_b = 1.059E6 \text{ psi} \qquad k_m = 1.063E7 \text{ psi} \qquad C = 0.090\,56 \tag{a}$$

4   The portions of the peak fluctuating load P felt by the bolt and the material and the resulting loads in bolt and material after the load is applied are also the same as in the previous example:

$$P_b = 181.12 \text{ lb} \qquad P_m = 1\,818.87 \text{ lb}$$

$$F_b = 4\,192.08 \text{ lb} \qquad F_m = 2\,192.08 \text{ lb} \tag{b}$$

5   Because these loads are fluctuating, we need to calculate the mean and alternating components of the force felt in the bolt. Figure 14-27 shows the load-deflection diagram for this problem, drawn to scale with the above forces applied. The shallow sine wave between the initial force line A and the maximum bolt force line B is the only fluctuating load felt by the bolt. The mean and alternating forces are then

$$F_{alt} = \frac{F_b - F_i}{2} = \frac{4\,192.08 - 4\,010.95}{2} = 90.56 \text{ lb}$$

$$F_{mean} = \frac{F_b + F_i}{2} = \frac{4\,192.08 + 4\,010.95}{2} = 4\,101.37 \text{ lb} \tag{c}$$

14

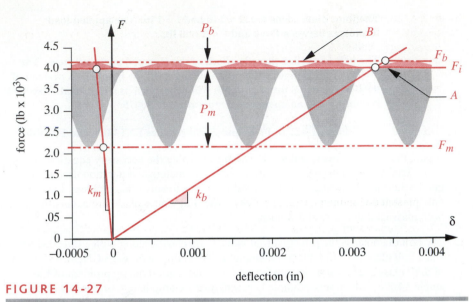

FIGURE 14-27

Dynamic Forces in Bolt and Material for Example 14-3, Drawn to Scale

Note how little of the 0-2 000 lb fluctuating force is felt by the bolt.

6   The mean and alternating stresses in the bolt are:

$$\sigma_{alt} = K_f \frac{F_{alt}}{A_t} = 3.0 \frac{90.56}{0.052\,43} = 5\,182 \text{ psi}$$

$$(d)$$

$$\sigma_{mean} = K_{fm} \frac{F_{mean}}{A_t} = 0.99 \frac{4\,101.37}{0.052\,43} = 77\,336 \text{ psi}$$

The fatigue stress-concentration factor K_f for rolled threads is taken from Table 14-8. The mean stress-concentration factor $K_{fm} = 0.99$ from equation 6.15 in this case.

7   The stress at the initial preload is

$$\sigma_i = K_{fm} \frac{F_i}{A_t} = 0.99 \frac{4\,010.94}{0.052\,43} = 75\,628 \text{ psi} \qquad (e)$$

8   An endurance strength must be found for this material. Using the methods of Section 6.6 we find

$$S_e' = 0.5 S_{ut} = 0.5(120\,000) = 60\,000 \text{ psi} \qquad (f)$$

$$S_e = C_{load}\, C_{size}\, C_{surf}\, C_{temp}\, C_{reliab}\, S_e'$$
$$= 0.70(1)(0.76)(1)(0.81)(60\,000) = 25\,837 \text{ psi} \qquad (g)$$

where the strength reduction factors are taken from the tables and formulas in Section 6.6 for, respectively, axial loading, the bolt size, a machined finish, room temperature, and 99% reliability.

9   The corrected endurance strength and the ultimate tensile strength are used in equation 14.16 to find the safety factor from the Goodman line.

$$N_f = \frac{S_e(S_{ut} - \sigma_i)}{S_e(\sigma_m - \sigma_i) + S_{ut}\sigma_a}$$

$$= \frac{25\,837(120\,000 - 75\,628)}{25\,837(77\,336 - 75\,628) + 120\,000(5\,182)} = 1.7 \qquad (h)$$

The modified-Goodman diagram for this stress state is shown in Figure 14-28.

10   The maximum static stress and the safety factor against yielding are the same as in the previous example:

$$N_y = \frac{S_y}{\sigma_b} = \frac{92\,000}{79\,043} = 1.2 \qquad (i)$$

11   The load required to separate the joint and the safety factor against joint separation are found from equations 14.14c and 14.14d.

$$N_{sep} = \frac{F_i}{P(1-C)} = \frac{4\,011}{2\,000(1-0.91)} = 2.2 \qquad (j)$$

12   The fatigue and separation safety factors are acceptable. The yielding safety factor is low. The *TKSolver* model was list-solved for the entire range of possible preloads from zero to 100 percent of proof strength and these three safety factors plotted

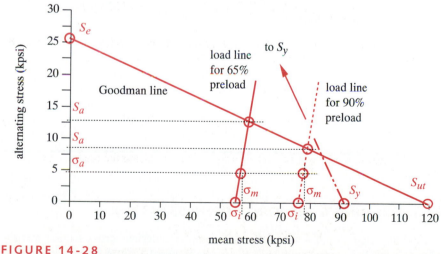

FIGURE 14-28

Modified-Goodman Diagram for Example 14-3 with Solutions for 65% and 90% Preload Shown

14

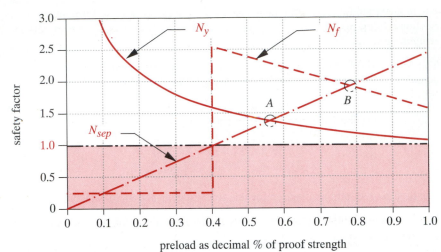

FIGURE 14-29

Safety Factors Versus Preload for Dynamically Loaded Bolt in Example 14-3

versus percent preload. The results are shown in Figure 14-29. Note that the fatigue and separation safety factors are < 1 below 40% preload. The optimum combination of fatigue and separation factors occurs at a preload of 84% of proof strength (point B). However, at that preload the bolt will yield from an overload before it separates. The optimum preload for dynamic yielding resistance and separation resistance is now 56% with a safety factor = 1.4 and the fatigue safety factor is 2.3 at point A. Bolt preloading typically has more beneficial effect against fatigue loading than against static loading. Note the sudden jump in the fatigue safety factor (at 40% preload in Figure 14-29) when the preload becomes effective at keeping the joint closed.

13 The preload required is found from

$$F_i = 0.56 S_p A_t = 0.56(85\ 000)(0.052\ 43) \cong 2\ 496 \ \text{lb} \qquad (k)$$

14 The recommended design is then a 5/16-18 UNC-2A, grade 5.2 bolt, 2.5 in long, preloaded to 56% of proof strength with a force of 2 496 lb. The safety factors against both static yielding and separation are optimized at 1.4 and the fatigue safety factor is 2.3. Note that this one, small, preloaded bolt will support a ton of fluctuating load!

14.8   DETERMINING THE JOINT STIFFNESS FACTOR

In the previous discussion, for simplicity, the clamped materials' cross section was assumed to be a small-diameter cylinder, as shown in Figure 14-23. A more realistic situation is depicted in Figure 14-21, in which the clamped materials are a continuum extending well beyond the region of the bolt's influence. In fact, most assemblies will have a number of bolts distributed in a pattern over the clamped surface. (When the bolt pattern is circular, the *circumference on which the bolt centerlines lie* is called the **bolt circle**.) The question then becomes, what amount of clamped material should be included in a calculation of the material stiffness k_m whose value is needed to find the joint stiffness factor C?

The stress distribution within the material under the bolt has a complex geometry. This problem has been studied by a number of investigators[7], [8], [9] and an accurate computation of the distribution of the stressed volume is quite complicated. The compressive stress in the material is (not surprisingly) highest directly under the bolt and falls off as you move laterally away from the bolt centerline (CL). At some lateral distance from the CL, the compressive stress at the joint interface goes to zero and beyond that point, the joint tends to separate since it cannot sustain a tensile stress.

Figure 14-30 shows the results of a finite-element analysis (FEA) study of the stress distribution in a two-part joint-sandwich clamped together with a single, preloaded bolt.[10], [11] Only one-half of the sandwich is analyzed because of its axisymmetry. The vertical bolt centerline is to the left of the left-hand edge of each diagram. The stress distribution around the bolt resembles a barrel shape, as seen in Figure 14-30a. Figure 14-30b shows the deformed geometry and greatly exaggerates the vertical dimension to show the very small deflections in the separation zone, which is clearly visible in the right-hand half of the clamped assembly. This study indicated that, if the two parts clamped were of equal thickness, the point of separation of the joint occurred at a dis-

Stress Distribution

Deformed Geometry and Stress Contour

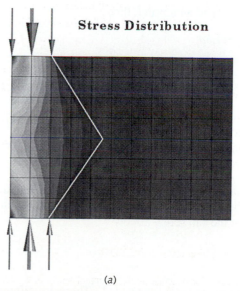
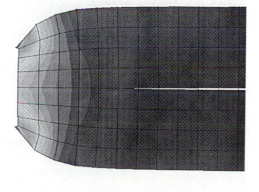
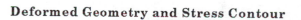

(a)

(b)

FIGURE 14-30

Finite-Element Analysis of the Stress Distribution and Deformation Within the Clamp Zone of a Bolted Connection

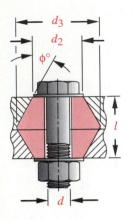

(a) Cone-frusta model

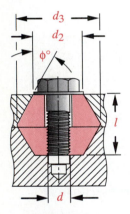

(b) Cap screw frusta

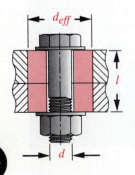

(c) Effective cylinder

FIGURE 14-31

Estimating the Material
Compressed by the Bolt

tance from the bolt centerline that subtended an angle ϕ of approximately $42°$ measured from the edge of the bolt head or washer, as shown in Figure 14-31. Even if the two parts joined are not of equal thickness, the stress distribution still extends outward to a point that can be approximated by the $42°$ cone shown in Figures 14-30 and -31.

Thus, an estimate of the effective clamped volume can be made with a double-cone-shaped "barrel," as shown in Figures 14-31a and 14-31b. This shape is actually two frusta of cones with the small-end diameter taken equal to the bolt head (or nut) diameter, which is approximately 1.5 times the bolt diameter. If hardened washers are used under either the head or nut, their standard diameters of $2d$ can be used instead. A cone angle of $42°$ is used to approximate the spreading out of the force pattern within the material.

If both clamped materials are the same, then the effective cross-sectional area of this cone is simple to calculate if we assume it to be a cylinder (see Figure 14-31c) whose volume equals the volume of the cone frusta. If the materials are different, they must be treated as springs in series, taking their different E's into account. (See Eq. 13.2b.)

For uniform materials in the joint, the effective cross-sectional area of compression A_m is the average cross-sectional area of the frustum-cone barrel:

$$A_m \cong \frac{\pi}{4}\left(D_{eff}^2 - d^2\right) \cong \frac{\pi}{4}\left[\left(\frac{d_2 + d_3}{2}\right)^2 - d^2\right]$$ (14.17a)

with the diameters as defined in Figure 14-31b.

$$d_3 = d_2 + l\tan\phi$$ (14.17b)

If no hardened washers are used under head and/or nut, then

$$d_2 = 1.5d$$ (14.17c)

to match the head or nut diameter.

If standard hardened washers are used under both head and nut, then:

$$d_2 = 2d$$ (14.17d)

The area A_m is then used in equation 14.11 to find k_m, and k_m is used in equation 14.13c to find the *joint stiffness constant C*.

The material spring constant for a machine screw or cap screw can be estimated in similar fashion. Figure 14-31b shows the length of the barrel defined to include only that portion of the clamped thickness that the cap screw affects.

Gasketed Joints

Gaskets are often used in joints where pressure seals are needed. There are different styles of gaskets, which can be divided into two general classes: **confined** and **uncon-**

Table 14-9   Young's Modulus for Some Gasket Materials
Source: Reference 7 with permission of McGraw-Hill, Inc., New York

Material	Modulus of Elasticity	
	psi	MPa
Cork	12.5E3	86
Compressed asbestos	70E3	480
Copper-asbestos	13.5E6	93E3
Copper (pure)	17.5E6	121E3
Plain rubber	10E3	69
Spiral wound	41E3	280
Teflon	35E3	240
Vegetable fiber	17E3	120

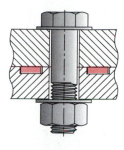

(a) Confined gasket

fined. Figures 14-32a and 14-32b show two variations of confined gaskets, one being an O-ring. All confined gaskets allow the hard faces of the mating parts to contact and this makes the joint behave as an ungasketed one in terms of its spring constant k_m. The approaches described above can be used to estimate k_m for confined gaskets.

Unconfined-gasket joints as shown in Figure 14-32c have the relatively soft gasket completely separating the mating surfaces. The gasket then contributes to the spring constant of the joint. The gasket's spring constant k_g can be combined with the spring constants of the mating parts in equation 13.2b to find an effective spring constant k_m for the assembly. The moduli of elasticity for several gasket materials are in Table 14-9.

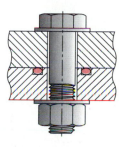

(b) Confined O-ring

With the exception of the copper and copper-asbestos gasket materials in Table 14-9, typical gaskets' moduli are so low that they will dominate equation 13.2b and essentially determine the joint stiffness. In those cases, it is not necessary to solve equation 13.2b and k_m can be set equal to k_g. With a copper-asbestos or copper gasket (or any other stiff, unconfined gasket), the gasket stiffness may be sufficiently high to warrant computing k_m from equation 13.2b. Use an outside diameter of gasket material consistent with the outside diameter of the cone in Figure 14-31 at the level of the gasket to estimate k_g. This will equal d_3 if the gasket is centered within the clamp zone, but will be smaller otherwise.

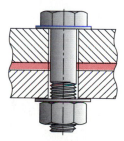

(c) Unconfined gasket

14

EXAMPLE 14-4

Determining the Material Stiffness and Joint Constant

Problem   A pressure chamber is sealed by a gasketed cap fastened with eight preloaded bolts. Find the material stiffness and joint constant for two designs of the assembly as shown in Figure 14-33, one with a confined

FIGURE 14-32

Confined and Unconfined Gaskets

gasket and one with an unconfined gasket. Also determine the loads felt by bolts and material.

Given   The cylinder diameter D_p = 4 in. Bolt circle diameter D_{bc} = 5.5 in. Outside flange diameter D_f = 7.25 in. The eight 3/8-16 UNC bolts are equispaced on the bolt circle. The clamped length l of the joint is 1.5 in. The gasket thickness t = 0.125 in. The pressure in the cylinder is 1 500 psi. All parts are steel.

Assumptions   The gasket material is rubber. Hardened, standard washers are used under the bolt heads and nuts.

Solution   See Figure 14-33 and *TKSolver* files EX14-04a and EX14-04b.

1   Figure 14-33 shows two alternate designs of gasket for the joint on the same view to save space. Whichever design is used, its gasket configuration will be present on both sides of the centerline. Don't be confused by the depiction of different gaskets top and bottom; only one or the other will be used in the final assembly. We will deal first with the confined gasket configuration.

2   The forces on the individual bolts can be found from the known pressure and cylinder dimensions assuming that all bolts share the load equally. The total force on the end cap is

$$P_{total} = pA = p\frac{\pi D_p^2}{4} = 1\,500\frac{\pi(4)^2}{4} = 18\,850 \text{ lb} \qquad (a)$$

and the applied force P on each bolt is

$$P = \frac{P_{total}}{N_{bolts}} = \frac{18\,850}{8} = 2\,356 \text{ lb} \qquad (b)$$

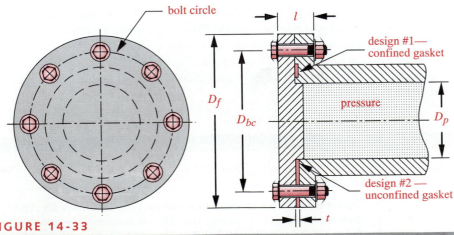

FIGURE 14-33

Pressure Vessel End Plate Secured with Preloaded Bolts on a Bolt Circle

14

3   First find the stiffness of one bolt. Its tensile stress area is found from Table 14-1 to be 0.077 in². The clamp length is given as 1.5 in. A 2-in bolt will have sufficient protrusion to grip the nut. The lengths of thread and shank of the bolt are then

$$l_{thd} = 2d + 0.25 = 2(0.375) + 0.25 = 1.0 \text{ in}$$

$$l_s = l - l_{thd} = 2.0 - 1.0 = 1.0 \text{ in} \tag{c}$$

from which we can find the length of thread l_t that is in the clamp zone:

$$l_t = l - l_s = 1.5 - 1.0 = 0.5 \text{ in} \tag{d}$$

4   Find the stiffness of the bolt from equation 14.11a.

$$\frac{1}{k_b} = \frac{l_t}{A_t E} + \frac{l_s}{A_b E} = \frac{0.5}{0.077(30E6)} + \frac{1.0(4)}{\pi(0.375)^2(30E6)}$$

$$k_b = 1.935E6 \text{ lb/in} \tag{e}$$

5   A confined gasket allows the metal surfaces to contact just as if there were no gasket present. So the analysis of the material stiffness can ignore the confined gasket. The effective area in the clamp zone around any one bolt with washers is found from equations 14.17.

$$d_2 = 2d = 2(0.375) = 0.750$$

$$d_3 = d_2 + l \tan\phi = 0.750 + 1.5 \tan 42° \cong 1.913$$

$$A_m \cong \frac{\pi}{4}\left[\left(\frac{d_2 + d_3}{2}\right)^2 - d^2\right]$$

$$\cong \frac{\pi}{4}\left[\left(\frac{0.750 + 1.913}{2}\right)^2 - (0.375)^2\right] \cong 1.093 \text{ in}^2 \tag{f}$$

6   The material stiffness from equation 14.11c is

$$k_m = \frac{A_m E_m}{l} = \frac{1.093(30E6)}{1.5} \cong 2.186E7 \text{ lb/in} \tag{g}$$

7   The joint stiffness factor for the confined-gasket design from equation 14.13c is

$$C = \frac{k_b}{k_m + k_b} = \frac{1.935E6}{2.186E7 + 1.935E6} \cong 0.081$$

and $$(1 - C) = 0.919 \tag{h}$$

8   The portions of the applied load P felt by the bolt and the material can now be found from equations 14.13.

$$P_b = CP = 0.081(2\,356) \cong 192 \text{ lb}$$

$$P_m = (1 - C)P = (1 - 0.089)(2\,356) \cong 2\,165 \text{ lb} \tag{i}$$

14

9   Now we will address the unconfined-gasket case. The bolt stiffness is not affected
    by the gasket but the material stiffness is. We now have two springs in series, the
    slightly shorter metal "barrel" of material and the gasket. These combine according
    to equation 13.2b. The area of the barrel shortened by the gasket thickness t is

$$d_3 = d_2 + (l - t)\tan\phi = 0.750 + (1.5 - 0.125)\tan 42° \cong 1.801$$

$$A_m \cong \frac{\pi}{4}\left[\left(\frac{d_2 + d_3}{2}\right)^2 - d^2\right]$$

$$\cong \frac{\pi}{4}\left[\left(\frac{0.750 + 1.801}{2}\right)^2 - (0.375)^2\right] \cong 0.986 \text{ in}^2 \qquad (j)$$

10 The stiffness of this shorter barrel by itself is

$$k_{m_1} = \frac{A_m E_m}{l} = \frac{0.986(30E6)}{1.5 - 0.125} \cong 2.151E7 \text{ lb/in} \qquad (k)$$

11 The diameter of the gasket subjected to the clamp force can be assumed to be the
    same as the largest diameter d_3 of the theoretical "barrel-shaped" clamp zone shown
    in Figure 14-31 and defined in equation 14.17d:

$$d_3 = d_2 + l\tan\phi = 0.750 + 1.5\tan 42° \cong 1.913 \qquad (l)$$

    The area of the clamped gasket is then

$$A_g = \frac{\pi\left(d_3^2 - d^2\right)}{4} = \frac{\pi\left(1.913^2 - 0.375^2\right)}{4} = 2.436 \text{ in}^2 \qquad (m)$$

12 The stiffness of this piece of gasket is

$$k_{m_2} = \frac{A_g E_g}{l_g} = \frac{2.436(10E3)}{0.125} \cong 1.949E5 \text{ lb/in} \qquad (n)$$

    The modulus of elasticity E_g of the gasket material is found in Table 14-9.

13 The combined stiffness of the gasketed joint (from equation 13.2b) is

$$\frac{1}{k_m} = \frac{1}{k_{m_1}} + \frac{1}{k_{m_2}} = \frac{1}{2.151E7} + \frac{1}{1.949E5}$$

$$k_m \cong 1.931E5 \text{ lb/in} \qquad (o)$$

    Note that the combined stiffness is essentially the same as that of the soft gasket
    since it dominates the equation. We could have used the gasket stiffness k_g to
    represent the joint stiffness k_m with little error.

14 The joint constant is now

14

$$C = \frac{k_b}{k_m + k_b} = \frac{1.935E6}{1.931E5 + 1.935E6} = 0.909$$

and $(1-C) = 0.091$ (p)

15 The portions of the applied load P felt by the bolt and the material with a soft, unconfined gasket in the joint can now be found from equations 14.13.

$$P_b = CP = 0.909(2\,356) \cong 2\,142 \text{ lb}$$

$$P_m = (1-C)P = (1-0.909)(2\,356) \cong 214 \text{ lb}$$ (q)

16 See what has happened as a result of introducing an unconfined soft gasket! Compare equation (i) to (q). The bolt has gone from feeling only 8% of the applied load with no gasket (or a confined gasket) to feeling 91% with a soft, unconfined gasket.

In effect, the roles of the bolt and the material have been reversed by the introduction of the soft gasket. The unconfined, soft gasket severely limits the ability of the bolt to accommodate fatigue loads, as was accomplished in the previous example.

Note that to obtain the benefits of high preloads in terms of protecting the fasteners from fatigue loading, it is necessary to have a material stiffness greater than the bolt stiffness. Soft, unconfined gaskets reduce the material stiffness so severely that they limit the effectiveness of preloading. For heavily loaded joints, unconfined gaskets should be made of high-stiffness material such as copper or copper-asbestos, or be replaced with confined gaskets.

Some rules of thumb regarding patterns of bolts such as used in Example 14-4 are

1 Bolt spacing around a bolt circle or in a pattern should not exceed about 6 bolt diameters between adjacent bolts for good force distribution.

2 Bolts should not be closer to an edge than about 1.5 to 2 bolt diameters.

3 The assumptions regarding the barrel-shaped load distribution within the clamped material shown in Figure 14-31 presume that the bolt is not closer to a material edge than half the diameter d_3 of the barrel from equation 14.17b.

14.9   CONTROLLING PRELOAD

The amount of preload is obviously an important factor in bolt design. Thus we need some means of controlling the preload applied to a bolt. The most accurate methods require that both ends of the bolt be accessible. Then the amount of bolt elongation can be directly measured with a micrometer, or an electronic length gage and the bolt stretched to a length consistent with the desired preload based on equation 14.10a. Ultrasonic transducers are sometimes used to measure change in bolt length when tightened and these only need access to the head end. These methods are not as useful in

14

high-production or field-service situations since they require time, care, precision instruments, and skilled personnel.

A more convenient but less accurate method measures or controls the torque applied to the nut or to the head of a cap screw. A torque wrench gives a readout on a dial of the amount of torque applied. Torque wrenches are generally considered to give an error in preload of up to ±30%. If great care is taken and the threads are lubricated (which is desirable anyway) this error can perhaps be halved, but it is still large. Pneumatic impact wrenches can be set to a particular torque level at which they stop turning. These give more consistent results than a manual torque wrench and are preferred.

The torque necessary to develop a particular preload can be calculated from equation 14.5a, developed for the power screw. Substitute equation 14.3 in 14.5a to get it in terms of lead angle λ:

$$T_i = F_i \frac{d_p}{2} \frac{(\mu + \tan\lambda\cos\alpha)}{(\cos\alpha - \mu\tan\lambda)} + F_i \frac{d_c}{2}\mu_c \tag{14.18a}$$

The pitch diameter d_p can be roughly approximated as the bolt diameter d and the mean collar diameter can be approximated as the average of the bolt diameter d and a standard head or nut size of 1.5d,

$$T_i \cong F_i \frac{d}{2} \frac{(\mu + \tan\lambda\cos\alpha)}{(\cos\alpha - \mu\tan\lambda)} + F_i \frac{(1+1.5)d}{2}\mu_c \tag{14.18b}$$

Factor out the force and bolt diameter to get

$$T_i \cong K_i F_i d \tag{14.18c}$$

where

$$K_i \cong \left[0.50 \frac{(\mu + \tan\lambda\cos\alpha)}{(\cos\alpha - \mu\tan\lambda)} + 0.625\mu_c \right]$$

K_i is called the *torque coefficient*.

Note that the friction coefficient μ_c between the head or nut and the surfaces as well as the thread friction coefficient μ contribute to the torque coefficient K_i. If we assume a friction coefficient of 0.15 for both locations, and calculate the torque coefficients K_i for all standard UNC and UNF threads, (using their correct pitch diameters d_p rather than the approximation of equation 14.18b) the value of K_i varies very little over the entire range of thread sizes, as shown in Table 14-10. Thus, the tightening torque T_i needed to obtain a desired preload force F_i **in lubricated threads** can be approximated (subject to the friction assumptions listed above) as

$$T_i \cong 0.21 F_i d \tag{14.18d}$$

The Turn-of-the-Nut Method

Another technique often used to control preload is called the *turn-of-the-nut* method. Since the lead of the fastener is known, turning the nut a specified number of turns will stretch the bolt a known amount provided that the starting point is such that all of the nut advance contributes to bolt stretch. The nut is first brought to a useful start point, called *snug-tight*, defined as the tightness obtained from a few strikes of an impact wrench or, if manually done, as tight as a person can make the nut with a standard wrench. Then the nut is turned (with a longer wrench) through an additional number of turns, or fractions thereof, calculated to stretch the bolt the desired amount based on equation 14.10*a*.

Torque-Limited Fasteners

The need for accurate preloads in high-strength bolts has caused bolt manufacturers to provide special "controlled-tension break-away" bolts, as shown in Figure 14-34. These bolts are provided with a splined extension on the end. This extension is designed with a shear area calculated to fracture when the proper torque is reached. Special sockets are provided that engage the spline as shown in Figure 14-35, which also details how they are used. These fasteners are often used in structural-steel construction, where their preload uniformity compared to a manual torque wrench or impact wrench is a big advantage in terms of minimizing operator error when tens of thousands of bolts must be properly installed to ensure that the skyscraper or bridge stays up.

FIGURE 14-34

Controlled Tension Break-Away Bolt - *Courtesy of Cordova Bolt Inc, Buena Park, CA 90621*

Load-Indicating Washers

Another aid to proper bolt tensioning involves using special washers under the bolt head that either control the tension load or indicate when it is correct. Belleville-spring washers are sometimes used under bolt heads. The Belleville spring is designed to give the desired bolt force when it is compressed flat (see Section 13.9). The bolt is simply tightened until the Belleville spring is flat.

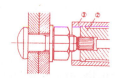

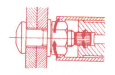

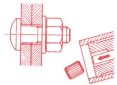

(*a*) Place bolt in hole with washer under nut and finger-tighten the nut

(*b*) Fit inner socket over spline and engage outer socket on nut

(*c*) Turn wrench on. The outer socket rotates tightening nut until the spline is sheared

(*d*) Remove socket from nut and eject sheared extension from the inner socket

FIGURE 14-35

Instructions for Use of Break-Away Bolts *Courtesy of Cordova Bolt Inc, Buena Park, Calif., 90621*

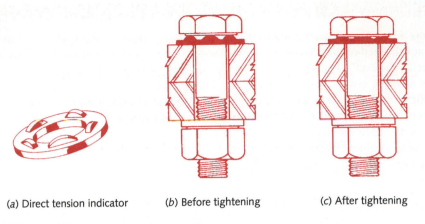

(a) Direct tension indicator   (b) Before tightening   (c) After tightening

FIGURE 14-36

Direct Tension Indicator Washers   *Courtesy of Cordova Bolt Inc., Buena Park, Calif., 90621*

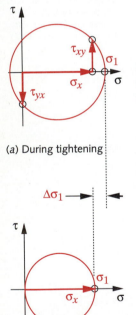

(a) During tightening

(b) After torsion is relaxed

FIGURE 14-37

Mohr's Circles for a Preloaded Bolt both During and After Tightening

Load-indicator washers (also called direct tension indicators) are made with protrusions that crush under the desired preload as shown in Figure 14-36. The bolt is tightened until the height of the washer has been reduced to the proper dimension, as shown in the figure.

Torsional Stress Due to Torquing of Bolts

When a nut on a bolt is torqued to a preload, a torsional load is applied to the bolt through the threads. If the friction in the threads is large, the torsion in the bolt can be appreciable. This is the principal reason for lubricating the threads before assembling fasteners. If there were no thread friction, the torsion load on the bolt would be close to zero. A dry lubricant such as graphite powder with molybdenum disulfide added works well, as will a petroleum oil.

A torsional stress is generated in the bolt shank during tightening, as defined in equation 14.9. This torsional stress combines with the axial tensile stress in the shank to create a principal stress larger than the applied tensile stress as shown in the Mohr's circle diagram of Figure 14-37a. If a reversed torque is applied to the nut after it is fully tightened, without actually loosening it, the torsional stress component can be relieved. Even if nothing is deliberately done to relieve the torsional stress after tightening, it will tend to relax over time, especially if there is any vibration present. When the torsional stress is relieved or goes away over time, the principal stress will be reduced by an amount $\Delta\sigma_1$, as shown in Figure 14-37b. This is the other reason, alluded to earlier, why a bolt that doesn't break when torqued to a high preload (near its proof strength) will probably not break under the applied loads for which it was designed.

14

EXAMPLE 14-5

Determining the Torque to Generate a Bolt Preload

Problem   Find the torque required to preload the bolt in Example 14-3.

Given   A 5/16-18 UNC-2A, grade 5.2 bolt, 2.5 in long, preloaded to 65% of proof strength with an axial force of 2 900 lb.

Assumptions   The threads will be lubricated. Assume a coefficient of friction of 0.15.

Solution   See *TKSolver* file EX14-05.

1   The required tightening torque can be estimated with equation 14.18*d*:

$$T_i \cong 0.21 F_i d = 0.21(2\ 900)(0.3125) = 190 \ \text{lb-in} \tag{a}$$

14.10 FASTENERS IN SHEAR

Bolts are also used to resist shear loads, as shown in Figure 14-38, though this application is more common in structural design than in machine design. Structural-steel building and bridge frames are frequently bolted together with high-strength, preloaded bolts. (Alternatively, they may be welded or riveted together.) The tensile preload in this case serves the purpose of creating large frictional forces between the bolted elements, which can resist the shear loading. Thus, the bolts are still loaded in tension with high preloads. If the friction in the joint is not sufficient to support the shear loads, then the bolt(s) will be placed in direct shear.

In machine design, where the dimensional relationships between parts typically require much closer tolerances than in structural work, it is not considered good practice to use bolts or screws in shear to locate and support precision machine parts under shear loads. Instead a combination of bolts or screws and **dowel pins** should be used with the screws or bolts serving to clamp the joint in compression and the hardened-steel dowel pins providing accurate transverse location and shear resistance. The joint friction developed from the bolt-clamp force should be expected to sustain the shear loads in combination with the dowel pins loaded in direct shear. In effect, the task is split between these different types of fasteners. *Dowel pins support shear loads but not tensile loads, and bolts/screws support tensile loads but not direct shear loads.*

There are several reasons for this approach, all centering around the typical need for accurate positional location of the machine's functional parts, (e.g., within ±0.005 in {0.13 mm} or closer in most machinery). There are exceptions to this of course, as in the case of a machine frame, which, other than its mounting surfaces, may be less accurately made and may even be an inherently inaccurate weldment.

14

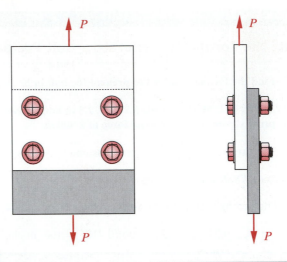

FIGURE 14-38

A Bolted Joint Loaded in Shear

Consider an assembly of two parts, loaded in shear as shown in Figure 14-38. There is a pattern of multiple bolts clamping the two parts together. Bolts and screws are not made to close tolerances. Holes for bolts or screws must be made oversize to provide some clearance for the bolt/screw insertion. Tapped holes for machine screws will have radial clearance versus the inserted screw, meaning that the concentricity of a screw in a tapped hole or a bolt in a clearance hole is not guaranteed. They will be eccentric.

The above observations are true for any one bolt/screw in any one hole. When a pattern of fasteners is used as in Figure 14-38, the required clearances between bolts/screws and holes becomes significantly greater than for one hole because of the tolerances on the dimensions between the hole centerlines in the two mating parts. For reliable and interchangeable assembly, the holes will have to be significantly larger than the removable fasteners in order to accommodate all the variation possible within customary manufacturing tolerances. Figure 14-39 shows an exaggerated mismatch between a pair of holes in mating pieces and indicates why the holes have to be larger in diameter than the fasteners to allow assembly.

Now consider what will happen if we rely on the four bolts in Figure 14-38 to both locate the parts and take the shear loads without tensile preloading. The positional location of one part to the other is severely compromised by the needed clearances in the holes and by the variations in diameter of commercial bolts. The ability of the four bolts to share the loads in direct shear is also compromised by the clearances. At best, two bolts will probably take all the shear load with the others not even contacting the appropriate sides of their holes on both parts in order to share the load.

So what is the solution to this problem? A better design is shown in Figure 14-40, which adds two hardened-steel dowel pins to the pattern of four bolts. More dowels

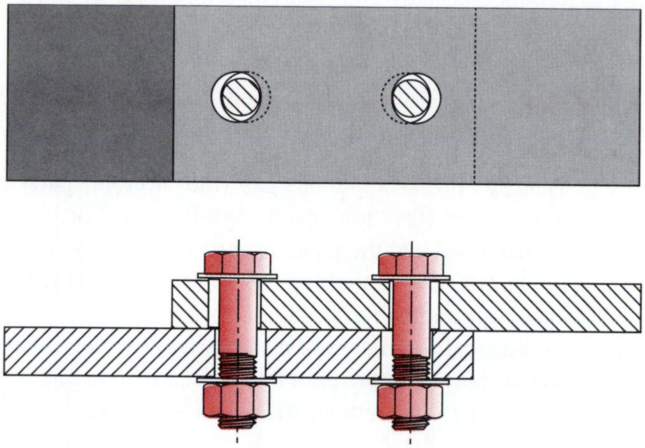

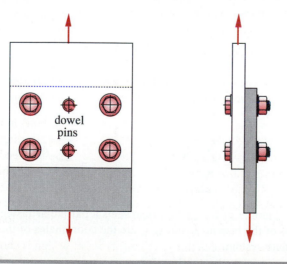

FIGURE 14-39

Clearance is Needed in Fastener Holes to Accommodate Manufacturing Tolerances

could be added, but two is the minimum number needed in order to withstand a couple in that plane and are usually sufficient. A brief digression on the proper application of dowel pins is necessary at this point.

Dowel Pins

Standard straight dowel pins[13] are made to extremely close tolerances (typically ± 0.0001 in variation in diameter), are ground to a fine finish, and are truly round. They

FIGURE 14-40

A Bolted and Doweled Joint Loaded in Shear

are available in low-carbon steel, corrosion-resistant (chrome) steel, brass, and alloy steel hardened to 40-48 HRC, and are purchased to the required length. They are relatively inexpensive. Tapered dowel pins are also available. Other varieties of grooved, knurled, and spring-roll pins are available that do not require as close-toleranced holes for a press-fit. We will limit our brief discussion to straight, solid dowel pins.

Dowel pins are typically press-fit into one part (the "bottom" part) and made to be a close slip fit in the other part (the "top" part). While tapped or clearance holes for the bolts or screws are machined into the separate parts prior to assembly, the dowel pin holes are not drilled until after the assembly has been bolted or screwed together and aligned to its proper configuration. Then pilot holes for the dowels, smaller than the dowel pin diameter, are drilled through both clamped parts in the specified locations. In some cases, the top part may have the pilot holes already in it, and it will be used as the jig to drill the holes through the bottom part, when assembled.

Once the (exactly concentric) pilot holes are in both parts, they can be reamed to the proper diameter for a press-fit with the dowel pin while still clamped together. It is then disassembled and the pilot holes in the "top" part are reamed slightly larger for a slip-fit with the dowel pin. The reamer will follow the hole, accurately keeping its center location. The dowels are then pressed into the bottom part and the top part is carefully fitted over the protruding dowels. The screw fasteners are then replaced and torqued to the appropriate preload.

We now have an assembly that is **accurately relocatable** when disassembled and reassembled and one in which we have **essentially zero radial clearance** between some number of hardened pins, which can, if necessary, resist shear loads in a shared fashion. Eccentric loading is not a problem since the two dowels can resist couples in the shear plane. Without dowels (or sufficient compressive preload to generate friction between the plates) applied couples will rack the bolts in their clearance holes, allowing relative motion between the top and bottom plate.

Centroids of Fastener Groups

When a group of fasteners is arranged in a geometric pattern, the location of the centroid of the fastener areas is needed for force analysis. With respect to any convenient coordinate system, the coordinates of the centroid are

$$\tilde{x} = \frac{\sum_1^n A_i x_i}{\sum_1^n A_i}, \qquad \tilde{y} = \frac{\sum_1^n A_i y_i}{\sum_1^n A_i} \qquad (14.19)$$

where n is the number of fasteners, i represents a particular fastener, A_i are the cross-sectional areas of the fasteners, and x_i, y_i are the coordinates of the fasteners in the selected coordinate system.

Determining Shear Loads on Fasteners

Figure 14-41a shows a shear joint with an eccentric load applied. Four bolts and four dowel pins are used to connect the two parts. Assume that the four dowels will take all the shear loading and will share the load equally. The eccentric load can be replaced with the combination of a force P acting through the centroid of the pin pattern and a moment M about that centroid, as shown in Figure 14-41b. The force through the centroid will generate equal and opposite reactions F_1 at each pin. In addition, there will be a second force F_2 at each pin, acting perpendicular to a radius from the centroid to the pin, due to the moment M.

The magnitude of the force component F_1 at each pin due to the force P acting through the centroid will be

$$\left| F_{1_i} \right| = \frac{P}{n} \tag{14.20a}$$

where n is the number of pins.

To determine how much force each pin feels from the moment M, assume that one part is allowed to rotate slightly about the centroid with respect to the other part. The displacement at any hole will be proportional to its radius from the centroid. The strain developed in the pin will be proportional to that displacement. Stress is proportional to strain in the elastic region, and force is proportional to stress for constant shear area. This means that the magnitude of the force component felt at any pin due to the moment M will be directly proportional to the pin's radius r_i from the centroid:

$$\left| F_{2_i} \right| = \frac{M}{r_i} = \frac{Pl}{r_i} \tag{14.20b}$$

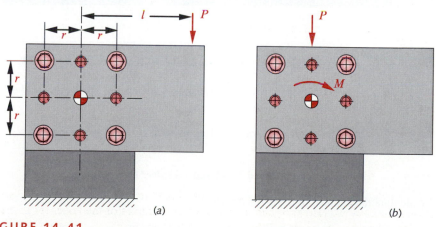

(a)   (b)

FIGURE 14-41

A Bolted and Doweled Joint Eccentrically Loaded in Shear

14

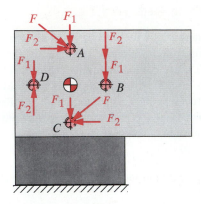

FIGURE 14-42

Pin Forces in a Joint Eccentrically Loaded in Shear

The total force F_i at each pin is then the vector sum of the two components F_{1_i} and F_{2_i} for that pin, as shown in Figure 14-42. The total force is greatest at pin B in the particular case shown.

The stress in the pin is found from equation 14.8c for direct shear stress. The shear yield strength can be estimated from the relationship in equation 5.9b, repeated here:

$$S_{ys} = 0.577 S_y \qquad (5.9b)$$

The minimum shear yield strengths S_{ys} for several common dowel-pin materials can be found in Table 14-11.

EXAMPLE 14-6

Fasteners in Eccentric Shear

Problem   Determine a size for the dowel pins in the bracket of Figure 14-41.

Given   The static force $P = 1\,200$ lb is applied at $l = 5$ in. The radius to the dowel pins is $r = 1.5$ in.

Assumptions All pins share load equally. Use 40-48 HRC alloy steel for the pin material.

Solution   See Figures 14-41 and 14-42 and *TKSolver* file EX14-06.

1   Calculate the moment of the applied force.

$$M = Pl = 1\,200(5) = 6\,000 \ \text{lb - in} \qquad (a)$$

2   Calculate the magnitude of the force due to this moment at each pin.

Table 14-11

Minimum Shear
Strengths for Dowel Pins.
Source: Drive-Lok, Inc.,
Sycamore, Ill.

Material	S_{ys} (kpsi)
Low-carbon steel	50
Alloy steel 40-48 HRC	117
Corrosion-resistant steel	83
Brass	40

14

$$F_M = \frac{M}{r} = \frac{6\,000}{1.5} = 4\,000 \text{ lb} \qquad (b)$$

3 Find the amount of the direct force P felt at each pin.

$$F_P = \frac{P}{n} = \frac{1\,200}{4} = 300 \text{ lb} \qquad (c)$$

4 Based on the vector diagram in Figure 14-42, pin B is the most heavily loaded and its resultant force is

$$F_B = F_P + F_M = 300 + 4\,000 = 4\,300 \text{ lb} \qquad (d)$$

5 Assume a trial pin diameter of 0.375 in and calculate the direct shear stress in pin B.

$$\tau = \frac{F_B}{A_B} = \frac{4\,300(4)}{\pi(0.375)^2} = 38\,933 \text{ psi} \qquad (e)$$

6 Find the shear yield strength of the material from Table 14-11 and calculate the safety factor against static shear failure.

$$N_s = \frac{S_{ys}}{\tau} = \frac{117\,000}{38\,933} = 3.0 \qquad (h)$$

14.11 CASE STUDIES

We will now address the design of preloaded fasteners in one of the Case Study assemblies that were defined in Chapter 8.

Designing Headbolts for an Air Compressor

The preliminary design of this device is shown in Figure 8-1, repeated here. The compressor cylinder and head are cast aluminum. The head is fastened to the cylinder block with a number of cap screws that insert in tapped holes arranged on a bolt circle. The pressure generated in the cylinder creates a force-time function on the head, as shown in Figure 8-2, repeated here.

CASE STUDY 7D

Design of the Headbolts for an Air Compressor

Problem   Design a set of fasteners to attach the compressor head to the cylinder in Figure 8-1 based on the loadings defined in Case Study 7A.

14

Given

The compressor bore is 3.125-in diameter. The dynamic force acting on the head fluctuates from 0 to 1 000 lb each cycle from the 130 psi cylinder pressure. A 0.06-in-thick, unconfined copper-asbestos gasket covers the entire head-cylinder interface. The head thickness at the attachment points (exclusive of cooling fins) is 0.4 in.

Assumptions

Infinite life. Use standard hex-head cap screws without washers. The operating temperature is less than 350°F. Use 99.9% reliability.

Solution

See Figures 8-1, 8-2, and the TKSolver files CASE7D-1 and CASE7D-2.

1 Choose a trial diameter d for the screws of 0.25 in. Use UNC threads to avoid stripping problems in the cast-aluminum cylinder. The trial fastener is then a 1/4-20 UNC-2A cap screw with rolled threads for fatigue resistance.

2 Choose a bolt circle and outside diameter based on the cylinder bore and the rule of thumb of *at least* 1.5d to 2d of distance between any bolt and an edge. We will use 2d because of the need for sealing area against the cylinder pressure.

$$d_{bc} = 3.125 + 2(2)(0.25) = 4.125 \text{ in}$$

$$d_o = 4.125 + 2(2)(0.25) = 5.125 \text{ in}$$

$$(a)$$

3 To get the recommended 6-bolt-diameters spacing between bolts we will need about 8 screws equispaced around the bolt circle. Calculate the spacing between screws in units of bolt diameters.

$$\Delta b = \frac{\pi D_p}{n_b d} = \frac{\pi(4.125)}{8(0.250)} = 6.5 \text{ bolt diameters} \qquad (b)$$

This is slightly larger than the recommended 6 bolt diameters, but we will nevertheless proceed with it on the basis that the head is quite stiff in bending due to its cooling fins and the cylinder pressure of 130 psi is relatively low. We will later calculate the gasket pressure to check for possible leakage.

4 Assume a trial bolt length of 1 in. The head thickness of 0.40 in at the screw holes plus the 0.1 in gasket will leave 0.5 in of thread penetration into the cylinder's tapped hole. This is 2x the screw diameter (10 threads), which is the minimum recommended length for a steel screw in aluminum threads. A 1-in-long 1/4 in screw has threads over 0.75 in of its length[2], which allows the desired penetration. The trial clamp length for the stiffness calculations is then 1 in, since the entire cap screw is engaged.

5 Try an SAE grade 7 bolt preloaded to 70% of its proof strength. Table 14-6 shows the proof strength of this bolt to be 105 kpsi. The tensile stress area from Table 14-1 is 0.0318 in². The required preload is then

$$F_i = 0.7 S_p A_t = 0.7(105\ 000)(0.0318) = 2\ 339 \text{ lb} \qquad (c)$$

6 Find the lengths of thread and shank of the bolt:

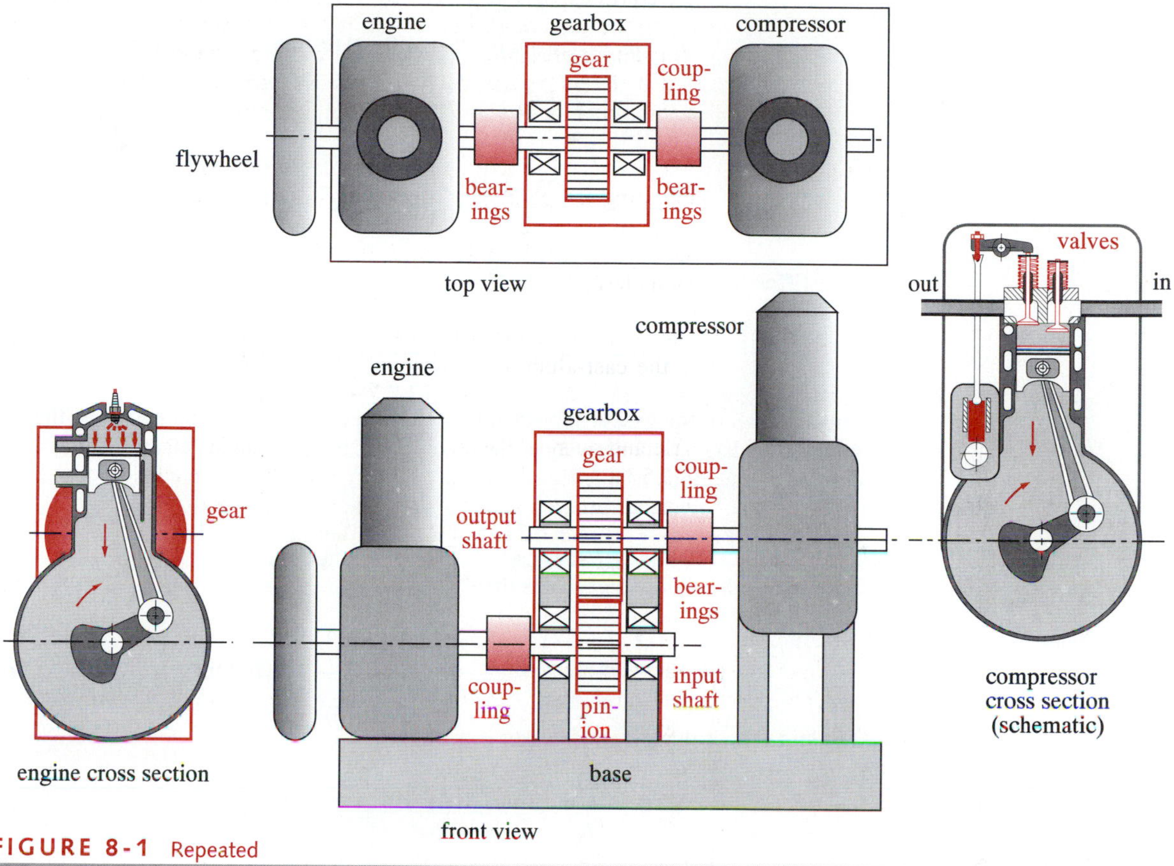

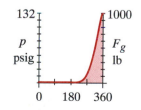

FIGURE 8-1   Repeated

Preliminary Design Schematic of Gasoline-Engine-Powered Portable Air Compressor, Gearbox, Couplings, Shafts, and Bearings

$$l_{thd} = 2d + 0.25 = 2(0.25) + 0.25 = 0.75 \text{ in}$$

$$(d)$$

$$l_s = l - l_{thd} = 1 - 0.75 = 0.25 \text{ in}$$

Because it is a cap screw, the entire thread is in the clamp zone.

$$l_t = l - l_s = 1.0 - 0.25 = 0.75 \text{ in}$$

$$(e)$$

7   Find the stiffness of the bolt from equation 14.11a.

$$\frac{1}{k_b} = \frac{l_t}{A_t E} + \frac{l_s}{A_b E} = \frac{0.75}{0.032(30E6)} + \frac{0.25(4)}{\pi(0.25)^2(30E6)}$$

$$k_b = 1.047E6 \text{ lb/in}$$

$$(f)$$

8   The material and gasket stiffness combine according to equation 13.2b. The area of a "barrel" of material shortened by the gasket thickness (from equation 14.17c) is:

Repeated

FIGURE 8-2

Pressure and Force Within Cylinder During One Cycle

$$d_2 = 1.5d = 1.5(0.250) = 0.375$$

$$d_3 = d_2 + l\tan\phi = 0.375 + (1.0 - 0.06)\tan 42° \cong 1.221 \text{ in}$$

$$A_m \cong \frac{\pi}{4}\left[\left(\frac{d_2 + d_3}{2}\right)^2 - d^2\right]$$

$$\cong \frac{\pi}{4}\left[\left(\frac{0.375 + 1.221}{2}\right)^2 - (0.250)^2\right] \cong 0.451 \text{ in}^2 \qquad (g)$$

9   The stiffness of the material by itself is

$$k_{m_1} = \frac{A_m E_m}{l} = \frac{0.451(10.4E6)}{1.0 - 0.06} \cong 4.993E6 \text{ lb/in} \qquad (h)$$

10   The diameter of the gasket subjected to the clamp force can be assumed to be the same as the largest diameter d_3 of the theoretical "barrel-shaped" clamp zone defined in equations 14.17:

$$d_3 = d_2 + l\tan\phi = 0.375 + (1.0 - 0.06)\tan 42° \cong 1.221 \text{ in} \qquad (i)$$

The area of the clamped gasket is then

$$A_g = \frac{\pi(d_3^2 - d^2)}{4} = \frac{\pi(1.221^2 - 0.250^2)}{4} = 1.123 \text{ in}^2 \qquad (j)$$

11   The stiffness of this piece of gasket is then

$$k_{m_2} = \frac{A_g E_g}{l_g} = \frac{1.123(13.5E6)}{0.06} \cong 2.562E8 \text{ lb/in} \qquad (k)$$

The modulus of elasticity of the gasket material is found in Table 14-9.

12   The combined stiffness of the gasketed joint (from equation 13.2b) is then

$$\frac{1}{k_m} = \frac{1}{k_{m_1}} + \frac{1}{k_{m_2}} = \frac{1}{4.993E6} + \frac{1}{2.562E8}$$

$$k_m = 5.052E6 \text{ lb/in} \qquad (l)$$

Note that the combined stiffness in this case is dominated by the aluminum because the copper-asbestos gasket is stiffer.

13   The joint constant is

$$C = \frac{k_b}{k_m + k_b} = \frac{1.047E6}{5.052E6 + 1.047E6} \cong 0.172 \qquad (m)$$

and $$(1 - C) = 0.828$$

14   The 1 000-lb load is assumed to be divided equally among the 8 bolts at 125 lb each.
The portions of the applied load felt by each bolt and material (equations 14.13) are:

$$P_b = CP = 0.172(125) \cong 21 \ \text{lb}$$

$$P_m = (1-C)P = 0.828(125) \cong 104 \ \text{lb}$$

(n)

15   The resulting peak loads in bolt and material are

$$F_b = F_i + P_b = 2\,339 + 21 = 2\,360 \ \text{lb}$$

$$F_m = F_i - P_m = 2\,339 - 104 = 2\,235 \ \text{lb}$$

(o)

16   The alternating and mean components of force on the bolt are

$$F_{alt} = \frac{F_b - F_i}{2} = \frac{2\,360 - 2\,339}{2} \cong 11 \ \text{lb}$$

$$F_{mean} = \frac{F_b + F_i}{2} = \frac{2\,360 + 2\,339}{2} \cong 2\,350 \ \text{lb}$$

(p)

17   The mean and alternating stresses in the bolt are

$$\sigma_a = K_f \frac{F_{alt}}{A_t} = 3.0 \frac{11}{0.031\,82} \cong 1\,011 \ \text{psi}$$

$$\sigma_m = K_{fm} \frac{F_{mean}}{A_t} = 1.23 \frac{2\,350}{0.031\,82} \cong 90\,924 \ \text{psi}$$

(q)

The fatigue stress-concentration factor K_f for rolled threads is taken from Table 14-8.
The mean stress-concentration factor K_{fm} is found from equation 6.17 and is 1.23.

18   The stresses at the initial preload and at the maximum bolt force are

$$\sigma_i = K_{fm} \frac{F_i}{A_t} = 1.0 \frac{2\,339}{0.031\,82} \cong 90\,509 \ \text{psi}$$

$$\sigma_b = K_{fm} \frac{F_b}{A_t} = 1.0 \frac{2\,360}{0.031\,82} = 91\,340 \ \text{psi}$$

(r)

19   The endurance strength for this material is found using the methods of Section 6.6:

$$S_e' = 0.5 S_{ut} = 0.5(133\,000) = 66\,500 \ \text{psi}$$

(s)

$$S_e = C_{load} C_{size} C_{surf} C_{temp} C_{reliab} S_e'$$
$$= 0.70(1)(0.74)(1)(0.75)(66\,500) = 26\,438 \ \text{psi}$$

(t)

14

where the strength reduction factors are taken from the tables and formulas in
Section 6.6 for, respectively, axial loading, the bolt size, a machined finish, room
temperature, and 99.9% reliability.

20 The corrected endurance strength and the ultimate tensile strength are used in equation 14.16 to find the safety factor from the Goodman line.

$$N_f = \frac{S_e(S_{ut} - \sigma_i)}{S_e(\sigma_m - \sigma_i) + S_{ut}\sigma_a}$$

$$= \frac{26\,438(133\,000 - 90\,509)}{26\,438(90\,924 - 90\,509) + 133\,000(1\,011)} \cong 7.7 \tag{u}$$

21 The safety factor against yielding is

$$N_s = \frac{S_y}{\sigma_b} = \frac{115\,000}{91\,340} = 1.3 \tag{v}$$

22 The load required to separate the joint and the safety factor against joint separation are found from equations 14.14c and 14.14d.

$$N_{sep} = \frac{F_i}{P(1-C)} = \frac{2\,339}{125(1 - 0.172)} \cong 23 \tag{w}$$

23 The joint will leak unless the clamping forces are sufficient to create more pressure at the gasket than exists in the cylinder. The minimum clamping pressure can be found from the total area of the gasketed joint and the minimum clamping force F_m.

$$P_{avg} = \frac{F_m}{A_j} = \frac{4F_m}{\pi\left(D_o^2 - D_i^2\right) - n_b A_b} = \frac{4(2\,235)}{\pi\left(5.125^2 - 3.125^2\right) - 8(0.049)} \cong 178 \ \text{psi} \tag{x}$$

$$N_{leak} = \frac{P_{avg}}{P_{cyl}} = \frac{178}{130} \cong 1.4$$

This ratio of clamp pressure over cylinder pressure makes the spacing of the screws acceptable. The above design is shown in the disk file CASE7D-1.

24 The preload force needed was found in step 5 and is

$$F_i = 2\,339 \ \text{lb} \tag{y}$$

25 The torque required to obtain that preload is

$$T_i \cong 0.21 F_i d = 0.21(2\,339)(0.250) \cong 120 \ \text{lb-in} \tag{z}$$

26 These safety factors are acceptable, but could be improved by increasing the diameter of the bolts to 0.312 in and reducing their number to 7. This second design is shown in the disk file CASE7D-2.   It uses seven 5/16-18 UNC-2A, grade 7 hex-head cap screws, 1.0 in long, preloaded to 80% of proof strength and equispaced on a 4.125-in-diameter bolt circle. This design increases the safety factor against leakage to 2, and the fatigue safety factor to 7, with a dynamic yielding safety factor of 1.2.

14

14.12 SUMMARY

This chapter has dealt with only a small sample of commercially available fasteners. An extremely varied collection of fasteners is produced by vendors. The "right" fastener can usually be found for any application and if not (and the required quantity is high enough) some vendor will make a new one for you. Many standards exist that define the configurations, sizes, strengths, and tolerances of fasteners. Threaded fasteners are made to one or another of these standards which provides good interchangeability. Unfortunately metric and English threads are not interchangeable and both are in wide use in the United States.

Power screws are threaded devices used primarily to move loads or accurately position objects. They have low efficiency due to their large friction losses unless the ball-screw variety is used, which lowers the friction significantly. However, low-friction screws give up one of their advantages, which is self-locking or the ability to hold a load in place with no input of energy (as in a jack). Back-drivable screws are the opposite of self-locking and can be used as a linear to rotary motion converter.

Threaded fasteners (bolts, nuts, and screws) are the standard means of holding machinery together. These fasteners are capable of supporting very large loads, especially if they are preloaded. Preloading tightens the fastener to a high level of axial tension before any working loads are applied. The tension in the fastener causes compression in the clamped parts. This compression has several salutary effects. It keeps the joint tightly together and thus able to contain fluid pressure and resist shear loads with its interfacial friction. The compressive forces in the clamped material also serve to protect the fastener from fluctuating fatigue loads by absorbing most of the applied load oscillations. The high clamping forces also guard against vibratory loosening of the fastener by creating high friction forces in the threads.

Threaded fasteners are also capable of resisting shear loads and are used extensively in that manner in structural applications. In machine design it is more common to rely on close-fitted dowel pins to take shear loads and let the threaded fasteners provide the tension to hold the joint together.

The interested reader is referred to the publications listed in the bibliography for more information on the diverse and fascinating world of fasteners.

Important Equations Used in this Chapter

See the referenced sections for information on the proper use of these equations.

Torque Required to Raise the Load with a Power Screw (Section 14.2):

$$T_u = T_{s_u} + T_c = \frac{Pd_p}{2} \frac{\left(\mu\pi d_p + L\cos\alpha\right)}{\left(\pi d_p \cos\alpha - \mu L\right)} + \mu_c P \frac{d_c}{2} \qquad (14.5a)$$

Self-locking of a Power Screw Will Occur if (Section 14.2):

$$\mu \geq \frac{L}{\pi d_p}\cos\alpha, \qquad \text{or} \qquad \mu \geq \tan\lambda\cos\alpha \qquad\qquad (14.6a)$$

Efficiency of a Power Screw (Section 14.2):

$$e = \frac{W_{out}}{W_{in}} = \frac{PL}{2\pi T} \qquad\qquad (14.7c)$$

Spring Constant of a Threaded Fastener (Section 14.7):

$$\frac{1}{k_b} = \frac{l_t}{A_t E_b} + \frac{l-l_t}{A_b E_b} = \frac{l_t}{A_t E_b} + \frac{l_s}{A_b E_b} \qquad\qquad (14.11a)$$

Spring Constant of the Clamped Material (Section 14.7):

$$k_m = \frac{A_m E_m}{l} \qquad\qquad (14.11c)$$

Approximate Effective Area of Clamped Material Around One Fastener (Section 14.8):

$$A_m \cong \frac{\pi}{4}\left(D_{eff}^2 - d^2\right) \cong \frac{\pi}{4}\left[\left(\frac{d_2+d_3}{2}\right)^2 - d^2\right] \qquad\qquad (14.17a)$$

Load Taken by a Preloaded Bolt and the Joint Constant C (Section 14.7):

$$P_b = \frac{k_b}{k_m + k_b}P$$

or
$$\qquad\qquad (14.13c)$$

$$P_b = CP \qquad \text{where} \quad C = \frac{k_b}{k_m + k_b}$$

Load Taken by the Preloaded Material (Section 14.7):

$$P_m = \frac{k_m}{k_b + k_m}P = (1-C)P \qquad\qquad (14.13d)$$

Minimum Load in Material and Maximum Load in Bolt (Section 14.7):

$$F_m = F_i - (1-C)P \qquad\qquad (14.14a)$$

$$F_b = F_i + CP \qquad\qquad (14.14b)$$

Load Required to Separate a Preloaded Joint (Section 14.7):

$$P_0 = \frac{F_i}{(1-C)} \qquad\qquad (14.14c)$$

14

Mean and Alternating Loads Felt by a Preloaded Bolt (Section 14.7):

$$F_{alt} = \frac{F_b - F_i}{2}, \qquad F_{mean} = \frac{F_b + F_i}{2} \qquad (14.15a)$$

Mean and Alternating Stresses in a Preloaded Bolt (Section 14.7):

$$\sigma_{alt} = K_f \frac{F_{alt}}{A_t}, \qquad \sigma_{mean} = K_{fm} \frac{F_{mean}}{A_t} \qquad (14.15b)$$

Preload Stress in a Bolt (Section 14.7):

$$\sigma_i = K_{fm} \frac{F_i}{A_t} \qquad (14.15c)$$

Fatigue Safety Factor for a Preloaded Bolt (Section 14.7):

$$N_f = \frac{S_e(S_{ut} - \sigma_i)}{S_e(\sigma_m - \sigma_i) + S_{ut}\sigma_a} \qquad (14.16)$$

Approximate Torque Needed to Preload a Bolt (Section 14.9):

$$T_i \cong 0.21 F_i d \qquad (14.18d)$$

Centroid of a Group of Fasteners (Section 14.10):

$$\tilde{x} = \frac{\sum_1^n A_i x_i}{\sum_1^n A_i}, \qquad \tilde{y} = \frac{\sum_1^n A_i y_i}{\sum_1^n A_i} \qquad (14.19)$$

Forces on Fasteners Eccentrically Loaded in Shear (Section 14.10):

$$\left| F_{1_i} \right| = \frac{P}{n} \qquad (14.20a)$$

$$\left| F_{2_i} \right| = \frac{M}{r_i} = \frac{Pl}{r_i} \qquad (14.20b)$$

14.13 REFERENCES

1 *Product Engineering*, vol. 41, p. 9, Apr. 13, 1970.

2 **H. L. Horton**, ed. *Machinery's Handbook*, 21st ed. Industrial Press, Inc.: New York. p. 1256, 1974.

3 **ANSI/ASME Standard B1.1-1989**, American National Standards Institute, New York, 1989.

4 **ANSI/ASME Standard B1.13-1983 (R1989)**, American National Standards Institute, New York, 1989.

5 **T. H. Lambert**, Effects of Variations in the Screw Thread Coefficient of Friction on Clamping Force of Bolted Connections, *J. Mech Eng. Sci.*, **4**: p. 401, 1962.

6 **R. E. Peterson**, *Stress-concentration factors*. John Wiley & Sons: New York, pp. 253, 1974.

7 **H. H. Gould and B. B. Mikic**, Areas of Contact and Pressure Distribution in Bolted Joints, *Trans ASME, J. Eng. for Industry*, **94**: pp. 864-869, 1972.

8 **N. Nabil**, Determination of Joint Stiffness in Bolted Connections, *Trans. ASME, J. Eng. for Industry*, **98**: pp. 858-861, 1976.

9 **Y. Ito, J. Toyoda and S. Nagata**, Interface Pressure Distribution in a Bolt-Flange Assembly, *Trans. ASME, J. Mech. Design*, **101**: pp. 330-337, 1979.

10 **J. F. Macklin and J. B. Raymond**, *Determination of Joint Stiffness in Bolted Connections using FEA*, Major Qualifying Project, Worcester Polytechnic Institute, Worcester MA, Dec. 31, 1994.

11 **B. Houle**, *An Axisymmetric FEA Model Using Gap Elements to Determine Joint Stiffness*, Major Qualifying Project, Worcester Polytechnic Institute, Worcester MA, Dec. 31, 1995.

12 **J. Shigley and L. Mitchell**, *Mechanical Engineering Design*, 4th ed. McGraw-Hill: New York, p. 389, 1983.

13 **ANSI Standard B18.8.2-1978 (R1989)**, American National Standards Institute, New York, 1989.

14.14 BIBLIOGRAPHY

American Institute of Steel Construction Handbook. AISI: New York.

ASME Boiler and Pressure Vessel Code. ASME: New York.

Industrial Ball Bearing Screw Catalog, Saginaw Steering Gear Division, General Motors Corp., Saginaw Mich.

Helpful Hints for Fastener Design and Application. Russell, Burdsall & Ward Corp.: Mentor, Ohio, 1976.

SAE Handbook. Soc. of Automotive Engineers: Warrandale Pa., 1982.

"Fastening and Joining Reference Issue," *Machine Design*, vol. 55, Nov. 17, 1983.

J. H. Bickford, *An Introduction to the Design and Behavior of Bolted Joints*, 2nd ed. Marcel Dekker, New York, 1990.

H. L. Horton, ed., *Machinery's Handbook*. 21st ed. Industrial Press.: New York, 1974.

R. O. Parmley, ed. *Standard Handbook of Fastening and Joining*. McGraw-Hill: New York. 1977.

H. A. Rothbart, ed., *Mechanical Design and Systems Handbook*. McGraw-Hill: New York, Sections 20, 21, 26, 1964.

14

14.15 PROBLEMS

14-1   Compare the tensile load capacity of a 5/16-18 UNC thread and a 5/16-24 UNF thread made of the same material. Which is stronger? Make the same comparison for M8 x 1.25 and M8 x 1 ISO threads. Compare them all to the strength of a 5/16-14 Acme thread.

*14-2   A 3/4-6 Acme thread screw is used to lift a 2 kN load. The mean collar diameter is 4 cm. Find the torque to lift and to lower the load using a ball-bearing thrust washer. What are the efficiencies? Is it self-locking?

14-3   A 1 3/8-4 Acme thread screw is used to lift a 1-ton load. The mean collar diameter is 2 in. Find the torque to lift and to lower the load using a ball-bearing thrust washer. What are the efficiencies? Is it self-locking?

*†14-4   The trailer hitch from Figure 1-1 (p. 12) has loads applied as shown in Figure P14-1. The tongue weight of 100 kg acts downward and the pull force of 4 905 N acts horizontally. Using the dimensions of the ball bracket in Figure 1-5 (p. 15), draw a free-body diagram of the ball bracket and find the tensile and shear loads applied to the two bolts that attach the bracket to the channel in Figure 1-1. Size and specify the bolts, their preload, and tightening torque for a safety factor of at least 1.7.

†14-5   For the trailer hitch of Problem 3-4, determine the horizontal force that will result on the ball from accelerating a 2 000-kg trailer to 60 m/s in 20 sec. Assume a constant acceleration. Size and specify the bolts, their preload, and tightening torque for a safety factor of at least 1.7.

*†14-6   For the trailer hitch of Problem 3-4, determine the horizontal force that will result on the ball from an impact between the ball and the tongue of the 2 000-kg trailer if the hitch deflects 1 mm on impact. The tractor weighs 1 000 kg. The velocity at impact is 0.3 m/s. Size and specify the bolts, their preload, and tightening torque for a safety factor of at least 1.7.

*14-7   A 1/2-in-dia UNC, class 7 bolt with rolled threads is preloaded to 80% of its proof strength when clamping a 3-in-thick sandwich of solid steel. Find the safety factors against static yielding and joint separation when a static 1 000-lb external load is applied. Use 99% reliability.

14-8   An M14 x 2, class 8.8 bolt with rolled threads is preloaded to 75% of its proof strength when clamping a 3-cm-thick sandwich of solid aluminum. Find the safety factors against static yielding and joint separation when a static 5 kN external load is applied. Use 99% reliability.

*14-9   A 7/16-in-dia UNC, class 7 bolt with rolled threads is preloaded to 70% of its proof strength when clamping a 2.75-in-thick sandwich of solid steel. Find the safety factors against fatigue failure, yielding and joint separation when a 5 000-lb (peak) fluctuating external load is applied. Use 99% reliability.

14-10   An M12 x 1.25, class 9.8 bolt with rolled threads is preloaded to 85% of its proof strength when clamping a 5-cm-thick sandwich of aluminum. Find the safety factors against fatigue failure, yielding, and joint separation when a 20 kN (peak) fluctuating external load is applied. Use 99% reliability.

*14-11   Find the tightening torque required for the bolt in Problem 14-7.

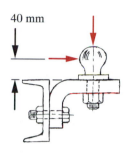

40 mm

FIGURE P14-1

Problems 14-4 to 14-6

* Answers to these problems are provided in Appendix G.

† These problems are based on similar problems in previous chapters with the same –number, e.g., Problem 14-4 is based on Problem 3-4, etc.

14-12   Find the tightening torque required for the bolt in Problem 14-8.

*14-13   Find the tightening torque required for the bolt in Problem 14-9.

14-14   Find the tightening torque required for the bolt in Problem 14-10.

14-15   An automobile manufacturer would like a feasibility study of the concept of building-in electric-motor powered screw jacks at each end of the car to automatically jack the car wheels off the ground for service. Assuming a 2-ton vehicle with a 60/40 front/rear weight distribution, design a self-locking screw jack capable of lifting either end of the car. The jack body will be attached to the car frame and the screw will extend downward to engage the ground. Assume a minimum installed clearance of 8 in under the retracted screw in the up position. It must lift the car frame at least 8 additional inches. Use rolling element thrust bearings. Determine a minimum screw size safe against column buckling. Determine its required lifting torque and efficiency and the power required to lift it to full height in 45 sec. What is your recommendation as to the feasibility of this idea?

14-16   Design a manual screw jack similar to that shown in Figure 14-4 for a 20-ton lift capacity and a 10-cm lift stroke. Assume that the operator can apply a 400-N force at the tip of its bar handle to turn either the screw or nut depending on your design. Design the cylindrical bar handle to fail in bending at the design load before the jack screw fails so that one cannot lift an overload and fail the screw. Use rolling element thrust bearings. Seek a safety factor of 3 for thread or column failure. State all assumptions.

*14-17   Determine the effective spring constant of the following sandwiches of materials under compressive load. All are uniformly loaded over their 10-cm^2 area. The first and third-listed materials are each 10 mm thick and the middle one is 1 mm thick, together making a 21-mm-thick sandwich.

(a)   aluminum, copper-asbestos, steel
(b)   steel, copper, steel
(c)   steel, rubber, steel
(d)   steel, rubber, aluminum
(e)   steel, aluminum, steel

In each case determine which material dominates the calculation.

*14-18   Repeat problem 14-17 with the assumption that the sandwich is compressed by a single 10-mm bolt in the center of the area. Determine the effective spring constant of the material sandwich based on the frustum-cone assumption.

14-19   What tightening torque would be needed to obtain a safety factor of 2 against joint separation in each case of Problem 14-18 if an external load of 35 000 lb is applied?

*14-20   A single-cylinder engine head sees explosive forces that range from 0 to 18.5 kN each cycle. The head is 10-cm-thick aluminum, the unconfined gasket is 1-mm-thick copper-asbestos, and the block is cast iron. The piston is 75 mm dia and the cylinder is 140 mm outside dia. Specify a suitable number, class, preload, tightening torque, and bolt circle for the cylinder head cap screws to give a minimum safety factor of 1.5 for any possible failure mode. Use fine-thread screws.

14-21   Repeat Problem 14-20 for an engine with cast-iron head and block.

14

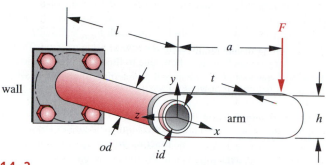

FIGURE P14-2

Problems 14-23 through 14-25

14-22   The forged steel connecting rod for the engine of problem 14-20 is split around the 38-mm-dia crankpin and fastened with two bolts and nuts that hold its two halves together. The total load on the two bolts varies from 0 to 28 kN each cycle. Design these bolts for infinite life. Specify their size, class, preload, and tightening torque.

*†14-23   (See also Problem 4-33.) The bracket in Figure P14-2 is fastened to the wall by 4 bolts equispaced on a 10-cm-dia bolt circle and arranged as shown. The bracket is subjected to a static force F, where F and the beam's other data are given in the row(s) assigned from Table P14-1. Find the forces acting on each of the 4 bolts due to this loading and choose a suitable bolt size and preload that will give a minimum safety factor of 2 for any possible mode of failure.

Table P14-1   Data for Problems 14-23 through 14-25

Use Only Data Relevant to the Particular Problem—Lengths in mm, Forces in N

Row	l	a	t	h	F	od	id	E
a	100	400	10	20	50	10	4	steel
b	70	200	6	80	85	12	6	steel
c	300	100	4	50	95	15	7	steel
d	800	500	6	65	250	25	15	alum
e	85	350	5	96	900	40	30	alum
f	50	180	4	45	950	30	25	alum
g	160	280	5	25	850	45	40	steel
h	200	100	2	10	800	40	35	steel
i	400	150	3	50	950	45	38	steel
j	200	100	3	10	600	30	20	alum
k	120	180	3	70	880	60	55	alum
l	150	250	8	90	750	45	30	alum
m	70	100	6	80	500	20	12	steel
n	85	150	7	60	820	25	15	steel

* Answers to these problems are provided in Appendix G.

† These problems are based on similar problems in previous chapters with the same –number, e.g., Problem 14-4 is based on Problem 3-4, etc.

14

*†14-24   (See also Problem 6-33.) The bracket in Figure P14-2 is fastened to the wall by 4
          bolts equispaced on a 10-cm-dia bolt circle and arranged as shown. The bracket is
          subjected to a sinusoidal force time function with $F_{max} = F$ and $F_{min} = -F$, where F
          and the beam's other data are given in the row(s) assigned from Table P14-1. Find
          the forces acting on each of the 4 bolts due to this fully reversed loading and choose
          a suitable bolt size and preload that will give a minimum safety factor of 1.5 for any
          possible mode of failure for $N = 5E8$ cycles.

*†14-25   (See also Problem 6-34.) The bracket in Figure P14-2 is fastened to the wall by 4
          bolts equispaced on a 10-cm-dia bolt circle and arranged as shown. The bracket is
          subjected to a sinusoidal force time function with $F_{max} = F$ and $F_{min} = 0$, where F
          and the beam's other data are given in the row(s) assigned from Table P14-1. Find
          the forces acting on each of the 4 bolts due to this fluctuating loading and choose a
          suitable bolt size and preload that will give a minimum safety factor of 1.5 for any
          possible mode of failure for $N = 5E8$ cycles.

†14-26   (See also Problem 6-42.) A cylindrical, steel tank with hemispherical ends is
          required to hold 150 psi of pressurized air at room temperature. The pressure cycles
          from zero to maximum. The tank diameter is 0̄.5 m and its length is 1 m. The
          hemispherical ends are attached by some number of bolts through mating flanges on
          each part of the tank. A 0.5-mm-thick, compressed asbestos, unconfined gasket is
          used between the 2.5-mm-thick steel flanges. Determine a suitable number, class,
          preload for, and size of bolts to fasten the ends to the tank. Specify the bolt circle
          and outside diameter of the flange needed to prevent leakage. A minimum safety
          factor of 2 is desired against leakage and a safety factor of 1.5 against bolt failure
          for infinite life.

†14-27   Repeat Problem 14-26 using a confined O-ring gasket.

* Answers to these problems
are provided in Appendix G.

† These problems are based
on similar problems in
previous chapters with the
same –number, e.g., Problem
14-4 is based on Problem
3-4, etc.

CLUTCHES AND BRAKES

A big book is a big nuisance.
CALLIMACHAS, 260 BC.

15.0 INTRODUCTION

Clutches and brakes are essentially the same device. Each provides a frictional, magnetic, hydraulic, or mechanical connection between two elements. If both connected elements can rotate, then it is called a clutch. If one element rotates and the other is fixed, it is called a brake. A clutch thus provides an interruptible connection between two rotating shafts as, for example, the crankshaft of an automobile engine and the input shaft of its transmission. A brake provides an interruptible connection between one rotating element and a nonrotating ground plane as, for example, the wheel of an automobile and its chassis. The same device may be used as either clutch or brake by fixing its output element to a rotatable shaft or by fixing it to ground.

Brakes and clutches are used extensively in production machinery of all types, not just in vehicle applications where they are needed to stop motion and allow the internal-combustion engine to continue turning (idling) when the vehicle is stopped. Clutches also allow a high inertia load to be started with a smaller electric motor than would be required if it were directly connected. Clutches are commonly used to maintain a constant torque on a shaft for tensioning of webs or filaments. A clutch may be used as an emergency disconnect device to decouple the shaft from the motor in the event of a machine jam. In such cases, a brake will also be fitted to bring the shaft (and machine) to a rapid stop in an emergency. To minimize injuries, many U. S. manufacturers require their production machinery to stop within one or fewer revolutions of the main driveshaft if a worker hits the "panic bar" that typically spans the length of the machine. This can be a difficult specification to achieve on large machines (10 to 100

Table 15-0   Variables Used in this Chapter

Symbol	Variable	ips units	SI units	See
a	length	in	m	Sect. 15.6
b	length	in	m	Sect. 15.6
c	length	in	m	Sect. 15.6
d	diameter	in	m	various
F	force	lb	N	various
F_a	applied force	lb	N	Sect. 15.6
F_f	friction force	lb	N	Sect. 15.6
F_n	normal force	lb	N	Sect. 15.6
R_x	reaction force	lb	N	Sect. 15.6
R_y	reaction force	lb	N	Sect. 15.6
K	arbitrary constant	none	none	various
l	length	in	m	various
M	moment	lb-in	N-m	various
N	number of friction surfaces	none	none	Eq. 15.2
P	power	hp	Watts	Ex. 15.1
p	pressure	psi	N/m^2	Sect. 15.4
P_{max}	maximum pressure	psi	N/m^2	Sect. 15.4
r	radius	in	m	various
r_i	inside radius of disk lining	in	m	various
r_o	outside radius of disk lining	in	m	various
T	torque	lb-in	N-m	various
V	linear velocity	in/sec	m/sec	Eq. 15.4
W	wear rate	psi-in/sec	Pa-m/sec	Eq. 15.4
w	width	in	m	various
θ	angular position	rad (deg)	rad (deg)	Sect. 15.6
μ	coefficient of friction	none	none	Eq. 15.2
ω	angular velocity	rad/sec	rad/sec	Ex. 15-1

feet long) driven by multi-horsepower electric motors. Manufacturers provide clutch-brake combinations in the same package for such applications. Applying power disengages the brake and engages the clutch making it fail-safe. Fail-safe brakes are engaged (typically by internal springs) unless power is applied to disengage them. Thus they "fail safe" and stop the load if the power fails. Highway truck and railroad-car air brakes are of this type. Air pressure releases the brake, which is normally engaged. If the railroad car or truck trailer breaks loose and severs its air hose connection to the engine or tractor, the brakes automatically engage.

15

Title page photograph courtesy of the Logan Clutch Corporation, Cleveland, Ohio.

This chapter will describe a number of types of commercially available clutches and brakes and their typical applications, and will also discuss the theory and design of a few particular types of friction clutches/brakes. Table 15-0 lists the variables used in this chapter and indicates the equation or section in which they can be found.

15.1   TYPES OF BRAKES AND CLUTCHES

Brakes and clutches can be classified in a number of ways, by the *means of actuation*, the *means of energy transfer* between the elements, and the *character of the engagement*. Figure 15-1 shows a flow chart depicting these characteristics. The means of actuation may be **mechanical**, as in the pushing of an automobile's clutch pedal, **pneumatic or hydraulic**, in which fluid pressure drives a piston to mechanically engage or disengage as with vehicle brakes, **electrical**, which is typically used to excite a magnetic coil, or **automatic** as in an anti-runaway brake that engages by relative motion between the elements.

POSITIVE CONTACT CLUTCHES   The means of energy transfer may be a **positive mechanical contact**, as in a toothed or serrated clutch, which engages by mechanical interference as shown in Figure 15-2. The character of engagement is mechanical interference obtained with jaws of square or saw-toothed shape, or with teeth of various shapes. These devices are not as useful for brakes (except as holding devices) because they cannot dissipate large amounts of energy as can a friction brake, and as clutches they can only be engaged at low relative velocities (about 60-rpm max for jaw-clutches and 300 rpm max for toothed-clutches). Their advantage is positive engagement and, once coupled, can transmit large torque with no slip. They are sometimes combined with a friction-type clutch, which drags the two elements to nearly the same velocity before the jaws or teeth engage. This is the principle of a **synchromesh clutch** in a manual automotive transmission.*

FRICTION CLUTCHES AND BRAKES   are the most common types used. Two or more surfaces are pressed together with a normal force to create a friction torque. The friction surfaces may be flat and perpendicular to the axis of rotation, in which case the normal force is axial (disk brake or clutch), as shown in Figure 15-3, or they may be cylindrical with the normal force in a radial direction (drum brake or clutch) as shown in Figures 15-8 through 15-9, or conical (cone brake or clutch). Cone clutches can tend to grab or refuse to release and are not now used very much in the United States, but are popular in Europe.[1]

At least one of the friction surfaces is typically metal (cast iron or steel) and the other is usually a high-friction material, referred to as the lining. If there are only two elements, there will be either one or two friction surfaces to transmit the torque. A cylindrical arrangement (drum brake or clutch) has one friction surface, and an axial arrangement (disk brake or clutch) has one or two friction surfaces depending on whether the disk is sandwiched between two surfaces of the other element or not. For higher torque capacity, disk clutches and brakes are often made with multiple disks to increase

* Automotive transmissions typically use helical gears, for quiet operation, as was noted in Chapter 12. The helical gears cannot be easily shifted in and out of engagement in manual transmissions because of their helix angle. So, they are all kept in constant mesh and clutched/declutched from the transmission shaft to engage a particular ratio. Each gear has a synchromesh clutch connecting it to its shaft. This clutch actually consists of conical friction surfaces that drag the two elements (shaft and gear) into near zero relative velocity before the teeth of its companion positive-contact clutch engage. The shift lever moved by the driver is shifting these synchromesh clutches into and out of engagement, rather than moving gears around in the transmission.

15

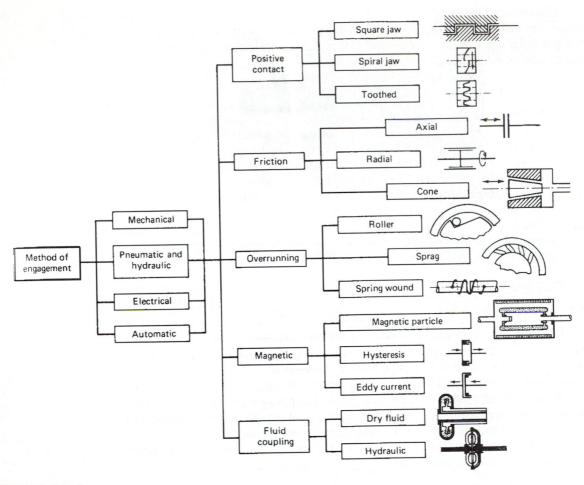

FIGURE 15-1

Classification of Clutches and Brakes   *Source: U. Hunhede et al., Machine Design Fundamentals, Prentice-Hall, 1983, with permission*

the number of friction surfaces (see Figure 15-3). A clutch or brake's ability to transfer the heat of friction generated can become the limiting factor on its capacity. Multidisk clutches are more difficult to cool so are appropriate for high-load, low-speed applications. For high-speed dynamic loads, fewer friction surfaces are better.[1]

Friction clutches may be operated either dry or wet, the latter running in an oil bath. While the oil severely reduces the coefficient of friction, it greatly increases the heat transfer. Friction coefficients of clutch/brake material combinations typically range from 0.05 in oil to 0.60 dry. Wet clutches often use multiple disks to make up for the lower friction coefficient. Automatic transmissions for automobiles and trucks contain many wet clutches and brakes operating in oil that is circulated out of the transmission for cooling. Manual transmissions for off-road vehicles, such as motorcycles, use sealed, oil-

15

FIGURE 15-2

Positive Contact Clutch   *Courtesy of American Precision Industries, Deltran Division, Amherst, N.Y., 14228*

filled, multi-disk wet clutches to protect the friction surfaces from dust, water, and dirt. Manual-transmission automobiles and trucks typically use single-disk, dry clutches.

OVERRUNNING CLUTCHES   (also called one-way clutches) operate automatically based on the relative velocity of the two elements. They act on the circumference and allow relative rotation in only one direction. If the rotation attempts to reverse, the internal geometry of the clutch mechanism grabs the shaft and locks up. These *backstop clutches* can be used on a hoist to prevent the load from falling back if power to the shaft is interrupted, for example. These clutches are also used as indexing mechanisms. The

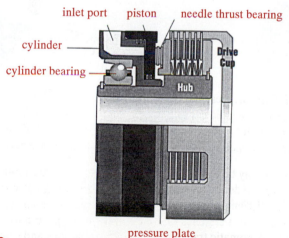

FIGURE 15-3

A Multiplate Disk Clutch Actuated by Fluid Pressure *Courtesy of Logan Clutch Corporation, Cleveland, Ohio*

input shaft can oscillate back and forth but the output turns intermittently in only one direction. Another common application of an overrunning clutch is in the rear hub of a bicycle to allow freewheeling when the wheel speed exceeds that of the drive sprocket.

Several different mechanisms are used in one-way clutches. Figure 15-4a shows a **sprag clutch**, which has an inner and outer race like a ball bearing. But, instead of balls, the space between the races is filled with odd-shaped *sprags,* which allow motion in one direction but jam and lock the races together in the other, transmitting a one-way torque. A similar result is obtained with balls or rollers captured in wedge-shaped chambers between the races, then called a **roller clutch**. Figure 15-4b shows another type of one-way or overrunning clutch called a **spring clutch**, which uses a spring wrapped tightly around the shaft. Rotation in one direction wraps the spring tighter on the shaft and transmits torque. Counterrotation unwinds the spring slightly allowing it to slip.

CENTRIFUGAL CLUTCHES engage automatically when the shaft speed exceeds a certain magnitude. Friction elements are thrown radially outward against the inside of a cylindrical drum to engage the clutch. Centrifugal clutches are sometimes used to couple an internal combustion engine to the drive train. The engine can idle decoupled from the wheels and when the throttle is opened, its increased speed automatically engages the clutch. These are common on go-karts. Used on chain saws for the same purpose, they also serve as an overload release that slips to allow the engine to continue running when the chain jams in the wood.

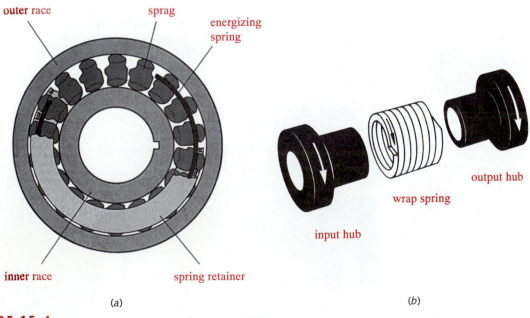

(a)   (b)

15 **FIGURE 15-4**

Overrunning Clutches (a) Sprag Clutch (b) Wrap-Spring Clutch *Courtesy of Warner Electric, South Beloit, Ill., 61080*

MAGNETIC CLUTCHES AND BRAKES   are made in several types. **Friction clutches** are commonly electromagnetically operated, as shown in Figure 15-5a. These have many advantages such as rapid response times, ease of control, smooth starts and stops, and are available powered on or powered off (fail-safe). Both clutch and brake versions are supplied as well as combined clutch-brake modules.

Magnetic particle clutches and brakes (not shown) have no direct frictional contact between the clutch disk and the housing and no friction material to wear. The gap between the surfaces is filled with a fine ferrous powder. When the coil is energized, the powder particles form chains along the magnetic field's flux lines and couple the disk to the housing with no slip. The torque can be controlled by varying the current to the coil and the device will slip when the applied torque exceeds the value set by coil current, providing a constant tension.

Magnetic hysteresis clutches and brakes (Figure 15-5b) have no mechanical contact between the rotating elements and so have zero friction when disengaged. The rotor, also called the drag cup, is dragged along (or braked) by the magnetic field set up by the field coil (or permanent magnet). These devices are used to control torque on shafts in applications such as winding machines, where a constant force must be applied to a web or filament of material as it is wound up. The torque on a hysteresis clutch is controllable independent of speed. These devices are extremely smooth, quiet, and long-lived as there is no mechanical contact within the clutch except in its bearings.

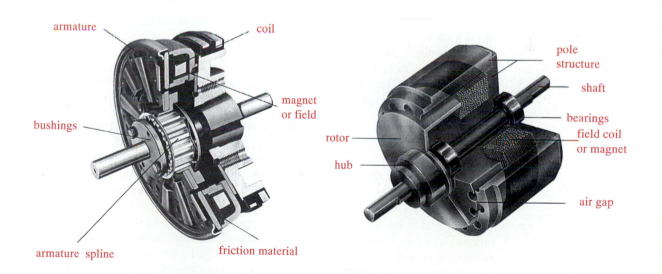

FIGURE 15-5

Magnetic Clutches (a) Magnetically Operated Friction Clutch (b) Hysteresis Clutch *Source: (a) Courtesy of Warner Electric, South Beloit, Ill., 61080, (b) Courtesy of Magtrol, Buffalo, N.Y., 14224*

15

Eddy current clutches (not shown) are similar in construction to hysteresis devices in that they have no mechanical contact between rotor and pole. The coil sets up eddy currents, which magnetically couple the clutch together. There will always be some slip in this type of clutch as there has to be relative motion between rotor and pole to generate the eddy currents that supply the coupling force; so an eddy current brake cannot hold a load stationary, only slow it from one speed to another. These have similar advantages to the hysteresis devices and are used for similar applications such as coil or filament winders, etc.

FLUID COUPLINGS   transmit torque through a fluid, typically an oil. An impeller having a set of blades is turned by the input shaft and imparts angular momentum to the oil that surrounds it. A turbine (or runner) with similar blades is attached to the output shaft and is turned by the moving oil impinging on it. The principle of operation is similar to placing two electric fans face to face and turning only one on. The airflow from the powered fan blades will cause the facing, unpowered blades to windmill, passing power without any mechanical contact. Using incompressible oil in a confined volume is much more efficient than two open-air fans, especially when the impeller and turbine blades are optimally shaped to pump the oil. A fluid coupling provides extremely smooth starts and absorbs shocks as the fluid simply shears when there is a speed differential, then gradually accelerates (or decelerates) the output turbine to nearly match the speed of the impeller. There will always be some slip, meaning the turbine can never reach 100% of the impeller's speed (0% slip), but it can operate at 100% slip when the turbine is stalled. All the input energy will then be converted to heat in shearing the oil. Heat transfer is an important consideration when designing a fluid coupling. The outside case is often finned to improve heat transfer. A fluid coupling transmits the input torque to the output shaft at any speed including stall, so can never be totally decoupled like a friction clutch. The output must be braked to hold it stationary when the input shaft is turning.[*] The horsepower rating of a fluid coupling varies as the fifth power of its diameter. A 15% increase in diameter doubles its power capacity. If used as a brake, a fluid coupling can only provide a drag to slow a device as in a dynamometer, but cannot hold a load stationary.

If a third, stationary, element with a set of curved blades, called a **reactor** or **stator**, is placed between impeller and turbine, additional angular momentum is imparted to the fluid and the device is then called a **torque converter**. Torque converters are used in vehicles to couple the engine to an automatic transmission. The engine can idle with the vehicle stopped (turbine stalled—100% slip). At stall the impeller and reactor blades create about a 2:1 torque multiplication, which is available to help accelerate the vehicle when the brakes are released and the engine speed increased. As the vehicle (and thus turbine) speed increases, the turbine will approach the impeller speed and the torque multiplication will decrease to essentially zero at a slip of a few percent. If more torque is needed momentarily (as in passing) the slip between impeller and turbine will automatically increase when the engine is sped up to provide more torque and power to accelerate the vehicle.

[*] This is why you must keep your foot on the brake when stopped at a traffic light in an automobile that has an automatic transmission if the engine is running and the transmission is in "drive." The fluid coupling between the engine and transmission is always transmitting torque and the car will creep forward at idle unless the brakes are applied.

15

15.2   CLUTCH/BRAKE SELECTION AND SPECIFICATION

Manufacturers of specialty clutches and brakes such as those described above provide extensive information on the torque and power capacities of their various models in catalogs, many of which are as informative as a textbook on the particular subject. They also define procedures for selection and specification, usually based on the anticipated torque and power for the application plus suggested service factors that attempt to accommodate different loading, installation, or environmental factors than those under which the products were tested. For example, the manufacturer's standard rating for a clutch model may be based on a smooth driver such as an electric motor. If the particular application uses an internal combustion engine of the same power, there will be impulse loads and a clutch or brake of larger capacity than dictated by the average power will need to be selected. This is sometimes referred to as *derating* the clutch (or brake), meaning that its actual capacity under the anticipated conditions is considered to be less than the rated capacity of the chosen device.

SERVICE FACTORS   According to many clutch manufacturers, a common cause of clutch trouble is the designer's failure to properly apply adequate service factors to account for the particular conditions of the application.[1] This may, in part, be due to confusion engendered by a lack of standardization for definitions of service factors. One manufacturer may recommend a service factor of 1.5 for a particular condition while another manufacturer recommends 3.0 for the same condition. Both will be correct for their particular clutch designs because, in one case, the manufacturer may have already built-in a safety factor while the other applies it in the service factor. The wise designer will carefully follow each manufacturer's recommended selection procedures for their products realizing that they are typically based on extensive (and expensive) testing programs as well as on field-service experience with that particular product.

A clutch even slightly too small for the applied loading will slip and overheat. A clutch too large for the load is also bad as it adds unnecessary inertia and may overload the motor that must accelerate it. Most manufacturers of machine parts are generous in providing engineering help to properly size and specify their products for any application. The machine designer's principal concern should be to accurately define the loading and environmental conditions that the device must accommodate. This may require extensive and tedious calculations of such things as the moments of inertia for all the elements in the drive train actuated by the clutch or brake. The load-analysis methods of Chapter 3 are applicable to such a task.

CLUTCH LOCATION   When a machine has both high- and low-speed shafts (as it will whenever a speed reducer is used, such as in Case Studies 7 and 8), and a clutch is needed in the system, the question immediately arises, should the clutch be placed on the low- or high-speed side of the gear reducer? Sometimes the answer is dictated by function. For example, it would make little sense to put an automotive clutch on the output shaft of the transmission instead of the input side since in this instance, the principle purpose of the clutch is to interrupt the connection between engine and transmission, *ergo*, it must go on the high-speed side. In other cases, function does not dictate

15

the clutch location as, for example, in Case Study 7, where the coupling on either shaft could be replaced by a clutch if it were desired to decouple the compressor from the engine. (See Figure 8-1 on p. 539.) The choice is less clear in these situations and there are two competing schools of thought.

The torque (and any shock load) is larger on the low-speed shaft than on the high-speed shaft by the gear ratio. The power is essentially the same at both locations (neglecting losses in the gear train), but the kinetic energy at the high-speed shaft is larger by the square of the gear ratio. A clutch on the low-speed side must be larger (and thus more expensive) in order to handle the larger torque. However, a smaller and cheaper clutch on the high-speed side must dissipate the greater kinetic energy at that location and thus may more readily overheat. Some manufacturers recommend always using the high-speed side for the clutch location if function allows, opting for its better initial economy. Other clutch manufacturers suggest that the higher initial cost of the larger, low-speed clutch will be counterbalanced by lower maintenance cost in the long run. The balance of expert opinion seems to tilt toward the high-speed location with the caveat that each situation should be individually evaluated on its own merits.[1]

15.3   CLUTCH AND BRAKE MATERIALS

Materials for the structural parts of brakes and clutches such as the disks or drums are typically made of gray cast iron or steel. The friction surfaces are usually lined with a material having a good coefficient of friction, and with sufficient compressive strength and temperature resistance for the application. Asbestos fiber was once the most common ingredient in brake or clutch linings, but is no longer used in many applications because of its danger as a carcinogen. Linings may be molded, woven, sintered, or of solid material. Molded linings typically use polymeric resins to bind a variety of powdered fillers or fibrous materials. Brass or zinc chips are sometimes added to improve heat conduction and wear resistance, and reduce scoring of drums and disks. Woven materials typically use long asbestos fibers. Sintered metals provide higher temperature resistance and compressive strength than molded or woven materials. Materials such as cork, wood, and cast iron are sometimes used as linings as well. Table 15-1 shows some frictional, thermal, and mechanical properties of a few friction-lining materials.

15.4   DISK CLUTCHES

The simplest disk clutch consists of two disks, one lined with a high-friction material, pressed together axially with a normal force to generate the friction force needed to transmit torque, as shown in Figure 15-6. The normal force can be supplied mechanically, pneumatically, hydraulically, or electromagnetically and is typically quite large. The pressure between the clutch surfaces can approach a uniform distribution over the surface if the disks are flexible enough. In such cases, the wear will be greater at larger

Table 15-1   Properties of Common Clutch/Brake Lining Materials

Friction Material Against Steel or CI	Dynamic Coefficient of Friction		Maximum Pressure		Maximum Temperature	
	dry	in oil	psi	kPa	°F	°C
Molded	0.25-0.45	0.06-0.09	150-300	1 030-2 070	400-500	204-260
Woven	0.25-0.45	0.08-0.10	50-100	345-690	400-500	204-260
Sintered metal	0.15-0.45	0.05-0.08	150-300	1 030-2 070	450-1 250	232-677
Cast iron or hard steel	0.15-0.25	0.03-0.06	100-250	690-720	500	260

diameters because wear is proportional to pressure times velocity (pV) and the velocity increases linearly with radius. However, as the disks wear preferentially toward the outside, the loss of material will change the pressure distribution to a nonuniform one and the clutch will approach a uniform wear condition of $pV = $ constant. Thus the two extremes are a **uniform pressure** and a **uniform wear** condition. A flexible clutch may be close to a uniform-pressure condition when new, but will tend toward a uniform-wear condition with use. A rigid clutch will more rapidly approach the uniform-wear condition with use. The calculations for each condition are different and the uniform-wear assumption gives a more conservative clutch rating, so is favored by some designers.

Uniform Pressure

Consider an elemental ring of area on the clutch face of width dr as shown in Figure 15-6. The differential force acting on this ring is

$$dF = 2\pi pr\, dr \tag{15.1a}$$

where r is the radius and p is the uniform pressure on the clutch face. The total axial force F on the clutch is found by integrating this expression between the limits r_i and r_o.

$$F = \int_{r_i}^{r_o} 2\pi pr\, dr = \pi p\left(r_o^2 - r_i^2\right) \tag{15.1b}$$

The friction torque on the differential ring element is

$$dT = 2\pi p\mu r^2\, dr \tag{15.2a}$$

where μ is the coefficient of friction. The total torque for one clutch disk is

$$T = \int_{r_i}^{r_o} 2\pi p\mu r^2\, dr = \frac{2}{3}\pi p\mu\left(r_o^3 - r_i^3\right) \tag{15.2b}$$

15

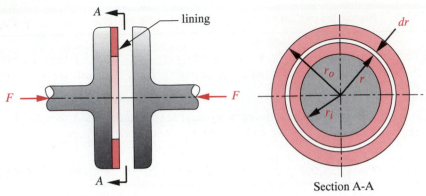

FIGURE 15-6

A Single-Surface Axial Disk Clutch

For a multiple disk clutch with N friction faces:

$$T = \frac{2}{3}\pi p\mu\left(r_o^3 - r_i^3\right)N \qquad (15.2c)$$

Equations 15.1b and 15.2c can be combined to give an expression for torque as a function of axial force.

$$T = N\mu F\,\frac{2}{3}\,\frac{\left(r_o^3 - r_i^3\right)}{\left(r_o^2 - r_i^2\right)} \qquad (15.3)$$

Uniform Wear

The constant wear rate W is assumed to be proportional to the product of pressure p and velocity V.

$$W = pV = \text{constant} \qquad (15.4a)$$

And the velocity at any point on the face of the clutch is

$$V = r\omega \qquad (15.4b)$$

Combine these equations assuming a constant angular velocity ω.

$$pr = \text{constant} = K \qquad (15.4c)$$

The largest pressure p_{max} must then occur at the smallest radius r_i.

$$K = p_{max}r_i \qquad (15.4d)$$

Combining equations 15.4c and 15.4d gives an expression for pressure as a function of the radius r:

$$p = p_{max} \frac{r_i}{r} \qquad (15.4e)$$

where the maximum allowable pressure p_{max} will vary with the lining material used. Table 15-1 shows recommended values of p_{max} and coefficients of friction for various clutch/brake lining materials.

The axial force F is found by integrating equation 15.1a for the differential force on the ring element of Figure 15-6 with equation 15.4e substituted for p.

$$F = \int_{r_i}^{r_o} 2\pi pr\, dr = \int_{r_i}^{r_o} 2\pi \left(p_{max} \frac{r_i}{r} \right) r\, dr = 2\pi r_i p_{max} (r_o - r_i) \qquad (15.5a)$$

The torque is found by integrating equation 15.2a with the same substitution:

$$T = \int_{r_i}^{r_o} 2\pi p\mu r^2\, dr = \pi \mu r_i p_{max} (r_o^2 - r_i^2) \qquad (15.5b)$$

Combine equations 15.2a and -b for an expression relating torque to axial force for the uniform wear case:

$$T = N\mu F \frac{(r_o + r_i)}{2} \qquad (15.6)$$

where N is the number of friction surfaces in the clutch.

From equation 15.5b, it can be shown that the maximum torque for any outside radius r_o will be obtained when the inside radius is:

$$r_i = \sqrt{1/3}\, r_o = 0.577 r_o \qquad (15.7)$$

Note that the uniform-wear assumption gives a lower torque capacity for the clutch than does the uniform-pressure assumption. The higher initial wear at the larger radii shifts the center of pressure radially inward, giving a smaller moment arm for the resultant friction force. Clutches are usually designed based on uniform wear. They will have a greater capacity when new but will end up close to the predicted design capacity after they are worn in.

EXAMPLE 15-1

Design of a Disk Clutch

Problem   Determine a suitable size and required force for an axial disk clutch.

Given   The clutch must pass 7.5 hp at 1 725 rpm with a service factor of 2.

Assumptions   Use a uniform-wear model. Assume a single-dry-disk with a molded lining.

15

Solution   See the *TKSolver* file EX15-01.

1   The service factor of 2 requires derating the clutch by that factor so we will design
    for 15 hp instead of 7.5. Find the torque required for that power at the design rpm.

$$T = \frac{P}{\omega} = \frac{15 \text{ hp}\left(6\,600\,\dfrac{\text{in - lb/sec}}{\text{hp}}\right)}{1725 \text{ rpm}\left(\dfrac{2\pi}{60}\,\dfrac{\text{rad/sec}}{\text{rpm}}\right)} = 548.05 \text{ lb - in} \tag{a}$$

2   Find the coefficient of friction and maximum recommended pressure for a dry,
    molded material from Table 15-1. Use the average of the ranges of values shown:
    $p_{max} = 225$ psi and $\mu = 0.35$.

3   Substitute equation 15.7 relating r_i to r_o for maximum torque into equation 15.5b to
    get

$$T = \pi\mu r_i p_{max}\left(r_o^2 - r_i^2\right) = \pi\mu(0.577r_o)p_{max}\left(r_o^2 - \frac{1}{3}r_o^2\right) = 0.3849 r_o^3 \pi\mu p_{max}$$

$$r_o = \left(\frac{T}{0.3849\pi\mu p_{max}}\right)^{\frac{1}{3}} = \left(\frac{548.05}{0.3849\pi(0.35)(225)}\right)^{\frac{1}{3}} = 1.79 \text{ in} \tag{b}$$

4   From equation 15.7:

$$r_i = 0.577r_o = 0.577(1.79) = 1.03 \text{ in} \tag{c}$$

5   The axial force needed (from equation 15.5a) is:

$$F = 2\pi r_i p_{max}(r_o - r_i) = 2\pi(1.034)(225)(1.792 - 1.034) = 1\,108 \text{ lb} \tag{d}$$

6   The clutch specification is then a single disk with 3.6-in outside diameter and 2-in
    inside diameter, a molded lining with $\mu_{dry} \geq 0.35$, and an actuating force $\geq 1\,108$ lb.

15.5   DISK BRAKES

The disk clutch equations also apply to disk brakes. However, disk brakes are seldom
made with linings covering the entire circumference of the face because they would then
overheat. While clutches are often used with light duty cycles (engagement time a small
fraction of total time), brakes often must absorb large amounts of energy in repeated
applications. Caliper disk brakes, such as those used on automobiles, use friction pads
applied against only a small fraction of the disk circumference, leaving the rest exposed
for cooling. The disk is sometimes ventilated with internal air passages to promote cooling. The caliper typically straddles the disk and contains two pads, each rubbing one
side of the disk. This cancels the axial force and reduces axial loads on the bearings.

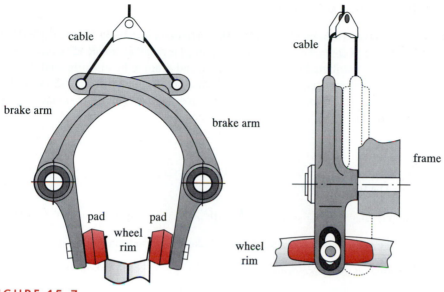

FIGURE 15-7

Bicycle Disk-Type Brake

The common bicycle caliper brake as shown in Figure 15-7 is another example in which the wheel rim is the disk and the calipers pinch against only a small fraction of the circumference. Disk brakes are now commonly used on automobiles, particularly on the front wheels, which provide more than half the stopping force. Some advantages of disk over drum brakes are their good controllability and linearity (braking torque is directly proportional to applied axial force).

15.6 DRUM BRAKES

Drum brakes (or clutches) apply the friction material to the circumference of a cylinder, either externally, internally, or both. These devices are more often used as brakes than as clutches. The part to which the friction material is riveted or bonded with adhesive is called the brake shoe, and the part against which it rubs, the brake drum. The shoe is forced against the drum to create the friction torque. The simplest configuration of a drum brake is the **band brake**, in which a flexible shoe is wrapped around a majority of the outer circumference of the drum and squeezed against it. Alternatively a relatively rigid, lined shoe (or shoes) can be pivoted against the outer or inner circumference (or both) of the drum. If the shoe contacts only a small angular portion of the drum, it is a **short-shoe brake**, otherwise a **long-shoe brake**. The geometry of the short versus long shoe contact requires a different analytical treatment in each case. We will examine the cases of external short-shoe and external long-shoe drum brakes to illustrate their differences and features, particularly in contrast to disk brakes. The principles are the same for internal-shoe brakes as well.

Short-Shoe External Drum Brakes

Figure 15-8a shows a schematic of a short-shoe external drum brake. If the angle θ subtended by the arc of contact between shoe and drum is small (< about 45°), then we can consider that the distributed force between shoe and drum to be uniform, and it can be replaced by the concentrated force F_n in the center of the contact area, as shown in Figure 15-8b. For any maximum allowable lining pressure p_{max} (Table 15-1) the force F_n can be estimated as

$$F_n = p_{max} r \theta w \tag{15.8}$$

where w is the width of the brake shoe in the z direction and θ is the subtended angle in radians. The frictional force F_f is

$$F_f = \mu F_n \tag{15.9}$$

where μ is the coefficient of friction for the brake lining material (Table 15-1).

The torque on the brake drum is then

$$T = F_f r = \mu F_n r \tag{15.10}$$

Summing moments about point O on the free-body diagram of Figure 15-8b and substituting equation 15.9 gives

$$\sum M = 0 = a F_a - b F_n + c F_f \tag{15.11a}$$

$$F_a = \frac{b F_n - c F_f}{a} = \frac{b F_n - \mu c F_n}{a} = F_n \frac{b - \mu c}{a} \tag{15.11b}$$

The reaction forces at the pivot are found from a summation of forces.

$$R_x = -F_f$$
$$R_y = F_a - F_n \tag{15.12}$$

SELF-ENERGIZING   Note in Figure 15-8b that with the direction of drum rotation shown, the friction moment $c F_f$ adds to the actuating moment $a F_a$. This is referred to as **self-energizing**. Once any application force F_a is applied, the friction generated at the shoe acts to increase the braking torque. However, if the brake drum rotation is reversed from that shown in Figure 15-8a, the sign of the friction moment term $c F_f$ of equation 15.11a becomes negative and the brake is then **self-deenergizing**.

This self-energizing feature of drum brakes is a potential advantage as it reduces the required application force compared to a disk brake of the same capacity. Drum brakes typically have two shoes, one of which can be made self-energizing in each direction, or both in one direction. The latter arrangement is commonly used in automobile brakes to aid in stopping forward motion at the expense of stopping backward motion, which is normally at lower speeds.

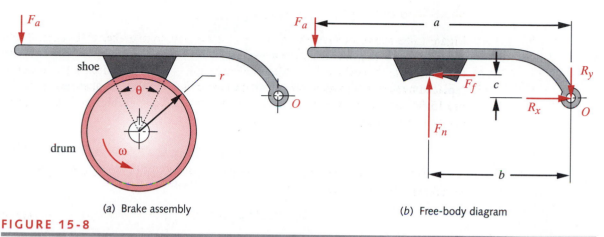

(a) Brake assembly   (b) Free-body diagram

FIGURE 15-8

Geometry and Forces for a Short-Shoe External Drum Brake

SELF-LOCKING   Note in equation 15.11 that if the brake is self-energizing and the product $\mu c \geq b$, the force F_a needed to actuate the brake becomes zero or negative. The brake is then said to be **self-locking**. If the shoe touches the drum, it will grab and lock. This is usually not a desired condition except in so-called *backstopping* applications as were described under overrunning clutches above. In effect, a self-locking brake can function as an overrunning clutch to backstop a load and prevent it from running away if power is lost. They are sometimes used in hoists for that purpose.

EXAMPLE 15-2

Design of a Short-Shoe Drum Brake

Problem   For the drum brake arrangement shown in Figure 15-8, determine the ratio c / r that will give a self-energizing ratio F_n / F_a of 2. Also find the c / r ratio that will cause self-locking.

Given   The dimensions are $a = b = 6$, $r = 5$.

Assumptions   Coefficient of friction $\mu = 0.35$

Solution   See Figure 15-8 and the *TKSolver* files EX15-02a and EX15-02b.

1   Rearrange equation 15.11 to form the desired ratio.

$$\frac{F_n}{F_a} = \frac{a}{b - \mu c} \qquad (a)$$

2   Substitute the desired self-energizing ratio, the given dimensions, and solve for c.

15

$$\frac{F_n}{F_a} = 2 = \frac{6}{6 - 0.35c}$$

$$c = \frac{-3}{-0.35} = 8.571 \qquad\qquad (b)$$

3   Form the $c\,/\,r$ ratio for a self-energizing ratio of 2 with the given brake geometry.

$$\frac{c}{r} = \frac{8.571}{5} = 1.71 \qquad\qquad (c)$$

4   For self-locking to begin, F_a becomes zero, making $F_n\,/\,F_a = \infty$ and $F_a\,/\,F_n = 0$. The second of these ratios will need to be used to avoid division by zero. Rearrange equation 15.11 to form the desired ratio and solve for c.

$$\frac{F_a}{F_n} = \frac{b - \mu c}{a} = \frac{6 - 0.35c}{6} = 0$$

$$c = 17.143 \qquad\qquad (d)$$

5   Form the $c\,/\,r$ ratio for self-locking with the given brake geometry.

$$\frac{c}{r} = \frac{17.143}{5} = 3.43 \qquad\qquad (e)$$

6   Note that these ratios are specific to the dimensions of the brake. The length a was set equal to b in this example in order to eliminate the effect of the lever arm ratio $a\,/\,b$, which further reduces the application force F_a required for any normal force F_n.

Long-Shoe External Drum Brakes

If the angle of contact θ between shoe and drum in Figure 15-8 exceeds about 45°, then the assumption of uniform pressure distribution over the shoe surface will be inaccurate. Most drum brakes have contact angles of 90° or more, so a more accurate analysis than the short-shoe assumption is needed. Since any real brake shoe will not be infinitely rigid, its deflections will affect the pressure distribution. An analysis that accounts for deflection effects is very complicated and is not really warranted here. As the shoe wears, it will pivot about point O in Figure 15-9 and point A will travel farther than point B because of its greater distance from O. The pressure at any point on the shoe will also vary in proportion to its distance from O. Assume that the drum turns at constant velocity and that wear is proportional to the friction work done, i.e., the product pV. Then, at any arbitrary point on the shoe such as C in Figure 15-9, the normal pressure p will be proportional to its distance from point O.

$$p \propto b \sin \theta \propto \sin \theta \qquad\qquad (15.13a)$$

Since the distance b is constant, the normal pressure at any point is just proportional to $\sin\theta$. Call its constant of proportionality K.

15

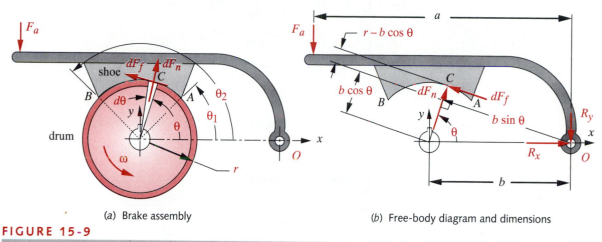

(a) Brake assembly   (b) Free-body diagram and dimensions

FIGURE 15-9

Geometry and Forces for a Long-Shoe External Drum Brake

$$p = K \sin \theta \tag{15.13b}$$

If the maximum allowable pressure for the lining material is p_{max} (Table 15-1) then the constant K can be defined as

$$K = \frac{p}{\sin \theta} = \frac{p_{max}}{\sin \theta_2} \tag{15.13c}$$

and

$$p = \frac{p_{max}}{\sin \theta_2} \sin \theta \tag{15.13d}$$

Equation 15.13d defines the normal pressure at any point on the shoe and it varies as $\sin\theta$ since p_{max} and θ_2 are constant for any particular brake. Thus, the friction force is small at small θ, is optimum at $\theta = 90°$, and diminishes at angles larger than $90°$. Little is gained by using $\theta_1 < 10°$ or $\theta_2 > 120°$.

To obtain the total force on the shoe, the pressure function must be integrated over the angular range of the shoe. Consider the differential element $d\theta$ shown in Figure 15-9. Two differential forces act on it, dF_n and dF_f. They have respective moment arms about point O of $b\sin\theta$ and $r - b\cos\theta$, as shown in Figure 15-9b. Integrating to create their moments about O for the entire surface gives, for the moment due to the normal force:

15

$$M_{F_n} = \int_{\theta_1}^{\theta_2} pwr\, d\theta\, b\sin\theta = \int_{\theta_1}^{\theta_2} wrbp\sin\theta\, d\theta$$

$$= \int_{\theta_1}^{\theta_2} wrb\,\frac{p_{max}}{\sin\theta_2}\sin^2\theta\, d\theta$$

$$M_{F_n} = wrb\,\frac{p_{max}}{\sin\theta_2}\left[\frac{1}{2}(\theta_2 - \theta_1) - \frac{1}{4}(\sin 2\theta_2 - \sin 2\theta_1)\right] \qquad (15.14a)$$

where w is the drum width in the z direction and the other variables are as defined in Figure 15-9. For the moment due to the frictional force:

$$M_{F_f} = \int_{\theta_1}^{\theta_2} \mu pwr\, d\theta\, (r - b\cos\theta)$$

$$= \int_{\theta_1}^{\theta_2} \mu wr\,\frac{p_{max}}{\sin\theta_2}\sin\theta\,(r - b\cos\theta)\, d\theta$$

$$M_{F_f} = \mu wr\,\frac{p_{max}}{\sin\theta_2}\left[-r(\cos\theta_2 - \cos\theta_1) - \frac{b}{2}(\sin^2\theta_2 - \sin^2\theta_1)\right] \qquad (15.14b)$$

Summing moments about point O gives

$$F_a = \frac{M_{F_n} \mp M_{F_f}}{a} \qquad (15.14c)$$

where the upper sign is for a self-energizing brake and the lower one for a self-deenergizing brake. Self-locking can only occur if the brake is self-energizing and $M_{Ff} > M_{Fn}$.

The torque for the brake is found by integrating the expression for the product of the friction force F_f and drum radius r.

$$T = \int_{\theta_1}^{\theta_2} \mu pwr\, d\theta\, r$$

$$= \int_{\theta_1}^{\theta_2} \mu wr^2\,\frac{p_{max}}{\sin\theta_2}\sin\theta\, d\theta$$

$$T = \mu wr^2\,\frac{p_{max}}{\sin\theta_2}(\cos\theta_1 - \cos\theta_2) \qquad (15.15)$$

The reaction forces R_x and R_y are found by summing forces in the x and y directions (see Figure 15-9b):

$$R_x = -F_n \cos\theta - F_f \sin\theta$$

$$= -\int_{\theta_1}^{\theta_2} wrp\, d\theta \cos\theta - \int_{\theta_1}^{\theta_2} \mu wrp\, d\theta \sin\theta$$

$$= -\int_{\theta_1}^{\theta_2} wr \frac{p_{max}}{\sin\theta_2} \sin\theta \cos\theta\, d\theta - \int_{\theta_1}^{\theta_2} \mu wr \frac{p_{max}}{\sin\theta_2} \sin^2\theta\, d\theta$$

$$R_x = -wr \frac{p_{max}}{\sin\theta_2} \left\{ \left(\frac{\sin^2\theta_2}{2} - \frac{\sin^2\theta_1}{2} \right) + \mu\left[\frac{1}{2}(\theta_2 - \theta_1) - \frac{1}{4}(\sin 2\theta_2 - \sin 2\theta_1) \right] \right\} \tag{15.16a}$$

$$R_y = F_a + F_f \cos\theta - F_n \sin\theta$$

$$= F_a + \int_{\theta_1}^{\theta_2} \mu wrp\, d\theta \cos\theta - \int_{\theta_1}^{\theta_2} wrp\, d\theta \sin\theta$$

$$= F_a + \int_{\theta_1}^{\theta_2} \mu wr \frac{p_{max}}{\sin\theta_2} \sin\theta \cos\theta\, d\theta - \int_{\theta_1}^{\theta_2} wr \frac{p_{max}}{\sin\theta_2} \sin^2\theta\, d\theta$$

$$R_y = F_a + wr \frac{p_{max}}{\sin\theta_2} \left\{ \mu\left(\frac{\sin^2\theta_2}{2} - \frac{\sin^2\theta_1}{2} \right) - \left[\frac{1}{2}(\theta_2 - \theta_1) - \frac{1}{4}(\sin 2\theta_2 - \sin 2\theta_1) \right] \right\} \tag{15.16b}$$

EXAMPLE 15-3

Design of a Long-Shoe Drum Brake

Problem   For the drum brake arrangement shown in Figure 15-9, determine the friction torque T, application force F_a, and reaction forces R_x, R_y.

Given   The dimensions are $a = 180$ mm, $b = 90$ mm, $r = 100$ mm, $w = 30$ mm, $\theta_1 = 30°$, $\theta_2 = 120°$.

Assumptions   Coefficient of friction $\mu = 0.35$, maximum lining pressure $p_{max} = 1.5$ MPa, and the brake is self-energizing.

Solution   See Figure 15-9 and the *TKSolver* file EX15-03.

1 Convert the given angles θ_1 and θ_2 to radians $\theta_1 = 0.524$ rad, $\theta_2 = 2.094$ rad.

2 Calculate the moment M_{F_n} about O due to the normal force using equation 15.14a.

15

$$M_{F_n} = wrb\frac{p_{max}}{\sin\theta_2}\left[\frac{1}{2}(\theta_2 - \theta_1) - \frac{1}{4}(\sin 2\theta_2 - \sin 2\theta_1)\right]$$

$$= 30(100)(90)\frac{1.5}{\sin(2.094)}\left[\begin{array}{l}\frac{1}{2}(2.094 - 0.524)\\[2mm]-\frac{1}{4}\left(\sin\{2(2.094)\} - \sin\{2(0.524)\}\right)\end{array}\right]$$

$$= 573.9 \text{ N-m} \tag{a}$$

3 Calculate the moment M_{F_f} about O due to the normal force using equation 15.14b.

$$M_{F_f} = \mu wr\frac{p_{max}}{\sin\theta_2}\left[-r(\cos\theta_2 - \cos\theta_1) - \frac{b}{2}(\sin^2\theta_2 - \sin^2\theta_1)\right]$$

$$= 0.35(30)(100)\frac{1.5}{\sin(2.094)}\left[\begin{array}{l}-100(\cos\{2.094\} - \cos\{0.524\})\\[2mm]-\frac{90}{2}\left(\sin^2\{2.094\} - \sin^2\{0.524\}\right)\end{array}\right]$$

$$= 209.0 \text{ N-m} \tag{b}$$

4 Find the application force from equation 15.14c.

$$F_a = \frac{M_{F_n} \mp M_{F_f}}{a} = \frac{573.9 - 209.0}{0.180} = 2\,013 \text{ N} \tag{c}$$

5 Find the friction torque from equation 15.15.

$$T_f = \mu wr^2 \frac{p_{max}}{\sin\theta_2}(\cos\theta_1 - \cos\theta_2)$$

$$= 0.35(30)(100)^2 \frac{1.5}{\sin(2.094)}(\cos\{2.094\} - \cos\{0.524\})$$

$$= 250 \text{ N-m} \tag{d}$$

6 The reaction forces are found from equations 15.16.

$$R_x = -wr\frac{p_{max}}{\sin\theta_2}\left\{\begin{array}{l}\left(\dfrac{\sin^2\theta_2}{2} - \dfrac{\sin^2\theta_1}{2}\right)\\[3mm]+\mu\left[\dfrac{1}{2}(\theta_2 - \theta_1) - \dfrac{1}{4}(\sin 2\theta_2 - \sin 2\theta_1)\right]\end{array}\right\}$$

$$= -30(100)\frac{1.5}{\sin(2.094)}\left\{\begin{array}{l}\left(\dfrac{\sin^2\{2.094\}}{2} - \dfrac{\sin^2\{0.524\}}{2}\right)\\[3mm]-0.35\left[\begin{array}{l}\frac{1}{2}(2.094 - 0.524)\\[2mm]-\frac{1}{4}\left(\sin\{2(2.094)\} - \sin\{2(0.524)\}\right)\end{array}\right]\end{array}\right\}$$

$$= 917 \text{ N} \tag{e}$$

$$
R_y = F_a + wr\frac{p_{max}}{\sin\theta_2}\left\{\mu\left(\frac{\sin^2\theta_2}{2} - \frac{\sin^2\theta_1}{2}\right) - \left[\frac{1}{2}(\theta_2 - \theta_1) - \frac{1}{4}(\sin 2\theta_2 - \sin 2\theta_1)\right]\right\}
$$

$$
= 2\,013 + 30(100)\frac{1.5}{\sin(2.094)}\left\{0.35\left(\frac{\sin^2\{2.094\}}{2} - \frac{\sin^2\{0.524\}}{2}\right) - \left[\frac{1}{2}(2.094 - 0.524) - \frac{1}{4}(\sin\{2(2.094)\} - \sin\{2(0.524)\})\right]\right\}
$$

$$
R_y = -3\,864 \text{ N} \tag{f}
$$

Long-Shoe Internal Drum Brakes

Most drum brakes (and virtually all automotive ones) use internal shoes that expand against the inside of the drum. Typically two shoes are used, pivoted against the ends of an adjusting screw and forced against the drum by a double-ended hydraulic cylinder. Light springs hold the shoes against the pistons of the wheel cylinder and pull the shoes away from the drum when not activated. Typically, one shoe is self-energizing in the forward direction and the other is self-energizing in the reverse direction of drum rotation. The automobile wheel attaches directly to the brake drum. The analysis of an internal-shoe brake is the same as that of an external-shoe one.

15.7   SUMMARY

Clutches and brakes are used extensively in all kinds of machinery. Vehicles all need brakes to stop their motion, as do many stationary machines. Clutches are needed to interrupt the flow of power from a prime mover (motor, engine, etc.) to the load so the load can be stopped (by a brake) and the prime mover continue running. A clutch and a brake are essentially the same device with the principal difference being that both sides of a clutch (input and output) are capable of rotation, but the output side of a brake is fixed to some nonrotating "ground plane," which itself may have some other motion, as in the case of an automobile chassis.

Many different styles of clutches/brakes are made but the most common style uses frictional contact between two or more surfaces to couple the input and output sides together. The friction surfaces can be moved into and out of engagement by any of several means including direct mechanical, electromagnetic, pneumatic, hydraulic, or combinations thereof. Other styles include direct magnetic (magnetic particle, hysteresis,

15

and eddy current), some of which have no mechanical contact (and thus zero drag) when disengaged, and fluid couplings, which are commonly used to couple vehicle engines to automatic transmissions.

Except for high-volume, specialized applications such as vehicle design, a machine designer seldom designs a clutch or brake from scratch. For the typical machine-design application, one usually selects a commercially available clutch or brake assembly from the many manufacturers' offerings. The design problem then becomes one of properly defining the torque, speed, and power requirements, and the character of the load, whether smooth or shocky, continuous or intermittent, etc. The inertia of the rotating elements to be accelerated by a clutch or decelerated by a brake can have a significant effect on the required size of that device and must be carefully calculated. Any gear ratios present in the system will cause the reflected or effective inertia to vary as the square of the gear ratio and this effect must be carefully included when calculating the inertia. (See Chapter 3 or any text on dynamics of machinery.)

The clutch/brake manufacturers' catalogs contain extensive engineering data that rates each device on its torque and power capacity and also suggests empirical derating factors for situations with shock loads, high duty cycles, etc. Once the loading is well defined, a suitable device can be specified using the manufacturers' rating data modified by their suggested service factors. The designer's (nontrivial) task then becomes that of proper load definition for the application, followed by proper use of the manufacturers' rating data. Engineering assistance is usually available from the manufacturer for the latter task, but the result can be only as suitable to the design requirements as the accuracy of the load analysis allows. The mechanical configurations of several clutch designs are briefly described within this chapter. Manufacturers' catalogs and applications engineers can provide more detailed information on the capabilities and limitations of the various clutch/brake styles.

Commercial friction clutches and brakes are most often made in a single or multiple disk configuration. Vehicle brakes are typically made in either disk or drum configurations. Disk configurations provide a friction torque that is linearly proportional to the applied actuating force and this can be an advantage from a control standpoint. Drum configurations can be designed to be self-energizing, meaning that once the brake or clutch is initially engaged, the friction force tends to increase the normal force, thus nonlinearly increasing the friction torque in a positive feedback fashion. This can be an advantage when braking large loads as it decreases the required application force but it makes control of braking torque more difficult. The analysis of both friction disk devices and friction drum devices is developed in this chapter.

Clutches and brakes are essentially energy transfer or energy dissipation devices and as such generate a great deal of heat in operation. They must be designed to absorb and transfer this heat without damage to themselves or their surroundings. Often, the heat-transfer ability of a device rather than its mechanical torque transmission ability limits its capacity. The thermal design of clutches and brakes is a very important consideration, but it goes beyond the scope of this text and space does not permit its treat-

15

ment here. Nevertheless, the designer needs to be aware of the heat-transfer aspect of clutch/brake design and take it into account. See any text on heat transfer for the theoretical background and see the references mentioned in the bibliography to this chapter as well as other manufacturers' catalogs for more specific information.

Friction clutches may be operated either dry or wet (typically in oil). Dry friction is obviously more effective, as the coefficient of friction is severely reduced with lubrication. However, running in oil can significantly improve the heat-transfer situation especially when the oil is circulated and/or cooled. More friction surfaces (e.g., multiple disks) are needed to achieve the same torque capacity wet that can be obtained with a single dry disk, but the trade-off can be positive because of enhanced cooling. Modern vehicle automatic transmissions use many internal clutches and brakes to interconnect or stop various members of their epicyclic (planetary) gear trains in order to shift among gear ratios. These are either multidisk clutches or band brakes and are run immersed in the transmission oil that is continuously circulated through a heat exchanger in the vehicle's radiator for cooling.

Important Equations Used in this Chapter

Torque in a Disk Clutch with Uniform Pressure (Section 15.4):

$$T = N\mu F \frac{2}{3} \frac{\left(r_o^3 - r_i^3\right)}{\left(r_o^2 - r_i^2\right)} \tag{15.3}$$

Torque in a Disk Clutch with Uniform Wear (Section 15.4):

$$T = N\mu F \frac{\left(r_o + r_i\right)}{2} \tag{15.6}$$

Forces and Torque on a Short-Shoe Drum Brake (Section 15.6):

$$F_n = p_{max} r \theta w \tag{15.8}$$

$$F_a = \frac{bF_n - cF_f}{a} = \frac{bF_n - \mu c F_n}{a} = F_n \frac{b - \mu c}{a} \tag{15.11b}$$

$$T = F_f r = \mu F_n r \tag{15.10}$$

Forces and Torque on a Long-Shoe Drum Brake (Section 15.6):

$$M_{F_n} = wrb \frac{p_{max}}{\sin\theta_2}\left[\frac{1}{2}(\theta_2 - \theta_1) - \frac{1}{4}(\sin 2\theta_2 - \sin 2\theta_1)\right] \tag{15.14a}$$

$$M_{F_f} = \mu wr \frac{p_{max}}{\sin\theta_2}\left[-r(\cos\theta_2 - \cos\theta_1) - \frac{b}{2}(\sin^2\theta_2 - \sin^2\theta_1)\right] \tag{15.14b}$$

15

$$F_a = \frac{M_{F_n} \mp M_{F_f}}{a} \tag{15.14c}$$

$$T = \mu w r^2 \frac{p_{max}}{\sin\theta_2}\left(\cos\theta_1 - \cos\theta_2\right) \tag{15.15}$$

$$R_x = -wr\frac{p_{max}}{\sin\theta_2}\left\{\left(\frac{\sin^2\theta_2}{2} - \frac{\sin^2\theta_1}{2}\right) + \mu\left[\frac{1}{2}\left(\theta_2 - \theta_1\right) - \frac{1}{4}\left(\sin 2\theta_2 - \sin 2\theta_1\right)\right]\right\} \tag{15.16a}$$

$$R_y = F_a + wr\frac{p_{max}}{\sin\theta_2}\left\{\mu\left(\frac{\sin^2\theta_2}{2} - \frac{\sin^2\theta_1}{2}\right) - \left[\frac{1}{2}\left(\theta_2 - \theta_1\right) - \frac{1}{4}\left(\sin 2\theta_2 - \sin 2\theta_1\right)\right]\right\} \tag{15.16b}$$

15.8   REFERENCES

1   **J. Proctor**, Selecting Clutches for Mechanical Drives, *Product Engineering*, pp. 43-58, June 19, 1961.

15.9   BIBLIOGRAPHY

Mechanical Drives Reference Issue. Penton Publishing: Cleveland, Ohio,

Deltran, *Electromagnetic Clutches and Brakes.* American Precision Industries: Buffalo, N.Y., 716-631-9800.

Logan, *Multiple Disk Clutches and Brakes.* Logan Clutch Co.: Cleveland, Ohio, 216-431-4040.

Magtrol, *Hysteresis Brakes and Clutches.* Magtrol: Buffalo, N.Y., 800-828-7844.

MCC, *Catalog 18e.* Machine Components Corporation: Plainview, NY, 516-694-7222.

MTL, *Eddy Current Clutch Design.* Magnetic Technologies Ltd.: Oxford, Mass., 508-987-3303.

W. C. Orthwein, *Clutches and Brakes: Design and Selection.* Marcel Dekker: New York, 1986.

Placid, *Magnetic Particle Clutches and Brakes.* Placid Industries Inc.: Lake Placid, N.Y., 518-523-2422.

15

Warner, *Clutches, Brakes, and Controls: Master Catalog.* Warner Electric: South Beloit, Ill., 815-389-2582.

15.10 PROBLEMS

*15-1   Find the torque that a 2-surface, dry disk clutch can transmit if the outside and inside lining diameters are 120 mm and 70 mm, respectively, and the applied axial force is 10 kN. Assume uniform wear and $\mu = 0.4$. Is the pressure on the lining acceptable? What lining material(s) would be suitable?

15-2   Repeat Problem 15-1 assuming uniform pressure.

*15-3   Design a single-surface disk clutch to transmit 100 N-m of torque at 750 rpm using a molded lining with a maximum pressure of 1 MPa and $\mu = 0.25$. Assume uniform wear. Find the outside and inside diameters required if $r_i = 0.577\ r_o$. What is the power transmitted?

15-4   Repeat Problem 15-3 assuming uniform pressure.

*15-5   How many surfaces are need in a wet disk clutch to transmit 120 N-m of torque at 1000 rpm using a sintered lining with a maximum pressure of 1.8 MPa and $\mu = 0.06$? Assume uniform wear. Find the outside and inside diameters required if $r_i = 0.577\ r_o$. How many disks are needed? What is the power transmitted?

15-6   Repeat problem 15-5 assuming uniform pressure.

*15-7   Figure P15-1 shows a single short-shoe drum brake. Find its torque capacity and required actuating force for $a = 100$, $b = 70$, $e = 20$, $r = 30$, $w = 50$ mm, and $\theta = 35°$. What value of c will make it self-locking? Assume $p_{max} = 1.3$ MPa and $\mu = 0.3$.

15-8   Repeat Problem 15-7 with the drum rotating clockwise.

15-9   Figure P15-1 shows a single short-shoe drum brake. Find its torque capacity and required actuating force for $a = 8$, $b = 6$, $e = 4$, $r = 5$, $w = 1.5$ in, and $\theta = 30°$. What value of c will make it self-locking? Assume $p_{max} = 250$ psi and $\mu = 0.35$.

15-10   Repeat Problem 15-9 with the drum rotating clockwise.

*15-11   Figure P15-2 shows a double short-shoe drum brake. Find its torque capacity and required actuating force for $a = 90$, $b = 80$, $e = 30$, $r = 40$, $w = 60$ mm, and $\theta = 25°$. What value of c will make it self-locking? Assume $p_{max} = 1.5$ MPa and $\mu = 0.25$. Hint: Calculate the effects of each shoe separately and superpose them.

15-12   Figure P15-2 shows a double short-shoe drum brake. Find its torque capacity and required actuating force for $a = 12$, $b = 8$, $e = 3$, $r = 6$ in, and $\theta = 25°$. What value of c will make it self-locking? Assume $p_{max} = 200$ psi, width = 2 in, and $\mu = 0.28$. Hint: Calculate the effects of each shoe separately and superpose them.

*15-13   Figure P15-3 shows a single long-shoe drum brake. Find its torque capacity and required actuating force for $a_X = 100$, $b_X = 70$, $b_Y = 20$, $r = 30$, $w = 50$ mm, and $\theta_1 = 25°$, $\theta_2 = 125°$. Assume $p_{max} = 1.3$ MPa and $\mu = 0.3$.

* Answers to these problems are provided in Appendix G.

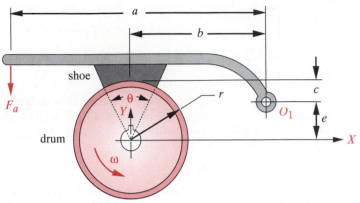

FIGURE P15-1

Geometry for a Short-Shoe External Drum Brake

15-14 Figure P15-3 shows a single long-shoe drum brake. Find its torque capacity and required actuating force for $a_X = 8$, $b_X = 6$, $b_Y = 4$, $r = 5$, $w = 1.5$ in, and $\theta_1 = 35°$, $\theta_2 = 155°$. Assume $p_{max} = 250$ psi and $\mu = 0.35$.

*15-15 Figure P15-4 shows a double long-shoe drum brake. Find its torque capacity and required actuating force for $a_X = 90$, $b_X = 80$, $b_Y = 30$, $r = 40$, $w = 30$ mm, and $\theta_1 = 30°$, $\theta_2 = 160°$. Assume $p_{max} = 1.5$ MPa and $\mu = 0.25$. Hint: Calculate the effects of each shoe separately and superpose them.

15-16 Figure P15-4 shows a double long-shoe drum brake. Find its torque capacity and required actuating force for $a_X = 12$, $b_X = 8$, $b_Y = 3$, $r = 6$, $w = 2$ in, and $\theta_1 = 25°$,

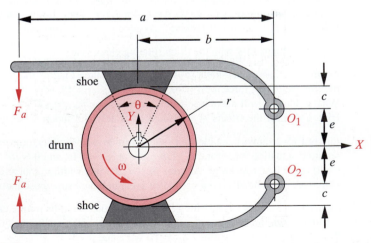

FIGURE P15-2

Geometry for a Double Short-Shoe External Drum Brake

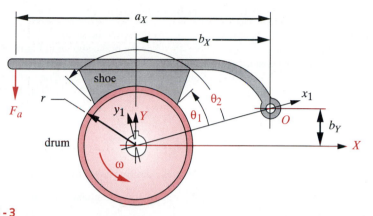

FIGURE P15-3

Geometry for a Long-Shoe External Drum Brake

$\theta_2 = 145°$. Assume $p_{max} = 200$ psi and $\mu = 0.28$. Hint: Calculate the effects of each shoe separately and superpose them.

*15-17   The short-shoe approximation is considered to be valid for brake shoes with an included angle of up to about 45°. For the brake shown in Figure P15-3, calculate its torque capacity and required force by both the short-shoe method and the long-shoe method and compare the results for the following data: $a_X = 90$, $b_X = 80$, $b_Y = 30$, $r = 40$, $w = 30$ mm. Assume $p_{max} = 1.5$ MPa and $\mu = 0.25$. Note that $\theta = \theta_2 - \theta_1$ for the short-shoe approximation.

(a) $\theta_1 = 75°$, $\theta_2 = 105°$.
(b) $\theta_1 = 70°$, $\theta_2 = 110°$.
(c) $\theta_1 = 65°$, $\theta_2 = 115°$.

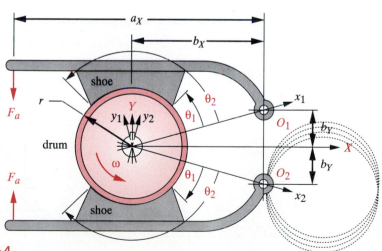

FIGURE P15-4

Geometry for a Double Long-Shoe External Drum Brake

* Answers to these problems are provided in Appendix G.

15

*15-18   Repeat Problem 15-17 for the brake design shown in Figure P15-4.

15-19   The short-shoe approximation is considered to be valid for brake shoes with an included angle of up to about 45°. For the brake shown in Figure P15-3, calculate its torque capacity and required force by both the short-shoe method and the long-shoe method and compare the results for the following data: $a_X = 8$, $b_X = 6$, $b_Y = 4$, $r = 5$, $w = 2$ in. Assume $p_{max} = 250$ psi and $\mu = 0.35$. Note that $\theta = \theta_2 - \theta_1$ for the short-shoe approximation.

(a) $\theta_1 = 75°$, $\theta_2 = 105°$.
(b) $\theta_1 = 70°$, $\theta_2 = 110°$.
(c) $\theta_1 = 65°$, $\theta_2 = 115°$.

15-20   Repeat Problem 15-19 for the brake design shown in Figure P15-4.

*15-21   Find the reaction forces at the arm pivot in the global XY system for the brake of Problem 15-11.

15-22   Find the reaction forces at the arm pivot in the global XY system for the brake of Problem 15-12.

*15-23   Find the reaction forces at the arm pivot in the global XY system for the brake of Problem 15-13.

15-24   Find the reaction forces at the arm pivot in the global XY system for the brake of Problem 15-14.

*15-25   Find the reaction forces at the arm pivot in the global XY system for the brake of Problem 15-15.

15-26   Find the reaction forces at the arm pivot in the global XY system for the brake of Problem 15-16.

Appendix A

CROSS-SECTIONAL PROPERTIES

A = area

C = centroid location

I_x = second moment of area about x axis = $\int x^2 dA$

I_y = second moment of area about y axis = $\int y^2 dA$

J_z = second polar moment of area about z axis through C

$$= \int r^2 dA = \int \left(x^2 + y^2 \right) dA = I_x + I_y$$

k_x = radius of gyration about x axis

k_y = radius of gyration about y axis

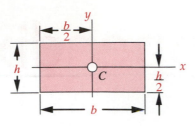

$$A = bh \qquad\qquad J_z = I_x + I_y$$

$$I_x = \frac{bh^3}{12} \qquad\qquad I_y = \frac{b^3 h}{12}$$

$$k_x = \sqrt{\frac{I_x}{A}} \qquad\qquad k_y = \sqrt{\frac{I_y}{A}}$$

(a) Rectangle

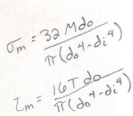

$\sigma_m = \dfrac{32 M}{\pi d^3}$

$\tau_m = \dfrac{16 T}{\pi d^3}$

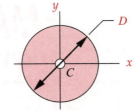

$$A = \frac{\pi D^2}{4} \qquad\qquad J_z = \frac{\pi D^4}{32}$$

$$I_x = \frac{\pi D^4}{64} \qquad\qquad I_y = \frac{\pi D^4}{64}$$

$$k_x = \sqrt{\frac{I_x}{A}} \qquad\qquad k_y = \sqrt{\frac{I_y}{A}}$$

(b) Circle

$\sigma_m = \dfrac{32 M d_o}{\pi (d_o^4 - d_i^4)}$

$\tau_m = \dfrac{16 T d_o}{\pi (d_o^4 - d_i^4)}$

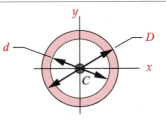

$$A = \frac{\pi}{4}\left(D^2 - d^2\right) \qquad\qquad J_z = \frac{\pi}{32}\left(D^4 - d^4\right)$$

$$I_x = \frac{\pi}{64}\left(D^4 - d^4\right) \qquad\qquad I_y = \frac{\pi}{64}\left(D^4 - d^4\right)$$

$$k_x = \sqrt{\frac{I_x}{A}} \qquad\qquad k_y = \sqrt{\frac{I_y}{A}}$$

(c) Hollow circle

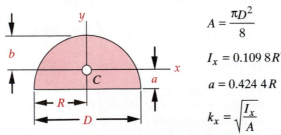

$$A = \frac{\pi D^2}{8} \qquad\qquad J_z = I_x + I_y$$

$$I_x = 0.109\,8 R^4 \qquad\qquad I_y = \frac{\pi}{8} R^4$$

$$a = 0.424\,4 R \qquad\qquad b = 0.575\,6 R$$

$$k_x = \sqrt{\frac{I_x}{A}} \qquad\qquad k_y = \sqrt{\frac{I_y}{A}}$$

(d) Solid semicircle

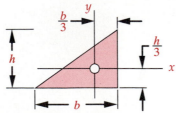

$$A = \frac{bh}{2} \qquad\qquad J_z = I_x + I_y$$

$$I_x = \frac{bh^3}{36} \qquad\qquad I_y = \frac{b^3 h}{36}$$

$$k_x = \sqrt{\frac{I_x}{A}} \qquad\qquad k_y = \sqrt{\frac{I_y}{A}}$$

(e) Right triangle

A

Appendix $\mathbf{B}$

MASS PROPERTIES

V = volume

m = mass

C_g = location of center of mass

I_x = second moment of mass about x axis = $\int \left(y^2 + z^2 \right) dm$

I_y = second moment of mass about y axis = $\int \left(x^2 + z^2 \right) dm$

I_z = second moment of mass about z axis = $\int \left(x^2 + y^2 \right) dm$

k_x = radius of gyration about x axis

k_y = radius of gyration about y axis

k_z = radius of gyration about z axis

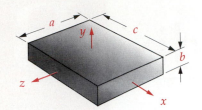

$$V = abc \qquad\qquad m = V \cdot \text{mass density}$$

$$x_{Cg} @\frac{c}{2} \qquad\qquad y_{Cg} @\frac{b}{2} \qquad\qquad z_{Cg} @\frac{a}{2}$$

$$I_x = \frac{m(a^2 + b^2)}{12} \qquad I_y = \frac{m(a^2 + c^2)}{12} \qquad I_z = \frac{m(b^2 + c^2)}{12}$$

$$k_x = \sqrt{\frac{I_x}{m}} \qquad\qquad k_y = \sqrt{\frac{I_y}{m}} \qquad\qquad k_z = \sqrt{\frac{I_z}{m}}$$

(a) Rectangular prism

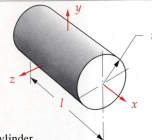

$$V = \pi r^2 l \qquad\qquad m = V \cdot \text{mass density}$$

$$x_{Cg} @\frac{l}{2} \qquad\qquad y_{Cg} \text{ on axis} \qquad z_{Cg} \text{ on axis}$$

$$I_x = \frac{mr^2}{2} \qquad\qquad I_y = I_z = \frac{m(3r^2 + l^2)}{12}$$

$$k_x = \sqrt{\frac{I_x}{m}} \qquad\qquad k_y = k_z = \sqrt{\frac{I_y}{m}}$$

(b) Cylinder

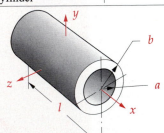

$$V = \pi(b^2 - a^2)l \qquad m = V \cdot \text{mass density}$$

$$x_{Cg} @\frac{l}{2} \qquad\qquad y_{Cg} \text{ on axis} \qquad z_{Cg} \text{ on axis}$$

$$I_x = \frac{m(a^2 + b^2)}{2} \qquad I_y = I_z = \frac{m(3a^2 + 3b^2 + l^2)}{12}$$

$$k_x = \sqrt{\frac{I_x}{m}} \qquad\qquad k_y = k_z = \sqrt{\frac{I_y}{m}}$$

(c) Hollow cylinder

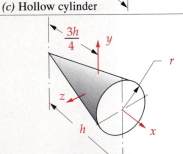

$$V = \pi\frac{r^2 h}{3} \qquad\qquad m = V \cdot \text{mass density}$$

$$x_{Cg} @\frac{3h}{4} \qquad\qquad y_{Cg} \text{ on axis} \qquad z_{Cg} \text{ on axis}$$

$$I_x = \frac{3}{10}mr^2 \qquad\qquad I_y = I_z = \frac{m(12r^2 + 3h^2)}{80}$$

$$k_x = \sqrt{\frac{I_x}{m}} \qquad\qquad k_y = k_z = \sqrt{\frac{I_y}{m}}$$

(d) Right circular cone

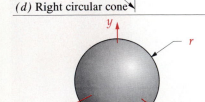

$$V = \frac{4}{3}\pi r^3 \qquad\qquad m = V \cdot \text{mass density}$$

$$x_{Cg} \text{ at center} \qquad y_{Cg} \text{ at center} \qquad z_{Cg} \text{ at center}$$

$$I_x = I_y = I_z = \frac{2}{5}mr^2$$

$$k_x = k_y = k_z = \sqrt{\frac{I_y}{m}}$$

(e) Sphere

A

Appendix C

MATERIAL
PROPERTIES

The following tables contain approximate values for strengths and other specifications of a variety of engineering materials compiled from various sources. In some cases the data are minimum recommended values and in other cases are data from a single test specimen. These data are suitable for use in the engineering exercises contained in this text, but should not be considered as statistically valid representations of specifications for any particular alloy or material. The designer should consult the materials' manufacturers for more accurate and up-to-date strength information on materials used in engineering applications, or conduct independent tests of the selected materials to determine their ultimate suitability to any application.

Table No.	Description
C-1	Physical Properties of Some Engineering Materials
C-2	Mechanical Properties for Some Wrought-Aluminum Alloys
C-3	Mechanical Properties for Some Cast-Aluminum Alloys
C-4	Mechanical Properties for Some Wrought- and Cast-Copper Alloys
C-5	Mechanical Properties for Some Titanium Alloys
C-6	Mechanical Properties for Some Magnesium Alloys
C-7	Mechanical Properties for Some Cast-Iron Alloys
C-8	Mechanical Properties for Some Stainless Steel Alloys
C-9	Mechanical Properties for Some Carbon Steels
C-10	Mechanical Properties for Some Alloy and Tool Steels
C-11	Mechanical Properties for Some Engineering Plastics

Table C-1   Physical Properties of Some Engineering Materials

Data from Various Sources.[*] These Properties are Essentially Similar for All Alloys of the Particular Material

Material	Modulus of Elasticity E		Modulus of Rigidity G		Poisson's Ratio ν	Weight Density γ	Mass Density ρ	Specific Gravity
	Mpsi	GPa	Mpsi	GPa		lb/in^3	Mg/m^3	
Aluminum Alloys	10.4	71.7	3.9	26.8	0.34	0.10	2.8	2.8
Beryllium Copper	18.5	127.6	7.2	49.4	0.29	0.30	8.3	8.3
Brass, Bronze	16.0	110.3	6.0	41.5	0.33	0.31	8.6	8.6
Copper	17.5	120.7	6.5	44.7	0.35	0.32	8.9	8.9
Iron, Cast, Gray	15.0	103.4	5.9	40.4	0.28	0.26	7.2	7.2
Iron, Cast, Ductile	24.5	168.9	9.4	65.0	0.30	0.25	6.9	6.9
Iron, Cast, Malleable	25.0	172.4	9.6	66.3	0.30	0.26	7.3	7.3
Magnesium Alloys	6.5	44.8	2.4	16.8	0.33	0.07	1.8	1.8
Nickel Alloys	30.0	206.8	11.5	79.6	0.30	0.30	8.3	8.3
Steel, Carbon	30.0	206.8	11.7	80.8	0.28	0.28	7.8	7.8
Steel, Alloys	30.0	206.8	11.7	80.8	0.28	0.28	7.8	7.8
Steel, Stainless	27.5	189.6	10.7	74.1	0.28	0.28	7.8	7.8
Titanium Alloys	16.5	113.8	6.2	42.4	0.34	0.16	4.4	4.4
Zinc Alloys	12.0	82.7	4.5	31.1	0.33	0.24	6.6	6.6

[*] *Properties of Some Metals and Alloys*, International Nickel Co., Inc., N.Y.; *Materials Handbook*, American Society for Metals, Materials Park, Ohio.

Table C-2   Mechanical Properties for Some Wrought-Aluminum Alloys

Data from Various Sources.[*] Approximate Values. Consult Material Manufacturers for More Accurate Information

Wrought-Aluminum Alloy	Condition	Tensile Yield Strength (2% offset)		Ultimate Tensile Strength		Fatigue Strength at 5E8 cycles		Elongation over 2 in	Brinell Hardness
		kpsi	MPa	kpsi	MPa	kpsi	MPa	%	-HB
1100	sheet annealed	5	34	13	90			35	23
	cold rolled	22	152	24	165			5	44
2024	sheet annealed	11	76	26	179			20	-
	heat treated	42	290	64	441	20	138	19	-
3003	sheet annealed	6	41	16	110			30	28
	cold rolled	27	186	29	200			4	55
5052	sheet annealed	13	90	28	193			25	47
	cold rolled	37	255	42	290			7	77
6061	sheet annealed	8	55	18	124			25	30
	heat treated	40	276	45	310	14	97	12	95
7075	bar annealed	15	103	33	228			16	60
	heat treated	73	503	83	572	14	97	11	150

[*] *Properties of Some Metals and Alloys*, International Nickel Co., Inc., N.Y.; *Materials Handbook*, American Society for Metals, Materials Park, Ohio.

A

Table C-3   Mechanical Properties for Some Aluminum Casting Alloys
Data from INCO.* Approximate Values. Consult Material Manufacturers for More Accurate Information

Aluminum Casting Alloy	Condition	Tensile Yield Strength (2% offset)		Ultimate Tensile Strength		Elongation over 2 in	Brinell Hardness
		kpsi	MPa	kpsi	MPa	%	-HB
43	permanent mold casting—as cast	9	62	23	159	10	45
195	sand casting—as cast	24	165	36	248	5	-
220	sand casting—solution heat treated	26	179	48	331	16	75
380	die casting—as cast	24	165	48	331	3	-
A132	permanent mold casting—heat treated + 340°F	43	296	47	324	0.5	125
A142	sand casting—heat treated + 650°F	30	207	32	221	0.5	85

* *Properties of Some Metals and Alloys*, International Nickel Co., Inc., New York.

Table C-4   Mechanical Properties for Some Wrought- and Cast-Copper Alloys
Data from INCO.* Approximate Values. Consult Material Manufacturers for More Accurate Information

Copper Alloy	Condition	Tensile Yield Strength (2% offset)		Ultimate Tensile Strength		Elongation over 2 in	Brinell or Rockwell Hardness
		kpsi	MPa	kpsi	MPa	%	
CA110—Pure Copper	strip annealed	10	69	32	221	45	40HRF
	spring temper	50	345	55	379	4	60HRB
CA170—Beryllium Copper	strip annealed plus age	145	1 000	165	1 138	7	35HRC
	hard plus age	170	1 172	190	1 310	3	40HRC
CA220—Commercial Bronze	strip annealed	10	69	37	255	45	53HRF
	spring temper	62	427	72	496	3	78HRB
CA230—Red Brass	strip annealed	15	103	40	276	50	50HB
	hard temper	60	414	75	517	7	135HB
CA260—Cartridge Brass	strip annealed	11	76	44	303	66	54HRF
	spring temper	65	448	94	648	3	91HRB
CA270—Yellow Brass	strip annealed	14	97	46	317	65	58HRF
	spring temper	62	427	91	627	30	90HRB
CA510—Phosphor Bronze	annealed	19	131	47	324	64	73HRF
	spring temper	80	552	100	689	4	95HRB
CA614—Aluminum Bronze	soft	45	310	82	565	40	84HRB
	hard	60	414	89	614	32	87HRB
CA655—High Silicon Bronze	annealed	21	145	56	386	63	76HRF
	spring temper	62	427	110	758	4	97HRB
CA675—Manganese Bronze	soft	30	207	65	448	33	65HRB
	half-hard	60	414	84	579	19	90HRB
Leaded-Tin Bronze	as cast	19	131	34	234	18	60HB
Nickel-Tin Bronze	as cast	20	138	50	345	40	85HB
	cast and heat treated	55	379	85	586	10	180HB

* *Properties of Some Metals and Alloys*, International Nickel Co., Inc., New York.

A

Table C-5   Mechanical Properties for Some Titanium Alloys

Data from INCO.* Approximate Values. Consult Material Manufacturers for More Accurate Information

Titanium Alloy	Condition	Tensile Yield Strength (2% offset)		Ultimate Tensile Strength		Elongation over 2 in	Brinell or Rockwell Hardness
		kpsi	MPa	kpsi	MPa	%	
Ti-35A	sheet annealed	30	207	40	276	30	135HB
Ti-50A	sheet annealed	45	310	55	379	25	215HB
Ti-75A	sheet annealed	75	517	85	586	18	245HB
Ti-0.2Pd Alloy	sheet annealed	45	310	55	379	25	215HB
Ti-5 Al-2.5 Sn Alloy	annealed	125	862	135	931	13	39HRC
Ti-8 Al-1 Mo-1 V Alloy	sheet annealed	130	896	140	965	13	39HRC
Ti-8 Al-2 Sn-4 Zr-2 Mo Alloy	bar annealed	130	896	140	965	15	39HRC
Ti-8 Al-6 V-2 Sn Alloy	sheet annealed	155	1 069	165	1 138	12	41HRC
Ti-6 Al-4 V Alloy	sheet annealed	130	896	140	13	2.5	39HRC
Ti-6 Al-4 V Alloy	heat treated	165	1 138	175	1 207	12	–
T1-13 V-11 Cr-3 Al Alloy	sheet annealed	130	896	135	931	13	37HRC
T1-13 V-11 Cr-3 Al Alloy	heat treated	170	1 172	180	1 241	6	–

* *Properties of Some Metals and Alloys*, International Nickel Co., Inc., New York.

Table C-6   Mechanical Properties for Some Magnesium Alloys

Data from INCO.* Approximate Values. Consult Material Manufacturers for More Accurate Information

Magnesium Alloy	Condition	Tensile Yield Strength (2% offset)		Ultimate Tensile Strength		Elongation over 2 in	Brinell or Rockwell Hardness
		kpsi	MPa	kpsi	MPa	%	
AZ 31B	sheet annealed	22	152	37	255	21	56HB
	hard sheet	32	221	42	290	15	73HB
AZ 80A	as forged	33	228	48	331	11	69HB
	forged and aged	36	248	50	345	6	72HB
AZ91A & AZ91B	die cast	22	152	33	228	3	63HB
AZ91C	as cast	14	97	24	165	2.5	60HB
	cast, solution treated and aged	19	131	40	276	5	70HB
AZ92A	as cast	14	97	25	172	2	65HB
	cast, solution treated	14	97	40	276	10	63HB
	cast, solution treated and aged	22	152	40	276	3	81HB
EZ33A	cast and aged	16	110	23	159	3	50HB
HK31A	strain hardened	29	200	37	255	8	68HB
	cast & heat treated	15	103	32	221	8	66HRB
HZ32A	cast - solution treated and aged	13	90	27	186	4	55HB
ZK60A	as extruded	38	262	49	338	14	75HB
	extruded and aged	44	303	53	365	11	82HB

* *Properties of Some Metals and Alloys*, International Nickel Co., Inc., New York.

A

Table C-7   Mechanical Properties of Some Cast-Iron Alloys

Data from Various Sources.* Approximate Values. Consult Material Manufacturers for More Accurate Information

Cast-Iron Alloy	Condition	Tensile Yield Strength (2% offset)		Ultimate Tensile Strength		Compressive Strength		Brinell Hardness
		kpsi	MPa	kpsi	MPa	kpsi	MPa	-HB
Gray Cast Iron—Class 20	as cast	–	–	22	152	83	572	156
Gray Cast Iron—Class 30	as cast	–	–	32	221	109	752	210
Gray Cast Iron—Class 40	as cast	–	–	42	290	140	965	235
Gray Cast Iron—Class 50	as cast	–	–	52	359	164	1 131	262
Gray Cast Iron—Class 60	as cast	–	–	62	427	187	1 289	302
Ductile Iron 60-40-18	annealed	47	324	65	448	52	359	160
Ductile Iron 65-45-12	annealed	48	331	67	462	53	365	174
Ductile Iron 80-55-06	annealed	53	365	82	565	56	386	228
Ductile Iron 120-90-02	Q & T	120	827	140	965	134	924	325

* *Properties of Some Metals and Alloys*, International Nickel Co., Inc., N.Y.; *Materials Handbook*, American Society for Metals, Materials Park, Ohio.

Table C-8   Mechanical Properties for Some Stainless Steel Alloys

Data from INCO.* Approximate Values. Consult Material Manufacturers for More Accurate Information

Stainless Steel Alloy	Condition	Tensile Yield Strength (2% offset)		Ultimate Tensile Strength		Elongation over 2 in	Brinell or Rockwell Hardness
		kpsi	MPa	kpsi	MPa	%	
Type 301	strip annealed	40	276	110	758	60	85HRB
	cold rolled	165	1 138	200	1 379	8	41HRC
Type 302	sheet annealed	40	276	90	621	50	85HRB
	cold rolled	165	1 138	190	1 310	5	40HRC
Type 304	sheet annealed	35	241	85	586	50	80HRB
	cold rolled	160	1 103	185	1 276	4	40HRC
Type 314	bar annealed	50	345	100	689	45	180HB
Type 316	sheet annealed	40	276	90	621	50	85HRB
Type 330	hot rolled	55	379	100	689	35	200HB
	annealed	35	241	80	552	50	150HB
Type 410	sheet annealed	45	310	70	483	25	80HRB
	heat treated	140	965	180	1 241	15	39HRC
Type 420	bar annealed	50	345	95	655	25	92HRB
	heat treated	195	1 344	230	1 586	8	500HB
Type 431	bar annealed	95	655	125	862	25	260HB
	heat treated	150	1 034	195	1 344	15	400HB
Type 440C	bar annealed	65	448	110	758	14	230HB
	Q & T 600F	275	1 896	285	1 965	2	57HRC
17-4 PH (AISI 630)	hardened	185	1 276	200	1 379	14	44HRC
17-7 PH (AISI 631)	hardened	220	1 517	235	1 620	6	48HRC

* *Properties of Some Metals and Alloys*, International Nickel Co., Inc., New York.

A

286.8   1034

Table C-9   Mechanical Properties for Some Carbon Steels
Data from Various Sources.* Approximate Values. Consult Material Manufacturers for More Accurate Information

SAE / AISI Number	Condition	Tensile Yield Strength (2% offset)		Ultimate Tensile Strength		Elongation over 2 in	Brinell Hardness
		kpsi	MPa	kpsi	MPa	%	-HB
1010	hot rolled	26	179	47	324	28	95
	cold rolled	44	303	53	365	20	105
1020	hot rolled	30	207	55	379	25	111
	cold rolled	57	393	68	469	15	131
1030	hot rolled	38	259	68	469	20	137
	normalized @ 1 650°F	50	345	75	517	32	149
	cold rolled	64	441	76	524	12	149
	Q&T @ 1 000°F	75	517	97	669	28	255
	Q&T @ 800°F	84	579	106	731	23	302
	Q&T @ 400°F	94	648	123	848	17	495
1035	hot rolled	40	276	72	496	18	143
	cold rolled	67	462	80	552	12	163
1040	hot rolled	42	290	76	524	18	149
	normalized @ 1 650°F	54	372	86	593	28	170
	cold rolled	71	490	85	586	12	170
	Q&T @ 1 200°F	63	434	92	634	29	192
	Q&T @ 800°F	80	552	110	758	21	241
	Q&T @ 400°F	86	593	113	779	19	262
1045	hot rolled	45	310	82	565	16	163
	cold rolled	77	531	91	627	12	179
1050	hot rolled	50	345	90	621	15	179
	normalized @ 1 650°F	62	427	108	745	20	217
	cold rolled	84	579	100	689	10	197
	Q&T @ 1 200°F	78	538	104	717	28	235
	Q&T @ 800°F	115	793	158	1 089	13	444
	Q&T @ 400°F	117	807	163	1 124	9	514
1060	hot rolled	54	372	98	676	12	200
	normalized @ 1 650°F	61	421	112	772	18	229
	Q&T @ 1 200°F	76	524	116	800	23	229
	Q&T @ 1 000°F	97	669	140	965	17	277
	Q&T @ 800°F	111	765	156	1 076	14	311
1095	hot rolled	66	455	120	827	10	248
	normalized @ 1 650°F	72	496	147	1 014	9	13
	Q&T @ 1 200°F	80	552	130	896	21	269
	Q&T @ 800°F	112	772	176	1 213	12	363
	Q&T @ 600°F	118	814	183	1 262	10	375

* *SAE Handbook*, Society of Automotive Engineers, Warrendale Pa.; *Materials Handbook*, American Society for Metals, Materials Park, Ohio.

A

Table C-10   Mechanical Properties for Some Alloy and Tool Steels

Data from Various Sources.* Approximate Values. Consult Material Manufacturers for More Accurate Information

SAE / AISI Number	Condition	Tensile Yield Strength (2% offset)		Ultimate Tensile Strength		Elongation over 2 in	Brinell or Rockwell Hardness
		kpsi	MPa	kpsi	MPa	%	
1340	annealed	63	434	102	703	25	204HB
	Q&T	109	752	125	862	21	250HB
4027	annealed	47	324	75	517	30	150HB
	Q&T	113	779	132	910	12	264HB
4130	annealed @ 1 450°F	52	359	81	558	28	156HB
	normalized @ 1 650°F	63	434	97	669	25	197HB
	Q&T @ 1 200°F	102	703	118	814	22	245HB
	Q&T @ 800°F	173	1 193	186	1 282	13	380HB
	Q&T @ 400°F	212	1 462	236	1 627	10	41HB
4140	annealed @ 1 450°F	61	421	95	655	26	197HB
	normalized @ 1 650°F	95	655	148	1 020	18	302HB
	Q&T @ 1 200°F	95	655	110	758	22	230HB
	Q&T @ 800°F	165	1 138	181	1 248	13	370HB
	Q&T @ 400°F	238	1 641	257	1 772	8	510HB
4340	Q&T @ 1 200°F	124	855	140	965	19	280HB
	Q&T @ 1 000°F	156	1 076	170	1 172	13	360HB
	Q&T @ 800°F	198	1 365	213	1 469	10	430HB
	Q&T @ 600°F	230	1 586	250	1 724	10	486HB
6150	annealed	59	407	96	662	23	192HB
	Q&T	148	1 020	157	1 082	16	314HB
8740	annealed	60	414	95	655	25	190HB
	Q&T	133	917	144	993	18	288HB
H-11	annealed @ 1 600°F	53	365	100	689	25	96HRB
	Q&T @ 1 000°F	250	1 724	295	2 034	9	55HRC
L-2	annealed @ 1 425°F	74	510	103	710	25	96HRB
	Q&T @ 400°F	260	1 793	290	1 999	5	54HRC
L-6	annealed @ 1 425°F	55	379	95	655	25	93HRB
	Q&T @ 600°F	260	1 793	290	1 999	4	54HRC
P-20	annealed @ 1 425°F	75	517	100	689	17	97HRB
	Q&T @ 400°F	205	1 413	270	1 862	10	52HRC
S-1	annealed @ 1 475°F	60	414	100	689	24	96HRB
	Q&T @ 400°F	275	1 896	300	2 068	4	57HRC
S-5	annealed @ 1 450°F	64	441	105	724	25	96HRB
	Q&T @ 400°F	280	1 931	340	2 344	5	59HRC
S-7	annealed @ 1 525°F	55	379	93	641	25	95HRB
	Q&T @ 400°F	210	1 448	315	2 172	7	58HRC
A-8	annealed @ 1 550°F	65	448	103	710	24	97HRB
	Q&T @ 1 050°F	225	1 551	265	1 827	9	52HRC

* *Machine Design Materials Reference Issue*, Penton Publishing, Cleveland Ohio; *Materials Handbook*, ASM, Materials Park, Ohio.

A

Table C-11   Properties of Some Engineering Plastics
Data from Various Sources.* Approximate Values. Consult Material Manufacturers for More Accurate Information

Material	Approximate Modulus of Elasticity E		Ultimate Tensile Strength		Ultimate Compressive Strength		Elongation over 2 in	Max Temp	Specific Gravity
	Mpsi	GPa	kpsi	MPa	kpsi	MPa	%	°F	
ABS	0.3	2.1	6.0	41.4	10.0	68.9	5 to 25	160-200	1.05
20-40% glass filled	0.6	4.1	10.0	68.9	12.0	82.7	3	200-230	1.30
Acetal	0.5	3.4	8.8	60.7	18.0	124.1	60	220	1.41
20-30% glass filled	1.0	6.9	10.0	68.9	18.0	124.1	7	185-220	1.56
Acrylic	0.4	2.8	10.0	68.9	15.0	103.4	5	140-190	1.18
Fluoroplastic (PTFE)	0.2	1.4	5.0	34.5	6.0	41.4	100	350-330	2.10
Nylon 6/6	0.2	1.4	10.0	68.9	10.0	68.9	60	180-300	1.14
Nylon 11	0.2	1.3	8.0	55.2	8.0	55.2	300	180-300	1.04
20-30% glass filled	0.4	2.5	12.8	88.3	12.8	88.3	4	250-340	1.26
Polycarbonate	0.4	2.4	9.0	62.1	12.0	82.7	100	250	1.20
10-40% glass filled	1.0	6.9	17.0	117.2	17.0	117.2	2	275	1.35
HMW Polyethylene	0.1	0.7	2.5	17.2	–	–	525	–	0.94
Polyphenylene Oxide	0.4	2.4	9.6	66.2	16.4	113.1	20	212	1.06
20-30% glass filled	1.1	7.8	15.5	106.9	17.5	120.7	5	260	1.23
Polypropylene	0.2	1.4	5.0	34.5	7.0	48.3	500	250-320	0.90
20-30% glass filled	0.7	4.8	7.5	51.7	6.2	42.7	2	300-320	1.10
Impact Polystryrene	0.3	2.1	4.0	27.6	6.0	41.4	2 to 80	140-175	1.07
20-30% glass filled	0.1	0.7	12.0	82.7	16.0	110.3	1	180-200	1.25
Polysulfone	0.4	2.5	10.2	70.3	13.9	95.8	50	300-345	1.24

* *Modern Plastics Encyclopedia*, McGraw-Hill, New York.; *Machine Design Materials Reference Issue*, Penton Publishing, Cleveland, Ohio.

Appendix D

BEAM TABLES

Loading, shear, moment, slope, and deflection functions for a selection of common beam configurations and loadings are presented in these tables. **Cantilever**, **simply supported**, and **overhung** beams with either a **concentrated load** at any point or a **uniformly distributed load** across any portion of the span are defined. A general set of equations is derived for each beam. Special cases, such as those with the load at center span, are accommodated by appropriate choice of dimensions in the general formulas. In all cases, singularity functions are used to write the beam equations, which gives a single expression for the entire span for each function. See Section 3.9 for a discussion of singularity functions. The equations for the beam cases in this appendix have been encoded in *TKSolver* files, which are provided on the disk included with this text. In some cases, the *TKSolver* files allow multiple loads to be applied at different locations on the beam, but the derivations in this appendix each accommodate only one load per beam. *Use superposition to combine various beam cases when more than one type of load is present on a beam.* For a more complete collection of beam formulas see Roark and Young, *Formulas for Stress and Strain*, 6th ed., McGraw-Hill, New York, 1989. A key to the Figures in this appendix and their related *TKSolver* files follows.

Figure No.	Case	TKSolver File
D-1*a*	Cantilever beam with concentrated load	CANTCONC
D-1*b*	Cantilever beam with uniformly distributed load	CANTUNIF
D-2*a*	Simply supported beam with concentrated load	SIMPCONC
D-2*b*	Simply supported beam with uniformly distributed load	SIMPUNIF
D-3*a*	Overhung beam with concentrated load	OVHGCONC
D-3*b*	Overhung beam with uniformly distributed load	OVHGUNIF

(a) Cantilever beam with concentrated loading

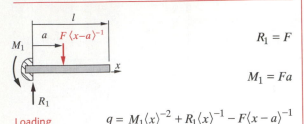

$$R_1 = F$$

$$M_1 = Fa$$

Loading

$$q = M_1 \langle x \rangle^{-2} + R_1 \langle x \rangle^{-1} - F \langle x - a \rangle^{-1}$$

(b) Cantilever beam with uniformly distributed loading

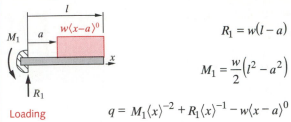

$$R_1 = w(l - a)$$

$$M_1 = \frac{w}{2}\left(l^2 - a^2\right)$$

Loading

$$q = M_1 \langle x \rangle^{-2} + R_1 \langle x \rangle^{-1} - w \langle x - a \rangle^{0}$$

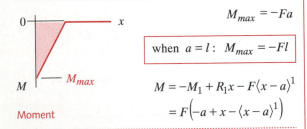

$$V_{max} = R_1 = F$$

$$V = M_1 \langle x \rangle^{-1} + R_1 - F \langle x - a \rangle^{0}$$
$$= F\left(1 - \langle x - a \rangle^{0}\right)$$

Shear

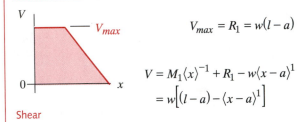

$$V_{max} = R_1 = w(l - a)$$

$$V = M_1 \langle x \rangle^{-1} + R_1 - w \langle x - a \rangle^{1}$$
$$= w\left[(l - a) - \langle x - a \rangle^{1}\right]$$

Shear

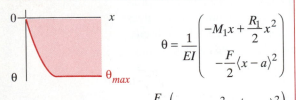

$$M_{max} = -Fa$$

when $a = l$: $M_{max} = -Fl$

$$M = -M_1 + R_1 x - F \langle x - a \rangle^{1}$$
$$= F\left(-a + x - \langle x - a \rangle^{1}\right)$$

Moment

$$M_{max} = M_1 = \frac{w}{2}\left(l^2 - a^2\right)$$

when $a = 0$: $M_{max} = \frac{wl^2}{2}$

$$M = -M_1 + R_1 x - w \langle x - a \rangle^{2}$$
$$= \frac{w}{2}\left[2(l - a)x - \left(l^2 - a^2\right) - 2\langle x - a \rangle^{2}\right]$$

Moment

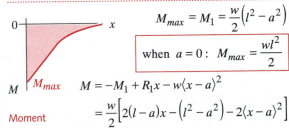

$$\theta = \frac{1}{EI}\begin{pmatrix} -M_1 x + \dfrac{R_1}{2}x^2 \\ -\dfrac{F}{2}\langle x - a \rangle^2 \end{pmatrix}$$

$$\theta = \frac{F}{2EI}\left(-2ax + x^2 - \langle x - a \rangle^2\right)$$

Slope

$$\theta = \frac{1}{EI}\begin{pmatrix} -M_1 x + \dfrac{R_1}{2}x^2 \\ -\dfrac{w}{6}\langle x - a \rangle^3 \end{pmatrix}$$

$$\theta = \frac{w}{6EI}\left(3(l - a)x^2 - 3\left(l^2 - a^2\right)x - \langle x - a \rangle^3\right)$$

Slope

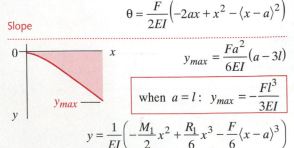

$$y_{max} = \frac{Fa^2}{6EI}(a - 3l)$$

when $a = l$: $y_{max} = -\frac{Fl^3}{3EI}$

$$y = \frac{1}{EI}\left(-\frac{M_1}{2}x^2 + \frac{R_1}{6}x^3 - \frac{F}{6}\langle x - a \rangle^3\right)$$

Deflection

$$= \frac{F}{6EI}\left(x^3 - 3ax^2 - \langle x - a \rangle^3\right)$$

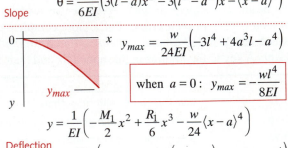

$$y_{max} = \frac{w}{24EI}\left(-3l^4 + 4a^3 l - a^4\right)$$

when $a = 0$: $y_{max} = -\frac{wl^4}{8EI}$

$$y = \frac{1}{EI}\left(-\frac{M_1}{2}x^2 + \frac{R_1}{6}x^3 - \frac{w}{24}\langle x - a \rangle^4\right)$$

Deflection

$$= \frac{w}{24EI}\left(4(l - a)x^3 - 6\left(l^2 - a^2\right)x^2 - \langle x - a \rangle^4\right)$$

FIGURE D-1

Cantilever Beams with Concentrated or Distributed Loading. Note: < > Denotes a Singularity Function

A

(a) Simply supported beam with concentrated loading

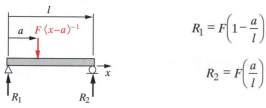

$$R_1 = F\left(1 - \frac{a}{l}\right)$$

$$R_2 = F\left(\frac{a}{l}\right)$$

Loading $q = R_1\langle x\rangle^{-1} - F\langle x-a\rangle^{-1} + R_2\langle x-l\rangle^{-1}$

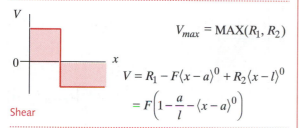

$$V_{max} = \text{MAX}(R_1, R_2)$$

$$V = R_1 - F\langle x-a\rangle^0 + R_2\langle x-l\rangle^0$$

$$= F\left(1 - \frac{a}{l} - \langle x-a\rangle^0\right)$$

Shear

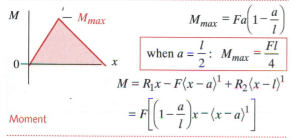

$$M_{max} = Fa\left(1 - \frac{a}{l}\right)$$

$$\boxed{\text{when } a = \frac{l}{2}: \ M_{max} = \frac{Fl}{4}}$$

$$M = R_1 x - F\langle x-a\rangle^1 + R_2\langle x-l\rangle^1$$

$$= F\left[\left(1 - \frac{a}{l}\right)x - \langle x-a\rangle^1\right]$$

Moment

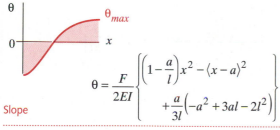

$$\theta = \frac{F}{2EI}\left\{\begin{array}{l}\left(1 - \frac{a}{l}\right)x^2 - \langle x-a\rangle^2 \\ + \frac{a}{3l}\left(-a^2 + 3al - 2l^2\right)\end{array}\right\}$$

Slope

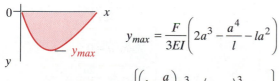

$$y_{max} = \frac{F}{3EI}\left(2a^3 - \frac{a^4}{l} - la^2\right)$$

$$y = \frac{F}{6EI}\left\{\begin{array}{l}\left(1 - \frac{a}{l}\right)x^3 - \langle x-a\rangle^3 \\ + \frac{a}{l}\left(-a^2 + 3al - 2l^2\right)x\end{array}\right\}$$

Deflection

(b) Simply supported beam with uniformly distributed loading

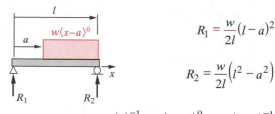

$$R_1 = \frac{w}{2l}(l-a)^2$$

$$R_2 = \frac{w}{2l}\left(l^2 - a^2\right)$$

Loading $q = R_1\langle x\rangle^{-1} - w\langle x-a\rangle^0 + R_2\langle x-l\rangle^{-1}$

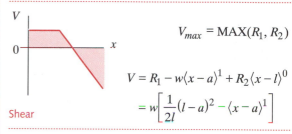

$$V_{max} = \text{MAX}(R_1, R_2)$$

$$V = R_1 - w\langle x-a\rangle^1 + R_2\langle x-l\rangle^0$$

$$= w\left[\frac{1}{2l}(l-a)^2 - \langle x-a\rangle^1\right]$$

Shear

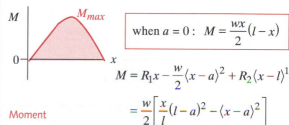

$$\boxed{\text{when } a = 0: \ M = \frac{wx}{2}(l-x)}$$

$$M = R_1 x - \frac{w}{2}\langle x-a\rangle^2 + R_2\langle x-l\rangle^1$$

$$= \frac{w}{2}\left[\frac{x}{l}(l-a)^2 - \langle x-a\rangle^2\right]$$

Moment

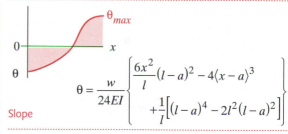

$$\theta = \frac{w}{24EI}\left\{\begin{array}{l}\frac{6x^2}{l}(l-a)^2 - 4\langle x-a\rangle^3 \\ + \frac{1}{l}\left[(l-a)^4 - 2l^2(l-a)^2\right]\end{array}\right\}$$

Slope

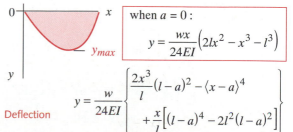

$$\boxed{\begin{array}{l}\text{when } a = 0: \\ y = \frac{wx}{24EI}\left(2lx^2 - x^3 - l^3\right)\end{array}}$$

$$y = \frac{w}{24EI}\left\{\begin{array}{l}\frac{2x^3}{l}(l-a)^2 - \langle x-a\rangle^4 \\ + \frac{x}{l}\left[(l-a)^4 - 2l^2(l-a)^2\right]\end{array}\right\}$$

Deflection

FIGURE D-2

Simply-Supported Beams with Concentrated or Distributed Loading. Note: < > Denotes a Singularity Function

A

(a) Overhung beam with concentrated loading

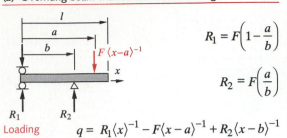

$$R_1 = F\left(1 - \frac{a}{b}\right)$$

$$R_2 = F\left(\frac{a}{b}\right)$$

Loading

$$q = R_1\langle x\rangle^{-1} - F\langle x - a\rangle^{-1} + R_2\langle x - b\rangle^{-1}$$

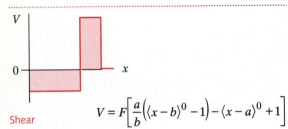

Shear

$$V = F\left[\frac{a}{b}\left(\langle x - b\rangle^0 - 1\right) - \langle x - a\rangle^0 + 1\right]$$

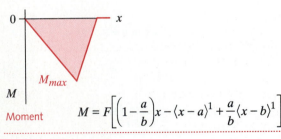

Moment

$$M = F\left[\left(1 - \frac{a}{b}\right)x - \langle x - a\rangle^1 + \frac{a}{b}\langle x - b\rangle^1\right]$$

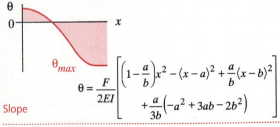

Slope

$$\theta = \frac{F}{2EI}\left[\left(1 - \frac{a}{b}\right)x^2 - \langle x - a\rangle^2 + \frac{a}{b}\langle x - b\rangle^2 + \frac{a}{3b}\left(-a^2 + 3ab - 2b^2\right)\right]$$

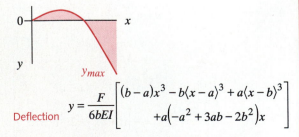

Deflection

$$y = \frac{F}{6bEI}\left[(b - a)x^3 - b\langle x - a\rangle^3 + a\langle x - b\rangle^3 + a\left(-a^2 + 3ab - 2b^2\right)x\right]$$

(b) Overhung beam with uniformly distributed loading

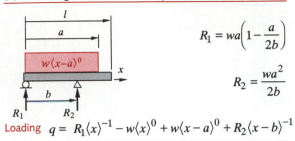

$$R_1 = wa\left(1 - \frac{a}{2b}\right)$$

$$R_2 = \frac{wa^2}{2b}$$

Loading

$$q = R_1\langle x\rangle^{-1} - w\langle x\rangle^0 + w\langle x - a\rangle^0 + R_2\langle x - b\rangle^{-1}$$

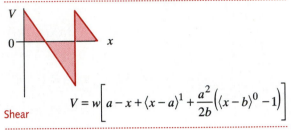

Shear

$$V = w\left[a - x + \langle x - a\rangle^1 + \frac{a^2}{2b}\left(\langle x - b\rangle^0 - 1\right)\right]$$

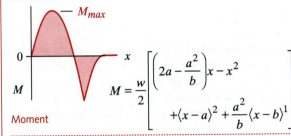

Moment

$$M = \frac{w}{2}\left[\left(2a - \frac{a^2}{b}\right)x - x^2 + \langle x - a\rangle^2 + \frac{a^2}{b}\langle x - b\rangle^1\right]$$

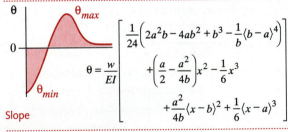

Slope

$$\theta = \frac{w}{EI}\left[\frac{1}{24}\left(2a^2b - 4ab^2 + b^3 - \frac{1}{b}\langle b - a\rangle^4\right) + \left(\frac{a}{2} - \frac{a^2}{4b}\right)x^2 - \frac{1}{6}x^3 + \frac{a^2}{4b}\langle x - b\rangle^2 + \frac{1}{6}\langle x - a\rangle^3\right]$$

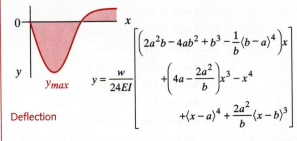

Deflection

$$y = \frac{w}{24EI}\left[\left(2a^2b - 4ab^2 + b^3 - \frac{1}{b}\langle b - a\rangle^4\right)x + \left(4a - \frac{2a^2}{b}\right)x^3 - x^4 + \langle x - a\rangle^4 + \frac{2a^2}{b}\langle x - b\rangle^3\right]$$

FIGURE D-3

Overhung Beams with Concentrated or Distributed Loading. Note: < > Denotes a Singularity Function

Appendix E

STRESS CONCENTRATION FACTORS

Stress-concentration factors for 14 common cases are presented in this appendix as listed below. All curves are taken from R. E. Peterson, "Design Factors for Stress Concentration, Parts 1 to 5." *Machine Design*, February - July, 1951, Penton Publishing, Cleveland, Ohio, with permission. Approximate equations for these curves have been fitted and are defined in each figure. These equations have been encoded as *TKSolver* functions (noted below) that can be incorporated in other models to allow automatic generation of approximate stress-concentration factors during calculations.

Figure	Case	TK File Name
E-1	Shaft with Shoulder Fillet in Axial Tension	APP_E-01
E-2	Shaft with Shoulder Fillet in Bending	APP_E-02
E-3	Shaft with Shoulder Fillet in Torsion	APP_E-03
E-4	Shaft with Groove in Axial Tension	APP_E-04
E-5	Shaft with Groove in Bending	APP_E-05
E-6	Shaft with Groove in Torsion	APP_E-06
E-7	Shaft with Transverse Hole in Bending	APP_E-07
E-8	Shaft with Transverse Hole in Torsion	APP_E-08
E-9	Flat Bar with Fillet in Axial Tension	APP_E-09
E-10	Flat Bar with Fillet in Bending	APP_E-10
E-11	Flat Bar with Notch in Axial Tension	APP_E-11
E-12	Flat Bar with Notch in Bending	APP_E-12
E-13	Flat Bar with Transverse Hole in Axial Tension	APP_E-13
E-14	Flat Bar with Transverse Hole in Bending	APP_E-14

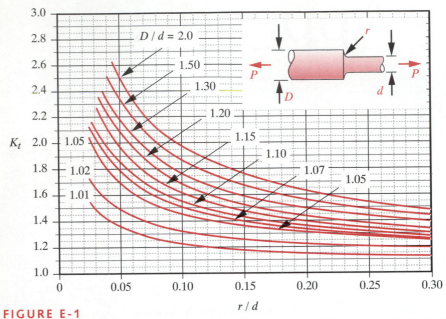

$$K_t \cong A\left(\frac{r}{d}\right)^b$$

where:

D/d	A	b
2.00	1.014 70	−0.300 35
1.50	0.999 57	−0.282 21
1.30	0.996 82	−0.257 51
1.20	0.962 72	−0.255 27
1.15	0.980 84	−0.224 85
1.10	0.984 50	−0.208 18
1.07	0.984 98	−0.195 48
1.05	1.004 80	−0.170 76
1.02	1.012 20	−0.124 74
1.01	0.984 13	−0.104 74

FIGURE E-1

Geometric Stress-Concentration Factor K_t for a Shaft with a Shoulder Fillet in Axial tension

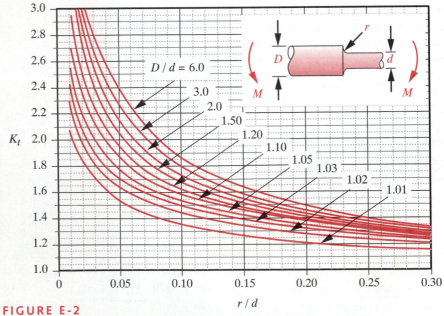

$$K_t \cong A\left(\frac{r}{d}\right)^b$$

where:

D/d	A	b
6.00	0.878 68	−0.332 43
3.00	0.893 34	−0.308 60
2.00	0.908 79	−0.285 98
1.50	0.938 36	−0.257 59
1.20	0.970 98	−0.217 96
1.10	0.951 20	−0.237 57
1.07	0.975 27	−0.209 58
1.05	0.981 37	−0.196 53
1.03	0.980 61	−0.183 81
1.02	0.960 48	−0.177 11
1.01	0.919 38	−0.170 32

FIGURE E-2

Geometric Stress-Concentration Factor K_t for a Shaft with a Shoulder Fillet in Bending

A

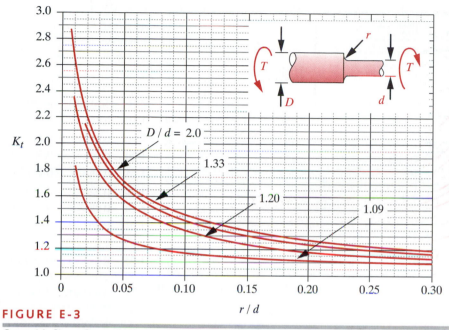

FIGURE E-3

Geometric Stress-Concentration Factor K_t for a Shaft with a Shoulder Fillet in Torsion

$$K_t \cong A\left(\frac{r}{d}\right)^b$$

where :

D/d	A	b
2.00	0.863 31	−0.238 65
1.33	0.848 97	−0.231 61
1.20	0.834 25	−0.216 49
1.09	0.903 37	−0.126 92

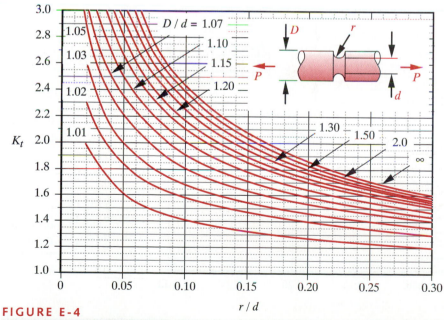

FIGURE E-4

Geometric Stress-Concentration Factor K_t for a Grooved Shaft in Axial Tension

$$K_t \cong A\left(\frac{r}{d}\right)^b$$

where :

D/d	A	b
∞	0.993 72	−0.393 52
2.00	0.993 83	−0.382 31
1.50	0.998 08	−0.369 55
1.30	1.004 90	−0.355 45
1.20	1.010 70	−0.337 65
1.15	1.026 30	−0.316 73
1.10	1.027 20	−0.294 84
1.07	1.023 80	−0.276 18
1.05	1.027 20	−0.252 56
1.03	1.036 70	−0.216 03
1.02	1.037 90	−0.187 55
1.01	1.000 30	−0.156 09

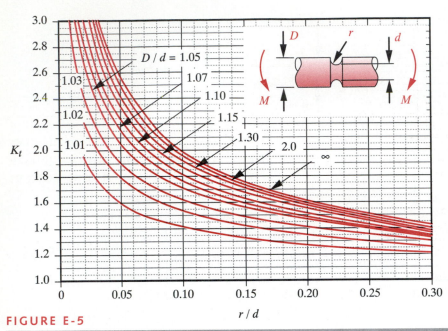

$$K_t \cong A\left(\frac{r}{d}\right)^b$$

where :

D / d	A	b
∞	0.948 01	−0.333 02
2.00	0.936 19	−0.330 66
1.50	0.938 94	−0.323 80
1.30	0.942 99	−0.315 04
1.20	0.946 81	−0.305 82
1.15	0.953 11	−0.297 39
1.12	0.955 73	−0.288 86
1.10	0.954 54	−0.282 68
1.07	0.967 74	−0.264 52
1.05	0.987 55	−0.241 34
1.03	0.990 33	−0.215 17
1.02	0.977 53	−0.197 93
1.01	0.993 93	−0.152 38

FIGURE E-5

Geometric Stress-Concentration Factor K_t for a Grooved Shaft in Bending

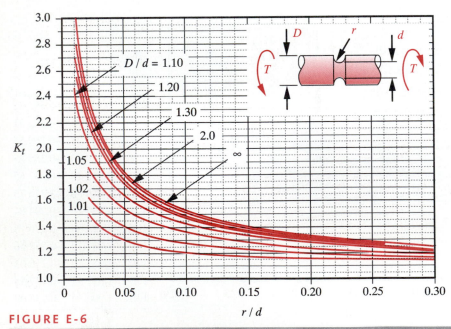

$$K_t \cong A\left(\frac{r}{d}\right)^b$$

where :

D / d	A	b
∞	0.881 26	−0.252 04
2.00	0.890 35	−0.240 75
1.30	0.894 60	−0.232 67
1.20	0.901 82	−0.223 34
1.10	0.923 11	−0.197 40
1.05	0.938 53	−0.169 41
1.02	0.968 77	−0.126 05
1.01	0.972 45	−0.101 62

FIGURE E-6

Geometric Stress-Concentration Factor K_t for a Grooved Shaft in Torsion

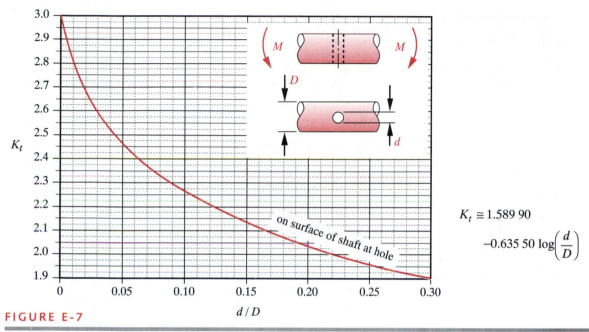

$$K_t \cong 1.589\ 90$$

$$-0.635\ 50 \log\left(\frac{d}{D}\right)$$

FIGURE E-7

Geometric Stress-Concentration Factor K_t for a Shaft with a Transverse Hole in Bending

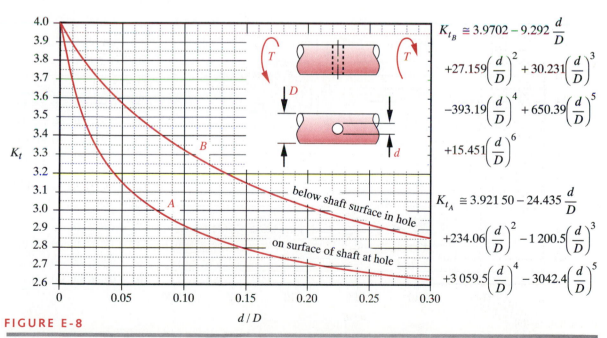

$$K_{t_B} \cong 3.9702 - 9.292\,\frac{d}{D}$$

$$+27.159\left(\frac{d}{D}\right)^2 + 30.231\left(\frac{d}{D}\right)^3$$

$$-393.19\left(\frac{d}{D}\right)^4 + 650.39\left(\frac{d}{D}\right)^5$$

$$+15.451\left(\frac{d}{D}\right)^6$$

$$K_{t_A} \cong 3.921\ 50 - 24.435\,\frac{d}{D}$$

$$+234.06\left(\frac{d}{D}\right)^2 - 1\ 200.5\left(\frac{d}{D}\right)^3$$

$$+3\ 059.5\left(\frac{d}{D}\right)^4 - 3042.4\left(\frac{d}{D}\right)^5$$

FIGURE E-8

Geometric Stress-Concentration Factor K_t for a with a Transverse Hole in Torsion

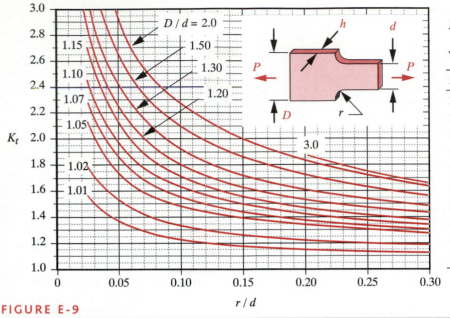

$$K_t \cong A\left(\frac{r}{d}\right)^b$$

where :

D/d	A	b
2.00	1.099 60	−0.320 77
1.50	1.076 90	−0.295 58
1.30	1.054 40	−0.270 21
1.20	1.035 10	−0.250 84
1.15	1.014 20	−0.239 35
1.10	1.013 00	−0.215 35
1.07	1.014 50	−0.193 66
1.05	0.987 97	−0.138 48
1.02	1.025 90	−0.169 78
1.01	0.976 62	−0.106 56

FIGURE E-9

Geometric Stress-Concentration Factor K_t for a Filleted Flat Bar in Axial tension

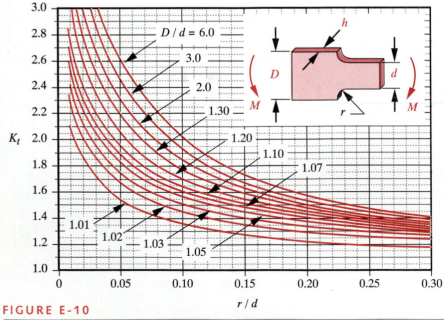

$$K_t \cong A\left(\frac{r}{d}\right)^b$$

where :

D/d	A	b
6.00	0.895 79	−0.358 47
3.00	0.907 20	−0.333 33
2.00	0.932 32	−0.303 04
1.30	0.958 80	−0.272 69
1.20	0.995 90	−0.238 29
1.10	1.016 50	−0.215 48
1.07	1.019 90	−0.203 33
1.05	1.022 60	−0.191 56
1.03	1.016 60	−0.178 02
1.02	0.995 28	−0.170 13
1.01	0.966 89	−0.154 17

FIGURE E-10

Geometric Stress-Concentration Factor K_t for a Filleted Flat Bar in Bending

A

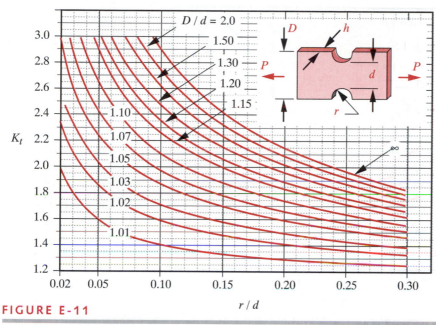

$$K_t \cong A\left(\frac{r}{d}\right)^b$$

where:

D/d	A	b
∞	1.109 50	−0.417 12
3.00	1.113 90	−0.409 23
2.00	1.133 90	−0.385 86
1.50	1.132 60	−0.365 92
1.30	1.158 60	−0.332 60
1.20	1.147 50	−0.315 07
1.15	1.095 20	−0.325 17
1.10	1.085 10	−0.299 97
1.07	1.091 20	−0.268 57
1.05	1.090 60	−0.241 63
1.03	1.051 80	−0.222 16
1.02	1.054 00	−0.188 79
1.01	1.042 60	−0.141 45

FIGURE E-11

Geometric Stress-Concentration Factor K_t for a Notched Flat Bar in Axial tension

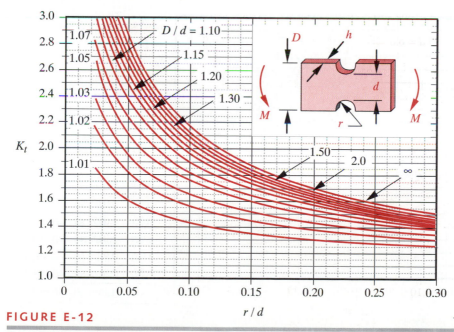

$$K_t \cong A\left(\frac{r}{d}\right)^b$$

where:

D/d	A	b
∞	0.970 79	−0.356 72
3.00	0.971 94	−0.350 47
2.00	0.968 01	−0.349 15
1.50	0.983 15	−0.333 95
1.30	0.982 88	−0.326 06
1.20	0.990 55	−0.313 19
1.15	0.993 04	−0.302 63
1.10	1.007 10	−0.283 79
1.07	1.014 70	−0.261 45
1.05	1.025 00	−0.240 08
1.03	1.029 40	−0.211 61
1.02	1.037 40	−0.184 28
1.01	1.060 50	−0.133 69

FIGURE E-12

Geometric Stress-Concentration Factor K_t for a Notched Flat Bar in Bending

A

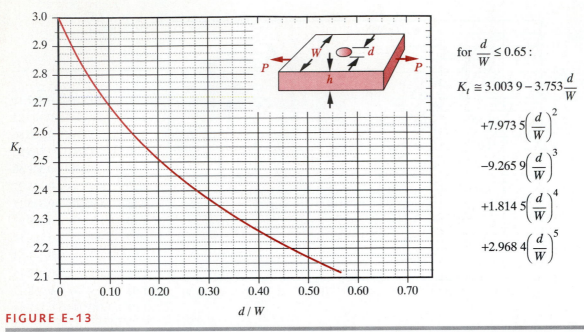

for $\dfrac{d}{W} \le 0.65$:

$$K_t \cong 3.003\,9 - 3.753\dfrac{d}{W}$$

$$+7.973\,5\left(\dfrac{d}{W}\right)^2$$

$$-9.265\,9\left(\dfrac{d}{W}\right)^3$$

$$+1.814\,5\left(\dfrac{d}{W}\right)^4$$

$$+2.968\,4\left(\dfrac{d}{W}\right)^5$$

FIGURE E-13

Geometric Stress-Concentration Factor K_t for a Flat Bar with Transverse Hole in Axial tension

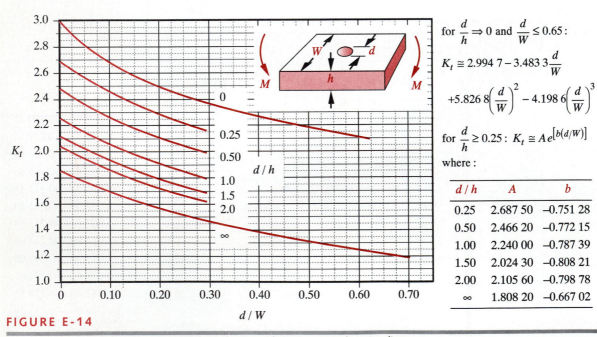

for $\dfrac{d}{h} \Rightarrow 0$ and $\dfrac{d}{W} \le 0.65$:

$$K_t \cong 2.994\,7 - 3.483\,3\dfrac{d}{W}$$

$$+5.826\,8\left(\dfrac{d}{W}\right)^2 - 4.198\,6\left(\dfrac{d}{W}\right)^3$$

for $\dfrac{d}{h} \ge 0.25$: $K_t \cong A\,e^{[b(d/W)]}$

where :

d/h	A	b
0.25	2.687 50	−0.751 28
0.50	2.466 20	−0.772 15
1.00	2.240 00	−0.787 39
1.50	2.024 30	−0.808 21
2.00	2.105 60	−0.798 78
∞	1.808 20	−0.667 02

FIGURE E-14

Geometric Stress-Concentration Factor K_t for a Flat Bar with Transverse Hole in Bending

A

Appendix F

CONVERSION FACTORS

Table F-1   Selected Units Conversion Factors

Note That These Conversion Factors (and Others) are Built Into the *TKSolver* Files UNITMAST and STUDENT

Multiply this	by	this	to get	this	Multiply this	by	this	to get	this
acceleration					**mass moment of inertia**				
in/sec^2	x	0.0254	=	m/sec^2	$lb\text{-}in\text{-}sec^2$	x	0.1138	=	$N\text{-}m\text{-}sec^2$
ft/sec^2	x	12	=	in/sec^2	**moments and energy**				
angles					in-lb	x	0.1138	=	N-m
radian	x	57.2958	=	deg	ft-lb	x	12	=	in-lb
					N-m	x	8.7873	=	in-lb
area					N-m	x	0.7323	=	ft-lb
in^2	x	645.16	=	mm^2					
ft^2	x	144	=	in^2	**power**				
					HP	x	550	=	ft-lb/sec
area moment of inertia					HP	x	33 000	=	ft-lb/min
in^4	x	416 231	=	mm^4	HP	x	6 600	=	in-lb/sec
in^4	x	4.162E–07	=	m^4	HP	x	745.7	=	watts
m^4	x	1.0E+12	=	mm^4	N-m/sec	x	8.7873	=	in-lb/sec
m^4	x	1.0E+08	=	cm^4					
ft^4	x	20 736	=	in^4	**pressure and stress**				
					psi	x	6 894.8	=	Pa
density					psi	x	6.895E-3	=	MPa
lb/in^3	x	27.6805	=	g/cc	psi	x	144	=	psf
g/cc	x	0.001	=	g/mm^3	kpsi	x	1 000	=	psi
lb/ft^3	x	1 728	=	lb/in^3	N/m^2	x	1	=	Pa
kg/m^3	x	1.0E–06	=	g/mm^3	N/mm^2	x	1	=	MPa
force					**spring rate**				
lb	x	4.448	=	N	lb/in	x	175.126	=	N/m
N	x	1.0E+05	=	dyne	lb/ft	x	0.08333	=	lb/in
ton (short)	x	2 000	=	lb					
					stress intensity				
length					$MPa\text{-}m^{0.5}$	x	0.909	=	$ksi\text{-}in^{0.5}$
in	x	25.4	=	mm	**velocity**				
ft	x	12	=	in	in/sec	x	0.0254	=	m/sec
					ft/sec	x	12	=	in/sec
mass					rad/sec	x	9.5493	=	rpm
blob	x	386.4	=	lb					
slug	x	32.2	=	lb	**volume**				
blob	x	12	=	slug	in^3	x	16 387.2	=	mm^3
kg	x	2.205	=	lb	ft^3	x	1 728	=	in^3
kg	x	9.8083	=	N	cm^3	x	0.061023	=	in^3
kg	x	1 000	=	g	m^3	x	1.0E+9	=	mm^3

A

Appendix G

ANSWERS TO SELECTED PROBLEMS

These problems have been solved using *TKSolver* wherever appropriate. A complete set of *TKSolver* problem-solution files is made available with the solutions manual for instructor use.

CHAPTER 1   INTRODUCTION TO DESIGN

1-4   220.5 lbf, 220.5 lbm, 6.854 slug, 0.571 blob, 980.8 N.

1-5   25.91 lbf.

1-6   1 000 lbf, 31.09 slug, 2.591 blob, 4 448.2 N.

CHAPTER 2   MATERIALS AND PROCESSES

2-6 E = 30 Mpsi, U_R = 60 psi, steel.

2-8 E = 45 GPa, U_R = 0.2 MPa, magnesium.

2-10 E = 16.5 Mpsi, U_R = 303 psi, titanium.

2-12 U_T = 12 000 psi, U_R = 60 psi.

2-14 S_{ut} = 170 kpsi, 360HV, 36HRC.

2-16   Iron and carbon, 0.95% carbon, can be through hardened or surface hardened without carburization.

CHAPTER 3   LOAD DETERMINATION

3-3 $T = 255$ N-m on sprocket, $T = 90$ N-m on arm, $M = 255$ N-m on arm.

3-6   55 044 N.

3-7   9 805 N.

3-8   99 rad/sec or 15.8 Hz.

3-10 $V = -1\ 821$ N @ 0.7 m, M = $-1\ 284$ N-m @ 0.7 m.

3-11   3 060-N dynamic force and 40.8-cm deflection. $V = -5\ 677$ N @ 0.7 m, $M = -4\ 003$ N-m @ 0.7 m.

3-15 $\mu = 0.03$.

3-18   Tipover at 19.0 to 20.3 mph, load slides at 16.4 to 19.5 mph.

3-22   (a) 898 N, (b) 3 592 N.

3-23   (a) $R_1 = 264$ N, $R_2 = 316$ N, $V = 264$ N, $M = 126$ N-m.

3-24   (a) $R_1 = 620$ N, $M_1 = 584$ N-m, $V = 620$ N, $M = -584$ N-m.

3-25   (a) $R_1 = -353$ N, $R_2 = 973$ N, $V = 580$ N, $M = -216$ N-m.

3-26   (a) $R_1 = 112$ N, $R_2 = 559$ N, $R_3 = -51$ N, $V = -426$ N, $M = 45$ N-m.

CHAPTER 4   STRESS, STRAIN, AND DEFLECTION

4-1 $\sigma_1 = 1\ 207$ psi, $\sigma_2 = 0$, $\sigma_3 = -207$ psi, $\tau_{max} = 707$ psi.

4-4   (a) $\sigma_1 = 6.28$ MPa, $\sigma_2 = 0$, $\sigma_3 = -7.67$ MPa, $\tau_{max} = 6.97$ MPa.

    (b) 9.93 MPa.

    (c) 6.79 MPa.

    (d) $\sigma = 8.65$ MPa, $\tau = 1.7$ MPa.

    (e) $\sigma_1 = 72.9$ MPa, $\sigma_2 = 0$, $\sigma_3 = 0$, $\tau_{max} = 36.4$ MPa.

4-6   (a) $\sigma_1 = 77.2$ MPa, $\sigma_2 = 0$, $\sigma_3 = -78.6$ MPa, $\tau_{max} = 77.9$ MPa.

    (b) 111.4 MPa.

    (c) 76.24 MPa.

    (d) $\sigma = 97$ MPa, $\tau = 1.7$ MPa.

    (e) $\sigma_1 = 635$ MPa, $\sigma_2 = 0$, $\sigma_3 = 0$, $\tau_{max} = 317$ MPa.

4-7   OD = 0.375 in, ID = 0.230 in.

4-8   19.9 cm.

4-10   24.5-MPa stress, −12.8-cm deflection.

A

4-11   76-MPa stress, –40-cm deflection.

4-15   6.38-mm-dia pin.

4-18   13 254-lb force, 0.36-in deflection.

4-19   2.1-in-dia pin, 3.1-in outside radius.

4-22   (a) 5.72 MPa.   (b) 22.87 MPa.

4-23   Row (a) $R_1 = 264$ N, $R_2 = 316$ N, $V = 264$ N, $M = 126$ N-m, $\theta = 0.001$ rad, $y = -3.25E-4$ m.

4-24   Row (a) $R_1 = 620$ N, $M_1 = 584$ N-m, $V = 620$ N, $M = -584$ N-m, $\theta = -0.0085$ rad, $y = -5.73E-3$ m.

4-25   Row (a) $R_1 = -353$ N, $R_2 = 973$ N, $V = 580$ N, $M = -216$ N-m, $\theta = -0.0026$ rad, $y = -8.6E-4$ m.

4-26   Row (a) $R_1 = 112$ N, $R_2 = 559$ N, $R_3 = -51$ N, $V = -426$ N, $M = 45$ N-m, $\theta = -0.0002$ rad, $y = -4E-5$ m.

4-29   Row (a) 307.2 N/m.

4-30   Row (a) 17.7 N/m.

4-31   Row (a) 110.6 N/m.

4-32   Row (a) 2 844 N/m.

4-33   Row (a) $\sigma_1 = 16$ MPa @ A, $\sigma_1 = 13$ MPa @ B.

4-34   Row (a) $\sigma_1 = 13$ MPa @ A, $\sigma_1 = 13$ MPa @ B.

4-35   Row (a) $\sigma_1 = 16$ MPa @ A, $\sigma_1 = 13$ MPa @ B.

4-37 $e = 0.84$ mm, $\sigma_i = 410$ MPa, $\sigma_o = -273$ MPa.

4-41   (a) 42 MPa, (b) 8.8 MPa.

4-49   Row (a) Johnson—part: (a) 247 MPa, (b) 273 MPa, (c) 287 MPa, (d) 106 MPa.

4-50   Row (a) Euler—part: (a) 33.8 MPa, (b) 67.7 MPa, (c) 135.2 MPa, (d) 8.46 MPa.

4-51   Row (a) Johnson—part: (a) 287 MPa, (b) 293 MPa, (c) 297 MPa, (d) 247 MPa.

4-52   Row (a) Secant—part: (a) 17.57 MPa, (b) 17.6 MPa, (c) 17.6 MPa, (d) 17.1 MPa.

CHAPTER 5   STATIC FAILURE THEORIES

5-1   Row (a) $\sigma_1 = 1\ 207$ psi, $\sigma_2 = 0$ psi, $\sigma_3 = -207$ psi, $\tau_{13} = 707$ psi, $\sigma' = 1\ 323$ psi.

        Row (h) $\sigma_1 = 1\ 140$ psi, $\sigma_2 = 250$ psi, $\sigma_3 = 110$ psi, $\tau_{13} = 515$ psi, $\sigma' = 968$ psi.

5-4   (a) $N = 24.8$, (b) $N = 30.2$, (c) $N = 25.5$, (d) $N = 34.7$, (e) $N = 4.1$.

5-6   (a) $N = 2.2$, (b) $N = 2.7$, (c) $N = 2.3$, (d) $N = 3.1$, (e) $N = 0.5$.

A

5-7   for $N = 3.0$, $OD = 0.375$ in, $ID = 0.222$ in.

5-8 $ID = 17.7$ cm.

5-10 $N = 4.1$.

5-11 $N = 1.3$.

5-15 $N = 1.0$.

5-17 $N = 3.5$.

5-19   2.2-in-dia pin and 3-in outside radius.

5-22   (a) $N = 40.3$ (b) $N = 10.1$.

5-23   Row (a)—part: (a) $N = 3.4$, (b) $N = 1.7$.

5-24   Row (a)—part: (a) $N = 0.73$, (b) $N = 0.37$.

5-25   Row (a)—part: (a) $N = 2$, (b) $N = 1$.

5-26   Row (a)—part: (a) $N = 9.5$, (b) $N = 4.8$.

5-27   (a) $a = 24$ cm, $b = 10$ cm, $N = 3.1$, (b) $a = 24$ cm, $b = 10$ cm, $N = 1.6$.

5-32   Modified-Mohr, $N = 1.58$.

5-33   Row (a) $\sigma' = 23$ MPa at point A, $\sigma' = 22$ MPa at point B.

5-34   Row (a) Distortion-energy theory: $N = 17.4$ at point A, $N = 17.8$ at point B, Max shear theory: $N = 15.2$ at point A, $N = 15.4$ at point B, Max normal stress theory: $N = 24.5$ at point A, $N = 30.8$ at point B.

5-35   Row (a) Coulomb-Mohr theory: $N = 17.7$ at point A, $N = 20.0$ at point B, Modified-Mohr theory: $N = 21.4$ at point A, $N = 27.0$ at point B.

5-37 $N = 1.0$ at inner fiber, $N = 4.4$ at outer fiber.

5-38 $N = 1.4$.

5-39   Crack half-width = 0.216 in.

5-41   (a) $N = 9.5$, (b) $N = 34.9$.

CHAPTER 6   FATIGUE FAILURE THEORIES

6-1   Row (a) $\Delta\sigma = 1\,000$, $\sigma_a = 500$, $\sigma_m = 500$, $R = 0$, $A = 1$.

   Row (c) $\Delta\sigma = 500$, $\sigma_a = 1\,000$, $\sigma_m = 500$, $R = 0.3$, $A = 0.5$.

   Row (e) $\Delta\sigma = 1\,500$, $\sigma_a = 750$, $\sigma_m = 250$, $R = -2$, $A = -3$.

6-3 $N_f = 0.28$.

6-6   (a) 2.3, (b) 2.8, (c) 2.4, (d) 3.2, (e) 0.5.

6-7   for $N_f = 1.5$, $OD = 0.375$ in, $ID = 0.240$ in.

6-8 $ID = 19.15$ cm.

6-10 $N_f = 2.5$.

6-11 $N_f = 0.8$.

6-15   Row (j) $a = 0.122$ in$^{0.5}$, $q = 0.804$, $K_f = 2.21$.

6-17 $N_f = 3.8$.

6-19   2.431-in-dia pin and 3-in outside radius.

6-22   (a) 32, (b) 8.

6-23   Row (a) Use a material with $S_{ut} = 538$ MPa.

6-24   Row (a) Use a material with $S_{ut} = 755$ MPa.

6-25   Row (a) Use a material with $S_{ut} = 635$ MPa.

6-26   Row (a) Use a material with $S_{ut} = 447$ MPa.

6-27   (a) $a = 21$ cm, $b = 10$ cm, $N = 2$, (b) $a = 27$ cm, $b = 10$ cm, $N = 2$.

6-29 $N_f = 2.3$.

6-31 $N_f = 1.7$.

6-33   Row (a) Use a material with $S_{ut} = 398$ MPa.

6-34   Row (a) Use a material with $S_{ut} = 318$ MPa.

6-37 $N_f = 2.0$.

6-39 $N_f = 2.3$

6-41   (a) $N_f = 4.5$, (b) $N_f = 16.4$.

6-47 $N_f = 2.8$.

CHAPTER 7   SURFACE FAILURE

7-1 $A_r = 0.333$ mm^2.

7-2 $\mu = 0.4$.

7-3 $N = 5E9$.

7-4 $\sigma_1 = -61$ kpsi, $\sigma_2 = -61$ kpsi, $\sigma_3 = -78$ kpsi.

7-8   6.438-cm total width.

7-10   0.11-mm total width.

7-13   (a) 1 177 strokes, (b) 589 strokes.

7-16   1-mm dia.

7-18   0.525-mm total width.

7-20   contact-patch half-dimensions: 0.933 x 0.713 mm, $\sigma_1 = -5.39$ GPa, $\sigma_2 = -5.81$ GPa, $\sigma_3 = -7.18$ GPa.

7-22   (a) $\sigma_1 = -41$ MPa, $\sigma_2 = -46$ MPa, $\sigma_3 = -48$ MPa, (b) $\sigma_1 = -65$ MPa, $\sigma_2 = -73$ MPa, $\sigma_3 = -76$ MPa.

7-23 $\sigma_1 = 0$ psi, $\sigma_2 = -777$ psi, $\sigma_3 = -1\,486$ psi.

CHAPTER 9   SHAFTS, KEYS, AND COUPLINGS

9-1   Row (a) $d = 1.069$ in.

9-2   Row (a) $d = 4.44$ cm.

9-4   Row (a) $y = 0.004$ in, $\theta = 0.22$ deg.

9-5   Row (a) $y = 2.947$ μm, $\theta = 0.21$ deg.

9-6   Row (a) 5/16-in square key, 0.563 in long, $N_f = 2.2$, $N_{bearing} = 2$.

9-8   Shaft $ID = 19.1$ cm.

9-9   Row (a) $d = 1.155$ in.

9-11   Row (a) 0.003 3 in.

9-13   Row (a) 3 510 rad/sec or 558.7 Hz, or 33 520 rpm.

9-15   Row (a) min = 0, avg = 11.9 hp, max = 23.8 hp.

9-16   Row (a) min = 0, avg = 10.4 kW, max = 5.2 kW.

9-17   Row (a) $N_f = 1.33$, $\theta = 0.63$ deg, $w_n = 5\,357$ rad/sec.

9-18   Row (a) $y = 30$ μm.

9-19   Row (a) $d = 1.21$ in.

CHAPTER 10   BEARINGS AND LUBRICATION

10-1   Row (a)—part (a) $d = 1.08$ in, $l = 1.35$ in, $C_d = 1.62$E–3 in, $R_L = 111$ lb, $R_R = 889$ lb, $\eta_L = 0.1$ μreyn, $\eta_R = 0.8$ μreyn, $p_{avgL} = 76$ psi, $p_{avgR} = 610$ psi, $T_L = 0.08$ lb-in, $T_R = 0.62$ lb-in, $\Phi_{l_L} = 0.002$ hp, $\Phi_{l_R} = 0.015$ hp.

Part (b) # 6306 bearings at each end give 9E10 cycles life on left bearing and 1.8E8 cycles on right.

10-3   267 cP.

10-5   0.355 in-lb.

10-6   10.125 μm.

10-7 $T_r = 3.765$ N-m, $T_0 = 2.181$ N-m, $T_s = 2.611$ N-m, $\Phi = 979$ W.

10-8 $d = 22$ cm, $l = 16.5$ cm, $C_d = 0.44$ mm, $R_L = R_R = 26\ 975$ N, $\eta = 63.7$ cP, $p_{avg} = 743$ kPa, $T = 6.1$ N-m, $\Phi_l = 31.6$ W.

10-10 $h_{min} = 4.94$ μm.

10-14 $\eta = 1.885$ μreyn, $T_s = 523.1$ N-mm, $T_0 = 327.1$ N-mm, $T_r = 704.2$ N-mm, $\Phi = 183$ W, $P = 19.22$ kN.

10-17   Row (a) - part (a) $d = 4$ cm, $l = 3$ cm, $C_d = 0.04$ mm, $R_L = 61$ N, $R_R = 1\ 189$ N, $\eta_L = 0.1$ cP, $\eta_R = 1.9$ cP, $p_{avgL} = 51$ kPa, $p_{avgR} = 991$ kPa, $T_L = 0.002$ N-m, $T_R = 0.049$ N-m, $\Phi_{lL} = 0.467$ W, $\Phi_{lR} = 9.1$ W,

Part (b) # 6308 bearings at each end give 1.5E14 cycles life on left bearing and 2.0E10 cycles on right.

10-20   Mixed boundary / EHD lubrication.

CHAPTER 11   SPUR GEARS

11-1 $d_p = 5.4$ in, addendum = 0.2 in, dedendum = 0.25 in, $OD = 5.8$ in, $p_c = 0.628$ in.

11-3   1.491.

11-5   6.667 in.

11-7   7.159.

11-9   96:14 and 96:14 compounded give 47.02:1.

11-11 $N_{ring} = 75$ t, ratio = −0.44.

11-14   7 878 in-lb on pinion shaft, 19 524 in-lb on gear shaft.

11-16 $p_d = 2.5$ gives $N_{pinion} = 6.1$ and $N_{gear} = 2.3$.

11-18 $p_d = 2.5$ gives $N_{pinion} = 2.8$ and $N_{gear} = 2.2$.

11-20   1 786 in-lb on sun shaft, 5 846 in-lb on arm shaft.

11-23 $p_d = 3$ gives $N_{pinion} = 5.4$ and $N_{gear} = 2.8$.

11-25 $p_d = 3$ gives $N_{pinion} = 2.8$ and $N_{gear} = 1.7$.

11-27 $T_1 = 8\ 921$ in-lb, $T_2 = 77\ 505$ in-lb, $T_3 = 673\ 322$ in-lb, $T_4 = 5\ 852\ 718$ in-lb.

11-28   Stage 1 $p_d = 2.5$, stage 2 $p_d = 1.25$, stage 3 $p_d = 1.0$.

11-29   106:12 and 106:12 compounded give 78.028:1.

CHAPTER 12   HELICAL, BEVEL, AND WORM GEARS

12-1 $d_p = 5.4$ in, addendum = 0.2 in, dedendum = 0.25 in, $OD = 5.8$ in, $p_c = 0.628$ in, $p_n = 0.544$ in $p_a = 1.088$ in.

A

12-3 $m_p = 1.49$, $m_F = 0.56$.

12-5 $\alpha_g = 83.66°$, $\alpha_p = 6.34°$, $d_g = 21$ in, $d_p = 2.33$ in, $W_{ag} = W_{rp} = 25$ lb, $W_{rg} = W_{ap} = 2.8$ lb.

12-7 $\alpha_g = 78.69°$, $\alpha_p = 11.31°$, $d_g = 11.43$ in, $d_p = 2.29$ in, $W_{ag} = 60.5$ lb, $W_{ap} = 169.6$ lb, $W_{rp} = 136.3$ lb, $W_{rg} = 209$ lb.

12-9 $l = 20$ mm, $\lambda = 7.26°$, λ per tooth = $3.63°$, $d_p = 140$ mm, $c = 95$ mm, self-locking.

12-11 $l = 5$ mm, $\lambda = 2.28°$, λ per tooth = $2.28°$, $d_p = 130.5$ mm, $c = 85.3$ mm, self-locking.

12-12 $T_w = 27.8$ N-m, $T_g = 495.7$ N-m rated, $W_t = 7\,028$ N, friction = 215 N, output power available = 2.34 kW, rated input power = 2.91 kW.

12-14   7 878 in-lb on pinion, 19 524 on gear.

12-16 $p_d = 3$ gives $N_{pinion} = 5.6$ and $N_{gear} = 2.0$.

12-18 $p_d = 3$ gives $N_{pinion} = 2.1$ and $N_{gear} = 1.7$.

12-20 $p_d = 16$ gives $N_{pinion} = 3.5$ in bending and 2.0 in surface failure, with $N_{gear} = 2.7$ in bending.

12-23 $p_d = 20$ gives $N_{pinion} = 1.8$ in bending and 1.6 in surface failure, with $N_{gear} = 1.4$ in bending.

12-27   Rated input power = 2.11 hp, output power available = 1.69 hp, rated output torque = 1 290 lb-in.

CHAPTER 13   SPRING DESIGN

13-1   1 600 N-m.

13-3 $S_{ys} = 110\,931$ psi, $S_{us} = 148\,648$ psi.

13-4 $S_{es'} = 86\,526$ psi.

13-6 $C = 10$, $k = 987.5$ N/m.

13-7   114 Hz or 6 840 rpm.

13-10 $k = 7\,614$ N/m, $f_n = 1.39$ Hz.

13-11 $d = 0.125$ in, $D = 0.94$ in, $L_f = 3.16$ in, $k = 36$ lb/in, 13.75 coils RH, music wire, squared and ground ends, unpeened, set.

13-13 $N_a = 19.75$, $D = 1.37$ in, $L_f = 7.84$ in, $L_{shut} = 6.79$ in, $k = 266.7$ lb/in, $y_{initial} = 0.19$ in, hole = 1.75 in.

13-17 $N_y = 1.6$ torsion in hook, $N_f = 2.44$ bending in hook, $N_{surge} = 5.6$.

13-19 $N_y = 1.3$ shut, $N_f = 1.2$, $N_{surge} = 12.3$.

13-21 $d = 5$ mm, $D = 40$ mm, $L_f = 115.5$ mm, $k = 8\,889$ N/m, 12.75 coils RH, music wire, S&G ends, unpeened, set.

13-22 $d = 16$ mm, $D = 40$ mm, $k = 1\ 600$ N/m/rev, 6.25 coils RH, 40-mm straight ends, music wire, relieved.

13-24 $d = 16$ mm, $D = 132.8$ mm, $k = 450$ N/m/rev, 6.25 coils RH, 40-mm straight ends, music wire, relieved.

13-26 $d_o = 39.6$ mm, $d_i = 19.78$ mm, $t = 0.76$ mm, $h = 1.075$ mm, $h/t = 1.414$, 1-mm working deflection, $S_{ut} = 1\ 700$ MPa, $N_s = 1.11$.

CHAPTER 14   SCREWS AND FASTENERS

14-2   Lifting torque = 42.68 lb-in, lowering torque = 18.25 lb-in, lifting efficiency = 27.95%, lowering efficiency = 65.36%, screw is self-locking.

14-4   M12 x 1.75 bolts, ISO class 8.8, $F_{preload} = 62.2\%$ of proof strength, $T = 80$ N-m, $N_y = 1.75$, $N_{sep} = 14.8$.

14-6   M12 x 1.75 bolts, ISO class 8.8, $F_{preload} = 55.6\%$ of proof strength, $T = 71$ N-m, $N_y = 1.75$, $N_{sep} = 1.2$.

14-7 $N_y = 1.4$, $N_{sep} = 12.6$.

14-9 $N_f = 2.4$, $N_y = 1.2$, $N_{sep} = 1.6$.

14-11   1 252 in-lb.

14-13   718 in-lb

14-17   (a) $K_{eff} = 5.052E9$ N-m, aluminum dominates.

       (b) $K_{eff} = 9.534E9$ N-m, no one material dominates.

       (c) $K_{eff} = 6.854E7$ N-m, rubber dominates.

       (d) $K_{eff} = 6.812E7$ N-m, rubber dominates.

       (e) $K_{eff} = 9.049E9$ N-m, no one material dominates.

14-18   (a) $K_{eff} = 2.034E9$ N-m, aluminum dominates.

       (b) $K_{eff} = 3.890E9$ N-m, no one material dominates.

       (c) $K_{eff} = 5.610E7$ N-m, rubber dominates.

       (d) $K_{eff} = 5.538E7$ N-m, rubber dominates.

       (e) $K_{eff} = 3.790E9$ N-m, no one material dominates.

14-20   Use 8 M6 x 1.0, ISO class 5.8 cap screws, torqued to 45% of proof strength on a 9.17-cm-dia bolt circle. $T = 4.37$ N-m, $N_f = 67$, $N_{sep} = 1.5$, $N_y = 1.56$ dynamic and 1.57 static.

14-23   Row (a) M8 x 1.25 bolts, class 8.8, $F_{preload} = 11\ 861$ N, (54% of proof), load on top bolts: 11 867 N, load on bottom bolts: 11 855 N, $N_y = 2$.

A

14-24   Row (a) M8 x 1.25 bolts, class 9.8, $F_{preload}$ = 11 660 N, (49% of proof), load on top bolts when force is maximum and on bottom bolts when force is minimum: 11 666 N, load on top bolts when force is minimum and on bottom bolts when force is maximum: 11 654 N, N_y = 1.5.

14-25   Row (a) M8 x 1.25 bolts, class 9.8, $F_{preload}$ = 11 660 N, (49% of proof), load on top bolts: 11 666 N, load on bottom bolts: 11 652 N, N_y = 1.5.

CHAPTER 15   CLUTCHES AND BRAKES

15-1 T = 382.7 N-m, p_{max} = 1.82 MPa, molded or sintered metal lining will work.

15-3 d_o = 138 mm, d_i = 79.6 mm, P = 10.5 hp.

15-5   N = 7, d_o = 102 mm, d_i = 58.6 mm, P = 16.7 hp.

15-7 T = 10.8 N-m, F_a = 798 N, will self-lock when c = 233 mm.

15-11 T = 30.7 N-m (15.8 top shoe, 14.9 bottom shoe), F_a = 1 352 N, will self-lock when c = 320 mm.

15-13 T = 31.9 N-m, F_a = 2 062 N.

15-15 T = 166.3 N-m (95.7 top shoe, 70.6 bottom shoe), F_a = 6 414 N.

15-17   (a)   short shoe: T = 9.49 N-m, F_a = 811.6 N,
                  long shoe: T = 9.72 N-m, F_a = 798.2 N.

         (b)   short shoe: T = 12.66 N-m, F_a = 1 082.1 N,
                  long shoe: T = 13.2 N-m, F_a = 1 073.6 N.

         (c)   short shoe: T = 15.82 N-m, F_a = 1 352.6 N,
                  long shoe: T = 16.9 N-m, F_a = 1 358.3 N.

15-18   (a)   short shoe: T = 18.41 N-m, F_a = 811.6 N,
                  long shoe: T = 17.37 N-m, F_a = 798.2 N.

         (b)   short shoe: T = 24.55 N-m, F_a = 1 082.1 N,
                  long shoe: T = 23.58 N-m, F_a = 1 073.6 N.

         (c)   short shoe: T = 30.68 N-m, F_a = 1 352.6 N,
                  long shoe: T = 30.17 N-m, F_a = 1 358.3 N.

15-21   Top pivot: R_x = 3 937 N, R_y = –218 N, bottom pivot: R_x = –369 N, R_y = 123 N.

15-23 R_x = 342 N, R_y = –854 N.

15-25   Top pivot: R_x = 2 339 N, R_y = –1 630 N, bottom pivot: R_x = –1 726 N, R_y = 478 N.

Appendix H

LIST OF SOFTWARE INCLUDED ON CD-ROM

A CD-ROM is included with this textbook that contains over 200 *TKSolver* model files which encode most of the Example and Case-Study solutions and a number of general solution files for beams, springs, shafts, fasteners, etc. A collection of rule, list, and procedure functions that can be merged into other files is also provided. Each of these *TKSolver* files is present on the CD-ROM in three versions, for use with DOS, Windows, and Macintosh computers, respectively. See the README file on the CD-ROM for instructions on installation and use of the software.

The fully capable student edition of *TKSolver* (on the CD-ROM in three versions to run on any of the computers listed above) and its student manual are available as an option, packaged with this text. This commercial equation solver package also comes with a library of general-purpose files to solve many common mathematical and engineering problems. If you purchased the book with the optional *TKSolver* program, see the attached *TKSolver* student manual for more detailed information on its use and for a listing of the library of standard functions included with it.

If you purchased this text without the optional *TKSolver* program, you will need to have that program to run the models provided on the CD-ROM. The *TKSolver* program is available separately from *Universal Technical Systems, Inc.*, 1220 Rock St., Rockford, Ill., 61101, (815) 963-2220.

This appendix lists the custom-written *TKSolver* files for this text that are on the CD-ROM-disk. Additional software in the form of independently executable programs for kinematic, dynamic, and stress analysis is also included on the CD-ROM.

H-1   TKSOLVER FILES - GENERAL

Beams   Path: TKFiles \ General \ Beams

BEAMFUNC   A collection of rule functions for various beam loadings and supports for use in programs. Can be combined for superposition of loads on any beam with consistent constraints. (See Figure 3-22.)

CANTCONC   Calculates the shear, moment, slope, and deflection functions for a cantilever beam with a concentrated load at any point along its length. Finds reactions, plots the beam functions, and finds their max and min values. (See Figure 3-22*b*.)

CURVBEAM   Calculates eccentricity of neutral axis and stresses for curved beams of various cross sections—ellipse, circle, square, rectangular, and trapezoidal. (See Figure 4-16.)

INDTUNIF   Calculates the shear, moment, slope, and deflection functions for an indeterminate beam with a uniformly distributed load over a portion of its length ending at one support. Finds reactions, plots the beam functions, and finds max and min values. (See Figure 4-22*d*.)

OVHGCONC   Calculates the shear, moment, slope, and deflection functions for an overhung beam with a concentrated load at any point along its length. Finds reactions, plots the beam functions, and finds their max and min values. (See Figure 4-22*c*.)

OVHGUNIF   Calculates the shear, moment, slope, and deflection functions for an overhung beam with a uniformly distributed load over a portion of its length beginning at one support and with an optional concentrated load at any point along its length . Finds reactions, plots the beam functions, and finds their max and min values. (See Figure 4-22*c*.)

SIMPCONC   Calculates the shear, moment, slope, and deflection functions for a simply supported beam with a concentrated load at any point along its length. Finds reactions, plots the beam functions, and finds their max and min values.

SIMPUNIF   Calculates the shear, moment, slope, and deflection functions for a simply supported beam with a uniformly distributed load over a portion of its length ending at one support. Finds reactions, plots the beam functions, and finds max and min values. (See Figure 4-22*a*.)

Bearings   Path: TKFiles \ General \ Bearings

BALL6200   A ball-bearing selection program that calculates the B10 life for 6200 series ball bearings under specified loads. Based on data from the SKF bearing catalog. (See Figure 10-17.)

BALL6300   A ball-bearing selection program that calculates the B10 life for 6300 series ball bearings under specified loads. Based on data from the SKF bearing catalog. (See Figure 10-17.)

EHD_BRNG   Solves for film pressure in general Elastohydrodynamic (EHD) contact between lubricated, nonconforming surfaces. Also finds the minimum oil-film thickness. (See Figure 10-4.)

SLEEVBRG   Calculates the film thickness, eccentricity, and oil pressure in a short (Ocvirk) sleeve bearing under hydrodynamic lubrication conditions. (See Figure 10-8.)

Clutch/Brake   Path: TKFiles \ General \ ClchBrak

DISKCLCH   Designs a disk clutch for uniform wear. Allows single or multiple disks. (See Figure 15-6.)

LONGDRUM   Designs a long-shoe drum brake. (See Figure 15-9.)

SHRTDRUM   Designs a short-shoe drum brake. (See Figure 15-8.)

Columns   Path: TKFiles \ General \ Columns

COLMNDES   A column design program that handles round, square, or rectangular concentric columns and uses both Johnson and Euler criteria to find critical load, weight and safety factor. Combines **ROUNDCOL** and **SQUARCOL** in one model. (See Figure 4-41.)

COLMPLOT   Calculates and plots the critical load versus slenderness ratio for Euler, Johnson, and short columns for various end conditions. (See Figure 4-42.)

ROUNDCOL   Calculates critical load for solid or hollow round concentric columns using both Johnson and Euler criteria—a simple model demonstrating the use of rule functions and conditional calls to functions. (See Figure 4-41.)

SECANT   An eccentric column design program that handles round, square, or rectangular concentric columns and uses the secant method and Johnson and Euler criteria to find critical load and safety factor. Plots critical load curves. (See Figure 4-43.)

SQUARCOL   Calculates critical load for solid or hollow square concentric columns using both Johnson and Euler criteria—a simple model demonstrating the use of rule functions and conditional calls to functions. (See Figure 4-41.)

Fastener   Path: TKFiles \ General \ Fastener

BLTFATIG   Calculates the safety factors for preloaded bolts with fluctuating tensile loads. Determines necessary tightening torque and plots the load-sharing, safety factor, and modified-Goodman diagrams. (See Figure 14-26.)

BOLTSTAT   Calculates the safety factors for preloaded bolts with static tensile loads. Determines necessary tightening torque and plots the load-sharing and safety factor diagrams. (See Figure 14-24.)

PWRSCREW   Calculates the torque and efficiency of an Acme-thread power screw. (See Figure 14-4.)

Fatigue   Path: TKFiles \ General \ Fatigue

GDMNPLTR   A plotting utility that creates and plots a modified-Goodman diagram for any set of supplied stresses and strengths. No calculations are done on the data. (See Figure 6-44.)

GOODMAN   Calculates the corrected endurance strength based on supplied data about finish, size, strength, etc. and draws a modified-Goodman diagram for supplied levels of alternating and mean stresses and material strengths. Also calculates safety factors. (See Figure 6-46.)

S_NDIAGM   Calculates the corrected endurance strength based on supplied data about finish, size, strength, etc. and draws an S-N diagram for supplied levels of alternating and mean stresses. Draws both log-log and semilog plots. (See Figure 6-33.)

S_NALUM   Calculates the corrected endurance strength based on supplied data about finish, size, strength, etc. and draws an S-N diagram for supplied levels of alternating and mean stresses for nonferrous material. Draws both log-log and semilog plots. Based on the file **S_NDIAGM.** (See Figure 6-33.)

S_NFCTRS   Calculates the coefficient and exponent of the S-N line for a material. (See Figure 6-33.)

SESAFTIG   Calculates and plots the variation on stress with phase in multiaxial fatigue based on the SESA algorithm. (See Figure 6-49.)

A

Flywheels   Path: TKFiles \ General \ Flywheel

FWDESIGN   Program to find the best combination of flywheel diameters and thickness to balance its weight against size, stress and safety factor. Calculates the maximum stress, outside diameter, weight, and safety factor as a function of the thickness of a flywheel. (See Figure 9-21.)

FWRATIO   Optimizes flywheel mass versus the ratio of radius to thickness and plots that function. (See Figure 9-25.)

FWSTDIST   Program to find the flywheel stress distribution over its radius. Calculates and plots the stresses across the radius of a flywheel. (See Figure 9-24.)

Fract Mech   Path: TKFiles \ General \ Fracture

STRSINTS   Plots the stress intensity around a crack tip. (See Figure 5-16.)

Gearing   Path: TKFiles \ General \ Gearing

BVLGRDES   Program for straight bevel gearset design. Finds bending and surface stresses in gear teeth and safety factors using AGMA methods. Requires I and J factors be manually looked up in AGMA Tables. (See Figure 12-4.)

GRTRNDES   Calculates bending and surface stresses for a two-stage spur gear train based on AGMA formulas and determines safety factors for supplied material strengths. Also determines kinematics for the two stages. (See Figure 11-14.)

HELGRDES   Program for helical gearset design. Finds bending and surface stresses in gear teeth and safety factors using AGMA methods. Requires I and J factors be manually looked up in AGMA Tables. (See Figure 12-2.)

SPRGRDES   Calculates bending and surface stresses for a single spur gearset (with or without an idler) based on AGMA formulas and determines safety factors for supplied material strengths. (See Figure 11-2.)

WORMGEAR   Worm and wormgear design based on AGMA formulas. (See Figure 12-9.)

Impact   Path: TKFiles \ General \ Impact

IMPCTHRZ   Calculates the impact force on a horizontal rod struck by a mass. (See Figure 3-18.)

IMPCTVRT   Calculates the impact force on a vertical rod struck by a mass. (See Figure 3-18.)

Linkages   Path: TKFiles \ General \ Linkages

4BARSTAT   Calculates the joint forces for a static fourbar linkage subjected to a known force applied to the coupler. (See Figure 3-13.)

DYNAFOUR   Calculates the kinematics and inverse dynamics of the fourbar linkage. Plots various parameters such as accelerations, forces, and torques. (See Figure 12-3 in reference 1.)

ENGINE   Calculates a slider crank's kinematics and the gas force and gas torque due to a specified explosion pressure at any position of the crank. It is a static force analysis. (See Figure 14-3 in reference 1.)

ENGNBLNC   Calculates the dynamic balance condition of an IC engine. Plots shaking forces, torques, and moments. (See Figure 14-12 in reference 1.)

FOURBAR   Calculates the kinematics of a fourbar linkage, position, velocity, and acceleration of various point for any range of motion. (See Figure 12-3 in reference 1.)

A

SLIDER   Calculates the offset slider crank's kinematics for any one position and input omega and alpha. Lists can be added for multiple position analysis. No forces are calculated. (See Figure 7-6 in reference 1.)

Shafts   Path: TKFiles \ General \ Shafts

HOLTZER   Find first natural frequency of a shaft with lumped masses using Holtzer's method. (See Figure 9-27.)

SHFTCONC   Program to design a simply supported shaft with concentrated load at any point along its length. Left support must be at $x = 0$, but right support can be anywhere. A fluctuating torque may be applied to the shaft. The moment is assumed to be fully reversed. Calculates and plots the shear, moment, and deflection functions. (See Figure P9-1.)

SHFTDESN   Program to design a simply supported shaft for fatigue in combined bending and torsion. A fluctuating torque may be applied to the shaft as well as a fluctuating moment. Also calculates stresses in a standard square key for the shaft diameter. (See Figure P9-3.)

SHFTUNIF   Program to design a simply supported shaft with uniform load over any portion of its length. Left support must be at $x = 0$ and right support is at length R_{2x}. A fluctuating torque may be applied to the shaft and the moment is assumed to be fully reversed. Calculates and plots the shear, moment and deflection functions. (See Figure P9-2.)

STATSHFT   Calculates the shear stress in a shaft subjected to a constant torque with no transverse loads or moments. (See Figure 4-28.)

STEPSHFT   Program to design a simply supported stepped shaft with uniform load over any portion. Calculates deflection for stepped shaft. (See Figure 9-5.)

Springs   Path: TKFiles \ General \ Springs

BELLEVIL   Calculates the load, deflection, and spring rate for a Belleville spring. Plots the nonlinear force-deflection curves for a family of springs. (See Figure 13-29.)

COMPRESS   Designs a helical coil compression spring for fatigue or static loading. Plots curves to allow an optimization of spring design. (See Figure 13-7.)

EXTENSN   Designs a helical coil extension spring for fatigue or static loading. Plots curves to allow an optimization of spring design. (See Figure 13-20.)

TORSION   Designs a helical coil torsion spring for fatigue or static loading. Plots curves to allow an optimization of spring design. (See Figure 13-26.)

Stress   Path: TKFiles \ General \ Stress

COULMOHR   Calculates factors for the Coulomb-Mohr diagram for brittle, uneven materials. (See Figure 5-9.)

ELLIPSE   Draws the distortion-energy ellipse for demonstration purposes. (See Figure 5-3.)

MOD_MOHR   Modified-Mohr theory calculator for brittle materials. Uses Dowling's method to find an effective stress for combined loading in brittle, uneven materials. (See Figure 5-11.)

STRSFUNC   Two rule functions, one for the calculation of 2-D principal and one for Von Mises stresses. Use for merging into other programs that need these functions. (See Figure 4-4.)

STRES_2D   Calculates the principal stresses, maximum shear stress, and Von Mises stress for any two-dimensional applied stress state specified. (See Figure 4-3.)

A

STRES_3D	Calculates the principal stresses and maximum shear stress for any three-dimensional applied stress state specified. It also plots the stress cubic function. (See Figure 4-1.)
VONMISES	Uses the distortion-energy method to find an effective stress for combined loading in ductile, even materials under static loading. (See Figure 5-3.)

Stress Conc.   Path: TKFiles \ General \ StrsConc

APP_E-01	Calculates stress-concentration factor for a shaft with shoulder fillet in tension. (See Appendix E, Figure E-1.)
APP_E-02	Calculates stress-concentration factor for a shaft with shoulder fillet in bending. (See Appendix E, Figure E-2.)
APP_E-03	Calculates stress-concentration factor for a shaft with shoulder fillet in torsion. (See Appendix E, Figure E-3.)
APP_E-04	Calculates stress-concentration factor for a shaft with a U groove in axial tension. (See Appendix E, Figure E-4.)
APP_E-05	Calculates stress-concentration factor for a shaft with a U groove in bending. (See Appendix E, Figure E-5.)
APP_E-06	Calculates stress-concentration factor for a shaft with a U groove in torsion. (See Appendix E, Figure E-6.)
APP_E-07	Calculates stress-concentration factor for a shaft with transverse hole in bending. (See Appendix E, Figure E-7.)
APP_E-08	Calculates stress-concentration factor for a shaft with transverse hole in torsion. (See Appendix E, Figure E-8.)
APP_E-09	Calculates stress-concentration factor for a flat bar with shoulder fillet in tension. (See Appendix E, Figure E-9.)
APP_E-10	Calculates stress-concentration factor for a flat bar with shoulder fillet in bending. (See Appendix E, Figure E-10.)
APP_E-11	Calculates stress-concentration factor for a flat bar with notch in axial tension. (See Appendix E, Figure E-11.)
APP_E-12	Calculates stress-concentration factor for a flat bar with notch in bending. (See Appendix E, Figure E-12.)
APP_E-13	Calculates stress-concentration factor for a flat bar with transverse hole in tension. (See Appendix E, Figure E-13.)
APP_E-14	Calculates stress-concentration factor for a flat bar with transverse hole in bending. (See Appendix E, Figure E-14.)
NTCHSENS	Plots the notch sensitivity curves for steels. (See Figure 6-36 part 1.)
Q_CALC	Calculates the notch sensitivity q of a material. (See Figure 6-36 part 2.)
SC_HOLE	Calculates and plots stress concentration at an elliptical hole in a semi-infinite plate. (See Figure 4-35.)

Surface Stress   Path: TKFiles \ General \ SurfStrs

SURFCYLX	Calculates the surface stresses for Hertzian contact of two cylinders, with or without a sliding component. Plots subsurface stress distributions across the contact-patch X-width at surface or at any Z-depth into material. (See Figure 7-17.)

A

SURFCYLZ   Calculates the surface stresses for Hertzian contact of two cylinders with or without a sliding component. Plots subsurface stress distributions from surface to any Z depth at any X-width across contact patch. (See Figure 7-17.)

SURFGENL   Calculates the surface stresses for Hertzian contact of two bodies of general shape. (See Figure 7-11.)

SURFSPHR   Calculates the surface stresses for Hertzian contact of two spheres, sphere-on-plane or sphere in bowl. Plots subsurface stress distributions. (See Figure 7-13.)

THCK_CYL   Calculates the stresses in walls of thick-cylinder pressure vessels. (See Figure 4-46.)

H-2   TKSOLVER FILES - EXAMPLES

Examples   Path: TKFiles \ Examples \ Chap_No

EX01-02   An introduction to TKSolver using the rule and variable sheets. (See Figure 1-7.)

EX01-03   An introduction to TKSolver showing how to switch variables from input to output on the variable sheet. (See Figure 1-7.)

EX01-04   An introduction to TKSolver showing how to solve simultaneous nonlinear equations using automatic iteration. (See Figure 1-7.)

EX01-05   An introduction to TKSolver showing how to solve for multiple conditions using lists and plot the results. Optimizes the solution. (See Figure 1-7.)

EX01-06   An introduction to TKSolver showing how to use lists with built-in library functions. (See Figure 1-7.)

EX01-07   An introduction to TKSolver showing how to write custom rule functions. (See Figure 1-7.)

EX01-08   An introduction to TKSolver showing how to write custom procedure functions. (See Figure 1-7.)

EX01-09   An introduction to TKSolver showing how to write custom list functions. (See Figure 1-7.)

EX01-10   An introduction to TKSolver showing how to use the unit conversion sheet and format sheet. (See Figure 1-7.)

EX03-01A   Example of impact of a mass against a horizontal rod in axial tension. Examines and plots the sensitivity of the impact force to the length/diameter ratio of the rod for a constant mass ratio. (See Figure 3-18.)

EX03-01B   Example of impact of a mass against a horizontal rod in axial tension. Examines and plots the sensitivity of the impact force to the mass ratio of the rod for a constant length/diameter ratio. (See Figure 3-18.)

EX03-02   Calculates the loading, shear, and moment functions for a simply supported beam with a uniformly distributed load over a portion of its length ending at one support. Finds reactions and plots the beam functions. (See Figure 3-22a.)

EX03-03   Calculates the loading, shear, and moment functions for a cantilever beam with a concentrated load at any point along its length. Finds reactions and plots the beam functions. (See Figure 3-22b.)

EX03-04   Calculates the loading, shear, and moment functions for an overhung beam with a moment load at any point along its length and with a ramp load over a portion of its length beginning at one support. Finds reactions and plots the beam functions. (See Figure 3-22c.)

A

EX04-01	Solves the stress cubic and finds the principal stresses and maximum shear for given values. Is same program as **STRESS3D** with data for this example. (See Figure 4-5.)
EX04-02	Solves the stress cubic and finds the principal stresses and maximum shear for given values. Is same program as **STRESS3D** with data for this example. (See Figure 4-6.)
EX04-03	Solves the stress cubic and finds the principal stresses and maximum shear for given values. Is same program as **STRESS3D** with data for this example. (See Figure 4-7.)
EX04-04	Calculates the shear, moment, slope, and deflection functions for a simply supported beam with a uniformly distributed load over a portion of its length ending at one support. Finds reactions, plots the beam functions, and finds max and min values. (See Figure 3-22*a*.)
EX04-05	Calculates the shear, moment, slope, and deflection functions for a cantilever beam with a concentrated load at any point along its length. Finds reactions, plots the beam functions, and finds their max and min values. Is the same program as **CANTCONC** with data for this example. (See Figure 4-22*b*.)
EX04-06	Calculates the shear, moment, slope, and deflection functions for an overhung beam with a concentrated load at any point along its length and with a uniformly distributed load over a portion of its length beginning at one support. Finds reactions, plots the beam functions, and finds their max and min values. (See Figure 4-22*c*.)
EX4-07	Calculates the shear, moment, slope, and deflection functions for a statically indeterminate beam with a uniformly distributed load over a portion of its length. Finds reactions, plots the beam functions, and finds their max and min values. (See Figure 4-22*d*.)
EX04-08	Determines the best cross-sectional shape for a hollow bar loaded in pure torsion. (See Figure 4-29.)
EX04-09	Calculates the stresses due to combined bending and torsional loading. (See Figure 4-30.)
EX04-10C	Designs columns in circular cross sections for concentric loading using both Johnson and Euler criteria to find critical load, weight, and safety factor. Is the same program as **COLMNDES** with data for this example. (See Figure 4-41.)
EX04-10S	Designs columns in square cross sections for concentric loading using both Johnson and Euler criteria to find critical load, weight, and safety factor. Is same program as **COLMNDES** with data for this example. (See Figure 4-41.)
EX05-01	Calculates the principal and von Mises stresses for a bracket made of ductile material and loaded in combined bending and torsion. Finds the safety factors based on the distortion-energy and maximum shear stress theories. (See Figure 5-7.)
EX05-02	Calculates the principal and von Mises stresses for a bracket made of brittle material and loaded in combined bending and torsion. Finds the safety factors based on the modified-Mohr theory. (See Figure 5-7.)
EX05-03	Calculates the fracture mechanics failure criteria for a cracked part. Compares the fracture-mechanics failure stress with a yield failure. (See Figure 5-15.)
EX06-01	Calculates the corrected endurance strength of ferrous metals based on supplied data about finish, size, strength, etc., and draws an estimated S-N diagram for supplied levels of alternating and mean stresses. Is the same program as **S_NDIAGM** with data for this example. (See Figure 6-34.)
EX06-02	Calculates the corrected endurance strength of nonferrous metals based on supplied data about finish, size, strength, etc., and draws an estimated S-N diagram for supplied levels of alternating and mean stresses. Is the same program as **S_NDIAGM** with data for this example. (See Figure 6-33.)

EX06-03	Finds the fatigue stress-concentration factor for a part of known material and geometry. (See Figure 4-36.)
EX06-04A	Design of a cantilever bracket for fully-reversed bending — part a: an unsuccessful design. (See Figure 6-41.)
EX06-04B	Design of a cantilever bracket for fully-reversed bending — part b: a successful design. (See Figure 6-41.)
EX06-05A	Design of a cantilever bracket for fluctuating bending — part a: an unsuccessful design. (See Figure 6-47.)
EX06-05B	Design of a cantilever bracket for fluctuating bending — part b: a successful design. (See Figure 6-47.)
EX06-06	Design of a cantilever bracket for multiaxial stresses in fatigue. (See Figure 5-7.)
EX07-01	Stresses in a ball thrust bearing. Uses **SURFSPHR** to calculate surface stresses in a spherical-flat contact. (See Figure 10-19.)
EX07-02	Stresses in cylindrical contact. Uses **SURFCYLZ** to calculate surface stresses in a wheel-on-rail contact. (See Figure 7-17.)
EX07-03	Stresses in general contact. Uses **SURFGENL** to calculate surface stresses in a crowned cam-follower contact. (See Figure 7-11.)
EX07-04A	Stresses in combined rolling and sliding in cylindrical contact. Uses **SURFCYLX** to calculate surface stresses in a nip-roller contact—part a. (See Figure 7-13.)
EX07-04B	Stresses in combined rolling and sliding in cylindrical contact. Uses **SURFCYLX** to calculate surface stresses in a nip-roller contact—part b. (See Figure 7-13.)
EX07-05	Safety factor in combined rolling and sliding in cylindrical contact problem of Example 7-04. Uses data from Table 7-8. (See Figure 7-13.)
EX09-01	Shaft design for steady torsion and fully reversed bending (parts a to d). (See Figure 9-5.)
EX09-02	Shaft design for repeated torsion combined with repeated bending (parts a to d). (See Figure 9-5.)
EX09-03	Designing a stepped shaft to minimize deflection. (See Figure 9-5.)
EX09-04	Designing shaft keys (parts a and b). (See Figure 9-5.)
EX09-05	Designing an interference fit. (See Figure 9-18.)
EX09-07	Designing a solid-disk flywheel. (See Figure 9-21.)
EX09-08	Determining the critical frequencies of a shaft. (See Figure 9-30.)
EX10-01A	Sleeve bearing design—part a. (See Figure 10-8.)
EX10-01B	Sleeve bearing design—part b. (See Figure 10-8.)
EX10-02	Lubrication in a crowned cam-follower interface. (See Figure 10-4.)
EX10-04	Selection of ball bearings for combined radial and thrust loads.
EX11-01	Determining gear-tooth and gearmesh parameters. (See Figure 11-8.)
EX11-03	Analyzing an epicyclic gear train. (See Figure 11-16.)
EX11-04	Load analysis of a spur-gear train. (See Figure 11-21.)
EX11-05	Bending stress analysis of a spur gear train. (See Figure 11-20.)
EX11-06	Surface stress analysis of a spur gear train. (See Figure 11-20.)

A

EX11-07	Material selection and safety factor for spur gears. (See Figure 11-20.)
EX12-01	Stress analysis of a helical gear train. (See Figure 12-2.)
EX12-02	Stress analysis of a bevel gear train. (See Figure 12-4.)
EX13-03	Design of a helical compression spring for static loading. (See Figure 13-7.)
EX13-04	Design of a helical compression spring for cyclic loading. (See Figure 13-7.)
EX13-05	Design of a helical extension spring for cyclic loading. (See Figure 13-20.)
EX13-06	Design of a helical torsion spring for cyclic loading. (See Figure 13-26.)
EX13-07	Design of a Belleville spring for static loading. (See Figure 13-29.)
EX14-01	Torque and efficiency of a power screw. (See Figure 14-4.)
EX14-02	Preloaded fasteners in static loading. (See Figure 14-24.)
EX14-03	Preloaded fasteners in dynamic loading. (See Figure 14-26.)
EX14-04	Determining material stiffness and the joint constant. (See Figure 14-31.)
EX14-05	Determining the torque needed to generate a bolt preload. (See Figure 14-31.)
EX14-06	Fasteners in eccentric shear. (See Figure 14-42.)
EX15-01	Design of a disk clutch. (See Figure 15-6.)
EX15-02	Design of a short-shoe drum brake. (See Figure 15-8.)
EX15-03	Design of a long-shoe drum brake. (See Figure 15-9.)

H-3   TKSOLVER FILES - CASE STUDIES

Case Studies   Path: TKFiles \ Cases \ CaseNo

CASE1A	Case study of the force analysis of a bicycle brake lever under static, 2-D loading. Finds reaction forces. See Chapter 3 and Figure 3-2.
CASE1B	Case study of the stress and deflection analysis of a bicycle brake lever under static, 2-D loading. Finds and plots the beam functions, shear, moment, and deflection and determines stresses at particular locations. See Chapter 4 and Figure 4-47.
CASE2A	Case study of the force analysis of a hand crimping tool under static, 2-D loading. Finds reaction forces. See Chapter 3 and Figure 3-4.
CASE2B	Case study of the stress and deflection analysis of a hand crimping tool under static, 2-D loading. Finds and plots the beam functions, shear, moment, and deflection and determines stresses at particular locations. See Chapter 4 and Figure 4-49.
CASE2C	Case study of the failure analysis of a hand crimping tool under static, 2-D loading. Finds the safety factors at particular locations. See Chapter 5 and Figure 5-19.
CASE3A	Case study of the force analysis of a scissors jack under static, 2-D loading. Finds reaction forces. See Chapter 3 and Figure 3-8.
CASE3B	Case study of the stress and deflection analysis of a scissors jack under static, 2-D loading. Finds and plots the beam functions, shear, moment, and deflection and determines stresses at particular locations. See Chapter 4 and Figure 4-52.
CASE3C	Case study of the failure analysis of a scissors jack under static, 2-D loading. Finds the safety factors at particular locations. See Chapter 5 and Figure 5-20.

A

CASE4A   Case study of the force analysis of a bicycle brake arm under static, 3-D loading. Finds reaction forces. See Chapter 3 and Figure 3-10.

CASE4B   Case study of the stress and deflection analysis of a bicycle brake arm under static, 3-D loading. Finds and plots the beam functions, shear, moment, and deflection and determines stresses at particular locations. See Chapter 4 and Figure 4-54.

CASE4C   Case study of the failure analysis of a bicycle brake arm under static, 3-D loading. Finds the safety factors at particular locations. See Chapter 5 and Figure 5-21.

CASE5AEN   Case study of the force analysis of a fourbar linkage under dynamic, 2-D loading. Finds theoretical reaction forces in English units. See Chapter 3 and Figure 3-13.

CASE5ASI   Case study of the force analysis of a fourbar linkage under dynamic, 2-D loading. Finds theoretical reaction forces in SI units. See Chapter 3 and Figure 3-13.

CASE6-X   Eight TK files (–0 through –7) for a case study of the fatigue analysis and redesign of a failed power loom laybar under dynamic, 2-D loading. See Chapter 6 and Figure 6-51.

CASE7A   Design of an engine-powered air compressor. This TK file sets up the design problem. See Chapter 8 and Figure 8-1.

CASE7B-X   Design of an engine-powered air compressor. These 3 TK files (-1, -2, -3) design the transmission shafts connecting the engine and compressor. See Chapter 9 and Figure 8-1.

CASE7C   Design of an engine-powered air compressor. These 2 TK files (-1, -2) design the spur gears connecting the engine and compressor. See Chapter 11 and Figure 8-1.

CASE7D   Design of an engine-powered air compressor. These 2 TK files (-1, -2) design the headbolts for the compressor. See Chapter 14 and Figure 8-1.

CASE8A   Design of a hay-bale lifter. This file sets up the design problem. See Chapter 8 and Figure 8-4.

CASE8B   Design of a hay-bale lifter. This TK file designs a worm and worm gear for the speed reducer. See Chapter 12 and Figure 8-4.

CASE9B   Design of a cam-test machine. This TK file designs the sleeve bearings for the camshaft. See Chapter 10 and Figure 8-8.

CASE9C   Design of a cam-test machine. This TK file designs the coil spring for the cam follower. See Chapter 13 and Figure 13-36.

H-4   TKSOLVER FILES - MASTERS

Masters Path: TKFiles \ Masters

FORMATS   This file contains only a format sheet that can be added to (merged into) any other *TKSolver* file without disturbing its other contents. The format sheet enables formatting of variables to any desired number of decimal places. See Example 1-10.

STUDENT   This file is blank except for the format sheet from **FORMATS** and the units sheet from **UNITMAST**. It is intended to be used by the student as a starter file for a new *TK* model to which rules, functions, variables, etc., can be added. Starting each model with this file eliminates the need to merge the **UNITMAST** or **FORMATS** files into your models and provides their advantages with minimal effort. Be sure to save the file with a new name to avoid overwriting the master file **STUDENT** each time it is used. Use SAVE AS from the FILE menu to provide a new filename.

Λ

UNITMAST	This file contains only a units sheet that can be added to (merged into) any other *TKSolver* file without disturbing its other contents. The units sheet enables units conversion of variables. See Example 1-10.
PROCDURS	This file contains a large number of rule, list, and procedure functions that are used in many of the other *TK* files provided. These functions can be imported and used in new *TK* models. See the functions' listings for documentation.

H-5   EXECUTABLE FILES - KINEMATICS, DYNAMICS, AND STRESS ANALYSIS

Kinematics   Path: EXEFiles \ Kinematics

FOURBAR.EXE[*†]	Program to compute position, velocity, and acceleration of any fourbar linkage.
FIVEBAR.EXE[*†]	Program to compute position, velocity, and acceleration of any geared fivebar linkage.
SIXBAR.EXE[*†]	Program to compute position, velocity, and acceleration of some inversions of Watt's and Stephenson's sixbar linkages.
SLIDER.EXE[*†]	Program to compute position, velocity, and acceleration of one inversion of a fourbar crank-slider linkage.

Dynamics   Path: EXEFiles \ Dynamics

DYNAFOUR.EXE[*†]	Program to compute position, velocity, and acceleration and dynamic forces of any fourbar linkage.
DYNACAM.EXE[*†]	Program to design and compute kinematics and dynamics of any plate cam with translating follower. (Some data files for this program that are mentioned in the text are also included, such as SPRAY.CAM.)
ENGINE.EXE[*†]	Program to compute kinematics and dynamics of any single or multicylinder internal-combustion engine of inline, vee, opposed, or W configuration.
MATRIX.EXE[*†]	Program to solve any linear system of up to 20 equations in 20 unknowns.
ASDEQ.EXE[*]	Program to solve any system of differential equations. (Contains a help manual.)
CASE8A.ASD	Input file for program ASDEQ that solves the differential equations of motion for Case Study 8A's force analysis.

Stress Analysis Path: EXEFiles \ Stress

MOHR.EXE[*]	Program to compute the 3-D stress cubic and plot the Mohr's circles for any stress state.
CONTACT.EXE[*]	Calculates and plots the stress distributions at and below the surface for Hertzian contact. Principal, shear, and Von Mises stresses are plotted for user-selected increments in x, y, and z directions. See Chapter 7.

---

[*] DOS version only.

[†] Instructions for using these programs can be found in R. L. Norton, *Design of Machinery*, McGraw-Hill, New York, 1992.

H-6   REFERENCE

1 **R. L. Norton**, *Design of Machinery*, McGraw-Hill, New York, 1992.

A

INDEX